SECOND EDITION

THE VACUUM INTERRUPTER

THEORY, DESIGN, AND APPLICATION

SECOND EDITION

THE VACUUM INTERRUPTER

THEORY, DESIGN, AND APPLICATION

PAUL G. SLADE

CRC Press is an imprint of the
Taylor & Francis Group, an **informa** business

Second edition published 2021
by CRC Press
6000 Broken Sound Parkway NW, Suite 300, Boca Raton, FL 33487-2742

and by CRC Press
2 Park Square, Milton Park, Abingdon, Oxon, OX14 4RN

First edition published by CRC Press 2007
CRC Press is an imprint of Taylor & Francis Group, LLC

Library of Congress Control Number: 2020942082

ISBN: 978-0-367-27505-1 (hbk)
ISBN: 978-0-429-29891-2 (ebk)

To my wife Dee-Dee
For her love, support and contribution

Contents

Preface and Acknowledgments for the First Edition xiii
Preface and Acknowledgments for the Second Edition xv
Author xvii
Introduction xix

PART 1 Vacuum Interrupter Theory and Design

Chapter 1 High Voltage Vacuum Interrupter Design 3

1.1 Introduction 3
1.2 The External Design 6
1.2.1 Electrical Breakdown in Gas 6
1.2.2 Creepage Distance 17
1.2.3 Insulating Ambients and Encapsulation 21
1.3 Electrical Breakdown in Vacuum 23
1.3.1 Introduction 23
1.3.2 The Electric Field 31
1.3.2.1 The Microscopic Enhancement Factor (β_m) 31
1.3.2.2 The Geometric Enhancement Factor (β_g) 34
1.3.3 Pre-Breakdown Effects 38
1.3.3.1 Field Emission Current 38
1.3.3.2 Anode Phenomena 44
1.3.3.3 Microparticles 51
1.3.3.4 Microdischarges 56
1.3.4 Vacuum Breakdown and the Transition to the Vacuum Arc 62
1.3.5 The Transition to a Self-Sustaining Vacuum Arc 75
1.3.6 Time to Breakdown 84
1.3.7 Conditioning 86
1.3.7.1 Spark Conditioning Using a High Voltage AC Power Supply 87
1.3.7.2 Spark Conditioning Using a High Voltage Pulse 88
1.3.7.3 Current Conditioning 91
1.3.7.4 Other Conditioning Processes 92
1.3.8 Puncture 94
1.3.9 Deconditioning 95
1.4 Internal Vacuum Interrupter Design 96
1.4.1 The Control of the Geometric Enhancement Factor, β_g 96
1.4.2 Breakdown of Multiple Vacuum Interrupters in Series for Contact Gaps Greater Than 2mm 103
1.4.3 Voltage Wave Shapes and Vacuum Breakdown in a Vacuum Interrupter 104
1.4.4 Impulse Testing of Vacuum Interrupters 105
1.4.5 Testing for High Altitude 113
1.5 X-Ray Emission 114

1.6 Arc Initiation When Closing a Vacuum Interrupter 125
References 126

Chapter 2 The Vacuum Arc 137

2.1 The Closed Contact 137
2.1.1 Making Contact, Contact Area, and Contact Resistance 137
2.1.2 Calculation of Contact Resistance 139
2.1.2.1 The Real Area of Contact a Small Disk of Radius "*a*" 139
2.1.3 Contact Resistance and Contact Temperature 141
2.1.3.1 The Calculation of Contact Temperature 142
2.1.4 Blow-Off Force 143
2.1.4.1 Butt Contacts 145
2.1.4.2 Contact Interface Melting During Blow-Off 146
2.2 The Formation of the Vacuum Arc during Contact Opening 147
2.3 The Diffuse Vacuum Arc 152
2.3.1 Cathode Spots 153
2.3.2 The Plasma between the Cathode Spot and the Anode 161
2.3.3 Current Chop 165
2.3.4 The Formation of the Low-Current and High-Current Anode Spot 169
2.4 The Columnar Vacuum Arc 173
2.5 The Transition Vacuum Arc 179
2.6 The Interaction of the Vacuum Arc and a Transverse Magnetic Field 181
2.6.1 The Diffuse Vacuum Arc and a Transverse Magnetic Field 181
2.6.2 The Columnar Vacuum Arc and a Transverse Magnetic Field 185
2.7 The Vacuum Arc and an Axial Magnetic Field 187
2.7.1 The Low-Current Vacuum Arc in an Axial Magnetic Field 188
2.7.2 The High-Current Vacuum Arc in an Axial Magnetic Field 193
2.8 Overview and Review of the Three Forms of Anode Spot 209
References 210

Chapter 3 The Materials, Design, and Manufacture of the Vacuum Interrupter 219

3.1 Introduction 219
3.2 Vacuum Interrupter Contact Materials 220
3.2.1 Introduction 220
3.2.2 Copper and Copper-Based Contact Materials That Have Been Developed Following the Initial Experiments on High Current Vacuum Arcs Using Copper Contacts 221
3.2.3 Refractory Metals Plus a Good Conductor 221
3.2.4 Semi-Refractory Metals Plus a Good Conductor 224
3.2.5 Copper Chromium Materials Plus an Additive 233
3.2.6 Chopping Current 233
3.2.7 Summary 242
3.3 The Contact Structures for the Vacuum Interrupter 245
3.3.1 Introduction 245
3.3.2 Disc- or Butt-Shaped Contacts 246
3.3.3 Contacts to Force the Motion of the High Current, Columnar Vacuum Arc 247

3.3.4 Contacts to Force the High Current, Columnar Arc into the Diffuse Mode 262
3.3.5 Summary 283
3.4 Other Vacuum Interrupter Design Features 283
3.4.1 The Insulating Body 283
3.4.2 The Shield 286
3.4.3 The Bellows 289
3.5 Vacuum Interrupter Manufacture 291
3.5.1 Assembly 291
3.5.2 Testing and Conditioning 298
3.5.3 Summary 304
References 306

PART 2 Vacuum Interrupter Application

Chapter 4 General Aspects of Vacuum Interrupter Application 321

4.1 Introduction 321
4.2 The Interruption of AC Circuits 323
4.2.1 The Interruption of the Diffuse Vacuum Arc for AC Currents Less Than 2 kA (rms.) with a Fully Open Contact Gap 323
4.2.2 The Interruption of the Vacuum Arc for AC Currents Greater than 2 kA (rms.) 337
4.2.3 The Interruption of High Current Vacuum Arcs 345
4.3 Interruption of AC Circuits When the Contacts Open Just Before Current Zero 361
4.3.1 Low Current Vacuum Arcs 361
4.3.1.1 Low Current Interruption of Inductive Circuits 362
4.3.1.2 Low Current Interruption of Capacitive Circuits 366
4.3.2 High Current Interruption 367
4.4 Contact Welding 369
4.4.1 Introduction 369
4.4.2 Welding of Closed Contacts 371
4.4.2.1 Cold Welding and Diffusion Welding 371
4.4.2.2 Welding Caused by the Passage of High Current 371
4.4.3 A Comparison of the Calculated "i_W" with Experimental Values 378
4.4.3.1 Simple Butt Contacts with One Region of Contact and a Short Current Pulse 378
4.4.3.2 Simple Butt Contacts with More Than One Region of Contact and a Short Current Pulse 379
4.4.3.3 Axial Magnetic, Large Area, Vacuum Interrupter Contacts 380
4.4.4 The Model to Determine the Threshold Welding Current for Closed Contacts with "n" Regions of Contact for Passage of Current of 1 to 4 Seconds 381
4.4.4.1 Closed Large Area Vacuum Interrupter Contacts Passing Fault Currents from 1 to 4 Seconds 383
4.4.5 Welding of Contacts That Slide 385
4.4.6 Welding when Contacts Close an Electrical Circuit 386
References 394

Chapter 5 Application of the Vacuum Interrupter for Switching Load Currents 401

5.1 Introduction 401
5.2 Load Current Switching 403
5.2.1 Switches Used at Distribution Voltages 403
5.2.2 Switches Used at Transmission Voltages 409
5.3 Switching Inductive Circuits 414
5.3.1 Voltage Surges When Closing an Inductive Circuit 414
5.3.2 Voltage Surges When Opening an Inductive Circuit 414
5.3.3 Surge Protection 415
5.3.4 Switching Three-Phase Inductive Circuits: Virtual Current Chopping 421
5.3.5 Transformer Switching 423
5.3.5.1 Tap Changers 424
5.3.5.2 Switching Off Unloaded Transformers 424
5.3.5.3 Switching Off an Unloaded Transformer's In-Rush Current 426
5.3.5.4 Switching Off Loaded Transformers 427
5.4 Vacuum Contactors 428
5.4.1 Introduction 428
5.4.2 Solenoid Operation 431
5.4.3 Sizing the Contact 433
5.4.4 The Shield 436
5.4.5 The Contact Material 436
5.5 Switching Capacitor Circuits 437
5.5.1 Inserting a Capacitor Bank 438
5.5.2 Disconnecting a Capacitor Bank 443
5.5.3 Switching Three-Phase Capacitor Banks 448
5.5.4 The Capacitor Switch Recovery Voltage, Late Restrikes, and NSDDs 450
5.5.5 Switching Cables and Overhead Lines 462
5.6 Vacuum Interrupters for Circuit Switching, Circuit Isolation, and Circuit Grounding 464
5.6.1 Background 464
5.6.2 Vacuum Interrupter Design Concepts for Load Switching and for Isolation 468
5.6.3 Vacuum Interrupter Design for Switching and Grounding 469
5.6.4 Vacuum Interrupter Design for Fault Protection, Isolation, and Grounding 473
5.7 Summary 474
References 475

Chapter 6 Circuit Protection, Vacuum Circuit Breakers, and Reclosers 481

6.1 Introduction 481
6.2 Load Currents 482
6.3 Short Circuit Currents 492
6.3.1 Introduction 492
6.3.2 The Short Circuit Current and Asymmetry 493
6.3.3 The Transient Recovery Voltage (TRV), for a Terminal Fault 496
6.3.3.1 First Pole-to-Clear Factor 499
6.3.4 The Terminal Fault Interruption Performance of Vacuum Interrupters 502

6.3.5 The Transient Recovery Voltage for Short Line Faults (SLF) 507
6.3.6 TRV from Transformer Secondary Faults 509
6.4 Late Breakdowns and Non-Sustained Disruptive Discharges (NSDDs) 509
6.5 Vacuum Circuit Breaker Design 515
6.5.1 Introduction 515
6.5.2 Closed Contacts 516
6.5.3 Mechanism Design 520
6.5.4 The Vacuum Interrupter Mounting and Insulation 532
6.5.5 The Vacuum Circuit Breaker's Electrical Life 540
6.6 Vacuum Circuit Breaker Testing and Certification 546
6.6.1 Developmental Testing of the Vacuum Interrupter 546
6.6.2 Certification Testing at an Independent High-Power Testing Laboratory 548
6.6.3 Fault Current Endurance Testing 549
6.7 Vacuum Circuit Breakers for Capacitor Switching, Cable and Line Switching, and Motor Switching 550
6.7.1 Introduction 550
6.7.2 Capacitor Switching 551
6.7.2.1 Capacitor Switching and NSDDs 554
6.7.3 Cable Switching and Line Dropping 555
6.7.4 Motor Switching 556
6.8 Application of Vacuum Circuit Breakers for Distribution Circuits (4.76 kV to 40.5 kV) 556
6.8.1 Indoor Switchgear 556
6.8.2 Outdoor Circuit Breakers 561
6.8.3 Vacuum Reclosers 561
6.8.4 The Ring Main Unit (RMU) for Secondary Distribution 564
6.8.5 Pad-Mount Secondary Distribution Systems 565
6.8.6 The Generator Vacuum Circuit Breaker 565
6.8.6.1 High Continuous Currents 568
6.8.6.2 Transformer/System Fed Faults 569
6.8.6.3 Generator Fed Faults 571
6.8.6.4 Out-of-Phase Switching 572
6.8.7 Transportation Circuit Breakers 572
6.8.7.1 Interrupting Fault Currents at Frequencies Less Than and Greater Than 50/60 Hz 573
6.8.8 Switching Electric Arc Furnaces (EAF) 575
6.9 Vacuum Interrupters in Series 576
6.10 Vacuum Interrupters for Subtransmission and Transmission Systems 582
6.11 Switching DC Circuits Using Vacuum Interrupters 589
6.11.1 DC Interruption Using the Natural Vacuum Arc Instability 589
6.11.2 DC Current Interruption Using an External Magnetic Field Pulse 590
6.11.3 Switching High Voltage DC Transmission Circuits Using a Current Counter Pulse 590
6.12 Development of Vacuum Interrupters for Low Voltage (< 1000V) Circuit Breakers 598
6.13 Concluding Summary 598
References 600

Author Index 613

Subject Index 627

Preface and Acknowledgments for the First Edition

Now as I was young and easy under the apple boughs
About the lilting house and happy as the grass was green

Dylan Thomas
(Fern Hill)

In the late 1960s, after completing my PhD, I joined a team of researchers at the Westinghouse R&D Center to work on a new, medium voltage, switching component called the vacuum interrupter. My assignment was to investigate and develop contact materials for this device. This was a daunting task for a new PhD who had, up to that time, only worked on low current, low voltage, dc relay contacts. Not only was there the shock of a new technology, but there was also the shock of a completely new vocabulary. The team that I joined seemed quite familiar with terms such as: vacuum arc, cathode spots, asymmetric fault currents, transient recovery voltage, rate of rise of recovery voltage (commonly referred to as triple-R, V) and so on. Both the science and the application of this new technology were completely new to me. I learned later that, although my colleagues' knowledge was certainly greater than mine, they too were only beginning their own understanding of this new switching technique as the subject was still developing. It took another decade of research and development before the vacuum interrupter became accepted as a viable component for medium voltage, circuit breakers, reclosers, and contactors. It took another 20 years before it became the dominant technology for these circuit switching and protection devices. Throughout this time period, there was an ever-increasing effort in research, development, and the application of the vacuum interrupter by both industrial R&D groups and university researchers. In fact, these R&D efforts have continued to this day. As this knowledge has accumulated so has my own understanding of the vacuum interrupter and the science that has driven its further development. So, for the past 40 years I have enjoyed gradually unraveling its mysteries. I did this first of all at the Westinghouse R&D Center and then, after Westinghouse sold its switching business to the Eaton Corporation, at the Eaton Corporation's Vacuum Interrupter Factory as the product's Technology Director. In my present position I have complete responsibility for R&D and new vacuum interrupter design as well as extensive consultation with customers on their application of this product.

One of the first things I did when the vacuum interrupter product became my full-time occupation was to develop an internal course on the science behind vacuum interrupter design and the engineering reasons behind vacuum interrupter application. Although this course was initially given for Eaton personnel, it soon became evident that there was a much wider audience for this information. This was especially true as more and more people were beginning to use vacuum interrupters for an ever-widening range of applications. In 1994, I began a two-day intensive course on vacuum interrupter design and application that has been offered up to three times a year. This course has been given in the United States, Europe, Africa, and Asia to a very wide audience. It has also undergone extensive revision as the attendees have offered constructive criticism and have made suggestions for changes and additions. Although the attendees are given course notes, a number of them have asked about a permanent source book on the subject that they can use once they have returned to their places of work.

This book has been written to partially satisfy this need, but it cannot be the complete comprehensive volume on the subject. The book does, however, present my biased understanding of this fascinating technology. While I have tried to cover a broad vision of vacuum interrupter research, development, and application, there will be aspects that I will have passed over too quickly or not even addressed at all. Sometimes you may find my personal interpretations of the scientific data somewhat controversial and surely in some cases it may well be quite mistaken: we are all fallible. For good or ill these interpretations are my own, as are any other mistakes you will find in the text. In spite of this, I hope that my discussions and my conclusions will prove useful. In developing this book, I owe a great debt of gratitude to the scientists and engineers who have contributed to the present body of technical literature on the subject. A book such as this rests upon the shoulders of those that have preceded it: Lafferty, J. (Editor) *Vacuum Arcs: Theory and Application*; Greenwood, A., *Vacuum Switchgear*; Mesyats, G., *Cathode Phenomena in a Vacuum Discharge: The Breakdown, the Spark and the Arc*; Latham, R., *High Voltage Vacuum Insulation: Basic Concepts and Technological Practice* and Boxman, R., Martin, P., Sanders, D. (editors) *Handbook of Vacuum Science and Technology: Fundamentals and Applications.* Each book covers different aspects of the subject, but its author(s) share a wide experience in vacuum interrupter technology.

Among my own colleagues I have had the good fortune to work with over the years, I would especially like to mention Drs. Clive Kimblin, Roy Voshall, Ken Davies, Michael Schulman, Jockel Heberlein, Kirk Smith, Erik Taylor, and Wang Pei Li. Each has given me different, but deep insights into the phenomena involved in developing and applying vacuum interrupters. I would also like to acknowledge the extensive body of experimental research conducted under Professor Manfred Lindmayer at the Technical University of Brunswick and under the late Professor Werner Rieder at the Technical University of Vienna. Both institutions have provided us with a wealth of experimental data, which over the years has proved invaluable. There are of course many other researchers who have silently contributed to this book through their publications. I have attempted in each chapter to give an extensive bibliography for those of you who wish to pursue one aspect of the subject in greater depth. I would especially urge you to look back over some of the seminal papers that were written more than 30 years ago.

I thank Marsha Coats who deciphered my handwriting and who typed the initial manuscript of this book. I am deeply grateful to Erik Taylor, Wang Pei Li, and Kirk Smith who read the chapters in manuscript form and offered up the constructive criticism that has enhanced the text. I am especially indebted to Erik Taylor, who has developed critical data and has supplied the many photographs of the vacuum arc used in this volume. I would also like to thank my father-in-law, Professor Ralph Armington, who as an expert in power engineering constructively read each chapter and offered his insights and suggestions from his many years of experience. My wife Dee-Dee has potentially lived with this book since its inception. She has been very supportive and tolerant of my retreats to my home office evenings, weekends, and holidays while I prepared and wrote the text and developed the figures. She also undertook the careful and constructive reading of the text, but I owe her so much more: this book is dedicated to her.

Paul G. Slade, 2007

Preface and Acknowledgments for the Second Edition

O may my heart's truth
Still be sung
On this high hill in a year's turning

Dylan Thomas
(Poem in October)

Since the publication of the first edition of this book in early 2008, there has been continued research in and development of vacuum interrupter science and vacuum interrupter application. Especially encouraging has been the work by a new generation of graduate students who have enhanced our understanding of the vacuum arc and vacuum breakdown phenomena.

Advances in vacuum interrupter science and application do not advance with the same hectic pace that we have come to expect from the application of the integrated circuit: e.g. in computer technology, software development, and communications technology. It sometimes seems that these systems bring obsolescence not in our lifetime, but even in our recent memory! That being said, the understanding of the science behind vacuum interrupter development has steadily advanced over the past 15 years. The vacuum interrupter is now the technology of choice for all distribution circuit breaker and switch applications (i.e., 3.6kV to 40.5kV) worldwide. Vacuum interrupters for the subtransmission voltage of 72.5kV have been successfully introduced by all the leading vacuum interrupter manufacturers. Significant single contact gap designs have also been tested for the transmission voltages 123kV to170kV. There have even been prototype design concepts for a single contact gap 242kV vacuum interrupter. These new vacuum interrupters have been developed into successful vacuum circuit breakers. They are gradually being introduced in place of the presently successful SF_6 puffer circuit breaker, now that SF_6 has been recognized as a major source of greenhouse gas.

I have written this second edition to bring the subject of the vacuum interrupter up to date. This new volume retains the essence of the original book, but it adds the contributions that reflect the continued diversity of the research from the international university and industrial R&D communities. I wish to express a personal note of thanks for the excellent work that has been introduced by scientists and engineers from China, Russia, Japan, and Europe. Again, although I have attempted to cover the broad subject of vacuum interrupter research, development, and application, there will be some aspects that I have passed over and some that I have not addressed at all. Some of my interpretations may sometimes be somewhat controversial and at times might even be incorrect. They are, however, my own and I hope that you will find these discussions and conclusions to be of some use.

I wish to acknowledge the help I received from Rene Smeets who has readily answered the many questions I have asked of him as I developed this volume. I thank Zhenxing Wang and Haruki Ejiri for helping me in further understanding vacuum breakdown phenomena. Also, I wish to thank Erik Taylor who continues to develop significant papers with me on contact welding and capacitor switching phenomena. Since the publication of the first edition, Andre Anders has published the book *Cathodic Arcs: From Fractal Spots to Energetic Condensation* and Pramod Naik has published the book *Vacuum: Science, Technology and Applications*. There are also many researchers past and present who have made significant contributions to this volume. I have thus retained all the references from the first edition so that new those of you who are new to the subject will have the opportunity to follow the development of the subject from the 1930s to the present day. Even Isaac Newton, who changed the world as it was seen in his time, famously said

If I have seen further it is by standing on the shoulders of giants.

He introduced the laws of motion, the concept of the gravitational force and, at the same time as Leibnitz, developed the calculus. He was so respected in his time that the eighteenth-century poet Alexander Pope wrote in his "An Essay on Man":

All nature and its law hid in night
God said: Let Newton Be: and all was light

I therefore strongly urge those of you who are new to the subject to look back over the seminal papers that have been written in the past 50 years.

My wife Dee-Dee has lived with this volume for over the past year and a half while I revised the text and developed additional figures. She has been extremely encouraging and supportive. She again undertook careful and constructive reading of the text. I dedicate this volume to her.

Author

Paul G. Slade received his BSc and PhD in Physics from the University of Wales, Swansea, UK, and an MBA from the University of Pittsburgh, Pennsylvania, USA. He joined a group of scientists at the Westinghouse R&D Center who, at that time, were developing a new electric circuit interruption technology; the vacuum interrupter. In his career at the R&D Center he held a number of technical management positions culminating in the position of Chief Scientist for the Westinghouse Industrial Group. When Westinghouse sold its Distribution Equipment Division to the Eaton Corporation, he joined Eaton's Vacuum Interrupter Product operation as manager of Vacuum Interrupter Technology, where he had complete responsibility for R&D and for new vacuum interrupter designs. Since his retirement from Eaton in 2007 he has been an independent consultant for vacuum interrupter technology, electrical contacts and circuit interruption. His email address is: paulgslade@verizon.net

Dr Slade has over 50 years' experience in the application of electric contacts for switching electric current. His research has covered many aspects of electric contact and arcing phenomena in vacuum, air and SF_6. He has published over 130 technical papers in archival journals and scientific conferences. He also holds 23 US patents. He is a major contributor and the editor of the second edition of the book *Electrical Contacts: Principles and Applications*. He is the recipient of the IEEE Ragnar Holm Scientific Achievement Award and of the German VDE Albert Keil Pries for his contributions to the science of electrical contacts. He is a fellow of the IEEE.

Introduction

All the world's a stage,
And all men and women merely players;
They have their exits and their entrances,
And one man in his time plays many parts……..

William Shakespeare
(As You Like It, Act II, Scene IV)

All experience is an arch, to build upon

Henry Brooke Adams
(The Education of Henry Adams)

The vacuum interrupter is deceptively simple. Figure I.1 shows a typical cross-section. The contacts are enclosed in a vacuum chamber originally manufactured from glass, but since the mid-1960s, increasingly from a high alumina ceramic. The contacts are closed by a mechanism connected to the moving terminal. The moving terminal is connected to the vacuum chamber by a stainless steel bellows that permits the contacts to open and close while maintaining the vacuum. Typical vacuum levels inside the device are about 10^{-4} Pa (~10^{-6}mbar). The often-stated advantages are:

- The contacts require no maintenance for the life of the vacuum interrupter
- The vacuum interrupter has long life. The electrical switching life usually equals the mechanical life. The mechanical life is determined by the fatigue life of the bellows. In most applications the vacuum interrupter's life will equal or even exceed the life of the mechanism in which it is housed
- A relatively low energy mechanism is required to operate it and the vacuum interrupter can work in any orientation
- It has a silent operation
- It is self-contained, it requires no extra supply of gases or liquids and it emits no flame or smoke
- The vacuum interrupter's performance neither affects nor is it affected by its surrounding ambient
- The vacuum interrupter can be used in adverse atmospheres and it is unaffected by temperature, dust, humidity, or altitude. It can be built from corrosion resistant materials
- When interrupting current the arc between the opening contacts is completely confined within the vacuum envelope; there is no danger of explosion in a potentially explosive environment such as a coal mine
- The vacuum interrupter can withstand high voltages with only a relatively small contact gap
- It usually interrupts at the first or second ac current zero after it opens
- It requires no resistors or capacitors to interrupt very fast transient recovery voltages such as those present with short line faults and transformer fault currents
- Its components are environmentally benign and can be easily recycled

In spite of these obvious advantages, the vacuum interrupter took a very long time to become a reliable and manufacturable device. It has also taken even longer to be accepted by the electric power industry and manufacturers of competing circuit interruption technologies.

The first patent for a vacuum switch was awarded at the end of the nineteenth century when Enholm received one in 1890 for a "Device for transforming and controlling electric currents" [1]. Thus, the concept of using vacuum as a medium to interrupt current is nearly as old as the distribution of electricity itself. The vacuum interrupter is an excellent example of technology development

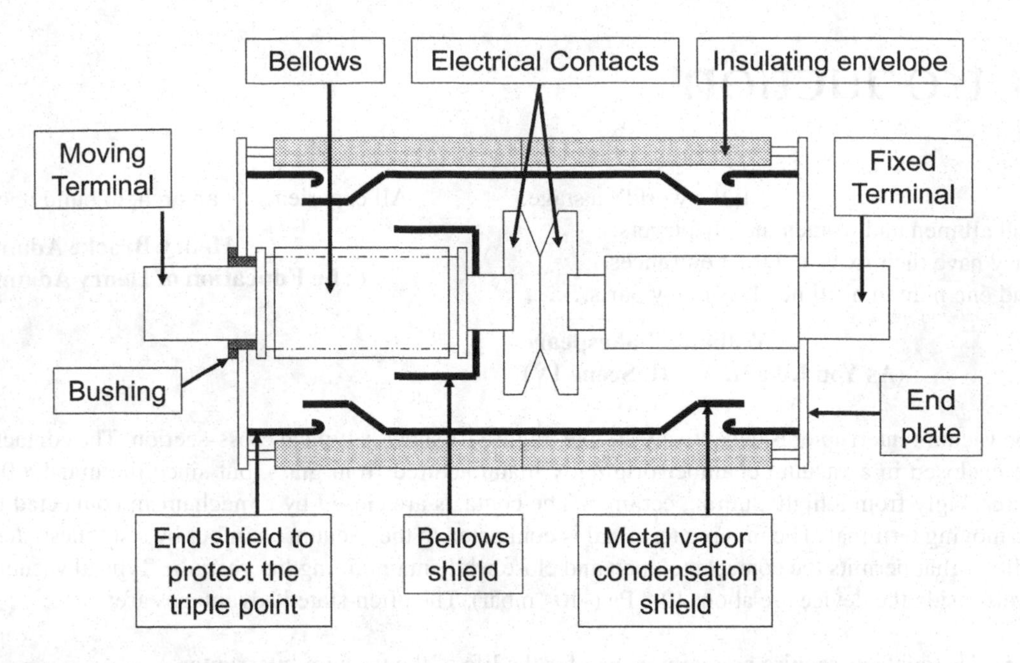

FIGURE I.1 The cross-section of a vacuum interrupter.

that has taken over 100 years from its first conception, to its first experimental verification and finally to its acceptance in the marketplace. As late as 1971, Hayes stated in an article in *Electrical World* [2]:

> *Vacuum switching at this stage is like the ingénue understudy waiting in the wings for the established star to tire – it has great potential, but it must be content with bit parts until its growing capability and appeal are too great to be denied.*

In 2011, 40 years later, *Electrical World* presented an editorial that showed the vacuum interrupter's "great potential" had indeed exceeded all expectations; see Chapter 6 in this volume, reference [246].

From 1890 to 1920, further patents followed at a rate of about one a year; see, for example, Moore [3]. Practical uses of the technology during these early years were confined to low power circuits. The first German patent was by Gerdien in 1920 [4]. In the early 1920s, a number of experimental vacuum switches were built and tested both in the United States and in Europe. These switches had contacts made from various metals and were enclosed in evacuated glass envelopes. They were operated both mechanically and magnetically. In 1926, considerable interest was aroused by the work of Sorensen et al. at the California Institute of Technology [5, 6]. They reported the interruption of currents up to 900A at 40kV with very little contact erosion. This work immediately stimulated both the Westinghouse Corporation and the General Electric Corporation (GE) in the United States to initiate research efforts to develop vacuum switches. They both encountered serious difficulties primarily from gases involved from the contacts, current chopping and contact welding. In spite of this, Tanberg, Berkey, Mason, and Slepian developed the first major insights into the interesting world of the vacuum arc and especially of the cathode spot [7–12]. Their work showed that the cathode spot moved in a retrograde direction when subjected to a transverse magnetic field. They also determined that the vapor stream moving away from the cathode spot was ionized to a considerable degree and had a velocity of ~ $1.5\text{x}10^4\text{ms}^{-1}$. Interestingly, this latter result was rediscovered in the 1960s by Davis et al. [13] and Plyutto et al. [14]. These research efforts led to a steady stream of patents, e.g., Escholz [15], Greenwood [16], and Rankin et al. [17]. Rankin's designs included field

FIGURE I.2 Commercially available power vacuum interrupters in 1961 [22].

coils for the development of transverse magnetic fields to control the high current vacuum arc. He claimed the ability to interrupt currents as high as 5kA in 12–15kV circuits. The Siemens Company also worked on a vacuum switch during this period [18]. The depression of the 1930s led to a hiatus of any real efforts to develop a practical high-power interrupter.

During World War 2 considerable advances were made in vacuum technology. The achievement of and the measurement of vacuum levels orders of magnitude lower than prewar practice were made possible. Also, the technology to permanently seal a vacuum chamber was developed. By the 1950s, Jennings [19] developed a practical vacuum interrupter for switching load current. Early models used W and W–Cu contacts, a glass shield to keep metal deposits off the glass insulating envelope and metal bellows for the mechanical operation. Ross et al. developed and commercialized these vacuum interrupters. He demonstrated series operation to obtain high voltage standoff [20, 21] and in 1955 installed such a switch at California Power and Light. By 1961, the Jennings Company had more than ten different types of vacuum interrupters in production and a large amount of accumulated experience in their application: See, for example, Figure I.2 [22]. Other companies were also actively pursuing vacuum switch technology; see, e.g., Schwager et al. [23], and Pearson et al. [24].

In 1952, the General Electric Corporation (United States) began a large R&D program to produce a practical high-power vacuum switch capable of interrupting high fault currents in distribution and transmission voltage circuits. They developed a gas free contact material that did not weld, had adequate chopping characteristics and satisfactory contact erosion. They also developed the spiral contact structure to control the high current vacuum arc. The resulting vacuum interrupters looked similar to the one shown in Figure I.1. This design has become the industry standard.

GE maintained secrecy on these extensive development efforts. From today's vantage point one wonders, "Why the secrecy?" Perhaps they believed that they could corner the circuit breaker market with the "new technology." If so, their market research was severely lacking. Even in 1979, my own extensive survey [25] showed that while the vacuum interrupter was gaining general acceptance, the minimum-oil interrupter [26] was still the dominant, worldwide technology for medium voltage circuit breaker applications. I concluded in that study that a premium price for the vacuum interrupter would not be accepted and therefore the technology would have to be continually developed to make the vacuum interrupter cost competitive with other interruption technologies. Figure I.3 shows how one company has, over time, continually reduced the size and hence the cost for the 15kV, 12kA function.

The technology shift toward the medium voltage vacuum circuit breaker has only occurred since 1980, about a century after the first patent was awarded; see Figure I.4. This is a very conservative market that resists rapid technical changes. Beginning in 1958, the GE team began to publish the

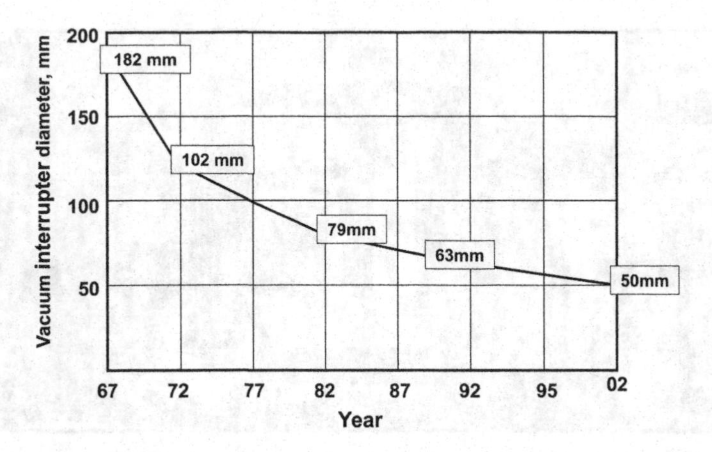

FIGURE I.3 The Westinghouse Corporation (now Eaton Corporation) experience in the reduction of the vacuum interrupter's diameter from 1968 for the 15kV, 12kA function.

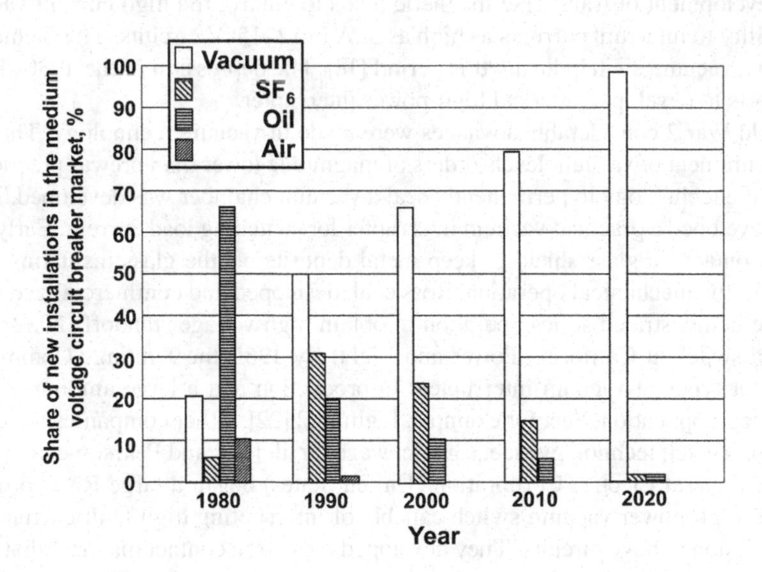

FIGURE I.4 The percentage share of the worldwide distribution circuit breaker market for air, oil, SF_6, and vacuum 1980–2020.

results of their work [27, 28], which together with the research by Reese in the United Kingdom (UK) [29] has served as the foundation for the modern vacuum interrupter design, for the emerging vacuum circuit breaker, the vacuum recloser and the vacuum contactor. By the mid-1960s GE had developed their first successful power vacuum interrupter rated at 15 kV (15.5 kV), 12kA, 1200A. In fact, in 1967 the magazine "Power" showed a photograph of a GE engineer looking lovingly at the 200mm diameter monster they had developed for this rating [30]. Other companies in the United States, Europe, and Japan began their development of the vacuum interrupter in the 1960s.

I joined the Westinghouse team in the mid 1960s. My task was to help develop a contact material that would overcome the problems associated with the Cu–Bi material developed by GE. The work at Westinghouse, in association with the English Electric Company (UK), culminated in the development of the Cu–Cr material that has found worldwide acceptance in present-day power circuit breaker and recloser, vacuum interrupters; see Section 3.2.4 in this volume. By the 1970s to

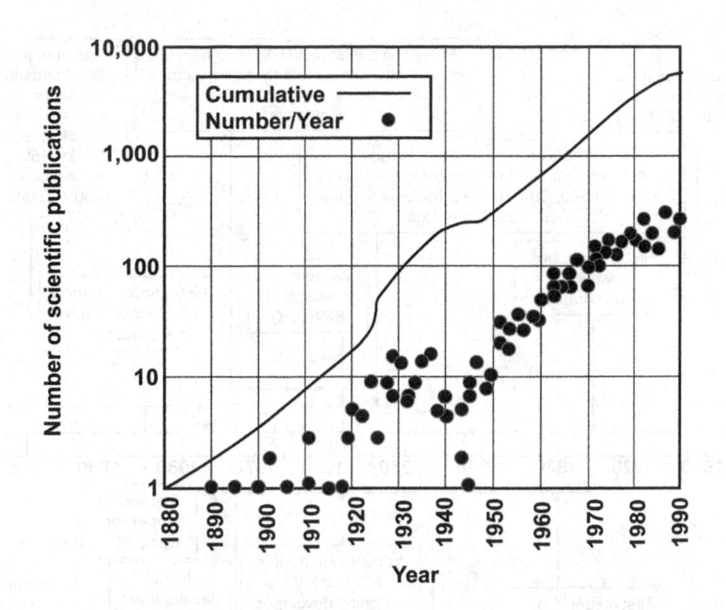

FIGURE I.5 The number of vacuum interrupter related publications 1880–1990 [32, 33].

the present there has been an extensive increase in issued patents worldwide, as well as a significant increase in the number of scientific papers published [31, 32]; see Figure 1.5 [33].

One of the major thrusts by the GE team was to develop a range of transmission circuit breakers, i.e., for voltages greater than 115kV. The initial assumption being that holding off very high voltages would not be a problem with a vacuum gap, but interrupting high currents would present a considerable challenge. Experience has shown, however, that the interruption of high current, while challenging, has proven to be attainable, but designing a single vacuum interrupter to withstand voltages above 242kV has been shown to be extremely challenging. The Associated Electrical Industries Company (UK) did demonstrate a 132kV, 15.3kA vacuum circuit breaker with eight vacuum interrupters in series in 1967 [34, 35]. Meidensha, in Japan, presented a 72kV, 20kA vacuum interrupter in 1975 [36]. Kurtz et al. and Shores et al. published the final GE papers on the transmission breaker in 1975 [37, 38]. These papers showed very large vacuum interrupters, 200mm outside diameter by 430 mm long, and 305mm outside diameter by 710mm long. This proved to be the swan song of the GE effort. After this, there was a decision by the GE corporate management to disband their vacuum interrupter team and curtail any more research on the subject. Thus, the pioneer company in the design and development of power vacuum interrupters no longer played a leading role in their further development and application.

GE did continue to manufacture vacuum circuit breakers using their obsolete technology which included tabulation exhaust and the Cu–Bi contact material. Eventually in 2006 they sold the distribution vacuum circuit breaker business to Powell Industries Incorporated, but GE retained their vacuum interrupter production business. Powell declined to offset GE's vacuum interrupter manufacturing costs and decided to upgraded the whole GE vacuum circuit breaker line using the latest vacuum interrupter technology. GE's vacuum interrupter production eventually ceased. GE also ceased to participate in the leading conference on vacuum interrupter phenomena and design, "The International Symposium on Discharges and Electrical Insulation in Vacuum" (ISDEIV) [39].

Since the first edition of this book, the isdeiv has continued to be the conference where the latest research on vacuum breakdown, the vacuum arc, vacuum interrupter design and its application to interrupt and control electric circuits is presented. There have been six conferences since 2006 where a total of about 800 papers have been published. Of these papers about 37% were from China, about 21% from Russia, about 13% from Germany, and about 10% from Japan. The final 20% were

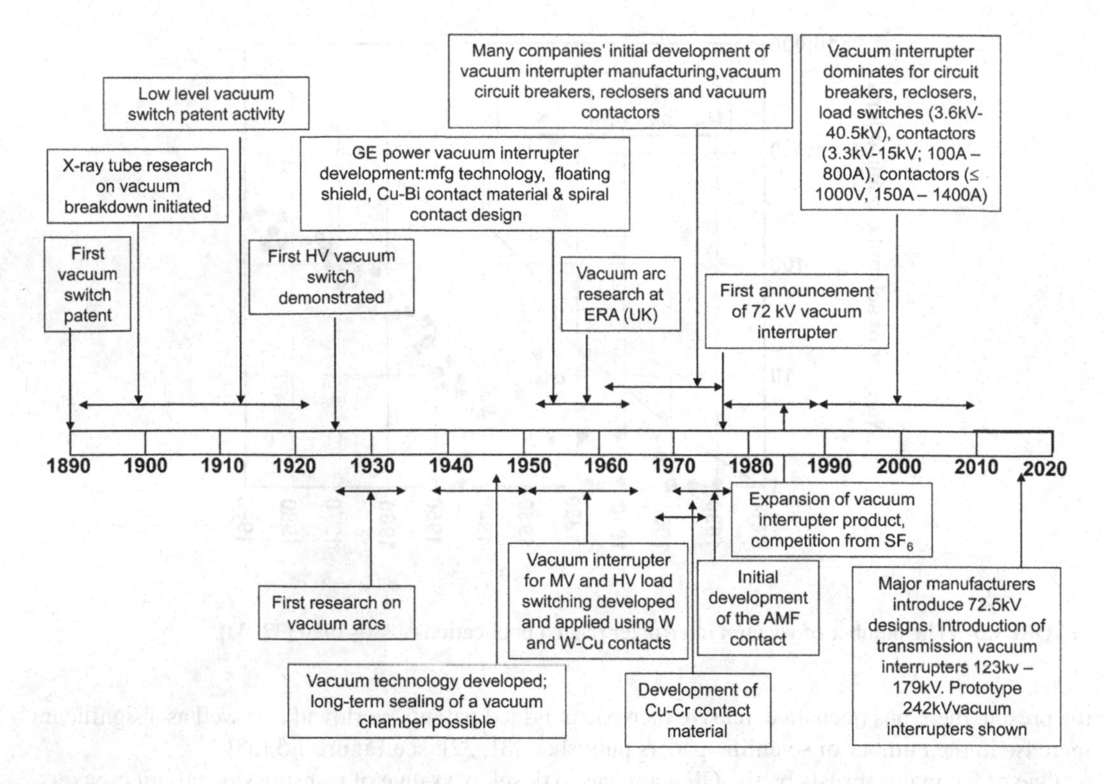

FIGURE I.6 A timeline of the significant developments in vacuum interrupter technology.

from 20 countries in Europe, Asia, and the Americas. As can be seen from this book, the research on the subject has continued to evolve and has led to a greater understanding of the criteria needed to design successful vacuum interrupters. It has also resulted in the development of cost-effective vacuum interrupters that are now dominant in switching and protecting distribution circuits and are now penetrating the sub-transmission and transmission markets. A timeline for the progression in the development of the vacuum interrupter is shown in Figure I.6.

Lafferty [40] edited the first book discussing vacuum interrupter design, with chapter authors from the original GE development team. Greenwood, a former member of the GE team, published his volume in 1994 [41]. The first chapter of his book has a very thorough historical review of the vacuum interrupter story. Mesyats, Latham, Boxman et al. (editors) and Anders [42–45] have published excellent volumes on the theory of vacuum arcs and vacuum breakdown.

In this book, I have concentrated on the design and the application of the modern vacuum interrupter and have divided it into two major sections. The first section has three chapters (1–3) that discuss the design and manufacture of the vacuum interrupter. The second section also has three chapters (4–6), which cover the general application of the vacuum interrupter.

In Chapter 1, I review the research to understand the vacuum breakdown process and what the vacuum interrupter designer has to consider when developing a design for a given high voltage application. I also present a challenge to the next generation of researchers for a potential research project to determine the exact interaction of the field emission current at the anode. Chapter 2 discusses the vacuum arc and how its appearance changes as a function of current. In Chapter 3, I present an overview of the contact materials that have been developed and summarize their advantages and disadvantages. I then present an analysis of vacuum interrupter contact design, which has proved to be essential along with the correct contact material in the development of successful, high current and high voltage, vacuum interrupters. I conclude this chapter with considerations for the manufacture of vacuum interrupters.

Chapter 4 begins the application section. Here I discuss the interruption process for low current and high current vacuum arcs. I also examine the voltage escalation event that can occur if the contact gap is very small at the ac current zero. The chapter concludes with a discussion on the phenomenon of contact welding. It presents a model of the threshold welding current as a function of closed contact force for contacts which have to remain closed while passing high short circuit currents. In Chapter 5, I review the application of vacuum interrupters to switch load currents only. Here I examine the effect of emission currents on the restriking of the open contact gap after switching a capacitor circuit. I also present another challenge for a potential research project to determine the exact relationship of the emission currents and the restrikes. Finally, in Chapter 6, I present a study of vacuum interrupters for circuit breaker and recloser applications. It includes a discussion on short circuit currents and mechanism design. It concludes with discussions of vacuum interrupters applied in series, vacuum interrupters for transmission voltages and the application of vacuum interrupters to switch high voltage dc circuits. Each chapter has an extensive bibliography for those readers who wish to delve further into the subjects presented.

Based upon my continued experience in vacuum interrupter development and design garnered over nearly five and a half decades, I would expect to see the range of applications for this technology to increase. Certainly, development at higher voltages will continue to be driven by the need for environmentally benign interrupting systems: i.e., reducing the need to apply circuit breakers that use the greenhouse gas SF_6. I also would expect that the development of precise mechanisms coupled with electronic control will permit an even wider range of applications.

When performing technology strategic planning for my own R&D team during my years with the Eaton Corporation 1993 to 2007, I had always projected the technology that might replace the vacuum interrupter ten years into the future: i.e., the "New Interrupter Technology." This category includes the use of Insulated-Gate Bipolar Transistors (IGBTs) to switch high currents and high voltages [46]. These devices continue to evolve they find increasing use in wind turbines, solar inverters, electric automobiles and have been proposed for transmission circuit breaker and high voltage dc circuit breaker applications. As of today, however, I still do not see them as a viable alternative to the vacuum interrupter even in 10 years' time. Also, considering the time it took for the vacuum interrupter to be accepted, if a new technology does appear, I suspect it may well take 50 years before it too becomes generally accepted by the electrical distribution and transmission industry.

REFERENCES

1. Enholm, O.A., "Device for transforming and controlling electric currents", *U.S. Patent No. 441,542*, November 1890.
2. Hayes, W.C., (T&D Editor), "An old technology is coming of age", *Electrical World*, (Transmission & Distribution), pp. 64–68, August 1971.
3. Moore, D.M., "Vacuum switches", *U.S. Patent Nos.: 593,230; 604,681-604,687; 672,452, and 702,318* from November 1897 to June 1902.
4. Gerdien, German Patent No. 351,809, 1920.
5. Sorensen, R., Millikan, R., "Electrical switches", *British Patent No. 291,815*, December 1926 and *Canadian Patent No. 275,004*, December, 1927.
6. Sorensen, R., Mendenhall, M., "Vacuum switching experiments at California Institute of Technology", *Transactions A.I.E.E.*, 45, pp. 1102–1105, 1926.
7. Tanberg, R., "The motion of an electric arc in a magnetic field under low pressure gas", *Nature*, 124(3123), pp. 371–372, September 1929.
8. Tanberg, R., "A theory of the electric arc drawn under low gas pressures", *Westinghouse Corporation Scientific Paper*, No. 408, December 1929.
9. Tanberg, R., "On the cathode of an arc drawn in vacuum", *Physical Review* , 35(9), pp. 1080–1089, May 1930.
10. Tanberg, R., Berkey, W., "On the temperature of the cathode in a vacuum arc", *Physical Review*, 38, p. 296, 1931.

11. Berkey, W., Mason, R., "Measurements on the vapor stream from the cathode of a vacuum arc", *Physical Review*, 38(5), p. 943, September 1931.
12. Slepian, J., Mason, R., "High velocity vapor jets at cathodes of vacuum arcs", *Physical Review*, 37(6), p. 779, March 1951.
13. Davis, W., Miller, H., "Analysis of the electrode products emitted by direct current arcs in a vacuum ambient", *Journal of Applied Physics*, 40, pp. 212–222, April 1969.
14. Plyutto, A., Ryzhkov, V., Kapin, A., "High speed plasma streams in vacuum arcs", *Soviet Physics-JETP*, 20, pp. 328–337, February 1965.
15. Escholz, O., "Vacuum circuit breaker", *U.S. Patent No. 1,819,914*, August 1931, and *U.S. Patent No. 1,901,639*, March 1933.
16. Greenwood, T., "Electrically operated circuit interrupter", *U.S. Patent No. 1,801,736*, April 1931.
17. Rankin, W., Hayward, C., "Vacuum switch", *U.S. Patent No. 2,027,836*, January 1936, and Rankin, W., "Vacuum circuit interrupter", *U.S. Patent No. 2,090,519*, August 1937.
18. Siemens Company, "Vacuum switches", *British Patent, No. 389,463*, March 1933.
19. Jennings, J.E., has many patents between, 1950 and the 1960's, including U.S. Patent Nos.: 2,740,867, April 1956; 2,794,885, June, 1957; 3,026,394, March, 1962.
20. Ross, H., "Vacuum switch properties for power switching applications", *Transactions A.I.E.E. Power Apparatus and Systems*, 77(3), pp. 104–117, April 1958.
21. Jennings, J., Schwagner, A., Ross, H., "Vacuum switches for power systems", *Transactions A.I.E.E., Power Apparatus and Systems*, 75, pp. 462–468, June 1956.
22. Ross, H., "Switching in vacuum – Interruption of high voltage power at moderate currents to 4000 amperes and possibly 10,000", *Proceedings of the International Research Symposium on Electric Contact Phenomena*, pp. 345–366, November 1961.
23. Schwager, H., "Encapsulated vacuum insulated circuit breaker", *US Patent, No. 2,870,298*, June 1959.
24. Pearson, J., Kowalczyk, M., "Improvements in vacuum electric switches", *British Patent. No. 839,083*, June 1960.
25. Slade, P., "The future of T&D circuit breaker technology volume 1: Methodology, summary and conclusions", *Westinghouse Report, No. 79-1C7-FORTA-R1*, December, 1979.
26. Lythall, R., *JSP Switchgear Book*, (pub. Newnes – Butterworth) 7th Edition, Chapter 13. "Small oil volume circuit breaker", pp. 418–432, 1989.
27. Cobine, J., "Research and development leading to the high-power vacuum interrupter – An historical review", *IEEE Transactions on Power Apparatus and Systems*, 82(65), pp. 201–217, April 1963.
28. Lee, T., Greenwood, A., Crouch, D., Titus, C., "Development of power vacuum interrupters, *A.I.E.E. Transactions Power Apparatus and Systems*, 81, pp. 629–639, February 1963.
29. Reece, M.P., "The vacuum switch, part 1: Properties of the vacuum arc" and "the vacuum switch, part II: Extinction of an a.c. vacuum arc", *Proceedings of the Institution of Electrical Engineers*, 110, pp. 796 – 811, April 1963.
30. Peach, N. (Assoc. Editor), "Vacuum interrupters", *Power*, pp. 189–196, September 1967.
31. Selzer, A., "Vacuum interruption – A review of the vacuum arc and contact functions", *IEEE Transactions on Industry Applications*, IA-8(6), pp. 707–722, November/December 1972.
32. Miller, H.C., "Bibliography: Electrical discharges in vacuum 1877–1979", *IEEE Transactions on Electrical Insulation*, 25, pp. 765–860, 1990: and "Bibliography: Electrical discharges in vacuum 1980-1990", *IEEE Transactions on Electrical Insulation*, 26, pp. 949–1043, 1991.
33. Boxman, R., Goldsmith, S., Greenwood, A., "Twenty-five years of progress in vacuum arc research and utilization", *IEEE Transactions on Plasma Science*, 25(6), pp. 1174–1186, December 1997.
34. "Vacuum circuit breaker at 132kV", *Engineering*, p. 996, June 1967.
35. Reece, M., "Improved electric circuit breaker comprising vacuum switches", *British Patent 1,149,413*, April 1969.
36. Umeya, E., Yanagisawa, H., "Vacuum interrupters", *Meiden Review*, 45(1), pp. 3–11, Series No, 1975.
37. Kurtz, D., Sofianek, J., Crouch, D., "Vacuum interrupters for high voltage transmission circuit breakers", *IEEE Power Engineering Society Winter Meeting*, Paper No. C 75 054-2, January 1975.
38. Shores, R., Phillips, V., "High voltage vacuum circuit breakers", *IEEE Transactions on Power Apparatus and Systems*, PAS – 94, pp. 1821–1831, 1975.
39. Miller, H., "History of the International symposium on discharges and electrical insulation in vacuum (ISDEIV)", *IEEE Electrical Insulation Magazine*, 21(2), pp. 30–37, March/April 2005.
40. Lafferty, J. (Editor), *Vacuum Arcs: Theory and Application*, (pub. John Wiley and Sons). ISBN 0 471 06506 4, 1980.
41. Greenwood, A., *Vacuum Switchgear*, (pub. IEE, London) ISBN 0 05296 855 8, 1994.

42. Mesyats, G., *Cathode Phenomena in a Vacuum Discharge: The Breakdown: The Spark and the Arc*, (pub. Nauka). ISBN 5 02 022567 3, 2000.
43. Latham, R., *High Voltage Vacuum Insulation: Basic Concepts and Technological Practice*, (pub. Academic Press). ISBN 0 12 437175 2, 1995.
44. Boxman, R., Martin, P., Sanders, D. (Editors), *Handbook of Vacuum Science and Technology: Fundamentals and Applications*, (pub. Noyes). ISBN 0 8155 1375 5, 1995.
45. Anders, A., *Cathodic Arcs: From Fractal Spots to Energetic Condensation*, (pub. Springer), ISBN 0 3877 9107 8, 2009.
46. https://en.wikipedia.org/wiki/Insulated-gate_bipolar_transistor.

Part 1

Vacuum Interrupter Theory and Design

1 High Voltage Vacuum Interrupter Design

We cannot pretend to offer proofs. *Proof* is an idol before whom the pure mathematician tortures himself. In *Physics* we are generally content to sacrifice before the lesser shrine of plausibility.

Arthur Eddington

1.1 INTRODUCTION

There are two major criteria to consider when designing a vacuum interrupter for high voltage:

1. The ability to support high voltage across the vacuum interrupter when the contacts are open
2. The ability to interrupt ac and or dc current over a wide range of circuit conditions

These two criteria are of course linked especially in the design for circuit interruption (this will be discussed in Chapters 3 and 4 in this volume). In this chapter, the high voltage design will be discussed without reference to interrupting current. What drives the high voltage design? The simple answer is that once a vacuum interrupter in an electrical circuit opens its contacts, it is expected to permanently withstand the voltage imposed across the contacts by that circuit. For the vacuum interrupter designer, however, the answer is somewhat more complicated. For a design to be successful, the vacuum interrupter, when placed in an operating mechanism, must pass a series of voltage tests. Committees around the world consisting of equipment users and equipment manufacturers have set the criteria for these high voltage tests. These committees have determined the voltage levels by consensus decisions over the last century. In this chapter I shall consider only the values given by the IEC (International Electro-Technical Commission), IEEE (American Institute of Electrical and Electronic Engineers) and the GB-DL (Chinese National Standard [GB] and Chinese Department of Electrical Power Standard [DL], which are similar to the Russian (Gost) standards) [1–4]. These now represent the major standards to which most countries conform even though some countries maintain their own individual variations [5]. These voltage standards therefore give vacuum interrupter designers a set of guidelines on which they must base their designs.

There are two major high voltage tests that are used to ensure that an open vacuum interrupter has a high probability of satisfactorily withstanding any voltage that may appear across it when installed in an ac circuit. The first is the ability to withstand an ac voltage for one minute across the open vacuum interrupter. The value of this withstand voltage is considerably higher than the maximum operating voltage in any given ac circuit. The withstand voltage level, set early in the last century by the American Institute of Electrical Engineers is 2.25 times the rated circuit voltage plus 2000 volts. Table 1.1 gives the present standards, which now show a considerable variation from this simple formula. This ac withstand requirement is very conservative [2–6] ranging from greater than three times the system voltage for lower circuit voltages to two times for higher circuit voltages. Garzon [1] speculates that the higher one minute withstand voltage values were originally adopted in lieu of switching surge tests.

The second test is used to simulate an overvoltage resulting from a lightning strike on the circuit. The voltage pulse has a rise time of 1.2μs to its peak value and a decay time to half this peak

TABLE 1.1
Voltage Ratings According To IEEE, IEC, and GB/DL Standards

Application Standards	Line-to-Line Voltage, kV (rms.)	1 Minute ac Withstand Voltage, kV (rms.)	1 Minute ac Withstand Voltage, kV (peak.)	Basic Impulse Voltage, kV (peak)	3 μsec, chopped-wave Impulse Voltage, kV (peak)	2 μsec, chopped-wave Impulse Voltage, kV (peak)
IEC	3.6	10	14.1	20		
IEC	3.6	10	14.1	40		
IEEE Indoor	4.76	19	26.9	60		
IEC	7.2	20	28.3	40		
IEC	7.2	20	28.3	60		
IEEE Indoor	8.25	36	50.9	95		
IEC	12	28	39.6	60		
IEC	12	28	39.6	75		
GB-DL	12	48	67.9	75		
GB-DL	12	48	67.9	85		
IEEE Indoor	15	36	50.9	95		
IEEE Outdoor	15.5	50	70.7	110	126	142
IEC	17.5	38	53.7	75		
IEC	17.5	38	53.7	95		
IEC	24	50	70.7	95		
IEC	24	50	70.7	125		
IEEE Indoor	27	60	84.8	125		
IEEE Outdoor	25.8	60	84.8	150	172	194
IEC	36	70	99.0	145		
IEC	36	70	99.0	170		
IEEE Indoor	38	80	113.1	150		
IEEE Outdoor	38	80	113.1	200	230	258
GB-DL	40.5	95	134.4	185		
IEEE Outdoor	48.3	105	148.5	250	288	322
IEC	52	95	134.3	250		

(Continued)

TABLE 1.1 (CONTINUED)
Voltage Ratings According To IEEE, IEC, and GB/DL Standards

Application Standards	Line-to-Line Voltage, kV (rms.)	1 Minute ac Withstand Voltage, kV (rms.)	1 Minute ac Withstand Voltage, kV (peak.)	Basic Impulse Voltage, kV (peak)	3 μsec, chopped-wave Impulse Voltage, kV (peak)	2 μsec, chopped-wave Impulse Voltage, kV (peak)
IEC	72.5	140	198.0	325		
IEEE Outdoor	72.5	160	226.3	350	402	452
IEC	72.5	160	226.3	350		
		175	247.5			
IEC	123	260	367.7	550		
		280	396			
IEC	145	310	438.4	650		
		325	459.6			
IEC	242	425	601	900		
		425	601			

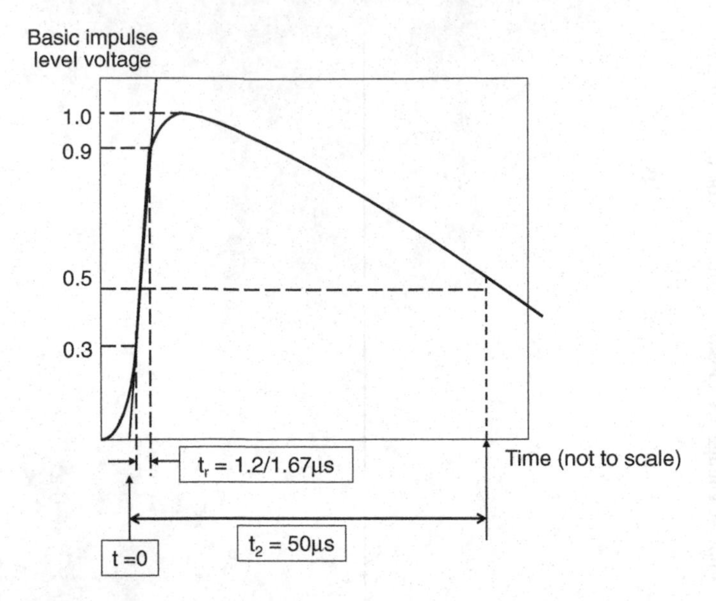

FIGURE 1.1 The basic impulse level (BIL) or lightning impulse withstand value (LIWV) voltage wave shape.

value in 50μs; see Figure 1.1. The 1.2μs value is defined as 1.67 times the time interval t_r that is the time for the pulse to go from 30% to 90% of its final peak value. The 50μs time begins where the straight line joining the 30% and the 90% voltage values intersects with the time axis and ends at the time where the voltage has declined to its 50% level. This Basic Impulse Level (BIL) or Lightning Impulse Withstand Voltage (LIWV) actually reflects the insulation coordination practices used in the design of electrical systems. Table 1.1 gives the peak voltages for the BIL tests. For some IEEE voltages two BIL levels are given. The lower one is for applications on a grounded wye distribution system equipped with surge arresters [2]. The IEC standards specify two values for all voltage classes up to 36kV [3]. In practice the vacuum interrupter designer uses the higher value at each voltage. In contrast with the conservative ac withstand voltage test, the BIL test only defines a limit for the switching system and defines the level of system coordination required when higher voltage pulses are impressed on a circuit by lightning strikes.

In the IEEE standards for outdoor circuit breakers another impulse test is required. This is known as the "chopped wave" test [1, 2]. Here the initial voltage pulse has the same shape as the BIL voltage but is chopped to zero 2μs or 3μs after t = 0 in Figure 1.1. This test reflects the coordination with a surge arrestor and has a peak voltage value higher than that for the BIL test, see also Table 1.1.

1.2 THE EXTERNAL DESIGN

1.2.1 Electrical Breakdown in Gas

There are two components to the vacuum interrupter's high voltage design. These are the internal and the external design. The voltage that is impressed across the open contacts in the vacuum is also impressed across the end plates and the terminals of the vacuum interrupter; see Figure 1.2. To a first approximation the outside of the vacuum interrupter consists of two metal plates, which sandwich an insulating cylinder. For most modern vacuum interrupters this insulating cylinder is high alumina content porcelain. The ambient surrounding the interrupter for a large number of applications is atmospheric air. There are also applications where the vacuum interrupter is surrounded by

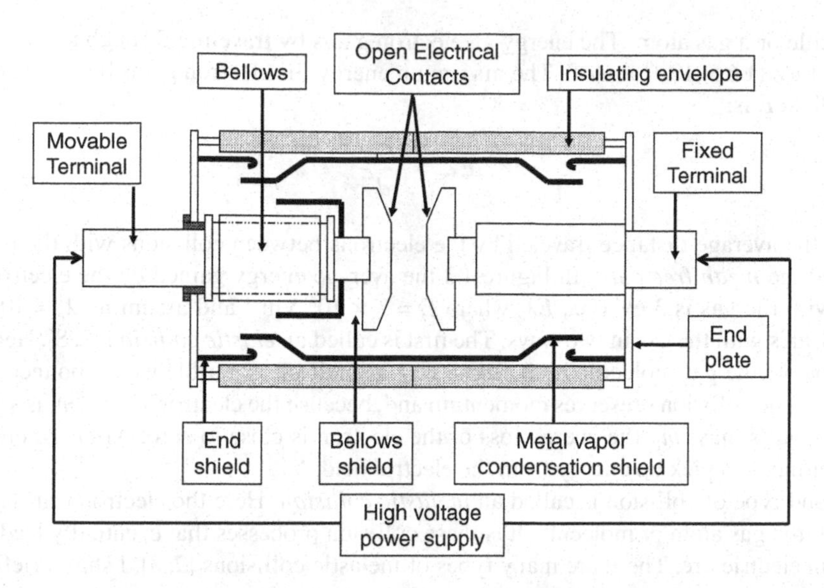

FIGURE 1.2 A cross-section of an open vacuum interrupter with a high voltage power supply connected to the terminals.

other dielectric media such as SF_6, oil and solid insulation. In this section, we shall first consider the external design for operation in air.

Let us consider Figure 1.3. It illustrates possible interactions of an electron with gas molecules and atoms, located between open contacts, which have a constant voltage U impressed across them and are a distance d apart in air at atmospheric pressure. The voltage or potential drop per unit distance i.e., U/d, is called the *electric field E*. An electron introduced into the gas between the contacts, being negatively charged, drifts towards the anode and gains energy before it collides with

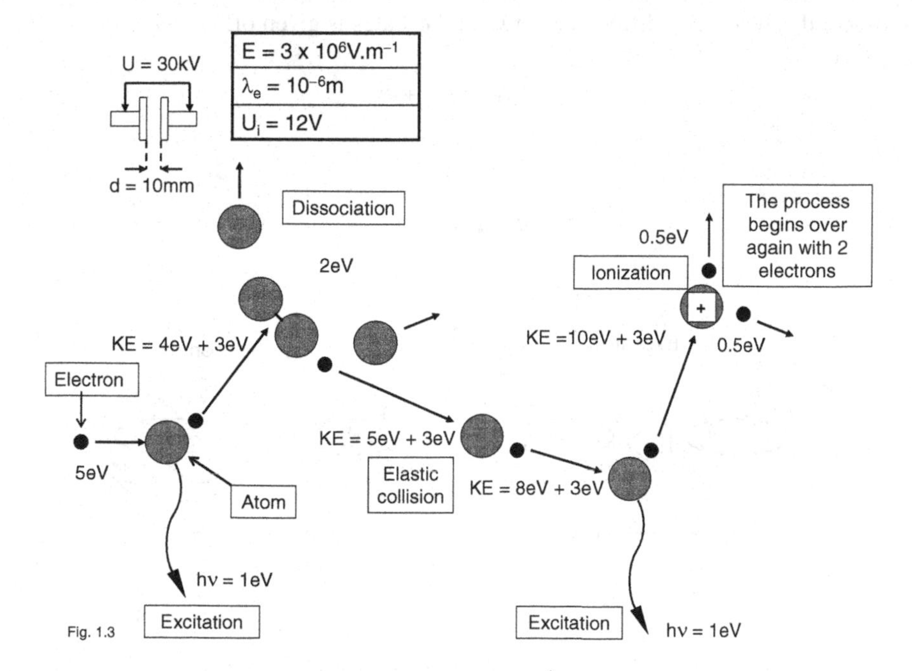

FIGURE 1.3 Possible interactions with a gas of an electron accelerated by an electric field.

a gas molecule or a gas atom. The energy an electron gains by traveling through a one-volt drop in potential is 1 eV (1.6×10^{-19} Joules). The maximum energy the electron gains between collisions in an electric field E is:

$$E\lambda_e = \frac{U\lambda_e}{d} \tag{1.1}$$

where λ_e is the average distance traveled by the electrons between collisions with the ambient gas and is called the *mean free path*. In Figure 1.3 the average energy gained by the electron between collisions with the gas is 3 eV (i.e., $E\lambda_e$ where $E = 3 \times 10^6$ Vm^{-1} and assuming $\lambda_e = 10^{-6}$ m). The electron collides with the gas in two ways. The first is called an *elastic collision* [7, 8]. Here the electron bounces off the gas molecule in the same way a small glass marble would bounce off a large bowling ball. The collision preserves momentum and, because the electron's mass m_e is significantly less than the gas's mass m_g, the energy loss of the electron is close to zero. After the collision, the electron continues to pick up energy from the electric field.

The second type of collision is called an *inelastic collision*. Here the electron transfers some of its energy to the gas atom or molecule. It is these collision processes that eventually lead to the formation of an electric arc. There are many types of inelastic collisions [7, 8]. I shall briefly describe three of them.

1. *Dissociation* is the process of splitting a molecule. For example, a nitrogen molecule *(N_2)* can be dissociated into two nitrogen atoms *(N)*.

$$e + N_2 \rightarrow e + N + N \tag{1.2}$$

Dissociation is possible if the electron energy is greater than the bond strength of the molecule.

2. *Excitation and relaxation* are the processes by which light is emitted from the gas. Figure 1.4 illustrates the process. The impacting electron forces an electron in an atom or a molecule to a higher energy level, leaving it in an excited state. The excited state is usually very unstable and the excited electron returns to its original energy state in less than a nanosecond. The energy difference between the states is given off as radiation.

$$e + N \rightarrow N^* + e \tag{1.3}$$

then

$$N^* \rightarrow N + hv \tag{1.4}$$

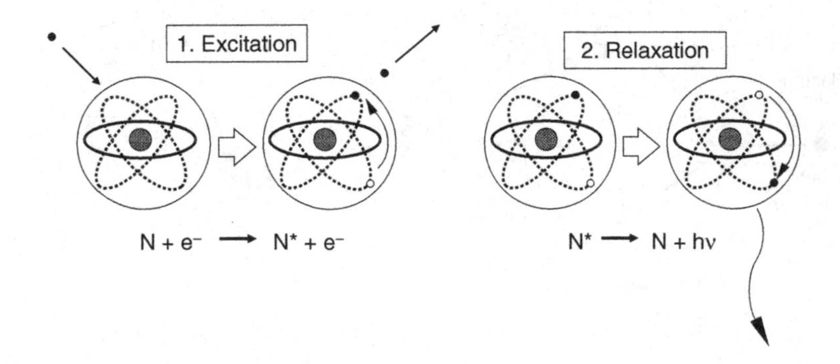

FIGURE 1.4 Excitation and relaxation produce radiation.

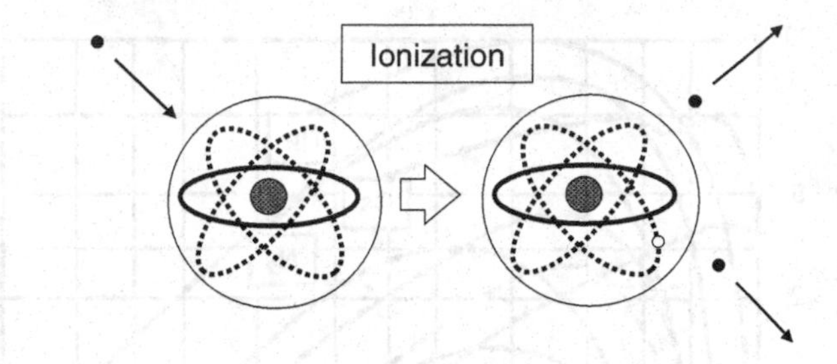

FIGURE 1.5 Ionization releases an electron from the atom leaving a positive ion.

where N^* is the excited state, h is Planck's constant, and ν is the radiation's frequency.

3. *Ionization* is the process that directly results in an electric arc. If the impacting electron has an energy great enough to remove electron with the weakest bound from the atom, there is a finite probability that it will do so, see Figure 1.5. This process results in a positive ion (N^+) and two electrons; the original impacting electron and the electron freed from the atom.

$$e + N \rightarrow 2e + N^+ \tag{1.5}$$

The minimum energy needed to remove the weakest bound electron from its normal state in the neutral atom to a distance beyond the sphere of influence of the nucleus is called the *ionization potential,* U_i. Values of ionization potentials, U_i, for some gases and for metal atoms that are used in vacuum interrupter contacts, are given in Table 1.2. Note that this table presents an important observation for the vacuum interrupter designer: U_i for metal atoms is usually much lower than that for gas atoms. An example of the ionization efficiency of electrons (i.e., the number of ions produced per unit distance per unit gas pressure) as a function of their energy is shown in Figure 1.6 [9]. Here we see that for a given gas there is a value of electron energy where a maximum ionization probability occurs. Also, we see that as the electron energy increases beyond 10^3eV the ionization efficiency decreases rapidly. The energy an electron receives depends upon the strength of the electric field E and upon the distance it is accelerated between collisions. Figure 1.3 illustrates one hypothetical

TABLE 1.2
Values of Ionization Potential

Gas	Ionization Potential (Volts)
Air	14
A	15.7
CO_2	14.4
H	13.5
N	14.5
O	13.5
C	11.3
Cu	7.7
Ag	7.6
Cr	6.8
W	8.0
Bi	8.0

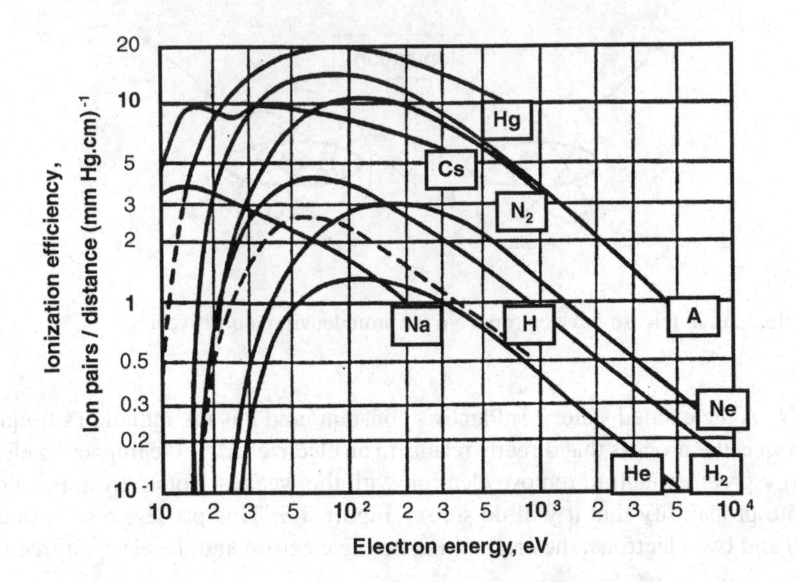

FIGURE 1.6 Ionization efficiency of electrons of uniform energy in a number of gases [9].

sequence of electron/gas interactions in an electric field. In this example an ionization event occurs after five collisions. At this time two electrons are produced and can proceed to pick up energy from the electric field. It only takes 30 such doublings to produce 10^9 electrons i.e., an avalanche of electrons. The number of new ions produced per meter of path by the accelerated electrons is inversely proportional to the mean free path λ_e. If α is the number of ionizing collisions per meter in the direction of the field, then:

$$\alpha = \frac{f(E\lambda_e)}{\lambda_e} \tag{1.6}$$

now $f(E\lambda_e)$ can be represented by [10]

$$f(E\lambda_e) = exp.\left(\frac{-U_i}{E\lambda_e}\right) \tag{1.7}$$

and

$$\frac{1}{\lambda_e} = Ap \tag{1.8}$$

where p is the gas pressure and A is a constant. Thus

$$\alpha = Ap.exp.\left(\frac{-AU_i}{E/p}\right) \tag{1.9}$$

α is called the first Townsend ionization coefficient [7, 8]. Values for λ_e are given in Table 1.3. If two contacts are separated by a distance d and if n_o electrons per square meter are liberated from the cathode initially then the number n_1 of electrons per square meter arriving at the anode is:

$$n_1 = n_0 e^{\alpha d} \tag{1.10}$$

TABLE 1.3
Mean Free Path of Molecules, Ions, and Electrons at Atmospheric Pressure and Room Temperature

Gas	Molecular mean free path, λ_g(nm)	Approximate ion mean free path, $\lambda_i \approx \sqrt{2}\,\lambda_g$ (nm)	Approximate electron mean free path, $\lambda_e \approx 4\sqrt{2}\,\lambda_g$ (nm)
Air	96	135	280–536
A	99	140	295–555
CO_2	61	86	185–340
H_2	184	260	550–1,030
He	296	418	890–1,655
H_2O	72.2	102	215–405
N_2	93.2	132	280–520
Ne	193	273	580–1,080
N_2O	70	99	210–390
O_2	99.5	141	300–560
SO_2	45.7	65	135–225

or, converting this to current [8]:

$$I_1 = I_0 e^{\alpha d} \tag{1.11}$$

However, as Figure 1.7 shows, if the ratio I_l/I_o exceeds a given value, the increase in I_l exceeds the value calculated from Equation (1.11). As the current increases by the ionization process, the ions, ever increasing in number, drift toward the cathode. When these ions reach the cathode, they help liberate more electrons and thus increase I_o. The extra electrons liberated give rise to a secondary

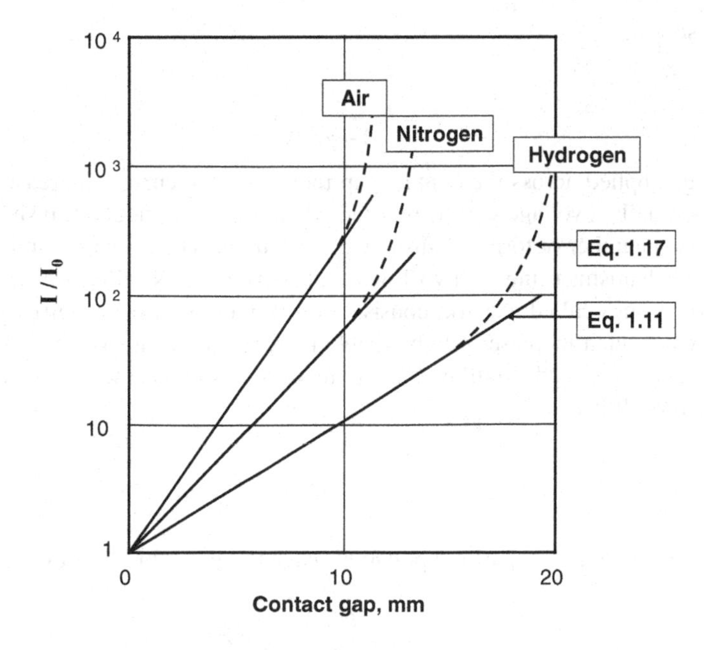

FIGURE 1.7 Typical $\log_e(I/I_0)$ vs. contact gap curves for Townsend breakdown in air, nitrogen and hydrogen [8].

ionization process that causes the current to increase faster than in Equation (1.11). There have been other secondary processes formulated for the further liberation of electrons from the cathode such as photoelectric emission as well as enhanced ionization in the gas itself [8].

If we assume that the extra electrons are liberated by ion bombardment, then suppose a number γ (called the second Townsend coefficient) of new electrons are emitted from the cathode for each of the incoming positive ions. Again, let us assume the following; that n_1 electrons per square meter reach the anode per second, that initially only n_0 electrons leave the cathode per square meter per second and that n_c is the total number of electrons per square meter that are liberated from the cathode from all the effects combined. The number of positive ions in the gas is equal to $(n_1 - n_c)$. The number of electrons leaving the cathode is:

$$n_c = n_0 + \gamma\left(n_1 - n_c\right) \tag{1.12}$$

$$n_c = \frac{n_0 + \gamma n_1}{1+\gamma} \tag{1.13}$$

Using Equation (1.10), the number of electrons reaching the anode is:

$$n_1 = \frac{n_0 + \gamma n_1}{1+\gamma} e^{\alpha d} \tag{1.14}$$

$$n_1 = \frac{n_0 e^{\alpha d}}{1-\gamma\left(e^{\alpha d}-1\right)} \tag{1.15}$$

or converting to current.

$$I_1 = I_0 \frac{e^{\alpha d}}{1-\gamma\left(e^{\alpha d}-1\right)} \tag{1.16}$$

Now exp(αd)>>1 so:

$$I_1 = I_0 \frac{e^{\alpha d}}{1-\gamma e^{\alpha d}} \tag{1.17}$$

Thus, as the voltage applied across the contact gap increases, the current increases as the electric field increases. Eventually a voltage will be reached when there is a sudden transition from a "dark discharge" to one of a number of forms of sustained discharges. The initiation of a discharge based directly upon the mechanisms using the two Townsend coefficients is called *Townsend breakdown.* This transition, sometimes called a spark, consists of a sudden increase of current in the gap and is accompanied by a sudden increase in light visible between the contacts; this spark initiates the arc. From Equation (1.17), I_1 tends to infinity, as γe^{d} tends to 1. Let us assume when the contact gap breaks down or sparks [10]:

$$\gamma e^{\alpha d} = 1 \qquad \text{or} \qquad log_e\left(\frac{1}{\gamma}\right) = \alpha d \tag{1.18}$$

If U_B is the breakdown voltage or sparking potential when the gap d breaks down i.e.:

$$E = \frac{U_B}{d} \tag{1.19}$$

From Equation (1.9), we know that:

$$\alpha = Ap.exp.\left(\frac{-AU_i}{E/p}\right) \tag{1.20}$$

Thus, at breakdown using Equation (1.19) and Equation (1.20):

$$\alpha = Ap.exp.\left(\frac{-AU_i pd}{U_B}\right) \tag{1.21}$$

Now, using Equation (1.18), and assuming breakdown occurs when *I* suddenly increases:

$$log_e\left(\frac{1}{\gamma}\right) = Apd.exp.\left(\frac{-AU_i pd}{U_B}\right) \tag{1.22}$$

$$U_B = \frac{AU_i\{pd\}}{log_e\left[\frac{-A\{pd\}}{log_e(1/\gamma)}\right]} \tag{1.23}$$

Thus, the breakdown voltage U_B for a given gas with an ionization potential U_i is a function of the gas pressure multiplied by the contact gap *(pd)* alone. This is known as Paschen's law, and was discovered in 1889 [11]. Now, $p = nkT$ where *n* is the number of gas molecules per cubic meter, *k* is Boltzmann's Constant (1.38×10^{-23} joules K^{-1}) and *T* is the absolute temperature, thus:

$$U_B = f(pd) or f(nkTd) \tag{1.24}$$

Figure 1.8 shows typical Paschen curves for air and SF_6 with a (pd) in the range of 1 mbar.mm to 10^5 mbar.mm for contacts shaped to have a uniform field between them [12, 13]. It is difficult to measure the true breakdown voltage to the left-hand side of the Paschen curve for contact gaps of a few millimeters (i.e., for *pd* values below about 3 mbar.mm), because the actual breakdown path is not easily controlled. By carefully controlling the breakdown gap, Schönhuber [13] has shown that a breakdown voltage of 20 kV occurs at about 1 mbar.mm. For a (pd) in air over the range 2 bar.mm to 50 bar.mm the breakdown voltage U_B for a contact gap with a uniform electric field is given by:

$$U_B \approx 3.8(pd)^{0.925} kV(\text{where pd is in bar.mm}) \tag{1.25}$$

Over the range of 1×10^2 to 4×10^4 Pa.m (i.e., 1 bar.mm to 4×10^2 bar.mm) from Equation 1.23, the breakdown voltage in air can be calculated from [14]:

$$U_B = \frac{B^*.pd}{log_e pd + C^*} \tag{1.26}$$

where *pd* is in Pa.m, $B^* = 0.3902$ and $C^* = 5.399$. Also shown in Figure 1.8 is the Paschen curve for SF_6 over the range 3 mbar.mm to 10^4 mbar.mm. Equation 1.26 can be used to also calculate the breakdown voltage for SF_6 over the range 3×10^1 to 1.2×10^4 Pa.m using $B^* = 1.48$ and $C^* = 10.71$ [14].

A qualitative description can be used to illustrate Paschen's law. For a given contact gap, at higher pressures (to the right of the minimum value) the electron's mean free path λ is smaller. The electrons, therefore, lose energy through more frequent collisions. In order to ensure a breakdown,

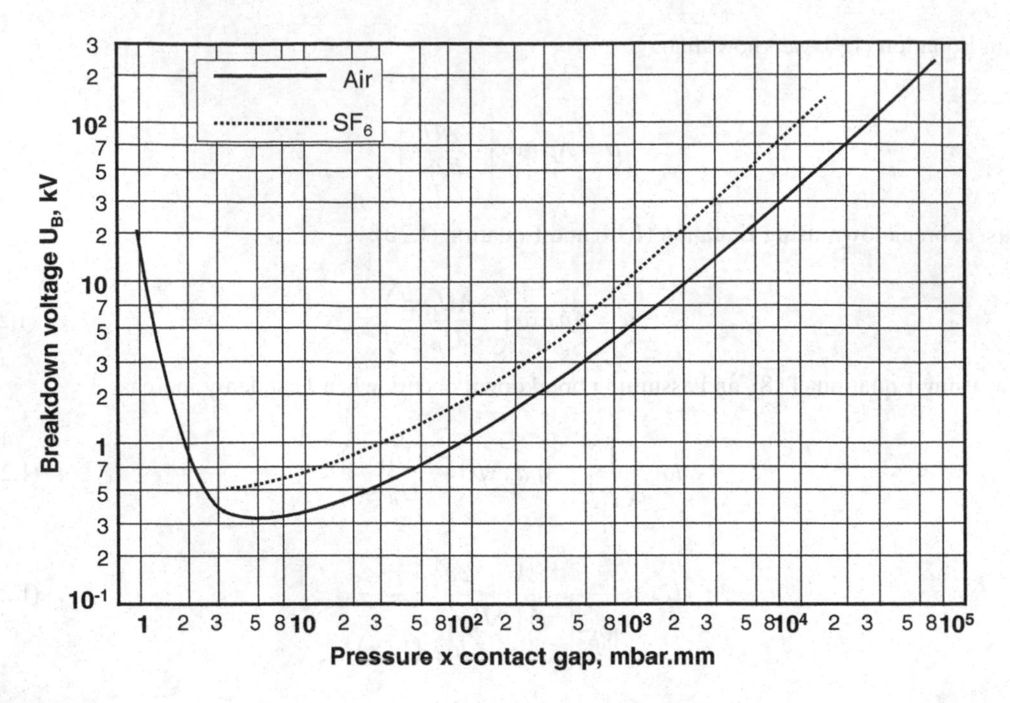

FIGURE 1.8 An example of Paschen curves in air and SF_6 for contacts with a uniform electric field between them [12–14].

the electric field must be high enough for the electrons to gain sufficient energy between collisions. As $E = U_B/d$ the U_B has to increase. At low values of pressure (to the left of the minimum value) the electron can now travel further before hitting an atom, but the probability of impact has decreased enough that each collision requires a higher probability for ionization. For this to occur the electrons must gain more energy from the electric field and thus U_B has to increase again. Table 1.4 gives the minimum breakdown voltage $(U_B)_{min}$ and minimum $(pd)_{min}$ values for various gases. Note that except at $(U_B)_{min}$ there are two values of pd for a given U_B. Thus, for a given U_B and gas pressure there are two possible electrode gaps at which breakdown can occur.

Under some circumstances, therefore, the intuitive answer to an unwanted breakdown of increasing the contact gap and thus increasing the breakdown distance will not always apply. Using Figure 1.8 it is possible to calculate U_B for a given contact gap at atmospheric pressure. It has been shown, however, that at contact gaps of less than 6μm Paschen's Law is no longer valid [15]. In this region

TABLE 1.4
Minimum Breakdown Voltage $(U_B)_{min}$ and $(pd)_{min}$ VALUES

Gas	$(U_B)_{min}$ (volts)	$(pd)_{min}$ (10^{-3} mbar.mm)
Air	327	7.0
A	137	5.1
N_2	251	8.8
H_2	273	1.9
O_2	450	9.3
SF_6	540	2.6

U_B is a linear function of the contact gap and its breakdown mechanism is similar to that of "vacuum breakdown," which will be discussed in Section 1.3.

For breakdown to occur in a gas, three things are necessary:

(i) An electric field to provide energy to the electrons
(ii) Electrons to initiate ionization and the subsequent electron avalanche
(iii) A gas to be ionized and provide the ions necessary for current conduction in the quasi-neutral plasma

In most careful experiments for measuring α and γ, photoemission or thermionic emission from the cathode is used to initiate the current I_0. In most vacuum interrupter applications, the initial electron current for a breakdown across the outside of the interrupter results from electrons that are liberated from the cathode end plate by field emission or by other random physical processes in the ambient gas. There may well be a period of time when U_B across an open contact is exceeded before breakdown begins, but once it is initiated, it can occur extremely quickly, see, for example, Figure 1.9.

After the application of a sufficiently high voltage, the time required for the establishment of an electrical discharge across the outside of an open vacuum interrupter has considerable variability. Once a voltage is impressed across an open contact gap, there are two important time periods that make up the time to breakdown t_B. The first is the statistical time lag, t_{st}. This depends upon the rate at which electrons are produced, the probability that one of these electrons will travel to a region where breakdown can occur and the probability that this electron will begin the avalanche process. In other words, when the first electron becomes available that can begin the breakdown process. The second is the formative time lag t_f. This is the time required for the discharge to become established. The t_f depends upon the overvoltage, the gas, the contact geometry, and the number of initiating electrons [16]. Thus:

$$t_B = t_{st} + t_f \tag{1.27}$$

If the threshold voltage for breakdown is U_B and a voltage $U > U_B$ is impressed across the contact gap, then it has been shown that

$$t_f \propto \theta^{-1} \tag{1.28}$$

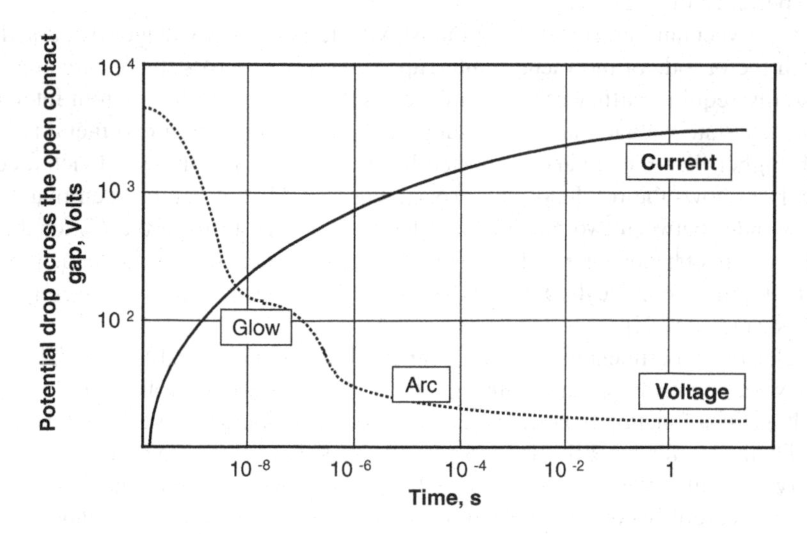

FIGURE 1.9 A schematic representation of time for the breakdown of a contact gap and the formation of an arc in air at atmospheric pressure.

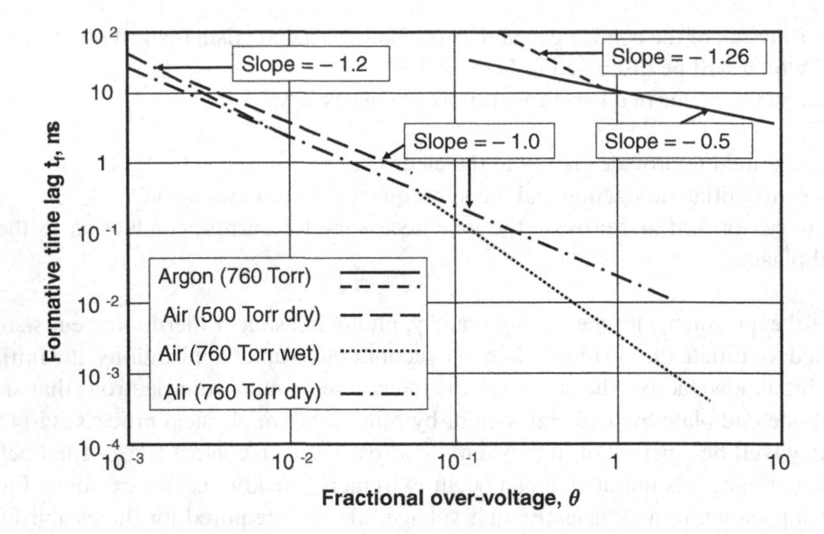

FIGURE 1.10 Plots of formative time lag, t_f, versus fractional over voltage ($\theta = \{U - U_B\} / U_B$) using data compiled by a number of researchers [16].

where $\theta = (U - U_B) / U_B$. Figure 1.10 shows the compilation of experimental data [16]. From this figure, it can be seen that t_f can range from about 4 ns when $U = 2\ U_B$, 400ns when $U = 1.1\ U_B$, 4μs when $U = 1.01\ U_B$ and 40μs when $U = 1.001\ U_B$. Therefore, once a voltage greater than or equal to U_B is impressed across the contacts and the first electron is initiated, the avalanche processes can occur very quickly.

The vacuum interrupter for the most part is applied in a power circuit. Once the breakdown across the outside of the vacuum interrupter is initiated, the potential drop across the resulting arc would be a few hundred volts at the most. Thus, the impedance of the external circuit would be the only limit placed on the value of the arc current. The vacuum interrupter's design, therefore, must be such that breakdown across its exterior surface never takes place. If it does, not only would the vacuum interrupter itself be damaged or even destroyed, but the mechanism in which it resides would also experience the same fate.

The successful vacuum interrupter design must withstand the test voltages discussed in Section 1.1 across both the outside of the vacuum interrupter as well as across the open contacts inside it. In air this usually requires sufficient length of the insulator between the vacuum interrupter's end plates. The steady state voltage that can be supported by a gap between two metal plates in air is usually much higher than the voltage supported by two metal plates that sandwich a ceramic cylinder. Figure 1.11 shows the results of one experiment [10]. Here it can be seen that U_B in air for the ceramic cylinder between two metal electrodes is about one-third of the U_B for the two metal electrodes alone. This phenomenon is the result of electron production at the triple point interface between the plate, the ceramic cylinder, and the air [17, 18] together with the charging effects along the cylinder's surface [19–21].

Li et al. [22] using experimental vacuum chambers shown in Figure 1.12 (One with two ceramics and the other with one ceramic) have determined the BIL breakdown voltage and the one-minute withstand voltage as a function of ceramic length. A compilation of their average data is shown in Figure 1.12. Their data for the BIL breakdown voltage is comparable to that for a ceramic cylinder between two metal plates seen in Figure 1.11. Their experimental vacuum interrupter design does not have the carefully controlled electrodes/contacts and the ceramic cylinder that produced the data shown in Figure 1.11. For example, the joint between the end plate and the ceramic is a braze material whose shape and electrical performance varies on a microscopic scale. In general,

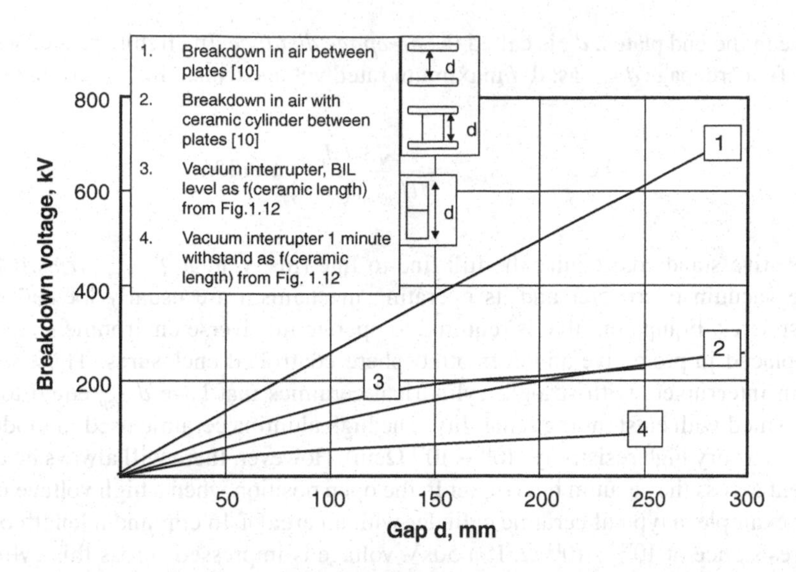

FIGURE 1.11 Breakdown voltage in atmospheric air as a function of contact gap for (1) open contacts with a uniform electric field between them (2) open contacts with a porcelain right cylinder between them [10]: (3) the BIL breakdown voltage across a vacuum interrupter as a function of ceramic length and (4) the one-minute rms withstand voltage across a vacuum interrupter in atmospheric air as a function of ceramic length [22].

Figure 1.12 shows that if the vacuum interrupter can support the BIL voltage externally, it will also be able to hold off the one-minute rms withstand voltage.

1.2.2 Creepage Distance

The ceramic lengths required to support different BIL voltages shown in Figures 1.11 and 1.12 are for clean cylinders in a relatively clean and low humidity environment. The distance along the

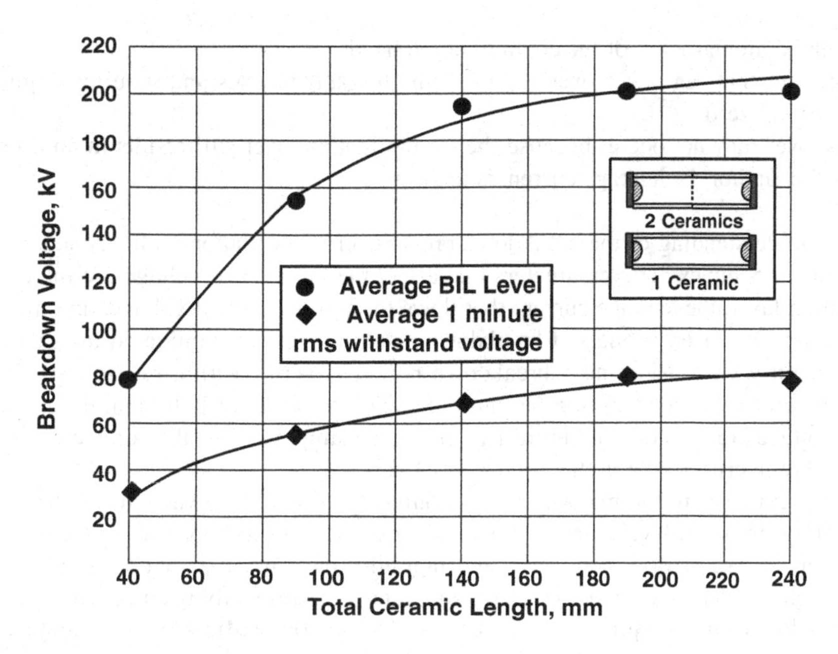

FIGURE 1.12 The average BIL withstand voltage and the average one-minute rms withstand voltage across vacuum interrupter ceramics as a function of the ceramic length [22].

ceramic between the end plates, d_c, is called the *creepage distance*. In circuit breaker standards it is usual to specify a creepage d_{creep} as: d_c / maximum rated voltage (rms.), line to ground (in mm/kV)

$$\text{i.e.,} \qquad d_{creep} = \frac{\sqrt{3} \times d_c}{U_{circuit}\left(rms\right)} \text{mm/kV} \tag{1.29}$$

Some conservative standards require the full line-to-line rms. voltage $[U_{circuit}\ (rms.)]$ for this calculation. The vacuum interrupter and its operating mechanism are usually located in relatively benign atmospheres. Equipment that is required to operate in adverse environments or outdoors is also usually placed in protective and even atmosphere-controlled enclosures. Thus, since the late 1960s vacuum interrupters with straight cylindrical ceramics that have d_{creep} equal to 12–14mm/kV have performed with outstanding reliability. The high alumina ceramic used in modern vacuum interrupters has a very high resistivity ($10^{11} - 10^{14}$ Ωcm). However, there will always be a very small leakage current across the vacuum interrupter in the open position when a high voltage is impressed across it. For example, a typical ceramic cylinder with an area of 16 cm^2 and a length of 16cm will have a total resistance of $10^{11} - 10^{14}$ Ω. If a 50kV voltage is impressed across this cylinder, then a leakage current of $5 \times 10^{-10} - 5 \times 10^{-7}$ A may be measured.

Outdoor high voltage equipment directly exposed to the environment, such as high voltage bushings, has long been concerned with creepage distance [6]. Here if adverse ambient conditions dominate, it is possible that the insulator's surface can have a "dust" deposit that becomes wetted by condensation from a high humidity atmosphere. This can result in a surface layer on the insulator that has a high enough conductivity to allow a measurable surface current to flow across the face of the insulator. In practice, however, the contaminating layer is not usually continuous. In most environments, outdoor equipment operates well with creepage distances of 18–22mm/kV. In heavily contaminated areas, however, such as those exposed to heavy industrial pollution, salt or fog, creepage distances of up to 50mm/kV are sometimes required.

Contaminated insulators with a high voltage impressed across them can exhibit three possible effects:

1) An immediate flashover of the contamination band
2) A flashover that can occur over several ac high voltage cycles and requires reignition at each current zero
3) A flashover may not occur, because the contamination layer still results in an acceptable insulation and/or the leakage current is too low

A qualitative understanding of the breakdown process across insulation with a conducting layer of contamination is complex. The parameters that affect the breakdown voltage are the conductivity of the pollution layer, the leakage current that flows for a given applied voltage, and the continuity of the contamination layer. Models of breakdown resulting from contaminated insulation between two plates begin with considering a breakdown region in series with a resistor, see Figure 1.13 [23–25]. The models proceed to become more complex as bands of pollution and the effect of a varying ac voltage are considered. Figure 1.14 gives an example of how the conductivity of the pollution layer plays a crucial role in these models [24, 25].

As stated previously, in the majority of environments in which the vacuum interrupter is used, consideration of pollution levels and increased creepage lengths are not necessary. In some countries, however, industrial pollution, dust, and humidity have been shown to affect even indoor equipment such as telephone exchanges [26], which have traditionally been considered very clean environments for electrical equipment. The China Electric Power Ministry, for example, has given a series of guidelines for creepage distances required for indoor equipment used in that country; see Tables 1.5, 1.6 and 1.7.

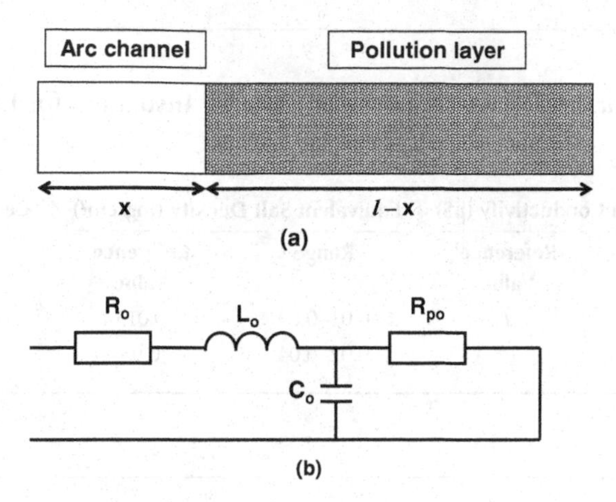

FIGURE 1.13 (a) Model used to analyze the growth of a discharge across a contaminated surface. (b) Electrical circuit model describing the whole discharge propagation [25].

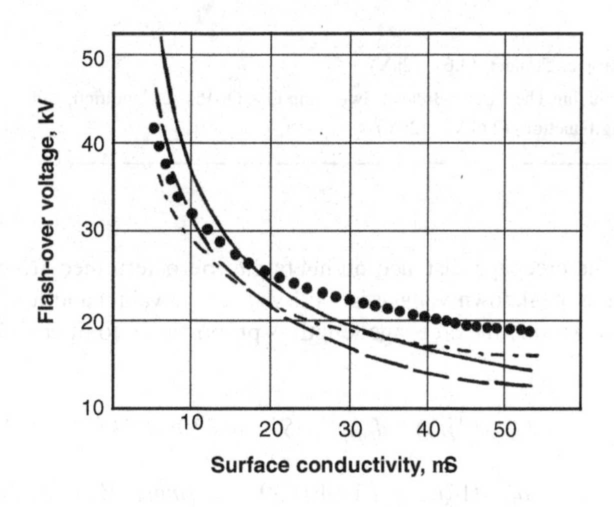

FIGURE 1.14 Flashover voltage vs. surface conductivity [25]: (i) dynamic ac model (––––), (ii) static model (–– ––), (iii) dynamic Sundararajan model (●●●●) [24] and (iv) static Sundararajan model (–– ––) [24].

TABLE 1.5
Creepage Distance for Pollution and Condensation Levels

C_0 little or no condensation	C_1 condensation	C_2 frequent condensation
P_0 no pollution	(≈ 1/month)	(> 2/month)
	P_l light pollution	P_h heavy pollution

Design Class	Condensation × Pollution	Minimum Creepage – mm/kV For Ceramic Insulation
0	C_0P_l	14
1	C_1P_l or C_0P_h	16
2	C_1P_h or C_2P_h or C_2P_h	18

TABLE 1.6
The Maximum Nominal Creepage Distance of External Insulation for Indoor Switchgear

Pollution Class	Pollution Conductivity (μS)		Equivalent Salt Density (mg/cm²)		The Minimum Nominal Creepage Distance For Ceramic Material, (kV/mm)
	Range	Reference Value	Range	Reference Value	
I	5–10	7	0.01–0.02	0.015	14
II	12–16	14	0.02–0.04	0.03	18

TABLE 1.7
Application Factors for the Minimum Nominal Creepage Distance

Applications to Insulator	Applications Factor
Phase to Ground	1
Phase to Phase	$\sqrt{3}$
Across Open Vacuum Interrupter Contacts (3.6 – 72kV)	1.0
Across Isolating Break (Including The Circuit Breaker Switching Out-Of-Phase Condition, and Switches With an Isolating Function) (2.6kV – 72kV)	1.15

The logic behind the creepage distance numbers can be determined from Figure 1.14. For a conductivity of 7 μS the breakdown voltage is ~40kV (average value) and for 14μS the breakdown voltage is ~30kV. If we assume the creepage length is proportional to the breakdown voltage, then from Table 1.6:

$$d_{creep}(14\mu S) = d_{creep}(7\mu S) \times 40/30 mm/kV \tag{1.30}$$

$$d_{creep}(14\mu S) = 14 \times 40/30 = 18.7 mm/kV \tag{1.31}$$

Because the buildup of contamination on the ceramic is a long-term effect, there is no adequate test for the vacuum interrupter. In practice, what has to be done is to take the worst-case creepage standard (i.e., 18mm/kV) and design the vacuum interrupter accordingly. This means for a 12kV (rms.) circuit voltage a minimum ceramic length of 125mm is required (or 216mm if *U(rms)* line-to-line is used in Equation 1.29), and for a 40.5kV circuit voltage a minimum ceramic length of 421mm is needed. If the straight cylindrical ceramic shown in Figure 1.2 were used, the vacuum interrupter would need to be very long in order to satisfy these creepage values. This would negate the advantage of the vacuum interrupter's compact internal design.

Vacuum interrupter designers have thus borrowed design concepts from bushing designers and produced a contour-wave ceramic, which produces a long creep length and yet maintains a compact overall ceramic length. Examples are shown in Figures 1.15 and 1.16. One interesting aspect is that a vacuum interrupter using a contour wave ceramic has a similar outer BIL performance, from end plate to end plate in air, to one that uses a clean straight cylinder ceramic with the same distance in air, end plate to end plate. Thus, while the creepage distance alleviates effects of pollution, the high voltage performance of the vacuum interrupter in air is still mainly determined by the distance between the end plates.

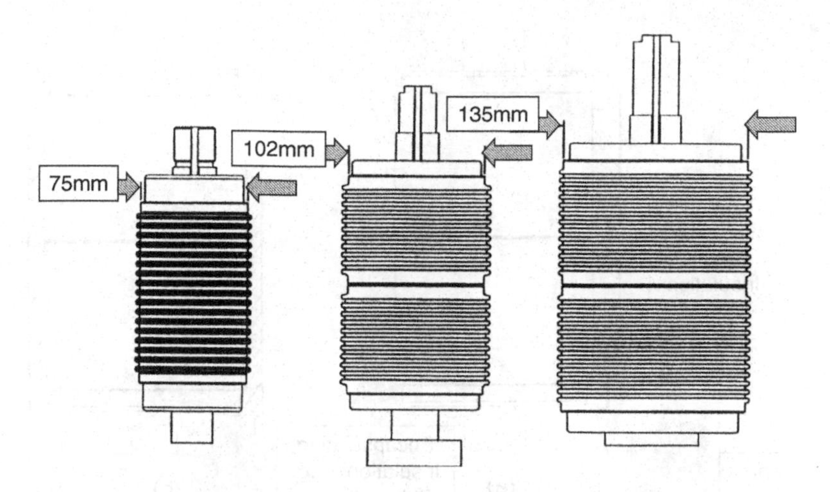

FIGURE 1.15 Contour wave ceramic vacuum interrupter designs for 12 kV application in air.

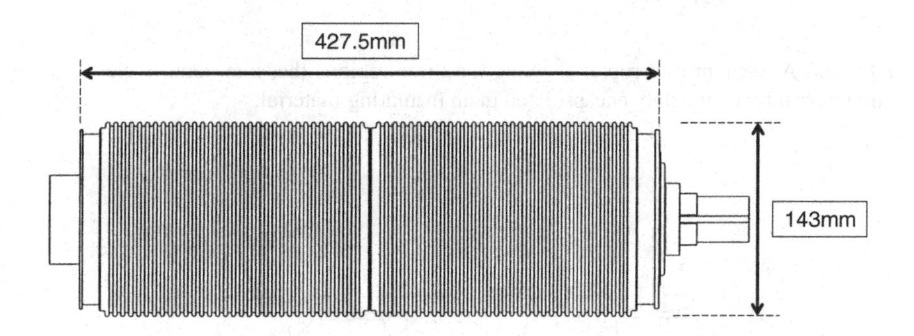

FIGURE 1.16 A contour wave ceramic vacuum interrupter design for 40.5 kV application in air.

1.2.3 Insulating Ambients and Encapsulation

The external BIL and one-minute withstand performance of a given ceramic length can be significantly improved by placing the vacuum interrupter in an ambient that has a higher dielectric strength than does air. Mineral oil, which has a dielectric strength about 5 times that of air, has been used for this purpose since the earliest application of vacuum interrupters in outdoor reclosers [27]. This insulating medium has been extremely reliable and has performed well. Mineral oil, however, has two disadvantages, one is environmental, and two, it is a potential fire hazard. The first disadvantage has now been alleviated with the development of biodegradable oils for high voltage insulation [27, 28]. In recent years vacuum interrupters have been placed in an SF_6 ambient. Usually the SF_6 is at a pressure of just about 1 atmosphere gauge i.e., one atmosphere above the ambient atmosphere. Figure 1.8 shows that for a 10mm contact gap the breakdown voltage in SF_6 at 2 atmospheres is about 5 times that of air at 1 atmosphere. This has permitted the use of relatively short vacuum interrupters in rather compact designs especially at voltages of 24kV, 36kV, and higher voltages [29]. Unfortunately, the realization that SF_6 is a potent greenhouse gas [30, 31] has led to a search for alternative methods of developing compact systems.

A third way of increasing the external high voltage performance of the vacuum interrupter is to encapsulate it in a solid dielectric. Figure 1.17(b) shows an example of a complete encapsulation using polyurethane, silicone rubber, or shrink-wrap. Here the external BIL for this ceramic length is now increased from 110kV to 180kV. In practice total encapsulation of the whole vacuum

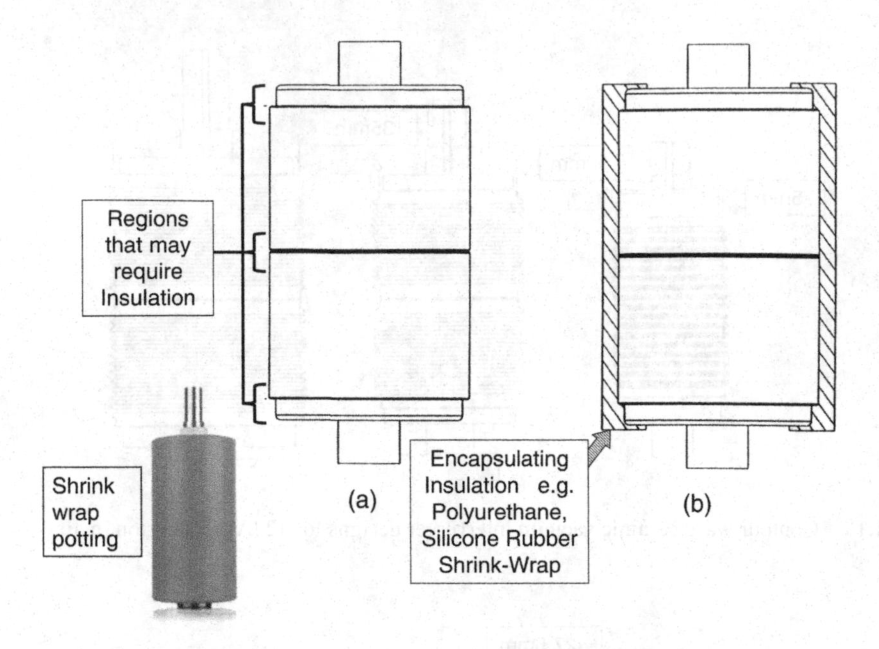

FIGURE 1.17 (a) A vacuum interrupter showing the three regions that may require electrical insulation. (b) A vacuum interrupter completely encapsulated in an insulating material.

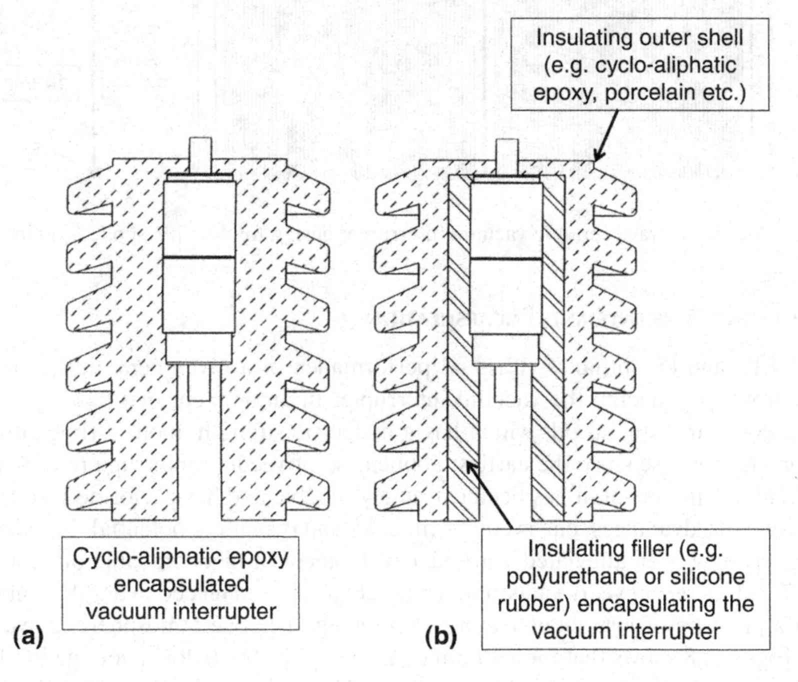

FIGURE 1.18 (a) A vacuum interrupter directly encased in an insulating housing and (b) a vacuum interrupter encapsulated inside an insulating housing with a secondary potting material.

interrupter is not really necessary. The same excellent external high voltage performance can be obtained with a thick band of insulating material at each end of and around the center braze joint of the vacuum interrupter as shown in Figure 1.17(a). Total encapsulation does give a more pleasing appearance and is easier to incorporate into a manufacturing operation. Other coatings have been used, e.g., cyclo-aliphatic epoxy [32] and thermoplastics [33]. It is also possible to mold this

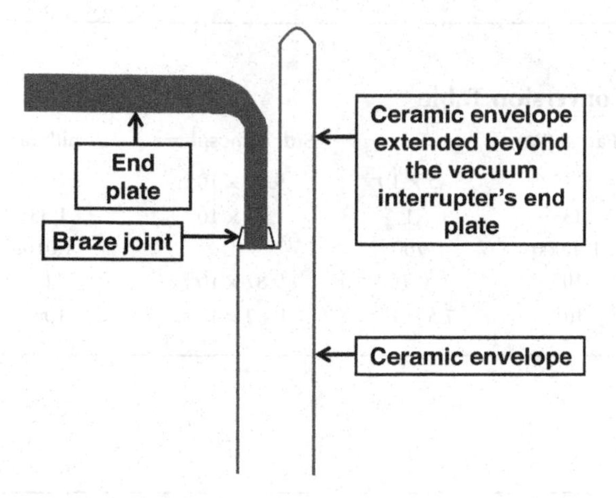

FIGURE 1.19 The extension of the ceramic envelope beyond the vacuum interrupter's end plate to improve its BIL withstand performance [35].

material with "sheds" for outdoor application, see Figure 1.18(a). Another way of using solid insulation is shown in Figure 1.18(b). Here a pre-molded bushing is used. The vacuum interrupter is fixed inside it and the space between the vacuum interrupter and the outer molding is filled with a second dielectric material.

When using a solid encapsulation, it is important to maintain the following, because a poor encapsulation can be worse than no encapsulation at all:

1) The solid encapsulation must form a hermetic seal against the vacuum interrupter. The best hermetic seal is one that is chemically bonded to the vacuum interrupter
2) There should be no voids or small holes in the insulation where high field stresses can be imposed when voltage is applied across an open vacuum interrupter
3) Care must be taken that the solid insulation design be compatible with any thermal expansion of the vacuum interrupter. This is especially important if the solid dielectric material encapsulates not only the vacuum interrupter but also the bus that connects it to the mechanism and to the electrical system [34]. Lamara et al. [35] suggest that extending the ceramic envelope beyond the end plate as shown in Figure 1.19 will also improve the vacuum interrupter's high voltage withstand performance. While this does have the desired effect, filling the space between the ceramic extension and the end plate with an insulating material would improve it even further [36].

1.3 ELECTRICAL BREAKDOWN IN VACUUM

Causa latet, vis est notissima.
[The cause is hidden, but the effect is clear.]

Ovid
(Metamorphoses I)

1.3.1 Introduction

Vacuum technology developed very rapidly in the 1950s with the advent of advances in vacuum pumps, improved sealing techniques, user-friendly leak detection equipment and a wide variety of vacuum compatible materials. These developments made it possible to consider producing a sealed chamber, resulting in a vacuum interrupter that would maintain a good vacuum over its

TABLE 1.8
Units of Pressure – Conversion Table

	Pascal (Pa)	Torr	Std. Atmosphere	millibar	dyne/cm²
1 Pascal = (N/m²)	1	7.5×10^{-3}	9.87×10^{-6}	10^{-2}	10
1 Torr = 1mm Hg	133	1	1.32×10^{-3}	1.33	1,330
1 std. atmosphere	101,000	760	1	1,010	1,010,000
1 millibar (mbar)	100	0.75	9.87×10^{-4}	1	1000
1 dyne/cm²	10^{-1}	7.5×10^{-4}	9.87×10^{-7}	10^{-3}	1

TABLE 1.9
Pressure Ranges in Vacuum Technology and Some Characteristic Features

	Rough Vacuum	Medium Vacuum	High Vacuum	Ultra-High Vacuum
Pressure (mbar)	1,013 1	$1–10^{-3}$	$10^{-3}–10^{-7}$	$< 10^{-7}$
Pressure (Pa)	$10^{5}–10^{2}$	$10–10^{-1}$	$10^{-1}–10^{-5}$	$< 10^{-5}$
Particle number density, n / m³	$10^{25}–10^{22}$	$10^{22}–10^{19}$	$10^{19}–10^{15}$	$< 10^{15}$
Gas mean free path (λ), cm	$< 10^{-2}$	$10^{-2}–10$	$10–10^{5}$	$> 10^{5}$
Monolayer formation time in secs.	$< 10^{-5}$	$10^{-5}–10^{-2}$	$10^{-2}–100$	> 100
Other features	Convection depends on pressure	Marked change in gas thermal conductivity	Marked reduction in volume related collision rate	Surface effects dominate

operating life. The historical development of the vacuum interrupter has already been discussed in the Introduction. You can read more details in a number of sources, e.g., [37–39]. The units by which vacuum pressure are given can be somewhat confusing. Table 1.8 gives the conversion factors. Most of these units will be used in this book. Table 1.9 presents some important physical data for various ranges of vacuum. The vacuum interrupter usually operates in the vacuum range, 10^{-2} to 10^{-4} Pa. Before we can discuss the internal design of the vacuum interrupter it is important to understand what causes a vacuum gap to break down when a high enough voltage is impressed across it.

At first glance a gap in vacuum should easily be able to withstand voltage and not breakdown. Of the three things necessary for breakdown presented in Section 1.2.1, only the first one is obviously available, i.e., the electric field. This results from the voltage impressed upon the open contact gap. Table 1.9 shows that the density of gas in a vacuum interrupter at 10^{-2} to 10^{-4}Pa is low enough that the chance of an electron/atom collision is very unlikely ($\lambda_e \approx 4\sqrt{2}\,\lambda_g$, where λ_g is the mean free path of the gas atom; see Table 1.3). Thus, there is no opportunity for the pre-breakdown electron avalanche to develop in the residual gas between the contacts. If, however, you take a given contact gap in vacuum and gradually increase the voltage, U, across it, breakdown will occur at some value (U_B). If this gap is in a power circuit a conducting arc can form whose current will only be limited by

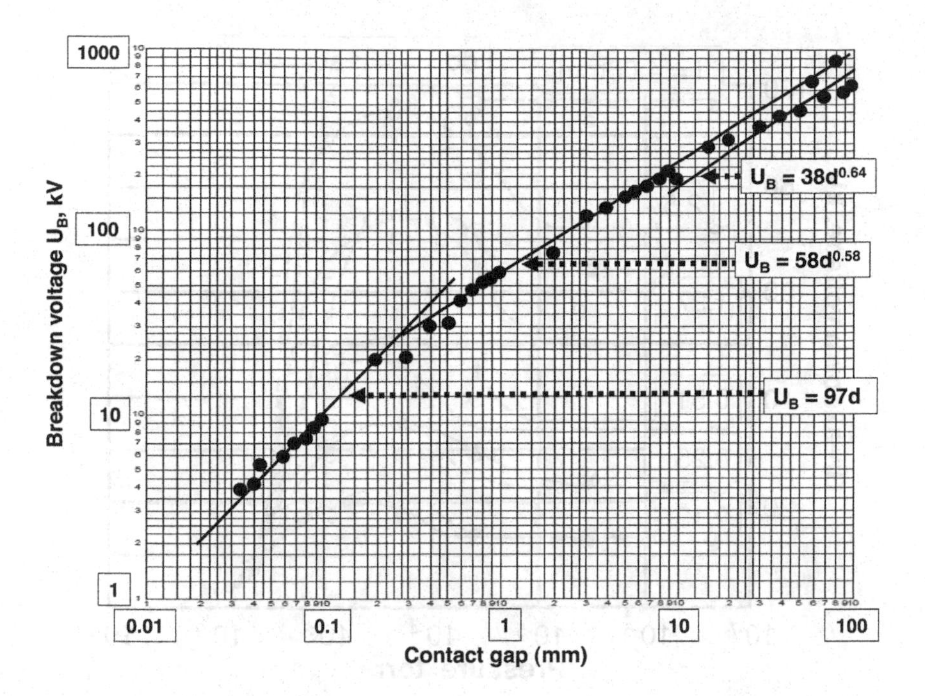

FIGURE 1.20 A log-log plot of the breakdown voltage, U_B, as a function of contact gap in vacuum [40].

the circuit's impedance. Figure 1.20 shows a compendium of the vacuum breakdown voltage, U_B, as a function of the contact gap d from many sources [40]. For small gaps $d < 0.2mm$, $U_B \propto d$ and for larger gaps, $0.3mm < d < 100mm$, $U_B \propto d^{\eta}$ (where η is between 0.5 and 0.7). In Figure 1.20, $\eta \cong 0.58$ for $0.4mm < d < 20mm$ and $\eta \cong 0.68$ for $20mm < d < 100mm$. The dependence of U_B on contact gap is more clearly seen in the linear plot shown in Figure 1.21. Here it can be observed that the value of

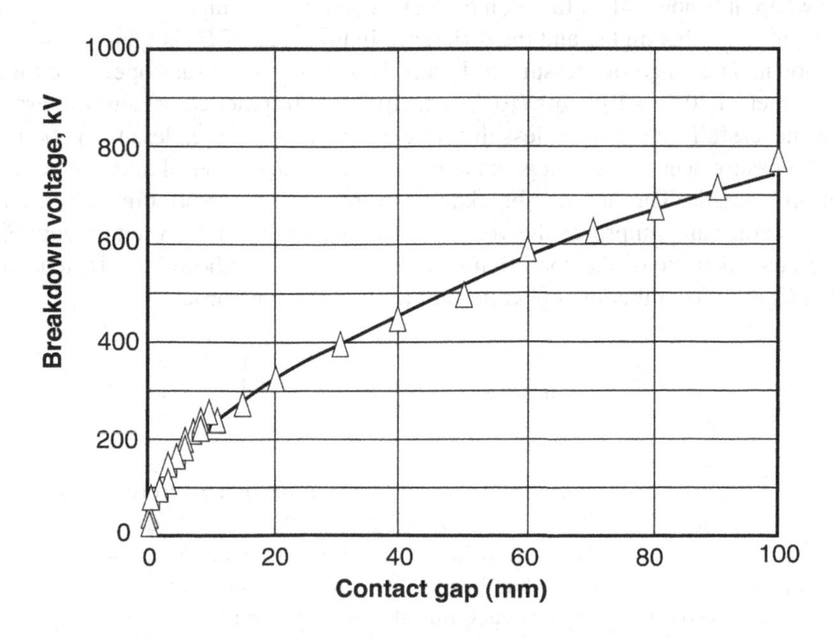

FIGURE 1.21 A linear plot of the breakdown voltage, U_B, as a function of contact gap in vacuum.

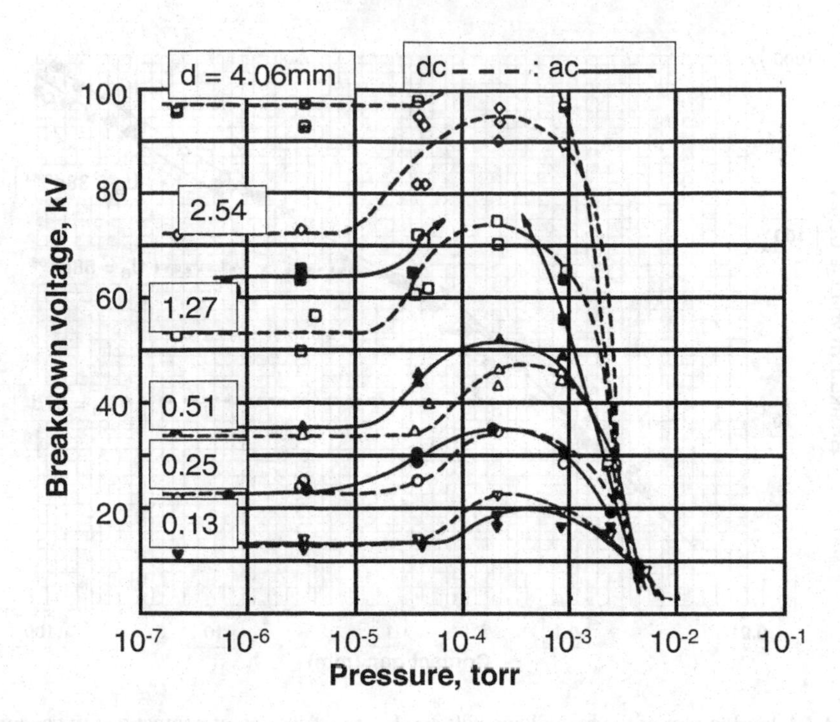

FIGURE 1.22 An example of the ac and dc breakdown voltages as a function of pressure and contact gap for Nb contacts [41].

U_B seems to be reaching a limiting value as the contact gap increases above 100mm. In general, the experiments that have contributed to these data sets use contacts that are carefully cleaned and polished, have a more or less uniform field between them and use a dc high voltage source across them.

The dependence of U_B on pressure or level of vacuum is shown in Figure 1.22 for plane-parallel Nb contacts [41]. It can be seen for both ac and dc voltages that U_B is independent of pressure below 3×10^{-5} torr where it is now only a function of contact gap. In the range 2×10^{-3} torr and 10^{-5} torr the value of U_B is somewhat higher and the difference in the value of U_B is more pronounced at the largest gap shown. The range of pressure in Figure 1.22 covers the usual operating range for the vacuum interrupter of 10^{-4} to 10^{-6} torr (10^{-2} Pa to 10^{-4} Pa). In practice, vacuum interrupters will operate quite successfully at pressures less than or equal to 10^{-3} torr (i.e., less than about 10^{-1} Pa).

The actual measurement of U_B for a given contact gap, contact material and contact preparation is somewhat problematic. The first few breakdown events have a conditioning effect on the open contacts. This conditioning improves the voltage withstand ability of the vacuum gap. Figure 1.23 illustrates the general shape of the conditioning curve [42]. It also shows that U_B tends to reach a limiting value U_L after ň number of applications of the conditioning process, i.e.,

$$U_B \rightarrow U_{BN} = U_1 + U_A \left(1 - e^{-(\check{n}-1)/C^{**}}\right) \tag{1.32}$$

where $U_A = U_L - U_l$ *and* C^{**} is a conditioning factor. This will be discussed further in Section 1.3.7.

The breakdown voltage U_B will have a statistical distribution [43]. The value most often chosen to represent the vacuum breakdown voltage for a particular contact gap is the 50% probability or mean value i.e., U_{B50}. There has been a long debate over which statistical distribution to use. Those of you who have designed vacuum interrupters and other high voltage components know from experience that when you impress a low enough voltage across a given contact gap,

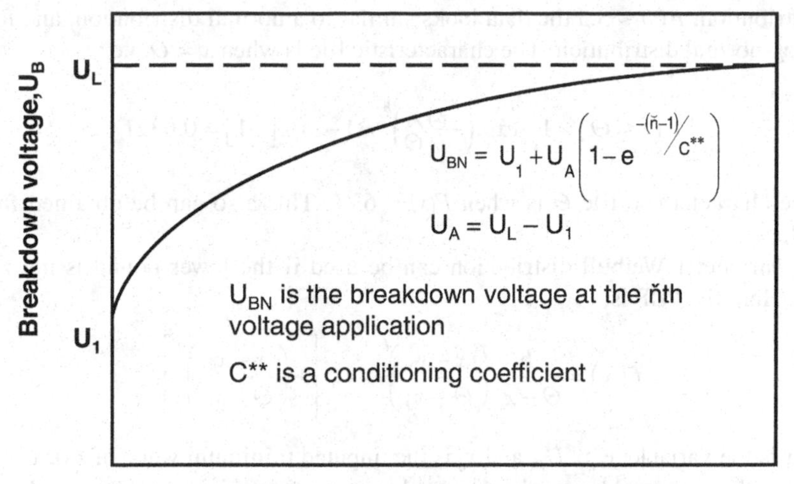

FIGURE 1.23 The typical characteristic of the conditioning curve [42].

the gap will never breakdown. The concept of "the impossibility of a breakdown" below a certain voltage level is, of course, usually quite repugnant to a traditional statistician. A distribution that does take this into account is one introduced by Weibull in the 1930s [44]. Engineers are increasingly using this distribution for failure analysis and for developing high voltage equipment [45–48] In Weibull's approach, the measured sample selects the distribution parameters of the population. The analysis can also include engineering experience and judgment. This is quite different from the traditional statistical analysis where it is assumed that the population is constant, but unknown. The sample statistics are then used as estimators and confidence intervals give errors in these estimates.

For the practicing engineer, the Weibull approach is very attractive. It enables the experienced engineer to use the cumulative years of knowledge together with a limited set of experimental data. Two forms of the Weibull distribution are commonly used: (a) the two-parameter and (b) the three-parameter. The mean and the standard deviation are not the only parameters considered in each of these forms. The two-parameter Weibull distribution assumes that the lower bound is zero probability: e.g., the probability of breakdown is zero when zero volts are applied across a vacuum gap. In this case the *probability density function* (PDF) is given by:

$$f(x)=\frac{b}{\Theta}\left(\frac{x}{\Theta}\right)^{b-1} exp.\left[-\left(\frac{x}{\Theta}\right)^{b}\right] \tag{1.33}$$

where x is the variable (e.g., U_B), b is the Weibull shape or "shape" factor and Θ is the characteristic breakdown value, or "scale" parameter. On integration of the PDF you obtain the *cumulative density function* (CDF):

$$F(x)=1-exp.\left[-\left(\frac{x}{\Theta}\right)^{b}\right] \tag{1.34}$$

The "shape" factor b is obtained from the plot of the cumulative probability vs. the $\log_e$ (of the variable). For $b < 1$ you have infant mortality, $b = 1$ you have random failures, $b > 1$ you have a statistical distribution. At $b \approx 3.5$ the data looks similar to a normal distribution, and for $b \approx 2$ it is similar to a log-normal distribution. The characteristic life is when $x = \Theta$, i.e.:

$$F(x=\Theta) = 1 - exp.\left(-\Theta/_{\Theta}\right)^b = 1 - exp.[-1] = 0.6321 \tag{1.35}$$

Therefore, the characteristic life Θ is when $F(x) = 63\%$. This also can be obtained from the plot described below.

The three-parameter Weibull distribution can be used if the lower bound is not zero, see for example [45]. Here the PDF is:

$$f(x) = \frac{b}{\Theta - x_0}\left(\frac{x - x_0}{\Theta - x_0}\right)^{b-1} exp\left[-\left(\frac{x - x_0}{\Theta - x_0}\right)^b\right] \tag{1.36}$$

where again x is the variable, e.g., U_B, and x_0 is the imputed minimum value of x or U_{B0} i.e., a value of U_B where breakdown would never be expected to occur for a given contact gap. Note, Equation 1.34 is a special case of Equation 1.36 with $x_0 = 0$. The CDF in this case is:

$$F(x) = 1 - exp.\left[-\left(\frac{x - x_0}{\Theta - x_0}\right)^b\right] \tag{1.37}$$

To determine the three parameters:

1. Sort the data in ascending order
2. Estimate x_0 usually between 0 and the lowest value in the sample. This is where experience comes in. It is also possible to estimate this when analyzing the raw data set to determine b and
3. Determine $x - x_0$ for all the sample values
4. Assign median ranks; $F(x)$
5. Determine $X = log_e\,(x - x_0)$ and $Y = log_e\,[-log_e(1/\{1 - F(x)\}]$
6. Perform linear regression fit to X and Y to determine b, Θ, and x_0
7. You can adjust the value of x_0 and repeat the analysis

A similar determination of the two parameters is performed with $x_0 = 0$. Figure 1.24 shows the analysis of a set of vacuum breakdown data [49, 50] using the two-parameter, Figure 1.24(a), and three-parameter, Figure 1.24(b), Weibull distributions. The results of this analysis are shown in Table 1.10 [50]. While the shape factors are different in each case, the scale factors, the mean and the standard deviation are similar. Figure 1.25 shows the probability of breakdown for the two cases. Here it can be seen that for this set of data it is reasonable to assume that the probability of failure is zero for an applied voltage across the contacts of 72kV (i.e., $\cong 0.5\ U_{B50}$). In an analysis of U_B for uniform contact gaps in vacuum, Shioiri et al. [43], show for BIL voltage pulses, U_{B0} could range anywhere from $0.5\ U_{B50}$ to $0.8\ U_{B50}$ depending upon the experimental approach, the contact design and the contact material used. Interestingly in their data the shape parameter "b" is the same for both Cu and stainless steel even though $F(U_B)$ for both materials is quite different. They also show that the way of performing the breakdown measurements also affects the results. Shioir et al. [43] and Hirose [48] use two methods [43, 48]:

1) *The up-and-down method*; here the initial applied voltage is set around the mean value i.e., U_{B50}. If breakdown does not occur then the voltage is increased to a higher value $U_{B50} + \Delta$. If breakdown does occur then the voltage is decreased to $U_{B50} - \Delta$. This up and down procedure continues for a proscribed number of times
2) *The step-up method*; here the initial applied voltage is set to a sufficiently low level e.g., U_{B0} where breakdown is unlikely to occur. The voltage is increased to a higher value $U_{B0} + \Delta$.

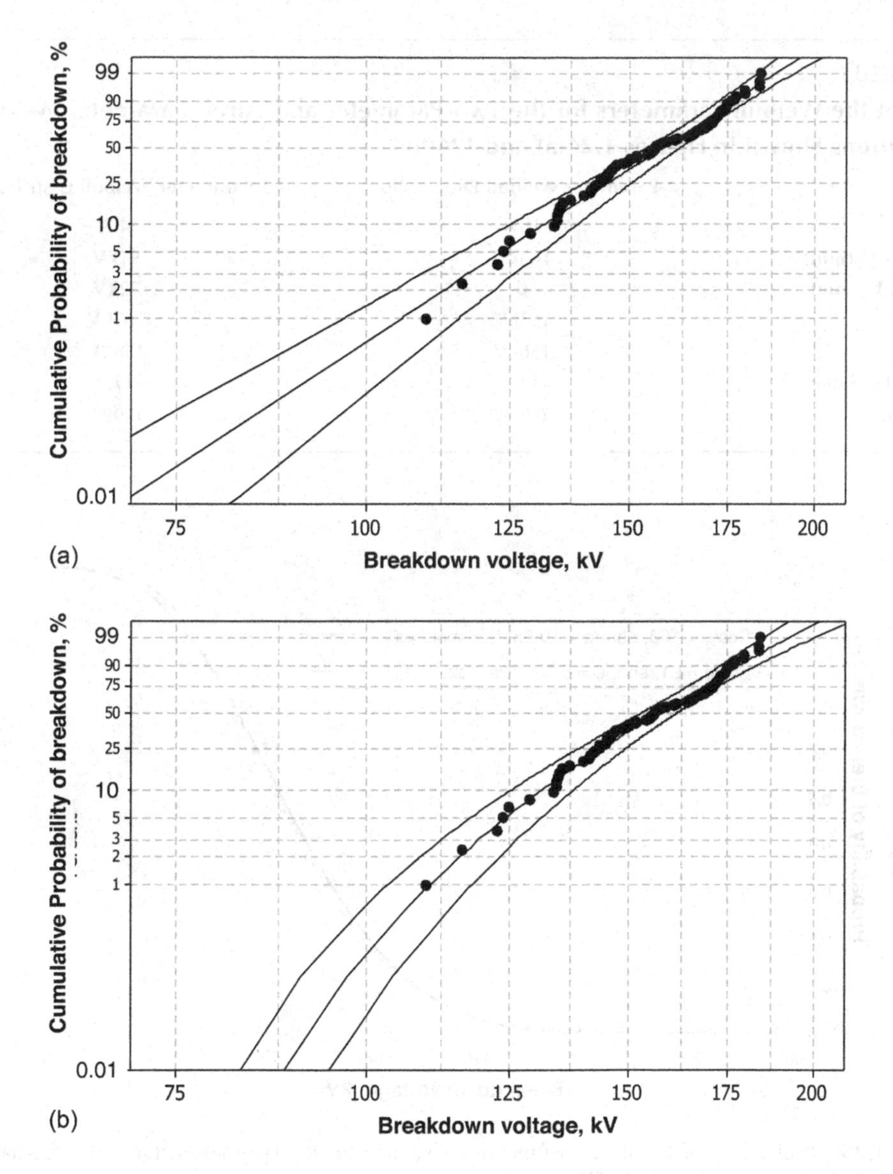

FIGURE 1.24 Cumulative frequency plots for (a) a two-parameter and (b) a three-parameter Weibull analysis of measured vacuum breakdown voltages for conditioned vacuum interrupter with Cu–Cr contacts and a 4mm contact gap [49, 50].

If breakdown does not occur in a set number of voltage applications then it is set at U_{B0} + 2Δ. This procedure is continued until the contact gap will no longer hold off the voltage for the required number of voltage applications.

For example, using a pair of stainless-steel contacts, an analysis of their data with the up-and-down voltage application method gives a $U_{B0} = 0.67\ U_{B50}$, but the step-up method gives a $U_{B0} = 0.52U_{B50}$. Thus, in practice, it would be wise to take a very conservative value for U_{B0} when applying a vacuum interrupter in a situation where the voltage withstand is of critical importance. Liu [51] has used the Weibull analysis to study the effect on vacuum breakdown of switching different levels of current in a practical vacuum interrupter. Sidorov et al. [52] have determined analytically the breakdown statistics for two vacuum gaps in series. They consider both an equal distribution and a non-uniform

TABLE 1.10
Values of the Weibull Parameters for the Two-Parameter and Three-Parameter Weibull Distributions Shown In Figures 1.24(a) and 124(b)

Factors	2-Parameter Weibull Distribution	3-Parameter Weibull Distribution
Shape	10.6	4.8
Scale [(Θ – U_B(min)]	163kV	91kV
Threshold U_B(min)	0	72kV
Θ	163kV	163kV
Mean	156kV	156kV
Standard Deviation	17.6	17.9
Correlation	0.992	0.994

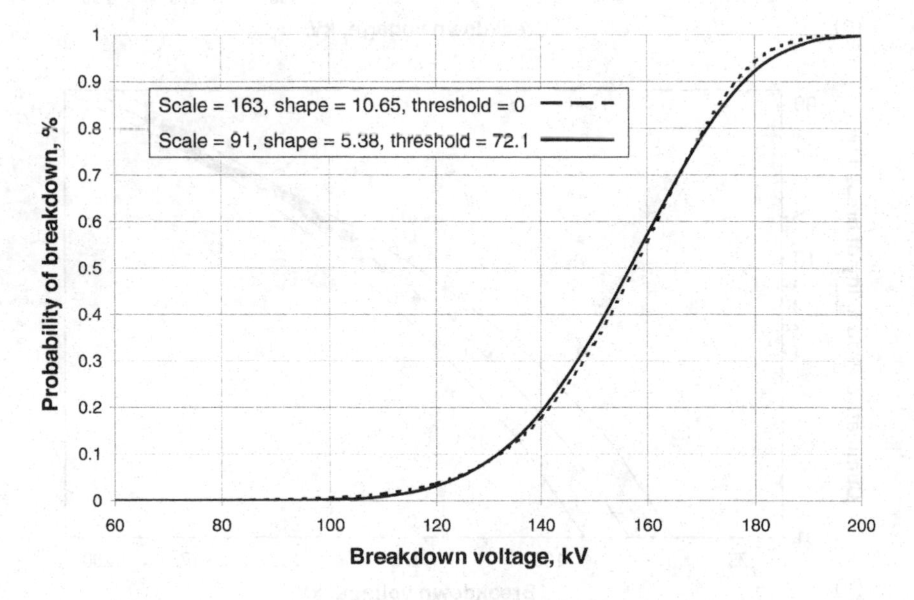

FIGURE 1.25 Probability of failure as a function of voltage for the two-parameter and three-parameter Weibull analysis shown in Figure 1.24 [50].

distribution of voltage across the two vacuum gaps. They show that the breakdown voltage follows the Weibull distribution and that the U_{B50} value for the two gaps in series is approximately equal to the U_{B50} value of gap 1 plus that of gap 2. Their analytical analysis seems to agree reasonably well with their experimental data.

The scientific investigation of vacuum breakdown has a long history starting at the end of the nineteenth century [53]. Farrall's historical review [54] gives a detailed and thorough account of the development of this subject. A thorough background summary of the subject can also be found in four excellent reviews [54–57]; the latest being Latham's excellent book [56]. When discussing vacuum breakdown phenomena and applying the information in the high voltage design of a vacuum interrupter, it is helpful to split the process into four parts:

(i) The electric field;
(ii) The conditions that lead up to the vacuum breakdown, i.e., the pre-breakdown effects;

(iii) The breakdown processes and the transition to the vacuum arc;
(iv) The transition to a self-sustained vacuum arc.

1.3.2 The Electric Field

In this section, I will discuss the electric field between the open contacts and the specific details that must be considered when discussing vacuum breakdown. When studying breakdown in air where the electron avalanche is initiated in the intercontact gas, researchers usually take care to ensure that there is a uniform field in the contact gap by profiling the contact surfaces [10, 58]. This frees the researcher from having to consider the effects of a nonuniform field structure between the contact faces. Thus, when breakdown occurs, the position of the final conducting channel is independent of its position between the contact faces. For studies of vacuum breakdown, it has long been recognized that consideration of the exact electric field is more complicated than just the inter-contact field. Alpert et al. [59] first suggested that the electric field, E, is not determined by U/d where U is the voltage impressed across a gap d, but is given by:

$$E = \frac{\beta U}{d} \tag{1.38}$$

Where β is an enhancement factor. Breakdown occurs when a critical field E_C is impressed across the contacts. Measured values of E_C for contact gaps < 0.2mm are given in Table 1.11 [55, 60, 61]. The enhancement factor has two components; (a) β_m is an enhancement factor resulting from microscopic surface projections and (b) β_g a geometric enhancement factory resulting from a nonuniform field distribution between the two open contacts. Thus:

$$E = \frac{\beta_m \beta_g U}{d} \tag{1.39}$$

where $\beta = \beta_m \times \beta_g$. The differentiation between β_m and β_g can be somewhat problematic, but it is very important for the vacuum interrupter designer to understand the effect of both enhancement factors.

1.3.2.1 The Microscopic Enhancement Factor (β_m)

It is well known that even a seemingly smooth contact surface is anything but smooth when observed on a microscopic level [62]. Figure 1.26 shows examples of surface profiles for three different contact finishes [63]. Care must be taken when interpreting these profiles. For example, the vertical

TABLE 1.11
Critical Vacuum Breakdown E_C at Short Contact Gaps (< 0.2mm) and Work Function φ for Various Metals

Metal	E_c ($\times 10^8$ V.m^{-1})	φ (eV)
Au	64	4.8
Cu	69	4.5
Cr	53	4.6
Ni	99	4.6
Mo	54	4.4
W	67	4.5
Stainless Steel	59	4.4

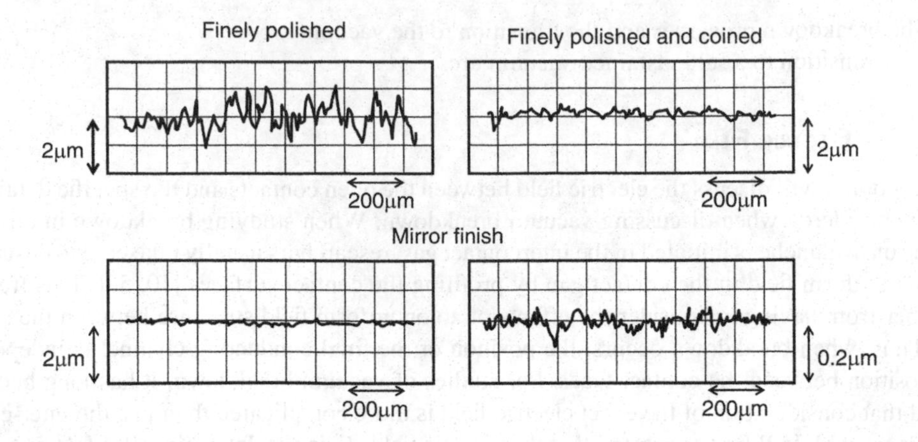

FIGURE 1.26 Measured microscopic surface profiles for fine silver contacts with different surface treatments [63].

scale for the finely polished cases is 0.01 times the horizontal scale. So, in this case the slopes to the peak values are quite gentle being about 3.5°. On the same scale the mirror finish shows an almost flat profile, but when the vertical scale is magnified, its complex structure is revealed. In fact, values of β_m of 100 ~ 300 for highly polished Cu contacts have been measured by Utsumi [64]. I will discuss the appearance of vacuum interrupter contact surfaces and the effect of conditioning on β_m in Section 1.3.7.

It is possible to calculate β_m values for hypothetical projections. Figure 1.27 shows the results of such a calculation [65, 66]. It also indicates the typical shapes needed to have the value of β_m in the experimentally observed range of 100–300. In the past there has been considerable controversy

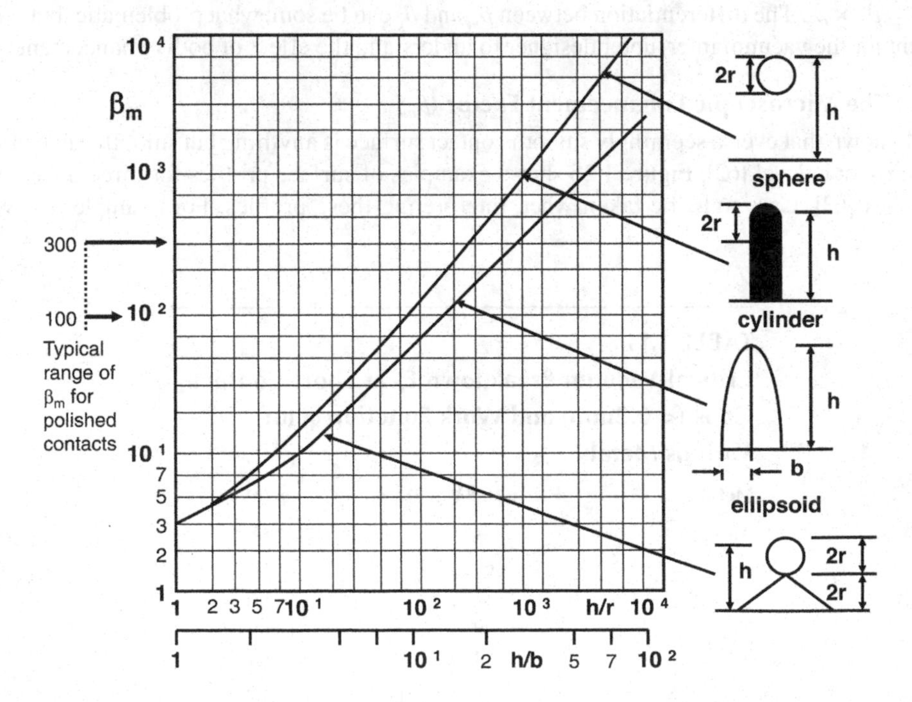

FIGURE 1.27 Calculated microscopic enhancement factors for various microscopic projections for contact gaps d much greater than the projection height h [65].

over whether or not such microscopic projections exist. Little [67] first showed a number of examples and it is now generally accepted that it is possible to have high β_m values even on polished contact surfaces [64, 68]. The vacuum interrupter designer has to assume that there will always be microscopic projections on the contact surfaces. Such projections will be left over after contact machining [69], contact cleaning, vacuum processing and initial conditioning. They will also occur as a result of normal operation of the vacuum interrupter used for switching and protecting electric circuits [70]. There are other types of surface imperfections [70, 71, 72] that can give rise to a β_m such as surface cracks, grain boundaries, imbedded particles, etc.: Figure 1.28 gives some examples. Batrakov et al. [73] indeed show photographs of electron emission sites on the surface of stainless steel from imperfections and from inclusions such as FeS in the steel's surface. Malucci [74] shows it is possible for a metal-insulator-metal particle on the cathode surface to give β_m values in the range 100 ~ 300.

In a series of experiments Sato et al. [75] have examined the effect of center-line-average roughness (R_{CLA}) [76] on the 50% vacuum breakdown (U_{B50}) using a BIL voltage pulse for Cu and Cu–Cr(25 wt.%) contacts. The R_{CLA} for the Cu contacts ranges from 0.03μm (a mirror finish) to 0.42µm. The range for the Cu–Cr contacts is 1.4μm to 4.42μm (typical for machined Cu–Cr contacts placed in a practical vacuum interrupter). Figure 1.29 shows their results. It is interesting that the U_{B50} for a given Cu–Cr surface roughness is considerably higher than that for pure Cu. Thus, for a 10mm contact gap the Cu–Cr contact with a R_{CLA} = 1.41µm has a similar breakdown voltage to the one measured for a Cu contact with a R_{CLA} = 0.47µm. For the vacuum interrupter designer this is an important observation. It means that even though the initial contact surface of a Cu–Cr contact may not be ideally smooth, it can still provide satisfactory voltage withstand values over the normal voltage range required; see Table 1.1. An empirical fit to their data gives:

$$U_{B50} = A^*\left(R_{CLA}\right)^{-n} \tag{1.40}$$

where $A^* = [12.8d + 14.5]$ and $n = [-0.0148d + 0.297]$ for the Cu contacts and $A^* = [19.4d + 44.1]$ and $n = [-0.0196d + 0.449]$ for the Cu–Cr contacts.

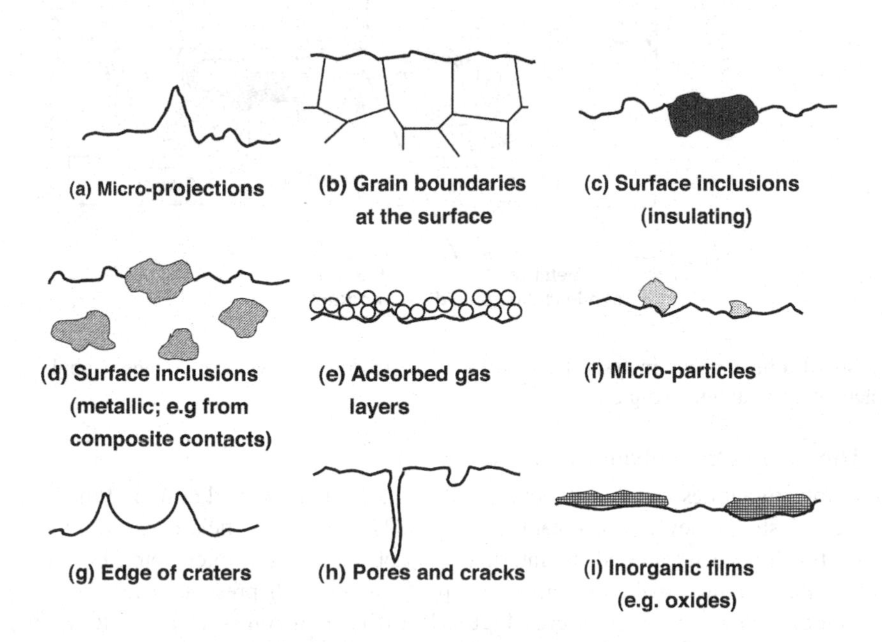

FIGURE 1.28 Examples of other surface effects that can give rise to a microscopic enhancement factor.

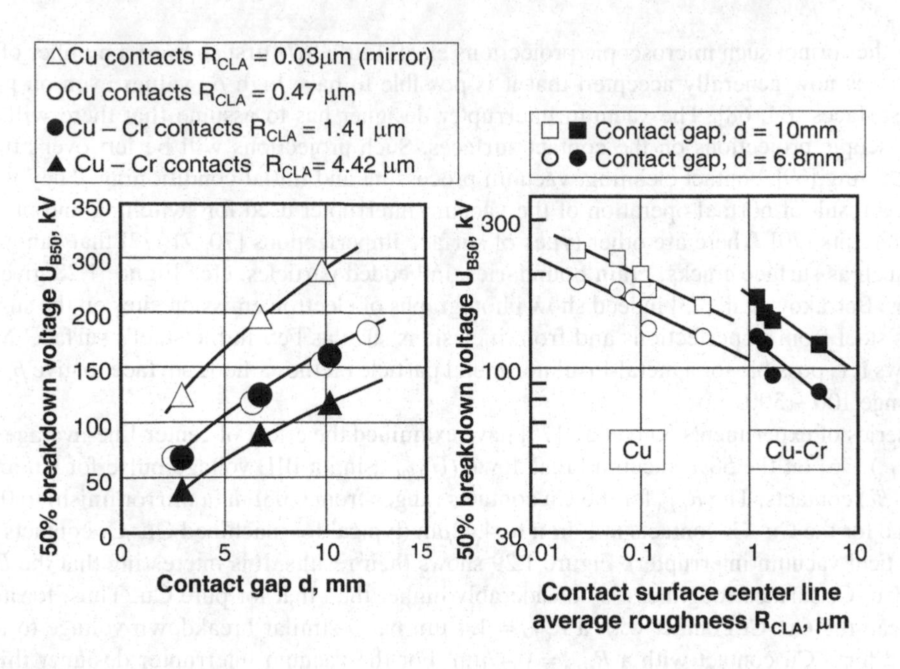

FIGURE 1.29 The 50% vacuum breakdown voltage U_{B50} as a function of contact gap and the contact's surface center-line average roughness for Cu and Cu–Cr contacts [75].

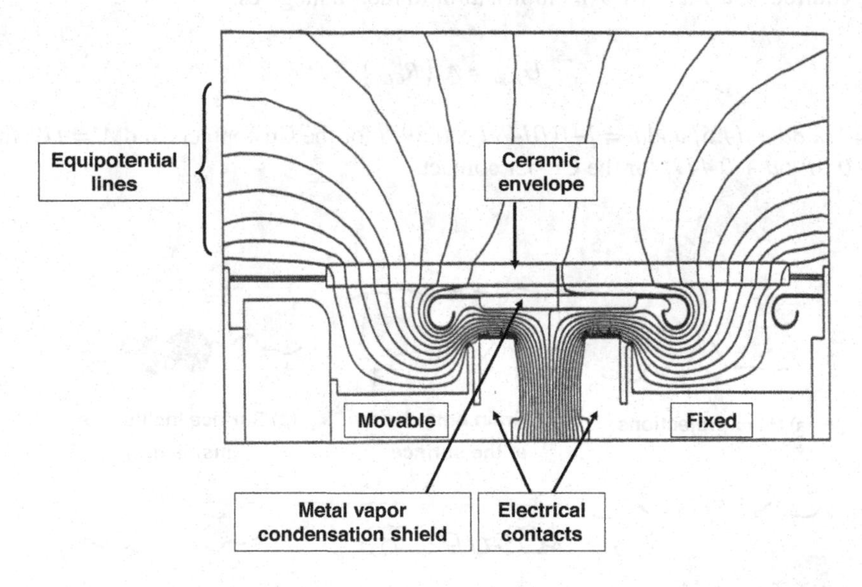

FIGURE 1.30 Finite element analysis showing the equipotential lines when a voltage is applied across the open contacts of a vacuum interrupter.

1.3.2.2 The Geometric Enhancement Factor (β_g)

The vacuum interrupter designer has direct control of the β_g. The internal design of the vacuum interrupter can be adjusted to develop acceptable β_g values. The geometric field is determined by the way the equipotential lines bend around the internal parts of the vacuum interrupter. Figure 1.30 shows an example for a floating shield vacuum interrupter. Farrall [60] presents a number of analytical approaches for certain geometric structures; Figure 1.31 [77] shows one such example. With the advent of high power, personal computers, it has now become possible for the vacuum interrupter designer

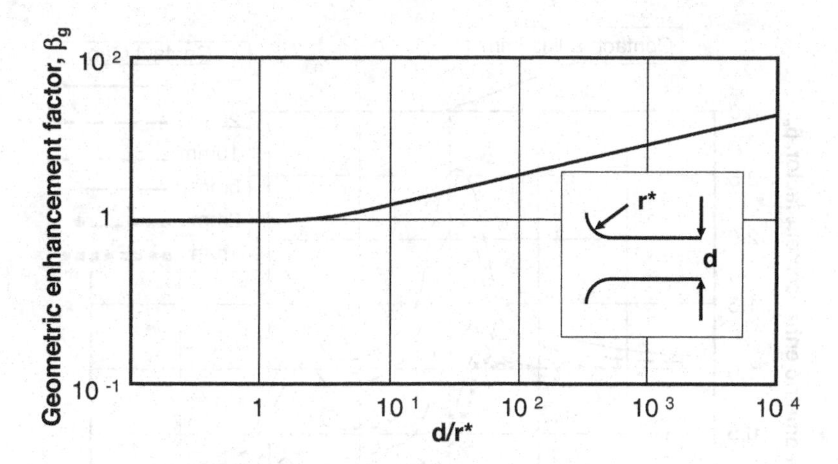

FIGURE 1.31 The geometric enhancement factor as a function of d/r for a pair of semi-infinite plane contacts with rounded edges [77].

to use a Finite Element Analysis (FEA) to analyze the electric field structure for any internal vacuum interrupter design. Using this technique, it is possible to determine the true geometric field at a contact's surface. Except for contact gaps where the ratio of the contact diameter to contact gap is large, it is always greater than the *U/d* value. Figure 1.32 illustrates the effect of geometry on the equipotential lines. Here the equipotential lines are shown for 30mm diameter, 10mm thick, butt contacts, with an edge radius of 2mm, for a contact gap of 10mm and with a 95kV voltage across them. Figure 1.33 shows how the geometric enhancement factor, β_g, varies across the contact radius and for the contact gap from 1mm to 30mm [50]. The highest β_g and hence the highest E_g at each contact gap is at the contact edges where the equipotential lines bend around the contact's face. It can be seen that the maximum β_g for the 10mm contact gap is about 1.75 at the contacts' edge. Thus, if the highest geometric field,

$$E_g = \frac{\beta_g U}{d} \tag{1.41}$$

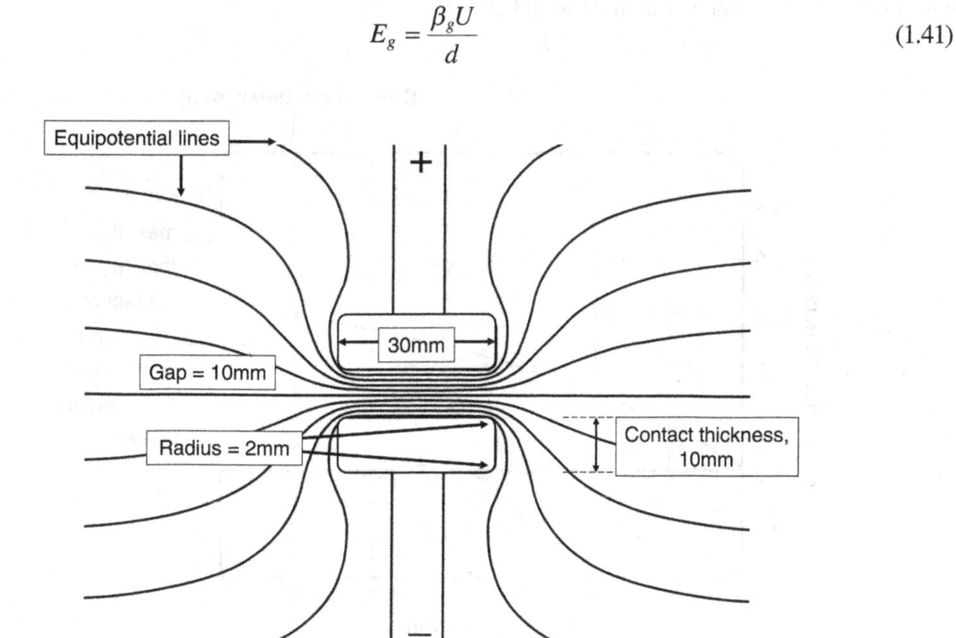

FIGURE 1.32 Finite element analysis showing the equipotential lines when a voltage is applied across the open contacts in a vacuum chamber with the chamber walls remote from the contacts [50].

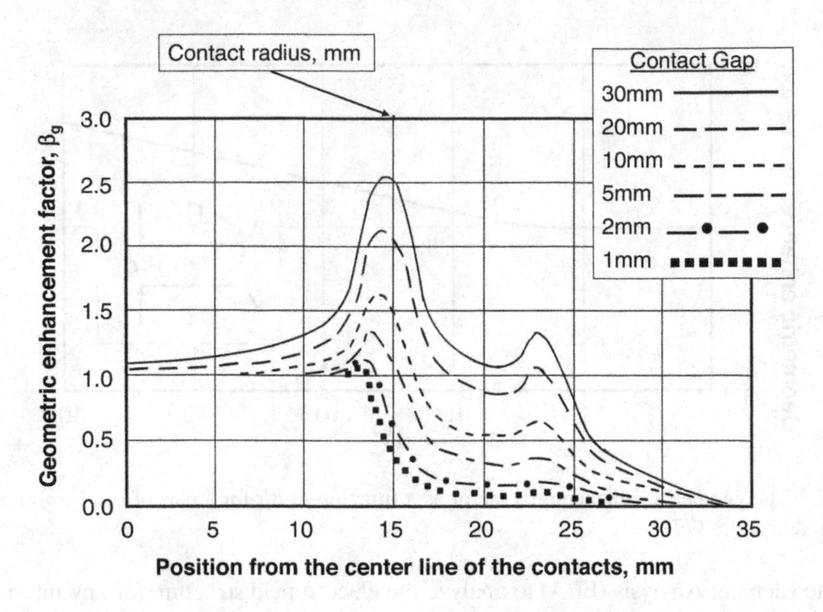

FIGURE 1.33 The variation of the geometric field enhancement factor β_g as a function of the distance from the center of the contact and the contact gap for the structure shown in Figure 1.32 [50].

Then

$$E_g = 1.75 \times 95 \times 10^3 / 10 \times 10^{-3}\,\mathrm{V.m^{-1}} \tag{1.42}$$

So, in this example, $E_g = 1.66 \times 10^7$ V.m^{-1}. The β_g for the maximum field as a function of contact gap for this contact structure is shown in Figure 1.34. Here we see that β_g is a function of the contact gap d for a given contact system. If we take:

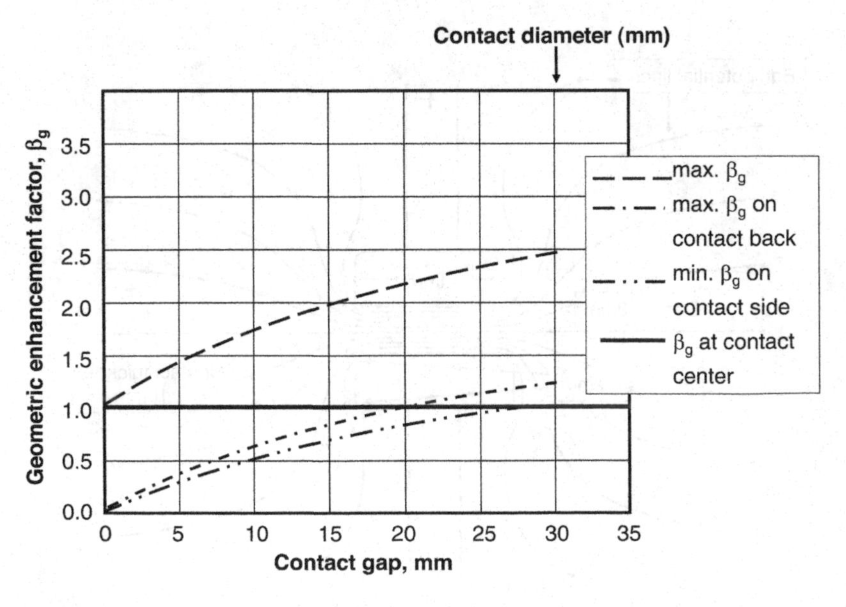

FIGURE 1.34 The geometric enhancement factors for four places on the contact surface for the contact structure shown in Figure 1.32 as a function of contact separation [50].

$$E(d) = \frac{\beta_m \beta_g(d).U(d)}{d} \tag{1.43}$$

and assume that breakdown occurs at a critical field E_c for a given breakdown voltage U_B then:

$$E_c(d) = \frac{\beta_m \beta_g(d).U_B(d)}{d} \tag{1.44}$$

Now from Figure 1.20:

$$U_B(d) = K_2 d^n \tag{1.45}$$

Thus:

$$E(d) = \beta_m \beta_g(d).K_2 d^{(n-1)} \tag{1.46}$$

Figure 1.33 shows that for a 1mm gap $\beta_g(1) \approx 1$, assuming β_m is constant then:

$$\frac{E_c(d)}{E_c(1)} = \beta_g(d).d^{(n-1)} \tag{1.47}$$

A plot of $E_C(d) / E_C(1)$ as a function of d for the contact structure shown in Figure 1.32 is shown in Figure 1.35. It is interesting to note that even though U_B / d decreases as the contact gap increases, the critical field $E_C(d)$ remains at a constant value for contact gaps greater than about 10mm for this contact geometry.

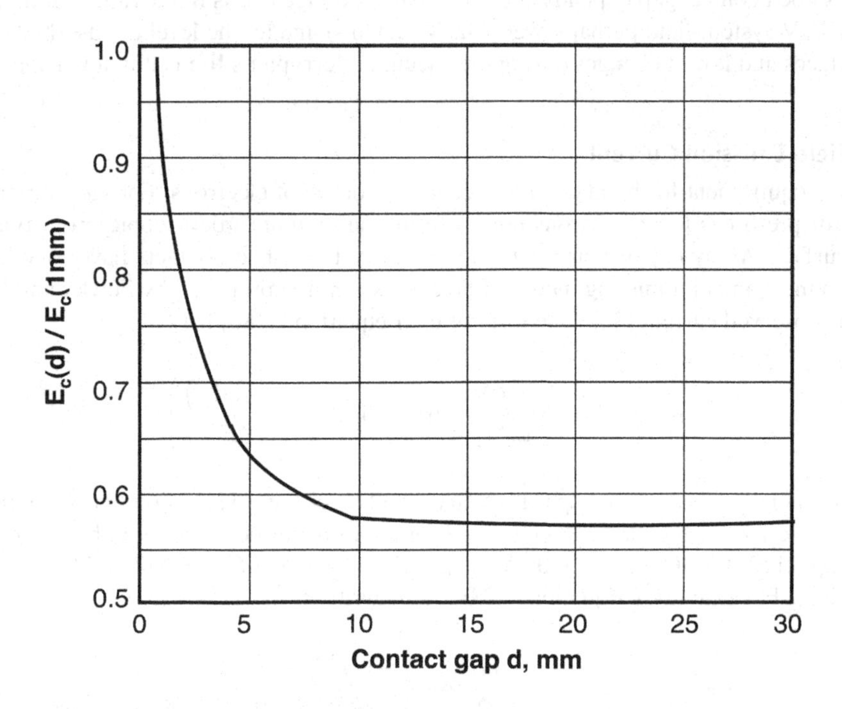

FIGURE 1.35 $E_C(d) / E_C(1)$ as a function of contact gap d [50].

1.3.3 Pre-Breakdown Effects

For the vacuum interrupter designer, the wealth of literature on pre-breakdown phenomena can to be somewhat confusing and sometimes even contradictory. The experimental research that examines these effects has generally been performed under carefully controlled vacuum conditions, with carefully prepared contacts, and at short gap spacings. More recent investigations, however, have developed vacuum breakdown information for the longer gaps used in vacuum interrupters (20mm–100mm). They have also used practical vacuum interrupter contact materials such as Cu–Cr.

In order to understand the pre-breakdown effects, it is best to separate the discussion into four convenient categories:

(i) Field emission from the cathode
(ii) Anode phenomena
(iii) Microparticles
(iv) Microdischarges

Even though there is continued controversy as to which of the pre-breakdown effects are dominant, there seems to be a general consensus that field emission electrons from the cathode are responsible for initiating the majority of these breakdown events. Experimental data allows for some generalizations. If the geometric field at the surface of a clean cathode in a good vacuum exceeds about 2×10^7 Vm^{-1} then field emission currents will be observed [55]. If the geometric field exceeds about 1×10^8 Vm^{-1}, then breakdown will occur [78]. If the contact surfaces have adsorbed gas on them then microdischarges may be observed [55]. Their probability of occurrence increases as the voltage impressed across the contact gap becomes closer to U_B. Microparticles can occur at any contact gap but their effect is more prevalent at larger gap spacings [79]. It is entirely possible that all four processes will occur in any given vacuum breakdown event inside a vacuum interrupter. This is especially true in a practical vacuum interrupter because: (a) the precise condition of even new contact surfaces is unknown and these surfaces change during the interrupter's life; (b) a vacuum interrupter's open contact gap depends upon the system voltage and is in the range 4mm to 100mm (3.6kV to 170kV systems and perhaps even 242kV systems) and (c) the level of adsorbed gas on the contact surfaces and how it changes during the vacuum interrupter's life is unknown and cannot be measured.

1.3.3.1 Field Emission Current

A necessary requirement for breakdown to occur is a source of electrons. For vacuum breakdown the source of pre-breakdown electrons comes from one or more microscopic projections on the cathode's surface. Many experiments with closely spaced polished contacts have established that the quantum mechanical tunneling theory of field emission first proposed by Fowler and Nordheim explains the observed current [80]. One form of their equation is [80] [81]:

$$j_e = 1.54x10^{-6} \frac{E^2}{\varphi t^2(y)} .exp. \left(-6.83x10^7 \frac{\varphi^{1.5} v(y)}{E} \right) \quad (1.48)$$

where j_e is in A.cm^{-2}, E is in V.cm^{-1}, φ is in eV and v and t are slowly varying functions of the dimensionless variable $y = 3.79 \times 10^{-4} E^{1/2}/\varphi$, $t(y) \approx 1$ for the range of fields likely to be found at micro-projection tips (i.e., $3 \times 10^9 \sim < E < \sim 10^{10}$ V.m^{-1}), $v(y) = 0.956 - 1.062y^2$. Miller [82] has tabulated these functions. For simplicity Equation (1.48) can be written:

$$j_e = \frac{B_1 E^2}{\varphi} exp. \left(\frac{-B_2 \varphi^{1.5}}{E} \right) \quad (1.49)$$

where B_1 and B_2 are constants. If we put $j_e = I_e / A_e$ where I_e is the emission current, A_e is the emission area and $E = \beta U / d$ and substitute for E and j_e in (Equ.1.49) and express the resulting equation in log form then:

$$\frac{I_e}{A_e} = \frac{B_1 (\beta U)^2}{\varphi d^2} exp.\left(\frac{-B_2 \varphi^{1.5} d}{\beta U}\right) \text{ or } log_{10}\left(\frac{I_e}{U^2}\right) = log_{10}\left(\frac{A_e B_1 \beta^2}{\varphi d^2}\right) - \frac{B_2 \varphi^{1.5} d}{2.303 \beta}\left(\frac{1}{U}\right) \tag{1.50}$$

Thus, if the current I_e measured as U is varied and a plot of $\log_{10} (I_e/U^2)$ vs. U^{-1} is made, the resulting graph will give a straight line whose slope and intercept will allow β and A_e to be calculated. An example of such a Fowler–Nordheim plot by Davis et al. [81] is shown in Figure 1.36(a). The slope of this line gives $\beta = 248$. As the contacts used in these experiments are profiled, the β_g value would be ≈ 1, so in this case $\beta = \beta_m = 248$. This illustrates a point made in the previous section that even carefully prepared, polished contacts can still have a microscopic surface profile that yields a high β_m. In this example $A_e = 4.8 \times 10^{-16}$ m^2 giving the emitter's radius $r_e = 12.4$ nm. Figure 1.36(b) shows the current level plotted as a function of U. Curve fitting gives:

$$I_e \approx 1.044x10^{-17} e^{0.692U} \text{A} \tag{1.51}$$

where U is in kV, thus at the breakdown voltage $U_B = 50\ kV$ the current is about 11.1mA.

Diamond [83] shows that the Fowler–Nordheim equation is very sensitive to small changes in E and φ. Figure 1.37 illustrates this sensitivity. Here we see that the exponential term in Equations (1.48 and 1.49) can change by many orders of magnitude for changes in E or φ of only a factor of 2. The effect of contact gap on the geometric enhancement factor is illustrated in Figure 1.38 [83]. Here I_e as a function of the electric field U/d (i.e., neglecting the effects of the enhancement factors β_m and β_g) is shown for contacts with a 2mm and a 4mm gap. An FEA analysis of the effect of β_g at the edge of the cathode easily accounts for this difference, assuming that β_m is the same in both cases. The main assumption in using the Fowler–Nordheim equation is that there is only one emitting site during these experiments. Many experiments have shown that even on carefully prepared broad area contact surfaces more than one emitting site is always present. In fact, Takahashi et al.

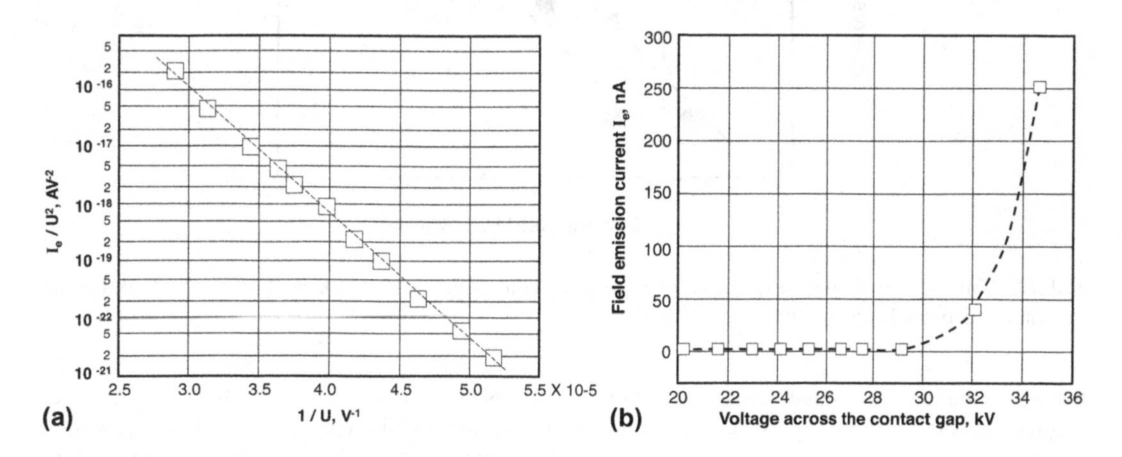

FIGURE 1.36 The variation of emission current as a function of a slowly rising dc voltage across open contacts in a vacuum, (a) a Fowler–Nordheim plot (□ experimental data, (-------) calculated from Equation 1.50 with $\varphi = 4.5$eV, $\beta = 248$ and $A_e = 4.8 \times 10^{-12}$ cm^2) and (b) emission current vs. applied voltage [81].

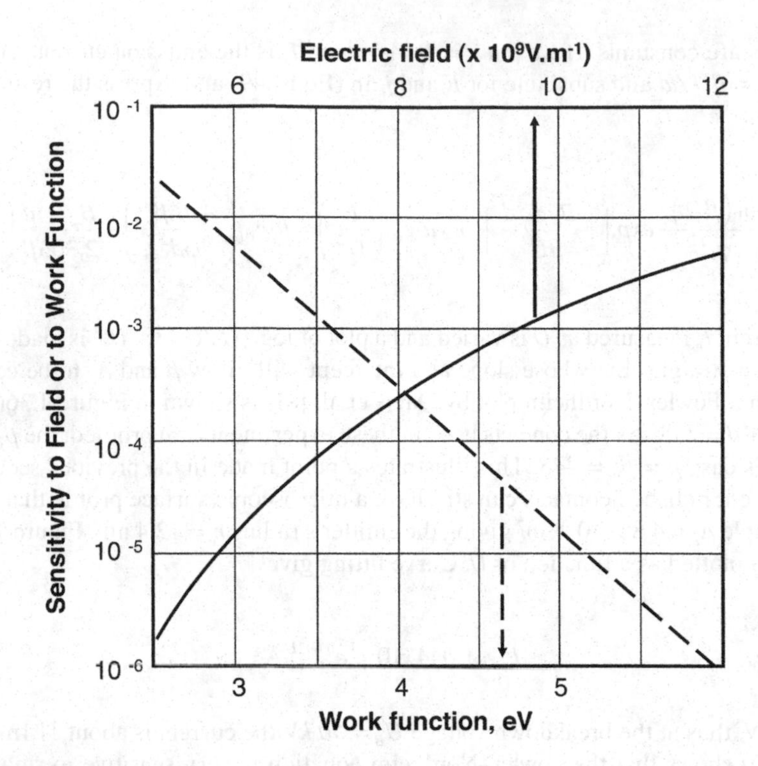

FIGURE 1.37 Sensitivity of the exponential term in Equation1.48 to changes in the electric field and the work function [83].

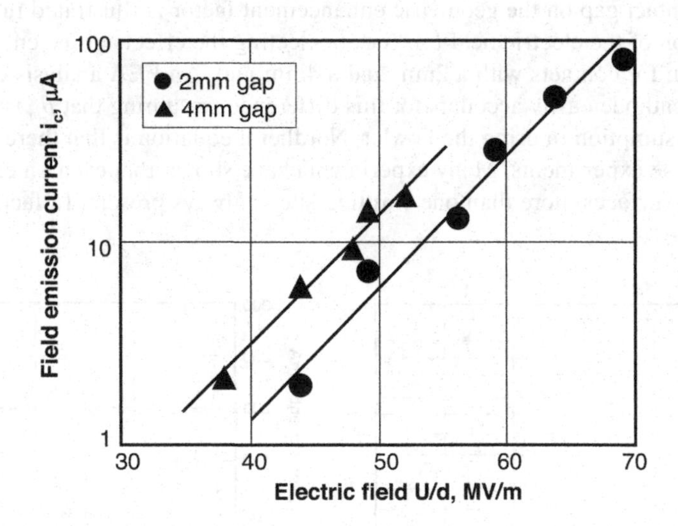

FIGURE 1.38 The emission current as a function of the contact gap field (U/d), gap (i.e., without taking into account the geometric enhancement factor) [83].

[84] show that as the voltage across the contact increases not only does I_e increase but the number of emitting sites increase as well, see Figure 1.39. An increase in voltage will permit potential emission sites with lower β_m values to reach a high enough field strength for electron emission to occur. Both Takahashi et al. [84] and Isono et al. [85] observe that while emitting sites occur at very low I_e (< 10^{-14}A), they are not stable and will change with time. Perhaps as I_e increases it anneals

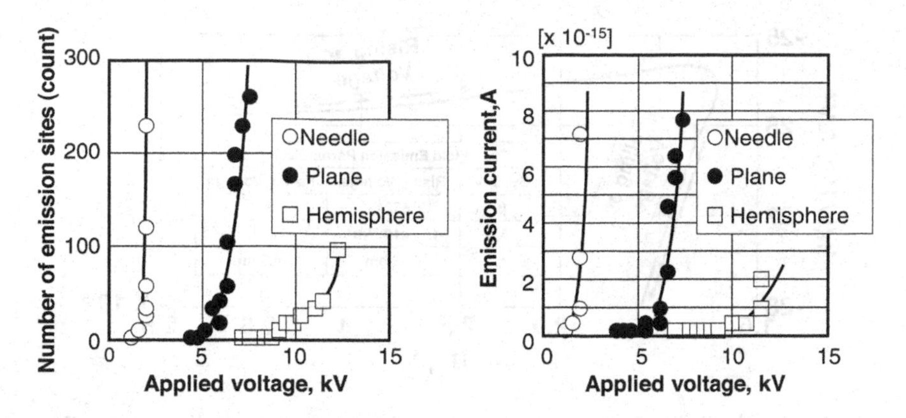

FIGURE 1.39 An observation of the change in the number of emission sites on the cathode contact as the voltage applied to the open contacts increases [84].

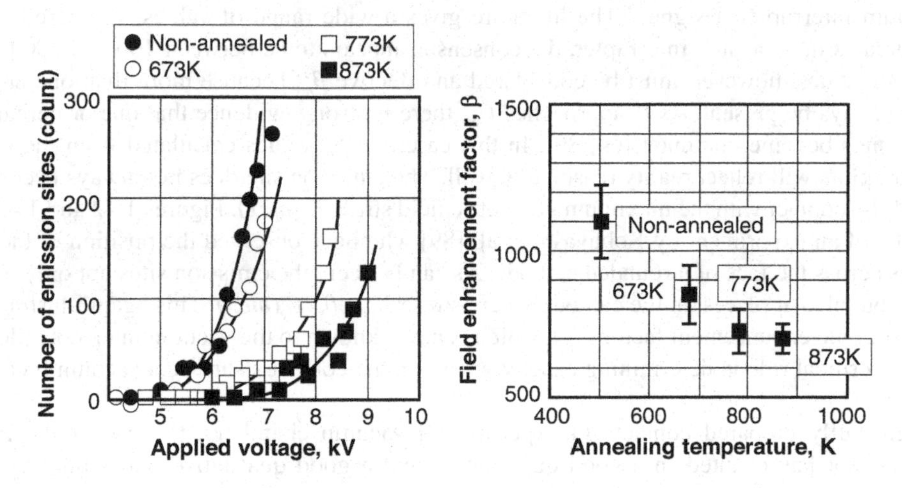

FIGURE 1.40 Effect of annealing on the number of emission sites and on the microscopic enhancement factor [84].

a microprojection on the cathode contact and thus it causes it to become less prominent than other sites around it. Certainly, annealing a cathode reduces the number of emission sites at a given applied voltage, see Figure 1.40 [84]. On the other hand, at higher currents softening or even melting of the projection in the presence of a strong E_g may well cause it to pull into a finer point with the result that β_m increases considerably [83, 84]. For the experiments described above, dc voltage is applied across the contact gap. Xu et al. [86] studied the emission currents with an ac voltage applied across a W needle cathode (tip radius $20\mu n$) and a 20mm diameter, plane, Cu anode. Figure 1.41 shows the Fowler–Nordheim plot when the W needle is the cathode and when the ac voltage is increasing and then decreasing. They attribute the difference to the annealing of the W needle as the emission current increases during the voltage rise, Yu et al. [87] show a similar result using a commercial vacuum interrupter.

The currents measured in some of these experiments are lower than the expected leakage current through a vacuum interrupter's ceramic body (see Section 1.2.2). Thus, in a practical vacuum interrupter it would not be possible to differentiate between this leakage current and a very low-level emission current. In practice the emission currents in a vacuum interrupter only become of

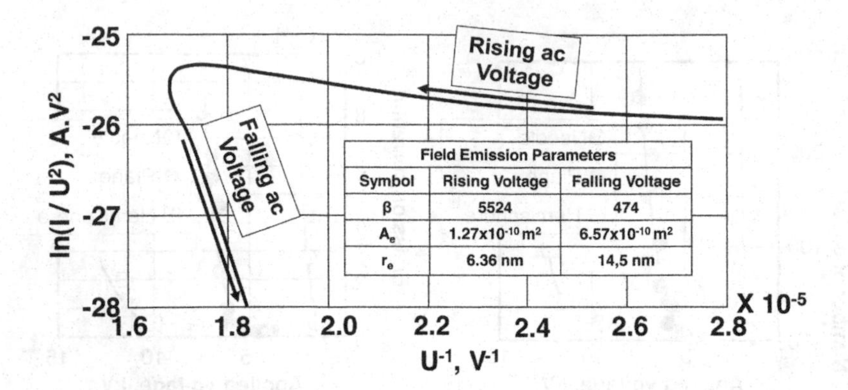

Field Emission Parameters		
Symbol	Rising Voltage	Falling Voltage
β	5524	474
A_e	1.27x10⁻¹⁰ m²	6.57x10⁻¹⁰ m²
r_e	6.36 nm	14,5 nm

FIGURE 1.41 Fowler-Nordheim plot for emission currents for an ac voltage applied across a W needle cathode and a plane Cu anode [86]

interest when they exceed a few tens of microamperes. What value of β_m should be considered for the vacuum interrupter designer? The literature gives a wide range of values. For carefully prepared contacts in a vacuum interrupter, the consensus appears to be between 100 and 300 [61, 64, 88]. These values, however, must be considered an effective β_m, because more than one emission site will always be present. As U approaches U_B, there is strong evidence that one or perhaps two emitting sites become dominant [64, 89]. In this case, the β_m values calculated from the Fowler-Nordheim plots will reflect reality reasonably well. The emission site does not always occur in the region of the contact with the maximum geometric field stress $E_g(max)$. Figures 1.42 and 1.43 show the results of an experiment by Kobayashi et al. [89] who have observed the position of the emission sites across the face of a rounded cathode. As can be seen, the emission sites not only occur at $E_g(max)$, but also at places on the cathode where stress $E_g = 0.4E_g(max)$. This again illustrates that the microscopic enhancement factor (β_m), which is never known to the vacuum interrupter designer can play a critical role in determining exactly where on the contact's surface a vacuum breakdown can occur.

For carefully prepared contacts in experimental vacuum chambers, the use of the Fowler-Nordheim plot has resulted in a good quantitative and a good qualitative understanding of the

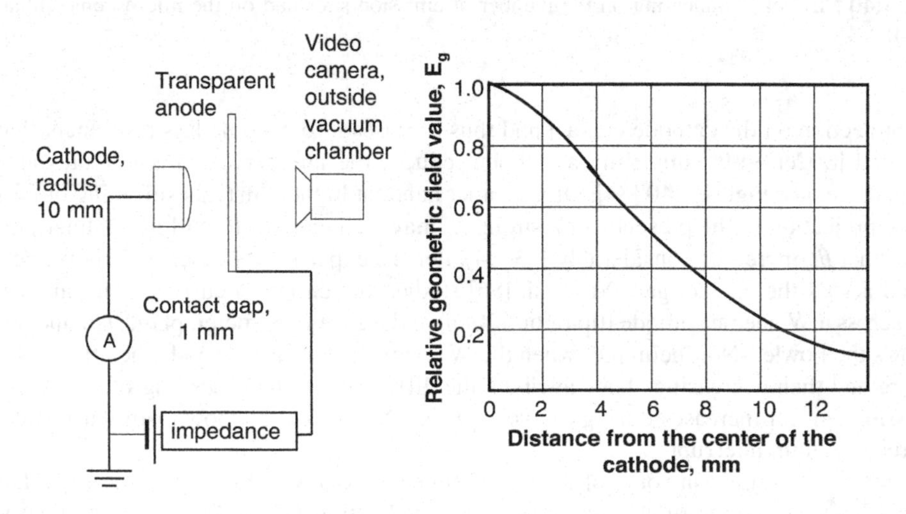

FIGURE 1.42 The experiment to observe electron emission sites [89].

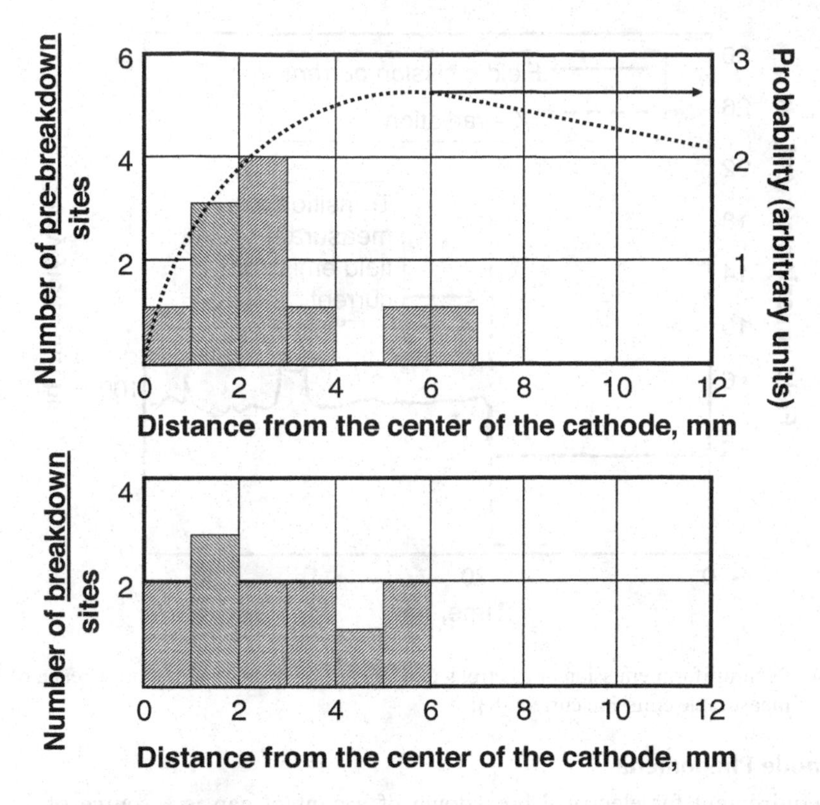

FIGURE 1.43 The number of prebreakdown sites and the number of breakdown sites observed as a function of the contact radius [89].

electron emission process. When considering it for application in vacuum interrupter design, care must be exercised. Equation (1.48) indicates for a given contact gap a smooth increase in I_e would be expected as U (and hence E_g and $E_m = \beta_m E_g$) is increased. Careful experiments measuring very low currents (down to 10^{-16}A) show that an I_e can be observed for very low value of E_g. Diamond [83], however, shows that a voltage of 55kV is required in his experiment with plane contacts and a 1mm gap before an appreciable I_e accompanied by an observable X-ray emission can be measured, see Figure 1.44. Here the current suddenly increases from a few picoamperes to a few microamperes as the voltage across the contact gap increases from about 50 kV to 55kV. It can also be seen that I_e is not necessarily constant for a given E_g. It can suddenly increase for a short period of time before settling back to its steadier value. The duration of this current pulse can range from a few microseconds to a few milliseconds (see later in this Section when I introduce *microdischarges*). Jüttner has observed similar instabilities in I_e [90]. The contact gap can also support quite high values of I_e without breakdown occurring. One practical value often measured is the voltage U_{-4} where $I_e = 10^{-4}$ A or 100 μA: a value that is proportional to the final breakdown voltage U_B [91], see Figure 1.45.

Finally, the work function φ in the Fowler–Nordheim plot is usually assumed to be that of the pure metal. The presence of oxide layers, other adsorbed gases, and variations in surface hardness will produce variable and uncontrolled changes in the work function's value. Diamond has shown in Figure 1.37 that a 0.75eV change in φ gives a decade change in the exponential term of the Fowler–Nordheim equation. Thus, in a practical situation, if the work function is not exactly known, the calculation of the enhancement factor from the slope of the Fowler–Nordheim plot is somewhat ambiguous.

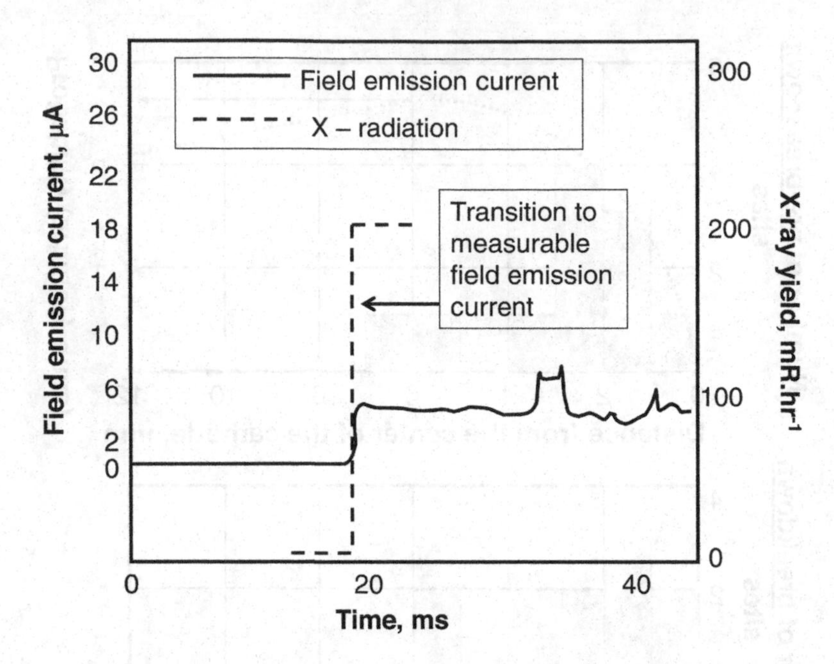

FIGURE 1.44 Non-uniform emission of electrons in a vacuum gap showing the occurrence of X-rays with the initiation of measurable emission current [83].

1.3.3.2 Anode Phenomena

The third requirement for electrical breakdown of a contact gap is a source of gas in which the production of ions would be possible. In a vacuum chamber the only source of gas could be adsorbed gases on the contact surfaces and/or metal vapor from the contacts themselves. The study of the source of gas has also had a long history. It now appears that for the vacuum

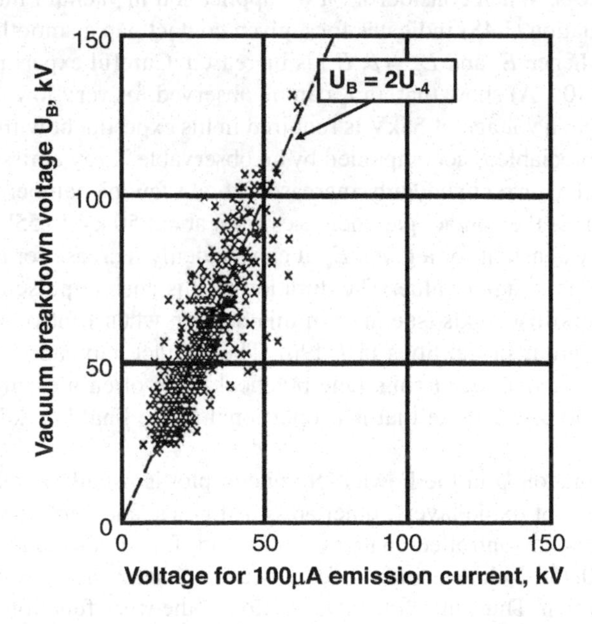

FIGURE 1.45 Final vacuum breakdown voltage, U_B, versus the onset voltage, U_{-4}, where the electron emission current is 10^{-4} A (100μA): Cu–Cr contacts with a 10mm contact gap [91].

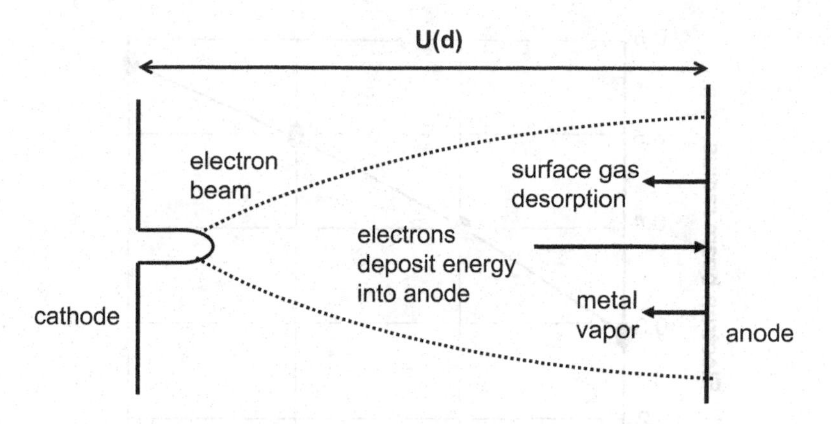

FIGURE 1.46 Schematic diagram of an electron beam from a cathode projection impinging on the anode.

interrupter designer most of these theories need to be considered, because they each can, depending upon the particular circumstance, play a role in the vacuum breakdown of a particular device operating in a specific electrical circuit. One of the earliest hypotheses for the source of gas originated from considering a narrow electron beam from the cathode impinging on the surface of a small anode area, heating it and releasing adsorbed gases and eventually evaporating metal into the contact gap, see Figure 1.46. Although Utsumi [64] thought this effect to be critical for contact gaps greater than a few millimeters, Davies et al. [92] have shown there are perhaps a number of problems with this concept. First of all, except for very small contact gaps, the electron beam is not exactly narrow; see Figure 1.47. It can be shown that for a contact gap *d* and an

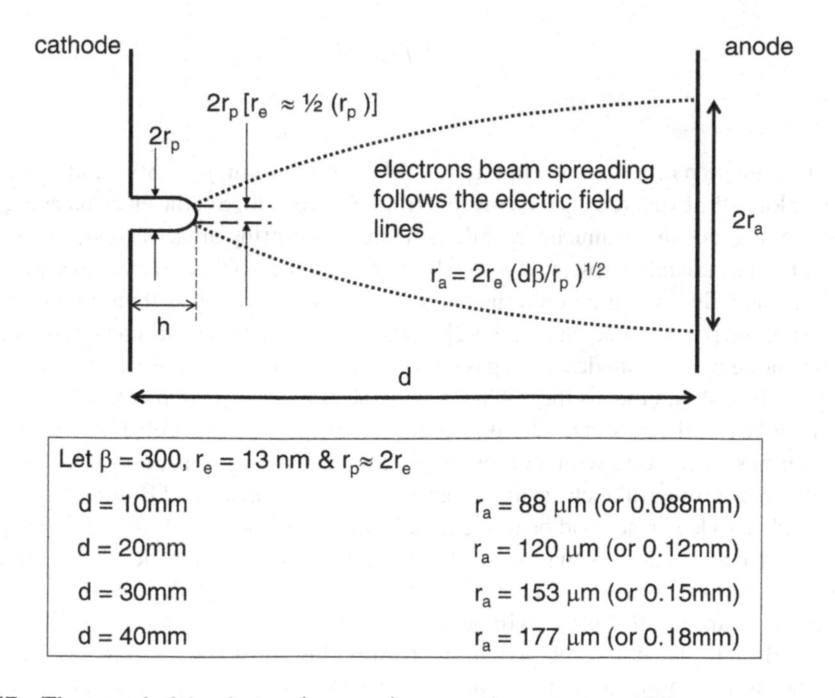

FIGURE 1.47 The spread of the electron beam as it crosses the contact gap [56].

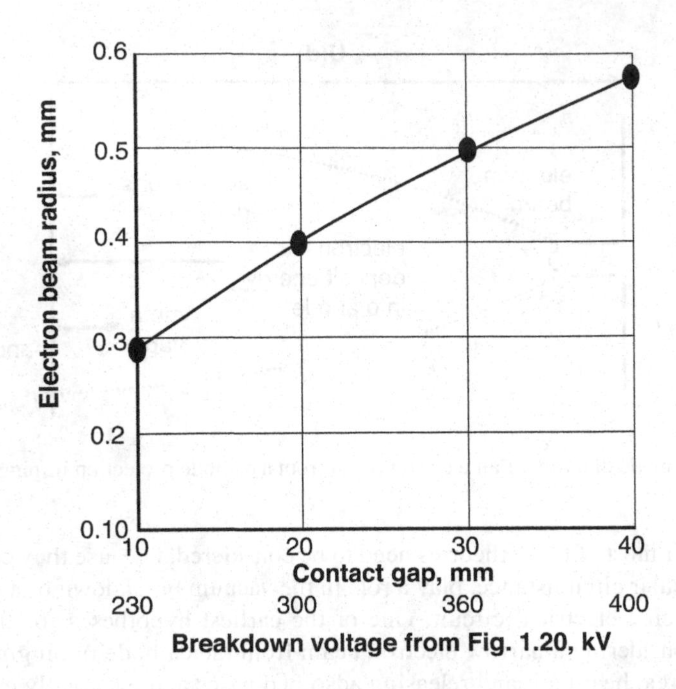

FIGURE 1.48 The electron beam radius at the anode as a function of contact gap for an emitter radius of 13 nanometers [56].

emitting radius r_e at the cathode micro-projection of radius r_p, the radius of the beam r_a when it reaches the anode is [56, 93]:

$$r_a = 2r_e \left(\frac{\beta_m d}{r_p} \right)^{\frac{1}{2}} \tag{1.52}$$

A plot of r_a as a function of d is shown in Figure 1.48 for $r_e = 13$ nm, $\beta_m = 300$, and $r_p = 2 \times r_e$. Also shown in this plot is the expected U_B for a given d from Figure 1.20. So, the electron energy is spread over an area at the anode that is much larger than the initial emitting area. Second, calculations have shown that for small contact gaps, with a slowly increasing dc voltage, the temperature rise from emission electrons only impinging onto the anode's surface is much less than the melting point of the anode during the pre-breakdown phase [92]. In the case of Cu this would result in a very low gas density at the anode either from desorbed gas or from evaporation of metal vapor. For a contact gap of 1mm, Figure 1.20 shows that as the voltage across the contact gap approaches U_B, electrons with energies up to 56keV will arrive at the anode. Figure 1.6 tells us that the probability of electrons with these high energies interacting with and ionizing a low-pressure gas just above the anode surface and beginning an electron avalanche is just about zero. In fact, Davies and Biondi [92] calculate that even if the whole anode surface had been evaporated in their experiments, the metal vapor density is eight orders of magnitude less than is needed for a Townsend avalanche type of breakdown to occur. Their calculations, however, did not consider the electron energy deposited below the anode's surface. The importance of this effect will be discussed later in this section.

Davies and Biondi [94, 95] developed an interesting, but unlikely, hypothesis of how an electron beam heating the anode could result in a process that would result in an electron avalanche being produced. This is illustrated in Figure 1.49. In this case, the region of the anode heated by the electron beam softens. This makes it possible for the electric field at the anode's surface to

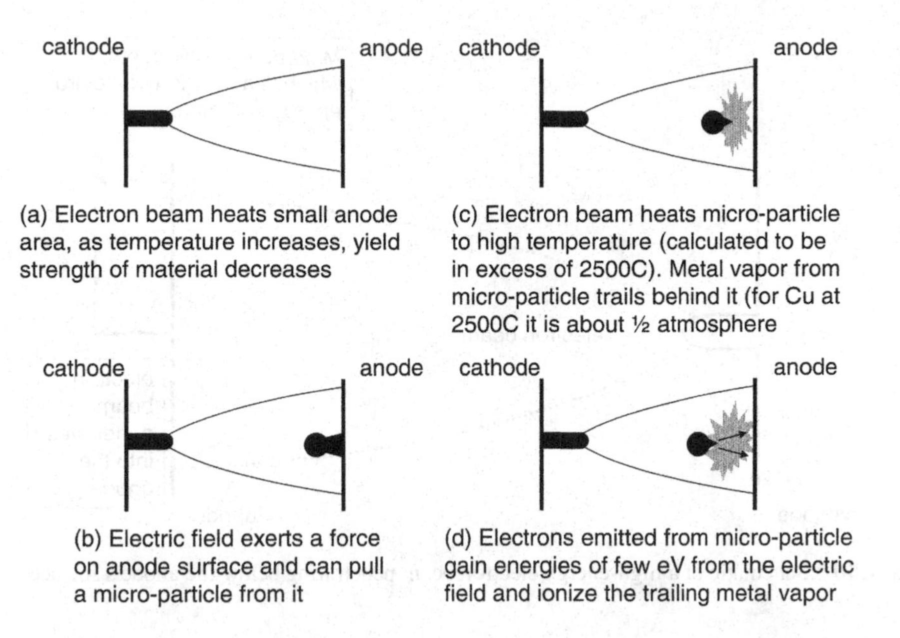
cathode
anode
(a) Electron beam heats small anode area, as temperature increases, yield strength of material decreases
cathode
anode
(c) Electron beam heats micro-particle to high temperature (calculated to be in excess of 2500C). Metal vapor from micro-particle trails behind it (for Cu at 2500C it is about ½ atmosphere
cathode
anode
(b) Electric field exerts a force on anode surface and can pull a micro-particle from it
cathode
anode
(d) Electrons emitted from micro-particle gain energies of few eV from the electric field and ionize the trailing metal vapor

FIGURE 1.49 Schematic diagram of electron beam heating of the anode, a resulting anode particle being produced, it being heated in transit across the contact gap, leaving a trail of metal vapor, which is ionized by electrons emitted by the particle [94, 95].

distort the metal in this region and eventually even to pull a microparticle of metal from its surface. This microparticle will initially have a positive charge. As this microparticle begins to move toward the cathode, it continues to be in the path of the electron beam. It will thus experience a charge neutralization, which will increase its transit time. It will also be heated by the electron beam to a temperature where it begins to evaporate and leaves a trail of metal vapor behind it. Electrons emitted from the microparticle itself with energies of a few 10s eV can now interact with this metal vapor and ionization can occur, see again Figure 1.6. If this process continues, an electron avalanche may develop which could eventually help to develop into a full breakdown of the contact gap.

Davies and Biondi [94, 95] ignored the effect of the electron beam penetrating below the anode's surface. I believe that heating the volume below the anode's surface by the emission, electron beam directly influences the vacuum breakdown. This is shown in Figure 1.50. Looking again at Figure 1.20, for polished contacts with contact gaps of > 10mm the breakdown voltage is > 200 kV. This means that for contact gaps > 10mm the electrons will reach the anode with energies > 200 keV as the voltage across the contact gap approaches the breakdown voltage, U_B. Even for practical vacuum interrupter contacts the breakdown voltage will be > 100 kV for a 10mm contact gap. When these electrons reach the anode, they will not be stopped at the anode's surface, but will penetrate below its surface and will thus deposit their energy there. Lamarsh [96] and Feldman et al. [97] give the maximum penetration depth, R_p, as:

$$R_p \approx 0.06 W^{\Upsilon} / \delta \tag{1.53}$$

where R_p is in μm, δ is in gm.cm^{-3}, W is in keV and Υ ranges from 1.2 to 1.7.
or:

$$R_p \approx 0.412 \mathrm{W}^{\gamma^*} / \delta \tag{1.54}$$

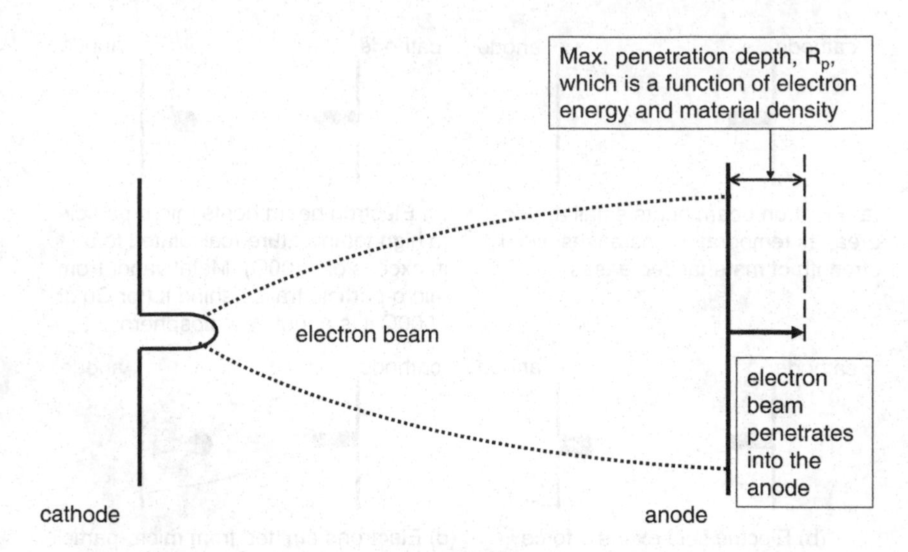

FIGURE 1.50 Schematic of a high-energy electron beam penetrating below the anode's surface.

where R_p is in cm, δ is in gm.cm^{-3}, W is now in MeV and $\Upsilon^* = \{1.265 - 0.0954 \log_e(W)\}$. The values for R_p are similar if $\Upsilon^* = 1.64$ for 50 kV $\leq V_B \leq$ 500 keV. Figure 1.51 shows R_p as a function of electron energy in a Cu anode [50]. Figure 1.52 gives a Monte-Carlo simulation of a thin, 20 keV electron beam interacting with an Fe anode [97]. Here it can be seen that although the maximum penetration depth as given by Equations (1.53) and (1.54) can be reached; most of the electron beam's energy is deposited to a depth of about $0.5\ R_p$. This example is for a very narrow electron beam. Figure 1.52 shows that once this thin, electron beam penetrates the anode's surface it spreads. Most of its energy is deposited in a cylinder 0.3μm in diameter. If we take into account the typical

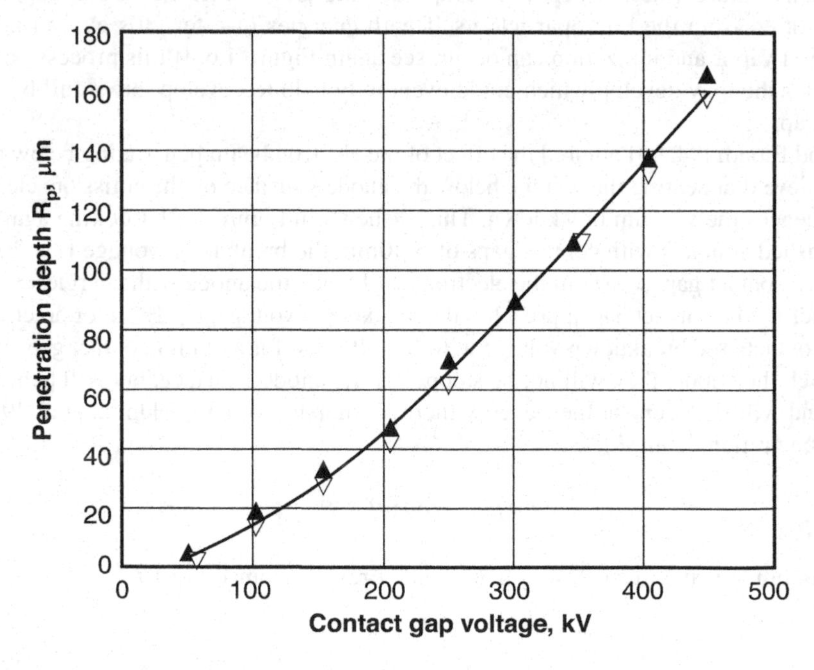

FIGURE 1.51 The maximum penetration of an electron beam into a copper anode as a function of incident electron energy (△ R_p from Equation1.53, ▲ R_p from Equation 1.54) [50].

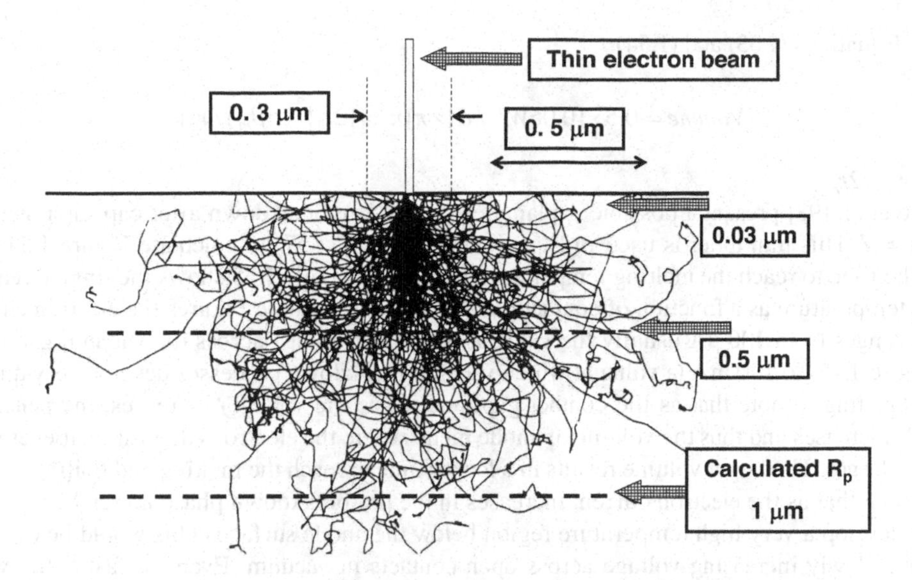

FIGURE 1.52 Monte Carlo simulation of a thin, 20 keV electron beam at normal incidence on to an Fe anode [97].

electron beam spreading shown in Figures 1.47 and 1.48, the spread of this beam inside the anode surface will be significant compared with its diameter at the surface. It is reasonable to assume that its spread inside the anode to be $n \times r_a$. It is now possible to calculate the temperature rise of this subsurface region by assuming all the electron energy being deposited below the anode surface in a cylindrical volume bounded by a diameter resulting from the spread of electron beam inside the anode and with a depth of $0.5\ R_p$ as is illustrated in Figure 1.53. If no conduction heat loss is assumed, then for an electron current I_e and a contact gap voltage U, the time t to reach a given temperature T in this region will be:

$$I_e U t = c_p \times Volume \times \delta \times \{T - T_0\} \quad (1.55)$$

where c_p is the specific heat, T_0 the initial temperature and the volume is given by:

$$Volume = 0.5 R_p \pi \{n r_a\}^2 \quad (1.56)$$

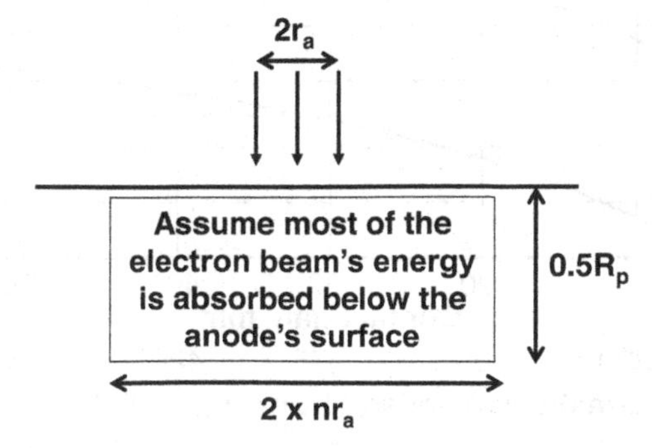

FIGURE 1.53 Schematic showing the volume below the anode's surface where most of the electron beam's energy is deposited.

or from Equations (1.53) and (1.54):

$$Volume = 0.5\times[0.06W^{\Upsilon} / \delta]\times\pi[n^2\times\{2r_e\}^2 (d\beta_m)/r_p] \tag{1.57}$$

Where $r_p \approx 2r_e$

Slade et al. [98] present a possible explanation for the late breakdown after capacitor switching using $n = 7$. This multiplier is used in Equations (1.55) and (1.57) to calculate Figure 1.54 which shows the time to reach the melting temperature for Cu and Figure 1.55 shows the time to reach the boiling temperature as a function of contact gap and current. In these Figures the electron emission current ranges from 100 μA usually the value at which the voltage across the vacuum gap is $\frac{1}{2}U_B$ (see Figure 1.45) to 100 mA (a value at which vacuum breakdown processes develop very quickly). It is interesting to note that as the contact gap (and hence the voltage) increases, the penetration depth R_p increases and thus the volume of anode metal where the electrons deposit their energy also becomes larger. The larger volume results in a longer time to reach the melting and boiling points. It can be seen that as the electron current increases in the pre-breakdown phase it then becomes possible to develop a very high temperature region below the anode surface. This would be especially true for a slowly increasing voltage across open contacts in vacuum. Even for fast-rising voltage pulse such as a BIL voltage the pre-breakdown current has been shown to follow the voltage pulse and reach value greater than 100mA. Even so, Figure 1.55 shows that the sub-anode temperature can reach the boiling point in less than 1μs. For faster rising voltages, however, there may be a time delay for the sub-anode temperature to increase to the boiling point. The expansion of this metal would create a large force on the anode's surface layer. Wang et al. [99] calculate that when a Cu anode reaches 2000C the flux of evaporated Cu atoms could become large enough to result in a

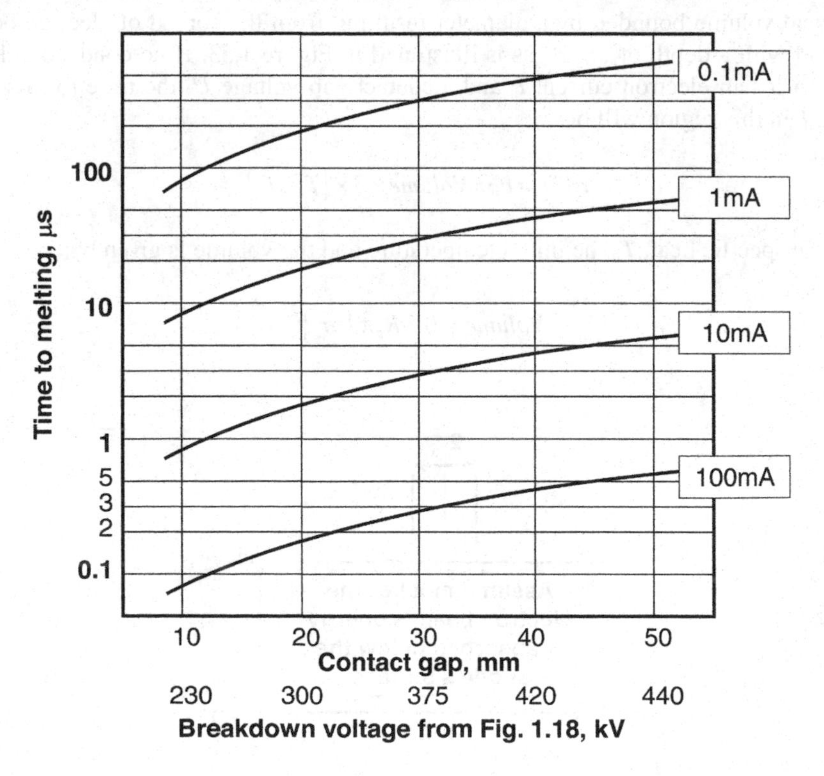

FIGURE 1.54 Time to reach the melting point of Cu for the volume below the anode surface shown in Figure 1.53 with n = 7 as a function of contact gap (i.e., breakdown voltage) and emission current.

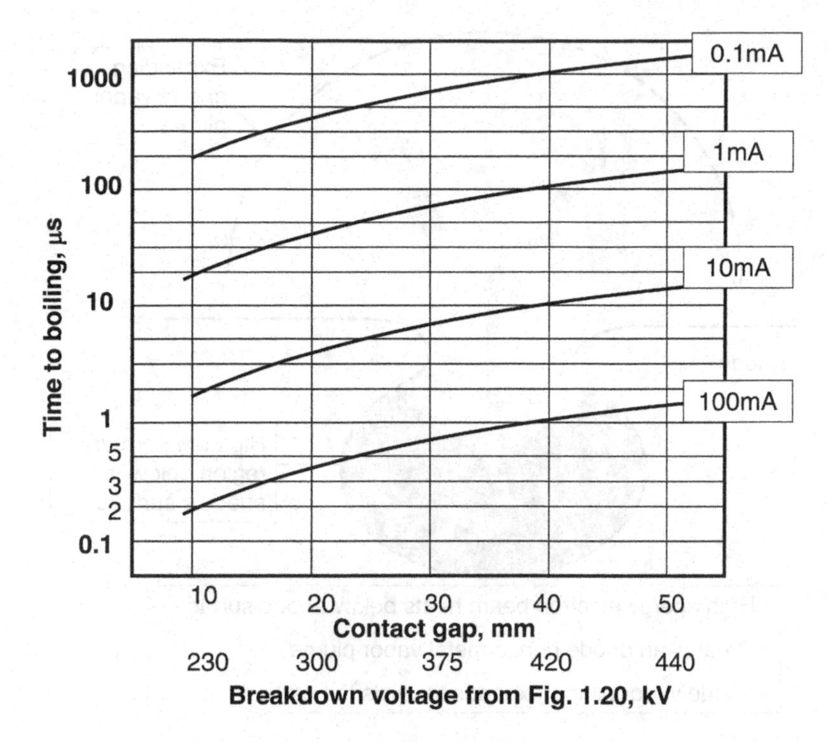

FIGURE 1.55 Time to reach the boiling point of Cu for the volume below the anode surface shown in Figure 1.53 with n = 7 as a function of contact gap and emission current.

micro-explosion of Cu vapor. It would then be quite possible for this subsurface metal to erupt from the anode as illustrated in Figure 1.56. A very dense cloud of metal vapor would then be established above the anode. This anode flare would then expand into the contact gap. This analysis indicates that the thermal conductivity of the anode would be an important parameter when considering the breakdown of practical vacuum interrupter contact gaps. Experimental evidence indeed shows that anodes with a low thermal conductivity have lower values of U_B[100].

I would like to challenge researchers who have an interest in studying vacuum breakdown to determine if my hypotheses of the electron beam causing the eruption of anode metal vapor into the contact gap is valid. An experiment could be established with a pin cathode to ensure only one emission region and a flat anode. A steady dc voltage should then be impressed across the contact gap and a steady emission current obtained. By observing the contact gap, the time for the eruption of the metal vapor from the anode can be measured. If the temperature of the metal vapor is assumed to be 2000C then the mass of the heated anode volume can be calculated. Knowing the penetration depth of the electron beam the diameter of the heated volume can be calculated and the spread of the electron beam inside the anode determined. The experiment can then be repeated for different values of emission current.

1.3.3.3 Microparticles

In a practical vacuum interrupter design, microparticles are always present on the contact surfaces. In one study Kamikawaji et al. [101] estimate there are between $5.7 \times 10^3 cm^{-2}$ and $8.7 \times 10^3 cm^{-2}$ particles on the contact's surface after contact machining. After the formation of the vacuum arc during contact opening more particles will be generated; see Figure 1.57 (this will be discussed further in Chapter 2 in this volume) and their effect on the breakdown of a vacuum gap may be felt long after the arcing event [102]. Thus, when considering the vacuum breakdown processes in a vacuum interrupter the effect of microparticles should be taken into account. The influence of microparticles

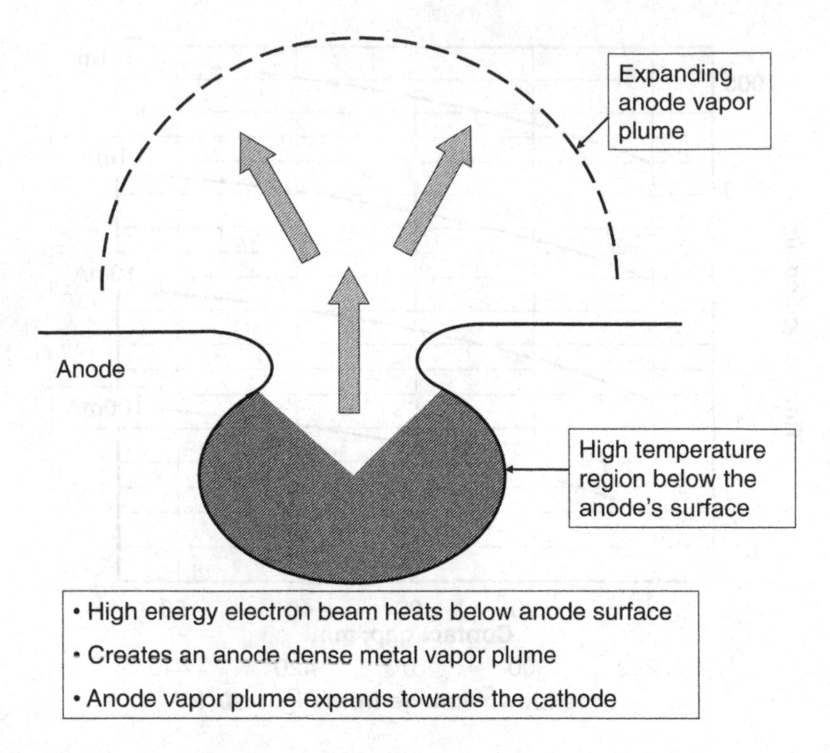

FIGURE 1.56 Schematic of a dense, metal vapor plume erupting from below the anode surface.

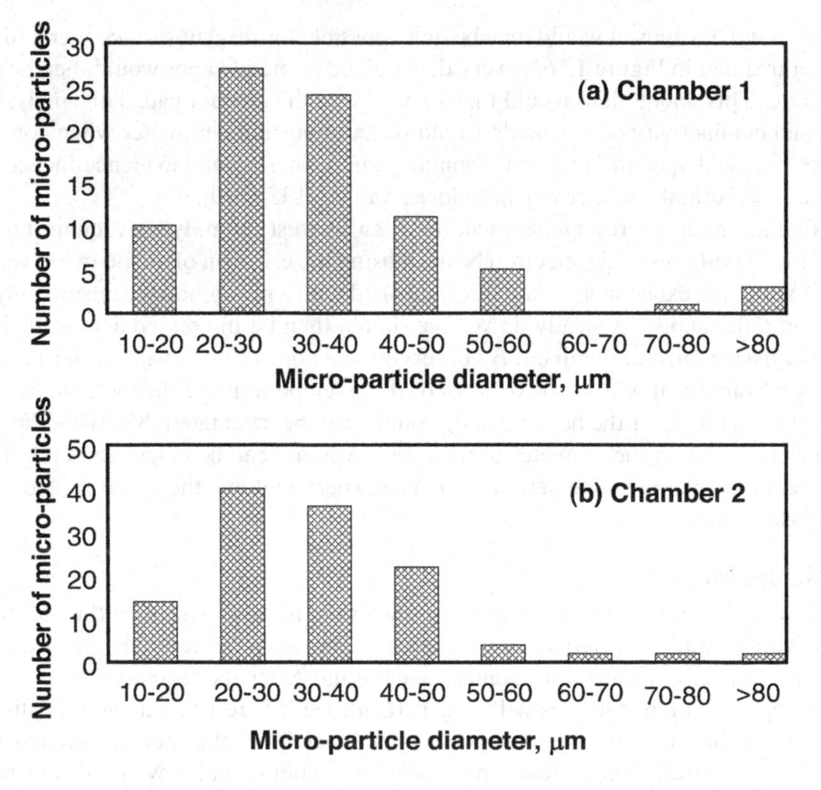

FIGURE 1.57 Distribution of microparticles left on machined Cu–Cr contact surface after switching 1000A, 100 times [101].

on pre-breakdown phenomena in vacuum has a long history having been first considered by Bennett in 1931 [103]. A detailed review of the subject can be found in Latham's book [56, 79]. Metallic microparticles can originate from a number of sources. As explained above, they can be pulled from the cathode and anode surfaces. They can be left behind on the contact surfaces after machining and after arc interruption [104]. They can evolve from the impact on a contact's surface of a microparticle driven across the contact gap by the electric field. They can also result from a previous breakdown event. Non-metallic microparticles can result from contact preparation processes such as machining, polishing and even perhaps foreign body contamination.

In the previous section, I discussed the experiments of Davies et al. where they imagine that a microparticle can be pulled from the anode [92,94,95]. Here the electron beam emitted from the cathode by field emission imparts its energy into a region on the anode surface. When this anode region reaches a high enough temperature, it is possible for the electric field at the anode to raise a microprojection in the softened contact metal. This microprojection can then neck down and a particle of anode material can break away from the surface. A similar process can also occur at the cathode microprojection [87] In this case, resistive heating of the cathode microprojection occurs when the field emission current passes through it. It has also been shown, however, that more than 95% of the microparticles originate at the anode [105]. Thus, most of them travel from anode to cathode. One explanation for this effect is that a microparticle on the cathode could become an emission site and, by losing electrons, be more tightly bound to the cathode contact [106]. Microparticles that are left on the contact surfaces after machining or arcing can adhere to these surfaces with a strong electrostatic force. It will, therefore, require a very strong electric field to lift these particles from the cathode contact surface. This means that microparticles may be held so strongly onto the contact surface that the normal fields resulting from vacuum interrupter design voltages will not cause them to lift off. These bound microparticles can themselves produce high β_m values [107]. For particles in the range of 1–10μm, however, the lift off field is usually less than that needed for breakdown to occur [107]. Kolyada et al. [108] have shown that a very small microparticle approaching close to a cathode surface can also greatly enhance the electron emission.

Let us examine the forces on a microparticle in a vacuum gap. If we assume the field at the contact surface is E, the charge Q on a spherical microparticle of radius r, and dielectric constant ε_0 is given by:

$$Q = 6.6\pi\varepsilon_0 E r^2 \tag{1.58}$$

where E is in Vm^{-1} and is given by $\beta U/d$ where $\beta = \beta_m \beta_g$. The force F on the particle at the contact surface will be given by:

$$F = QE = Q\,\beta_m\,\beta_g\,U\,/\,d \tag{1.59}$$

Once the particle has lifted off the contact surface the force on it will be $F = QU/d$. The average acceleration f of the particle will be this force divided by the particle's mass i.e.,

$$f = 3QU\,/\,4\pi r^3 \delta d \tag{1.60}$$

where δ is the density of the microparticle. The impact energy W_i is:

$$W_i = QU = 6.6\pi\,\varepsilon_0\,\beta_m\,\beta_g U^2 r^2\,/\,d \tag{1.61}$$

Velocity V_i of a particle with charge Q in a potential drop U, when the particle impacts the opposite contact is:

$$V_i = \{(3QU)\,/\,2\pi r^3 \delta\}^{0.5} \tag{1.62}$$

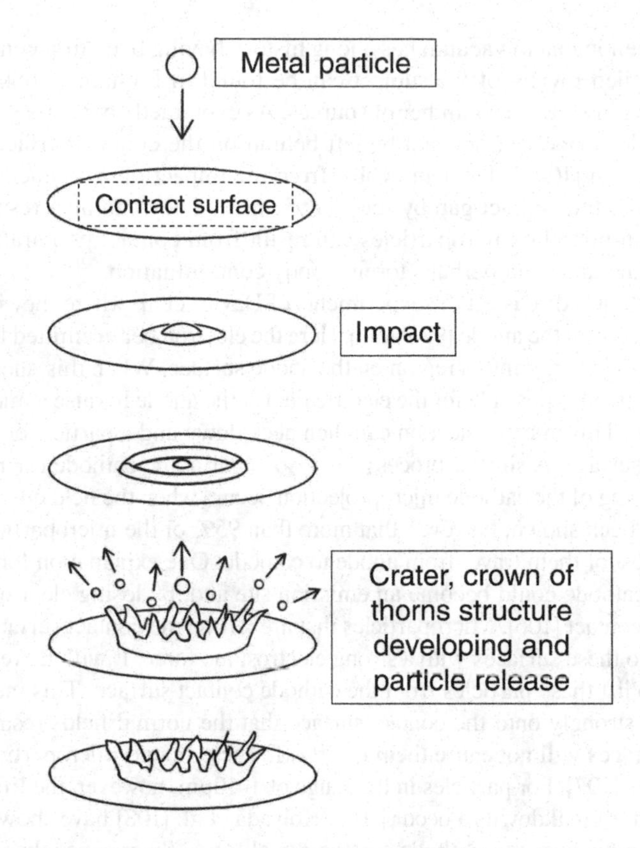

FIGURE 1.58 Schematic of a particle impact resulting on a crown of thorns structure on the contact surface plus the ejection of secondary microparticles.

If we assume that Q remains constant during the particle's transit across the vacuum gap, the impact velocity V_i depends upon $U^{0.5}$ and the impact energy W_i depends upon U^2 and upon the particle's radius r^2. It also depends inversely upon the contact gap d. Latham [79] gives a qualitative assessment of the particle's effect on the contact's surface when it impacts as a function of its impact velocity V_i compared to a critical impact velocity V_c, where $V_c \approx [8\sigma/\delta]^{1/2}$, σ is the yield strength of the material:

1) Low-impact velocities ($V_i < V_c$) result in plastic deformation of the contact surface
   a) For Cu, $V_c \approx 200$m/s, for stainless steel $V_c \approx 500$m/s
   b) V_i will always be greater than V_c for particles with a radius $< 0.7\mu$m
   c) The particle will either stick to the contact's surface or it will bounce off.
2) Intermediate impact velocities ($V_c < V_i < 5\ V_c$).
   a) Energy results in a permanent deformation of the contact's surface
      (i) Crater with a crown of thorns structure, Figure 1.58
      (ii) Imbedded particle enhances local β_m at the cathode
      (iii) The generation of secondary particles, Figure 1.58
3) High-impact velocities ($V_i > 5\ V_c$)
   a) Complete evaporation of the particle releases a cloud of metal vapor
      (i) Ionization possible
      (ii) Can develop into a microdischarge or even a full breakdown

Thus, there are a number of ways the microparticles can affect the pre-breakdown conditions of the vacuum gap:

1) The microparticles can be completely evaporated on impact releasing a cloud of metal vapor
2) The microparticles on impact can give rise to secondary particles and produce a crater on the contact's surface. This crater can have the characteristic "crown of thorns" structure on its edges; see Figure 1.58. These sharp points will result in microscopic regions of high field which, if on the cathode, will in turn produce field emission currents
3) When the microparticle approaches close to the cathode contact surface it is possible to obtain an increase in the local microscopic field enhancement (β_m) factor. This could be as much as 10 times. The result would be an increase in the field emission current [79, 109]
4) It is possible, but unlikely, that the microparticle may also be vaporized during flight by the field emission electron beam with ionization occurring in the resulting metal vapor [92,94,95]

It is possible for a particle in an ac field not to reach the opposite contact during one half cycle. It would then reverse its direction and move toward the other contact [79, 109] thus perhaps pick up more energy with each reversal. This would continue until impact with a contact occurs or the particle leaves the intercontact gap [108], see Figure 1.59. Ejiri et al. have observe the motion of Cu and alumina (Al_2O_3) microparticles between Cu planar contact in 10^{-3} Pa vacuum [110–112]. The particles are introduced onto the lower contact and a 50kV ac voltage is applied. The motion of the particles is recorded using a high-speed camera using the scattered light from a laser illuminating the contact gap. Figure 1.60 presents examples of Cu particle motion at the cathode contact [112]. The typical bouncing behavior of both the Cu and the alumina microparticles is shown in Figure 161 [112]. At the peak of the ac voltage the microparticles have a speed of $10 ms^{-1}$. At this speed the microparticles either stick on the contact surface or else they bounce off it. Also, at the ac voltage peak an emission current of 2–3mA is measured. When a vacuum interrupter is an integral part of

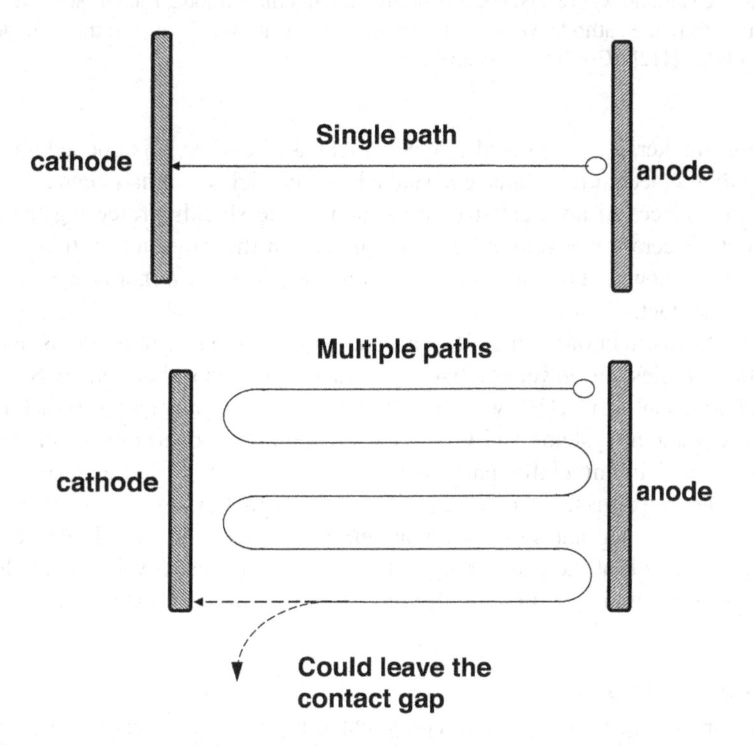

FIGURE 1.59 Schematic of a particle making an oscillatory passage across the contact gap.

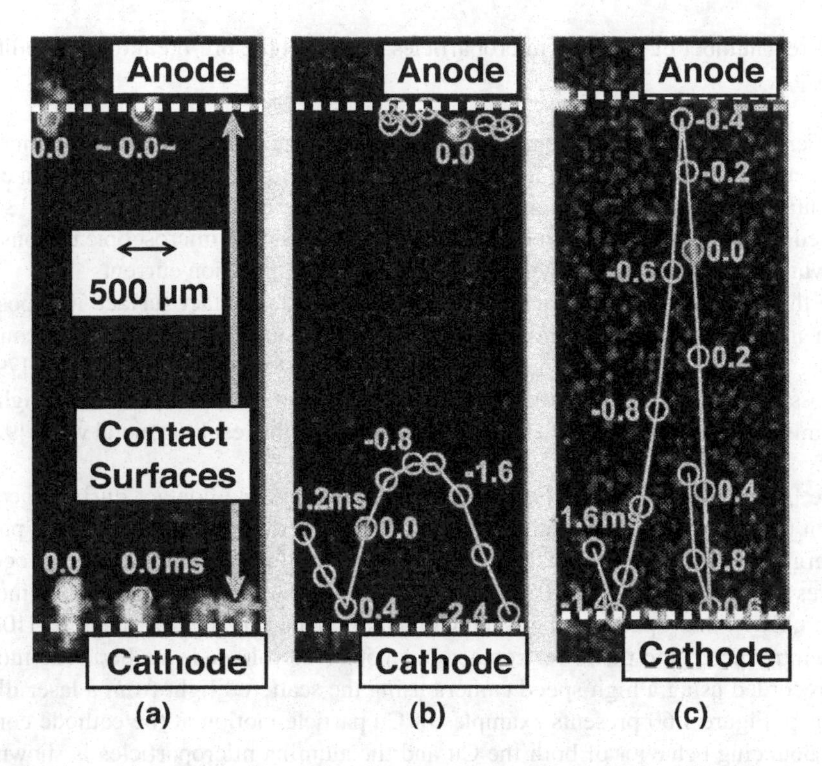

FIGURE 1.60 Observed motions of bouncing Cu and alumina microparticles with diameters 36–212 μm between planar Cu contacts (a) Attached at the cathode (b) Bouncing at the cathode and (c) Bouncing between cathode and anode. Examples of a discharge (BD) between a particle and the cathode: (a) The particle travels to the cathode where a BD occurs between it and the cathode, then it travels away from the cathode (b) The particle travels to the cathode where a BD occurs between it and the cathode, then attaches to the cathode and (c) The particle travels to the cathode where a BD occurs from the particle's tip and the cathode, then it travels away from the cathode [112] (Courtesy Haruki Ejiri).

a vacuum circuit breaker, it is subjected to mechanical shock when it is opened and closed. It has been suggested that a mechanical shock can shake loose particles that had been attached to contact surfaces or from surfaces of adjacent structures such as the shields protecting the interior of the vacuum interrupter's ceramic envelope (see Section 3.4.1 in this volume). Juttner [113] and Juttner et al. [114], however, show that mechanical shock can also result in an increase in electron emission from the cathode contact.

The pre-breakdown effect of nonmetallic particles is not obvious. There seems to be a consensus that nonmetallic particles can lower U_B, but experimental data is rather sparse. Studies by Farrall et al. [115] and Johnson et al. [116] with heavily contaminated surfaces shows lower U_B values. However, in most vacuum systems and vacuum interrupters this condition is unlikely to occur. It might be expected that a nonmetallic particle sitting on a cathode would give rise to a triple point where enhanced electron emission would be possible [73]. However, Kraft et al. [117] show that even if the vacuum discharge originates near the nonmetallic particle, the overall effect of the particles on the breakdown strength of the gap is marginal; see Table 1.12. Pokrovskaya-Soboleva et al. show that there is even some evidence that not all non-metallic particles enhance the electron emission [118].

1.3.3.4 Microdischarges

The breakdown process in vacuum is also complicated by the phenomenon of the microdischarge [55]. The microdischarge is a unique phenomenon which can be observed when a voltage is applied

TABLE 1.12
Effect Of Insulating Inclusions Incorporated into Iron Contacts During Manufacture by Powder Metallurgy [117, 118]

Inclusion Material	Inclusion Size (μm)	Vacuum Breakdown Voltage for a 0.05mm Contact Gap (kV)	Vacuum Breakdown Voltage for a 2.0mm Contact Gap (kV)
None, pure Fe	0	10.0	38
FeO	< 30	10.0	37
SiO_2	< 30	10.0	36
FeS	< 30	10.0	34
Al_2O_3	50–500	6.5	31
Alumina Silicate	50–500		28
SiO_2	50–500	5.5	26

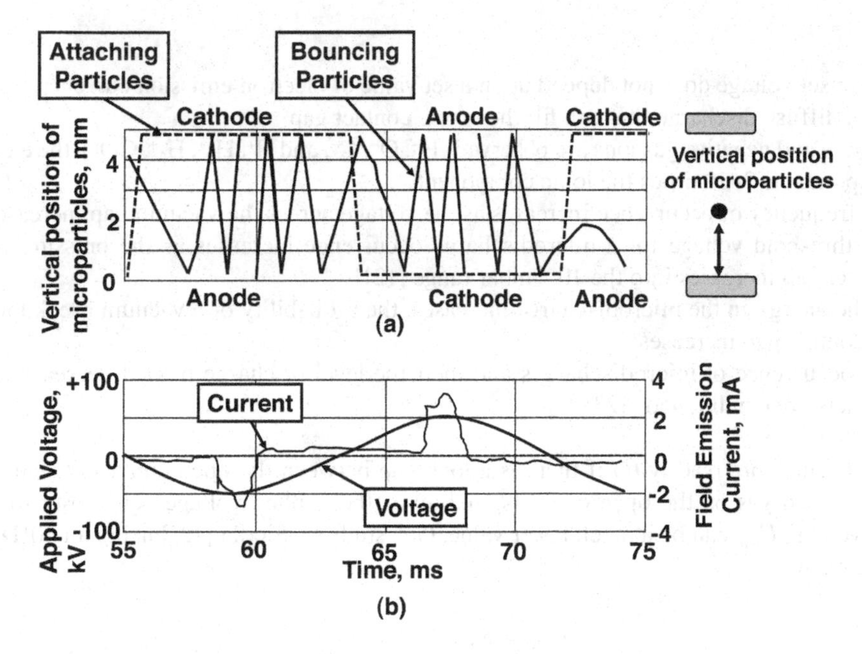

FIGURE 1.61 The motion of bouncing Cu and Alumina microparticles with diameters 5–50 μm between planar Cu contacts during the application of 50 Hz, 50kV, ac voltage (a) The bouncing and attachment patterns (b) the emission current at the peak of the ac voltage [112] (Courtesy Haruki Ejiri).

across open contacts in a vacuum. It allows a passage of a self-limiting current pulse, but does not usually result in the full breakdown of the vacuum gap with the passage of the full circuit current. The duration of the current pulse in the microdischarge can vary from 50μs to 100 ms [119–121] and its peak value usually ($\leq$ 0.01A) can exceed a typical pre-breakdown emission current by perhaps one or two orders of magnitude. A typical current waveform taken by König et al. [122] is shown in Figure 1.62. The shape of the pulse depends upon the area of the contacts and perhaps the measuring system. Some general observations about microdischarges can be stated [55]:

(i) The energy dissipated in a microdischarge is close to the total energy stored by the capacitance in the contacts plus some stray capacitance to ground

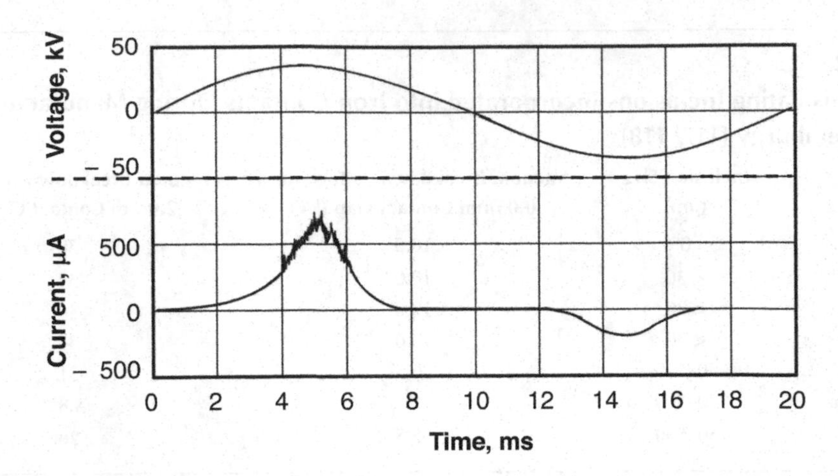

FIGURE 1.62 A current pulse seen during a microdischarge across a vacuum gap [122].

(ii) The onset voltage does not depend upon a set value of electron emission current
(iii) It is a diffuse discharge that can fill the whole contact gap
(iv) Positive and negative gas ions are observed (H^-, O^-, C^-, and H^+, H_2^+, H_3^+, CO^+), the electron component is 3–10 times the ionic component
(v) The frequency of occurrence increases as the voltage across the vacuum gap increases
(vi) The threshold voltage for a microdischarge occurrence increases as the pressure in the contact gap increases into the 10^{-4} mbar range [121]
(vii) As the energy in the microdischarge increases, the probability of a vacuum breakdown of the contact gap increases
(viii) The occurrence of microdischarges and their the level of charge passed decrease as the contacts are conditioned [123]

They also occur more readily (a) if there is a long gap between the open contacts (>8mm); (b) if there is absorbed gas on the open contacts, and (c) as the applied voltage gets closer to U_B. The inception voltage U_{mdi} can be a much lower value. One study by Siodla [121] using an 50Hz ac voltage source shows:

$$U_{mdi} = 5.3d^{0.67} kV\left(rms.\right) \tag{1.63}$$

where d (mm) ranges from 4mm to 15mm. This compares with U_B from Figure 1.20, where:

$$U_B = 58d^{0.58} kV\left(dc\right) \tag{1.64}$$

Aoki et al. [124] in an interesting set of experiments have studied the effect of adsorbed gas on the contact surfaces on the measured current in a microdischarge. They use OHFC Cu contacts with a 5mm and a 10mm contact gap at two vacuum pressures 10^{-6}Pa and 10^{-2}Pa. They apply a negative pulse voltage (–30/100μs) across the open contacts. They begin by applying a voltage pulse with a 16.8 kV voltage peak. If a microdischarge less than 30mA is observed three times they raise the voltage peak in 5.6kV steps. Eventually when breakdown occurs, they lower they next voltage pulse. The peak of the microdischarge does not necessarily occur at the peak voltage as can be seen in Figure 1.63. Figure 1.64 shows that at a pressure of 10^{-6}Pa they observe that as the number of discharges increase the peak current in the microdischarges decreases. However, at a pressure of

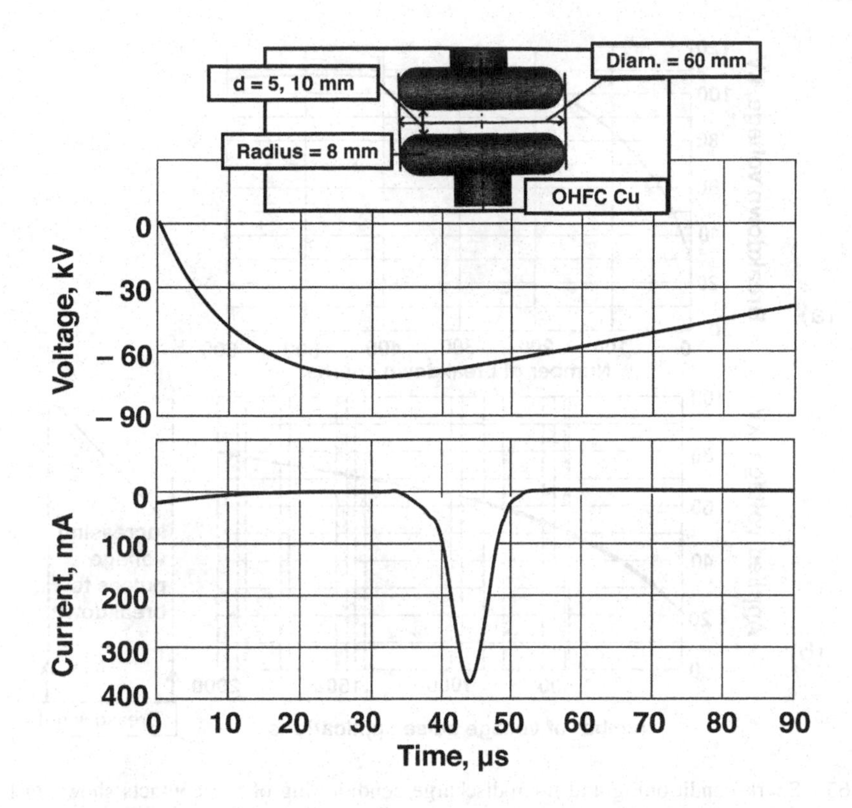

FIGURE 1.63 An example of the –30/100µs voltage pulse and corresponding microdischarge current waveforms [124].

10^{-2}Pa the peak current increases. They conclude that at 10^{-6}Pa there is a decrease of adsorbed gas on the contacts, resulting from an increase in the adsorbed gas on the walls of the vacuum chamber. Thus, the current level continues to decrease as the number of voltage pulses goes up. At 10^{-2} Pa the opposite occurs and as the adsorbed gas on the contacts increases, the observed milli-ampere current increases.

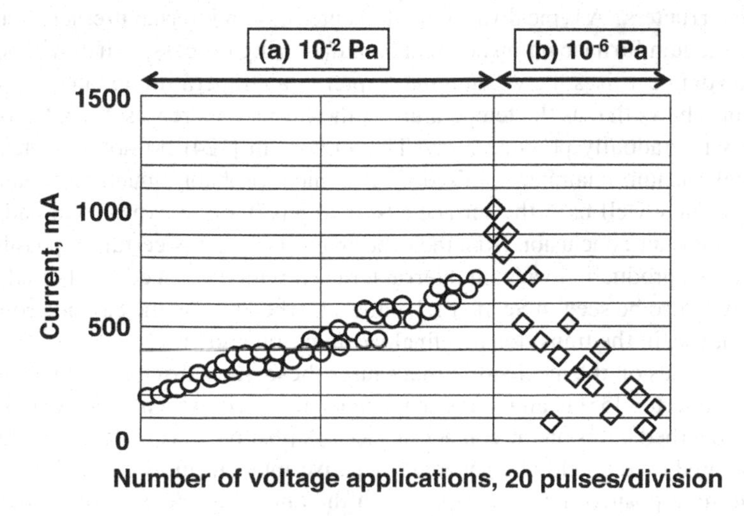

FIGURE 1.64 The change in the microdischarge current with the gradual increase of the applied pulse voltage for vacuum pressures of 10^{-2} Pa and 10^{-6} Pa [124]

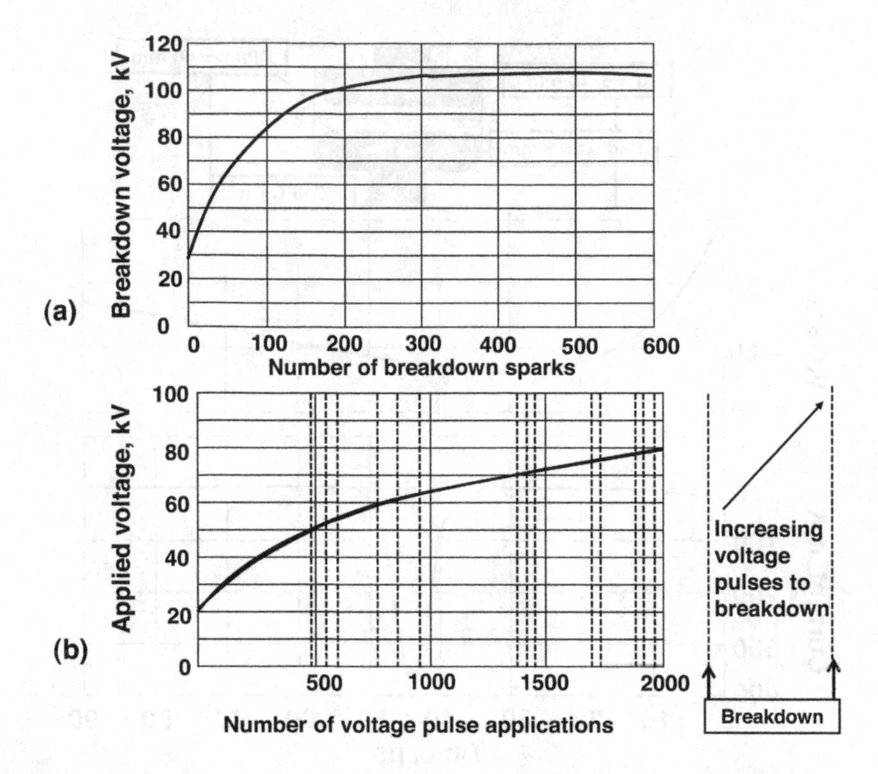

FIGURE 1.65 Spark conditioning and microdischarge conditioning of the contacts shown in Figure 163 [124].

The authors also compare the effect of spark conditioning of the contacts with conditioning by the microdischarge: See, for example, Figure 1.65. The spark conditioning is more effective, because it reduces contact surface defects such as micro-projections and also smooths, to some extent, the contact's surface roughness. The microdischarge conditioning only has a minor effect on the metal surface. Its main effect is to reduce the adsorbed gas on the contact's surface and this, in turn, helps to raise the breakdown voltage. How are these experiments relevant to the usual production of vacuum interrupters? A typical vacuum interrupter has an internal pressure between 10^{-4} Pa and 10^{-2} Pa. The vacuum furnace pump and seal manufacturing process, which will be discussed in Chapter 3 in this volume, raises the vacuum interrupter to a temperature greater that 800C. Figure 3.83 in this volume shows that as the temperature of the furnace increases, the adsorbed gas on the contacts' surfaces is gradually pumped away. The authors in [124] do not say what temperature their experimental vacuum chamber is raised to. It is most probably much less than 800C. So, a vacuum interrupter may well have the surface adsorbed gas that the experiment had at a pressure of 10^{-6} Pa. The important conclusion from the experiment is that it is certainly possible to observe a microdischarge in a production vacuum interrupter even with a reduced level of adsorbed gas on the contacts. As will also be seen in Section 1.3.4, the adsorbed gas on the cathode contact's surface plays an important role in the transition to a final vacuum breakdown.

There is no consensus on the mechanism that causes these microdischarges to occur. One explanation given by Chatterton [55] is an ion exchange mechanism, in which positive and negative ions are ejected from adsorbed gases on the contact surfaces. A positive ion created at random in the contact gap is accelerated by the field and generates negative ions on impact with the cathode. These, in turn, generate more positive ions on impact with the anode. If "A" negative ions are produced per positive ion impact on the cathode and "B" positive ions are produced per negative impact on the anode, and $A \times B > 1$, then the subsequent multiplication of charge leads to a microdischarge.

This theory is supported by the fact that H^-, O^-, C^-, and H^+, H_2^+, H_3^+, CO^+, are gases observed in an analysis of gas pressure during the pre-breakdown phase [55, 125]. Another explanation by Diamond centers on a subcritical explosive emission of electrons from a cathode microprojection [126]. This process could produce enough charge carriers for a short duration discharge to occur across the contact gap. The explosive emission may be caused by a high current density in the projection itself or as the result of a particle impacting the cathode. A third explanation by Diamond is similar to Chatterton's, but is initiated by the release of gas trapped in a contact's subsurface [126]. While there is an electron component in these microdischarges, a self-sustaining electron emission from the cathode is not achieved and the discharge ceases. In summary, a microdischarge can be initiated by:

(i) Particles removed from one contact impacting the other contact
(ii) Explosive destruction of a cathode microprojection
(iii) A mutual secondary emission of positive and negative ions from gases adsorbed on the contact surfaces
(iv) Removal of absorbed gases from the cathode projection or the anode region of electron impact
(v) Gases released from micropores in the contact surfaces
(vi) A microburst of gas that supports a local electron avalanche

The occurrence of a microdischarge does not necessarily indicate a vacuum gap in stress. In some earlier work no correlation is found between the presence of microdischarges and the eventual vacuum breakdown voltage U_B [122]. Work by Ziomek et al, however, indicate that, as the charge passed in a microdischarge increases, so the probability of breakdown of the contact gap also increases [120]. These authors also showed the charge Q passed in a microdischarge varies with the contact gap d as well as the applied voltage U:

$$Q = exp.\left(-Ad^{-0.7}\right).U^{M[d]^{-0.45}} \tag{1.65}$$

where Q is the total charge (pC) for one period of ac voltage, U (kV), $0.13 \le A \le 0.36$, $0.2 \le M \le 0.55$, d (m). Figure 1.66 shows one set of data with Q being measured using a partial discharge meter

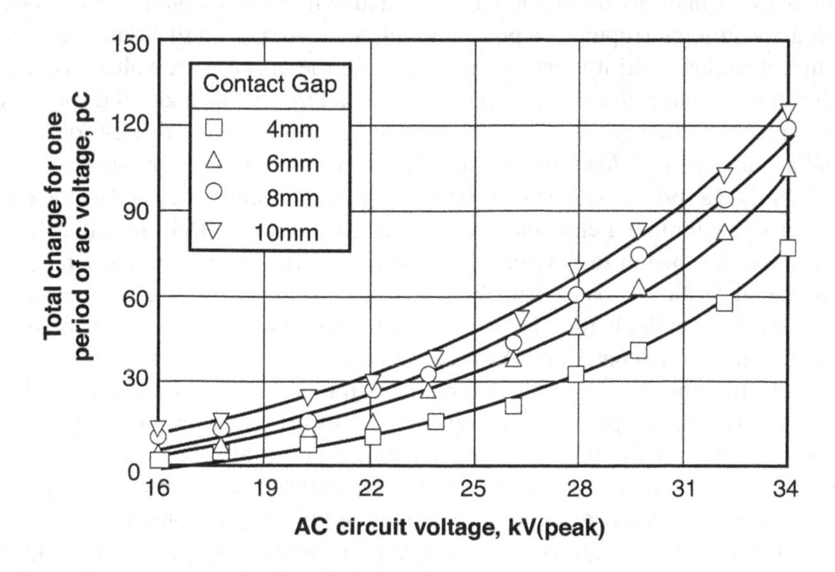

FIGURE 1.66 The charge passed by a microdischarge [120].

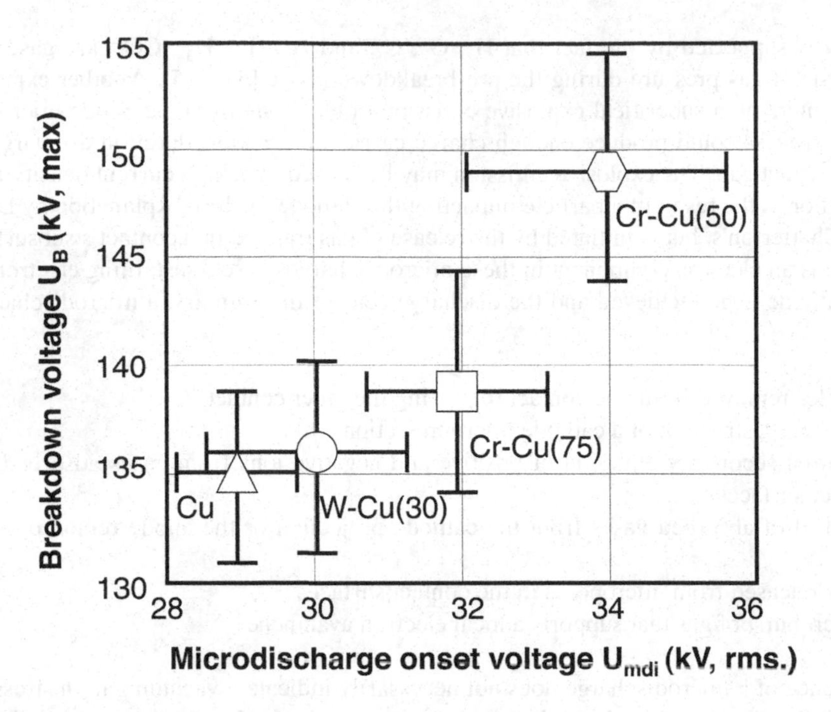

FIGURE 1.67 Correlation between the microdischarge initiation voltage and the ultimate vacuum breakdown voltage for different contact materials [123].

[120]. In a further set of experiments Mościk-Grzesiak et al. [123] show in Figure 1.67 an interesting correlation of the eventual breakdown voltage U_B, with the microdischarge initiation voltage U_{mdi} for conditioned contact materials.

1.3.4 Vacuum Breakdown and the Transition to the Vacuum Arc

As the voltage across a contact gap in vacuum is raised one or more of the pre-breakdown phenomena discussed in Section 1.3.3 will be observed. On continuing to increase the voltage, these various events can increase in intensity and eventually a breakdown of the vacuum gap will occur. If this happens with a vacuum interrupter in a power circuit, a vacuum arc will result which will permit the circuit current to flow until it is interrupted. Because the vacuum arc voltage is usually much lower than the circuit voltage, the circuit impedance will determine the magnitude of this current. It is important to note that the initiation of the vacuum breakdown process differs considerably from the Townsend avalanche breakdown in atmospheric air (or in other high-pressure gases) discussed in Section 1.2. In the gas breakdown, once a high enough field is impressed across the open contact gap, only one or a few initiating electrons are required to begin the electron avalanche in the gas. Once the avalanche has begun to develop, the transition to the arc is extremely rapid ($< 1\mu s$). If the current in the arc in air is greater than about 400 milli-amperes, the arc will be self-sustaining [127] and will develop a cathode root during the breakdown phase that is capable of supplying the electrons to maintain current continuity. During the avalanche process the development of a neutral plasma (for conducting the current) in the inter-contact region occurs as a direct result of the avalanche process where ions are produced as copiously as are the electrons. The electrons, ions and neutral atoms in this region are in local thermodynamic equilibrium [127]: i.e., their temperatures are approximately equal $T_e = T_i = T_n$. In the vacuum breakdown case, it is necessary to first of all establish a cathode region that can supply a continuous supply of electrons. Second, it is also necessary to establish an ionization process that will supply enough ions to produce and maintain a quasi-neutral plasma in the contact gap. If these two conditions are not met, then the full breakdown of the contact gap in vacuum will not occur. Instead, a microdischarge will be observed which will soon self-extinguish and the current will cease to pass between the contacts.

Each of the pre-breakdown phenomena discussed above has influence on the vacuum breakdown. Shioiri et al. [128], using voltage pulses, observe that I_e can increase to greater than 20mA as the voltage increases without the contact gap breaking down and then can drop to a low value as the voltage pulse decreases to zero. They show that this current increase and decrease follows the Fowler–Nordheim equation and that a value of β can be determined for each voltage pulse. In their experiments when I_e increases to about 30mA vacuum breakdown takes place. These authors also show that the absorbed power at breakdown (i.e., $P_B = U_B \times I_B$) is approximately constant for smooth Cu spheres ($\approx$ 5000W) with contact gaps in the range 2mm to 4mm. From Figure 1.55 it can be seen that for an I_e = 30mA, the subsurface anode material could have reached the boiling temperature in less than 5μs. Visual measurements of the contact gap using high-speed cameras show light first from the cathode and then much stronger and more dispersed light from the anode [129, 130]. The macroscopic field at breakdown is $E_g \approx 4-8 \times 10^7\ Vm^{-1}$ [54]. For Cu–Cr contacts with $\beta = 100-300$, the microscopic field for breakdown is $E_m \approx 8 \times 10^9 - 11 \times 10^9\ \text{Vm}^{-1}$ [61, 131]. The thermal stability and mechanical properties of the anode also influence the breakdown voltage. Anodes with low thermal conductivity and low vapor pressure for a given temperature tend to have lower values of U_B [99, 132]. Anodes with high tensile strength (note: tensile strength $\approx$ ⅓(hardness)]) have higher values of U_B [54, 83]. Zhou et al. [133] show in Figure 1.68 an example of the voltage breakdown in vacuum between a needle cathode and a plane anode.. Here a very fast-rising voltage pulse (> 70ns) shows a breakdown initiating at t_{BD}. In this case the breakdown results entirely from field emission electrons from the cathode. The anode flare is only initiated about 0.4μs later at t_{FB}. This indicates that for a very fast-rising voltage pulse, the anode sub-surface heating is delayed.

Particles can certainly influence the breakdown voltage for a given contact gap. Kamikawaji et al. [101], for example, show the effect that different diameter particles have on reducing the breakdown strength of a carefully prepared, polished vacuum contact system; see Figure 1.69. It is interesting to note that the smallest particles lower U_B the most. Figure 1.70 shows another data set taken by Sato et al. [134] for clean, polished Cu contacts and for Cu–Cr contacts with a practical roughness expected in a manufactured vacuum interrupter. The influence of microparticles on the U_B value for a rough surface is much less than that on a polished surface. The dependence of U_B as a function of

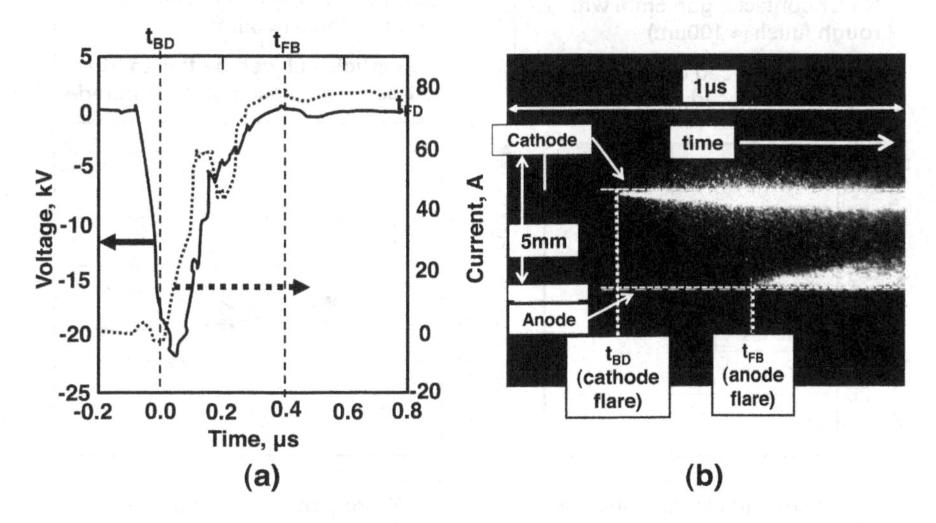

FIGURE 1.68 A streak image for the initial stage of a vacuum breakdown process: cathode material W, anode material Cu, gap length 5mm, for a very fast voltage rate of rise to breakdown (~ 70ns) for a total voltage pulse width of 5 μs and the steady state arc current 80 A (Courtesy Zhenxing Wang).

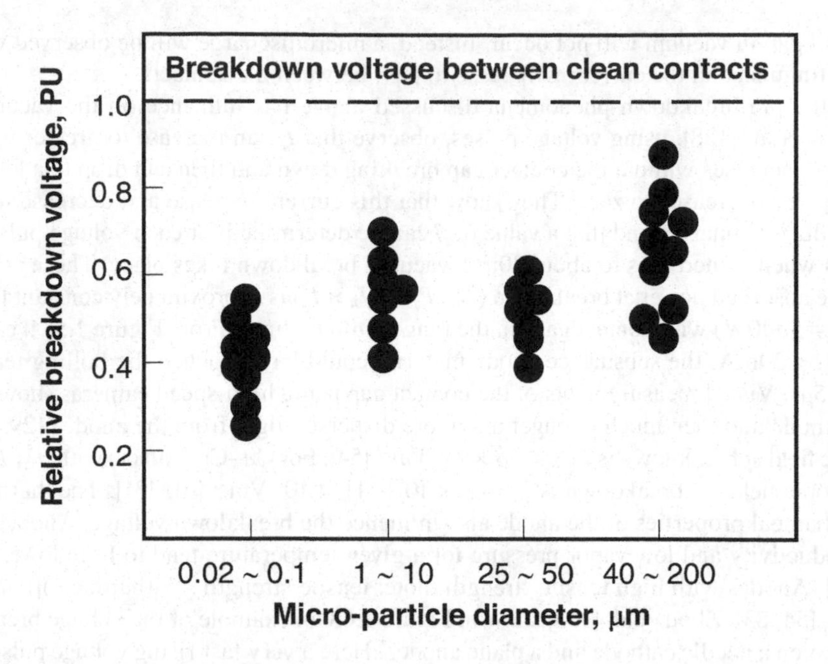

FIGURE 1.69 The effect on the vacuum breakdown voltage of introducing particles on to polished contacts as a function of particle diameter [101].

the spacing between the rough-finish, Cu–Cr contacts for two particle sizes is shown in Figure 1.71. These data are of great importance to a vacuum interrupter designer who has little control over the microscopic roughness of the contact surfaces during the vacuum interrupter's operating life. These data suggest that the effect of particles on a rough surface is minimized by the roughness. So, if the vacuum interrupter designer takes into account an initial contact surface roughness, then it is possible for the contact structure to maintain its high voltage performance in spite of particles being

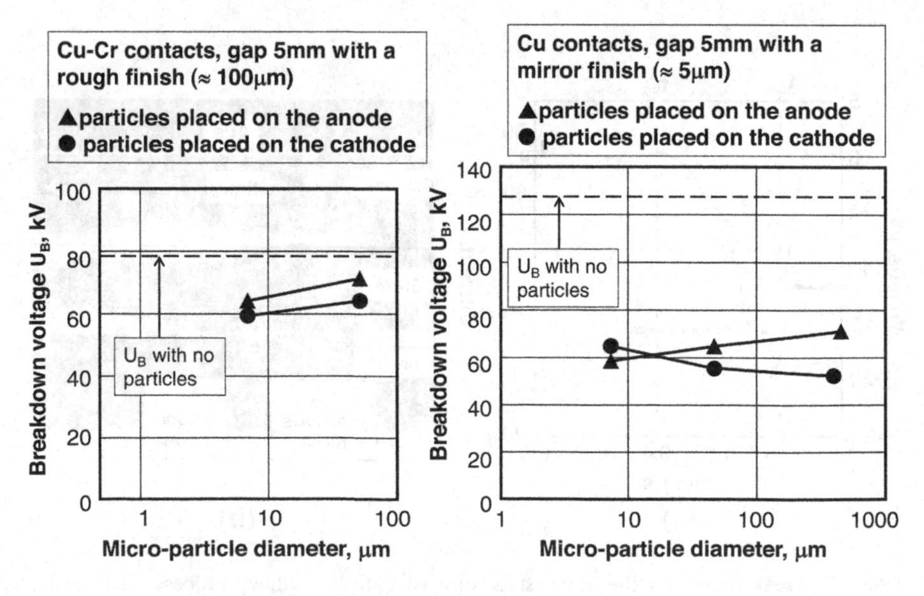

FIGURE 1.70 Effect of the introduction of microparticles on the vacuum breakdown voltage for Cu–Cr contacts with a rough finish and for Cu contacts with a mirror finish [134].

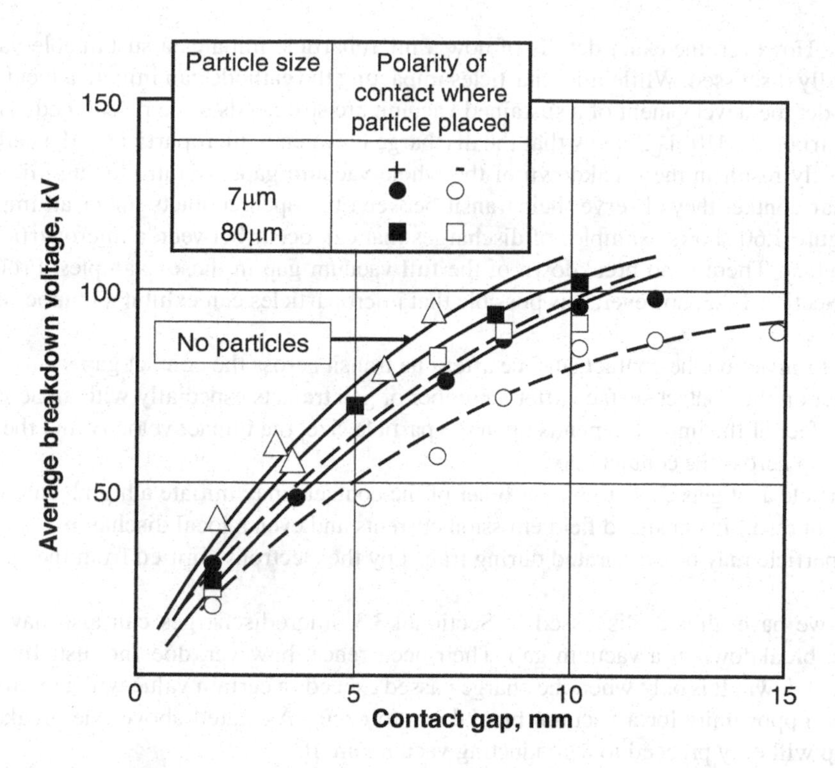

FIGURE 1.71 Average breakdown voltage between machined Cu–Cr contacts for clean contacts and after the introduction of 7 μm and 80 μm microparticles onto the contact surface [134].

produced each time the vacuum interrupter switches current. Microparticles for the most part travel from anode to cathode [135]. Indeed for a slowly rising voltage, anode material is usually present on the cathode after vacuum breakdown [95].

One compelling argument for particle initiation of vacuum breakdown is Cranberg's analysis [136]. He has proposed that once a microparticle obtains a critical energy W_c from the electric field, the impact of this particle on a contact surface would result in the vacuum breakdown of the contact gap. Thus from Equation 1.60, the impact energy per unit area from a particle of radius r can be approximated by:

$$W_c / \pi r^2 = 6.6 \varepsilon_0 \beta_m \beta_g U_B{}^2 / d \tag{1.66}$$

Thus:

$$U_B = K_c d^{1/2} \tag{1.67}$$

where K_c is a constant. Further refinement of the exponent for the contact gap d for the case of a dc voltage has it ranging from 0.5 to 0.625 [137]. Experimental measurements by Sato et al. [134] show even greater variations in the exponent, from 0.3 to 1.1, as is also seen in Figure 1.20. They also show that the exponent has a strong dependence upon particle size and contact gap. Farrall [138] has shown that for fast impulse voltages a single transit particle could result in breakdown voltage $U_B \propto r^{1/3}d^{5/6}$ for and a breakdown voltage $U_B \propto r^{1/3}d^{5/2}$ for pulses with a constant rise time. The fact that $U_B \propto d^n$ and that a relationship of this nature can be explained by a microparticles effect has given rise to an acceptance that microparticles do play an important role in vacuum breakdown especially in contact gaps greater than a few millimeters, i.e., the range of contact gap common in vacuum interrupters. However, the exact details of how a microparticle initiates a sustainable vacuum arc are not usually discussed. While microparticles impacting the cathode can initiate a local discharge at the cathode, the development of a sustained vacuum arc still needs to be researched. The experiments by Ejiri et al. [110–112] show that the discharge between a microparticle and a cathode does not necessarily result in the breakdown of the whole vacuum gap. By introducing microparticles onto a planar contact they observe their transit between the open contacts under an impressed ac voltage. Figure 1.60 shows examples of discharges that can occur between a microparticle and the cathode contact. There is no breakdown of the full vacuum gap in these examples. From the discussion in Section 1.3.3, however, it is possible that microparticles can exhibit a number of effects:

1. Direct impact on the contact surface after one transit across the contact gap
2. Impact on the contact surface after a number of gap transits especially with an ac voltage
3. The effect of the impact depends upon the particle size, the impact velocity and the potential drop across the contact gap
4. A particle as it gets close to one or other of the contacts may initiate a high local field and this can result in enhanced field emission currents and even a local discharge
5. The particle may be evaporated during transit by the electrons emitted from the cathode

Finally, as we have already discussed in Section 1.3.3, microdischarges can also have an influence on the breakdown of a vacuum gap. Their occurrence, however, does not usually result in a vacuum breakdown. It is only when the charge passed exceeds a certain value will a microdischarge enhance the opportunity for a vacuum breakdown to occur. As stated above, the breakdown of a vacuum gap will only proceed to a conducting vacuum arc if:

1. The cathode region develops into an efficient source of electrons
2. There is sufficient gas for the electrons to ionize
3. A region for the long-term production of ions is established

Let us first of all consider the effects of the electron emission from a cathode micro-projection. As we have already shown in Figure 1.36(b), the current I_e from this cathode region increases exponentially as the voltage across the vacuum gap is increased. At very low values of I_e the temperature of the projection will slowly increase. When the current density j_e reaches a value of about 2×10^{12} Am^{-2}, Fursey [139] has shown, with a two-dimensional analysis of a long projection whose tip radius is 0.4μm, the very rapid growth in temperature shown in Figure 1.72. In performing this analysis, Fursey considers two major energy sources into the microprojection. First, the current flow through the bulk of the metal produces the usual Joule heating effect and, second, it can initiate the Nottingham effect. [56, 140, 141]. The Nottingham effect results from the fact that the average energy of an emitted electron will be different from the average energy supplied by its replacement electron from the metal lattice inside the projection. For low temperature field emission, electrons will be emitted from states below the Fermi level, so there will be a heating effect. At high temperatures the electrons will tunnel from populated states above the Fermi level and so will give rise to a cooling effect. The transition between heating and cooling depends upon the metal's work function and the microscopic electric field at the microprojection; a typical value is approximately 1500 C. As the electron density in the projection increases and the projection increases in temperature, it is important to note that the highest temperature is not at the surface of the projection, but inside it; see for example, Figure 1.73. Other calculations show that at high current densities, the electron temperature exceeds the phonon temperature in the metal lattice. This difference can be very significant at current densities greater than 5x10^{12} Am^{-2}. In fact, at these current densities the microprojection will be destroyed in $\approx$ 3ns [139]. Is it possible to achieve these current densities in a cathode microprojection? Let us take the information given in Figures 1.36(a) and (b). If we extrapolate the data in Figure 1.36(b) for I_e to the breakdown voltage of 50 kV for this

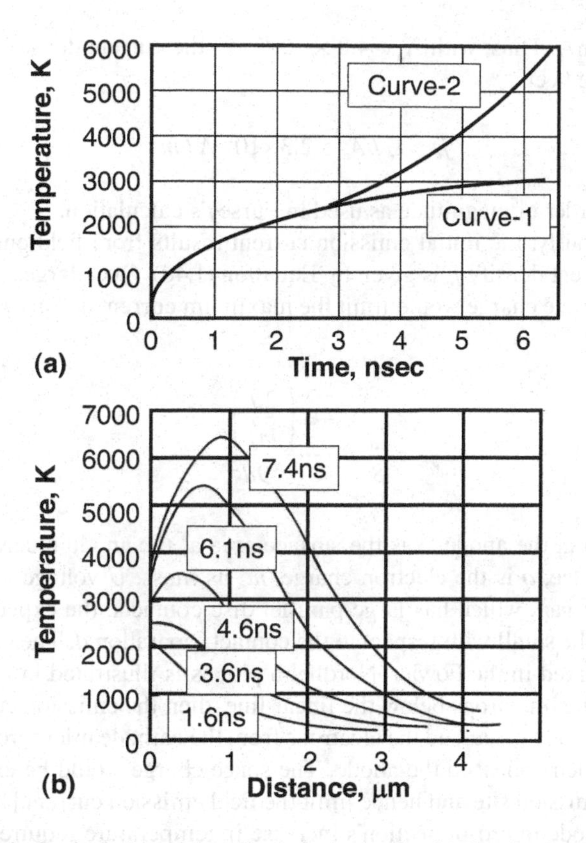

FIGURE 1.72 A numerical calculation of the effect of microprojection heating by a high density, field emission current (j_{FE} = 2x10^{12} A/m^2); (a) the temperature as a function of time: Curve-1, the microprojection's surface temperature and Curve-2, the maximum temperature below the microprojection's surface and (b) the temperature distribution along the microprojection's axis (note, the "0" on the abscissa is at the microprojection's tip and the distance is in μms below the tip into the microprojection) [139].

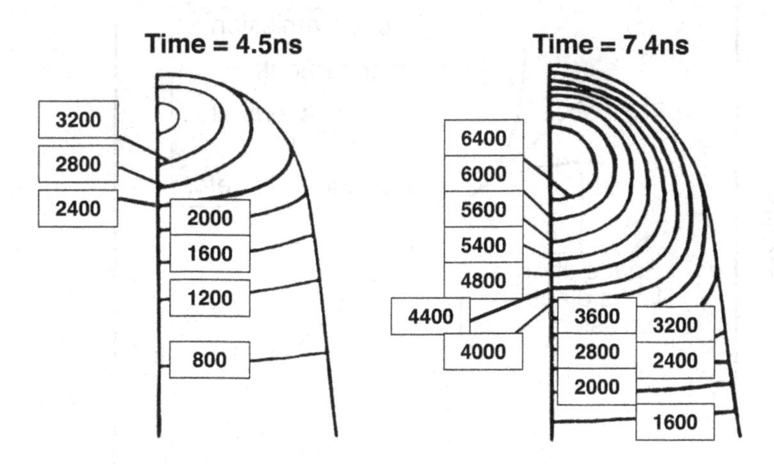

FIGURE 1.73 A cross-section of an electron emitting microprojection on the cathode used to develop Figure 1.72 [139].

system, then $I_e \cong 11.1$mA. Thus, with $A_e = 4.8 \times 10^{-16}$ m^2, the current density at the cathode in this experiment would have been:

$$j_e = I_e / A_e \cong 2.3 \times 10^{13} A/m^2 \tag{1.68}$$

which is of the same order of magnitude as used in Fursey's calculation.

As discussed previously, the initial emission current results from field emission at the cathode projection and the current density j_e is given by Equation (1.48). Once larger currents begin to flow the effect of electron space charge would limit the maximum current density j_c by Child's law [142].

$$j_e = \frac{I_a}{S} = \frac{4\varepsilon_0 \left(\frac{2q}{m_e}\right)^{0.5} U^{1.5}}{9d^2} \tag{1.69}$$

where I_a is the current at the anode, S is the surface area at the anode receiving current, ε_0 is the permittivity in free space, q is the electron charge, m_e its mass, U voltage across the contact gap d. In a normal contact gap, which has large parallel disc contacts, the expected roughness of the contact surface would be small with respect to the contact separation d. The effects of space charge limitation can be observed in the Fowler–Nordheim plot as is illustrated in Figure 1.74. As can be seen in this figure, if the line drops below the linear line, then the emission current is space charge limited. As we have already discussed the electrons from the cathode microprojection would remain in a narrow cone on their transit to the anode. The space charge would be expected to reduce the effective field at the emission site and hence limit the field emission current [143]. Fursey's compelling model of the cathode micro-projection's increase in temperature requires a current density in the range $j_e = 10^{12} - 10^{13}$ A.m^{-2}. When this current density is reached there is a violent rupture of the cathode micro-projection, because surfaces forces holding it together can no longer do so. Before this current density can be reached by pure field emission, Dyke et al. [144–147] have shown that

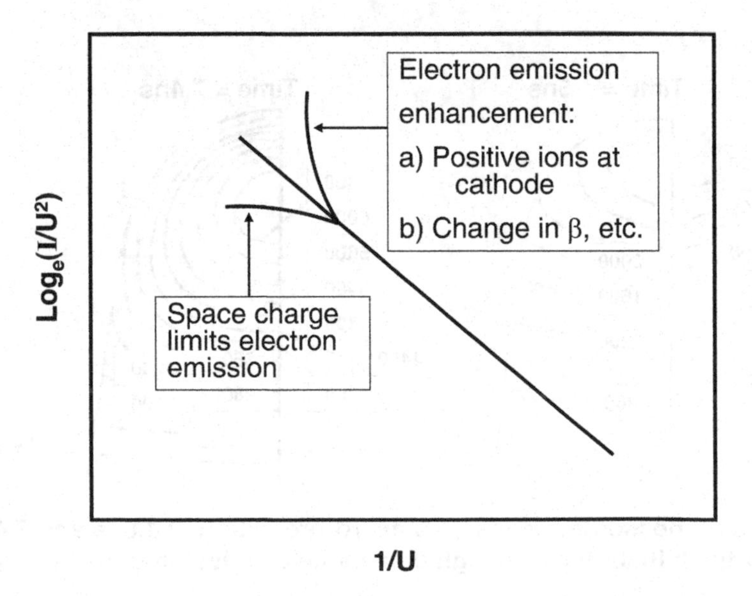

FIGURE 1.74 Schematic graph showing how the slope of a Fowler–Nordheim plot can be modified by space charge effects and by emission enhancement effects.

it would be space charge limited. Thus, the electric field near the emitting site would also decrease once the current becomes space charge limited. This in turn would lower j_e and the temperature of the emitting site.

How then is it possible to continue increasing j_e? One answer results from the interaction of the emitted electrons and the adsorbed gas on the cathode's surface which Aoki et al. [124] demonstrate exists even at a vacuum pressure as low as 10^{-6}Pa. A proposal by Schwirzke et al. [148] is illustrated in Figure 1.75. They suggest that the increase of j_e to a value greater than the space charge limited value requires the presence of ions at the cathode-emitting site. In their model adsorbed gas at the cathode-emitting site is desorbed by the joule heating of the site. This could release a monolayer of 2×10^{19} molecules.m^{-2} and an expanding gas cloud would appear across the cathode. For a typical vacuum interrupter contact gap, the breakdown voltage is greater than the BIL value, so for an 8mm gap U_B may be 100kV. Thus, emitted electrons will have energies of less than 100eV at less than 8μm from the contact surface. Figure 1.6 indicates that electrons with this energy will ionize this gas efficiently. The resulting ions would return to the cathode. The heating of the projection would then involve the field emission current plus the returning ion current. The total current density j_e in the projection would be:

$$j_e = j_- + j_+ \tag{1.70}$$

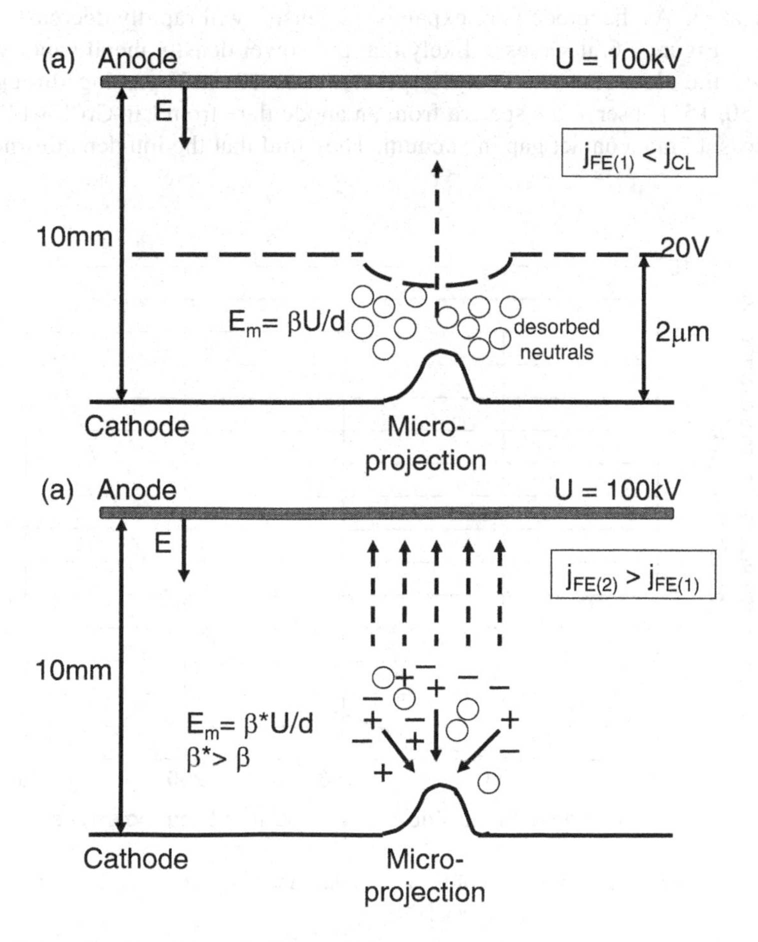

FIGURE 1.75 The ionization of desorbed monolayer at the cathode reduces space charge limitations on electron emission from a cathode microprojection [148].

The returning ion bombardment not only allows a rapid increase of j_e, but also could lead to further gas desorption and could even result in initiating an evaporation of the cathode metal. As more ions are produced, the electric field at the cathode will be enhanced, j_e will increase and the microprojection will quickly become unstable and explode. The subsequent release of metal vapor above the cathode provides a medium for the production of metal ions and a plasma flare will develop in the region of the cathode's microprojection. This cathode flare would then expand toward the anode. Uimanov [149] has developed a model that shows the space charge may not be the limiting factor for the emitted electrons close to the surface of a cathode microprojection. For emitters with a tip radius less than 100nm the Fowler–Nordheim current density can reach about 10^{10} A.cm^{-2} for applied voltages 250-300kV. In Figure 136a, the emitter area is 4.8×10^{-12} cm^2. In this example the emission current can reach 4.8×10^{-2} A or 48 mA. This current is well above the value of 11.1 mA calculated from Figure 136b at the breakdown voltage of 50 kV.

Another source of ions could be the anode. As I have discussed in Section 1.3.3.2, the field emission current penetrates the anode. It heats a well-defined volume below the anode's surface. When this volume reaches a temperature close to the metal's boiling point a vapor flare will erupt into the inter-contact gap. At this temperature thermal radiation from the anode flare will emit an intense white light generated by the thermal motion of the metal atoms. The initial volume of the anode at the anode's surface will have a very high density. At this stage there could be a small, but finite probability that even high energy electrons from the cathode flare may achieve some degree of ionization. As the anode flare expands its density will rapidly decrease. As discussed in Section 1.2.1, Figure 1.6, it seems unlikely that this lower density metal vapor would be ionized directly by the electrons with energies greater than 10s keV passing through it. Indeed, Nagai et al. [150, 151] observe the spectra from an anode flare from Cu–Cr(35wt.%) contacts as it expands across a 2mm contact gap in vacuum. They find that the ion densities near the anode

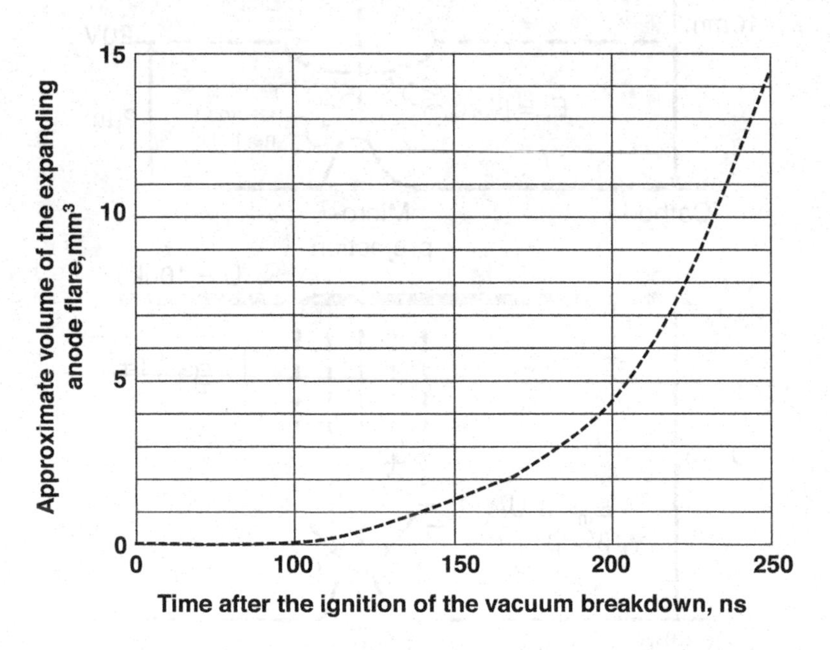

FIGURE 1.76 The approximate expansion of the anode flare volume from a Cu–Cr(30wt.%) across a 2mm contact gap calculated from figure 4 in reference [151].

are higher at 100ns after the flare's initiation than at 200ns. Also, the ratio of Cu to Cr atoms in the anode flare is similar to that in the contact material. Figure 1.76 shows the approximate expansion of the anode flare from their experiment. Bochkarev et al. [152] also observe the spectra during the breakdown of the vacuum gap between a Cu rod cathode and a Mo planar anode. They conclude that the anode flare shows mainly neutral metal spectral lines while the cathode flare shows strong spectral lines from singly and doubly ionized Cu atoms. Another possible, but unlikely, source of ions could result from the model adopted by Davies et al. [93, 94] in which the electron beam would evaporate a particle pulled from the anode and the region behind the particle would be ionized by lower energy electrons emitted from the particle itself, Figure 1.49. The ions produced by this process would then travel to the cathode and alleviate the space charge limiting effects. These metal ions would then also enhance j_e and give rise to the instability of the cathode-emitting site.

If the line in Figure 1.74 is above the Fowler–Nordheim linear line at high current densities, then two effects can be occurring. First of all, ions returning to the cathode will reduce the space charge limitation and even enhance I_e. Second, as the cathode projection reaches a higher temperature, its metal will soften. Emelyanov et al. [153] show that this can result in the local electric field pulling the projection into a finer point; hence increasing β_m and E_m. Some experiments even suggest that the tip of the cathode projection is in the liquid phase just before the explosive rupture of the electron emitting microprojection [154]. Mesyats et al. [155] have developed the idea that the ultra-high field at a microprojection will result in a decrease in the metal's potential barrier below its work function. This can result in a substantial increase in the electron density near the emitter's surface. They show that if this effect does occur, saturation of the field emission current does not happen and the current density increases more or less linearly with the increase with the electric field. Once the j_e exceeds 10^{12} to 10^{13} A.m^{-2}, the projection will explode and it would be possible to form a plasma in the dense metal vapor cloud. Interestingly enough this can occur at the same time at more than one site [77, 154]. Figure 1.77 shows an example of multiple cathode flares [150, 151].

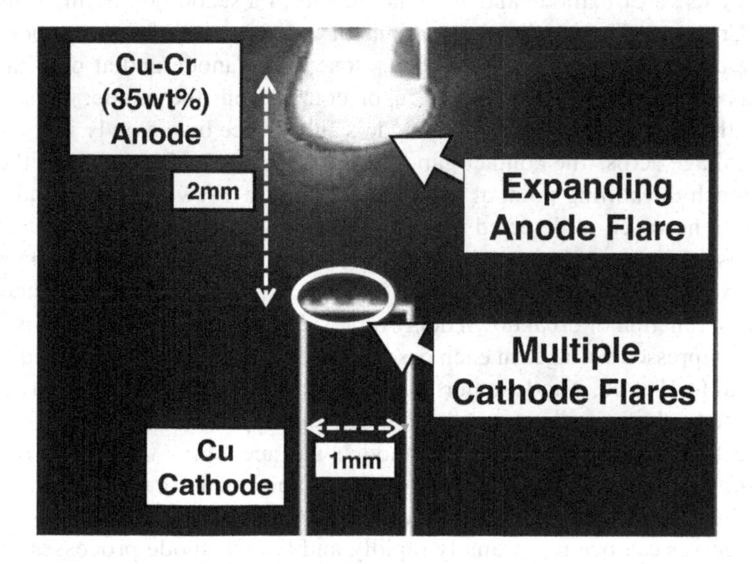

FIGURE 1.77 Multiple cathode flares during the initial stages of vacuum breakdown between a Cu–Cr(30wt.%) anode and a Cu cathode [150, 151].

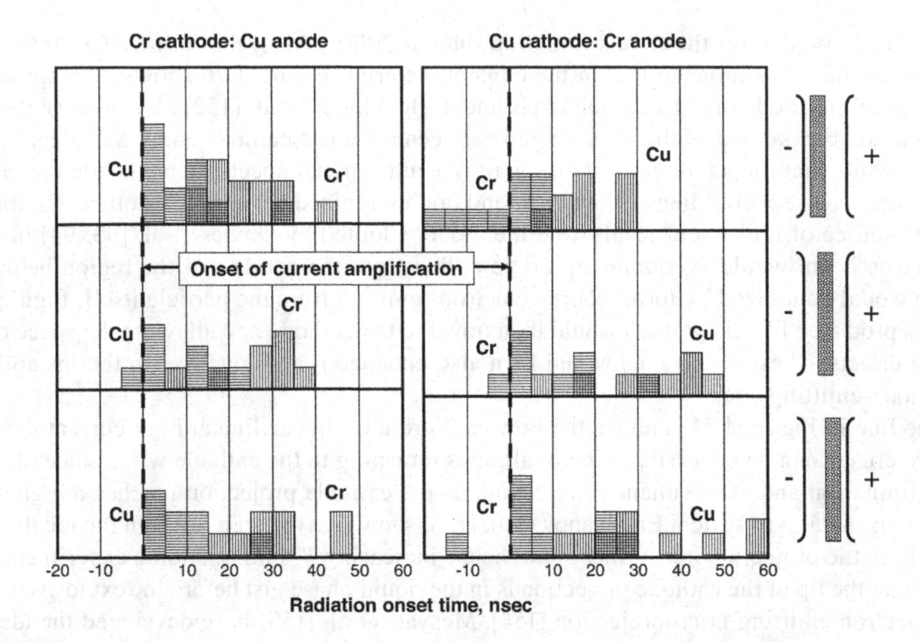

FIGURE 1.78 The position of metal vapor from the cathode and anode contacts after the initial observation of emission current with a slowly increasing dc voltage across the contacts as a function of time and contact gap position [95]; anode material ▒ and cathode material ▒

In an interesting series of experiments, Davies et al. [94] have explored the origin of metal vapor during the breakdown phase in the intercontact gap using a slowly increasing dc voltage across it. To do this they use different metals on the cathode and the anode. In one experiment, Figure 1.78, they use a Cr cathode and a Cu anode and in a second experiment they used a Cu cathode and a Cr anode. They make observations of the presence of metal vapor as a function of time in three positions: close to the cathode, close to the anode and at mid-gap. They show that in all cases the first observable metal vapor comes from the anode. This observation is consistent with the expected heating of the anode's subsurface by a slowly increasing emission current as the voltage across the contact gap increases. Figure 1.55 shows that the interior of a Cu anode can reach the boiling point at a current of 1mA in a few micro-seconds. Metal vapor from the cathode, however, is detected close to the cathode within 5 to 10 ns after the initial detection of emission current. In a subsequent series of experiments [156] using a rectangular-shaped, pulsed voltage source, they recorded whether anode material or cathode material is observed first as a function of breakdown delay time t_B, see Figure 1.79. Here t_B is clearly shown as a function of impressed voltage. In each case the cathode material is observed first if the t_B is less than or equal to 10 μsec. Anode material is only observed first for longer breakdown times. The inferences from these studies are: (a) although the hypothesis presented by Davies et al. [93, 94] of a particle being pulled from the anode's surface which then gives rise to a plasma cloud is intriguing, a more likely explanation of their experimental results is that they observed an anode flare that resulted from the emitted electrons heating the subsurface of the anode; and (b) cathode processes can occur extremely rapidly and (c) the anode processes take more time to develop. Certainly, Figure 1.68 shows that for a rapidly rising voltage pulse the cathode flare

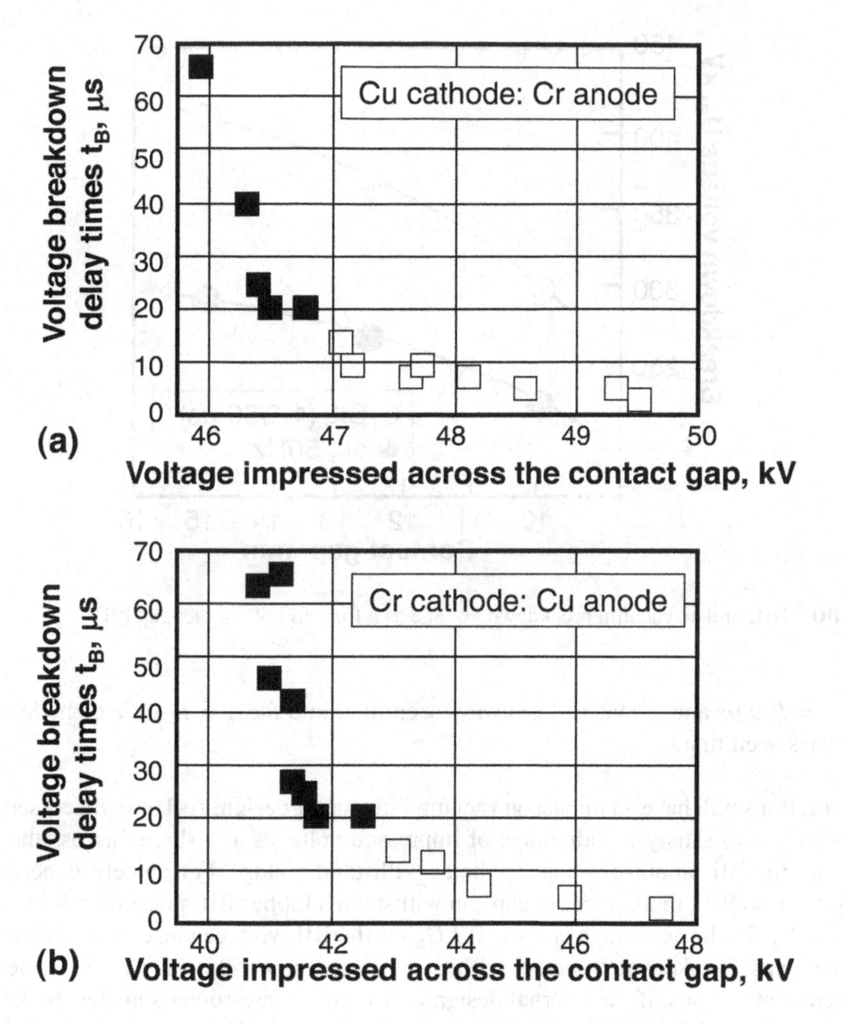

FIGURE 1.79 Breakdown delay times as a function of breakdown voltage for a rectangular pulse applied to the open contacts; (a) Cu cathode–Cr anode and (b) Cr cathode–Cu anode; ■ breakdown events in which radiation from anode vapor is emitted first and □ breakdown events where radiation from cathode vapor is emitted first [156].

is initiated well before the anode flare. For a slowly applied dc voltage across a 1mm contact gap [156]:

1) At mid gap the anode material is seen before cathode material
2) Radiation is seen initially in the cathode region
3) As the voltage U across the contacts increases the power input to the cathode goes as j_e^2, but the power input to the anode goes as $j_e \times U$

and for a fast-rectangular voltage pulse across a 1mm contact gap [156]:

1) $j_e^2 \times t_B = constant$ (for times 1ns < t_B < 4µs)
2) If k* = applied field/ dc breakdown field (k* ≤ 1.2 for $t_B > 10\,\mu s$. And k *≥ 1.3 for $t_B < 10\,ns$)

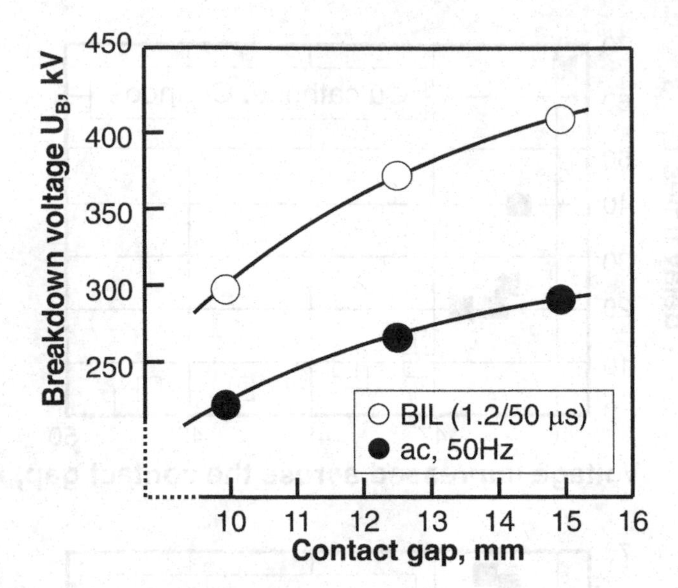

FIGURE 1.80 BIL and ac vacuum breakdown voltage as a function of contact gap [91].

3) For t_B = *160 μs* anode material is always seen first; and for t_B = *14 μsec* cathode material is always seen first.

These observations will have an impact on vacuum interrupter design. As I have discussed in Section 1.1 the design has to satisfy a wide range of impressed voltages and shapes across the open contact gap; from the BIL impulse voltage to the ac withstand voltage. Fortunately, experiments have shown, Figure 1.80 [91], that a vacuum gap can withstand a higher BIL pulse than it can a 50 Hz ac voltage. Here U_B for the ac voltage is about 0.7 U_B for the BIL voltage pulse. This indicates that for these contact gaps, breakdown processes with times greater than 50μs play a role in their eventual vacuum breakdown. Thus, if the internal design of a vacuum interrupter satisfies the BIL requirements given in Table 1.1, it will also pass the ac withstand test, but not vice versa. I discuss this further in Section 1.4.2. While this is true for the internal design of the vacuum interrupter, it is not necessarily valid for the external dielectric design. So, when developing a new vacuum interrupter design, the internal and external high voltage withstand capabilities must be considered separately.

Another example of an experiment to show the spatial distribution of light from a contact gap in vacuum during the breakdown process is shown in Figure 1.81 [129]. The original streak photograph clearly shows an initial intermittent light from the cathode, which only becomes steady once light from the anode appears. This observation is similar to that shown in Figure 1.68 for a very fast rising voltage pulse across the open contact gap. Other photographic observations of the vacuum gap show simultaneous radiation from the cathode and from the anode [77, 129, 150, 151, 153, 157]. In general, the thermal radiation from the anode tends to be more intense than that from the cathode. This has led some researchers to conclude that the anode plays the dominant role in establishing the vacuum arc. While the anode region can initially develop some ions [151], it is the development of an electron emitting cathode spot that determines whether or not a sustained vacuum arc is formed. One problem with the photographic observation of the cathode is that the cathode spot's initial formation is very small [126, 151, 153]. The high-pressure plasma that forms directly above it moves away at very high speeds. Thus, its density decreases rapidly and the radiation from it is undetectable at distances of about 0.1mm. The thermal radiation from the anode flare, on the other hand, can have a much broader luminous volume, which can overwhelm the detection of radiation from the cathode; see for example, Figure 1.77.

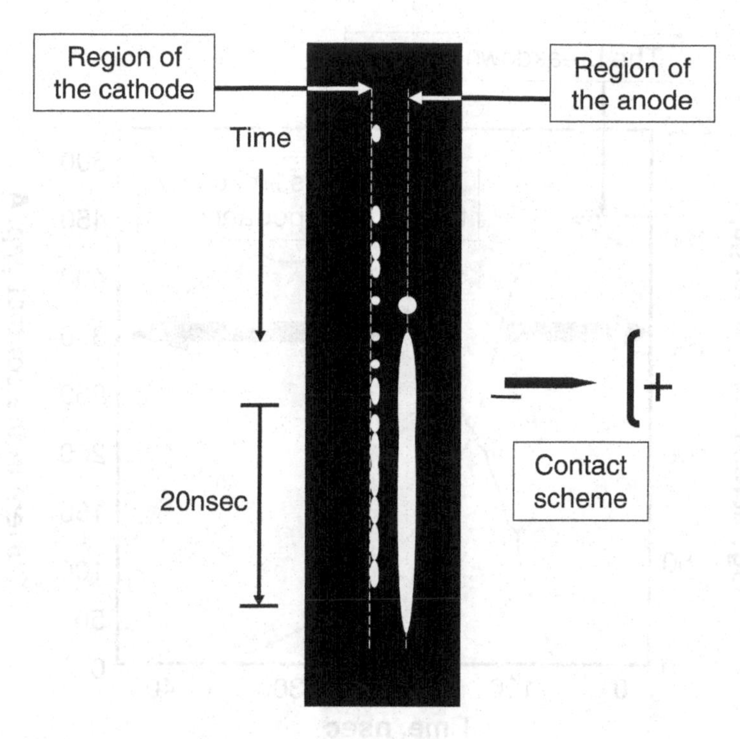

FIGURE 1.81 The pre-breakdown, spatial distribution of light in a contact gap showing light first from the cathode [129].

1.3.5 The Transition to a Self-Sustaining Vacuum Arc

As I have already discussed, in order for the pre-breakdown emission current to develop, the macroscopic field at the cathode has to be greater than 4×10^7 V/m. This field can result from the potential drop across the open contacts, or perhaps it can result from the field between the cathode and a charged particle close to the cathode surface. Once a discharge begins to develop the voltage across the contact gap will begin to collapse. Djogo et al. [158, 159] show the voltage and current waveforms across a 7mm contact gap as it breaks down in vacuum; see Figure 1.82. Figure 1.83 shows their measurements for the time of voltage collapse to be 10% of U_B for contact gaps 2mm to 10mm [158]. As can be calculated from Equation (1.48), the field resulting from such a reduction in voltage will result in a drastic drop in emission current, perhaps by three or four orders of magnitude. Therefore, once the transition to a vacuum arc begins, the cathode region has to develop a self-sustaining electron production process in an extremely short time. If this new cathode emission process does not develop to replace the field emission process, then the discharge will self-extinguish and a microdischarge will be observed between the open contacts.

To illustrate this let us consider a series of experiments by Cobine et al. [160]. They establish an arc by drawing contacts apart in vacuum (see Chapter 2 in this volume) and observe how long the arc will remain burning. Figure 1.84(a) shows their results for currents up to 20A for a variety of contact metals. At these currents an arc established in vacuum can be unstable and self-extinguish even though the power circuit voltage is much greater than that needed to satisfy a stable vacuum arc. For example, a 10A arc between Cu contacts (a current at which an arc in air would remain quite stable [127]) here has an average lifetime of only $10^4 \mu$s or 0.01s. The curves shown in Figure 1.84(a) seem to correlate with the vapor pressure of the contact materials, Figure 1.84(b), i.e., the higher the vapor pressure of the contact material, the longer the arc duration for a given current. Thus, even though a vacuum arc is established it may not be stable. For the case of an open contact gap, I have already described the microdischarge phenomena where a short duration current pulse

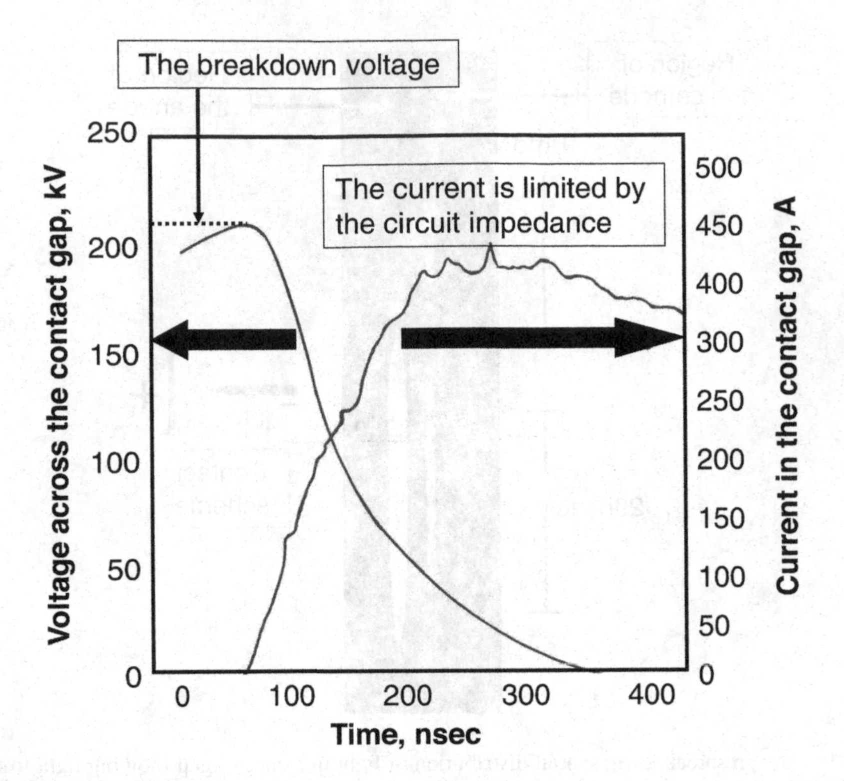

FIGURE 1.82 The change in the voltage drop across the open contacts as the vacuum breakdown process develops [158].

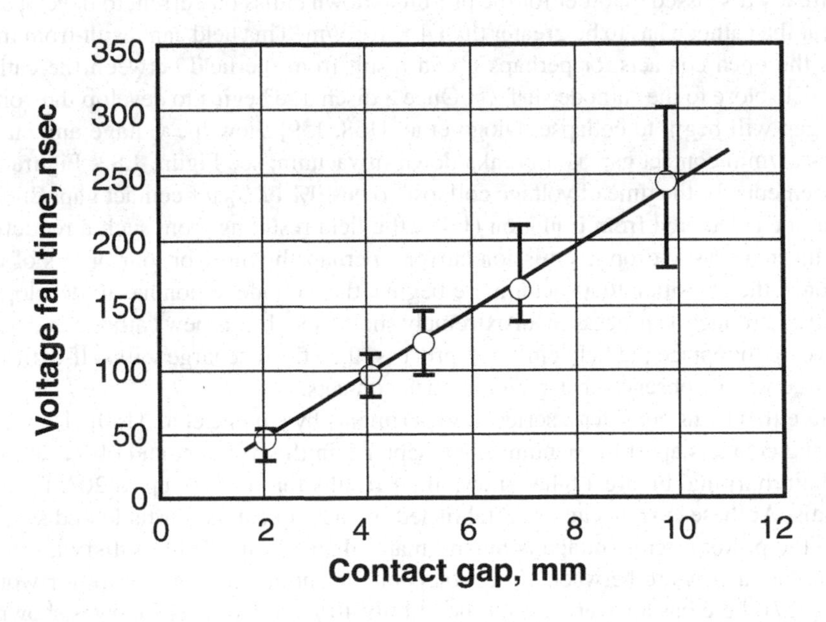

FIGURE 1.83 The voltage fall time (i.e., the time the voltage across the vacuum gap goes from 90% of is prebreakdown value to 10% of that value) for copper contacts as a function of contact gap [158].

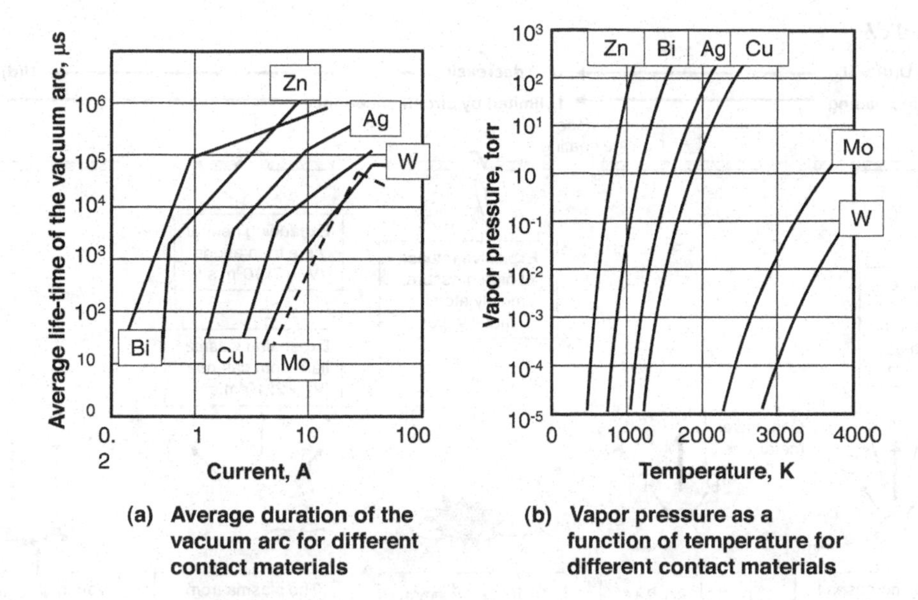

FIGURE 1.84 (a) The average lifetime of a drawn vacuum arc as a function of current and contact material. (b) The vapor pressures of some contact materials [160].

passing currents in the milliampere range can develop, but not result in breakdown of the contact gap. Figure 1.60 also shows that while a discharge can develop between a microparticle and a cathode, breakdown of the whole vacuum gap does not necessarily occur. Now we see that it is even possible to develop a conducting plasma, which can permit quite high currents (perhaps > 10A) to flow between the open contacts, for up to seconds and still not establish a long-term conducting path between the contacts in a vacuum ambient.

A microdischarge seems to be sufficient to discharge the contact capacitance and some stray circuit capacitance to ground. However, this discharge does not necessarily lead to a sustained arc in a power circuit. It is entirely possible for the current flow to discontinue and the contact gap to hold off the circuit voltage impressed across it. A microparticle's interaction at the cathode may release a momentary pulse of electrons which cross the contact gap and gives rise to a flow of current as much as a few milliamps, but not cause a complete breakdown of the vacuum gap. How then is the vacuum interrupter designer to use the wealth of vacuum pre-breakdown data and vacuum breakdown data to identify the processes that can occur in the vacuum interrupter and how can this information be used to achieve a better high voltage design?

The development of a self-sustaining electron emission from a cathode spot at the moment of vacuum breakdown is thus of crucial importance. There has been considerable discussion on how the self-sustaining discharge begins. First of all, for research work on vacuum breakdown that has been performed using high voltage supplies with very limited current capacity, it is entirely possible that what has been called a vacuum breakdown may well have been a long microdischarge. For a truly sustainable vacuum arc to develop from the pre-breakdown phenomena, we must concentrate upon the two conditions:

1. For current continuity at the cathode surface there must be a sustainable electron emission.
2. In order to support a high current flow across the inter-contact gap, there must be a generation of ions that together with the electrons results in a quasi-neutral, inter-contact plasma.

Jüttner [157] has given a good explanation of the formation of the development of the vacuum arc. This explanation is based upon the pioneer work by Mesyats and his coworkers [77]. It is based also

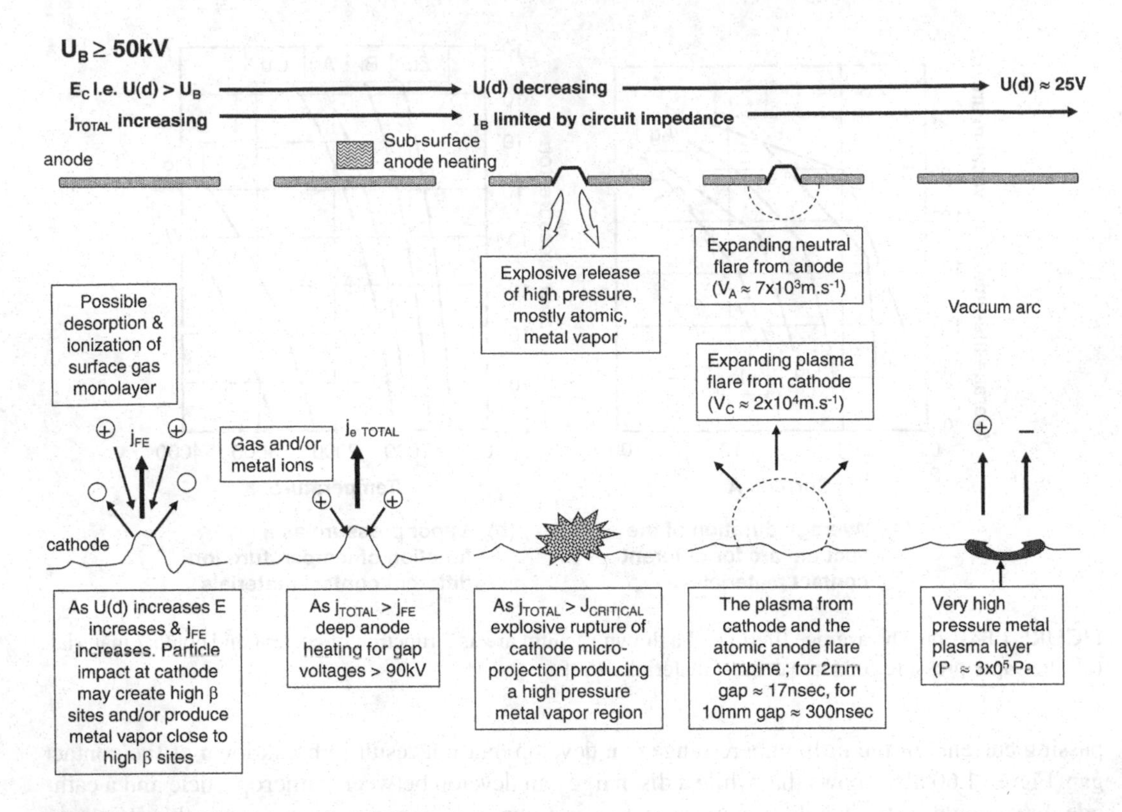

FIGURE 1.85 The vacuum breakdown sequence.

on results from other research on explosive emission from cathode projections discussed above [77, 139, 154, 155]. Let us consider the vacuum breakdown sequence illustrated in Figure 1.85. From Figure 1.20 for contact gaps less than 0.2mm the breakdown voltage is a linear function of the contact gap. The breakdown voltage for contact gaps less than 0.2mm is less than 20kV. The energy of the emission electrons as they impact the anode may be insufficient to cause the temperature inside the anode to increase enough for the eruption of a metal vapor cloud. Thus, the breakdown process will only depend upon the cathode's emission currents. The critical breakdown field is shown in Table 1.11 for this linear dependence. For contact gaps greater than about 0.2mm the breakdown voltage U_B is proportional to $d^{0.58}$. In this region the effect of the anode metal vapor flare affects the breakdown process. As the emission current increases, space charge effects are limited by gaseous ions returning to the emission site. At the same time the emission electrons begin to heat the subsurface anode material. At a critical current (or current density) the cathode microprojection explodes and an expanding plasma flare is developed at the cathode. As the electron current increases, it is possible for this subsurface region to erupt through the anode surface and give rise to an anode flare. This can occur a few hundred nanoseconds after the rupture of the cathode microprojection for rapidly rising voltage pulses, but it can develop at the same time as the initiation of the cathode flare for slower rising voltage applications. Once the cathode flare reaches the anode, or it meets the anode flare, conduction can occur across the contact gap with a low applied voltage. The current will continue to pass only if the cathode spot (see Chapter 2 in this volume) that maintains a supply of electrons is established. This process has been well documented by Mesyats [77] in a series of photographs, Figure 1.86. Here for a contact gap of 0.35mm, the cathode flare is initiated first and then is followed by the anode flare. In this example the current rises from a very low value to about 230A in about 30nsec.

Let us now examine the processes that occur after the initiation of the cathode and anode flares. Here Jüttner's analysis will be used [157]. Figure 1.87 shows the vacuum gap after the initiation of

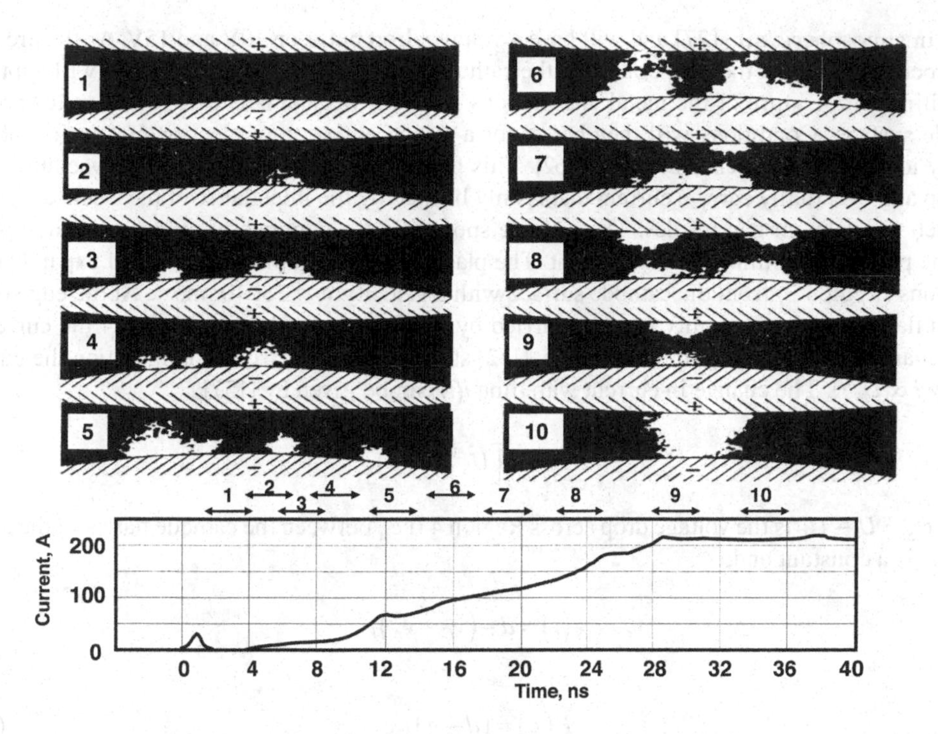

FIGURE 1.86 Photographic observation of the light from a vacuum breakdown of a 0.35 contact gap. Note, these photographs are taken over a series of breakdown events and are not a sequence for one breakdown [77].

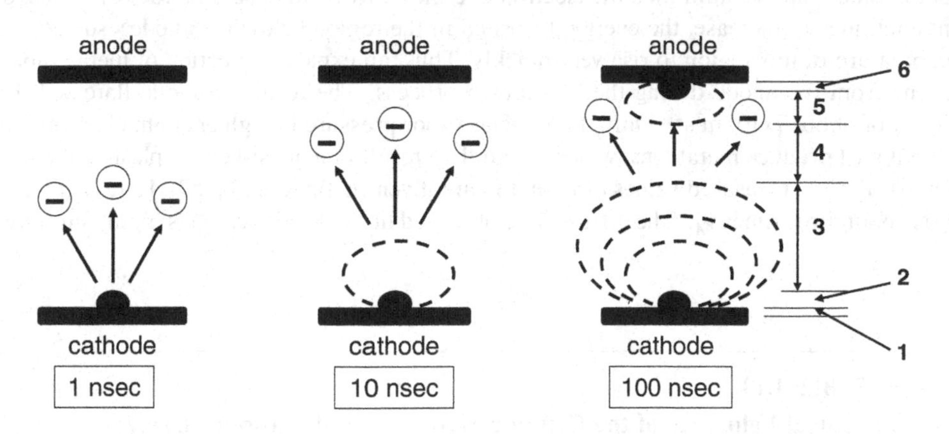

FIGURE 1.87 Schematic of the discharge zones in a vacuum gap as the vacuum breakdown process develops: (1) Cathode fall region, (2) cathode spot plasma, (3) expanding cathode plasma flare, (4) vacuum zone, (5) expanding anode flare [157].

the cathode and anode flares. After the explosion of the cathode projection a small ball of dense metal vapor is located in the region of the cathode projection. From our discussions in Section 1.3.3 and 1.3.4, this first ignition site could also be assisted by, or perhaps result from (a) the heating of the microprojection by the field emission current, (b) the effect of a close encounter with a microparticle, (c) a microparticle impact, or (d) the evaporation of a microparticle close to the cathode. It could well result from a combination of these events. This high-density gas quickly becomes ionized and a plasma flare is formed. In region 1 in Figure 1.87(c) a space charge sheath U_C (~10^{-6} m) forms between the plasma and the cathode. This region is similar to the cathode fall region in an arc in atmospheric air [127] and will have a voltage drop between 10V and 15V. As the arc ignition process continues, the temperature of the cathode surface under the plasma flare will approach the boiling temperature of the metal and the ions from the plasma ball will be driven toward the cathode surface. Conditions will then be set for a strong emission of electrons from the cathode spot by ion assisted T-F emission [161, 162]. This flow of electrons will permit the vacuum arc to develop and eventually carry a current that is only limited by the impedance of the electrical circuit in which the vacuum gap resides. The cathode spot will also continue to produce metal vapor to feed the plasma flare immediately above it. The plasma flare (or cathode flare) itself expands in all directions (Region 3) from the cathode surface with a velocity $v_C \approx 2 \times 10^4$ m.s^{-1}. At the edge of the plasma flare, current continuity must be carried by the electrons. Thus, in Region 4 the current is space charge limited and again Child's Law [142] states for any electrode configuration the current density $j \propto U_S^{3/2}$. The change in current with time $i(t)$ can be given by [157];

$$i(t) = A\left[U_S^{3/2} F(g)\right] \tag{1.71}$$

where $U_S = U - U_C$ is the voltage drop across Region 4 (i.e., between the cathode flare and the anode flare), A is a constant and:

$$g(t) = d - (v_C + v_A)t \tag{1.72}$$

$$F(g) = (d - g)/g \tag{1.73}$$

As $U_C << U_S$, $U_S \approx U$, so the electrons in Region 4 enter the anode region with a high energy. If we consider contact gaps > 8mm then the electron energies will be in excess of 100keV. Thus, as the current continues to increase, the energy deposited in the region below the anode's surface causes the temperature in this region to rise very quickly. Thus, an explosive ejection of metal vapor will also occur from the anode during the breakdown process. The resulting anode flare will have a velocity v_A of about 7×10^3 m.s^{-1}. Only if the metal vapor pressure is high enough, electron interaction with it will produce metal ions, which would then result in a quasi-neutral plasma at the anode (Region 6). Typical measured values of v_C and v_A are given in Table 1.13 [157]. Here it can be seen that v_C is about three times v_A., The diffuse flare observed in the anode region is highly luminous. If,

TABLE 1.13
Typical Velocities of the Cathode Flare v_C and the Anode Flare v_A During the Vacuum Breakdown Process [157]

Contact Material	Cathode Flare Velocity, Km.s^{-1}	Anode Flare Velocity, Km.s^{-1}
Al	18–26	7–9
Cu	17–26	5–8
Mo	18–26	
Nb	30	
Ni	19	
Stainless Steel		4–5
Ta	35	
W	19–27	5.3

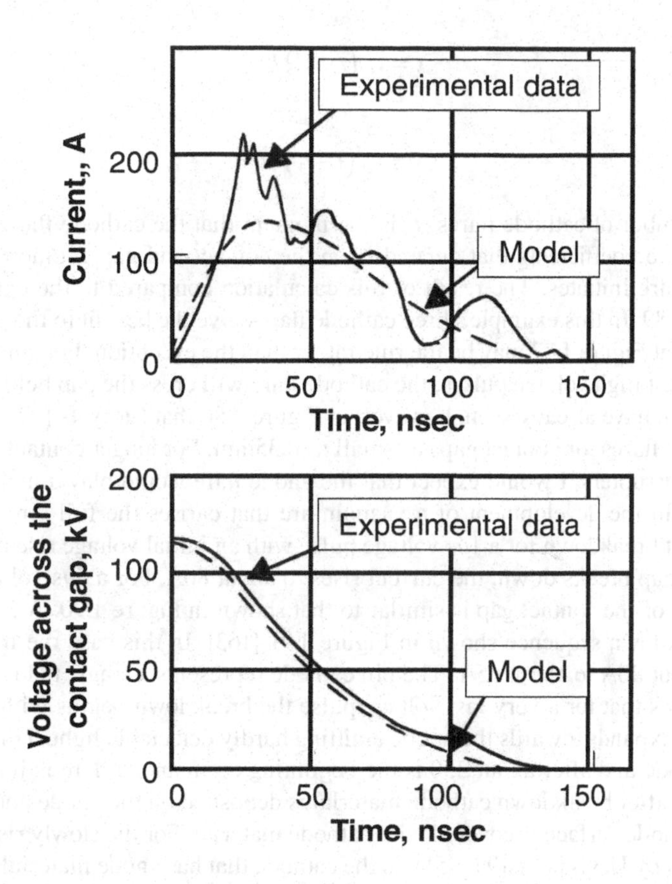

FIGURE 1.88 The current and voltage waveforms of a capacitor discharging into a vacuum gap compared to Equation (1.74): Mo contacts, 4mm contact gap, 121.5pF capacitor and U_B = 112 kV [157].

in a first approximation, the voltage drop in Regions 2 to 6 is neglected, and putting $g \cong d - v_C \times t$ the current growth *i (t)* can be given by:

$$i(t) = \frac{A\left[U(t)\right]^{3/2} v_C t}{d - v_C t} \tag{1.74}$$

Jüttner uses Equation (1.74) to explain the observed current and voltage waveforms for a capacitor discharging into a vacuum gap; see Figure 1.88. The voltage drop across Region 4 will eventually become smaller as Regions 3 and 5 expand. Once they overlap, a vacuum arc can form. In a power circuit where the voltage across the vacuum arc is much less than the circuit voltage, the current will only be limited by the circuit voltage and the circuit impedance. Djago et al. [158] use a similar analysis to calculate the change in the voltage across the vacuum gap during the breakdown process assuming a number of plasma clouds initiating at the cathode. They begin by assuming the current in the vacuum gap is given by:

$$i = P_t U^{3/2} \tag{1.75}$$

where the time varying perveance, P_t is given by:

$$P_t = n\left(ax^2 + bx + c\right) / \left(1 - x - y\right) \tag{1.76}$$

$$x = v_C(t - t_0)/d \tag{1.77}$$

$$y = v_A(t - t_1)/d \tag{1.78}$$

where n is the number of cathode flares, t_0 is the moment that the cathode flares initiate and a, b, and c are perveance coefficients that depend upon the contact and gap geometries. The time t_1 is when the anode flare initiates. The result of this calculation compared to the experimental data is shown in Figure 1.89. In this example, three cathode flares give the best fit to the experimental data, which indicates that Figure 1.77 may be the rule rather than the exception. For small contact gaps, or perhaps for very fast high voltage pulses, the cathode flare will cross the gap before the slower moving anode flare. We have already seen, however, in Figure 1.86 that Mesyats [77] has observed both cathode and anode flares for contact gaps as small as 0.35mm. For longer contact gaps that are typical in vacuum interrupters, I would expect that the anode flare would play a major role as it meets the cathode flare in the development of a vacuum arc that carries the full circuit current. Figure 1.90 shows the fast breakdown for a 1μs voltage pulse with an initial voltage rise of 70ns [163]. Here after the vacuum gap breaks down, the current rises to about 80A. For a 5μs voltage pulse the time to full breakdown of the contact gap is similar to that shown in Figure 1.90. A 5μs voltage pulse is used for the breakdown sequence shown in Figure 1.91 [163]. In this case the final circuit current declines from about 80A to about 65A. The pin cathode represents a single field emission site. This figure clearly shows that for a very fast voltage pulse the breakdown begins with the cathode flare. The cathode flare expands towards the anode emitting hardly detectable light. This current heats the interior of the anode and after about 250ns the beginning of an anode flare is initiated. Zhou et al. [163] observe that after breakdown cathode material is deposited on the anode and as the number of breakdowns the anode surface is covered with cathode material. For the slowly rising voltage pulses in the experiments by Davis et al. [94, 95] it is the cathode that has anode material deposited upon it. The beginning of an anode flare is seen at 250ns when the current has reached close to its maximum value. The vacuum arc is initiated after 3μs and the intercontact brightness decreases. At 5μs, toward the end of the current pulse, cathode spots on the cathode show that the diffuse vacuum arc has been

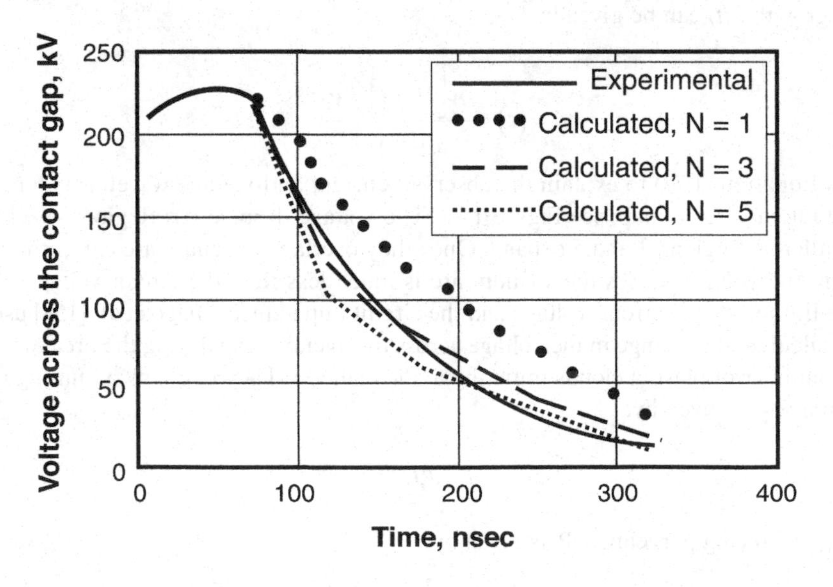

FIGURE 1.89 Comparison of calculation and the experimental measurement of the voltage collapse across a 7mm vacuum gap during the breakdown process as a function of the number of cathode plasma flares (n) [158].

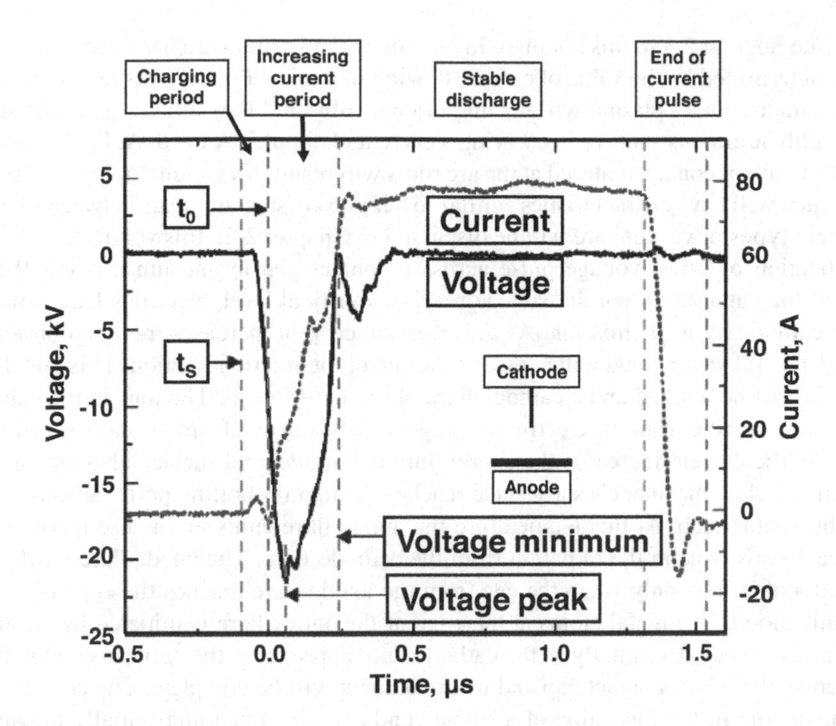

FIGURE 1.90 Typical waveforms of the voltage and the current during a vacuum breakdown for cathode pin to anode plane contacts with a 5mm gap. The voltage pulse width is 1 μs, t_s is the initiation of the voltage pulse and t_0 is the beginning of the rise in current [163] (Courtesy Zhenxing Wang).

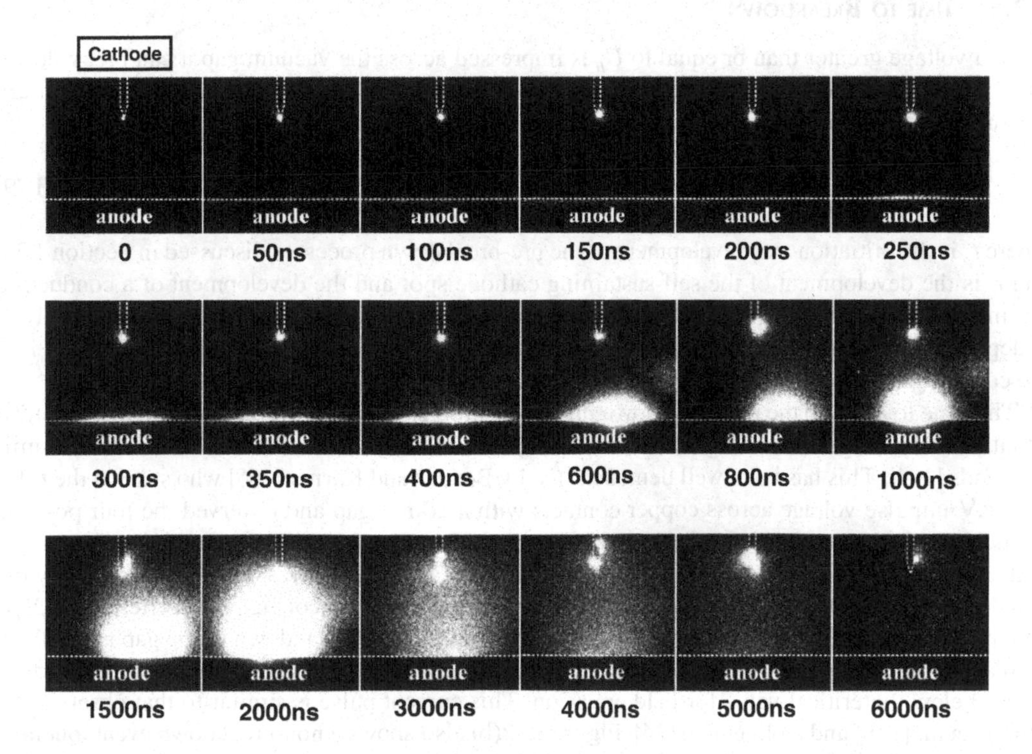

FIGURE 1.91 Time-resolved light emission of a vacuum breakdown process for cathode pin to anode plane contacts with a 5mm gap. For a very fast rate of voltage rise and a total voltage pulse width of 5 µs. The frame width is 50ns [163] (Courtesy Zhenxing Wang).

established (see Section 2.3 in this volume). In fact, once the vacuum arc has developed, its appearance will be determined by the value of current flowing in the arc. For currents less than about 5kA, the luminous inter-contact plasma will decline dramatically and the cathode spot will split into a number of highly luminous spots each carrying a current of about 50A to 100A. For higher currents, the evaporation rate of contact material at the arc roots will result in a columnar, high-pressure metal vapor arc, which will have characteristics similar to a high-pressure arc in air between two contacts. These different types of vacuum arc will be discussed in Chapter 2, in this volume.

The application of a fast voltage pulse across a contact gap in vacuum initiates the vacuum breakdown at the cathode. When the voltage reaches a critical level, electrons leave microprojections on the cathode by field emission. As the emission current increases the microprojection will vaporize and a metal vapor cloud will appear at the site of the microprojection. This metal gas close to the cathode will be ionized and a cathode flare plasma will result. The ions in the cathode flare will cancel the screening from the electronic charge and the current from the cathode will continue to increase. As the current increases the energy into the anode's subsurface also increases. When the temperature below the anode's subsurface reaches the metal's boiling point, an anode flare will erupt into the contact gap. At this temperature the anode flare emits an intense thermal radiation which can easily overwhelm the radiation from the cathode flare. The anode flare mostly contains neutral metal atoms. It is only when the gas from the anode flare reaches the edge of the cathode flare that ionization of the metal atoms at the edge of the anode flare is initiated by electrons with energies of a few 10s eV. Eventually as the cathode flare approaches the anode a conducing plasma will exist across the whole contact gap and its breakdown will be complete. The cathode therefore plays the major role in the formation of a stable conducting channel and initially the anode only plays a secondary role. The intense light from the anode flare results from thermal radiation and it decays once the vacuum breakdown is complete.

1.3.6 Time to Breakdown

Once a voltage greater than or equal to U_B is impressed across the vacuum gap it can break down. The time to breakdown, t_B, of the vacuum gap and the formation of a sustainable vacuum arc depends upon two factors:

$$t_B = t_p + t_c \tag{1.79}$$

where t_p is the initiation and development of the pre-breakdown processes discussed in Section 1.3.3 and t_c is the development of the self-sustaining cathode spot and the development of a conducting plasma path between the contacts. As discussed in Section 1.3.5, t_c can occur extremely quickly. It is dependent upon the contact gap d, and the expansion speed of the cathode and anode flares into the contact gap.

The time to develop the pre-breakdown process is problematical, however, because in a practical situation it can range from being almost instantaneous to a time period that can be hundreds of milliseconds [164]. This has been well demonstrated by Bender and Kärner [165] who studied the t_p for a 250kV impulse voltage across copper contacts with a 20mm gap and observed the four possible breakdown situations shown in Figure 1.92. Figure 1.92(a) shows the case where the vacuum gap withstands the voltage. Note that the first current peak of several amperes ($t < 3\mu s$) is caused by the gap displacement current (this depends upon the capacitance of the contact gap). After about $25\mu s$ an emission current pulse is observed. This does not result in the breakdown of the gap and falls to zero as the decrease in voltage across the gap causes the macroscopic field at the cathode surface to fall below the critical value for field emission. This current pulse is similar to that observed by Shioiri et al. [128] and Aoki et al. [124]. Figure 1.92(b) also shows a non-breakdown event, but here the current pulse has two peaks. In Figure 1.92(c) the delayed current pulse leads to a breakdown and the current rapidly increases after $35\mu s$. In 1.92(d) the current pulse is initiated immediately and

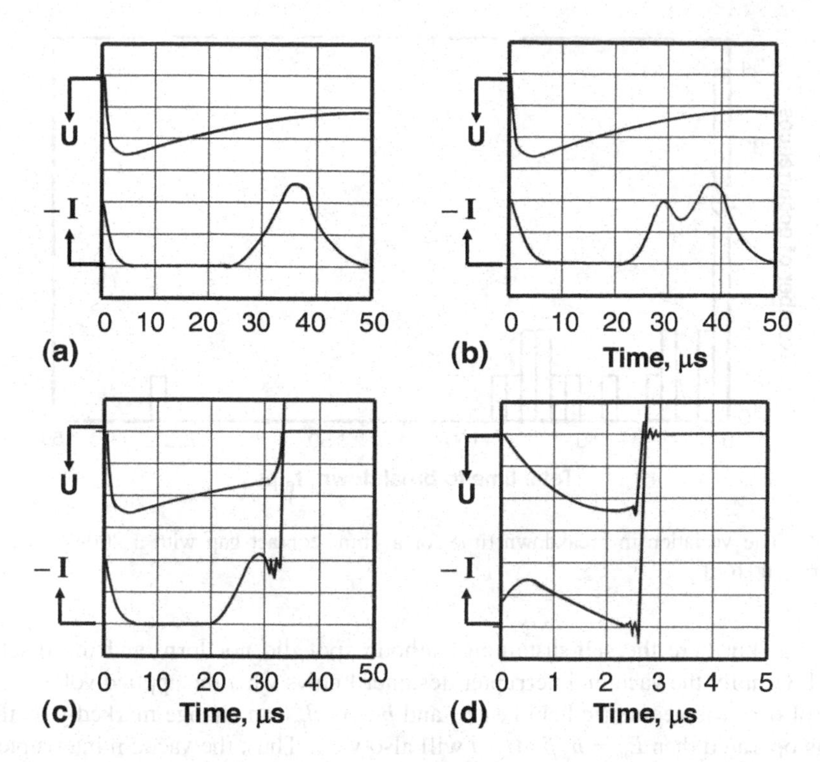

FIGURE 1.92 The variation in the time to breakdown for a pulse voltage with a peak value of 250 kV across a 20mm copper, contact gap (I = 2A/div, U = 100 kV/div): (a) and (b) no breakdown, (c) a breakdown after 30 μsec and after the peak of the voltage pulse and (d) an immediate breakdown once the voltage exceeds a given value [165].

eventually leads to the breakdown as the pre-breakdown current decreases to zero. These examples show the difficulty of assuming breakdown will occur even if you obtain a pre-breakdown pulse of current.

In Figures 1.92(a) and (b) a considerable current pulse of about 5A peak flows for between 10μs and 20μs and the gap does not breakdown. This is perhaps the result of a microparticle impact at the cathode or a microdischarge event discussed in Section 1.3.3. Anders et al. [166] have observed a similar variability in breakdown time. In a series of experiments with a 200kV impulse voltage across a 4mm gap (a gap and voltage at which breakdown would be expected), they observe a considerable variation to t_B; see Figure 1.93. While the majority of breakdowns occur in less than 10μs some occur with delays up to 200μs. Ritskaya [167] attributes the late breakdown events (Figures 1.92(c)) in vacuum interrupters to the effect of metal vapor deposited on their interior ceramic walls. The metal vapor deposit results from 60 switching operations with currents ranging for 17kA to 60kA. He also states that free metal particles are not responsible for these late breakdowns.

These experiments highlight the difficulty faced by the vacuum interrupter designer. In Figure 1.92 the maximum voltage impressed across the vacuum gap is the same in all four cases. The exact field, however, at the contact surfaces is quite unknown and obviously changes from one application of the voltage pulse to another: i.e., the microscopic enhancement factor (β_m) changes from one voltage application to another. This, in turn, means that the value of k^* = applied field / dc breakdown field (see Section 1.3.4), is not known from one application of the voltage pulse to another. When the contact gap breaks down immediately, Figure 1.92(d), k^* must be greater than 1.3, so t_B is of the order of tens of nanoseconds. In the example shown in Figure 1.92(c), k^* must be less than 1.2, so the breakdown time is tens of microseconds. For the no breakdown cases, Figures 1.92 (a) and (b),

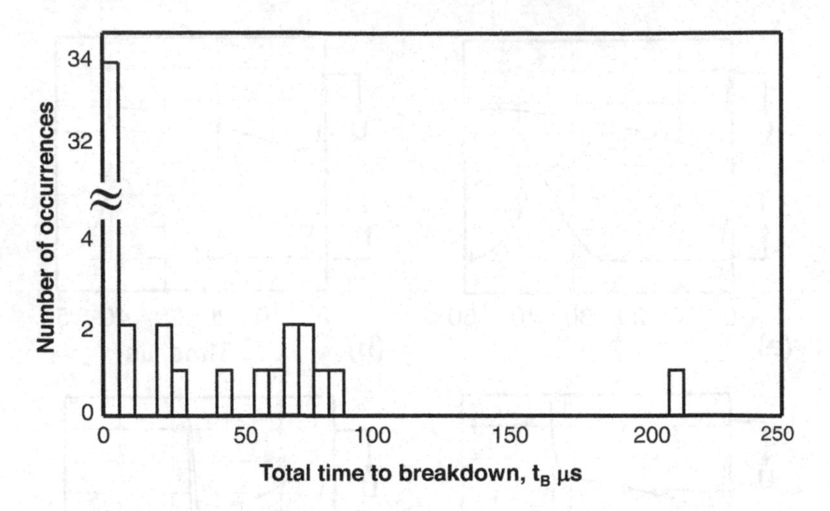

FIGURE 1.93 The variation in breakdown time for a 4mm contact gap with a 200kV impulse voltage impressed across it [166].

again $k^* < 1.2$, but here the self-sustaining cathode spot did not form and the discharge self-extinguished. Usually the vacuum interrupter designer knows what the applied voltage will be and he has control over the geometric field i.e., E_g and β_g. As β_m can change markedly as the vacuum interrupter is operated then $E_m = \beta_g \beta_m U / d$ will also vary. Thus, the vacuum interrupter designer has to carefully control the geometric field at the open contact surface, i.e., control β_g. The design then should be such that no matter how β_m varies during the vacuum interrupter's life, there will be minimal field emission of electrons.

Another conclusion from these experiments is that for the IEEE, outdoor, chopped wave, impulse voltage tests (see Section 1.1), the vacuum interrupter must be designed to pass the normal BIL impulse level for the higher chopped wave peak voltage levels. The fact that the voltage pulse is chopped to zero 2 or 3 microseconds after the application of the peak voltage would have no influence if the early breakdown event shown in Figure 1.92(d) occurs during this test. Thus, a vacuum interrupter designed for IEEE outdoor use would have to pass a BIL level about 29% higher than the normal rated outdoor BIL level if it is to also pass the chopped wave tests (see Table 1.1).

1.3.7 Conditioning

From the earliest research on vacuum breakdown, it has been observed that the high voltage withstand ability increases to a limiting value as the number of breakdown events increases. It appears as if this "conditioning" process eliminates a number of the sources of pre-breakdown current such as the surface roughness [65] microprojections, microparticles and sharp edges left on the contact surfaces after machining [68, 168]. After conditioning, the "stability" of the vacuum gap is considerably increased. For any vacuum gap, improving the voltage hold-off ability is an obvious advantage; therefore, much research has been performed on developing conditioning processes. Indeed, conditioning is the only practical way that a vacuum interrupter manufacturer can exert some control over β_m.

Experiments on conditioning have demonstrated that the major impact of conditioning is on the cathode contact [169]. This is quite consistent with the fact that most breakdowns in vacuum are initiated at the cathode. A vacuum interrupter manufacturer must also consider conditioning other components such as the interior shields. Certainly, changes in the contact surfaces by reduction of microprojections and the elimination of particles residing on the surfaces have been well demonstrated [101]. Changes in the surface structure of the contact surfaces down to the atomic level have also been shown to be part of the conditioning practice [170]. For large area contacts such as those found in vacuum interrupters it is possible to condition in three ways:

1.3.7.1 Spark Conditioning Using a High Voltage AC Power Supply

Here high voltage ac circuits like the one shown in Figure 1.94 are used. The voltage on the primary of the transformer is gradually increased until breakdown events are recorded. Typically, the current in the breakdown discharge is severely limited to a few tens of milliamps and the discharge will self-extinguish. Kojima et al. [171] show that as the breakdown charge increases the conditioning effect increases. They also show that when the charge becomes too large it damages the contact surfaces and the conditioning effect decreases. The voltage is held at this level until the gap holds it off before continuing to increase it. This process is continued until the contact gap holds off the required peak ac voltage. A typical conditioning curve developed by Ballat et al. [172] is shown in Figure 1.95 for Cu–Cr contacts with a 3mm contact gap. As the conditioning events occur so the value of U_B increases to a limit set by the contact material and the contact gap. It follows the general conditioning characteristic given in Figure 1.23. Figure 1.95 shows that U_B increases from about 20kV to between 85kV and 120kV after about 100 breakdowns. After about 100 conditioning sparks the contacts are fully conditioned. In fact, Figure 1.95 also shows that the lower bound of the measured breakdown voltage can decrease if the process continues further. Seki et al. [173] have recorded changes in the contact surfaces as they became conditioned. They conclude that the loss of preferential emission sites such as Cr particles and the homogenization of field intensification sites by sputtered particles contribute to the increase in U_B. It is interesting to note, however, even when the contacts are conditioned these researchers show that the contact surfaces still contain many imperfections and irregularities. Ballat et al. [172] in a series of experiments also show that you can obtain an optimum conditioning process by creating a circuit that supplies a very high frequency, quickly decaying, high current pulse when the contact gap breaks down. Figure 1.96 shows an optimum-conditioning curve for both positive and negative voltages. The current flowing after breakdown has a peak value of 500A with a frequency of 17 MHz and it decays to zero in about 400 ns. The effect on the enhancement factor is shown in Figure 1.97.

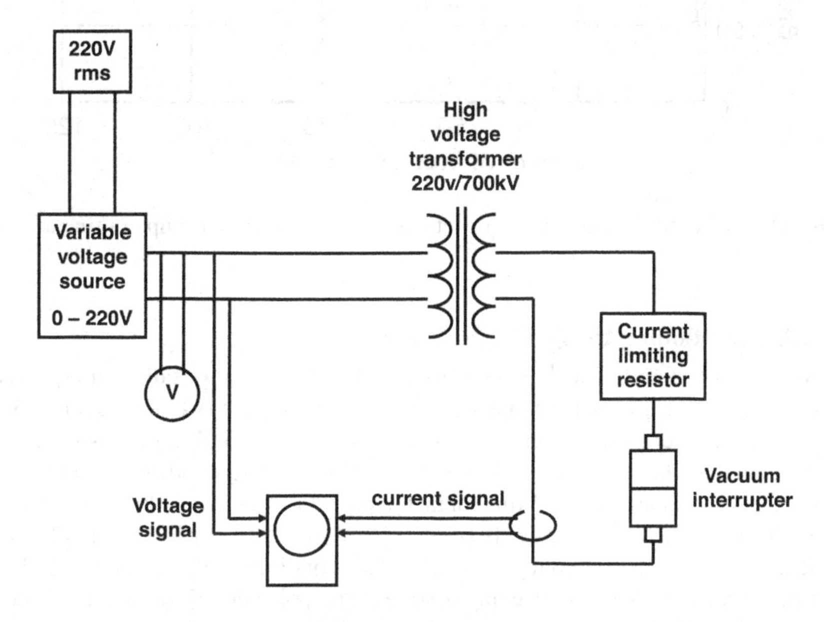

FIGURE 1.94 A typical high voltage ac circuit for spark conditioning the contacts in a vacuum interrupter.

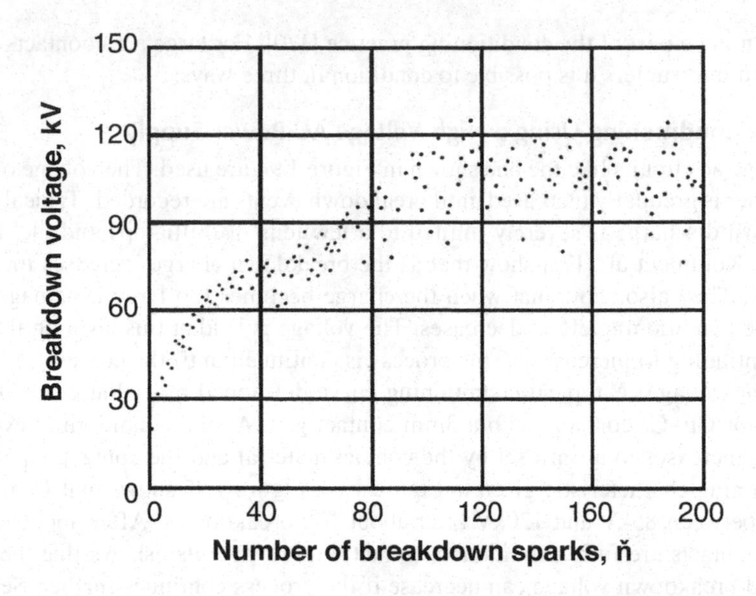

FIGURE 1.95 A typical high voltage spark conditioning curve for Cu–Cr contacts with a 3mm contact gap [172]

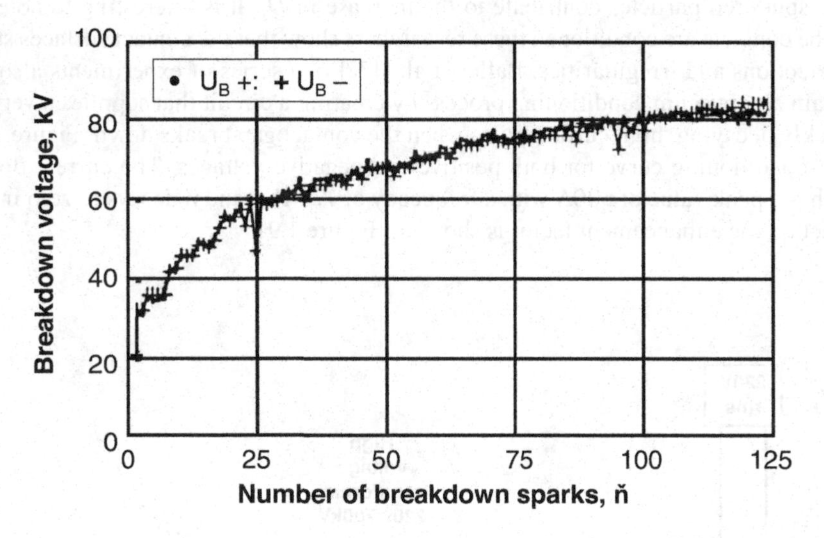

FIGURE 1.96 Optimal conditioning of Cu–Cr contacts inside a vacuum interrupter using a 3.5mm contact gap [172].

1.3.7.2 Spark Conditioning Using a High Voltage Pulse

In this process a Marx generator capable of supplying a BIL type voltage pulse replaces the ac transformer in Figure 1.94. This is a useful conditioning mechanism when BIL testing a new vacuum circuit breaker as will be discussed in Section 1.4.4. This method of conditioning is not as easy to use on a routine basis in manufacturing as is the ac method. A BIL pulse-conditioning curve is shown in Figure 1.98 for Cu–Cr contacts with a 3mm contact gap [172]. In this case the gap initially holds off a BIL pulse of 100kV, which increases to between 140kV and 240kV after about 120 breakdowns. As with conditioning using an ac voltage wave, further conditioning operations result in a wider dispersion of the breakdown voltage. A comparison of BIL pulse conditioning and ac conditioning is shown in Figure 1.99 [166]. In this experiment, U_B from ac conditioning is only marginally lower

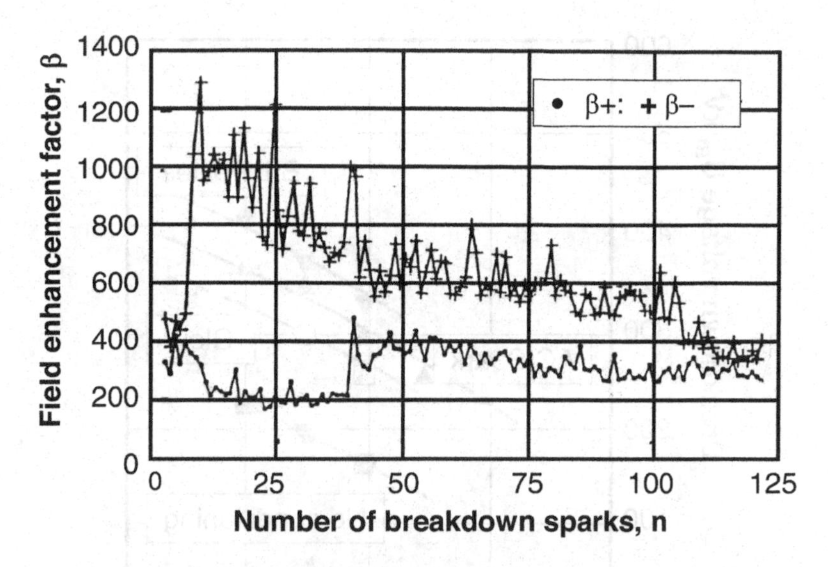

FIGURE 1.97 The effect on the enhancement factor as the conditioning shown in Figure 1.96 proceeds; the *β*+ corresponds the positive voltage pulse and the β- corresponds to the negative [172].

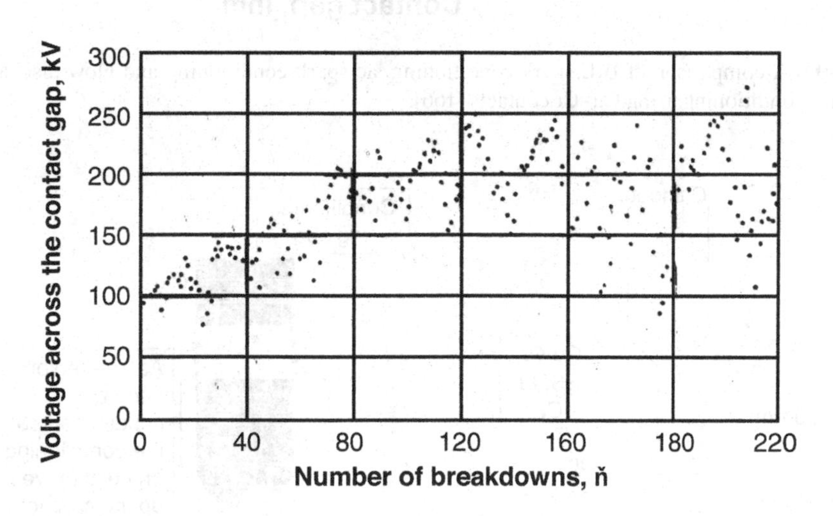

FIGURE 1.98 A typical BIL impulse-conditioning curve for Cu–Cr contacts with a 3mm contact gap [172].

than that from spark conditioning. Both, however, show a marked improvement over no conditioning. Interestingly, the ratio between the breakdown voltages afterglow conditioning and BIL pulse conditioning is similar to the difference between microdischarge conditioning and pulse conditioning shown in Figure 1.65 [124].

Miyazaki et al. [174] have investigated the BIL pulse conditioning of the contact arrangement shown in Figure 1.100(a). Table 1.14 shows their ultimate conditioning U_{B50}, for Cu–Cr, for Cu cathodes and for stainless steel. In their experiment the region of highest E_g; i.e., the tip of the cathode is initially conditioned. After this region is conditioned the conditioning effect moves up the sides of the cathode, gradually removing high spots and imperfections, see Figure 1.100(b), thus generally smoothing the sides of the cathode. Again, as Seki et al. [173] observe, the cathode surface still contains a major roughness even after conditioning. Miyazaki et al. divide their cathodes into 20 regions (Figure 1.100(b)) and show that as the E_g in each region exceeds 1 to 2 × 10^7V.m^{-1} the

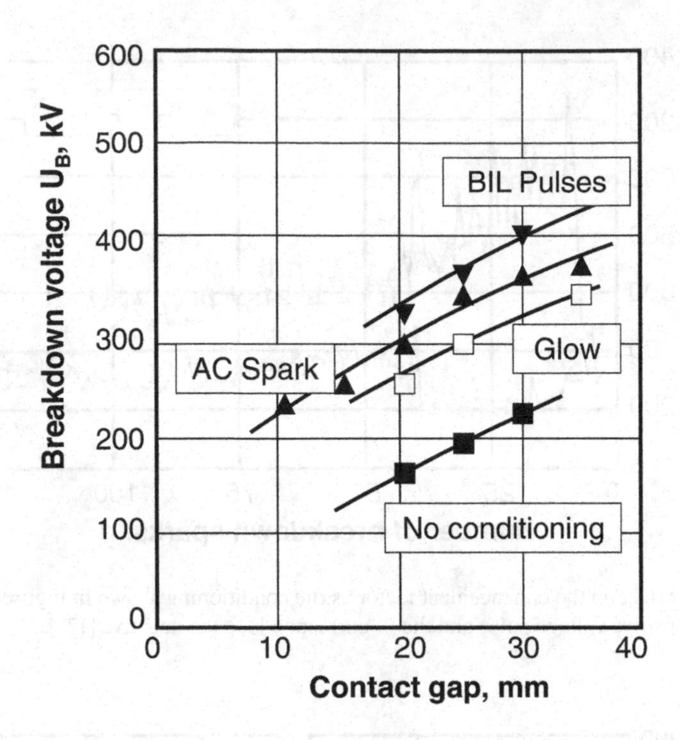

FIGURE 1.99 A comparison of BIL spark conditioning, ac spark conditioning and glow discharge conditioning and no conditioning using Cu–Cr contacts [166].

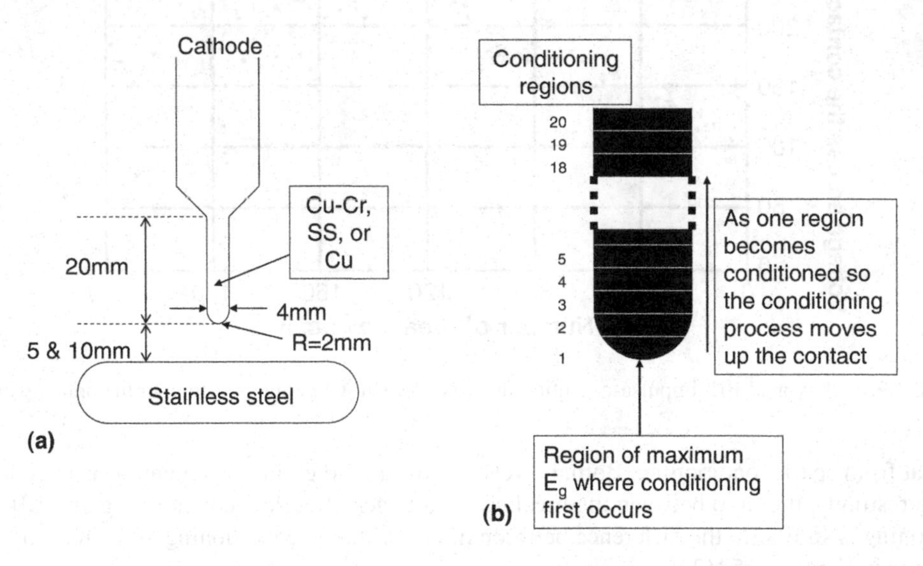

FIGURE 1.100 Experimental arrangement to investigate the conditioning of a finger shaped cathode [174].

conditioning moves to the next higher region. The saturation U_{B50} is reached when the non-uniform field and the "conditioning degree" are balanced. The "conditioning degree" in each region is proportional to the nonuniform electric field for each of the cathode materials investigated. The U_{B50} at the region of the maximum E_g is highest after each region of the cathode has been conditioned. The importance of this experiment for the vacuum interrupter manufacturer is that not only the facing surfaces of the vacuum interrupter contacts and the initial high E_g regions need to be conditioned,

TABLE 1.14
Conditioning of Rod Cathodes with a Large Area Plane Anode [174]

	Material			
Parameter	**Cu–Cr**		**OHFC Cu**	**304 Stainless Steel**
Contact gap, mm	5	10	5	5
Initial U_B, kV	25	80	60	60
Number of breakdown operations until U_B reaches conditioning plateau	135	105	90	135
Plateau, U_{B50}, kV	177	201	95	330
U_{B50} Standard deviation, kV	21	15	36	27

but also the conditioning process has to be continued until lower E_g regions have also been subjected to the smoothing effect of the conditioning process. Kulkani et al. [175] emphasize, that for the commercial vacuum interrupter, it is important to condition not only the contacts, but other internal metal structures such as the shields that protect the ceramic envelope.

1.3.7.3 Current Conditioning

In this method the interrupter is placed in a power circuit capable of supplying currents of a few hundred amperes. The contacts are then opened under load and a diffuse vacuum arc is formed (see Section 2.3 in this volume). This arc form of vacuum arc has numerous cathode spots that move over the cathode in retrograde motion. These moving cathode spots reduce microprojections and eliminate surface particles and leave a somewhat smoother contact surface. This process has to be performed on both contact surfaces. The contacts should be closed with no load after each opening operation. This process is continued until the desired U_B is achieved. A typical conditioning sequence for a vacuum interrupter [49] with Cu–Cr contacts follows the characteristic curve shown in Figure 1.23. A fully conditioned contact gap can be achieved after 60 to 80 switching operations. The disadvantage of this conditioning method is the time, the power and the inconvenience of also measuring U_B after each set of conditioning events. It also only conditions the vacuum interrupter's contact faces. Schneider at al [176] have successfully conditioned Cu–Cr(30wt.%) contacts inside a vacuum interrupter using a combination of a high-frequency current arc plus a lightning pulse voltage. During this conditioning process the surface structure of the Cu–Cr contact will also change (see Section 3.2.4 in this volume). The contacts surface becomes smoother thus increasing U_B.

Current conditioning is of great practical importance to the vacuum interrupter designer and user. In normal use the vacuum interrupter will continually be called upon to switch currents. In doing so it will be subjected to conditioning events throughout its life. Thus, its high voltage performance will also be maintained at a high level throughout its life. In fact, there is a good chance that its performance may even improve as the interrupter is switched during service. Figure 1.101 compares the high voltage performance of Cu–Cr after high current conditioning at different levels of current and after spark conditioning [101]. Here an axial magnetic field is used across the contacts to maintain a diffuse vacuum arc (see Chapter 2 in this volume) at 20 kA. In another experiment, Pursch et al. [88] have shown the effect of spark conditioning and arc conditioning on the enhancement factor, β, and the emitting area A_e. Figure 1.102 shows that there is a general reduction in both of these parameters with both conditioning methods. Godechat et al. [177] show that current conditioning of Cu–Cr contacts results in a thin surface Cu–Cr layer whose structure is different from the original contact material: See Section 3.2.4 in this volume. In their experiments a half wave of 8kA current is more effective than a dc current of 500A, because the greater number of cathode spots cover the cathode,

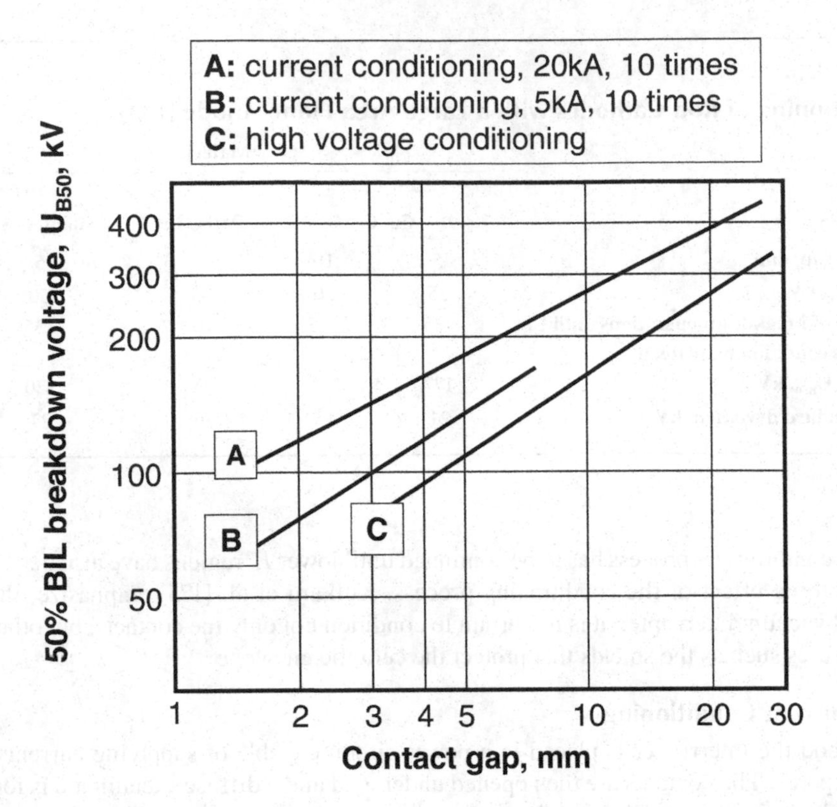

FIGURE 1.101 A comparison of ac spark conditioning and current conditioning using Cu–Cr contacts [101].

contact surface more effectively. This is true even if the dc current has a higher charge quantity. Contact gaps > 10mm allow the cathode spots to travel more easily away from previously conditioned contact surfaces. They show it is thus possible to condition > 90% of a cathode's surface.

1.3.7.4 Other Conditioning Processes

An experimental study by Zhang et al. [178] is shown in Figure 1.103. Here two vacuum interrupters close on no current. The first opens and passes 450A dc current for 500ms. The second opens and passes 750A dc also for 500ms. The vacuum interrupters then receive up to 650 BIL conditioning pulses. Figure 1.103 compares the results for these two vacuum interrupters to one that had no current operations. The current operations improved the high voltage withstand. Again, the authors attribute the improved performance to the surface melting of the contacts during the current operation. Another interesting technique for conditioning polished contacts is an experimental system that uses the emission current to blunt the microprojections [179]. In this technique, the emission current is monitored as the voltage across the open contacts is increased. At a certain level of emission current, the voltage is held steady until the value of the emission current decreases. The voltage is again increased. While interesting, this technique is not practical for commercial vacuum interrupter manufacture. A technique that is quite useful in experimental systems is to develop a glow discharge between the open contacts with perhaps 2 torr of an inert gas such as Argon [180, 181]. This process works very well for cleaning adsorbed gases and for blunting micro-projections, but Figure 1.99 indicates that this process is not as effective as ac or pulse conditioning. Proskurovsky et al. [182] show that the electric field strength of vacuum gaps can also be increased considerably by electropolishing the contacts and following this with a pulsed electron beam treatment and then implantation of a high dose of metal ions. Again, these are not practical conditioning methods for commercial vacuum interrupter manufacture.

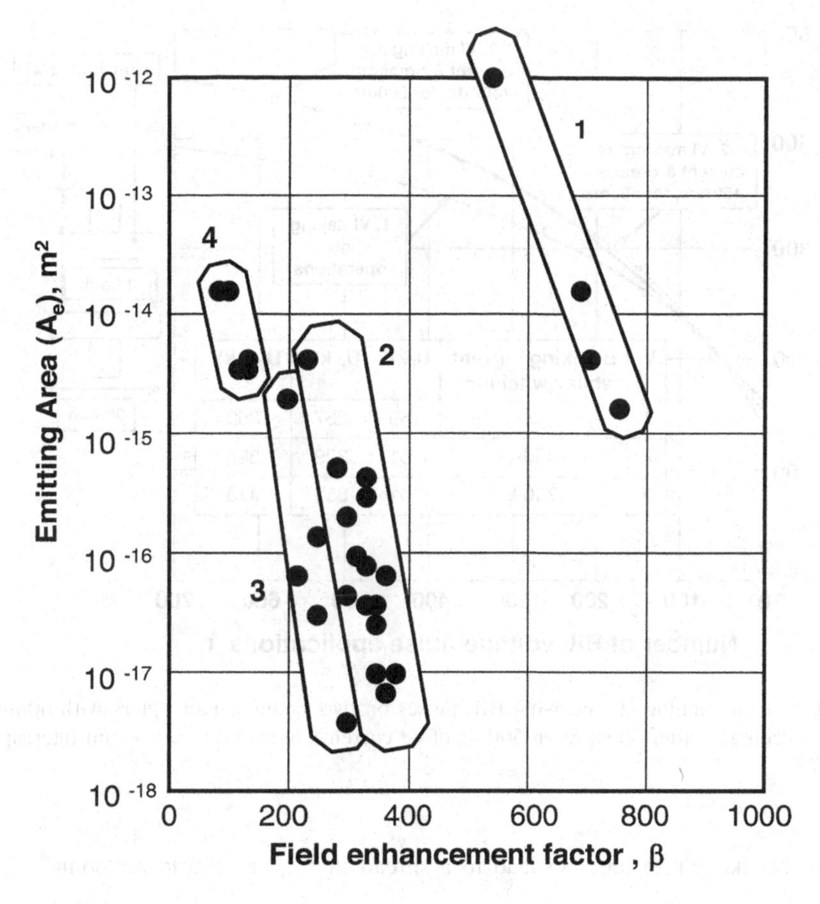

FIGURE 1.102 The enhancement factor β and the emitting area A_e for a number of Cu–Cr contact surface conditions. (1) After opening closed contacts without current after the passage of a high current pulse that resulted in contact welding, (2) New contacts, after heating to 400 C, (3) New contacts, after applying several hundred dc breakdowns (microdischarges?). (4) Contacts after arc conditioning (20A for many operations to a few operations at 10kA to 20kA) [88].

Research continues to develop a practical, efficient conditioning process for vacuum interrupter production. For example, Fink et al. [183] have extended the conditioning method developed by Ballat et al. [172]. In their method they follow a high voltage spark breakdown of the contact gap with a high frequency current, which is supplied by parallel resonance circuits. The main current has an initial peak of about 10kA and a frequency of about 7kHz. This main current has superimposed on it a current with a frequency of about 100kHz. When this current is periodically interrupted the high voltage is again seen across the contacts. If the contact gap again breaks down, the high frequency currents flow again. The sequence continues until the contact gap withstands the voltage after it has interrupted the current: one sequence lasts about 1ms. Fink et al. show that it is possible to condition a vacuum interrupter with Cu–Cr (25 wt.%) contacts after about six or seven operations of this circuit.

Evaluation of the studies on contact conditioning in vacuum results in the following summary:

1) The major impact of conditioning is on the cathode
2) Breakdown events "clean" the contact surfaces
3) After 100–150 breakdown events, U_B reaches its maximum value

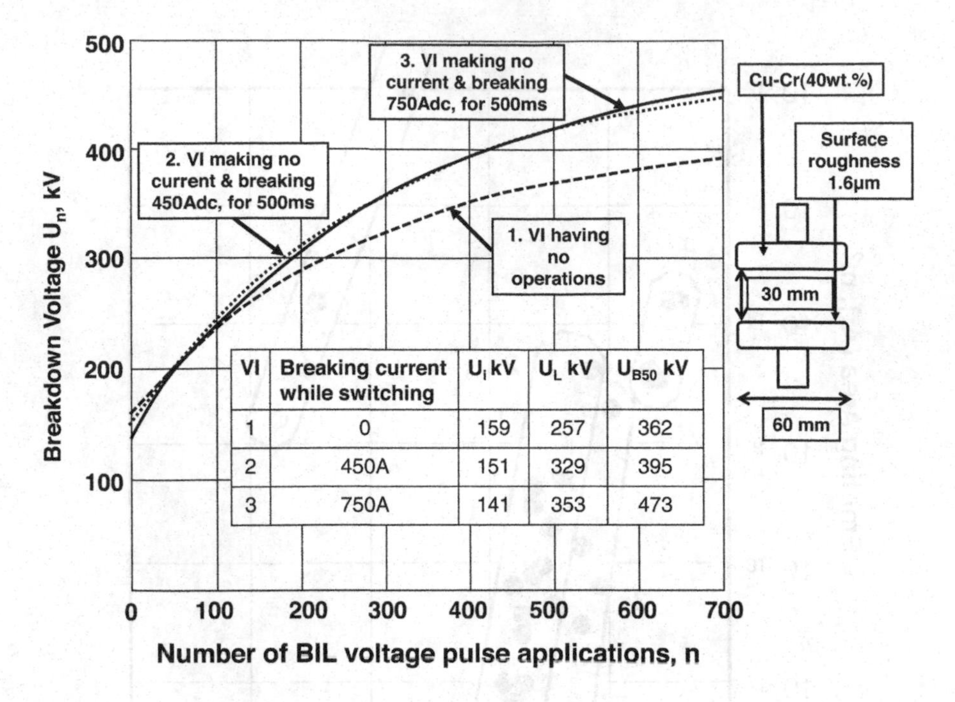

VI	Breaking current while switching	U_i kV	U_L kV	U_{B50} kV
1	0	159	257	362
2	450A	151	329	395
3	750A	141	353	473

FIGURE 1.103 Conditioning curves using BIL pulses on two vacuum interrupters with 60mm diameter, Cu–Cr(40wt.%) contacts after opening on 500ms of dc current compared to a vacuum interrupter with no operations [176].

4) Further breakdown events can lead to a spread in U_B, i.e., the lower bound of U_B can decrease
5) If the current in the breakdown pulses is too high ($> \approx$ 100mA), de-conditioning may occur
6) For very low currents (2 to 5μA) deconditioning occurs with the formation of microneedles on the contact surfaces
7) Conditioning removes adsorbed gases, forms melted and resolidified layers and smooths microprojections
8) β_m decreases as conditioning proceeds
9) The optimum i_B for conditioning depends upon the contact material and the surface roughness. Also, i_B should be increased as the material becomes smoother and as its melting point increases
10) Conditioning eliminates microflakes and microparticles left over from machining
11) Conditioning increases U_B for all contact gaps, all contact diameters, all contact materials, and all intact surface preparations
12) Electropolishing of rough surfaces allows for rapid U_B improvement
13) Conditioning begins at high E_g regions and then migrates to regions with lower values of E_g

1.3.8 Puncture

One practical, but negative, aspect of spark conditioning using a high voltage ac power supply should be considered; i.e., the possibility of puncturing the vacuum interrupter's ceramic envelope during the conditioning procedure [184]. This can occur if an intense enough electron emission occurs outside the central contact region and these electrons are permitted to impact the ceramic directly. Such an electron beam is illustrated in Figure 1.104. Here the electrons are emitted from

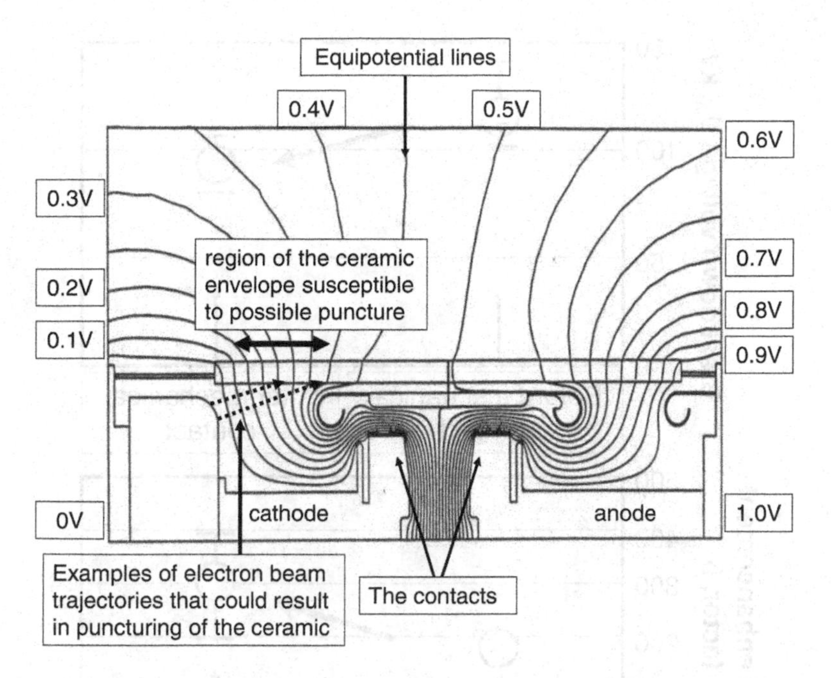

FIGURE 1.104 Cross-section of a vacuum interrupter illustrating a possible electron path that could result in a puncture of the ceramic envelope and the regions of the ceramic that would be vulnerable to puncture.

the upper part of the negative terminal and are driven by the electric field toward a region in the ceramic. Their energy will now be deposited below the ceramic's surface (see Figures 1.54 and 1.55). Eventually the ceramic can melt and a small crater can form [185]. They will also create a region of high electric charge. If this electron beam is allowed to continue, it is quite possible for the ceramic envelope itself to rupture in that region, which, in turn, will create a passage to the ambient. The result of this will be a loss of vacuum inside the interrupter. Figure 1.104 shows an example of a region on the ceramic's wall that may be susceptible to this puncture phenomenon. The probability of puncture occurring depends not only upon the magnitude of the conditioning voltage, but also upon the internal design of the vacuum interrupter and the components (e.g., the materials and their surface preparation) employed outside the contact region. One way of preventing puncture from occurring is to closely monitor the emission current and reduce the applied ac voltage once the current reaches a value of a few tens of milliamperes and does not drop to zero in a few milliseconds. Once the voltage has been reduced, it can be increased again. This procedure can be repeated until the emission current and its duration are reduced to an acceptable level. In my experience for a well-designed vacuum interrupter this phenomenon occurs only occasionally during the high voltage ac conditioning procedure. It never occurs during any of the certification high voltage tests or in the practical application of the vacuum interrupter in an ac circuit.

1.3.9 Deconditioning

When two contact surfaces that have been processed in a vacuum chamber are forced together, the micro-contact spots can form a cold weld [186]. As the contacts are parted these small welds will break and microprojections will appear on the contact surfaces. These microprojections can act as electron emission sites and a lowering of the breakdown voltage across the vacuum gap can result. Tsutsumi et al. [187] show the effect of mechanical contact on the breakdown voltage and the field enhancement factor (β); Figure 1.105. When contacts are closed in a high current circuit and opened with no current, the effect on β and hence on U_B can be quite dramatic. Ballat et al. [188]

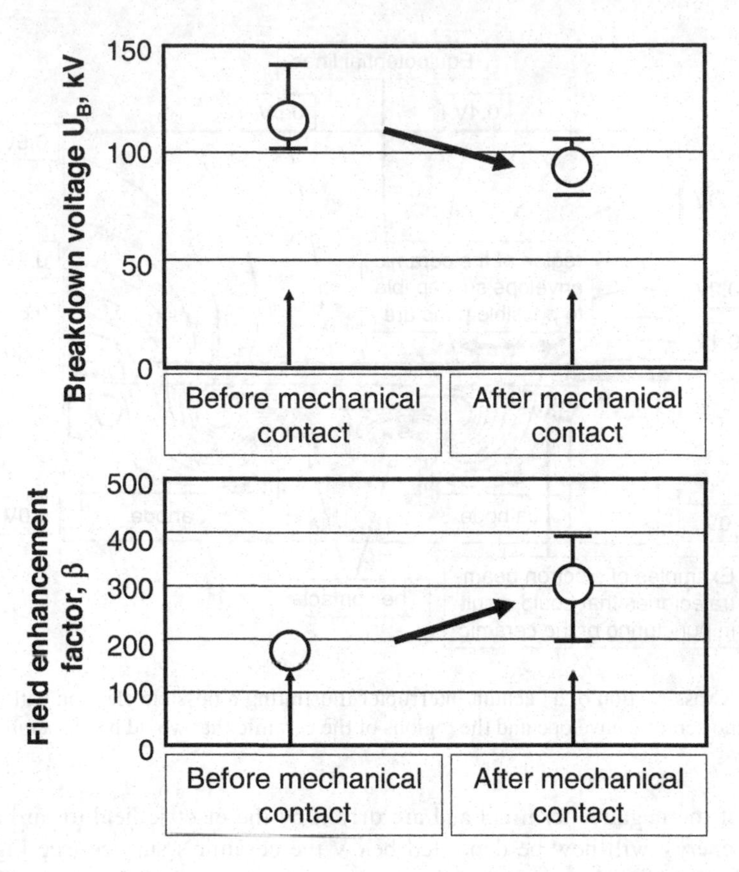

FIGURE 1.105 The effect of mechanical contact on the vacuum breakdown voltage and on the field enhancement factor [187].

show for Cu–Cr contacts that β increases from close to 200 for conditioned contacts to about 500 after closing on 1 kA, to about 1000 after closing on 7.65 kA and, finally, about 1500 after closing on 15 kA. Slade et al. [189] show the microscopic enhancement factor, β_m, increases as the number of operations closing and latching on 50kA peak current increases. They show that the contact gap at which prestrike occurs increases as the number of close and latch operations increase: i.e., the duration of the prestrike arc increases. Even passing a high current through closed contacts can also have a similar effect on β and on A_e; see Figure 1.102. In practice it is possible to minimize the deconditioning effect by the choice of contact material and the hardness of the contact surfaces. As I have discussed in Section 1.3.3, work hardening of the contact surface can affect the value of the work function φ and can even decrease the pre-breakdown field emission current. Again, it must be noted that, in general, vacuum interrupters are required to switch load currents and as they do, the surfaces of their contacts will be continuously conditioned as can also be seen in Figure 1.102.

1.4 INTERNAL VACUUM INTERRUPTER DESIGN

1.4.1 The Control of the Geometric Enhancement Factor, β_G

By taking into account the effects discussed in Section 1.3, the designer can develop vacuum interrupters that will satisfy the high voltage standards given in Table 1.1. The primary consideration for the designer is the geometric electric field that can occur across the open contacts. The vacuum interrupter's compactness and the presence of components such as the shielding complicate an analysis of the electric fields inside it. The designer, however, has almost no control over the microscopic fields at the surface of the contacts; hence there is almost no control over β_m in the primary design. As discussed in Section 1.3.7, the effects of β_m will be reduced by contact conditioning during manufacture. It will also be controlled during the vacuum interrupter's life by switching load currents. Knowledge and control of β_g, however, is essential for the control of electron emission and for minimizing the effects of β_m. In former times the designer had to rely upon approximate analytical tools, electrolytic baths and trial and error to determine the geometric electric field structure of the vacuum interrupter. Today, the advent of powerful personal computers and user-friendly finite elements analysis (FEA) software has given us a powerful vacuum interrupter design tool [190]. As shown in Figure 1.1, the vacuum interrupter is rotationally symmetric. Therefore, the designer usually only has to perform a two-dimensional FEA analysis on half the vacuum interrupter. Figure 1.106(a) shows a typical FEA giving the equipotential lines for a vacuum interrupter designed with a floating shield [50]. From this initial analysis, the software can then calculate the maximum fields at various places inside the vacuum interrupter. It can even indicate where the maximum and minimum fields occur; see Figure 1.106(b).

The first item the designer should ensure is that the triple points have very low field stress. These are the regions at the ends of the vacuum interrupter where there is a junction between, metal, ceramic and air (A, in Figure 1.106(a)) or metal, ceramic and vacuum (B, in Figure 1.106(b)). These regions, if on the cathode, and under field high stress, can be electron sources [17, 18, 191]. In this example, the triple points have approximately the same potential as the end plates, thus the fields at the triple points are relatively low and little problem would be expected from them. Venna et al. [192] show, however, that the ceramic's edge shape and the angle it makes with a metal flange can have a significant effect on the field at the triple junction. The next thing to consider is the geometric electric field. Here in this design the highest value is between the contacts. Thus, it would be expected that any breakdown event would occur there and the resulting vacuum arc would operate between the contacts where it is designed to take place. The maximum field at this point for a 12kV device is $1.02 \times 10^7 \text{Vm}^{-1}$ for the peak BIL value of 95kV, which is a value where a low level of field

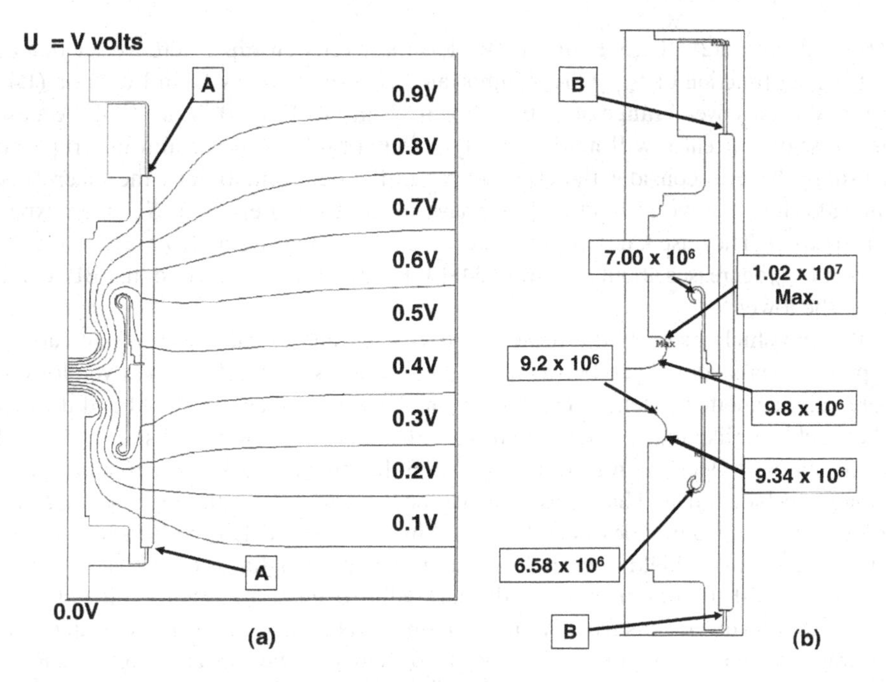

FIGURE 1.106 (a) The equipotential lines and the external triple point "A" and (b) the corresponding electric fields and the internal triple point "B" for one vacuum interrupter design [50].

emission current could be expected to occur. The BIL voltage pulse is of short duration, so unless there is a major microprojection, the BIL will be easily withstood. This type of analysis can also be applied to vacuum interrupters in a three-phase arrangement [193] and to evaluate the effects of the external insulation discussed in Section 1.2.3.

The breakdown voltage would also be expected to be dependent upon the area of the surfaces inside the vacuum interrupter. This results from the fact that the areas of the contacts and the shields are usually high. In typical vacuum interrupters the contact areas, for example, can range from 10^3mm^2 to 10^4mm^2. It is impossible to know precisely the microscopic surface structure over such a large surface. It is reasonable to assume, however, that only the areas close to high geometric field stress regions will have an impact on the vacuum breakdown. This, however, is not always true. As we have already seen in Section 1.3.2, Figures 1.42 and 1.43, the breakdown can occur in regions where $E_g < E_g$ (maximum). This can be true especially at close contact spacings where E_g is nearly constant across the whole contact surface, numerous field emission sites are possible. As discussed in Section 1.3.2, these sites can change position and seem to be randomly distributed. At the longer contact gaps, that are usual in vacuum interrupters, E_g is certainly not uniform and it is important to consider the region on the contact surface where breakdown is likely to occur. This has resulted in considering an effective area S_{eff}, which covers the contact area where $E_g > 0.9 \times E_g$ (maximum) [131, 194]. Three empirical important relationships have been derived as functions of S_{eff} for various practical geometries:

$$E_g \propto S_{eff}^{\ -\alpha(1)} \tag{1.80}$$

$$\beta_m \propto S_{eff}^{\ \alpha(2)} \tag{1.81}$$

$$E_c = E_g \times \beta_m \propto S_{eff}^{\ \alpha(2)/\alpha(1)} \tag{1.82}$$

where $\alpha(1) \sim 0.24$ and $\alpha(2) \sim 0.26$. Figure 1.107 plots these relationships. Here we see that E_c is only a slowly changing function of S_{eff}. This is important because we know from Equation (1.48) j_e can change by six decades over a range of E_c from 2×10^9 V.m^{-1} to 10×10^9 V.m^{-1} [83]. Because E_c and hence the emission current I_e will not be greatly affected by S_{eff}, the vacuum interrupter designer does not usually have to consider the effect of S_{eff} in the determination of the macroscopic field E_g. Having said this, however, Liu et al. [195] have found a small effect of S_{eff} in an experimental vacuum interrupter. They use Cu–Cr(50wt.%) contacts S_{eff} with ranging from 211mm^2 to 550mm^2. They find that U_{B50} decreases about 11% from 248kV to 221kV. They also find that BIL conditioning is slower for the lower S_{eff}.

While the end shield protects the triple points at each end of the vacuum interrupter, it does present a potential breakdown gap between it and the center shield. Also, when you look at Figure 1.106, there can be another potential breakdown gap between the center shield and the copper terminal. The shields, unlike the contacts, are usually made from a pure metal such as Cu or from an alloy such as stainless steel. With proper preparation the shields can be manufactured with a very low roughness surface finish. Thus, it is possible for them to exhibit the much higher breakdown voltage for contacts with a mirror finish shown in Figures 1.26 and 1.70. Shümann et al. [196] have evaluated the effect of the shield's curl radius and the shields' roughness on the vacuum breakdown strength. They show that for an increase of the curl radius from 2mm to 5mm the withstand ability increases for all breakdown gaps from 5mm to 20mm. Larger values of curl radius up to 8mm show no further improvement; see Figure 1.108. They also show that the surface roughness R_{CLA} [75], as expected, is an important parameter. They show for the center shield and end shield arrangement shown in Figure 1.108 with a breakdown gap, d, that:

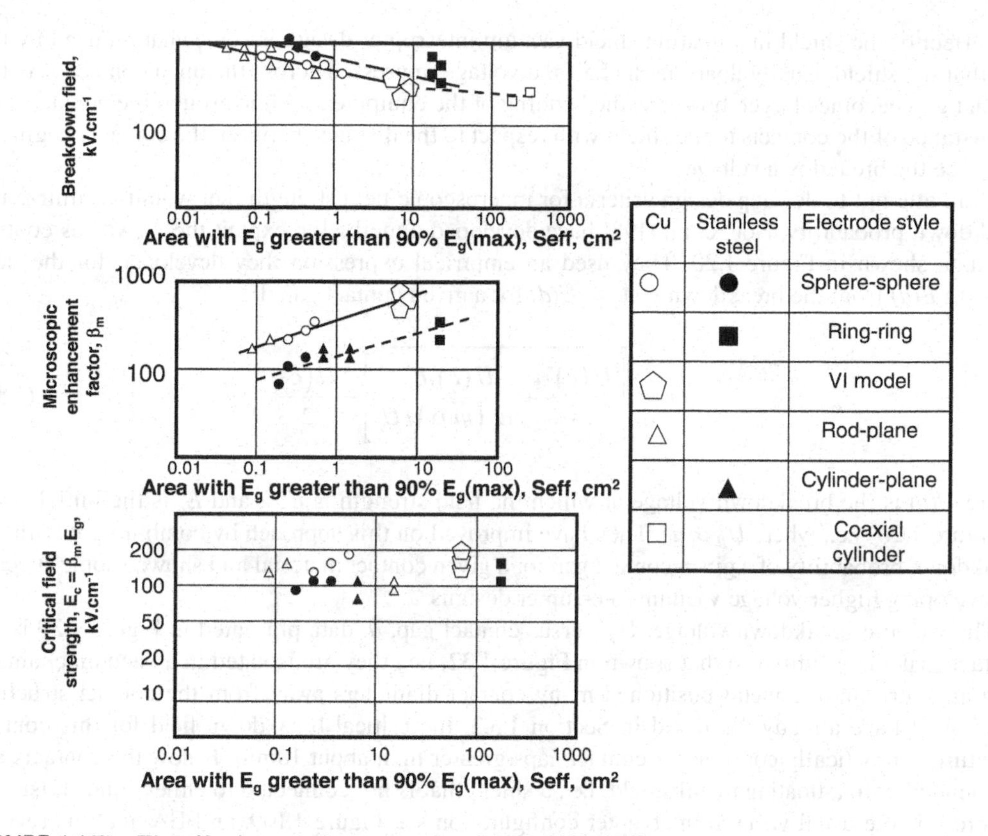

FIGURE 1.107 The effective area for vacuum breakdown [131, 194].

$$U_B(R_{CLA}, d) = 72.6d^{0.35}R_{CLA}^{-0.15} \tag{1.83}$$

They also show that over the range of breakdown gaps they have investigated the effect of S_{eff} is not significant for a given curl radius. Shioiri et al. [197] show using stainless steel contacts with an electro-polished surface with an $R_{CLA} = 0.04\mu m$ a 10% improvement in U_B over a machined contact with an $R_{CLA} = 0.16\mu m$ even though both surfaces have an R_{CLA} of about $0.22\mu m$ after conditioning.

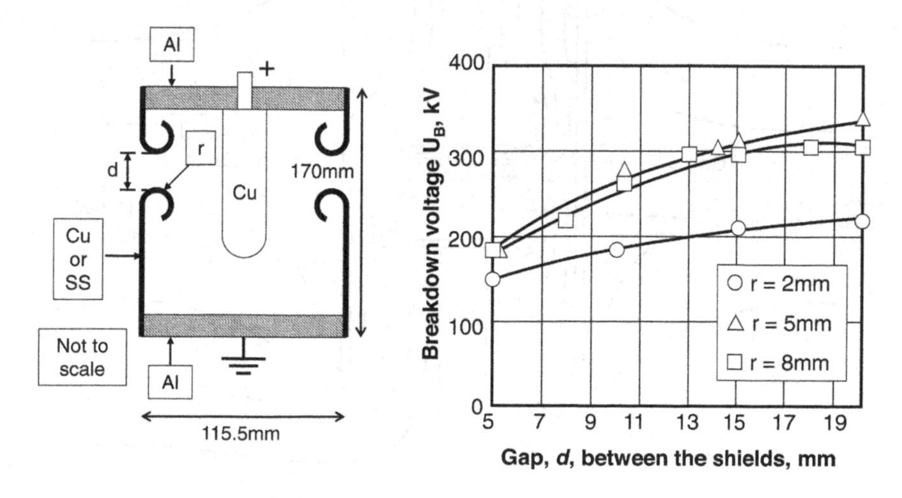

FIGURE 1.108 The vacuum breakdown voltage between a stainless-steel end shield and a stainless-steel center shield as a function of shield radius, *r*, and the shield gap, *d*, 196].

The effect of the shield in a floating shield vacuum interrupter design is somewhat reduced by the fact that the shield usually floats at one half the voltage impressed across the open contacts. As the contact gap becomes larger, however, the bending of the equipotential line around the contacts and the distance of the contacts to the shield with respect to the distance between the contacts begins to influence the breakdown voltage.

In an attempt to develop design criteria for macroscopic field strength that would minimize the breakdown probability, Noe, et al. [198] have developed a method to exploit the U_B versus contact gap data shown in Figure 1.20. They used an empirical expression they developed for the field strength *E(d)* from the breakdown voltage *U(d)* for a given contact gap d:

$$U(d) = \sqrt{\left[\frac{U(c)^2}{4} + \frac{U(c).E_0}{E_g(max)/U}\right]} - \frac{U(c)}{2} \tag{1.84}$$

where *U(c)* is the breakdown voltage at which the field strength is $E_o/2$ and E_o is the initial slope of Figure 1.20 (i.e., where $U_B \propto d$). They have improved on this approach by combining it with the breakdown probability of a given contact gap for a given contact material and showed some success at developing higher voltage vacuum interrupter designs.

The vacuum breakdown voltage, U_B, versus contact gap, *d*, data presented in Figure 1.20 is for contact structures similar to that shown in Figure 1.32, i.e., they are isolated in a vacuum chamber with any surrounding metal positioned many contact diameters away from the contact structure itself. As I have already discussed in Section 1.3.2, the critical breakdown field for this contact structure is practically constant for contact gaps greater than about 10mm. If now the contacts are surrounded with a floating metal shield, i.e., a shield that is not connected to either contact, (such a structure is the usual vacuum interrupter configuration see Figure 1.109, an FEA analysis reveals that the bending of the equipotential lines around the contacts becomes more severe. Figure 1.110

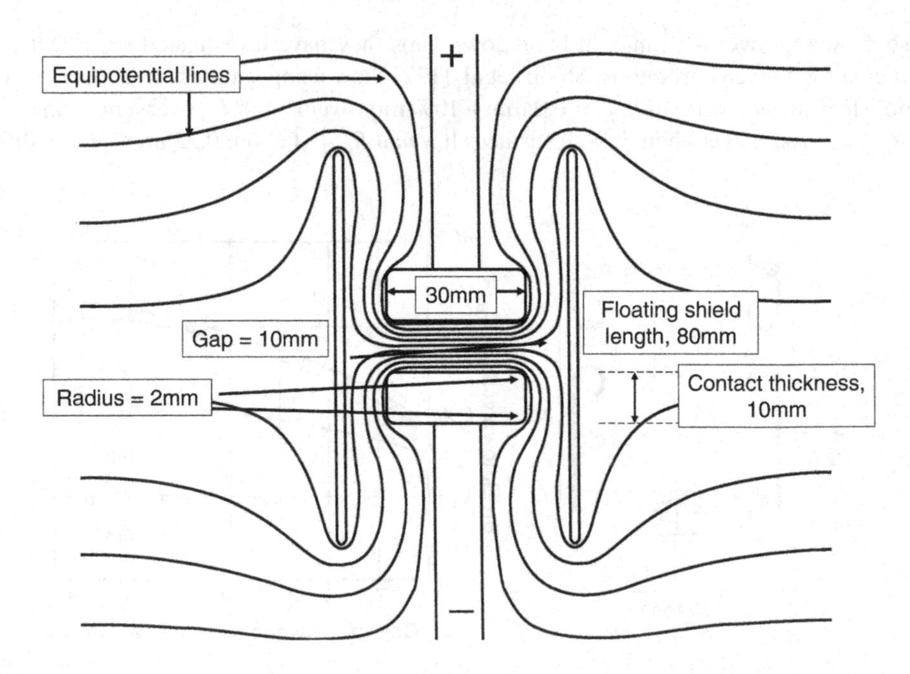

FIGURE 1.109 A contact structure with a surrounding metal shield showing the equipotential lines from an FEA field analysis [50].

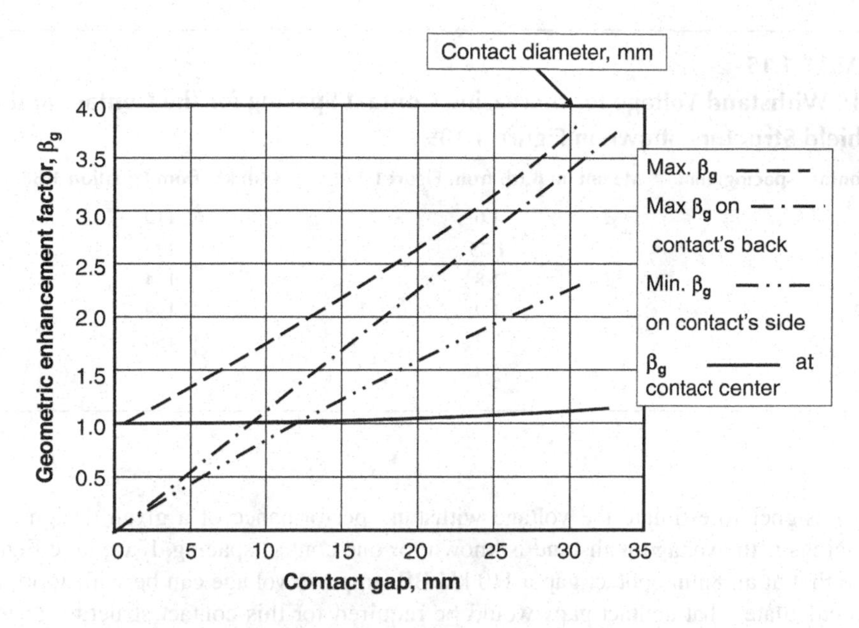

FIGURE 1.110 The geometric enhancement factors for four places on the contact surface for the contact structure shown in Figure 1.109 as a function of contact separation [50].

shows the effect on the geometric enhancement factor, β_g, around the periphery of the contact and as a function of the contact gap. Here the maximum β_g (and hence the maximum macroscopic electric fields E_g) is at the front and back edges of the contacts [50]. Figure 1.111 shows the effect of the contact gap on β_g values. If you compare this figure with Figure 1.33 you can see that the effect of the surrounding metal shield is to nearly double the β_g. This analysis can be used by the vacuum

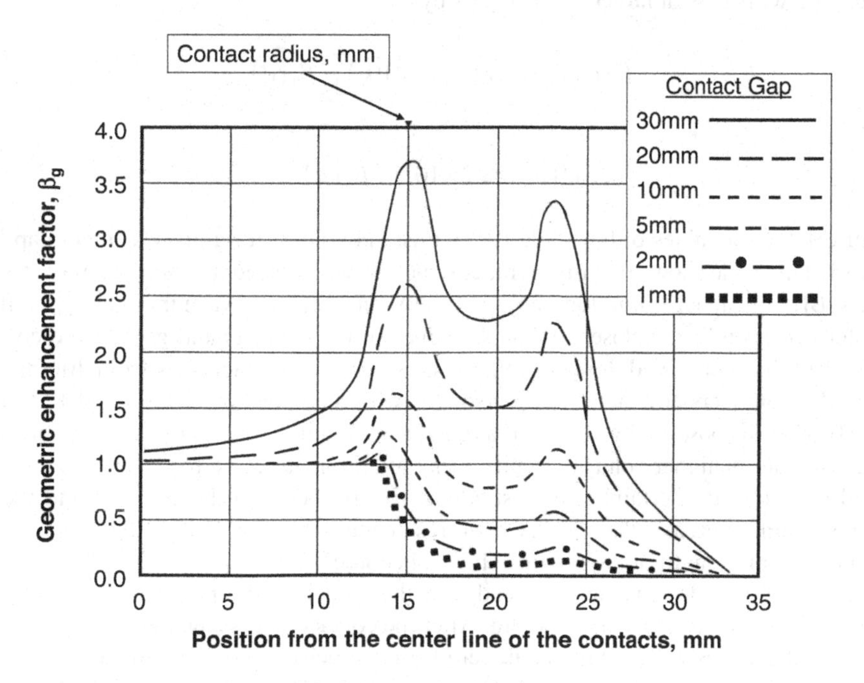

FIGURE 1.111 The geometric enhancement factor as a function of the distance from the center of the contact and the contact gap for the structure shown in Figure 1.109 [50].

TABLE 1.15
BIL Withstand Voltage for Increasing Contact Spacing for the Contact and Shield Structure Shown in Figure 1.109

Contact spacing, mm	Maximum $\beta_g(d)$ from Figure1.111	U(d) kV, from Equation 1.88
8	1.6	110
10	1.75	126
11	1.8	134
12	1.9	139
15	2.2	150
20	2.7	163

interrupter designer to estimate the voltage withstand performance of a given design at various contact spacings if the voltage withstand is known for one contact spacing. If we take Figure 1.111 and assume that at an 8mm contact gap a 110 kV BIL impulse voltage can be withstood, then it is possible to calculate what contact gaps would be required for this contact structure to withstand 125 kV and 150 kV. The macroscopic field at 8mm E_g(8mm) with 110 kV peak voltage impulse is:

$$E_g(8mm) = \beta_g(8mm) \times \left[110 \times 10^3\right] / \left[8 \times 10^{-3}\right] \text{V.m}^{-1} \tag{1.85}$$

$$E_g(8mm) = 1.6 \times \left[110 \times 10^3\right] / \left[8 \times 10^{-3}\right] = 2.2 \times 10^7 \text{V.m}^{-1} \tag{1.86}$$

If we assume that this macroscopic field is a conservative value then we can calculate the BIL voltages that can be withstood at larger contact gaps by:

$$U(d) = E_g(8mm)\left[d \times 10^{-3}\right] / \beta_g(d) \tag{1.87}$$

$$U(d) = 2.2 \times 10^3 d / \beta_g(d) \tag{1.88}$$

Using Figure 1.111 for values of β_g(d), the BIL withstand voltage for a given contact gap is shown in Table 1.15. This table shows that this contact structure would need a contact gap of about 10mm for a 125 kV BIL and about 15mm for a BIL of 150 kV. In a practical vacuum interrupter, the shield has a greater effect on E_g the closer it is to the edges of the contacts and greater the contact gap. Kahl et al. [199] have observed, however, that over a range of contact gaps from 10mm to 20mm for a 75mm diameter copper contact, copper shields with diameters of 100mm and 120mm have a minimal effect on the observed vacuum breakdown; see Figure 1.112. As vacuum interrupters are designed to operate in higher voltage circuits, strict attention has to be paid to the geometry of all the internal components. Certainly, the insulation strength between the contacts and the floating shield at BIL voltages greater than 250kV requires special attention [200, 201]. Also, the design, spacing, diameter and length of the floating shields become critical.

Of course, all of these design criteria depend upon the vacuum interrupter going through some conditioning process at the end of its manufacture. After the contact processing, machining, cleaning, and so on the contact surfaces will contain a number of surface defects. Conditioning to give the vacuum interrupter the highest possible U_B so that it can easily pass the standard tests discussed in Section 1.1 can minimize the effects of these practical surfaces. As discussed in Section 1.3.4, ac conditioning

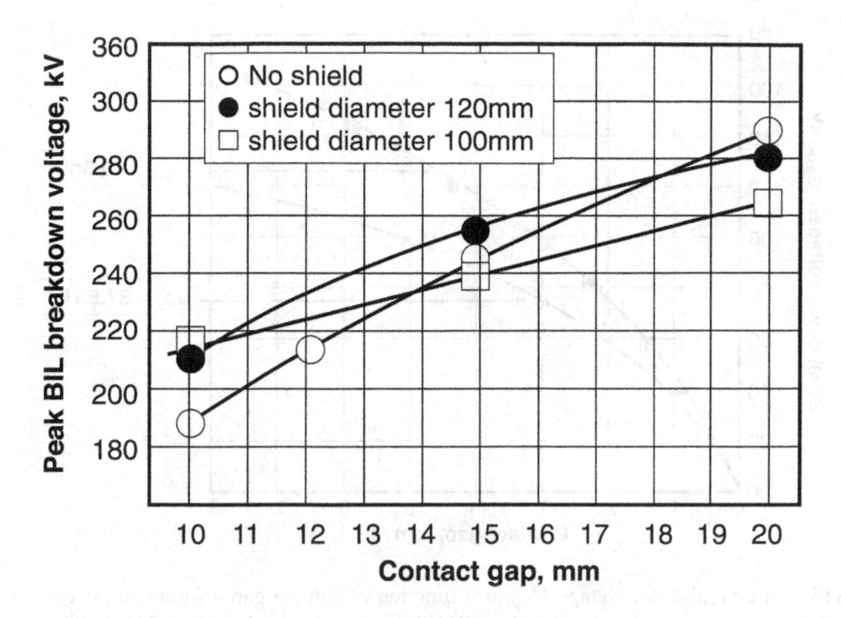

FIGURE 1.112 The BIL breakdown voltage as a function of contact gap for 75mm diameter Cu contacts with no surrounding shield and with shields of different diameters [199].

works well. If the vacuum interrupter is ac conditioned somewhat above the highest peak voltage required of it, e.g., 10% greater than its required BIL peak value), then it will satisfy both the BIL levels and the ac withstand levels needed to pass the standard ratings. As stated previously, once in service the vacuum interrupter contacts will experience continual conditioning from switching the load current.

1.4.2 Breakdown of Multiple Vacuum Interrupters in Series for Contact Gaps Greater Than 2mm

When the breakdown voltage $U_{B(1)}$ as a function of contact gap d for a single vacuum interrupters is determined the relationship is in the form:

$$U_{B(1)} = K_1^* d^\eta \tag{1.89}$$

where η can range from 0.3 to 0.7 and K_1^* is a constant. Now if two identical vacuum interrupters are connected in series and each has the same contact gap d, then we would expect that the breakdown voltage $U_{B(2)}$ across the two to be:

$$U_{B(2)} = 2U_{B(1)} = 2K_1^* d^\eta \tag{1.90}$$

Now for a single vacuum interrupter with a contact gap of $2d$:

$$U_{B(1)} = K_1^* (2d)^\eta \tag{1.91}$$

An improvement factor P_2^* for the two vacuum interrupters in series each with a contact gap of d compared to one vacuum interrupter with a contact gap of $2d$ can be given as:

$$P_2^* = \frac{U_{B(2)}}{U_{B(1)}} = \frac{2K_1^* d^\eta}{K_1^* (2d)^\eta} = 2^{(1-\eta)} \tag{1.92}$$

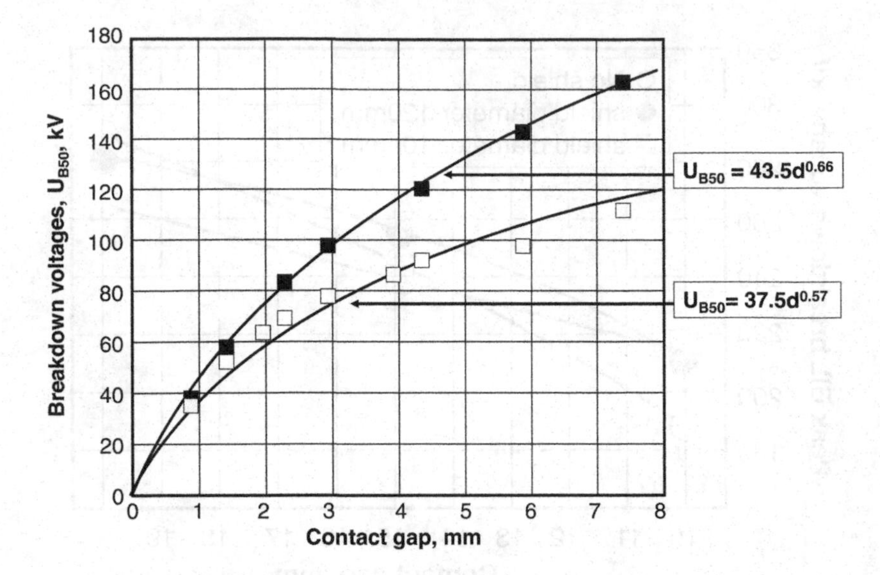

FIGURE 1.113 The breakdown voltage $U_{B(1)}$ as a function of contact gap d for a single vacuum interrupter (□) and for three vacuum interrupters in series (■)$U_{B(3)}$ with the same total contact gap (i.e., each vacuum interrupter has a contact gap *d/3*) [202].

Liao et al. [202] in Figure 1.113 give an example of $U_{B(1)}$ for a single interrupter as a function of contact gap d and $U_{B(3)}$ as a function of the same total contact gap d for three interrupters vacuum in series (i.e., each vacuum interrupter has a contact gap of *d/3*). The measured performance factor P_3^* for *d=7.5mm* is:

$$P_3^* = \frac{U_{B(3)}}{U_{B(1)}} = \frac{43.5 \times 7.5^{0.66}}{37.5 \times 7.5^{0.57}} = 1.39 \tag{1.93}$$

In this case with η = 0.57, the maximum expected value for three vacuum interrupters in series would be:

$$P_3^* = \frac{U_{B(3)}}{U_{B(1)}} = 3^{(1-0.57))} = 1.60 \tag{1.64}$$

The difference in the two performance factors most probably results from the non-uniform voltage division across vacuum interrupters in series. I will discuss how the voltage distributes across vacuum interrupters in series in Chapter 5, Section 5.2.2, in this volume. Even so, this example does show that three vacuum interrupters in series with have a higher voltage withstand capability than a single vacuum interrupter with the same total contact gap.

1.4.3 Voltage Wave Shapes and Vacuum Breakdown in a Vacuum Interrupter

The number and variety of voltage wave-shapes that can be impressed across the open contacts complicate the life of the vacuum interrupter designer. In this section, we will discuss the most common ones and then show how the internal high voltage design decision can be simplified. I would expect that the most common voltage experienced across the open contacts would be the system ac voltage. For a three-phase ungrounded system this will be:

$$U(peak) = \frac{\sqrt{2}.U(rms)}{\sqrt{3}} \quad (1.95)$$

Another voltage that can appear from switching a capacitor bank is approximately *(2.5 to 3) × U(peak)* and is unidirectional; this will be discussed in Chapters 5 and 6 in this volume. The switching of inductive circuits and short circuit currents will impose, after the first phase to clear, a peak value of the transient recovery voltage which is about *1.7 × U (system, rms.)* with a rate of rise *(dU/dt)* of between 0.5 and 1 kV.μs^{-1}; see Chapters 4, 5, and 6 in this volume. As discussed in Section1.1, the vacuum interrupter must also satisfy the standard one-minute withstand voltage test and the BIL impulse voltage test. Table 1.16 presents the peak values of these voltages for a range of system voltages. The ratios of the BIL voltage to the other peak voltage values are also given. It can be seen that the BIL value is between 1.5 and 1.9 times greater than the peak value of the one-minute withstand voltage. The ratios are even greater for the peak open circuit voltage (5 to 8 times), the inductive transient recovery voltage (2 to 3 times) and the peak capacitor switching voltage (1.8 to 3 times). These voltages have markedly different rates of rise, i.e., *dU/dt*. The BIL has a *dU/dt* of about 10^{2}kV.μs^{-1}and the inductive circuit transient recovery voltage is between 0.5 and 1kV.μs^{-1}, although in some special applications it can be greater than 3kV.μs^{-1}. The open circuit and the ac withstand voltages reach their peak values in 4 to 5 ms and the capacitor switching voltage reaches its peak value in 8 to 10 ms.

König et al. [203] have studied the effect of *dU/dt* on the breakdown voltage (U_B) for conditioned Cu–Cr 3ontacts; an example of their data is given in Figure 1.114. The first thing that strikes you about this data is the very wide range of U_B for each value of *dU/dt*. For the sake of this discussion, I will use only the lowest band of this data set; i.e., the minimum value of breakdown voltage $U_B(min)$ for each *dU/dt* value. For a contact gap of 3mm, *dU/dt* has little effect on $U_B(min)$. We would thus expect the same $U_B(min)$ for an ac voltage waveform as for a BIL pulse voltage. As the contact gap increases, however, the *dU/dt* has a marked effect on $U_B(min)$. For example, for a contact gap of 14mm, $U_B(min)$ at 1 kV/μs is about 130 kV while at 10^2 kV/μs it is about 170 kV. The ratio between these two values is 170/130 = 1.3, which is similar to the ratio that can be calculated from the data given in Figure 1.80. Thus, if a vacuum interrupter with a 14mm contact gap were designed to withstand a 125 kV BIL voltage (i.e., for use in a 24kV system), then it would be able to withstand a peak ac voltage of:

$$U_{peak}(min) = 125/1.3\,kV = 96\,kV \quad (1.96)$$

Table 1.16 shows us that this value is higher than any of the peak ac voltages that a vacuum interrupter has to withstand in a 24kV circuit. The BIL withstand level is thus of paramount importance for the vacuum interrupter designer. From the above discussion we can see that once the BIL level is satisfied, all other voltage waveforms that the vacuum interrupter will be subjected to in a practical ac system will be easily withstood.

The vacuum interrupter's contacts should be conditioned to hold off impulse voltages in excess of those required by the testing standards. If the contact surfaces become de-conditioned during operation then a voltage breakdown across an open vacuum gap during BIL testing may occur. If this were to happen it is likely that the breakdown will condition the contact surfaces and the vacuum interrupter will continue to operate successfully in the electrical circuit. Experience has shown that when a vacuum interrupter has been designed conservatively for BIL impulse voltages, it will easily withstand all the voltages an electric circuit can impress across its open contacts.

1.4.4 Impulse Testing of Vacuum Interrupters

When a vacuum interrupter is newly installed in a mechanism, it is usual to perform a number of no-load operations to test the mechanism's operation. As discussed in Section 1.3.7, this switching without

TABLE 1.16
Peak Voltages That Will Be Experienced by a Vacuum Interrupter in an Electrical System

A	B	C	D	E	F				
System Voltage (kV)	BIL Voltage (kV)	Peak Open Circuit Voltage for 3 Phase Ungrounded System (kV)	Maximum Peak Voltage for 3 Phase Capacitor Switching (kV)	Peak, One Minute Withstand Voltage (kV)	Peak Transient Recovery Voltage, First Phase to Clear, 3 Phase Ungrounded System (kV)	B/C	B/D	B/E	B/F
12	75	9.9	24.5	40	20.6	7.6	3.1	1.9	3.6
15	95	12.2	30.6	51	28	7.8	3.1	1.9	3.4
17.5	95	14.3	35.7	54	30	6.6	2.7	1.8	3.2
24	125	19.6	49.0	71	41.2	6.4	2.6	1.8	3
27	125	22	55.1	85	51	5.7	2.3	1.5	2.5
36/38	170	31	77.6	113	71	5.5	2.2	1.5	2.4

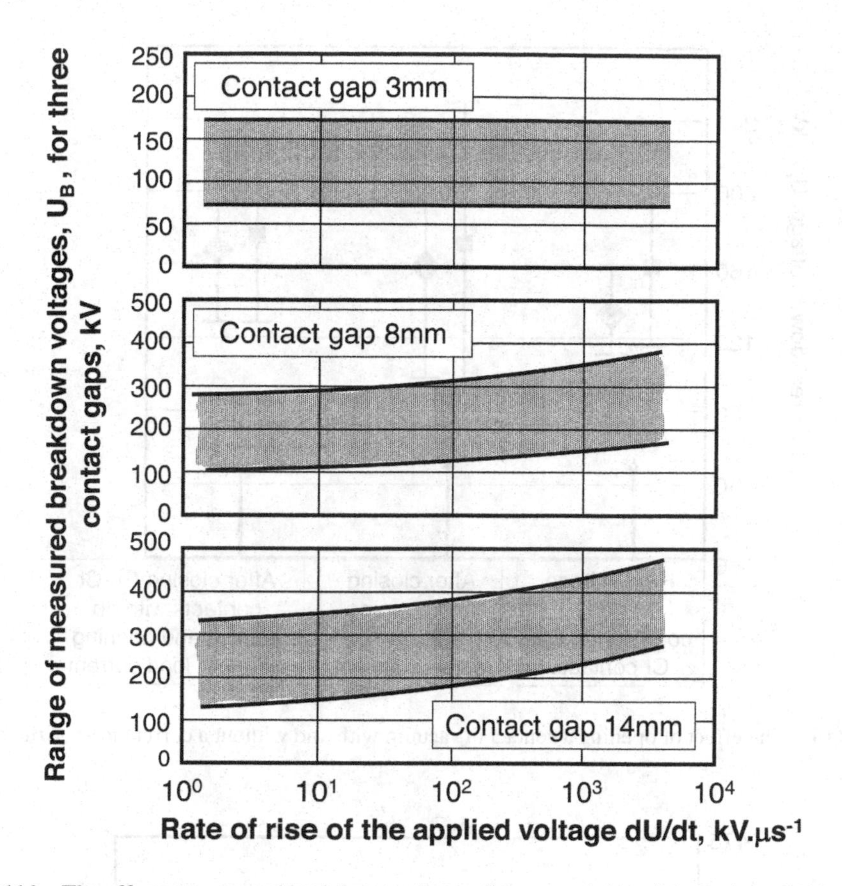

FIGURE 1.114 The effect on vacuum breakdown voltage of the rate of rise of that voltage [203].

current can have a temporary effect on the BIL high voltage withstand ability of the new vacuum interrupters. Figure 1.115 compares the effect of opening under high load current and opening under no load [49]. Zhang et al. [204] show similar results for closing with no load and then opening with and without current: See Figure 1.116. Under no load, however, there can be a wide variation in the breakdown voltage of the contact gap. The breakdown voltage can be much less than the conditioned value. This results from the microscopic cold-welding that can take place at the microscopic contact spots. When the contacts part those small welded regions can pull microprojections, which can enhance the breakdown effects i.e., increase the value of β_m. Smith [205] has recognized this effect when BIL testing of new vacuum interrupters installed in new circuit breaker mechanisms. He describes a testing procedure that both assures the reconditioning of the new contacts as well as enabling the true BIL level of the new vacuum circuit breaker to be determined. In fact, it is even possible for the BIL level to be much higher. One reason for this is that the contact surfaces will be work hardened as they make contact. The work hardening of the contact surfaces will help to increase their voltage withstand ability.

For circuit breakers and other switching devices, the open vacuum interrupter contacts must withstand any voltage impressed across them for the following reasons:

(a) To prevent phase-to-ground breakdowns
(b) To prevent phase-to-phase breakdowns
(c) To prevent line-to-load breakdowns

Here (a) and (b) represent outright failures of the vacuum circuit breaker's insulation design. The design's primary objective would be to prevent these from occurring. Line-to-load breakdowns,

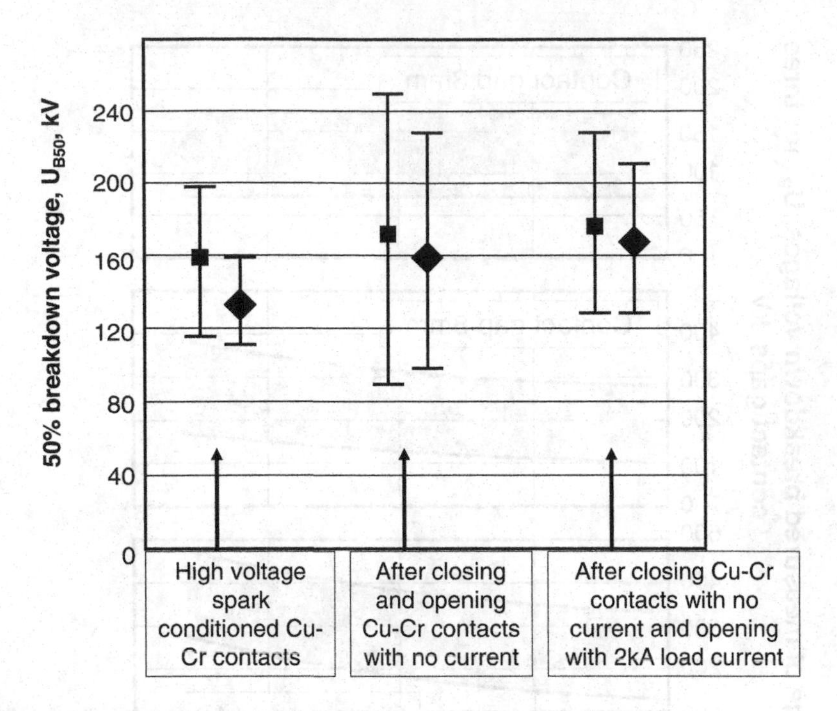

FIGURE 1.115 The effect of opening a contact in vacuum with and without a current load on the breakdown voltage [49].

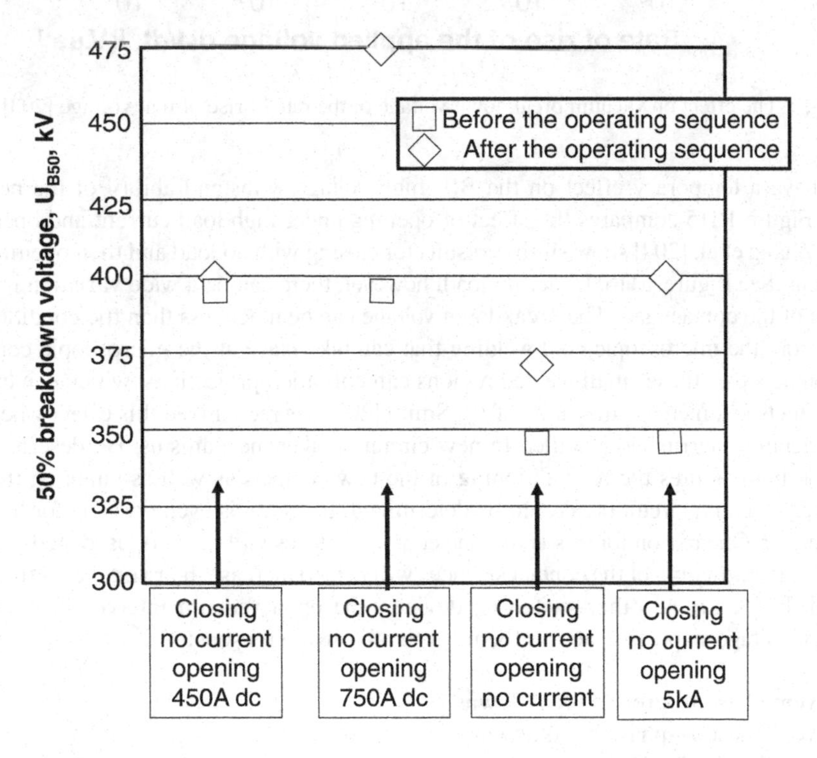

FIGURE 1.116 Change in BIL performance with no load closing and opening with different current options and with no load opening [204].

however, must be viewed in terms of the effect on the electrical system of the resulting current that flows through the vacuum interrupter. For example, a breakdown followed by an uninterrupted flow of current through the vacuum interrupter would represent a failure of the circuit breaker to maintain the open status of the power circuit and must not be allowed to occur. However, if the breakdown results in the passage of a very short duration current pulse, which is quickly interrupted, the circuit breaker can then be considered to have essentially preserved the open status of the power circuit. Vacuum interrupters are rather unique in being able to interrupt current even when the contacts are in the fully open position. Arc interruption in vacuum is not dependent on the motion of the contacts. As will be discussed in Chapter 3 in this volume, the arc is controlled by the geometry of the contact structure. Even if a breakdown does occur between open vacuum interrupter contacts, and results in a flow of power frequency current, it will be interrupted at the first current zero; see Chapter 4 in this volume. This behavior is rather unique to vacuum interrupters and is not seen in other techniques such as minimum oil, air magnetic or SF_6 puffer interrupters.

For newly installed vacuum interrupters in a circuit breaker or switch mechanism, pre-certification BIL testing should consider the following effects:

- The vacuum interrupter contacts may undergo deconditioning during installation as they are closed and opened with no current flow
- While pre-certification BIL testing, the new vacuum gap may break down at a value below the required value given in the testing standards (I shall call this "a disruptive discharge")
- If a breakdown occurs in a power circuit and a half cycle of current flows through the vacuum interrupter, it will be interrupted at the first current zero
- The breakdown event and the current flow will recondition the vacuum interrupter's contacts

For 12 kV and 15 kV vacuum interrupters where the impulse withstand rating is 95kV, the probability of the interrupter to withstand the BIL, impulse voltage is very high. In fact, as I have discussed in Section 1.4.1, the same designs with a slightly larger contact gap can be applied at 24kV with a 125kV impulse rating. The occasional disruptive discharge at a voltage that is less than the vacuum interrupter's full capability is usually more probable at a voltage greater than the 95kV rating. As designs for 36 to 40.5kV and higher voltages have become more commonplace and the demands for compact designs more insistent, the design capability of the interrupter and the rating have become much closer. As a result, an occasional disruptive discharge has been observed at less than rated BIL voltage.

The ac withstand test should always be done before impulse tests are performed. The ac withstand test is a prudent step which requires little time to perform. It is first of all a means for checking that the vacuum interrupters and the complete circuit breaker insulation system are in good condition. It will also identify any vacuum interrupters that have been damaged during transit during assembly. For example, the initial mechanical operations of the circuit breaker may cause a weak braze joint to leak. It is suggested that the ac withstand test be applied at 80% of the rated power frequency withstand voltage for 1 to 3 minutes and then 100% of the rated power frequency withstand voltage for 1 minute. The duration of the test at 100% of the rated power frequency withstand voltage test should be limited to 1 minute since other circuit breaker insulation paths are subjected to this same voltage and should not be stressed beyond what is required by the certification standards.

A small amount of reconditioning of the vacuum interrupter may occur during the ac power frequency withstand test, but this is very unlikely. The voltage magnitudes used in the ac power frequency withstand test are much lower than those used in the manufacturing conditioning operation. As shown in Table 1.16 for a 36/38kV vacuum circuit breaker the peak voltage of 113 kV applied during the ac power frequency withstand voltage test is only 66% of the 170kV impulse withstand. I would expect that any reconditioning provided by this voltage level would be minimal. Preliminary impulse tests should then be performed on each interrupter by applying both positive and negative impulse voltages with two purposes in mind:

1. Reconditioning the interrupter's ability to withstand impulse voltages
2. Reversing the trapped charge on the internal floating shield whenever the polarity is changed

Each interrupter in a vacuum circuit breaker should be tested in the following manner. First of all, preliminary tests need to be performed starting at a fraction of the rated impulse withstand voltage and then proceed to 110% of the rated impulse voltage. These preliminary tests provide some reconditioning of vacuum interrupter's contacts to smooth sharp spots produced by mechanical touching and cold-welding of the contacts. In addition, the preliminary tests are especially important whenever changing from one polarity to the opposite one. It is important to remove and reverse the charge that builds up on the floating shield during the testing at the first polarity. Any disruptive discharges that occur in the preliminary trials are not counted in the statistics for pass/fail determination at the rated impulse voltage.

The sequence of tests in Table 1.17 is recommended. This sequence is based on the IEEE and IEC impulse voltage test methods described in the relevant switchgear standards. IEEE has used a method called the 3 × 3 method for many years while IEC has used a 2 × 15 method. A compromise method called the 3 × 9 method is now the standard method in more recent revisions of IEEE standards and is an acceptable alternative method in IEC standards. These test methods are explained below.

1. 3 × 3 Impulse Voltage Test Method [206]:

Step 1: Apply three impulses of a desired crest voltage:

- If all thee impulses are withstood, then the device has passed the test
- If two disruptive discharges are observed in the three impulse trials, then the device has failed the test.
- If one disruptive discharge is observed in the three impulse trials, then perform three more tests in Step 2 at the same crest voltage.

Step 2: Apply three additional impulses of the same crest voltage:

- If all three additional impulses are withstood for a total of one disruptive discharge in six tests, then the device has passed the test.
- If a second disruptive discharge is observed, then the device as failed the test.

2. 3 × 9 Impulse Voltage Test Method:
The 3 × 9 method is the same as the 3 × 3 method except for two changes:

- The number of additional impulse trials to perform is nine if there is one disruptive discharge in the first three impulse trials, and
- If all 9 additional impulses trials are withstood for a total of 1 disruptive discharge in 12 tests, then the device has passed.

3. 2 × 15 Impulse Voltage Test Method [207]:
Apply 15 impulse trials of a desired crest voltage:

- If no more than two disruptive discharges are observed for a total of two disruptive discharges in 15 impulse trials, then the device has failed the test. Table 1.18 gives an example for a high voltage vacuum interrupter whose performance at 346kV and 360kV gave satisfactory results, but shows that it cannot achieve 380 kV [208].

TABLE 1.17
Recommended Voltage Steps for Impulse Testing

Test condition	Voltage Polarity	Test voltage applied	No. Of Trials
Initial	**Polarity**	**% of rated**	
Preliminary	Positive	50%	3 Note 1
	Positive	75%	3 Note 1
	Positive	90%	3 Note 1
	Positive	110%	3 Note 1
Certification	Positive	100%	N Note 2
Reverse	**Polarity**		
Preliminary	Negative	50%	3 Note 1
	Negative	75%	3 Note 1
	Negative	90%	3 Note 1
	Negative	110%	3 Note 1 and Note 3
Certification	Negative	100%	N Note 2

Note 1:
If a disruptive discharge occurs in one of these trials,
the use the 3 × 3 method at this voltage or, for more conditioning, perform additional trials at the same voltage until three to five impulses are withstood in a row.

Note 2:
The number of trials performed at the rated impulse
withstand voltage depends on the standard used.

- For IEC tests to IEC standard 56 and 694 and 60: N = 15 and Pass ≤ 2 breakdowns in 15 trials.
- For IEEE tests to C37.09 and IEEE Standard 4: N = 3 or 6 and Pass ≤ 1 breakdown in 6 trials.
- For both IEEE and IEC Standards, recent revisions: N = 3 or 12 and Pass ≤ 1 breakdown in 12 trials.

Note 3:
If the external insulation will not support the 110% BIL value then the voltage should be limited to the 100% value.

TABLE 1.18
BIL Conditioning of a High Voltage Vacuum Interrupter with a 40mm Contact Gap from 346 KV to 380 KV

BIL Peak	Pulse Sequence														
kV	1	2	3	4	5	6	7	8	9	10	11	12	13	14	15
346	✓	✓	✓	✓	✓	✓	✓	✓	✓	✓	✓	✓	✓	X	✓
360	✓	✓	✓	✓	✓	✓	✓	✓	✓	✓	X	✓	X	✓	✓
380	X	X	X	X	X	X									

(✓Success, **X** Breakdown)

TABLE 1.19
An Example of Reconditioning A 38 KV Vacuum Interrupter after Shipping with the Contacts Closed Using BIL Impulse Voltages and a 14mm Contact Gap

Voltage	Positive Pulse	Negative Pulse
50% (87 kV)	✓✓✓	✓✓✓
75% (130 kV)	✓✓✓	X✓✓✓✓✓
90% (157 kV)	✓✓✓	✓✓
100%	(175 kV) ✓✓✓	(176 kV) ✓ X
		(173 kV) ✓✓✓

Required BIL Voltage 170 kV (✓success, **X** breakdown)
Continued testing showed that this vacuum interrupter could pass the 200kV BIL level

It is interesting to note that all of these standards recognize that no interruption device will pass a rated BIL value every time: they accept that disruptive discharges can occur during any sequence of BIL testing. The three certification pass criteria presented above for impulse voltage withstand tests provide that 1 of 6 (17%), or 1 of 12 (8%) or 2 of 15 (13%) breakdowns may well occur at the rated BIL voltage during any test sequence. It would be expected that the occurrence of breakdowns at less than the rated BIL value would be quite infrequent. An example of a test on a 36kV vacuum interrupter for a BIL level of 170kV is shown in Table 1.19. Here the vacuum interrupter had been conditioned during manufacture to satisfy a BIL level of 185kV. It was then shipped closed and placed in a mechanism and operated with no current. After the lower voltage conditioning breakdowns had occurred, the vacuum interrupter easily passed the required BIL level. In fact, continued testing showed this vacuum interrupter capable of operating at a 200kV BIL level.

The fact that an open vacuum interrupter gap occasionally has a breakdown event during pre-certification BIL tests presents no problem for vacuum circuit breakers in service on power distribution systems. There are four main reasons for this:

1. During normal operation, switching load current, the vacuum interrupter's contacts are continually conditioned
2. Surge protection keeps impulse voltages low
3. Breakdown is a statistically rare event
4. Vacuum interrupters can interrupt the power follow-through current even when sitting open

Also, in service few impulse voltages as high as the rated value actually reach most circuit breakers or reclosers. Systems are designed using insulation coordination techniques to avoid overstressing individual pieces of equipment. In addition, protective devices are used, such as lightning arresters, ground wires above lines, ground mats around substations and spark gaps at various locations to limit the impulse voltages that reach the location of circuit breakers or reclosers. Moreover, circuit breakers and reclosers spend most of their lives in the closed position carrying current to feed loads. During lightning storms, these closed, circuit breakers and reclosers are there to protect the circuit. For example, they will open and protect the circuit when phase-to-ground or phase-to-phase faults result from lightning induced breakdown of the system insulation or by water, ice and wind-blown actions. When open for extended periods, circuit breakers and reclosers are usually isolated with disconnect switches, especially during line repair work. So, the application of impulse voltages to open vacuum interrupters is limited in magnitude and very infrequent.

If a breakdown does occur and results in a power frequency current, the vacuum interrupter will clear the circuit at the next current zero. Assuming that there are no faults in the system, the follow current is likely to be rather small and easily interrupted by the vacuum interrupter. In ungrounded systems, it would take breakdowns in two interrupters simultaneously to result in a power frequency current, and this is an unlikely event. More typical in such systems would be a breakdown in only one phase followed by a high frequency current, which is also easily interrupted by vacuum interrupters. Many such incidents have probably occurred over the years and not been noted since circuit isolation is quickly restored by the vacuum interrupters with an insignificant effect on the system.

1.4.5 Testing for High Altitude

When circuit breakers and other switching devices are applied at high altitudes the effects of the reduction in atmospheric pressure has to be taken into account. Table 1.20 gives the atmospheric pressure, gas density and temperature at different altitudes. For high altitudes the right-hand side of Paschen's Law can be written as:

$$U_B = 2440\left(\frac{293pd}{T}\right) + 61\left(\frac{293pd}{T}\right)^{1/2} \tag{1.97}$$

where T is the absolute temperature in degrees Kelvin, p is in atmospheres and the breakdown gap d is in meters. From Equation 1.97 and Table 1.20 for $d = 10\text{mm}$ the ratio of the breakdown voltages U_B *@ 5000m /* U_B *@ sea-level* is:

$$U_B @ 5000m \,/\, U_B @ sea-level = 0.64 \tag{1.98}$$

Now, if you wish to test at sea level for the external withstand ability in air at an altitude of 5000 m, the voltage levels across the vacuum interrupter would have to increase by about 57%. As most certification laboratories are located close to sea level, a special series of high voltage tests is required for certification for use at high altitudes. The most expedient way to test is to increase the test voltage at sea level and if the equipment passes the test, the use at high altitudes will be confirmed. This type

TABLE 1.20
Average Hypothetical Atmospheric Pressure, Temperature, and Air Density as a Function of Height Above Sea Level

Height above Sea Level (m)	Pressure (mbar)	Temperature (C)	Density (Kg.m^{-3})
0	1014	23.1	1.19
500	957	20.5	1.14
1000	902	18.0	1,07
1500	850	15.5	1.03
2000	802	12.9	1.01
2500	755	9.8	0.93
3000	710	6.6	0.88
4000	627	0.5	0.80
5000	554	−5.7	0.72
6000	487	−10.0	0.65

of test is realistic for the phase-to-ground and the phase-to-phase insulation. If the vacuum interrupter occasionally breaks down internally during this over-voltage test, this internal breakdown should be ignored. The internal high voltage withstand is of course not affected by the external gas pressure. Thus, if a vacuum interrupter passes the standard tests internally at sea level, it will also do so at high altitude. So, when performing over-voltage stress tests to check if the external performance of the device is suitable for use at high altitudes, any internal breakdowns at voltages in excess of the vacuum interrupter's internal design voltage should be ignored. A more realistic test for a vacuum interrupter and for the device would be to place them in a pressure chamber and perform the standard voltage tests at reduced air pressure. If the vacuum circuit breaker is enclosed in a hermetically sealed chamber with SF_6 gas insulation or it is completely encapsulated in a solid dielectric material, the altitude will have no effect on the voltage performance and the high voltage tests performed at sea level will also be valid at any altitude.

1.5 X-RAY EMISSION

If a high enough voltage is applied across the open contacts of a vacuum interrupter then it is possible to generate X-rays. In fact, the detection of X-ray emission has been used as a diagnostic tool in the investigation of pre-breakdown phenomena [83, 129]. The X-rays result from the electrons emitted from the cathode by field emission, gaining energy from the potential drop between the contacts, and giving up that energy on impact with the anode. The intensity and energy of the X-rays is dependent upon the level of electron emission and the electric field (βU/d; see Section 1.3.2) across the contact gap. As the electrons impact the anode contact most of their energy is converted into heat. Less than 1% results in the production of X-rays. For a Cu–Cr anode the approximate efficiency Є for producing X-rays is:

$$\epsilon = U \times I_e \times Z \times 10^{-6} \tag{1.99}$$

Where U is the applied voltage, I_e the emission current (usually less than 1mA), $U \times i_e$ is in keV, and Z is the atomic number of the anode metal. For Cu with an atomic number of 29 the efficiency of X-ray production for 100keV, 200keV, and 300keV electrons is 0.29%, 0.58% and 0.86% respectively. The electron–anode interaction produces two types of X-radiation: (a) The slowing and direction change of the electrons by the atom's nucleus produces *Bremsstrahlung* (braking of deceleration radiation) X-ray photons and (b) the impact of the incoming electrons with electrons in the atom's shells produced *characteristic* X-ray photons. As an electron passes close to an atomic nucleus it is deflected and slowed by the attractive force of the nucleus. The energy lost by the electron is converted into X-radiation. Very few of the X-ray photons have energies close to that of the impinging electrons. Figure 1.117 illustrates the production and energy spectrum of the *Bremsstrahlung* production. The *characteristic* radiation only occurs if the oncoming electron has a kinetic energy greater than the binding energy of an electron in an inner shell of the anode's atoms. When this electron is removed it leaves a vacancy which is filled by an electron from a higher energy level. The energy this electron gives up is released in the form of an X-ray photon with a specific energy. For vacuum interrupters, the interaction that produces most X-rays is the *Bremsstrahlung* process. The measured X-ray quantity is inversely proportional to the square of the distance from the vacuum interrupter. For a detectable X-ray dose three requirements have to be met:

1. An open contact gap
2. A voltage across the open contact gap that results in a high enough microscopic field at the cathode for the field emission of electrons and also that gives these electrons a high enough energy when they impact with the anode.
3. A long enough voltage pulse

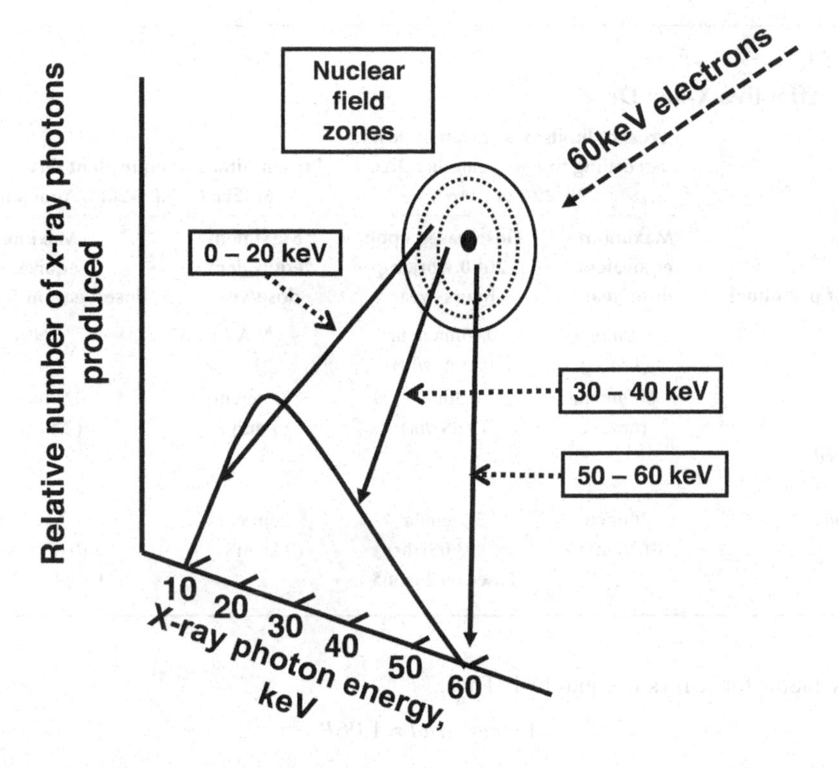

FIGURE 1.117 An illustration of the production of Bremsstrahlung X-rays with an image of the expected X-ray photon energy spectrum.

A vacuum interrupter does not produce X-rays, or if they are produced, the dose is less than the background radiation level under the following conditions:

- Sitting in a box
- Sitting on the shelf
- In a deenergized switch
- In an energized switch when the interrupter contacts are closed
- In an energized switch at the normal operating distribution circuit voltages (up to 145 kV, rms)
- In a breaker switching either load currents or faults currents
- In a breaker undergoing impulse (BIL) voltage withstand testing

The unit of X-ray exposure is the *roentgen (R)*. A dose of *1R* in living tissue corresponds to about 95ergs/gm (~0.01J/kg). The unit of absorbed X-ray dose is the *rad* (radiation absorbed dose), by definition:

$$1rad = 0.01J.kg^{-1} \approx 1.06\,R \tag{1.100}$$

The SI unit for absorbed dose is the *gray* (Gy) and 1Gy = 100rads. The actual dose absorbed by living tissue is called the dose equivalent. Its conventional unit is the *rem* and is defined by:

$$\text{Dose equivalent} = \text{absorbed dose} \times \text{quality factor}$$

TABLE 1.21
Limits on Effective X-Ray Dose

Category of personnel	Present limits of equivalent dose according to European directive 96/29/Euratom		Present limits of equivalent dose according to IEEE C37.85-2000 Appendix A	
	Maximum equivalent dose/year	Dose based upon 2000 working hours /year	Maximum equivalent dose/year	Maximum equivalent dose/year1m from the VI
General Public	100mrem (1 mSv)	0.05mrem/hr (0.5 μSv/hr)	N/A	N/A
Maximum exposure In supervised area	600mrem (6mSv)	0.3mrem (3 μSv/hr)	500mrem (5 mSv)	025 mrem/hr (2.5 μSv/hr)
Background radiation exposure	26mrem (0.26 mSv)	3 μrem/hr (0.03 μSv/hr) based on 24x365	26mrem (0.26 mSv)	3μrem/hr (0.03 μSv/hr) based on 24x365

The quality factor for X-rays is equal to 1. Thus:

$$1\,rem = 1\,rad \approx 1.06R \tag{1.101}$$

The SI unit for dose equivalent is the sievert (Sv), 1Sv = 100rem. Typical allowed exposure limits are given in Table 1.21.

We are all exposed to a general background radiation. It mainly results from four major sources:

1) The inhalation of air – e.g., breathing radon and its daughters: $^{226}Ra \rightarrow {}^{222}Rn \rightarrow {}^{210}Pb$
2) The ingestion of food and water – e.g., ^{40}K and ^{14}C
3) Terrestrial and ground radiation – e.g., soil and building materials
4) Cosmic radiation – e.g., γ rays. The dose level depends upon altitude

This background radiation varies greatly by region. In the United States, for example, the lowest values are found on the Atlantic and Gulf coastal plains, the middle range is found in the northeastern, eastern, central, and western areas and the high range is in the Colorado Plateau area. Table 1.22 shows the average human exposure to background radiation for different regions of the world.

The X-radiation from the open contacts of a vacuum interrupter is mostly concentrated perpendicular to the contact gap [209–212]. The power of the X-radiation P is proportional to the emission current I_e and the square of the applied voltage U [209, 212]:

$$P \sim I_e^2 \times U^2 \tag{1.102}$$

TAVLE 1.22
Average Annual Human Exposure to Environmental Ionizing Radiation in Millisieverts (mSv) Per Year

World	United States	China	Japan	Germany
2.4	3	2.3	1.5	0.5–1.5

The imported energy ΔE_D is defined by [213]

$$E_D = E_{in} - E_{out} \quad (1.103)$$

The absorbed dose D is given by:

$$D = \frac{\Delta E_D}{\Delta m} \quad (1.104)$$

Where m is the mass. The conventional unit of absorbed dose is the "rad":

$$1\,rad = 0.01J\,/\,kg = 100\,ergs\,/\,gm \quad (1.105)$$

The SI unit for absorbed dose is the Gray (Gy):

$$1Gy = 1j\,/\,kg = 100rads \quad (1.106)$$

The dose equivalent H is given by:

$$H = D \times Q \quad (1.107)$$

Where Q = 1 for X-radiation. The conventional unit for absorbed dose is the rem:

$$1\,rem = 1\,rad\;and\;in\;SI\;units\;1\,Sv = 100\,rem \quad (1.108)$$

The dose rate *dH/dt* is therefore:

$$\frac{dH}{dt} = \frac{dD}{dt} \quad (1.109)$$

Now *dD/dt* can be given by: [210, 213]

$$\frac{dD}{dt} = \frac{dH}{dt} = K^* I_e \times U^2 \quad (1.110)$$

Where K^* is a constant and I_e is given by the Fowler–Nordheim equation (see Equation 1.50) Geire et al. [212] show that dH/*dt* can be calculated using:

$$\frac{dH}{dt} = \frac{a_1}{R^2} \times \beta^2 \times \frac{1}{d^2} \times U^4 \times e^{-\left[\frac{a_2 d}{\beta U}\right]} \quad (1.111)$$

Where a_1 and a_2 are constants dependent upon the composition of the cathode contact, β is the cathode's enhancement factor, d is the contact gap and R is the distance from the vacuum interrupter to the X-ray detector. Equation (1.111) indicates, that for a given voltage impressed across the contact gap, the absorbed rate should be a function of the contact gap and also it should depend upon the cathode's enhancement factor. Yan et al. [210, 211] observe the X-ray 3-minute dose 1 meter from a 126kV vacuum interrupter do indeed show a dependence upon the contact gap: See Figure 1.118. Geire et al. [212] measure the X-ray dose rate 1 meter from a 145kV vacuum interrupter with a 60mm contact gap, then calculate using Equation (1.111) the expected rate for a shorter contact and a longer contact gap: See Figure 1.119. The effect of the enhancement factor is reduced during the manufacturing process by conditioning the contacts. Yan et al. demonstrate the effect of this reduced enhancement factor on the X-ray dose rate: See Figure 1.120.

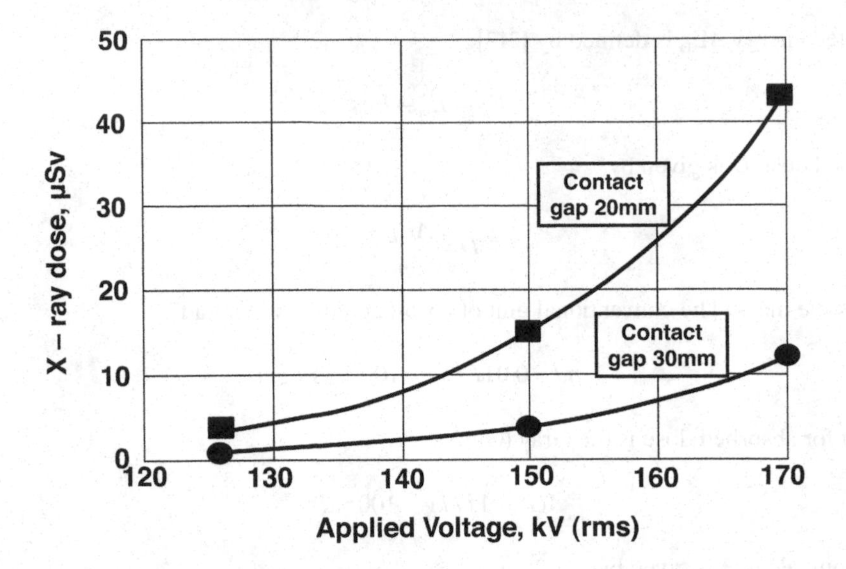

FIGURE 1.118 Average 3-minute X-ray dose as a function of applied power-frequency voltage [210, 211].

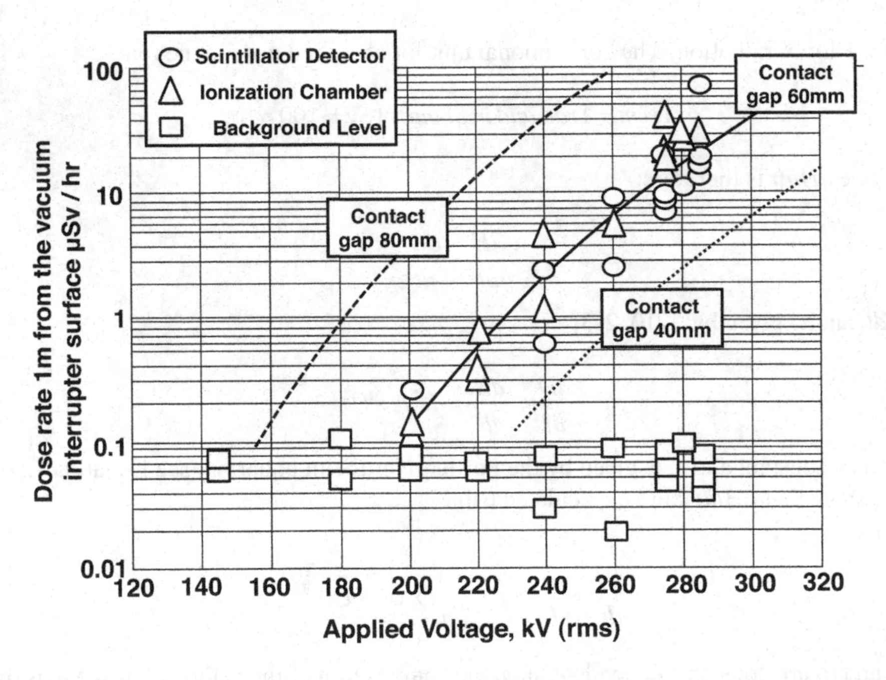

FIGURE 1.119 X-ray dose rate 1m from a 145kV vacuum interrupter as a function of power-frequency voltage for 60mm contact gap compared to calculated values from Equation (1.111) for a smaller and a larger contact gap [212].

What is the effect of these X-ray dose levels on manufacturing personnel and users of vacuum interrupters? Table 1.23 shows the expected voltages experienced by increased vacuum interrupter ratings. Manufacturing personnel will be exposed to the highest voltages when conditioning a new vacuum interrupter with the high-voltage power-frequency voltage. They will also be exposed to a relatively high voltage when performing the one-minute high-voltage withstand tests on new vacuum interrupters. The yearly X-ray dose for these workers will depend upon the frequency of these

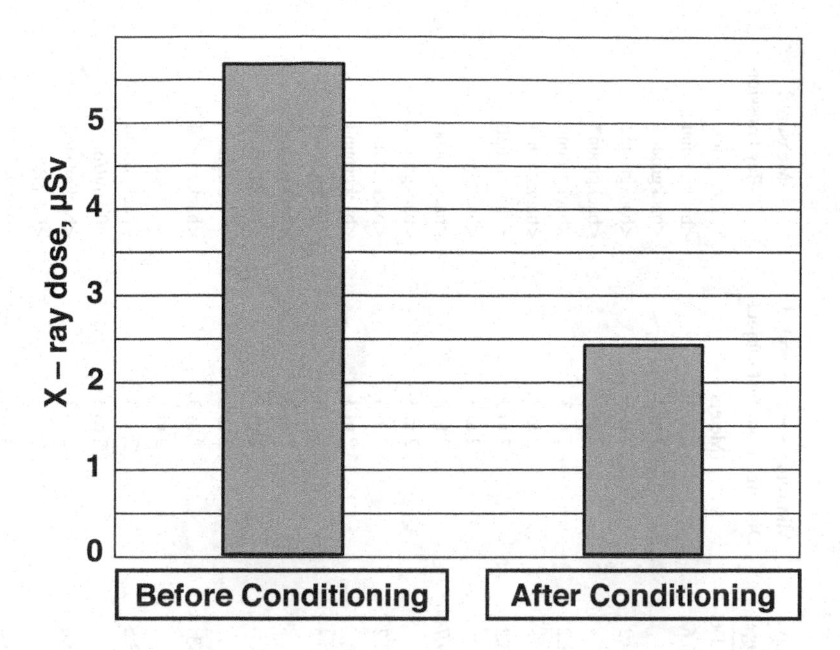

FIGURE 1.120 Effect of the enhancement factor "β" on the average 3-minute X-ray dose [210].

tests being performed. For a high-volume vacuum interrupter production line, perhaps 300–600 tests per day. Thus, the accumulated dose could be substantial. The high voltage BIL test is of such short duration that the X-ray dose from it will be very low. And the personnel will be a substantial distance from the high voltage source and from the vacuum interrupter being tested. The vacuum interrupter user will experience the highest voltage when performing the one-minute high-voltage withstand test. Renz et al. [209] show in Figure 1.121, that for vacuum interrupter ratings up to 36kV with their rated contact gaps, the expected X-ray dose rate will be equivalent or less than the background radiation rate. Of course, the X-ray does rate has to be added to the background rate. Geire et al. [212] also show similar data for a 145kV vacuum interrupter: see Figure 1.122. Interestingly in their experiments. the X-ray dose rate for new and used vacuum interrupters is similar. The new vacuum interrupters would have been conditioned to lower the microscopic enhancement factor. The vacuum interrupters in these experiments would have had axial magnetic contact structures. Thus, the diffuse vacuum arc formed when opening the contacts to switch current would continuously condition the contacts. So, the result shown in Figure 1.122 might have been expected as the applied voltage increases the X-ray dose rate seems to vary considerably.

As reported in Appendix A of the IEEE Standard, C37.85-2001 [212]; ac or dc high voltage withstand testing is the only condition under which a vacuum interrupter with its contacts open can have a sustained voltage applied that is high enough in magnitude to possibly produce measurable X-rays. As Renz et al. [209] and Geire et al. [212] show this seems to be true for vacuum interrupters with ratings up to 145kV with their rated contact gaps. Therefore, the people that need to be concerned about X-ray exposure are:

- Personnel performing high voltage ac withstand tests at the vacuum interrupter factory
- Personnel conditioning manufactured vacuum interrupters using high-voltage power frequency voltage in the factory
- Personnel performing similar high voltage tests at the vacuum circuit breaker and at the switchgear factories
- Maintenance personnel performing similar high voltage tests in the field

TABLE 1.23
X-Ray Emission at Recommended Distance from a Vacuum Interrupter According to IEEE and IEC Standards

Applicable Standard	Rated System Voltage	Rated Power Frequency Withstand Voltage	Factory Test Voltage at 100% of Rated Withstand Voltage	Field Test Voltage at 75% of Rated Withstand Voltage		Minimum Recommended Distance from Test Object	Measured X-Ray Emission
	kV, rms.	kV, rms.	kV, rms.	kV, rms.	kV, DC	Meters	Millirem/hour
IEC 694	3.6	10		7.5	10.6	2 to 3	<background
		10	10			2 to 3	<background
IEEE C37.06	4.76	19		14.3	20.2	2 to 3	<background
		19	19			2 to 3	<background
IEC 694	7.2	20		15	21.2	2 to 3	<background
IEC 694		20	20			2 to 3	<background
IEEE C37.06	8.25	36		27	38.2	2 to 3	<background
		36	36			2 to 3	<background
IEC 694	12	28		21	29.7	2 to 3	<background
		28	28			2 to 3	<background
IEEE C37.06	15	36		27	38.2	2 to 3	<background
		36	36			2 to 3	<background
IEC 694	17.5	38		28.5	40.3	2 to 3	<background
		38	38			2 to 3	<background
IEEE C37.06	15.5	50		37.5	53	2 to 3	<background
		50	50			2 to 3	<0.5mR/hr
IEC 694	24	50		37.5	53	2 to 3	<background
		50	50			2 to 3	<0.5mR/hr
IEEE C37.06	25.8	60		45	63.6	2 to 3	<background
		60	60			2 to 3	<0.5mR/hr
IEEE C37.06	27	60		45	63.6	2 to 3	<background
		60	60			2 to 3	<0.5mR/hr
IEC 694	36	70		52.5	74.2	3 to 4	<0.5mR/hr

(Continued)

TABLE 1.23 (CONTINUED)

X-Ray Emission at Recommended Distance from a Vacuum Interrupter According to IEEE and IEC Standards

Applicable Standard	Rated System Voltage	Rated Power Frequency Withstand Voltage	Factory Test Voltage at 100% of Rated Withstand Voltage	Field Test Voltage at 75% of Rated Withstand Voltage		Minimum Recommended Distance from Test Object	Measured X-Ray Emission
		70	70			3 to 4	<3.5mR/hr
IEEE C37.06	38	80		60	84.8	3 to 4	<0.5mR/hr
		80	80			3 to 4	<3.5mR/hr

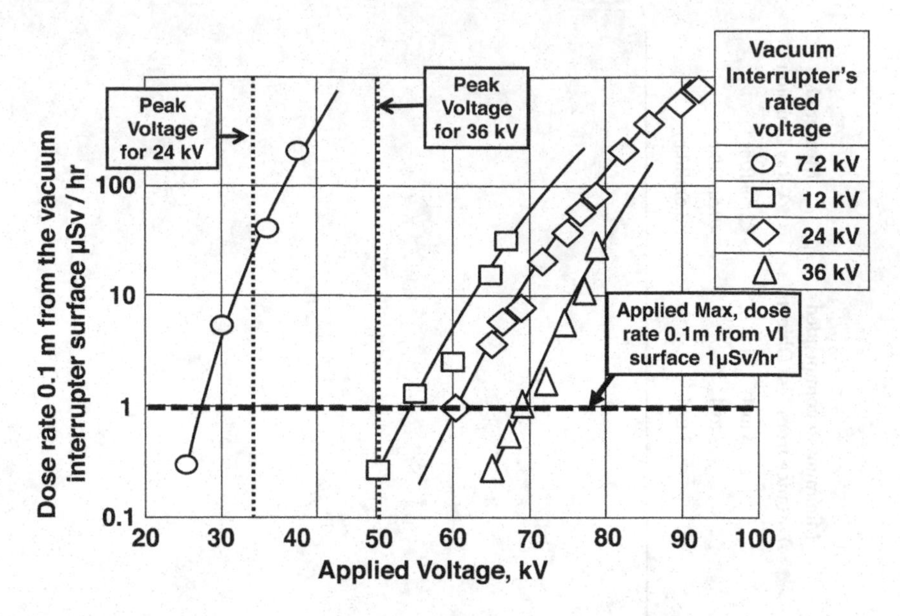

FIGURE 1.121 X-ray dose rate measured 0.1m from four vacuum interrupters of different ratings as a function of applied voltage [209].

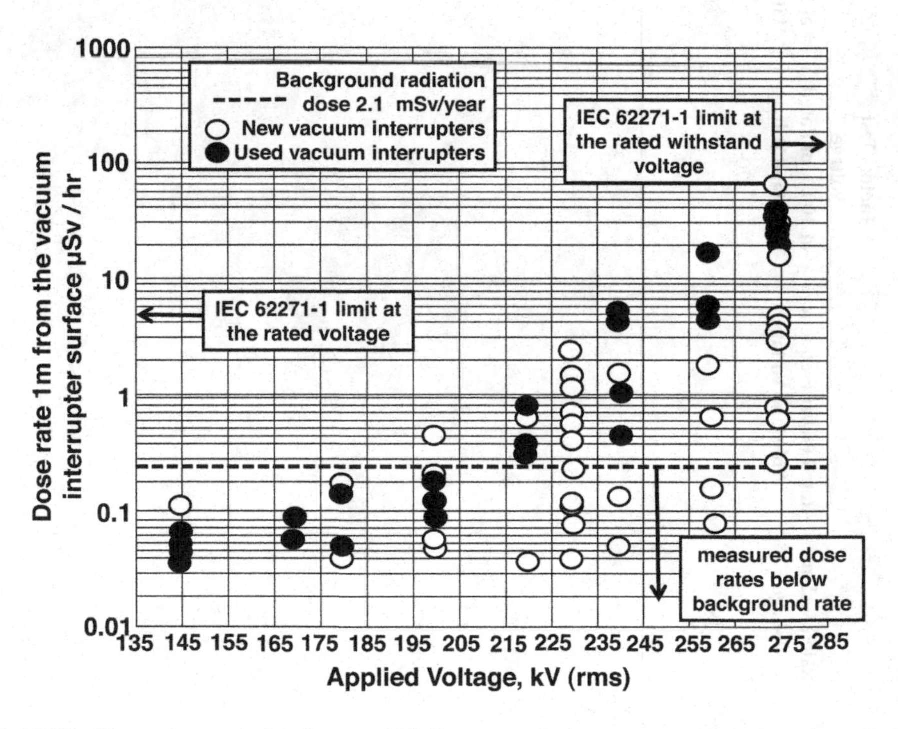

FIGURE 1.122 X-ray dose rate 1m from a 145kV vacuum interrupter as a function of applied power-frequency voltage for new vacuum interrupters and those that have operated mechanically and have been used to switch current for a 60mm contact gap [212].

The IEEE standard states that a vacuum interrupter will be in compliance with the standard if the X-radiation that is emitted when measured 1 meter from the tested vacuum interrupter does not exceed:

1. 0.5 mR/hour at the maximum operating system voltage shown in Column 2 of Table 1.23 (Table 1 of IEEE C37.85-2001 [212]). Note that the IEEE standard does not use the *rem* or *sievert*, but instead uses the radiation exposure measurement, the *roentgen*; this can easily be measured with an X-ray exposure meter
2. 15.0 mR/hour, 1m from the vacuum interrupter, at the ac frequency, one-minute, voltage withstand test voltage shown in column 3 of Table 1.23. This corresponds to 0.25 mR/test

As reported in Appendix A of C37.85-2001 [214]:

> *The manufacturers of vacuum interrupters have concluded that neither the general public nor the users will be subjected to harmful X-radiation due to normal application and operation of 15.5 kilovolt-rated vacuum interrupter devices when applied within their assigned ratings and when the voltage applied across the open contacts of these interrupters is 15.5 kilovolts or less. The manufacturers also concluded that at the permissible user dielectric withstand test voltage of 37.5kV radiation levels are negligible for vacuum interrupters rated 15.5kV. Normal electrical safety precautions require the user to be at a distance from the interrupters that provides sufficient protection. ... A minimum distance of 2 to 3 meters can normally be expected to be used for reasons of electrical safety.*

With an X-ray level of 15mR/hr or 0.25mR/test a user would have to perform over 2000 one-minute tests per year standing 1 meter from a vacuum interrupter to exceed the 500 mrem/year cumulative dose limit. This number of tests is at least 100 times greater than a normal Utility worker would be expected to perform in a year. Data from tests conducted at Eaton Corporation's vacuum interrupter factory are summarized in column 8 of Table 1.23. For 15kV to 27kV vacuum interrupters the X-ray exposure 1m to 3m away from the interrupter for system voltages up to 40kV ac (rms.) is indistinguishable from the background level of about 3 μrem/hr. It is just above background but less than 0.5 mR/hr (5 μSv/hr) at a distance 2m to 3m at 60kV ac (rms.). Here the maximum exposure rate is 0.5 mR/hr or 8 μR/test. On tests with 38kV circuit breakers, the radiation levels do not exceed 0.5 mR/hr at a distance of 3 meters from the device when the applied voltage is 60kV ac rms. (the end user's test voltage) with all three phases of the device energized at the same time.

Under normal testing conditions performed by the circuit breaker manufacturer, the test voltage does not exceed 80kV ac rms. Here the maximum instantaneous radiation level does not exceed 3.5 mrem/hr or 60 μR/test a distance of 3 meters from the device (the IEEE limit is 15 mrem/hr). Even at this level an operator would have to perform over 8000 such tests in a year. Thus, the potential X-ray exposure in the field is rather low. This becomes especially true when it is remembered that such field tests are normally performed at 75% of the voltage rating. The potential for X-ray exposure, however, can be significant during tests in a factory since these tests are not only performed at 100% of the rated, power frequency, withstand voltages but are also performed more frequently during vacuum interrupter and during vacuum circuit breaker manufacture. Thus, there is a greater potential for an operator to accumulate an X-ray exposure over time.

It is extremely prudent for the manufacturers to take appropriate precautions to protect personnel and meet local codes by using monitoring devices and safety measures such as shielding. The effects of extra shielding have also been evaluated. A 3mm thick steel shield and a separation of 3m results in the X-ray exposure level at 60kV ac rms that is again indistinguishable from the background level. At 80kV ac rms the exposure level is only about 0.8mR/hr. which is a negligible dose for testing personnel and only about 1/20th of the 15mR/hr. limit set in IEEE C37.8-2001. A 1.5mm thick portable lead shield and a separation of 3 meters results in X-ray levels at 60kV ac rms that are again indistinguishable from the background level. At 80 kV ac rms, it is now only about 0.4 mR/hr, which again is a negligible dose for testing personnel and only about 1/40th of the 15mR/hr

TABLE 1.24
Practical Shielding for Protecting Personnel from X-Ray Exposure During High Voltage Testing During Vacuum Interrupter Manufacture

Barrier Material	Barrier Thickness	X-ray level at 3 meters reduced to:
Steel	3mm	Less than ¼ of the unshielded case
Lead	1.5mm	Less than ⅛ of the unshielded case

limit set in IEEE C37.85 (2001). Thus, for a vacuum interrupter manufacturer or a vacuum circuit breaker manufacturer, it is a good practice to ensure that X-ray levels are within safe limits for test personnel by providing:

1. At least the minimum recommended separation between the test personnel and the test object
2. Shielding between the test personnel and the test object for extra protection. The shielding enclosure should have a lid, because X-rays scattered by air molecules can travel over walls
3. X-ray cumulative-dose monitoring of all test personnel
4. Training of all test personnel in the hazards of X-rays and safe test practices

The first priority is to establish a high voltage test area where a minimum clearance between the vacuum interrupter under test and the test personnel is 3 to 4 meters. In this manner, switchgear of all voltage ratings can be tested in the same location and no special consideration between one voltage and another is needed since the largest separation distance is employed. A second, but equally important priority is to provide an enclosure for the tested device or a barrier between the test personnel and the switchgear device under test made from sheet metal walls and possibly a lead lining or a lead screen that will provide some additional protection from X-ray exposure. An example of the reduction in the exposure level expected from such barriers is given in Table 1.24.

In addition, all high voltage test personnel or others that are regularly in close proximity to the test should wear film badges that are checked monthly for accumulated radiation exposure. This practice records the actual accumulated exposure experienced by the people involved and provides a warning when the exposure exceeds allowable limits in a 1-month time period. This practice has been followed at the Eaton Corporation's vacuum interrupter factory where separation and shielding are in place. In their experience over 40 years, no accumulated exposure has ever been detected beyond what would have been expected from the background levels experienced in the Southern Tier of New York State.

Personnel who perform high voltage testing in the field should receive training outlining the potential hazards of X-rays from vacuum interrupters and should follow the proper precautions to take to keep the risk of exposure and the cumulative dose to minimum levels. As the voltage performance of vacuum interrupters increases from 72kV to 170kV and higher, the X-ray precautions such as shielding will be required even if normal system voltage is impressed across open vacuum interrupter contacts. It will certainly be required for the one-minute voltage withstand levels greater than 135kV rms (IEC) and 140kV rms (IEEE). Fortunately, good high voltage practice requires that it is prudent to stand more than 5m from such equipment when it is being tested.

As I have discussed above, X-ray radiation is seen as a health and safety issue. As higher voltage vacuum interrupter designs are introduced, higher X-radiation production may become another component in the initiation of vacuum breakdown. Photons produced by the high energy emission current impacting the anode would cross the contact gap and release secondary electrons from the

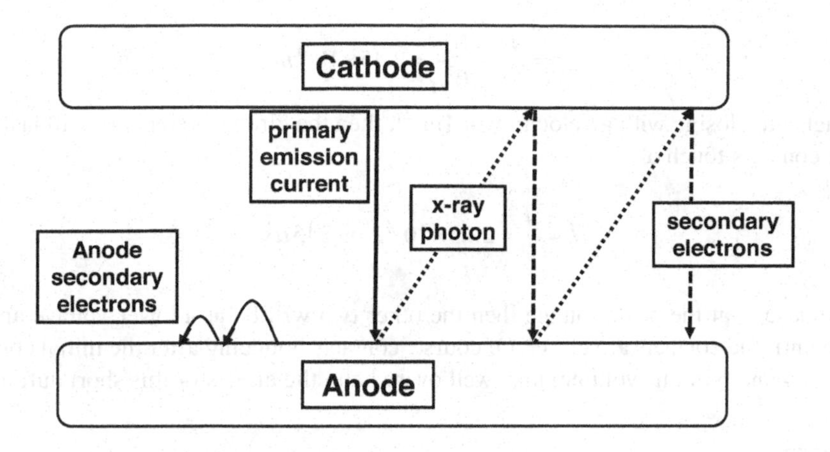

FIGURE 1.123 The X-ray enhanced current in a vacuum gap [215].

cathode. This would increase the current flowing in the contact gap during the initial phase of the breakdown process. Back scattered electrons may also enhance the X-ray yield, see Figure 1.123. Henken [215] has analyzed the effect of the production of secondary electrons for voltages up to 45kV. He shows that they have very little effect at these voltages. However, as vacuum interrupters are developed to higher voltages these secondary electrons will certainly be a factor in the pre-breakdown parameters of vacuum breakdown.

1.6 ARC INITIATION WHEN CLOSING A VACUUM INTERRUPTER

As a vacuum interrupter's contacts close in a power circuit, there will be a period just before they touch when the electric field across them will be high enough for breakdown to occur. This will give rise to a "prestrike" arc, which will form between the contacts and will carry the power circuit's current. If the current is high enough, e.g., in a fault current situation (see Chapter 4 in this volume), it is possible that the contact surfaces will melt. Where they touch a small weld can form. In a single-phase circuit and a three-phase grounded circuit, once a breakdown occurs in a single vacuum interrupter the circuit current will flow. In a three-phase ungrounded circuit two phases must break down at the same time for the circuit current to flow. The designer of the vacuum interrupter equipment must take this phenomenon into account when developing the mechanism. When this is done correctly, welding of vacuum interrupter contacts is never a problem.

As discussed in Section 1.3.6, the time to the breakdown of a vacuum gap is:

$$t_B = t_p + t_c \tag{1.112}$$

where t_p has great variability and t_c is very fast compared with the closing speed of a circuit breaker's contacts. For example, let us consider a 12kV, three-phase grounded circuit. The peak voltage that will be impressed across a vacuum interrupter's closing contacts will be:

$$U_p = 12 \times 10^3 \sqrt{\frac{2}{3}} \tag{1.113}$$

Assume a vacuum breakdown will occur when the contacts are close enough (d m) to give a macroscopic field of $E_g = 4 \times 10^7$ Vm^{-1}, then at U_p and assuming that t_p is a few microseconds (i.e., small compared with the closing speed of the contacts):

$$E_g = 4 \times 10^7 = \frac{U_p}{d} \tag{1.114}$$

$$d = \frac{9.8 \times 10^3}{4 \times 10^7} = 2.45 \times 10^{-4} m \quad (1.115)$$

If the contacts are closing with a velocity $\underline{v}$ of $1 ms^{-1}$, then the circuit current arc will last for a time t before the contacts touch of:

$$t = \frac{d}{\underline{v}} = 2.25 \times 10^{-4} s \text{ or } 245 \mu s \quad (1.116)$$

If this one phase is at the peak voltage then the other two will be at a lower voltage and will not break down until the contacts are closer. Of course, contacts bouncing after the initial contact touch (see Chapters 4 and 6 in this volume) may well overwhelm the effects of this short duration arc.

REFERENCES

1. Garzon, R., *High Voltage Circuit Breakers; Design and Application*, (pub. Marcel Dekker, New York), Chapter 7, pp. 225–255, 1997.
2. *IEEE C37.06-2000*, "Voltage ratings for electric power systems and equipment".
3. *IEC Publication, IEC 62771–1*, "High-voltage switchgear and control gear". These standards are undergoing continuous change at the present time in order to (a) update them and (b) to harmonize them with other standards such as those in [2].
4. Hazel, T., Norris, A., Safin, R., "Don't get lost in Gost standards", *IEEE Industry Applications Magazine*, pp. 53–60, November/December 2004.
5. Hammons, T., McQuin, N., Ibuki, K., Wei, L., Ananthakrishnan, S., "How national labs use standards to test electrical apparatus", *IEEE Power Engineering Review*, pp. 6–14, July 1997.
6. Wagner, C. L., "Circuit breaker application", In: Browne, T. E. (Editor), *Circuit Interruption: Theory and Techniques*, (pub. Marcel Dekker, New York), Chapter 3, Sections 3.3 and 3.4 and Section 3.4.1, pp. 47–48 1984.
7. Meek, J. M., Craggs, J. D. *Electrical Breakdown of Gases*, (pub. Wiley, NY, 1978).
8. Llewellyn–Jones, F., *Ionization and Breakdown in Gases*, (pub. Methuen, London, Wiley, New York), 1957.
9. Von Engel, A. *Encyclopedia of Physics*, XXI, "Electron emission, gas discharges I", p. 508, 1956.
10. Cobine, J. D., *Gaseous Conductors*, (pub. Dover, New York), 1958.
11. Paschen, F., "Über die zum Funkenübergang in Luft, Wasserstoff und Kohlensäuer bei verschiedenen Drucken erforderliche Potentialdifferenz", *Wied Annalen*, 37, ser. 3, pp. 69–96, March 1889.
12. For an excellent compilation of Paschen Curve data see: Dakin, T., Luxa, G., Oppermann, G., Vigreux, J., Wind, G., Winkelnkemper, H., "Breakdown of gases in uniform fields: Paschen Curves for nitrogen, air and sulfur hexafluoride", *Electra*, 32, pp. 61–82, 1974.
13. Schönhuber, M., "Breakdown of gas below the Paschen minimum: Basic design data of high voltage equipment", *IEEE Transactions Power Apparatus and Systems*, PAS-88(2), pp. 100–107, February 1969.
14. Heylan, A., "Sparking formulae for very high-voltage Paschen characteristics of gases", *IEEE Electrical Insulation Magazine*, 22(3), pp. 25–35, May/June 2006.
15. Slade, P. G., Taylor, E. D., "Electrical breakdown in atmospheric air between closely spaced (0.2μm–40μm) electrical contacts", *IEEE Transactions Components and Packaging Technology*, 25(3), pp. 390–396, September 2002.
16. Maier, W., Kadish, A., Buchenauer, C., Robiscoe, R., "Electrical discharge initiation and a microscopic model for formative time lags", *IEEE Transactions Plasma Science*, 26(6), pp. 676–683, December 1993.
17. Bektas, S., Farish, O, Hizal, M., "Computation of the electric field at a solid/gas interface in the presence of surface and volume charges", *IEE Proceedings*, 133, Pt. A(9), pp. 577–586, December 1986.
18. Schächter, L., "Analytical expression for triple point electron emission from an ideal edge", *Applied Physics Letters*, 72(4), pp. 421–423, January 1998.
19. Geisselman, M., Pfieffer, W., 'Influence of solid dielectrics upon breakdown voltage and predischarge development in compressed gases", In: *Gaseous Dielectrics IV*, (pub. Pergamon Press), pp. 431–436, 1984.

20. Sudarshan, T. S., Dougal, R. A., "Mechanisms of surface flashover along solid dielectrics in compressed gases: A review", *IEEE* Transactions Dielectrics *and* Electrical Insulation, EI–21(5), pp. 727–746, October1986.
21. Miller, H. C., "Surface flashover of insulators", *IEEE Transactions Dielectrics and Electrical Insulation*, 24(5), pp. 765–786, October 1989.
22. Li, W., Campbell, L., Balasubramaman, G., Loud, L., "Measuring external dielectric strength of the vacuum interrupter envelope", *Proceedings 28th International Symposium on Discharges and Electrical Insulation in Vacuum*, pp. 531–534, September 2018.
23. Gosh, P. S., Chatterjee, N., "Polluted insulator flashover model for ac voltage", *IEEE Transactions Dielectrics and Electrical Insulation*, 2(1), pp. 128–136, February 1995.
24. Sundararajan, R., Gorur, R. S., "Role of non-soluble contaminants on the flashover of porcelain insulators", *IEEE Transactions Dielectrics and Electrical Insulation*, 3(1), pp. 113–118, February 1996.
25. Dhahbi-Megriche, N., Beroual, A.,"Flashover dynamic model of polluted insulators under ac voltage", *IEEE Transactions Dielectrics and Electrical Insulation*, 7(2), pp. 283–289, April 2000.
26. Goa, J. C., Zhang, J, G, "Measurement of electrical charges carried by dust", *Proceedings 48th IEEE Holm Conference on Electrical Contacts*, pp. 191–196, October 2002.
27. McShane, C. P., "Relative properties of the new combustion-resistant vegetable-oil-based dielectric coolants for distribution and power transformers", *IEEE Transactions on Industry Applications*, 37(4), pp. 1132–1139, July/August 2001.
28. Oommen, T. V., Claiborne, C. C., "Electrical transformers containing insulation fluids comprising high oleic acid oil compositions", U. S. Patent 5,949,017, September 1999 and "An agriculturally based biodegradable dielectric fluid", *Proceedings IEEE T&D Conference*, New Orleans, LA, April 1999.
29. Masaki, N., Matsuzawa, K., Yoshida, T., Ohshima, I., "72/84 kV cubicle style SF_6 gas–insulated switchgear with vacuum circuit breaker", *Proceedings IEE 2nd International Conference on Developments in Distribution Switchgear*, (IEE publication # 261), 1981.
30. Christophorou, L. G., Olthoff, J. K., Brunt, R. J., "Sulfur hexafluoride and the electric power industry", *IEEE Electrical Insulation Magazine*, 13(5), pp. 20–24, September/October 1997.
31. See, for example, *The Proceedings International Conference on SF_6 and the Environment: Emission Reduction Strategies*, November 2002.
32. Leusenkamp, M., Hilderink, J., Lenstra, K., "Field calculations on epoxy resin insulated vacuum interrupters", *Proceedings 17th International Symposium on Discharges and Electrical Insulation in Vacuum*, pp. 1065–1069, July 1996.
33. New.abb.com/medium-voltage.
34. Bestel, F., "Encapsulated vacuum interrupter and method of making same", US Patent 5,917,167, June 1999.
35. Lamara, T., Tricarico, C., Fischer, B., "Improvement of external dieletric performance of vacuum interrupter", *Proceedings 27th International Symposium on Discharges and Electrical Insulation in Vacuum*, pp. 671–674, September 2016.
36. Ramesh, M., Summer, R., Singh, Serdyuk, Y., Gubanski, S., Kumara S., "Application of streamer criteria for calculations of flashover voltages of gaseous insulation with solid dielectric barrier", *Proceedings 18th International Symposium on High Voltage Engineering*, pp. 1258–1263, August 2013,
37. Selzer, A., Vacuum interruption–a review of the vacuum arc and contact functions", *IEEE* Transactions Industry Applications, IA–8, pp. 707–722, November/December 1972.
38. Greenwood, A., *Vacuum Switchgear*, (pub. IEE, London, UK), 1994.
39. Boxman, R. L., Goldsmith, S., Greenwood, A., "Twenty-five years of progress in vacuum arc research and utilization", *IEEE Trans Plasma Science*, 25(6), pp. 1174–1181, December 1997.
40. These data are taken from many sources. The points are average values.
41. Hackham, R., Altcheh, L., "AC and DC electric breakdown of vacuum gaps with variation of air pressure in the range 10^{-9} to 10^{-2} torr using OFHC copper, nickel and niobium parallel plane electrodes", *Journal of Applied Physics*, 46, pp. 627–636, 1975.
42. Shioiri, T., Kamikawaji, T., Kaneko, E., Homma, M., Takahashi, H., Ohshima, I., "Influence of electrode area on the conditioning effect in vacuum", *IEEE Transactions Dielectrics and Electrical Insulation*, 2(2), pp. 317–320, April 1995.
43. Shioiri, T., Kamikawaji, T., Yokokura, K., Kaneko, E., Ohshima, I., Yanabu, S., "Dielectric breakdown probabilities for uniform field gap in vacuum", *Proceedings 19th International Symposium on Discharges and Electrical Insulation in Vacuum*, pp. 17–20, September 2000.
44. Weibull, W. A., "A statistical theory of the strength of materials", *Ingeniors Ventenskaps Academien Handlinger*, 151, pp. 5–45, 1939.

45. Duffy, S., Powers, L., Starlinger, A., "Reliability analysis of structural ceramic components using a 3-parameter Weibull distribution", *The Journal of Engineering for Gas Turbines and Power*, 115(1), pp. 109–116, 1993.
46. Bruning, A., Kasture D., Ascher, H., "Absolute vs. comparative end-of-life age", *IEEE Transactions Dielectrics and Electrical Insulation*, 3(3), pp. 567–576, August 1996.
47. A discussion of [46] published in *IEEE Transactions Dielectrics and Electrical Insulation*, 4(2), pp. 241–247, April 1994.
48. Hirose, H., "More accurate breakdown voltage estimation for the new step-up test method", *IEEE Transactions Dielectrics and Electrical Insulation*, 10(3), pp. 475–482, June 2003.
49. Osmokrovic, P., "Influence of switching on dielectric properties of vacuum interrupters", *IEEE Transactions Dielectrics and Electrical Insulation*, 1(2), pp. 340–347, April 1994.
50. Taylor, E., "Private communication", 2005.
51. Liu, S-C, "Statistical properties of steady state impulse breakdown for commercial vacuum interrupters", *IEEE* Transactions *on Electrical Insulation*, EI-18(3), pp. 325–331, June 1983.
52. Sidorov, V., Alferov, D., Alferova, E., "Dielectric strength of series connected vacuum gaps", *IEEE Transactions Dielectrics and Electrical Insulation*, 13(1), pp. 18–25, February 2006.
53. Wood, R. W., "A new form of cathode discharge and the production of X-rays together with some notes on diffraction", *Physics Review*, 5(1), pp. 1–10, July 1897.
54. Farrall, G. A., "Electrical breakdown in vacuum", *IEEE* Transactions *on Electrical Insulation*, EI–20(5), pp. 815–841, October 1985.
55. Chatterton, P. A., "Vacuum breakdown", Chapter 2 in Reference [7].
56. Latham, R. V., *High Voltage Vacuum Insulation: Basic Concepts and Technological Practice*, (pub. Academic Press), 1995.
57. Litvinov, E., Mesyats, G., Proskurovskii, D., "Field emission and explosive electron emission process in vacuum discharges", *Soviet Physics Uspekhi*, 26(2), pp. 138–157, February 1983.
58. Rogowski, W., Rengier, H., "Plane spark gaps with correct edge development", *Archiv Fur Elektrot*, 16, pp. 73–75, April 1926.
59. Alpert, D., Lee, D., Lyman, E., Tomaschke, H., "Initiation of electrical breakdown in ultra high vacuum", *Journal of Vacuum Science and Technology*, 1(2), pp. 35–50, November/December 1964.
60. Farrall, G. A., Chapter 2, "Electrical breakdown in vacuum", In: Lafferty, J. M. (Editor), *Vacuum Arcs: Theory and Application*, (pub. John Wiley & Sons, New York), 1980, pp. 20–80.
61. Vries, L. M., Damstra, G. C., "Prebreakdown emission current measurements in a 24kV vacuum interrupter with butt contacts", *IEEE Transactions on Electrical Insulation*, 23(1), pp. 97–100, January 1988.
62. Timsit, R. S., "Electrical contact resistance: Fundamental principles", In: Slade, P. G. (Editior), *Electrical Contacts: Principles and Applications 2nd Edition*, (pub. CRC Press), 2014, pp. 1–2 and 69–79.
63. Keil, A., Merl, W. A., Vinaricky, E., *Elektrische Kontakte und ihre Werkstoffe*, (pub. Springer–Verlag, Berlin) p. 3, 1984.
64. Utsumi, T., "Cathode and anode induces electrical breakdown in vacuum", *Journal of Applied Physics*, 38(7), pp. 2969–2997, June 1967.
65. Rohrbach, F., "CERN Report, 71–5/TC-L, 1971. See also, Xu, N", In: Latham, R., (Editor), *High Voltage Vacuum Insulation: Basic Concepts and Technological Practice*, (pub. Academic Press), Chapter 4, p. 126, 1995.
66. Lathan, R. Op. Cit. chapter 4, p. 126.
67. Little, R. P., "Electrical breakdown in vacuum", *IEEE Transactions on Electron Devices*, ED-12, pp. 77–83, February 1965.
68. Zhang, Y., Xu, X., Jin, L., An, Z., Zhang, Y., "Fractal-based electric field enhancement modeling of vacuum gap electrodes", *IEEE Transactions on Dielectrics and Electrical Insulation*, 24(3), pp. 1957–1963, June 2017.
69. Sato, K., Yamano, Y., Asari, N., Shioiri, T., Icikawa, T., "Characteristics of vacuum breakdown field and comparison of field enhancement factor β in CuCr electrodes with slit configuration", *Proceedings of the 27th International Symposium on Discharges and Electrical Insulation in Vacuum*, pp. 60–63, September 2016.
70. Li, S., Geng, Y., Lui, Z., Wang, J., "Influence of arc-melted cathode layer depth on vacuum insulation", *IEEE Transactions on Dielectrics and Electrical Insulation*, 24(6), pp. 3327–3332, December 2017.
71. Cox, B., "The nature of field emission sites", *Journal of Physics D: Applied Physics*, 8, pp. 2065–2073, 1975.
72. Niedermann, P., Fischer, O., "Application of a scanning tunneling microscope to field emission studies", *IEEE* Transactions *on Electrical Insulation*, EI-24(6), pp. 905–910, December 1989.

73. Batrakov, A., Onischenko, S., Proskurovsky, D., Johnson, D., "On priorities of cathode and anode contaminations in triggering the short-pulsed breakdown in vacuum", *IEEE Transactions Dielectrics and Electrical Insulation,* 13(1), pp. 41–51, February 2006.
74. Malluci R., "The impact of micro-protrusions on field emission electrons", *Proc. 65th IEEE Holm Conference on Electrical Contacts*, pp. 64–67, 2019
75. Sato, S., Koyama, K., "Relationship between electrode surface roughness and impulse breakdown voltage in vacuum gap of Cu and Cu-Cr electrodes", *IEEE Transactions on Dielectrics and Electrical Insulation*, 10(4), pp. 576–582, August 2003.
76. ISO 4288: Geometrical product specification (GPS) –Surface texture: Rules and procedures for assessment of surface texture, 1996.
77. Mesyatt, G. A., *Cathode Phenomena in a Vacuum Discharge: The Breakdown, the Spark and the Arc*, (pub. Nauka, Moscow), 2000.
78. Kranjec, P., Ruby, L., "Test of the critical theory of electrical breakdown in vacuum", *Journal of Vacuum Science Technology*, 4(2), pp. 94–96, 1967.
79. Microparticle Phenomena, *High Voltage Vacuum Insulation: Basic Concepts and Technological Practice*, Lathan R., (Editor), (pub. Academic Press), Chapter 7, 1995.
80. Fowler, R., Nordheim, L., "Electron emission in intense electric fields", Proceedings Royal Society, A119, pp. 173–181, May 1928.
81. Davies, D., Biondi, M., "Vacuum electrical breakdown between plane–parallel copper plates", *Journal of Applied Physics*, 37(8), pp. 2969–2977, July 1966.
82. Miller, H., "Values of the Fowler-Nordheim emission functions v(y), t(y) and s(y)", *The Journal of the Franklin Institute*, 252(6), pp. 382–388, December 1966.
83. Diamond, W., "New perspectives in vacuum high voltage insulation. 1. The transition to field emission", *Journal of Vacuum Science and Technology A*, 16(2), pp. 707–719, Mar/April 1998.
84. Takahashi, E., Sone, M., "Observation of field emission sites and study of heat treatment effects", *IEEE Transactions Dielectrics and Electrical Insulation*, 5(6), pp. 929–934, December 1998.
85. Isono, H., Kojima, A., Sone, M., Mitsui, H., "Effect of minute projections on field emission electron", *Proceedings 14th International Symposium on Discharges and Electrical Insulation in Vacuum*, pp. 122–126, 1990.
86. Xu, S., Kumada, A., Hidaka, K., Ikeda, H., Kaneko, E., "Observation of pre-discharge phenomena with point-to-plane electrodes in vacuum under ac", *IEEE Transactions on Dielectrics and Electrical Insulation*, 22(6), pp. 3633–3640, December 2015.
87. Yu, Y., Wang, J., Yang, H., Geng, Y., Liu, Z., "Asymmetrical ac field emission current characteristics of a vacuum interrupter subjected to inrush current", *IEEE Transactions on Dielectrics and Electrical Insulation*, 23(1), pp. 49–56, February 2016.
88. Pursch, H., Siemroth, P., Juttner, B., "On the nature of prebreakdown emission currents in vacuum interrupters", *IEEE Transactions on Electrical Insulation*, 24(6), pp. 917–920, December 1989.
89. Kobayashi, S., Xu, N. S., Saito, Y., Latham, R. V., "Distributions of prebreakdown emission sites on broad area rounded shaped copper cathode of a vacuum gap", *Proceedings 18th International Symposium on Discharges and Electrical Insulation in Vacuum*, pp. 56–59, 1998.
90. Juttner, B., "Instabilities of prebreakdown currents in vacuum, II: The nature of emission sites", *Journal of Physics D: Applied Physics*, 32, pp. 2544–2577, 1999.
91. Fröhlich, K., Kärner, H. C., König, D., Lindmayer, M., Möller, K. Rieder, W., "Fundamental research on vacuum interrupters at Technical Universities in Germany and Austria", *IEEE Transactions on Electrical Insulation*, 28(4), pp. 592–606, August 1993.
92. Davies, D. K., Biondi, M. A., "The effect of electrode temperature on vacuum electrical breakdown between plane–parallel copper plates", *Journal of Applied Physics*, 39(7), pp. 2979–2990, June 1968.
93. Chatterton, P. A., "Further calculations on field emission initiated vacuum breakdown", *Proceedings Physical Society*, 88, pp. 231–245, 1966 and 89, pp. 178–180, 1966.
94. Davies, D. K., Biondi, M. A., "Mechanism of dc electrical breakdown between extended electrodes in vacuum", *Journal of Applied Physics*, 42(8), pp. 3089–3107, July 1971.
95. Davies, D. K., Biondi, M. A., "Emission of electrode vapor resonance radiation at the onset of dc breakdown in vacuum", *Journal of Applied Physics*, 48(10), pp. 4229–4233, October 1977.
96. Lamarsh, J. R., *Introduction to Nuclear Engineering*, (pub. Addison Wesley), 2nd Edition, pp. 91–94, 1983.
97. Feldman, L. C., Mayer, J. W., *Fundamentals of Surface Science and Thin Film Analysis*, (pub. Prentice Hall, NJ), Chap. 6, pp. 141–145, 1986.

98. Slade, P., Taylor, E., "Calculations on the potential role of emission currents on restrikes after capacitor switching", *Proceedings of the 28th International Symposium on Discharges and Insulation in Vacuum*, pp. 177–180, September 2018.
99. Wang, L., Jia, S., Yang D., Liu K., Su, G., Shi, Z., "Modelling and simulation of anode activity in high-current vacuum arc", *Journal of Physics D: Applied Physics*, 42(14), June 2009.
100. Balachanndra, T. C., Nagabhushana, G. R., "Anode hot spot temperature estimation in vacuum gaps under 50Hz alternating excitations", *IEEE Transactions on Electrical Insulation*, 28(3), pp. 392–401, June 1993.
101. Kamikawaji, T., Shioiri, T., Okawa, M., Kaneko, E., Ohshima, I., "Generation of micro-particles from copper-chromium contacts and their influence on insulating performance in vacuum", *Proceedings 23rd International Conference on Electrical Contact Phenomenon*, pp. 895–902, July 1994.
102. Bernauer, C., Kny, E., Rieder, W., "Restrikes in vacuum circuit-breakers within nine seconds after current interruption", *Proceedings 14th International Symposium on Discharges and Electrical Insulation in Vacuum*, pp. 512–5168, September 1990.
103. Bennett, W. H., "*Cold emission from unconditioned surfaces*", *Physics Review*, 37, p. 582, 1931.
104. Yan, W., Wang, Z., Liu, J., Li, Y., Sun, L., Liu, Z., Geng, Y., Wang, J., "The effect of contact material on particle behavior in high-current vacuum arcs", *Proceedings 28th International Symposium on Discharges and Electrical Insulation in Vacuum*, pp. 19–22, September 2018.
105. Eastham, D. A., Chatterton, P. A., "An investigation of micro-particle induced breakdown using a twin beam laser scattering system", *Proceedings 19th International Symposium on Discharges and Electrical Insulation in Vacuum*, pp. 17–21, 1982.
106. Shioiri, T., Kamikawaji, T., Kaneko, E., Okawa, M., Ohshima, I., "Breakdown characteristics with vacuum gaps contaminated by metallic particles", *Proceedings 2nd International Conference on Electrical Contacts, Arcs, and their Apparatus*, pp. 74–78, May 1993.
107. Menon, M. M., Srivastava, K. D., "The nature of micro-particles and their role in vacuum breakdown", *Proceedings 6th International Symposium on Discharges and Electrical Insulation in Vacuum*, pp. 3–10, September 1974.
108. Kolyada, Y., E., Bulanchuk, O. N., Fedun, V. I., Onishchenko, I. N., "Aerosol micro-particles and emission characteristics of the pulsed high-current vacuum diode in a microsecond range", *Proceedings 19th International Symposium on Discharges and Electrical Insulation in Vacuum*, pp. 68–74, September 2000.
109. Latham, R. V., "Micro-particles charge acquisition and reversal at impact", *Journal of Physics D: Applied Physics*, 5, pp. 2044–2054, 1972.
110. Abe, K., Ejiri, H., Matsuoka, S., Kumada, A., Hidaka, K., Donen, T., Tsukima, M., "Optical measurement for particle production in vacuum interrupter' *Proceedings 27th International Symposium on Discharges and Electrical Insulation in Vacuum*, pp. 21–24, September 2016.
111. Ejiri, H., Abe, K., Matsuoka, S., Kikuchi, Y., Kumada, A., Hidaka, K., Donen, T., Tsukima, M., "Motion and production of microparticles in vacuum interrupter", *IEEE Transactions on Dielectrics and Electrical Insulation*, 24(6), pp. 3375–3380, December 2017.
112. Ejiri, H., Kumada, A., Hidaka, K., Donen, T., Kokura, K., "Motion and breakdown related to microparticle in vacuum gap", *Transactions IEEE on Plasma Science*, 47(6), pp. 3384–3392, August 2019.
113. Jüttner, B., "Surface migration as a possible cause for late breakdowns", *Proceedings 18th International Symposium on Discharges and Electrical Insulation in Vacuum*, pp. 408–411, 1998.
114. Jüttner, B., Lindmayer, M., Düring, G., "Instabilities of pre-breakdown currents in vacuum 1: Late breakdowns", *Journal of Physics D: Applied Physics*, 32, pp. 2537–2543, 1999.
115. Farrall G. A., Hudda, F. G., "The effect of glass debris on electron emission and electrical breakdown of vacuum interrupters", *IEEE Transactions on Electrical Insulation*, 15(4), pp. 61–67, April 1981.
116. Johnson, D., Savage, M., Sharpe, R.,Batrakov, A., Proskurovsky, D., "Pulsed HV breakdown of polished, powder coated and e-beam treated large area stainless steel electrodes with 0.5 to 7mm gaps", *IEEE Transactions Dielectrics and Electrical Insulation*, 13(1), pp. 52–64, February 2006.
117. Kraft, V. V., Stuchenkov, V. M., "Effect of nonmetallic inclusions in cathodes on vacuum breakdown", *Soviet Physics Technical Physics*, 17(1), pp. 66–70, July 1972.
118. Pokrovskaya-Soboleva, A., Kraft, V., Borisova, T., Mazurova, L., "The effect of nonmetallic inclusions on electrical breakdown in vacuum", *Proceedings 5th International Symposium on Discharges and Electrical Insulation in Vacuum*, pp. 105–109, September 1972.
119. Slivkov, I. N., "Initiation of electrical breakdown in a vacuum in the presence of micro-discharges", *Soviet Physics Technical Physics*, 13(5), pp. 663–666, November 1968.
120. Ziomek, W., Moscicka-Grzesiak, H., "Relation of breakdown voltage and prebreakdown micro-discharge parameters in vacuum", *IEEE Transactions on Electrical Insulation*, 28(4), pp. 481–487, August 1993.

121. Siodla, K., "Micro-discharges–the prebreakdown phenomenon in the vacuum interrupter", *Proceedings 14th International Symposium on Discharges and Electrical Insulation in Vacuum*, pp. 80–83, September 1990.
122. König, D., Heinemeyer, R., "Prebreakdown currents of vacuum tubes with increased pressure, stressed with ac voltage", *IEEE Transactions Electrical Insulation*, 24(6), pp. 937–941, December 1989.
123. Mośck-Grzesiak, H., Ziomek, W., Siodla, K., "Estimation of properties of contact materials used in vacuum interrupters based on investigations of the microdischarge phenomenon", *IEEE Transactions Components, Materials and Packaging–Part A*, 18(2), pp. 344–347, June 1995.
124. Aoki, K., Nishimura, R., Kojima, H., Homma, M., Shioiri, T., Okubo, H., "Enhancement of breakdown strength of microdischarge under impulse voltage applications in vacuum", *Proceedings 10th International Symposium on Discharges and Electrical Insulation in Vacuum*, pp. 32–35, September 2010.
125. Tsuruta, K., Ootaka, M., "Gases released at prebreakdown and breakdown stage in vacuum and its mass analysis", *Proceedings 19th International Symposium on Discharges and Electrical Insulation in Vacuum*, pp. 25–28, September 2000.
126. Diamond, W. T., "New perspectives in vacuum high voltage insulation. 2. Gas desorption", *Journal of Vacuum Science and Technology A*, 16(2), pp. 720–735, March/April 1998.
127. Slade, P., "The arc and interruption", In: Slade, P. G., (Editor), *Electrical Contacts: Principles and Applications, 2nd Edition*, (pub. CRC Press), pp. 578–584 and 585–588, 2014.
128. Shioiri, T., Ohshima, I., Honda, M., Okumura, H., Takahashi, H., Yoshida, H., "Impulse voltage field emission characteristics and breakdown dependency upon field strength in vacuum gaps", *IEEE Transactions on Power Apparatus and Systems*, PAS-101(10), pp. 4187–4184, October 1982.
129. Mazurek, B., Nowak, A., Tyman, A., "X-ray emission accompanying cathode micro-discharges", *IEEE Transactions on Electrical Insulation*, 28(4), pp. 488–493, August 1993.
130. Kaljatsky, I., Kassirov, G., Sekisov, F., "On the impulse electrical breakdown of centimeter vacuum gaps", *IEEE Transactions on Electrical Insulation*, 20(4), pp. 701–703, August 1985.
131. Shioiri, T., Ohshima, I., Honda, M., Okumura, H., Matsumoto, K., "Area effect of intensification factor in vacuum gap breakdown", *Proceedings 11th International Conference on Electrical Contact Phenomena*, pp. 321–325, June 1982.
132. Srinivasa, K. V., Nagabhushana, G. R., "Prebreakdown conduction in vacuum gaps under switching impulse excitations", *IEEE* Transactions *on Electrical Insulation*, EI-20(6), pp. 691–695, August 1985.
133. Zhou, Z., Kyritsakis, A., Wang, Z., Li, Y., Geng, Y., Djurabekova, F., "Direct observation of vacuum arc evolution with nanosecond resolution", www.nature,com/scientific reports, Article 1814, published online May 24, 2019, 53.
134. Sato, S., Koyama, K., Fujii, H., "Behavior of conductive microparticles under electric field in vacuum and their influence on breakdown characteristics", *Proceedings 19th International Symposium on Discharges and Electrical Insulation in Vacuum*, pp. 22–28, September 2000.
135. Eastham, D. A., Chatterton, P. A., "An investigation of micro-particle induced breakdown using a twin beam laser scattering system", *Proceedings 10th International Symposium on Discharges and Electrical Insulation in Vacuum*, pp. 17–21, September 1982.
136. Cranberg, L., "Initiation of electrical breakdown in vacuum", *Journal of Applied Physics*, 23(5), pp. 518–522, May 1952.
137. Slivkov, I. N., *Soviet Physics Technical Physics*, vol. 2, pp. 1928–1934, 1957.
138. Farrall, G. A., "Cranberg hypothesis of vacuum breakdown as applied to impulse voltages", *Journal of Applied Physics*, 33, pp. 96–99, 1962.
139. Fursey, G. N., "Field emission and vacuum breakdown", *IEEE* Transactions *on Electrical Insulation*, EI-20(4), pp. 659–670, August 1985.
140. Charbonnier, F. M., Strayer, R. W., Swanson, L. W., Martin, E. E., "Nottingham effect in field and T-F emission: Heating and cooling domains, and inversion temperature", *Physics Review Letters*, 13(13), pp. 397–401, September 1964.
141. Sun, J., Liu, G, "Numerical modeling of thermal response of thermofield electron emission leading to explosive electron emission", *IEEE Transactions Plasma Science*, 33(5), pp. 1487–1490, October 2005.
142. Child, C. D., "Discharge from hot CaO", *Physics Review*, 32, p. 492, 1911.
143. Batrakov, A., Pegel, I., Proskurovsky, D., "On the screening of the E field at the cathode surface by electron space charge at intense field emission", *IEEE Transactions Dielectrics and Electrical Insulation*, 6(4), pp. 436–440, August 1999.
144. Dyke, W. P., Trolan, J., Martin, E., Barbour, J., "The field emission initiated vacuum arc. 1, Experiments on arc initiation", *Physics Review*, 91(9), pp. 1043–1054, September 1953.

145. Dolan, W., Dyke, W., Trolan, J., "*Physics Review*, 91(9), pp. 1054–1057, September 1953.
146. Dyke, W., Dolan, W., "Field emission: Large current densities, space charge and the vacuum arc", *Physics Review*, 89(4), pp. 799–808, 1953.
147. Barbour, J., Dolan, W., Martin, E., Trolan, J., J., P. Dyke, W., "Space charge effects in field emission", *Physics Review*, 92(1), pp. 45–51, 1953.
148. Schwirzke, F., Hallal, M. P., Maruyama, X. K., "Onset of breakdown and formation of cathode spots", *IEEE Trans Plasma Science*, 21(5), pp. 410–415, October 1993.
149. Uimanov, I., "The dimensional effect of the space charge on the self-consistent electric field at the cathode surface", *IEEE Transactions on Dielectrics and Electrical Insulation*, 18(3), pp. 924–928, June 2011.
150. Nagai, H., Inada, Y., Kumada, A., Ikeda, H., Hidaka, K., Tateyama, C., Niwa, Y., Shiori, T., Ichikawa, T., "High-speed spectroscopy of vacuum breakdown process between CuCr electrode", *Proceedings of the 28th International Symposium on Discharges and Electrical Insulation in Vacuum*, pp. 23–26, September 2018.
151. Nagai, H., Kikuchi, R., Inada, Y., Matsuoka, S., Shiori, T., Kumada, A., Hidaka, "Initiation process of vacuum breakdown between Cu and CuCr electrodes", *IEEE Transactions on Plasma Science*, 47(11), November 2019.
152. Bochkarev, M., Zemskov, Yu. A., Uimanov, I., "High speed and spectroscopic investigation of 300kV pulsed vacuum spark in centimeter gap", *Proceedings of the 24th International Symposium on Discharges and Electrical Insulation in Vacuum*, pp. 72–75, September 2010.
153. Emelyanov, A., Emelyanov, E., Safonova, T., Serikov, I., "Estimation of cathode-initiated breakdown in vacuum", *IEEE Transactions Dielectrics and Electrical Insulation*, 13(1), pp. 26–33, February 2006.
154. Batrakov, A., Proskurovsky, D., Popov, S., "Observation of the field emission from the melting zone occurred just before the explosive electron emission", *IEEE Transactions Dielectrics and Electrical Insulation*, 6(4), pp. 410–417, August 1999.
155. Mesyats, G., Uimanov, I., "On limiting density of field emission current with metals", *IEEE Transactions Dielectrics and Electrical Insulation*, 13(1), pp. 105–110, February 2006.
156. Yen, Y. T., Tuma, D. T., Davies, D. K., 'Emission of electrode vapor resonance radiation at the onset of impulsive breakdown in vacuum", *Journal of Applied Physics*, 55(9), pp. 3301–3307, May 1984.
157. Jüttner, B., "Vacuum arc initiation and applications", pp. 516–519, in Reference [56].
158. Djogo, G., Cross, J. D., "Dependence of gap voltage collapse during vacuum breakdown on geometry and plasma dynamics", *IEEE Transactions Dielectrics and Electrical Insulation*, 4(6), pp. 848–853, December 1997.
159. Djogo, G., Cross, J. D., "Circuit Modeling of a vacuum gap during breakdown", *IEEE Trans Plasma Science*, 25(4), pp. 617–624, August 1997.
160. Cobine, J. D., Farrell, G. A., "Experimental study of arc stability", *Journal of Applied Physics*, 31, pp. 2296–2304, 1960.
161. Spatami, C., Teillet-Billy, D., Gauyacq, J., Teate, Ph., Chabrier, J., "Ion assist emission from a cathode in an electric arc", *Journal of Physics D: Applied Physics*, 30, pp. 1135–1145, 1997.
162. Lee, T. H., Greenwood, A., "theory of the cathode mechanism in metal vapor arcs", *Journal of Applied Physics*, 32, pp. 916 –923, 1961.
163. Zhou, Z., Kyritsakis, A., Wang, Z., Li, Y., Geng, Y., Djurabekova, F., "Spectroscopic study of vacuum arc plasma expansion", *Proceedings 8th International Workshop on Mechanism of Vacuum Arcs*, September 2019. Published in *Journal of Physics D: Applied Physics*, 53, pp. 1–11, January 2020. As "Spectroscopic study of vacuum arc plasma expansion".
164. Srinivasa, K., Nagabhushana, G., "Prebreakdown conduction in vacuum gaps under switching impulse excitation", *IEEE* Transactions *on Electrical Insulation*, EI-20(4), pp. 691–695, August 1985.
165. Bender, H. G., Kärner, H. C., "Breakdown of large vacuum gaps under lightning impulse stress", *IEEE Transactions on Electrical Insulation*, 23(1), pp. 37–41, January 1988.
166. Anders, S., Juttner, B., Lindmayer, M., Rusteberg, C., Pursch, H., Unger-Weber, F., "Vacuum breakdown with microsecond delay time", *IEEE Transactions on Electrical Insulation*, 28(4), pp. 461–466, August 1993.
167. Ritskaya, L., "Breakdowns vacuum interrupters behind front of lightning pulse", *Proceedings of the 24th International Symposium on Discharges and Electrical Insulation in Vacuum*, pp. 72–75, September 2012.
168. Lagotzky, S., Schellekens, H., Papillon, A., Müller, G., "Voltage conditioning effect on Cu-Cr contacts", *Proceedings of the 27th International Symposium on Discharges and Electrical Insulation in Vacuum*, pp. 49–53, September 2016.

169. Miller, H. C., Farrall, G. A., "Polarity effect in vacuum breakdown electrode conditioning", *Journal of Applied Physics*, 36, p. 1138–1343, April 1965.
170. Ohira, K., Iwai, A., Saito, Y., Kobayashi, S., "Changes of parameters influencing breakdown characteristics of vacuum gaps during spark conditioning", *Proceedings 10th International Symposium on Discharges and Electrical Insulation in Vacuum*, pp. 105–108, September 1998.
171. Kojima, H., Takahashi, T., Hayakawa, N., Hasegawa, K., Saito, H., Sakaki, M., "Dependence of spark conditioning on breakdown charge and electrode material under non-uniform electric field in vacuum", *IEEE Transactions on Dielectrics and Electrical Insulation*, 23(5), pp. 3224–3230, October 2016.
172. Ballat, J., König, D., Reininghaus, U., "Spark conditioning procedures for vacuum interrupters in circuit breakers", *IEEE Transactions on Electrical Insulation*, 28(4), pp. 621–627, August 1993.
173. Seki, T., Yamamoto, A., Okutomi, T., "A metallurgical study of Cu-Cr contact surface in breakdown", *Proceedings 17th International Conference on Electrical Contact Phenomena*, pp. 879–885, July 1994.
174. Miyazaki, F., Inagawa, Y., Kato, K., Sakahi, M., Ichikawa, H., Okubo, H., "Electrode conditioning characteristics in vacuum under impulse application in non-uniform electric field", *IEEE Transactions on Dielectrics and Electrical Insulation*, 12(1), pp. 17–23, February 2005.
175. Kulkarni, S., Masilamani, H., Chaudhari, S., Shanker, P., "Lightning impulse voltage conditioning of different electrode gaps in high voltage vacuum interrupters", *Proceedings of the 27th International Symposium on Discharges and Insulation in Vacuum*, pp. 585–588, September 2016.
176. Schneider, A., Dubrovskaya, E., Onischenko, S., "Effect of high-frequency conditioning of the electrodes on electric strength of vacuum insulation", *Proceedings of the 28th International Symposium on Discharges and Electrical Insulation in Vacuum*, pp. 81–84, September 2018.
177. Godechot, X., Chakraborty, S., Papillon1, A., Berthon, B., Triaire C., "Investigation of current conditioning process for vacuum interrupters", *Proceedings of the 27th International Symposium on Discharges and Electrical Insulation in Vacuum*, pp. 577–580, September 2016.
178. Zhang, Y., Liu, Z., Geng, Y., "Influence of no-load operation and current switching on breakdown characteristics of high-voltage vacuum interrupters at contact gap 30mm", *Proceedings of the 25th International Symposium on Discharges and Electrical Insulation in Vacuum*, pp. 60–63, September 2012.
179. Gruszka, H., Mościcka-Grzesiak, H., "Conditioning process of electro-polished electrodes by field emission current in vacuum", *IEEE* Transactions *on Electrical Insulation*, EI-20(4), pp. 705–708, August 1985.
180. Störi, H., "An in-situ glow discharge cleaning method for the LEP vacuum system", *Vacuum*, 33(3), pp. 171–m178, 1983.
181. Hayashi, T., Toya, H., "Glow discharge cleaning of vacuum switch tubes", *IEEE Trans Plasma Science*, 19(5), pp. 740–742, October 1991.
182. Proskurovsky, D., Batrakov, A., 'Treatment of the electrode surface with intense charged particle flows as a new method for improvement of the vacuum insulation", *Proceedings 19th International Symposium on Discharges and Electrical Insulation in Vacuum*, pp. 9–15, September 2000.
183. Fink, H., Gentsch, D., Heil, B., Humpert, C., Schnettler, A., "Conditioning of series vacuum interrupters (VI) for medium voltage by applying high-frequency (HF) current to increase the dielectric strength", *IEEE Transactions on Plasma Science*, 35(4), Part 2, pp. 873–878, August 2007.
184. Shea, J., "Punch-through of ceramic insulators", *Proceedings* IEEE *Conference on Electrical Insulation and Dielectric Phenomena*, Library of Congress No. 79-649806, pp. 441–450, October 1990.
185. Vine, J., Einstein, P., "Heating effect of an electron beam impinging on a solid surface", *Proceedings IEEE*, 111(5), pp. 921–930, May 1964.
186. Slade, P., "An investigation into the factors contributing to welding of contact electrodes in high vacuum", *IEEE Transactions Parts, Materials and Packaging*, PMP-7(1), pp. 23–33, March 1971.
187. Tsutsumi, T., Shioiri, T., Okubo, H., Yanabu, S., "The effect of mechanical contact on breakdown characteristics in vacuum", *IEEE Transactions on Electrical Insulation*, 24(6), pp. 921–924, December 1989.
188. Ballat, J., König, D., "Insulation characteristics and welding behavior of vacuum switch contacts made from various Cr-Cu alloys", *IEEE Transactions on Electrical Insulation*, 28(4), pp. 628–634, August 1993.
189. Slade, P., Kirkland Smith, R., Taylor, E., "The effect of contact closure in vacuum with fault current on prestrike arcing time, contact welding and the field enhancement factor", *Proceedings IEEE Holm Conference on Electrical Contacts*, pp. 32–36, September 2007.
190. For Example: "ANSOFT 2D or 3D Electromagnetic Software", Ansoft Corporation, Pittsburgh, PA., USA.

191. Yamano, Y., Ito, S., Kato, K., Okubo, H., "Charging characteristics and electric field distribution on alumina as affected by triple junctions in vacuum", *IEEE Transactions Dielectrics and Electrical Insulation*, 9(2), pp. 173–177, April 2002.
192. Venna, K., Schramm, H., "Simulation analysis on reducing the electric field stress at triple-junctions & on the insulator surface of the high voltage vacuum interrupters", *Proceedings of the 26th International Symposium on Discharges and Electrical Insulation in Vacuum*, pp. 53–56, October 2014.
193. Yungdong, C., Xiaoming, L., Erzhi, W., Licheng, W., "Numerical analysis of electric field of three phase outdoor vacuum circuit breaker", *Proceedings 19th International Symposium on Discharges and Electrical Insulation in Vacuum*, pp. 131–134, September 2000.
194. Okawa, M., Shioiri, T., Okubo, H., Yanabu, S., "Area effect on dielectric breakdown of copper and stainless-steel electrode in vacuum", *IEEE Transactions on Electrical Insulation*, 23(1), pp. 77–81, February 1988.
195. Liu, Z., Cheng, S., Zhang, Y., Yang, L., Geng, Y., Wang, J., "Influence of contact contour on breakdown behaviors in vacuum under uniform field", *Proceedings of the 23ed International Symposium on Discharges and Electrical Insulation in Vacuum*, pp. 25–28, September 2008.
196. Shümann U., Gere, S., Kurrat, M., "Breakdown voltage of electrode arrangements in vacuum circuit breakers", *IEEE Transactions Dielectrics and Electrical Insulation*, 10(4), pp. 557–562, August 2003.
197. Shioiri, T., Kamikawaji, T., Kaneko, E., Okawa, M., Ohshima, I., "Conditioning effect of stainless-steel electrodes in vacuum", *Proceedings 8th International Symposium on HV Engineering*, pp. 473–477, August 1993.
198. Noe, S., Böhm, H., Pilsinger, G., Schmidt, B., "Empirical criterion for electrical breakdown in high-voltage vacuum insulations", *Proceedings 17th International Symposium on Discharges and Electrical Insulation in Vacuum*, pp. 28–31, July 1996.
199. Kahl, B., Kärner, H., C., "The influence of a cylindrical shield electrode on the dielectric strength of large vacuum gaps", *IEEE Transactions on Electrical Insulation*, 28(4), pp. 473–480, August 1993.
200. Okubo, H., "development of electrical insulation techniques in vacuum for higher voltage vacuum interrupters", *Proceedings 22nd International Symposium on Discharges and Electrical Insulation in Vacuum*, pp. 7–12, September 2006.
201. Wang, j., Liu, Z., Xiu, S., Wang, Z., Yuan, S., Jin, L., Zhou, H., Yang., R., "Development of high voltage vacuum circuit breakers in China", *Proceedings 22nd International Symposium on Discharges and Electrical Insulation in Vacuum*, pp. 247–252, September 2006.
202. Liao, M., Zou, J., Duan, X., Fan, X., Sun, H., "Dielectric strength and statistical property of single and triple-break vacuum interrupters in series", *Proceedings 22nd International Symposium on Discharges and Electrical Insulation in Vacuum*, pp. 157–160, September 2006.
203. König, D., Schmidt, H., "Breakdown voltage of conditioned 24 kV vacuum tubes as a function of ramp rate", *IEEE* Transactions *on Electrical Insulation*, EI-20(4), pp. 715–720, August 1985.
204. Zhang, Y., Liu, Z., Geng, Y., "Influence of no-load operation and current switching on breakdown characteristics of high voltage vacuum interrupters at contact gap 30mm", *Proceedings 25th International Symposium on Discharges and Electrical Insulation in Vacuum*, pp. 60–63, September 2012.
205. Smith, R., K., "vacuum interrupter impulse voltage testing procedures should recognize initial breakdowns as reconditioning events", *CIRED 99*, (Nice, France), 1999.
206. *IEEE Standards C37.09-1999*, "Test procedures for ac high-voltage circuit breakers on a symmetrical current basis".
207. IEC Standards 60060–1: 2010.
208. Zou, J, He, J., Cheng, L., "Improvement of voltage withstand level of vacuum interrupters", *Proceedings IEE 2nd International Conference on Advances in Power System Control and Management*, pp. 338–341, December 1993.
209. Renz, R., Gentsch, D., "Permissible x-ray radiation emitted by vacuum–interrupters–devices at rated operating conditions", *Proceedings 24th International Symposium on Discharges and Electrical Insulation in Vacuum*, pp. 133–137, August 2010.
210. Yan, J., Liu, Z., Geng, Y., Zhang, S., zhang, Y., "X-radiation of a 126kV vacuum interrupter", *Proceedings 25th International Symposium on Discharges and Electrical Insulation in Vacuum*, pp. 461–464, September 2012.
211. Yan, J., Liu, Z., Zhang, S., Geng, Y., Zhang, Y., He, G., "Investigation on X-radiation for 126kV vacuum interrupters", *Plasma Science and Technology*, 18(5), pp. 577–582, May 2016.

212. Geire, S., Heinz, T., Lawail, A., Steiler, C., Taylor, E., Wenze;, N., Wethekam, S., "X-radiation emission of high-voltage vacuum interrupters: Dose rate control under testing and operation conditions", *Proceedings 28th International Symposium on Discharges and Electrical Insulation in Vacuum*, pp. 523–526, September 2018.
213. Lamarsh, J., "Radiation units", In: *Introduction to Nuclear Engineering*, (pub. Addison Wesley), 2nd Edition, pp. 399–406, 1983.
214. IEEE Standards C37.85-2001.
215. Henken, K., "Investigation of the role of x-ray photons in the pre-breakdown current in vacuum interrupter gaps", *Proceedings 25th International Symposium on Discharges and Electrical Insulation in Vacuum*, pp. 5–8, September 2012.

2 The Vacuum Arc

Felix qui portuit rerum cognoscere causas.
Happy is the man who could search out the causes of things.

Virgil
(Goegrics I)

2.1 THE CLOSED CONTACT

The vacuum interrupter's contacts are the heart of its performance. When closed they permit the flow of current in an electrical circuit. The current must flow through the contacts without overheating them. In some applications the contacts may stay closed for many years: in others, frequent opening and closing may be normal. When contacts open under load the vacuum arc that forms between them must be stable to the natural current zero of an ac circuit or to the forced current zero in a dc circuit. The region of the contacts where this electric arc attaches is subjected to very high temperatures. Even under these severe conditions, the contacts must resist excessive erosion and maintain their mechanical integrity. The contact surfaces must also maintain a reasonably high electrical conductivity so that, when they again close, current can flow without excessive heating. When the contacts are opened under fault current conditions (such as short-circuit currents), the designer has to consider how to control the resulting high-current vacuum arc. When current zero is reached in an ac or dc circuit and the arc extinguishes, the gap between the electrical contacts has to recover its dielectric properties very rapidly.

In all vacuum interrupters the properties of the electric contacts are vital for a successful dielectric recovery. For example, the contacts must not exhibit severe field distortion or stay hot long enough to liberate large numbers of electrons or too much metal vapor. In some applications the contacts are required to close and latch on very high short-circuit currents. Here they not only have to withstand the mechanical forces involved but also the opening mechanism must be designed to break any contact welds that form. Thus, it is apparent that the electric contact participates in every phase of a vacuum interrupter's operation. No matter what the voltage range or current level, the choice of the contact material and the correct design of the electrical contact are essential (see Chapter 3, this volume).

In a power system an understanding of electric contacts is not just confined to the design of the vacuum interrupter. It is also required for the proper connection of the vacuum interrupter into its switching mechanism and into the electrical circuit: e.g., how the vacuum interrupter's fixed terminal is connected to the circuit breaker's bus, how the connection is made to the moving terminal and how the circuit breaker itself is joined to the electrical circuit. These contacts are usually not affected by arcing, but their design is an integral part of the successful operation of the vacuum interrupter; see Section 6.2. I will therefore begin this chapter with a brief discussion of electrical contact theory. A more thorough review of this subject is presented in my book *Electrical Contacts* (CRC Press, 2014) [1] and in the seminal work by Ragnar and Else Holm [2]. I will then continue with how the vacuum arc forms between two opening contacts. After this I will present the different kinds of vacuum arc. I will conclude with a review of how these different modes of vacuum arc are affected by the imposition of external magnetic fields.

2.1.1 MAKING CONTACT, CONTACT AREA, AND CONTACT RESISTANCE

If the two cylinders shown in Figure 2.1 are butted together and the resistance between points a and b is measured, the resistance will be found to be:

$$\text{Total resistance a} \leftrightarrow \text{b} = \left(\text{bulk resistance of the cylinders}\right) + \left(\text{contact resistance}\right)$$

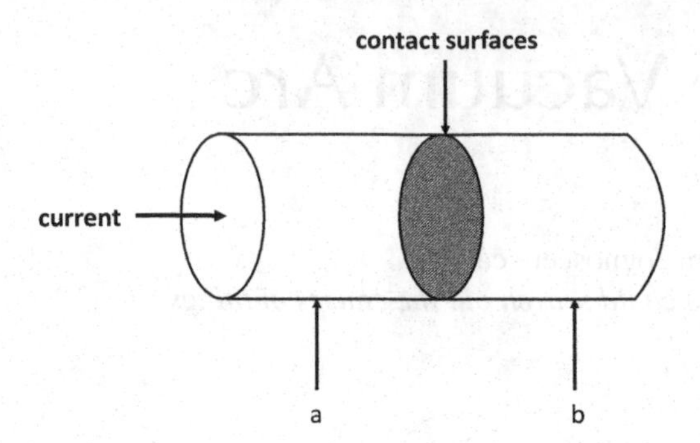

FIGURE 2.1 The electrical contact.

or

$$R_T = R_B + R_C \tag{2.1}$$

The reason for this is that no matter how carefully the cylinders' faces are prepared; they will never be perfectly flat. Indeed, they will make contact only at a number of discrete points on these flat surfaces [2–4]. You have already seen in Figure 1.26 that a high-magnification picture taken of a smooth metal surface reveals a number of microscopic peaks and valleys. Thus, when two such surfaces are brought together, they initially touch at the two highest micro-peaks. Even under light loads the pressures at these peaks will be very high [2, 3] so the peaks will deform plastically. As the first two micro-peaks deform, more micro-peaks will come into contact and they in turn will deform plastically. This process will continue until the force on the contacts is fully supported by a small number of microscopic contact spots. This is shown conceptually in Figure 2.2, and the process can be represented by [2, 3].

Contact closing force ≈ [contact material's hardness] × [$\sum$ real microscopic areas of contact] or:

$$F = \xi \times H \times \sum A_i \tag{2.2}$$

where ξ is a constant (It is ≈ 1 for plastic deformation). This state is usual for the forces typically used in the mechanisms that operate vacuum interrupters, i.e., greater than 10N) and "i" is the "i^{th}" microscopic contact spot. It can be seen that Equation (2.2) implies that the actual area of contact depends only on the contact force and the contact's material properties; it does not depend upon the total area of the contact face. This is true for contacts carrying very low currents (e.g., in electronic

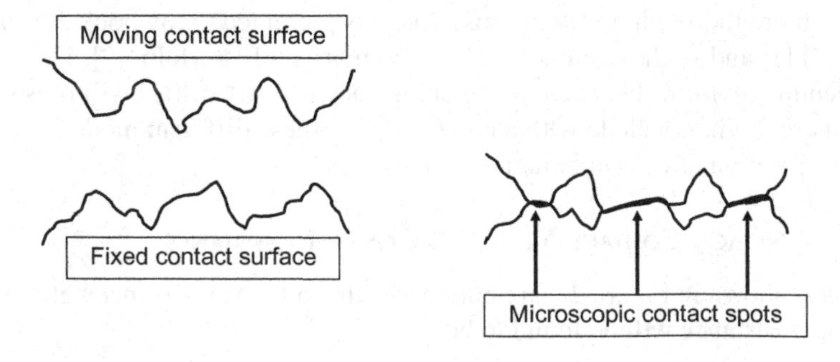

FIGURE 2.2 Plastic deformation and the real area of contact.

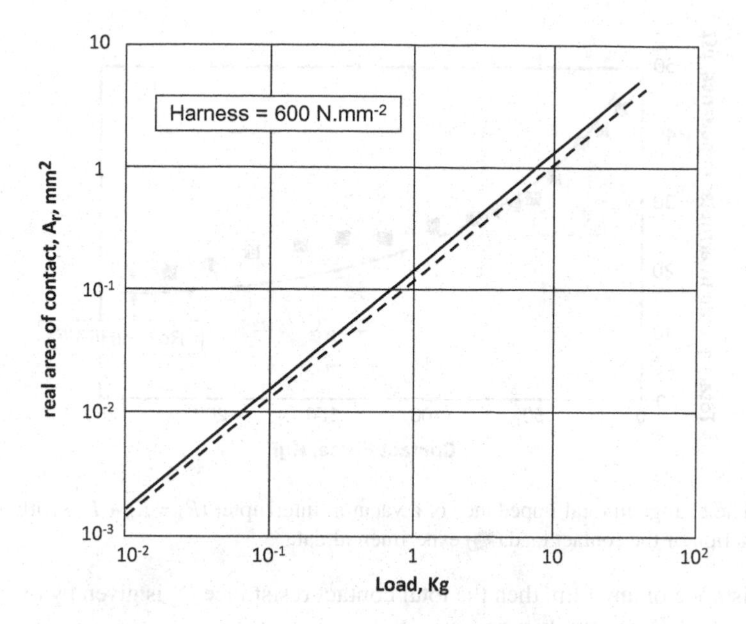

FIGURE 2.3 Example of the real area of contact vs. contact load for two contact sizes: (a) solid line, 10cm^2 nominal area: (b) dashed line, 1cm^2 nominal area [5].

circuits) as well as for contacts carrying the range of currents usually found in vacuum interrupters (e.g., 50A to 80kA). Figure 2.3 illustrates this for a 10-fold change in contact face area [5].

2.1.2 Calculation of Contact Resistance

2.1.2.1 The Real Area of Contact a Small Disk of Radius "a"

Consider as a first approximation of Figure 2.4, where a disk-shaped area A_r of radius "a" is achieved after the contacts have been forced together. The flow of current from one conductor to the other would then be constrained to flow through this area. The constriction resistance is given by [2, 3, 6]:

$$R_K = \frac{\rho}{2a} \tag{2.3}$$

Where ρ is the resistivity of the contact material. Substituting from Equation (2.2) and noting that $A_r = \pi a^2$:

$$R_K = \frac{\rho}{2}\sqrt{\frac{\pi H}{F}} \tag{2.4}$$

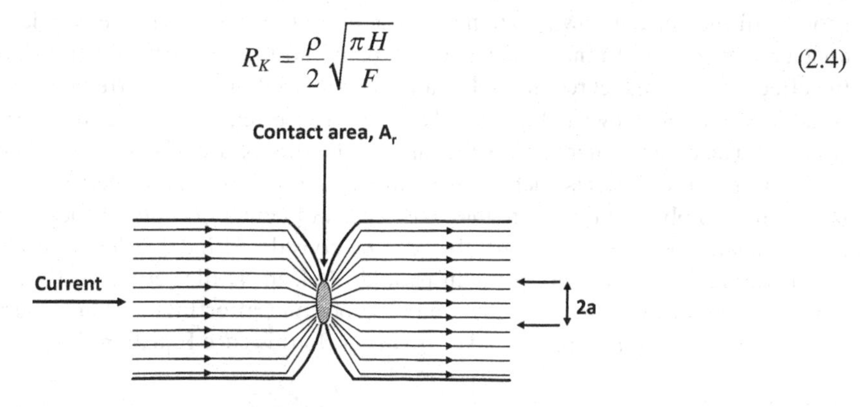

FIGURE 2.4 The average area of contact, A_r, showing how the lines of current flow are constricted to flow through it.

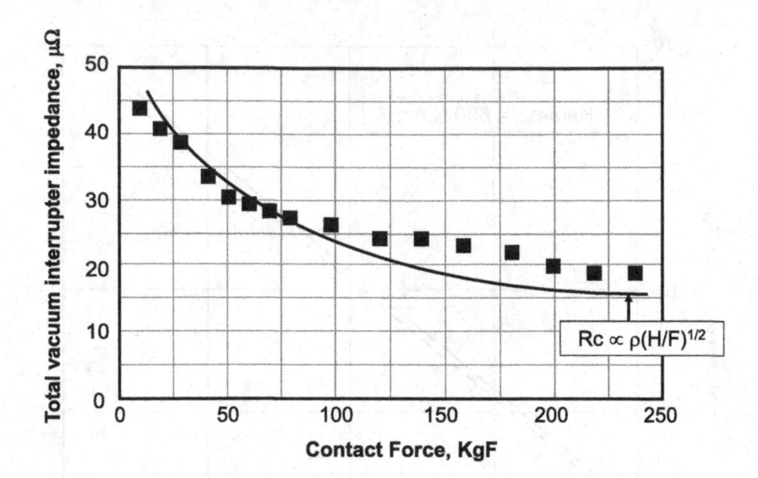

FIGURE 2.5 The change in total impedance of a vacuum interrupter ($R_T = R_B + R_C$) with annealed Cu–Cr contacts as a function of the contact load;(■) experimental data.

If R_F is the resistance of any film, then the total contact resistance R_C is given by ($R_C = R_K + R_F$). In a vacuum interrupter R_F is usually zero, so:

$$R_C = R_K \tag{2.5}$$

Thus, the contact resistance results from the current being forced to flow from the bulk of the conductor into a small area radius "*a*." Figure 2.5 presents data for fully annealed Cu–Cr (25 wt% Cr) contacts inside a vacuum interrupter showing how ($R_T = R_B + R_C$), where R_B is the bulk resistance of the Cu terminals plus the bulk resistance of the contact material (see Figure 1.2) varies with F and how closely Equation (2.4) describes the data. The actual contact spot is usually not just one spot as shown in Figure 2.4. In most practical contact systems and certainly with vacuum interrupter contacts the region of actual contact is made up of a number of microscopic contact spots distributed within an overall contact region. This is illustrated in Figure 2.6 [4]. Fortunately, calculations on randomly arrayed contact spots with a practical distribution of diameters [2–4], show that the microscopic effect of these spots gives a similar relationship to those given in Equations (2.3) and (2.4) with an average microscopic radius of contact like the one shown in Figures 2.4 and 2.6. For practicing engineers Equations (2.3) and (2.4) give values of constriction resistance, which are close enough to the real value (within 20%). Thus no one is really interested in the actual individual contact spots. The reason why these two equations satisfy most practical situations is that the constriction of the current as it travels to the region of contact does not recognize that there are individual microscopic contact spots until the current flow is extremely close to each contact's surface. While the individual micro contact spots do determine the final current path, most of the current constriction effect, and hence the effect on the contact resistance has already occurred. It is also fortunate that the average contact constriction radius shown in Figures 2.4 and 2.6 gives an equivalent area of plastic deformation of the contact surface that is more or less the same as the sum of areas of all the individual contact spots.

For large area contacts such as those found in power vacuum interrupters (see Section 3.3), it is common to observe that more than one region of contact occurs. If these regions of contact are located a sufficient distance apart, they can be considered independent of each other [7, 8]. These are therefore similar current paths in parallel. Indeed, Dullni et al. [9] show that for large area, Cu–Cr, vacuum interrupter contacts both before and after high-current interruption testing that three regions of contact are most likely to occur. In this case Equation (2.4) becomes:

$$R_C = \frac{\rho}{2n}\sqrt{\frac{\pi H}{F/n}} \tag{2.6}$$

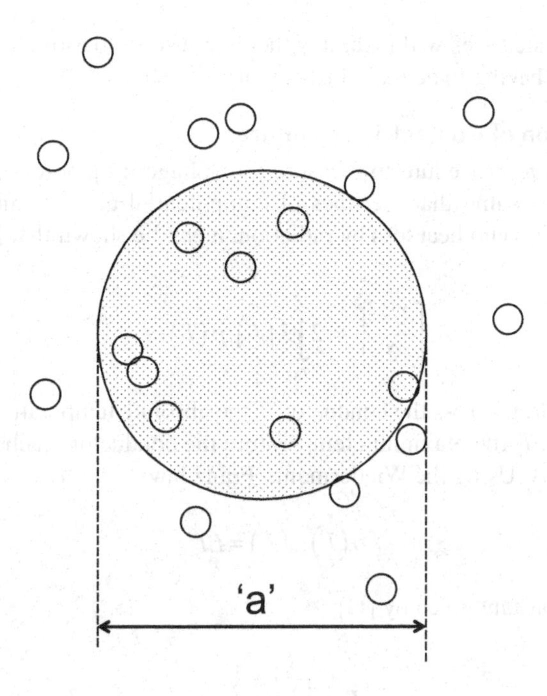

FIGURE 2.6 A random distribution of contact spots, giving the equivalent single contact area shown in Figure 2.4 [4].

where $n = 3$. They also show the top 100–200 μm of the contact surfaces change their composition: see Figure 3.6, this volume. This surface layer has a higher hardness and a higher resistivity than the original contact material: see Table 2.1. This results in a higher contact resistance after the series of high-current interruption tests. They conclude that the higher hardness contributes to most of the increase in the contact resistance and the increase in the surface resistivity only has a small effect. This observation is consistent with the fact that the contact resistance results from the flow of current from the bulk of the contact into the small areas of contact. Thus, the thin surface later only has a minor effect on its value. Taylor et al. [10] also observe a similar effect in the contact resistance of Cu–Cr contacts after high-current interruption testing.

2.1.3 Contact Resistance and Contact Temperature

As R_C increases so does the power input at the contact junction. This results in a temperature increase at the contact junction. In extreme cases, the resulting increase in temperature can be high

TABLE 2.1
Surface Constants Cu–Cr Contacts before and after High-Current Interruption Tests [9]

	Before the Tests	After the Tests
Surface resistivity, ρ μΩ.m	0.036	0.043
Surface hardness, N.mm^{-2}	900	2000
Number of contacting regions	3	3
$R_B + R_C$ for vacuum interrupter 1, μΩ	4.5	8.5
$R_B + R_C$ for vacuum interrupter 2, μΩ	3.4	7.4

enough that the contact interfaces will melt. It is therefore useful to estimate the temperature of the contact junction without having to actually measure it.

2.1.3.1 The Calculation of Contact Temperature

It is a very straightforward procedure to measure the voltage drop across a closed contact. If we take any conductor and assume that the lines of equipotential are the same as the line of equi-temperature and that there is no heat loss by radiation, it can be shown that [2, 3, 6]:

$$\left[\frac{U_C}{2}\right]^2 = 2\int_{T_0}^{T_p} \lambda(T).\rho(T)\,dt \tag{2.7}$$

where U_C is the voltage drop across the conductor, T the absolute temperature in degrees Kelvin, T_0 the ambient temperature, T_p the maximum temperature the conductor reaches, ρ the resistivity, and λ the thermal conductivity. Using the Wiedermann–Franz law:

$$\rho(T)\lambda(T) = ŁT \tag{2.8}$$

where $Ł$ is the Lorenz constant given by [11]:

$$Ł = \frac{\pi^2}{3}\left[\frac{k}{e}\right]^2 \tag{2.9}$$

where k is Boltzmann's constant, e is the electronic charge, $Ł = 2.45 \times 10^{-8}$ W Ω /(degrees Kelvin)^2. Integrating Equation (2.7) gives:

$$U_C = \left[4Ł\left\{T_p^2 - T_0^2\right\}\right]^{1/2} \tag{2.10}$$

If $T_p >> T_0$ (Kelvin) this can be reduced to:

$$T_p = U_C / [2\sqrt{Ł}] \tag{2.11}$$

Figure 2.7 presents a comparison of the measured voltage [1] across a conductor at its melting point with Equation (2.11). It can be seen that it is possible to obtain a very good estimate of the temperature of the actual contact junction by just measuring the voltage across the contacts. It is interesting to note that a very small voltage drop represents a very high temperature (e.g., if the voltage across a Cu contact is only 0.43V, the microscopic contact spots will be molten).

Another important consideration is the softening temperature of metals. This temperature, usually from about 25 to 45% of the melting temperature, is where the metal softens and can deform plastically very easily. Equation (2.10) then allows us to define a softening voltage U_S as:

$$U_S = \left[4Ł\left\{(ST_m)^2 - T_0^2\right\}\right]^{1/2} \tag{2.12}$$

where S is a constant between 0.25 and 0.45 and T_m is the melting temperature. Table 2.2 shows experimentally determined softening and melting voltages for a few contact materials. An extensive list can be found in reference [1], Chapter 13. The effect of the softening temperature on the contact resistance is shown in Table 2.3. The data for two different contact forces certainly show the effect that heating the contact spot has on the total impedance of the vacuum interrupter. At ambient temperature the impedance in both cases is 17μΩ. Immediately after the passage of the 20kA rms current for 1 second the impedance is 22μΩ. However, after allowing the system to cool for

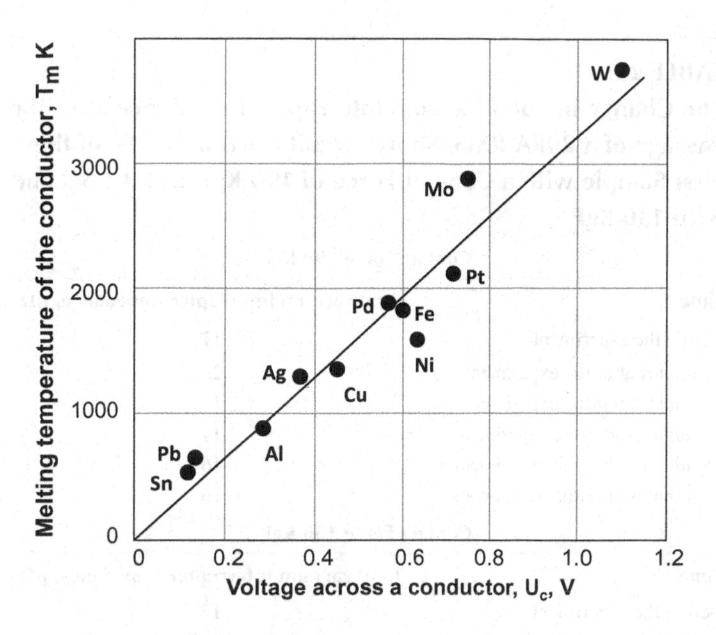

FIGURE 2.7 Comparison of the measured melting voltages for various metals with those calculated from Equation 2.10.

TABLE 2.2
Comparison of Softening Temperatures and Melting Temperatures and the Measured Softening and Melting Voltage

Metal	Melting Temp. T_m K	Softening Temp. T_s K	T_m / T_s	Melting Voltage, U_m, V	Softening Voltage, U_s, V	U_m / U_s
Ag	1233	453	0.37	0.37	0.09	0.24
Cr	1875			≈ 0.67		
Cu	1356	463	0.34	0.43	0.12	0.28
Ni	1726	793	0.46	0.65	0.16	0.25
Fe	1810	773	0.43	0.6	0.19	0.32
W	3683	1273	0.35	1.1	0.4	0.36

20 minutes, the resistance is now 16μΩ, lower than the initial value. This is a strong indication that the lowering of the hardness at the maximum temperature resulted in a larger area of contact Ac and a corresponding lowering of R_C: see Equation (2.4).

2.1.4 Blow-Off Force

From Section 2.1.1 we know that the current flow through a pair of contacts is constricted to flow through a very small area. This is illustrated in Figure 2.8 The flow of the current along the contact surfaces gives rise to a force F_B that tends to blow the contacts apart. This force F_B is given by [2, 12, 13]

$$F_B = \frac{\mu_0 i^2}{4\pi} log_e\left(\frac{R}{a}\right) \tag{2.13}$$

TABLE 2.3.
The Change in Total Vacuum Interrupter Impedance after the Passage of a 20kA RMS, Short-Circuit Current for 1s, of the First Sample with a Contact Force of 190 KgF and the Second with 130 KgF

Contact Force 190 KgF	
Time	**Total vacuum Interrupter Impedance, μΩ**
Before the experiment	17
0 minutes after the experiment	22
5 minutes after the experiment	19
10 minutes after the experiment	17
15 minutes after the experiment	16
20 minutes after the experiment	16
Contact Force 130 KgF	
Time	**Total vacuum Interrupter Impedance, μΩ**
Before the experiment	17
0 minutes after the experiment	22
5 minutes after the experiment	18
10 minutes after the experiment	18
15 minutes after the experiment	17
20 minutes after the experiment	16

where μ_o is the magnetic permeability in vacuum, i is the instantaneous current, R is the measured contact radius, and a is the average radius of the constriction. Using Equations (2.4) and (2.13) and noting that [$log_e (R/a) = 1/2 \, log_e (R/a)^2$]

$$F_B = \frac{\mu_0 i^2}{8\pi} log_e \left(\frac{HA}{F} \right) \tag{2.14}$$

where A is the total area of the contact face. The flaw in using Equation (2.14) is that as the blow-apart force on the contacts increases, the total contact force F_T decreases. Barkan [14] has considered this problem. Given that the total force on the contacts F_T is:

$$F_T = F_S \pm F_A - F_B \tag{2.15}$$

where F_S represents the design spring force holding the contacts together and F_A represents a force resulting from the current flow in the whole contact structure which can be either a blow-on or a blow-off force. This force is very often arranged to be a blow-on force that can counteract F_B and can usually be represented by:

$$F_A = \pm K * i^2 \tag{2.16}$$

where $K*$ is a constant. Barkan showed that the minimum spring force required to counter-balance the blow-off force is:

$$F_B = \frac{\mu i^2}{8\pi} \left[1 + log_e \frac{8\pi Ha}{\mu i^2} - \frac{8\pi}{\mu i^2} F_A \right] \tag{2.17}$$

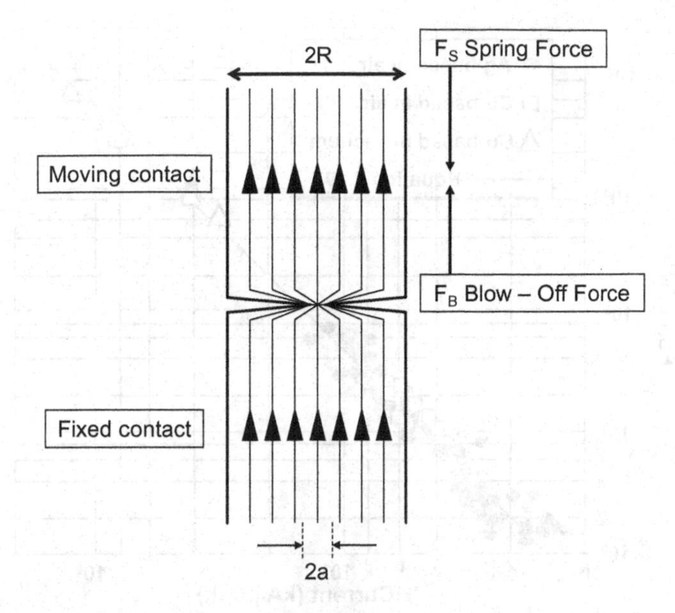

FIGURE 2.8 Current path in mating contacts constricted to flow through the contact area A_r. Opposing currents in the contact surfaces tend to force the contacts apart.

2.1.4.1 Butt Contacts

Let us take the contact shown in Figure 2.8, with one region of contact. Here $F_A = 0$, so Equation (2.17) reduces to:

$$F_B = \frac{\mu i^2}{8\pi}\left[1 + log_e \frac{8\pi Ha}{\mu i^2}\right] \tag{2.18}$$

Figure 2.9 shows data for the contact force F_S required to balance the high-current blow-off force F_B for butt contacts in high vacuum. It can be seen that Equation (2.18) correlates well with the

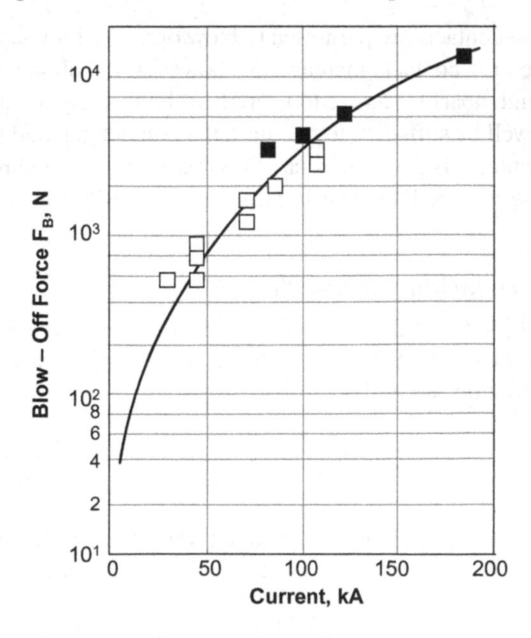

FIGURE 2.9 Contact force required to balance the blow-off force during the passage of high current (■, □ experimental data) [14].

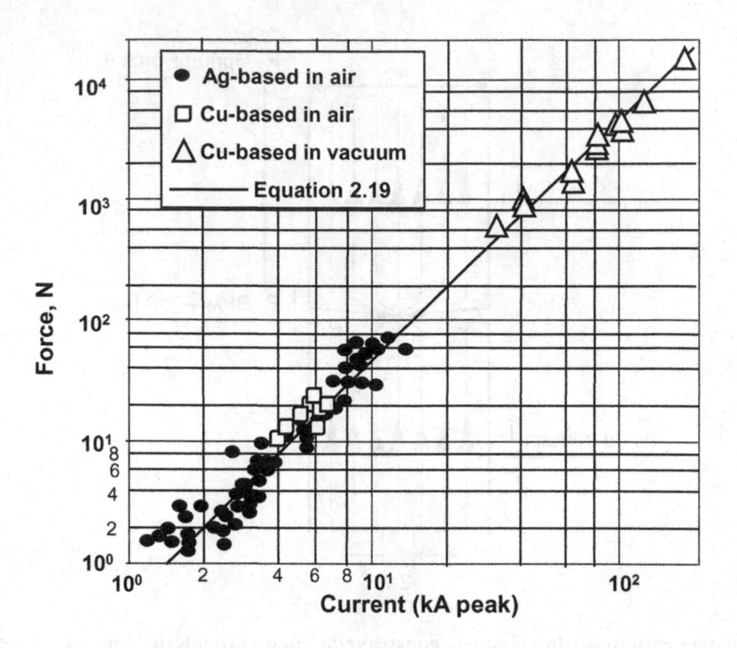

FIGURE 2.10 Plot of the blow-off force of Ag-based and Cu-based contact materials used in vacuum or air, compared to the empirical formula Equation 2.19 [15].

experimental data. Figure 2.10 shows blow-off data compiled by Taylor et al. [15] from many sources including that shown in Figure 2.9. The curve in Figure 2.10 shows that Equation (2.14) can be represented (within experimental error) by:

$$F_B = 4.8\times10^{-7} i^2 \text{Newtons} \tag{2.19}$$

where i is in amperes. If the contacts are permitted to blow open the blow-apart force will rapidly be reduced as the gap between the contacts increases. If, however, the blow-apart force is sufficient to force the contacts far enough apart for an arc to form then the forces generated by the heated gas in the region of the arc may well be sufficient to maintain the contact gap and the arc [16, 17]. This arc between closely spaced contacts is sometimes called a floating arc and may have values of voltage below the minimum arcing voltage [18]. This floating arc can result in severe contact erosion and/or severe contact welding.

2.1.4.2 Contact Interface Melting During Blow-Off

If the total contact force ($F_T = F_S - F_B$) is reduced enough during a blow-off event, then the contact interfaces can melt. If the melting point of the contact metal is T_m then the voltage U_m across the closed contacts when melting occurs will be:

$$U_m = \left(10^{-7}\left[T_m^2 - T_0^2\right]\right)^{1/2} \tag{2.20}$$

Holm [2] shows that if the bulk temperature of a contact remains at T_0 K and the contact spot is at a temperature T_1 K then the effective resistivity for the contact resistance shown in Equation (2.4) is:

$$\rho_T = \rho_0\left(1 + \frac{2}{3}\alpha\left[T_1 - T_0\right]\right) \tag{2.21}$$

Where α is the temperature coefficient of resistivity. Slade [19] has shown for a single contact region that it is possible to calculate the current at which the contact region melts for a given contact material and a given design spring force. He assumes adiabatic heating in the contact region, the hardness of the contacting spot for a temperature T_1 is $0.1H_0$ (where H_0 is the hardness at T_0) and the temperature of the contact spot is $(T_1 = T_m - 50C)$, The temperature T_1 is chosen, because once the contact spot melts the resistivity doubles and the hardness has no meaning. Using Equations (2.4), (2.19), and (2.21):

$$U_m = i_m \times \frac{\rho_{T_1}}{2}\sqrt{\frac{\pi H_{T_1}}{F_T}} \tag{2.22}$$

$$i_m = \frac{2U_m}{\rho_0\left(1+\frac{2}{3}\left[T_1 - T_0\right]\right)}\sqrt{\frac{F_T}{\pi \times 0.1H_0}} \tag{2.23}$$

Using Equation (2.19), the true force on the contact is, to a first approximation:

$$F_T = F_S - 4.8\times10^{-7} i^2 \tag{2.24}$$

$$i_m = \frac{2U_m}{\rho_0\left(1+\frac{2}{3}\left[T_1 - T_0\right]\right)}\sqrt{\frac{F_S - 4.8\times10^{-7} i_m^2}{\pi \times 0.1H_0}} \tag{2.25}$$

Rearranging Equation (2.25)

$$i_m = \frac{2U_m\sqrt{F_S}}{\left[\left\{\rho_0\left[1+\frac{2}{3}\alpha\left(T_1 - T_0\right)\right]\right\}^2 \pi\left(0.1H_0\right) + 4.8\times10^{-7}\times 4U_m^2\right]^{1/2}} \tag{2.26}$$

if ρ is in ohm-mm, H in N.mm^{-2}, and F_S is Newtons then i_m is in amperes. So, for a given design spring force F_S there is a current at which the contacts will melt. I have found from experience that the following expression gives a quick way of making a conservative estimate for the required design spring force needed to keep vacuum interrupter contacts together:

$$F_S = 8\times10^{-5} i^{1.54}\text{Newtons} \tag{2.27}$$

2.2 THE FORMATION OF THE VACUUM ARC DURING CONTACT OPENING

If the circuit current is greater than a minimum value and the voltage that appears across the opening contacts is also greater than a minimum value, an arc will always form between them. The arc formation depends entirely upon the properties of the contact material and the arc always initiates in metal vapor from the contacts themselves. From Equation (2.4) we know that the contact resistance R_C is given by:

$$R_C \propto \left(F_T\right)^{-1/2} \tag{2.28}$$

As the contact begins to open F_T tends to 0 and R_C increases, the voltage across the contact junction, $(U_C = I \times R)$ increases. Using Equation (2.26) for a constant current i_C the contact region melts when $(U_C = U_m)$, that is, when the spring force F_S reduces to:

$$F_{S(melt)} = \left[\frac{\left\{ \rho_0 \left[1 + \frac{2}{3}\alpha\left(T_1 - T_0\right) \right] \right\}^2 \pi\left(0.1H_0\right) + 4.8\times10^{-7} \times 4U_m^2}{4U_m^2} \right] \times i_C^2 \tag{2.29}$$

Once the contact spot melts and the contacts continue to part, they will draw a molten metal bridge between them even when they open in vacuum [20, 21]. Slade [22] describes the transition from the initial formation of the molten metal bridge to the development of the arc between opening contacts both in air and vacuum. A typical change in the voltage drop across the contacts during this molten metal bridge stage is shown in Figure 2.11. This figure shows a jump in the voltage as the metal as the contacting region melts [23] and its resistivity doubles. Its voltage then increases as the contacts continue to open and the molten metal bridge lengthens. Initially the molten metal bridge is quite stable with a slow increase of the voltage drop across the bridge between the opening contacts. As the contacts continue to separate, a stage will be reached when the molten metal bridge becomes unstable and ruptures. There are a number of physical reasons for this instability, including surface tension effects, boiling of the highest temperature region, convective flows of molten metal resulting from the temperature variation between the bridge roots and the high-temperature region [6]. In the vacuum environment an additional instability mechanism results from the evaporation of metal from the bridge's hot surface. When the bridge ruptures, the breaking voltage U_B is greater than the calculated boiling voltage U_{bl} of the contact material [from Equation (2.10)] and metal vapor is released into the contact gap. The bridge rupture results in some metal particles being expelled from its vicinity. In the few tens of nanoseconds after the rupture of the molten metal bridge this entire metal vapor is inertially confined to a region close to the bridge's original position. Thus, there is a confined region of metal vapor in the small gap between the contacts. This region of metal vapor will also be very dense (i.e., have a high pressure) and will have a high temperature.

At the moment of bridge rupture, the metal vapor between the contacts is essentially an insulator. The contact with this metal vapor between them can be considered a very small capacitor. As the circuit inductance will prevent a rapid change in current, charge flows from the circuit's inductance

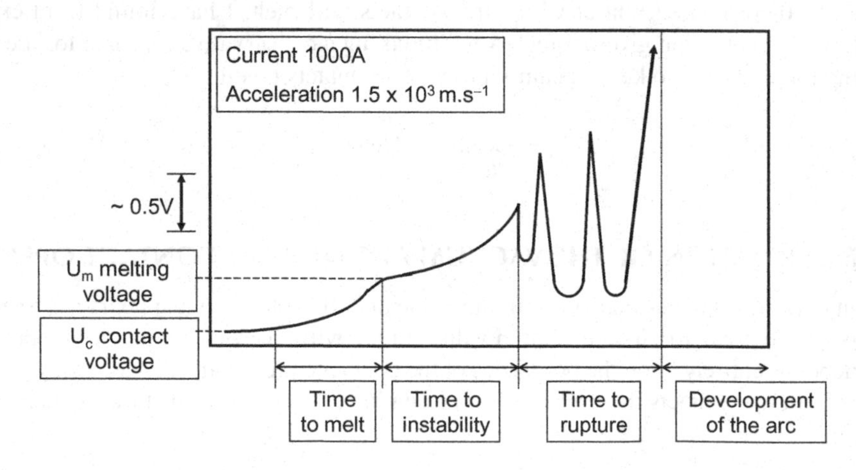

FIGURE 2.11 The voltage drop across an opening Cu contacts showing the formation and rupture of the molten metal bridge.

into this very small capacitor, causing a very rapid voltage rise across the opening contacts [22]. Figure 2.12 shows the expected voltage across the contacts after the rupture of the molten metal bridge. Initially, the voltage across the contacts rises very rapidly to a value somewhat greater than the minimum arcing voltage expected for the contact metal [22, 24, 25] before dropping to a value close to the metal's minimum arcing voltage [20, 22, 26]. This initial voltage peak can have a duration from tens of nanoseconds to a few hundred nanoseconds depending upon the level of current at contact part and the contact material. During the time of rapid voltage rise, the pressure of the metal vapor between the contacts decreases to perhaps a few atmospheres [22]. Thus, a compact, high-pressure region of metal vapor will exist between the contacts with a voltage of a few tens of volts across it. This hot metal gas initially released into the space between the contacts establishes a condition where a dense, high-pressure, non-ideal plasma/glow discharge described by Puchkarev et al. [27] and Korolev et al. [28] can exist. This type of discharge operates in a metal vapor of high density (greater than about 10^{25} m^{-3}). At high densities/pressures, the coulomb interaction of the outer bound electrons of the neutral and ionized metal vapor with the surrounding charged particles (the ions and free electrons) leads to a substantial shift in energy levels and a lowering of the electrons' binding energy. The result can be a sharp increase in the ionization state of the plasma; see, for example, Anders [28] and Ebeling et al. [29]. The initial plasma can have an ion current density ~ 10^{10} Am^{-2}, which will drop to ~ 10^{7} Am^{-2} as an efficient electron emission mechanism is established at the cathode. During this initial pseudo-arc phase, the electrons from the cathode that are required for current continuity will result from secondary emission caused by ion impact at the cathode. As the metal vapor in this region expands a normal arc will be established, which will have a cathode region generating electrons and the usual neutral plasma between the contacts enabling current to flow between them. Thus, conditions are such just after the rupture of the molten metal bridge that ionized plasma forms (pseudo-arc or a high-density, low-voltage glow discharge) that will maintain the flow of current across the contact gap. This initial pseudo-arc will quickly transition into the usual arc discharge with a voltage impressed across the contacts whose value is about that of the minimum value U_{min} (≈ 15 – 25V) for the contact metal. The whole sequence is illustrated in Figure 2.12 for currents up to a few thousand amperes.

Slade et al. [25] have taken high-speed streak photographs of contacts opening in vacuum. Figure 2.13 shows the typical opening sequence of Cr contact carrying 1000A. The molten metal bridge is observed once its temperature is high enough for incandescence. After the bridge ruptures, the voltage across the contacts increases to about 18V as the high-pressure, non-ideal plasma forms. The voltage then drops to about 17V after perhaps 100ns and an arc is formed. Initially, there is sufficient metal vapor in the region of the molten metal bridge for the arc to have a columnar structure. I shall call this

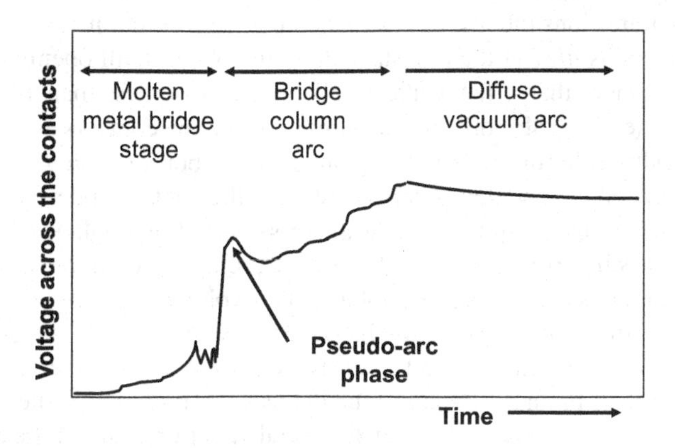

FIGURE 2.12 The voltage drop across opening contacts in a low-current circuit showing the rupture of the molten metal bridge and the formation of the vacuum arc.

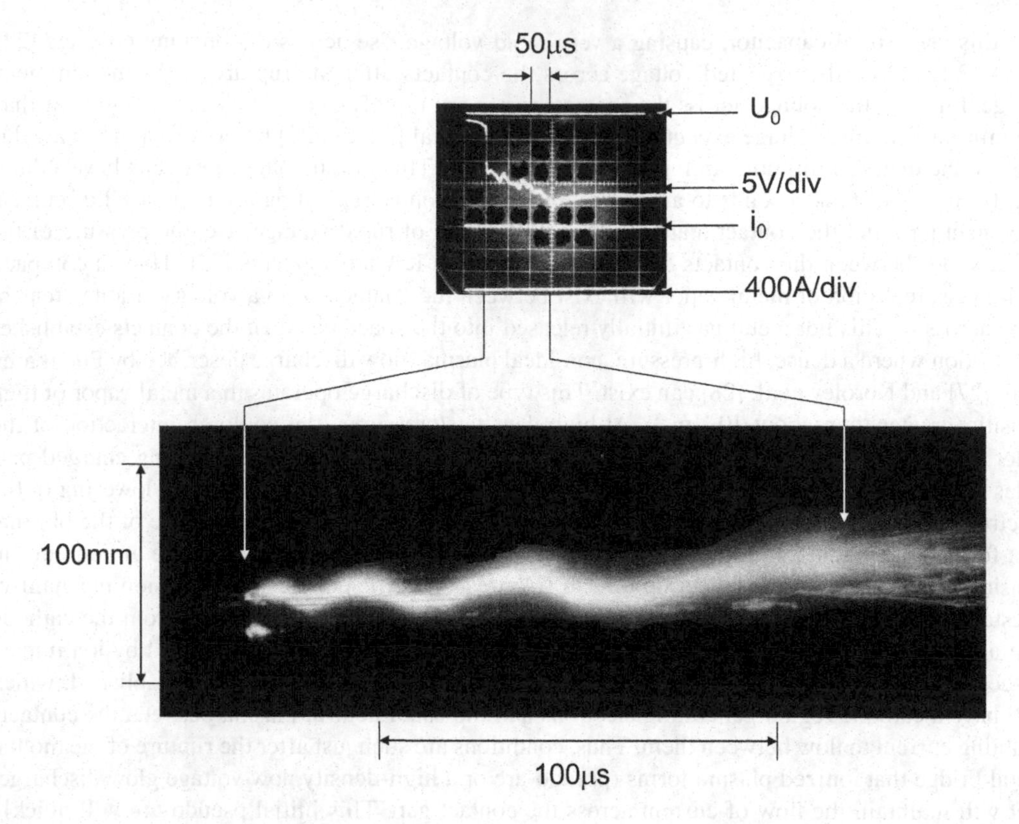

FIGURE 2.13 Streak photograph of the formation of the bridge column arc and then the diffuse vacuum arc between opening Cr contacts [25].

arc *the bridge column arc*. This arc has a pressure of about one atmosphere and has a cathode region similar to that of an arc in air at atmospheric pressure. In an important publication Logachev et al. [31] presented the opening sequence for Cu–Cr contacts for rectangular current pulses 0.5kA to 15kA. They showed that the bridge column arc always formed after the rupture of the molten metal bridge before it transitioned into one of the forms of true vacuum arc. Figure 2.14 shows the voltage across the opening Cu–Cr contacts for a dc current of 4.3kA. After the rupture of the molten metal bridge the characteristic voltage rise of the bridge column arc as shown in Figure 2.13 is seen. In this example, the bridge column arc transitions into the diffuse vacuum arc (see Section 2.3). It is important to note that the bridge column arc is *always* the first stage of arcing between all opening contacts in circuits with a circuit voltage of more than a few volts, for all currents from less than 1A to very high values and for all atmospheres (e.g.,air, SF_6, oil, and vacuum) in which the contacts are placed [24].

In vacuum, this bridge column arc is initially quite stable, but it eventually transitions to a true "vacuum arc" after a period of time that is dependent upon the contact material and the current. The stability of the bridge column arc requires that the arc roots provide enough metal vapor to replace the metal vapor continually being lost from the high-pressure arcing region into the surrounding vacuum and to accommodate the arc's ever-increasing volume. This volume increase has two components: (a) the continued parting of the contacts gradually increases its length and (b) the gradual radial expansion of the current flowing in the cathode and anode regions from the initial confined passage limited by the molten metal bridge. Figure 2.15 shows the decrease in pressure of the drawn arc shown in Figure 2.13. In this case it can be seen that when the metal vapor pressure in the arc decreases below about one half of an atmosphere, the column arc can no longer be sustained. Therefore, at 1000A the evaporation of material from the arc roots is not enough to maintain the high-pressure column.

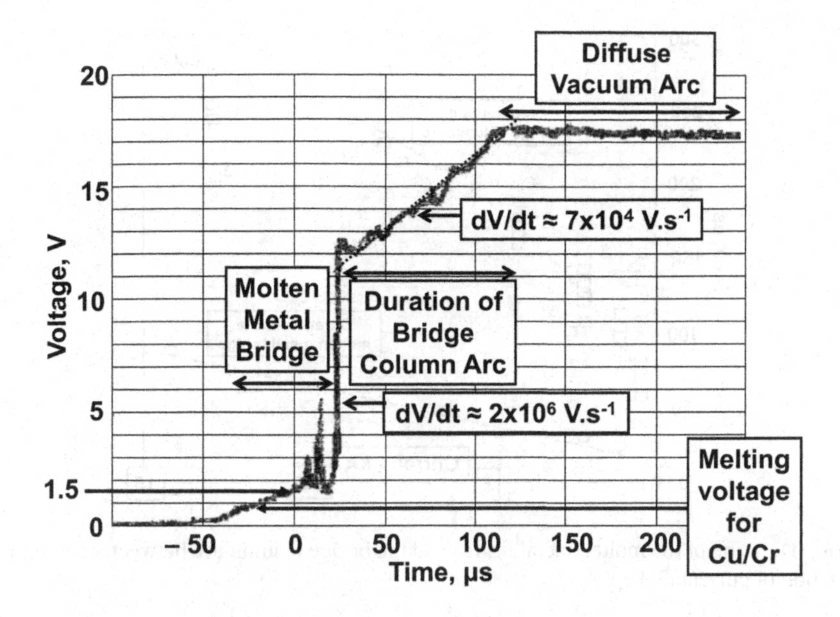

FIGURE 2.14 The voltage across Cu–Cr contacts as they open a 4.3kA current showing the formation of the bridge column arc and its transition into the diffuse vacuum arc [30].

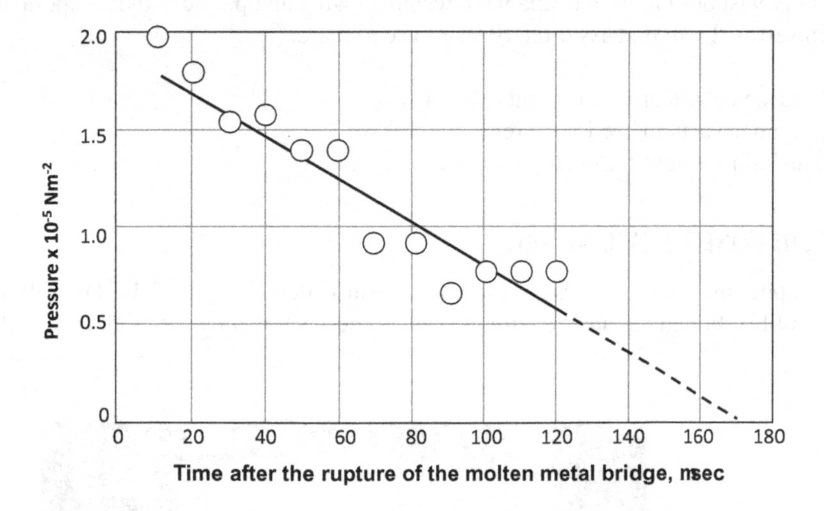

FIGURE 2.15 The pressure as a function of time of the bridge column arc shown in Figure 2.13 [25].

Figure 2.16 shows the duration of the molten metal bridge stage and the duration of the bridge column arc as a function of current [31]. As the opening current increases so the molten bridge stage column duration increases. Also, the bridge column arc is stable for longer time periods. Above about 10kA the bridge column arc transitions into the high-current columnar arc (see Section 2.4). Therefore, when the current at the contact parting is high enough, the high-pressure columnar arc will be maintained, no matter what the contact gap. It is critical for the vacuum interrupter designer to understand the contact opening sequence and the initial formation of the bridge column, because the subsequent development of the different vacuum arc formations is highly dependent upon it.

All arcs that initiate between opening contacts form in metal vapor from the molten metal bridge and are thus metallic arcs. They all eventually transition into arcs that operate in the ambient

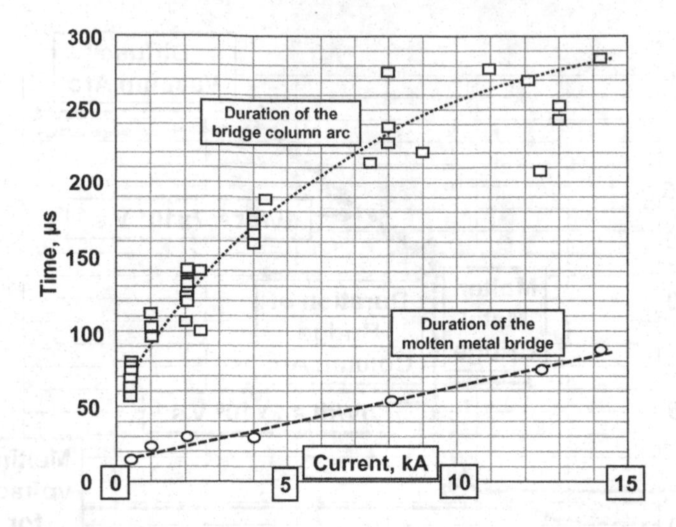

FIGURE 2.16 Duration of the molten metal bridge and the bridge column arc between opening Cu–Cr contacts as a function of current [31].

atmosphere surrounding the contacts. Thus, a contact opening in air will transition from an arc burning in metal vapor to one burning in air [22]. Similarly, an arc burning between contacts opening in SF_6 will transition to an SF_6 arc. In vacuum, the "vacuum arc" continues to be a metal vapor arc, i.e., the only gaseous medium available in which the arc can operate is that evaporated from the contacts themselves. I will discuss three types of vacuum arc:

(a) The diffuse vacuum arc for currents ≤ ~ 6kA
(b) The columnar vacuum arc for currents ≥ ~ 10kA
(c) The transition vacuum for currents ≥ ~ 6kA ≤ ~ 10kA

2.3 THE DIFFUSE VACUUM ARC

The general appearance of the diffuse vacuum arc is illustrated in Figure 2.17. The cathode surface has a number of bright regions named *cathode spots*. These spots move at random over the cathode

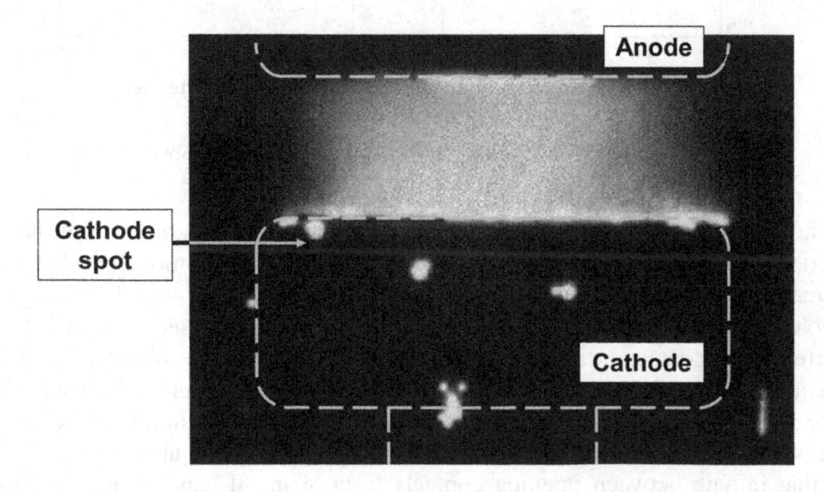

FIGURE 2.17 The diffuse vacuum arc.

surface and they generally repel each other: i.e., move in a retrograde direction. Between the top surface of a cathode spot and the anode contact there exists a diffuse neutral plasma that permits the circuit current to flow with a very small potential drop. The anode contact is generally passive and serves to collect electrons over its entire surface. In this section I will first discuss the properties of cathode spots. I will then describe the neutral plasma that exists between the cathode spot and the anode contact. Following this the stability of the cathode spot at low currents and the phenomenon of current chop will be examined. Finally, I will present a brief overview for the formation of two anode spot types.

2.3.1 Cathode Spots

In this section I will present a discussion of the interesting world of the cathode spot. For a very detailed review of our present knowledge of this phenomenon I refer the reader to the excellent surveys by Jüttner et al. [32, 33], Mesyats [34], Hantzsche [35], and Anders [36]. It is generally agreed that there are two characteristic cathode spots: Type I and Type II. Type I cathode spots are found on oxidized or contaminated contacts [25, 32]. Even thick oxide layers (perhaps a few μm) will be removed by cathode spots moving randomly over them [25, 37]. The contact surfaces in a vacuum interrupter are cleaned during the manufacturing process and by the vacuum arc each time they switch current. Thus, for the vacuum interrupter designer only the Type II cathode spot needs to be considered. Table 2.4 gives the typical characteristics of Type II cathode spots. Cathode spots have a maximum current level they can support before another companion cathode spot is formed: for Cu contacts this is about 100A. Table 2.5 gives values of the current per cathode spot for a number of contact metals. This table shows that the current per cathode spot is not an exact value. Beilis et al. [38], for example, show for a 350A vacuum arc between Cu contacts, the number of spots can range

TABLE 2.4
Characteristics of Type II Cathode

Contact Condition	Cleaned by arcing
Velocity	$1ms^{-1}$–$100ms^{-1}$
Spot Current (Dependent upon the Vacuum Interrupter's contact materials)	30A – 300A
Lifetime	~0.1ms
Crater Appearance	Overlapping
Crater Radius	40μm–100μm
Erosion Rate (Vacuum Interrupter Contact Materials)	$10\mu gC^{-1}$–$100\mu gC^{-1}$

TABLE 2.5
The Average Current per Cathode Spot

Cathode Material	Spot Current
Bi	3–5
C	200
Cr	30–50
Cu	75–100
Ag	60–100
W	250–300

from 2 to 6. However, the number with the highest probability of occurring is 2 or 3. Even though at 1000A you would expect about 10 cathode spots on Cu contacts, their data show only 5 to 6. The cathode spot current on the Cu matrix of a Cu–Cr contact is tens of amperes, but when it is located on a Cr grain the cathode spot current is only a few amperes [39, 40]. For a complete physical understanding of the cathode spot processes an explanation of this statistical distribution of current per cathode spot is important. For the development of a new vacuum interrupter design, however, the exact current per cathode spot is of little interest. Most designers assume that all contact materials have a cathode spot current of about 100A.

A cathode spot on a contact operated in a vacuum interrupter environment has a gross diameter between 40μm and 100μm, giving a current density of between 10^{10} Am^{-2} and 10^{11} Am^{-2} for a cathode spot carrying 100A [41]. Many researchers have observed several subspots, or cells within the cathode spot [42, 43]. The current density in these cells can be extremely high, $(1–10) \times 10^{12}$ $A.m^{-2}$. It is most probably the interaction of these cells within each cathode spot that determines the exact current that a given cathode spot will carry. That is, as the number of cells within a cathode spot varies, so will the current that it carries. These cells have been observed to move extremely rapidly over a short distance with speeds up to 60 ms^{-1} [44]. The cathode spot has a finite lifetime of up to 100μs. The individual cells, however, can extinguish and reappear in a time duration as short as 10ns to 20ns. As a cathode spot extinguishes, a new one is formed. For typical vacuum interrupter contact materials, the new cathode spot forms on the rim of the one it is replacing [45]. A single cathode spot moves with a "random-walk" motion over the contact's surface with a speed of about 1 ms^{-1} [46]. One model of the cathode spot that has been used to explain the spot motion is the continuous extinction and reforming of the individual cathode spot cells. To explain the spot's overall motion, the explosive cathode processes used as a model in Chapter 1, this volume, to explain the vacuum breakdown phenomenon, can also be evoked; see Mesyats [34, 47] and Barengolts et al. [48, 49]. In this model, as the cathode spot momentarily resides in one place, there is a gradual increase in the cathode crater size and an associated decrease in current density. The result is an increase in arc voltage and the initiation of a new spot on a high field region, usually on the rim of the original crater. For an individual cell within a cathode spot this occurs typically in a time frame of about 10ns. Certainly, measurements of arc voltage of low-current vacuum arcs show rapid variations in arc voltage within this time frame and high-speed photographs of cathode spots show corresponding changes in the luminosity of the cathode spot region [43, 50, 51]. Also, the ejection of metal particles from the cathode spots primarily in the plane of the cathode [52–56] is quite consistent with the explosive formation of the cathode spot. Siemroth et al. [56] show in Figure 2.18 that while

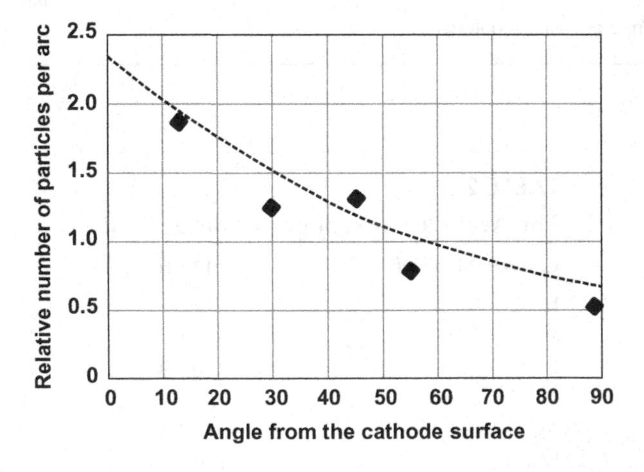

FIGURE 2.18 The average angular distribution of ejected metal particles from the cathode spots [56].

the peak of the metal particle flux is in the plane of the cathode contact, a good number are ejected in a forward direction.

For an ac current, the number of spots will gradually increase as the current increases to its maximum value and then decrease as the current decreases toward zero. Sometimes the number of cathode spots increase by the splitting of existing spots, but their number can also increase by the spontaneous generation of new cathode spots. These multiple cathode spots repel each other (this is called "retrograde motion") and move rapidly over the cathode surface with speeds of up to about $100 ms^{-1}$ for a 1000A arc. In a typical vacuum interrupter, the contacts are the same; i.e., the anode and the cathode have the same shape and the same diameter. With such contact pairs the cathode spots are frequently observed to move down the side of the cathode contact. Here they may extinguish. If this happens, new cathode spots are generated on the contact's face to take their place. In fact, after testing vacuum interrupters in practical switching circuits, it is common to observe cathode spot tracks running down the contact supports and even on the shield that surrounds the open contacts. In experimental systems, however, it is possible to limit the cathode spot excursion and to prevent most of them from traveling over the edge of the cathode contact. To achieve this the anode must have a diameter appreciably smaller than that of the cathode. Heberlein et al. [57] have observed with such an experimental arrangement that the cathode spots still repel each other, but their travel is limited to a circle on the cathode that has a diameter only slightly larger than that of the anode. When the cathode spots reach the edge of this circle they continue to move, but now they travel the periphery of this circle.

There is no consensus on why the cathode spots move with a retrograde motion. Many models have been proposed, but as of this date, none has been universally accepted by the research community. Jüttner et al. [32] give a nice review of these models pre-1995. Benelov et al. [58], in a more recent survey, analyze three possible mechanisms that have been proposed to explain the retrograde motion. They conclude that a first principle understanding of the retrograde motion remains elusive. They do, however, offer a phenomenological description of the observed motion. The appearance of the contact surface after arcing shows many fine cathode tracks spread over the whole cathode surface [42, 44, 54]; see Figure 2.19(a). After many operations in a practical vacuum interrupter the erosion of the contacts is typically very uniform. High magnification of these tracks shows that these consist of overlapping craters similar to those seen with a single cathode spot; i.e., as the cathode spot moves it reforms on the crater of the old cathode spot [59, 60]: see Figures 2.19(b) and 2.19(c). Thus, the retrograde motion of multiple cathode spots is also achieved by many small jumps, which results in new cathode spots forming on what appears to be the rim of the old cathode spot crater.

Quantitative modeling of the cathode spot is an extremely complex problem. A successful model of the cathode spot requires knowledge of:

- The energy balance at the contact surface
- The electrical and thermal conductivities as a function of temperature in the contact region under the cathode spot
- The effect of metal evaporation
- The change in the contact metal's work function
- The effect of electron emission by the space charge region established very close to the cathode surface
- The pressure of the evaporated metal atoms
- The ionization of the metal atoms and the flow of material from the high-pressure region close to the contact surface into the surrounding vacuum

Each of these effects will be nonlinear at the extreme temperatures and pressures expected within the micrometer dimensions of the cathode spot. Hantzsche and Beilis [61] and Bellis [62] present an excellent review of the cathode spot model with the governing analytical equations for each region within it. Fortunately for the vacuum interrupter designer a detailed quantitative understanding of

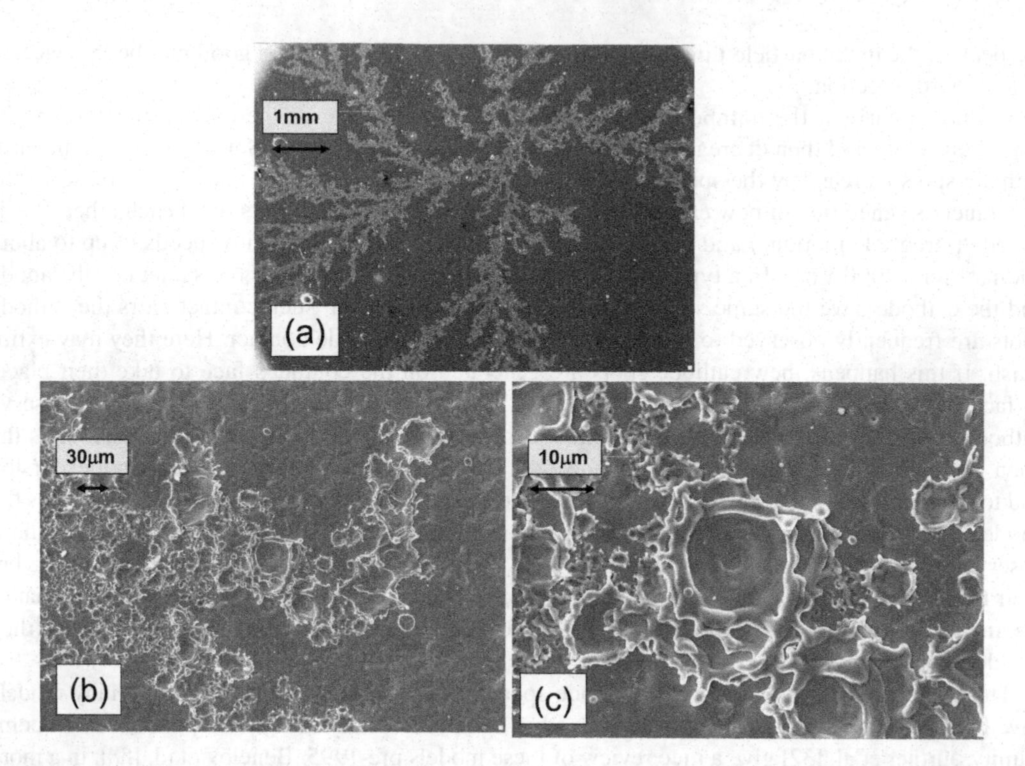

FIGURE 2.19 A cathode spot track showing the variation in crater diameter and the formation of new craters on the rim of existing craters.

the cathode spot operation is not necessary. A general qualitative picture of a mature cathode spot's operation is, however, useful.

A quasi-stationary model of the cathode spot is illustrated in Figure 2.20 [61–64]. Of course, the boundaries shown in Figure 2.20 are just for illustration. In practice each region will overlap adjacent regions. The contact surface immediately below the cathode spot is molten. Its temperature will be in the region of 3500K to 5500K [39], depending upon the metal from which the contact is made. The energy to sustain this temperature results mainly from the flow of ions from the dense plasma just above the cathode surface and from Joule heating by the high-density circuit current flowing into the cathode spot. The actual energy balance at the cathode surface under the cathode spot is, in fact, extremely complex. Cobine [65] has developed a guide to this complexity, which is illustrated in Figure 2.21. The conditions are such at the contact surface that metal atoms will be readily evaporated and move away from it in a substantially uniform direction perpendicular to it. When a positive space charge is established just above the cathode surface, the resulting high field (in the range of 2×10^9 Vm^{-1}) and the high temperature of the cathode surface result in the copious emission of electrons by an ion enhanced, T– F emission [63]. The electrons and the neutral metal atoms travel through this cathode fall region without making collisions, i.e., it is a ballistic zone with a thickness in the range of 7.5×10^{-9}m. The strong electric field accelerates the electrons away from the cathode and also accelerates metal ions back toward the cathode. The metal atoms will collide with each other and with the returning metal ions in the metal gas relaxation region. The electrons will also begin to collide with the metal atoms in this metal gas relaxation region, but they will eventually move into a zone where considerable ionization of the metal atoms occurs. This metal atom relaxation region and ionization region just above the cathode has a very high pressure (perhaps five atmospheres) and is extremely thin between 10^{-6}m to 10^{-7}m. In this zone it is generally assumed that 100% of the neutral metal atoms are ionized. Some of them will be doubly charged or

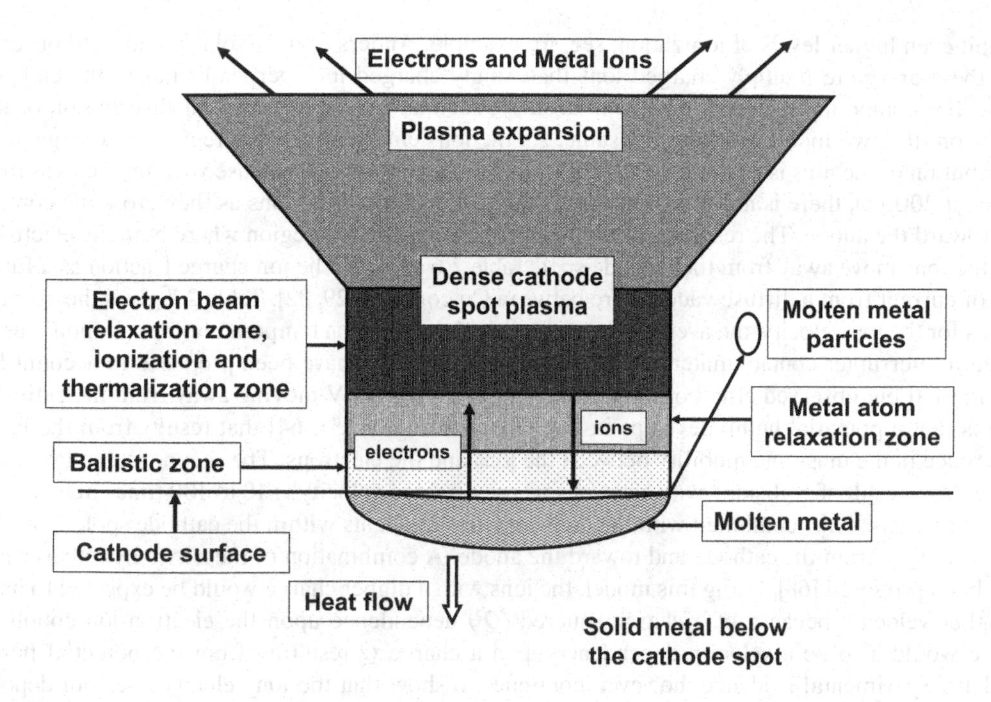

FIGURE 2.20 A model of the cathode spot showing its complexity.

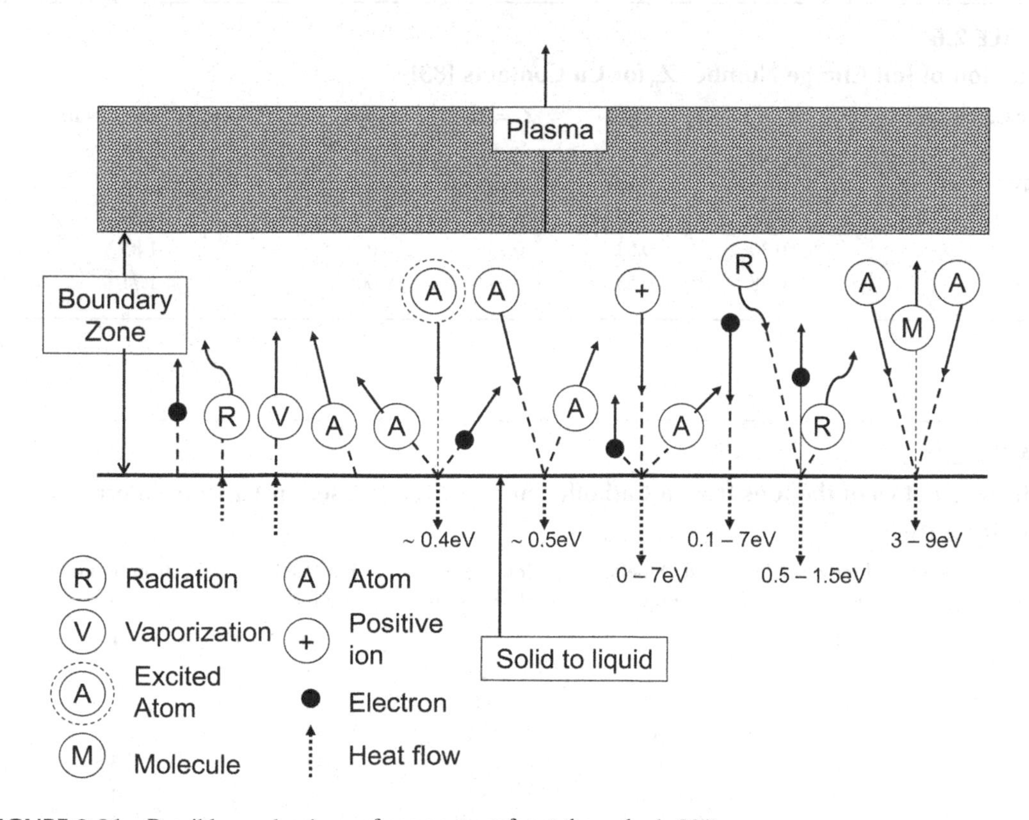

FIGURE 2.21 Possible mechanisms of energy transfer at the cathode [65].

exhibit even higher levels of ionization; see, for example, Anders [29]. Nikolaev et al. [66] observe that there are more multiply-charged ions than singly charged ions perpendicular to the cathode spots. For compound materials with similar mass, such as Cu–Cr, the angular distribution of the ions from the two metals is about the same. As the ions emerge from this region the charge state distribution of the ions is essentially "frozen": i.e., for the low-current diffuse vacuum arc (less than or about 2000A), there is no change in the charge distribution of the ions as they cross the contact gap toward the anode. The resulting plasma exits into an expansion region where both the electrons and the ions move away from the cathode spot. Table 2.6 presents the ion charge fraction as a function of current from a diffuse vacuum arc between Cu contacts [29, 32]. Table 2.7 gives the average values for the ion velocity, the average charge state and the electron temperature for commonly used vacuum interrupter contact materials [29, 67]. Two hypotheses have been proposed to account for the unusual but observed effect of ions with energies up to 100eV moving away from the cathode. One is that a potential hump develops in the ionization region [53, 64] that results from the large difference in the mass and mobility between the ions and the electrons. The second is that ions gain energy as a result of collisions with electrons whose directed velocity is 10 to 100 times higher than that of the ions. This, combined with the high-pressure gradients within the cathode spot, drive the ion flow away from the cathode and toward the anode. A combination of these two hypotheses has also been proposed [68]. Using this model, the ions with a higher charge would be expected to have a higher velocity, because of a charge squared (Q^2) dependence upon the electron-ion coupling. There would also be a velocity dependence upon a charge Q resulting from the potential hump model. Experimental evidence, however, continues to show that the ion velocity does not depend strongly upon the ion charge. Tsuruta et al. [69] have observed little dependence of ion velocity on ion charge for vacuum arcs between Cu and Ag contacts. Yushkov et al. [70] have shown a similar

TABLE 2.6
Fraction of Ion Charge Number Z_n for Cu Contacts [83]

Arc Current A	$Z_1 = 1$	$Z_2 = 2$	$Z_3 = 3$	$Z_4 = 4$	Average Charge Number
100	0.16	0.63	0.2	0.01	2.06
200	0.44	0.42	0.14		1.7
200	0.26	0.49	0.25		1.99
100	0.30	0.53	0.15	0.004	1.85
100	0.38	0.55	0.07	0.005	1.71

TABLE 2.7
Characteristics of the Ions from a Cathode Spot for Metals Used in Vacuum Interrupter Contacts [67]

Metal	Ion velocity, $\times 10^4$ ms^{-1}	Ion kinetic energy, eV	Ion average charge state	Ion Mach number	Electron temperature, eV
Bi	0.42	19	1.2	3.3	1.8
Ag	1.04	61	2.1	3.9	4
C	2.97	54	1	5.2	2
Cr	1.94	101	2.1	5.5	3.4
Cu	1.28	54	2	4.0	3.5
W	1.05	106	3.1	5.0	4.3

result for nine further metals. Byon et al. [71] have shown, in studies of the ion velocity distribution coming from a cathode spot for 20 metals, that there is a strong peak at one velocity. While there seem to be some minor peaks at higher velocities these cannot be interpreted by the measured ion distribution or by surface contamination of lighter ion species such as oxygen. Figure 2.22 shows their ion velocity distribution for a Cr cathode. It seems as if the ion acceleration does not result from a potential hump within the cathode spot. The potential hump hypothesis does, in fact, contain a fundamental problem [67]. Such a potential hump within the cathode spot would become a potential well for electrons. Electrons with an energy less than kT_e (where T_e is the electron temperature) would quickly fill the potential well and thus reduce its depth to $U_{well} \approx kT_e/e$. The maximum value of U_{well} would then be only a few volts. This is far too low to explain ions with energies up to 100eV. Thus, it seems probable that the ion acceleration is caused by a force on the ions that results from the very high-pressure gradients within the cathode spot coupled to some extent with the electron-ion friction effect. Yushkov et al. [70] have shown for this model the limit on velocity of the ions (v_i^{lim}) leaving a cathode spot is:

$$v_i^{lim} = \mathcal{M} v_s \tag{2.30}$$

where $\mathcal{M}$ is the Mach number (i.e., the ion sound velocity [67]), $v_s = [\gamma \{Z\} T_e / m_i]^{1/2}$, $\gamma = 5/3$ is the adiabatic index, $\{Z\}$ is the average ion charge and m_i is the ion mass. The Mach number for a distance greater than 1mm from the cathode is its ultimate value $\mathcal{M} \approx 3.5$. Certainly, this is an acceptable distance from the cathode spot and a distance where an ion would have achieved v_i^{lim}, thus:

$$v_i^{lim} = \mathcal{M}_{lim} v_s \approx \left[20\{Z\} T_e / m_i\right]^{1/2} \tag{2.31}$$

Figure 2.23 shows their comparison of measured ion velocities for a wide range of metals and values of $\{Z\}$, which seems to present strong support for this model. Thus, while the potential hump

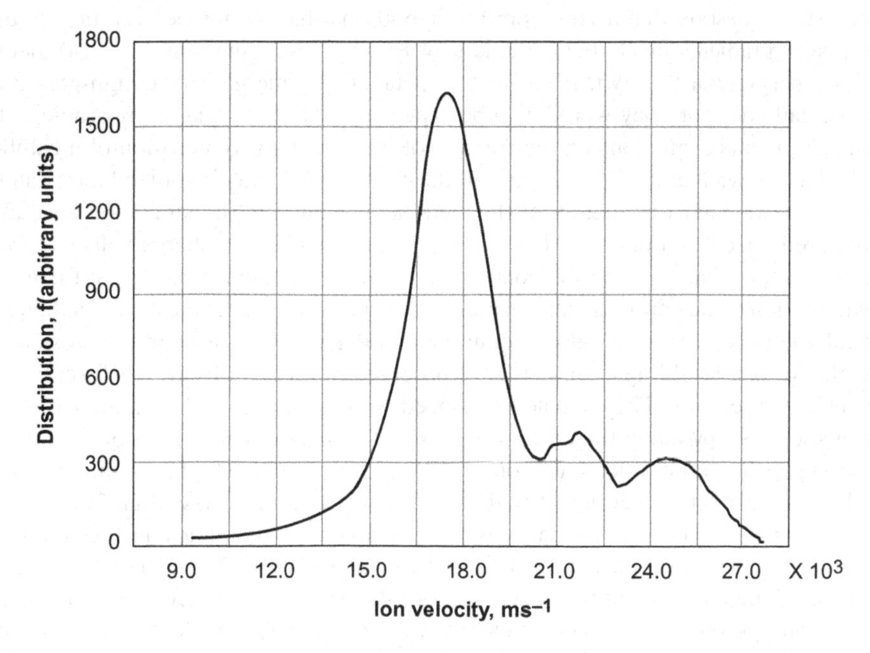

FIGURE 2.22 Distribution of ion velocities from the cathode spot for a vacuum arc with a Cr cathode for a current of 100A [71].

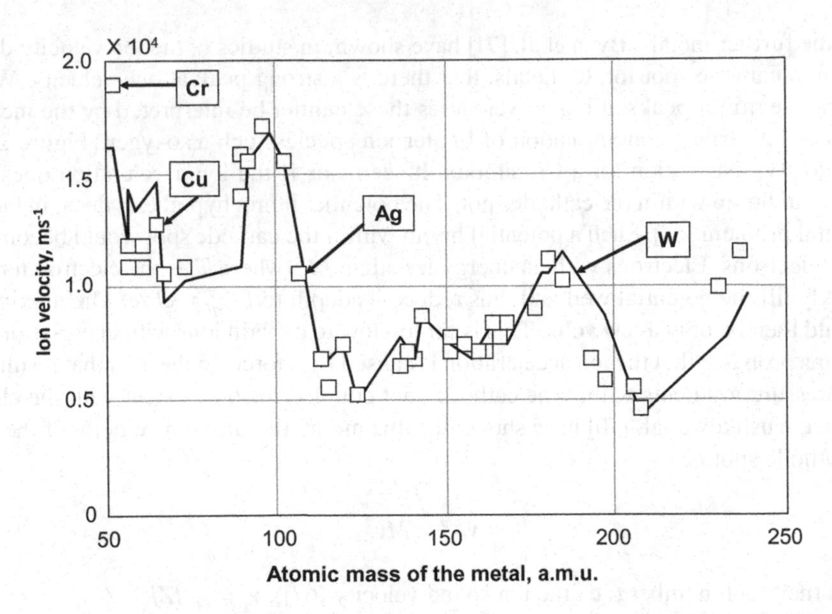

FIGURE 2.23 The experimental (□) and calculated (————) values from the gas-dynamic theory for the velocity of the ions from a cathode spot for different metals [70].

hypothesis gives a qualitative explanation of the unexpected high-energy flux of ions moving away from the cathode, the experimental evidence does not really support it.

Models of the cathode spot region continue to be developed. For example, Shmelev et al. [72] differentiate between a quasi-stationary cathode spot with a radius of about 100μm and a fast-moving cathode spot of radius about 1μm. They show that strong current instabilities create conditions for the acceleration of ions both toward the cathode and toward the anode.

One interesting question that arises from the hypothesis that the ion acceleration results from the high-pressure gradients inside the cathode spot is: why do the measured neutral metal atoms have such low energies (≈1 to 2 eV)? If the neutral metal atoms emerge from a high-pressure region just above the cathode spot, they would also be expected to have energies comparable to those of the ions. Indeed, from calculations on the fraction of ions from a given erosion of a cathode spot, Kimblin [73] has shown that for Cu only 55% of the eroded Cu results in ionized metal atoms. The answer to this question must be that all of the metal atoms entering the ionization zone above the cathode spot are ionized. Not only are they ionized, but they also have charge values Q from +1 to +4. The neutral vapor that exists in the contact gap above the cathode spots must therefore come from the evaporation of the particles that can account for up to 80% of the measured cathode erosion. Kimblin's calculations do not take into account the dominance of particles in the erosion products from the cathode spot. Methling et al. [74] have observed the spectra from a diffuse 2kA vacuum arc between Cu–Cr contacts. They couple high-speed, video imaging with spatially-resolved, optical emission spectroscopy along the contact gap axis. The lateral distribution of ionic lines has a maximum mid gap. The ratio of ionic to atomic spectral lines is much higher in mid gap than close to the anode. The calculated energies of both the Cu and Cr atoms is less than 2eV. As might be expected, the plasma layer close to the anode is not in local thermodynamic equilibrium (LTE). The ratio of electron temperature to neutral atom temperature is about 1.3. Anders [75] has suggested that the potential hump theory can be reconciled with the gas-dynamic theory by considering traveling potential humps. Each of these being associated with an explosive electron emission event.

You can see from this description of the cathode spot that there can be a huge variability of its measured parameters. Among the examples that have been identified by Anders [36] are:

1. The cathode spot's random walk
2. The resulting trace of the random walk on the cathode's surface
3. The nonlinear ion charge state
4. The a-periodic arc voltage fluctuations
5. The chaotic velocities of the cathode spots
6. The power law distribution of the microparticle flux from the cathode spots

He has shown that these many varying features of the cathode spots can be analyzed and can be described by using the theory of fractals [76, 77]. This interesting insight into the behavior of the cathode spot will certainly result in a more comprehensive understanding of all the processes involved with its formation and operation.

There has also been some discussion (see, e.g., [28, 30]) concerning the early phase of a cathode spot's life being in the form of the pseudo-arc that I discussed earlier in the transition from the molten metal bridge to the arc between opening contacts. Again, for designers of practical vacuum interrupters complete analysis of the chaotic nature of the cathode spot and knowledge of its exact description while of considerable interest will not usually affect the overall performance of their designs.

2.3.2 The Plasma between the Cathode Spot and the Anode

At first sight, the plasma above the cathode spot in the intercontact space appears as a diffuse, barely luminous, neutral plasma that serves solely to conduct the arc current from the anode to the cathode. The anode acts as a passive collector of electrons. If we examine this region more closely, we find that the physical processes occurring within it do, in fact, affect the nature of the discharge. A general understanding of these effects and their impact on vacuum interrupter design as the circuit current increases to values of a few kiloamperes is useful. Goldsmith [78] gives a comprehensive review of this plasma. For close contact spacings of 1mm–2mm there is practically no voltage drop across the plasma, i.e., all the arc voltage is concentrated across the cathode spot, as is illustrated in Figure 2.24. In fact, for the large diameter contacts usually found in vacuum interrupters, the total arc voltage of the diffuse vacuum arc is only in the range of ~20V to ~45V for currents up to 3150A and for typical contact spacing of 8mm to 20mm.

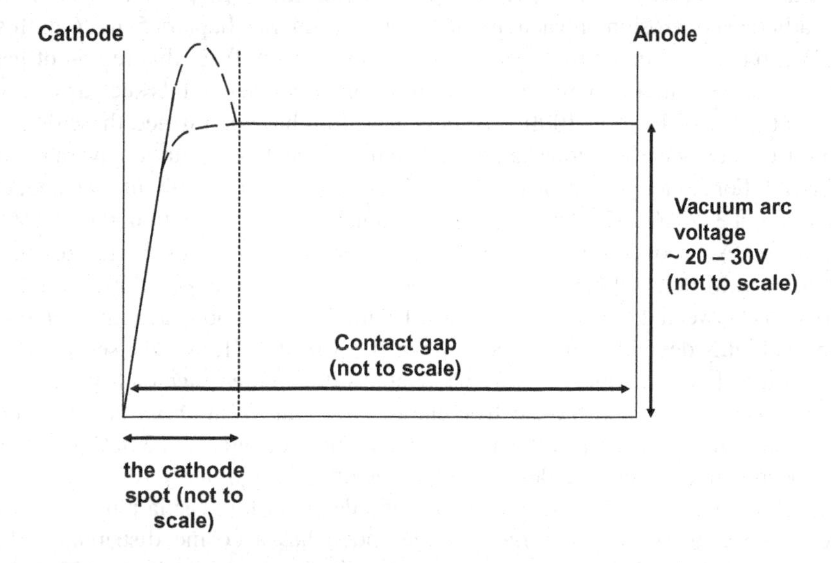

FIGURE 2.24 The arc voltage across a diffuse vacuum arc with closely spaced contacts.

TABLE 2.8
Cathode Spot Erosion Rates for Metals Used in Vacuum Interrupter Contacts

Material	Erosion Rate μgC^{-1}
Ag	140, 150
Cu	115, 130
Cr	22–27, 40
W	55, 62, 64

As I have discussed above, all the material in the intercontact space comes from the cathode spots. The absolute erosion from a cathode spot has been measured by many investigators. Table 2.8 gives the erosion rates for commonly used pure metals that are found in vacuum interrupter contacts. The commonly used parameter for vacuum arc erosion is "grams per coulomb," i.e., the number of grams lost by the cathode as a function of the current times the arcing time; i.e., $\int i(t)dt$. The bulk of the erosion has been shown to be particles [52, 53, 56, 79], perhaps as much as 80%. This is most probably the result of the violent formation of the cathode spots as they move over the cathode surface [34, 49, 55]. These metal particles have diameters that range from less than a micrometer to a few tens of micrometers. They travel with a velocity from 10 to 800 m/sec and they have a temperature above the melting point, but below the boiling point of the cathode material. Thus, for Cu contacts the metal particles have a temperature about 2500K. The remainder of the eroded cathode material is metal vapor, which becomes completely ionized in the ionization zone immediately above the cathode. Figure 2.25 shows an example for W–Cu (10 wt%) contacts with a 600A vacuum arc [80]. In general, most of the metal particles are ejected horizontally in the plane of the cathode. In a vacuum interrupter they are deposited on the vacuum interrupter's shield opposite the open contact gap [79]: see Figure 2.26. Theoretical modeling of such particles moving through the high-pressure plasma region immediately above the cathode spot shows that significant and perhaps even total evaporation of such a particle is possible. Particles from a high-temperature material such as W can evaporate even faster than those from lower melting metals [81]. I will discuss the implications of the cathode spot erosion on vacuum interrupter design in Chapters 5 and 6 in this volume.

Kimblin [82] has found an approximately constant ion current from this region of between 7% and 12% of the total current for common vacuum interrupter contact metals such as Cu, Ag, Cr, and W in the current range of 100A to 1000A. Anders et al. [83] have confirmed these data. As shown in Table 2.6 for Cu contacts these ions can have a charge as high as 4, but, on average, they have a charge of about 2. For Ag and Cr, the average ion charge number is also about 2 [29, 53, 66, 84–86] and for W it is about 3 [29, 67, 87]. These ions have a much higher energy than the ~2 eV one would expect from a 1000A arc operating between Cu contacts in air at atmospheric pressure [88]. For example, Davis and Miller [89] have measured ion energies in the range of 50eV for Cu contacts. The ion velocity is between 0.4×10^4 ms^{-1} and 3×10^4 ms^{-1} for common vacuum interrupter contact materials and is highly dependent upon the mass of the ions [67, 71, 90, 91]; see Figure 2.23. The electrons have much lower energies, e.g., for Cu contacts it is in the range of 1eV–4eV [29, 67], but other researchers have measured the electron energies ranging from about 2eV to about 6eV for some contact materials. For example, the average value for Cu contacts is 3.5eV [67]. The electron density is proportional to the current density in this region; approximately 10^{21} at 10 A/mm^2 [61].

The angular distribution of the ions above the cathode spot (the plasma plume) for low-current vacuum arcs (consisting of two or three cathode spots) has a cosine distribution [57, 90]. At higher currents (with 12 to 20 cathode spots) this distribution becomes flatter [57, 90]. However,

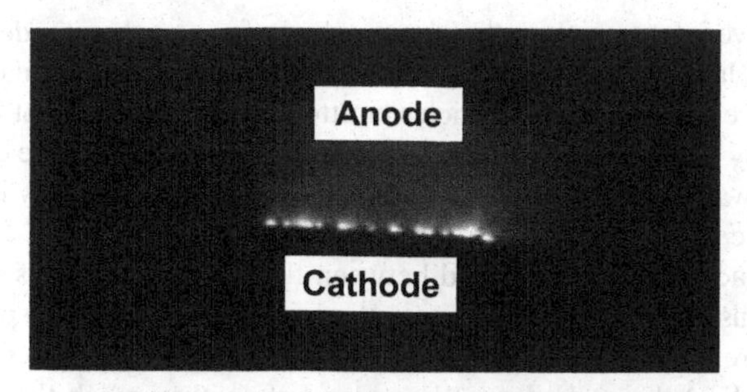

(a) A 0.1ms exposure of a 800A, diffuse vacuum arc, with a neutral density filter, showing the cathode spots

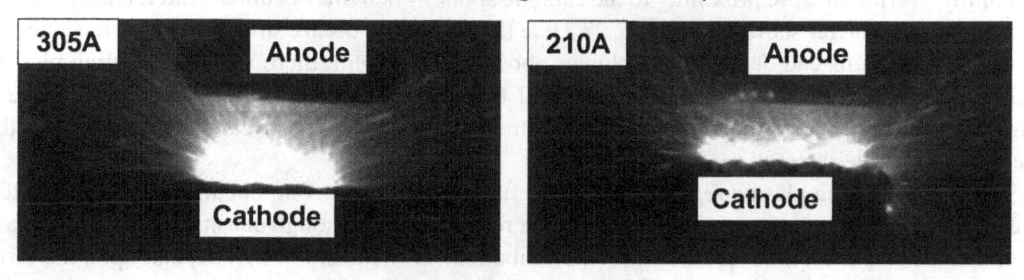

(b) A 0.1ms exposure of the diffuse vacuum arc, with no density neutral density filter, showing the tracks of incandescent W particles

FIGURE 2.25 Incandescent W particles emitted from the cathode spots of a diffuse vacuum arc between W–Cu (10 wt%) contacts [80].

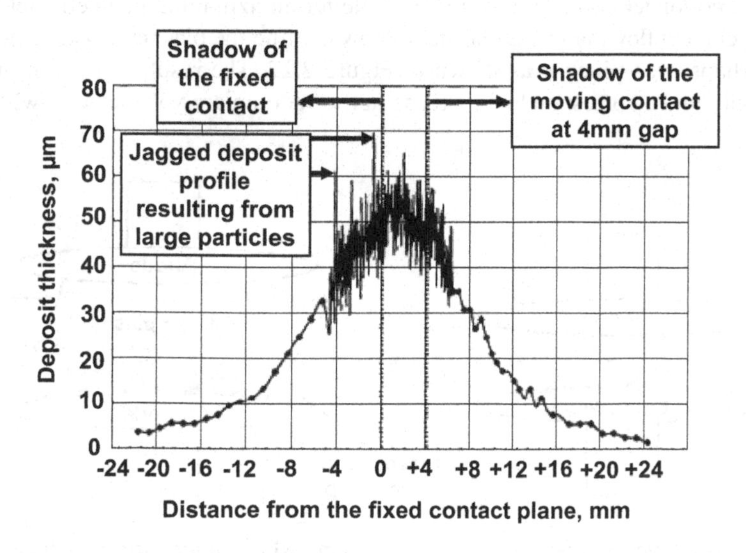

FIGURE 2.26 Thickness profile of the deposit on the shield of a vacuum interrupter after interrupting 400A ac, 52K times. The center jagged portion of the profile is from tracing a collage of optical images of the actual deposited layer [79].

for simplicity it is convenient to picture the plasma plume from each cathode spot located within a solid angle perpendicular to the cathode surface, subtended by a cone with angle of about 70° [66, 91]. As shown in Figure 2.18, the microparticles emitted by the cathode spot have a completely different distribution; they are mostly ejected at an angle less than 30° from the cathode surface. Some of them do absorb forward momentum from collisions with the high-energy ions. Most of the mass flow is carried by these heavier particles.

Once the diffuse vacuum arc has formed between the opening contacts of a practical vacuum interrupter, there will usually be more than one cathode spot. Each of these spots will have a plasma plume above it. If there is a high enough density of cathode spots (i.e., the current is high enough for a given contact radius) these plumes will overlap before they reach the anode. As the current increases, the cathode spot density will increase. The result will be that the plasma plumes will eventually overlap in close proximity to the cathode spots. When this occurs the intercontact plasma will have the character shown in Figure 2.27. Here the cathode spots are still spread over the cathode surface, but now instead of individual plumes above each of them, there is an overlapping plasma region that stretches to the anode. The nature of this plasma will be different from that above a single cathode spot where the ions and electrons making up the plasma stream away from the cathode with only little interaction.

In this overlapping plasma region collisions between the plasma components would be expected as the plasma density increases. When the plasma reaches a high enough density, elastic and inelastic collisions will affect the local ionization distribution. This transition will depend upon the level of the arc current and the diameter of the contacts. Some of the microparticles ejected by the cathode spots will be evaporated in this plasma region. The metal vapor released can also become ionized and this too will affect the properties of the intercontact plasma. Rusteberg et al.'s data [92] in Figure 2.28 show one effect of this change in the nature of the intercontact plasma. Here the peak of the ion energy distribution decreases and the probability of the emergence of very high-energy ions also decreases as the arc current increases. In this experiment, using 60mm diameter contacts, the transition between a collisionless plasma and one where the plasma density increases enough for collisions to take place occurs at a current of about 1.8kA. As the plasma density increases and the plasma stream is no longer collisionless, it is possible for an azimuthal magnetic field, B, resulting from the circuit current flowing in the plasma, to now confine the plasma to a specific well-defined column and perhaps even pinch it as shown in Figure 2.27. The result can be an increase in the plasma's resistivity and a corresponding increase in the vacuum arc voltage. This will be discussed

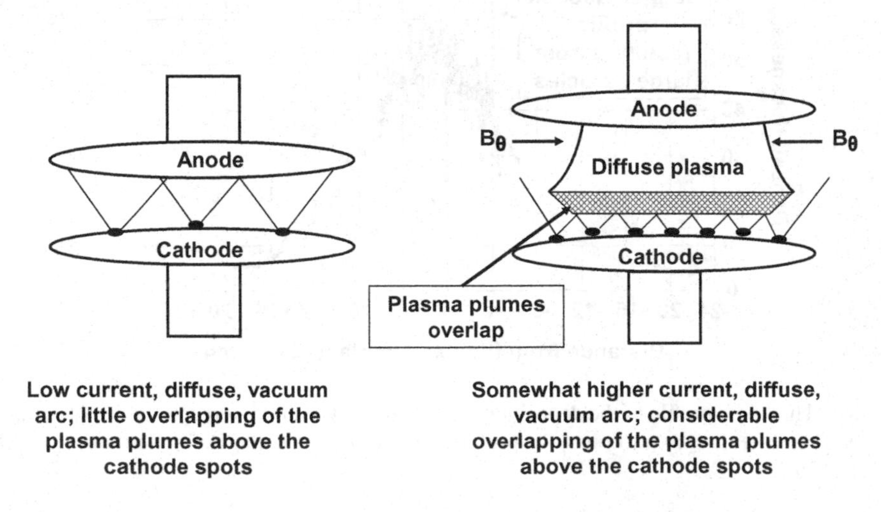
Anode
Cathode
Low current, diffuse, vacuum arc; little overlapping of the plasma plumes above the cathode spots
Anode
B_θ
B_θ
Diffuse plasma
Cathode
Plasma plumes overlap
Somewhat higher current, diffuse, vacuum arc; considerable overlapping of the plasma plumes above the cathode spots

FIGURE 2.27 The effect of the overlapping plasma plumes from the cathode spots as the current increases for a given contact diameter.

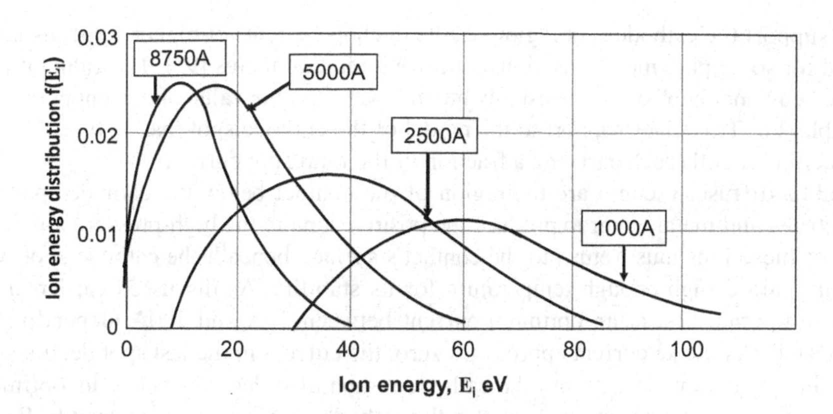

FIGURE 2.28 Ion energy distribution functions f(E_i) for different rms currents at 2ms before an ac current zero [92].

again in Section 2.6.3. As the current continues to increase to values between 6kA and 10kA, the interaction of the plasma plumes results in the development of the transition vacuum arc (see Section 2.5) where the cathode spots are confined to a well-defined region on the cathode and the plasma is a well-defined, low-pressure column.

2.3.3 Current Chop

For a practical vacuum interrupter in an ac circuit, the typical load breaking current ranges from about 100A (rms.) to 3150A (rms.). In Section 2.2 1 have already discussed the formation of the vacuum arc as the vacuum interrupter's contacts begin to part. Initially, a molten metal bridge is formed at the last point of contact. Once this bridge ruptures, a high-pressure bridge column arc develops whose duration depends upon the circuit current (see Figure 2.16). As the contacts continue to part and the erosion of the contact material at the arc roots can no longer keep up with the loss of metal vapor from the expanding column, the vacuum arc transforms into the diffuse mode, with multiple cathode spots. As the ac current increases to its maximum value, the number of cathode spots also increases and the intercontact plasma also changes its character, as has been discussed above.

As the current then decreases toward zero the number of cathode spots also decrease. When a cathode spot extinguishes, the energy input into its last location on the cathode contact ceases immediately. The thermal time constant in the cathode spot extinction region is so fast (in the order of microseconds) that its temperature very quickly falls to that of the surrounding contact metal. The region of the extinguished cathode spot therefore no longer participates in the vacuum arc process. The elimination of the cathode spots as the ac current half cycle falls to its natural current zero is thus extremely rapid when compared to the millisecond time scale for the change in current. In fact, the cathode spot easily adapts to current changes up to 100A/μs [93]. Interestingly enough Rusteberg et al. [92] have shown that as the current decreases so too the ion energy distribution increases, again indicating that changes in the vacuum arc's intercontact plasma also occur on a time scale that is much shorter than the changes in the ac current. When the current is below a certain value, which is dependent upon the contact material, only one cathode spot will remain.

It has long been observed that as the current in any arc drops to a low enough value the arc will suddenly extinguish. When this happens, the plasma channel between the contacts no longer permits the passage of the circuit current and that current drops to zero in a very short time. This phenomenon is termed *current chop.* In air the minimum current that closely spaced contacts will support is less than about 0.5A [94]. It has been observed, however, that there is a probability for a very short duration arc for currents down to 50mA in air [24, 94]. In vacuum the energy level required to support the cathode spot region results in chop currents of higher value, usually greater than 1A and for some pure metals, as high as a few hundred amperes [95]. The value of the current at which the "current chop" occurs is usually much lower than the value of current per cathode spot given in Table 2.5. This gives support to the model of the cathode spot that pictures it being made from a collection of cells each carrying a fraction of the total spot current.

For a stable, diffuse, vacuum arc the region of the contact below the cathode spot must emit enough electrons and metal vapor to produce the positive ions in the high-pressure ionization zone. About half of these ions must return to the contact's surface beneath the cathode spot where they will help maintain a high enough temperature for its stability. As discussed earlier, the cathode spots seem to operate best at an optimum current between 30A and 100A (depending upon the contact material). As the ac current approaches zero, the current in the last spot declines below the optimum value. Thus, the energy into the cathode region also declines below its optimum value. When this happens, the conditions are such that the cathode spot can become unstable. For example, the electron emission and metal evaporation can fall sharply. This, in turn, will reduce the ion flow necessary to maintain the temperature of the cathode spot for efficient electron emission as well as the ions required to maintain charge neutrality in the intercontact region.

Another way to look at this phenomenon is to study the stability of a vacuum arc at low currents, i.e., the lifetime of a vacuum arc before it chops out when it is formed at currents below the contact metal's optimum cathode spot value. See, for example, Figure 1.84 developed by Cobine et al. [96]. Here you can see, for a given current, metals with a higher vapor pressure are more stable, i.e., have longer lifetimes. This correspondence seems to hold for the ac vacuum arc: In general, the average value of the chopping current decreases as the vapor pressure of the metal increases, see Table 2.9.

In fact, it is generally agreed that the higher the contact material's vapor pressure and the lower its thermal conductivity, the lower will be its chopping current in an ac circuit [97]. More controversial are the experimental and theoretical data that shows a higher work function for the contact metal and a higher ionization potential for the metal atoms will also result in a lower chopping current [98]. Mixing of metals to form practical vacuum interrupter contacts profoundly affects the value of the chopping current. This will be discussed in more detail in Chapter 3 in this volume, as will be its effect on vacuum interrupter design. The effect of current chop in ac circuits will be investigated in Chapter 5 in this volume.

Figure 2.29 illustrates the observed phenomena of the ac current and the vacuum arc voltage as the current drops close to zero; see Smeets [99–101]. An experimental record taken by Ding et al. [102] is shown in Figure 2.30. First, an initial current instability is observed. Yamamoto et al. [103] have shown that this first current instability has a typical value up to six times that of the final chopping current for a given metal. For example, the initial instability for Cu contacts is three times that for Ag contacts. As Ag is alloyed with Cu the value of the first instability drops to the value of pure Ag as the Ag content is increased. The initial current instability has a corresponding fluctuation in the arc voltage. They also show that during this increase in voltage the number of metal ions in the

TABLE 2.9
Chopping Current for Some Pure Metals

Metal	Average Chop Current, A	Maximum Chop Current, A
Ag	3.5	6.5
Cu	15	25
Cr	7	16
Ni	7.5	14
W	16	350

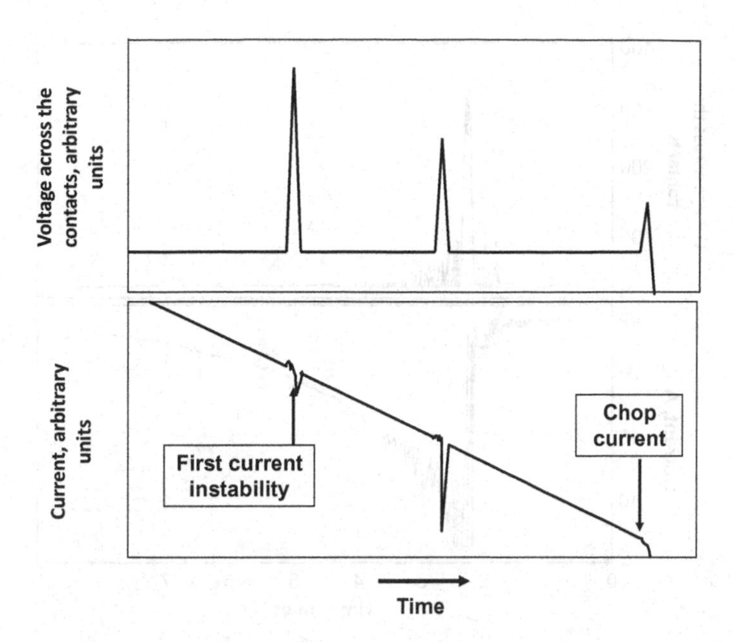

FIGURE 2.29 An illustration of the vacuum arc's current instabilities and current chop as an ac current goes to current zero.

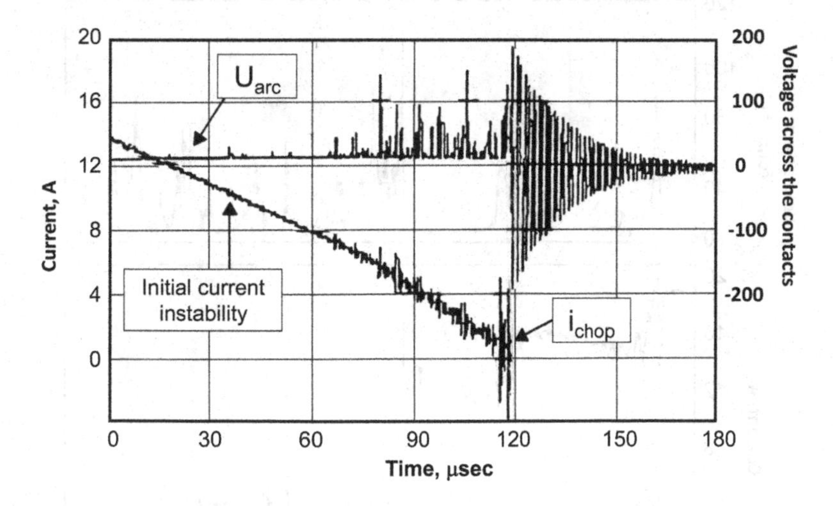

FIGURE 2.30 Experimental voltage and current trace showing the first current instability and the current chop for WC–Ag (33 wt %) contacts, 60A peak current. [102].

vacuum arc also increases. As the current continues to drop, the instabilities become more frequent. Smeets [100, 101] shows that as the arc current falls, during the initial instability, the arc voltage increases and the current recovers to its initial value in about 7μs: see Figure 2.31. As the current decreases further, however, it becomes harder for the current level to recover. If the current does not fully recover before the next instability occurs, the current proceeds to chop to zero: see Figure 2.32 [100, 101]. Thus, it appears that the final current instability leading to the "spontaneous" vacuum arc extinction results from a succession of current instabilities that do not recover to the full value of the circuit current. In this case the total time for the final chop event is of the order of 8μs, which gives an average di/dt in the range of 10^6 As^{-1}.

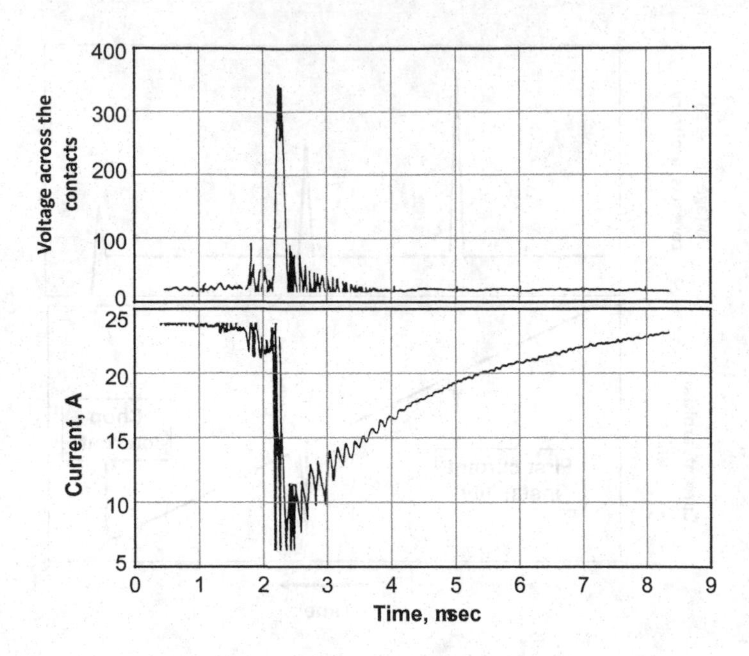

FIGURE 2.31 An example of the first current instability showing its recovery to its original value [101].

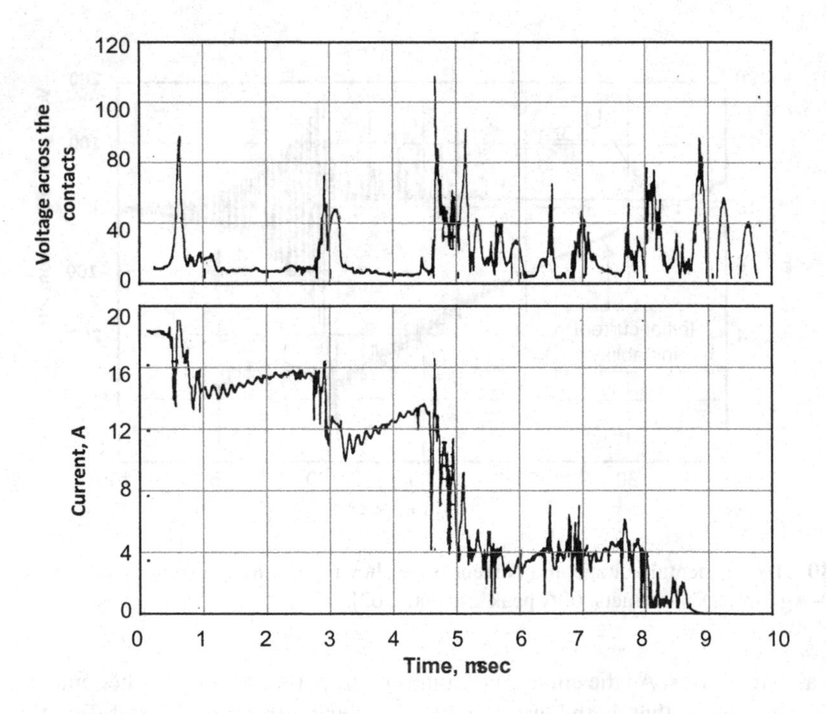

FIGURE 2.32 Detailed view of the final current chop showing its demise over 8μs [101].

The initial instability in the vacuum arc occurs at a level of current much higher than the final chopping current. In Figure 2.31, after the current decreases by about 20A in about 0.2μs, the voltage increases momentarily to a value of about 300V before returning to the normal vacuum arc voltage of 20V as the current recovers to its original value. In fact, as the current approaches zero, the current instabilities become more frequent and are accompanied by voltage fluctuations of less than 1μs duration, and with values up to ten times the normal vacuum arc voltage; see for example, Figure 2.32. At higher currents when the vacuum arc is stable, its voltage also has rapid fluctuations. As I have presented previously, these fluctuations have been shown to correspond to changes in the radiation output from the cathode spot [50, 51]. Thus, during the normal operation of the cathode spot, there is a continual adjustment within it to maintain the flow of ions and electrons for current continuity between the open contacts. As the current decreases to a value close to the chop current level, there is no correspondence between the radiation intensity of the cathode spot and the level of arc voltage [100, 101]. This has led researchers to believe that the final instability is the result of ion starvation in the intercontact plasma that in the end prevents the flow of current across the contact gap. Thus, the description of the chop phenomenon given by Smeets seems to be reasonable [100, 101].

A certain level of current is imposed on the vacuum arc by the external ac circuit. This requires that a certain minimum electron density must be maintained in the intercontact plasma to guarantee the flow of the circuit current. This minimum electron density requires a certain density of ions to maintain the neutral plasma above the cathode spot. As these ions are produced within the cathode spot, a minimum flow of evaporated metal atoms and the emission of electrons into the cathode spot's ionization region is required. When the current decreases below an optimum value for a stable cathode spot this production decreases. At currents greater than the final chopping current, however, the energy into the cathode spot increases and it is possible for the cathode spot to recover from the momentary instability. It then provides enough electrons and ions for current continuity between the contacts even though the circuit current continues to fall. As the current further decreases, a stage will be reached when the cathode spot can no longer provide the necessary conditions for the plasma charge neutrality in the intercontact region. The voltage in the plasma increases to force the electrons across the contact gap. At this time, however, the vacuum arc ceases to function and its current chops to zero. Thus, small contact gaps and a small plasma volume would favor arc stability. Hence the value of the actual chop current is not a fixed value but has a statistical distribution. The value usually given for a contact material is its most probable value: I shall discuss this further in Chapter 3 in this volume.

We also have to remember from our discussion of cathode spots that a cathode spot itself is not a fixed, stable structure. Once it has formed, it grows in size and then forms a new spot on the rim of the old spot's crater. This is usually accompanied by an increase in arc voltage and fluctuation of current flow within the cathode spot [43, 46]. Thus, the chop phenomenon is not only a failure of the cathode spot to produce the required charge carriers for current continuity, but it is also the unsuccessful continuation of repetitive cathode spot formation that is so characteristic of a stable, diffuse vacuum arc [34]. One interesting consequence of this is that the vacuum arc may well be more stable on contacts that have a high density of field-enhancing micro-protrusions. A second consequence may be that the vacuum arc may also be more stable in the presence of an axial magnetic field where the cathode spot's speed of motion is slower; see Section 2.6.3.

2.3.4 The Formation of the Low-Current and High-Current Anode Spot

Kimblin [104] has shown that the anode spot can form from a diffuse vacuum arc at currents of only a few hundred amperes if certain conditions are satisfied. This phenomenon, while of scientific interest, has little relevance for the vacuum interrupter designer. The phenomenon, however, still is of considerable academic interest [105, 106]. It almost never occurs during the opening of a vacuum interrupter's contacts where the initial bridge column arc and the ac circuit current determine the final structure and appearance of the vacuum arc. It may occasionally develop as a transient mode in a transition from a diffuse arc to a high-current column arc during vacuum breakdown with a high circuit current follow-through. A good example is given by Zhang et al. [107]. They use a high-voltage pulse to break down a 10mm contact gap between Cu contacts. At 2.8kA (rms) the vacuum arc remains diffuse,

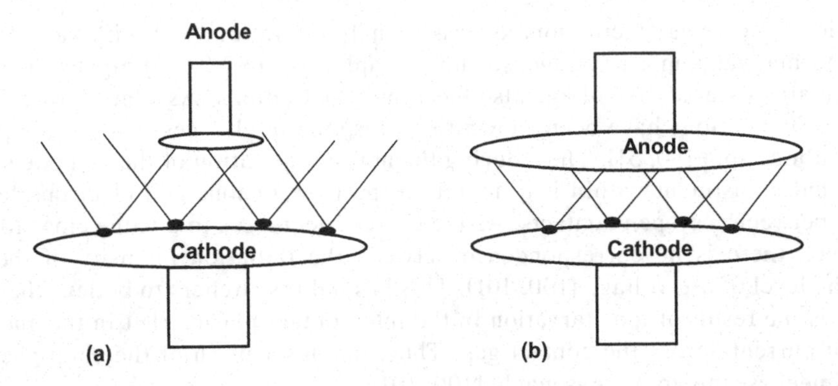

FIGURE 2.33 The diffuse vacuum arc between contacts with different diameter anodes.

but at 7.5kA (rms) an anode spot forms. A different form of anode spot can form with high-current vacuum arcs with axial magnetic fields imposed across them: see Section 2.7.2. Figure 2.33 illustrates opening contacts: (a) one with a small diameter anode and (b) the other with a large diameter anode [104], each operating against a cathode with a constant diameter. In 2.33(a) as the gap "*d*" increases less plasma is incident on the anode contact and more escapes to the surrounding shield. What fraction is captured depends upon the solid angle subtended by the anode. Thus, in this illustration the smaller anode receives less plasma from the cathode spots than does the larger anode as the contact gap "*d*" increases. When the contacts separate beyond a given distance, a point will be reached when the voltage of the vacuum arc must increase for the anode to attract a high enough flow of electrons to maintain the circuit current. This will happen sooner for a small diameter anode *(2r)*, i.e., when *2r/d* is smaller. The voltage gradient close to the anode produces an anode fall. The energy of the electrons striking the anode will increase as they fall through the anode fall region and this energy when deposited onto the anode's surface will result in its heating. As the temperature of the anode increases an ever-increasing mass of metal atoms will evaporate from it. Once the metal vapor above the anode exceeds a threshold pressure it will become ionized and a visible anode plume will be observed. This is called an *anode spot*. Once the anode spot forms experimental observation shows that the cathode spots tend to stay in a region just below it and no longer spread over the whole cathode surface. This results in a concentrated cathode region being heated. Eventually a cathode plume may also develop which will form a column arc once it joins with the anode plume: see, for example, Figure 2.34 [107].

Figure 2.35 [104] gives an example of the voltage versus current characteristics for a 50mm diameter cathode and for two anode contact diameters (12.5mm and 50mm), for dc currents

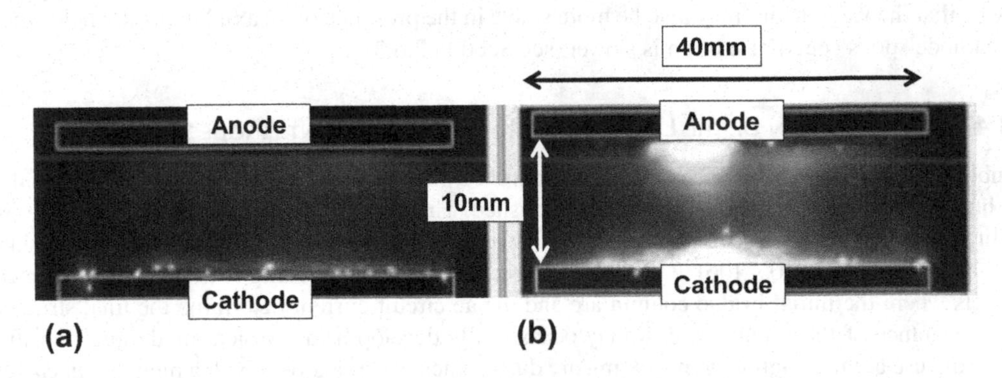

FIGURE 2.34 The vacuum arc appearance between Cu contacts after the breakdown of the contact gap: (a) at the peak current 4kA and (b) at peak current 10.6kA [107].

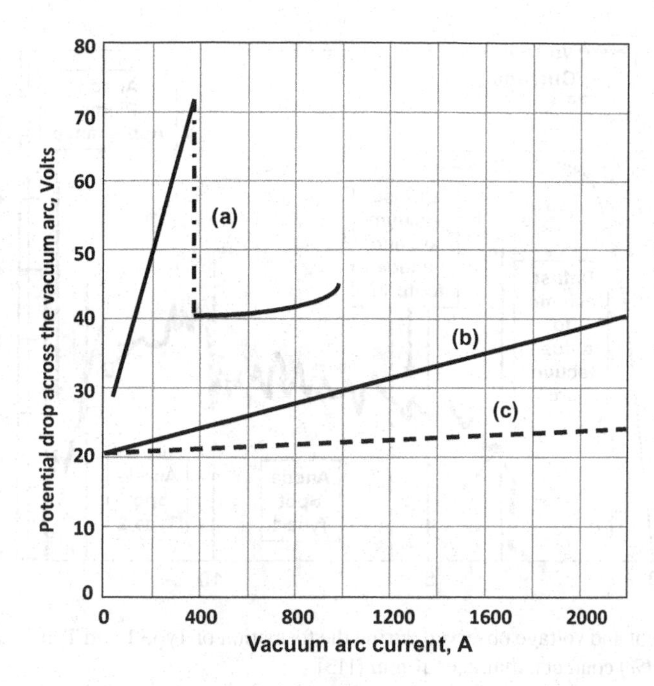

FIGURE 2.35 The arc voltage, arc current characteristics for vacuum arcs between contacts with a 25mm gap; (a) 12.5mm diameter anode and a 50mm cathode (b) both anode and cathode 50mm (c) the characteristic for ≈ 1mm contact gap for both anodes [104].

100A–2100A. At a small contact spacing (few millimeters) the vacuum arc voltage increases from 21V to 24V as the current increases from 100A to 2100A for both diameter contact structures. For the smaller diameter anode contact at a gap of 25mm the voltage increases rapidly as a function of current until at ≈ 400A an anode spot forms and the voltage drops. For the larger diameter contacts with a contact gap of 25mm, the arc voltage slowly increases as a function of current from 21V to 40V and anode spot activity is observed to begin only at 2100A.

Modeling this form of anode spot usually assumes an open contact gap. This does not take into account the initiation of the vacuum arc as the contacts open [108–110]. The effect of the bridge column arc (see Section 2.2) is ignored as are the transition stages to the columnar vacuum arc (see Section 2.4). Certainly, even a low-current, diffuse vacuum arc given time will cause the anode temperature to increase. Beilis et al. [111] show a titanium anode's temperature increases to 1300K after being exposed for 50s to a 175A dc diffuse vacuum arc and to 1500K for a current of 225A dc. Once a Cu anode temperature reaches about 2200K the evaporated Cu atom flux can erupt explosively [108].

This form of anode spot can occur between opening contacts under specific experimental conditions. Khakpour et al. [112–114] and Franke et al. [115] in a series of papers have explored the formation of this type of anode spot between opening Cu–Cr (25wt%) contacts. The anode diameter in these experiments is 10mm and the contact gap varies from 10mm to 20mm. The contacts open both a simulated 10ms, dc current pulse (2kA to 2.75kA) and a 50Hz current (4.4kA peak). The small anode diameter ensures that anode spots develop from the initial diffuse arc even at the low currents used in these experiments. They identify two forms of this type of anode spot while opening the simulated dc current pulse. A Type 1 anode spot and a Type 2 anode spot. For a 20mm contact gap these are identified from the difference in the arc voltage for contacts opening a 3.7kA 50Hz ac current: see Figure 2.36 [113]. The Type 1 anode spot forms about 3ms after contact opening and the Type 2 anode spot forms about 5ms after the contacts open. The light intensity increases dramatically in the transition from the Type 1 anode spot to the Type 2 anode spot when the arc

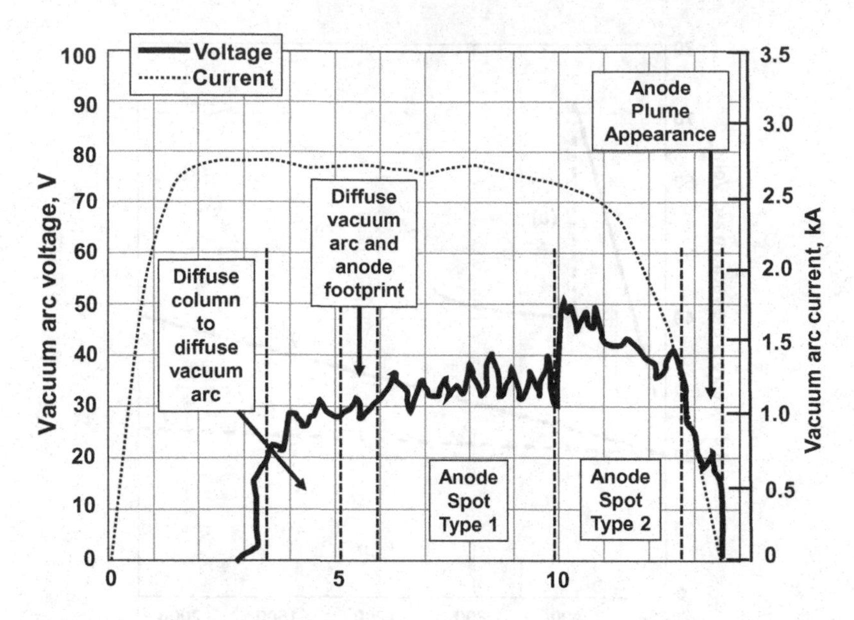

FIGURE 2.36 Current and voltage observed during the formation of Type 1 and Type 2 anode spots between opening Cu–Cr (25 wt%) contacts, diameter 10mm [113].

voltage increases: see Figure 2.37 [115]. The transition occurs as the contact gap increases as can be seen from Figure 2.37 [115]. The intensities of the CuI, CuII, and CuIII spectral lines increase near the anode for both types of anode spot. Their intensity, however, is much greater for the Type 2 anode spot [115]. As the current falls to zero the Type 2 anode spot transitions into a low intensity anode plume [115]. Batrakov [116] and Gortschakow [117] show similar results for opening Cu–Cr (25wt%) contacts. Interestingly the Type 1 anode spot shows an increase in conductivity before the Type 2 anode spot forms. When the Type 2 anode spot forms there is a substantial drop in the arc conductivity: see Figure 2.38 [116]. Kong et al. [118] observe the formation of anode spots between opening Cu–Cr (25wt%) and Cu–Cr (50wt%) contacts with a 50Hz ac currents ranging from 2.29kA to 4.15kA. Their experiments use two contact diameters, 12mm and 25mm. They show for opening

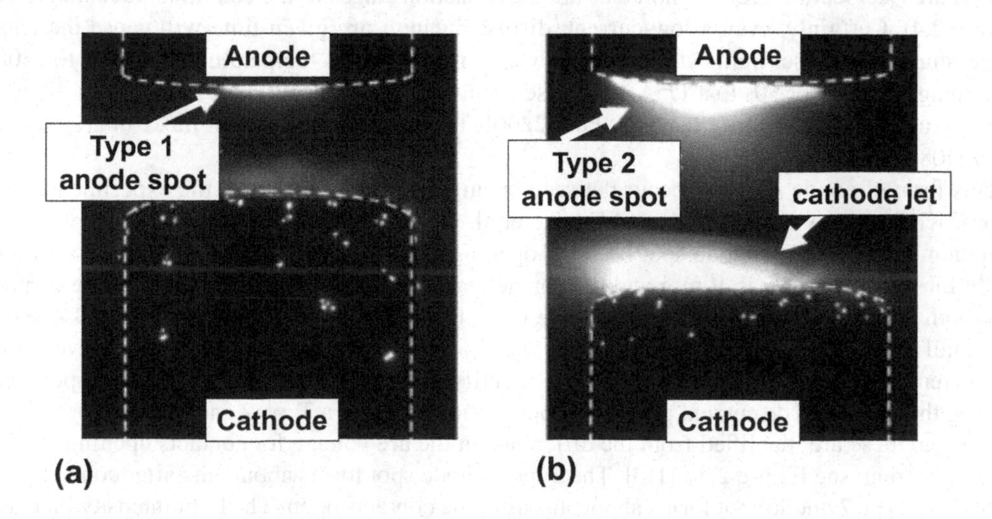

FIGURE 2.37 Photographs of (a) the Type 1 and (b) Type 2 low-current anode spots [113].

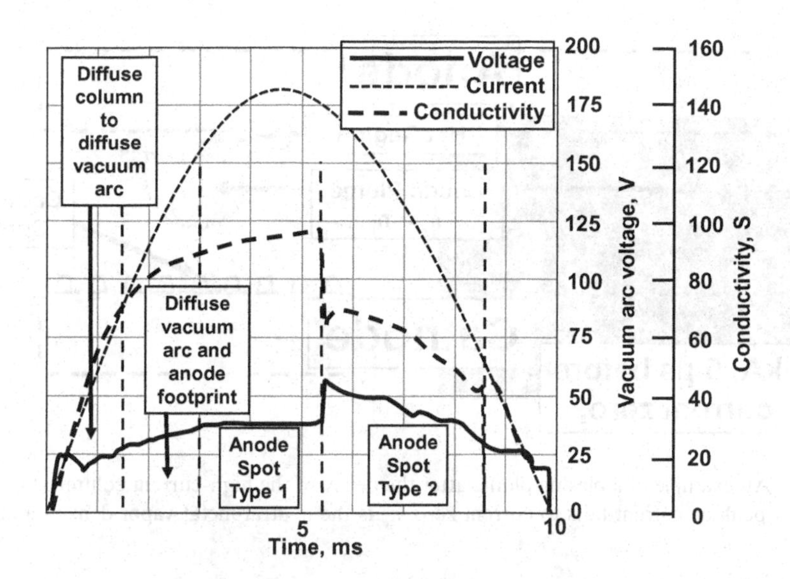

FIGURE 2.38 The change in the vacuum arc conductivity during the transition from the Type 1 anode spot to the Type 2 anode spot [116].

speeds of $1.3ms^{-1}$ and $1.8ms^{-1}$ that the anode spot formation depends upon the contact gap for both contact diameters and for both contact compositions. The formation of this type of low-current anode spot depends upon a combination of:

1) The diameter of the anode contact: the anode spot forms more readily on a smaller diameter anode than on a larger one
2) The contact gap: the anode spot forms more readily as the contact gap increases
3) The current level: for a given contact gap and anode diameter

For the designer of commercial, power vacuum interrupters, where the diameter of the contacts for a given fault current are generally larger than those used in the experiments discussed above, it is unlikely that this form of anode spot plays a role in their current interruption performance.

A second form of anode spot is observed by Batrakov et al. [119]. Their experiment used Cu–Cr (25wt%) to interrupt 50Hz currents of 10kA to 14kA. As the contacts open with a velocity of $1ms^{-1}$, a columnar arc forms (see Section 2.4). Photographs of the contact gap 0.5ms before the ac current zero show that the columnar arc transitions to a decaying anode plume. As the current continues to just before the zero, the vacuum arc becomes diffuse with multiple cathode spots and an anode hot spot: see Figure 2.39 [119].

2.4 THE COLUMNAR VACUUM ARC

In the development of vacuum interrupters for interrupting high short-circuit currents, understanding the formation and the control of the columnar vacuum arc is of great importance. When the contacts in vacuum initially part while carrying high currents, the transition from the molten metal bridge to the bridge column arc still occurs. If the current is high enough, the power into the contact surfaces at the arc roots is high enough to fully compensate for the slow expansion of the radius of the bridge column arc and for the material being lost to the surrounding vacuum. The bridge column will then transition into a high-pressure columnar arc, which has properties similar to those of an arc in air at atmospheric pressure: see Figure 2.40. The stationary arc roots cause considerable erosion of the contact. Figure 2.41 shows erosion data taken by Mitchell [120] for an arc between

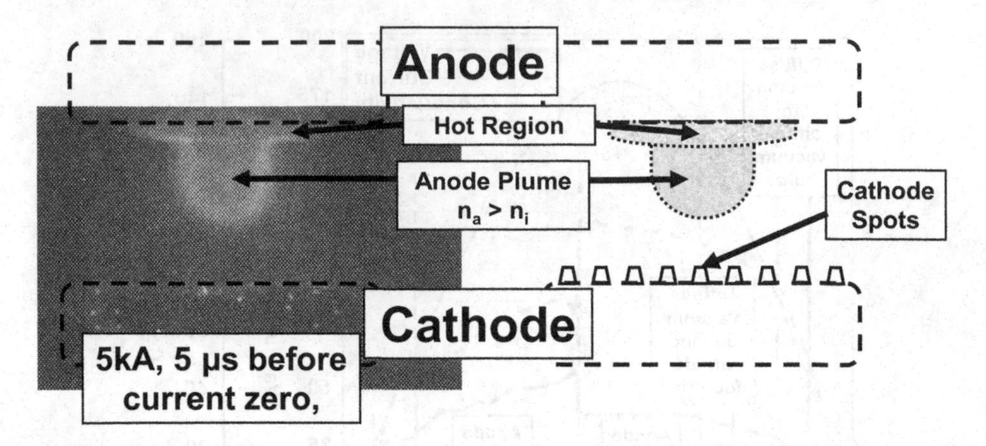

FIGURE 2.39 An example of a plasma plume after the decay of the high-current columnar vacuum arc, 5µ before the 14kA peak ac current falls to current zero: n_a is the neutral metal vapor density and n_i is the ion density [119].

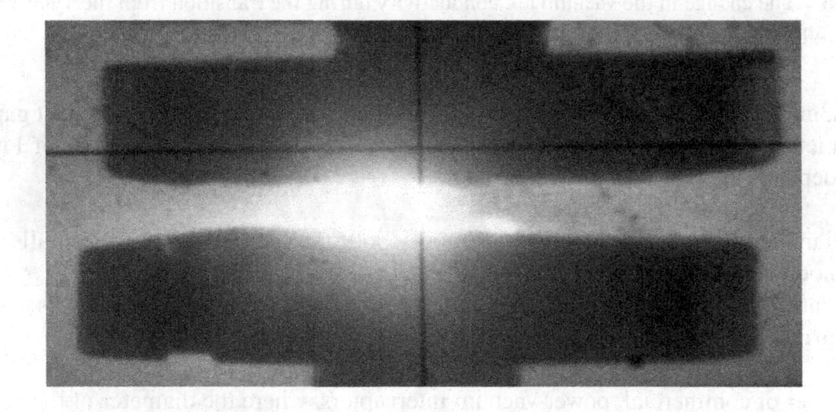

FIGURE 2.40 The columnar vacuum arc.

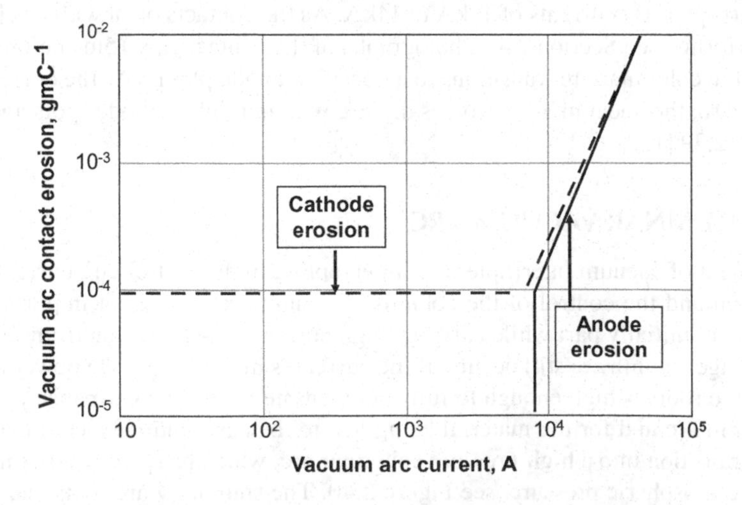

FIGURE 2.41 Vacuum arc erosion as a function of the arc current for disc-shaped (butt), Cu contacts [120].

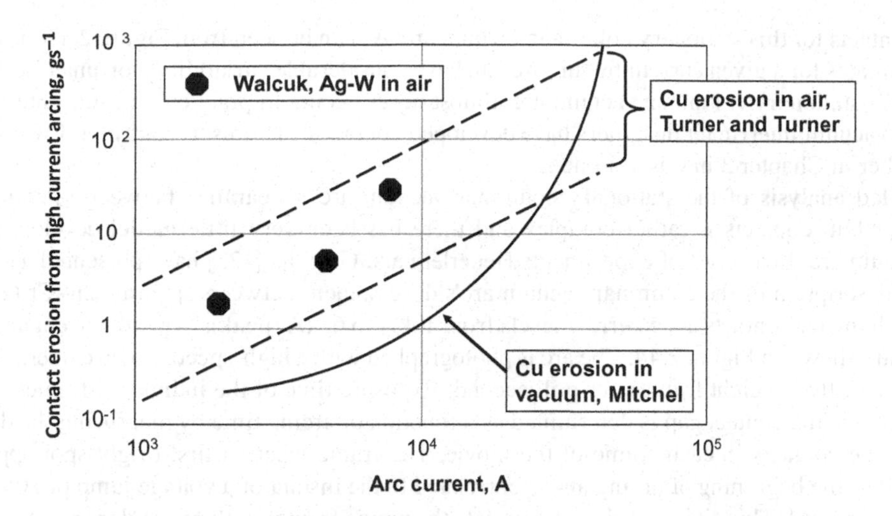

FIGURE 2.42 Comparison of arc erosion of butt contacts in air and vacuum [120, 121, 122].

disc-shaped or butt Cu contacts. For currents below ≈10kA the erosion is entirely from the cathode spots and gives the usual value of about 100 μgC^{-1}. Once the columnar arc forms there is a rapid increase in erosion from both the cathode contact and the anode contact i.e., both contacts now have similar erosion rates. Figure 2.42 compares the erosion of these Cu contacts in vacuum with the erosion rate given by Turner & Turner [121] for arcs in air and for the actual, high-current, arc erosion of W–Ag contacts also operating in air [122]. Figure 2.43 presents a summary of the energy balance

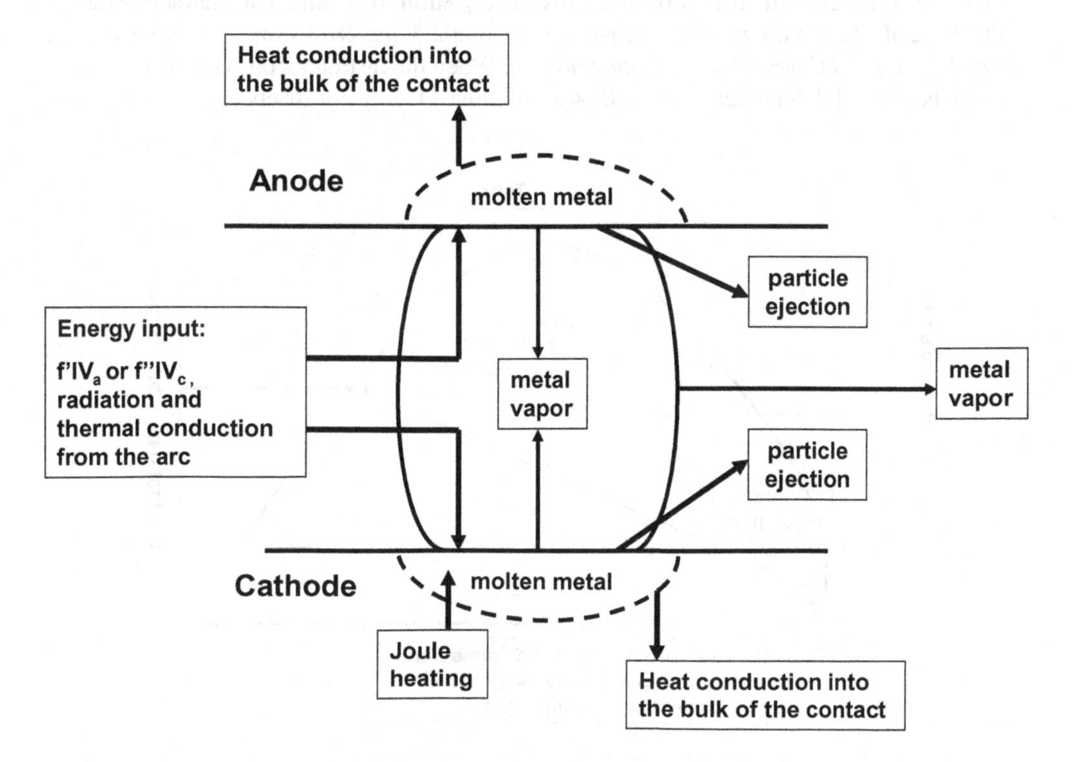

FIGURE 2.43 Energy balance for the columnar vacuum arc.

at the contacts for this stationary columnar vacuum arc. As can be seen from Figure 2.42 the erosion of the contacts for a given current by this arc can have considerable variability. Fortunately, the condition of a stationary columnar vacuum arc almost never occurs in practical vacuum interrupters, because vacuum interrupter designers have developed contact structures to control it. I will discuss this further in Chapter 3 in this volume.

Detailed analysis of the stationary columnar vacuum arc appearance between opening disc-shaped or butt contacts is rather complex and there has been very little model development for this vacuum arc. In a series of experiments Heberlein and Gorman [123] have presented a comprehensive description of the columnar vacuum arc's development between opening Cu–Cr (25wt%), 100mm diameter contacts at ac current levels from 10kA to 67kA (peak). Typical current and travel records are shown in Figure 2.44. The arc is photographed with a high-speed movie camera running at a speed of five to eight frames per millisecond. Exposure time of the individual frames is 50μs. In each movie the contact gap is determined as a function of arcing time by measuring the distance between the contacts in each frame of the movie. The frame where a first bright spot appears is identified as the beginning of arcing and is correlated to the instant of a voltage jump of 20V on the arc voltage record. Thus, it is possible to correlate the movie frames with particular arc phenomena, changes in arc voltage, current level and contact gap. By photographing many opening operations and a large number of ac current levels the *"Appearance Diagram"* shown in Figure 2.45 is developed. Here the arc appearance from the photographs is given as a function of contact gap and circuit current. They developed the following descriptive names for the observed columnar arcs, which are illustrated in Figures 2.45 and 2.46.

1. *The Bridge Column Arc*: I have described this in Section 2.2. i.e., for currents less than about 5kA the bridge column gradually increases in diameter until the diffuse arc forms
2. *The Diffuse Column Arc*: When contacts are separated at current levels between 7kA and 15kA the bridge column transitions into a diffuse column. The diffuse column diameter increases linearly with current from approximately 8mm to 20mm. The characteristics of this type of vacuum arc will be discussed in Section 2.5, the *Transition Vacuum Arc*
3. *The Constricted Column Arc or Columnar Arc*: When the instantaneous current exceeds a value between 10kA and 20kA, the diffuse column suddenly constricts and becomes very

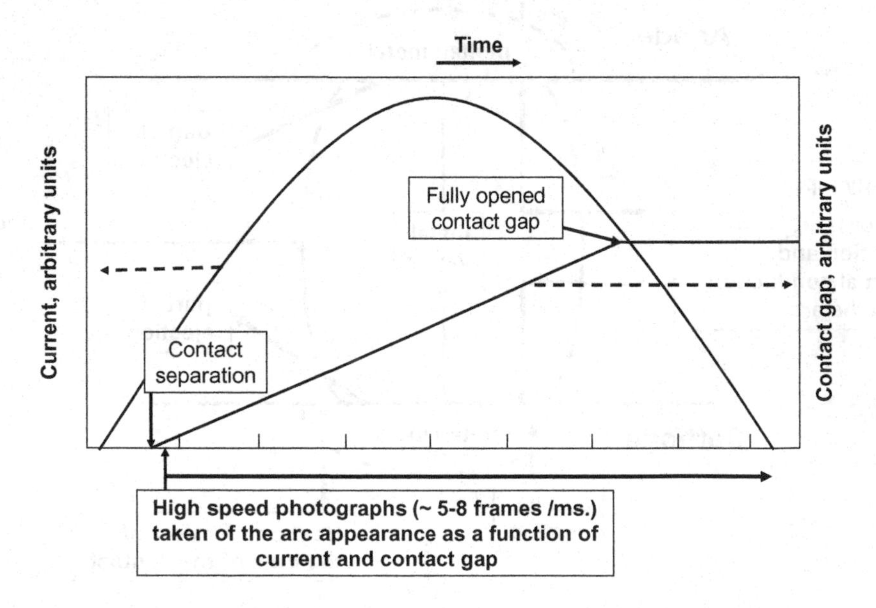

FIGURE 2.44 Current and contact opening curves as the vacuum is photographed with a high-speed camera [123].

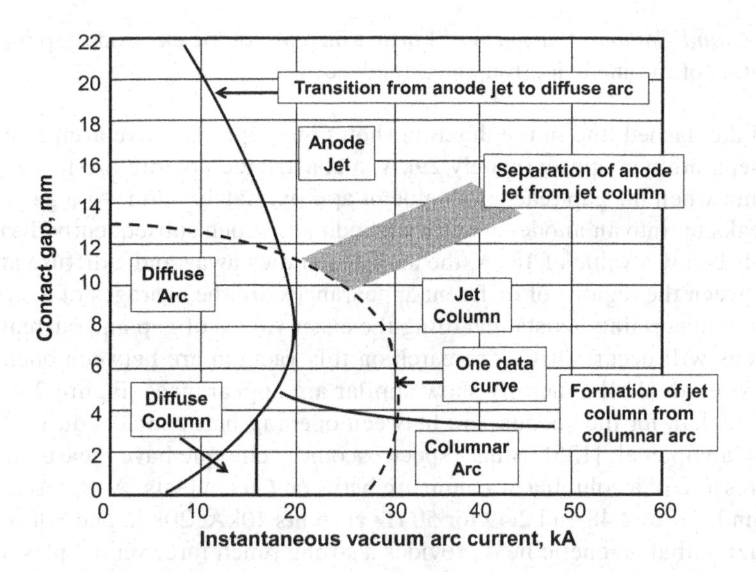

FIGURE 2.45 The "*Appearance Diagram*" for the columnar vacuum arc at various currents and contact spacings [123].

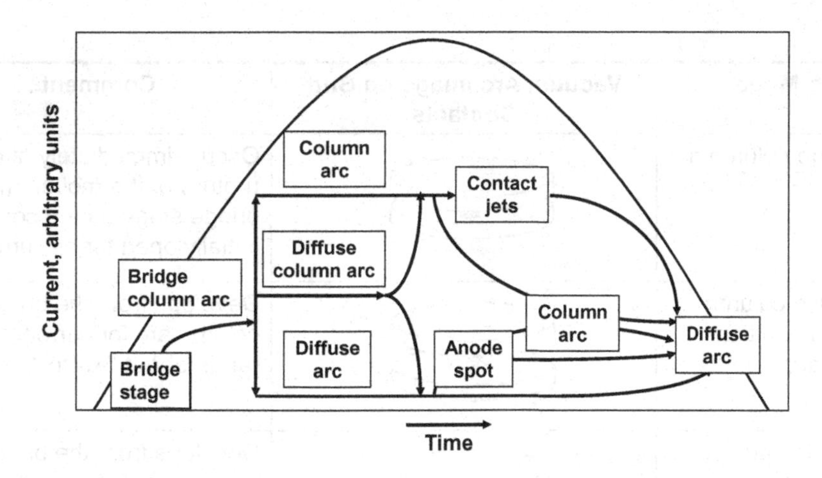

FIGURE 2.46 The appearance of the high-current columnar vacuum arc

luminous with well-defined boundaries (constricted column). The strong radiation intensity gradient at the column boundaries is demonstrated by the fact that the column diameter remains essentially constant for a wide variation of the photographic film exposures using neutral density filters. The column is cylindrical with a slight constriction in front of the anode; column diameter is typically 10mm and is relatively insensitive to the value of the instantaneous current. When contacts are separated at current values larger than 15kA, the columnar arc transitions directly from the bridge column arc

4. *The Plasma Jet Column Arc*: This arc is comparable to the constricted column in intensity, but it is wider where it attaches to the contact and it has a constriction in its center section. Its resulting appearance is thus two cones meeting in their apexes. Typical dimensions for the column diameters are 12mm at the anode, 10mm at the narrowest point, and 20mm at the cathode, but the diameters increase for currents above 36kA

5. *The Anode and Cathode Plasma Jets*: Further increase of the electrode gap finally results in separation of the anode jet from the cathode jet

In Figure 2.45 the dashed line in the diagram shows one representative arcing sequence. Here after contact separation at approximately 29kA, a constricted column arc forms; this changes into a jet column when the gap reaches a value of approximately 4mm. At a gap of 9.5mm, the jet column breaks up into an anode jet and a cathode jet, which subsequently disappear. When the current falls below a value of 18kA, the anode spot dies away and a diffuse arc forms. The boundaries between the regions of different appearances are the averages of a large number of points, and some uncertainty exists regarding the exact values of gap and current at which the actual transitions will occur. Further research on this vacuum arc between opening butt contacts by Zalucki et al. [124] generally show similar arc appearances. Figure 2.47 summarizes their appearance data for the vacuum arc between opening butt contacts during an ac current half cycle. Abplanalp et al. [125] using a spectroscopic technique have measured the temperature and the pressure of a columnar vacuum arc between Cu contacts. A representation of their data is shown in Figures 2.48 and 2.49 for 50 Hz currents 10kA, 20kA, and 30kA. As might be expected the azimuthal magnetic field provides a strong pinch force on the plasma column. At 30kA this results in a higher arc pressure than at 10kA. Surprisingly the average temperature for the three currents are similar. Both the pressure and the arc temperature decline as the current approaches zero.

Arc Mode	Vacuum Arc Image on Butt Contacts	Comments
The bridge column arc		Occurs immediately after the rupture of the molten metal bridge stage as the contacts initially open for all currents
The diffuse column arc (the transition vacuum arc)	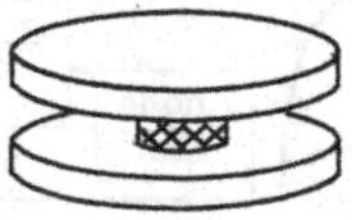	Develops from the bridge column arc for currents in the range about 6kA to 15kA.
The constricted column vacuum arc	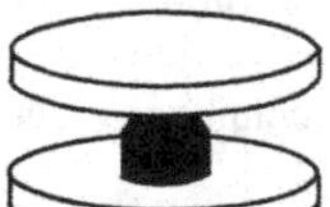	Develops from the bridge column arc from a diffuse column arc for currents > 15kA.
The plasma jet column vacuum arc	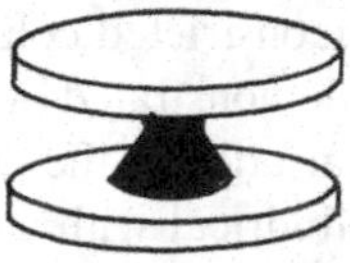	Similar to the constricted column arc with a constricted center region and a cathode foot wider than the anode foot
The anode anode and cathode jet vacuum arc	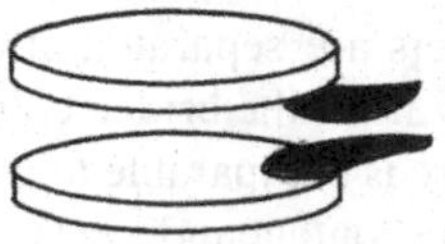	As the current increases above 20kA, the contact gap increases and the arc column moves to the contacts' edge

FIGURE 2.47 Observed sequences of vacuum arc appearance modes for a high-current vacuum arc on butt contacts [124].

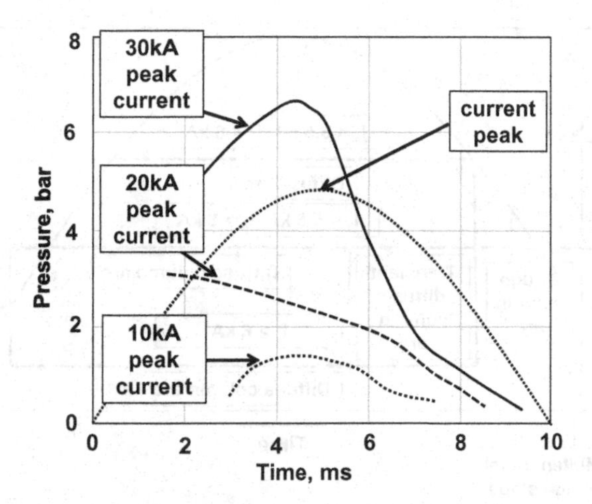

FIGURE 2.48 The arc pressure of the columnar vacuum arc during the passage of the ac current 10kA, 20kA, and 30kA [125].

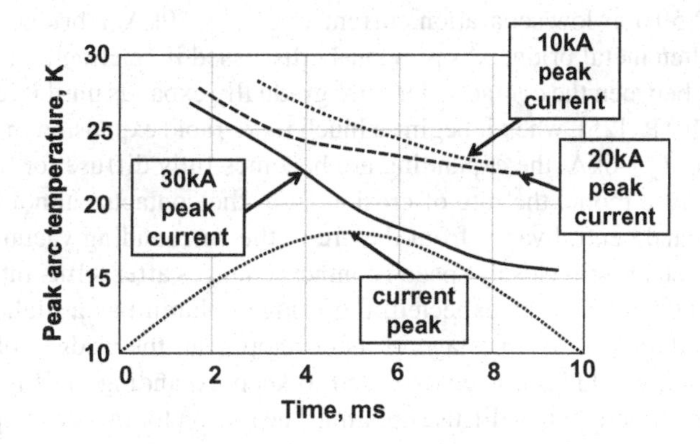

FIGURE 2.49 The arc temperature of the columnar vacuum arc during the passage of the ac current 10kA, 20kA, and 30kA [125].

For the vacuum interrupter designer, the "*Appearance Diagram*" gives a good qualitative understanding of expected vacuum arc modes between large area contacts switching high ac currents in vacuum. As this columnar vacuum arc is generally stationary on butt contacts, the arc root regions will be strongly eroded. At an ac current zero the contact regions close to the arc roots would remain at a high temperature and metal vapor would continue to evaporate into the contact gap. Above a certain current level, therefore, the performance of a vacuum interrupter with butt contacts would be degraded. I will discuss this in Section 4.2.2. Thus, in a practical vacuum interrupter it is important to control this arc by either forcing the arc roots to move over the contact surfaces or developing a way to ensure a diffuse vacuum arc at high currents. Contact designs for the control of the high-current vacuum arcs will be discussed in Chapter 3 in this volume.

2.5 THE TRANSITION VACUUM ARC

Schulman and Slade [126] have given a full description of this vacuum arc mode. In their experiments the arc appearance has been photographed for disc-shaped, Cu–Cr (25 wt%) butt contacts opening an ac current for peak current values from 2.8kA to 15.7kA. The observed arc modes are

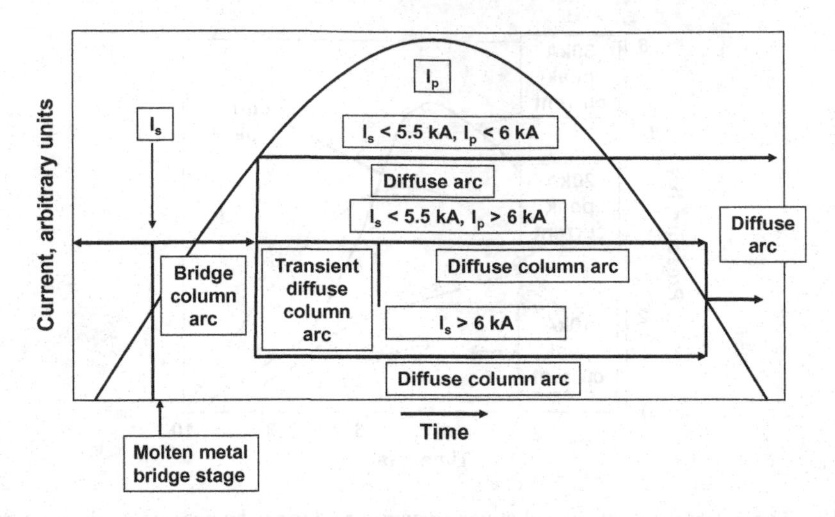

FIGURE 2.50 Observed sequences of vacuum arc appearance modes for a transition vacuum arc on butt contacts [126].

illustrated in Figure 2.50. For low separation currents, i.e., $I_s \leq 10kA$, a bridge column exists after the rupture of the molten metal bridge. As previously discussed in Section 2.2, this column begins as a high-pressure arc between the contacts. This arc gradually expands until it reaches a pressure of less than about 0.5×10^5 Pa [25], when it begins a much more rapid expansion into a diffuse vacuum arc. If the peak current $I_p \leq 6kA$, the expanding arc becomes fully diffuse for the remainder of the half cycle. For the diffuse mode, the rate of erosion from the contacts cannot keep pace with the rate of loss of plasma and neutral vapor from the arc to the surrounding vacuum. If $I_p \geq 6kA$, the diffuse-type arc between the still closely spaced contacts changes after ~1ms into a diffuse column.

For opening current $I_p \geq 6kA$, as expected, the bridge column has a higher pressure when it reaches the end of its duration. At this stage of its development, the bridge column will not form a diffuse arc, although it would have been expected to keep expanding until it did so. Instead, the bridge column changes directly into a diffuse column. The reason for this development is apparently related to the instantaneous current level, the contact diameter and the overlapping plasma plumes above the cathode spots. When the diffuse vacuum arc collapses into the diffuse columnar mode, the diffuse column begins to form at the position of the original bridge column arc roots, or at the position of a particularly intense cathode spot. This could be expected to result from a continued heating of the localized hot area of the anode.

This diffuse column mode is characterized by its limited current range ($I \leq \approx 15kA$) and the high rate of vapor loss through its effective surface area. This new equilibrium condition can only be maintained over a limited range of pressure, which is roughly 0.2×10^5 Pa to 2×10^5 Pa. The contact erosion pattern of the diffuse column arc is quite unique and reveals lightly melted craters. There is no gross erosion or jets of material from the contacts. Analysis of the cathode crater shows that it consists of many cathode spots, which have been confined within it. Thus, for the transition vacuum arc, the cathode spots do form and initially move apart after the bridge column arc, but then reach a position where they no longer repel each other. Since this occurs at currents greater than several kiloamperes, the effects discussed in Sections 2.3.1 and 2.3.2, are forcing the development of this transition arc. Now the intercontact plasma has a high enough density to experience many inelastic collisions and metal particles that enter this plasma column can be evaporated. The resulting metal vapor will, in turn, become ionized and contribute to an increase in the plasma density. Conditions are achieved where the azimuthal magnetic field that results from the circuit current flowing in the plasma confines the plasma to a well-defined column. The cathode spots now are restricted from expanding beyond the boundaries of this plasma column. The work by Kimblin [127] may also give a clue to the structure of the transition vacuum arc. He shows that as the ambient pressure in an arc chamber increases from about 10^{-4} Pa to above 10^4 Pa, the cathode spots tend to remain within a fixed location, the cathode spot erosion decreases and the ion current collected on a surrounding shield biased to the cathode potential also decreases. At present, however, there is no adequate analytical model of the transition to this type of vacuum arc.

As the circuit current goes to zero in an ac circuit, the plasma plumes cease to strongly overlap at a given current and a given contact diameter. When this occurs the character of the plasma changes back to be essentially collisionless and the vacuum arc returns to the diffuse mode. Experimental observations of the diffuse column vacuum arc show that it always returns to the diffuse mode as the current goes to zero.

2.6 THE INTERACTION OF THE VACUUM ARC AND A TRANSVERSE MAGNETIC FIELD

2.6.1 THE DIFFUSE VACUUM ARC AND A TRANSVERSE MAGNETIC FIELD

In the literature on the vacuum arc, the term "Transverse Magnetic Field" really refers to an "Transverse Magnetic Flux." In order to prevent confusion, I will generally adopt the commonly used expression "Transverse Magnetic Field" instead of the more correct "Transverse Magnetic Flux" (interestingly, the abbreviation TMF can be used interchangeably for both expressions). The observation that the cathode spots from a low-current diffuse vacuum arc move away from each other in a retrograde (i.e., an anti-Amperian) motion has a long history, This has led to experimental studies of a single cathode spot subjected to an external transverse magnetic field "B_T" impressed across it: see, for example, Fang [128] and Persky et al. [129]. Wang et al. [130] show that a single cathode spot with a current of 40A and a contact gap of 4mm moves in a retrograde motion faster for $B_T = 180mT$ than for $B_T = 20mT$: see Figure 2.51. Figure 2.52 shows that the speed "v" of a single cathode spot on both Cu and Cu–Cr (25wt%) cathodes increases as a function of B_T:

$$v = P^* B_T^{1/2} \tag{2.32}$$

where P^* is a constant and B_T ranges from 20mT to 200mT. Nemchinsky [131] modeled the retrograde motion of a cathode spot in the presence of a transverse magnetic field. He explains the

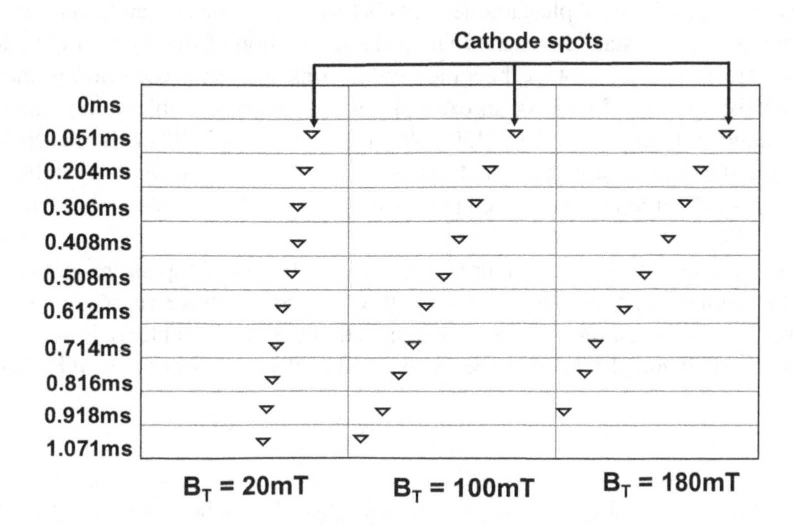

FIGURE 2.51 Single, 40A, cathode spot's retrograde motion on a Cu cathode with a contact gap of 4mm, for three values of external transverse magnetic fields [130]

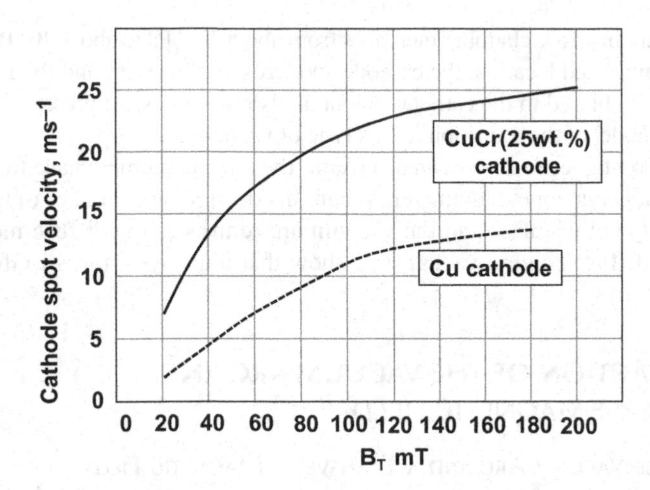

FIGURE 2.52 Retrograde velocity of single 40A cathode spot on Cu and Cu–Cr (25 wt%) cathodes as a function of the transverse magnetic field [130].

retrograde motion in terms of the voltage drop fluctuations, theHall field created by surface positive charges at the retrograde side of the cathode spot and the probability of the spot jumping. His model gives a cathode spot velocity for a spot current of 30A and a transverse field of 80mT of about 6–7 ms^{-1}. This velocity is similar to that given by Wang et al. [130] for a cathode spot on Cu with a 40A current shown in Figure 2.52. Shi et al. [132] present a similar relationship for the B_T range 20mT to 1250mT. They also develop a random-walk model of a single cathode's spot motion in a B_T. The relationship between the external B_T and the ignition probability of a new cathode spot is analyzed. They assume that the ignition probability is proportional to the magnetic pressure around the former cathode spot. Their model uses Beilis' hypothesis [133] that the ignition probability of the new cathode spot is highest at the retrograde side of the cathode crater where the magnetic pressure is the highest. Chaly et al. [134] also show that cathode spots move in a retrograde manner when subjected to a transverse magnetic field B_T.

Alferov et al. [135, 136] show that a dc current can be interrupted by applying an external B_T across a diffuse vacuum arc. They open Cu–Cr (50wt%) contacts in dc circuits ranging from 100A to 200A with B_T values from 100mT to 200mT. Initially as the contacts begin to open a low voltage diffuse vacuum arc with multiple cathode spots is formed. As the contacts continue to open the contact gap increases and a stage is reached when the interaction of the B_T with the cathode spots begins to take effect. The arc voltage becomes wildly unstable with oscillations that can reach greater than 1000V. Figure 2.53 shows an example of the highly variable voltage across a single cathode spot subjected to an external B_T [130]. When the B_T is applied across the diffuse vacuum arc in a dc circuit, the high voltages cause a decrease in the dc current until the unstable vacuum arc extinguishes and the dc circuit is interrupted [135, 136]. Figure 2.54 illustrates this effect for an 80A vacuum arc in a 500V circuit [134]

Consider the open contacts shown in Figure 2.55(a) [137]. The cathode spots are spread over the contact surface. They emit the ions and electrons that make up the neutral plasma carrying a current density in the intercontact space. If a uniform transverse magnetic field, B_T, is suddenly impressed across the contacts as is shown in Figure 2.55(a), then the plasma will be dominated by the Hall Field E_H given by:

$$E_H = \frac{j \times B_T}{n_e e} \tag{2.33}$$

where n_e is the electron density and e the electron charge. E_H is the transverse field required to move electrons across the magnetic lines of force. Its direction is shown in Figure 2.55a. At a distance (see Figure 2.55b) from a cathode spot emitting an electron current i_e and an ion current

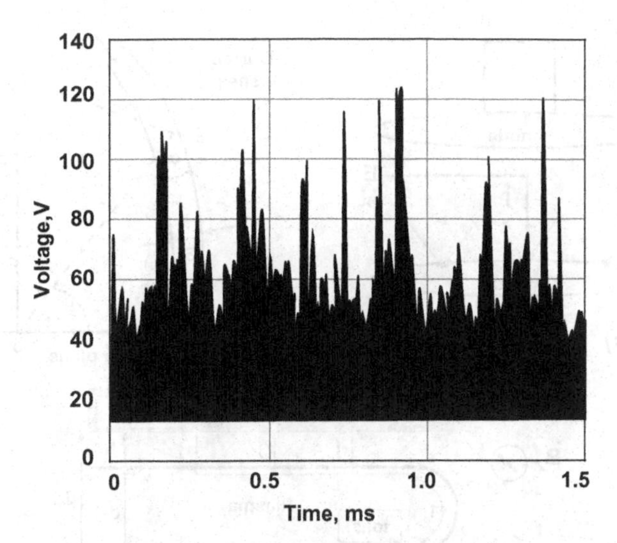

FIGURE 2.53 The voltage excursions shown by a single 40A cathode spot subjected to an external, transverse magnetic field, B_T [130].

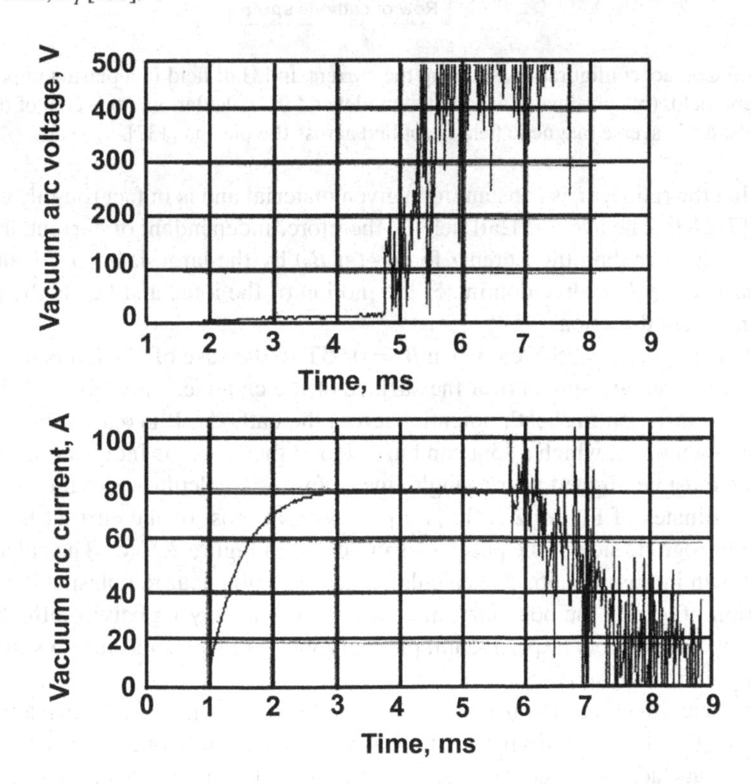

FIGURE 2.54 The effect of an external magnetic field on cathode spots showing the interruption of a dc current [136].

i_i we have $j = -i_e/(2\pi r^2)$ and $n = i_i/(2\pi r^2 e v_i)$, where v_i is the mean ion velocity. The mean Hall field is therefore:

$$E_H = \left(\frac{i_e}{i_i}\right) v_i B_T \tag{2.34}$$

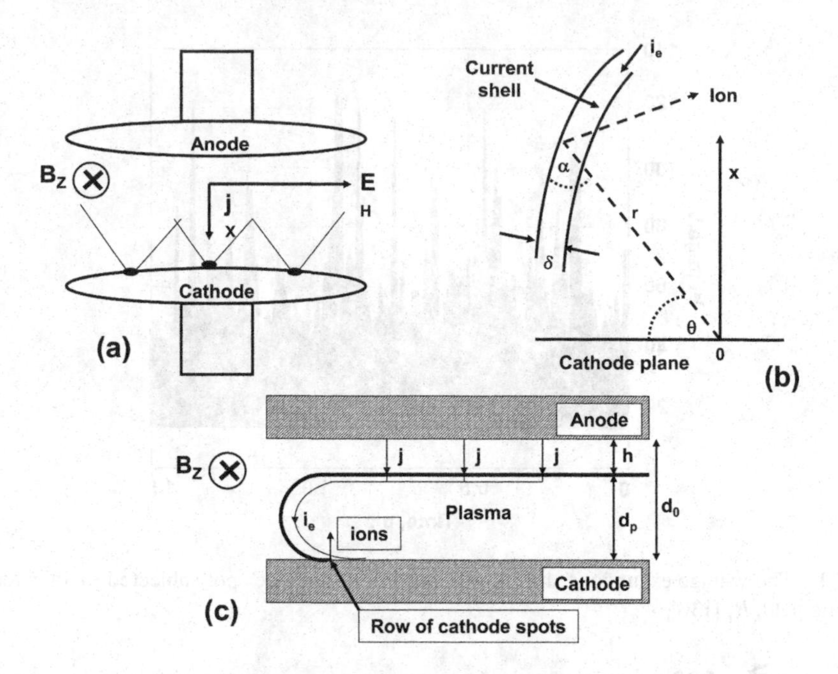

FIGURE 2.55 (a) Contact configuration, showing the current and Hall field in a plasma unperturbed by the transverse magnetic field: (b) coordinates used in the model and the calculation; (c) sketch of the plasma configuration as a pulsed transverse magnetic field is applied across the plasma [137].

Now we know that the ratio i_e / i_i is constant for a given material and is in fact roughly equal to 10 for most materials [73, 83]. The average Hall field is, therefore, independent of current. The Hall force on the ions eE_H is greater than the Lorentz force $-(ev_iB_T)$ by the large factor i_e / i_i and is opposite in sense. For that reason that force dominates the motion of the ions, and bends the plasma in the "forward" or Amperian direction.

The Hall field is large, $E_H = 35$ V/cm when $B_T = 0.05$T in the case of Cu and is transverse. If the cathode spots were to remain spread over the surface of the cathode, as in Figure 2.55(a), the Hall field would cause a variation in sheath potential across the cathode. It is reasonable to assume that there is only one potential at which a spot can burn stably; therefore, as indicated in Figure 2.55(c), the cathode spots must be aligned near a single line of force. A calculation of the plasma structure that uses the coordinates of Figure 2.55(b) [136] shows that most of the current is restricted to a thin shell on the retrograde side of the plasma, as indicated in Figure 2.55(c). The calculated plasma structures are shown in Figure 2.56. The calculated dense region near the plasma boundary results from the reflection of ions at the boundary, and all reflected ions lying between the boundary and the internal reflection envelope. Experimental photographs of this plasma show a similar structure, but it is markedly more diffuse [137].

Once the ions head in a direction parallel to the anode surface, in order to drive the circuit current into the anode, the electrons will have to flow across the gap 'd' between the neutral plasma and the anode. The voltage across the contact gap will increase to a value given by the Child's Law for space charge limited current [138]:

$$U = Kj^{2/3}d^{4/3} \tag{2.35}$$

where $K = 5690\text{V}/(\text{A})^{2/3}$. This effect has been used, as I have discussed above, by Alferov et al. to interrupt dc currents up to 400A in 200V circuits. It has also been used to successfully develop a

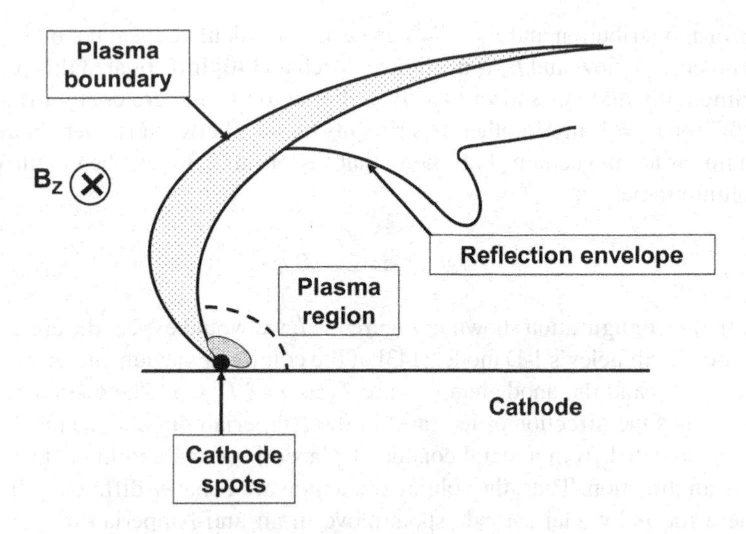

FIGURE 2.56 A sketch of the calculated plasma structure [137].

high-voltage, dc, circuit breaker that can interrupt currents up to 15kA for circuit voltages up to 80kV [139]; see Section 6.11.

2.6.2 The Columnar Vacuum Arc and a Transverse Magnetic Field

I will discuss the practical application of the transverse magnetic field (TMF) to control the columnar vacuum arc in Chapter 3 in this volume (Section 3.3.3), when I discuss practical vacuum interrupter contact structures. Here, let us consider a columnar vacuum arc that has formed between the two rails shown in Figure 2.57. If an external TMF of strength, B_T, is impressed on the columnar vacuum arc then a force F_T in the direction shown will be imposed upon the column arc of length L_a, carrying a current i, which is given by:

$$F_T = i \times L_a \times B_T \tag{2.36}$$

In a practical contact design, the current flow in the rails is shown in Figure 2.57. Here the current flowing in the rails supplies the TMF, B_T. The actual value of B_T in the region of the columnar arc is not obvious. At a position 'P' a good distance behind the arc column, the current in the rails has a

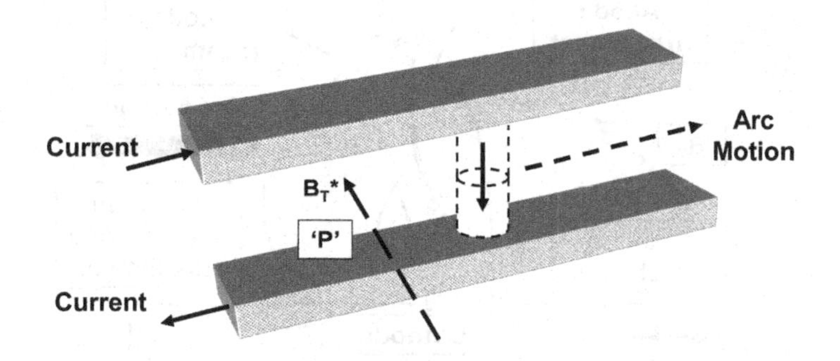

FIGURE 2.57 A columnar vacuum arc between conducting rails.

more or less uniform distribution and $B_T = B_T^*$ can easily be calculated. On the other side of the arc column there is no current flow and B_T is negligible. Michal [140] investigated this problem analytically and experimentally and has shown that B_T in the region of the arc column is about half that at the position "P" for $L_a = 3$mm. In other experiments on arcs between runners in air [141] and an analysis of a columnar arc in vacuum [142] shows that Equation (2.36) can be modified for the force F_T on the arc column to be:

$$F_T = n^* \times i \times L_a \times B_T^* \tag{2.37}$$

where $n^* \approx 0.5$. In the configuration shown in Figure 2.57, we would expect the arc column to move in the direction shown. Shmelev's 1-D model [143] of the columnar vacuum arc gives the ratio of the cathode temperature T_C and the anode temperature T_A as $T_C / T_A \approx 1$. The columnar plasma in the contact gap determines the direction of its travel in the Amperian direction. This direction is now one that would be expected from a metal conductor placed between and in contact with the rails; i.e., in the Amperian direction. Thus, the columnar vacuum arc behaves differently from the diffuse vacuum arc where the individual cathode spots move in an anti-Amperian motion or retrograde motion. It has been found experimentally that when an arc forms between opening contacts it does not immediately move under the influence of this B_T. The arc roots initially dwell at the location where they are first established. It is only after a certain dwell time (or perhaps after a minimum value of arc length is reached) that the arc column begins to move [141, 144, 145]. Once the arc does begin to move, there will be opposing forces on the column arc from the arc roots attached to the cathode and to the anode. The arc roots cannot move instantaneously and thus will limit the speed of the arc motion. Delachaux et al. [146] and Shmelev et al. [147] have developed models of the high-current columnar vacuum arc with a B_T impressed across it. They use a 2-D MHD model together with a radiation transfer approximation. The attachment at the cathode is assumed to be a solid uniform area. Both models also assume that the arc roots at both contacts have a temperature in excess of 3000K. This is somewhat higher than experimental data that gives anode temperature of a columnar vacuum arc to be between 2000K and 3000K [148, 149].

A schematic of the arc column motion is shown in Figure 2.58 [147]. It illustrates a possible sequence for its travel. The plasma jet from the anode advances beyond the main arc column and heats the cathode area in the Amperian direction. Once this surface reaches a high enough temperature (perhaps greater than 2200K), there will be a copious emission of metal vapor that will allow the cathode attachment to jump to its new position. The anode attachment will then follow a

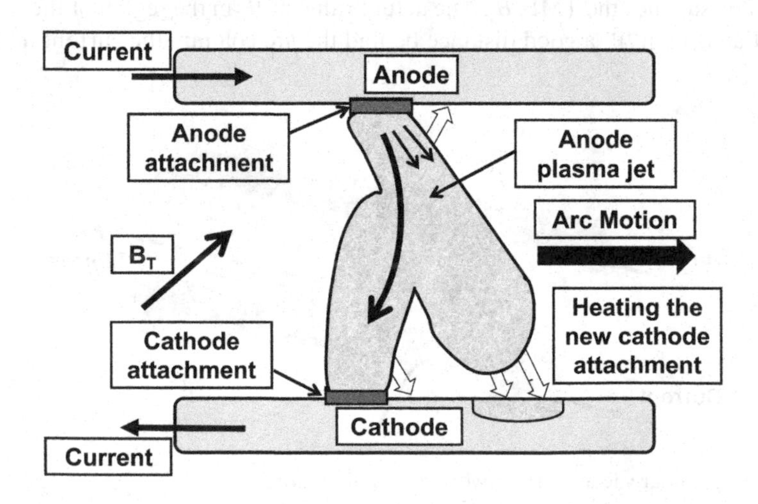

FIGURE 2.58 The motion of the columnar vacuum arc subjected to a transverse magnetic field [147].

few microseconds later. In vacuum there is no drag force on the arc that results from the arc motion through an ambient atmosphere. There are, however, drag forces resulting from momentum lost by neutral atoms escaping from the arc column as it moves [145] and the reluctance of the arc roots on the contacts to move to a new location. When the driving force from the magnetic field is greater than the drag forces the arc will begin to move.

2.7 THE VACUUM ARC AND AN AXIAL MAGNETIC FIELD

In the literature on the vacuum arc the term "Axial Magnetic Field" really refers to an "Axial Magnetic Flux." In order to prevent confusion, I will generally adopt the commonly used expression "Axial Magnetic Field" instead of the more correct "Axial Magnetic Flux" (interestingly, the abbreviation AMF can be used interchangeably for both expressions). Let us consider the single cathode spot in an axial magnetic field (AMF), B_A, shown in Figure 2.59. An electron with a charge e, a velocity v_e, a mass m_e on a plane perpendicular to B_A will experience a force F_R given by:

$$F_R = ev_eB_A \tag{2.38}$$

Thus

$$ev_eB_A = \frac{m_e v_e^2}{r} \tag{2.39}$$

or

$$r = \frac{m_e v_e}{eB_A} \tag{2.40}$$

i.e., an electron traveling into the intercontact gap above the cathode spot will form a helix around the magnetic field line with a maximum radius given by Equation (2.40). If B_A = 100mT and the electron energy is 6eV, Equation (2.40) gives a radius of about 0.08mm. Thus, the electrons traveling from a cathode spot in the presence of an AMF will be confined to a helix around the magnetic flux line. Once the electrons are confined in this manner, the ions attempting to travel away from the cathode spot will be subjected to very high electric fields that will also force them to remain with the electrons. As both the electrons and the ions are confined, it would be expected that more collisions would occur than if they were allowed to move freely in the intercontact space. This would result in

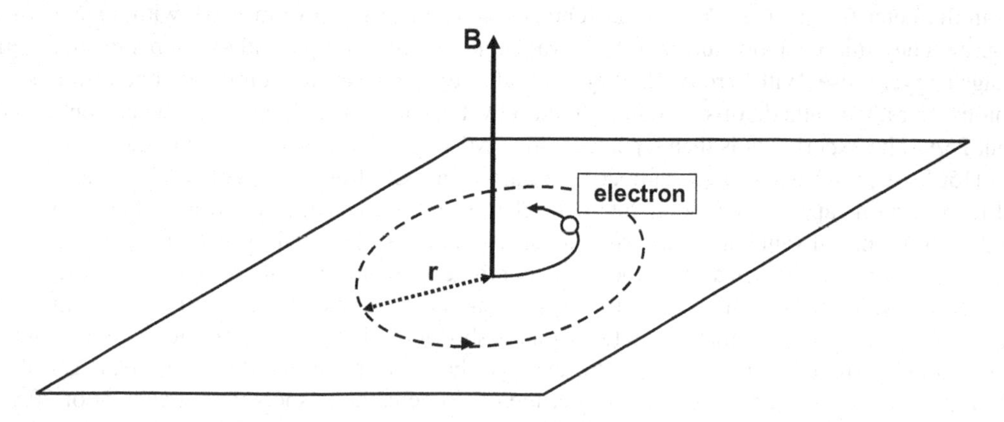

FIGURE 2.59 An electron emerging from a single cathode spot perpendicular to the direction of an applied axial magnetic field.

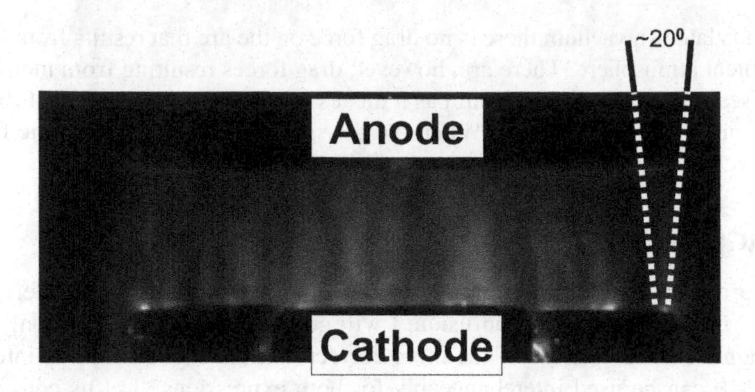

FIGURE 2.60 Image of a 7kA diffuse vacuum arc with an 93mT axial magnetic field showing the divergence of a cathodic plasma plume as it crosses the contact gap [150].

an increase in arc voltage. The radius of the electron helix would also be expected to expand as the electron travels toward the anode.

2.7.1 The Low-Current Vacuum Arc in an Axial Magnetic Field

For the low-current vacuum arc, once the diffuse vacuum arc forms from the initial bridge column arc, the cathode spots will spread over the cathode surface, but now the plasma above each cathode spot will be confined to a much narrower cone. In fact, observations of a cathode spot with its confined plasma plume show that it resembles a "search-light." If the AMF is high enough, say greater than 80 mT, then the ion current that can be measured on a surrounding shield biased to cathode potential drops by 60%: see Heberlein et al. [57]. The ion distribution cone is now very forward directed with most of the ions confined to a cone with an angle of less than 25° normal to the cathode surface [57]: see, for example, Figure 2.60 [150]. The plume radius is difficult to determine. Nemchinsky [151] and Gundlach [152] have modeled the radius of the plasma plume at a distance from the cathode spot. Their data together with experimental points from the work of Morimiya et al. [153] and Rondeel [151] is shown in Figure 2.61. The cathode spots are also less mobile [148, 151] and are more content to remain on the cathode surface under the shadow of the anode [57]. Measurements of the expansion of a ring of cathode spots from a central breakdown point show a marked dependence upon the axial magnetic field; see Figure 2.62 [152].

The AMF affects the current in the cathode spot. Wu et ai [156] have investigated this effect for 30mm diameter Cu and Cu–Cr (25wt%) contacts in an external, uniform AMF with a contact gap of 6mm. They initiate a low-current diffuse vacuum arc with multiple cathode spots using a high-voltage trigger pulse. With zero AMF, they gradually reduce the current until only one cathode spot remains. For a Cu cathode this current is about 65A, but for the Cu–Cr cathode the current is only about 25A. The experiment is then repeated with AMFs up to 250mT: their data is shown in Figure 2.63 [156]. At 20mT the average cathode spot current for Cu drops from about 60A to about 35A and then remains approximate level for B_A levels up to 220mT. The Cu–Cr data shows a similar trend with the initial value dropping from about 22A to about 18A and the remains constant as B_A increases. The researchers in a new experiment continue to reduce the current, with zero AMF, until they reach a value below which the cathode spot can no longer exist. They call this the threshold current. The effect of B_A on the threshold current is shown in Figure 2.64 [156]. It is interesting that the threshold current for Cu is similar to the average chop current observed for that metal: see Table 3.7. Wang et al. [157] record the average lifetime of single cathode spots with a current of 20A as a function of B_A. Figure 2.65 shows that the lifetime of the cathode spot increases significantly for values of B_A up to 60mT then slowly decreases in the range 60mT to 220mT.

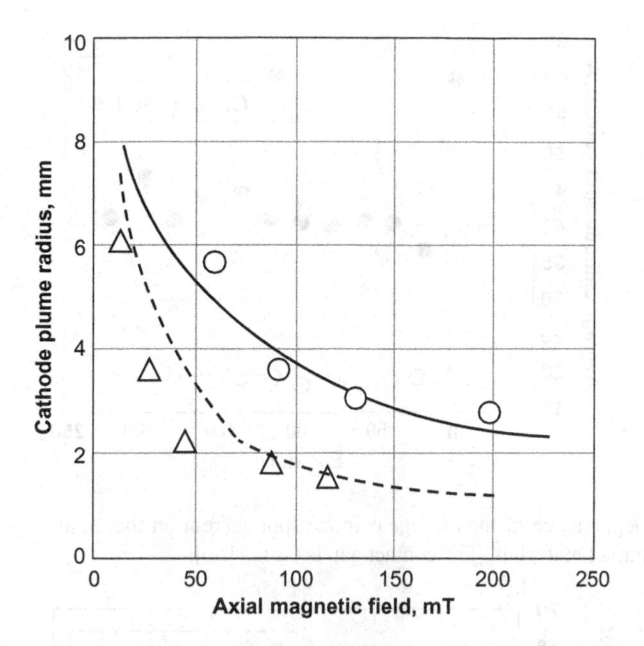

FIGURE 2.61 Radius of the plasma plume a few millimeters above a the cathode spot as a function of the axial magnetic field flux; (– – – –) [152], (————) [151], ○[154], △[153].

Kimblin et al. [158] first established that the AMF has an interesting effect on the arc voltage of the vacuum arc for currents greater than several kiloamperes once the diffuse vacuum arc has been stabilized and the cathode spots have spread over the cathode's surface. Figure 2.66 illustrates this for a 4.2kA dc arc between 70mm diameter contacts with a floating metal shield surrounding them (i.e., not connected to the anode or cathode contact). When the AMF is zero, the vacuum arc has

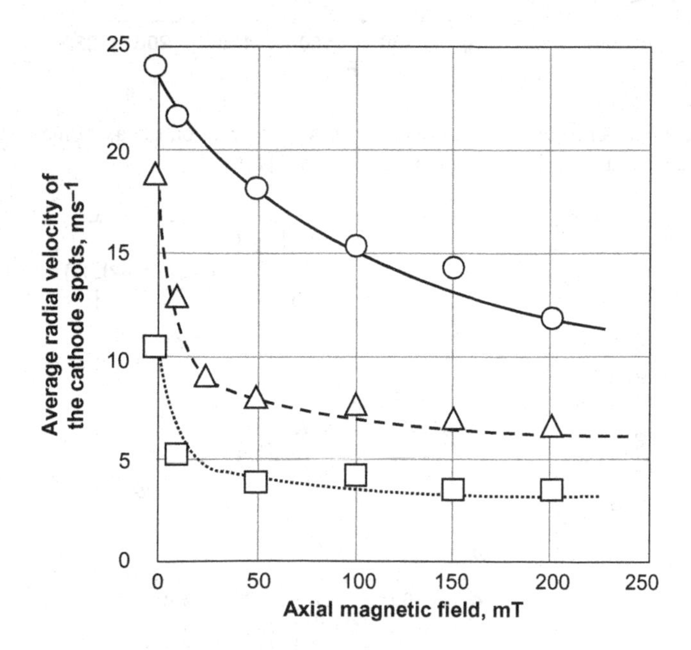

FIGURE 2.62 Average radial velocity of cathode spots on Cu as a function of the axial magnetic field and current: □1kA, △3kA, ○5kA [152].

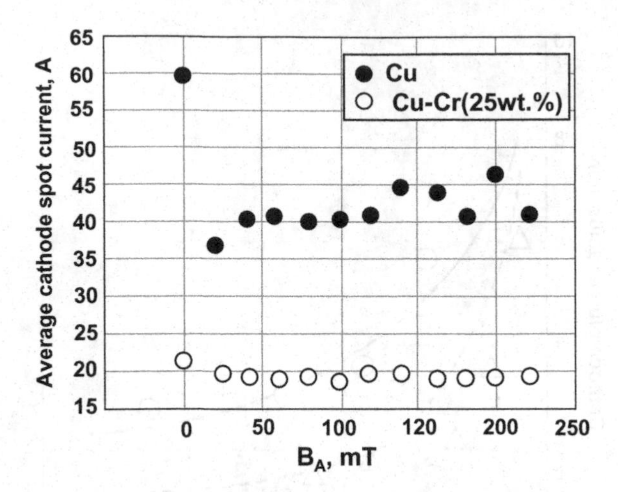

FIGURE 2.63 The dependence of the average cathode spot current on the axial magnetic field, B_A, for Cu and C–Cr (25 wt%) contact materials. The contact gap is 6mm [156].

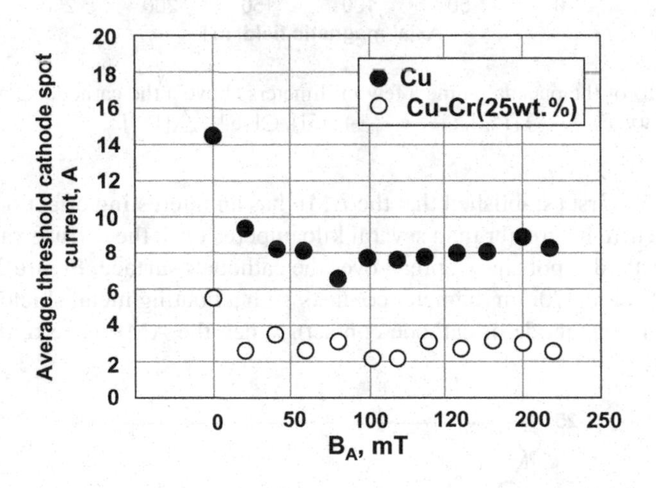

FIGURE 2.64 The dependence of the threshold cathode spot current on the axial magnetic field B_A, for Cu and C–Cr (25 wt%) contact materials. The contact gap is 6mm [156].

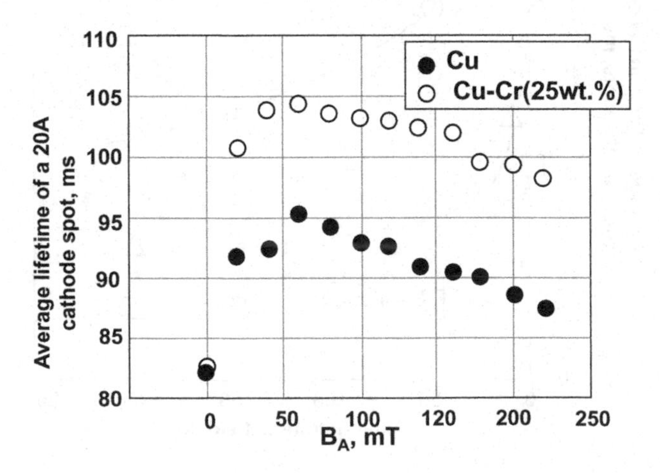

FIGURE 2.65 The average lifetime for a 20A cathode spot on Cu and Cu–Cr (25 wt%) cathodes as a function of axial magnetic field. The contact gap is 6mm [157].

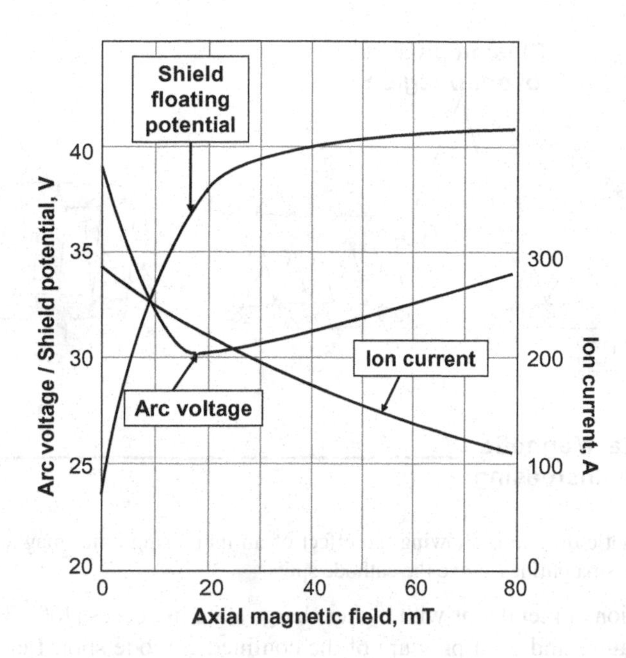

FIGURE 2.66 Effect of axial magnetic field on a 4.2kA a vacuum arc between 70mm diameter Cu contacts 19mm apart [158].

a voltage of about 38V and the shield floats at about 24V (i.e., the potential just above the cathode spot). As the AMF increases, so the arc voltage decreases, until a minimum value of about 30V is reached for an AMF of about 20mT. At the same time, the potential of the surrounding, floating shield also gradually increases. As the AMF increases further, the arc voltage increases; so, for an AMF of 80mT, the arc voltage is about 33V. The shield voltage rises to a value of 40V (i.e., a potential greater than the anode potential). When the shield is biased at cathode potential there is a gradual decrease in the collected ion current as the magnetic flux increases. These phenomena can be discussed in terms of the effect of the AMF on the cathode spots and on the intercontact plasma [151, 152]. As presented in Section 2.3.2, for the case of zero AMF, if the density of cathode spots is high enough, the plasma plumes above the cathode spots will overlap. This results in an increase in the plasma density and an increase in collisions between the plasma constituents. The azimuthal magnetic field from the circuit current flowing though the plasma confines it more and its resistivity increases. This results in a higher arc voltage. Figure 2.67 illustrates the effect that an increasing AMF has on the structure of a multicathode spot diffuse vacuum arc. As soon as the AMF is applied the confinement of the electrons will be initiated and they will form a helix along the magnetic flux lines. The plasma plume above the cathode spot will also be confined with the electrons. As the AMF increases the position where the plasma plumes initially overlap moves toward the anode. The AMF thus initially forces the electrons to travel to the anode in a much more direct path and the region of plasma plume overlap is reduced. Thus, the arc voltage will show a gradual decrease as the AMF increases. The motion of the cathode spots is reduced as the AMF is increased. In fact, the reduced motion of the cathode spots not only results in a slightly higher erosion rate at the cathode but also a buildup of material at the anode [154]. Table 2.10 shows erosion data measured by Anders et al. [159] for the cathode erosion rate with and without an AMF. As the magnetic field continues to increase, the electrons will form tighter helixes around the magnetic flux lines and the plasma plumes will eventually separate. Consequently, each cathode spot will operate independently and in parallel. The decreasing plume diameter results in an increase in the current density and some energy loss from collisions between the ever-increasing confinement of the plasma components. As a result, the arc voltage will increase as the AMF increases. One other consequence of an increasing

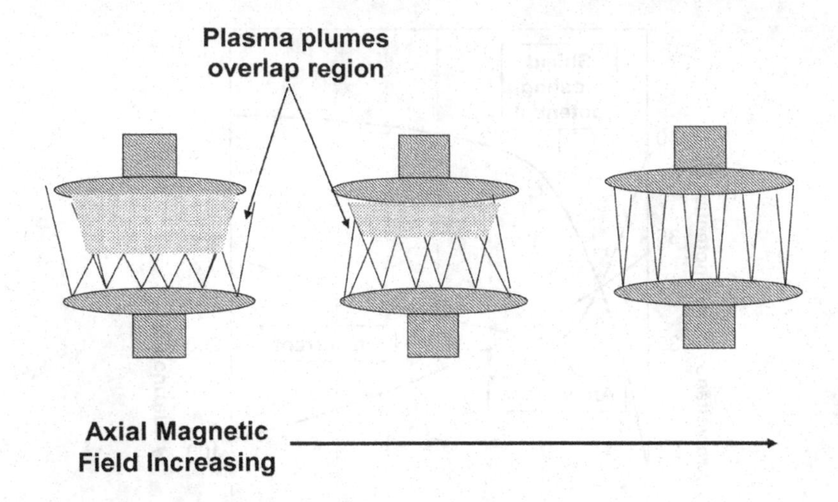

FIGURE 2.67 Schematic diagrams showing the effect of an increasing axial magnetic field reducing the overlap region of the plasma plumes above the cathode spots.

AMF is that the fraction of metal ions with higher charge state increases [160]. This may result from an increased temperature and even pressure of the confined cathode spot. Indeed, Galonska et al. [161] have shown that the electron temperature can increase from a value of about 2 to 3eV for $B_A = 0$ to about 8 to 12eV for a $B_A = 60mT$. Simulations by Oks et al. [162] have shown that a 1eV increase in electron temperature can explain the increase in the charge state. Finally, at high AMF values only the most energetic ions can escape from the intercontact space, which results in the surrounding shield now floating above the anode potential. Also, the confined arc roots at both the anode and the cathode can now result in material loss from both contacts [163].

In experiments using a triggered low current (3kA to 8kA), Song et al. [164] show that the periphery of the expanding cathode spot ring expands slower as an AMF increases: see Figure 2.68. They analyze the effect of the overall azimuthal magnetic field B_θ that surrounds the expanding cathode spot ring interacting with the AMF. This causes them to divert from a straight radial expansion observed for $B_A = 0$. The individual cathode spots are deflected an angle θ, the Robson angle [165] and follow a longer curved path: see Figure 2.69. This would be seen as a slower expansion of the periphery of the expanding cathode spot ring. The Robson angle is given by:

$$\theta = \acute{\eta} \times \varphi \tag{2.41}$$

TABLE 2.10
Cathode Erosion Rate with and without a 170mT AMF for Thin Film Cathods and 250μs Current Pulses [159]

Metal	Erosion rate without an AMF (μg.C^{-1})	Erosion rate with a 170mT AMF (μg.C^{-1})	Increase (%)
Ag	76.3	83.7	10
Cu	51.1	56.0	10
Au	86.2	100.9	17
Ni	38.4	49.1	28
Pt	111.1	116.4	5
Mo	30.9	38.8	26
W	44.4	50.0	13

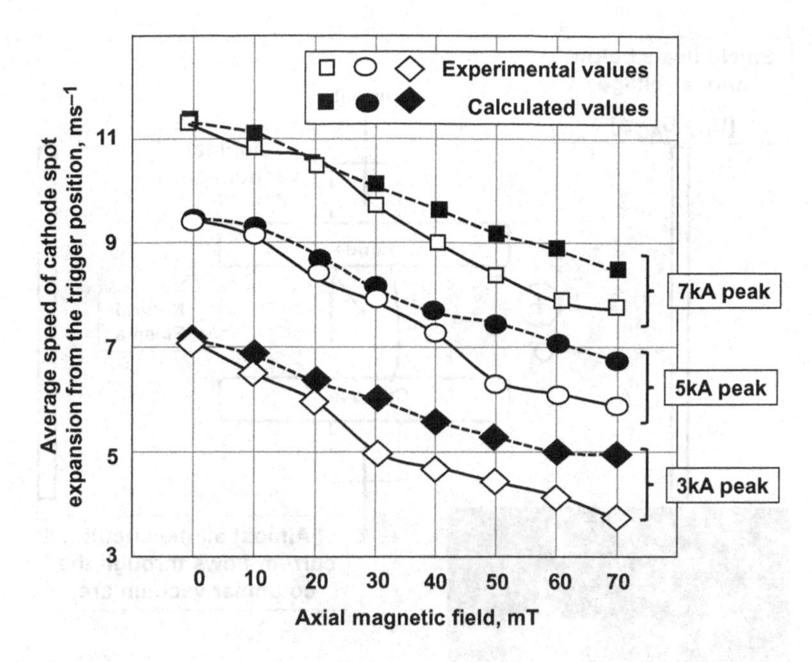

FIGURE 2.68 A comparison between the calculated cathode spot periphery ring expansion speeds with experimental measurements for Cu–Cr (25 wt%) cathodes as a function of peak ac current and uniform axial magnetic fields [164].

where $\varphi = arctan(B_A/B_\theta)$, $\acute{\eta} \approx 0.8$ for Cu–Cr contacts [134] and the new ring expansion speed is:

$$v_{AMF} = v_0 cos\theta \tag{2.42}$$

Song et al's calculation is shown in Figure 2.68, It can be seen that the bending of and the lengthening of the cathode spot paths does not totally explain the slower expansion of the periphery of the expanding cathode spot ring as B_A increases.

2.7.2 The High-Current Vacuum Arc in an Axial Magnetic Field

The opening of contacts carrying high currents has been discussed in Section 2.4. Figure 2.70 can represent the resulting columnar arc. Here the metal vapor required to maintain the high-pressure

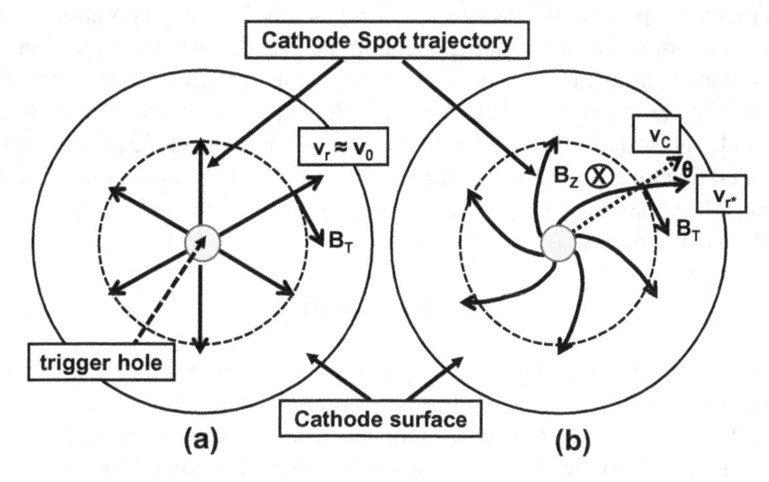

FIGURE 2.69 Schematic diagrams of the expansion of the expanding cathode spot ring's periphery (a) without an AMF and (b) with an AMF: v_r expansion speed without an AMF: v_{r*} absolute magnitude of cathode spot retrograde speed with an AMF [164].

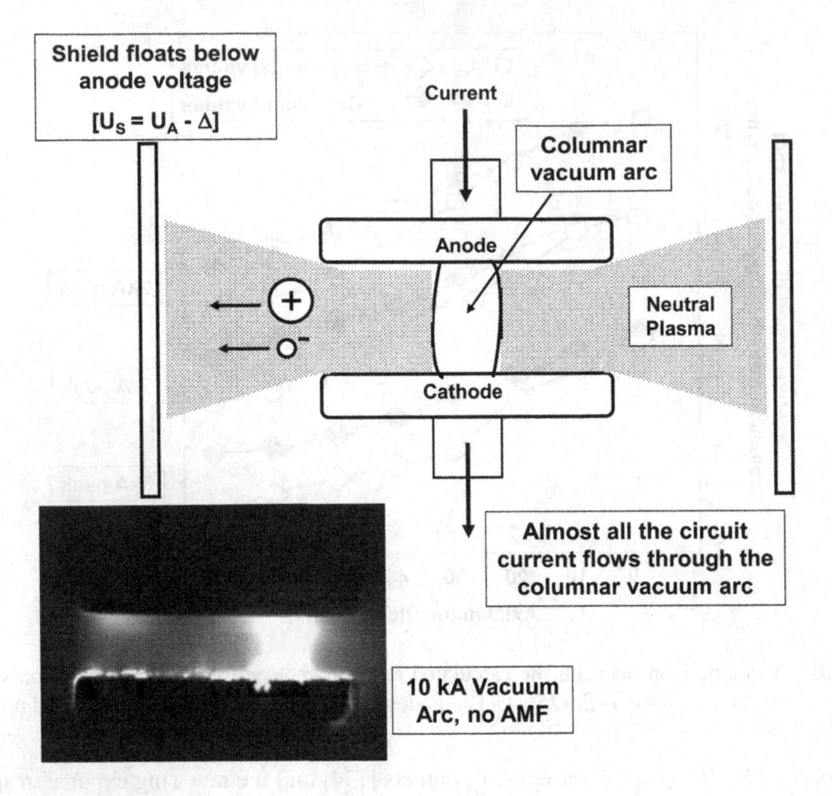

FIGURE 2.70 The high-current columnar vacuum arc with no axial magnetic field.

plasma column is supplied by the evaporation from the arc roots on the cathode and the anode. Any electrons and ions leaving this column will be free to escape the intercontact region and a surrounding floating shield will have a potential close to that of the potential just above the cathode region.

Figure 2.71 illustrates the effect of an AMF when it is applied to this columnar arc [166]. The electrons that escape from the arc column will be confined to helixes around the magnetic flux lines. When they do this, they also confine any escaping ions, thus a plasma region will form surrounding the main arc column, i.e., a parallel current path will develop. When the current in the main column is no longer sufficient to provide the energy into the contacts for the evaporation of enough metal vapor to replace that lost to the surrounding vacuum, the vacuum arc will transition into a diffuse mode. For the complete transition to the fully diffuse mode at high currents a threshold AMF is required; Figure 2.72 illustrates this. If the AMF is below the shaded line, the arc will not be fully diffuse. Here anode spots can be observed, which can result in rapid anode erosion. When the AMF is above the shaded line for a given current, then the high-current vacuum arc is fully diffuse. From these experiments Schulman et al. [166] have established the concept of the Critical Axial Magnetic Field, B_{crit}. In their experiments:

$$B_{crit} = 3.2\left(i_p - 9\right) \tag{2.43}$$

where i_p is the peak current in kiloamperes. They also show in Figure 2.73, for an AMF greater than B_{crit}, the time required for the arc to go diffuse after contact separation.

Observations by Taylor et al. [167] on vacuum arc behavior from the initial contact separation until the formation of the high-current diffuse arc in an AMF show that it follows a three-step process. This is illustrated in Figure 2.74. The first step occurs when the bridge column forms after the rupture of the molten metal bridge connecting the two contacts. The high outward plasma pressure

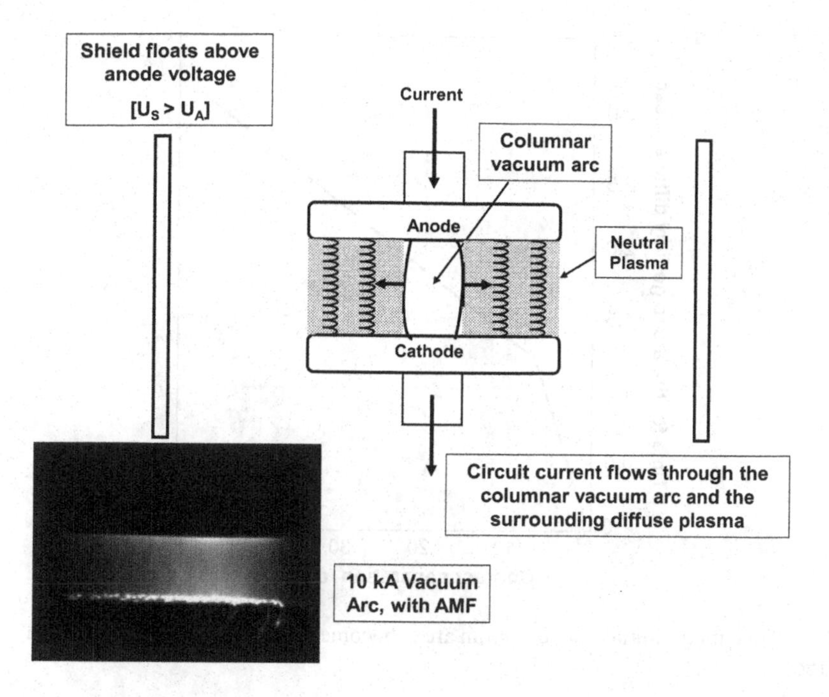

FIGURE 2.71 The high-current columnar vacuum arc in the presence of an axial magnetic field [166].

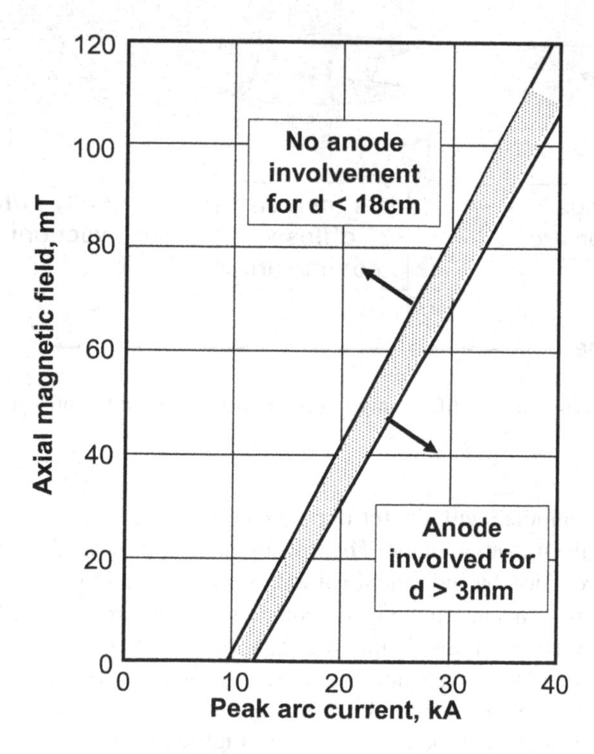

FIGURE 2.72 Anode involvement observed in high-current vacuum arcs with an applied axial magnetic field showing the influence of a critical axial magnetic field B_{crit} [166].

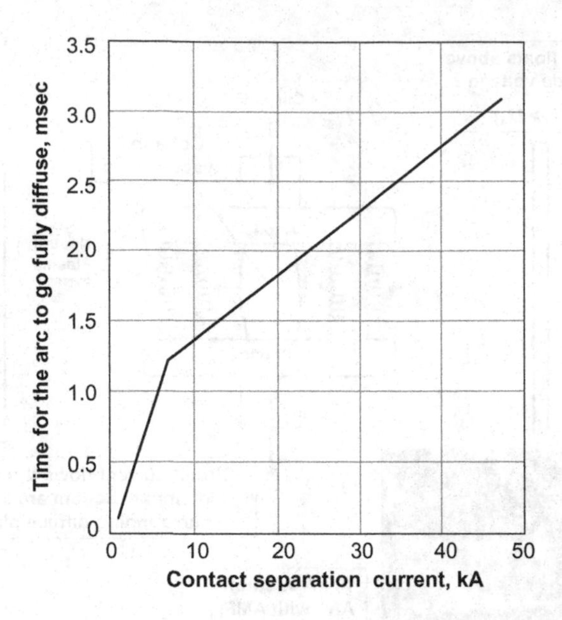

FIGURE 2.73 Time for the initial bridge column arc to become diffuse as a function of the current at contact separation [166].

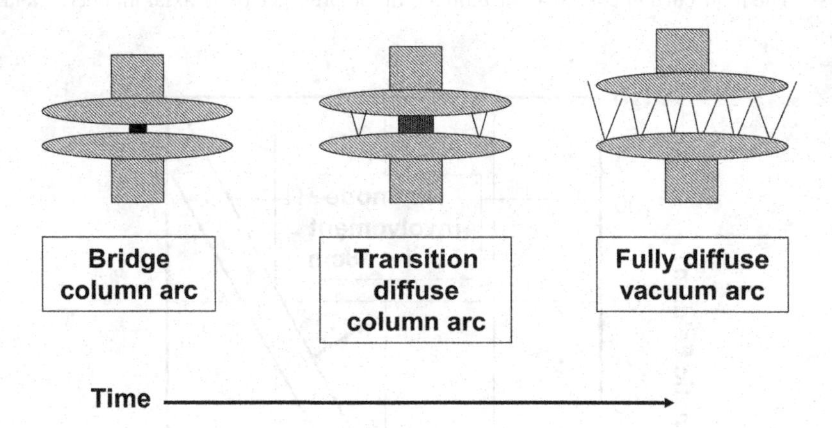

FIGURE 2.74 The transition to the diffuse high-current vacuum arc in the presence of an axial magnetic field.

in the bridge column, combined with the retrograde motion of the cathode spots, drives the expansion into the next step, the transition mode. The transition mode possesses a bright central core with a stable diameter, surrounded by individual cathode spots. The proximity of the contacts during the contact separation may encourage the formation of a concentrated arc [168], as well as increase the influence of metal vapor from anode melting [105, 106]. The constriction of the plasma in the central core increases the plasma resistance, and thereby increases the arc voltage. This produces a transient peak in the arc voltage during the transition mode; see Figure 2.75. The duration of the transient diffuse arc depends upon the magnetic field. Figure 2.75 shows the effect of doubling the magnetic field. Photographs of the change in the arc appearance for the two cases can clearly be seen in Figure 2.76. Further experiments by Taylor [169] and modeling by Taylor et al. [170, 171] show that the transition mode can only exist if the arc voltage is above a critical value. Observations of the transition mode show that it can exist in two forms; (a) a central diffuse column with few or

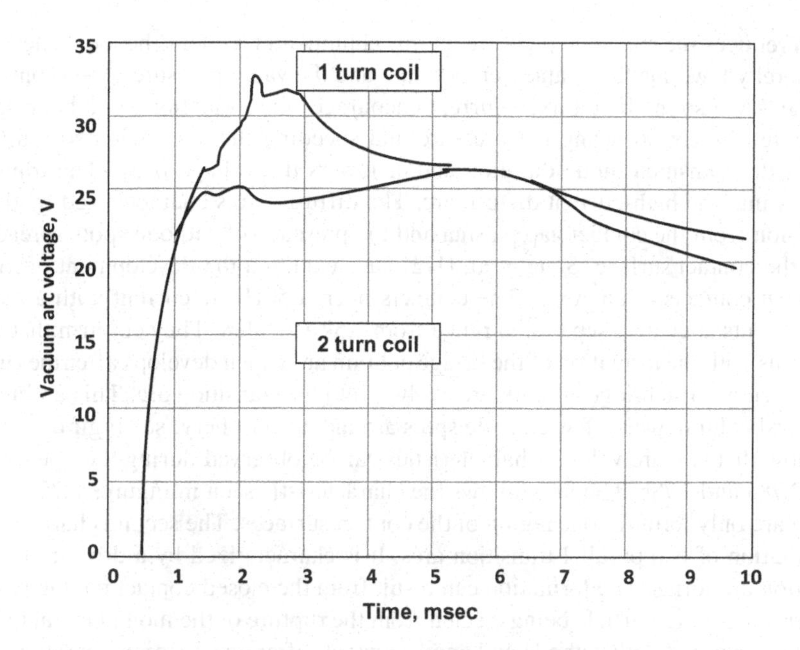

FIGURE 2.75 Example of the variation in the arc voltage across opening Cu–Cr contacts when the axial magnetic field doubles [167]

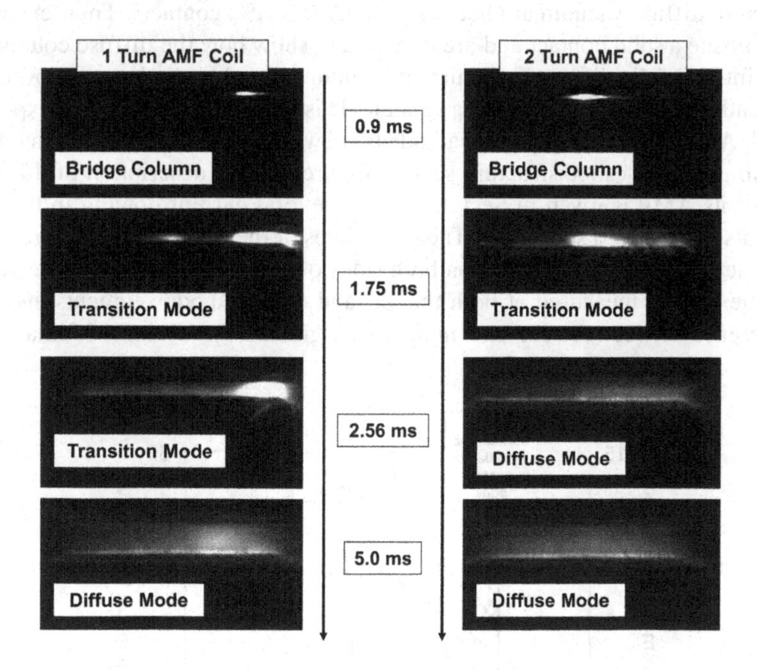

FIGURE 2.76 Photographs of the developing vacuum arc between opening contacts that correspond to the axial magnetic fields and arc voltage traces shown in Figure 2.75 [167].

no cathode spots outside the main column and (b) a central column with a number of cathode spots outside it. When the arc voltage drops below the critical value, the arcs are now in the fully diffuse mode. The model also shows that the fraction of the current flowing through the central column decreases with time during this transition to the diffuse vacuum arc. Thus, over time, more cathode spots appear outside the central core, diverting current away from the core. Lowering the current in the core both reduces the magnetic pressure on this column and reduces the local energy input into the anode, thereby lowering the contact temperature and the vapor pressure of the contact material. Increasing the AMF strength at a fixed current encourages the formation of cathode spots outside the central core, thereby lowering the pressure and speeding the conversion to a diffuse mode. This shortens the transition mode duration and/or lowers the voltage peak. The transition mode finally evolves into the high-current diffuse arc. The diffuse arc is characterized by the reduction in light emission from the contact gap plasma and the presence of cathode spots spread fairly uniformly over the contact surface. Song et al. [172] have explored the development of the transition arc between Cu contacts in an AMF. The contacts open a 50Hz ac current with a peak value of 12kA. The currents at contact separation range from 1.5kA to 9kA. They confirm that the opening sequence begins with the formation of the bridge column arc which develops after the rupture of the molten metal bridge. The bridge column arc evolves into the transition arc. This arcing mode consists of a central column with a few cathode spots around its periphery: see Figure 2.77. Song et al. [173] also show that two arc voltage characteristics can be observed during the opening sequence: see Figures 2.78a and 2.78b. The first follows the characteristic seen in Figures 2.75 and 2.76. Here the transition arc only forms in one region of the contact surfaces. The second characteristic occurs with the formation of two parallel transition arcs. It is characterized by a drop in the arc voltage as the transition arc forms. The formation can result from the closed contact having two regions of contact or perhaps from a particle being ejected from the rupture of the molten metal bridge. Figure 2.79 shows the surface of both cathode and anode contacts after one opening operation at 16kA and 25kA. It clearly shows the shallow craters formed by the transition diffuse vacuum arc [174]. In an important paper, Li et al. [175] have studied the effect of the axial magnetic field on the development of the high-current diffuse vacuum arc between Cu–Cr (50wt%) contacts. Their experiments use a split contact opposite a solid contact and are designed to show how the diffuse columnar transition arc transforms into the fully diffuse high-current vacuum arc with cathode spots covering the whole surface of the cathode. In one experiment a split anode is used and in a separate experiment a split cathode is used. An example using a split cathode is shown in Figure 2.80. The axial magnetic field is generated using the horseshoe structure seen in Figure 2.80. A description of this structure and the formation of its AMF is given in Section 3.3.4. The total current flowing in the left and right cathode segments is measured separately. The steel plates of the horseshoe structure are not split so they affect the actual currents measured on both sides of the split cathode. Figure 2.80 shows the total current, the current measured at both the left and right cathode segments and, most importantly, the difference between the currents in the two segments. An example of images observed in

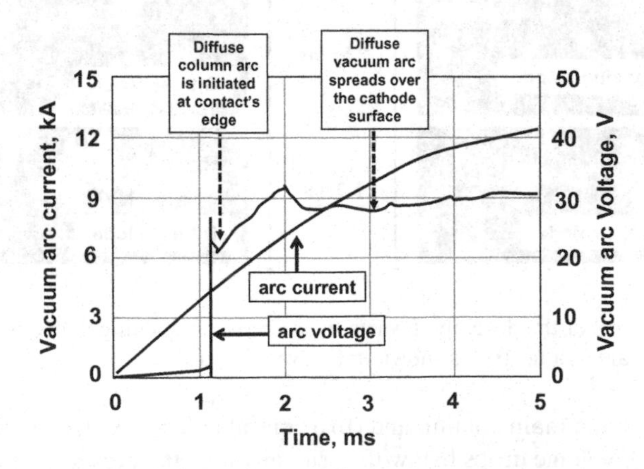

FIGURE 2.77 Initial stages of the vacuum arc current and voltage for opening Cu contacts in an axial magnetic field with the formation of the diffuse column transition arc developing into the fully diffuse vacuum arc. [172].

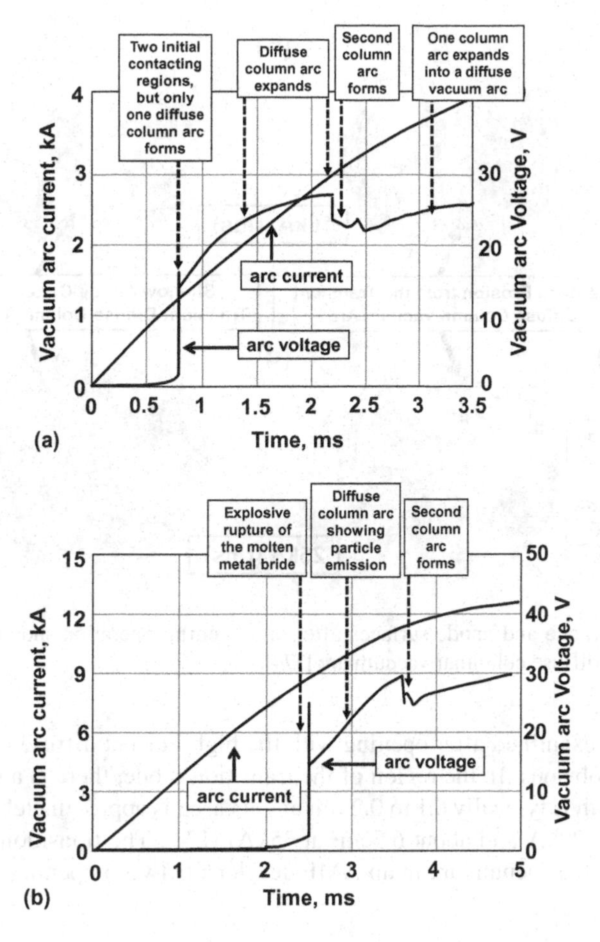

FIGURE 2.78 Initial stages of the vacuum arc current and voltage for opening Cu contacts in an axial magnetic field (a) the formation of a second diffuse column transition arc 2ms after the first formed from the molten metal bridge rupture and (b) the formation of a second diffuse column transition arc 1ms after the first resulting from a liquid droplet emitted from the molten metal bridge rupture [173].

an experiment using a split cathode is shown in Figure 2.81(a–f) for a 19.3kA peak current. Figure 2.80 shows that when the contacts are closed approximately equal currents are measured for both cathode segments. Once the contacts part, Figure 2.81(a) shows that a diffuse column arc forms on the left cathode segment. As would be expected, Figure 2.80 shows that the current measured in the left cathode segment is significantly greater than that flowing in the right segment. In Figure 2.81(b) the diffuse column arc still dominates, but cathode spots are beginning to be seen forming on the right cathode segment confirming the hypothesis presented in Figure 2.71 given by Schulman et al. [166]. The current flowing to the left cathode segment remains greater than the current flowing in the right segment. In Figure 2.81(c) only a residual cathode plume remains and the cathode spots have covered most of the surfaces of the two cathode segments. Figure 2.80 shows that the current difference between the two segments is gradually becoming less. There is a short period when the current to both segments is equal. Images for Figures 2.81 (d) to (f) show more intense light coming from the cathode spots on the right cathode segment. This is reflected by a greater current measured in that segment until current zero is reached. Figure 2.82 summarizes the possible modes of vacuum arc appearance between two butt contacts in the presence of an AMF during a half cycle of ac current.

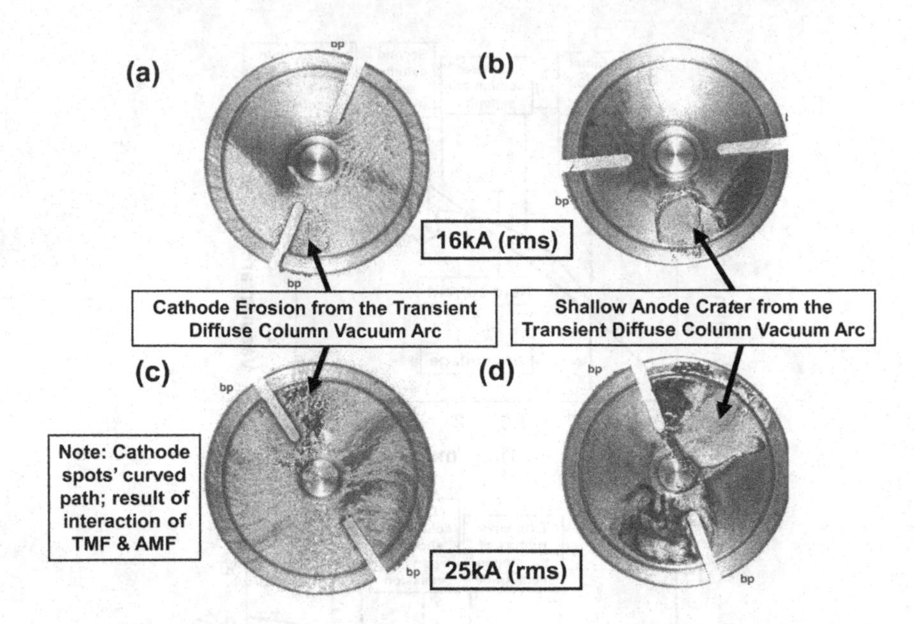

FIGURE 2.79 The cathode and anode surfaces after one opening operation showing the shallow craters formed by the transient diffuse columnar vacuum arc [174].

If the contacts are examined after opening with the high-current diffuse arc, the effects of this two-step process are obvious. In the region of the transition mode, there is a shallow crater on the anode contact. Its depth is typically 0.1 to 0.2mm and its area is approximately proportional to current (about 2.75cm^2 at 20kA and about 6.5cm^2 at 35kA) [174]. The transition mode can last up to a few milliseconds as the vacuum arc in an AMF develops between opening contacts. Chaly et al.

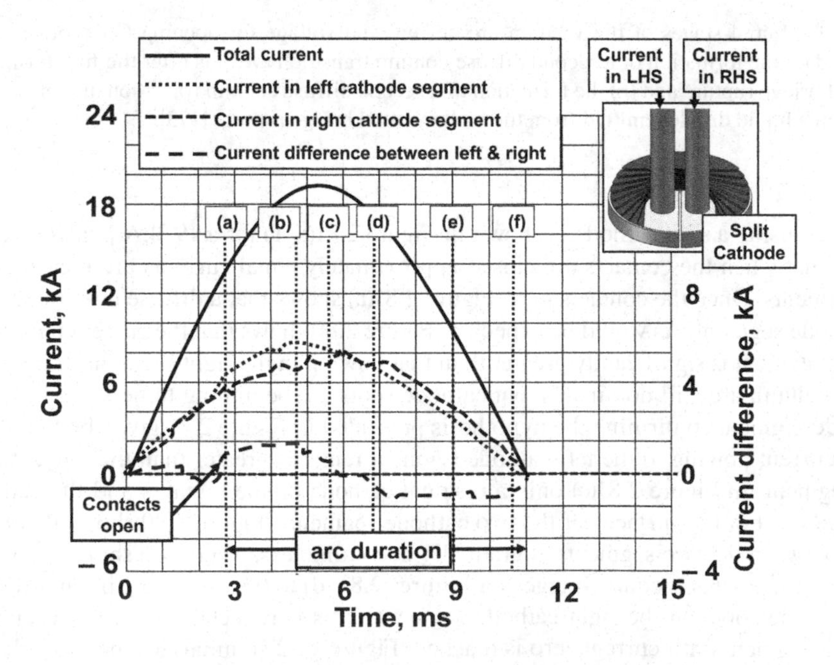

FIGURE 2.80 Arc current and cathode current measured for each cathode segment and the difference in the segment cathode currents. (difference between two arc columns for a 19.3kA peak current) [175].

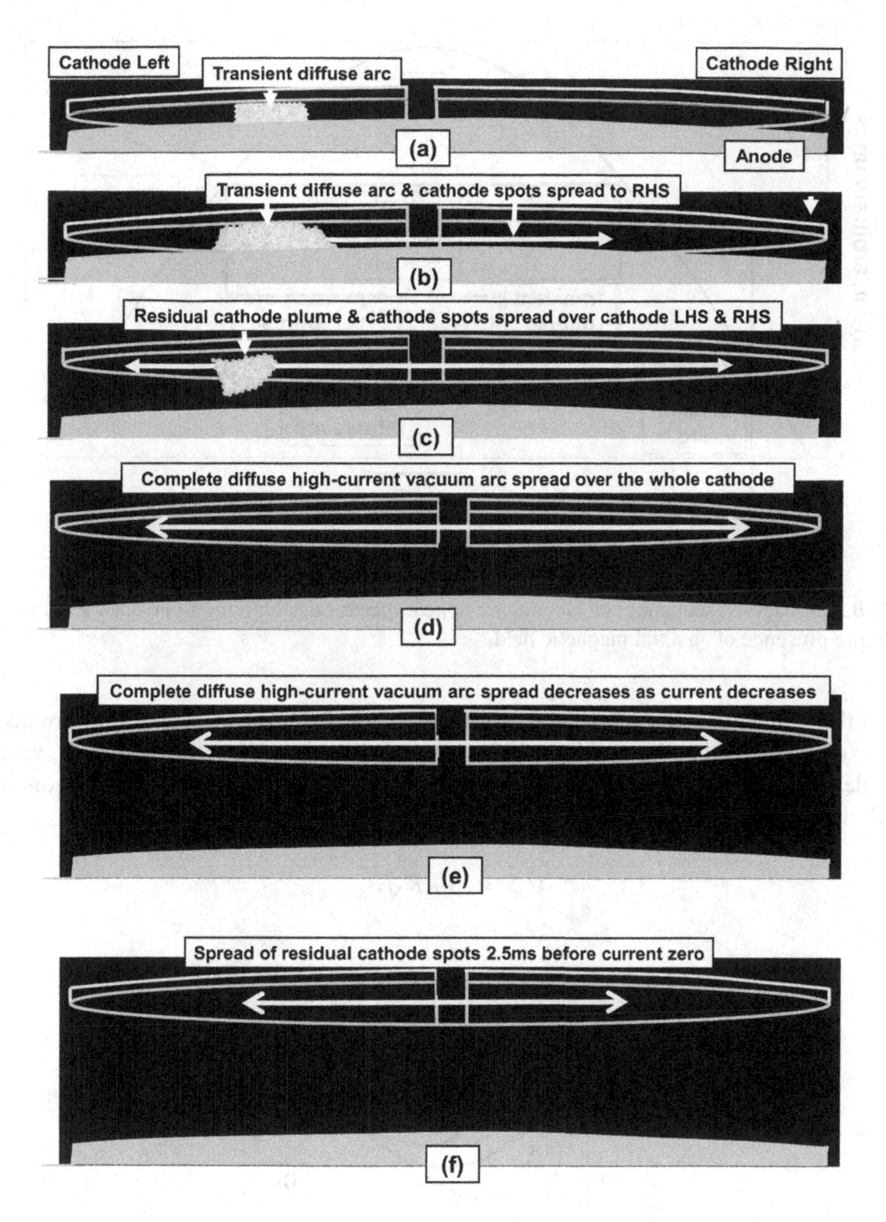

FIGURE 2.81 The development of a diffuse high-current vacuum arc using a split cathode in an axial magnetic field for a 19.6kA peak current from Figure 6 of [175].

[176] consider that it is a characteristic of cathode spots to group together at short contact gaps. The effect of the transition mode on the cathode is a confinement of the cathode spots to an area perhaps 50% smaller than that of the anode crater. These erosion patterns are very similar to those observed for the transition vacuum arc discussed in Section 2.5. Once the high-current vacuum arc becomes fully diffuse, the cathode spots spread over the cathode's surface and there is minimal erosion of the anode [167]. Jia et al. [177] and Chaly et al. [134] give precise overviews of the models of the vacuum arc in an AMF and relate them to experimental data.

Once the diffuse, high-current, vacuum arc in the presence of an AMF has fully developed, Gundlach [152] has shown that at high currents the arc voltage has a similar relationship with the

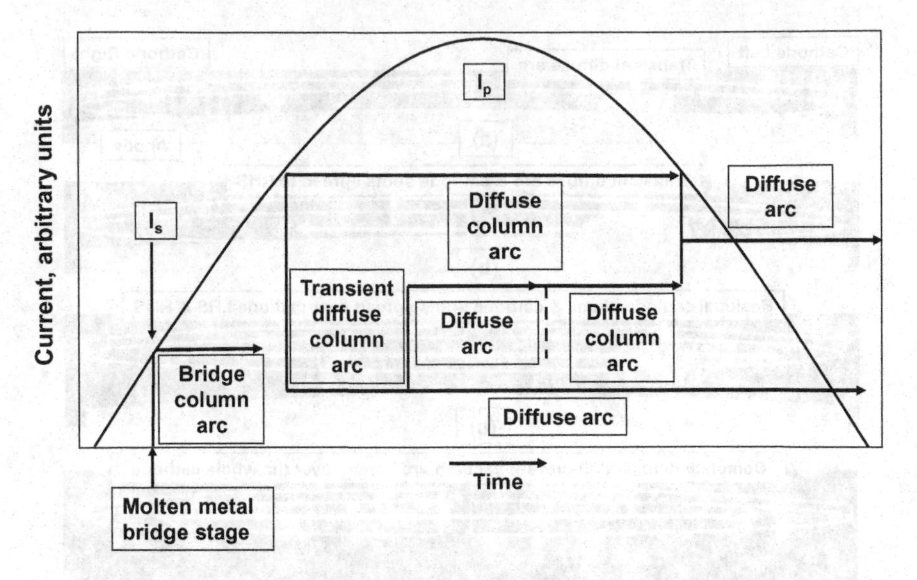

FIGURE 2.82 Observed sequences of vacuum arc appearance modes for a high-current vacuum arc on butt contacts in the presence of an axial magnetic field.

strength of the AMF as it does at lower currents; see Figure 2.83. His measurements are taken for a contact gap of 10mm and at the peak of a 50 Hz current after the diffuse, high-current, vacuum arc had fully developed. At 100A (i.e., a single cathode spot) the arc voltage increases monotonically with AMF and can be represented by the relationship:

$$U_{arc} = 20 + K\sqrt{B_A} \tag{2.44}$$

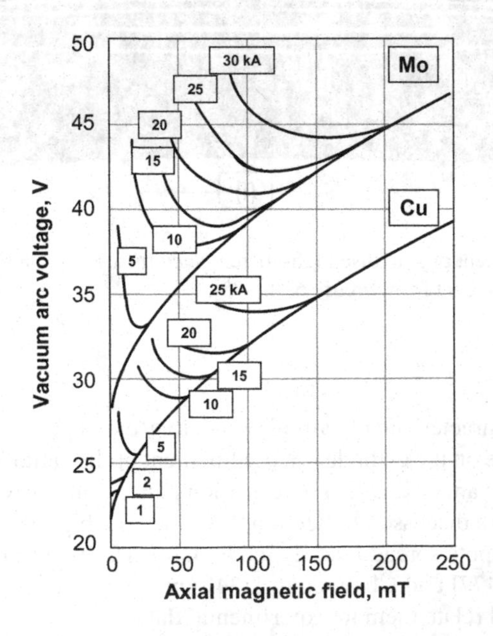

FIGURE 2.83 The vacuum arc voltage as a function of a uniform axial magnetic field and as a function of current for Cu and Mo contact materials each with a 60mm contact diameter and a 10mm contact gap [152].

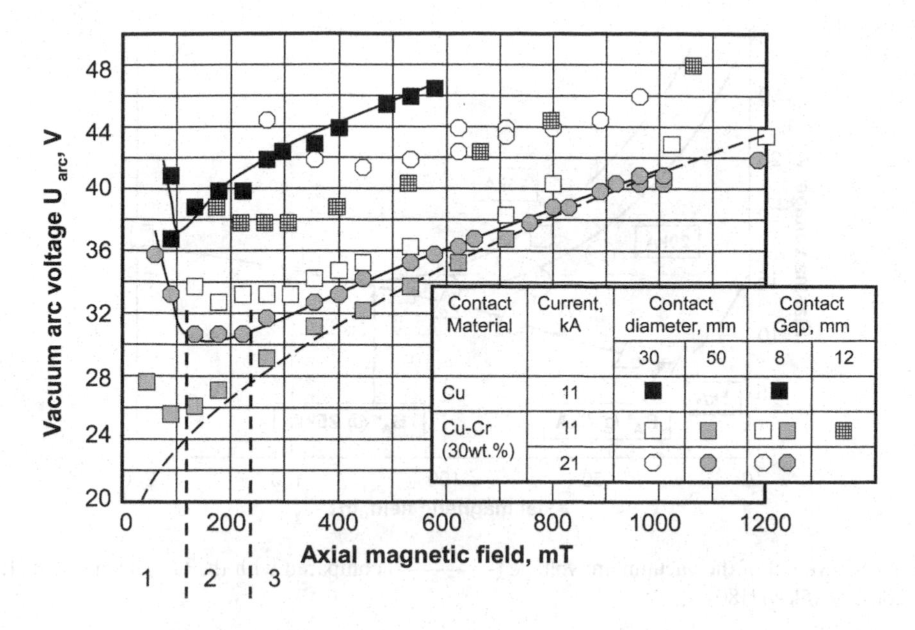

FIGURE 2.84 The vacuum arc voltage as a function of axial magnetic field for Cu, and Cu–Cr (30 wt%) contacts passing a rectangular, 6ms current pulse [178].

where $K = 30.5$ for Cu contacts, U_{arc} is in volts and B_A is in Teslas. Chaly et al. [178] have developed similar data using a rectangular, 6ms current pulse that compares Cu and Cu–Cr (30wt%) contacts; see Figure 2.84. The data for Cu contacts is similar to that of Gundlach. They show that there is a strong dependence on the level of current, contact diameter, and contact gap. It is interesting to note that the data for the Cu–Cr contact converge if the AMF is strong enough. Chaly [179] has also related U_{arc} to B_A and the contact gap d for a Cu–Cr contact with the relationship:

$$U_{arc} = 17 + \left(7.7\sqrt{B_A} + 0.47\right) \times log_e 20d \tag{2.45}$$

where U_{arc} is in volts and B_A is in Teslas and d is in cm. At 5kA Gundlach shows that U_{arc} as $f(B_A)$ has a characteristic similar to that shown in Figure 2.66; i.e., the vacuum arc voltage initially decreases as the applied AMF increases until it reaches a minimum value (region 1 in Figure 2.84), after which the arc voltage increases with an increase in AMF. In this region the arc voltage decreases, because more ions are confined to the intercontact region, see Figure 2.66. At the region of minimum arc voltage (region 2 in Figure 2.84) the electrons are increasingly confined to form their characteristic helix around the magnetic flux lines [178]. Figure 2.84 shows that the minimum value of the arc voltage increases with (a) an increasing contact gap for a constant current and contact diameter and (b) an increasing current for a constant contact diameter and contact gap. The minimum value of the arc voltage decreases with an increasing contact diameter for a constant current and contact gap.

As will be shown in Chapter 3 in this volume (Section 3.3.4), the AMF developed by practical AMF contact structures is usually in the range 100mT to 400mT. The arc voltage eventually converges to the relationship given in Equation (2.44) in region 3 of Figure 2.84. At higher currents the arc voltage as a function of AMF has a similar shape. Now, however, at the higher currents the initial arc voltage at low values of AMF has a higher value. Above a certain AMF (which is a function of the current), the arc voltage is independent of current, and also converges to satisfy Equation (2.44). Gundlach concludes that this condition is reached when the multiple cathode spots each act independently as a collection of single cathode spots operating in parallel. Keidar et al. [180] explored this hypothesis further using a two-dimensional model of the free-boundary,

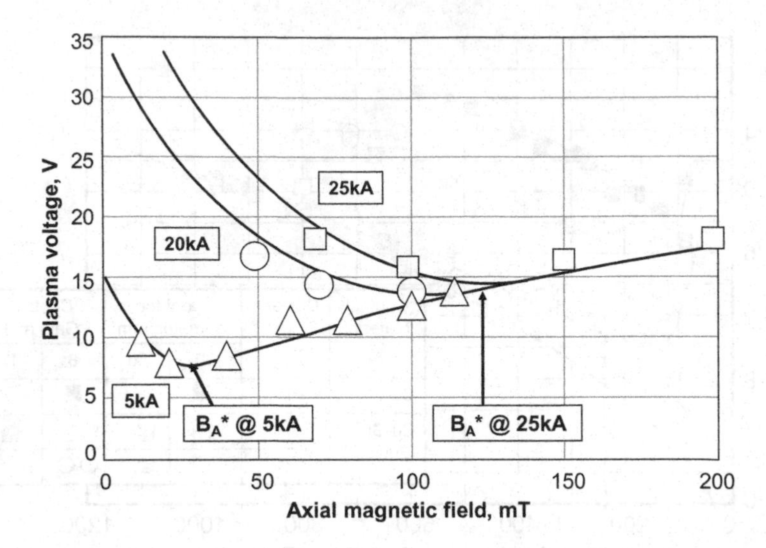

FIGURE 2.85 Model of the vacuum arc voltage (————) compared with the Gundlach's data [152], (□ 25kA; ○20kA; △5kA) [180].

intercontact, plasma flow in an AMF, which had been previously developed by Keidar et al. [181, 182]. This model includes the effects of plasma density gradients and the self-magnetic field in such a manner that the plasma expansion and current distribution can be calculated. Using this model, they calculate the arc voltage U_{arc} for the parameters used in Gundlach's experiment [152] (60mm diameter, Cu contacts, and a contact gap of 10mm). Figure 2.85 compares the result of this model with Gundlach's experimental data for 5kA, 20kA, and 25kA. They find that the best fit with the experimental data occurs when the cathode spots expand over only 50% of the contact area. In fact, it is commonly observed that the cathode spots do not always spread completely over the whole cathode's surface in the presence of an AMF [183]. The model permits them to determine an AMF (B_A*) where the U_{arc} has its minimum value. Their model predicts that B_A* will increase as the arc current increases for a given contact diameter. This would be expected because the density of cathode spots would increase and a higher AMF would be required before the plasma plumes would no longer overlap. They also show that B_A* has to increase if the contact gap is increased. Again, this is expected, because the plasma plumes expand radially as they cross the contact gap. The longer the contact gap, the more the plumes have to be confined to prevent them from overlapping before they reach the anode. Their analysis also compares well with the experimental data reported by Chaly et al. [168]. In general, they show that $B_A^* < B_{crit}$, see Equation (2.43). This indicates that the condition where the plasma plumes are nonoverlapping at the anode does not necessarily indicate that the heat input to the anode is low enough to prevent the occurrence of gross melting. Thus, the vacuum interrupter designer should consider both the experimental observation for B_{crit} where gross melting of the anode no longer occurs as well as the value B_A* where the U_{arc} is at its minimum value.

Jia et al. [184] have studied the temperature rise at the anode for a fully diffuse, high-current, vacuum arc in an axial magnetic field. They initially determine the cathode spot dispersion over a 50mm diameter Cu cathode, for a contact gap of 10mm and a current 15kA (rms). Figure 2.86 shows that the cathode spots cover the whole cathode surface at the peak current of 21kA. From this it is possible to assume that the cathode plumes will also be spread over the whole anode surface. Each cathode spot carries the expected current of between 75A and 100A: see Table 2.5. They use the cathode spot distribution to calculate the energy into the anode and to calculate its temperature rise during the passage of a half cycle of ac current. This is shown in Figure 2.87. The maximum temperature rise is about 1160K, which is well below the melting point of Cu (1350K). It occurs at

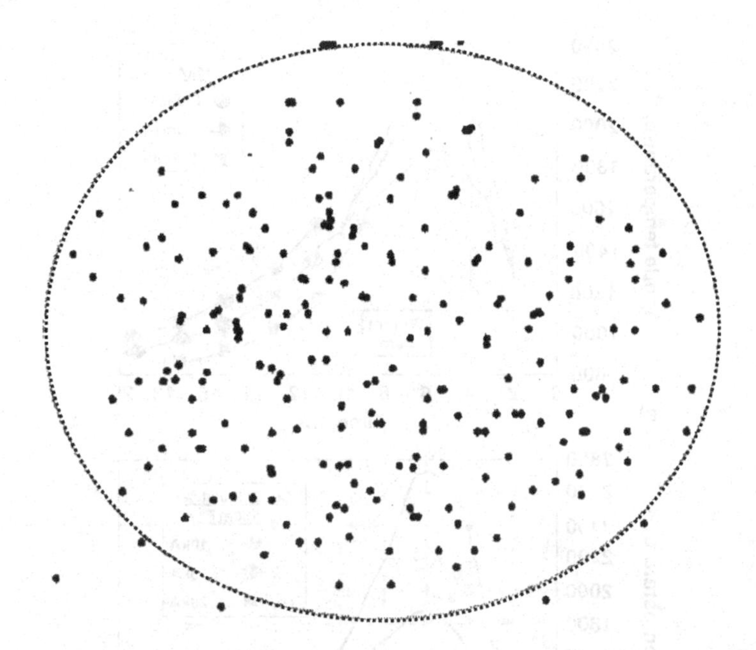

FIGURE 2.86 The distribution of cathode spots over a 58mm diameter Cu cathode with a 10mm contact gap at 21kA current in an axial magnetic field [184].

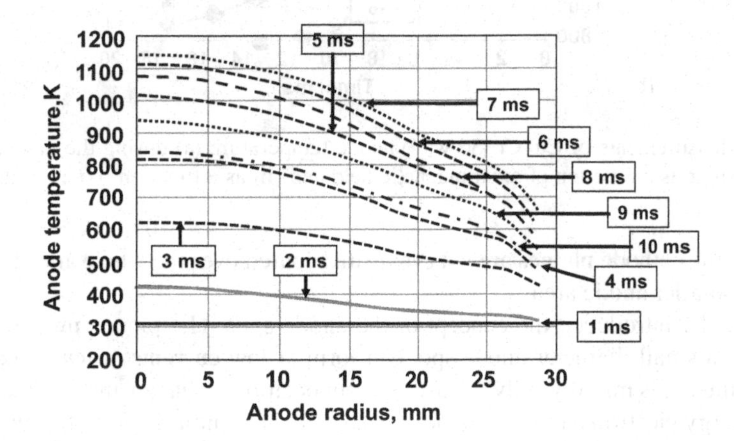

FIGURE 2.87 The calculated temperature of the Cu anode during the passage of the half cycle of 15kA (rms) current [184].

7ms, which is after the passage of the peak current. A steep decline in the anode temperature is seen as the current drops to zero. At 10ms (close to current zero) the anode temperature is only about 850K. Experimental observation of the anode after interrupting the 15kA (rms) current shows no obvious melting. So, the experiment seems to confirm the calculated result. The implication is that if the diffuse vacuum arc keeps the anode temperature below the melting point of the Cu–Cr contact material. and the cathode plumes are also equally dispersed over the anode, the anode will be a passive receiver of the electron current. Methling et al. [185] use pyroscopic and optical spectra in the near infrared measurements to determine the anode temperature for Cu–Cr (50wt%) contacts. They could only accurately measure the anode's temperature after the vacuum arc extinguishes at current zero, Figures 2.88a and 2.88b show their results for currents 10kA, 20kA, and 30kA and for B_A values 120mT, 150mT, and 180mT. Their data seems to show that only the 10kA (rms) with a $B_A = 18mT$ has a temperature below the melting point of the Cu–Cr contact material. The other

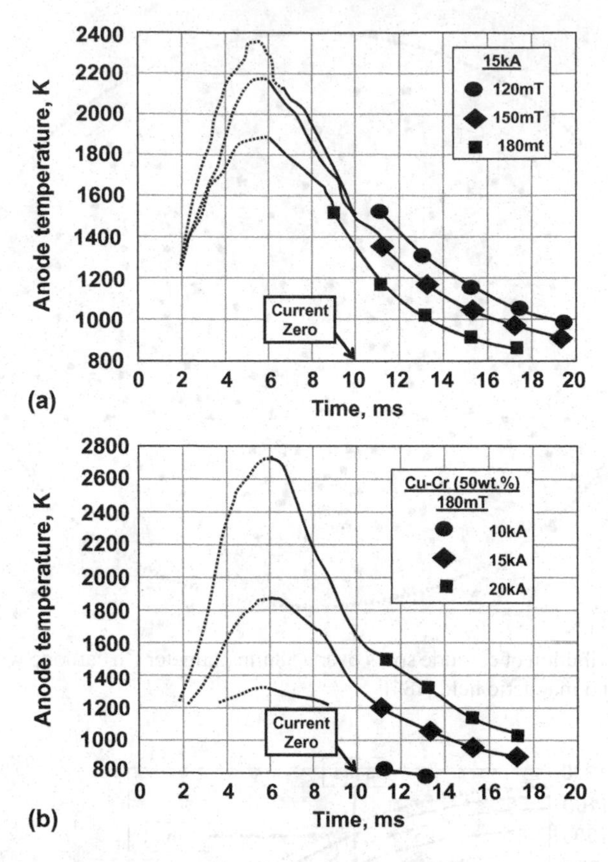

FIGURE 2.88 Measurements of Cu–Cr (50 wt%) anode temperature (a) during the passage of half cycle of 15kA (rms) current as a function of axial magnetic field and (b) as a function of (rms) current for a $B_A = 180mT$ [185].

data implies that the cathode plumes are not evenly distributed over the whole anode surface but are constricted to a smaller anode area.

In Section 2.3.4 I introduced the concept of the anode spot. The pioneering work by Kimblin (104) shows that a small diameter anode spot can form at low currents (a few kiloamperes) even though the vacuum arc is initially fully diffuse. The supposition is that an anode fall develops which drives higher energy electrons into the anode's surface. This eventually heats the anode to its melting point and an anode plume of vaporized anode material results. Also, introduced in Section 2.3.4, an anode spot can form after the ac current in a high-current columnar vacuum arc approaches a current zero. The residual hot spot at the anode can continue to evaporate anode metal and produce an anode plume [119]. A third type of anode spot has been observed when a high-current vacuum arc forms in the presence of an axial magnetic field (AMF). Figures 2.60 and 2.89a show the ideal form of diffuse vacuum arc in an AMF with multiple cathode spots distributed over the whole cathode surface and the cathode plumes are separate from each other and are equally distributed over the whole anode surface. In Figure 2.89b when the current increases for the same AMF, contact gap and contact diameter the cathode plasma plumes begin to overlap toward the anode. Also, when the contact gap increases for the same AMF, the same current and contact diameter the cathode plasma plumes begin to overlap toward the anode. When the plasma density in these overlapping plume reaches a critical value the azimuthal magnetic field "B_θ" begins to pinch it to a smaller diameter. The energy into the anode is now concentrated in a smaller area. As the effect of the pinching develops, the anode can reach the melting point of the anode material. When this occurs, an anode spot develops that will provide a high density of metal vapor into the vacuum arc column. Usually

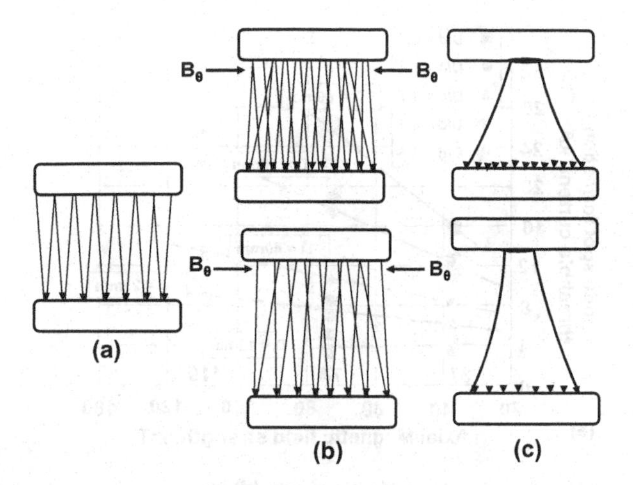

FIGURE 2.89 Vacuum arc modes in an axial magnetic field (a) the fully diverse vacuum arc, (b) Overlapping cathode plumes and (c) the formation of the anode spot.

in this condition the cathode spots tend to concentrate below the anode spot, but some still remain distributed over the cathode's surface: Figure 2.89(c). Figure 2.90 shows an example of the pinched plasma and the anode spot. The formation of the anode spot is a function of the current, the axial magnetic field strength and the contact gap [186]. Kong et al. [187], Zhang et al. [188], and Liu et al. [189] have studied the critical AMF and the critical current for the formation of a anode spot as a function of contact gap *"d"* and contact diameter *'D'* for Cu–Cr contacts. For a 24mm contact gap the current threshold for the formation of an anode spot is a linear function of AMF values 37mT to 110mT and for contact diameters ranging from 12mm to 80mm: see Figure 2.91 (a) [187]. For a constant contact diameter of 60mm, however, the dependence of the threshold current on contact gap as a function of AMF is much weaker: see Figure 2.91(b) [187]. The formation of the anode spot is also dependent on the contact materials. Zhang et al. [188] show that the contact metal's melting point and its vapor pressure at its melting point are critical. They determine that the threshold current for anode spot formation on a pure Cr anode is about 15kA (rms) while for a Pure Mo anode it is about 22.5kA (rms). Liu et al. [189] show that there is only a small difference in threshold current for the formation anode spots with Cu, Cu–Cr (25wt%) and Cu–Cr (50wt%) 60mm diameter anodes with a contact gap of 24mm. This might be expected as the material properties of these three materials

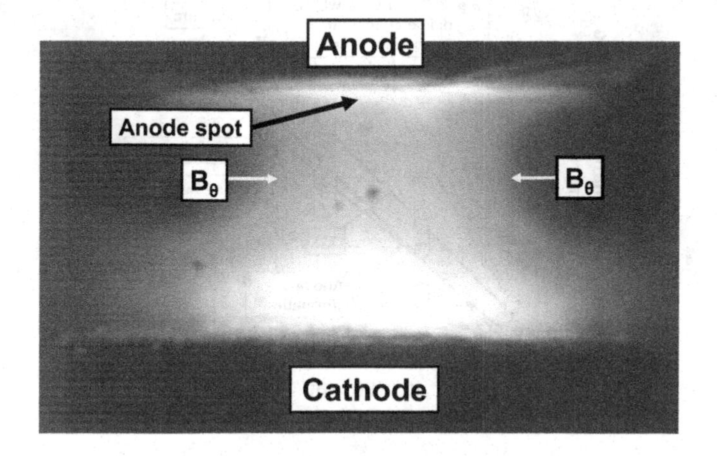

FIGURE 2.90 A photograph of a vacuum arc in an axial magnetic field with cathode spots covering the cathode, a constricted plasma above them and the formation of an anode spot.

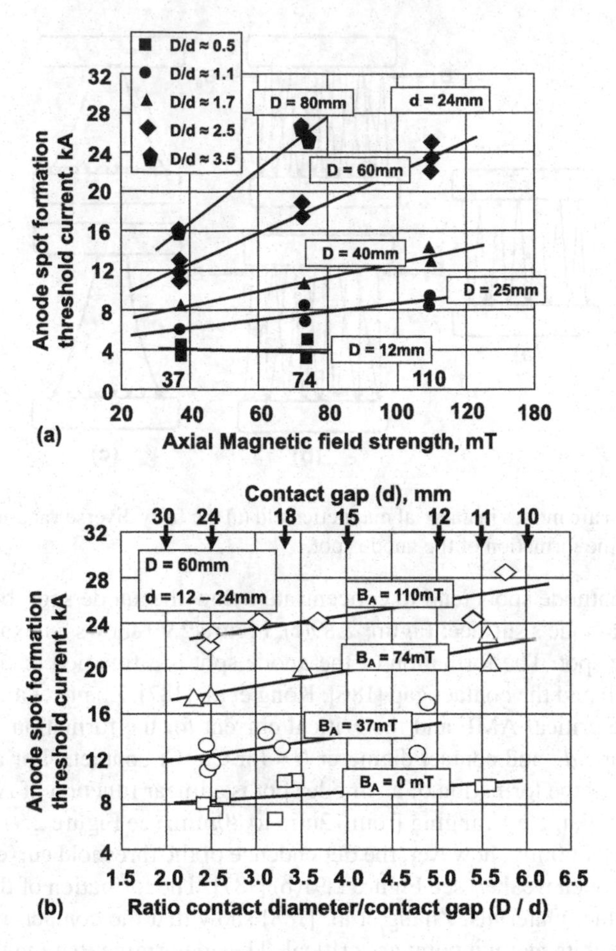

FIGURE 2.91 The threshold current for the formation of an anode spot on Cu–Cr (25wt,%) contacts for a vacuum arc as a function of axial magnetic field, (a) as a function of contact diameter for a constant contact gap and (b) as a function of contact diameter / contact gap for a constant contact diameter of 60mm [188].

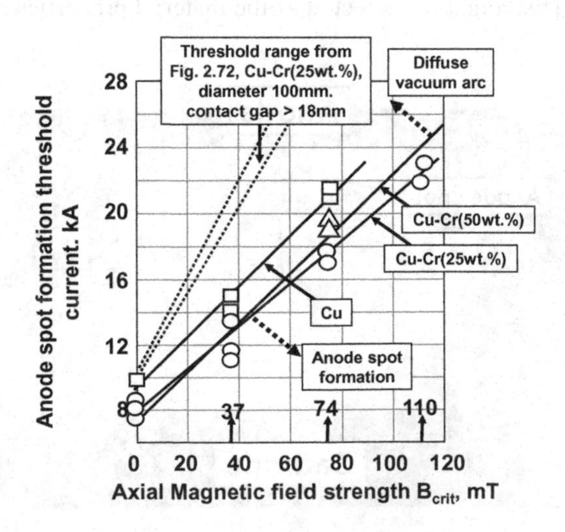

FIGURE 2.92 The critical magnetic field above which a fully diffuse vacuum arc forms and below which an anode spot can form [166,189].

are quite similar. Figure 2.92 [189] illustrate this small difference and compares their data with the 100mm diameter Cu–Cr (25wt%) contact given by Heberlein et al. in Figure 2.72.

Most of the experimental observations described in this section have been performed on contacts opening in a uniform AMF. External Helmholtz coils placed above and below the contact structure usually supply this AMF. In Chapter 3, I will discuss practical contact designs to supply the AMF where the AMF is not uniform across the contact gap. Also, in a practical vacuum interrupter design the contact diameter is restricted in order to make the overall diameter of the vacuum interrupter as small as possible and yet still interrupt the short-circuit current for which it is designed. This sometimes results in such a high cathode spot density that the plasma plumes always overlap before they reach the anode for any practical AMF that can be applied. It is thus very important when designing a commercial vacuum interrupter using the AMF concept, to understand the limitations that are imposed by the formation of the anode spot. After the passage of a high short-circuit current, too high a level of residual metal vapor in the contact gap can result in a failure to interrupt the circuit. This will be discussed in Chapter 4, this volume.

2.8 OVERVIEW AND REVIEW OF THE THREE FORMS OF ANODE SPOT

The vacuum interrupter designer must consider all aspects of the complex formation and development of the vacuum arc. As can be seen from the description in this chapter, it can exist in many different forms. These depend upon the contact's opening sequence and the magnitude of the circuit's current. As the contacts begin to open the vacuum arc always begins as a high-pressure confined columnar arc: the bridge column. At low currents this transitions into the diffuse vacuum arc. Here cathode spots that move rapidly over the cathode contact's surface and provide the electrons and ions for the neutral intercontact plasma are dominant. At high currents the vacuum arc can continue as a high-pressure column. This columnar arc is similar in form to a high-current arc in air at atmospheric pressure. The high-current vacuum arc can be controlled by applying a transverse magnetic field (TMF) or an axial magnetic field (AMF). The formation of the *anode spot* has been shown to occur in three forms:

1) At low current (less than 4kA) with small diameter anode. An anode fall can develop which accelerates electrons onto the anode. This increases the energy input to the anode and its temperature can increase. When the temperature reaches the anode's melting point, metal vapor is emitted from the anode and an anode spot is observed. This effect is unlikely to occur in practical vacuum interrupters whose contact diameters are equal and generally greater than 24mm
2) As the current falls to zero after the formation of the high-current columnar vacuum arc. Here the emission of metal vapor from the cathode and the anode can no longer maintain the columnar structure. The residual temperature at the anode, however, can continue to emit metal vapor which can be seen as an anode spot. The decay of this anode plume plays a critical role in practical vacuum interrupters when interrupting high fault currents using transverse magnetic field contacts; see Section 4.2.3
3) In a high-current vacuum arc controlled by an axial magnetic field. When the dispersed cathode plasma plumes overlap to produce a uniform plasma density the azimuthal magnetic field causes the plasma column to contract. The energy into the anode then is deposited in a smaller area and its temperature increases. When the temperature reaches the melting point an anode spot develops which emits anode metal vapor into the arc column

The design of a vacuum interrupter that only has to switch load currents (less than about 4kA) where the diffuse vacuum arc is dominant will be quite different from one that must switch both load currents and fault currents ($\geq$ 10kA). The designer must also consider the desired current switching life. This will depend upon the arc erosion of the contacts, which, in turn, will depend upon the vacuum arc mode. While the erosion of the contacts is essential to sustain the vacuum arc, the designer must control it so that the high-voltage performance of the vacuum interrupter is maintained throughout its life. The development of successful vacuum interrupter, contact materials, and the design of practical contact structures will be discussed in Chapter 3 in this volume. The use of these structures in practical vacuum interrupter designs to interrupt ac circuits will be discussed in Chapters 4, 5, and 6 in this volume.

REFERENCES

1. Slade, P. (Editor), *Electrical Contacts: Principles and Applications: 2nd Edition*, (pub. CRC Press), 2014.
2. Holm, R., Holm, E., *Electrical Contacts: Theory and Application* (pub. Springer-Verlag), 1967 (reprinted 2000).
3. Timsit, R., "Electrical contact resistance: Fundamental principles," In: Slade, P. (Editor), *Electrical Contacts: Principles and Applications: 2nd Edition* (pub. CRC Press), 2014, pp. 5–18.
4. Greenwood, J., "Constriction resistance and the real area of contact," *British Journal of Applied Physics*, 17(12), pp. 1621–1632, 1966.
5. Greenwood, J., Williamson, J., "Contact of nominally flat surfaces," *Proceedings of the Royal Society*, 295A, pp. 300–319, 1966.
6. Llewellyn-Jones, F., *The Physics of Electrical Contacts* (pub. Clarendon Press, Oxford), 1957.
7. Kawase, Y., Mori, H., Ito, S., "3-D finite element analysis of electro-dynamic repulsion forces in stationary electric contacts taking into account asymmetric shape," *IEEE Transactions on Magnetics*, 33(2), pp. 1994–1999, 1997.
8. Malucci, R., "The effects of current density variations in power contact interfaces," *Proceedings 57th IEEE Holm Conference on Electrical Contacts*, pp. 55–61, 2011.
9. Dullni, E., Gentsch, D., Shang, W., Delachaux, T., "Resistance increase of vacuum interrupters due to high-current interruptions," *IEEE Transactions on Dielectrics and Electrical Insulation*, 23(1), pp. 1–7, February 2016.
10. Taylor, E., Baus, S., Lawall, A., "Increase in contact resistance of vacuum interrupters after short-circuit testing," *Proceedings 27th International Conference on Electrical Contacts*, pp. 203–206, 2014.
11. Wilson, A. H., *The Theory of Metals* (pub. Cambridge), 1953.
12. Snowdon, A., "Studies of electrodynamic forces occurring at electrical contacts," *AIEE Transection*, 80, pp. 24–28, March 1961.
13. Piccoz, D., Tetse, Ph., Andlauer, R., Leblanc, T., Chabrerie, J., "The repulsion of electrical contacts crossed by short circuit currents," *Proceedings 44th IEEE Holm Conference on Electrical Contacts*, pp. 129–135, October 1999.
14. Barkan, P., "A new formulation of the electromagnetic repulsion phenomenon in electrical contacts at very high currents," *Proceedings 11th International Conference on Electrical Contact Phenomena*, pp. 185–188, June 1982.
15. Taylor, E., Slade, P., "The repulsion or blow-off force between closedcontacts carrying current," *Proceedings 62nd IEEE Holm Conference on Electrical Contacts*, pp. 159–162, October 2016.
16. Shea, J., de Vault, B., Chen, Y., "Blow open forces on double break contacts," *IEEE Transactions on Components, Hybrids and Manufacturing Technology*, 17(1), pp. 32–38, March 1994.
17. Zhou, X., Theissen, P., "Investigation of arcing effects during contact blow open process," *IEEE Transactions on Components and Packaging Technology*, 23(2), pp. 271–277, June 2000.
18. Huber, B., "An oscilloscope study of the beginning of a floating arc," *Proceedings Holm Seminar on Electrical Contacts*, pp. 141–152, November 1967.
19. Slade, P., "The current level to weld closed contacts," *Proceedings 59th Holm Conference on Electrical Contacts*, pp. 123–128, September 2013.
20. Slade, P., Nahemow, M., "Initial separation of electrical contacts carrying high, Currents," *Journal of Applied Physics*, 49(9), pp. 3290–3297, 1971.
21. Hauh, R., Kouahou, T., Doremieux, J., Phenomena preceding arc ignition between opening contacts: Experimental study and theoretical approach," *Proceedings 36th IEEE Holm Conference on Electrical Contacts*, pp. 543–549, August 1990.
22. Slade, P., "Opening contacts: The transition from the molten metal bridge to the electric arc," *IEICE Transactions on Electronics*, E93-C(9), p. 13801386, September 2010.

23. Wakatsuki, N., Kudo, T., "Melting phenomena of electric contacts during current interruption," *Proceedings 60th IEEE Holm Conference on Electrical Contacts*, pp. 185–191, October 2014.
24. Slade, P., "The consequences of arcing," p. 509 and "The arc and interruption," In: Slade, P. (Editor), *Electrical Contacts: Principles and Applications: 2nd Edition* (pub. CRC Press), 2014, p. 463.
25. Slade, P., Hoyaux, M., "The effect of electrode material on the initial expansion of an arc in vacuum," *IEEE Transactions on Parts, Hybrids and Packaging*, 8(1), pp. 35–47, March 1972.
26. Slade, P., "The minimum arc current and the minimum arc voltage. In: *Electrical Contacts: Principles and Applications: 2nd Edition* (pub. CRC Press), 2014, p. 585.
27. Puchkarev, V., Bochkarev, M., "High current density spotless vacuum arc as a glow discharge," *IEEE Transactions on Plasma Science*, 25(4), pp. 593–597, August 1997.
28. Korolev, Y., Frants, O., Landl, V., Kasyanov, A., Bolotov, A., Shemyakin, L., "Investigation of the high current stages in pseudospark discharge," *Proceedings 25th International Symposium on Discharges and Electrical Insulation in Vacuum*, pp. 95–98, September 2012.
29. Anders, A., "Ion charge distributions of vacuum arc plasmas: The origin of the species," *Physical Review E*, 55(1), pp. 969–981, January 1997.
30. Ebeling, W., Förster, A., Radtke, R., *Physics of Non-Ideal Plasmas* (pub. Teubner Verlagsgesellschaft, Stuttgart), 1991.
31. Logachev, A., Tenitskiy, P., Vykhodtsev, A., "Analysis of voltage oscillograms at disconnecting contacts," *Proceedings 27th International Symposium on Discharges and Electrical Insulation in Vacuum*, pp. 225–228, September 2014.
32. Juttner, B., Puchkarev, V., Hantzsche, E., Beilis, I., "Cathode spots," In: Boxman, R. L., Martin, P. J. and Sanders, D. M., (Editors), *Handbook of Vacuum Arc Science and Technology* (pub. Noyes), 1995.
33. Juttner, B., "Cathode spots of electric arcs," *Journal of Physics. Part D: Applied Physics*, 24(17), pp. R103–R123, 2001.
34. Mesyats, G., *Cathode Phenomena in a Vacuum Discharge* (pub. Nauka), 2000.
35. Hantzsche, E., "Mysteries of the arc cathode spot; a retrospective glance", *IEEE Transactions on Plasma Science*, 31(5), pp. 799–808, October 2003.
36. Anders, A., "The fractal nature of vacuum arc cathode spots", *IEEE Transactions on Plasma Science*, 33(5), pp. 1456–1464, October 2005.
37. Li, W., Shi, Z., Wang, C., Shi, F., Jia, S., "The motion characteristics of a single cathode spot in removing oxide layer on metal surface by vacuum arc", *IEEE Transactions on Plasma Science*, 45(1), pp. 106–112, January 2017.
38. Beilis, I., Djakov, B., Juttner, B., Pursch, H., "Structure and dynamics of high–Current arc cathode spots in vacuum", *Journal of Physics: Part D: Applied Physics*, 30(1), pp. 119–130, 1997.
39. Almeida, N., Benilov, M., Benilov, L., Hartmann, W., Wenzel, N., "Near-cathode plasma layer on Cu-Cr contacts of vacuum arcs", *IEEE Transactions on Plasma Science*, 41(8), pp. 1938–1949, August 2013.
40. Benilov, M., Cunha, M., Hartmann, W., Kosse, S., Lawall, A., Wenzel, N., "Space-resolved modeling of stationary spots on copper vacuum arc cathodes and on composite Cu-Cr cathodes with large grains", *IEEE Transactions on Plasma Science*, 41(8), pp. 1950–1958, August 2013.
41. Porto, D., Kimblin, C., Tuma, D., "Experimental observations of cathode spot surface phenomena in the transition from a metal vapor arc to a nitrogen arc", *Journal of Applied Physics*, 53(7), pp. 4740–4749, July 1982.
42. Kesaev, I., *Soviet Physics, Technical Physics*, 9, pp. 1146–1154, 1964.
43. Juttner, B., "The dynamics of arc cathode spots in vacuum: Part III, measurements with improved resolution and UV radiation", *Journal of Physics. Part D: Applied Physics*, 31(14), pp. 1728–1736, 1998.
44. Daalder, J., "Erosion structures on cathodes arced in vacuum", *Journal of Physics. Part D: Applied Physics*, 12(10), pp. 1769–1779, 1979.
45. Beilis, I., "Continuous transient cathode spot operation on a micro-protrusion: Transient cathode drop", *IEEE Transactions on Plasma Science*, 39(6), pp. 1277–1283, June 2011.
46. Daalder, J., "Random walk of cathode arc spots in vacuum", *Journal of Physics. Part D: Applied Physics*, 16(1), pp. 17–27, 1983.
47. Mesyats, G., Uimanov, I., "Semiempirical model of the microcrater formation in the cathode spot of a vacuum arc", *IEEE Transactions on Plasma Science*, 45(8), pp. 2087–2092, August 2017.
48. Barengolts, S., Mesyats, G., Shmelev, D., "Structure and time behavior of vacuum arc cathode spots", *IEEE Transactions on Plasma Science*, 31(5), pp. 809–816, October 2003.
49. Barengolts, S., Shmelev, D., Uimanov, I., "Pre-explosion phenomena beneath the plasma of a vacuum arc cathode spot", *IEEE Transactions on Plasma Science*, 43(8), pp. 2236–2240, August 2015.

50. Smeets, R., Schulpen, F., "Fluctuations of charged particle and light emission in vacuum arcs", *Journal of Physics. Part D: Applied Physics*, 21(2), pp. 301–310, 1988.
51. Anders, S., Anders, A., Jüttner, B., "Brightness distribution and current density of vacuum arc cathode spots", *Journal of Physics. Part D: Applied Physics*, 25(11), pp. 1591–1599, 1992.
52. Tuma, D., Chen, C., Davies, D., "Erosion products from the cathode spot region of a copper vacuum arc", *Journal of Applied Physics*, 49(7), pp. 3821–3831, July 1978.
53. Plyutto, A., Ryzhkov, V., Kapin, A., "High speed plasma streams in vacuum arcs", *Soviet Physics JETP*, 20, pp. 328–337, February 1965.
54. Daalder, J., "Cathode spots and vacuum arcs", *Physica*, 104C(1–2), pp. 91–106, 1981.
55. Mesyats, G., Uimanov, I., "Aerodynamics of the molten metal during the crater formation on the cathode surface in a vacuum arc", *IEEE Transactions on Plasma Science*, 43(8), pp. 2241–2246, August 2015.
56. Siemroth, P., Laux, M., Pursch, H., Sachteben, J., "Diameters and velocities of droplets emitted from Cu cathode in vacuum arc", *IEEE Transactions on Plasma Science*, 47(8), pp. 3470–3477, August 2019.
57. Heberlein, J., Porto, D., "The interaction of vacuum arc ion currents and axial magnetic fields", *IEEE Transactions, on Plasma Science*, PS-11, pp. 152–159, September 1983.
58. Benelov, M., Kaufmann, H., Hartmann, W., Benelov, L., "Revisiting theoretical description of the retrograde motion of cathode spots of vacuum arcs", *IEEE Transactions, on Plasma Science*, 47(8), pp. 3434–3441, August 2019.
59. Beilis, I., "Transient cathode spot operation at a microprotusion in a vacuum arc", *IEEE Transactions on Plasma Science*, 35(4), pp. 966–972, August 2007.
60. Mesyats, G., "Ecton mechanism of the cathode spot phenomena in a vacuum arc", *IEEE Transactions on Plasma Science*, 41(4), pp. 676–694, April 2013.
61. Hantzsche, E., "Theories of cathode spots" and Beilis, I., "Theoretical modeling of cathode spot phenomena," pp. 151–256, in reference [23].
62. Bellis, I., "Vacuum arc cathode spot theory: History and evolution of the mechanisms", *IEEE Transactions on Plasma Science*, 47(8), pp. 3412–3433, August 2019.
63. Lee, L., Greenwood, A., "Theory for the cathode mechanism in metal vapor arcs", *Journal of Applied Physics*, 32(5), pp. 916–923, 1961.
64. Harris, L., "A mathematical model for cathode spot operation", *Proceedings 8th International Symposium on Discharges and Electrical Insulation in Vacuum*, pp. F1–F18, September 1978.
65. Cobine, J., "Introduction to vacuum arcs," In: Lafferty, J. M. (Editor), *Vacuum Arcs, Theory and Application*, (pub. John Wiley & Sons), p.11, 1980.
66. Nikolaev, A., Yu, G., Savkin, K., Oks, E., "Angular distribution of ions in a vacuum arc plasma with single-element and composite cathodes", *IEEE Transactions on Plasma Science*, 41(8), pp. 1923–1928, August 2013.
67. Yushkov, G., Anders, A., Oks, E., Brown, I., "Ion velocities in vacuum arc plasmas", *Journal of Applied Physics*, 88(10), pp. 5618–5622, November 2000.
68. Hantzsche, E., "Two dimensional models of expanding vacuum arc plasmas", *IEEE Transactions on Plasma Science*, 23(6), pp. 893–898, December 1995.
69. Tsuruta, K., Sekiya, K., Watanabe, G., "Velocities of copper and silver ions generated from an impulse vacuum arc", *IEEE Transactions on Plasma Science*, 25(4), pp. 603–608, August 1997.
70. Yushkov, G., Bugaev, A., Krinberg, I., Oks, E., "On a mechanism of ion acceleration in vacuum arc-discharge plasma", *Doklady Physics*, 46(5), pp. 307–309, 2001.
71. Byon, E., Anders, A., "Ion energy distribution functions of vacuum arc plasmas", *Journal of Applied Physics*, 93(4), pp. 1899–1906, February 2003.
72. Shmelev, D., Barengolts, S., Tsventoukh, M., "Numerical simulation of near the cathode spot of vacuum arc", *IEEE Transactions on Plasma Science*, 45(11), pp. 3046–3053, November 2017.
73. Kimblin, C., "Erosion and ionization in cathode spot regions of vacuum arcs", *Journal of Applied Physics*, 44(7), pp. 295–302, July 1973.
74. Methling, R., Gorchacov, S., Lysniak, M., Franke, S., Khakpour, A., Popov, S., Batrakov, A., Uhrlandt, D., Weltmann, K.-D., "Spectroscopic investigation of a Cu-Cr vacuum arc", *IEEE Transactions on Plasma Science*, 43(8), pp. 2303–2309, August 2015.
75. Anders, A., "Ion energies in vacuum arcs: A critical review of data and theories leading to travelling potential humps", *Proceedings 26th International Symposium on Discharges and Electrical Insulation in Vacuum*, pp. 201–204, September 2014.
76. Mandelelbrot, B., *The Fractal Geometry of Nature*, (pub. Freeman, NY), 1983.

77. Gleik, J., *Chaos: Making a New Science*, (pub. Penguin Books), 1987: this is an excellent introduction to the subject.
78. Goldsmith, S., "The inter-electrode plasma", In: Boxman, R., Martin, P. and Sanders, D. (Editors), *Handbook of Vacuum Arc Science and Technology*, (pub. Noyes), 1995.
79. Slade, P., Lee, W., Loud, L., Haskins, R., "The unusual electrical erosion of high tungsten content, W-Cu contacts switching load current in vacuum", *IEEE Transactions Components, Manufacturing and Packaging Technology*, 24(3), pp. 320–330, September 2001.
80. Taylor, E., "Cathode spot behavior on tungsten-copper contacts in vacuum and the effect on erosion", *Proceedings 51st IEEE Holm Conference on Electrical Contacts*, pp. 135–138, September 2005.
81. Kozyrev, A., Shishkov, A., "Movement, heating and evaporation of droplets in near cathode region of vacuum arc", *Proceedings 21th International Symposium on Discharges and Electrical Insulation in Vacuum*, pp. 229–232, September/October 2004.
82. Kimblin, C., "Vacuum arc ion currents and electrode phenomena", *Proceedings of the IEEE*, 59(4), pp. 546–555, April 1971.
83. Anders, A., Oks, E., Yushkov, G., Savin, K., Brown, I., Nikolaev, A., "Measurement of total ion flux from vacuum arc cathode spots", *IEEE Transactions on Plasma Science*, 33(5), pp. 1532–1536, October 2005.
84. Brown, I., Galvin, J., MacGill, R., West, M., "Multiply charged metal ion beams", *Nuclear. Instruments and Methods*, Vol. 43, No. 3, pp. 455–458, 1989.
85. Lunev, V., Padalka, V., Khoroshikh, V. M., *Soviet Physics, Technical Physics*, 22, pp. 858–861, 1977.
86. Brown, I., Godechot, X., "Vacuum arc ion charge-state distributions", *IEEE Transactions on Plasma Science*, 19(5), pp. 713–717, 1991.
87. Browns, I., Feinberg, B., Galvin, J., "Multiply stripped ion generation in metal vapor vacuum arc", *Journal of Applied Physics*, 63(10), pp. 4889–4898, 1988.
88. Slade, P., Schulz-Gulde, E., "Spectroscopic analysis of high current, free burning, ac arcs between copper contacts in argon and air", *Journal of Applied Physics*, 44(1), pp. 157–162, 1973.
89. Davis, W., Miller, H., "Analysis of electrode products emitted by dc arcs in a vacuum environment", *Journal of Applied Physics*, 40(5), pp. 2212–2221, April 1969.
90. Anders, A., Yushkov, G., "Angularly resolved measurements of ion energy of vacuum arc plasmas", *Journal of Applied Physics*, 80(14), pp. 2457–2459, April 2002.
91. Aksenov, I., Khoroshikh, V., "Angular distributions of ions in a plasma steam of a steady-state vacuum arc", *Proceedings 18th International Symposium on Discharges and Electrical Insulation in Vacuum*, pp. 211–214, September 1998.
92. Rusteberg, C., Lindmayer, M., Juttner, B., Pursch, H., "On the ion energy distribution of high current arcs in vacuum", *IEEE Transactions on Plasma Science*, 23(6), pp. 909–914, December 1995.
93. Paulus, I., Holmes, R., Edels, H., "Vacuum arc response to current transients", *Journal of Physics. Part D: Applied Physics*, 16, pp. 17–27, 1983.
94. Ben Jemma, N., "Short arc duration laws and distributions at low current (< 1A) and voltage (14–42 V dc)", *IEEE Transactions on Components and Packaging Technologies*, 24(3), pp. 358–362, September 2001.
95. Taylor, E., Slade, P., Li, W.-P., "High chop currents observed in vacuum arcs between tungsten contacts", *Proceedings 23rd International Conference on Electrical Contacts*, pp. 1–6, June 2006.
96. Cobine, J., Farrell, G., "Experimental study of arc stability. I", *Journal of Applied Physics*, 31(12), pp. 2296–2304, 1960.
97. Greenwood, A., *Vacuum Switchgear*, (pub. I.E.E.), pp. 81–89, 1994.
98. Czarnecki, L., Lindmayer, M., "Experimental and theoretical investigations of current chopping in vacuum with different contact materials", *Proceedings 13th International Conference on Electrical Contact Phenomena*, pp. 128–134, 1986.
99. Smeets, R., "Stability of low current vacuum arcs", *Journal of Physics. Part D: Applied Physics*, 19(4), pp. 575–587, 1986.
100. Smeets, R., "The origin of current chopping in vacuum arcs", *IEEE Transactions on Plasma Science*, 17(2), pp. 303–310, April 1989.
101. Smeets, R., "An experimental study of vacuum arc instability for conventional and low-surge contact material" *Proceedings 15th International Symposium On Discharges and Electrical Insulation in Vacuum*, pp. 518–522, September/October 1992.
102. Ding, C., Yanabu, S., "Effect of parallel circuit parameters on the stability of a low-current vacuum arc", *IEEE Transactions on Plasma Science*, 31(5), pp. 877–883, October 2003.

103. Yamamoto, A., Kusano, T., Okutomi, T., Homma, M., Okawa, M., "Effect of composition on current chopping characteristics in Ag-base contacts", *Proceedings 17th International Conference on Electrical Contact Phenomena*, pp. 887–892, 1994.
104. Kimblin, C., "Anode voltage drop and anode spot formation in dc vacuum arcs", *Journal of Applied Physics*, 40(4), pp. 1744–1752, March 1969.
105. Miller, C., "Anode modes in vacuum arcs", *IEEE Transactions on Dielectrics and Electrical Insulation*, 4(4), pp. 382–388, August 1997.
106. Miller, H., C., "Anode modes in vacuum arcs: Update", *IEEE Transactions, on Plasma Science*, 45(8), pp. 2366–2374, August 2017.
107. Zhang, Y., Yi, X., Zhang, Z., Liu, Z., Geng, Y., Wang, J., Wang, W., Li, G., Wang, X., "Electron temperature and electron density in diffuse arc mode and anode spot mode", *Proceedings International Symposium on Discharges and Electrical Insulation in Vacuum*, pp. 298–302, September 2016.
108. Wang, L., Jia, S., Yang, D., Liu, K., Su, G., Shi, Z., "Modelling and simulation of anode activity in high-current vacuum arc", *Journal of Physics. Part D: Applied Physics*, pp. 1–13, 2009.
109. Huang, X., Wang, L., Zhang, X., Jia, S., Shi, Z., "Numerical simulation of high-current vacuum arcs with considering the microprocess of anode vapor", *Proceedings International Symposium on Discharges and Electrical Insulation in Vacuum*, pp. 263–266, September 2016.
110. Tian, Y., Wang, Z., Jiang, Y., Ma, H., Liu, Z., Geng, Y., Wang, J., "Simulation of surface erosion of anode under high-current vacuum arcs", *Proceedings International Symposium on Discharges and Electrical Insulation in Vacuum*, pp. 329–332, September 2016.
111. Beilis, I., Koulik, Y., Boxman, R., "Anode temperature evolution in a vacuum arc with a black body electrode configuration, *Proceedings International Symposium on Discharges and Electrical Insulation in Vacuum*, pp. 295–297, September 2016.
112. Khakpour, A., Gortschakow, S., Uhrlandt, D., Methling, R., Franke, S., Popov, S., Batrakov, A., Weltmann, K.-D., "Video spectroscopy of vacuum arcs during transition between different high-current modes", *IEEE Transactions on Plasma Science*, 44(10), pp. 2462–2469, October 2016.
113. Khakpour, A., Franke, S., Methling, R., Uhrlandt, D., Gortschakow, S., Popov, S., Batrakov, A., Weltmann, K.-D., "Optical and electrical investigation of transition from anode spot type 1 to anode spot type 2", *IEEE Transactions on Plasma Science*, 45(8), pp. 2126–2134, August 2017.
114. Khakpour, A., Franke, S., Gortschakow, S., Uhrlandt, D., Imani, M., "An improved mode; for vacuum arc regarding anode spot modes", *IEEE Transactions on Dielectrics and Electrical Insulation*, 26(1), pp. 120–127, February 2019.
115. Franke, S., Khakpour, A., Methling, R., Gortschakow, S., Uhrlandt, D., "Optical emission spectroscopy during the formation of an anode plume", *Proceedings International Symposium on Discharges and Electrical Insulation in Vacuum*, pp. 227–230, September 2019.
116. Batrakov, A., "Vacuum-arc anode phenomena: New findings and new applications", *Proceedings International Symposium on Discharges and Electrical Insulation in Vacuum*, pp. 163–168, September 2018.
117. Gortschakow, S., Methling, R., Popov, S., Schneider, A., Franke, S., Khakpour, A., Uhrlandt, D., "Cu and Cr density Determination during high-current discharge modes in vacuum arcs", *Proceedings International Symposium on Discharges and Electrical Insulation in Vacuum*, pp. 181–184, September 2018.
118. Kong, G., Liu, Z., Wang, D., Rong, M., "High-current vacuum arc: The relationship between anode phenomena and the average opening velocity of vacuum interrupters", *IEEE Transactions on Plasma Science*, 39(6), pp. 2370–1378, June 2011.
119. Batrakov, A., Popov, S., Schneider, A., Sandolache, G., Rowe, S., "Observation of the plasma plume at the anode of high-current vacuum arc", *IEEE Transactions on Plasma Science*, 39(6), pp. 1291–1295, June 2011.
120. Mitchell, G., "High current vacuum arcs: Part I and Part II", *Proceedings IEE*, 117, pp. 2315–2332, 1970.
121. Turner, H., Turner, C., "Discontinuous contact erosion", *Proceedings 3rd International Conference on Electrical Contact Phenomena*, pp. 309–320, June 1966.
122. Walcuk, E., "Arc erosion of high current contacts in the aspect of CAD of switching devices", *Proceedings 38th IEEE Holm Conference on Electrical Contact Phenomena*, pp. 1–16, October 1992.
123. Heberlein, J., Gorman, J., "The high current metal vapor arc column between separating electrodes", *IEEE Transactions on Plasma Science*, PS-8(4), pp. 283–289, December 1980.
124. Zalucki, Z., Janiszewski, J., "Transition from constricted to diffuse vacuum arc modes during high arc current interruption", *IEEE Transactions on Plasma Science*, 27(4), pp. 991–1000, August 1999.
125. Abplanalp, M., Menzel, K., Delachaux, T., Sutterlin, R.-P., Kassubek, F., "Optical investigation of constricted vacuum arcs", *Proceedings International Symposium on Discharges and Electrical Insulation in Vacuum*, pp. 279–282, September 2016.

126. Schulman, M., Slade, P., "Sequential modes of drawn vacuum arc between butt contacts for currents in the range 1kA to 16kA", *IEEE Transactions on Components, Hybrids and Manufacturing Technology*, 18(1), pp. 417–422, March 1995.
127. Kimblin, C., "Cathode spot erosion and ionization phenomena in the transition from vacuum to atmospheric arcs", *Journal of Applied Physics*, 45(12), pp. 5235–5244, December 1974.
128. Fang, D., "Cathode spot velocity of vacuum arcs", *Journal of Physics. Part D: Applied Physics*, 15(5), pp. 833–844, May, 1982.
129. Persky, N., Sysun, S., Khromoi, Y., "the dynamics of cathode spots in vacuum discharge", *High Temperature*, 27(6), pp. 832–839, 1989.
130. Wang, C., Shi, Z., Wu, B., Jia, S., Wang, L., "Experimental investigation on the dynamic of cathode spot of vacuum arc in external transverse magnetic field", *Proceedings International Symposium on Discharges and Electrical Insulation in Vacuum*, pp. 239–242, September 2016.
131. Nemchinsky, V., "Model of the retrograde motion of a vacuum arc: comparison to experiment", *Ieee Transatctions on Plasma Science,* Vol. 48, N0. 4, pp. 1151–1161, April 2020.
132. Shi, Z., Wang, C., Song, X., Jia, S., Wang, L., "Stepwise simulation on the motion of a single cathode spot of vacuum arc in external transverse magnetic field", *IEEE Transactions on Plasma Science*, 43(1), pp. 427–479, January 2015.
133. Beilis, I., "Vacuum arc cathode spot grouping and motion in magnetic fields", *IEEE Transactions, on Plasma Science*, 30(6), pp. 2124–2132, December 2002.
134. Chaly, A., Shkol'nik, S., "Low-current vacuum arcs with short arc length in magnetic fields of different orientations: A review", *IEEE Transactions on Plasma Science*, 39(6), pp. 1311–1318, June 2011.
135. Alferov, D., Yevsin, D., Londer, Y., "Studies of the stable stage of the electric arc at the contact separation in a vacuum gap with a transverse magnetic field", *IEEE Transactions on Plasma Science*, 35(4), pp. 953–958, August 2007.
136. Alferov, D., Belkin, G., Yevsin, D., "DC vacuum arc extinction in a transverse axisymmetric magnetic field", *IEEE Transactions on Plasma Science*, 37(8), pp. 1433–1437, August 2009.
137. Emtage, P., Kimblin, C., Gorman, J., Holmes, F., Heberlein, J., Voshall, R., Slade, P., "Interaction between vacuum arcs and transverse magnetic fields with application to current limitation", *IEEE Transactions on Plasma Science*, PS-8(4), pp. 314–319, December 1980.
138. Child, C. D., "Discharge from hot CaO", *Physiological Reviews*, 32(5), p.492, 1911.
139. Gorman, J., Kimblin, C., Voshall, R., Wein, R., Slade, P., "The interaction of vacuum arcs with magnetic fields and applications", *IEEE Transactions on Power Apparatus and Systems*, PAS-102(2), pp. 257–266, February 1983.
140. Michal, R., "Theoretical and experimental determination of the self-field of an arc", *Proceedings 26th Holm Conference on Electrical Contacts*, pp. 265–270, September 1980.
141. Lee, A., Chien, Y., Koren, P., Slade, P., "High current arc movement in a narrow insulating channel", *IEEE Transation Components, Hybrids and Packaging Technologies*, CHMT 5(1), pp. 51–55, March 1982.
142. Boxman, R., "High current vacuum arc column motion on rail electrodes", *Journal of Applied Physics*, 48(5), pp. 1885–1889, May 1977.
143. Shmelev, D., "Kinetic model of short vacuum arc with hot evaporating anode", *IEEE Transactions on Plasma Science*, 41(8), pp. 1969–1973, August 2013.
144. Lindmayer, M., "Medium to high current switching: low voltage contactors, circuit breakers and vacuum interrupters " Chapter 14, p 788-790, In: Slade, P. (Editor), *Electrical Contacts: Principles and Applications: 2nd Edition*, (pub. CRC Press), 2014.
145. Teichmann, J., Romheld, M., Hartman, W., "Magnetically driven high current switching arcs in vacuum and low-pressure gas", *IEEE Transactions on Plasma Science*, 27(4), pp. 1021–1025, August 1999.
146. Delachaux, T., Fritz, O., Gentsch, D., Schade, E., Shmelev, D., "Numerical simulation of a moving high-current vacuum arc driven by a transverse magnetic field (TMF)", *IEEE Transactions on Plasma Science*, 35(4), pp. 905–911, August 2007.
147. Shmelev, D., Delachaux, T., "Physical modeling and numerical simulation of constricted high-current vacuum arcs under the influence of a transverse magnetic field", *IEEE Transactions on Plasma Science*, 37(8), pp. 1379–1385, August 2009.
148. Pieniak, T., Kurrat, M., Gentsch, D., "Surface temperature analysis of transverse magnetic field contacts using thermography camera", *IEEE Transactions on Plasma Science*, 45(8), pp. 2157–2163, August 2017.
149. Methling, R., Franke, S., Gortschacow, S., Abplanalp, M., Sutherlin, R.-P., "Anode surface temperature determination in high-current vacuum arcs by different methods", *IEEE Transactions, on Plasma Science*, 37(8), pp. 2099–2107, August 2009.

150. Yang, D., Jia, S., Shi, Z., Wang, L., "Influence of axial magnetic field on cathode plasma jets in high-current vacuum arc", *Proceedings International Symposium on Discharges and Electrical Insulation in Vacuum*, pp. 247–250, September 2016.
151. Nemchinsky, V., "Vacuum arc in axial magnetic field", *Proceedings 14th International Symposium on Discharges and Electrical Insulation in Vacuum*, pp. 260–262, 1998.
152. Gundlach, H., "Interaction between a vacuum arc and an axial magnetic field", *Proceedings 8th International Symposium on Discharges and Electrical Insulation in Vacuum*, pp. A2-1–A2-11, 1978.
153. Morimiya, O., Sohma, S., Sugarwara, T., Mizutani, H., "High current vacuum arcs stabilized by axial magnetic fields", *IEEE Transactions on Power Apparatus and Systems*. PAS-92, pp. 1723–1732, 1973.
154. Rondeel, W., "The vacuum arc in an axial magnetic field", *Journal of Physics. Part D: Applied Physics*, 8(8), pp. 934–942, 1975.
155. Agarwal, M., Holmes, R., "Cathode spot motion in high-current vacuum arcs under self-generated azimuthal and applied magnetic fields", *Journal of Physics. Part D: Applied Physics*, 17(4), pp. 743–756, 1984.
156. Wu, B., Shi, Z., Wang, C., Jia, S., Wang, L., "Experimental investigation on the current of cathode spot of vacuum arc in axial magnetic fields", *Proceedings International Symposium on Discharges and Electrical Insulation in Vacuum*, pp. 220–222, September 2016.
157. Wang, C., Shi, Z., Wu, B., Jia, S., Wang, L., "Experimental investigation on the influence of axial magnetic field on the lifetime of cathode spot in vacuum arc", *Proceedings International Symposium on Discharges and Electrical Insulation in Vacuum*, pp. 235–238, September 2016.
158. Kimblin, C., Voshall, R., "Interruption ability of vacuum interrupters subjected to axial magnetic fields", *Proceedings IEE*, 119(12), pp. 1754–1758, December 1972.
159. Anders, S., Anders, A., Yu, K., Yao, X., Browne, I., "On the microparticle flux from vacuum arc cathode spots", *IEEE Transactions on Plasma Science*, 21(5), pp. 440–446, October 1993.
160. Oks, E., "Generation of multiply-charged metal ions in vacuum arc plasmas", *IEEE Transactions on Plasma Science*, 30(1), pp. 202–207, February 2002.
161. Galonska, M., Hollinger, R., Krinberg, I., Spaedtke, P., "Influence of an axial magnetic field on the electron temperature in a vacuum arc plasma", *IEEE Transactions on Plasma Science*, 33(5), pp. 1542–1547, October 2005.
162. Oks, E., Anders, A., Brown, I. G., Dickinson, M. R., MacGill, R. A., "Ion charge state distributions in high current vacuum arc plasmas in a magnetic field", *IEEE Transactions on Plasma Science*, 24, pp. 1174–1183, June 1996.
163. Foosnaes, J., Rondeel, W., "The vacuum arc subjected to an axial magnetic field", *Journal of Physics. Part D: Applied Physics*, 12(11), pp. 1867–1871, 1975.
164. Song, X., Shi, Z., Liu, C., Jia, S., Wang, L., "Influence of AMF on the expansion speed of cathode spots in high-current triggered vacuum arc", *IEEE Transactions, on Plasma Science*, 41(8), pp. 2061–2067, August 2013.
165. Robson, A., "The motion of a low-pressure arc in a strong magnetic field", *Journal of Physics. Part D: Applied Physics*, 11(13), pp. 1917–1923, September 1978.
166. Schulman, M., Slade, P., Heberlein, J., "Effect of an axial magnetic field on the development of the vacuum arc between opening electric contacts", *IEEE Transactions on Components, Hybrids and Manufacturing Technology*, 16(1), pp. 180–189, March 1993.
167. Taylor, E., Keider, M., Schulman, M., "Transition to the diffuse mode for high-current drawn arcs with an axial magnetic field", *IEEE Transactions on Plasma Science*, 31(5), pp. 909–917, October 2003.
168. Chaly, A., Logatchev, A., Shkol'nik, S., "Cathode processes in free burning and stabilized by axial magnetic field vacuum arcs", *IEEE Transactions on Plasma Science*, 27(4), pp. 827–835, August 1999.
169. Taylor, E., "Visual measurements of plasma arc modes in a high-current vacuum arc with an axial magnetic field", *Proceedings 49th IEEE Holm Conference on Electrical Contacts, Washington, DC, U.S.A.*, pp. 70–75, September 2003.
170. Taylor, E., Keider, M., "Transition mode of the vacuum arc in an axial magnetic field: Comparison of experimental results and theory", *IEEE Transactions on Plasma Science*, 33(5), pp. 1527–1531, October 2005.
171. Keidar, M., Taylor, E., "Model for the transition to the diffuse column vacuum arc based on an arc voltage criteria ", *Proceedings 22nd International Symposium on Discharges and Electrical Insulation in Vacuum*, Matsue, Japan, September 2006.
172. Song, X., Shi, Z., Qian, Z., Jia, S., Liu, C., Wang, L., "The influence of ignition position on arc characteristics in the transition mode of a drawn vacuum arc in cup-shaped AMF contacts", *IEEE Transactions on Plasma Science*, 40(8), pp. 2051–2055, August 2012.

173. Song, X., Shi, Z., Jia, S., Qian, Z., Liu, C., Wang, L., "Experimental investigation on the initial expansion process in a drawn vacuum arc and the influence of axial magnetic field", *IEEE Transactions on Plasma Science*, 40(2), pp. 528–534, February 2012.
174. Taylor, E., Slade, P., Schulman, M., "Transition to the diffuse mode of the drawn arc in vacuum with an axial magnetic field", *IEEE Transactions on Plasma Science*, 31(5), pp. 909–1917, October 2003.
175. Li, H., Wang, Z., Liu, S., Geng, Y., Liu, Z., Wang, J., Uhrlandt, D., "Current balance characteristics of twin vacuum arc columns of horseshoe-type electrode", *Proceedings International Symposium on Discharges and Electrical Insulation in Vacuum*, pp. 291–293, September 2018.
176. Chaly, A., Logatchev, A., Shkol'nik, S., "Cathode dynamics on pure metals and composite materials in high-current arc", *IEEE Transactions on Plasma Science*, 25(4), pp. 564–570, August 1997.
177. Jia, S., Shi, Z., Wang, L., "Experimental investigation and simulation of vacuum arc under axial magnetic field: A review", *Proceedings 22nd International Symposium on Discharges and Electrical Insulation in Vacuum*, pp. 267–272, September 2006.
178. Chaly, A., Logatchev, A., Zabello, K., Shkol'nik, S., "High-current vacuum arc in a strong axial magnetic field", *Proceedings 22nd International Symposium on Discharges and Electrical Insulation in Vacuum*, pp. 309–312, September 2006.
179. Chaly, A., "Magnetic control of high-current vacuum arcs with the aid of of an axial magnetic field: A review", *IEEE Transactions on Plasma Science*, 33(5), pp. 1497–1503, October 2005.
180. Keider, M., Schulman, M. B., "Modeling the effects of an axial magnetic field on the vacuum arc", *IEEE Transactions on Plasma Science*, 29(5), pp. 684–689, October 2001.
181. Keider, M., Beilis, I. I., Boxman, R. L., Goldsmith, S., " 2-D expansion of the low-density of the interelectrode vacuum arc plasma jet in an axial magnetic field", *Journal of Physics. Part D: Applied Physics*, 29(7), pp. 1973–1983, 1996.
182. Keider, M., Beilis, I. I., Boxman, R. L., Goldsmith, S., "Theoretical study of plasma expansion in a magnetic field in a disc anode vacuum arc", *Journal of Applied Physics*, 83(2), pp. 709–707, 1998.
183. Henon, A., Altimani, T., Picot, P., Schellekens, H., "3-D finite element simulation and synthetic tests of vacuum interrupters with axial magnetic field contacts", *Proceedings 20th International Symposium on Discharges and Electrical Insulation in Vacuum*, pp. 463–466, 2002.
184. Jia, S., Zhang, L., Wang, L., Chen, B., Shi, Z., Sun, W., "Numerical simulation of high-current vacuum arcs under axial magnetic fields with consideration of current density distribution at the cathode", *IEEE Transactions on Plasma Science*, 39(11), pp. 3233–3243, November 2011.
185. Methling, R., Franke, S., Gortschakow, S., Aplanalp, M., Sutterlin, R.-P., Delachaux,T., Menzel, K., "Anode surface temperature determination in high-current vacuum arcs by different methods", *IEEE Transactions on Plasma Science*, 45(8), pp. 2099–2197, August 2017.
186. Zhai, X., Kulkarni, S., Yao, X., Acharya, V., Hemachander, M., Zhang, W., "Drawn arc behaviors of ½ coil-type axial magnetic field contact", *Proceedings International Symposium on Discharges and Electrical Insulation in Vacuum*, pp. 226–229, September 2016.
187. Kong, G., Liu, Z., Geng, Y., Ma, H., Xue, X., "Anode spot formation threshold current dependent on dynamic solid angle in vacuum subjected to axial magnetic fields", *IEEE Transactions on Plasma Science*, 41(8), pp. 2051–2060, August 2013.
188. Zhang, Z., Ma, H., Zhang, Y., Yi, X., Liu, Z., Geng, Y., Wang, J., "Anode spot threshold current for four pure metals subjected to uniform axial magnetic field in high current vacuum arcs", *IEEE Transactions on Plasma Science*, 45(8), pp. 2135–2143, August 2017.
189. Liu, Z., Kong, G., Ma, H., Geng, Y., Wang, J., "Estimate of critical axial magnetic field to prevent anode spots in vacuum interrupters", *IEEE Transactions on Plasma Science*, 42(9), pp. 2277–2283, September 2014.

3 The Materials, Design, and Manufacture of the Vacuum Interrupter

Three things are to be looked to in a building: that it stands on the right spot; that it be securely founded; that it be successfully executed.

Goethe
(Elective Affinities)

3.1 INTRODUCTION

The modern development of the vacuum interrupter began in the 1950s. There has been continuous development of its performance and application since that time. Figure 3.1 illustrates my own experience from 1966 to 2002 while working for the Westinghouse Electric Corporation and its successor the Eaton Corporation in the vacuum interrupter business. In 1966, we see that in order to achieve a 12.5kA interrupting ability in a 15kV circuit, the vacuum interrupter had to have a diameter of at least 182mm. Since 1966 there has been continuous development in vacuum evacuation technology, vacuum interrupter materials and computer design software. There has also been improved understanding of vacuum arc physics and high voltage, vacuum breakdown phenomena. The result is that the vacuum interrupter's size has steadily decreased until years later the same 15kV, 12.5kA performance can be achieved in a vacuum interrupter with a diameter of only 50mm.

The internal components of a typical vacuum interrupter are shown in Figure 3.2. The contacts are housed within an evacuated envelope in which the ambient gas pressure is between $10^{-2} \sim 10^{-4}$ Pa. When the contacts touch, the vacuum interrupter is in the closed condition, its natural state. The contacts open by withdrawing the contact attached to the moving terminal from the contact attached to the fixed terminal. The moving terminal is attached to a bellows; consequently, the vacuum is maintained inside the vacuum interrupter. When the contacts separate arcing is established within the interrupter. As I have discussed in Chapter 2 in this volume, this arc burns in the metal vapor evaporated from the contact surfaces. The metal vapor continually leaves the intercontact region and condenses on the contact surfaces and the surrounding metal vapor shield. The latter is isolated from one or from both contacts and serves to protect the insulating envelope from excessive metal vapor deposition. At current zero, contact vapor production ceases and the original vacuum condition is rapidly approached. The dielectric strength of the interrupter also increases, and the circuit is interrupted. The vacuum arc interruption process and the high voltage recovery of the contact gap will be discussed in Chapter 4 in this volume. As I have discussed in Chapter 1 in this volume, the contacts in the open position isolate the circuit voltage internally by the intercontact gap and externally by the insulating envelope.

In this chapter I will first present the development of vacuum interrupter contact materials: their properties will be discussed, as well as their strengths and weaknesses for particular applications. I will then discuss the contact structures that have been developed to control the high-current, columnar vacuum arc (see Section 2.4). Finally, I will present a general description of vacuum interrupter design, the components that are incorporated in it and outline the manufacturing techniques that have evolved.

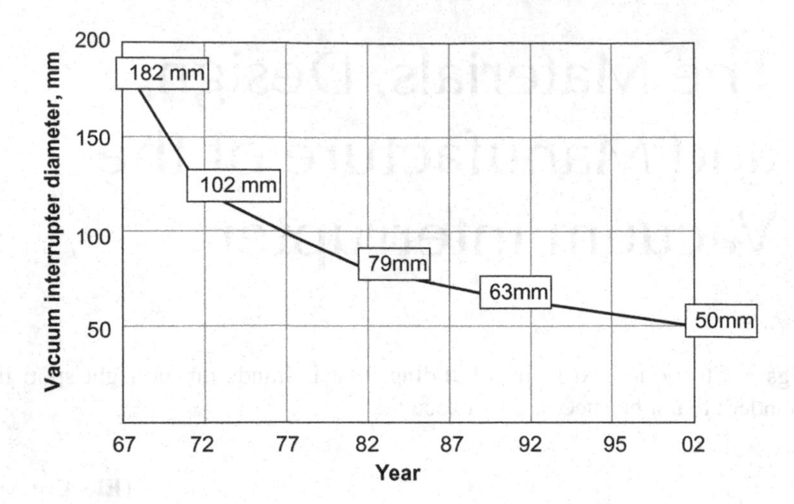

FIGURE 3.1 The Westinghouse (now the Eaton) experience in the reduction of the vacuum interrupter's diameter from 1968 for the 15kV, 12.5kA function.

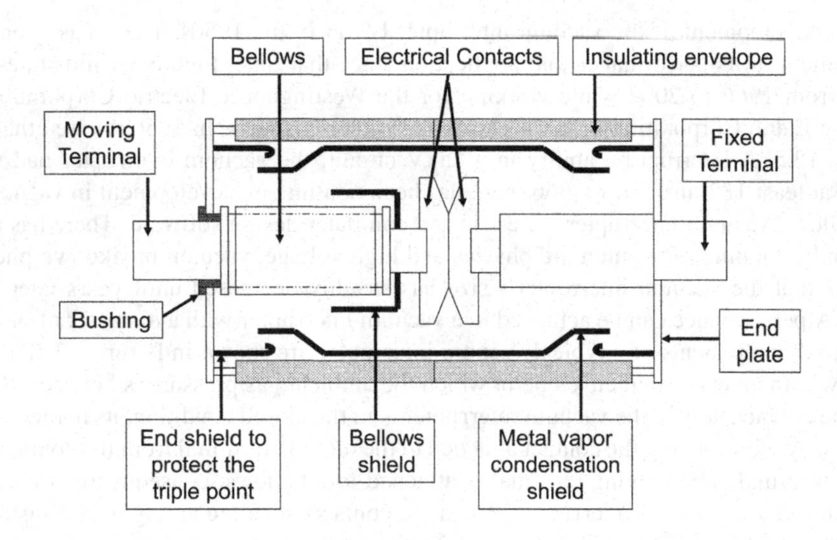

FIGURE 3.2 The cross-section of a vacuum interrupter.

3.2 VACUUM INTERRUPTER CONTACT MATERIALS

3.2.1 Introduction

The vacuum environment offers definite advantages to the developer of contact materials for use in the vacuum interrupter since there is no ambient gas to contaminate the contact surfaces. Thus, mixtures of materials that cannot be contemplated for application in gaseous environments such as air or SF_6 can be considered. Also, changes in the contact surface after arcing are only affected by the interaction of the contact materials themselves, and not by complex oxides that can form, for example, in air [1]. Thus, once the contact surface has stabilized, the contact resistance will be steady and consistent throughout the vacuum interrupter's life. The lack of ambient gas also allows for a high-voltage withstand across a small contact gap and therefore permits the creation of relatively compact vacuum interrupter designs. Even with these advantages, the development of practical vacuum interrupter contact materials continues to be limited by the traditional compromise between the desired electrical and mechanical properties and the contact material's own limitations [2]. Table 3.1 presents a matrix showing the complex interaction between the contact material properties and a typical vacuum interrupter's performance requirements.

3.2.2 Copper and Copper-Based Contact Materials That Have Been Developed Following the Initial Experiments on High Current Vacuum Arcs Using Copper Contacts

Much of the initial experimental research on high current, vacuum arcs has been performed on contacts made from pure Cu [3]. While valuable information on the nature of the vacuum arc was obtained from these studies, the tendency of Cu to form strong welds when closing on high currents in vacuum prevented its use in practical vacuum interrupter designs. This welding property, in fact, has prevented the use of all pure metals except perhaps the limited application of W contacts.

The first successful high current contact material developed for vacuum interrupters used in vacuum circuit breaker is a Cu-based material with other metals added to increase its weld resistance by reducing its mechanical strength. This material combined Cu with a small percentage of Bi. The addition of a small percentage of Bi to molten Cu results in the Bi migrating to the grain boundaries of the Cu during solidification. This makes the resulting contact more brittle than pure Cu. The inherent defects of this material are its high erosion rate, mechanical weakness, and its high level of chop current. Further development of this concept led to a whole class of Cu–Bi type of binary alloys [4, 5] (e.g. Cu–Sn, Cu–Pb, Cu–Sb, Cu–Zn, etc.). A very high percentage of Bi is required (> 5%) to obtain a very low chopping current. This amount of Bi made the resulting contact material mechanically very weak, its current interruption ability and its high voltage-withstand ability undesirably low, and its erosion undesirably high.

An effective contact material evolved when a very small amount, about 0.5% Bi, was added to Cu. This material had satisfactory current interruption ability, reasonably high voltage withstand capability and did not readily weld [4]. This class of materials was later expanded to include all materials where the major constituent had a boiling point of less than 3500K and where the minor constituent had a freezing temperature lower than that of the major constituent. The minor constituent of these materials also had to have a substantial solubility in the liquid state of the major constituent and little solubility in the solid state of the major constituent [6]. In fact, when reading the old patents for these materials, it seems as if the authors were trying to patent the whole periodic table. Research continued to improve this class of contact material using ternary systems [7, 8]. In the end, this research effort proved to be a dead end.

With the advent of the Cu–Cr contact material all research on the Cu–Bi type of material eventually ceased and the Cu–Cr contact material superseded its use in practical, high current vacuum interrupters. Even so, research continued on high Cu content, contact alloys such as Cu–Co–Ta [9], which showed some promise, but did not seem to compete with the Cu–Cr contact materials. Table 3.2 gives the material properties for this class of vacuum interrupter contact material.

3.2.3 Refractory Metals Plus a Good Conductor

One of the earliest contact materials applied in low current vacuum switches depended upon the use of a refractory material. The most common type of material in this class is W–Cu or Mo–Cu [10] and variations of them, e.g., W–Cu–Ti–Bi, W–Cu–Ti–Sn [11], W–Cu–Ti, W–In–Cu [12], W–Cu–Zr [13, 14], W–Zr [15], and of course W–Cu–Bi [16]. In these materials the refractory W is usually a sintered matrix (> 50% by volume), which is infiltrated with the good conductor. The infiltrate is usually Cu or Cu alloy. The elements like Ti, Zr, and In are used to aid the vacuum infiltration of the W and the elements Bi and Sn are there to aid the chopping and also the anti-welding capability.

TABLE 3.1
Contact Performance Needs and Contact Properties

Contact Performance	Material Properties								
	Gas Content	Melting Point	Vapor Pressure	Work Function & Electron Emission	Ionization Potential	Electrical & Thermal Conductivities	Residual Gas Gettering	Structural Quality	Smooth Surface, no Cracks and Pits
Circuit Interruption & Dielectric Recovery	✓	✓	✓	✓	✓	✓	✓	✓	✓
Endurance & Resistance to Erosion		✓	✓			✓		✓	✓
Dielectric Strength	✓			✓		✓	✓	✓	✓
Current Carrying Capability		✓				✓		✓	✓
Chopping Current	✓	✓	✓	✓	✓	✓	✓		
Resistance to Welding		✓				✓		✓	✓
Ability to interrupt high frequency currents	✓	✓	✓	✓	✓	✓	✓	✓	✓

TABLE 3.2
Material Properties of Copper-Based Contact Materials

	Material Properties of Cu Based Contacts			
Contact Material	**Weight %**	**Electrical Conductivity $MS.m^{-1}$**	**Hardness $x10^2$ Nmm^{-2}**	**Gas Content ppm**
Cu	100	60	4–6	5, O_2
Cu–Bi	99.5–0.5	~ 55	7	~ 5, O_2
Cu–Pb	99–1	~ 50	7	
Cu–Co–Ta	90–5–5		7.1	10, O_2
	80–15–5	13	8.9	20, N_2
	77–18–5	12	9.3	
	75–20–5	11	10.2	

The W–Cu material will readily interrupt a low current, diffuse vacuum arc when switching an ac circuit. It does not, however, work well at currents where the plasma plumes from the cathode spots overlap close to the cathode or when the columnar vacuum arc forms. This class of material does have very good high voltage withstand ability, a resistance to welding and an acceptable level of chopping current, so it has found use in load-break switches and in capacitor switches with circuit currents less than about 2kA (see Chapter 5 in this volume).

Another commonly used contact material in this class is manufactured from WC and Ag. This material has found almost universal acceptance for application in vacuum interrupters for vacuum contactors over the range of circuit voltages 400V $\leq U \leq$ 15kV; i.e., this material is used in motor switching applications. This material's properties of its low-surge capability, which results from its inability to interrupt high frequency currents, its low chopping current, and its resistance to welding make it particularly attractive for this application. I will discuss this further in Chapters 4 and 5 in this volume. The WC–Ag material has the same limitation on interrupting higher currents, as does the W–Cu material: in ac circuits up to 7.2kV the limit is usually about 4.5kA. It has been found capable, however, of interrupting much higher currents ($\approx$ 25kA rms.) if the contacts have a large enough diameter and if the high current vacuum arc is forced into the diffuse mode by the application of a high enough axial magnetic field (AMF) [17, 18]. It can also interrupt somewhat higher currents if the transverse magnetic field spirals are cut into its face. This material has been shown to be extremely resistant to welding, even when the circuit current is high enough for the blow-off force (see Section 2.1.4 in this volume) to open the contacts and for an arc to be established. This property is very desirable for contactor applications where the contact holding forces are generally low.

The cost of the Ag in the WC–Ag contacts has resulted in a search for a lower cost alternative. Behrens et al. [19, 20] report work on WC–Ag, WC–Cu, W–Cu, and W–Cu–Sb for low voltage contactor applications. They evaluate chop current values, susceptibility of re-ignition and contact resistance change during operating life. An example of their data is shown in Figure 3.3. In general, all the contact materials have acceptable chop currents (i.e., $I \leq$ 6A), the lowest being for the WC–Ag contacts. The Cu containing materials have lower re-ignition rates than the Ag containing materials. As I will discuss in Chapter 4 in this volume, I believe this will result in materials that do not have the same low-surge property of WC–Ag. Temborius et al. [21] show that the exact performance of these materials is strongly affected by the particle size and the size distribution of the W powders: an example is shown in Figure 3.4. They also show that contacts with a low chopping current also have a low current interruption capability. Pure W contacts have found a limited application in vacuum interrupters used for switching low dc currents (a few amperes) in medium voltage circuits. The use of this material relies upon the property of W shown in Figure 1.84. At currents less than 20A the vacuum arc between W contacts is very unstable and will chop to zero within a

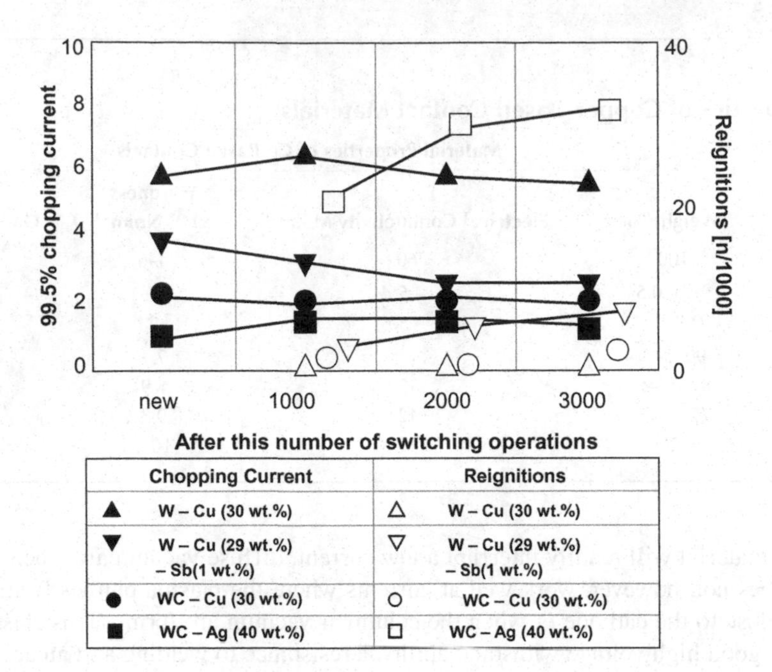

FIGURE 3.3 Comparison of the 99.5% chopping current and reignition percentage for WC–Cu (30 wt.%), WC–Cu (40 wt.%), WC–Ag (40 wt.%) and WC–Cu (29 wt.%)–Sb (1 wt.%) [19, 20].

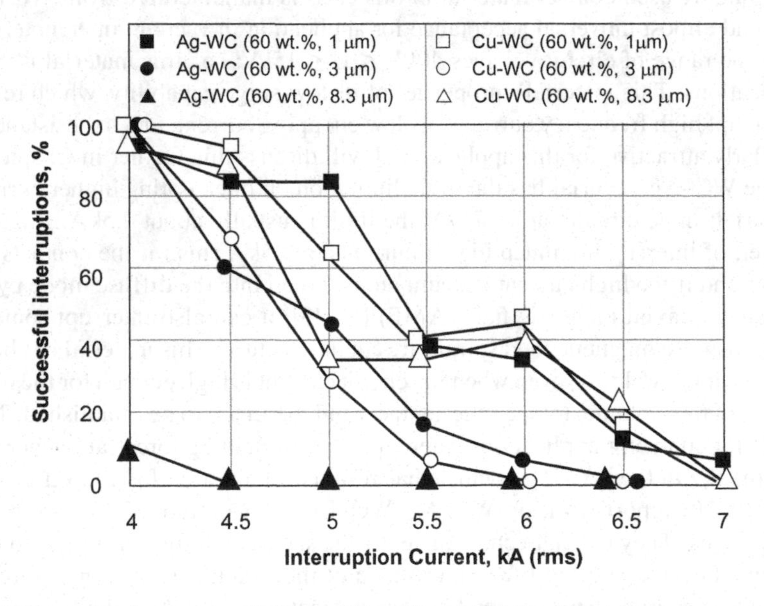

FIGURE 3.4 Interruption frequency of WC–Ag (40wt.%) and WC–Cu (40wt.%) contacts manufactured with different WC grain sizes [21].

few milliseconds even in a medium voltage circuit (5kV–15kV). Table 3.3 gives a general overview of the material properties for this class of vacuum interrupter contact material.

3.2.4 Semi-Refractory Metals Plus a Good Conductor

Since it has been introduced, this class of contact material has assumed dominance in vacuum interrupters used for vacuum circuit breakers and vacuum reclosers. In 1972 one material in this class,

TABLE 3.3
Material Properties of Refractory-Based Materials

		Material Properties of Refractory Based Contacts				
				Gas Content ppm		
Contact Material	Weight %	Electrical Conductivity $MS.m^{-1}$	Hardness $\times 10^2$ Nmm^{-2}	H_2	O_2	N_2
W	100	17.7	12–40			
Cu–W	10–90	~ 13	~ 30	2	35	3
	15–85	~ 17–18	27–31	3	40	2
	20–80	~ 17–20	25	3	35	2
	25–75	~ 20	19	3	40	3
	30–70	~ 20	22			
Cu–W–Sb	29–70–1	~ 14	26	2	35	20
Cu–Mo	25–75	24	19	2	150	20
	45–55	29	16	2	120	20
Cu–WC	40–60	15	27.5	2	50	20
	30–70	14	33	3	40	21
Ag–WC	40–60	~ 24	22	2	50	20
	50–50	~ 25	18			
Ag–WC–Co	30–68–2	15	35	3	70	60
Ag–WC–Zr	39.5–60–0.5	18	31.5	2	280	65

Cu–Cr, was developed as a possible contender to the then widely accepted Cu–Bi class of material [22]. Since that time, extensive research into the properties and the performance of Cu–Cr contacts in medium voltage, high-current vacuum interrupters has confirmed their early promise. In fact, this material is now the preferred contact for this class of vacuum interrupter. Research comparing the properties of other material combinations in this class, such as Co–Cu and V–Cu, continues to confirm the superior performance of Cu–Cr [23]. While there continues to be research [24] and patent activity in additives to the basic Cu–Cr material (e.g. [25–28], none has found general use as a vacuum interrupter contact material.

This material class has an interesting history. Robinson (English Electric Ltd., UK) [29] proposed it in the mid-1960s as a possible replacement for Cu–Bi. It was identified as a material combination with the potential to have the high-voltage withstand and the arc erosion properties of W–Cu together with the high-current interruption ability of Cu–Bi. Robinson called it "C" material. In its first embodiment, it consisted of Cr powder coated with a very thin Ni layer, which was pressed, vacuum sintered and then vacuum infiltrated with Cu–Zr alloy in a carbon crucible. The carbon crucible was thought to be necessary to reduce the chromium oxide, but it introduced some chromium carbide into the Cr matrix. The Ni coating tended to form a ternary alloy with the infiltrating Cu–Zr. This combined with the chromium carbide, resulted in a high-resistivity contact material. While English Electric pioneered this material, the ability to consistently manufacture a uniformly reliable contact for the commercial production of vacuum interrupters was pioneered by Westinghouse when the two Corporations formed a technical partnership at the end of the 1960s. Under the leadership of Bill Platt new embodiments followed with: (1) the CLR material, a vacuum sintered Cr matrix (eliminating the Ni coating) and infiltrating in vacuum with high purity Cu. This resulted in a lower resistivity contact and (2) a material with a pressed and vacuum sintered Cu–Cr powder with higher Cu content, called LR. The industrial development and manufacture of this contact material was begun at the Westinghouse Horseheads factory by Bob Thomas and used in commercial vacuum interrupters in the early 1970s.

Universal recognition that Cu–Cr could be used as a vacuum interrupter contact material was slow to occur. Even in 1980, only Siemens and Mitsubishi followed Westinghouse (now Eaton) and English Electric (This company no longer exists, but a distant descendant is Vacuum Interrupters Ltd.) in using Cu–Cr contacts for commercial interrupters [30, 31]. At that time most other manufacturers still used the Cu-alloy class of material. Since 1980, however, Cu–Cr contacts have become so widely accepted that now they are used in all the commercial, high-current, distribution and transmission voltage vacuum interrupters.

Table 3.4 shows the techniques that are used to produce a reliable Cu–Cr contact material. The initial production of the Cu–Cr material uses a vacuum sintered matrix of Cr, which is then vacuum infiltrated with high purity, oxygen free Cu. This technique limits the resulting contact to a composition with about 50% by weight of Cr. To obtain a higher percentage of Cu, a mixture of Cu and Cr powders is pressed together and then solid-state or liquid-phase sintered in a high vacuum. This produces a highly successful, cost effective, contact structure with a low enough gas content that is still widely used. Müller [32] introduced the arc melted Cu–Cr contact material. In his technique the Cu and the Cr are arc melted in a low-pressure, inert gas ambient. The resulting vacuum arc molten mixture (VAM) is deposited in a water-cooled Cu crucible where a rapid solidification results in a fine dispersion of Cr dendrites in the Cu matrix. A consequence from Müller's work on low-pressure arc melting of Cr and Cu to form Cu–Cr ingots has been the development of the new Cr-Cu phase diagram shown in Figure 3.5. This diagram shows a monotectic system with a flat miscibility gap between Cr mass fractions of 40% and 94.5%. The monotectic temperature of 1750C is 280C higher than previously reported [33]. The solubility of Cu in Cr at the eutectic temperature (1076C) with a Cu mass fraction of 1.28% is higher than the usually accepted value of 0.1%. Successful Cu–Cr contact materials have also been manufactured using vacuum induction melting (VIM) [34, 35] and vacuum arc melting [36] of Cu–Cr pressed powder mixtures followed by rapidly freezing the melt in a cooled mold. Each of these techniques has produced contacts that have had successful application in vacuum interrupters [37]. Miao et al. [38] present a detailed description of the vacuum cast Cu–Cr(25wt.%) contact material.

There is a difference in the final structure of the VAM and VIM materials. In experiments to show the interruption endurance at 25kA/12kV the VAM material eroded less rapidly [39]. In a

TABLE 3.4
Manufacturing Techniquest for Cu–Cr Contact Materials

a) Densification of Cu and Cr powder shaped into contact blanks
   (i) Solid phase sintering under vacuum
   (ii) Solid phase sintering under Hydrogen
   (iii) Liquid phase sintering under vacuum
b) Infiltration of Cu into a Cr matrix shaped contact blank
   (i) Mold-free infiltration
   (ii) Infiltration of a matrix with variable porosity
   (iii) Infiltration with a cast Cu backing
c) Pressed layer of Cu–Cr powder on top Cu powder in a stainless-steel cup and vacuum sinter
d) Arc melting of Cu and Cr in low pressure Argon and rapid freezing in a cooled mold
e) Vacuum induction melting of Cu and Cr mixture and rapid freezing in a cooled mold
f) Molten Cu and Cr atomized into small particles then pressed and sintered into contact blanks
g) Taking a Cu–Cr body manufactured in the processes above and hot extruding it.
h) Taking a Cu–Cr body manufactured in the processes above and cold extruding it

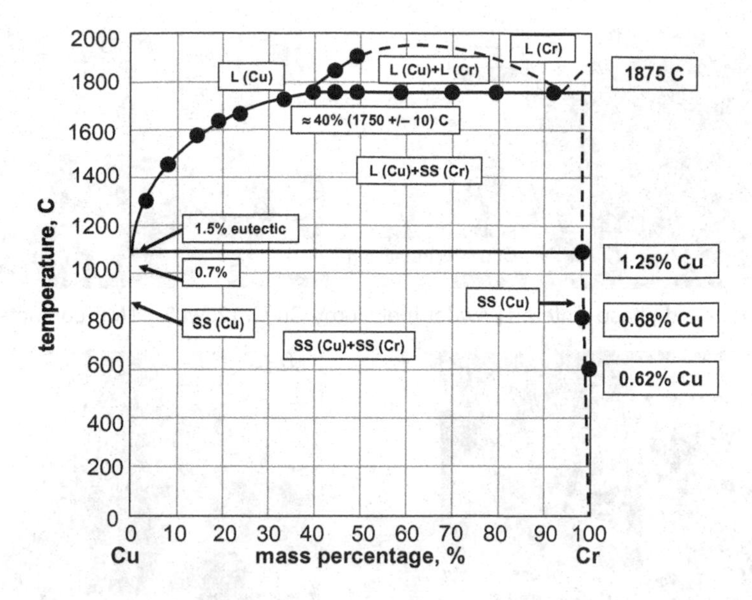

FIGURE 3.5 Müller's phase diagram for the Cr–Cu system [32]

study of the surface layer after the short circuit interruptions, however, all Cu–Cr materials have a similar structure that is shown in Figure 3.6. The melting of the Cu and Cr mixtures and their rapid cooling in a mold results in a 100% dense contact material. Another way of obtaining a 100% dense Cu–Cr contact is to form a Cu–Cr body using the press and vacuum sinter technique and then to cold extrude or hot extrude the resulting Cu–Cr body [40, 41].

More complicated manufacturing methods have also been suggested. In one example a mixture of molten Cu and Cr is atomized into small particles. These particles are then pressed and sintered into a suitable contact material [42]. A second example by Xiu et al. [43] is to melt the surface of a contact structure made with the powder metallurgy method discussed above with an arc or a laser. This certainly results in a fine, surface grain structure, a higher hardness and a lower surface porosity. A third method presented by Schneider et al. [44] is to melt the surface of a sintered Cu–Cr contact using an electron beam to produce a thin layer that has a similar structure to the induction melted or vacuum arced material, which is similar to that shown in Figure 3.6. Yu et al. [45] suggest using nano-crystalline Cr in a Cu matrix. The resulting Cu–Cr (25wt.%) material has Cr particles with an average dimension of less than 125nm, which compares to about 30μm in the induction melted material. The current per cathode spot on the resulting material is about 34A ± 10A.

The usual erosion limit given by most vacuum interrupter manufacturers is 3mm; i.e., about 1.5mm per contact. This has led to the suggestion that only a thin layer of Cu–Cr material is really required. Fink et al. [46] have developed one such contact structure, which results in a multilayered contact (MLC). Here a layer of Cu–Cr powders is placed on top a layer of Cu powder in a stainless-steel cup. After a vacuum sintering process the contact shape can be machined. This results in a Cu–Cr surface, a Cu sub layer and a stainless-steel backing for support. The resulting contact is thinner and has a slightly greater electrical conductivity than a pure Cu–Cr contact. They demonstrate that this MLC has a similar short circuit performance to that of the Cu–Cr contact. Durakov et al. [47] have manufactured Cu–Cr contacts by feeding a Cu–Cr powder mixture onto a Cu substrate and melting it under the action of a scanning electron beam. The resulting structure has Cr particles with a dimension of about 1.8μm.

The question naturally arises: Which technique produces the contact material with the best performance? In a practical application a vacuum interrupter designer must also consider the cost of producing the material versus its performance for a particular application. Devismes et al. [48] compare

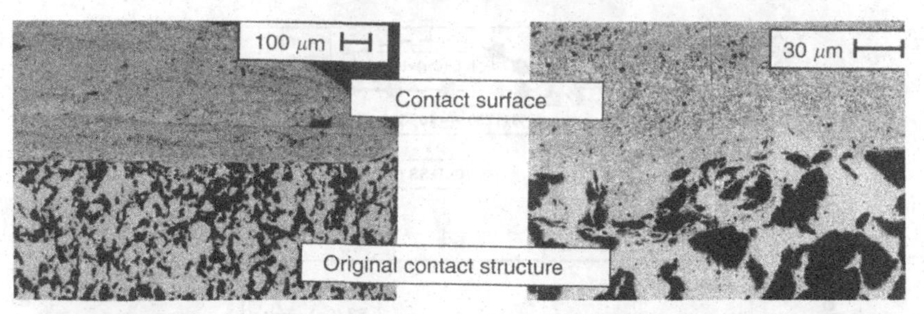

(a) Liquid-phase sintered powder metallurgy Cu-Cr (25wt.%) TMF contacts

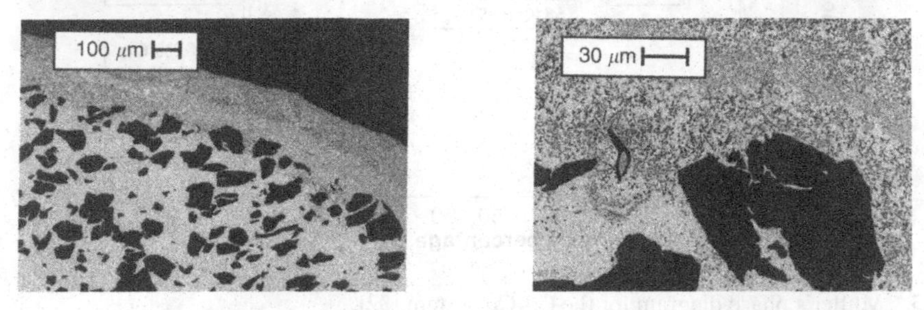

(b) Solid-phase sintered powder metallurgy Cu-Cr (25wt.%) AMF contacts

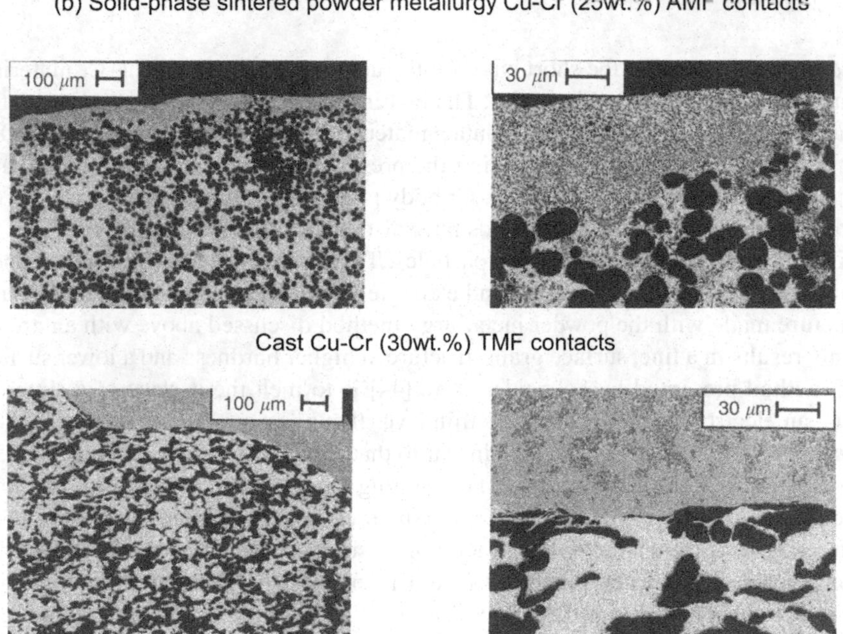

Solid- phase sintered powder metallurgy Cu-Cr (25wt.%) TMF contacts

FIGURE 3.6 Cross-section of a Cu–Cr (25 wt.%) contact's surface layers after fault current switching operations (totaling > 4000A.s) for TMF and AMF contact structures and different manufacturing methods.

three Cu–Cr (25 wt.%) contact materials produced by (a) solid-state sintering, (b) liquid-phase sintering, and (c) vacuum arc melting and rapid cooling in a mold. They show that the solid-state sintered materials have a larger variability in interruption performance compared to the liquid-phase sintered and the arc melted materials. They attribute this to the variability of the trapped gas in the closed porosity of the solid-state sintered material. They state, however, that any trapped gas can be controlled by the processing technique and the size distribution of the Cr powder. In practice, each of the manufacturing techniques shown in Table 3.4 produces contact structures that are successfully being used in commercial, fully certified, vacuum interrupters. Therefore, the choice of material for a particular application really depends upon the performance required and the cost of the material structure to obtain that performance; there is no universal criterion that makes one manufacturing method always the best choice. The vacuum interrupter designer should be careful if a contact manufacturer claims that its materials perform "better" than that of another manufacturer's. Table 3.5 gives the material properties for the Cu–Cr contact material used in vacuum interrupters.

In an experiment to determine the conductivity of solid phase, press sintered Cu–Cr contacts in vacuum, Klinski-Wetzel et al. [50] measure its value in the two directions shown in Figure 3.7. The transverse direction is the one in which the powders are pressed and the longitudinal direction is at right angles to it. Figure 3.8 shows that the conductivity in the transverse direction is about 14% less than that in the longitudinal direction. The transverse values agree with those given in Table 3.5. Assuming that the resistivity from impurities in both cases is equal, the authors conclude that a significant contribution of the phase boundaries of the composite is present. This may result from either segregation of the contaminants towards the grain boundaries or from internal stresses caused by a mismatch of thermal expansion coefficients. In practice the transverse value is the conductivity for practical Cu–Cr contacts. I would expect that this difference in conductivities will not be present in liquid phase, press sintered Cu–Cr contacts in vacuum or in the VIM and VAM Cu–Cr contacts.

Early work on a comparison of the interruption ability of contact materials [22] indicated that the Cu–Cr class of materials would be superior to the Cu–Bi contact. All published literature since that time has confirmed this initial indication. One illustrative example is the arc erosion of the contact surface and the ejection of metal droplets during arcing. Dullni et al. [51], quoting data from investigations using a laser shadow technique developed by Gellert et al. [52, 53] on the effect of high

TABLE 3.5
Material Properties of Cu–Cr Contact Materials

		Material Properties of Cu–Cr Contacts		
Manufacturing Methods	**Weight %**	**Electrical Conductivity MS.m^{-1}**	**Hardness ×10^2 Nmm^{-2}**	**Gas Content ppm**
Press Cu–Cr mixture, vacuum sinter	75–25	22–30	7–11	400–1600, O_2
	60–40	20–24	8–12	10–160, N_2
	50–50	17–21	10–13	
Press, vacuum sinter Cr matrix, vacuum infiltrate with Cu	50–50	15–17	8–13	400–1000, O_2 10–60, N_2
Arc melting Cu–Cr and rapid freezing	75–25	≥ 25	≥ 8	150–500, O_2
	70–30	≥ 21	≥ 9	8–30, N_2
	50–50	≥ 16	≥ 11	
Vacuum induction melting Cu–Cr and rapid freezing [49]	90–10	42–45	7 ~ 8	140–740, O_2
	75–25	32–34	8 ~ 9	90–150, N_2
	70–30	29–31	8 ~ 9	
	60–40	23–35	9 ~ 10	

Material Properties of Cu and Cr

Metal	**Melting Point, K**	**Boiling Point, K**	**Hardness, (×10^2) N.mm^{-2}**	**Conductivity, MS.m^{-1}**	**Latent Heat of Melting J.Kg^{-1}**	**Latent Heat of Vaporization J.Kg^{-1}**
Cu	1355	1868	4–9	59.9	2x10^5	4.8×10^7
Cr	2133	2953	7–13	6.7	4x10^5	6.6×10^7

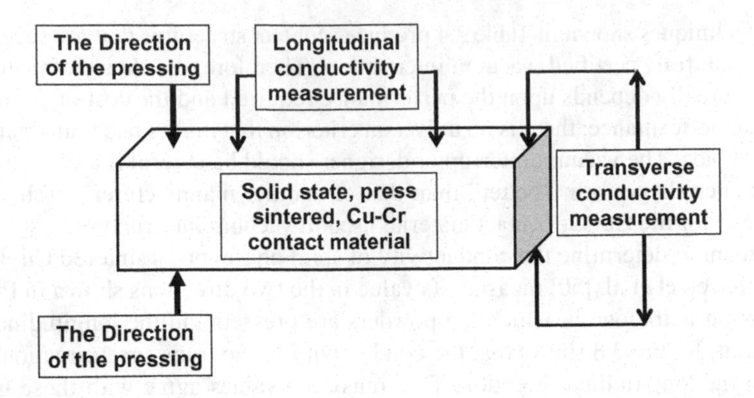

FIGURE 3.7 Measuring the electrical conductivity of pressed, solid state, sintered, Cu–Cr (25wt.%) contacts [50].

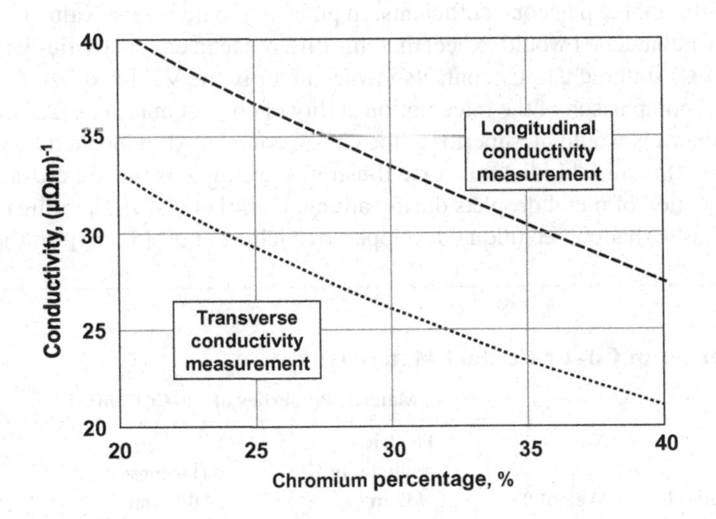

FIGURE 3.8 The average conductivities of pressed, solid state, sintered, Cu–Cr (25wt.%) contacts measured in the two directions shown in Figure 3.7 as a function of Cr content [50].

current arcing using disc (or butt) contacts, have shown that there is a major difference on the type of erosion, i.e. in the surface deformation and the particle ejection, for Cu and Cu–Cr contacts. The Cu–Cr data shows far fewer particles of smaller diameter than are observed from an arc between Cu contacts at the same current. They explain that this difference results from both the smaller melt depth on the surface of the Cu–Cr contacts (about 150μm versus about 1mm for Cu) caused by the lower bulk thermal conductivity and the presence of solidified Cr particles in the molten region. These two effects combine to hinder liquid motion at the edge of the molten pool, which, in turn, hampers the formation of large droplets and also decreases the number of particles that can form. Fewer metal droplets with smaller diameters (<12μm for Cu–Cr versus 50μm–450μm for Cu) after current zero results in much lower metal vapor densities in the contact gap during the appearance of the transient recovery voltage across the open contacts. The small melt depth and the fine dispersion of Cr particles in the Cu also leaves the Cu–Cr contact with a very smooth, crack-free surface. Figure 3.9 illustrates the smooth surface of a spiral Cu–Cr(25wt.%) contact after 31 high-current interruptions. This thin, arc-melted surface layer, shown in Figure 3.6 is also harder and more brittle than the original contact [54]. Dullni et al. [55] show that this layer has a higher resistivity, but that any increase in the material's contact resistance is primarily the result of the increase in its

FIGURE 3.9 The relatively smooth surface on a spiral Cu–Cr(25wt.%) contact after 31 high-current ac interruptions.

hardness: see Equation (2.4). The arced surface layer also seems to improve the material's lightning impulse withstand voltage rating (BIL rating): see Rylskaya [56]. There is also some evidence that the frozen melted Cu–Cr surface improves the material's current interruption performance [57]. This seemingly advantageous surface structure occurs with weight percentages of Cr from 25wt.% to 50wt.%, For contacts with Cr content greater than 60wt.%, Gentsch et al. [58] show that the surface changes from the desirable smoothness to a cracked and brittle structure. While this surface makes weld breaking easier it does not bode well for the contact's long-term viability.

The interruption capability of Cu–Cr contacts is not only superior at high currents, but they also perform well after switching low currents (up to a few thousand amperes). M. Glinkowski et al. clearly show [59] that the contact gap recovers its ability to withstand high voltage after switching a capacitor circuit much faster for Cu–Cr than for Cu–Bi; i.e., the probability of breakdown after capacitor switching is lower for Cu–Cr contacts. This confirms the research by Ohshima et al. [60] who show two effects: (1) Cu–Cr contacts have a lower probability of restriking after switching capacitor banks than do Cu-alloy contacts; and (2) this property is maintained over a wide range of currents. After low-current switching (isolated shunt capacitor banks), the Cu–Cr material is clearly superior. When switching high inrush currents (back-to-back capacitor banks), the contrast between the probability of restriking for Cu–Cr contacts and Cu-alloy contacts is even greater (capacitor switching will be discussed in more detail in Chapters 4, 5, and 6 in this volume). These data are consistent with experiments showing that Cu–Cr contacts have a lower probability of particle ejection after high-current arcing and the smoother surface after arc erosion [51, 52]. In experiments covering a range of currents from 800 to 4800 A, it has also been shown that, in this current range, the rate-of-rise of transient recovery voltage (TRV, see Chapters 5 and 6 in this volume) after arc extinction has little effect on successful interruption [61, 62]. Figure 3.10 shows the percentage of successful interruptions as a function of contact material for a positive polarity of the TRV under line-operating conditions, Heyn et al. [61]. The interruption ability does not differ greatly from that shown for high frequency current, Lindmayer et al. [62]. They conclude that droplet and particle-initiated processes influence the reignition process. It is likely, therefore, that the lower particle emission from the Cu–Cr vacuum arc is a major contributor to its superior interruption performance at all current levels.

In the 1960s considerable effort was placed upon obtaining extremely low gas content contact materials (e.g. Hässler et al. [63]). It was thought that any trapped gas in the contact structure

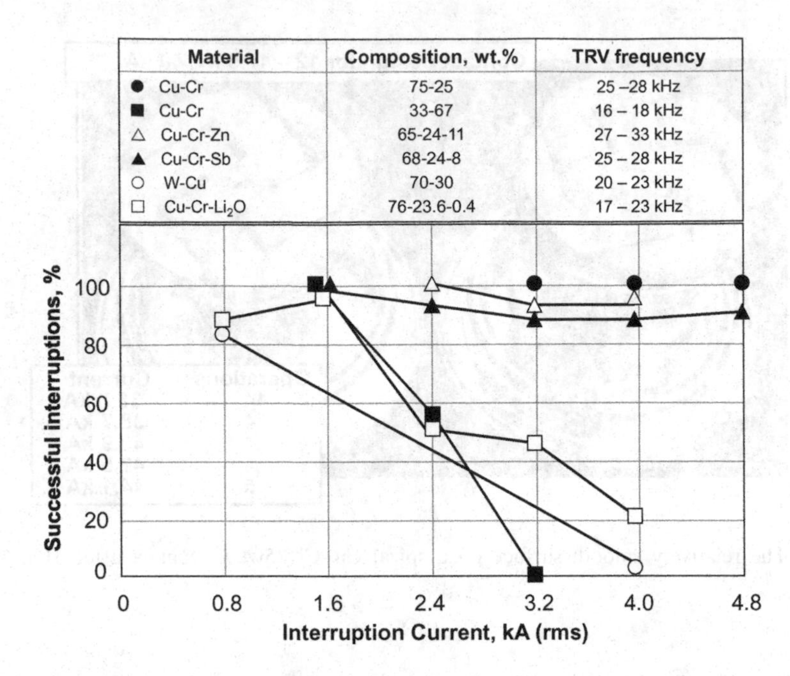

FIGURE 3.10 Percentage of successful interruptions of a 50 Hz current for different contact materials using a positive transient recovery voltage (TRV) that covers a frequency range from 10 kHz to 33 kHz and a peak value of 34.5 kV [61].

would severely degrade the vacuum and lead to interruption failure. Table 3.5 shows that Cu–Cr contacts can have a relatively high O_2 content. Experimental data continue to show that in spite of the high O_2 content, Cu–Cr contacts have very good interruption ability. This apparent contradiction has been explained by the experiments performed by Frey et al. [64], who show that after arcing, the deposited Cr effectively getters any residual O_2. It is so efficient, in fact, that the pressure of O_2 in the post arc period could not be measured. This has led the authors to speculate that an O_2 content in excess of 2000 ppm can easily be tolerated for the complete switching life of Cu–Cr contacts. This finding has been reinforced by a study comparing the interruption ability of Cu contacts with Cu–Cr contacts where it is found that the Cu–Cr material has excellent current interruption ability for a gas content as high as 2000 ppm, Wu et al. [65]. It is still important, however, to prevent excessive gas from becoming trapped in any closed porosity inside the contact during manufacture [48].

The Cu–Cr contact contains other impurities such as Fe, Al, and Si that can result from the aluminothermic, manufacturing process of the Cr powder. They can be found as high as 2000 ppm in the final contact structure. Cu–Cr contacts made with Cr powder manufactured using the electrolytic process have very low ppm of these impurities. Papillon et al. [66] have studied pressed, solid-state sintered Cu–Cr contacts. The contacts are sintered in either vacuum or in a hydrogen atmosphere. The surprising results show that the contacts sintered in a hydrogen atmosphere not only have a lower oxygen content but also they have a better current interruption performance. They also show that the contacts manufactured with the purer Cr using the electrolytic process have a lower current interruption ability than those manufactured with the Fe, Al and Si impurities. In fact, they observe that those contacts with the highest level of these impurities performed as well as those with lower levels. They speculate that contacts manufactured with the aluminothermic Cr powder perform better because the Fe impurity results in a higher surface thermal conductivity that ensures a smoother surface layer after arc interruption (Figure 3.6). A similar result has been observed by Klinski-Wetzel et al. [67]. Their data are only relevant to the solid-state sintered contact material.

The dielectric recovery of the Cu–Cr material is aided not only by the reduced level of particles after current interruption, but also by the smooth surface structure left on Cu–Cr contacts after arcing. After arcing, the contact surface exhibits shallow pools of melted metal, which solidify to form a smooth surface finish [22, 32, 53, 67, 68]. The thin molten layer solidifies very quickly, and the smooth surface allows a rapid reestablishment of the dielectric strength of the contact gap after arcing. The continued presence of the smooth surface with no peaks, troughs, or loose surface material ensures a continued high-voltage withstand ability for the Cu–Cr contacts throughout their lives. Certainly, the high voltage performance of vacuum interrupters after switching short circuit currents has been clearly demonstrated [67–70]. There is no lack of experimental evidence to show that Cu–Cr can interrupt very high short circuit currents over the whole range of distribution voltages from 4.6 kV to 72 kV and for transmission voltages up to 170 kV.

The optimum ratio of Cu to Cr has not been universally accepted for all vacuum interrupter designs and for all vacuum interrupter applications. There have been studies that show a Cu–Cr (50wt.%) works well [31]. Indeed, a contact with a high Cr content will have a good resistance to welding [71]. On the other hand, increasing the Cr content and the decreasing Cu content will lower the contact's thermal and electrical conductivities. For example, increasing the Cr content from 25 wt.% to 50wt.% reduces the contact's thermal conductivity between 40% and 50% (there is also a further decrease resulting from (a) the mutual solubility of the Cu and the Cr and (b) the contact's final density from the powder metallurgy process). Yanabu et al. [72] report that the threshold interruption current i_{th} at which the post-arc current (see Chapter 4 in this volume) dramatically increases depends upon the Cr content of the contacts: i.e. i_{th} at 25 wt.% Cr > i_{th} at 40 wt.% Cr > i_{th} at 50 wt.% Cr. In fact, Sato et al. [73] seem to show that the interruption ability of the Cu–Cr contact continues to improve as the Cr content decreases below 20 wt.%. In another study Li et al. [74] show that the best performance occurs for a Cr content between 20 wt.% Cr and 40 wt.% Cr. Their study certainly shows a major loss of performance if the Cr content goes above 50wt.% or below 10 wt.%. Contacts with 50 wt.% Cr seem to have somewhat better high voltage performance. I suspect that there will never be agreement on the optimum ratio of Cr to Cu for all vacuum interrupter contact structures and for all vacuum interrupter applications. At present, a review of the available commercial literature on this class of contact material shows that Cu–Cr (25 wt.%) is the most popular composition. Thus, it seems that in the future a material with a high Cu content will be more commonly used.

3.2.5 Copper Chromium Materials Plus an Additive

A subclass of Cu–Cr contact materials has been developed in order to enhance a particular performance characteristic. Table 3.6 presents examples of such additives [2]. The addition of a small percentage of Te to Cu–Cr(25wt.%) contacts gives a marginal improvement in its anti-welding characteristics [58, 75]. Contacts with added Nb {Cu–Cr(6.65wt.% – Nb(5.85wt.%)} do not perform better that Cu–Cr(25wt.%) or Cu–Cr(50wt.%) [76]. Delachaux et al. [77] show that the addition of a small percentage of Zr, Hf, Nb, C, and Mo to Cu–Cr(25wt.%) does not improve its current interruption performance. Thus, none of these contact structures has found general use in practical vacuum interrupter designs.

3.2.6 Chopping Current

I have already presented a full discussion of the current chop (i_c) phenomenon in Section 2.3.3 in this volume. As the current in the ac circuit falls towards current zero, a point will be reached, where the vacuum arc will no longer be able to sustain itself, and the current will suddenly "chop" to zero. If the value of chopping current is high enough, this rapid change in current interacting with the circuit's surge impedance can result in high over-voltages. A compendium of experimentally determined values of chop current is presented in Table 3.7. Note that the value of i_c is not an exact number, but it has a statistical distribution. The average value is much less than the maximum value.

TABLE 3.6
Suggested Additives to Cu–Cr Contact Materials

Material	Weight %	Purpose
W	2	Increase the high-voltage withstand
C	0.18–1.8	Reduction of the O_2 content and of the welding force
Te	0.1–4	Reduction of the welding force
Bi	2.5–15	Reduction of the chopping current and of the welding force
Si, Ti, Zr	1	Increase of the high-voltage withstand and interruption ability
Mo + Ta	High	Increase of the high-voltage withstand with good interruption ability
Mo + Nb	High	
Sb	2–9	Reduction of chopping current
WC	5	Reduction of chopping current
Ta, Te, Se	1, 1	Reduction of chopping current
Fe	15	Reduction of chopping current, arc erosion and the welding force. Increase the high-voltage withstand
Co		Increase the high-voltage withstand

For example, Cu has an average value of about 15A and a maximum value of greater than 21A. This has given rise to the impression that the "chop current" is a major factor when using a vacuum interrupter.

As presented in Section 3.2.2, the earliest commercial contact material for vacuum interrupters used in high current circuits was Cu with about 0.5% Bi. The vacuum arc chops when the current in the last cathode spot falls below a value to sustain it. Now if the last cathode spot resides on Cu, which is highly likely, given the very small percentage of Bi present in this contact structure, then the chop current can be greater than 21A (see Table 3.7). If, on the other hand, the last cathode spot is on a site where Bi is prevalent the highest chop current will be less than 0.5A. Thus, in the early development of vacuum interrupters using Cu–Bi (0.5 wt%) contacts, very high chop currents have been frequently observed. This, in turn, has led to high over-voltages in high surge impedance circuits and has resulted in the perception that chopping current is a problem with the vacuum interrupter. Interestingly enough Filip et al. [78] have demonstrated a similar effect with new Cu–Cr contacts. The distribution of their i_c values is shown in Figure 3.11(a). Here it can be seen that for new contacts about 80% of the measured chop currents have a mean value and a standard deviation of a few amperes. However, this figure shows that there is a second distribution of higher values of i_c. These researchers speculate that for new contacts there are islands of pure Cu between the Cr particles. When the last cathode spot resides on this region of pure Cu a higher i_c will result. After switching current the surface layer of the Cu–Cr contact becomes more homogeneous as a result of the interaction of the cathode spots and the surface layer of contact material. This surface takes the form of a Cu matrix with a fine dispersion of very fine Cr particles; thus, the single i_c distribution shown in Figure 3.11(b) results.

Exact comparison of data between different experimenters is not precise because i_c is dependent upon a circuit's surge impedance [$Z_s = f(L,C,R)$] and also upon the level of current being switched [79]. In Figure 2.30 I show the voltage and current characteristics for a Ag–WC contact as the ac current falls towards zero [80]. In looking at this figure, you will quickly notice that to measure the actual value of i_c is somewhat difficult. After the current chops, however,

TABLE 3.7
Chopping Currents for Vacuum Interrupter Contact Materials

Material	Average Chopping Current, A	Standard Deviation	Maximum Chopping Current, A
Ag	3.5–4.1		6.7
Cu	15		> 21
Cr	7		16
W	14–50		350
Cu–Cr (25 wt.%)	4	0.5	4.5–8
After 1000 ops. At 1kA	3.5		7.5
Cu–Cr (50 wt.%), switching 100A	4.5		10
Switching 950A, Surge Impedance, 240Ω	2		6
Switching 1550A, Surge Impedance, 240Ω	1.5		6
Cu–Cr (40 wt.%)	5.5		4.5
Cu–Cr (55 wt.%)			8
Cu–Cr (75 wt.%)	4.75		15
Cr (25 wt.%) Cu–Bi (5 wt.%)	1.1		13.1
Cr (25 wt.%) Cu–Bi (15 wt.%)	0.78		11.1
After 1000 ops at 600A	0.8		13.8
After 10,000 ops at 600A	1.45		12.6
Cr (25 wt.%) Cu–BiO_2 (5 wt.%)	1.2		7.5
Cr (25 wt.%) Cu–BiO_2 (15 wt.%)	0.75		10.5
After 1000 ops at 600A	0.76		9
After 10,000 ops at 600A	0.8		8
Cr (25 wt.%)–Cu–Sb (2 wt.%)	5.2		7.5
Cr (25 wt.%)–Cu–Sb (9 wt.%)	4		10.5
After 1000 ops. At 1kA	5		9
Cr (~23 wt.%)–Cu(~23 wt.%)–WC (~5 wt.%)	1		
Cr (~24 wt.%)–Cu (~24 wt.%)–Te/Se (~2wt.%)	2.5		
Cr (~44 wt.%)–Cu (~54 wt.%)–Ta (~2wt.%)	2.5		
W–Cu (10 wt.%)	5		
W–Cu (20 wt.%)	5		
W–Cu (30 wt.%)	5	5.3	8
WC–Cu (25 wt.%)			
WC–Cu (30 wt.%)	1		1.5
WC–Cu (40 wt.%)	1.9–2.7		4
WC–Cu (50 wt.%)	1.5		
WC–Ag (25 wt.%)	0.5	0.2	1.1
WC–Ag (40 wt.%)	0.6–1.1	0.2–0.4	1.4–1.5
WC–Ag (50 wt.%)	1.1	0.3	2.2
WC–Ag (60 wt.%)	1.5		4.0
WC (1μm grain)–Co (5 wt.%)–Ag (25 wt.%)	0.42		
WC (1μm grain–Co (5 wt.%)–Ag (30 wt.%)	0.27		
WC (6μm grain)–Co (5 wt.%)–Ag (30 wt.%)	0.38		
WC (1μm grain)–Co (0.5 wt.%)–Ag (30 wt.%)	0.58		
WC (1μm grain)–Co (5 wt.%)–Ag (40 wt.%)	0.67		
Mo–Cu (25 wt.%)			5.5
Mo–Cu (45 wt.%)			6.5

(Continued)

TABLE 3.7 (CONTINUED)
Chopping Currents for Vacuum Interrupter Contact Materials

Material	Average Chopping Current, A	Standard Deviation	Maximum Chopping Current, A
Cu (30 wt.%) W–Sb (1 wt.%)	3	5.3	5.5
After 10^5 ops AC4, 1500A	2.5	4.7	
Cu (30 wt.%) W–Sb (4 wt.%)	1.6	3.8	
After 105 ops AC4, 1500A	1.6	3.8	
Cu (30 wt.%) W–Li (0.24 wt.%)	22.1	63	
Cu (30 wt.%) W–SbBi (10 wt.%)	0.8	1.8	
Cu–Bi (0.5 wt.%)	2		> 21
Cu–Bi–Pb (1 wt.%)	5		11
Cu–Bi–Pb (13 wt. %)	1		2
After 1000 ops at ~ 400A	1.2		2.2
After 10^5 ops at ~ 400A	4.5		9
Co–Ag–Se	0.4		0.8
After 1000 ops at ~ 400A	0.4		0.8
After 10^5 ops at ~ 400A	0.4		0.8

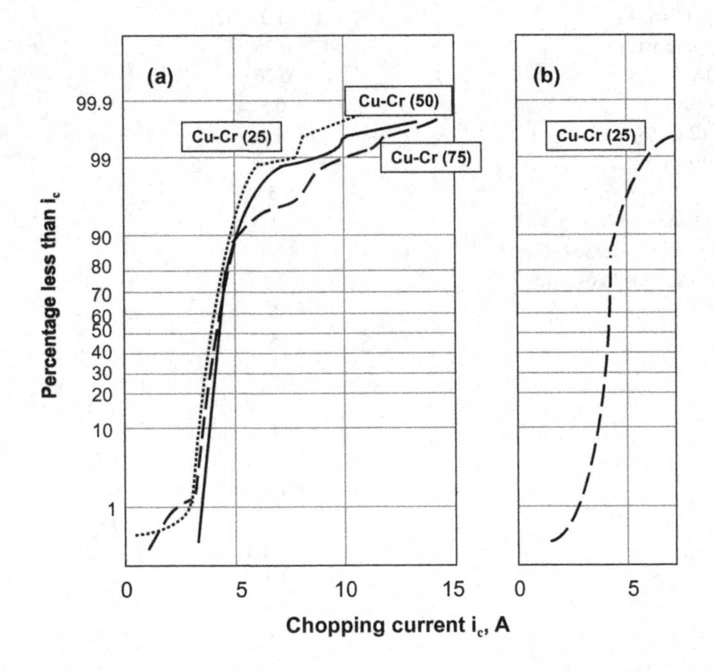

FIGURE 3.11 (a) Cumulative distribution of chopping currents for new Cu–Cr (25 wt.%), Cu–Cr (50 wt.%) and Cu–Cr (75 wt.%) contacts. (b) Cumulative distribution of chopping currents for Cu–Cr (25 wt.%) contacts after many switching operations [78].

the voltage across the contacts has a well-defined peak value and a well-defined frequency of oscillation. Holmes [81] used this peak voltage and its oscillation to accurately determine the actual chopping current. As can be seen from Figure 2.30 the voltage transient that occurs after the current chop is approximately a damped sinusoidal wave. Thus, it is possible to develop a simple analysis for this voltage transient using a simple, lumped parameter model shown in Figure 3.12. This model results in good agreement between measurements of chopping current

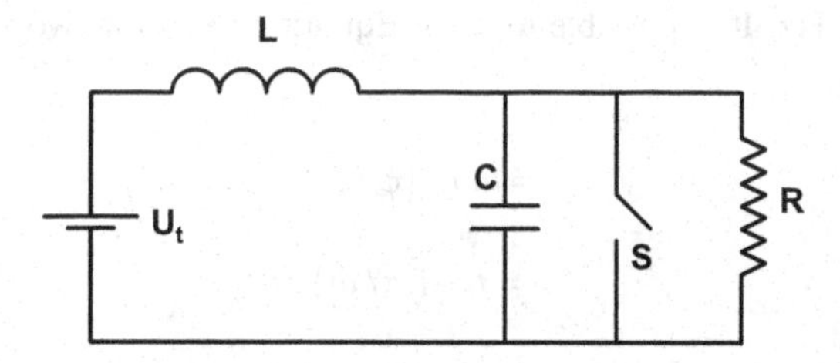

FIGURE 3.12 Lumped parameter circuit for analyzing the effect of chopping current on subsequent voltage oscillations.

and the calculation of the chopping current from the voltage transient that occurs after the current chops [82, 83].

The circuit most susceptible to the effects of current chop is an inductive circuit. Thus in Figure 3.12 the inductance L causes the current zero to occur when the circuit voltage and the transient recovery voltage, U_R, is a maximum and in the opposite polarity (I will present a complete explanation of the development of U_R in Section 4.3.1.1) This peak voltage remains essentially constant for the time period under consideration: up to about 200μs. For example, after 200 μs the voltage will have changed by less than 0.3%. Therefore, U_R can be represented by a constant emf in Figure 3.12. All of the effective circuit inductance is lumped into L and the capacitance placed across the switch is lumped into C. Switch S is an ideal switch that is perfectly conducting when closed and perfectly insulating when open (this switch represents the vacuum arc as it changes from a conductor to an insulator during the current chop). All of the circuit resistive or energy loss effects are lumped into R. In practice, the value for R can be calculated from the decrement of the voltage transient. The differential equation for the circuit shown in Figure 3.12 when the switch S is opened at the time the current equals i_c is:

$$L\frac{d^2q}{dt^2}+\frac{L}{RC}\frac{dq}{dt}+\frac{q}{C}=U_R \tag{3.1}$$

where q is the charge on the capacitor C at the time following the current chop. The voltage across the switch is:

$$U=\frac{q}{C} \tag{3.2}$$

The initial conditions are:

$$U=0\, at\, t=0 \tag{3.3}$$

$$\frac{dU}{dt}\, at\ t=0\ is -\frac{i_c}{C} \tag{3.4}$$

Using Equations (3.2), (3.3), and (3.4) results in a solution to Equation (3.1):

$$U=U_R-exp\left(-\frac{t}{2RC}\right)\left[\frac{i_c}{\omega C}+\frac{U_t}{2R\omega C}sin\omega t+U_Rcos\omega t\right] \tag{3.5}$$

where

$$\omega=2\pi f=\left[\frac{1}{LC}-\frac{1}{(2RC)^2}\right]^{1/2}\approx\left(\frac{1}{LC}\right)^{1/2} \tag{3.6}$$

and f is the frequency in Hz. It is possible to solve Equation (3.5) for two values of t where U can be a maximum

$$t_1 = tan^{-1}(\varphi / \omega) \tag{3.7}$$

$$t_2 = t_1 + (\pi / \omega) \tag{3.8}$$

and

$$\varphi = \frac{2RC\omega\left(\frac{i_c}{C} - \frac{U_R}{RC}\right)}{\frac{i_c}{C} + \frac{U_R}{2RC} + 2RC\omega^2 U_R} \tag{3.9}$$

The maximum voltage, U_p, occurs at t_1 when the chopping current is particularly severe (i.e. for large values of i_c) or where U_R is very small; for all other conditions the maximum value U_p occurs at t_2. In distribution circuits (5kV to 40.5kV), U_p will usually occur at t_2. The peak value of the first phase to clear for an ungrounded, inductive ac circuit is about 1.8 times the rms system voltage. In Holmes' experiment the maximum U_p is about 8kV for a chopping current of 8A in a high surge impedance circuit [81]. Thus, even in a 5kV circuit with $i_c = 0$, the peak of the recovery voltage will be 9kV. The voltage swing resulting from any current chop will take this voltage to a somewhat higher value. Also, for advanced vacuum interrupter contact materials a chopping current of 8A is a rare event. Holmes occasionally observes the highest U_p occurring at t_1, because he performs his experiment in a circuit with a peak system voltage $U_t = 620$V and one of the contact materials he investigates is Cu–Bi (0.5 wt.%).

In order to use these equations, you first must determine the value of U_p, the peak of the system's ac voltage. The stray capacitance, C, in parallel with the contacts and the circuit inductance, L, can be determined by observing the effect on the frequency of the transient voltage by adding additional parallel capacitance. The effective resistance, R, can be determined from the values of successive peaks, U_n and U_{n+1} of the transient voltage. The log-decrement "δ" is given by:

$$\delta = log_e\left[\frac{U_n}{U_{n+1}}\right] = \frac{\pi}{RC\omega} = \frac{1}{2RCf} \tag{3.10}$$

One useful parameter that directly influences the voltage transient during the current chop event is the circuit's surge impedance, Z_s, which is given by the relation:

$$Z_s = \frac{1}{\omega C} = \sqrt{\frac{L}{\left(C - \frac{L}{4R^2}\right)}} \tag{3.11}$$

When Equation (3.11) is substituted into Equation (3.5),

$$U = U_R - exp\left(\frac{-t}{2RC}\right)\left[\left(i_c Z_s + \frac{U_r Z_s}{2R}\right)sin\omega t + U_R cos\omega t\right] \tag{3.12}$$

In order to determine the validity of this analysis Holmes measures the actual chopping current, i_c, and the corresponding peak of the transient recovery voltage, U_R. He does this with zero additional capacitance across the contacts, for circuit currents ranging from 380A to 1550A, and for circuit inductances ranging from 0.75mH to 3.09mH. One representative data set is shown in

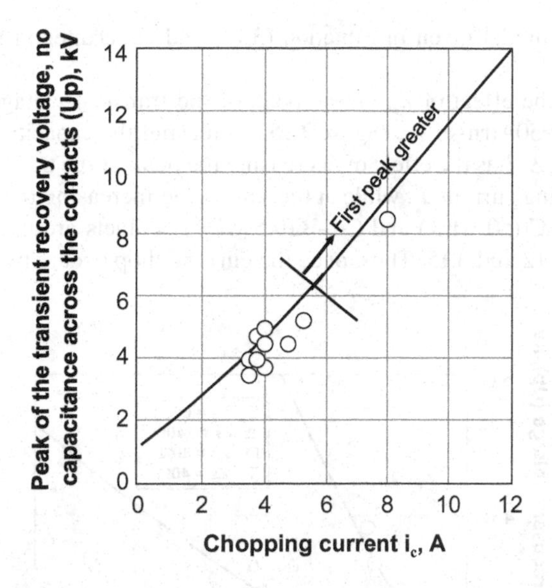

FIGURE 3.13 The relationship between peak of the transient recovery voltage and the chopping current for Zs = 1320Ω and circuit current i = 950A (rms), {○ experimental data, _______ calculated from Equation (3.12)} [81].

Figure 3.13 for one circuit current and for one surge impedance. For most of these data the peak voltage transient, U_p, occurs at t_2, but because his system voltage has a peak value of only 620V, he observes some peak values occurring at t_1 for higher chopping currents. This figure shows excellent agreement between the experimental data and the predicted value from Equations (3.5) and (3.12). Interestingly, the calculated curve also correctly predicts where the peak of the voltage transient occurs. Thus, it is possible to accurately calculate the true value of the chopping current from the transient recovery voltage using Equation (3.5) without having to rely upon an imprecise measurement of a current trace such as that shown in Figure 2.30. Taylor et al. [82] show in Figure 3.14

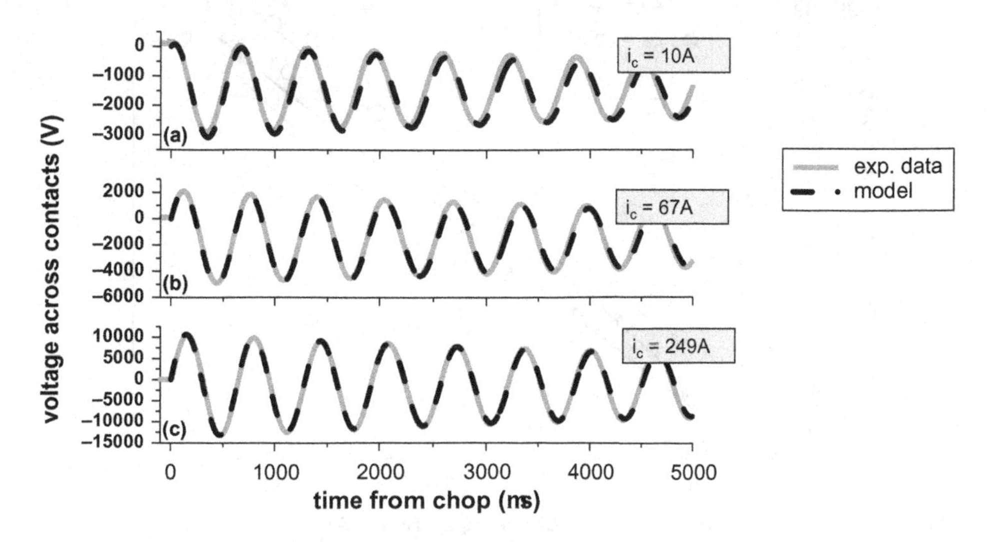

FIGURE 3.14 Comparison of the measured voltage transient after a current chop and the calculated voltage transient using Equation 3.12 [82].

agreement between the model given in Equation (3.12) and experimental values of U for a wide range of chop currents.

Holmes determined the effect of Z_s, on the peak of the transient voltage and on the chopping current for a current of 950A(rms); see Figure 3.15. Increasing the capacitance in parallel with the contacts, i.e., decreasing Z_s, has the effect of decreasing the peak of the transient voltage, U_p, associated with a given chopping current, i_c, while at the same time increasing the value of i_c; see Figures 3.16 and 3.17. Both Cu–Cr(50 wt.%) and Cu–Bi(0.5 wt.%) contacts are used for the experimental data shown in Figures 3.12 and 3.15. Thus, once the current chop occurs (which is dependant upon

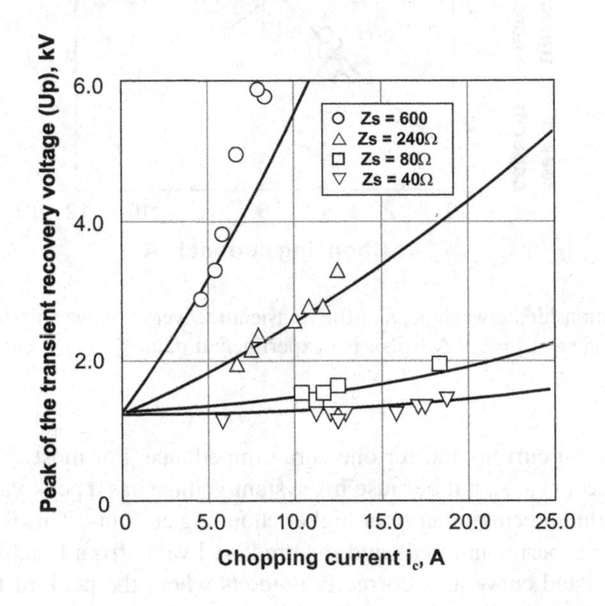

FIGURE 3.15 Comparison of the measured peak of the transient recovery voltage and that calculated from Equation (3.12) for a circuit current of 950A(rms) {O, □, △, ▽ experimental data, ______ calculated from Equation (3.12)} [81].

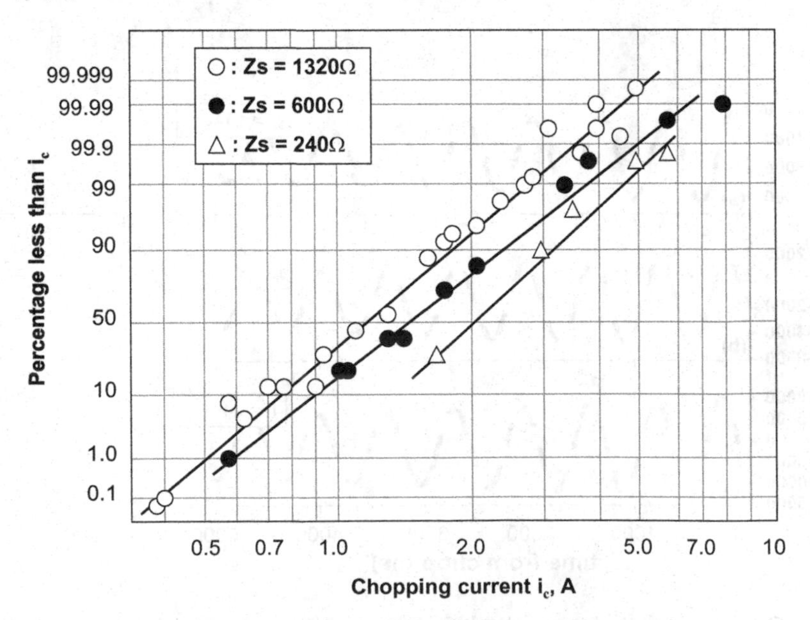

FIGURE 3.16 Distribution of chopping currents for Cu–Cr (50 wt.%) contacts [81].

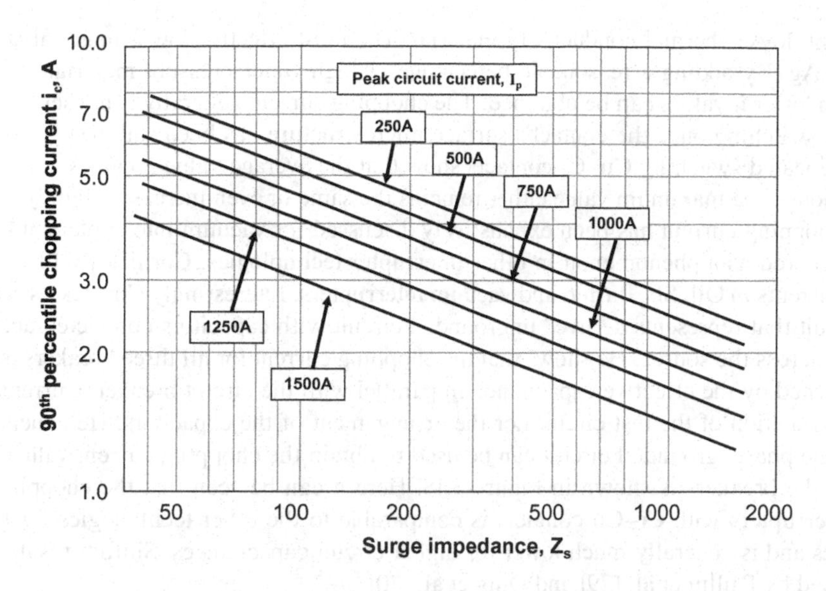

FIGURE 3.17 The 90th percentile value of chopping current for Cu–Cr (50 wt.%) contacts [81].

the contact material), the value of U_p, is determined only by the circuit's parameters and is independent of the contact material.

The current chop data in Holmes' experiment has a log-normal distribution, i.e., the $\log_e(i_c)$ is normally distributed, see Figure 3.16. Figure 3.17 shows the expected 90th-percentile i_c (i.e. the value at which the measured values of i_c have this value or less) as a function of circuit current (i) and the circuit's surge impedance (Z_s) for Cr–Cu(50 wt.%). Here we see that i_c ranges from 8A (for i= 250A, Z_s = 50Ω) to 1.1A (for i = 1500A, Z_s = 2000Ω). It is interesting to note how low these i_c values are. Table 3.7 gives an average i_c value for this material of 4.5A for current interruption levels of about 100A. This indicates that the chopping current also depends upon the peak current interrupted; i.e., the di/dt as the current falls to zero. The distribution of the i_c values in Holmes' experiments does not depend upon where on the ac current wave the contacts open. This confirms that the dependence on current level must be more related to the rate of change of current just before the current zero.

Most vacuum interrupters switch load currents in the range 100A to 3150A. It is therefore important to realize that the practical values of i_c may well be lower than those values given in the literature where the switched current is less than 100A [82]. From a practical point of view current chopping decreases to a negligible value when a vacuum interrupter is interrupting short circuit currents. Thus, when determining the chopping current level in a practical vacuum interrupter it is not only important to know the exact composition of the contact material, but it is also important to know the level of current being interrupted and the surge impedance of the circuit in which the vacuum interrupter will be applied. Other experiments have been performed by Ding et al. [80] and Smeets et al. [84] with a variable, parallel, capacitive (C) + inductive (L) + resistive (R) network across the opening vacuum contacts. Here i_c has been shown to increase as C increases (for a constant L and R), and to decrease as L or R increases (for a constant C). Finally, as is shown in Figure 3.11(a) and (b) the distribution of i_c values also depends upon the state of the contact's surface after switching electric current.

Table 3.7 shows that pure metals Cu, Ag, Cr, and W have a high value of i_c, but mixtures of these metals, in general, have much lower values. Thus, when Cu is mixed with Cr or with W, the chopping current drops to values that are considerably below those of the pure metal constituents. This is a result of the interaction between the lower melting point, high-vapor pressure material and the higher melting point, lower thermal conductivity material. The most effective low i_c material is a mixture of WC and Ag. By adding a reasonable percentage of high-vapor pressure materials to Cr–Cu or W–Cu, even lower i_c values can be obtained. The chopping current generally is not severely affected by repeated switching once the contact's surface microstructure has been stabilized. Experimental data after repeated switching Cu–Cr contacts show that the average value i_c can even decrease [79, 85, 86], although the maximum value either remains the same or even increases slightly.

While chopping current has been exhaustively discussed for vacuum interrupters, it has in general been ignored as a phenomenon in other interrupter technologies. Cornick [87] has compared chopping currents in Oil, SF_6 Puffer, and vacuum interrupters. Interestingly Cornick uses an experimental circuit that represents a "live" ungrounded circuit with capacitors connected across a load reactor and across the source. He shows that the chopping current for all three breakers is predominantly governed by the effective capacitance in parallel with the circuit breaker. Current chopping is neither a function of the test circuit nor the arrangement of the capacitors. He concludes that a simple, single phase, grounded circuit can be used to obtain the chopping current values. The data for each of the breakers is shown in Figure 3.18. Here it can be seen that the chopping level for vacuum interrupters with Cr–Cu contacts is comparable to the other technologies for low circuit capacitances and is generally much lower for higher circuit capacitances. Similar results have also been obtained by Dullni et al. [79] and Ding et al. [80].

3.2.7 Summary

Tables 3.8 and 3.9 present a qualitative assessment of the material properties and the expected performance of vacuum interrupter contact materials together with comments concerning their application. Table 3.10 presents an overview of the advantages and disadvantages of the three most commonly used contact materials. The Cr–Cu class of contact material has clearly emerged as the best material to use for vacuum interrupters that have to interrupt fault (i.e. short circuit) currents. This is based upon the published research on contact materials and experimental work performed since the early 1970s. It combines good electrical properties with outstanding circuit interruption ability and excellent arc erosion and welding resistance. The W–Cu contact material is popular for

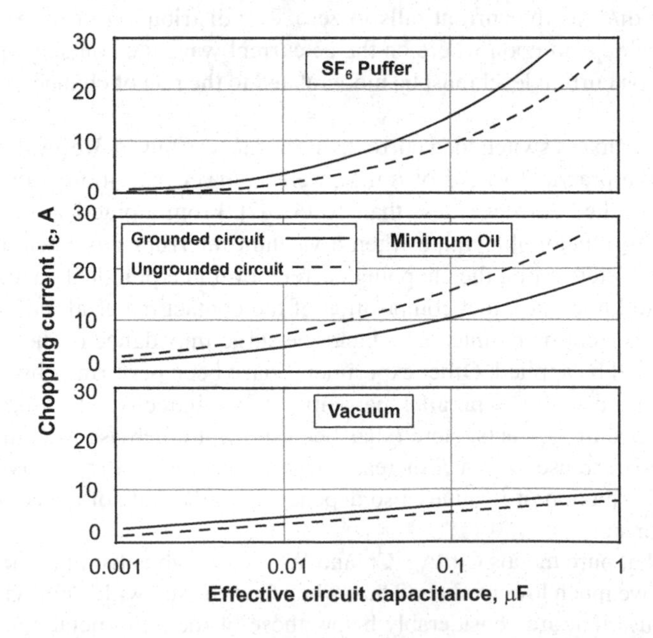

FIGURE 3.18 Chopping currents for an Oil Circuit Breaker, an SF_6 Puffer Breaker, and a Vacuum Breaker [87].

TABLE 3.8
An Assessment of the Material Properties for Vacuum Interrupter Contact

	Material Properties								
Contact Material Class	Gas Content	Melting Point	Vapor Pressure	Work Function & Electron Emission	Ionization Potential	Electrical & Thermal Conductivities	Residual Gas Gettering	Structural Quality	Smooth Surface, no Cracks and Pits
Copper Alloy (e.g. Cu–Bi)	++	+	+	++	+	++	++	–	–
Ag–WC (Diffuse vacuum arc only)	+	+	–	–	+	+	+	++	++
W–Cu (Diffuse vacuum arc only)	+	+	+	–	+	+	+	++	++
Cu–Cr	+	+	+	++	+	++	++	++	++
Cu–Cr + Additives	+	+	+	+	+	+	+	+	+
W–Cu + Additives	+	+	+	–	+	+	+	+	+

++ Excellent, + Acceptable, – Unacceptable

TABLE 3.9
Performance and Application of Vacuum Interrupter Contact Material

	Expected Contact Material Performance and its Application						
	Circuit interruption and recovery	Endurance and resistance to erosion	Dielectric strength	Current carrying capacity	Chopping current	Welding resistance	
Contact Material	Up to 80kA, 15 kV; 40kA, 38kV; & fault current interruption at 72kV and 140kV	Contactors (Up to 1M ops) Circuit Breakers: (Rated current ≥30K ops) (Short circuit current ≥ 30 ops)	7.2kV, 60kV BIL ; 38kV, 200kV BIL; 140kV 600kV BIL	Up to 3150A & for special needs up to 6000A	With modern contact materials the chopping current is low enough	Weld strength must be lower than the drive device's opening force	Application
Cu–Cr	●●●+	●●●+	●●●+	●●●	●●●	●●	Distribution and Transmission voltage power circuit breakers, reclosers and switches.
Cu–Cr + additives	●●	●●	●●	●●●	●●●+	●●●+	Distribution and Transmission voltage contactors and welding resistant switches
Ag–WC	●●● (up to 4.5kA)	●●●+	●●●	●●●	●●●+	●●●+	Contactors and switches that interrupt currents < 4.5kA. AMF vacuum interrupters for low surge, circuit breakers
Cu–W	●●● (up to 2kA)	●●●+	●●●+	●●●	●●	●●●	Distribution switches
Cu–W + additives	●● (up to 2kA)	●●●	●●	●●●	●●●+	●●●+	Specialty switches that require outstanding welding resistance
Cu alloy (e.g., Cu–Bi)	●●	●●	●	●●●+	●/●●	●●●	Originally developed for distribution circuit breaker applications, now superceded by the Cu–Cr contact material

● poor, ●● adequate, ●●● good, ●●●+ excellent

TABLE 3.10
General Advantages of Commonly Used Vacuum Interrupter Contact Material

Contact Material	Advantages	Disadvantages
Cu–Cr	• After high current interruption the contact surface remains smooth • The surface of the contact changes to very finely disperse Cr in a Cu matrix • Excellent electrical endurance • Excellent mechanical strength • Excellent high voltage performance • Withstands high speed transient recovery voltages • Low cost • Adequate resistance to contact welding in circuit breaker applications • Excellent high, short circuit interruption performance • Butt contacts interrupt up to 10kA • Maximum chop current < 6A	• Excellent high frequency current interruption can result in voltage escalation in high inductance circuits • Welding tendency too high for contactor applications
Ag–WC	• Excellent load current electrical endurance (> 10^6 electrical operations possible) • Poor high frequency current interruption, limits high voltage escalation effects, i.e. it is a "low surge" contact material (see Section 4.3.1.1) • Excellent resistance to contact welding • Maximum chop current <2A	• High cost • Butt contact interruption limit 4 to 4.5kA • Requires an AMF design to reach higher short circuit interruption values • High voltage performance generally limited to 15kV or lower systems • Use generally limited to contactor applications
Cu–W	• Excellent load current electrical endurance • High frequency current interruption ability between that of Cu–Cr and Ag–WC • Excellent high voltage performance • Good resistance to contact welding • Low cost • Maximum chop current <8A	• Butt contact interruption limit 2–3 kA • Use generally limited to load break switches

load break switches and for circuit currents below about 2000A. The WC–Ag contact material is now widely used for contactor vacuum interrupters. There are niche applications where very low surge voltages are needed so more complex materials such as Cu–Cr with high vapor pressure additives can be used. Also, as vacuum interrupters are developed for circuit voltages ≥ 72kV, additives may be added to Cu–Cr contacts to enhance their high voltage withstand performance. For all practical purposes, the maximum value of the chop current for modern, vacuum interrupter, contact materials is low enough that the phenomenon no longer presents a problem in the electrical circuits they are designed to switch and/or protect. I shall discuss this further in Chapter 4.

3.3 THE CONTACT STRUCTURES FOR THE VACUUM INTERRUPTER

3.3.1 Introduction

The vacuum interrupter's contact is one of the most important components for ensuring the successful operation of the vacuum interrupter. In the previous section I discussed the choice of materials and showed the considerations that have to be taken into account when choosing one over another for a particular vacuum interrupter application. The choice of material, however, is not the only concern when deciding upon a vacuum interrupter's contact. Equally important, some might even argue more important, is to consider the actual structure, or design, of the contact itself. As presented in Chapter 2 in this volume, there are three forms of vacuum arc:

1) The diffuse vacuum arc (currents < ≈ 6kA)
2) The high current column vacuum arc (currents > ≈ 10kA)
3) The transition vacuum arc (6kA ≈ < currents < ≈ 10kA)

The possible arc modes on a butt contact during a current half cycle are illustrated in Figures 2.45, 2.46, and 2.47. As the contacts part they are separated by a molten metal bridge, which forms at the last point of contact. Once the bridge has ruptured, there is a brief period during which an arc column is confined to the vicinity of the bridge, call the *bridge column arc*. Now, depending upon the current level, the vacuum arc will progress into one of three possible forms. For currents of less than about 6kA, a diffuse arc will develop. For currents between about 6kA and about 14–15kA, the arc can form a transition diffuse column, which depending upon the level of current, will either remain a diffuse column or develop into a more constricted column. At currents higher than about 14–15kA, the vacuum arc immediately develops into a constricted column from the bridge column.

At low currents, the diffuse arc will stay diffuse all the way to current zero. The transition vacuum arc will return to the diffuse mode just before current zero. At the highest currents, the contact regions of a stationary constricted column exhibit intense activity with jets of material being ejected from the contact faces. After severe contact activity such as this, even if the arc returns to the diffuse mode just before current zero, there will be copious residual metal vapor residing in the intercontact region; see Section 4.2.3 in this volume.

Electrical contacts that have a diffuse vacuum arc burning between them many milliseconds before current zero will have excellent dielectric recovery properties after current zero: See Section 4.2 for a complete description of the interruption process. The reason for this results from the properties of the cathode spots. First, the ions are ejected from the spot with very high velocities (about $10^4 \mathrm{ms}^{-1}$, Section 2.3) and so will clear the contact region in a few microseconds after current zero. Second, the spots themselves are extremely small (about 50μm). They will therefore cool very rapidly after current zero (in approximately microseconds or less and will cease to liberate metal vapor into the intercontact space. During the arcing period small particles of contact material are ejected from the cathode spots tangentially to the contact face, and moving away from the intercontact space in the shortest possible time. In contrast, if you have an uncontrolled columnar arc, the contact faces can be grossly heated and there can be liberal evaporation of metal vapor after the vacuum attempts to extinguish at current zero. This metal vapor prevents rapid dielectric recovery of the contact gap by providing a possible breakdown path [88].

It is important therefore to minimize the arc heating of the vacuum contacts during the current half cycle and to maximize the time for which the arc remains diffuse. In order to do this many electric contact designs have been proposed. It is possible to split these designs into three major groupings [89]:

1) disc or butt shaped contacts
2) contacts that cause the roots of the columnar arc to move rapidly over their surfaces
3) contacts that force the high current vacuum arc into a high current diffuse mode.

3.3.2 Disc- or Butt-Shaped Contacts

The earliest vacuum interrupters used simple butt contact structures, Figure 3.19(a). Indeed, for low current, diffuse vacuum arc switching (of less than about 6kA) these contacts worked very well. At higher currents, however, the intense arcing activity caused severe erosion of the contacts and

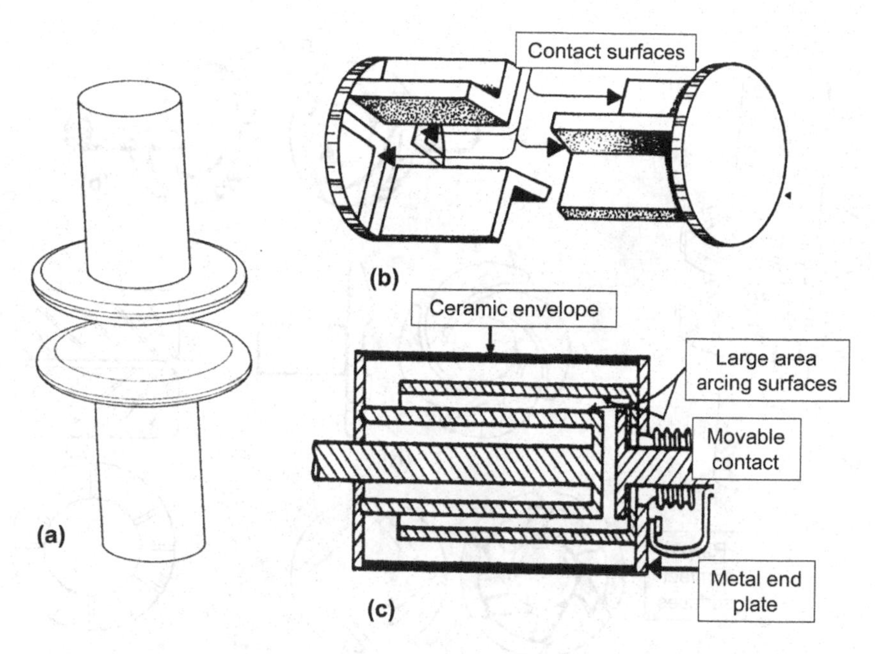

FIGURE 3.19 (a) The disc or butt contact for vacuum interrupters, (b) and (c) some examples of compact, "large area" contact structures.

failure to interrupt the current. Research into one form of contact activity, showed that an anode spot would form from a diffuse arc if, at a given contact gap, the anode area is too small [90]; see Section 2.3.4 in this volume. This led some researchers to develop the ultimately short-lived concept of using large area contacts for high-current vacuum interrupter applications. Examples of large area contacts are shown in Figure 3.19(b) and (c) [91, 92]. Many other designs based upon this concept can be found in the patent literature [93]. Unfortunately, this type of contact structure proved to be impractical because the inventors did not take into account the initial formation of the bridge column arc after the rupture of the molten metal bridge; see Section 2.2. When the arc current is high enough, this bridge column arc transitions directly into the columnar vacuum arc with the resulting gross erosion and overheating of the contacts. The large area does not come into play at all and has no impact on controlling the high current, columnar vacuum arc.

3.3.3 Contacts to Force the Motion of the High Current, Columnar Vacuum Arc

As I have discussed in Section 2.6.2 in this volume, the high current, columnar vacuum arc moves in the Amperian manner when a transverse magnetic field (TMF) is impressed across it [94]. This finding has led to some very practical high current contact designs for vacuum interrupters. Each of these designs accepts the occurrence of the columnar vacuum arc, but this arc is forced to move across the contact face through the interaction between the current flowing in the arc and a transverse magnetic field (TMF) resulting from the current flowing in the contact itself. The earliest design for this type of contact, the spiral contact, was patented by H. N. Schneider [95] in 1960 and is illustrated in Figure 3.20(b). Also, in 1960 Smith patented a cup TMF design shown in Figure 3.20(c) [96]. Other typical designs are shown in Figures 3.20(a), (d), and (e). The mode of operation with a columnar vacuum arc is illustrated in Figure 3.21(a) and (b). When opening usual load currents (i.e., < 4kA), however, the vacuum arc will be in the diffuse mode with cathode spots running over the whole spiral cathode's surface.

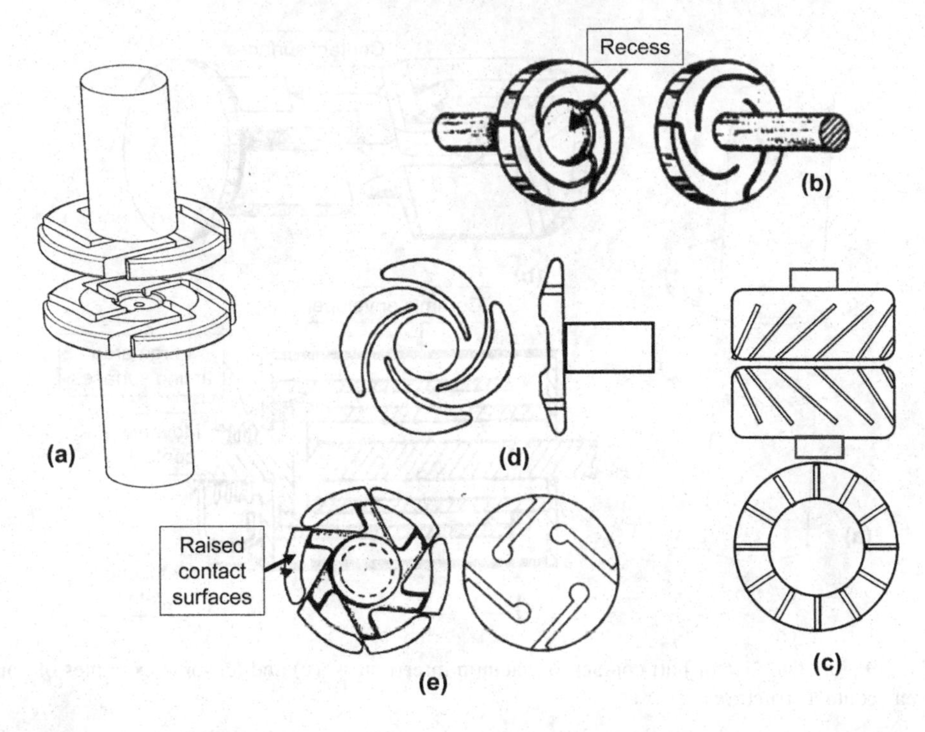

FIGURE 3.20 (a) An example of the "Transverse Magnetic Field" (TMF) contact structure for vacuum interrupters, (b) H.N. Schneider's original spiral contact design [95], (c) and (d) variations on the TMF design, (e) Smith's cup TMF design.

Feng et al. [97] have investigated the performance of the shape of the spirals shown in Figure 3.22 when interrupting currents up to 24kA peak. Seven arc modes are observed in these experiments: the bridge column, the diffuse column arc, the constricted column arc, the plasma jet arc, the anode jet arc the high and low current diffuse arcs. Only minor differences are observed in current interruption performance between the two designs for contact diameters of 40mm and 46mm. The

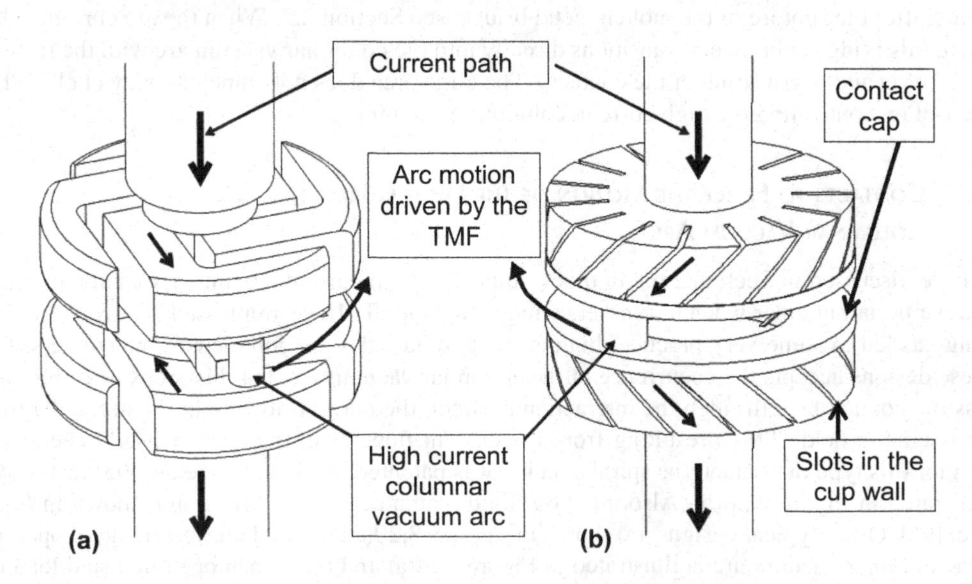

FIGURE 3.21 The motion of the high current columnar vacuum arc on the TMF contact structure; (a) spiral-shaped contacts and (b) the contrate cup shaped contacts.

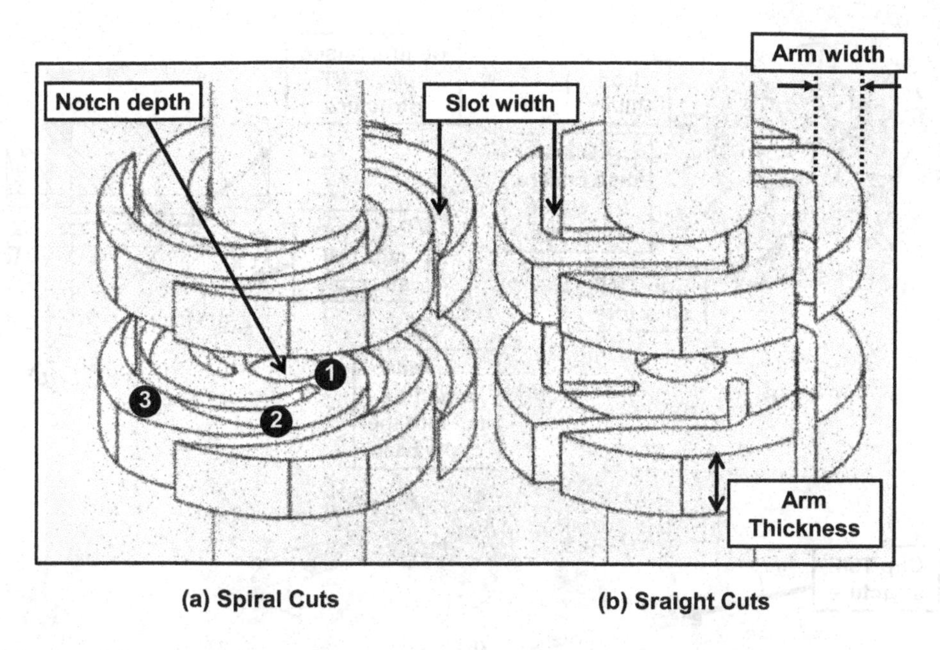

FIGURE 3.22 The spiral and straight cut Transverse Magnetic Field (TNF) contact designs.

effective contact area eroded by the spiral contacts is between 73% and 79% and between 91% and 97% for the straight cut design.

Figure 3.22 shows the design parameters for the spiral contact. They are, the width and thickness of the spiral arms, the width of the gap between the spiral arms and the notch depth in the contact's center. For most designs the contacts make closed contact at position (1). The spiral arms radiate from this region. As will be shown in Chapter 6 in this volume, the closing velocity and closing force are very high for vacuum interrupters that are designed to interrupt high fault currents. Thus, when the contacts touch the arms will have built up considerable inertia. The spiral arms experience high stress each time the contacts open and close. It is important to control the mass of these arms. This is done by optimizing the slot width and the arm thickness and width. Feng et al. [98] have investigate the TMF design in Figure 3.20(a) for slot width (1.5mm to 2.5mm), arm thickness (5mm to 7mm), contact diameter 40mm and initial column arc position (1), (2), and (3) when interrupting ac currents (5kA to 20kA). They conclude that when the initial arc column is at position (1) (i.e., at the contact's center), magnetic force is much lower that at positions (2) and (3). When the contact gap increases the magnetic force on the column arc increases. However, when the contact thickness increases the magnetic force decreases. The notch depth has little influence.

A second contact structure that uses a TMF to drive the columnar vacuum arc is shown in Figures 3.20(c) nd 3.21(b). This design is the contrate cup contact [96 ,99, 100]. It can be seen from Figures 3.21(b) that the slanted slots cut into the cup's side provide a transverse component to the magnetic field, which drives the arc around the cup's surface. It has been found that the arc runs best in this design if the slots do not extend all the way to the rim, i.e. there is a rim of solid contact material. When the slots do extend through the contact's surface a more or less continuous surface develops after a few mechanical operations. The impact from these mechanical operations will cause the slots at the contact's surface to collapse and touch each other. One way of preventing the slots from collapsing is suggested by Wang et al. [101] and is shown in Figure 3.23(a). Liu et al. [102] in Figure 3.23(b) show another way of enhancing the magnetic field strength. Lamara et al. [103] propose the raised central spiral and a fixed contrate cup design shown in Figure 3.23(c). The high current column vacuum arc is initiated on the central spiral contacts and migrates to the fixed cup

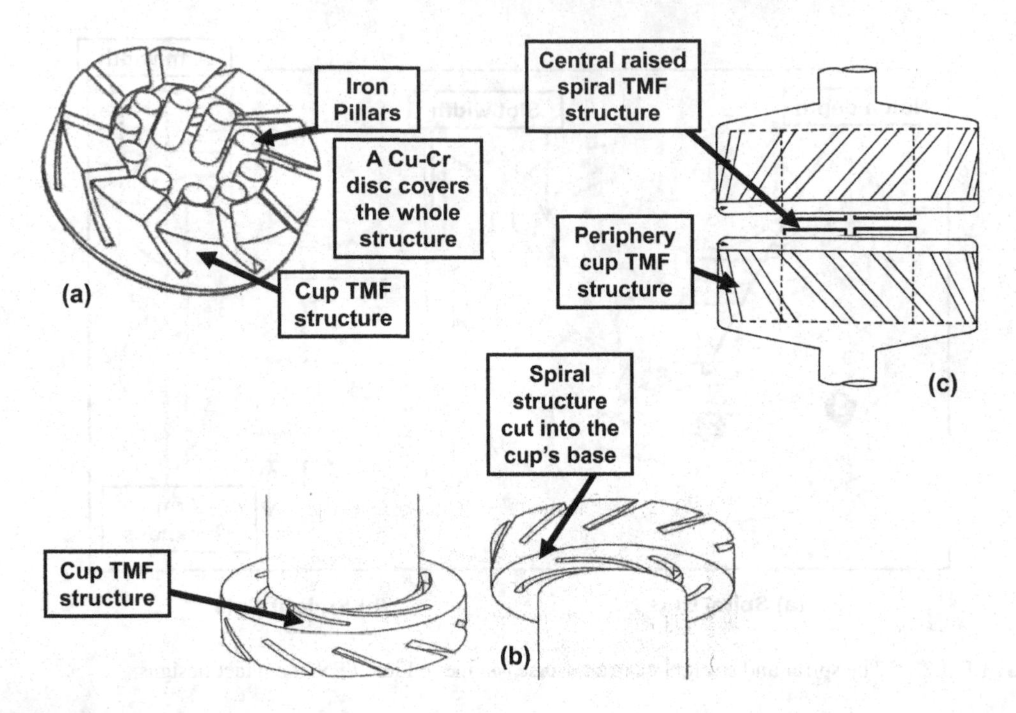

FIGURE 3.23 Examples of extensions to the cup TMF contact: (a) iron pillars inside the cup design [101], (b) Spiral cuts into the cup's base [102] and (c) spiral moving contacts inside a fixed cup [103].

structure. The authors of these designs each claim a somewhat improved interruption performance over the simpler designs shown in Figure 3.20.

Many variations of the TMF contact design have been proposed. For example, Figure 3.24 illustrates some TMF design variations proposed by Altof [104]. Most of the vacuum interrupter manufacturers who use this contact design to control the high current columnar arc have developed their own "optimum" designs. The optimum design will depend upon the shape of the spiral or the cup, the width of the slots, the length of the slots, the number of slots, the width of the spiral, the shape of the cup wall, and the slot angle in the cup's wall as well as the contact material used. Each manufacturer has developed its own optimized design and a benchmark review of available vacuum interrupters shows that there are as many variations of the TMF contact structure as there are manufacturers. Finally, the cost to manufacture a particular design to achieve a given interruption performance rating is, in practice, an important criterion for a commercial product. The spiral TMF contact has traditionally been produced by machining a disc of C–Cr material. Kowanda et al. [105] show that it is possible to produce a net-shape contact directly with the press and solid phase sinter technique given in Table 3.4(a).

Figure 3.21(a) shows a columnar arc between the spiral arms. As I have already discussed in Section 2.6.2 in this volume, the current path generates a transverse magnetic field with the magnetic flux B_T perpendicular to the page. This TMF, together with the current, produces a Lorentz force F_T, which is directed towards the right and tends to enlarge the area of the loop, thus making the arc move. Thus, when the arc is in the columnar mode, the self-magnetic field generated by the current flowing in the spiral arms interacts with the current flowing in the arc to force the arc to move between the spirals. When the arc roots reach the end of the spiral arm, they are forced to jump the gap to the next spiral arm by the arc column continuing to move around the circumference of the contacts. Figure 3.25 shows the typical arc voltage characteristic for spiral, TMF contacts opening a high-current ac circuit. After the initial column arc formation there is a dwell time when the column arc does not move. Once the contacts open far enough (in this example, after 2ms)

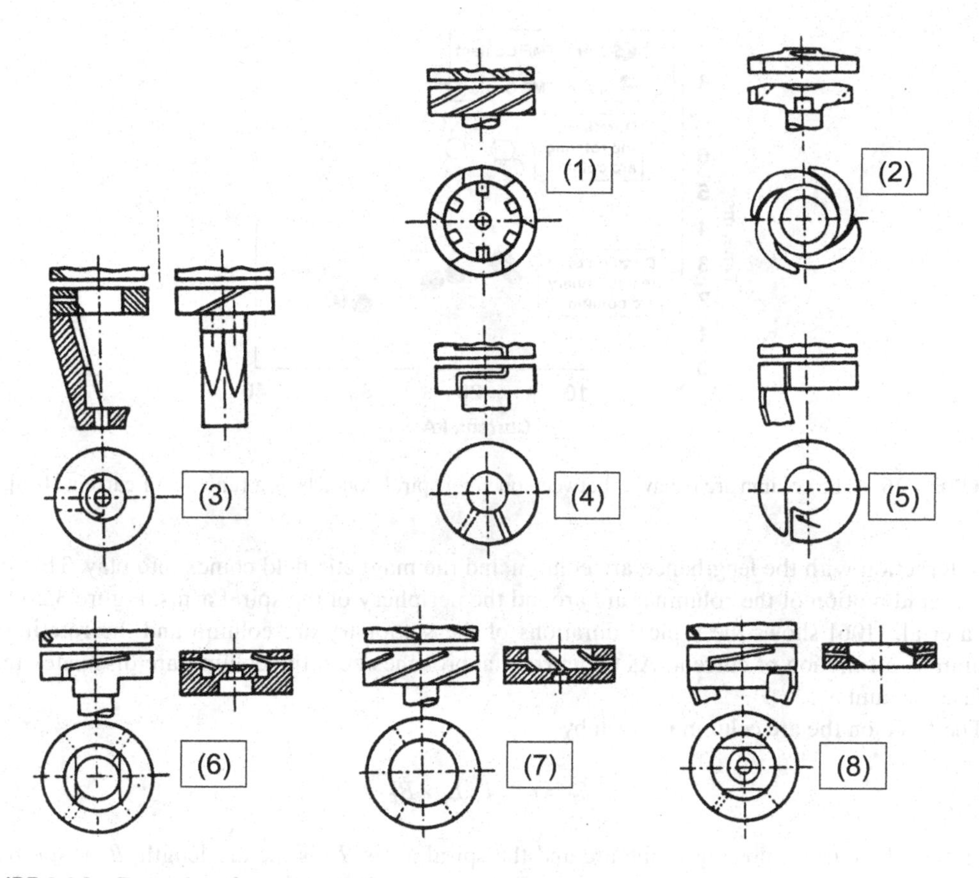

FIGURE 3.24 Examples of possible TMF structures for power vacuum interrupters [104].

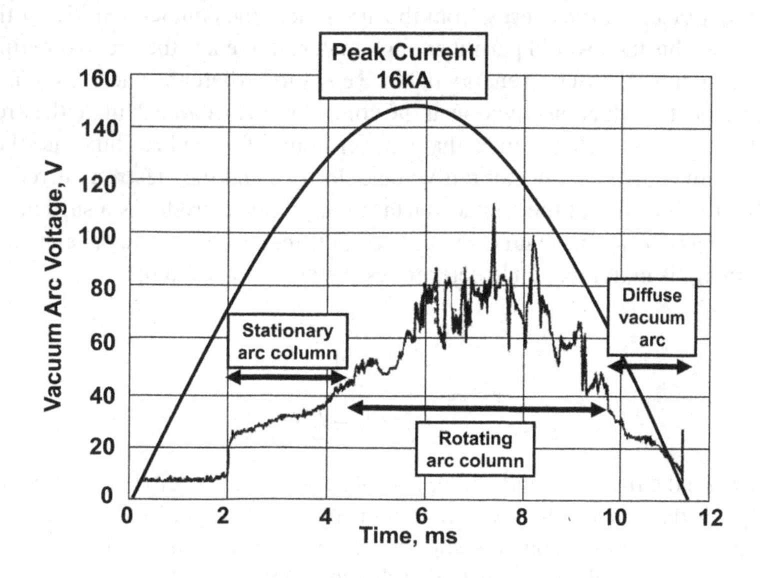

FIGURE 3.25 The voltage across spiral contacts as they open a high-current circuit showing the stages of the columnar arc motion and transition to a diffuse vacuum arc as the current approaches zero.

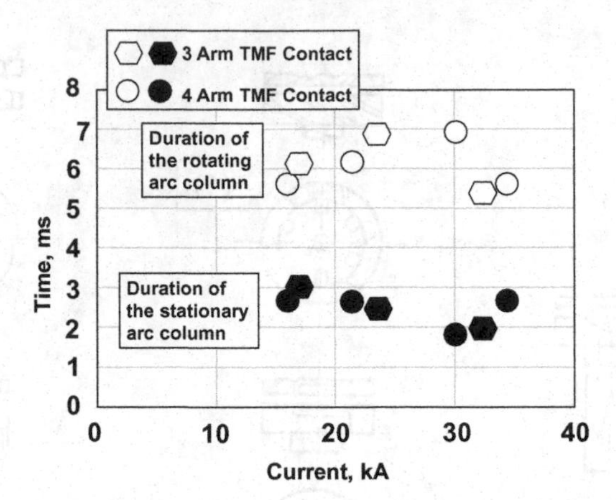

FIGURE 3.26 The column arc behavior between opening spiral contacts as a function of current [106].

the interaction with the lengthened arc column and the magnetic field comes into play. This initiates a rapid motion of the columnar arc around the periphery of the spiral arms. Figure 3.26 from Niwa et al. [106] shows the typical durations of the stationary arc column and the rotating arc column as a function of current. As the current approaches zero the column arc dissipates into a diffuse vacuum arc.

The force on the arc column is given by:

$$F_T = n^* \times i \times L_a \times B_T \tag{3.13}$$

where i is the current flowing in the arc and the spiral arms, L_a is the arc length, B_T is the transverse magnetic flux between the spiral arms is some distance behind the arc, and $n^* \approx 0.5$ (see Section 2.6.2 in this volume). The arc's velocity will be governed by this driving force and the forces that oppose the arc's motion. Dullni [107] assumes that the mechanism limiting the columnar arcs motion is the time it takes for the energy from the arc to heat the contact material at the arc's roots to the boiling point. This then would permit enough material to enter the arc and maintain its high-pressure columnar nature. He uses an energy balance equation where the energy loss is described by a one-dimensional heat conduction equation at the contact's surface to calculate the arc velocity v_{arc} as a function of the current i. He assumes that a lower bound for the heat flux into the contact can be estimated from an equivalent contact fall voltage drop U_f and the electron current density j. He also assumes that the increase of the contact surface temperature produces a sufficient evaporation rate to achieve the gas density necessary for the current transport in the arc plasma column, i.e., the columnar vacuum arc is maintained. He determines the arc velocity to be:

$$v_{arc} = \frac{8U_f^2 \left({}^{j}\!/_{\pi} \right)^{3/2} \sqrt{i}}{\kappa \zeta c \left(T_b - T_0 \right)^2} \tag{3.14}$$

where K, ζ, and c are the thermal conductivity, mass density, and specific heat of the contact material, respectively; i is the total discharge current in the plasma column, $j = i/\pi r^2$ (r is the mean radius of the arc column), T_b is the final contact temperature, and T_0 is the initial surface temperature. The resulting dependence of v_{arc} on the arc current is shown in Figure 3.27 for an equivalent contact fall voltage drop of $U_f = 17$ V. Except in a narrow range of currents the model overestimates the arc velocity that has been measured by experiment. One drawback of this model is that the energy balance equation is not solved for the arc column.

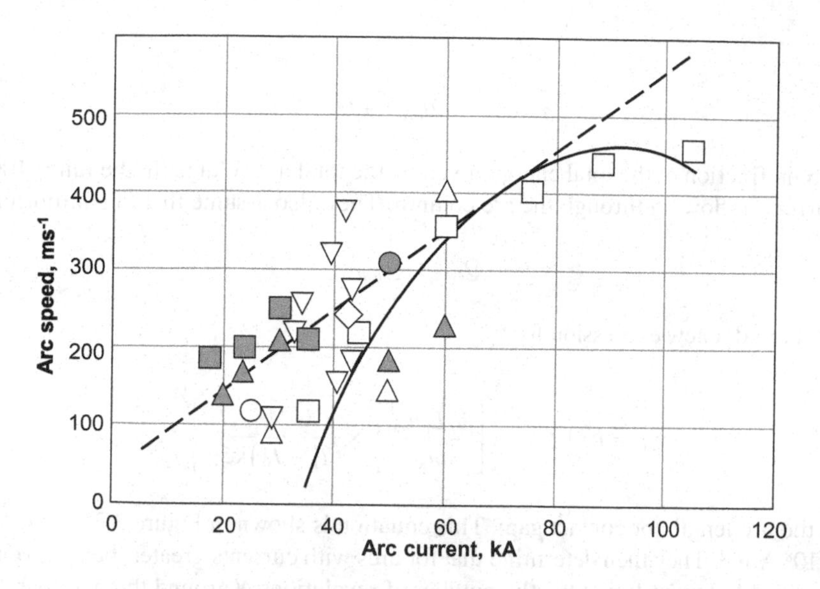

FIGURE 3.27 The speed of the columnar vacuum arc between TMF contacts (experimental data, from Westinghouse △, ○ from [111], ▽ from [112], □ from [109]; ● from [113] ▲ from [108], and ■ from [106]), ——— calculated from Teichmann et al's model [109] and – – – calculated from Dullni's model [107].

Experimental measurements show that the total arc voltage of a 40kA rms, columnar vacuum arc moving between spiral (TMF) contacts is 100V–200V [108]. Approximately 40% of the total arc power is dissipated in each contact. This corresponds to an equivalent anode fall voltage drop (equivalent particle energy) of the order of 40V. Teichmann et al. have further developed this model [109] by including the momentum loss from neutral metal vapor leaving the arc column by diffusion. The arc experiences a permanent loss of momentum, F_N:

$$F_N = \Upsilon_N m_N v_{arc} \tag{3.15}$$

Where Υ_N is the neutral vapor flux out of the plasma column in units of atoms.s^{-1} and m_N is the corresponding atomic mass. (It is assumed that the mass loss is balanced by a corresponding evaporation rate of contact material, Υ_{ev}). The energy balance equation solved simultaneously with the momentum balance equation (Lorentz force = momentum loss rate) yields a relation for the arc velocity, which is then solved numerically:

$$jU_{eq} = \tfrac{1}{2}\left(\frac{\pi j}{i}\right)^{1/4}\left(T_b - T_0\right)\left(\frac{\pi\kappa\zeta c v_{arc}}{2}\right)^{1/2} + \left(\frac{h_{ev} L_a b_T i j}{m_N v_{arc}}\right) \tag{3.16}$$

Here, j is the total current density, h_{ev} is the enthalpy of evaporation, b_T is the normalized B-field ($B_T^* / i \approx 5$ µT/A), and L_a is the arc length. The result provides a good agreement between experimental values and theory assuming a realistic equivalent voltage drop U_{eq} for the contact heating term of 40V. They conclude that the propagation of columnar vacuum arcs between spiral contacts is dominated by neutral vapor loss from the arc column. As can be seen from Figure 3.27, this model agrees quite well with experimental data.

Dullni et al. [110] have developed their model further by including in the total energy balance an evaporation term:

$$Q_{arc} = 2Q_{heat} + Q_{evap} + Q_{plasma} \tag{3.17}$$

where,

$$\alpha U_{arc} i = 2Q_{heat} + Q_{evap} \tag{3.18}$$

where α is some fraction of the total energy given by the total arc voltage (in the range 100V–150V) times the current, i, flowing through the arc column. They also assume that in equilibrium:

$$Q_{evap} \cong Q_{heat} \tag{3.19}$$

They then obtained a new expression for v_{arc}:

$$v_{arc} = i^{5/6}\left(\frac{j}{\pi}\right)^{1/2}\left(\left[\frac{b_T L_a h_{ev}}{m_N}\right]^2 \times \frac{4}{(T_b - T_0)\kappa\xi c}\right)^{1/3} \tag{3.20}$$

where L_a is the arc length (or contact gap). This equation is shown in Figure 3.28 for L_a =8mm and for j= 1.5 x 10^8 Am^{-2}. They then determine that for arcs with currents greater than a given value and for L_a greater than a few millimeters, the number of revolutions N around the periphery of a spiral contact, diameter D, can be given by:

$$N = \Delta t_{arc}\frac{v_{arc}}{\pi D} = \Delta t_{arc}\frac{const.i^{5/6}}{\pi D} \tag{3.21}$$

where Δt_{arc} is the time during which the columnar arc exists and the contacts have a large enough gap [114,115]. Comparison of their model with experimental data is shown in Figure 3.29. While there is a lot of scatter in the experimental data, the model does seem to fit, to some extent, with the maximum value of N for a given contact diameter. It also shows that at lower currents when the arc is not necessarily in the full columnar mode, it will not move as quickly. They conclude that the maximum rms current i_{max} that could be interrupted by a given spiral contact with a diameter D is:

$$i_{max} \propto D^{6/5} \tag{3.22}$$

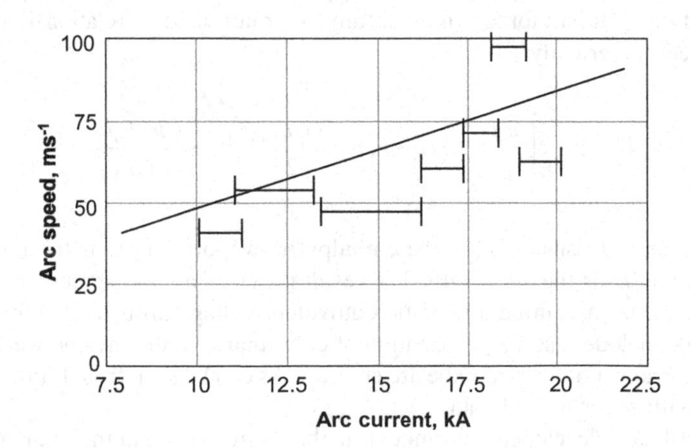

FIGURE 3.28 The speed of the columnar vacuum arc between spiral, TMF contacts for a contact gap of 8mm and for a current density of 1.5 × 10^8 $A.m^{-2}$ (––––– calculated from Equation (3.20) [110] and I––––I experimental data [110].

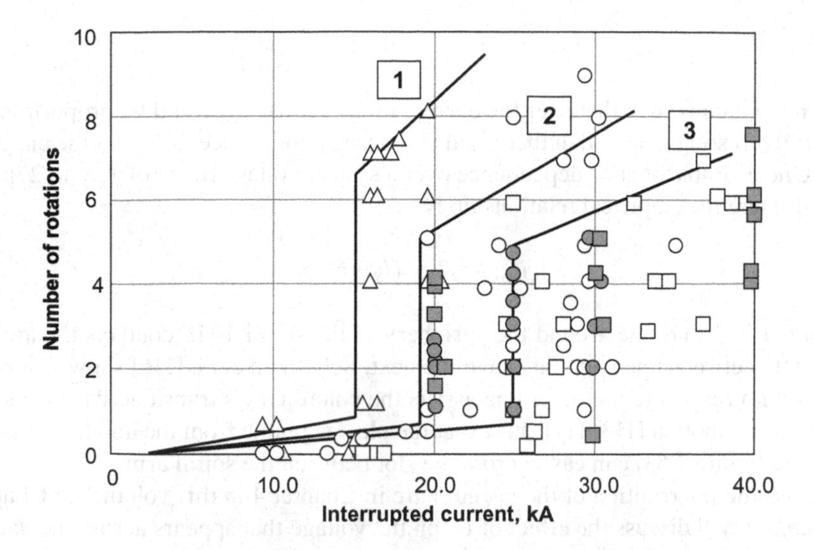

FIGURE 3.29 The number of rotations of a high current, columnar vacuum arc between spiral, TMF, vacuum interrupter, contacts opening at 1.5 ms^{-1} and a 50 Hz rms current, (1. Δ 40mm contact diameter, 2. ○ 62mm contact diameter and 3. □ 82mm contact diameter) [110] and (● 68mm contact diameter and ■ 90mm contact diameter) [108].

Their data as well as my own data for i_{max} as a function of contact diameter is given in Figure 3.30. The relationship given in Equation (3.22) is close to what might be an intuitive expectation for a columnar vacuum arc traveling around the periphery of a TMF contact: i.e., the i_{max} would depend upon the contact's circumference πD. Looking at the data given in Figure 3.30:

$$i_{max} = 0.64D \quad or \quad i_{max} = 0.26D^{6/5} \quad (@\ 12kV) \tag{3.23}$$

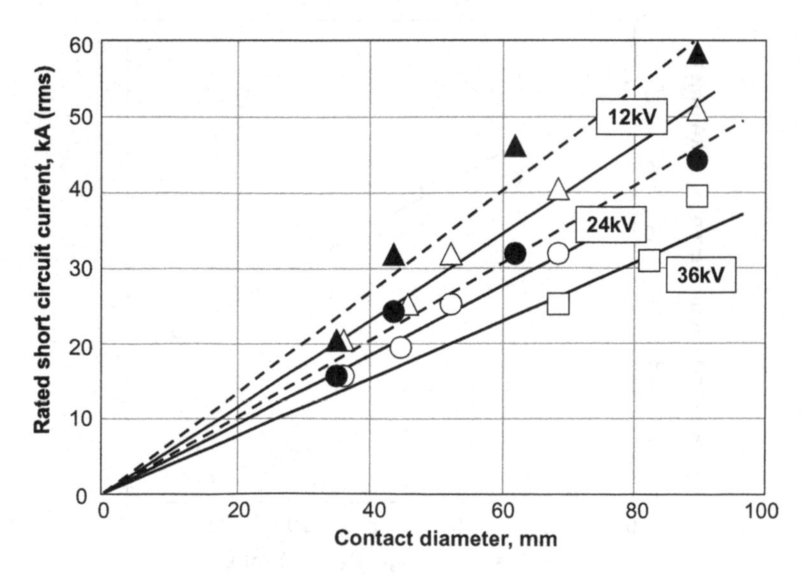

FIGURE 3.30 An example for two vacuum interrupter designs of the interruption ability of TMF contacts as a function of contact diameter for three-distribution circuit, voltage values (Δ 12kV, ○ 24kV and □ 36kV [110]) and (▲ 12kV, ● 24kV [Eaton Corporation]).

$$i_{max} = 0.48D \quad or \quad i_{max} = 0.21D^{6/5} \quad (@\ 24kV) \tag{3.24}$$

The scatter in the data is such that over the current range shown i_{max} could be proportional to either D or $D^{6/5}$. It is interesting to note that there is also a strong dependence of i_{max} on the circuit's system voltage, U_S. One example of such dependence over a system voltage range of 1kV to 27kV is shown in Figure 3.31. Here the empirical relationship is:

$$i_{max} = -1.33U_S + 61.5 \tag{3.25}$$

For the columnar arc to rotate around the periphery of the spiral TMF contacts the arc roots have to cross the slots between one spiral arm and the next. Schulman et al. [116] show that rotating one spiral contact with respect to the other one assists the column arc's transit across the slots. Models of the columnar arc motion [113, 117] show that the plasma plume from the anode that precedes the arc motion (See Figure 2.58) can easily cross the slot between the spiral arms.

I will discuss the interruption of the vacuum arc in Chapter 4 in this volume. In Chapters 5 and 6 in this volume, I will discuss the effect of U_S on the voltage that appears across the vacuum interrupter's open contacts when interrupting high fault currents. Care must be taken when interpreting Figures 3.30 and 3.31. It does show an interesting correlation between two sets of experimental data and a model of the arc motion in determining the interruption ability of a vacuum interrupter with a TMF contact.

However, the interruption ability of a given vacuum interrupter design does not solely depend upon the contact structure. It is also dependent upon the exact composition of the contact material and the overall internal design of the vacuum interrupter. Thus, the two data sets shown in Figure 3.30 should not be assumed to be the absolute limit of interruption for all vacuum interrupters with the contact diameters given there. Figure 3.32 shows a sequence of photographs of a 30kA rms columnar vacuum arc between 64mm diameter spiral contacts. Here the arc velocity is about 320m.s^{-1}, which is about what you would expect from Figure 3.27. The break-up of the arc column just before current zero is shown in Figure 3.33. At 1ms before the current zero the arc column shows

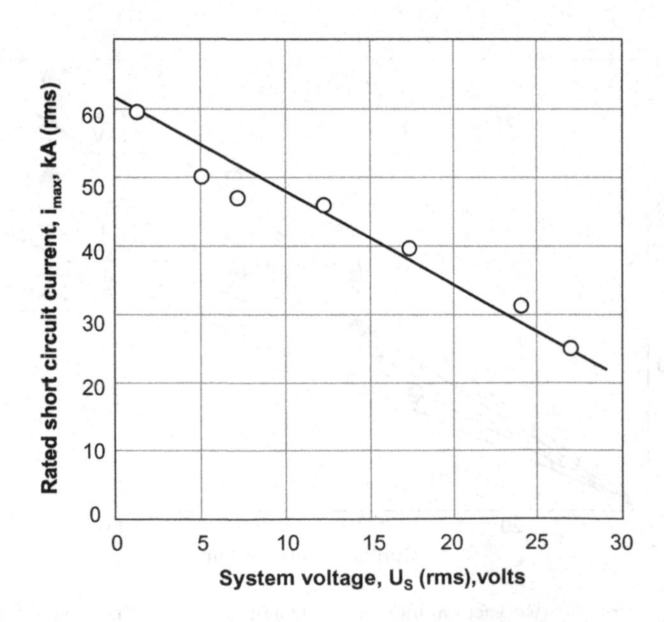

FIGURE 3.31 An example of the interruption ability of a TMF contact (62mm diameter) as a function of the system voltage.

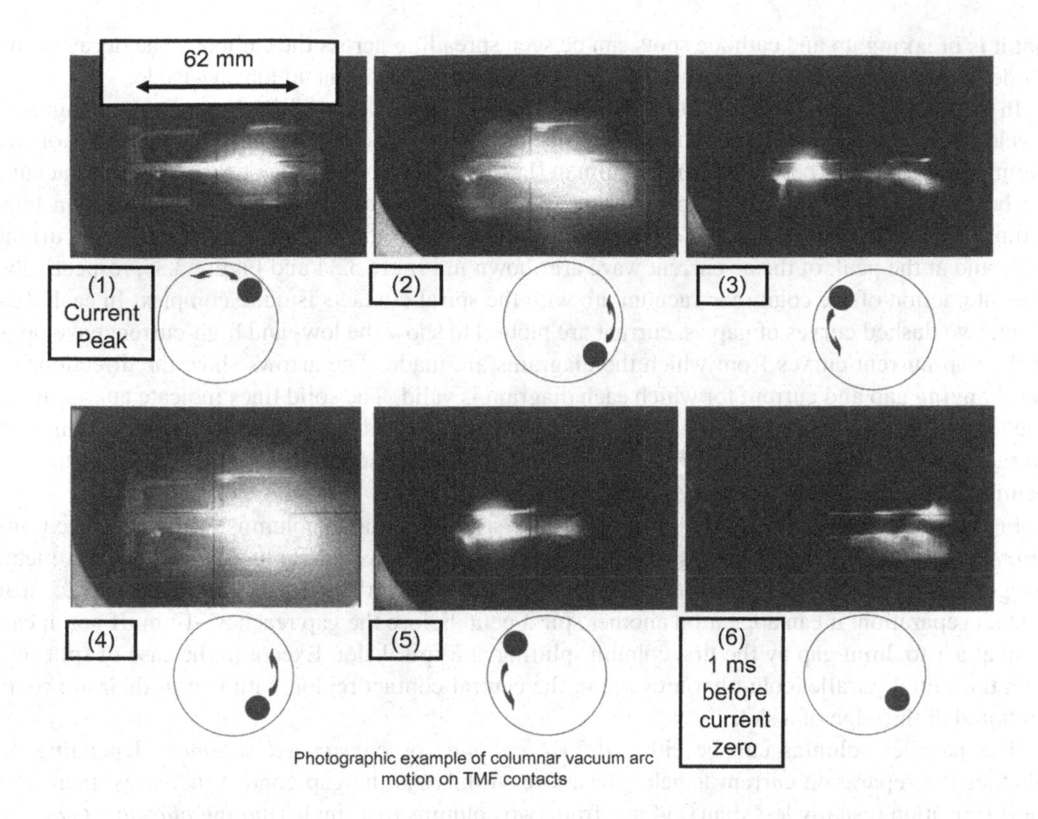

FIGURE 3.32 The columnar vacuum arc between 62mm diameter, spiral contacts with a circuit current of 30kA (rms).

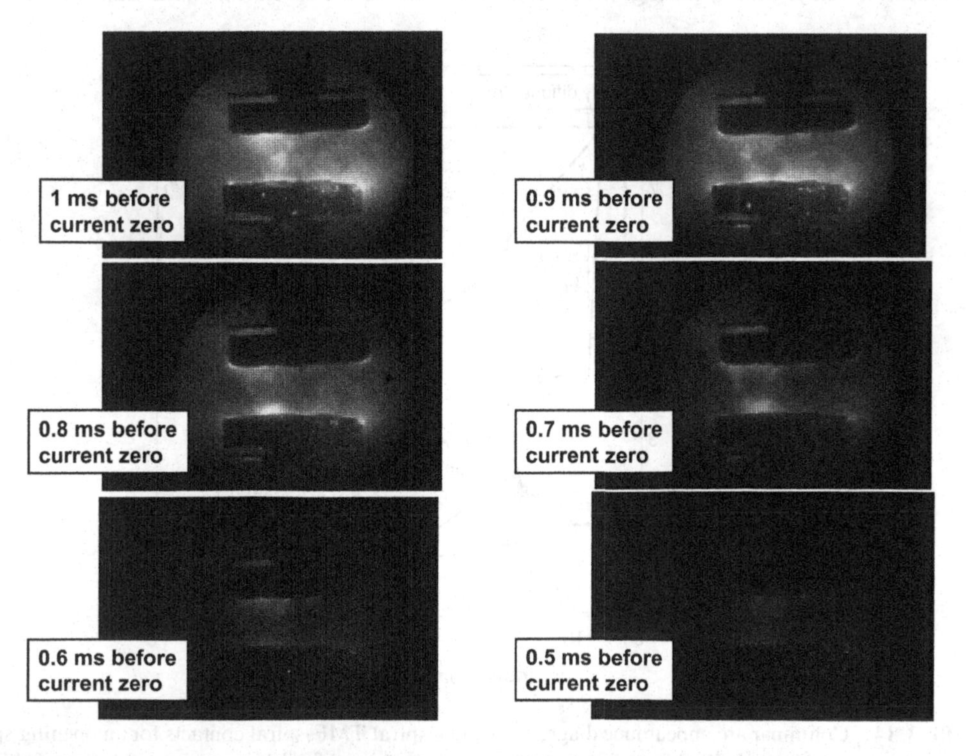

FIGURE 3.33 The transition from the moving columnar vacuum arc on spiral contacts to the diffuse vacuum arc as the circuit current approaches zero.

that it is breaking up and cathode spots can be seen spreading across the cathode. The decay of the anode spot continues until 0.5ms before current zero after which it is no longer visible.

In Section 2.4 in this volume, I have already presented an account of the "Appearance Diagram" development (Figure 2.45) for butt contacts opening with an ac current and the formation of the columnar vacuum arc between them. Schulman [114, 115] has studied the appearance of the vacuum arc between opening spiral contacts using a similar experimental technique and has shown how complex it can be. His "Appearance Diagrams" for contacts opening on the rise of an ac current wave and at the peak of the ac current wave are shown in Figure 3.34 and Figure 3.35, respectively. The interaction of the columnar vacuum arc with the spiral contacts is quite complex. In each diagram, two dashed curves of gap vs. current are plotted to show the low- and high-current envelopes of the gap-current curves from which the diagrams are made. The arrows show the directions of the changing gap and current for which each diagram is valid. The solid lines indicate approximate boundaries between arc modes. There are no observable changes in the diagrams over the ranges of average contact opening speed of 1.6 ms^{-1} to 2.1 ms^{-1}. I will describe these arcing modes using the terminology presented in Section 2.4 in this volume.

Figures 3.34 and 3.35 have some similarities. Firstly, two parallel columns form before the transition to a single column. Perhaps this indicates the formation and rupture of two molten metal bridges and the formation of parallel bridge column arcs. When the second column does not form at contact separation, it can appear on another spiral petal before the gap reaches ~1mm. If not, it can form at a 1 to 2mm gap by the first column splitting at a spiral slot. Except in the case of splitting, both the initial parallel columnar arcs are in the central contact region with one of their arc roots anchored at the edge of a slot.

The parallel columns can be either *diffuse columns* or *constricted columns,* depending on whether the separation current is below or above ~15kA. As the gap contact increases, there is a short transition (usually less than 0.14 ms) from two columns to a single running *plasma jet column* for instantaneous currents above ~20kA. The instantaneous gap at this point depends on current and the separation delay, which, for a given current, falls within a 1 to 1.5mm wide transition region.

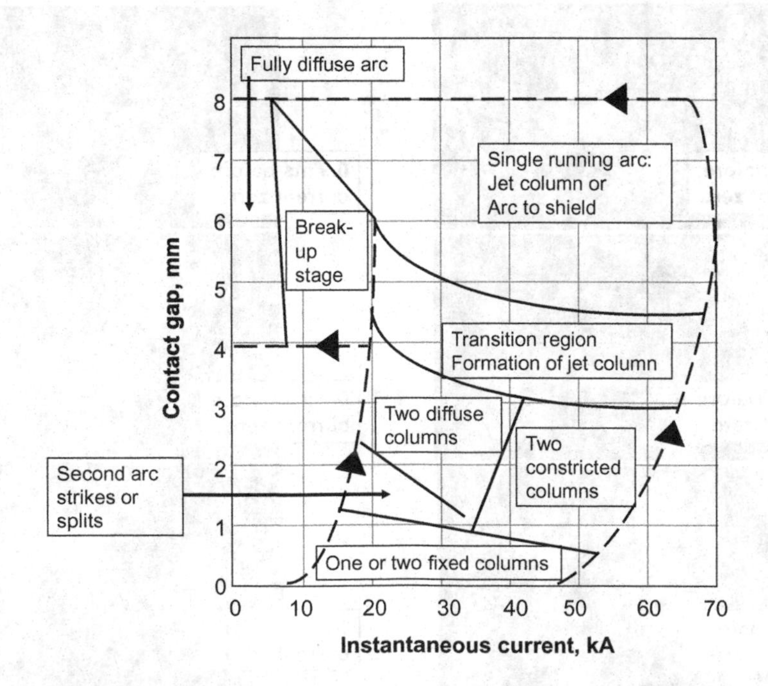

FIGURE 3.34 Columnar arc appearance diagram between spiral TMF, spiral contacts for an opening speed 1.6 ms^{-1}, opening delays of 0.48–2.3 ms into the current half cycle and final contact gaps of 4–8mm [114].

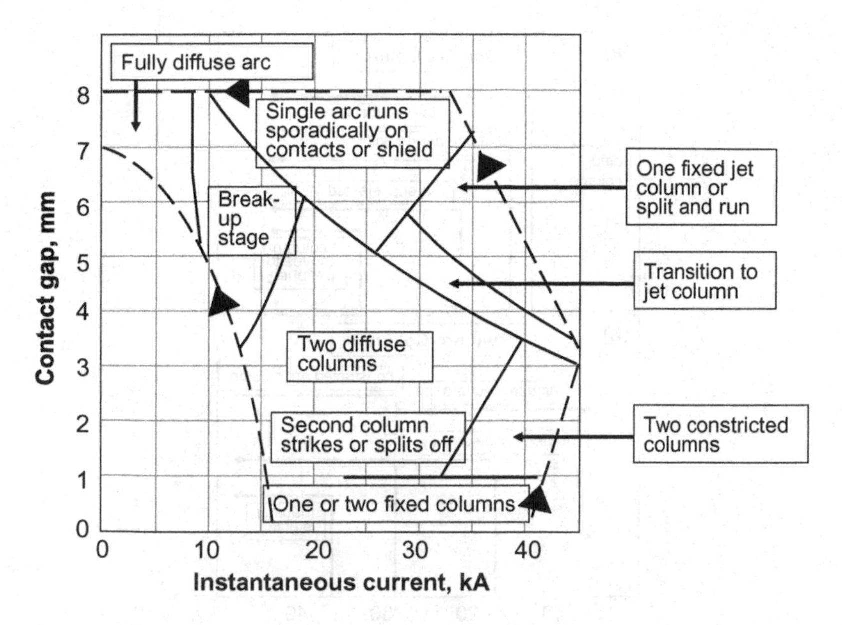

FIGURE 3.35 Columnar arc appearance diagram between spiral TMF, spiral contacts for an opening speed 2.0 ms^{-1}, opening delays of 2.4–4.4 ms into the current half cycle and final contact gaps of 7–8mm [114].

The threshold gap for formation of the single jet column approaches a range of 3 to 4mm for both diagrams at high currents. Once the single jet column forms, it quickly moves outward and begins running on the spiral arms along the periphery of the contacts. As the current decreases, the arc enters the *break-up stage* as it passes over the spiral slots. Anode footprints [118] are still present throughout the *break-up stage*. The final mode of *fully diffuse arcing* (no anode involvement seen in Figure 3.33) at the end of the half-cycle is indicated in the region at the top left of the diagrams, corresponding to the current approaching zero at maximum gap.

The differences in the arc behavior shown in Figures 3.34 and 3.35 result from the initiation and maintenance of the columnar vacuum arc's motion. In these experiments, when the contacts part early in the current half cycle, the running arcs make from 2.5 to 5 vigorous rotations (i.e., a speed between 150 and 300 ms^{-1}) along the periphery of the contacts before going diffuse. They also rarely become anchored long enough to develop intense erosion at the arc roots. Above ~40kA, the transition to single arc motion depends only on gap, with a threshold of 3.6 ± 0.8mm. At a gap of 8mm, the running column transforms to a diffuse arc as the ac current drops to zero without sticking or breaking up. When the contacts open close to the peak current, there is a marked effect on the motion of the columnar vacuum arc. At the current peak, the initial stationary arc results in copious melting and vapor production at the arc roots. This results in a longer contact gap before the arc begins to run around the spiral contact structure. Once the arc does move, it has a lower speed. There is also a greater tendency for the arc roots to stick briefly at the spiral tips. Sometimes the arc splits into two columns at the spiral tips for contact gaps greater than 4mm. The vacuum arc still goes fully diffuse before the current zero. The effect on interruption performance of a vacuum interrupter when it opens at the maximum of a high ac current will be discussed in Section 6.3.4 in this volume.

One obvious design criterion when applying this type of contact structure is that a minimum contact gap is required before a vigorous motion of the columnar arc is achieved. Schulman [115] has studied the effect of the interaction of columnar vacuum arcs with spiral contacts and small contact gaps (≤3mm). For the range of currents investigated (7.7kA peak to 36kA peak), the appearance diagram shown in Figure 3.36 shows that the columnar arcs, if they move at all, only move a minimum distance of one half of the contacts perimeter. He observes that one or two *bridge column*

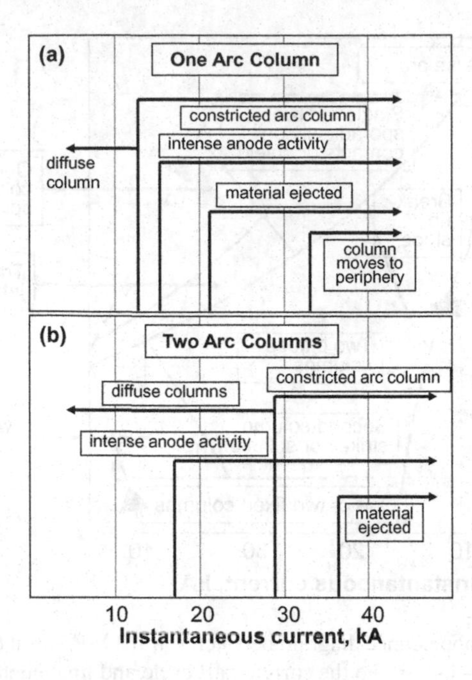

FIGURE 3.36 (a) Ranges of peak current for observed appearance of various arcing modes with a full gap of 2mm and a single, high current, vacuum arc column, (b) Ranges of peak current for observed appearance of various arcing modes with a full gap of 2mm and two parallel, high current, vacuum arc columns [115].

arcs form on contact separation and then become anchored at nearby spiral slots. If only one column forms initially, a second column will sometimes appear and stick at another slot before the gap reaches about 1mm. In other instances, part of the arc will expand across the slot, split off, and move away from the original column to the next slot. His results show that for contact gaps up to 3mm the spiral contacts interact with the arc essentially as butt-type contacts, and the spiral petals do not induce the column instability required for the arc to maintain motion over the slots. When the current is high enough, there is intense erosion at the arc roots. Thus, the conclusion from these experiments is that when applying a TMF contact design, a contact gap of at least 4mm is required and 6mm is preferable.

One advantage of the spiral TMF design is that the contacts are structured such that the last point of contact and hence the point of arc initiation as they open is towards the center of the contacts. Thus, the initial bridge column and the period of little or no arc motion when the contacts are close is away from the periphery of the contact where the arc will eventually run. Once the arc has moved to the periphery of the contact and begins to run around it, the region of arc initiation will have a chance to cool down and evaporation of metal vapor into the contact gap from that region will cease.

The arc voltage for the columnar vacuum arc between spiral contacts has some distinctive features that are illustrated in Figure 3.37. Schulman [114] correlates the arc voltage with the arc appearance at different stages of the contact opening sequence. Thus, he is able to correlate the arc voltage characteristics to the modes of arc behavior that he observed. It is interesting to note that while the arc is in motion, the vacuum arc voltage is generally about 100V. I would expect that the sooner the columnar vacuum arc breaks up and transitions to the diffuse mode, the better the high current interruption performance of a vacuum interrupter would be. Liu et al. [119] have shown that the residual axial magnetic field from the spiral contacts plays a role in this transition. Spiral contacts similar to those shown in Figure 3.22(a) with the slots continuing close to the contacts' center experience the transition to the diffuse mode sooner than the slot structure shown in Figure 3.22(b). Li et al. [120] have studied the high current interruption performance of TMF contacts with two,

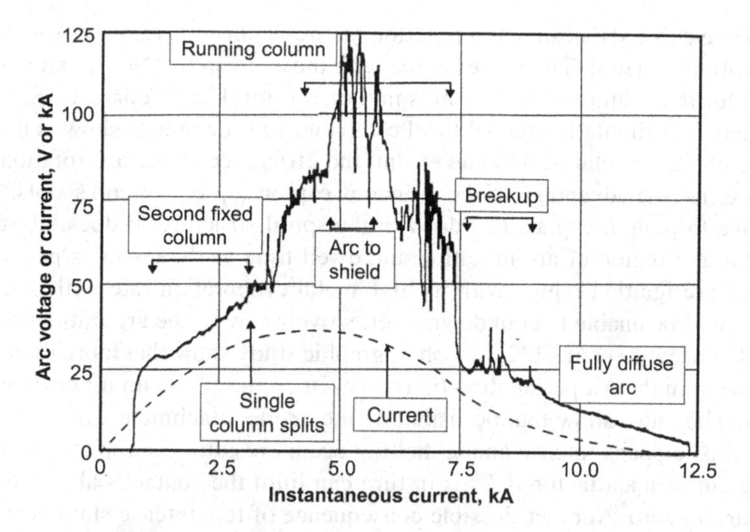

FIGURE 3.37 Arc voltage and current characteristics for an opening TMF contact with a 6mm full contact gap, opening with a speed of ~ 1.5 ms^{-1} and a peak current of 35kA [114].

three, four, five, and six spiral arms. Their experiment used two interruption circuits; 36kV/40kA and 17.5kV/63kA. They conclude that the four-arm spiral contact appears to be the best compromise between mechanical robustness and current interruption ability.

Paulus [121] has investigated the motion of the columnar arc with the contrate cup TMF contact structure shown in Figures 3.20(c) and 3.21(b). In his experiments he shows that the columnar arc does not run on the cup's rim for contact gaps ≤ 2mm. He also observes that, after the initial arc ignition, parallel, multiple constricted columnar arcs can occasionally occur. Again, these eventually merge into one columnar arc that proceeds to run intermittently or stay motionless on the cup's rim. He observes running speeds of up to 200ms^{-1}, but the typical speed is 20-60ms^{-1}. Wolf et al. [108] also show the initial parallel arc columns. They also show that the high-current arc can be both diffuse column as well as constricted column. This is quite different from the appearance of the arc with the spiral contacts, where once the constricted column forms, it alone moves across the contact surfaces. Figure 3.38 shows the constricted column arc

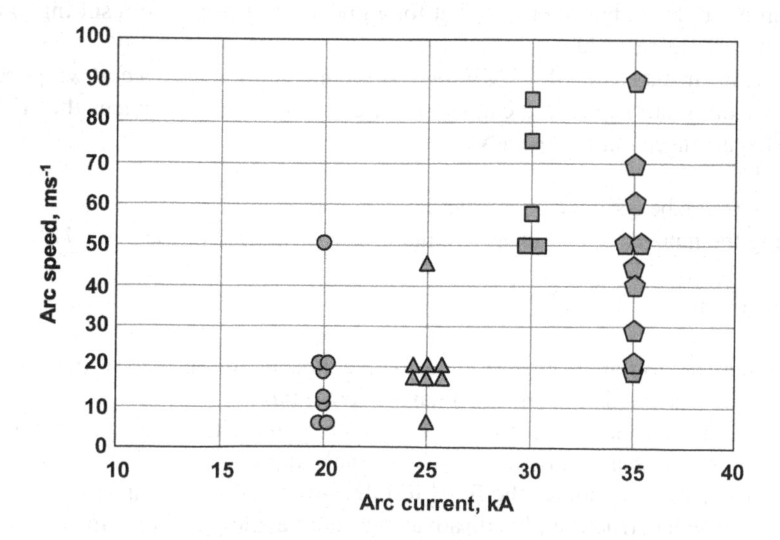

FIGURE 3.38 Column arc speed on a contrate cup TMF design as a function of circuit current [108].

speed on a contrate cup TMF design as a function of circuit currents ranging from 25kA to 35kA presented by Wolf et al. [108]. The values agree with those given by Paulus and are much slower than those seen for the columnar vacuum arcs motion on spiral TMF contacts: See, for example, Figure 3.27. Their 3-D simulation model for the cup and spiral contacts showed that the Lorentz force on the arc column explains, to some extent. the difference in the arc rotation speeds. This contact structure has the advantage of a continuous cap on top of the cup's slots. Thus, the arc roots do not have to jump a gap as they do with the spiral structure. It does, however, have the disadvantage that the region of arc initiation and dwell time as the contacts part is also on this cap. This melted arc ignition region with its high metal evaporation rate will be exposed to the running arc and will be unable to cool down as effectively as will the arc initiation region for the spiral contact. Janiszewski et al. [122] in a photographic study show that more than one columnar vacuum arc moving in the Amperian direction can exist on the cup's rim for currents in the range 7.2kA to 24kA. They also show that occasionally the anode attachment can form at the base of the cup. When this happens intense anode melting occurs resulting in the evaporation of the Cu from which the cup is manufactured. This in turn can limit the contact's ability to interrupt the ac current at current zero. Another possible consequence of this intense stationary anode is that the molten metal inside the cup can flow into the slots cut into its side walls thereby reducing the effect of the TMF.

3.3.4 Contacts to Force the High Current, Columnar Arc into the Diffuse Mode

In Chapter 2, Section 2.7.2, in this volume, I discussed the effect of an axial magnetic field (AMF) on the high current columnar arc, which forms between opening contacts. A typical sequence is illustrated in Figure 2.82. After the rupture of the molten bridge, the bridge column forms. As I have discussed in Section 2.7.2 in this volume, in the presence of a high enough AMF, this column expands into a diffuse column and then into a fully diffuse arc. The cathode spots spreading out over the cathode's surface characterizes this arc. Electrons from these cathode spots are confined by the magnetic flux lines in the inter-contacts region and, because of the associated creation of radial electric fields, the ions are also confined to the intercontact region. An example of a 50kA diffuse arc is shown in Figure 3.39. Under ideal conditions this high current, diffuse vacuum arc distributes the arc energy over the whole cathode and anode contact surfaces and thus greatly reduces the opportunity for gross erosion of the contacts. As I have also discussed in Section 2.7.2 in this volume, any practical contact design that develops an AMF for the purpose of forcing a high current diffuse vacuum arc to form, has to ensure that for a given arc current, the resulting axial magnetic field is higher than a critical value.

The earliest experiments using the AMF for current interruption used a coil wrapped around the outside of the vacuum interrupter that carries the circuit current and produces the AMF [123], see Figure 3.40. This arrangement has the advantages of:

1) The contacts can be a simple butt structure
2) Increasing the number of coil turns can increase the strength of the AMF [124]

The disadvantages are:

1) The coil must be insulated because it has to be connected to one end of the vacuum interrupter (usually the fixed end), and will have its potential
2) At high currents the coil is under a high "loop" force, thus either the coil itself will need to be very strong or else it will require extra external support
3) The coil and its connection to the fixed terminal will add extra impedance to the whole vacuum interrupter structure (this disadvantage can be alleviated to some extent by clever design [125])

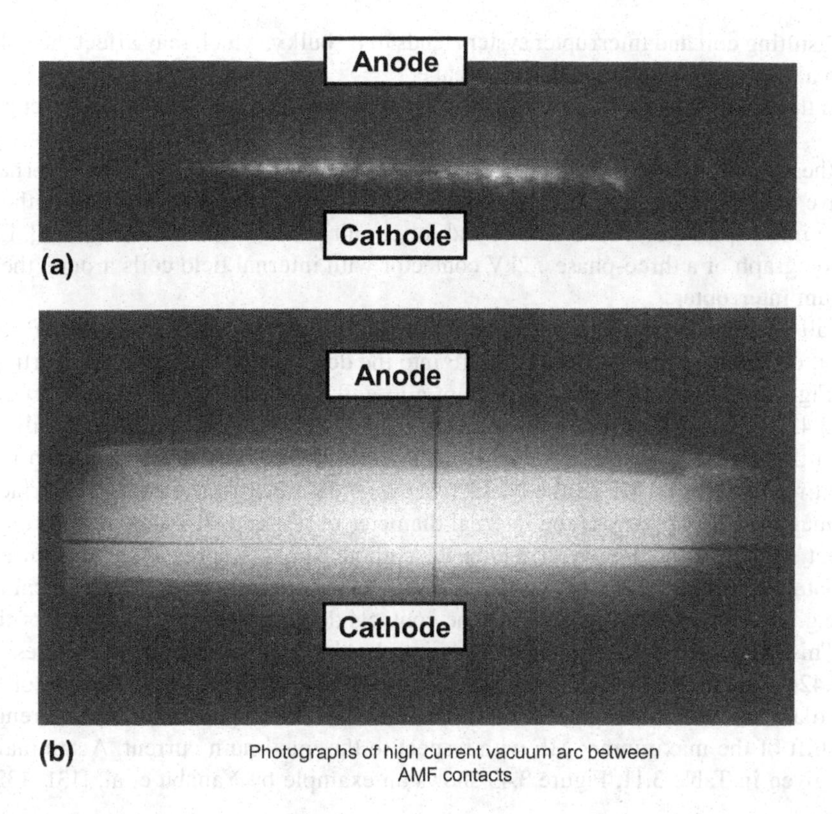

FIGURE 3.39 Examples of high current diffuse vacuum arcs between Cu–Cr(25wt.%) contacts in an axial magnetic field: (a) a current of 50kA for a contact diameter of 50mm and (b) a current of 101kA for a contact diameter of 86.5mm.

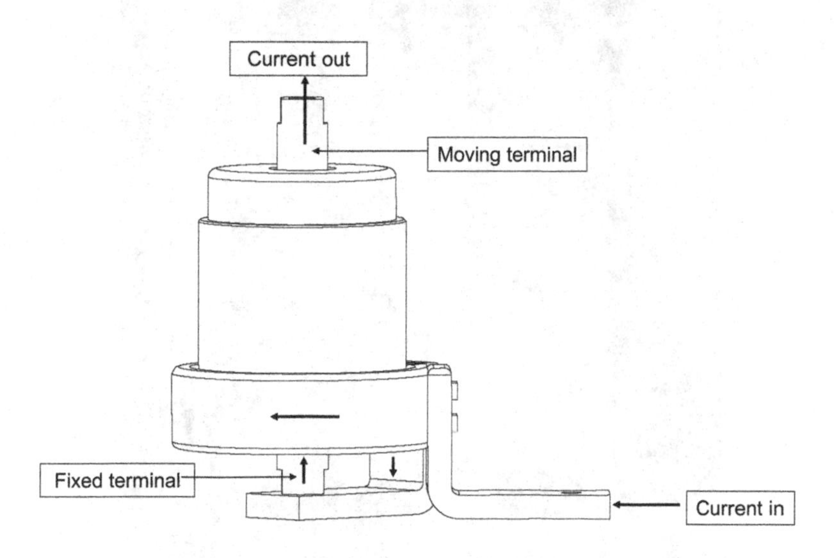

FIGURE 3.40 A vacuum interrupter arrangement with an external magnetic field coil to provide the axial magnetic field.

4) The resulting coil and interrupter system tends to be bulky, which may affect the pole spacing in a three-phase switch or circuit breaker
5) When the fault current occurs, it takes time for the AMF to penetrate the contact gap

In spite of these disadvantages, there have been a few practical applications of an external coil. All of these have been with vacuum interrupter designs where the shield is attached to the fixed terminal of the interrupter (see Section 3.4.2) and for circuit voltages ≤ 12kV [126, 127]. Figure 3.41 shows a photograph of a three-phase 7.2kV contactor with internal field coils around the fixed end of the vacuum interrupter.

Almost all vacuum interrupter manufacturers, who utilize the effect of the AMF to form the high current, diffuse arc, incorporate the AMF into the design of the structure supporting the contact itself. Figure 3.42(a) shows one example of a ½-coil, two-arm, AMF contact structure [128]. In Figures 3.42(a) and (b) the current flows up the copper terminal and splits into a half coil before passing into the contact [129]. The passage of the current through the half coils on the back of each contact results in an AMF in the contact gap. The diameter of the contact's surface through which the magnetic flux passes is the internal diameter of the coil. It is important to note that in these contact structures the area available for the cathode spots to spread can be significantly less than the contact's total area. There have been many designs described in the technical and patent literature, e.g., [128–141]. The coils behind the contacts do increase the impedance of the contact structure. This impedance can be reduced by using the three- or four-segment coil designs shown in Figure 3.42(c) and (d). Wang et al. [130] have analyzed the effect of the number of coil segments, the contact diameter, the coil height, the coil width and the contact gap on the AMF strength and on the phase shift of the maximum AMF with respect to the maximum current. A summary of their findings is given in Table 3.11. Figure 3.43 shows an example by Yanabu et al. [131, 132]. Here, a

FIGURE 3.41 A three-phase vacuum contactor with external axils magnetic field coils around the vacuum interrupters (courtesy of Eaton Corporation).

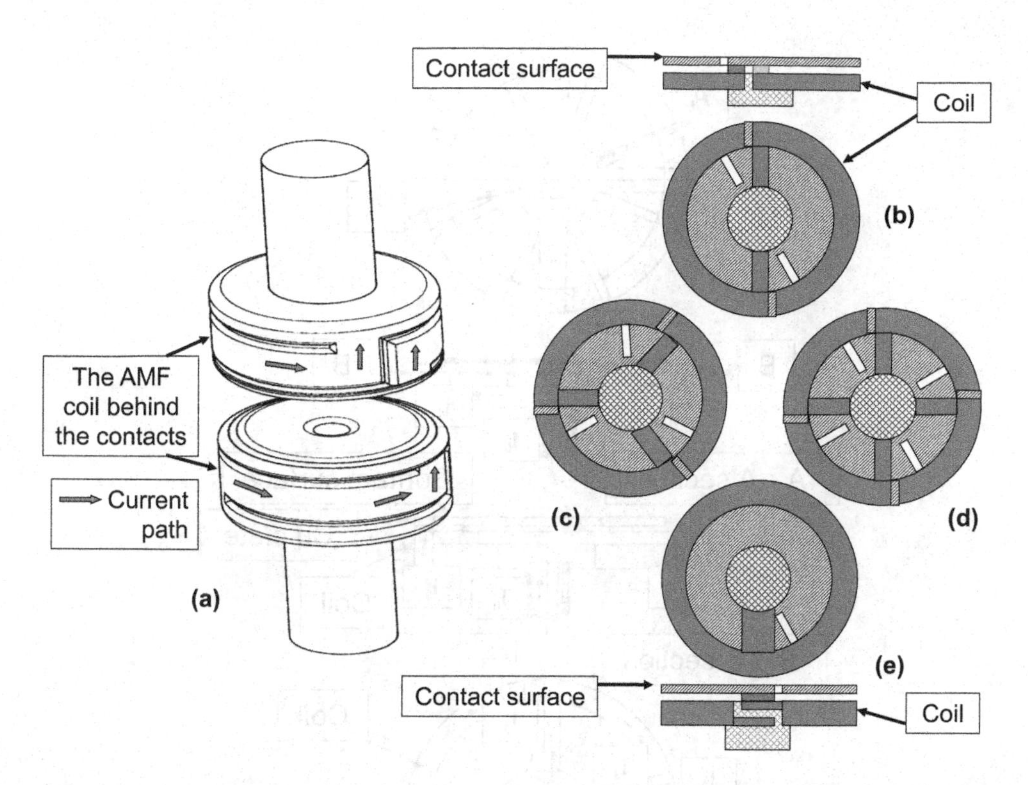

FIGURE 3.42 (a) An example of a contact structure for a vacuum interrupter with the magnetic field coil behind the contact face [128], (b) a two-segment coil design or ½-coil, two-arm design, (c) a three-segment coil design or ⅓-coil, three-arm design (d) a four-segment coil design or ¼-coil,four-arm design, and (e) a single coil design.

TABLE 3.11
The Effect of AMF Coil Parameters and Contact Gap on the AMF and the Phase Shift of the AMF with Respect to the ac Current [130]

Design Parameter		AMF strength
Number of coil segments (2–4)	↑	↓
	↓	↑
Contact diameter	↑	↓
(30mm–100mm)	↓	↑
Coil height	↑	↓
(4mm–18mm)	↓	↑
Coil width	↑	0
(5mm–15mm)	↓	0
Contact gap	↑	↓
(6mm–12mm)	↓	↑

↑ Increasing, ↓ Decreasing, 0 Little effect

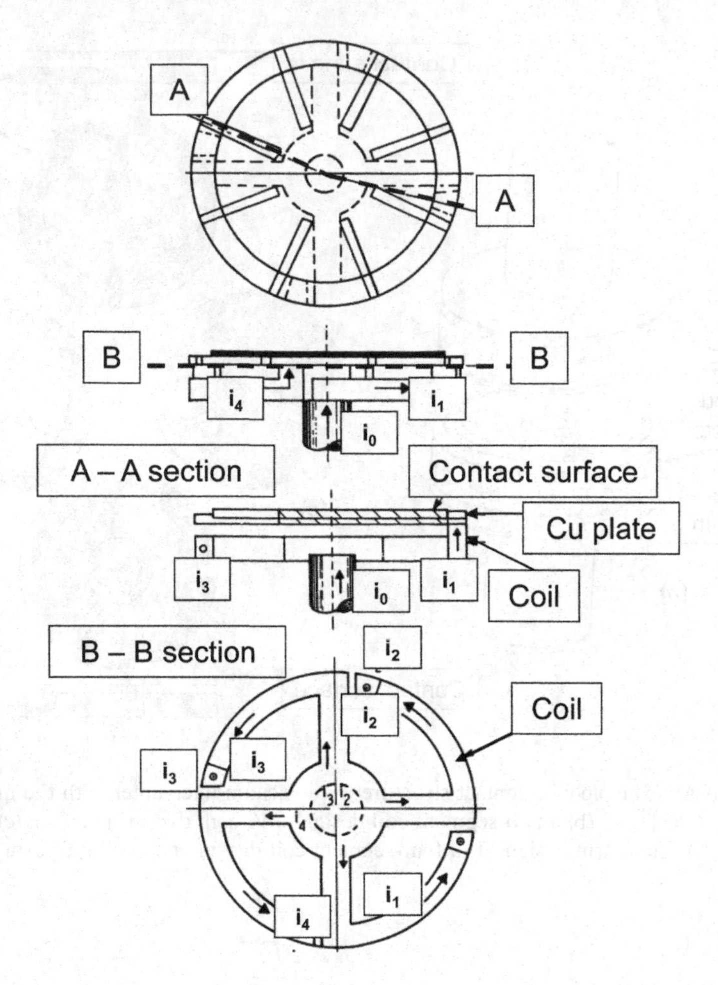

FIGURE 3.43 Four segment axial magnetic field contact structure proposed by Murano et al. [131] and Yanabu et al. 132].

four-segment coil is placed behind the contact. The slots in the contact faces are to reduce eddy currents, which result in reducing the effective AMF in the contact gap.

Another design by Kurasawa et al. [133] gives an interesting magnetic field structure that changes direction across the contact face (Figure 3.44). This type of design was first developed at the end of the 1970s and has been resurrected as the "new" development in AMF contact structures in the late 1990s [134]. Another structure that has also had practical success is shown in Figure 3.45. Here the "coil" is a cup with slots cut into it in the way shown [135–138]. The six-slot design shown in Figures 3.45(a) and (b) find the most common usage. Wang [139] has shown that a four-slot design gives an enhanced *AMF* that is useful at the longer contact gaps needed for single vacuum interrupters operating in 126kV circuits. While this design gives a lower AMF than that experienced in the coil designs shown in Figure 3.42, it also works well if slots are placed in the contact surface to disrupt the induced eddy currents. It also has a lower impedance than the coil designs. Figure 3.46 shows data from Xiu et al. [140] who have compared the AMF for the two-arm coil design, see Figure 3.45(b) and the six-slot cup design, Figure 3.45(a).

The major advantage of this class of contact structure is its ability to keep the arc diffuse during very high currents because the magnetic field increases as the current increases. It has been used to interrupt very high currents. Yanabu et al. [132] report interrupting currents as high as

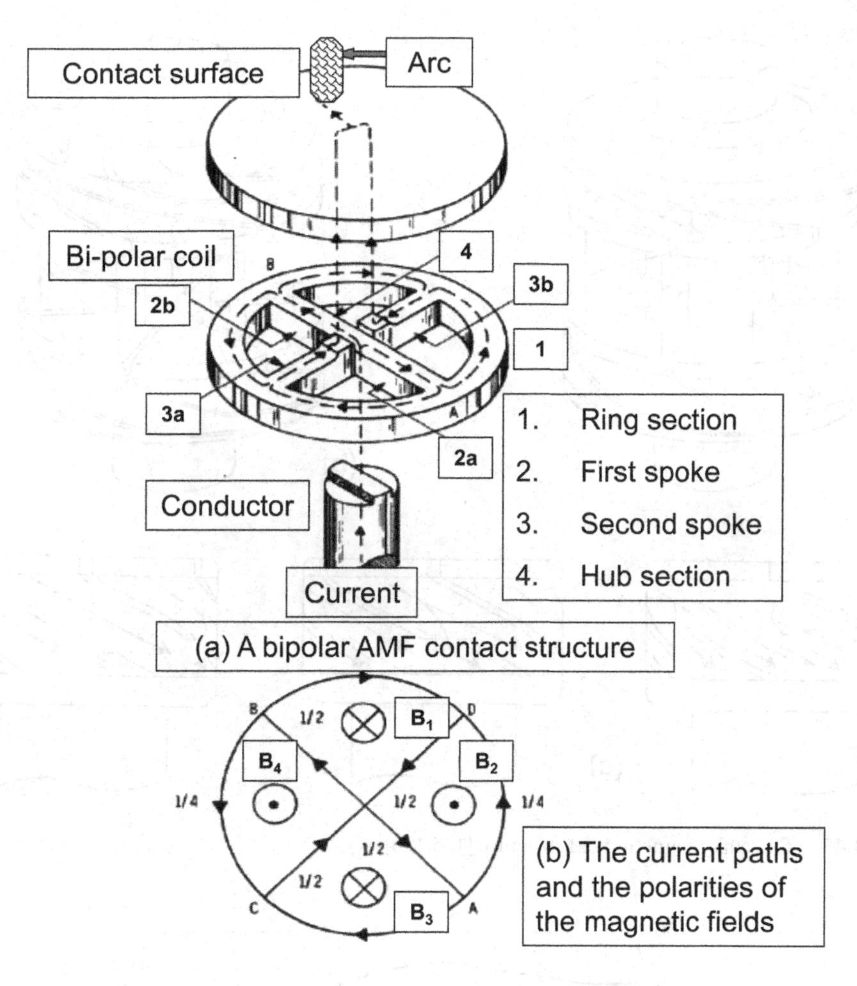

FIGURE 3.44 A bipolar axial magnetic field contact structure proposed by Kurosawa et al. [133].

200kA in 12kV circuits and Voshall et al. [68] report currents up to 63kA in 72kV circuits. The major disadvantage is that the contact design increases the impedance between closed contacts. Thus, the vacuum interrupter must dissipate a higher energy when passing high steady state currents. Mayo [141] has shown that the increase in impedance can be mitigated to some extent by using a coil with a trapezoid cross section. Liu et al. [142] have proposed another interesting design, which is illustrated in Figure 3.47. This uses a central Cu post that is forced to make contact to the underside of the main contact when the vacuum interrupter closes and deforms the cup. This allows the bulk of the circuit current to bypass the AMF coil. When the contacts are parted the cup restores itself to its original dimensions (this may be aided by some contact welding). It is also aided by the Lorenz force between the adjacent segments in the cup. The central pillar breaks contact with the underside of the main contact and the current flows through the cup and the AMF is generated.

One other proven AMF design that separates the generation of the AMF from the current carrying path is shown in Figure 3.48 [143]. Here a series of steel "horseshoe" plates is placed behind a butt contact. The magnetic field that surrounds the conductor from the current flowing through the conductor is trapped in the steel horseshoes. By judiciously designing the gap in the horseshoes, it is possible for this magnetic flux to preferentially jump the gap between the contacts to a horseshoe behind the opposite contact rather than jump the gap to its own horseshoe. When this happens, a

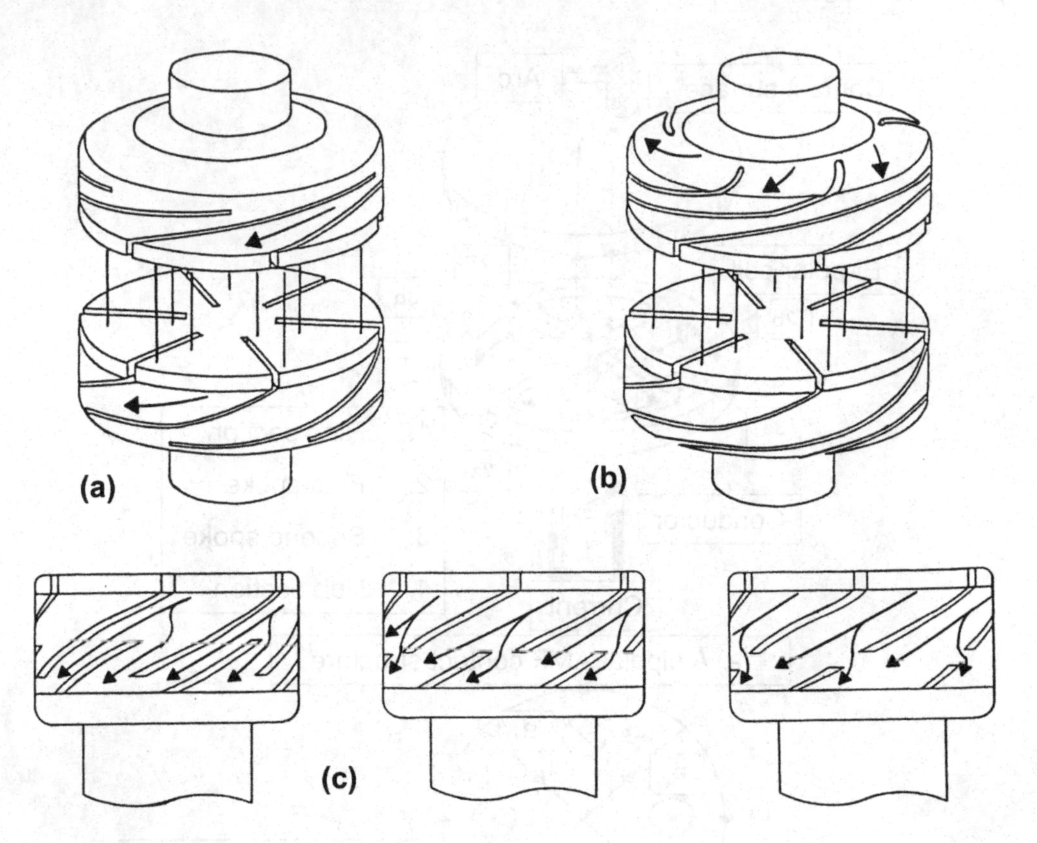

FIGURE 3.45 The cup, magnetic field structure [135,136].

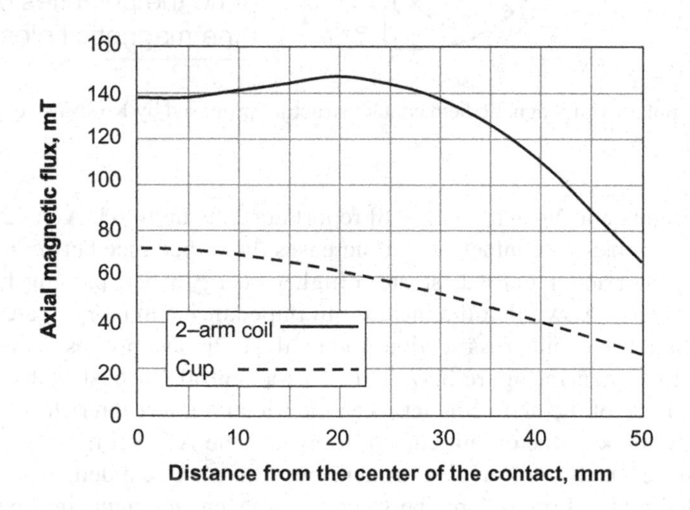

FIGURE 3.46 A comparison of the axial magnetic flux for a two-segment AMF coil design with a six-cut cup design, with a 100mm contact diameter and a current of 40 kA. [140].

bipolar, AMF results. The major disadvantage of this structure in the past has been that for contact gaps greater than about 12mm, the design of the horseshoe structure becomes more difficult. That being said, vacuum interrupters are being developed with long contact gaps [144] and for use in circuits with a system voltage of 126kV [145]. This structure has the major advantage that its total impedance can be lower than the coil or cup designs.

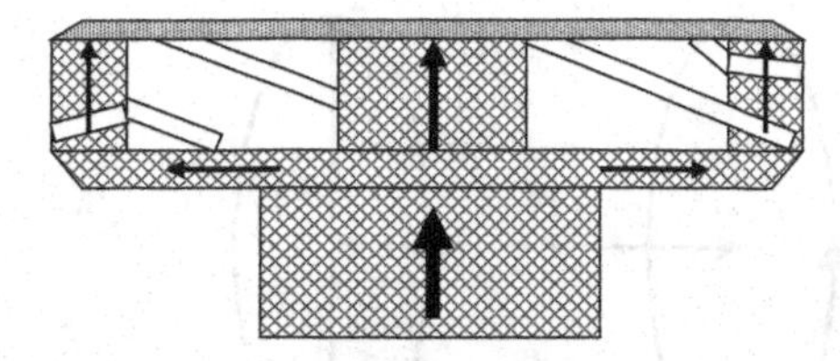

(a) With the contacts closed the central Cu pillar is forced to make contact with the under side of the main contact and most of the circuit current flows as shown

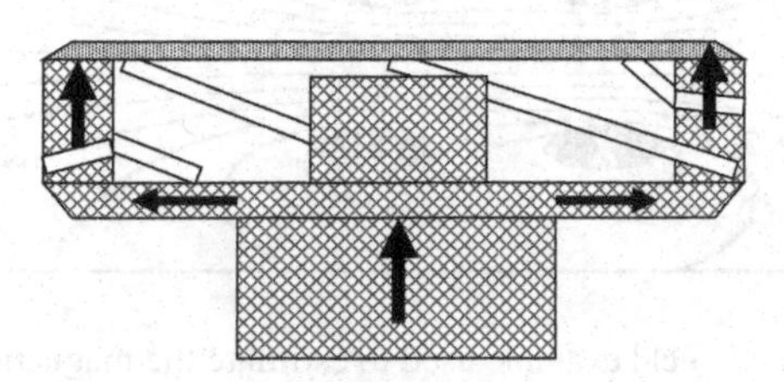

(b) With the contacts open the central Cu pillar disconnects from the under side of the main contact and circuit current flows through the AMF structure

FIGURE 3.47 One proposed AMF cup contact design to give a low impedance contact structure when carrying rated continuous current and to give the AMF when interrupting fault current [142].

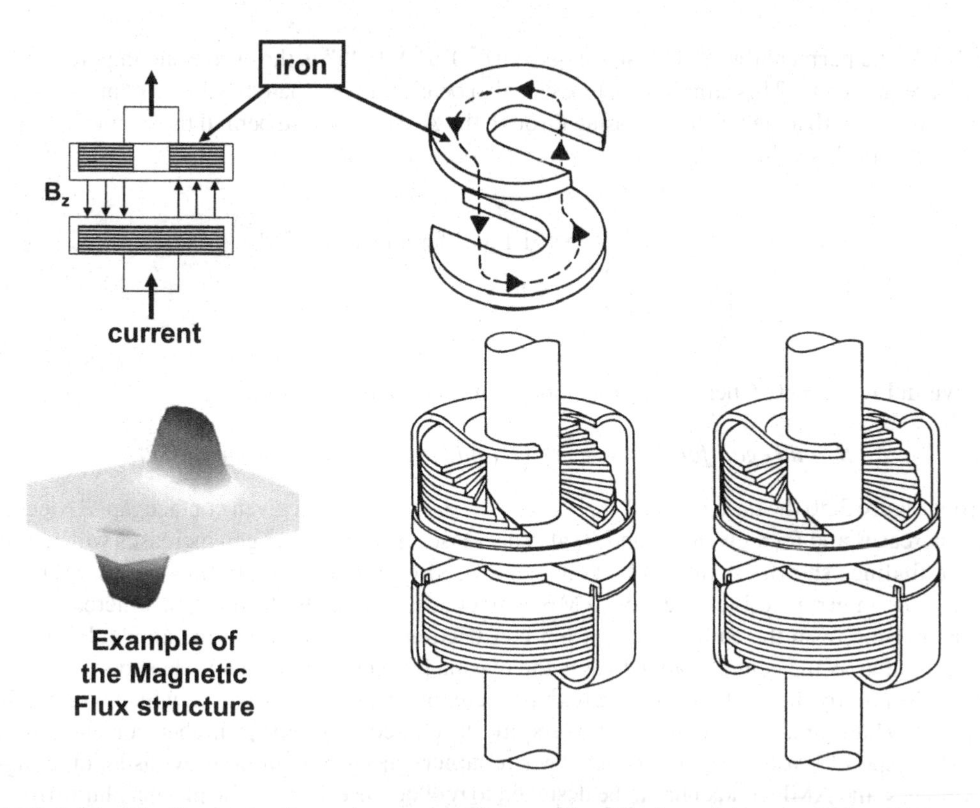

FIGURE 3.48 The "horse-shoe" contact structure that provides a bipolar axial magnetic flux [143].

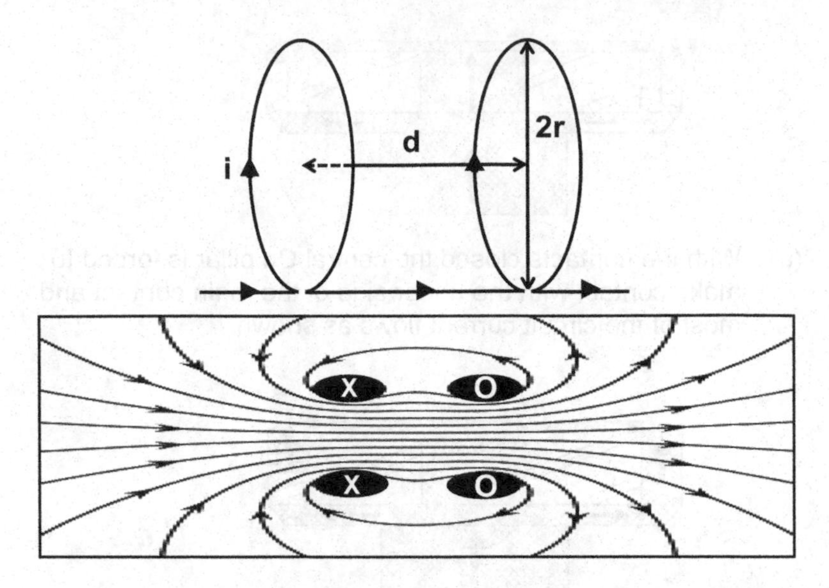

FIGURE 3.49 A simple axial magnetic field example used to estimate the magnetic flux.

An easy way to estimate the value of the axial magnetic field is shown in Figure 3.49. Here two rings of radius "*r*" a distance "*d*" apart give the magnetic field B_A at mid-gap:

$$B_A(mid-gap) = \frac{\mu_0 \times i \times r^2}{2\left(r^2 + \left[\frac{d}{2}\right]^2\right)^{3/2}} \tag{3.26}$$

Where μ_0 is the permeability in free space ($4\pi \times 10^{-7}$ T.m.A^{-1}), "*i*" is the current in amperes and "*r*" and "*d*" are in meters. This simple structure can also be used to estimate the added contact force for closed contacts with a high current passing through the AMF structure behind the contact faces: see Section 4.4.2 in this volume. A plot of:

$$\frac{\mu_0 \times i \times r^2}{\left(r^2 + \left[\frac{d}{2}\right]^2\right)^{3/2}} \text{ vs 'd' for different values of 'r'} \tag{3.27}$$

Is shown in Figure 3.50. Chen [146] gives for a 100mm diameter the ratio of:

$$\left(B_A \; mid\text{-}gap \; for \text{'d'} = 40 \; mm\right) / \left(B_A \; mid\text{-}gap \; for \text{'d'} = 60 \; mm\right) = 0.78 \tag{3.28}$$

From Figure 3.50 the ratio is also 0.78. Figure 3.50 shows that for a given contact gap, B_A decrease as "*r*" increases and for a given "*r*," and B_A also decreases as the contact gap increases which agrees with the findings shown on Table 3.11. I have discussed in Section 2.7.2 in this volume that as the current to be interrupted increases, the AMF contact area and/or the B_A must also increase to prevent the plasma from the cathode spots from constricting and overheating the anode. Figure 2.89 also shows that as the contact gap increases the plasma plumes from the cathode spots also tend to overlap. Eventually they will also form a region where the plasma is constricted which will overheat the anode. Therefore, as vacuum interrupters are developed to interrupt higher currents and are being developed for use at higher voltages, where larger gaps are required to withstand the higher BIL voltages, the AMF contact has to be designed to reduce the effects of the plasma plume overlap.

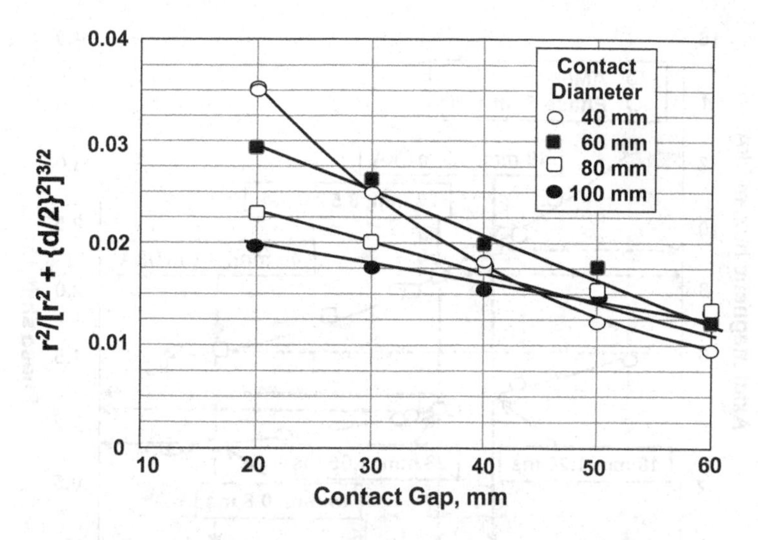

FIGURE 3.50 The calculated effect of the contact gap as a function of the coil radius.

Zhang et al. [147], Ryu et al. [148], Yao et al. [149] and Zhang et al. [150] have proposed the two-layer, three-arm coil design shown in Figure 3.51. The effect of the contact gap on the magnetic flux for this contact design is shown in Figure 3.52 [149]. Zhang et al. [151] propose a 3/4-coil design (see Figure 6.3), which also provides adequate field strengths at longer contact gaps.

As I have discussed in Section 2.4 in this volume, to be most effective the AMF must be greater than a minimum value. In an ac circuit the sinusoidal current flowing in the coil structure behind the

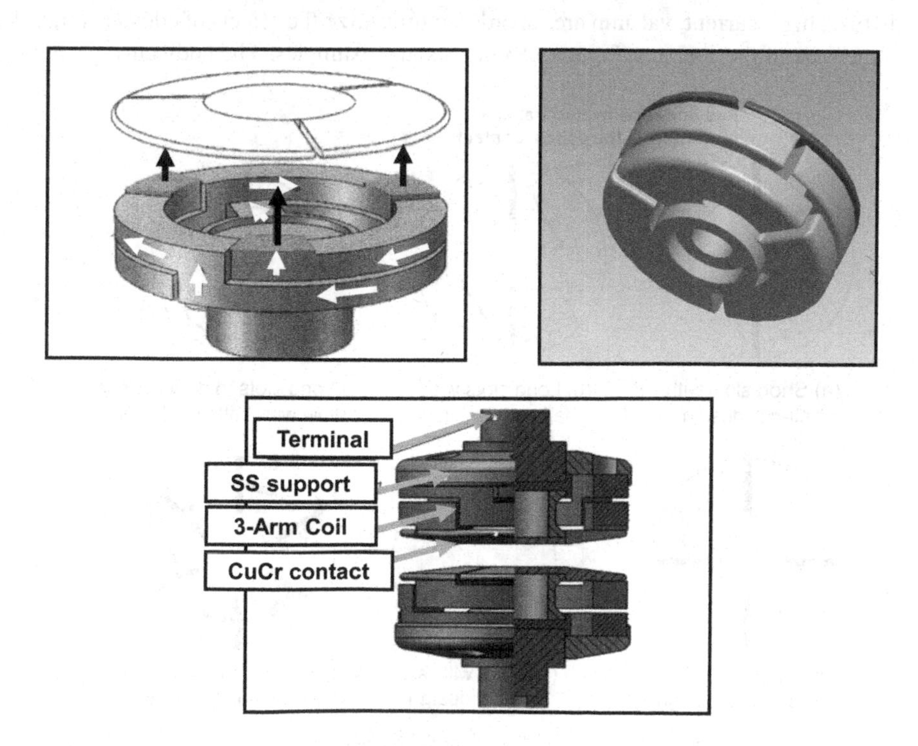

FIGURE 3.51 The two-layer, three-arm coil proposed for high voltage and high current vacuum interrupters [147–150].

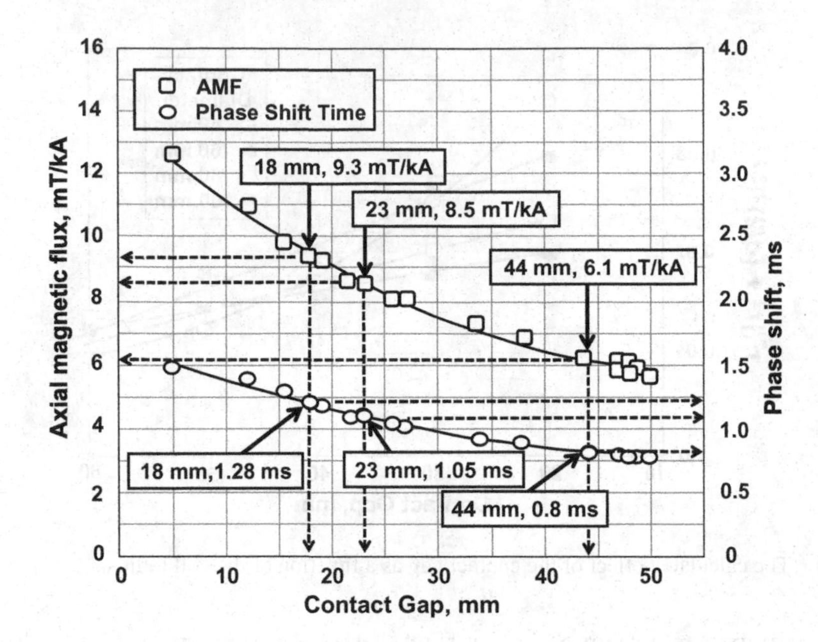

FIGURE 3.52 The magnetic flux as a function of contact gap for 100mm diameter, two-layer, three-arm coil design [149].

contact face results in an AMF that also has a sinusoidal structure. This varying AMF will in turn develop eddy currents in the contact structure's face and the base (for cup-shaped designs). These eddy currents produce a counter AMF that reduces the effect of the AMF from the coil structure behind the contact and hence the AMF in the contact gap itself. A reduced AMF can hinder the development of the fully diffuse, high-current, vacuum arc. In order to minimize the effects of eddy currents slots can be cut into the face of the contacts. Figure 3.53 gives some examples. The eddy currents also change

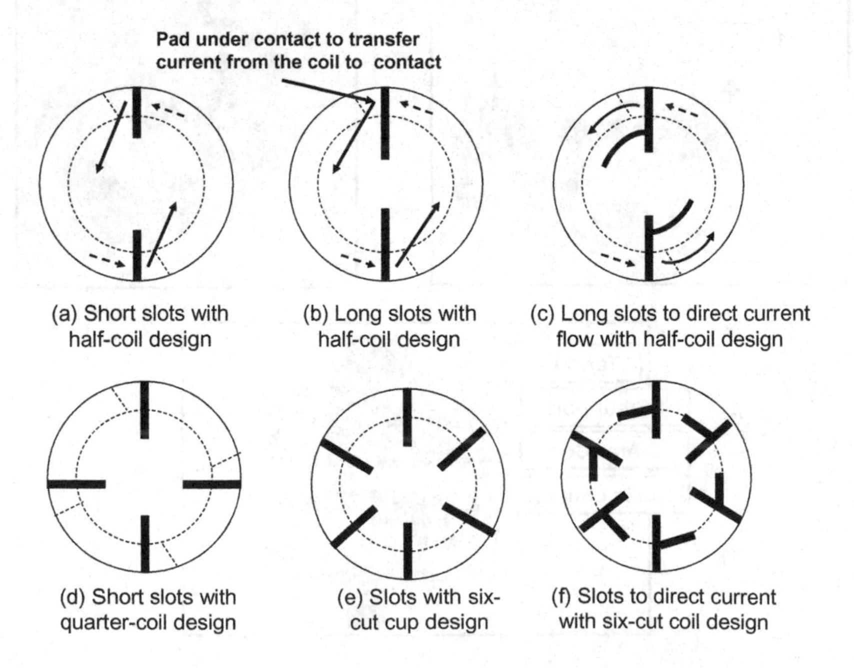

FIGURE 3.53 Examples of slots cut into the face of the vacuum interrupter's contact to reduce the effects of eddy currents in the AMF contact structure.

the phase of the AMF with respect to the ac current by as much as 20° to 30°, see for example Figure 3.52. The maximum flux from the AMF does not occur at the maximum current through the contacts. This is also an important design consideration during the opening of the contacts and the development of the high current diffuse vacuum arc. This phase difference between the AMF and the ac current results in a significant AMF across the open contacts when the ac current reaches zero.

The development of the AMF type of contact structure has been greatly assisted by the advent of engineer friendly, three-dimension, finite element software and high-speed, high capability personal computers. Stoving et al. [152] give an excellent example of FEA analysis of a number of AMF contact designs. This type of analysis has proven extremely useful in analyzing the effect of current flow through the coil and contacts on the distribution of AMF across the face of the contact [153]. It has also shown the advantage of placing slots in the contact face to alleviate the effects of the eddy currents induced by the AMF; see, for example, Figure 3.54 [154]. FEA analysis, other magnetic structure analysis and computer modelling are now the design tools used by all engineers involved with advanced contact design for vacuum interrupters: see, for example [147–150]

The vacuum interrupter designer has limitations placed upon the contact design. First of all, there is a cost constraint. As larger diameter ceramics are more expensive than ones of smaller diameter, the contact diameter is usually restricted. That is, the contacts diameter must be optimized for a given short circuit current interruption performance. Thus, the need to limit the vacuum interrupter's diameter will be satisfied. Another complication in the design is the spacing between the contact and the shield. In order to maintain the vacuum interrupter's high voltage performance this spacing also has to be optimized. Thus, it is possible, even if the AMF is high and the cathode spots spread out across the cathode's surface, the plumes above the cathode spots will begin to overlap if the current is high enough. Schulman et al. [155] describe three vacuum arc modes in an AMF after the bridge column stage of the opening contacts. These vacuum arc modes that depend upon the current and the contact's diameter are:

1) *The multi cathode spot vacuum arc*: Here the cathode spots spread over the cathode surface with their associated plasma plumes. The space between the cathode spots is great enough that their associated plasma plumes do not overlap. This results in a low level

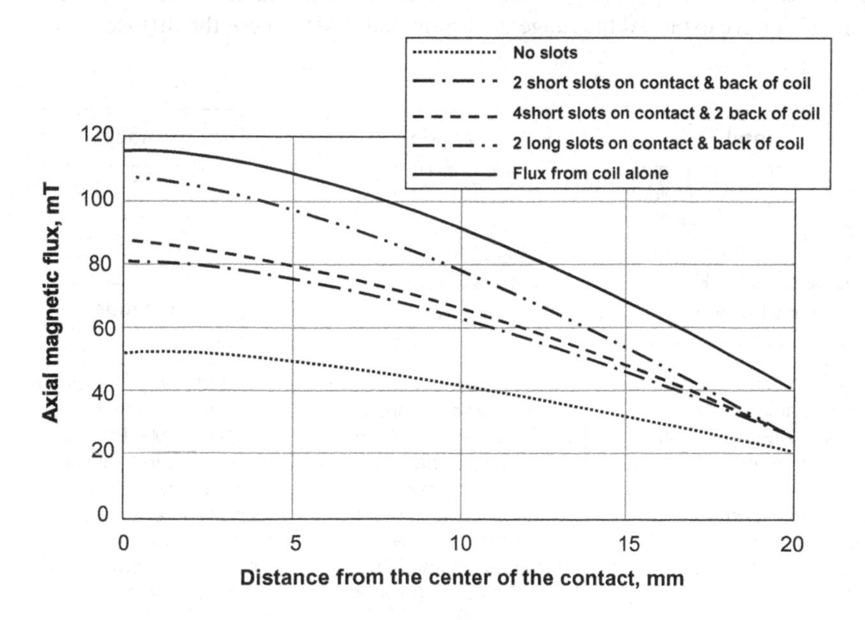

FIGURE 3.54 The effect of slots in the contact face and in the cup structure of the coil shown in Figure 3.42(a) on the magnetic flux in the center of the contact gap as a function of the radial distance from the center of the contact [154].

of light intensity from the inter-contact gap. As the cathode spots spread over the whole cathode surface, there is a more or less uniform erosion rate of the cathode, which can be measured as a (constant) gC^{-1}. The anode remains a passive receiver of electrons.

2) *The high current diffuse vacuum arc*: Here the cathode spots still spread over the cathode's surface, but now there is overlapping of the plasma plumes and visible plasma now fills the intercontact gap with a more or less uniform light intensity. The individual cathode spots can still be observed, but the individual plasma plumes can no longer be resolved.
3) *The high current diffuse column vacuum arc*: Here there is a central plasma core in the intercontact gap that joins cathode to anode. This plasma core is much brighter than the cathode spots that can no longer be resolved. It is also much brighter than the peripheral plasma that surrounds it. This diffuse column arc tends to constrict as it approaches the anode. The plasma constriction at the anode can then develop an anode spot. This anode spot reaches a high temperature which, in turn, releases metal vapor into the inter-contact gap: see Sections 2.7.2 and 2.8 in this volume.

The current level, the contact diameter, the contact gap and the strength and structure of the AMF generally determine these vacuum arc modes. Figure 3.55 shows a schematic of the local light intensity from the vacuum arc as the current increases. Chen et al. [156] present photographs of vacuum arcs in an AMF that illustrate these effects nicely. Schulman et al. [155] have used observations of such changes in the vacuum arc's luminosity to develop the "Appearance Diagram" shown in Figure 3.56. Here the distinctive plasma core develops at 35kA. They also have developed an analysis based upon Gundlach's research (see Figure 2.83 and Equations (2.44) and (2.45) that, if the arc voltage, U_{arc}, is greater than a value given by:

$$U_{arc}\left(i_{arc}(t),d(t)\right)=a\left[d(t)\right]\times\sqrt{i_{arc}(t)}+b\left[d(t)\right] \tag{3.29}$$

where U_{arc} as a function of time, t, is a function of $i_{arc}(t)$ and the contact gap $d(t)$, $a[d(t)]$ is a constant that is a function of $d(t)$ and $b[d(t)]$ is a constant that includes a term for the increase in U_{arc} with arc length $d(t)$ ($b \approx U_{arc}[B_A \rightarrow 0,\ i_{arc} \rightarrow 0]$)), then the vacuum arc would be in the diffuse column mode. Matsui et al. [157] have explored the range of current and AMF where the diffuse column results in

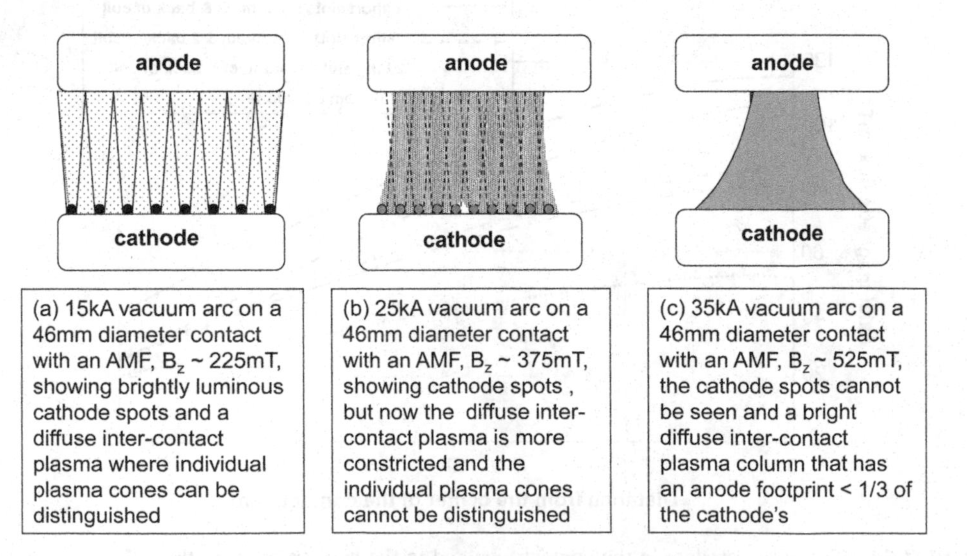
anode
anode
anode
cathode
cathode
cathode
(a) 15kA vacuum arc on a 46mm diameter contact with an AMF, Bz ~ 225mT, showing brightly luminous cathode spots and a diffuse inter-contact plasma where individual plasma cones can be distinguished
(b) 25kA vacuum arc on a 46mm diameter contact with an AMF, Bz ~ 375mT, showing cathode spots , but now the diffuse inter-contact plasma is more constricted and the individual plasma cones cannot be distinguished
(c) 35kA vacuum arc on a 46mm diameter contact with an AMF, Bz ~ 525mT, the cathode spots cannot be seen and a bright diffuse inter-contact plasma column that has an anode footprint < 1/3 of the cathode's

FIGURE 3.55 Schematics (not to scale) of the intercontact plasma, showing the increasing luminosity and the constriction to a bright diffuse plasma column as the arc current increases; contact gap about 8mm [155].

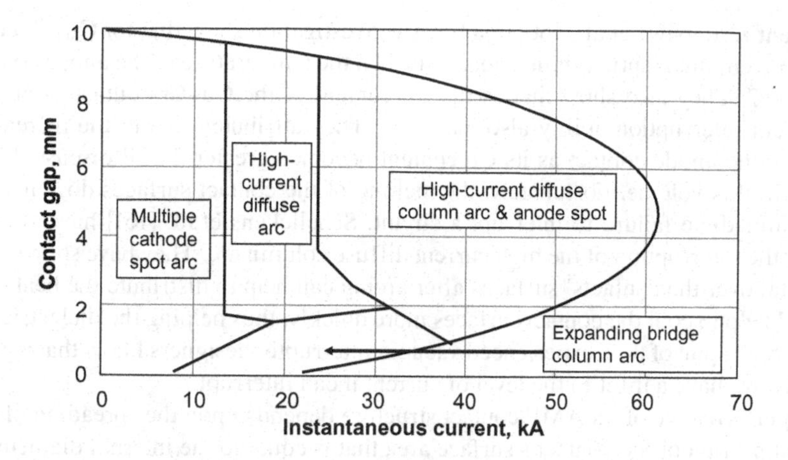

FIGURE 3.56 The appearance diagram for a high current, vacuum arc in an axial magnetic field [155].

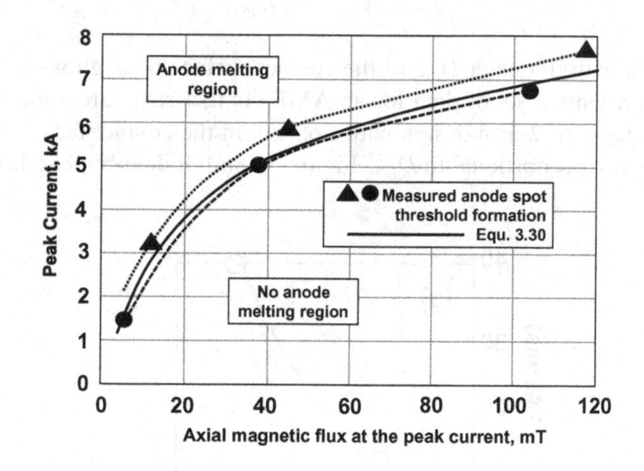

FIGURE 3.57 The threshold anode melting current as a function of axial magnetic field for 22mm diameter Cu–Cr contacts with a contact gap of 6mm [157].

anode melting. Figure 3.57 shows the range of currents above which anode melting occurs for 22mm diameter Cu–Cr contacts with a 6mm contact gap. They establish an equation:

$$i_{th} = 1.85 \times ln(B_A) - 1.6 \tag{3.30}$$

where i_{th} kA is the threshold current and B_A mT is the magnetic field. Keidar et al. [158] have shown that once U_{arc} of the diffuse column vacuum arc exceeds a given value, it is difficult for a cathode spot to exist outside it. It is thus possible for the spread of the cathode spots from a high current vacuum arc to be restricted to an area less than that of the total contact's area even in the presence of an AMF. At high currents, for a given contact diameter, the energy input into the cathode and the anode will be mostly confined to the limited area. This can result in melting of the contact surfaces and a potential limitation on the level of short circuit current that the vacuum interrupter containing this contact can handle. Watanabe et al. [159] show for a Cu–Cr (50 wt.%) contact the anode reaches a temperature of about 1750K at its current interruption limit. At this temperature the vapor density of evaporated contact material reaches about 3 × 10^{20} atoms/m^3 or a pressure of about 8 Pa. They conclude that the density of metal vapor in the contact gap dominates the reignition of the vacuum arc after current zero. Niwa et al. [160] in a further investigation show that for Cu–Cr contacts with an AMF the current interruption limit is reached when the temperature of the anode contact reaches 2000C to 2500C. They also show that as the Cu content of the Cu–Cr contacts increases, so the contact's current interruption ability also increases. They attributed this to the increased thermal conductivity of the anode contact as its Cu content becomes greater. I will discuss this further in Section 4.2.3 in this volume. Fortunately, the melting of the contact surfaces does not necessarily result in an immediate failure to interrupt a circuit. Schellekens et al. [161] have studied contact surfaces after the interruption of the high current diffuse column arc. They have shown that the flow of molten metal over the contacts' surfaces after arcing can help to distribute the heat energy more efficiently and help to cool the contact surfaces more quickly, thus helping the dielectric recovery of the contact gap. In spite of this, experienced vacuum interrupter designers know that a given contact design will always have a limit in the level of current it can interrupt.

Since the performance of an AMF contact structure depends upon the spreading of the cathode spots over that portion of the contact's surface area that is equal to the internal diameter of the coil structure, one would expect that the maximum current that can be interrupted i_{max} (rms) is given by:

$$i_{max} \propto A_{B(z)} or \propto D_{B(z)}^2 \tag{3.31}$$

where $A_{B(z)}$ is the area with diameter $D_{B(z)}$ of the contact surface through which the magnetic flux passes. Figure 3.58 presents a set of data for an AMF contact structure similar to that shown in Figure 3.42(a) (i.e., a ½–coil, 2-arm design with slots cut in the contact faces). The dependence of i_{max} (rms) is certainly not proportional to $D_{B(z)}$, Figure 3.58(a). It does appear, however, that it does

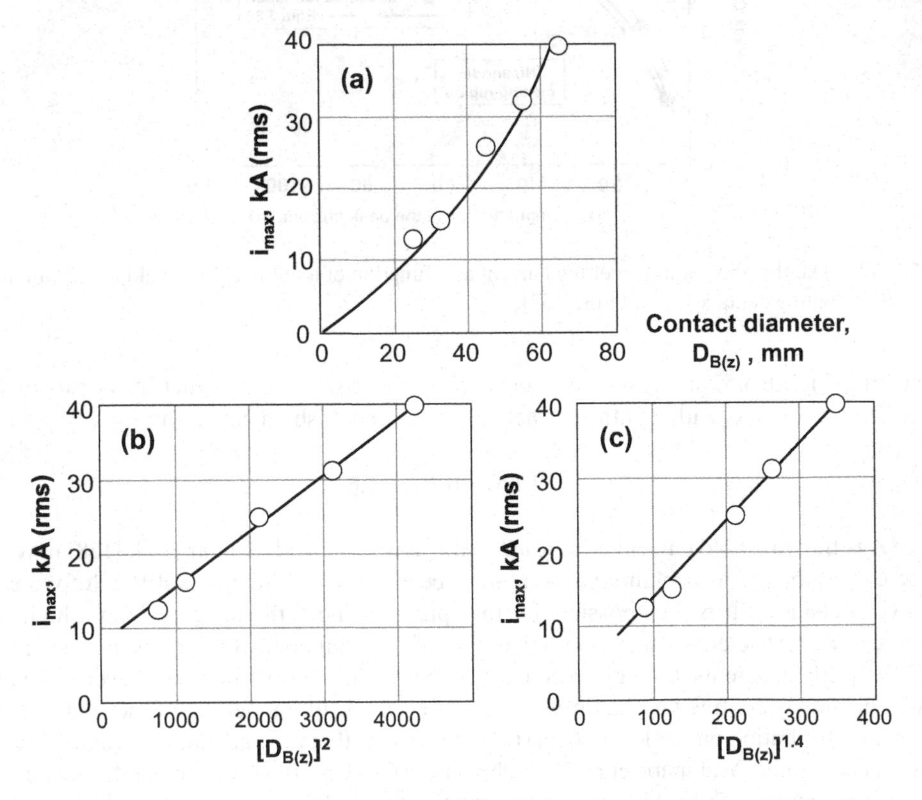

FIGURE 3.58 An example of the interruption ability of an AMF contact structure with Cu–Cr contacts as a function of $D_{B(z)}$ for a 38kV system voltage (a) Interruption current as f(contact diameter) (b) interruption data from (a) as f(contact diameter)^2 (c) interruption data from (a) as f(contact diameter)$^{1.4}$.

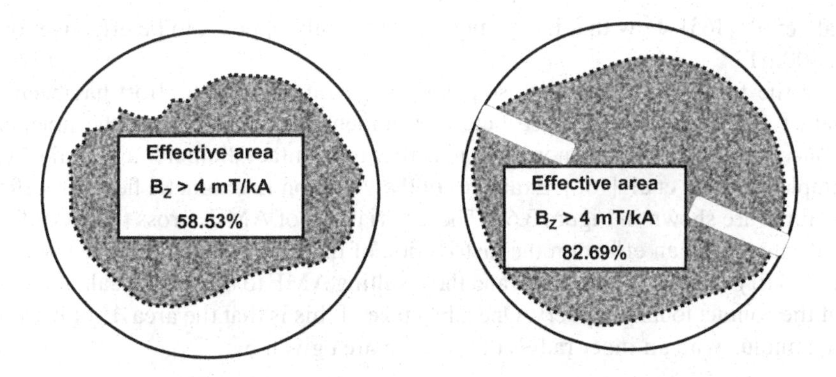

FIGURE 3.59 The contact surface area affected by the magnetic flux for a ½-coil design without and with slots in the contact face [162].

satisfy Equation (3.31); i.e., there is dependence on $D^2_{B(z)}$, Figure 3.58(b). Del Rio et al. [162] show in Figure 3.59 that the B_A does not cover the whole contact face. Henon et al. [163] suggest that the actual effective area on the contact's surface is where $B(z) \geq 4mT.kA^{-1}$. They also show in their study that:

$$i_{max} \propto D^{1.4}_{B(z)} \tag{3.32}$$

This dependence for the same set of data is shown in Figure 3.58(c). The scatter in the data is such that both Equations (3.31) and (3.32) seem to be satisfied. Again, the interruption data in Figure 3.58 should not be considered the absolute interruption limit for AMF contacts. The ultimate limit will depend upon many variables including; the exact composition of the contact material, the design of the AMF structure and the internal geometry of the vacuum interrupter. These data show that the contact diameter is an important design parameter. Its importance results from the greater area available for the expansion of the cathode spots from the high current diffuse vacuum arc. The limit on interruption comes when this area confines too high a density of cathode spots. When designing the AMF contact structure, the designer also needs to consider the required contact gap. The strength of the AMF for a given contact structure is a function of the contact gap. Also, the expanding plasma plumes above the cathode spots have a greater opportunity to overlap as the contact gap is increased. Now the AMF, B_Z is a function of the contact gap d. Liu et al. [164] propose that the maximum rms current, i_{max}, that can be interrupted by a given AMF contact design can be given by:

$$i_{max} = K^* \times D^{\alpha_1}_{B(z)} \times B^{\alpha_2}_z \tag{3.33}$$

They show that over the range of 45mm $\leq D_{B(z)} \leq$ 70mm and 3mm $\leq d_z \leq$ 15mm that:

$$i_{max} = 3.079 \times 10^{-3} \times D^2_{B(z)} \times B^{0.4}_z \tag{3.34}$$

where B_Z is the maximum AMF given in mT.kA^{-1}. Again, the vacuum interrupter designer should treat these data with care. Equations (3.33) and (3.34) seem to imply that for a given contact diameter the i_{max} can be increased simply by the ever increasing B_z. It is certainly true that before the high current vacuum arc can be fully diffuse, the B_z has to be greater than a critical value B_{crit} as I have already discussed in Section 2.7.2 in this volume. However, the density of cathode spots for a given contact diameter certainly limits the overall current interruption performance of the contact structure. Most AMF designs seem to show maximum values of B_z in the range 100mT to 400mT.

In fact, Chaly et al. [165] show that increasing the AMF only appears to be effective for B_z values of less than 400mT.

Because of the formation of the diffuse column vacuum arc, much effort has been applied to develop contact geometries, which distribute or fragment the AMF. All of the practical contact designs produce an AMF, which is nonuniform across the contact diameter and in the contact gap: see, for example Wootten et al. [166]. Examples of the variation in magnetic field for different AMF contact structures are shown in Figure 3.60. These variations of AMF across the face of the contact in practical designs have an effect on the distribution of the cathode spots. Some researchers have attempted to modify the contact structure and the resulting AMF to move the peak of the AMF from the center of the contact to its periphery. One advantage of this is that the area A^* of high $B(z)$ is now moved to an annulus with an inner radius of r_I and an area given by:

$$r_I = r - w \quad \text{and} \quad A^* = \pi w(2r - w) \tag{3.35}$$

where r is the contact radius and w is the annulus width. Indeed, it has been demonstrated by Chaly et al. [167], using a "magnetic barrier" to shape the AMF distribution to that shown in Figure 3.61 does effectively distribute the cathode spots more uniformly across the cathode's surface. They then show that the anode erosion resulting from an 11kA arc using a similar AMF with a 30mm diameter contact is much less pronounced than that obtained by using the profile shown in Figure 3.60(a). This work, however, uses only limited values of current, AMF and contact diameter. Homma et al. [168] propose such a design, which they call a SADE contact (self-arc diffusion by electrode). Their patent [169] presents the details of the desired AMF: it must have its lowest value in the center of the contact and have its highest value toward the contact's edge. They use a value of B_A in the center of the contact that is 75–90% of the $B_{U(min)}$ value given by Gundlach for a given current (see Figure 2.83). The B_A then gradually increases to the $B_{U(min)}$ value at a distance from the contact's center of between 20% and 40% of the contact's radius. It then peaks to values between 1.4 to 2.4 times $B_{U(min)}$

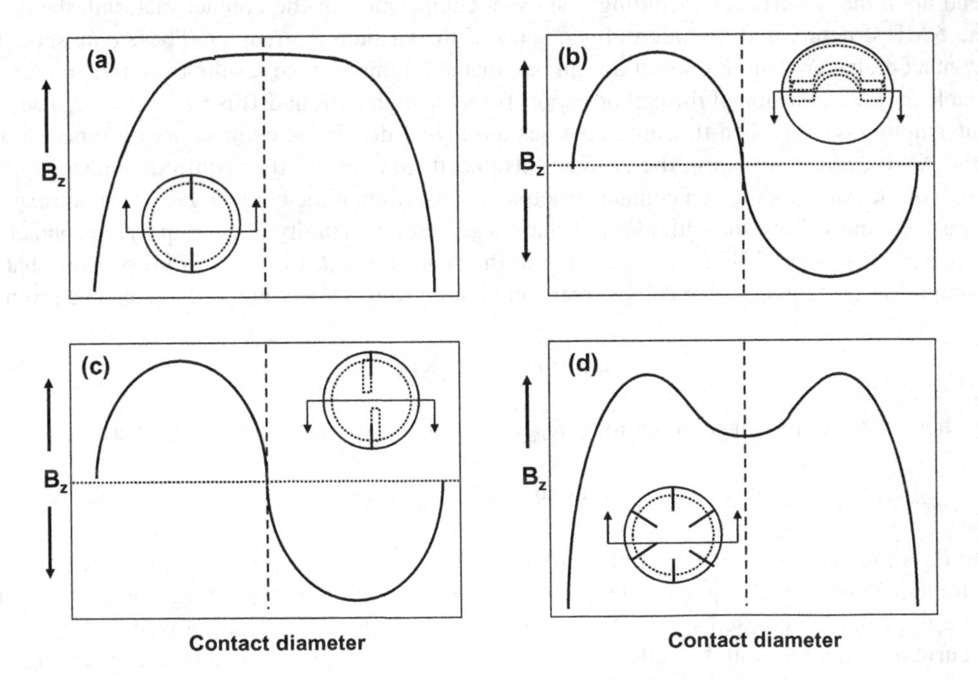

FIGURE 3.60 Examples of the variations of magnetic flux across the diameter of the contact gap, (a) the unipolar structure, Figure 3.42(a), (b) the horseshoe structure, Figure 3.48, (c) the bipolar structure, Figure 3.44 and (d) the structure shown in Figure 3.45(a).

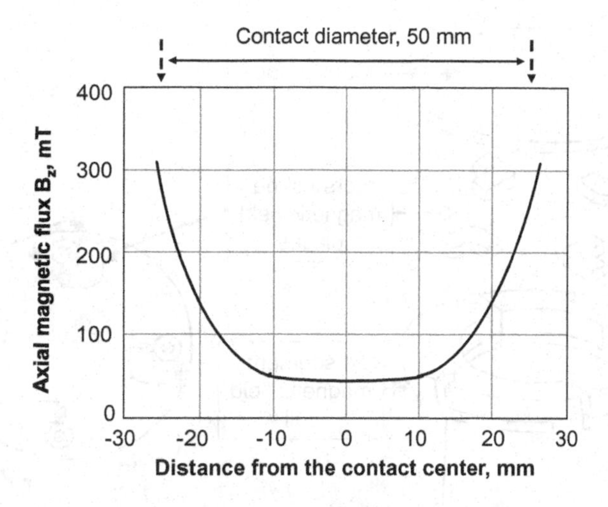

FIGURE 3.61 Extreme distribution of the magnetic flux to the edges of the contact [167].

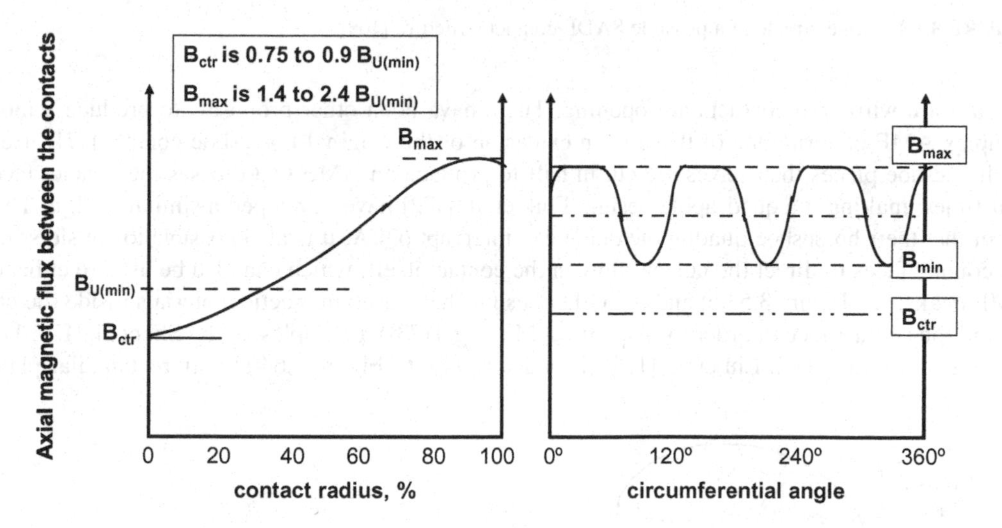

FIGURE 3.62 The distribution of the magnetic flux for the SADE, vacuum interrupter, contact structure [169].

at 70 to 100% of the contact's radius. Figure 3.62 taken from [169] gives a representative example of the profile of the B_A as a function of the contact's radius. One way of achieving such an AMF profile is shown in Figure 3.63. Here horseshoe steel plates are placed around the support posts between the field coil and the contact itself. Current passing through the support posts induces a magnetic flux through the steel. If the contact gap is chosen judiciously, the magnetic flux can be forced to cross the contact gap, giving rise to an enhanced AMF at the contact's edge. This is similar to the principle of the horseshoe contact described earlier, Figure 3.48. The major problem with this design is that in spite of the publications citing its virtues, there is very little real data on its practical design or of its actual performance in a power circuit. Thus, it is difficult to evaluate whether or not this contact does indeed perform in a manner superior to other AMF contacts. Experiments by Shkol'nik et al. [170] do seem to show that there is some advantage in using such AMF structures. Here the interruption data, however, should be judged warily, because they open their contacts with a low current and allow the cathode spots to spread over their contacts before applying the full current pulse. In doing so, their contacts do not experience the full effect of the formation of the high current diffuse

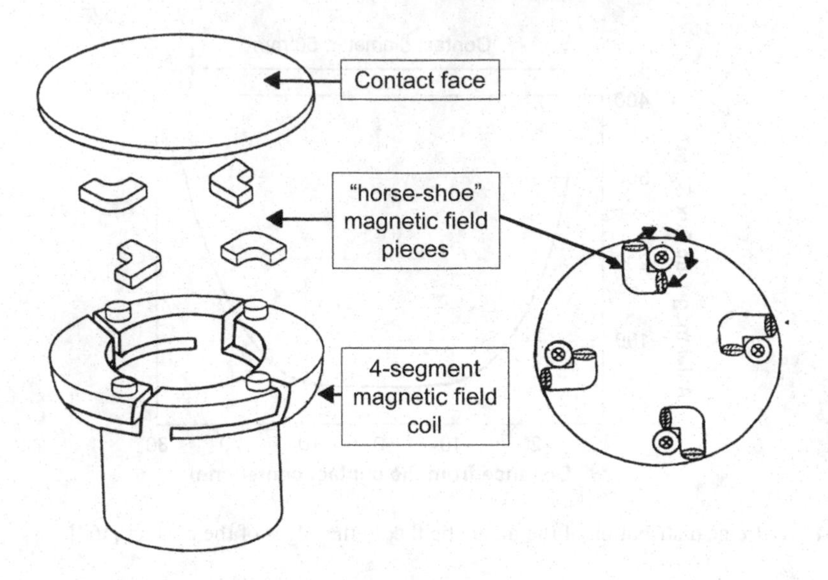

FIGURE 3.63 An example of a possible SADE contact structure [169].

vacuum arc while the contacts are opening. There have been other proposals to produce a more complex AMF pattern. One of these is an extension of the original horseshoe contact [171]. Here the horseshoe pieces themselves are cut in half to produce an AMF that crosses the contact faces four times; making it a quadrapole design. Fink et al. [172] have developed a similar design. They claim that their horseshoe quadrapole deign can interrupt 63kA. It is also possible to cut slots into the contact faces to direct the current flow in the contact itself, which can also be used to enhance AMF peaks; see Figure 3.53(c) and (f). Other designs have used magnetic materials inside the cup behind the contact face in order to shape the AMF (e.g. [173]). Examples are by Shi et al. [174, 175] are given in Figure 3.64. Liu et al. [176] show a variation of Figure 3.64(a) with rectangular plates

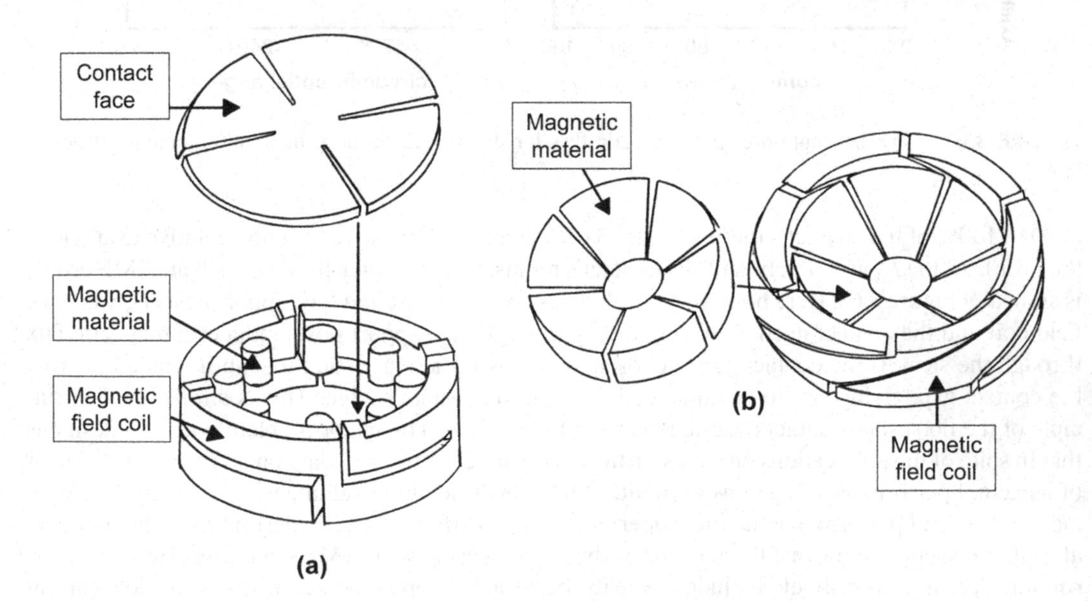

FIGURE 3.64 The use of magnetic materials inside a cup-shaped magnetic field structure to shape the AMF in the gap between the contacts [174, 175].

in place of the cylinders. They claim that the addition of the plates increases the B_A by as much as 125% over the design without plates for the contact diameter and contact gap they used. Zheng et al. [177] show the effect of placing an iron ring inside the cup AMF design: See Figure 3.65. At a 2mm contact gap the effect of the iron ring is much greater than at 10mm. At 2mm, however, the vacuum arc is developing into the fully diffuse form from the transition vacuum arc, see Section 2.7.2 in this volume. Therefore, the AMF structure formed with the iron ring would greatly assist the cathode spots to distribute to the outer periphery of the cathodes surface. Even though the outer edge of the AMF decreases as the contacts open, I would expect that the AMF would maintain the cathode spots in their widely dispersed pattern. Niwa et al. [178] also claim that a split iron ring placed inside the cup AMF design effectively produces a more uniform AMF across the contact gap. Liu at al. [179] have analyzed the iron plate structure shown in Figure 3.64(b). They show that the AMF is not greatly affected by the number of slots in the iron plate, its inner radius, its outer radius, its height and its thickness. Kulkani et al. [180] orientated the two contact structures with the ½-coil, two-arm design by 90° with respect to each other: see Figure 3.66. They show that the 90° orientation gives a more uniform B_A across the face of the contacts. This results in less severe erosion of the contacts after interrupting high currents.

It has been claimed that the bipolar designs and quadrapolar designs discussed above also fragment the magnetic field and split the plasma up across the contacts. Indeed, photographs of vacuum arcs between these contact structures [133, 134, 172] have shown multiple concentrated arc regions operating in parallel. Within each of these regions the vacuum arc mode is that of the diffuse

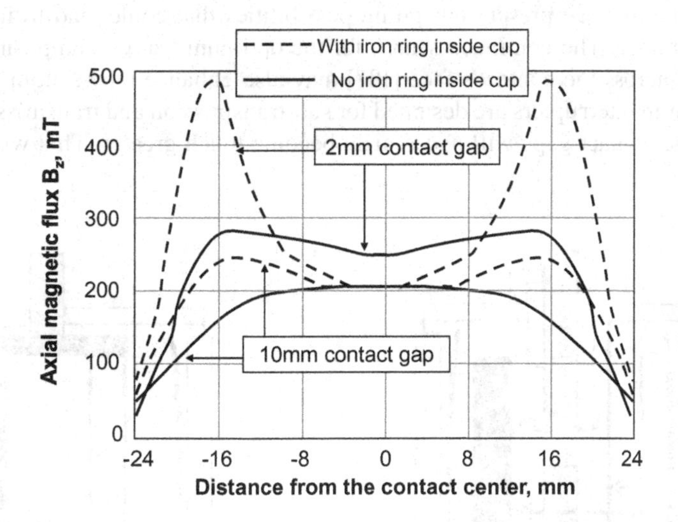

FIGURE 3.65 Comparison of the AMF for a six-cut, 48mm diameter, cup AMF contact with and without an iron ring inside the cup for a 2mm and a 10mm contact gap and a current of 20kA (rms) [177].

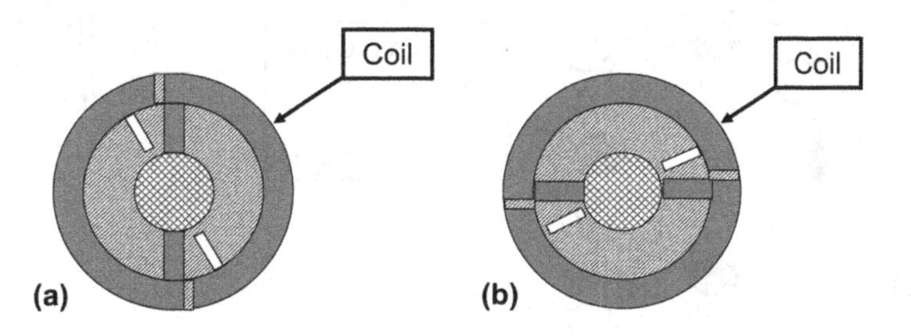

FIGURE 3.66 The orientation of the two AMF contacts 90° to each other [180].

vacuum arc. If, however, the arc is restricted to one segment (i.e. at the position of the initial bridge column arc), then an intense diffuse column will develop in that quadrant with a resulting overheating, perhaps gross melting of the contacts and the potential for the vacuum interrupter to fail to interrupt the circuit current. The effect of limited arc structure is well illustrated in Figure 4 of reference [134]. Here the erosion pattern for a bipolar magnetic field contact clearly shows that the vacuum arc is confined to only one portion of the contact. There is also some indication that the lower AMF, for a given current, produced by the quadrapole structure results in more constriction of the plasma towards the anode and a lowering of its current interruption ability [181]. Both the coil and cup AMF designs are usually manufactured from OFHC copper. After manufacture these structures will have been annealed and can easily be deformed. It is therefore necessary to braze a support structure inside these designs. This structure will prevent the AMF structures from collapsing or from expanding if the contact faces weld. Li et al. [182] show that a stainless-steel support does not affect the distribution of the original magnetic flux. A novel interrupter design first described by Reese [183] and developed by Alferov [184] is the multiple rod structure shown in Figure 3.67(a). A high-current diffuse arc forms between the metal rods during interruption of high currents. A triggered version of this design illustrated in Figure 3.67(b) by Wang et al. [185] showed the potential to interrupt 100kA. Lamara et al. [186] propose other designs using a combination of AMF and TMF structures, which separate the normal current conduction and the high-current interruption functions.

It is obvious from the technical literature that each of the vacuum interrupter manufacturers is actively developing their new generation of AMF contact designs. Certainly, the AMF modification ideas I have discussed above present intriguing possibilities that could lead to improved vacuum interrupter performance. The continued search for the optimum Cu–Cr composition or even multiple compositions across the contact's face [187] may also enhance the vacuum interrupter's performance. As vacuum interrupters are designed for sub-transmission and transmission voltages; i.e., 72kV to 242kV, the contact gaps will necessarily become much greater. This will place a greater

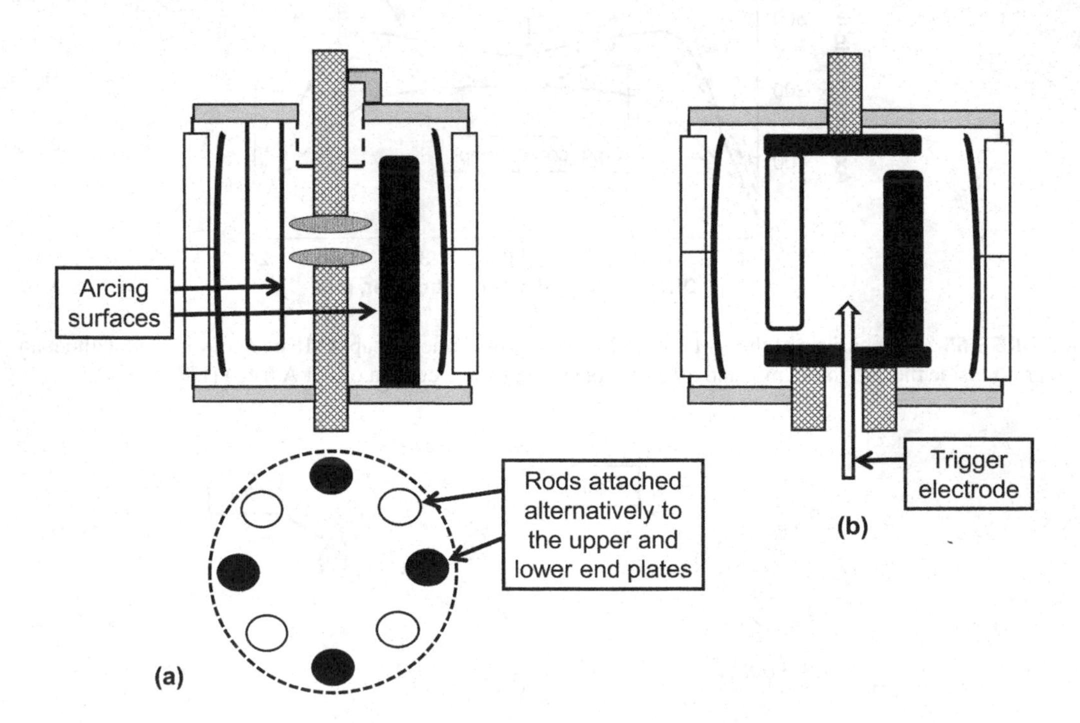

FIGURE 3.67 The rod vacuum interrupter structure [183,185].

burden on the vacuum interrupter designer to maintain the AMF at a value greater than B_{crit}. Xui et al. [188] Zhang et al. [151] and Yao et al. [149] present examples of a double coil design to achieve this at large contact gaps. The challenge will be to limit the contact structure's impedance to an acceptable level. As I will discuss in Chapter 6 in this volume, the vacuum interrupter placed in a vacuum circuit breaker does not open as soon as a fault or short circuit current occurs. There is a delay that results from the mechanism itself and from the need for coordination with the other protection devices in the electrical system. This delay allows time for the AMF to be somewhat out of phase with the fault current. Thus, in a practical situation the maximum AMF will never occur at the maximum current flowing through the vacuum arc and there will be a nontrivial AMF present at current zero. At present, there is no consensus on the best contact design or even the best contact material structure. Each proponent for a particular position has presented limited experimental data to support their claim for superiority. Meanwhile the AMF contact designs presently in production continue to be successfully and economically applied in the widest range of circuit switching applications.

3.3.5 Summary

The three types of vacuum interrupter contact structure have each found its application in practical vacuum interrupter design. The answer to the obvious question, "Which is the best design?" brings the obvious answer, "It all depends upon the application!" The butt contact structure is the easiest to manufacture, but can only really be used effectively for those applications where the vacuum arc is in the diffuse mode. Of course, if you can pass a certification test at higher currents using a butt contact, then it can be used. You do see in some manufacturers' literature an offering of vacuum interrupters with a butt contact for interruption of currents greater than 10 kA. While this is possible, it is not the best contact structure to use for this level of fault current.

Most manufacturers of power vacuum interrupters offer vacuum interrupter designs that use TMF or AMF contacts for interrupting currents greater than about 10 kA. There are many claims in the technical and commercial literature that the TMF contact structure is superior to the AMF structure and vice versa. For example, Renz [189] states that to interrupt fault currents greater than 31.5kA, the AMF contact is preferred. Since TMF contacts are routinely used in vacuum generator circuit breakers with interrupting ability up to 80kA, his conclusion is contradicted by practical experience. Indeed, Gentsch et al. [190] have shown that both TMF and AMF contact structures distribute the arc energy from a 30kA (rms) vacuum arc into the contacts' surfaces very effectively. Interestingly they also show that a surface melted layer of between 100μm and 150μm is similar for both contact structures and is similar to that shown in Figure 3.6.

Table 3.12 presents my qualitative assessment of the advantages and disadvantages of the two structures. For ac circuits up to and including 27 kV, the TMF contact performs with true reliability. At higher voltages, equal to and greater than 36 kV, the AMF contact has been shown to be superior. Although 36kV vacuum interrupters using both the spiral and contrate cup TMF designs have satisfied certification tests. At lower ac circuit voltages, the only real advantage of the AMF contact over the TMF design for distribution circuits less than 27kV is its lower erosion rate when interrupting high short circuit currents. For the very few practical applications where this is necessary, the AMF structure would be essential.

3.4 OTHER VACUUM INTERRUPTER DESIGN FEATURES

3.4.1 The Insulating Body

Two materials have been successfully used commercially for the vacuum interrupter's body: (a) a glass cylinder; and (b) a high alumina content porcelain ceramic. Another material that has been proposed by Parashar et al. [191] is a glass-ceramic envelope. It is the ceramic envelope that has

TABLE 3.12
A Qualitative Comparison for the Performance of TMF and AMF Contact Structures

Contact Structure	Advantages	Disadvantages
TMF	• Smaller diameter for a given short circuit current at 12kV–24kV • Good for very long arcing times i.e. low frequency and generator breaker applications • Good for high voltage up to 27kV, but some success has been achieved at 36–38kV especially with the cup design • Higher arc voltages during short circuit arcing (generator switching applications) • Adequate high current arc erosion endurance • Lower cost vacuum interrupter	• Not recommended for high voltage circuits > 38kV • Lower high current arc erosion endurance
AMF	• Excellent high current arc erosion endurance • Good high voltage performance up to 170kV • Required for long contact gaps used for high voltage performance (20–25mm for 250kV, BIL; 30–40mm for 350kV, BIL; and 60–80mm for 500–650kV, BIL)	• Requires bigger diameter contacts and larger diameter vacuum interrupters • Higher cost vacuum interrupters • More complicated contact structure • Higher closed contact impedance • Long arcing times result in more severe contact melting • Lower arc voltage (not recommended for generator breaker applications)

emerged since the late 1960s, as the dominant material for commercial vacuum interrupters. The reasons for this are:

1. It permits a much higher temperature to be used during the vacuum processing
2. It allows the use of a vacuum furnace exhaust and braze operation for batch vacuum interrupter production
3. With modern metallization techniques, it is easy to form very long-life vacuum tight braze seals with the metal components of the vacuum interrupter
4. The long-term permeation of He through the envelope is prevented and the permeation of H_2 is limited
5. Its use results in a robust vacuum interrupter structure that does not require delicate handling

Typical vacuum interrupter ceramics are cylinders manufactured from a mixture of alumina (about 94 wt%) and silica (about 6 wt%). The cylinders are formed in a "green" state from a paste of this mixture. They are then placed in an air furnace where they are fired at a high temperature. The resulting ceramic cylinder is ideal for the application, because of its mechanical strength, its low porosity, its low out-gassing, its ability to form vacuum-tight seals, and its excellent high voltage withstand characteristics. The finished ceramic cylinder is then usually coated with a glaze material and refired. The glazed surface permits easy handling and greatly facilitates the final cleaning of the finished vacuum interrupter product. In order to form a vacuum tight seal with the vacuum interrupter's metallic end cups, the end surfaces of the ceramics must first be metallized. This will then allow the metal end cap to be brazed to the ceramic with a hermetic, vacuum tight joint.

The metallization is a two-step process [192]. The ends of the ceramic are ground to a certain level of roughness. A paste of Mo–Mn (~ 10 wt%) with some SiO_2 particles in an organic binder is applied to these surfaces. This layer is dried in a hot air furnace. The ceramic is then placed in a furnace with a cracked ammonia atmosphere (a nitrogen–hydrogen gas with a well-controlled dew point). It is important to minimize the relative humidity, because water vapor can affect the high voltage withstand ability of the ceramic [193]. This firing removes the organic binder and allows the Mn to migrate into the body of the ceramic. Once there, it lowers the melting point of the small percentage of SiO_2 that resides in-between the alumna grains. The spaces between the comparatively small particles of Mo draw this molten glass into the body of the metallization through capillary action. When the ceramic is cooled, the glass solidifies and the Mo layer is tightly bonded to it. As a final step, the metallized layer is plated with Ni. It is now possible to braze metal parts to the ends of the ceramic cylinder and form vacuum tight seals; see Figure 3.68. I will briefly discuss this in Section 3.5, when I describe the manufacturing sequence.

The metallization process has proved to be highly successful and has allowed the production of vacuum interrupters that satisfy the "sealed for life" concept. The metallization does, however, add considerable cost to the production of the ceramic envelope: perhaps an increase of 15% to 40%, depending upon the ceramic's diameter. This has led to research on ways to braze the metal parts to the ceramic directly using a so-called "active braze" material. For example, this will permit the end caps of the vacuum interrupter to be sealed to the ceramic during the final exhaust and braze cycle (see Section 3.5.2) without the need for the Mo–Ni metallization layer. The braze alloy that shows some promise to do this is made from a mixture of Ag, Cu, and Ti. This material can be produced in the form of a solid foil and also in the form of a paste. The composition is Ag-rich, but depends upon manufacturer: e.g. Ag (~ 60 to 70 wt.%), Cu (~ 30 to 40 wt.%) and Ti (~ 1 to 3 wt.%). When melted (at about 800C) the Cu and Ti form a thin Cu–Ti–O layer close to the Al_2O_3 ceramic and which bonds to it. Above this thin layer there is a thicker layer of mixed Ag–Ti and Cu–Ti. Finally, there is an Ag–Cu layer that bonds to the metal part. Whether or not this active braze material can be incorporated into the manufacture of reliable vacuum interrupters remains to be seen. It is an interesting concept, but as of 2020 it has not been used in the production of commercial vacuum interrupters.

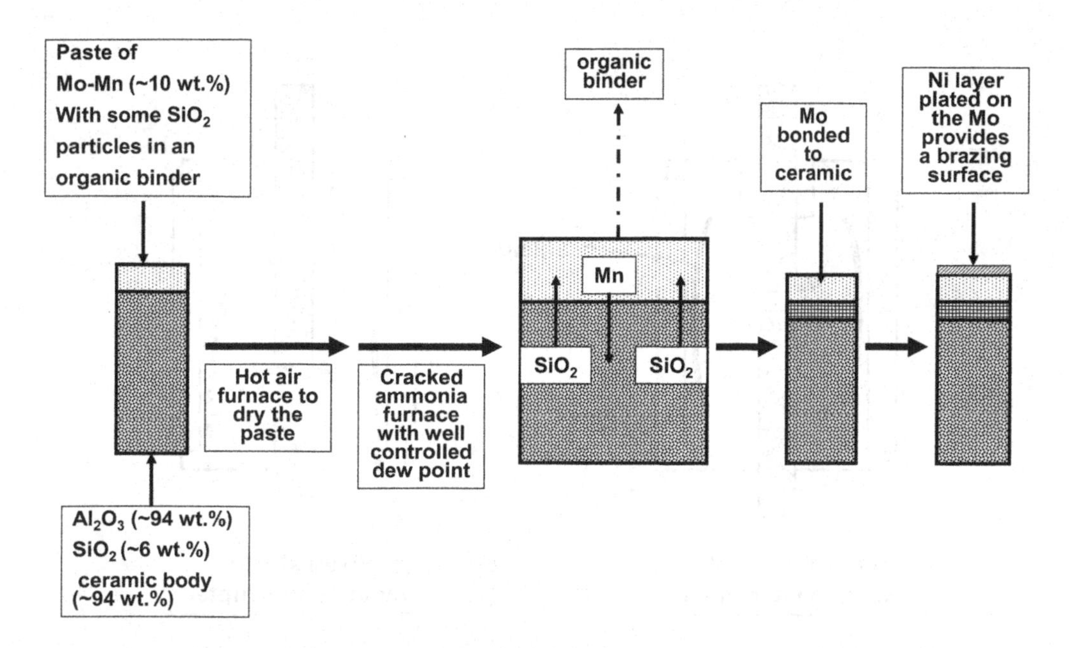

FIGURE 3.68 The metallization process to permit brazing of the metal end caps to the ceramic cylinders.

3.4.2 THE SHIELD

As I have presented in Section 1.4.1 in this volume, the shield is an essential component of the vacuum interrupter's design. It has multiple purposes:

1. To capture metal vapor from the vacuum arc and thus maintain the internal insulation of the ceramic cylinders
2. To shape the voltage distribution inside and outside the VI and to maintain this distribution throughout the vacuum interrupter's life
3. To protect the ceramic from the radiation and thermal energy produced by the vacuum arc. In extreme cases, to protect the ceramic from direct contact with the high current, columnar, vacuum arc

The shield is either electrically floating (i.e., it is not attached to either end of the vacuum interrupter, see Figure 3.69(a) [194]) or it is fixed (i.e., it is attached to one end of the vacuum interrupter, usually the end with the fixed terminal, see Figure 3.69(b)). Figure 3.70 shows four possible methods of attachment for the floating shield inside the ceramic cylinder. Figure 3.70(a) shows a flange attached to the shield, sandwiched between two ceramics and brazed in place during the manufacture of the vacuum interrupter. Second, Figure 3.70(b) [195] shows a flange attached to the shield captured in a groove in the ceramic's inner wall. Third, Figure 3.70(c) [196] shows the shield supported on a raised portion in the ceramic's inner wall. This can be achieved in a number of ways: e.g., it can be done by using a two-piece shield, or it can be done by shaping a soft metal-like Cu over the raised portion in the ceramic once the shield is in place. Some manufacturers even metallize the raised portion so that they can braze the shield to it. A fourth method is to make the floating shield part of the vacuum interrupter's envelope. I will call this a "belly-band" vacuum interrupter. An example is shown in Figure 3.70(d). This shield is manufactured from oxygen-free Cu.

Most manufacturers use stainless-steel or oxygen-free Cu for the shield material. Stainless steel is a very good high voltage material, but is unforgiving if exposed to too much energy from the high current vacuum arc. Copper has very good heat sinking properties and resists the effects of the high current vacuum arc better than does stainless steel. Stainless-steel has a low thermal conductivity

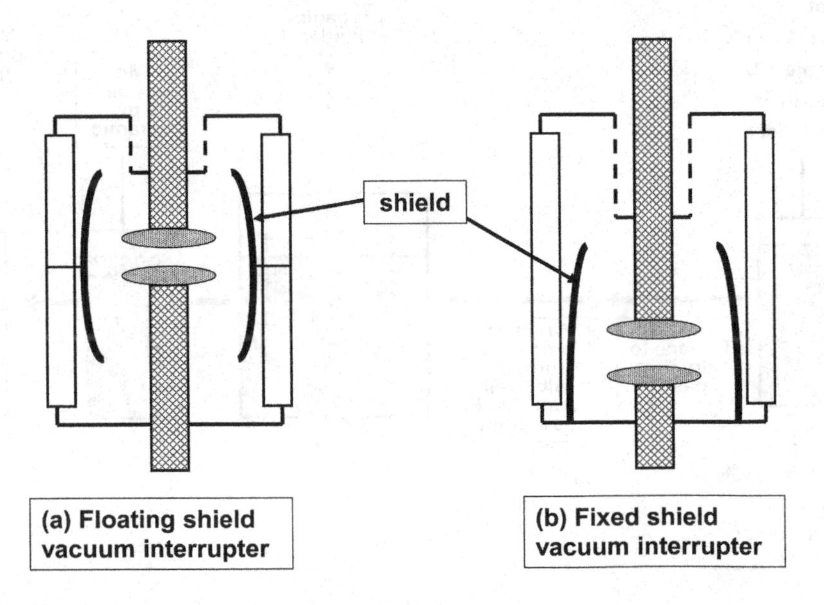

FIGURE 3.69 The floating shield and the fixed shield vacuum interrupters.

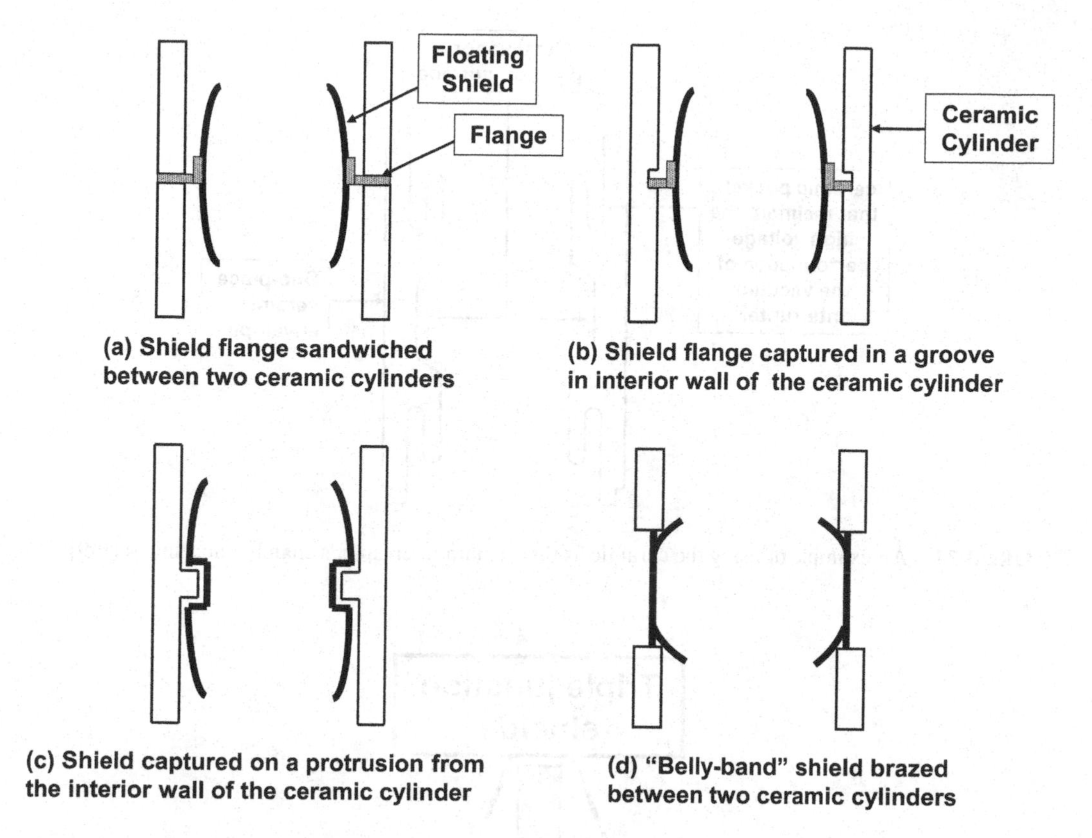
Floating Shield
Flange
Ceramic Cylinder
(a) Shield flange sandwiched between two ceramic cylinders
(b) Shield flange captured in a groove in interior wall of the ceramic cylinder
(c) Shield captured on a protrusion from the interior wall of the ceramic cylinder
(d) "Belly-band" shield brazed between two ceramic cylinders

FIGURE 3.70 Examples of attaching the floating shield to the ceramic cylinders [194, 195, 196].

and is therefore not able to dissipate the thermal energy from the high current vacuum arc as well as Cu. It thus will form "hot spots" that will melt the material and even create holes in the shield. Stainless-steel shields are more commonly used with AMF contacts where the vacuum arc tends to be confined to a region between the contacts.

Shield designs have been proposed with a Cu center portion facing the vacuum arc and stainless-steel end curls for high voltage withstand [197]. Some manufacturers use cylinders of Cu–Cr as a shield material with stainless-steel end curls [198]. The Cu–Cr material also resists the effects of the high current vacuum arc, but is expensive to manufacture. Falkingham has even proposed using the ceramic itself as the shield [199]. Figure 3.71 shows one way of achieving this. The exposed ceramic opposite the contacts gradually becomes coated with metal from the vacuum arc, but the internal high voltage withstand ability is maintained from the protection provided by the pocket regions shown in Figure 3.71. This concept, although interesting, has found only limited use.

The other shielding required in the vacuum interrupter are the end shields. These are shown schematically in Figure 3.72. They are positioned to produce a low field or even a field free region at the triple junction of the end plate, the ceramic and the vacuum. The need for shielding the triple junctions has been presented by Del Rio Etaya [200] and Venna et al. [201]. The design considerations for the end shields with respect to the center shield has been discussed in Section 1.4.1 and Figure 1.108 in this volume.

The floating shield is incorporated in vacuum interrupters used in circuits with a voltage of 12kV or higher. The fixed shield design is usually used on vacuum interrupters for circuits at or below 7.2kV. There are some uses of the fixed shield vacuum interrupter at 12kV/15kV that can be used for interruption conditions that do not have to experience excessive recovery voltages or where the manufacturer wishes to use an external coil to supply an axial magnetic field [126, 127]. By using

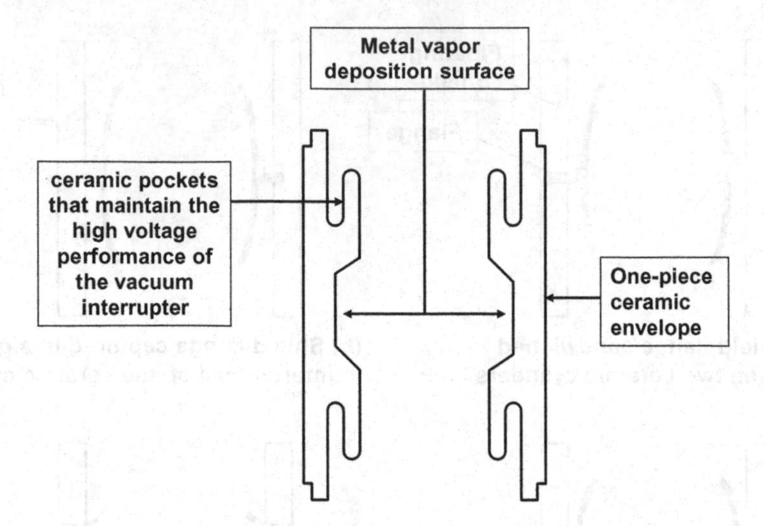

FIGURE 3.71 An example of using the ceramic as the vacuum interrupter's chamber and shield [199].

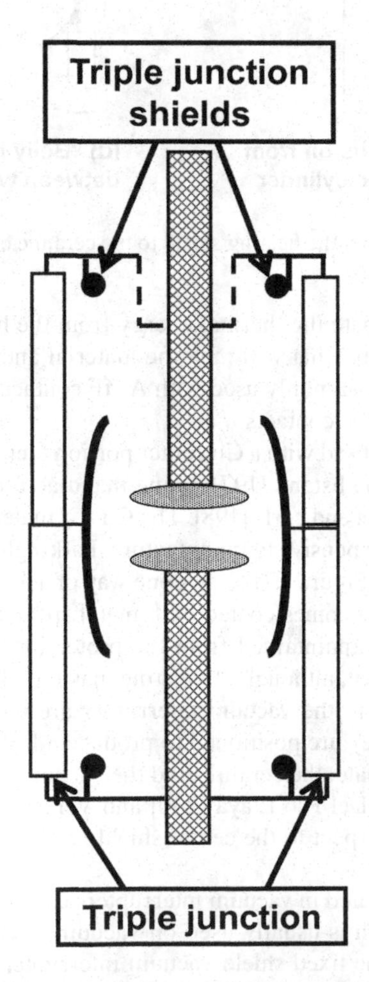

FIGURE 3.72 The end shields that provide low field or field free regions at the vacuum interrupter's triple junctions.

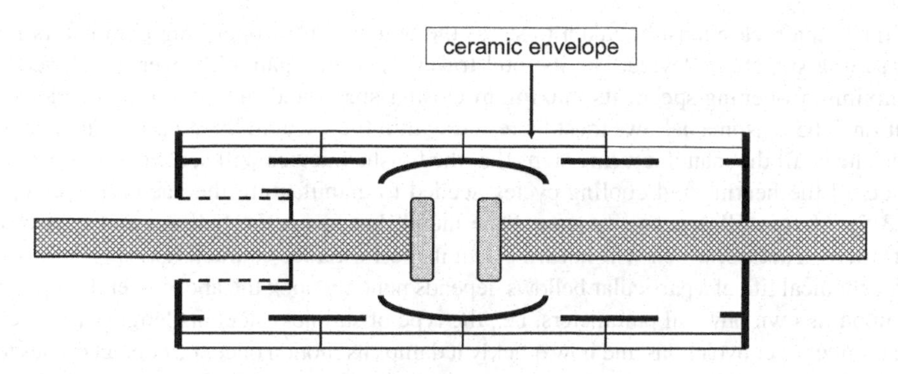

FIGURE 3.73 A schematic example of a three-shield vacuum interrupter for BIL voltages greater than 250kV [203].

a floating shield, it is possible to design vacuum interrupters with BIL requirements for circuit voltages up to 170kV [148, 202]. In order to distribute the voltage gradient more evenly across the exterior of the ceramic envelope, three floating shields have been suggested [203–206], see Figure 3.73. It is possible, however, to design vacuum interrupters with a single floating shield for voltages 72kV and 145kV [202]

3.4.3 The Bellows

The bellows permits the motion of the moving contact while maintaining a vacuum tight seal. The bellows is made from stainless steel and usually has a thickness of about 150μm. Three styles of bellows have been successfully applied in vacuum interrupter design:

(1) Seamless – hydro-formed bellows
(2) Seam-welded – hydro-formed bellows
(3) Bellows manufactured from edge welded, thin stainless steel, washers

General and detailed information on bellows design and performance can be found in the EJMA Standards [207]. As can be seen from Figure 3.2, one end of the bellows is fixed, being brazed to the end plate of the vacuum interrupter. The other end is brazed to the moving terminal and moves with that terminal as the contacts are opened and closed. In a vacuum interrupter the bellows is subjected to an impulsive motion as the contacts are made to open and close: the opening speed of the moving contact can go from zero ms^{-1} to as much as 2 ms^{-1} in a time of less than 100μs. At the end of the contact stroke, the moving end of the bellows suddenly stops. It also comes to a sudden stop when the contacts close. For some duty cycles, this open and close operation can occur many times, but in others it will occur only on rare occasions. The motion imparted to the bellows is thus anything but uniform and it is quite possible for the bellows to oscillate many times during one open or one close operation. For those readers interested in an analysis of this bellows motion, Barkan [208] has developed a general analytical approach for determining the dynamic stresses that can be experienced by a bellows operating under impulsive motion.

Most vacuum interrupter manufacturers purchase their bellows from reputable bellows manufacturers and work with them to obtain the required bellows life. This is usually done by building the bellows into a practical vacuum interrupter and performing a mechanical life test on a statistically significant number of vacuum interrupter samples. A specified mechanical life can then be assigned to the vacuum interrupter with that bellows by using the Weibull analysis described in Section 1.3.1 in this volume. Usually the limit of a vacuum interrupter's mechanical life is the number of operations the bellows can perform before it experiences fatigue failure. When mechanically testing a vacuum interrupter, care must be taken to stress the bellows at the operating parameters it will be exposed to in a switching device, i.e., its total travel (operating gap, plus over travel, see Chapter 6), its maximum opening speed, its maximum closing speed and the effects of acceleration and deceleration. The reason a bellows must be tested within the vacuum interrupter is to ensure that it has experienced all the manufacturing steps that the finished device will see. For example, it should experience all the heating and cooling cycles needed to manufacture the vacuum interrupter (see Section 3.5). These will necessarily anneal the metal that forms the bellows and thus change its granular microstructure, which will in turn affect its performance characteristics.

The mechanical life of a particular bellows depends not only upon the above operating parameters, but also upon its own physical parameters, i.e., the type of stainless steel, its length, diameter, thickness, the number of convolutions and how quickly it damps its motion once the contact ceases to move. It is possible to design bellows that will perform reliably for the normal 30,000 operations required for most vacuum circuit breakers and vacuum reclosers, and in excess of 10^6 operations required for vacuum contactors. While vacuum interrupter manufacturers have been diligent in designing their vacuum interrupters to meet the specified mechanical life of the various switching devices, most of the vacuum interrupters never see their stated mechanical life when placed in the field.

The vacuum interrupter designer has to be careful that the user cannot twist the bellows when the vacuum interrupter is placed into a mechanism. The mechanical life of a twisted bellows is severely reduced; it can be less than 1% of its design life. The torque that can be applied to the thin walled bellows used in a vacuum interrupter before it is permanently twisted is quite low, perhaps about 8.5–11.5mN. In order to prevent the bellows from becoming twisted the designer must place an antitwisting bushing into it. This bushing can be locked in place by a suitable attachment to the end plate of the interrupter. The inner surface of the bushing is shaped or has a key-way that prevents any rotation of the moving, Cu terminal that is attached to the bellows; see Figure 3.74. The bushing material can be metal or it can be a plastic material such as Nylatron. Care must be exercised when using plastic materials such as Nylatron and Valox. These materials can only be used in applications

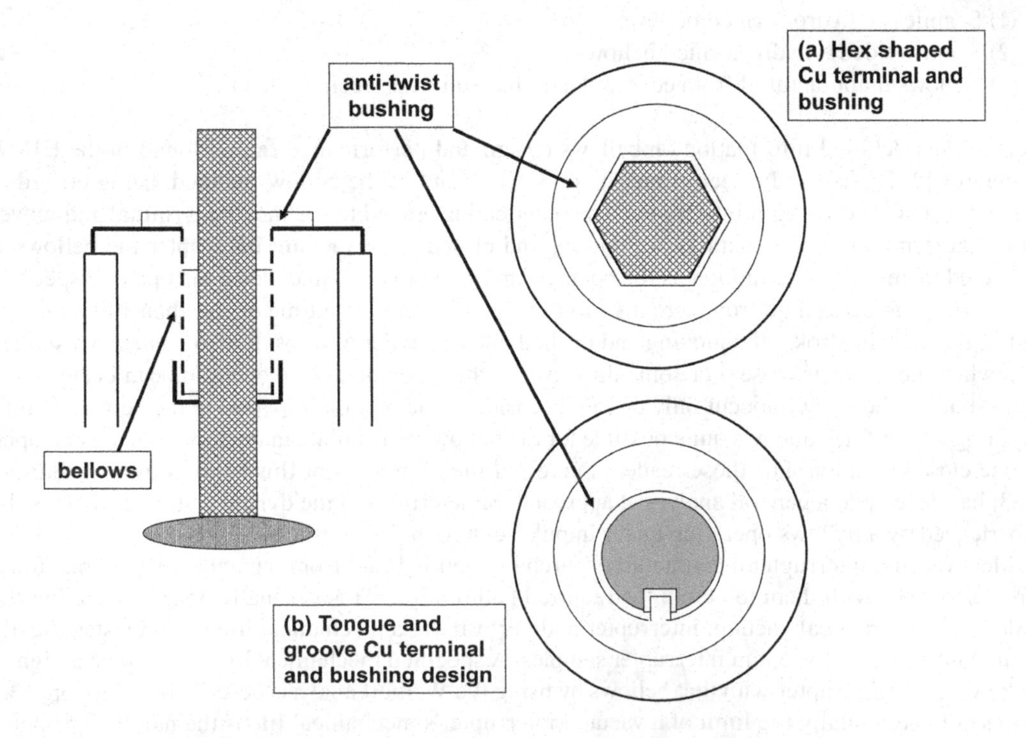

FIGURE 3.74 Examples of antitwist bushings to protect the bellows from damage.

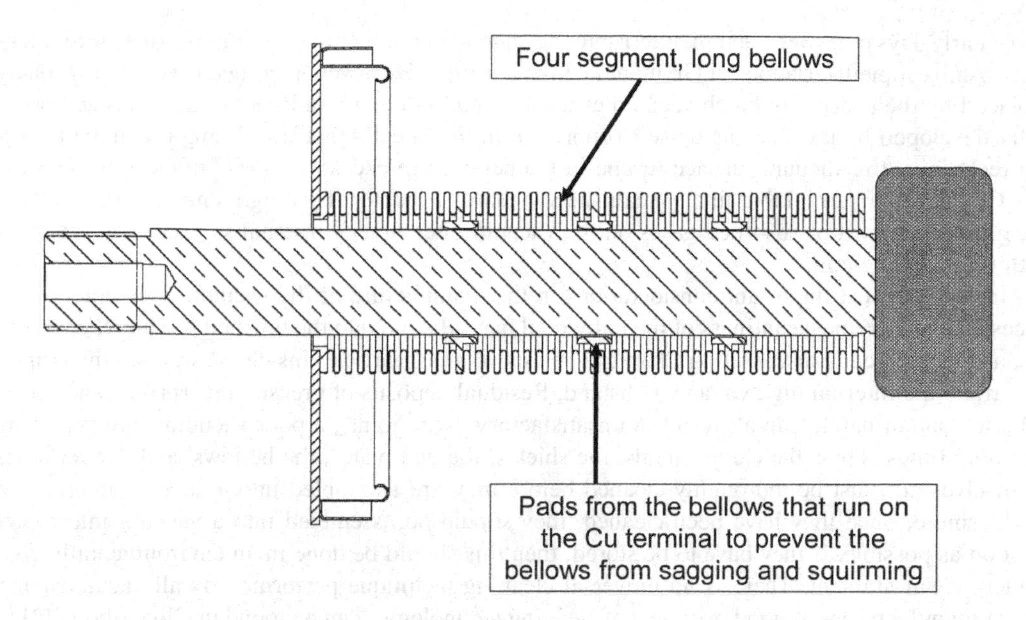

FIGURE 3.75 An example of a long bellows with pads that ride on the Cu terminal to prevent the bellows from sagging or squirming [210].

where there is a limit to the maximum permissible temperature the bushing material will experience (i.e., the temperature at which the tensional strength will be reduced to 50% after 100,000 hours). For Nylatron, this will occur at about 125°C (it does, however withstand a higher temperature for short periods without deformation, because of its glass fibre content); for Valox DR48 it occurs at about 140°C. There are other higher temperature and more expensive plastics such as "Ultem 2310 R." For this material the maximum permissible temperature is about 180°C, but this temperature can be exceeded for a short time (about 1 hour) without serious deformation occurring.

For vacuum interrupters that are used at higher circuit voltages longer strokes are required, e.g., at 72kV and 126kV strokes of about 40mm and 60mm respectively. To achieve this stroke the bellows must become correspondingly longer. A very long bellows will not close and open uniformly, but will squirm as it moves: See Sallappan et al. [209]. The squirm is caused by the circumferential membrane stress placed on the bellows by the opening and closing of the vacuum interrupter. When it does this the inner convolutions may well rub against the Cu terminal. This can result in considerable reduction to the bellows' life. In order to prevent this, bellows have been developed with internal pads that ride along the Cu terminals; an example is shown in Figure 3.75. Tsutsumi et al. [210] have used bellows of this type for strokes up to 60mm.

3.5 VACUUM INTERRUPTER MANUFACTURE

3.5.1 Assembly

The earliest manufacturing technique developed by the General Electric Corporation (GE) for power vacuum interrupters for circuit breaker use is well described by Lafferty [211] and Greenwood [212]. Initially, subassemblies of the fixed end and the moving end are brazed in a hydrogen furnace. These ends, the shield, and the insulating outer cylinder are then assembled and placed in a hydrogen furnace. The finished vacuum interrupter would have a tube attached to one of its end plates. This tube is attached to a vacuum system and the vacuum is then created inside the vacuum interrupter. The temperature of the vacuum interrupter is usually elevated to about 400°C while it is being pumped. Once a suitable vacuum level is achieved inside the vacuum interrupter the tubulation is sealed.

In the early days of power, vacuum interrupter development number of corporations such as Siemens and Toshiba copied variations of GE manufacturing system. However, it has been almost universally replaced by the process of batch vacuum evacuation and brazing in a large vacuum furnace originally developed by the Westinghouse Corporation in the late 1960s [213]. Using this manufacturing technique, the vacuum furnace reaches a temperature in excess of 800°C in order to melt the Ag–Cu brazing material that finally seals the vacuum interrupter. This high temperature results in the greater removal of adsorbed gases on the vacuum interrupter's internal surfaces and creates a better-quality vacuum.

One of the most important considerations in the manufacture of the vacuum interrupter is the necessity for extreme cleanliness of the individual parts that go into the finished product. For example, as I have already discussed in Chapter 1 in this volume, particles inside the vacuum interrupter can affect the internal high voltage withstand. Residual deposits of grease, fingerprints, and atmospheric contamination can all result in unsatisfactory braze joints, a poor vacuum or in very long pumping times. Thus, the Cu terminals, the shields, the end plates, the bellows, and the ceramics themselves all must be thoroughly cleaned before they are assembled into a vacuum interrupter. Furthermore, once they have been cleaned, they should be assembled into a vacuum interrupter as soon as possible. If they have to be stored, then this should be done in an environmentally controlled, clean ambient. There is no universal cleaning technique performed by all vacuum interrupter manufacturers. A good overview of cleaning technologies can be found in Cline's book [214]. Cleaning techniques that are used include: aqueous cleaning, bright dipping Cu in an acid solution, electropolishing (mostly of stainless steel and Cu parts), ultrasonic cleaning in a liquid bath (Freon, alcohol, TCE, and even deionized water are being used). Once the parts have been cleaned, they must *never* again be touched by a bare hand.

All the assembly stages of the vacuum interrupter are performed in a clean room [215]. This room is not only environmentally controlled for temperature and humidity, but is also controlled for the particulate make-up of the ambient air. Figure 3.76 illustrates a typical clean room structure for manufacturing vacuum interrupters. Clean rooms are given a classification according to the Federal Standard 209; see Table 3.13. For vacuum interrupter assembly, most manufacturers like to claim their clean room is Class 1000, but in reality, at the work bench the Class is somewhere between Class 1000 and Class10,000. Since 1970 this has proved to be a quite satisfactory level for the

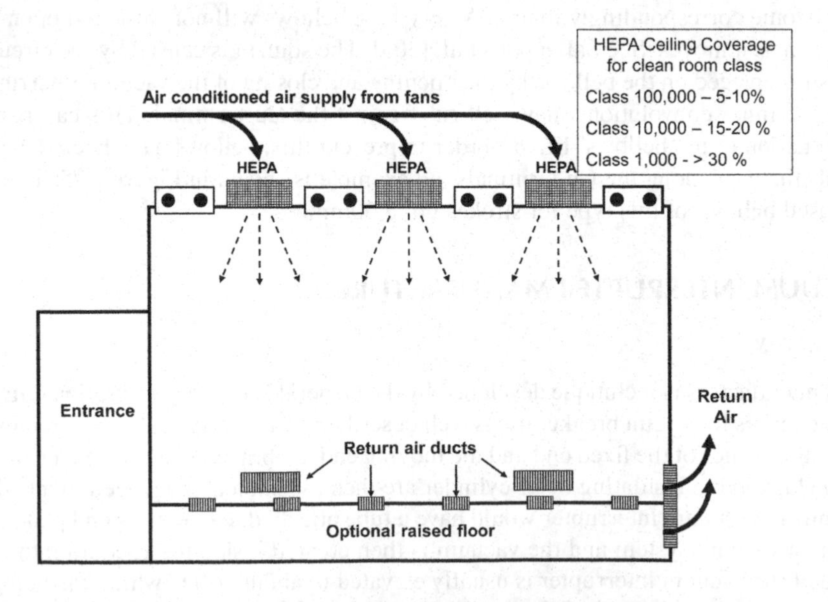

FIGURE 3.76 The clean room for manufacturing vacuum interrupters.

TABLE 3.13
Clean Room Classifications

Clean–Room Class	Number of Particles (≥ 0.5μm) per Cubic Foot (~ 0.028 m³)	Typical Tasks
1	1	Integrated circuits with submicron geometries
10	10	Integrated circuits with line widths 1μm < w< 2μm
100	100	Manufacture of bacteria: particulate free injectable medicines; integrated circuit manufacture; transplant operations
1,000	1,000	High-quality optical equipment; precision gyroscopes; miniature bearings and vacuum interrupter manufacture
10,000	10,000	Vacuum interrupter manufacture; precision hydraulic equipment; high-grade gearing
100,000	100,000	General optical work; assembly of electronic equipment; hydraulic and pneumatic equipment assembly

manufacture of high-quality vacuum interrupters. One of the reasons for the lowering of the clean-room class is the introduction of particulates by the workforce that has to assemble the vacuum interrupters and by the general activity of manufacturing; see Figure 3.77. Table 3.14 presents some examples of how this activity can affect the clean room ambient. In the typical vacuum interrupter assembly clean room, the personnel are required to wear special clean room smocks, caps, boots and gloves. Some manufacturers also require the use of facemasks that cover the mouth and nose; this in my experience is somewhat overly cautious.

There are two assembly techniques. The first is to manufacture the sub-assemblies in a vacuum or hydrogen furnace and then assemble the complete vacuum interrupter for brazing and sealing in a vacuum furnace. Naik [216] gives an excellent introduction to vacuum technology including vacuum pumping, vacuum furnaces, cleaning, vacuum brazing, etc. The second is to do the whole assembly in one step. There are, of course, variations of these assembly techniques where parts of the fixed and moving ends are put together (for example, the terminal and the contact), and then the whole vacuum interrupter is assembled. Figure 3.78 shows examples of the fixed contact

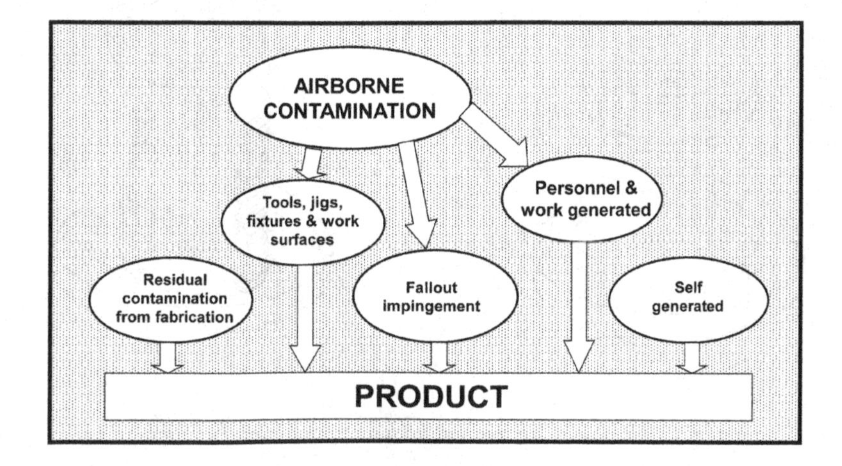

FIGURE 3.77 Activities that can reduce the clean-room classification.

TABLE 3.14
The Effect of Activity in a Clean Room

Activity	Times increase over the ambient particle levels (0.2 μm to 50 μm)
Personnel Movement	
Three of four people in one location	1.5 to 3
Normal walking	1.2 to 2
Sitting quietly	1 to 1.2
Personnel Protective Clothing	
Brushing sleeves	1.5 to 3
Stamping on floor with protective shoe covering	1.5 to 3
Removing handkerchief from pocket	3 to 10
Personnel–Other	
Breath of a smoker	2 to 5
Sneezing	5 to 20
Rubbing skin on hands or face	1 to 2
Other Sources of Particles	Typical Particle Size
Rubbing painted surfaces	90 μm
Sliding metal surfaces (nonlubricated)	75 μm
Crumpling paper	65 μm
Tightening screws and/or bolts	30 μm
Writing with a ballpoint pen on paper	20 μm
Handling passivated metals	10 μm
Rubbing skin	4 μm

sub-assembly and the moving contact subassembly. These are placed on plates and loaded into a vacuum furnace or into a hydrogen furnace. Once established inside the furnace, internal heaters begin to raise the temperature of the assemblies. This temperature is continually raised until the braze material placed between the components melts [217]. The furnace temperature is then lowered until the furnace cools to room temperature. The rate of temperature increase and the rate of cooling are usually a closely guarded manufacturing secret: each manufacturer determines it from experience. It is affected by the out-gassing rate of the components, the braze material's melting

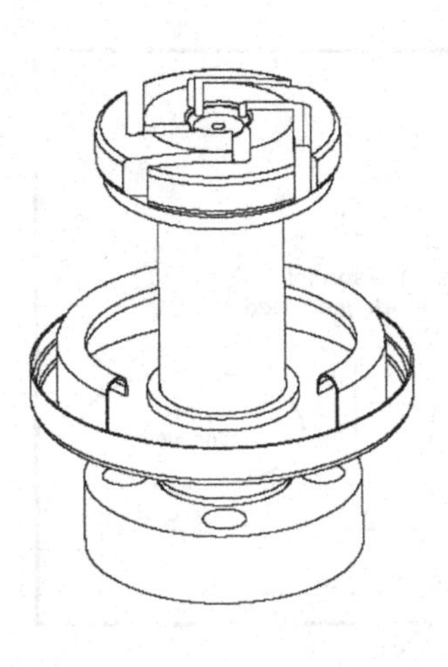
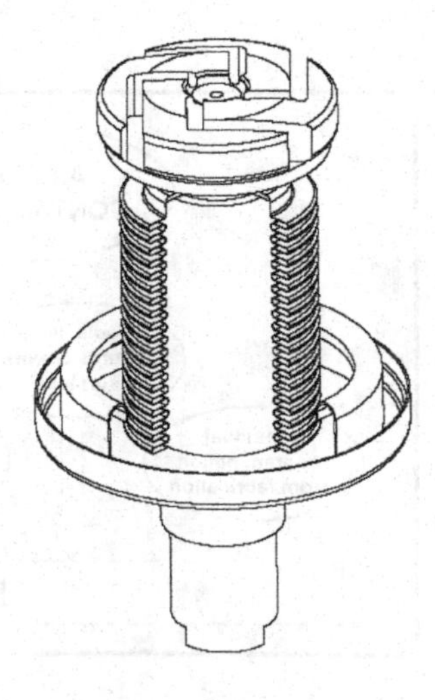

FIGURE 3.78 The fixed and movable subassemblies.

temperature and the necessity to ensure that all the components reach the final temperature at the same time. The subassemblies are then leak checked with a helium leak checker [218, 219] that can detect leaks down to about 3×10^{-11} standard cubic centimeters per second.

The final assembly is shown schematically in Figure 3.79. Here the sub-assemblies, the shield, and the ceramics are placed together with a suitable braze material between them. The final assembly requires either self-alignment of the components or external fixturing. Individual manufacturers have developed fixturing schemes best suited to their manufacturing production. The braze material most used is a Ag–Cu(28wt.%), which has a liquidus temperature of 779C. Hilderbrandt et al. [220] have developed a Cu–Ag(40(wt.%)–Ga(10wt.%) braze material whose solidus temperature is 726°C and its solidus temperature is 831°C. They show that this material wets stainless steel and forms a good vacuum seal between it and the ceramic body. It also has a cost advantage over the Ag–Cu braze material, because it uses less Ag. The braze material chosen for the final assembly should have a melting temperature that is lower than that used in the sub-assembly manufacture. During the final assembly a non-evaporable getter (NEG) [221] is attached to a suitable surface inside the vacuum interrupter. This surface is chosen to be as far away as possible from the contacts and the vacuum arc that forms when the contacts part.

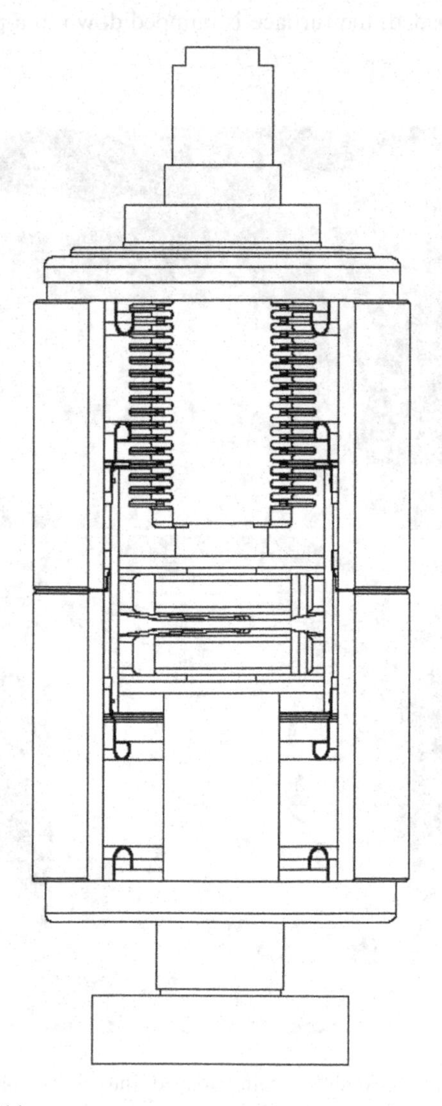

FIGURE 3.79 The final assembly.

Thus, the NEG is usually found somewhere in the vicinity of an end plate or seal cup. The purpose of the NEG is to provide a continuous pumping of a low level of H_2; hydrogen being the only gas that readily diffuses through the ceramic and metal walls of the vacuum interrupter. The H_2 is not only available from the ambient atmosphere, but is also available from the dissociation of H_2O molecules. This can occur in the presence of an active metal such as Cu or Ag when it oxidizes and leaves hydrogen close to the vacuum interrupter's surface. The getter also adsorbs nitrogen. Nitrogen and hydrogen can also be released by the out-gassing of the internal components inside the vacuum interrupter once it has been sealed. Hydrogen can also result from the dissociation of C_xH_y, a residual gas frequently observed inside sealed vacuum interrupters. The C_xH_y will be dissociated in the vacuum arc releasing the H_2. The NEG is a deposit of a material such as Zr–Al, Zr–V or Zr–V–Fe onto a strip of stainless steel. A deposit in a powdered form can have a large surface area as a result of many internal paths and voids to which the gettered gas can attach. Actual surface areas as high as 70–100 cm^2 of deposited area can be achieved. Getters must be activated at an elevated temperature. Care must be taken not to activate them before a good vacuum is achieved inside the vacuum interrupter. If they are heated in the presence of even a few millibars of nitrogen, they will lose some of their gettering capability.

The final assemblies are placed on a furnace plate and loaded into the vacuum furnace, see Figure 3.80. Once they have been loaded, the furnace is pumped down to a pressure below 10^{-4} Pa. The

FIGURE 3.80 A batch of final assemblies being loaded into a vacuum furnace (courtesy of Eaton Corporation).

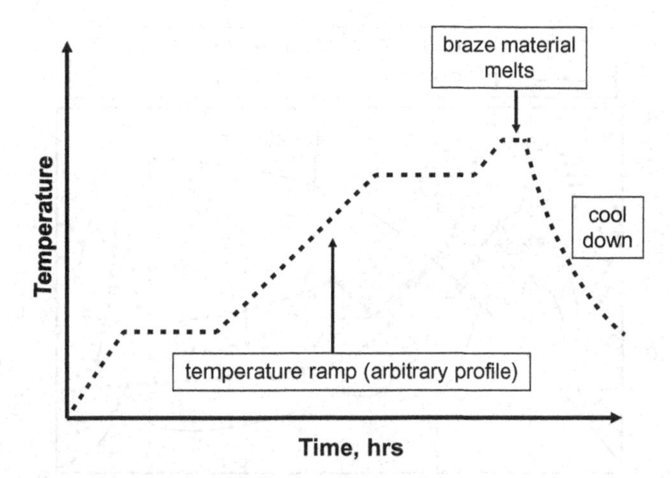

FIGURE 3.81 An example of a vacuum furnace temperature – time ramp for the final vacuum bake and seal of vacuum interrupter batch.

internal heaters are placed close to the polished faces of the furnace in order to reflect the radiation. The temperature of the furnace and of the assemblies is raised at a specific rate. Figure 3.81 gives a schematic example of a furnace temperature/time cycle. The plateaus in this example are required for temperature regimes where copious out-gassing may occur as well as to allow time for the entire vacuum interrupter batch to reach a given temperature. The total time of each brazing cycle depends upon the load size, the out-gassing rates, and the braze material's melting temperature.

When loading the vacuum furnace, it is important to ensure that all the assembled vacuum interrupters reach the final temperature at the same time. Aranaga et al. [222] have demonstrated that it is possible to simulate the heating cycle for a sample vacuum interrupter load and show correspondence to the measured temperatures for an actual furnace load. Figure 3.82 shows their measured and simulated temperature rise as a function of time compared to the profile set by the furnace operator. The simulation and the measured temperature give lower values up to about 400°C, but

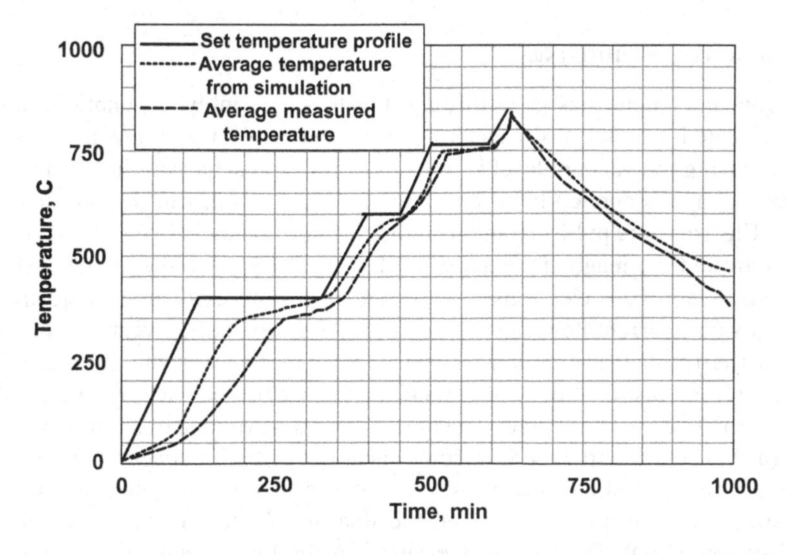

FIGURE 3.82 A Comparison of the set temperature profile, the simulation profile, and the measured profile as a function of time [222]

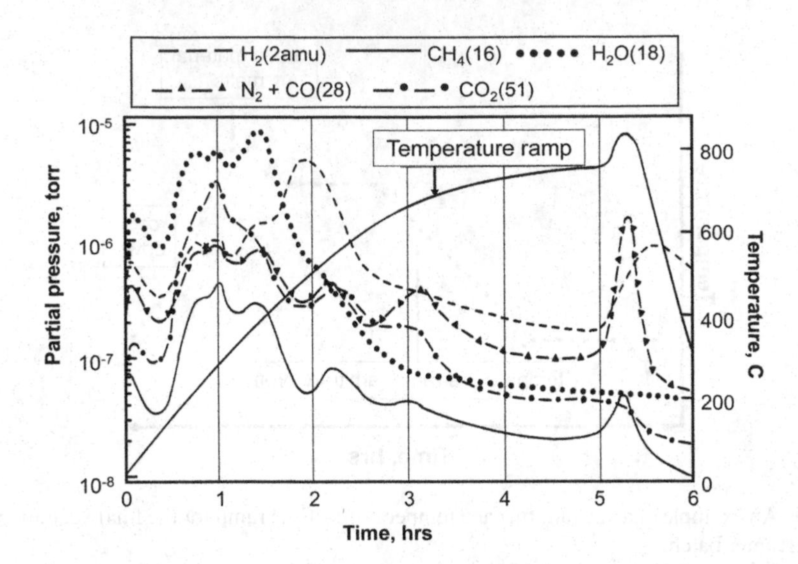

FIGURE 3.83 An example of the gas species from a vacuum interrupter as a function of temperature inside a vacuum furnace [223].

then follow the set profile until at the braze temperature they are all the same. This model provides a useful tool for optimizing loading of the vacuum furnace. Figure 3.83 gives one example of the gaseous species detected from a vacuum interrupter as the temperature is raised [223]. The temperature is increased to the melting point of the braze material (e.g., if it is Ag–Cu, the final temperature greater than 279°C). Once the braze has melted, the furnace begins its cool-down phase. As the braze cools and solidifies the components are joined with a vacuum tight seal and a good vacuum is trapped inside the vacuum interrupter. In the single step process, all the components are assembled at one time. These are loaded into the furnace only once for the final evacuation and braze. By using this technique, the total time in the vacuum furnace is reduced. The disadvantage is that more time is required for the assembly process and there can be no leak check of the subassembly joints before the final vacuum interrupter is manufactured.

3.5.2 Testing and Conditioning

Once the vacuum interrupter has gone through the final assembly, evacuation and braze processes, the completed product has a vacuum lower than 10^{-2} Pa sealed inside it. The usual method for measuring the vacuum level is to use a Penning discharge; see Figure 3.84(a) [224]. Here a high dc voltage is applied between two suitable electrodes in the presence of an axial magnetic field. Electrons are pulled from the cathode electrode by field emission, and then spiral back and forth around the magnetic flux lines. Their travel path to the anode is thus considerably lengthened. When these electrons collide with a residual gas atom/molecule, ionization can occur and a small current flow can be measured. The level of current measured is directly proportional to the residual gas pressure. One way of performing this measurement using a vacuum interrupter is illustrated in Figure 3.84(b) [225, 226]. The vacuum interrupter is placed inside a magnetic field coil. The contacts are opened and a dc voltage of about 10kV is impressed across them. In this case the trapped electron swarm is captured between the upper contact (in this example, the anode) and the floating shield; i.e., an inverted magnetron discharge (IMD). An electron swarm is also trapped between the floating shield and the lower cathode; i.e., a magnetron discharge (MD). The current measured in the high voltage circuit thus flows from the anode to the shield and then to the cathode. A typical current profile after the application of

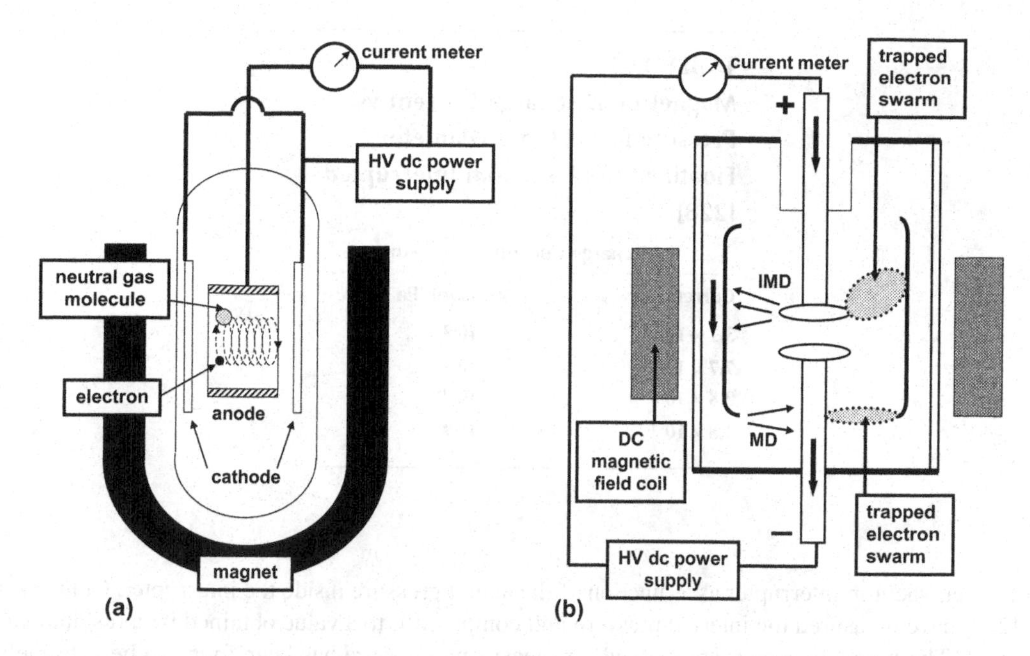

FIGURE 3.84 (a) The Penning discharge and (b) the magnetron pressure measurement technique for vacuum interrupters [223, 224].

the magnetic field and the voltage is shown in Figure 3.85. The current is related to the pressure inside the vacuum interrupter by [227]:

$$I_{(peak)} = k \times (U - U_e) \times P^a \tag{3.36}$$

Where "$i_{(peak)}$" is the peak value of the discharge current, "U" is the applied voltage "U_e" is the extinction voltage, "P" is the pressure, and $a \approx 1$. Typical values of the pressure for a given peak current in a 60mm diameter, floating shield vacuum interrupter are given in Table 3.15 [228]. Once the current pulse reaches its maximum value, the Penning discharge pumps the vacuum interrupter to a lower pressure [229]. Of course, the actual current/pressure relationship for a particular vacuum interrupter style depends upon its internal geometry. Thus, before the magnetron method can be used routinely as a manufacturing test device, it has to be calibrated for each vacuum interrupter style.

One way of performing this calibration is to add a tubulation to a vacuum interrupter and then attach it to a pressure gauge and to a vacuum system. The magnetron current can then be measured

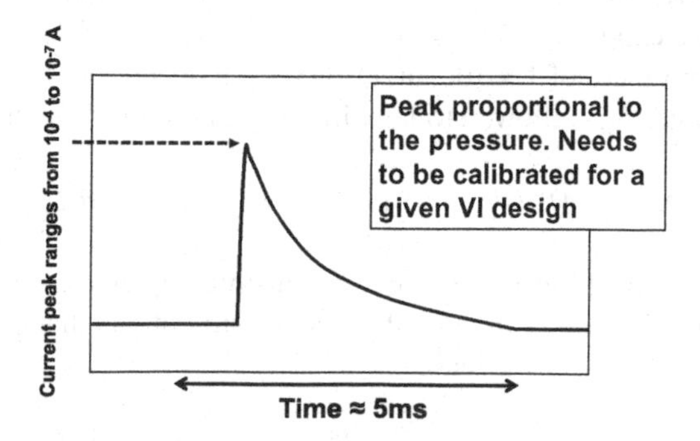

FIGURE 3.85 Typical vacuum interrupter's Magnetron current trace.

TABLE 3.15
Magnetron Discharge Current vs. Pressure for a 60mm Diameter Floating Shield Vacuum Interrupted [228]

Discharge Current vs. Pressure	
Current, A	**Pressure, Pa**
3.1×10^{-5}	10^{-2}
7.7×10^{-6}	10^{-3}
3.8×10^{-6}	10^{-4}
1.8×10^{-7}	10^{-5}

for a given vacuum interrupter as a function of the actual pressure inside the interrupter. Gentsch et al. [229] have measured the internal pressure and compared it to a value obtained by a residual gas analysis. The magnetron pressure method for measuring pressure has been found to be extremely reliable and is easily incorporated into the vacuum interrupter manufacturing process.

Over the years there have been many suggestions for measuring the pressure level inside the sealed vacuum interrupter both during its manufacture and while inside a mechanism in the field. Examples of some of these can be found in references [230–239]. However, none of these, except the Penning and magnetron methods described above and an ac high voltage withstand test, have found practical application.

Unlike other enclosed interrupting systems such as SF_6 puffer interrupters, the vacuum interrupter must be constructed to have a zero-leak rate. There have been some publications [240] that discuss a leak rate to give a 30-year vacuum life (i.e., reach an internal pressure of 10^{-1} Pa). However, these reports do not take into consideration the unique nature of vacuum seals and vacuum leaks in an external ambient such as atmospheric air. What really matters is determining what size leak can be tolerated. No one really knows the true length of time that commercial vacuum interrupters can retain a vacuum below 10^{-1} Pa. In my own experience with vacuum interrupters manufactured by Eaton Corporation (formerly Westinghouse Corporation) the field experience goes back to 1968. As far as I know, the vacuum interrupters we manufactured in 1968 are still operating reliably. In fact, an analysis in 1998 of a Westinghouse vacuum interrupter that had been in the field since 1970, showed an internal pressure of 8.7×10^{-4} Pa. Reeves et al. [241] show a similar residual pressure in vacuum interrupters manufactured 50 years before being retested. A 30- to 50-year life expectancy does impose a very strict requirement on the leak-tightness of the vacuum interrupter envelope and all the braze joints that are exposed to the ambient. If we conservatively assume a maximum allowable vacuum interrupter pressure of 1×10^{-3} mbar or 10^{-1}Pa (a commonly accepted criterion), then the maximum amount of gas N_{max} allowed to leak into a vacuum interrupter of volume V_{int} is:

$$N_{max} = \Delta P \times V_{int} = \left(0.1 - P_0\right) \times V_{int} \approx 0.1 V_{int} Pa.L_t \tag{3.37}$$

where P_0 is the initial vacuum level inside the vacuum interrupter after final assembly, which is typically orders of magnitude smaller than 0.1Pa. The amount of gas that can leak into the vacuum interrupter depends upon the size of the leak.

$$N_{max} = \int^{30yrs} Q dt \tag{3.38}$$

Where Q is the leak rate. The Q_{max} can be calculated for a given time t from the following expression [223]:

$$Q_{max} = \frac{(\Delta P) \times V_{int}}{t} = \frac{\left[10^{-1} - P_0\right]}{t} \text{Pa.L}_t.\text{s}^{-1} \quad (3.38)$$

Using Equation (3.38) it is possible to calculate the Q_{max} for a 30-year life:

$$Q_{max} = N_{max} / 30 years = 0.1\ X\ V_{int} / 30 \times 365 \times 24 \times 3600\ Pa.L_t.s^{-1} \quad (3.40)$$

Table 3.16 lists the maximum leak rate allowed for the internal pressure of vacuum interrupters with volumes ranging from 0.5 to 4 Liters (L_t) to reach 10^{-1} Pa as a function of time. Thus, to obtain a 30-year lifetime, the Q_{max} allowed is about 10^{-10} Pa. $L_t.s^{-1}$. This is one or two orders of magnitude beyond the detection limit of the typical helium leak detector [242, 219]. In fact, as stated above, the subassembly braze joints are leak checked with the helium leak detector for their leak tightness down to a level of about 5×10^{-8} Pa. L_t s^{-1}. Table 3.16 shows that leak rates of this value would only guarantee a lifetime of about one-month. Why then do vacuum interrupter manufacturers experience lifetimes of their product in excess of 30 years? Fortunately, nature works in their favor. Leak rates of about 1×10^{-6} to 1×10^{-8} Pa L_t s^{-1} may require a bake out to become detectable. Leak rates of about 1×10^{-8} to 1×10^{-9} Pa L_t s^{-1} or less are most likely plugged permanently when exposed to open air at one atmosphere [243]. Minute leaks smaller than leak rates of about 1×10^{-8} Pa L_t s^{-1} are of no practical importance for the vacuum life of the typical vacuum interrupter and a leak check level to that sensitivity is enough to guarantee its vacuum integrity over its required lifetime. From the data given in Table 3.16, it can safely be said that if a vacuum interrupter with an internal volume of 1 L_t does not show a pressure of 10^{-1} Pa within the one month, it does not have a leak larger than 3.9×10^{-8} Pa L_t s^{-1} and will, therefore be unlikely to leak further during its entire application life. This, of course, presupposes that the vacuum interrupter is not subjected to abusive handling or corrosive ambients that can generate unexpected leaks.

This analysis leads to a straightforward and effective method for assessing vacuum integrity and the expected life of a vacuum interrupter by measuring its high voltage withstand level. An ac high voltage tester is the best instrument if it can supply about 8–10 mA. The vacuum interrupter should be opened to its design contact gap and the high voltage withstand value can be measured. The process can be illustrated using the Paschen curve shown in Figure 4.16. For this vacuum interrupter the voltage withstand value is 50kV(rms) for a contact gap of 10mm and a pressure inside

TABLE 3.16
Maximum Leak Rate for Vacuum Interrupters Achieve a Given Lifetime

Internal Vacuum Interrupter Volume (L_t, Liters)		0.5	1	2	4
Maximum leak rate Q_{max}	30 years	0.5×10^{-10}	1.1×10^{-10}	2.1×10^{-10}	4.2×10^{-10}
in	10 years	1.6×10^{-10}	3.2×10^{-10}	6.3×10^{-10}	1.3×10^{-9}
Pa. L_t.s−1	5 years	3.2×10^{-10}	6.3×10^{-10}	1.3×10^{-9}	2.5×10^{-9}
for a life of:	1 years	1.6×10^{-9}	3.2×10^{-9}	6.3×10^{-9}	1.3×10^{-8}
	1 month	1.9×10^{-8}	3.8×10^{-8}	7.6×10^{-8}	1.5×10^{-7}
	1 week	8.2×10^{-8}	1.6×10^{-7}	3.3×10^{-7}	6.6×10^{-7}
	1 day	5.8×10^{-7}	1.2×10^{-6}	2.3×10^{-6}	4.6×10^{-6}

the vacuum interrupter of 2 × 10⁻² mbar. Thus, a worst case leak rate (or the maximum leak rate , $L_{r(max)}$ would be :

$$L_{r(max)} = \frac{2\times10^{-2}}{y_1}\text{mbar.year}^{-1} \quad (3.41)$$

where y_1 is the age in years since the vacuum interrupter's manufacture. Figure 4.16 shows that this vacuum interrupter would have an unacceptably high voltage performance at a pressure of 3.0×10^{-2} mbar. Thus the minimum time to reach this pressure would be:

$$y_2 = \frac{3.0\times10^{-2}}{L_{r(max)}} = \frac{3.0\times y_1}{2.0}\text{years} \quad (3.42)$$

Now if the vacuum interrupter is tested after a time $(y_2 - y_1)$ and it still withstands 50kV(rms) then Equations (3.41) and (3.42) can be used to calculate a new minimum life for this vacuum interrupter. Table 3.17 shows that it only takes seven test sequences to ensure the vacuum interrupter's integrity for 30 years. In fact in my experience if a vacuum interrupter has been in the field for 5 years without showing any sign of a small leak, then it will never leak. A dc high voltage tester can also be used for this measurement if the test voltage is set at $U_{ac} \times \sqrt{2}$. When a dc voltage is applied across the open contacts, a high field emission current from a micoscopic projection on the cathode can be interpreted as a sign of an unacceptable pressure inside the vacuum interrupter. To avoid this problem the open interrupter should always be subjected to a dc voltage test using both polarities. This will ensure that each contact in turn is the cathode contact. An interrupter with an unacceptable pressure will have a similar leakage current in both polarities. An interrupter with an acceptably low pressure may have a high leakage current, but this will usually only occur with one polarity. Some vendors offer special purpose "vacuum testers" that use a dc voltage. Usually they only provide a gauge that reads " good/ not good" for a measured current in the range of about 300μA. As we have seen in Chapter 1 in this volume, this level of emission current is quite possible from a cathode with an acceptably low level of vacuum. Again when using such a device it is imperative that both polarities are measured. There are three advantages of using this high voltage test method:

1) It is relatively easy to perform
2) The mechanism only has to be isolated from the circuit and the voltage can be applied across each vacuum interrupter in turn. The vacuum interrupter does not have to be removed from its mechanism

TABLE 3.17
Lifetime Calculation Integrity of a Vacuum Interrupter Using a High Voltage Withstand Test

Withstands 50kV high voltage after y_1 years after manufacture (pressure $\leq 2 \times 10^{-2}$ m bar)	Leak rate mbar.year^{-1} to reach 2×10^{-2} mbar since manufacture	Time in years to 3×10^{-2} mbar after the HV withstand test	Minimum years after manufacture to reach 3×10^{-2} mbar	Years after manufacture for the next HV withstand test
1	2.0×10^{-2}	0.5	1.5	2
2	1.0×10^{-2}	1.0	3	3
3	6.7×10^{-3}	1.5	4.5	5
5	4.0×10^{-3}	2.5	7.55	8
8	2.5×10^{-3}	4	12	12
12	1.7×10^{-3}	6	18	20
20	1.0×10^{-3}	10	30	30

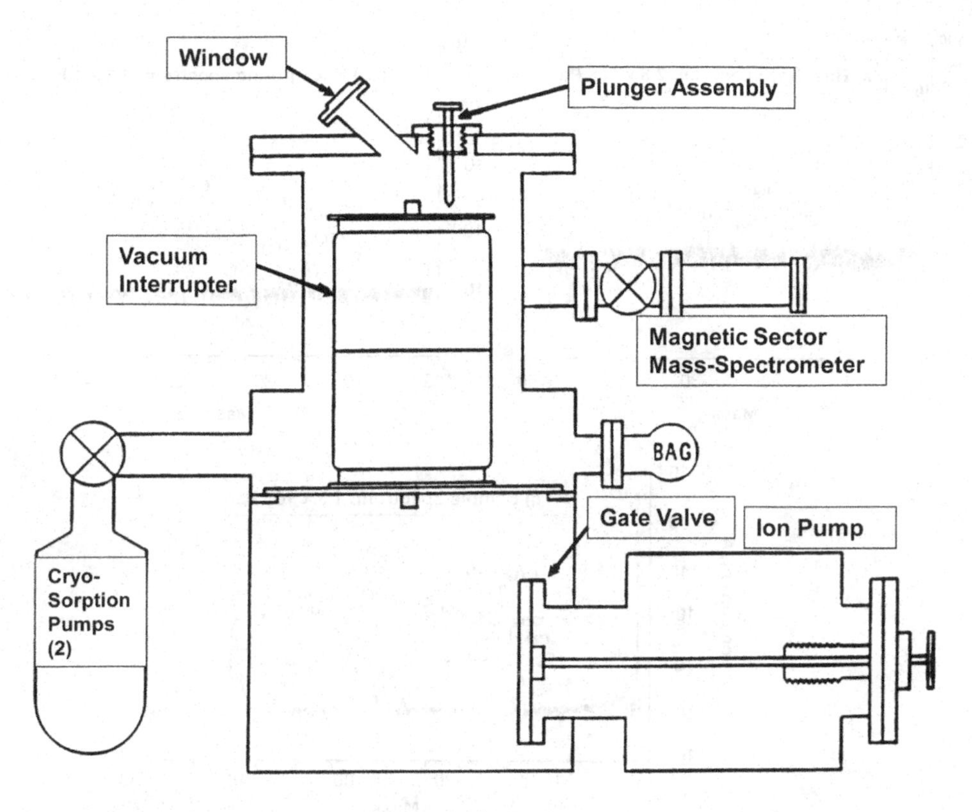

FIGURE 3.86 The residual gas analysis apparatus [244].

3) The test also measures the insulation integrity of the cicuit breaker/switch system and not just the vacuum intergrity of the vacuum interrupter

Another method used to assess the vacuum condition inside a manufactured vacuum interrupter is to perform a residual gas analysis [229, 244]. The experimental residual gas analysis (RGA) apparatus is illustrated in Figure 3.86. The principle of operation is to puncture a vacuum interrupter inside a closed chamber and detect the difference of the chamber gas contents before and after puncturing the interrupter. The chamber must have a much better vacuum than the interrupter sample to minimize the influence of any background gases. An all-metal bakeable ultrahigh vacuum system is therefore used. The gas content is measured with a magnetic sector mass-spectrometer AeroVac™ (Vacuum Technology Inc., Oak Ridge, TN) with an electron multiplier and a 250 amu capability. To avoid any oil contamination, a pumping system consisting of sorption pumps and an ion pump is employed. For the system to be able to test commercially available vacuum interrupters of various sizes, the entire vacuum interrupter sample is placed inside the chamber. This requires the interrupter to be thoroughly cleaned, i.e. any organic substances such as bushings, glues and paints must be removed. With an extended baking the chamber typically reaches a vacuum level of 5×10^{-5} Pa under constant pumping.

Figure 3.87 shows a typical set of mass-spectra, including the base-line spectrum of the continuously pumped chamber, the background spectrum of the valved-off chamber before puncturing the interrupter, and the sample spectrum after puncturing the sample interrupter. The difference between spectra (a) and (b) represents the gas content coming from inside the sample interrupter. The significant amount of helium observed in Figure 3.87(a) and (b) is indicative of the negligible pumping effect of the ion pump for inert gases. This particular mass-spectrometer has various sensitivity factors for different amu peaks, with that for H_2 about 20 times the values for N_2, CO, and

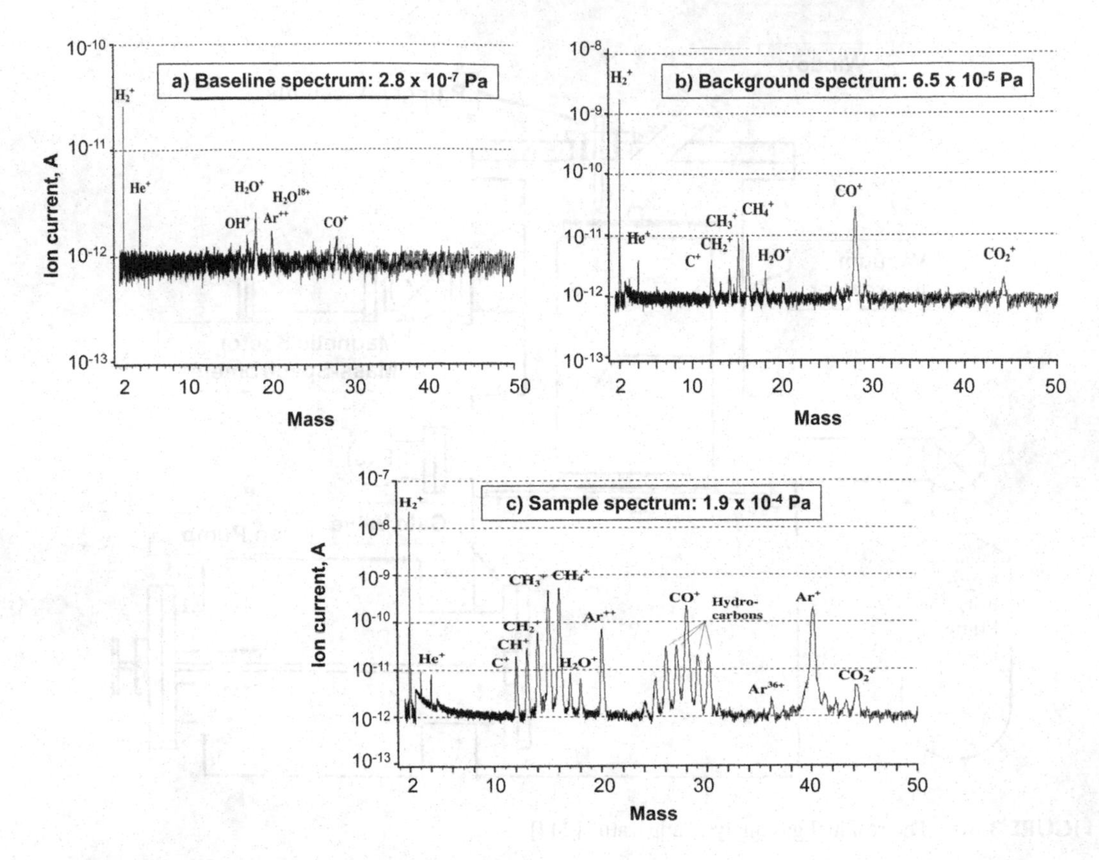

FIGURE 3.87 Data from a typical residual gas analysis of a processed vacuum interrupter [244].

CH_4. Because this method destroys the vacuum interrupter, it cannot be used during the normal vacuum interrupter manufacturing sequence. The technique is, however, extremely powerful for assessing changes made in the manufacturing process such as changes to the operating time of the vacuum furnace. It is also valuable for determining the effect of design and material changes in new vacuum interrupter developments.

Once a vacuum interrupter has passed the initial vacuum integrity test, it is then conditioned to bring its high voltage performance up to its expected level. I have described the various techniques to perform this function in Section 1.3.7 in this volume. Most manufacturers use either the high voltage ac or the current conditioning methods. The vacuum interrupter is then stored for a period to determine if there are any small leaks in the braze joints or if the ceramic has been punctured during the high voltage conditioning process. A final magnetron pressure test and usually a one-minute high voltage withstand test are performed on the vacuum interrupter before it is ready for the final assembly operations such as the Ag plating of the terminals, the addition of the anti-twist bushing and, for some manufacturers, the painting of the end plates.

3.5.3 Summary

When studying experiments of the high-current performance of TMF and AMF contact structures it is important to ask, "Does the experimenter open the contacts somewhere on the high-current ac cycle?" If the experimenter does not and uses, for example, a spark breakdown or opens on a low current before the high-current cycle is applied, then the formation of the vacuum arc between the contacts will be different, because the vital bridge column arc (see Section 2.2 in this chapter) will

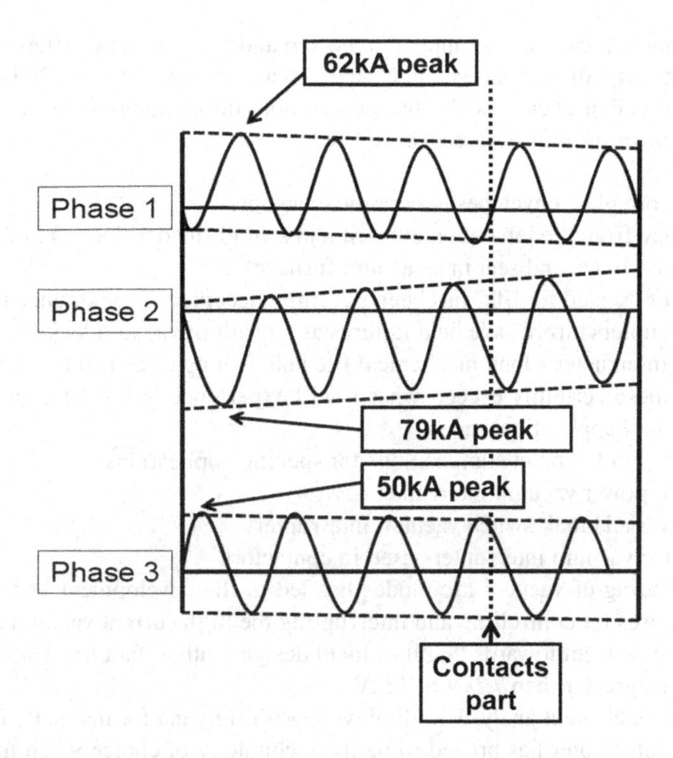

FIGURE 3.88 A 31.5kA short circuit current showing the asymmetric peak current values and the gradual decay of the to the symmetric current values. An example of the contact opening of the three vacuum interrupters in a three-phase vacuum circuit breaker occurring a few cycles after the short circuit current has been detected.

not precede the observed vacuum arc formations. Care should then be taken when interpreting these results for the performance of the contact structures in a practical vacuum interrupter.

Almost all studies of both TMF and AMF contact designs open the contacts on the current rise of an ac half cycle: See, for example, Figure 3.25. The high-current vacuum interrupter is the current interruption component of a three-phase, vacuum circuit breaker. The current phases in a three-phase circuit are 120^0 with respect to each other. Also, when a short circuit occurs, a decaying dc component is superimposed on the symmetrical currents. The result is that the initial peak current can be much greater than the peak value of the symmetrical rms current. Figure 3.88 shows an example of a 31.5KA (rms peak current of 44.5kA) short circuit current. Here the maximum asymmetric peak current is 79kA on phase 2. As the three currents phases are 120° with respect to each other the other two phases have lower peak current values: see Section 6.2 in this volume. As the dc component decays in perhaps 5 to 10 cycles the short circuit current returns to its symmetrical form. Figure 3.88 shows an example of the contacts parting a few cycles after the short circuit has been detected. Here only the phase 1, vacuum interrupter's contacts open on the initial rise of the ac current. Phase 2 opens close to the current maximum and phase 3 opens as the current begins its progress to current zero. Phase 3 has the first opportunity to reach current zero and interrupt the short circuit current. Therefore, when evaluating vacuum interrupter contact structures, experiments should be performed with contact parting not only on the rise of the ac current, but also parting at the peak of the ac current and perhaps as the current is decreasing towards current zero. As Schulman shows [114] (see Figures 3.34 and 3.35), the arc characteristics can be quite different depending upon where the contacts open in the ac cycle.

The vacuum interrupter has been successfully manufactured since the 1950s and the power vacuum interrupter since the 1960s. Since that time there has been continuous improvement in its operating performance. It can now be applied to switch and protect many different distribution circuits: these applications will be discussed in Chapters 5 and 6 in this volume. Indeed its application range continues to broaden There has also been continuous development in its design and manufacture. Among the innovative improvements have been:

1) The change from glass envelopes to ceramic envelopes
2) The move away from the tabulation exhaust and seal to the development of the batch process for vacuum braze and seal in a vacuum furnace
3) The concept of "sealed for life" has been generally accepted as experience has shown that vacuum interrupters rarely have field failures as a result of vacuum leaks [245–247]
4) The vacuum interrupter's long mechanical life and its long electrical life. In fact, for most applications these certainly exceed what it will experience in the field and, in fact, will exceed the actual application's need
5) The development of contact compositions for specific applications:
   a. Cu–Cr for power vacuum interrupters
   b. Cu–W for load break switch vacuum interrupters
   c. Ag–WC for vacuum interrupters used in contactors
6) The understanding of vacuum arc modes has led to the development of highly effective contact structures for controlling and interrupting the high current vacuum arc
7) The general movement towards floating shield designs rather than fixed shield designs for circuit voltages greater than 7.2kV to 15kV
8) The use of finite element analysis for high voltage design and for magnetic field design
9) The vacuum interrupter has proved to be the technology of choice when married to new mechanism developments and to precise electronic control (see Chapters 5 and 6 in this volume)
10) The move to apply a single vacuum interrupter to circuit voltages up to transmission voltages of greater than 170kV is moving the technology beyond the traditional distribution circuit applications.

Also, as can be inferred from the discussion of the materials used in the vacuum interrupter, another of its major advantages that it is environmentally friendly, i.e.:

- All its components are environmentally safe
- All its components are recyclable
- It emits no green-house gases
- It produces no toxic biproducts
- It produces minimum noise when it operates
- It has a low total lifetime cost

REFERENCES

1. Slade, P., "Variations in contact resistance resulting from oxide formation and decomposition in Ag-W and Ag-WC-C contacts passing steady currents for long time periods", *IEEE Transactions on Components, Hybrids and Manufacturing Technology*, CHMT-9, pp. 3–16, March 1986.
2. Slade, P., "Advances in material development for high power, vacuum interrupter contacts", *IEEE Transactions on Components, Packaging and Manufacturing Technology, Part A*, 17(1), pp. 96–106, March 1994.
3. Reece, M., "The vacuum switch Part I and Part II,", *Proceeding IEE*, 110(4), pp. 793–811, April 1963.
4. Barkan, P., Lafferty, J., Lee, T., Talento, J., "Development of contact materials for vacuum interrupters", *IEEE Transactions on Power Apparatus and Systems*, PAS-90, pp, 975, May/June 1971.
5. Lee, T., Cobine, J., "Vacuum type circuit interrupter", *U.S. Patent 2,975,256*, March 1961.

6. Lafferty, J., Barkan, P., Lee, T., Talento, J., "Vacuum circuit interrupter contacts", *U.S. Patent 3,246,979*, April 1969.
7. Horn, F., Porter, J., Talento, J., "Vacuum-type circuit interrupter with contact material containing a minor percentage of aluminum", *U.S. Patent 3,497,652*, February 1970.
8. Nakajima, Y., "Contact alloys for vacuum circuit interrupters", *U.S. Patent 3,502,465*, March 1970.
9. Xiu, S., Wang, J., Liu, Z., "Research on properties of Cu-Co-Ta contact materials for vacuum circuit breaker ", *Proceedings 22nd International Symposium on Discharges and Electrical Insulation in Vacuum*, pp. 319–323, September/October 2004.
10. Lafferty, J., Cobine, J., Burger, E., "Electrodes for vacuum circuit interrupters and methods of making some", *U.S. Patent 3,125,441*, March 1964.
11. Krock, R., Zdanuk, E., "Tungsten powder bodies infiltrated with copper-titanium-bismuth or copper-titanium-tin", *U.S. Patent 3,305,324*, February 1967.
12. Zdanuk, E., Krock, R., "Tungsten bodies infiltrated with copper titanium alloys" U.S. Patent 3,353,953, April 1968; also "Tungsten-indium powder bodies infiltrated with Cu," *U.S. Patent 3,353,931*, November 1967.
13. Zdanuk, E., Krock, R., "Tungsten powder bodies infiltrated with copper zirconium alloy", *U.S. Patent 3,340,022*, September 1967.
14. Wood, A., Brown, K., "Electrodes for electrode devices operable in vacuum", *U.S. Patent 3,379,846*, April 1968.
15. Krock, R., Zdanuk, E., "Method of producing tungsten powder bodies infiltrated with zirconium", *U.S. Patent 3,449,120*, June 1969.
16. Talento, J., "Vacuum type circuit interrupter with contacts containing a refractory metal", *U.S. Patent 3,596,025*, July 1971.
17. Satoh, Y., Kaneko, E., Yokokura, K., Okutomi, T., Homma, M., Tamagawa, T., Oshima, I., Yanabu, S., "Development of new type, low surge and high current vacuum interrupters", *Proceedings 14th International Symposium on Discharges and Electrical Insulation in Vacuum*, pp. 446–449, September 1990.
18. Kaneko, E., Yokokura, K., Homma, M., Satoh, Y., Okawa, M., Okutomi, T., Oshima, I., Yanabu, S., "Possibility of high current interruption of vacuum interrupter with low surge contact material: Improved Ag -WC", *IEEE Transactions on Power Delivery*, 10(2), pp. 797–830, April 1995.
19. Behrens, V., Honig, T., Kraus, A., "Comparisons of different contact materials for low voltage vacuum applications", *Proceedings 19th International Conference on Electrical Contact Phenomena*, pp. 247–251, September 1998.
20. Behrens, V., Honig, T., Kraus, A., "Tungsten and tungsten carbide based contact materials used in low voltage vacuum contactors", *Proceedings 45th IEEE Holm Conference on Electrical Contacts*, pp. 105–110, October 1999.
21. Temborius, S., Lindmayer, M., Gentsch, D., "Properties of WC – Ag and WC – Cu for vacuum interrupters", *IEEE Transactions on Plasma Science*, 31(5), pp. 945–952, October 2003.
22. Slade, P., "Contact materials for vacuum interrupters", *IEEE Transactions on Parts, Hybrids and Packaging*, PHP-10(1), pp. 43–47, March 1974.
23. Schellekens, H., "Development of materials for electrical contacts with improved high temperature and arcing properties", *Proceedings 13th International Symposium on Discharges and Electrical Insulation in Vacuum*, pp. 362–363, June 1988.
24. Wang, J., Zhang, C., Zhang, H., Yang, Z., Ding, B., "Cu-Cr(25) W(1) Ni(2) contact material of vacuum interrupter", *The Transactions of Nonferrous Metals Society*, 11(2), pp. 226–230, April 2001.
25. Watson, W.G., Breedis, J.F., "Copper base alloy and process", *U.S. Patent 4,224,066*, September 1980.
26. Kato, M., "Electrical contact composition for a vacuum type circuit breaker", *U.S. Patent 4,372,783*, February 1983.
27. Kashiwagi, Y., Noda, Y., Kitakizaki, K., "Contact of vacuum interrupter and manufacturing process", *U.S. Patent 4,554,425*, November 1985.
28. Naya, E., Okumura, M., "Contact material for vacuum circuit breaker", *U.S. Patent 4,626,282*, December 1986.
29. Robinson, A.A., "Vacuum type circuit interrupting device with contacts of infiltrated matrix material", *U.S. Patent 3,818,163*, June 1974.
30. Hässler, H., Kippenberg, H., Schreiner, H., "Demands on contact materials for HV vacuum circuit breakers", *Proceedings 10th International Conference on Electric Contacts*, pp. 219–230, August 1980.
31. Kippenberg, H., "Cr-Cu as a contact material for vacuum interrupters", *Proceedings 13th International Conference on Electric Contacts*, pp. 140–144, September 1986.

32. Müller, R., "Arc-melted Cu-Cr alloys as contact materials for vacuum interrupters", *Siemens Forsch-u. Entwickl.-Ber.*, 17(3), pp, 105–111, 1988. Springer-Verlag.
33. Hansen, M., Anderko, K., *Constitution of Binary Alloys*, 2nd Edition, (pub. McGraw-Hill, New York), pp. 524–525, 1958.
34. Zhang, C., Wang, J., Zhang, H., Yang, Z., Ding, B., "Deoxidization of Cu Cr(25) alloys prepared by vacuum induction melting", *The Transactions of Nonferrous Metals Society*, 11(3), pp. 338–339, June 2001.
35. Miao, B., Zang, Y., Zhoa, Y., Lui, G.,Ding, S., Hong, Y., "Two new Cu – Cr alloy contact materials", *Proceedings 20th International Symposium on Discharges and Electrical Insulation in Vacuum*, pp. 729–732, June 2002.
36. Zhou, F., Xu, H., Yang, Z., Ding, B., "Preparation of Cu Cr (25) alloys through vacuum arc smelting and their properties", *The Transactions of Nonferrous Metals Society*, pp. 73–75, February 2000.
37. Miao, B., Zang, Y., Lui, G., "Current status and developing trends of Cu-Cr contact materials", *Proceedings 21st International Symposium on Discharges and Electrical Insulation in Vacuum*, pp. 311–314, September/October 2004.
38. Miao, B., Guo, H., Zhang, Y., Liu, G., "Effects of interface characteristics of Cu/Cr phases on the contact performance of Cu-Cr25 alloy contact material", *Proceedings 23rd International Symposium on Discharges and Electrical Insulation in Vacuum*, pp. 185–188, September 2008.
39. Li, P., Yao, X., Zhang, S., Wang, X., Gang, Li, Liu, Z., "Comparison of electrical performance of Cu-Cr contact material manufactured by two methods", *Proceedings 28th International Symposium on Discharges and Electrical Insulation in Vacuum*, pp. 607–610, September 2018.
40. Hauner, F., Rolle, S., Lietz, A., "Method of producing a contact material for contact pieces for vacuum switching devices, and a contact material and contact pieces therefore", *US Patent 6,524,525, B2*, February, 2003.
41. Rolle, S., Lietz, A., Amft, D., Hauner, F., "Cr-Cu contact material for low voltage vacuum contactors", *Proceedings 20th International Conference on Electrical Contacts*, pp. 179–186, June 2000.
42. Noda, Y., Yoshioka, N., Suzuki, N., Fukai, T., Yoshihara, T., Koshiro, K., "Method for forming an electrical contact material", *U.S. Patent, 5,480,472*, January 1996.
43. Xiu, S., Jai, S., Wang, J., "Arc remelting experimental results and analysis on Cr–Cu contact materials", *Proceedings 20th International Symposium on Discharges and Electrical Insulation in Vacuum*, pp. 523–525, July 2002.
44. Schneider, A., Popov, S., Duracov, V., Dampilon, B., Dehonova, S., Batracov, A., "On breaking capacity of the CuCr25 composite material produced with electron beam cladding", *Proceedings 25th International Symposium on Discharges and Electrical Insulation in Vacuum*, pp. 269–271, September 2012.
45. Li, Yu., Wang, J., Geng, Y., Kong, G., Liu, Z., "High-current vacuum arc phenomena of nanocrystalline CuCr25 contact material", *IEEE Transactions on Plasma Science*, 39(6), pp. 1418–1425, June, 2011.
46. Fink, H., Gentsch, D., Heimbach, M., "multilayer contact material based on copper and chromium material and its interruption ability", *IEEE Transactions on Plasma Science*, 31(5), pp. 973–976, October 2003.
47. Durakov, Y., Gnyusov, S., Dampilon, B., Dehonova, S., Ubiennykh, B., "Microstructure and properties of vacuum electron beam facing CuCr25 contact material", *Proceedings 25th International Symposium on Discharges and Electrical Insulation in Vacuum*, pp. 525–528, September 2012.
48. Devisme, M.-F., Schellekens, H., Picot, P., Olive, S., Henon, A., "The influence of Cu-Cr (25) characteristics on interruption capability of vacuum interrupters", *Proceedings 22nd International Symposium on Discharges and Electrical Insulation in Vacuum*, pp. 359–363, September/October 2004.
49. Lin, R., Wang, L., Shi, W., Deng, J., Jia, S., "Experimental investigation on triggered vacuum arc and erosion behavior under different contact materials", *IEEE Transactions on Plasma Science*, 46(8), pp. 3047–3056, August 2018.
50. Klinski-Wetzel, K., Kowanda, C., Böning, M., Heilmaier, M., Müller, F., "Parameters influencing the electrical conductivity of Cu-Cr alloys", *Proceedings 25th International Symposium on Discharges and Electrical Insulation in Vacuum*, pp. 392–395, September 2012.
51. Dullni, E., Plessl, A., Reininghaus, U., "Research for vacuum circuit breakers", *ABB Review*, pp. 11–18, March 1989.
52. Gellert, B., Schade, E., Dullni, E., "Measurement of particles and vapor density after high-current vacuum arcs by laser techniques", *IEEE Transactions on Plasma Science*, PS-15, pp. 546–551, October 1987.

53. Gellert, B., Schade, E., "Optical investigation of droplet emission in vacuum interrupters to improve contact materials", *Proceedings 24th International Symposium on Discharges and Electrical Insulation in Vacuum*, pp. 450–454, September 1990.
54. Frey, P., Klink, N., Michal, R., Saeger, K.E., "Metallurgical aspects of contact materials for vacuum switching devices", *IEEE Transactions on Plasma Science*, 17(5), pp. 734–740, October 1989.
55. Dullni, E., Gentsch, D., Shang, W., "Resistance increase of vacuum interrupters due to high-current interruptions", *IEEE Transactions, on Dielectrics and Electrical Insulation*, 23(1), pp. 1–7, February 2016.
56. Rylskaya, "Breakdown of vacuum interrupters behind front of lightning impulse", *Proceedings 25th International Symposium on Discharges and Electrical Insulation in Vacuum*, pp. 72–75, September 2012.
57. Feilbach, A., Hauf, U., Böning, M., Hinrichsen, V., Heilmaier, M., Müller, F., "Investigation of current breaking capacity of vacuum interrupters with focus on contact material properties with the help of a reference model vacuum circuit breaker ", *Proceedings 27th International Symposium on Discharges and Electrical Insulation in Vacuum*, pp. 121–124, September 2016.
58. Gentsch, D., Gorlt, K., "Welding behavior of vacuum interrupters equipped with CuCr contact material caused by making and breaking operations under short circuit current interruption", *Proceedings 26th International Symposium on Discharges and Electrical Insulation in Vacuum*, pp. 481–484, September 2014.
59. Glinkowski, M., Greenwood, A., Hill, J., Mauro, R., Varneckes, V., "Capacitance switching with vacuum circuit breakers", *IEEE Transaction Power Delivery*, 6, pp. 1088–1095, July 1991.
60. Ohshima, I., Yokokura, K., Matsuda, M., Otobe, K., Satoh, K., Yanabu, S., "Shunt capacitor switching performance of vacuum interrupters", *Presented at the IEEE Power Engineering Society Summer Meeting*, 1990.
61. Heyn, D., Lindmayer, M., Wilkening, E.D., "Effect of contact material on the extinction of vacuum arcs under line frequency and high frequency conditions", *IEEE Transactions on Components, Hybrids and Manufacturing Technology*, 14(1), pp. 65–70, March 1991.
62. Lindmayer, M., Wilkening, E.D., "Influence of contact material on the reignition of high frequency vacuum arcs", *RGE*, 6, pp. 36–40, June 1989.
63. Hässler, H., Kuhl, W., Rüttel, R., Schlenk, W., "Measurements of the oxygen content of chromium-copper contacts", *Proceedings 12th International Conference on Electric Contacts*, pp. 357–361, September 1984.
64. Frey, P., Jäger, K., Klink, N., Saeger, K., "Investigations on the release of gas from contact materials for vacuum switching devices during operation", *Proceedings 11th International Conference on Electric Contacts*, pp. 317–320, June 1982.
65. Wu, W., Yuan, F., "Discussion on process of gas evolution and absorption in vacuum interrupter during arcing", *Proceeding Proceedings 19th International Symposium on Discharges and Electrical Insulation in Vacuum*, pp. 273–277, September 2000.
66. Papillon, A., Schellekens, H., Route, S., "Influence of impurities on the interruption performance of solid state sintered Cu-Cr(25)", *Proceedings 28th International Symposium on Discharges and Electrical Insulation in Vacuum*, pp. 623–626, September 2018.
67. Klinski-Wetzel, K., Kowanda, C., Rettenmaier, T., Heilmaier, M., Müller, F., Hinrichsen, V., "Correlation between microstructural features of the melt zone and switching behavior in CuCr contact material", *Proceedings 27th International Conference on Electrical Contacts*, pp. 197–202, June 2014.
68. Voshall, R., Kimblin, C., Slade, P., Gorman, J., "Experiments on vacuum interrupters in high voltage 72kV circuits", *IEEE Transactions on Power Apparatus and Systems*, PAS-99, pp. 658–666, March/April 1980.
69. Yanabu, S., Okawa, M., Kaneko, E., Tamagawa, T., "Use of axial magnetic fields to improve high current vacuum interrupters", *IEEE Transaction Plasma Science*, PS-15, pp. 524–532, October 1987.
70. Slade, P., Voshall, R., Wayland, P., Bamford, A., McCracken, G., Yeckley, R., Spindle, R., "The development of a vacuum interrupter retrofit for the upgrading and life extension of 121kV–145kV oil circuit breakers", *IEEE Transactions on Power Delivery*, 6(3), pp. 1124–1131, July 1991.
71. Ballat, J., König, D., "Insulation characteristics and welding behavior of vacuum switch contacts made from various Cu Cr alloys", *IEEE Transactions on Electrical Insulation*, 28(4), pp. 628–634, 1993.
72. Yanabu, S., Sato, J., Homma, M., Tamagawa, T., Kaneko, E., "Post arc current in vacuum interrupters", *IEEE Transactions on Power Delivery*, PWRD-1(4), pp. 209–214, October 1986.
73. Sato, J., Watanabe, K., Somei, H., Homma, M., "Effect of Cr content in Cu Cr contact material on the interrupting ability of VCB", *Proceedings 3rd International Conference on Electrical Contacts, Arcs, Apparatus and Their Applications*, pp. 249–253, May, 1997.

74. Li, W.-P., Thomas, R., Smith, K., "Effects of Cr content on the interruption ability of Cu Cr contacts", *IEEE Transactions on Plasma Science*, 29(5), pp. 744–748, October 2001.
75. Miao, B., He, J., Liu, G., Wang, W., Wang, X., Liu, K., "Microstructure, tensile strength and anti-welding property of Cu-25CrTe alloy contact materials", *Proceedings 26th International Symposium on Discharges and Electrical Insulation in Vacuum*, pp. 445–448, September 2014.
76. Li, W.-P., Leusenkamp, M., Ellis, D., "Testing a Cu-8Cr-4Nb contact material in vacuum interrupters", *Proceedings 26th International Symposium on Discharges and Electrical Insulation in Vacuum*, pp. 409–412, September 2014.
77. Delachaux, T., Rager, F., Simon, R., Schmoelzer, T., Boehm, M., Gentsch, D., "Development of a simple procedure for screening current interruption capability of vacuum interrupter contact materials", *Proceedings 26th International Symposium on Discharges and Electrical Insulation in Vacuum*, pp. 413–416, September 2014.
78. Filip, G., Rieder, W., Schussek, M., "Current chopping due to new and eroded vacuum contacts", *Proceedings 13th International Symposium on Discharges and Electrical Insulation in Vacuum*, pp. 224–228, September 1986.
79. Dullni, E., Lindell, E., Liljestrand, L., "Dependence of the chopping current level of a vacuum interrupter on parallel capacitance", *IEEE Transactions on Plasma Science*, 45(8), pp. 2150–2156, August 2017.
80. Ding, C., Yanabu, S., "Effect of parallel circuit parameters on the instability of a low current vacuum arc", *IEEE Transactions on Plasma Science*, 31(5), pp. 877–883, October 2003.
81. Holmes, F., "An empirical study of current chopping by vacuum arcs", *Presented at the IEEE Power Engineering Society Winter Meeting*, PES paper C-74-088-1, January, 1974.
82. Taylor, E., Slade, P., Li, W., "High chop currents observed in vacuum arcs between tungsten contacts", *Proceedings 23rd International Conference on Electrical Contacts*, pp. 1–6, June 2006.
83. Taylor, E., Niemeyer, K., Pietsch, C., "Generation of overvoltages by chop current on Ag-WC and Cu-W/WC contacts in vacuum", *Proceedings 64th IEEE Holm Conference on Electrical Contacts*, pp. 242–245, October 2018.
84. Smeets, R., Kaneko, E., Ohshima, I., "Experimental characterization of arc instabilities and their effect on current chopping in low-surge vacuum interrupters", *IEEE Transactions on Plasma Science*, 20(4), pp. 439–446, October 1992.
85. Kurosawa, Y., Iwasita, K., Watanabe, R., Andoh, H., Takasuma, T., Watanabe, H., "Low surge vacuum breakers", *IEEE Transactions on Power Apparatus and Systems*, PAS-104, pp. 3634–3642, December 1985.
86. Rieder, W., Schussek, M., Glätzle, W., Kny, E., "The influence of composition and Cr particle size of Cu/Cr contacts on chopping current, contact resistance, and breakdown voltage in vacuum interrupters", *IEEE Transactions on Components, Hybrids and Manufacturing Technology*, 12(2), pp. 273–283, June 1989.
87. Cornick, K., "Current chopping performance of distribution breakers", *Proceedings IEE 2nd International Conference on Developments in Distribution Switchgear*, p. 12-i6, May 1986.
88. Frind, G., Carroll, J., Goody, C., Tuoky, E., "Recovery times of vacuum interrupters which have stationary anode spots", *Presented at IEEE PAS Summer Meeting*, 81 SM 395-4, July 1981.
89. Slade, P., "The vacuum interrupter contact", *IEEE Transactions on Components, Hybrids and Manufacturing Technology*, CHMT–7 (1), pp. 25–32, March 1984.
90. Rich,J., "A means of raising the threshold current for anode spot formation in metal vapor arcs", *Proceedings of the IEEE*, 59(4), pp. 539–554, 1971.
91. Streater, A., "Vacuum-type circuit interrupter with arc-voltage limiting means", *U.S. Patent 3,321,598*, May 23, 1967.
92. Emmerich, W., Voshall, R., "Contact structures for vacuum-type circuit interrupters", *U.S. Patent 3,632,928*, January 4, 1972.
93. A sample of many patents: Hundstad, R., *U.S. Patent 3,889,081*, June 1975, *U.S. Patent 3,667,871*, June 1972*; Rich, J., U.S. Patent 3,471,733*, October 1969, *U.S. Patent 3,471,736*, October 1969; *U.S. Patent 3,509,405*, April 1970; *U.S. patent 3,509,406*, April 1970 Mitchell, G., *U.S. Patent 3,858,076*, December 1974; Geneguard, P., *U.S. Patent 3,705,144*, December 1972.
94. Schellekens, H., "50years of TMF design considerations", *Proceedings 23rd International Symposium on Discharges and Electrical Insulation in Vacuum*, pp. 95–98, September 2014.
95. Schneider, H., "Contact structure for an electric circuit interrupter", *U.S. Patent 2,949,520*, August 1960.
96. Smith, S., "Contact structure for an electric circuit interrupter", *U.S. Patent 3,089,936*, February 1960.

97. Feng, D., Xiu, S., Li, M., Yuan, B., Zhang, Y., Zhao, Li, "Experimental investigation on transition characteristics of arcing stages under transverse magnetic field", *IEEE Transactions on Plasma Science*, 46(3), pp. 622–630, March 2018.
98. Feng, D., Xiu, S., Liu, G., Sun, Z., Zheng, J., "Simulation of magnetic field spiral-type contacts", *Proceedings 26th International Symposium on Discharges and Electrical Insulation in Vacuum*, pp. 165–168, September 2014.
99. Reece, M., "Development of the vacuum interrupter", *Philosophical Transactions of the Royal Society of London Series. Part A*, 275, p. 121, 1973.
100. Reese, M., "Improvements to Vacuum switch contacts", *U.K. Patent 1,100,159*, January 1968.
101. Wang, T., Xui, S., Long, Y., Wang, X., Zhao, L., Zhang, Y., "Simulation and experimental investigation on a new type of cup-shaped contacts with iron cores", *IEEE Transactions on Plasma Science*, 46(8), pp. 3057–3064.
102. Liu, Z., Xiu, S., Yuan, B., Li, M., Wang, X., Long, Y., "Numerical and experimental investigation of an improved vacuum interrupter cup-shaped contact", *IEEE Transactions on Plasma Science*, 46(8), pp. 3040–3046, August 2018.
103. Lamara, T., Hencken, K., Gentsch, D., "A novel vacuum interrupter contact design for improved high current interruption performance based on a double-TMF arc control system", *IEEE Transactions on Plasma Science*, 43(5), pp. 1798–1805, August 2018.
104. Althoff, F., "Forschungsarbeiten auf dem Gebiet Elektrischer Vakuumschalter fur grosse Strome", *E.T.Z.-a*, 92(9), p. 538, 1971.
105. Kowanda, C., Hochstrasser, M., Schwaiger, A., Müller, F., Plakensteiner, A., Grohs, C., "Net shape manufacturing of CuCr for vacuum interrupters", *IEEE Transactions, on Dielectrics and Electrical Insulation*, 18(6), pp. 2131–2137, December 2011.
106. Niwa, Y., Palad, R., Sasage, K., Sakaguchi, W., "vacuum arc behavior in transverse magnetic field electrode of vacuum interrupter", *Proceedings 25th International Symposium on Discharges and Electrical Insulation in Vacuum*, pp. 505–508, September 2012.
107. Dullni, E., "Motion of high current vacuum arcs on spiral-type contacts", *Transaction IEEE on Plasma Science*, 17(6), pp. 875–879, 1989.
108. Wolf, C., Kurrat, M., Lindmayer, M., Gentsch, D., "Arcing behavior on different TMF contacts at high-current interrupting operations", *IEEE Transactions on Plasma Science*, 39(6), pp. 1284–1290, June 2011.
109. Teichmann, J., Romheld, M., Hartman, W., "Magnetically driven high current switching arcs in vacuum and low-pressure gas", *Transactions IEEE on Plasma Science*, 27(4), pp. 1021–1025, August 1999.
110. Dullni, E., Schade, E., Shang, W., "Vacuum arcs driven by cross-magnetic fields (RMF)", *Transactions IEEE on Plasma Science*, 31(5), pp. 902–908, October 2003.
111. Pavelescu, D., Pavelescu, G., Gherendi, F., Nitu, C., Dimitrescu, G., Nitu, S., "Investigation of the rotating arc plasma generated in a VCB, *Transactions IEEE on Plasma Science*, 33(5), pp. 1504–1509, October 2005.
112. Haas, W., Hartmann, W., "Investigation of arc roots of constricted high current vacuum arcs", *IEEE Transactions on Plasma Science*, 27(4), pp. 954–960, August 1999.
113. Delachaux, T., Fritz, O., Gentsch, D., Schade, E., Shmelev, D., "Numerical simulation of a moving high-current vacuum arc driven by a transverse magnetic field (TMF)", *IEEE Transactions on Plasma Science*, 35(4), pp. 905–911, August 2007.
114. Schulman, M.B., "Separation of spiral contacts and motion of vacuum arcs at high currents", *IEEE Transactions on Plasma Science*, 21(5), pp. 484–488, October 1993.
115. Schulman, M., "The behavior of vacuum arcs between spiral contacts with small gaps", *IEEE Transactions on Plasma Science*, 23(6), pp. 915–918, December 1995.
116. Schulman, M., Slade, P., "Multiple electrode structure for vacuum interrupter", *U.S Patent 5,444,201*, August1995.
117. Shmelev, D., Delachaux, T., "Physical modeling and numerical simulation of constricted high-current vacuum arcs under the influence of a transverse magnetic field", *IEEE Transactions, on Plasma Science*, 37(8), pp. 1379–1385, August 2009.
118. Miller, H., "Vacuum arc anode phenomena", *IEEE Transactions on Plasma Science*, PS-11, pp. 76–89, June 1983.
119. Liu, Z., Xiu, S., Wang, X., Yuan, B., Li, R., "Influence of axial self-magnetic field component on vacuum arc transition to diffuse mode of transverse magnetic field (TMF) contacts", *Proceedings 28th International Symposium on Discharges and Electrical Insulation in Vacuum*, pp. 251–254, September 2018.

120. Li, W., Smith, K., Leusenkamp, M., "Optimum number of petals for a spiral TMF contact", *Proceeding 26th International Symposium on Discharges and Electrical Insulation in Vacuum*, pp. 659–662, September 2016.
121. Paulus, I., "The short vacuum arc–Part 1; Experimental investigations", *IEEE Transactions on Plasma Science*, 16(3), pp. 342–347, June 1988.
122. Janiszewski, J., Zulucki, Z., "Photographic study of discharge development and high current arc modes in vacuum", *Proceeding Proceedings 17th International Symposium on Discharges and Electrical Insulation in Vacuum*, pp. 220–225, July 1996.
123. Ito, T., Ookura, T., Takani, T., "Vacuum type circuit interrupter with axial magnetic field", *UK Patent 1,258,015*, December 1971.
124. Dan, S., Hu, D., Li, R., Jiang, S., Meng, Z., Xia, X., "External coils for generation of axial magnetic field in vacuum interrupters", *Proceedings 28th International Symposium on Discharges and Electrical Insulation in Vacuum*, pp. 575–578, September 2018.
125. Leusenkamp, M., Hilderink, J., "Vacuum circuit breaker with coaxial coil for generating an axial magnetic field in the vicinity of the contact members of the circuit breaker", *Int'l Patent Classification H01U33/66, Int'l Patent No. WO 03/056591 A1*, May 2003; *US Patent 7,038,157 B2*, 2006.
126. Schellekens, M., "Arc behavior in axial magnetic field vacuum interrupters equipped with an external coil", *Proceedings 18th International Symposium on Discharges and Electrical Insulation in Vacuum*, pp. 514–517, 1998.
127. Bolongeat-Mobleu, R., Schellekens, H., "Vacuum electrical switch", *U.S. Patent 5,861,597*, January 1999.
128. Smith, R.K., "Cylindrical coil and contact support for vacuum interrupter", *U.S. Patent 5,804,788*, September 1998.
129. Wayland, P.O., Gorman, J.G., Voshall, R.E., "Vacuum type circuit interrupter with an improved contact with axial magnetic field", *U.S. Patent 4,260,864*, April 1981.
130. Wang, Z., Liu, Z., Zhang, X., Zheng, Y., Wang, J., "Analysis of axial magnetic field characteristics of coil type axial magnetic field vacuum interrupters", *Proceedings 22nd International Symposium on Discharges and Insulation in Vacuum*, pp. 344–347, September 2006.
131. Murano, M., Yanabu, S., Sodeyama, H., "Vacuum interrupter", *U.S. Patent 3,935,406*, 1976.
132. Yanabu, S., Kaneko, E., Okumura, H., Aiyoshi, T., "Novel electrode structure of vacuum interrupter and its application", *IEEE Transactions on Power Apparatus and Systems*, PAS-100, pp. 1966–1974, March/April 1981.
133. Kurosawa, Y., Sugawara, H., Kawakubo, Y., Abe, N., Tsuda, H., "Vacuum circuit breaker electrode generating multi-pole axial magnetic field and its interruption ability", *IEEE Transactions on Power Apparatus and Systems*, PAS-99(6), pp. 2079–2085, November/December1980.
134. Fenski, B., Lindmayer, M., Heimback, M., Shang, W., "Characteristics of a vacuum switching contact based on bipolar axial magnetic field", *IEEE Transactions on Plasma Science*, 27(4), pp. 949–953, August 1999.
135. Gorman, J.G., Roach, J.F., Wayland, P.O., "Vacuum type circuit interrupter with a contact having integral axial magnetic field means", *U.S. Patent 4,117,288*, September 1978.
136. Paul, B.-J., Renz, R., "Contact arrangement for vacuum switches", *European Patent 0155376, US Patent 4,620,074*, 1987 and *US Patent 4,532,391*, July 1985.
137. Nash, W., Bestel, E., "Vacuum interrupter", *U.S. Patent 4,839,481*, February 1988.
138. Akira, N., Hidemitsu, T., Yoshihiko, M., Takaaki, F., "Contact for vacuum interrupter and vacuum interrupter using same", *U.S. Patent 6,686,552 B2*, Feb.2004 and "Contact for vacuum interrupter and vacuum interrupter using the contact", *U.S. Patent 6,639,169 B2*, October 2003.
139. Wang, Z., "AMF electrode simulation for HV vacuum interrupter", *Proceedings 26th International Symposium on Discharges and Electrical Insulation in Vacuum*, pp. 411–444, September 2016.
140. Xiu, S., Pang, L., Wang, J., Lin, J., He, G., "Analysis of axial magnetic field electrode applied to high voltage vacuum interrupters", *Proceedings 22nd International Symposium on Discharges and Insulation in Vacuum*, pp. 317–320, September 2006.
141. Mayo, S., "Axial magnetic field coil for vacuum interrupter", *US Patent 5,777,287*, July 1998.
142. Liu, Z., Wang, Z., Wang, J., "A new slot type axial magnetic field contact with low resistance", *Proceedings 22nd International Symposium on Discharges and Insulation in Vacuum*, pp. 293–296, September 2006.
143. Schellekens, H., Shang, W., Lenstra, K., "Vacuum interrupter design based on arc magnetic field interaction for horse-shoe electrodes", *IEEE Transactions on Plasma Science*, 21(5), pp. 469–473, October 1993.

144. Li, H., Wang, Z., Geng, Y., Wang, J., Liu, Z., "high-current vacuum arc mode transition of a horse-shoe axial magnetic field contact with long contact gap", *IEEE Transactions on Plasma Science*, 45(8), pp. 2164–2171, August 2017.
145. Li, H., Wang, Z., Geng, Y., Liu, Z., Zhang, Y., Zhao, F., "Design parameters optimization of a 126kV horse shoe type axial magnetic field vacuum interrupter", *Proceeding Proceedings 27th International Symposium on Discharges and Electrical Insulation in Vacuum*, pp. 473–476, September 2016.
146. Cheng, S., "Study on influence of self-generated axial magnetic field upon high-current vacuum arc behavior at a long contact gap in high-voltage interrupter", *IEEE Transactions on Plasma Science*, 39(3), pp. 911–917, March 2011.
147. Zhang, Y., Geng, Y., Yu, L., Yan, J., Liu, Z., Hu, C., Yao, J., "Axial magnetic field strength needed for a 126kV single break vacuum interrupter", *Proceedings 24th International Symposium on Discharges and Electrical Insulation in Vacuum*, pp. 320–323, September 2010.
148. Ryu, J., Kim, Y.-G., Choi, J., Park, S., "The experimental research of 170kV VCB using single-break vacuum interrupter", *Proceedings 25th International Symposium on Discharges and Electrical Insulation in Vacuum*, pp. 493–496, September 2012.
149. Yao, X., Wang, J., Geng, Y., Yan, J., Liu, Z., Yao, J., Liu, P., "Development and type test of a single-break 126-kV/40kA-2500A vacuum circuit breaker", *IEEE Transactions on Power Delivery*, 31(1), pp. 182–190, February 2016.
150. Zhang, Y., Yao, X., Liu, Z., Geng, Y., Liu, P., "Axial magnetic field strength needed for a 126-kV single-break vacuum circuit breaker during asymmetrical current switching", *IEEE Transactions on Plasma Science*, 41(8), pp. 2034–2041, August 2013.
151. Zhang, X., Wang, X., Guan, Q., Li, M., "A new axial magnetic field contact three-quarters of coil for high-voltage vacuum interrupter and its properties", *Proceeding Proceedings 27thh International Symposium on Discharges and Electrical Insulation in Vacuum*, pp. 541–544, September 2016.
152. Stoving, P.N., Bestel, E.F., "Finite element analysis of AMF vacuum contacts", *Proceedings 18th International Symposium on Discharges and Electrical Insulation in Vacuum*, pp. 522–529, 1998.
153. Fenski, B., Lindmayer, M., "Vacuum interrupters with axial field contacts –3D element simulations and switching experiments", *IEEE Transactions on Dielectrics and Electrical Insulation*, 4(4), pp. 407–412, August 1997.
154. Taylor, E., "Private communication".
155. Schulman, M., Schellekens, H., "Visualization and characterization of high current diffuse arcs on axial magnetic field contacts", *IEEE Transactions on Plasma Science*, 28(2), pp. 443–452, April 2000, Corrections to this paper, *IEEE Transactions on Plasma Science*, 28(4), p. 1050, June 2000.
156. Chen, S., Xiu, S., Wang, J., Li, X., "Influence of gap distance on the arc modes transition of cup type AMF electrode", *Proceedings 22nd International Symposium on Discharges and Insulation in Vacuum*, pp. 384–387, September 2006.
157. Matsui, Y., Sana, A., Komatsu, H., Satau, H., Saito, H., "Vacuum arc phenomena under various axial magnetic field and anode melting", *Proceedings 24th International Symposium on Discharges and Electrical Insulation in Vacuum*, pp. 324–327, September 2010.
158. Kiedar, M., Schulman, M., Taylor, E., "Model of a diffuse column vacuum arc as cathode jets burning in parallel with a high-current plasma column", *IEEE Transactions on Plasma Science*, 32(2), pp. 909–917, April 2004.
159. Watanabe, K., Sato, J., Kagenaga, K., Somei, H., Homma, M., Kaneko, E., Takahashi, H., "The anode surface temperature of Cu-Cr contacts at the limit of current interruption", *Proceedings 17th International Symposium on Discharges and Electrical Insulation in Vacuum*, pp. 291–295, July 1996.
160. Niwa, Y., Sato, J., Yokokura, K., Kusano, T., Kaneko, E., Ohshima, I., Yanabu, S., "The effect of contact material on temperature and melting of anode surface in vacuum interrupters", *Proceedings 19th International Symposium on Discharges and Electrical Insulation in Vacuum*, pp. 524–527, September 2000.
161. Schellekens, H., Schulman, M.B., "Contact temperature and erosion in high current diffuse vacuum arcs on axial magnetic field contacts", *IEEE Transactions on Plasma Science*, 29(3), pp. 452–461, June 2001.
162. Del Rio, L., Barrio, S., Izcara, J., Aranaga, S., "Development of optimized vacuum interrupters through virtual experimentation", *Proceedings 25th International Symposium on Discharges and Electrical Insulation in Vacuum*, pp. 321–323, September 2014.
163. Henon, A., Altimani, T., Schellekens, H., "3-D finite element simulation and synthetic tests of vacuum interrupters with axial magnetic field contacts", *Proceedings 20th International Symposium on Discharges and Electrical Insulation in Vacuum*, pp. 463–466, June 2002.

164. Liu, Z., Cheng, S., Zhang, X., Wang, J., Wang, Q., He, G., "An interrupting capacity model for axial magnetic field vacuum interrupters with slot type contacts", *Proceedings 22nd International Symposium on Discharges and Insulation in Vacuum*, pp. 297–300, September 2006.
165. Chaly, A., Poluyanova, I., Poluyanova, V., "Maximum interrupting capacity of Cu-Cr contacts under the effect of uniform axial magnetic field (AMF)", *Proceedings 22nd International Symposium on Discharges and Insulation in Vacuum*, pp. 341–343, September 2006.
166. Wooton, R.E., Voshall, R.E., "Vacuum interrupter with a specially modulated axial magnetic field", *U.S. Patent 4,401,868*, April 30, 1983.
167. Chaly, A., Logatchev, A., Zabello, K., Shol'nik, S., "High current arc appearance in non-homogeneous axial magnetic field", *IEEE Transactions on Plasma Science*, 31(5), pp. 884–889, October 2003.
168. Homma, M., Somie, H., Niwa, Y., Yokokura, K., Ohshima, I., "Physical and theoretical aspects of a new vacuum arc control technology–Self-arc diffusion by electrode; SADE", *IEEE Transactions on Plasma Science*, 27(4), pp. 961–968, August 1999.
169. Okutomi, T., Seki, T., Ohshima, I., Homma, M., Somie, H., Uchiyama, K., Niva, Y., Watanabe, K., "Electrode arrangement of vacuum circuit breaker with magnetic member for longitudinal magnetization", *US Patent 6,080,952*, June 2000.
170. Shkol'nik, S., Afanas'ev, V., Barino, Y., Chaly, A., Logatchev, A., Malakhovsky, S., "Distribution of cathode current density and breaking capacity of medium voltage vacuum interrupters with axial magnetic fields", *IEEE Transactions on Plasma Science*, 33(5), pp. 1511–1518, October 2005.
171. Shang, W., Schellekens, H., Hilderink, J., "Experimental investigations into the arc properties of vacuum interrupters with horse-shoe electrode, four pole electrodes, and their application", *IEEE Transactions on Plasma Science*, 21(5), pp. 474–477, October 1993.
172. Fink, H., Heimbach, M., Wenkai, S., "Vacuum interrupters with axial magnetic field contacts based on bipolar and Quadrapolar design", *IEEE Transactions on Plasma Science*, 29(5), pp. 738–743, October 2001.
173. Glinkowski, M., "Non-linear magnetic field distribution in vacuum interrupter contacts", *US Patent 6,747,233*, June 2004.
174. Shi, Z., Jia, S., Fu, J., Wang, Z., "Axial field contacts with non-uniform distributioned axial magnetic fields", *IEEE Transactions on Plasma Science*, 31(2), pp. 289–294, April 2003.
175. Shi, Z., Jia, S., Wang, L., Wang, Z., "Experimental investigation of high current vacuum arc under non-uniformly distributed axial magnetic fields", *Proceedings 22nd International Symposium on Discharges and Electrical Insulation in Vacuum*, pp. 221–224, September/October 2004.
176. Liu, Z., Wang, D., Rong, M., Wang, J., Wang, Q., "Comparison of vacuum arc behaviors for slot-type axial magnetic field contacts with and without iron plates", *IEEE Transactions, on Plasma Science*, 37(8), pp. 1458–1467, August 2009.
177. Zheng, Y., Wang, Z., Liu, Z., Hao, M., Wang, J., "Influence of irons on magnetic field characteristics of vacuum interrupter with cup type axial magnetic field contacts", *Proceedings 22nd International Symposium on Discharges and Electrical Insulation in Vacuum*, pp. 482–484, September 2006.
178. Niwa, Y., Asari, N., Sakaguchi, W., Daibo, A., Sekimori, Y., "Fundamental research of uniform vacuum arc control by magnetic field and its application to VCB", *Proceedings 27th International Symposium on Discharges and Electrical Insulation in Vacuum*, pp. 667–670, September 2016.
179. Liu, Z., Hu, Y., Wang, J., Wang, Q., Bi, D., He, G., "Analysis of axial magnetic field contacts with iron plates", *Proceedings 22nd International Symposium on Discharges and Electrical Insulation in Vacuum*, pp. 289–292, September 2006.
180. Kulkarni, S., Rajan, S., Andrews, L., Rajhans, R., Thomas, J., "Influence of uniformity of the axial magnetic field in a vacuum interrupter on performance of the contacts", *Proceedings 23rd International Symposium on Discharges and Electrical Insulation in Vacuum*, pp. 422–425, September 2008.
181. Steinke, K., Lindmayer, M., "Influence of contact distance on the appearance of vacuum arcs with different AMF contacts", *IEEE Transactions on Plasma Science*, 33(5), pp. 1600–1604, October 2005.
182. Li, M., Wang, X., Li, W., Zhang, X., "The effect of contact support material on magnetic field and breakdown of vacuum interrupter", *Proceedings 27th International Symposium on Discharges and Electrical Insulation in Vacuum*, pp. 517–520, September 2016.
183. Reece, M., "Electrical switching apparatus", *US Patent 2,897,222*, July 1959.
184. Alferov, D., Ivanov, V., Sidorov, V., "High-current vacuum switching devices for power energy storages", *IEEE Transactions on Magnetics*, 35(1), No. 1, pp. 323–327, January 1999.
185. Wang, Y., Lin, F., Dai, L., Zhang, J., "Research on the self-magnetic field distribution of the triggered vacuum switch with multirod system", *IEEE Transactions, on Plasma Science*, 44(11), pp. 2886–2892, November 2016.

186. Lamara, T., Gentsch, D., "High current vacuum arc investigation with new innovative TMF-AMF contacts", *Proceedings 25th International Symposium on Discharges and Electrical Insulation in Vacuum*, pp. 173–176, September 2012.
187. Kusano, T., Seki, T., Yamamoto, A., Sato, J., Homma, M., Oshima, I., "Control of vacuum arc motion by three-layer contacts in concentric circles", *Proceeding 19th International Conference on Electric Contacts*, pp. 253–257, September 1998.
188. Xiu, S., Lei, P., Wang, J., Lin, J., He, G., "Analysis of axial magnetic field electrode applied to high voltage vacuum interrupters", *Proceedings 22nd International Symposium on Discharges and Insulation in Vacuum*, pp. 317–320, September 2006.
189. Renz, R., "On criteria of optimization of AMF and TMF contact systems in vacuum interrupters", *Proceedings 19th International Symposium on Discharges and Electrical Insulation in Vacuum*, pp. 176–179, September 2000.
190. Gentsch, D., Shang, W., "High speed observations of arc modes and material erosion of TMF and AMF contacts", *IEEE Transactions on Plasma Science*, 33(5), pp. 1605–1610, October 2005.
191. Parashar, R., "Improved glass-ceramic envelope for vacuum interrupters", *Proceedings 25th International Symposium on Discharges and Electrical Insulation in Vacuum*, pp. 449–452, September 2012.
192. Kohl, W.H., "Ceramic-to-metal sealing", In: *Handbook of Materials and Techniques*, (pub. Reinhold, NY), Chapter 15, 1967, pp. 446–449.
193. Jaitly, N., Sudarshan, T.S., Dougal, R.A., Miller, H., "Degradation due to wet hydrogen firing of the high-voltage performance of alumina insulators in vacuum applications", *IEEE Transactions on Electrical Insulation*, EI-22(4), pp. 447–452, August 1987.
194. Greenwood, A., Schneider, H., Lee, T., "Vacuum type circuit interrupter", *US Patent 2,892,912*, 1959.
195. Mayo, S., Slade, P., Rosenkrans, B., Burmingham, D., Dunham, B., "Method and apparatus for mounting vapor shield in vacuum interrupter and vacuum interrupter incorporating the same", *U.S. Patent 6,417,473*, July 2002.
196. Schellekens, H., Hilderink, J., Ter Hennepe, J., "Process for fixing a metal screen in the housing of a vacuum switch, screen, therefore, and vacuum switch provided with such screen", *European Patent Pub. No. 0 406 944 B1*, December 1995 and *US Patent 5,077,883*, January 1992.
197. Cherry, S., "Two material vapor shield for vacuum type circuit interrupter", *US Patent 4.020,304*, April, 1977.
198. Wayland, P., "Arc resistant vapor condensing shield for vacuum type circuit interrupter", *US Patent 4,553,007*, November 1985.
199. Falkingham, L., "Design and development of the shieldless vacuum interrupter concept", *Proceedings 22nd International Symposium on Discharges and Electrical Insulation in Vacuum*, pp. 430–433, September/October 2004.
200. Del Rio Etaya, L., "Analysis of a 38kV vacuum interrupter based on a triple shielding research", *Proceedings 25th International Symposium on Discharges and Electrical Insulation in Vacuum*, pp. 114–116, September 2012.
201. Venna, K., Schramm, H., "Simulation analysis on reducing the electric field stress at the triple junctions & on the insulation surface of high voltage vacuum interrupters", *Proceedings 26th International Symposium on Discharges and Electrical Insulation in Vacuum*, pp. 53–56, September 2014.
202. Saitoh, H., Ichikawa, H., Nishijima, A., Matsui, Y., Sakaki, M., Honma, M., Okubo, H., "Research and development on 145kV/40kA one break vacuum circuit breaker", *Proceedings of the IEEE/T&D Conf. and Exhibition 2002*, 2, pp. 1465–1468, 2002.
203. Voshall, R., "Vacuum interrupter for high voltage application", *US Patent 3,792, 214*, February 1974.
204. Wang, Z., Wang, J., "Theoretical research and design on high voltage vacuum interrupter with long electrode distances", *Proceedings 17th International Symposium on Discharges and Electrical Insulation in Vacuum*, pp. 258–262, 1988.
205. Shioiri, T., Homma, M., Miyagawa, M., Kaneko, E., Ohshima, I., "Insulation characteristics of vacuum interrupter for a new 72/84 kV C-GIS", *IEEE Transaction Dielectrics and Insulation*, 6(4), pp. 486–490, August 1999.
206. Zhoa, Z., Zou, H., Wen, H., Sun, H., "A utility method of a high voltage vacuum interrupter design", *Proceedings 22nd International Symposium on Discharges and Insulation in Vacuum*, pp. 189–192, September 2006.
207. *Standards of the Expansion Joint Manufacturer's Association, Inc.*, 10th Edition, 2003, ejma@ejma.org.
208. Barkan, P., "A study of the influence of dynamic overstressing and annealing on the fatigue life of convoluted bellows", *Israel Journal of Technology*, 9, pp. 571–578, 1971.

209. Sellappan, A., Chandrasekharan, M., More, S., Kamble, D., Jaghel, D., "Reliable design of bellows and end components for vacuum interrupters having longer stroke length", *Proceedings 26th International Symposium on Discharges and Electrical Insulation in Vacuum*, pp. 477–480, September 2014.
210. Tsutsumi, T., Kanai, Y., Okabe, N., Kaneko, E., Kamikawaji, T., Homma, M., "Dynamic characteristics of high-speed operated long-stroke bellows for vacuum interrupters", *IEEE Winter Power Meeting*, 92 WM 240-2 PWRI, New York, NY, 1992.
211. Lafferty, J. (Editor), *Vacuum Arcs, Theory and Application*, (pub. John Wiley and Sons), p. 327, 1980.
212. Greenwood, A., *Vacuum Switchgear*, (pub. I.E.E.), Chapter 9, p. 197, 1994.
213. Bereza, A., "Methods of sealing and evacuating vacuum envelopes", *US Patent 3,656,255*, April 1972, and revisions; 27,773, August 1973.
214. Cline, C. (Editor), "Technology handbook, precision cleaning", *Precision Cleaning*, pp. 15–41, December 1996.
215. Whyte, W. (Editor), *Clean Room Design*, (pub. John Wiley and Sons), 1991.
216. Niak, P., *Vacuum Science and Technology*, (pub. CRC Press), 2018.
217. Fabian, R. (Editor), *Vacuum Technology, Practical Heat Treating and Brazing*, (pub. ASM International), Chapter 4. In: "Furnaces and Equipment" by J. G. Conybear.
218. User's Manual, ASM 142/ASM 142D Helium Leak Detector, Alcatel Vacuum Products, Hingham, MA, pp. A800-801, Ed. 06-03/2005.
219. *Varian Vacuum Products Catalog*, (pub. Varian Vacuum Products, Lexington, MA), 1996, p. 201.
220. Hildebrandt, S., Wiehl, G., "A new coper based sealing material: A comparison with AgCu(28) GeCo(0.3)", *Proceedings 27th International Symposium on Discharges and Electrical Insulation in Vacuum*, pp. 587–590, September 2016.
221. Giorgi, T., Ferrario, B., Storey, B., "An updated review of getters and gettering", *Journal of Vacuum Science and Technology A*, A3(2), pp. 417–422, March/April 1985.
222. Aranaga, S., Izcara, J., Del Río, L., Vallejo, H., Seco, M., "Thermal modelling of a brazing process of vacuum interrupters", *Proceeding 27th International Symposium on Discharges and Electrical Insulation in Vacuum*, pp. 425–428, September 2016.
223. Li, W.-P., "Private communication".
224. Penning, F., "Ein neues Manometer für niedrige gasdrucke, insbesondere zwischen 10^{-n} und 10^{-u} mm", *Physica*, 4(2), pp. 71–75, 1937.
225. Kageyama, K., "Magnetically confined low-pressure gas discharge generated in a vacuum switch", *Journal of Vacuum Science and Technology A*, 1(3), pp. 1522–1528, July–September 1983.
226. Kageyama, K., "Properties of a series cross-field discharge in a vacuum switch", *Journal of Vacuum Science and Technology A*, 1(3), pp. 1529–1532, July-September 1983.
227. Godechot, X., Nicolle, C., Hairour, M., Olive, S., Picot, P., "Investigation and optimization of magnetron discharge in a vacuum switch", *Proceedings 25th International Symposium on Discharges and Electrical Insulation in Vacuum*, pp. 465–469, September 2012.
228. Acharya, V., Kulkarni, S., Rajhans, R., Paulzagade, D., "Uses of magnetron discharge to assess the vacuum level of vacuum interrupter", *Proceedings 28th International Symposium on Discharges and Electrical Insulation in Vacuum*, pp. 93–96, September 2018.
229. Gentsch, D., Fugal, T., "Measurements by residual gas analysis inside vacuum interrupters", *IEEE Transactions on Plasma Science*, 37(8), pp. 1484–1489, August 2009.
230. Cobine, J., "Vacuum device gas measurement apparatus and method", *U.S. Patent 3,495,165*, February 1970.
231. Cherry, S., "Vacuum type circuit interrupters with condenser shield at a fixed potential relative to the contacts", *U.S. Patent 4,002,867*, January 1977.
232. Howe, F., "Leak sensor and indicating system for vacuum circuit interrupter", *U.S. Patent 4,103,291*, July 1978.
233. Frontzek, F., König, D., Heinemeyer, R., "Electrical methods for verifying internal pressure of vacuum interrupters after long-time service", *IEEE Transactions on Electrical Insulation*, 28(4), pp. 635–641, August 1993.
234. Damstra, G., Smeets, R., Poulussen, H., "Vacuum state estimation of vacuum circuit breakers", *SPIE*, 2259, pp. 266–269, February 1994.
235. Kitamura, T., Kobayashi, T., Osawa, Y., Usui, N., Yamada, M., "Method and apparatus for detecting a reduction in the degree of vacuum of a vacuum valve while in operation", *U.S. Patent 5,399,973*, May 1995.

236. Merk, W., Damstra, G., Bouwmeester, C., Gruntjes, R., "methods for estimation of the vacuum status in vacuum circuit breakers", *IEEE Transactions on Dielectrics and Electrical Insulation*, 6(4), pp. 400–404, August 1999.
237. Walczak, K., "Method for vacuum switch evaluation based on analysis of dynamic changes of electron emission current and X-radiation in time", *Proceedings 20nd International Symposium on Discharges and Electrical Insulation in Vacuum*, pp. 231–234, July 2002.
238. Zhang, X., Liu, Z., Fan, X., Huang, Z., Fan, J., Liang, C., Shi, W., "A high accurate sensor research and its application for VCBs' internal pressure on-line condition monitor", *Proceedings 25th International Symposium on Discharges and Electrical Insulation in Vacuum*, pp. 477–480, and 485–488, September 2012.
239. Schellekens, H., "Continuous vacuum monitoring in vacuum circuit breakers", *Proceedings 26th International Symposium on Discharges and Electrical Insulation in Vacuum*, pp. 465–468, September 2014.
240. Okawa, M., Tsutsumi, T., Alyoshi, T., "Reliability and field experience of vacuum interrupters", *IEEE Transactions on Power Delivery*, PWRD 2(3), pp. 799–804, July 1987.
241. Reeves, R., Falkingham, L., "The return of permanent gas pressurein sealed vacuum interrupters", *Proceedings 27th International Symposium on Discharges and Electrical Insulation in Vacuum*, pp. 615–618, September 2016.
242. User's Manual, ASM 142/ASM 142D Helium Leak Detector, Alcatel Vacuum Products, Hingham, MA, pp. A800-801, Ed. 06-03/2005.
243. Wilson, N., Beavis, L., *Handbook of Vacuum Leak Detection*, (pub. American Vacuum Soc.), 1976, p. 55.
244. Li, W.-P., Thomas, R., Slade, P., "Residual gas analysis of vacuum interrupters", *Proceedings 3rd International Conference on Electrical Contacts, Arcs, Apparatus and Their Applications*, pp. 491–498, May, 1997.
245. Slade, P., Li, W., Mayo, S., Smith, R., Taylor, E., "The Vacuum interrupter, the high reliability component of distribution switches, circuit breakers and contactors", *International Conference on Reliability of Electrical Components & Electrical Contacts*, pp. 335–341, March 2007.
246. Renz, R., Gentsch, D., Fink, H., Slade, P., Schlaug, M., "Vacuum interrupters, sealed for life", *19th International Conference on Electrical Distribution*, May 2007.
247. Taylor, E., Lawall, A., Gentsch, D., "Long term integrity of vacuum interrupters", *Proceedings26th International Symposium on Discharges and Electrical Insulation in Vacuum*, pp. 433–436, September 2014.

Part 2

Vacuum Interrupter Application

4 General Aspects of Vacuum Interrupter Application

In our description of nature, the purpose is not to disclose the real essence of the phenomena, but only to track down, so far as it is possible, relations between the manifold aspects of our experience.

Niels Bohr
(Atomic Theory and the Description of Nature)

4.1 INTRODUCTION

In this chapter, I will first discuss the current interruption process that occurs inside a vacuum interrupter operating in an ac circuit. I will then present a general review of contact welding. Vacuum interrupters have been used primarily to switch and/or protect alternating current (ac) circuits. There has also been an increasing interest in using the vacuum interrupter to interrupt direct current (dc) circuits. I will discuss the development of dc applications in Chapter 6. In Chapter 2 in this volume, I described the contact opening process and the formation of the various modes of vacuum arc. I also began a discussion on the interaction of the vacuum arc with transverse magnetic fields (TMF) and axial magnetic fields (AMF). In Chapter 3 in this volume, I reviewed the contact materials that have been successfully developed for vacuum interrupter application and I also discussed the TMF and the AMF contact structures that control the high current vacuum arc. The effectiveness of these contact structures for interrupting ac current will be analyzed in this chapter.

When a vacuum interrupter is called upon to open in an ac circuit, its contacts first separate at a random point on the ac current wave. For example, if the contacts open just after a current zero the subsequent vacuum arc will pass close to a half cycle of current and there will be a large contact gap at the next current zero. If they separate close to the maximum current, the effects of the initial bridge column arc will be greatest. When the contacts separate just before a current zero there is an opportunity for the very short, recovering contact gap to reignite the vacuum arc and allow another half cycle of arcing. I will discuss this particular case in Section 4.3. Once the contacts have parted and the vacuum arc has formed, the arc will continue to the natural current zero of the ac current. At current zero, there is a very brief time when the contact gap changes from a relatively good conductor (with the arc present) to a good insulator that can hold off the voltage impressed across the open contacts by the ac circuit. In this chapter I will discuss the interruption process for the forms of vacuum arc presented in Chapter 2, in this volume; i.e., the diffuse vacuum arc, the columnar vacuum arc and briefly, the transition vacuum arc. I will also show the importance of the TMF and AMF contact designs for the successful interruption of high current vacuum arcs in ac circuits.

If the ac circuit contains only resistive elements (i.e., the inductance and capacitance are both negligible), the ac current wave will be in phase with the circuit voltage (see Figure 4.1). Once the contacts part an arc voltage, U_A, appears across the contacts. In an ideal case, when the current, i, reaches zero, the vacuum arc extinguishes and the transient recovery voltage (TRV), $U_R(t)$, is impressed across the contacts. In this case the impressed voltage is the same as the circuit voltage, which reaches its maximum value (i.e., the peak circuit voltage) a quarter of a cycle later: 5ms for a 50Hz circuit and 4.3ms for a 60Hz circuit. During this relatively slow increase in voltage the contact gap can continue to increase to its maximum value. It should be noted that the anode contact during the arcing period now becomes the new cathode during the recovery stage.

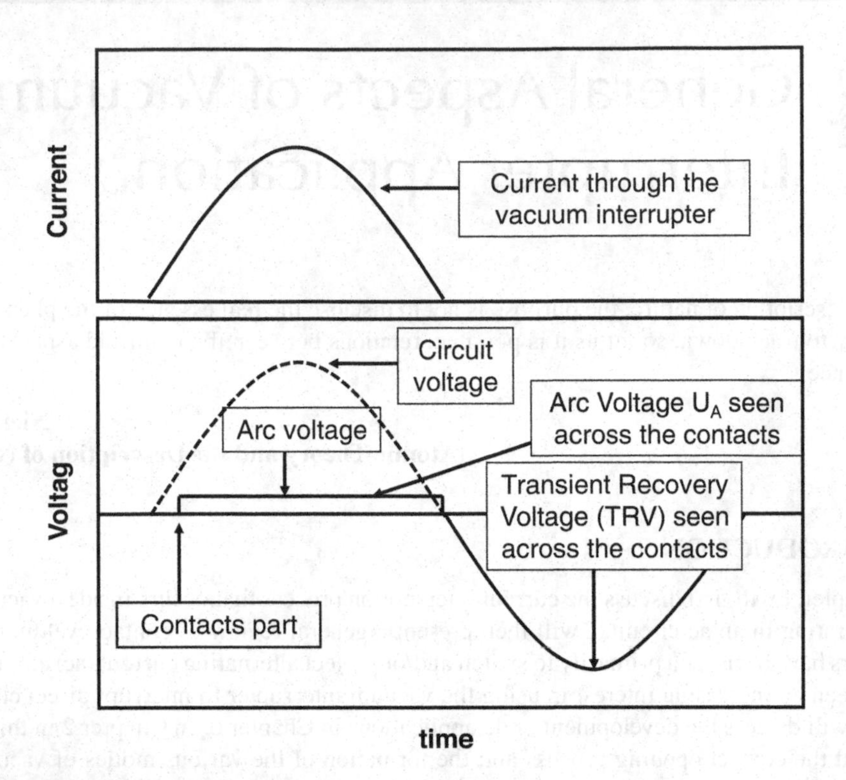

FIGURE 4.1 Schematic to show the transient recovery voltage (TRV) across the open vacuum interrupter contacts after the interruption of ac current in a resistive circuit.

In an inductive circuit (i.e., the resistance and capacitance are now negligible), the situation at current zero is quite different. In this case, when the contacts are closed, the circuit voltage leads the current by 90° (see Figure 4.2). When the contacts open, the vacuum arc forms with an arc voltage, U_A, which continues to the natural ac current zero. At current zero, when the arc extinguishes, the circuit voltage will be close to its peak value. The shape of the recovery voltage that now appears across the open contacts is complicated by the small capacitance that is present in all inductive circuits. This comes, for example, from the capacitance between the turns of the inductance and from a stray capacitance to ground. This small capacitance can be considered a parallel element to the arc. At the current zero it will be charged to a voltage approximately equal to the maximum value of the supply voltage plus the value of the arc voltage at the extinction of the vacuum arc. Once the vacuum arc has extinguished at current zero, the circuit voltage appears at both contacts, on the line side from the source and on the load side from the stray capacitance (C_L) to ground. Thus, the voltage across the contacts is initially zero and then begins to increase with a $1\text{-}cos(\omega t)$ shape where, $\omega = 2\pi/[1/C_L L]^{1/2}$. This voltage across the contacts rises to about 1.7 times the circuit voltage in about 10 to 70 μ sec. There is now a transfer of electro-magnetic energy stored in the inductance to electrostatic energy in the capacitor and vice versa. This results in a high frequency voltage that is superimposed on and oscillates around the circuit voltage. The shape of this recovery voltage is shown in Figure 4.2. It oscillates above and below the circuit voltage until the resistive damping leaves the open circuit voltage across the contact gap; usually in less than a quarter cycle.

Understanding the inductive circuit case is important. First, it is intuitively obvious that this rapidly increasing recovery voltage after current zero would be more difficult to withstand than the slower resistive recovery voltage discussed above. Second, all ac circuits have some level of inductance. Common circuits including electric motors can have a high level of inductance. Again, it should be noticed that the anode contact during the arcing phase before current zero is now the new cathode during the recovery phase.

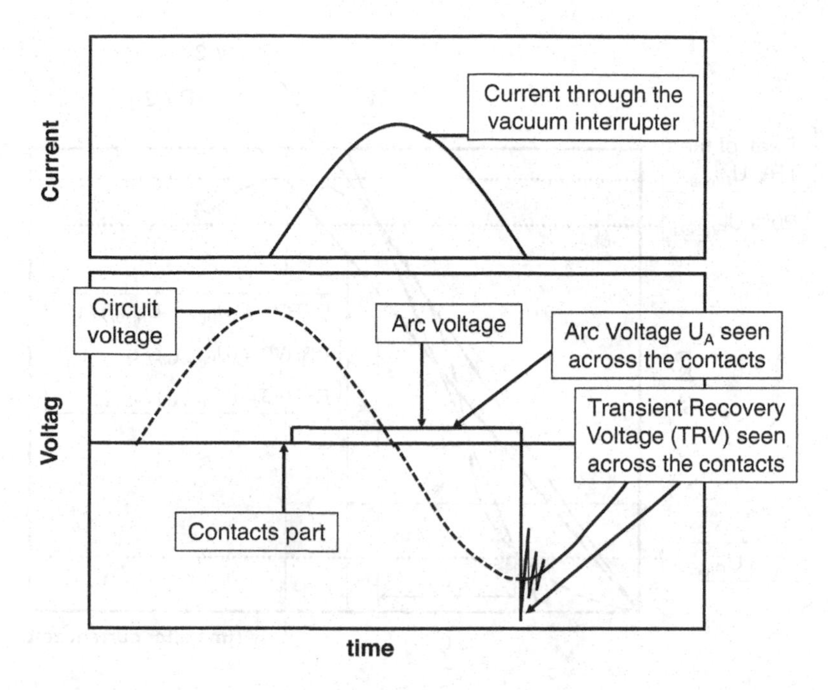

FIGURE 4.2 Schematic to show the transient recovery voltage (TRV) across the open vacuum interrupter contacts after the interruption of ac current in an inductive circuit.

Thus, any discussion of the changes in the contact gap after current zero that describes the change to a good insulating gap should include the effects of a rapidly rising voltage pulse whose peak value is above the peak value of circuit voltage. There are three terms that are important to future discussions:

(1) RRRV, the Rate of Rise of Recovery Voltage. Unfortunately, the definition of the value of RRRV varies with each researcher. Figure 4.3 shows the three most common ways of defining it. Some researchers also use an equivalent frequency. For a vacuum interrupter designer, each way can be used without a problem. I will discuss the TRV definition given by the IEEE and the IEC standards in Section 6.3.3 in this volume
(2) TRV, the Transient Recovery Voltage, $U_R(t)$, usually is the description of the whole recovery voltage pulse to its peak value or perhaps just after its peak value
(3) The peak recovery voltage, $U_{R(peak)}$, is the peak of the TRV, shown in Figure 4.3

Interrupting a purely capacitive circuit is a special case, which I will discuss in Sections 4.3.1.2 and 5.5, in this volume.

4.2 THE INTERRUPTION OF AC CIRCUITS

4.2.1 The Interruption of the Diffuse Vacuum Arc for AC Currents Less Than 2 kA (rms.) with a Fully Open Contact Gap

In this section, a fully open contact gap can mean a contact gap of at least several millimeters. At current zero in an ac circuit the rate of change of current is given by:

$$\frac{di}{dt} = -\omega i_{peak} \tag{4.1}$$

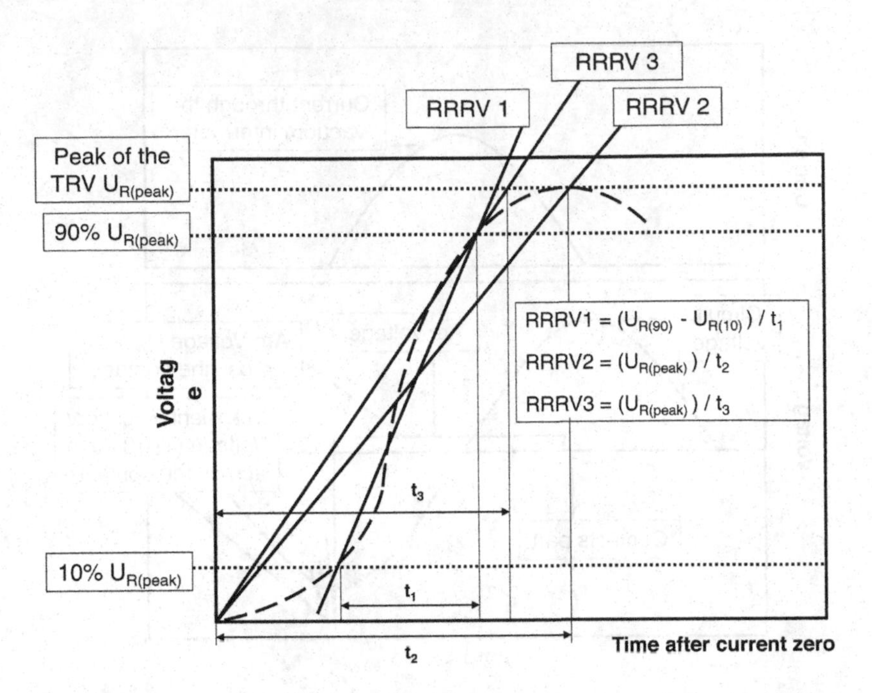

FIGURE 4.3 The transient recovery voltage (TRV) showing three ways of measuring the rate of rise of recovery voltage (RRRV).

i.e.,

$$\frac{di}{dt} = -2\pi \times 50 \times i_{peak} \left(\text{for 50Hz circuits}\right) \tag{4.2}$$

$$\frac{di}{dt} = -2\pi \times 60 \times i_{peak} \left(\text{for 60Hz circuits}\right)$$

For ac currents of about 2000A (rms.), the rate of change of current at current zero is thus approximately 1 A/μs. We already know from Section 2.3.3 in this volume that the cathode spots in a diffuse vacuum arc can react almost instantaneously to changes in current up to 100A/μs. Thus, as the current approaches current zero, these vacuum arcs will be characterized by a gradual extinction of the cathode spots, until just before current zero when only one spot remains. Also, from the discussion of current chop in Sections 2.3.3 and 3.2.6 in this volume, this final cathode spot will self-extinguish just before the true current zero. As I discussed in Chapter 2, the anode contact during the arcing phase of the diffuse vacuum arc is a passive collector of electrons over most of its surface. In a vacuum interrupter with its usual contact gap of 6mm to 15mm, there will be little or no excessive heating of the anode contact at this level of current. Once the arc extinguishes, the old anode now becomes the new cathode during the recovery phase. Experience has shown that in all practical vacuum interrupter designs it is almost impossible NOT to interrupt a current of 2000A (rms.) or less, in most ac circuits.

The above statement, at first sight, seems to be rather a bold one. Let us examine some of the extensive experimental evidence that supports it. One powerful tool that has been in use since the early days of vacuum interrupter development has been to study the recovery rate of the contact gap's dielectric strength after the extinction of the vacuum arc. At this current level the experiments usually involve a variation of the "free recovery" method that was first developed to investigate the recovery of arcs in air [1]. A vacuum arc is established between the opening contacts. The current can be either ac or dc. Once the contacts have reached a given separation and the vacuum arc has been established for a given length of time, the current is forced to zero using an auxiliary circuit in a time of 0.5μs to 3μs [2, 3]. At a given delay following the arc extinction, the dielectric strength of the recovering contact gap is investigated by applying a step function voltage pulse. The magnitude of this pulse can be varied for a given delay until a dielectric breakdown of the contact gap is obtained. By varying the delay time of the reapplied voltage, a free recovery reignition voltage characteristic as a function of time after current zero can be obtained. Of course, the value of the breakdown voltage for a given time delay can have some considerable variation. The 50% value, or mean value, is quite often used.

A variation of this technique is to apply a fast ramp voltage pulse (say 250kV/μs) after a given time delay and then monitor the voltage at which the dielectric breakdown of the contact gap occurs [4]. Figure 4.4 shows typical free recovery curves for contact gaps of a few millimeters after the forced reduction to zero from arc currents of about 200A. Here it can be seen that (a) the recovery rate does depend upon the contact material and (b) the contact gap recovers to its full strength very rapidly even though the *di/dt* at arc extinction is in excess of 100A/μs. From Equation (4.2), this *di/dt* is about two orders of magnitude greater than would be expected from the peak value of a 2 kA (rms.) ac current: a current at which a fully diffuse vacuum arc would occur.

In a series of very instructive experiments Lins [5–7] has measured the metal vapor density in the contact gap after a natural ac current zero for arcing currents in the range 500A (rms.) to 2000A (rms.). Figure 4.5 shows the Cu vapor density before and after current zero of a 500A (rms.) vacuum arc for both metal vapor evaporated from the cathode and also from the molten copper particles exiting the contact gap at $300ms^{-1}$. It can be seen that at current zero the total Cu density in mid-contact gap from the cathode spots and the molten particles is about 10^{17} atoms.m^{-3}. Relating these data to Table 4.1 it can be seen that this corresponds to a pressure of less than about 10^{-2} Pa even if the contact surfaces are at 2150K. At this pressure the electron mean free path is significantly greater than the contact gap. So, clearly any breakdown of the gap has to be considered a "vacuum breakdown" event and will not have been caused by a Townsend avalanche resulting from electron ionization of the residual metal vapor in the contact gap. In further work using Cu–Cr contacts and currents of 1kA (rms.) and 2kA (rms.), Lins has developed Figure 4.6(a) and (b). Note here the data begin 40μs after the current zero. Even so, the residual neutral metal vapor density is still too low

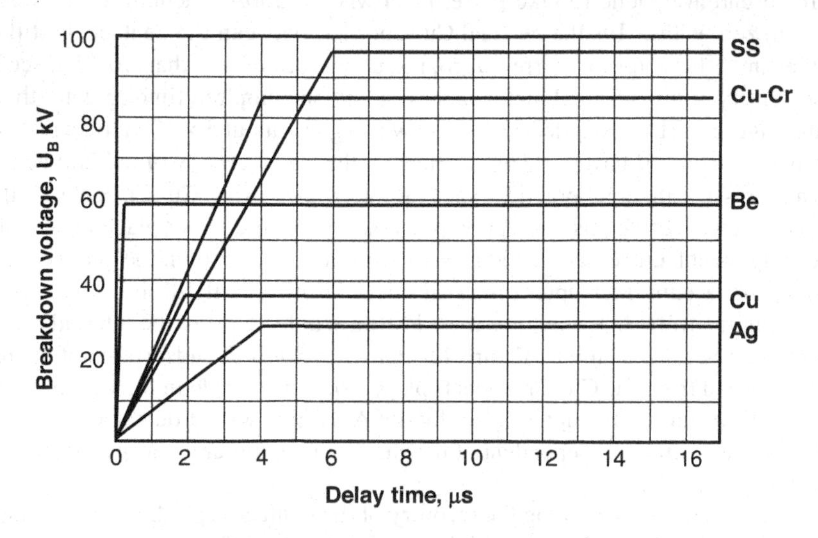

FIGURE 4.4 The free recovery of the vacuum contact gap after a current of about 200A has been ramped to zero in less than 2μs for a diffuse vacuum arc; the Cu–Cr data from [4], the other contact materials [2].

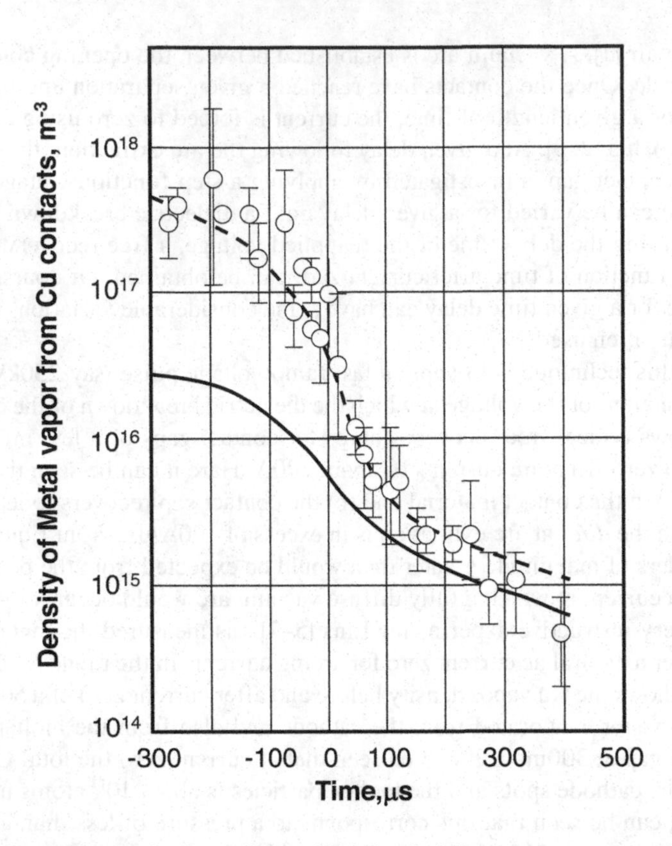

FIGURE 4.5 The Cu vapor density for a 50 Hz, 500A (rms) diffuse, vacuum arc at the center of a 14mm contact gap before and after current zero: [O] experimental data, [– – – –] calculated assuming the cathode has a temperature of 2000 K and an effective erosion rate of 3μg/C, [———] calculated vapor contribution from Cu molten droplets in flight, which begins with a temperature of 2000 K, a diameter of 10 μm and a velocity of $300ms^{-1}$ [6].

to allow a Townsend avalanche to take place. Even when a 200A, vacuum arc between Cu contacts is forced to zero within 1μs the residual Cu vapor left between the contacts is still only about 1.2×10^{18} $atoms.m^{-3}$, 1μs after the current zero, i.e., a pressure of less than 10^{-1} Pa; see Figure 4.7. Although the density of this residual metal vapor is about one hundred times greater than that for a 2kA (rms) arc, Figure 4.6(b), and it decays very slowly (e.g., to about 5×10^{17} $atoms.m^{-3}$ after 200μs) the residual metal vapor will have little or no effect on the free recovery of the 2mm contact gap.

The internal pressure inside a practical vacuum interrupter using Cu–Cr contacts after current interruption is affected by two gas adsorption processes. First, the getter placed inside the vacuum interrupter during manufacture: see Section 3.5.1 in this volume. Second, the gettering effect of the Cr deposited on the vacuum interrupter's internal shield. Weuffel et al. [8] show that for dc arcs with durations (t_{arc}) ranging from 3ms to 960ms there is always a decrease in the internal pressure of the vacuum interrupter: See, for example, Figure 4.8. For $t_{arc} \geq 30ms$ the adsorption of the residual gas by the metal deposited from the Cu–Cr contacts plays a significant role in the reduction of the internal pressure of the vacuum interrupter. Also, for 460A dc arcs with a duration greater than 30ms the internal pressure is almost independent of the vacuum interrupter's internal pressure before the arcing occurred.

Another powerful way of examining the recovery of the contact gap's dielectric strength is to use the Weil–Dobke "synthetic-circuit" approach [9]. A typical circuit that has been used by Wilkening et al. [10] is shown in Figure 4.9. It consists of a low-voltage, high-current, ac power supply on the

TABLE 4.1

An Estimation of the Gas Density, Electron Mean Free Path and Metal Vapor Mean Free Path as a Function of Pressure and Temperature

Pressure		Pressure × 10mm contact gap		Temperature, K					
mbar	Pa	mbar.mm	Pa.m	300			1350		
				Number density, $n.m^{-3}$	Approx. electron mean free path,mm	Approx. metal atom mean free path,mm	Number density, $n.m^{-3}$	Approx. electron mean free path,mm	Approx. metal atom mean free path,mm
10^3	10^5	10^4	10^3	2.4×10^{25}	4×10^{-4}	3×10^{-5}	5.3×10^{24}	1.8×10^{-3}	1.4×10^{-4}
10^2	10^4	10^3	10^2	2.4×10^{24}	4×10^{-3}	3×10^{-4}	5.3×10^{23}	1.8×10^{-2}	1.4×10^{-3}
10	10^3	10^2	10	2.4×10^{23}	4×10^{-2}	3×10^{-3}	5.3×10^{22}	1.8×10^{-1}	1.4×10^{-2}
1	10^2	10	1	2.4×10^{22}	0.4	3×10^{-2}	5.3×10^{21}	1.8	0.14
10^{-1}	10	1	10^{-1}	2.4×10^{21}	4	0.3	5.3×10^{20}	18	1.4
10^{-2}	1	10^{-1}	10^{-2}	2.4×10^{20}	40	3	5.3×10^{19}	1.8×10^2	14
10^{-3}	10^{-1}	10^{-2}	10^{-3}	2.4×10^{19}	4×10^2	30	5.3×10^{18}	1.8×10^3	1.4×10^2

Pressure		Pressure × 10mm contact gap		Temperature, K					
mbar	Pa	mbar.mm	Pa.m	1700			2150		
				Number density, $n.m^{-3}$	Approx. electron mean free path,mm	Approx. metal atom mean free path,mm	Number density, $n.m^{-3}$	Approx. electron mean free path,mm	Approx. metal atom mean free path,mm
10^3	10^5	10^4	10^3	4.3×10^{24}	2.2×10^{-3}	1.7×10^{-4}	3.4×10^{24}	2.8×10^{-3}	2.1×10^{-4}
10^2	10^4	10^3	10^2	4.3×10^{23}	2.2×10^{-2}	1.7×10^{-3}	3.4×10^{23}	2.8×10^{-2}	2.1×10^{-3}
10	10^3	10^2	10	4.3×10^{22}	2.2×10^{-1}	1.7×10^{-2}	3.4×10^{22}	2.8×10^{-1}	2.1×10^{-2}
1	10^2	10	1	4.3×10^{21}	2.2	0.17	3.4×10^{21}	2.8	0.21
10^{-1}	10	1	10^{-1}	4.3×10^{20}	22	1.7	3.4×10^{20}	28	2.1
10^{-2}	1	10^{-1}	10^{-2}	4.3×10^{19}	2.2×10^2	17	$3.4x 10^{19}$	2.8×10^2	21
10^{-3}	10^{-1}	10^{-2}	10^{-3}	4.3×10^{18}	2.2×10^3	1.7×10^2	$3.4x 10^{18}$	2.8×10^3	2.1×10^2

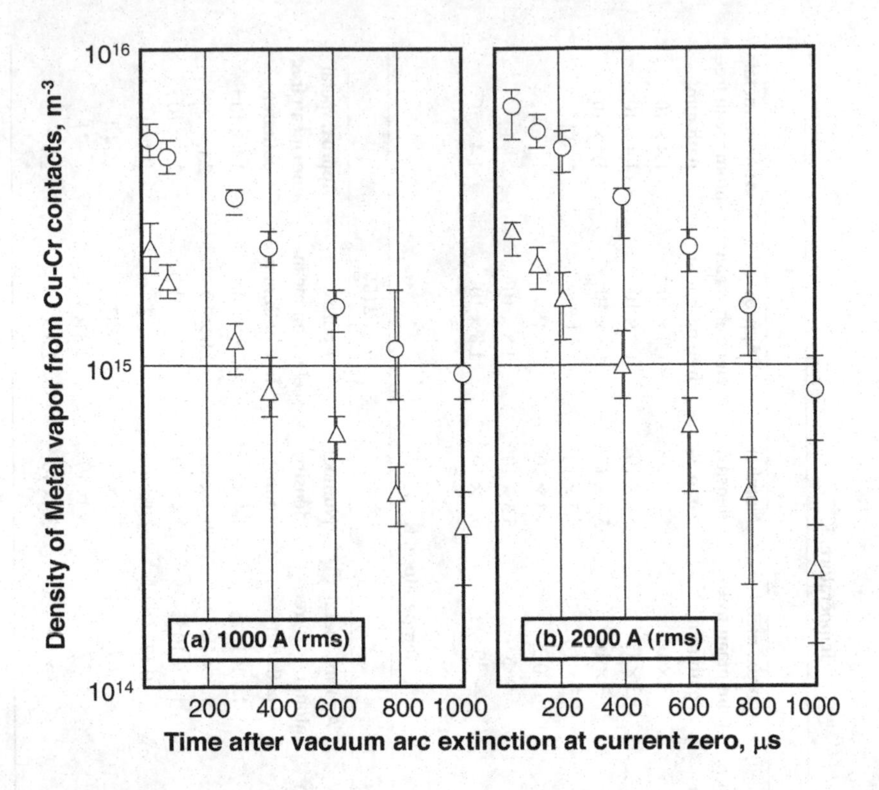

FIGURE 4.6 The vapor density of Cu and Cr from Cu–Cr contacts after current zero after interrupting 50 Hz, ac currents of 1000A (rms) and 2000A (rms): [O] Cu data, [△] Cr data [7].

left-hand side and a high-voltage, low current, high frequency power supply on the right-hand side. The main current flow, a half cycle (50Hz) ac wave, begins at time t_0 from a low voltage circuit. At this time, the high voltage power supply is isolated from this circuit. At time t_1 the contacts of the vacuum device being tested open and the vacuum arc is initiated. This can then give up to about 9ms of arcing. Longer arcing times allow a larger contact gap at current zero. An isolation switch in

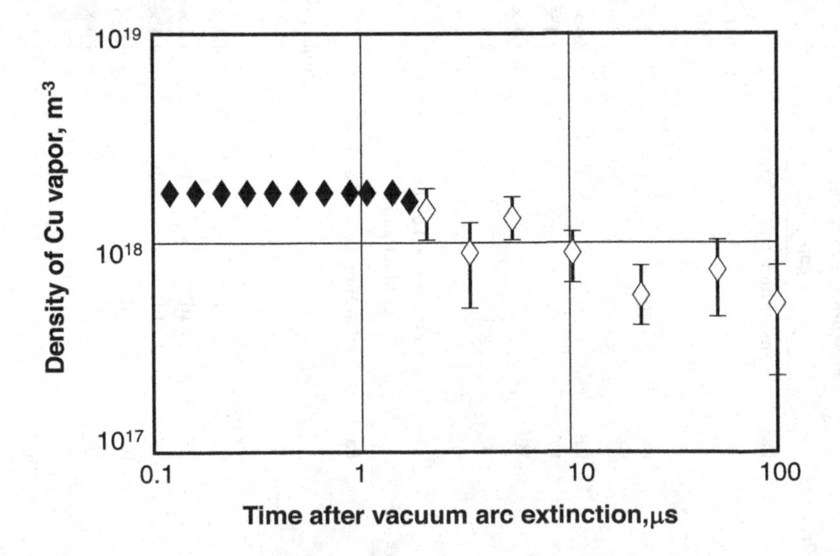

FIGURE 4.7 The Cu vapor density in a 2mm contact gap after the current for a 200A vacuum arc is ramped to zero in less than 2 μs [4].

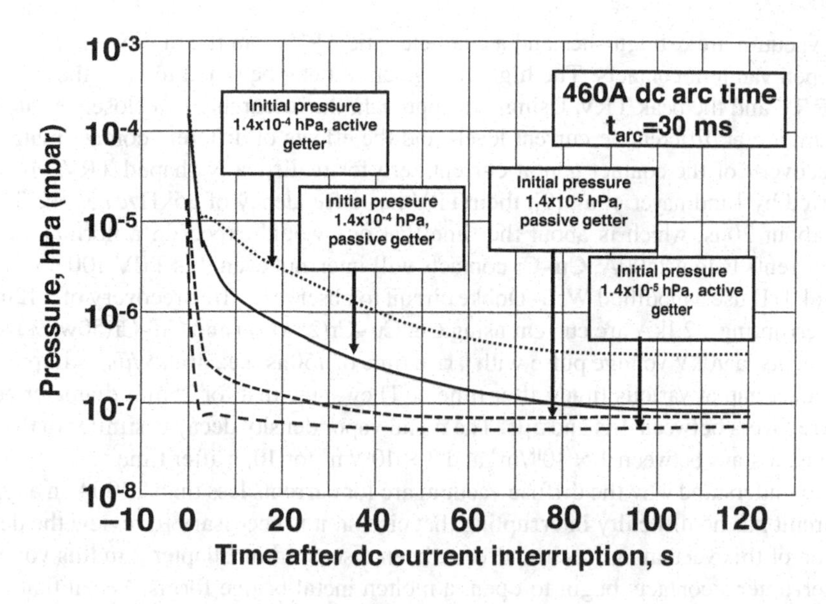

FIGURE 4.8 An example of the change in the internal pressure inside a vacuum interrupter with Cu–Cr(25wt.%) contacts with active and passive getters after interrupting 460A dc current at t = 0 [8]

series with the vacuum contacts opens at time t_2. This switch interrupts the main current at time t_4. At time t_3 a high frequency current from the parallel high voltage power supply is injected through the open vacuum contacts. Then the vacuum arc continues operating to the natural current zero of this high frequency current at time t_5. During the time interval $(t_5–t_4)$ the isolation switch fully recovers to its design withstand voltage and thus isolates the main current source from the high voltage, high frequency power supply. When the high frequency current passing through the vacuum contacts goes

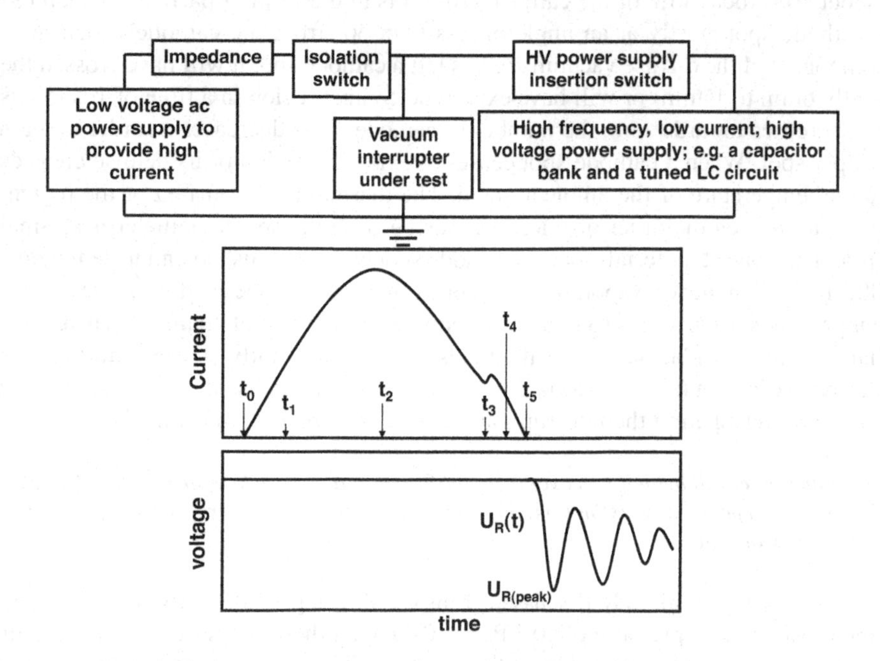

FIGURE 4.9 The Weil–Dobke synthetic circuit for evaluating the interruption performance of a vacuum interrupter [10].

to zero, the vacuum arc extinguishes and a characteristic TRV from the high voltage power appears across the open vacuum contacts. The high voltage circuit can be tuned to vary the di/dt at current zero, the RRRV and the peak TRV. Using this approach, the researcher can closely relate the effects of the vacuum arc at different ac current levels and the effects of different contact materials on the dielectric recovery of the contact gap at current zero for realistically shaped TRVs. In the experiments reported by Lindmayer et al. [9], their TRV has a frequency of 25kHz, i.e., the TRV peak is reached in about 20μs, which is about the time that one would expect in a normal, ac, inductive circuit. At currents below 2500A, Cu–Cr contacts will interrupt even this TRV 100% of the time.

Wang et al. [11] use a modified Weil–Dobke circuit to observe the free recovery of a 12mm contact gap after interrupting a 2.1kA arc current using Cu, Cu–Cr(25wt.%) and Cu–Cr(50wt.%) contacts. In their experiments, a 90kV voltage pulse with a rise time of 150ns (i.e., 460kV/μs) is impressed across the open contact gap at various times after time t_5. They show that for 25mm diameter contacts the contact gap recovers between 4μs and 5μs. The metal vapor density decay is similar to that shown on Figure 4.7: i.e., it stays between $1 \times 10^{18}/\text{m}^3$ and $3 \times 10^{18}/\text{m}^3$ for 10μs after time t_5.

In order to understand why the diffuse vacuum arc for currents less than 2000A in a typical 50 or 60 Hz ac circuit has no difficulty interrupting that circuit, it is necessary to review the development and operation of this vacuum arc as has already been discussed in Chapter 2 in this volume. As the vacuum interrupter's contacts begin to open, a molten metal bridge forms. When that bridge ruptures, a high-pressure bridge column arc develops in the region previously occupied by the molten metal bridge. This high-pressure arc will endure until the evaporation of metal vapor from the arc roots is no longer enough to replace the metal vapor lost to the surrounding vacuum and is also no longer enough to maintain the required arc pressure as the contacts continue to part and the total arc volume increases. When this happens, the diffuse vacuum arc forms with cathode spots (each with a current 50A–100A) moving in a retrograde motion over the cathode's surface with speeds up to 10^2m.s^{-1}. The cathode spots produce electrons, ions, neutral metal vapor and metal particles. For currents below 2000A, the ions have energies up to about 50eV and have speeds up to 10^4m.s^{-1}. The electrons have energies in the range of 1eV–4eV and have speeds of the same order as the ions.

The ions and electrons, for the most part, leave the cathode spot in a cone whose cross-sectional angle is about 70°. About 80% of the cathode erosion is in the form of particles, which move away from the cathode spot mostly at an angle of less than 30° from the cathode's surface. Once the plasma component of the diffuse vacuum arc has left a cathode spot it will have crossed the contact gap (typically 6mm to 15mm) or will have exited the contact region in a time of 1μs to 2μs. As the ac current decreases towards zero, the number of cathode spots decreases in order to maintain the 50A–100A per spot. When a cathode spot ceases to exist, the region of its demise cools extremely rapidly to the temperature of the ambient metal. The thermal time constant of the region close to the spot's location is on the order of microseconds because the spot's diameter is so small. Thus, for most practical contact materials, once a cathode spot has extinguished and its temperature drops below 1000°K the continued evaporation of contact material will be negligible; see Table 4.2. As the current proceeds to current zero in an ac circuit, where the rate of change of current is so much slower than the time for plasma dispersal and also very significantly slower than the after effects of cathode spot extinction (e.g., at 2000A rms. di/dt at current zero is about 1A/μs), the vacuum arc itself seems to be anticipating the interruption process. To paraphrase Farrall [12]:

> *a vacuum interrupter will begin to recover from a diffuse vacuum arc while the arc is still burning, just after the sinusoidal peak. ... At extinction, the contact gap only "remembers" a small period of arcing just preceding extinction.*

At current zero, as I have already discussed, Lins's and Wang et al.'s experiments show that the residual metal vapor has a pressure of 10^{-1} Pa to 10^{-2} Pa. At these pressures the electron mean free path is much greater than the contact gap. Thus, you would not expect the residual metal vapor to play a role in the re-establishment of the vacuum arc between the contacts. For a diffuse vacuum

TABLE 4.2
Vapor Pressure of Contact Metals at Increasing Temperatures

	Temperature, K							
	300K	1000K	1360K	1500K	1750K	2000K	2150K	2300K
Metal	Vapor Pressure, Pa	Vapor Pressure, Pa	Vapor Pressure, Pa	Vapor Pressure, Pa	Vapor Pressure, Pa	Vapor Pressure, Pa	Vapor Pressure, Pa	Vapor Pressure, Pa
Bi	$< 10^{-8}$	5.3	1.3×10^3	6.7×10^3	$> 10^5$	$> 10^5$	$> 10^5$	$> 10^5$
Cu	$< 10^{-8}$	2×10^{-6}	5.3×10^{-2}	9.3×10^{-1}	27	4×10^2	1.3×10^3	4×10^3
Cr	$< 10^{-8}$	1.2×10^{-8}	2×10^{-3}	6.7×10^{-2}	5.3	133	6.7×10^2	2.7×10^3
Ag	$< 10^{-8}$	5.3×10^{-4}	2.7	40	1.3×10^3	6.7×10^3	2.1×10^4	10^5
W	$< 10^{-8}$	$< 10^{-8}$	$< 10^{-8}$	$< 10^{-8}$	$< 10^{-8}$	$< 10^{-8}$	$< 10^{-8}$	2.7×10^{-7}

arc the ions carry about 10% of the current and the electrons carry the remainder. Just before current zero both the electrons and the ions are moving away from the last cathode spot as shown in Figure 4.10(a). Just after current zero, the former anode becomes the new cathode and the former cathode becomes the new anode. The ions thus continue their motion towards the new cathode, but the electrons very quickly reverse their direction towards the new anode, see Figure 4.10(b). So, the stage is set to discuss the changes that occur between the contacts after the current zero as the TRV appears across them.

The discussion begins with the work by Johnson et al. [13] who have analyzed what would happen when an identical floating double probe is placed in an ionized plasma and is separated from the source that created the plasma. When a difference in potential is applied between the probes, the floating system rapidly adjusts itself so that the more positive probe will be close to the plasma potential. That is, neither probe becomes greatly positive with respect to the plasma nor are large electron currents drawn to them. Also, the potential difference applied to the probes does not penetrate the main body of the plasma but appears across a space charge sheath adjacent to the negative probe. As the current in a vacuum arc approaches very close to the ac current zero, the intercontact plasma in the vacuum interrupter adjusts itself to maintain quasi charge neutrality. In fact, for

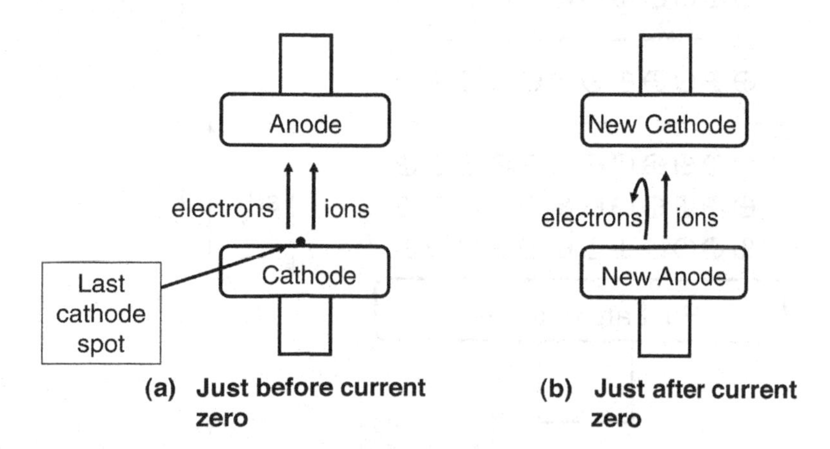

FIGURE 4.10 A schematic showing the motion of the electrons and ions from the last cathode spot before and after current zero and the initial rise of the transient recovery voltage.

currents of less than 2000A (rms.), the vacuum arc chops out just before current zero (see Section 2.3.3 in this volume). All emission of electrons and ions from the last cathode spot ceases. A low impedance plasma now bridges the intercontact gap as the restored voltage appears across the open contacts. The ions, which have considerable inertia, will continue to move toward the new cathode (the former anode). This residual current is termed the *post-arc current* (i_{pac}). The electrons, however, will rapidly decelerate, come to a halt (e.g., if they are moving at $10^{-3}ms^{-1}$, it will take a potential of only $10^{-5}V$ to achieve this) and try to reverse direction. The conditions will thus be created for the ions and electrons to rapidly form a stationary, quasi-neutral plasma between the contacts. This is similar to the double floating probe discussed above. During the brief adjustment period (typically about 1μs) there will be little voltage drop across the positive space charge sheath that appears at the new cathode. Soon, however, the potential drop from the TRV will appear across the sheath. There is practically zero voltage drop across the rest of the plasma, i.e., the plasma has the same potential as the new anode, see Figure 4.11. Figure 4.12 shows an example of a typical post-arc current pulse (i_{pac}) that would be observed in a vacuum interrupter with a floating shield [14]. The current crosses zero at time $t = 0$, then reaches a peak post-arc current, $\hat{\imath}_{pac}$, in the opposite polarity. The di_{pac}/dt during this initial stage is similar to that of the vacuum arc just before the current zero. This results from the ion current flowing to the new cathode, which has an initial velocity of up to 10^4ms^{-1} before the neutral plasma is established and the positive space charge sheath is established. During the initial period t_d there is a delay in the appearance of the TRV across the contacts. After an initial sharp decrease in i_{pac} there follows a flatter region until t_k. Interestingly, up to this time the shield potential has remained zero, i.e., the shield has been connected to the anode by way of the conducting plasma. At the time t_k, when there is a distinct drop in i_{pac}, the shield voltage begins to follow the capacitive distribution.

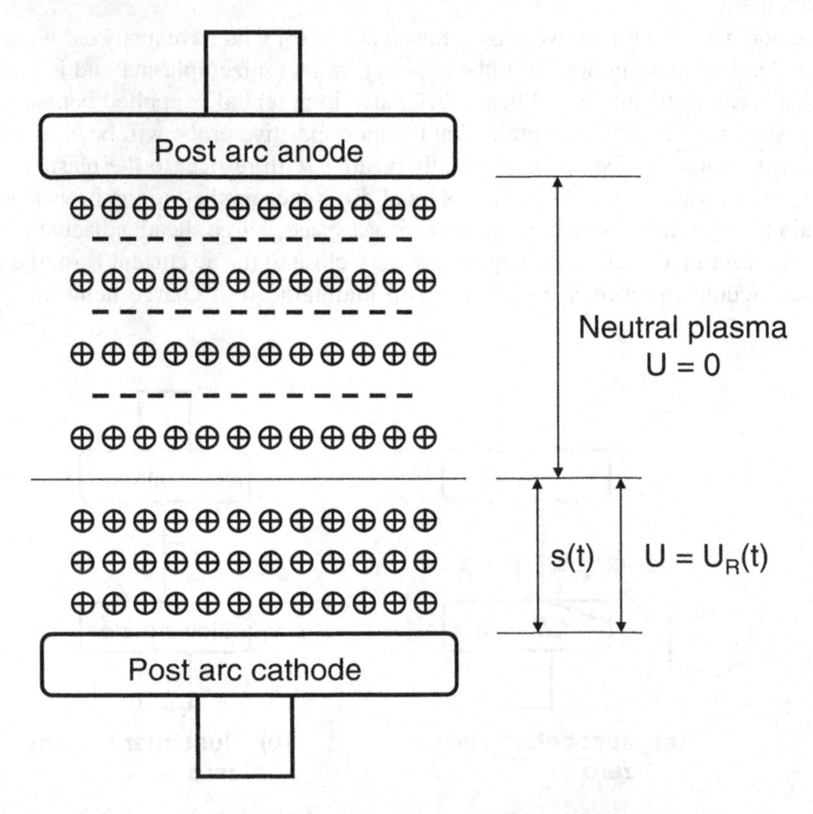

FIGURE 4.11 A schematic illustrating the sheath growth model during the recovery phase of intercontact region after the interruption of a diffuse vacuum arc and the initial rise of the transient recovery voltage [14].

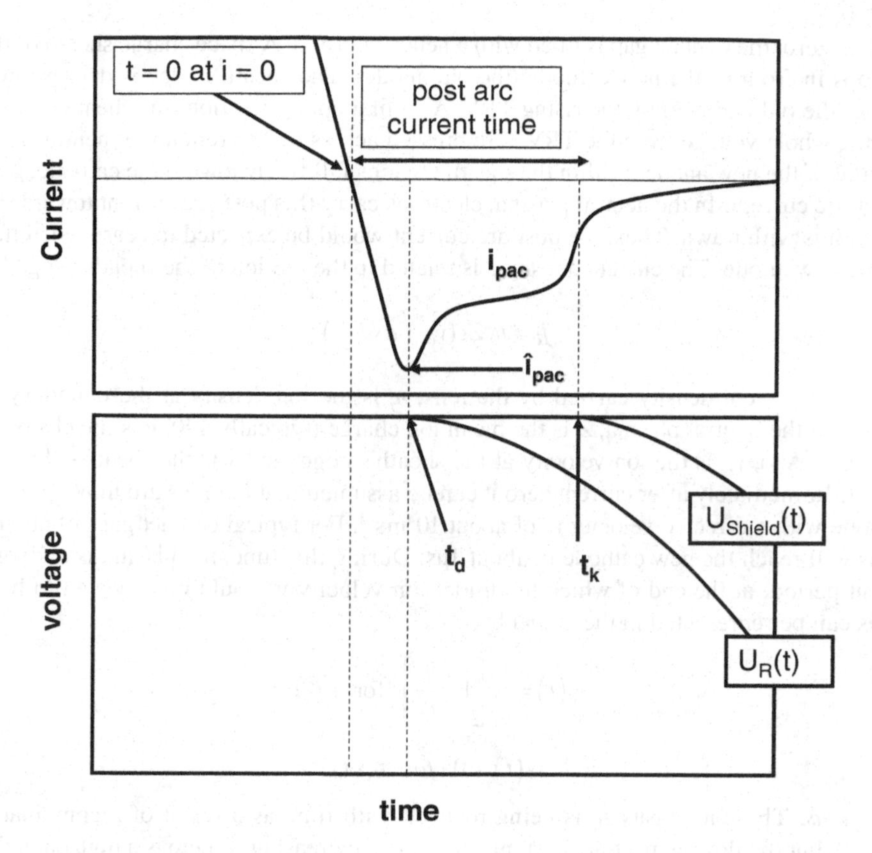

FIGURE 4.12 A schematic showing the current zero region and the development of the post arc current during the rise of the transient recovery voltage [14].

The post-arc current results from the ions in the sheath region falling freely towards the new cathode. In the plasma, the electrons carry this post-arc current as they withdraw towards the new anode. As they do this, they increase the thickness of the space charge sheath at the cathode. For the most part, the circuit's TRV appears only across the sheath. It is only when the sheath reaches the new anode that TRV is impressed across the full contact gap. Also, at this time the post-arc current ceases. At current levels less than 2000 A, discussed in this section, the contact gap can reignite only after current zero if a cathode spot forms on the new cathode (the former anode). As this former anode is quite passive during the arcing phase, its surface will be at a comparatively low temperature during the post-arc current phase. Thus, the only way to initiate a cathode spot at the new cathode is to develop a field at the new cathode high enough to develop the vacuum breakdown process discussed in Chapter 1 in this volume. The free recovery experiments (see Figures 4.4–4.7) show that the *dU/dt* after current zero has to be extremely rapid so that the field across the sheath $U_R(t)/s(t)$ is high enough to initiate a reignition of this contact gap. In most practical ac circuits, the *dU/dt* is usually about an order of magnitude (i.e., 10 to 20 times) lower than those given in Figures 4.4–4.7. The work by Wilkening et al. [10] and others has shown that low current vacuum arcs do not reignite even in the face of an extremely fast TRV across an open contact gap after a current zero.

There has been considerable modeling of the post-current zero period and of the post-arc current phenomena [14–25]. This modeling has been useful in providing a predictive quality to the above qualitative explanation of the physical changes between the electrical contacts after current zero. In order to outline the general analysis, I will use the model of the post current zero period presented by Fenski and Lindmayer [14] and their co-workers at the Technische Universität Braunschweig. Just after current zero, the contact gap is filled with a neutral plasma. A space charge sheath of thickness *s(t)* develops in front of the new cathode (the old anode). The sheath's edge is driven towards the new anode (the old cathode) by the rising TRV. As a first approximation this sheath contains only ions and the whole voltage from the TRV is impressed across it. The remaining neutral plasma has the potential of the new anode. Within the sheath the ions fall freely toward the cathode, giving rise to the post-arc current. In the neutral plasma, electrons carry this post-arc current toward the anode as the sheath is withdrawn. Thus, the post-arc current would be expected to cease when the sheath reaches the new anode. The current density j_i is related to the motion of the sheath's edge by:

$$j_i = n_i Ze(v_{ie} + ds / dt) \tag{4.3}$$

where j_i is the current density carried by the ions, n_i is the ion density at the boundary between the sheath and the neutral plasma, Z is the mean ion charge (typically 1.8), e is the electron charge (1.602×10^{-19} A.s), v_{ie} is the ion velocity at the sheath's edge, and s is the sheath's thickness, see Figure 4.11. Immediately after current zero it can be assumed that the ions are moving towards the new cathode with a directed velocity v_0 of about 10^4ms^{-1}. For typical contact gaps of about 10mm, these ions will reach the new cathode in about 1μs. During this time, the plasma is undergoing the adjustment period, at the end of which the initial ion velocity v_i would be expected to be close to zero. This can be represented in the model by:

$$\begin{aligned} v_i(t) &= v_0\left[1 - \frac{t}{t_d}\right] \quad \text{for } t \le t_d \\ v_i(t) &= 0 \quad for \ t > t_d \end{aligned} \tag{4.4}$$

with $t_d \cong 1\ \mu s$. The ion density n_i is being reduced with time as a result of recombination ($n_i = n_0 \exp\{t/\tau_R\}$), but the flowing post-arc current can also be increasing if there is a high enough density of metal vapor in the contact gap. This, however, is unlikely for ac arc currents of 2000A or less. This can be represented by:

$$\frac{dn_i}{dt} = -\left[\frac{n_i}{\tau_R} + \frac{i_{pac}}{eZV_p}\right] \tag{4.5}$$

where i_{pac} is the post-arc current and V_p is the plasma volume. The value of the ion decay time constant τ_R depends upon the arc current and the internal volume of the vacuum interrupter. For freely recovering arcs τ_R is 0.5μs for a 2mm contact gap and 10μs for a 10mm contact gap [26]. For a practical vacuum interrupter switching an ac current, values between 20μs and 100μs are more reasonable. Thus, for low currents where the i_{pac} is over in a few microseconds the ion density can be assumed to be constant. The second term on the right-hand side of Equation (4.5) results from the ions moving toward the cathode from the edge of the sheath. It is also possible that there will be an increase in the charge density as a result of the production of new charge carriers during the rise of the TRV. One process that has been postulated [27] is the secondary emission of electrons as a result of positive ions with energy $e\,Z\,U(t)$ (where $U(t)$ is the voltage drop across the sheath, i.e., the TRV, $U_R(t)$) impinging on the cathode. If $\gamma(U)$ is the yield at voltage $U(t)$ then the total current i_{pac} is given by:

$$i_{pac} = i_i\left[1 + \gamma(U)\right] \tag{4.6}$$

$\gamma(U)$ has been shown to rise linearly with voltage so that at 20kV, $\gamma(U)=3$. Using Equations (4.3) and (4.6):

$$\frac{ds}{dt} = \frac{i_{pac}}{An_iZe\left[1 + \gamma(U)\right]} - v_{ie}(t) \tag{4.7}$$

where, to a first approximation, A is the area of the contacts. A limitation to the actual post-arc current that flows between the recovering contacts is the space charge limit imposed by the Child-Langmuir equation, i.e.:

$$j_i = \frac{i_i}{A} = \frac{4}{9} \frac{\varepsilon_0 (2e)}{m_i} \frac{U(t)^{3/2}}{s^2} \tag{4.8}$$

where m_i is the ion mass and ε_o is the permittivity in vacuum, 8.85×10^{-12} A.s.V^{-1}.m^{-1}. This equation is not quite as accurate as that used by Andrews and Varley [16], but it is valid over most of the post-arc sheath [17] and thus is perfectly acceptable as a first approximation. For a simulation to proceed, a first-order differential equation is required that describes the electrical circuit and another that describes the relationship between the TRV and the current at the interrupter's terminals.

At current zero $s = 0$. As discussed above, for a short time after current zero when the polarity across the contacts reverses, the voltage across them stays nominally at zero. The total current i_{pac} during this time follows the pre-current zero di/dt, because of the inertia of the ions moving with a velocity v_i, so:

$$i_{pac} = t_d \frac{di}{dt} \tag{4.9}$$

This equation implies that the simulation should begin at time t_d in Figure 4.12 where the sheath formation begins, once the electrons have reversed and the electron current is zero, i.e., for $(t > t_d)$

$$i_{pac} = i = An_i eZ \left[v_{ie} - \frac{ds}{dt} \right] \tag{4.10}$$

where v_{ie} is the velocity of the ions moving from the edge of the space charge sheath to the new cathode. This can be the starting point for the post-arc current model. As stated above, the post-arc current should cease when the sheath reaches the new anode, i.e., $ds/dt = 0$ and at t_k in Figure 4.12. There always is, however a residual current that remains after t_k, i.e., after the sheath reaches the new anode. The cause of this is most probably the slow decay of the charges remaining in the intercontact gap. Van Lanen et al.'s analytical model gives the post-arc current during the rise of the TRV as [22]:

$$i_{pac} = en_i v_B A_{eff} \tanh\left(\frac{eV}{2kT_0}\right) + C_{sh} \frac{dV}{dt} \tag{4.11}$$

Where e electron charge, n_i ion density, $v_B = (kT_e/m_i)$, m_i ion mass, T_e electron temperature, k Boltzmann's constant, A_{eff} effective cathode area and C_{sh} sheath capacitance. Mo et al. [24] and Takahashi et al. [21] use the PIC–MCC software (Plasma in Cell with Monte Carlo Collisions software [28]) to model the post-arc current. The PIC–MCC code is a simulation model for plasmas that includes kinetic effects. The simulation requires a lot of computer memory and its run time can be rather long.

For ac currents up to 2kA (rms.), the value of the post-arc current is very small. Its duration is a function of the final contact gap at current zero. As the density of metal vapor in the intercontact gap at current zero is so low the usual concept of the post-arc current "heating" the residual gas (see, for example, in gas circuit breakers as reported by Frost et al. [29]), does not apply in low current vacuum arcs. Also, at a fully open contact gap, it is unlikely that a new cathode spot will form from almost all normal TRVs, so the possibility of a reignition after current zero is extremely remote. It is now possible to give an explanation for the rapid free recovery data shown in Figure 4.4. In this

Figure, the Cu–Cr contact with a 2mm contact gap recovers to its full withstand strength of 90kV in about 4μs after the low current (200A) vacuum arc has been ramped down to zero in just under 1.5μs; see Lins [4]. The ion density in this contact gap as measured by Lins [19] is about $3 \times 10^{17}m^{-3}$ just before and just after the fast-current ramp to zero. The ion density drops to below $10^{15}m^{-3}$ in about 3.5μs: the decay constant, τ_R, being 0.5μs. Lins speculates that the decay of the ion density and the recovery of the contact gap during this time must, in some way, be related. The ions crossing the contact gap during the free recovery period would impact the new cathode and liberate electrons that could contribute to the gap's breakdown during the 4μs to 5μs period. As the fastest ions from the last cathode spot would cross the contact gap in 1μs, the question remains why does it take 4μs to 5μs for the contact gap to fully recover? Wang et al. [11] conclude that during the initial breakdown period (0μs to 4μs or 5μs) the probability of breakdown drops with the decay of the ion current during the free recovery period. The collision of the ions with the residual metal vapor slows their decay so that it only drops below a critical density after 4μs or 5μs: see Figure 4.13. There seems to be a direct relationship between the decay of the ion density and the free recovery of the contact gap for a 2kA vacuum arc. In order for the vacuum arc to reignite after the current is ramped to zero the cathode has to develop a cathode spot capable of sustaining the resulting vacuum arc. Because the neutral gas density (Figure 4.7) and the ion density are so low, there will be no interaction of any electrons that may be liberated from the new cathode. Thus, any reignition must develop from the vacuum breakdown process described in Chapter 1 in this volume enhanced by the ion bombardment at the new cathode.

Up to about 1μs after the current zero, no voltage appears across the contacts. This is the plasma adjustment period. After this period the sheath is formed at the cathode: all the TRV voltage $U_R(t)$ appears across the sheath. The sheath moves toward the new anode with a velocity given by Equation (4.7). If $U_R(t)/s(t)$ has a value greater than $4.5 \times 10^7 Vm^{-1}$, then a high enough electron current could be liberated by field emission from the new cathode, which, in turn, could initiate the vacuum breakdown process. Equation (4.7) shows that ds/dt is inversely proportional to the ion density n_i. The ds/dt is slower just after current zero than 3μs later when the n_i is 0.01 of its value at current zero. Thus $dU_R(t)/dt$ has to be extremely rapid to create an electric field at the cathode that is high enough to result in the vacuum breakdown of the contact gap after the 200A arc, even though, in this case, the

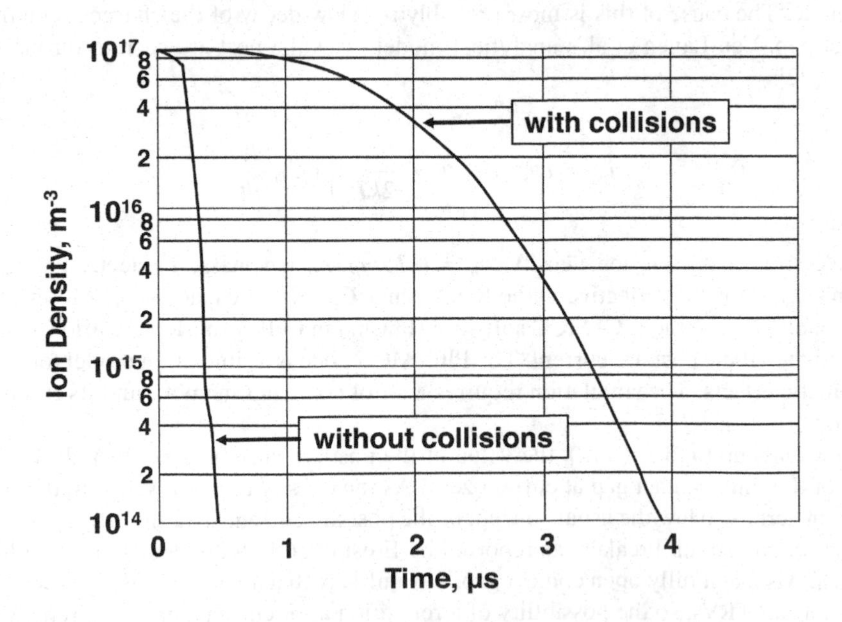

FIGURE 4.13 The decay of the ion density for 25mm diameter Cu contacts [11]

di/dt of the current just before the current zero is higher than 100A.μs^{-1}. In Figure 4.4 the average $dU_R(t)/dt$ for Cu–Cr contacts is about 21kV.μs^{-1}, which means that the average ds(t)/dt has to be less than 0.5mm.μs^{-1}.

4.2.2 The Interruption of the Vacuum Arc for AC Currents Greater than 2 kA (rms.)

In Chapter 2, I discussed the changes in the appearance of the vacuum arc, between butt contacts, as the arcing current is increased. The diffuse vacuum arc is certainly maintained for currents up to about 5kA. At currents above about 2kA, however, there is a distinct change in the distribution of ion energies, see Figure 2.28. Above about 5kA, the arc develops into the transition arc with a well-defined, low-pressure column; see Section 2.5 in this volume. This arc always returns to a diffuse vacuum arc as the current goes to zero. At currents greater than about 15kA the vacuum arc forms a constricted high-pressure column. When the stationary, columnar vacuum arc forms, there is also gross melting where the arc roots attach to the contacts. The effects of this gross heating and contact melting in these contact regions will linger after the current zero even if there is little visible evidence of the arc column at that time. Certainly, from my experience, butt contacts made from Cu–Cr (25 wt.%)) can interrupt ac currents up to 10kA (rms.) in a normal 12–15kV circuit, but at 25kA, the effect of the columnar vacuum arc will result in a reignition of the arc at current zero. In fact, the arc will continue to burn until a back-up device interrupts the ac circuit. Let us now examine how the increase in current affects the recovery process of the vacuum arc at higher currents.

As I have already shown in Chapter 2, Figures 2.45 and 2.46, even a very high current ac vacuum arc can change into the diffuse mode as the ac current decreases to zero. If, for example, a contact opens at the beginning of a 20kA (peak) ac current and it is assumed the vacuum arc is fully diffused at 5kA, there is still about 0.7ms before the current zero. This would result in enough time for the regions on the contacts where the columnar arc roots are attached to cool and also for the metal vapor density in the intercontact gap to dissipate. Frind et al. [30] have studied the effect of the higher arc currents on interruption using a dc current pulse that could be varied from 0.3ms to 9.5ms. They open the contacts with a dc low current of about 150A and let the resulting low current diffuse vacuum arc burn for about 10ms. The contacts are thus fully open (9.5mm), before a high dc current pulse is introduced. This high current pulse has a rise time and a ramp down time of 20μs. In these experiments there is no effect of the initial bridge column on the development of the high current vacuum arc mode. Also, the high current, columnar vacuum arc has to develop via the anode spot process discussed in Section 2.3.4. The 20μs ramp down of the current will have the result that most of the metal vapor from the column arc will remain in the contact gap during the free recovery period. Figure 4.14 shows that the recovery time of the contact gap to withstand a voltage of 38kV for a current pulse of 250A is only 7 μs. The 250A vacuum arc will be in the fully diffuse mode. This recovery time increases slowly to about 40μs for a 4kA current pulse. After this, however, the recovery time increases sharply with current until at 12kA it is about 700μs. Even though this series of experiments uses current pulses that are far removed from a normal ac current loop, they do illustrate the effects of a higher density of metal vapor in the contact gap once the recovery voltage is applied after a forced current zero.

In a more realistic set of experiments, Binnedijk et al. [31] have used a Weil–Dobke synthetic circuit similar to that described in Section 4.2.1; see Figure 4.9. They provide a 36.4Hz ac current wave with peak currents of 2.5kA to 20kA from the low voltage circuit and a TRV with a RRRV of 1.9kV/μs and a peak value of 25kV from the high voltage circuit. The experimental vacuum interrupters use 25mm diameter Cu, Cu–Cr (50 wt.%) and Ag–WC butt contacts. The di/dt at current zero from the synthetic circuit is 11.7A/μs. This di/dt at current zero is equivalent to that from a 37kA peak 50Hz current. The RRRV of 1.9kV/μs is somewhat faster than would be experienced from interrupting a normal short circuit current; see Chapter 6 in this volume. Their interesting results are, however, relevant to the description of current interruption in an ac circuit. First of all, the overall intensity of the light coming from the vacuum arc between Cu contacts is low at 4.1kA peak. It increases for a 5.7kA peak-current vacuum arc and then increases considerably for a

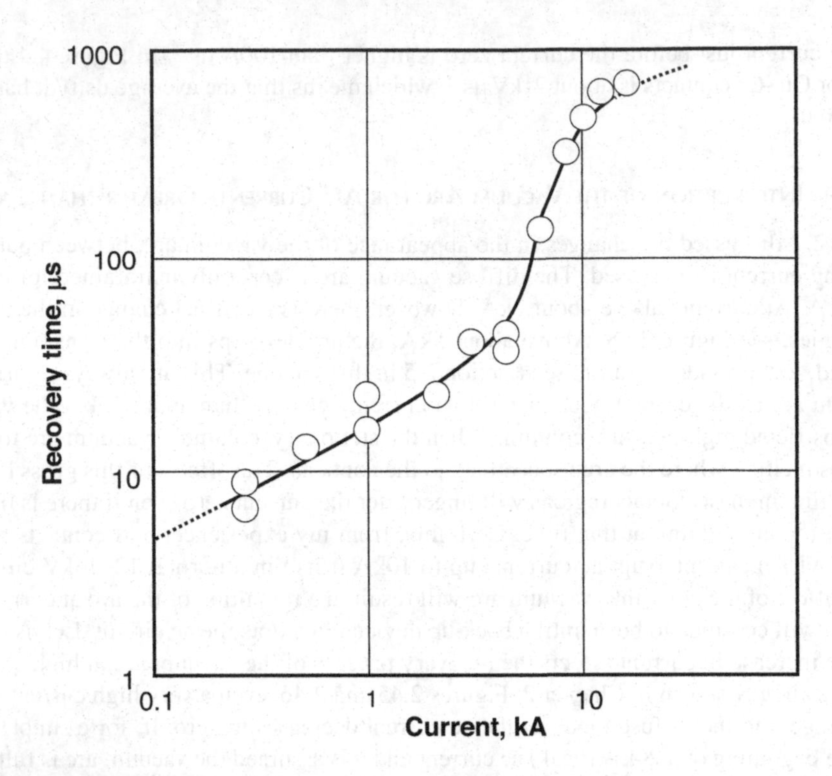

FIGURE 4.14 The free recovery time to a voltage of 38kV as a function of vacuum arc current for 1" (2.54mm) diameter, Cu–Bi(0.5 wt.%) contacts with a 9.5mm gap after the interruption of a trapezoid shaped arc current with a rise time and a ramp-down time of 20μs and a duration of 4.5ms [30].

10.4 kA peak-current vacuum arc. At 4kA, there is some overlapping of the plasma plumes above the cathode spots and a reduction of the ion energy distribution. The vacuum arc, however, still returns to an essentially diffuse intercontact plasma as it approaches current zero. At 6.7kA the researchers begin to observe melting of the Cu contact surfaces. This results in an increase in metal vapor density in the intercontact gap, more interaction with the energetic electrons and more opportunity for excitation of this metal vapor and hence, a greater light output. At 10.4kA, intense melting of the contacts is observed, which results in an ever-greater density of the Cu vapor in the contact gap, a greater opportunity of excitation and hence, a much greater light output from the arc. Interestingly enough, Cu–Cr (50 wt.%) contacts show much less melting of the contact surfaces than do the Cu contacts at the same current, with a resulting lower light output. For both the Cu and the Cu–Cr contacts the reignitions generally become more frequent as the main circuit current increases. Bennedijk et al. define the reignitions as dielectric reignitions when the TRV peak voltage is greater than 4kV and thermal reignitions if the TRV peak voltage is less than 4kV. I will discuss dielectric and thermal reignitions in more detail later in this chapter and again in Chapter 6, Section 6.3.4, in this volume. Thermal reignitions occur more frequently as the current increases and as the contact gap increases. Ag–WC contacts show a tendency for thermal reignitions for all contact gaps and all currents. This contact material is useful where surge free interruption is required; see Section 4.3.1. However, using it to interrupt high currents is a challenge. The Cu contacts tend to show thermal reignitions more frequently at lower currents than do those made from Cu–Cr; see Table 4.3. These data emphasize the advantage of using the Cu–Cr contact material over using a Cu material for switching high currents in vacuum.

In order to appreciate the effect of these high current vacuum arcs on the interruption of an ac circuit, it is instructive to examine the research that has been performed by Dullni et al. [27]. They

TABLE 4.3
Interruption of the High Current Vacuum Arc Showing Differences in the Type of Reignition for Cu, Cu–Cr and Ag–WC Butt Contacts [31]

Contact Material	Contact Gap, d mm	Current, i kA	Most probable reignition type
Cu	$d < 5$	$i \leq 10$	Dielectric
	$5 < d < 10$	$i \leq 15$	Dielectric
		$i > 15$	Thermal
	$d > 10$	$i > 10$	Thermal
Cu–Cr(50 wt.%)	$d < 5$	$i < 10$	Dielectric
	$5 < d < 10$	$i \leq 20$	Dielectric
	$d > 10$	$i \leq 20$	Dielectric
		$i > 20$	Thermal
Ag–WC	$d < 5$	$i < 18$	Thermal & Dielectric
	$5 < d < 10$	$i \leq 15$	Dielectric
		$i > 15$	Thermal
	$d > 10$	$i \leq 15$	Dielectric
		$i > 15$	Thermal

use 25mm diameter Cu butt contacts in a circuit similar to one used by Frind et al. [30] described above. The contacts open with a 50A dc current, which lasts for about 40ms. During this time, the contacts become fully opened to 7.5mm. An essentially dc current pulse is introduced to the arc. This pulse has a rise-time and a ramp-down time of 0.7ms, has a current level that can be varied from 0.9 to 11kA and lasts for 5.5ms. After the current zero they use both the free recovery method with a high *dU/dt* to determine the dielectric recovery strength of the contact gap and also the imposition of a TRV with a slower *dU/dt* right after current zero to determine the post-arc current. Although the experiments have a much greater *di/dt* at current zero than you would obtain from an ac current wave, it is slower than that used by Frind et al. For example, it is 1.4 $A.\mu s^{-1}$ for the 0.9kA arc, which is equivalent to the *di/dt* for a 4.6kA (peak) 50Hz, ac current at current zero and it is 17.6$A.\mu s^{-1}$ for the 11kA arc, which is equivalent to the *di/dt* for a 56kA, 50Hz ac current. Their data do, however, give some insight into the recovery process of high current, essentially stationary, transition and columnar vacuum arcs. At 9kA, where a transition arc is certainly present, the free recovery experiments still show that the contact gap can withstand a voltage of 30kV after 6μs and reaches a withstand value of 90kV in about 50μs. Interestingly, they show that after an 11kA arcing current pulse, it takes about 400μs for the contact gap to recover to its cold withstand voltage and this time period is essentially independent of *dU/dt*. This recovery time is somewhat shorter than that measured by Frind et al. [30].

Dullni et al. also show that at current zero, there is approximately 45 times the charge between the contacts for a 9kA arc than there is for a 0.9kA arc. In fact, in these experiments, there again seems to be a sharp distinction between the stationary vacuum arcs below about 4kA and those above 6kA. Below 4kA, the residual charge at current zero in the contact gap is about 1.82μC/kA, but above 6kA, the charge is 8μC/kA. Also, the decay of the charge after a 0.9kA arc is exponential with a time constant of 8.4μs, but after a 9kA arc, the decay time constant is 23.8μs. These observations show the major effect that the diffuse column, transition arc has on the intercontact gap after current zero. For example, the 8.4μs decay corresponds to a speed of 1200ms^{-1} (temperature of 4300K), and the 23.8μs decay corresponds to a speed of 408ms^{-1} (temperature of 500K) [27]. These measurements suggest a much higher density of both ions and neutral metal vapor in the post-arc contact gap once the diffuse column arc has formed.

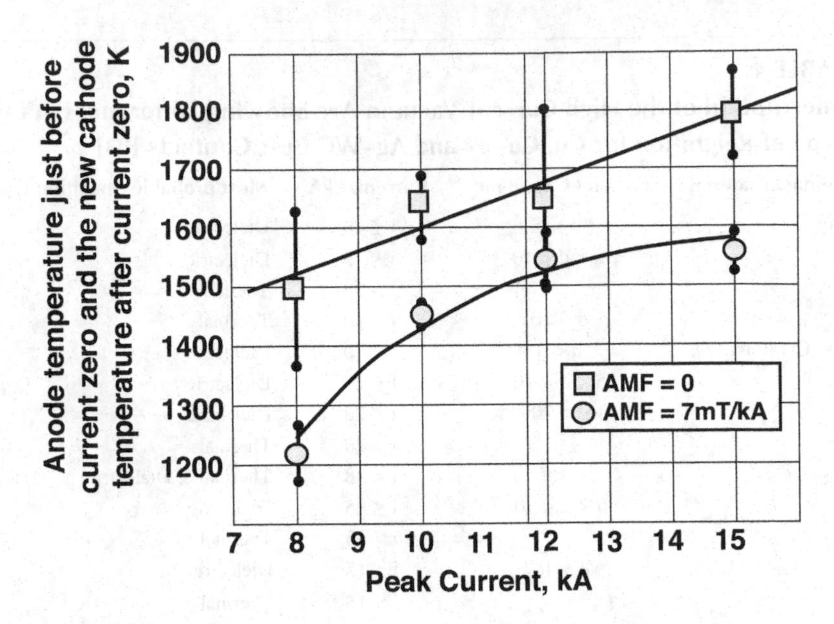

FIGURE 4.15 Temperature of the new cathode after current zero as a function of peak ac current between opening, butt Cu–Cr(25wt.%) contacts, diameter 20mm with and without an axial magnetic field [32]

In Section 4.2.1, I have discussed the production of secondary electrons at the new cathode from ion bombardment. It is also possible that electrons can be pulled from the remaining hot spot on the new cathode (former anode) at a much lower field than would be required for purely field emission. Electron emission from only field emission would be the case if the new cathode were only subjected to the purely diffuse vacuum arc. That is, its surface temperature would be relatively cool at current zero, because for the duration of the diffuse vacuum arc, at the anode, it would have been a passive collector of electrons over its whole surface; see Section 2.3.2 in this volume. As the vacuum arc current increases with butt contacts, an anode spot can form which can persist after current zero [32]. The temperature at the new cathode as a function of current is shown in Figure 4.15. With a further increase in current a stationary columnar vacuum arc will be formed. If this arc persists close to the ac current zero, the contact gap will experience a greatly increased density of metal vapor and a greatly increased ion density. Both of these will be slow to decay. The post-arc current model of the ion sheath moving across the contact gap is still valid, but now the electric field at a hot spot on the new cathode can be much lower for the production of electrons, e.g., by T-F emission [33]. These electrons, together with the secondary electrons emitted via ion bombardment interacting with the residual metal vapor (if its density is high enough) and residual density of ions, can lead to a reignition of the vacuum arc and a failure of the vacuum interrupter to interrupt the circuit current.

While no vacuum interrupter designer would use butt contacts to interrupt short circuit currents greater than about 10kA (rms), the experiments at higher currents with this contact design give some interesting insights into the eventual failure mechanism of all vacuum interrupter contact structures. Let us first consider the effect of the metal vapor after the current zero. Figure 4.16 shows a Paschen curve for a practical vacuum interrupter. Here the minimum breakdown voltage is $U_{B(min)} \approx 460V$ at a pressure times contact gap (*pd*) value of 7mbar.mm (or 0.7Pa.m). When this Figure is compared to Figure 1.8, for the contacts with a uniform electric field in air, you can see that the minimum value of the breakdown voltage is $U_{B(min)} \approx 330V$ and this occurs at about a similar *pd* value [34, 35]. The difference in the values of $U_{B(min)}$ most probably reflects the nonuniform electric fields typical inside the usual vacuum interrupter; see Section 1.3.2 in this volume. It is difficult to measure the true breakdown voltage to the left-hand side of the Paschen curve for contact gaps of a few millimeters (i.e., for *pd* values below about 1mbar.mm), because the actual breakdown path is not easily controlled [35]. By carefully controlling the breakdown gap, Schönhuber [35] has shown that a breakdown voltage of 10kV occurs at about 1.0mbar.mm. Figure 4.16 shows a similar value of about 0.35 mbar.mm; i.e., for a 10mm contact gap, a pressure of 3.5×10^{-2} mbar and a breakdown voltage of 10kV. The breakdown voltage rises rapidly to 40kV for $pd = 0.2$mbar.mm. The true vacuum ac breakdown voltage for this vacuum interrupter is greater than 50kV (rms). Table 4.1 shows the expected number density of gas and the estimated electron mean free path at 300K.

The consideration of the electron mean free path is important. For small contact gaps at atmospheric pressure, it has been shown that once the gap is less than about 10 to 20 electron mean free paths (i.e., less than about 6 to 7μm) the Townsend avalanche does not occur [36] and the U_B versus contact gap follows the $U_B = k \times d$ for short vacuum gaps shown in Figure 1.18. Table 4.1 shows that for a 10mm contact gap this occurs at gas pressures below 10^2Pa. There the number density of the gas is about 10^{22}m^{-3}. Immediately after an ac current zero inside a vacuum interrupter the residual gas is mostly metal vapor from the contacts. It is reasonable to assume that a similar metal atom number density would be required at the transition between gas breakdown and vacuum breakdown. Once the contact surfaces reach temperatures close to the melting point of Cr (~2150K) the vapor pressure from a Cu–Cr contact can reach a value of 10^3 to 2×10^3Pa; see Table 4.2. Table 4.1 shows that at this temperature and at this pressure the gas density is greater than 3.4×10^{22}m^{-3}, and the electron mean free path is such that about 40 collisions could be expected by an electron crossing a 10mm contact gap; i.e., a Townsend avalanche or a gas assisted breakdown could well be initiated. It is not until the pressure drops to below 10^2Pa that the vapor density falls to a value (below 10^{22}m^{-3}) where the electron mean free path falls below the 10 to 20 collisions in a 10mm contact gap. Under these conditions it is unlikely that a gas-assisted breakdown would occur. It is interesting to note that at the melting point of Cu (~1360K) the vapor pressure of Cu and even of Ag is too low to affect the breakdown of the contact gap. However, the vapor pressure does increase rapidly over the next 700K. Certainly, in my experience the short circuit vacuum arc melts both the Cu and the Cr in a Cu–Cr contact when the vacuum interrupter is called upon to switch high short circuit currents. So, the contact surface during at least part of the arcing time will be at or above the melting temperature of Cr. The pressure times gap data can be converted into density (n) times gap (d) data using Equation (4.12):

$$nd = \frac{pd}{kT}\,\mathrm{m}^{-2} \tag{4.12}$$

where p is in Pa, d is in meters, T degrees Kelvin, and k Boltzman's constant (1.38×10^{-23} JK^{-1}). Dullni et al. [37] give a value of $nd = n_o d$ for the $U_{B(min)}$ to be 3×10^{20}m^{-2}. The value of $n_o d$ from Figure 4.17 where the minimum is 7mbar.mm gives a similar value of 1.7×10^{20}m^{-2}. Dullni et al. then propose that gas breakdown will not occur for $nd \leq 2.7 \times 10^{19}$ m^{-2} where $U \approx 10kV$. The data for the vacuum interrupter shown in Figure 4.16 give $U_B \approx 10kV$ for $pd \approx 3.5 \times 10^{-1}$ mbar.mm or $nd \approx 8.5 \times 10^{18}$ m^{-2}. If we look at Table 4.1, the electron mean free path would be about 10mm and $n = 8.5 \times 10^{20}$ m^{-3} for a 10mm contact gap; i.e., the breakdown would most probably not be in the gas breakdown regime because there would be a low probability of even one electron/gas collision for one electron transit of the contact gap.

Figure 4.17 from Dullni et al. [37, 38] show the expected vapor density in the contact gap immediately after the interruption of a 50Hz ac current as a function of the current for Cu–Cr butt shaped contacts. Here the vapor density between the contacts is less than 10^{21}m^{-3} for currents less than or equal to 8kA. At 8kA the vacuum arc would be expected to be in the "transition arc" regime (see Section 2.5 in this volume) and Figure 4.15 shows a surface temperature of 1500K at current zero. Thus, even though a low pressure, diffuse column would exist as the current passes through its peak value, it would only erode broad shallow craters on the Cu–Cr contacts and the arc would return

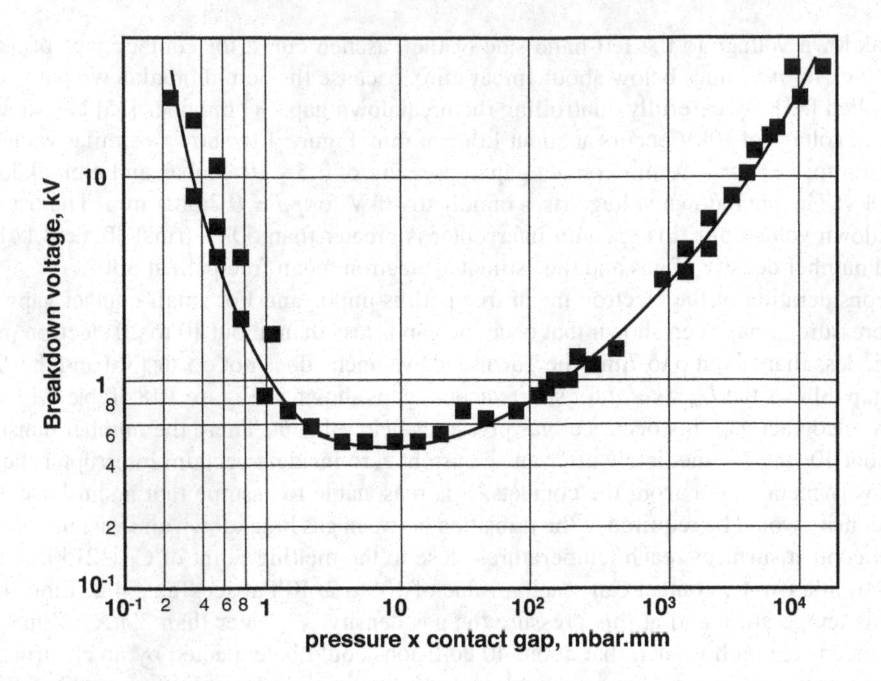

FIGURE 4.16 Paschen curve for a commercial vacuum interrupter with an AMF contact, the contact diameter is 75mm and the floating shield has an internal diameter 110mm.

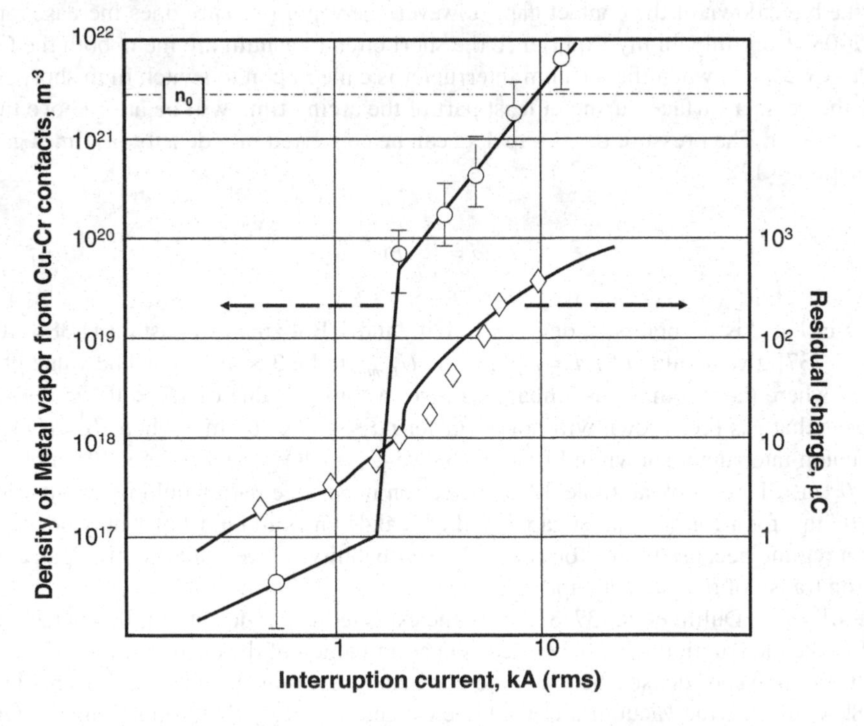

FIGURE 4.17 The metal vapor density and the residual electric charge between Cu–Cr contacts after the interruption of ac currents, [37, 38].

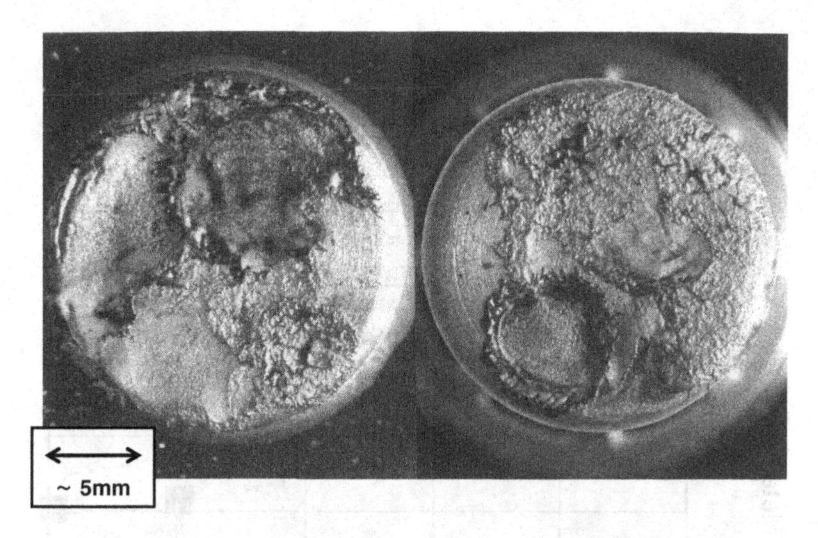

FIGURE 4.18 Contact erosion of a Cu–Cr(40 wt.%) contact after the interruption of 8kA (rms) transition vacuum arc 50 times, showing the shallow erosion craters.

to the diffuse mode before the current zero. Figure 4.18 shows an example of a Cu–Cr (40wt.%) contact after 50 operations at 8kA(rms). This Figure shows the effects of these shallow craters on the total erosion and illustrates that there is no gross erosion at this current level. From my discussion in Section 2.5 in this volume, the transition vacuum arc could persist up to about 11kA rms. Only at a greater current level would the true columnar vacuum arc give rise to the intense erosion of the contact surfaces. Dullni et al. [37], in fact, show even with their trapezoid shaped current pulse with a peak value of 11kA, where the arc between Cu–Cr contacts would produce density of metal vapor greater than $10^{22}m^{-3}$, the contact gap recovers immediately to greater than 20kV and then reaches its full recovery value of about 80kV in about 250ms. Thus, in a true 12kV ac circuit where the peak recovery voltage would be 20.4kV (i.e., 12 × 1.72, according to IEC, see Section 6.3.3 in this volume) and in a 15kV 28.3kV (i.e., 15 × 1.88, according to IEEE; see Section 6.3.3 in this volume), an interruption could well be expected for a 10kA(rms) short circuit current between the butt shaped Cu–Cr contacts.

It is only when the vacuum arc is permitted to remain in the columnar mode close to a current zero that the vapor density in the contact gap will be high enough for a Townsend avalanche or a gas assisted breakdown to be initiated and a failure to hold off the transient recovery voltage resulting in a reignition of the current. From the discussion above, I would expect this to occur with butt contacts at currents in the range of 11kA to 15kA. In fact, Lindmayer et al. [9] give a good example of butt shaped, Cu–Cr (25 wt.%) contacts successfully interrupting a 10kA(rms) current even with a very rapid RRRV of about $4kV.\mu s^{-1}$ and a peak value of the TRV of 160kV. For a current of 15kA, the contact gap reignites when the TRV reaches 100kV. Binnendijk et at [31] also show an immediate recovery of 20kV for butt shaped, Cu–Cr (50 wt.%) contacts for ac currents up to about 10kA and for contact gaps above 10mm. A decline in the value of recovery voltage is observed from about 15 to 20kA where the columnar vacuum arc is beginning to develop. From 20kA to 30kA, where an intense columnar arc will definitely form, the contacts will not interrupt the circuit. It is thus imperative for the vacuum interrupter designer to ensure that the metal vapor density in the center contact gap at current zero be less than about $10^{22}m^{-3}$. Below this value electrons crossing the contact gap would have a very low probability of interacting with the residual vapor and the breakdown voltage of the contact gap would be approaching the true vacuum breakdown value for the gap.

In order to achieve this low vapor density a diffuse vacuum arc has to form as the current approaches current zero. For ac currents above about 10kA (rms) this is only possible if one of the

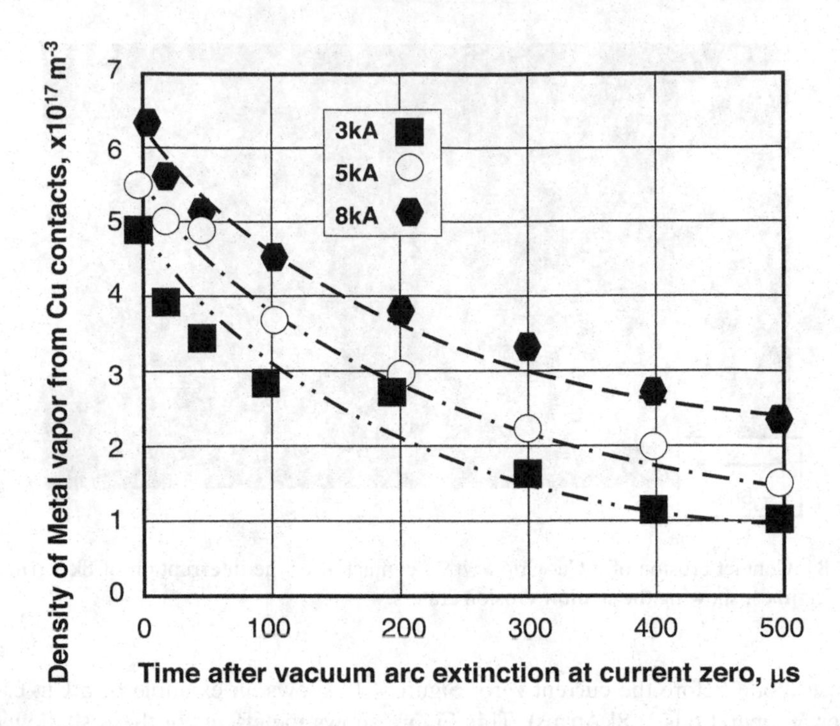

FIGURE 4.19 The Cu vapor density after interrupting currents in the range 3kA to 8kA using Cu contacts with an AMF contact structure [39].

contact structures, discussed in Sections 3.3.3 and 3.3.4 in this volume, that control the high current vacuum arc is used. Takahashi et al. [39] have presented a good example of how an AMF can affect the metal vapor density at current zero. They use an AMF contact structure similar to that shown in Figure 3.40(c) (with 3 arms), 40mm diameter Cu contacts and a contact gap of 30mm. They open 50Hz ac currents in the range 3kA (peak) to 10kA (peak). During the opening operation they observe a more or less uniform plasma column between the contacts 40mm diameter and 30mm long: i.e., the AMF has forced a fully diffuse vacuum arc between the contacts over the range of currents investigated. Figure 4.19 shows the measured metal vapor density at current zero as well as its decay at a later time. The metal vapor density at current zero is proportional to the peak current over the range 3kA to 10 kA. This is expected, as the number of cathode spots generating metal vapor would also be proportional to the current, and they are confined by the AMF to the cathode's diameter. The metal vapor density after the 3kA vacuum arc is similar to that shown by Dullni et al. in Figure 4.17; i.e., about $5 \times 10^{17}m^{-3}$. At this current the fully diffuse vacuum arc is expected in both experiments and thus the metal vapor densities at current zero would be the similar.

At higher currents the effect of the AMF forcing a fully diffuse vacuum arc that spreads the cathode spots over the whole cathode's surface becomes evident. The metal vapor density, while increasing with current, now reaches only between $6 \times 10^{17}m^{-3}$ and $7 \times 10^{17}m^{-3}$ at 8kA peak. The more constricted transition vacuum arc at this current (Figure 4.17) shows a metal vapor density of between $10^{20}m^{-3}$ and $10^{21}m^{-3}$. Figure 4.17 shows that as the current increases above 2 kA, the density of the metal vapor in the contact gap continues to increase. Figures 2.34(b) and 2.37 show that when an anode spot forms it melts the contact material beneath it. This surface will then evaporate even more metal vapor into the contact gap. This is especially true if the anode plume as shown in Figure 2.39 exists after the current has gone to zero. Wang et al. [40] in their model of the dielectric strength after vacuum arc extinction show: (a) there is a positive correlation of the metal vapor density in the contact gap with the anode temperature and (b) a higher anode temperature results in a higher probability of the contact gap breaking down.

Dullni et al. [37, 38] also present an interesting perspective on the influence of metal droplets on the recovery of the contact gap after the interruption of the high current, vacuum arc between Cu–Cr contacts. From their previous research they have shown that the flux of droplets from Cu–Cr contacts is low and that the droplets themselves have small diameter distribution; see the discussion in Section 3.2.4 in this volume. They conclude that droplets do not have the ability to cause the breakdown of the contact gap during the recovery voltage stage (at least up to the 100kV level in their experiments). One reason for this is that the velocity of the droplets is less that $100ms^{-1}$. As discussed in Section 1.3.3 in this volume, velocities would have to exceed about $2 \times 10^3 ms^{-1}$ to totally evaporate on impact. They also state that in other experiments, when they observe metal droplets impacting at the cathode after the vacuum arc has extinguished, they see no resulting breakdown of the contact gap.

4.2.3 The Interruption of High Current Vacuum Arcs

As will be discussed in Chapter 6 in this volume a vacuum circuit breaker has to successfully pass a series of certification tests before it can be offered for sale. The test series are identified in the IEEE C67 and the IEC 62271 publications. The tests include successful interruption of high, short-circuit currents. The individual vacuum interrupters in vacuum circuit breakers have to be designed to interrupt these high currents. This section discusses the performance criteria that have an impact on a vacuum interrupter's design that will give it an acceptable high-current interruption performance. I will firstly discuss the effect of the post-arc current that occurs immediately after the vacuum arc has extinguished at the ac current zero as the TRV appears across the open contacts. I will then explore the impact this post-arc current has on the expected TRV values experienced in an ac circuit. Finally, the effect of contact temperature at current zero and the effect of the residual metal vapor on interruption performance will be examined. It will be shown that the study of post-arc current in vacuum interrupters is largely of academic interest while the discussion of the residual metal vapor density in Section 4.2.2 is highly relevant.

In order to interrupt the high current vacuum arc, it is necessary to ensure that it returns to the diffuse mode long enough before current zero that all the effects of the column mode have been forgotten. In Chapter 3, Sections 3.3.3 and 3.3.4, in this volume, I discussed two contact structures that control and/or dissipate the energy of the columnar, high current, vacuum arc. The first contact design, the transverse magnetic field structure (TMF contact), forces the columnar vacuum arc to rotate around the periphery of the contact at quite high velocities. This effectively limits the energy per unit time that can be deposited into the arc roots on the contacts so that the necessary volume of metal vapor to sustain the columnar arc is not achieved all the way to current zero. Once the arc returns to the diffuse mode, the hot spots on the cathode and anode contacts cool on a millisecond time scale. For Cu–Cr, TMF contacts, the contact faces, show fairly uniform and smooth contact erosion after high current arcing.

The second contact design, the axial magnetic field structure (AMF contact), forces the columnar vacuum arc into a diffuse mode, even at very high currents. Pieniak et al. [41] have measured the surface temperature of Cu–Cr(25wt.%) spiral, TMF contacts during the passage of high currents. Figure 4.20 shows their data for a 19kA (rms) current. It shows the characteristic voltage of a high current vacuum arc between spiral, TMF contacts. As the contacts open the columnar vacuum arc remains in one position until the contacts are about 3–4mm apart. It then begins to rotate around the periphery of the spiral arm: region (a). During this time, the contacts in positions 1 and 2 show an increase in temperature well above the melting temperatures of Cu and Cr. At the beginning of region (b) the columnar arc begins to break up and transition to the diffuse mode. At position 5, the temperature has now decreased to 1863K just above the melting point of Cu. If we assume that the temperature decay rate for position 4 to 5 (360 $K.ms^{-1}$) continues to current zero, then the contact temperature at current zero will be about 1200 K. Looking at Table 4.2, the pressure of metal vapor in the contact gap would be less than 10^{-2}Pa (or 10^{-4}mbar). This number density of less than $10^{-18}m^{-3}$ is much lower than would be needed for the contact gap to breakdown and reignite the vacuum arc.

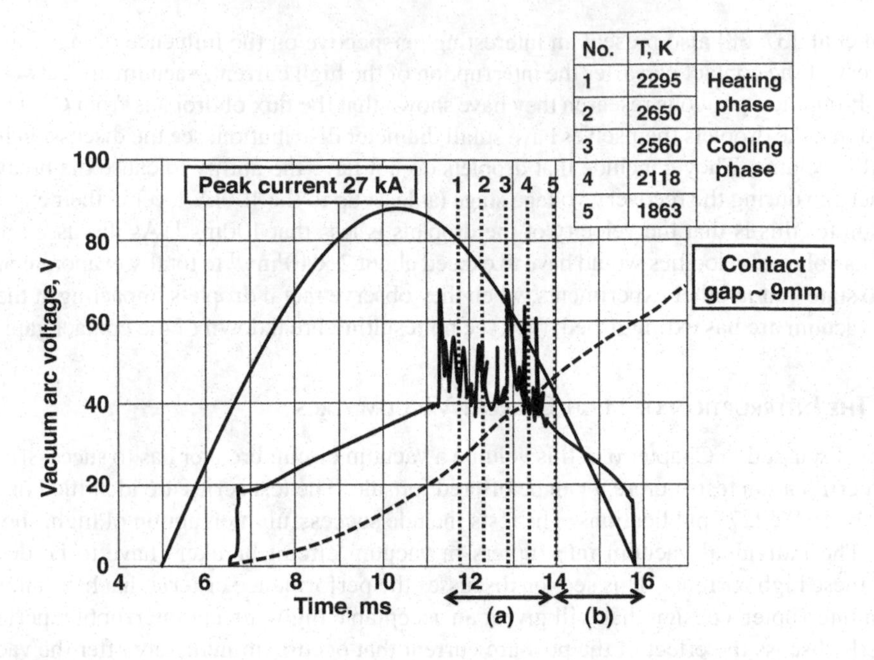

FIGURE 4.20 The surface temperature of TMF contacts during the columnar arc running phase (a) and as the column breaks up and the diffuse phase begins (b) [41].

This data confirms the hypothesis that the transition phase (b) in Figure 4.20 does indeed result in the successful interruption of high, short-circuit currents. Again, the arc energy is spread over the contact faces, so the contact erosion is also quite uniform. Gentsch et al. [42] show that the distribution of the energy from a 30kA(rms) vacuum arc is equally effective for both the TMF and the AMF contacts. Once the arc goes into the diffuse mode, it remains diffuse until the current goes to zero. Each of these structures has proven to be highly effective in permitting the vacuum interrupter to control and interrupt very high currents. As will be discussed below, however, a contact with a given diameter will eventually reach a limiting level of current above which it cannot interrupt the ac circuit.

Fenski et al. [14] have investigated the effect of a 50Hz half-cycle vacuum arc between AMF and TMF contacts for currents in the range 5kA (rms) to 40kA (rms). Figure 4.21 shows typical traces for a 25kA (rms.) current for both AMF and TMF contact structures. Both contact structures have a diameter of 90mm and are made from a Cu–Cr(25wt.%) material. In each case, the contacts open about 1ms after current initiation where the current is greater than 10kA. Thus, a well-established bridge column will exist immediately after the rupture of the molten metal bridge. In Figure 4.21 it can be seen that once the AMF contacts open, the arc voltage increases to a peak value at about 1ms. From the discussion in Section 3.3.4 in this volume, it is only after this period that the arc goes into the fully diffuse mode. As the arc voltage here remains at quite a low value, about 30V, it can be assumed (also from the discussion in Sections 2.6.3 and 3.3.4 in this volume) that there is sufficient cathode area for the cathode spots to spread with very little overlapping of the plasma plumes above them. This fully diffused arc would be expected to distribute its energy over most of the cathode and anode contact areas.

The arc voltage for the TMF contact in Figure 4.21, by contrast, is quite different. It does have, however, the exact structure that would be expected from my discussion in Section 3.3.3 in this volume. There is a dwell time of about 3ms during which time the contact gap opens to about 4.5mm. At this time, the columnar vacuum arc begins to move rapidly around the periphery of the contacts with the characteristic arc voltage spikes of up to 200V. About 1ms before current zero, the arc

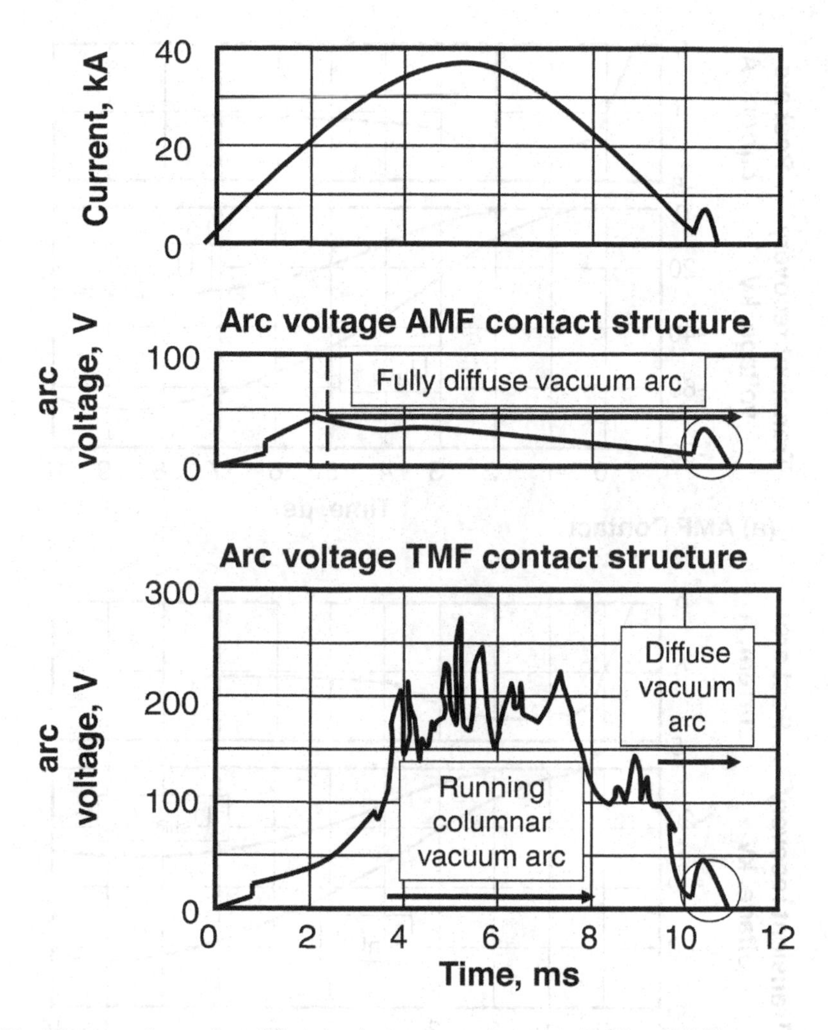

FIGURE 4.21 The arc voltages for a 25kA (rms) vacuum arc between AMF and TMF contact structures [14].

motion ceases. This is when the vacuum arc would be expected to return to the diffuse mode. The slight bumps in the voltage traces (circled), result from the injected current from the Weil–Dobke synthetic circuit. It is interesting to note that for the 500μs before this voltage "bump," both voltage traces have similar structures. Since we know that the vacuum arc between the AMF contacts is fully diffuse during this period before the current zero, it is reasonable to assume that this is good evidence that the arc between the TMF contacts is also fully diffuse at current zero.

It is now instructive to extend the post-arc current discussion that I began in Section 4.2.1 when I described the interruption of the diffuse vacuum arc. In fact, Fenski et al. [14] show that the post-arc currents (i_{pac}) for both the AMF and the TMF contacts after the 25 kA (rms) vacuum arcs shown in Figure 4.21 are almost identical; see Figures 4.22(a) and (b). Here you can see that the shapes and peaks of the post-arc currents are nearly equal, even though an extremely fast rising TRV is used (60kHz, 8.5kV/μs). An abrupt reduction in the post-arc current occurs in both cases about 3μs after current zero: i.e., when the ion sheath reaches the new anode. After this, there is a much slower decay, which results from the decay of the residual, low-density plasma remaining in the intercontact gap. I believe that these data present a very strong confirmation that in this experiment the columnar vacuum arc between the TMF contacts has transitioned into the diffuse mode long enough before current zero that all memory of the high current mode is forgotten during the post-arc

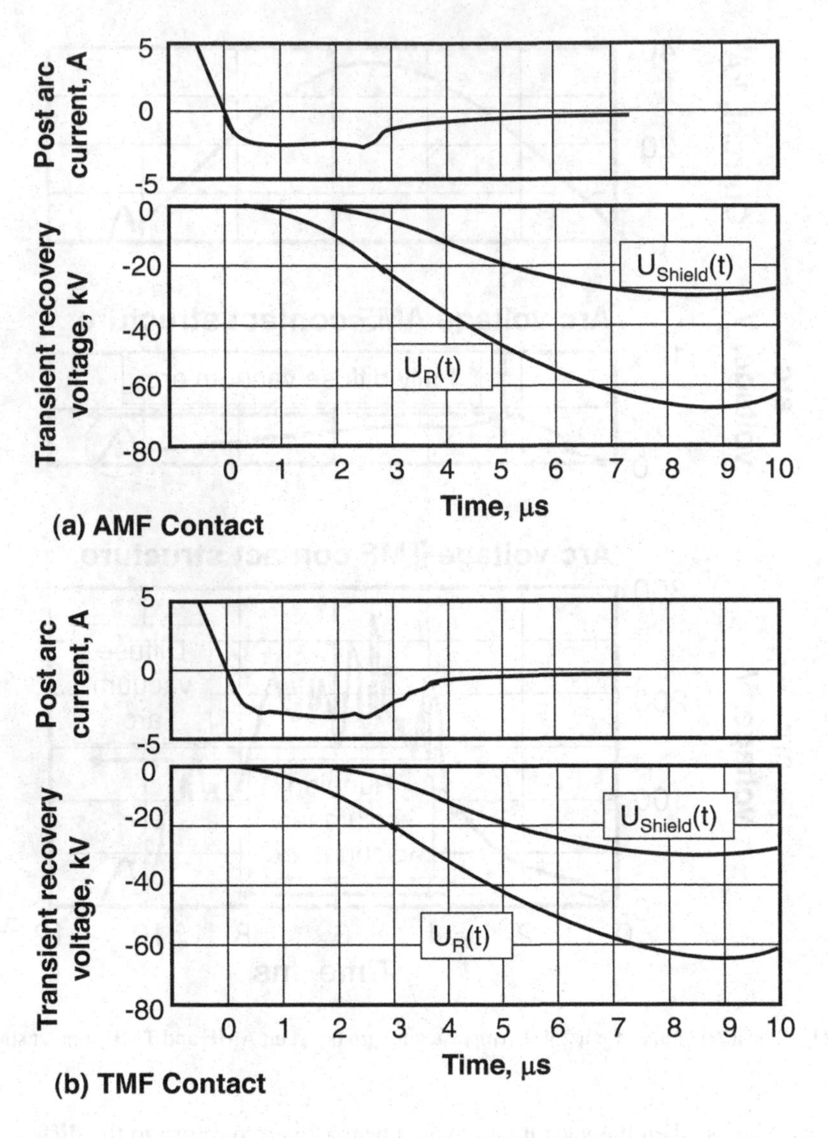

FIGURE 4.22 The post arc current after the interruption of a 25kA (rms) current with about 9 ms of arcing for (a) AMF contacts and (b) TMF contacts [14].

period. Figure 4.23 presents a compilation of high current interruption performance for TMF and AMF contacts as a function of vacuum interrupter diameter, which is generally proportional to the contact diameter, and also the circuit voltage. Here it can be seen that, in general, the greater the vacuum interrupter's diameter, the higher the current it can interrupt; also, the higher the circuit voltage, the greater the required vacuum interrupter's diameter to interrupt the same current.

Let us return to the main circuit current and see what effect it has on the level of post-arc current. In an interesting set of experiments, Hakamata et al. [43] have taken large diameter contacts (170mm) with a multi-pole AMF structure behind them (see Figure 3.44). These open at the beginning of a 52Hz, half-cycle of current from the primary circuit. A Weil–Dobke synthetic circuit injects a high-frequency current just before current zero and then supplies the TRV. The di/dt of the injected current just before current zero, is related to the peak TRV and its RRRV. An oscillogram of a typical post-arc current is shown in Figure 4.24. The delay of 1μs–2μs in the critical rise of the TRV after current zero is clearly seen. Post-arc current data for a 57kA(rms.) vacuum arc are shown in Figure 4.25; here the influences of the *di/dt*, $dU_R(t)/dt$ and $U_{R(peak)}$ on the peak post-arc current are

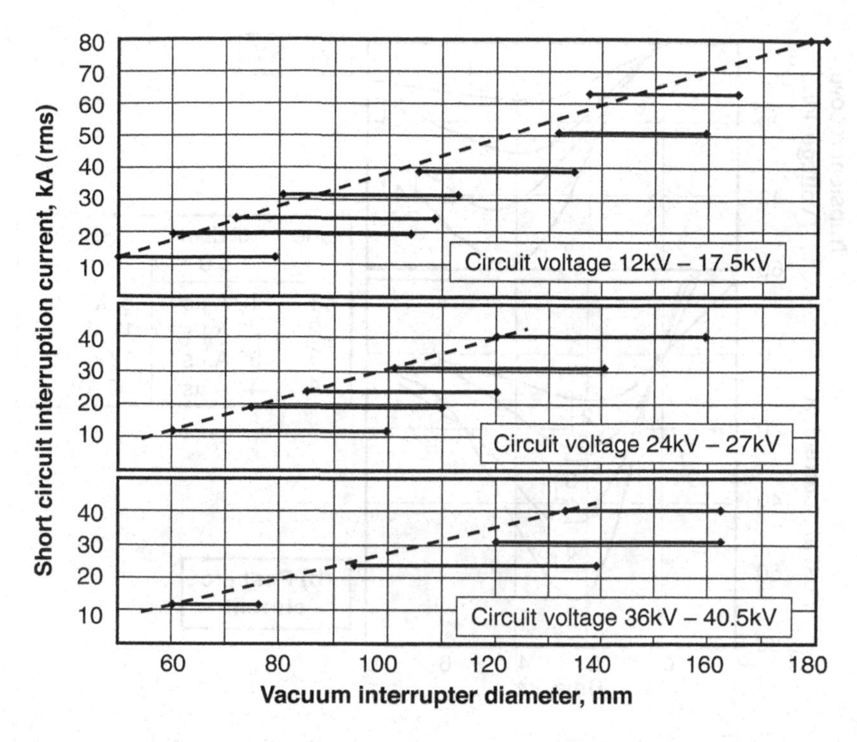

FIGURE 4.23 A compilation of data taken from 12 vacuum interrupter manufacturers, for both TMF and AMF contacts, of the interruption limit as a function of current, circuit voltage and vacuum interrupter diameter (i.e., generally proportional to contact diameter).

presented. The most interesting data, however, are shown in Figure 4.26. Here the post-arc current is shown to strongly depend upon the *di/dt* just before the current zero and the $dU_R(t)/dt$ and $U_{R(peak)}$ just after the current zero, but seems to have little memory of the high current arc that occurred before it. These results are of paramount importance to the vacuum interrupter designer. They mean that with a properly designed contact structure and with a properly sized contact to ensure that the vacuum arc is diffuse at current zero, the vacuum interrupter will successfully interrupt the ac circuit. In fact, contact structures have been suggested for the interruption of currents as high as 200kA [44].

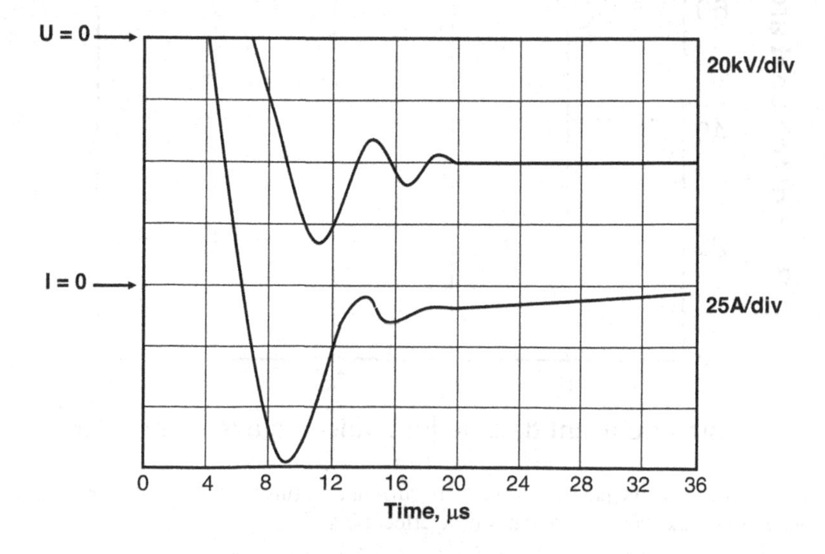

FIGURE 4.24 The post arc current and the TRV after the interruption of a 57kA (rms) [43].

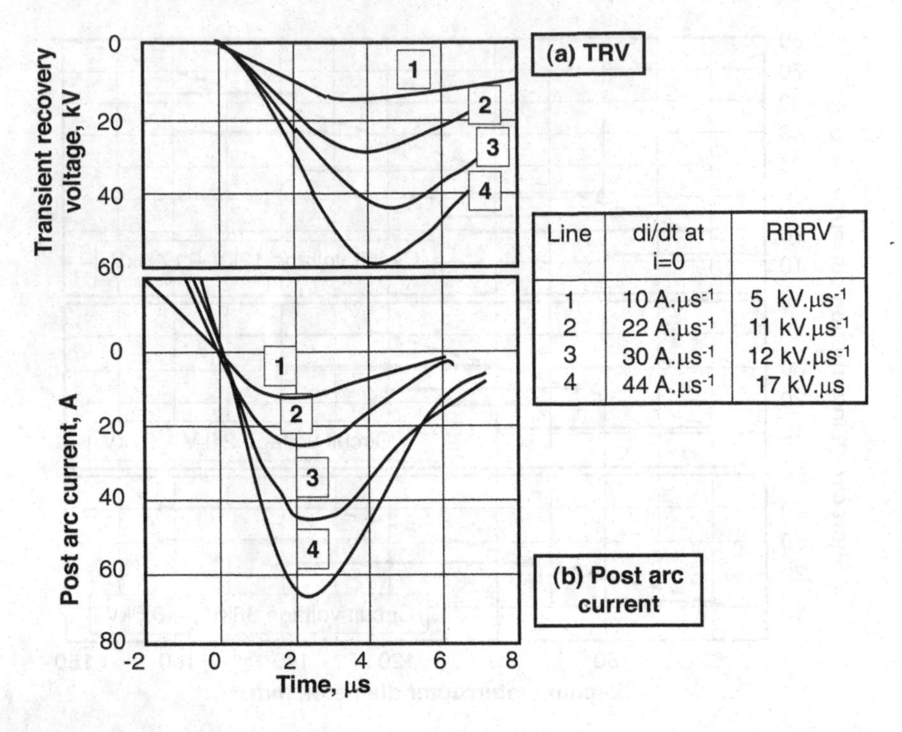

Line	di/dt at i=0	RRRV
1	10 A.μs⁻¹	5 kV.μs⁻¹
2	22 A.μs⁻¹	11 kV.μs⁻¹
3	30 A.μs⁻¹	12 kV.μs⁻¹
4	44 A.μs⁻¹	17 kV.μs

FIGURE 4.25 The effect of di/dt and TRV on the post arc current [43].

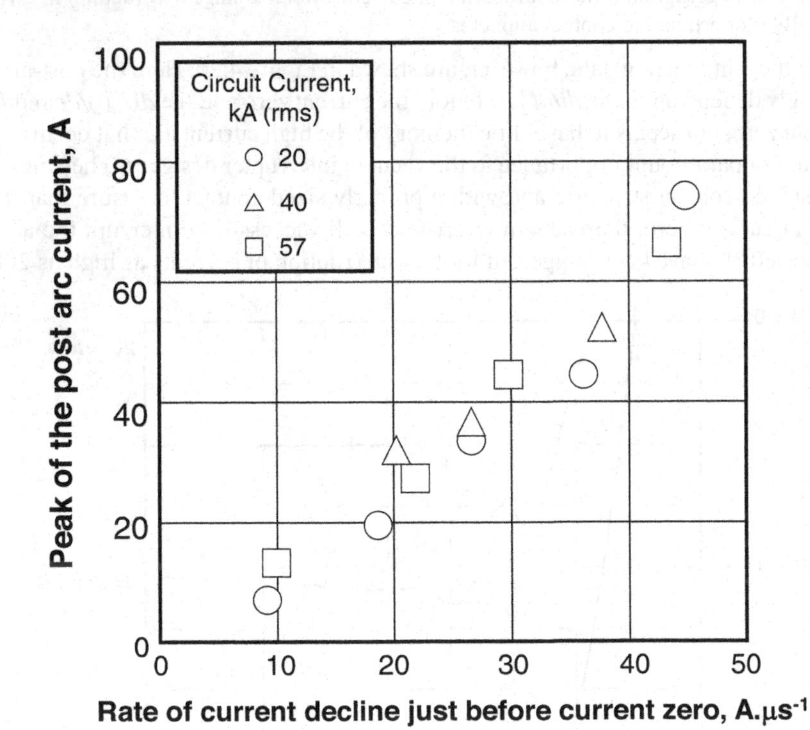

FIGURE 4.26 The effect on the peak of the post arc current as a function of di/dt before current zero, showing that memory of the peak arcing current has little effect [43].

Another aspect of these data that is important for the vacuum interrupter designer to consider is that a lower value of di/dt before current zero and a lower value of the TRV peak give a lower post-arc current. So, it might be expected that a contact with a given diameter would be able to interrupt a higher current in a 5kV circuit than it would in a 38kV circuit. Certainly in Figures 3.30, 3.58 and 4.22 the data for both TMF and AMF contacts seem to confirm this statement. A typical interruption characteristic for a well-designed vacuum interrupter, with either TMF or AMF Cu–Cr contacts, interrupting a symmetrical ac current is shown in Figure 4.27. In my experience this characteristic is valid for all circuit voltages and all symmetrical currents. Up to a limiting short circuit current ($1.0 \times i_{lim}$) the probability of interrupting a given short circuit current is 100%. After this i_{lim} the probability of successful interruption rapidly falls to zero. Figure 4.28 shows the variation that

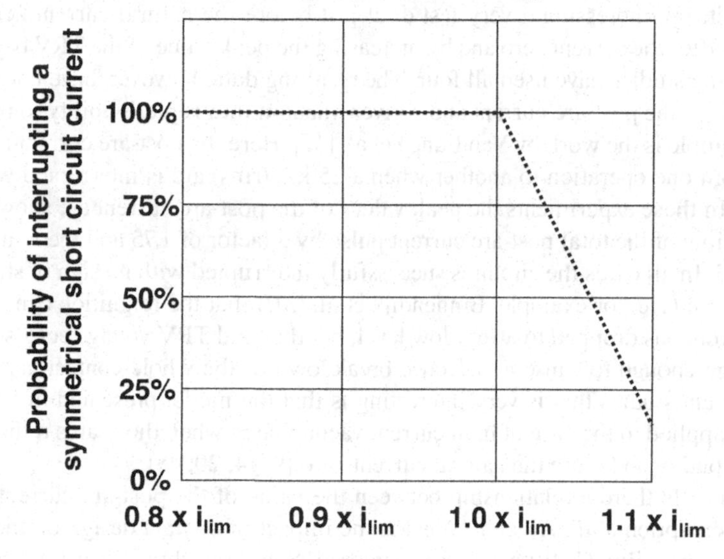

FIGURE 4.27 The probability of interruption of a symmetrical short circuit current as a function of a vacuum interrupter's limiting current (i_{lim}).

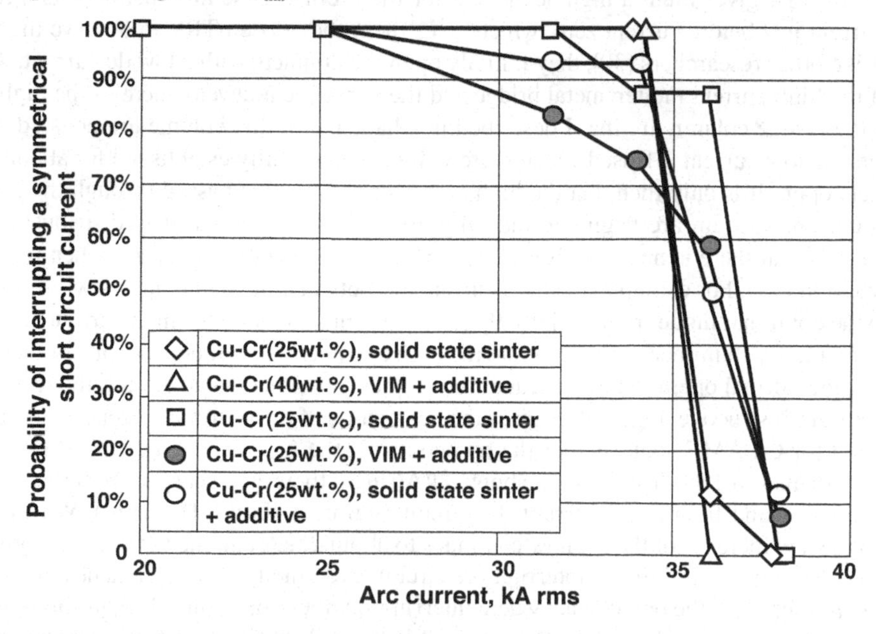

FIGURE 4.28 Variation in the probability of interruption of a symmetrical short circuit current as a function of a vacuum interrupter's contact material (note: VIM is vacuum induction melting) [45].

can result in this characteristic. Here Godechot et al. [45] have evaluated the performance for Cu–Cr contact material with different compositions and manufacturing methods.

There is no clear relationship between post-arc current and the high current interruption performance of a given vacuum interrupter. That being said, some researchers have tried to find a direct correlation between the post-arc current and a vacuum interrupter's ability to interrupt a given ac circuit [9, 46, 47]. For example, it might be expected that the greater the post-arc current, the greater will be the probability that the vacuum arc will reignite after current zero. However, in order to study the post-arc phenomenon in practical vacuum interrupters most researchers have had to investigate the vacuum interrupter's performance beyond its natural design limit. They have done this in a number of ways; e.g., by increasing the current interrupted beyond the vacuum interrupter's designed current interruption limit, by impressing a very fast di/dt just before the natural current zero, by having a very fast RRRV after the current zero and by increasing the peak value of the TRV beyond the design limit. In fact, some studies have used all four. The resulting data, however, have not given a definite correlation between the post-arc current and a given vacuum interrupter's ability to interrupt the circuit. A good example is the work by Van Lanen et al. [47]. Here the post-arc current is shown to vary considerably from one operation to another when a 25 kA (rms) arc is interrupted with a RRRV of about $6kV.\mu s^{-1}$. In these experiments the peak values of the post-arc current are shown to differ by a factor of 2, the time of the total post-arc current pulse by a factor of 1.75 and the total charge passed by a factor of 2.5. In all cases the circuit is successfully interrupted with no sign of stress. Also, most studies have shown (see, for example, Binnendijk et al. [31]) that the reignition can occur well after the post-arc current has dropped to a very low level, but the total TRV voltage across the contact gap has become large enough to cause a dielectric breakdown of the whole contact gap. One aspect of the post-arc current studies that is very interesting is that the model presented in Section 4.1.2 can be successfully applied to the case of high current vacuum arcs when those arcs return to the diffuse mode a millisecond or so before the natural current zero [9, 14, 20, 48].

The questions, "Is there a relationship between the value of the post-arc current and a vacuum interrupter's interruption ability?" and "What is the impact of contact design on the vacuum interrupter's interruption ability?" still need to be explored. Some insight to answer these questions can be found in the work of Yanabu et al. [49]. These researchers have worked with a synthetic experimental circuit that gives them a high ac current for the arcing phase and then injects a high-frequency current just before current zero, which gives them the needed TRV to observe the post-arc current. Like other researchers [30], they initially open the contacts with a low dc current. Thus, all effects of the high current molten metal bridge and the subsequent development of the high current arc from the bridge column arc that I described in Chapter 2 in this volume are negated. In these experiments, a low current diffused vacuum arc will have been fully established for about 10ms as the contacts open. It is only then that the high ac current (~50Hz) is passed through the arc. Thus, this high current vacuum arc begins in the fully diffused mode with widely distributed cathode spots. The fact that there is no initial bridge column arc is especially important when interpreting the interruption data. For example, the high current arc between the butt contacts will take time to form into the columnar mode from the initially diffuse vacuum arc and the high current arc between the AMF contacts will immediately form into the high current diffuse mode. In spite of these differences from the normal operation of a vacuum interrupter in a practical ac circuit, the results of the experiments are instructive. Figure 4.29 gives the peak value of the post-arc current ($\hat{\imath}_{pac}$) for Cu butt contacts and for Cu AMF contacts. For the butt contacts, the $\hat{\imath}_{pac}$ only increases slowly for circuit currents 4kA(rms.) to 18kA(rms.). After about 18kA(rms.), there is a sharp increase as the circuit current increases and failure to interrupt the circuit begins at about 21kA(rms.). With the AMF contacts, the slow increase of the $\hat{\imath}_{pac}$ now continues to about 25kA and then increases more rapidly. In this case there are no failures to interrupt the circuit, even though the $\hat{\imath}_{pac}$ reaches higher values than those measured for the butt contact where interruption does not occur. Thus, just measuring the $\hat{\imath}_{pac}$ alone does not necessarily give an indication of how easily a given contact structure interrupts a given high current circuit.

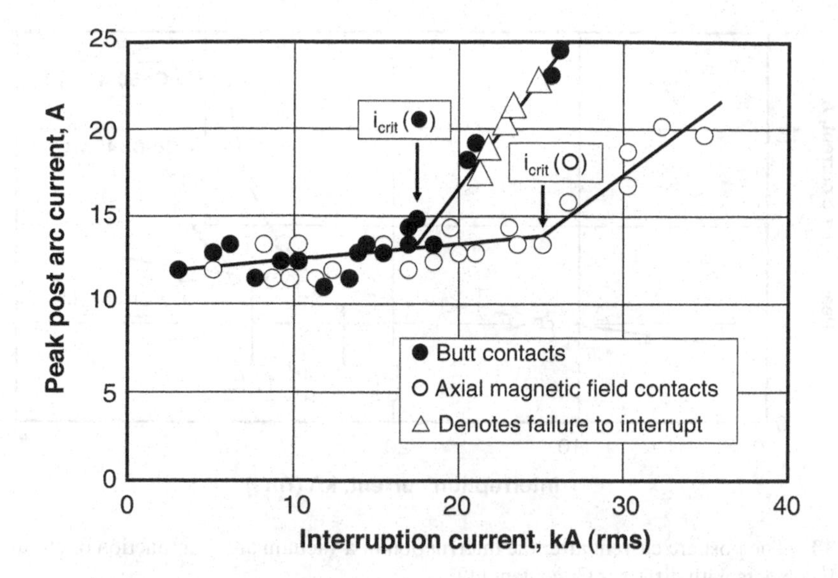

FIGURE 4.29 The post arc current after the interruption of a vacuum arc as a function of circuit current for Cu butt and Cu AMF contacts [49].

It is also important to understand the importance of the contact structure. These data can be interpreted using the discussion presented in Section 3.3 in this volume and our knowledge of the butt and the AMF contact structures. The vacuum arc between the butt contacts will develop into the columnar mode once the circuit current exceeds 14–15 kA(rms). As this columnar vacuum arc burns through current maximum, it will release a copious quantity of metal vapor into the contact gap in order to maintain the high-pressure arc column. It will also heat the contacts at the arc's roots to a temperature higher than the melting point of the Cu. At current zero this stationary columnar vacuum arc will have well-developed residual hot spots on the new cathode and anode as well as an increase in the residual metal vapor and plasma in the intercontact gap. Thus, not only will the contacts continue to supply metal vapor to the intercontact region, but it will also be possible to have electron emission from the new cathode even for relatively low fields impressed on it by the rising TRV. In these experiments, the columnar arc on the butt contacts remains in place until just before the current zero for a circuit current of about 20kA(rms) when the density of the metal vapor in the contact gap is high enough for the gap to break down after the current zero. The Cu contacts with an AMF impressed across them seem to have a large enough diameter for a fully diffuse high current vacuum arc to be developed for the whole current half cycle; even for currents as high as 35kA(rms). As I have discussed in Section 2.6.3 in this volume, the AMF does trap the plasma between the contacts, but for the fully diffuse high vacuum arc, the energy into the contact is spread over the faces of the contacts and is not concentrated in one small area. The AMF contact, with its diffuse arc at current zero, will have no residual hot spots. Consequently, even though there may be the same plasma density between the contacts at current zero, there is a much lower probability of electron emission from the new cathode and any failure to interrupt the circuit will occur at much higher currents. Figure 4.30 presents the effect of Cu content for Cr–Cu contacts on the $\hat{\imath}_{pac}$ for increasing arcing current: here the highest Cu content (75 wt.%) gives the lowest $\hat{\imath}_{pac}$. Again, no interruption failures are recorded even with a $\hat{\imath}_{pac} \approx 18A$.

While Hakamata et al. [43] have shown that for a well-designed contact only the di/dt of the ac current before current zero influences the $\hat{\imath}_{pac}$ and there is little memory of the peak of the arcing current, it is possible to demonstrate the effects of the vacuum arc's current especially if butt contacts are used. Of course, in a normal ac circuit the di/dt just before the current zero is directly proportional to the peak of the ac current; see Equation (4.1). For experiments with butt contacts

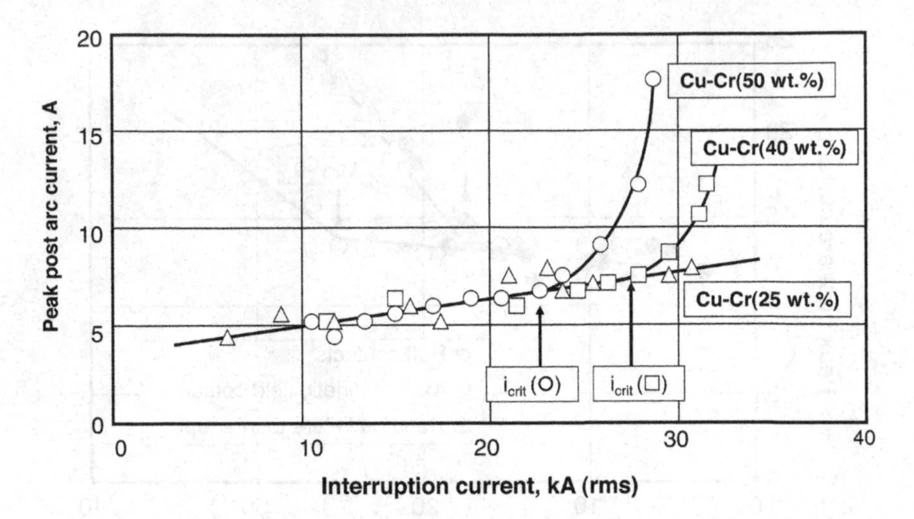

FIGURE 4.30 The post arc current after the interruption of a vacuum arc as a function of circuit current for Cu–Cr AMF contacts with different Cr content [49].

opening a 50Hz circuit and interrupting a 10kA current Lindmayer et al. [9] show that the i_{pac} pulse has a total time of about 10μs and a peak value of about 2.5A. At this time, in their experiment, the recovery voltage, $U_R(t)$, has a value of about 50kV and it is increasing, at a rate $U_R(t)/dt$ of about 5kV/μs, to its peak value of 160kV. After a half cycle of arcing with a 15kA (rms) current the i_{pac} continues to increase. When $U_R(t)$ reaches 75kV, the contact gap breaks down and the circuit is not interrupted. An example of the continued increase of the i_{pac} and the breakdown of the contact gap is shown in Figure 4.31. Here the contrast between the i_{pac} values after the passage of a 10kA current and the passage of a 15kA current is well illustrated. At 15kA, the high current stationary columnar vacuum arc would have formed. At current zero, there would have been hot spots on the new

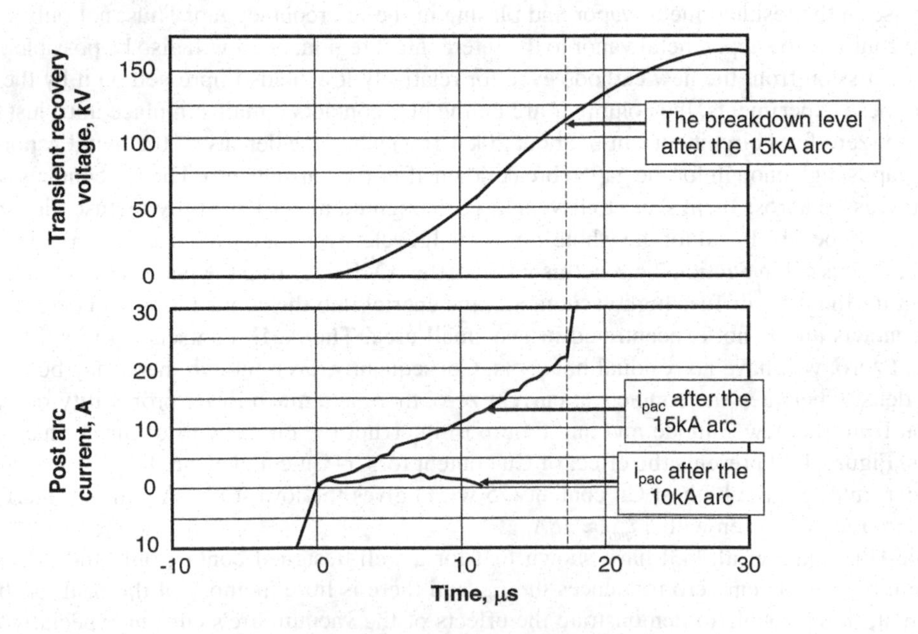

FIGURE 4.31 The post arc current after the interruption of a vacuum arc as a function of circuit current for Cu–Cr butt contacts for two circuit currents, 10kA and 15kA [9].

cathode and a high enough density of metal vapor in the contact gap to initiate a gas breakdown. Smeets et al. [50] show that the post-arc current on its own is not a measure of a vacuum interrupter's high-current interruption performance. While observing certification testing of vacuum circuit breakers, they observe the following instances of reignition and failure to interrupt high-currents:

1. Immediate reignition during the rising edge of the post-arc current. This type of reignition is extremely rare and is only observed when the current to be interrupted is much higher than the value for which the vacuum interrupter had been designed
2. Early reignition during the post-arc current stage. Again, this rarely occurs and indicates that the current is beyond the vacuum interrupter's design current
3. Delayed reignition or dielectric reignition which occurs after the post-arc current has ceased and the TRV across the open contacts increases to its peak value. This is the most common form of reignition in vacuum interrupters and infers that the current to be interrupted is greater than the i_{lim} shown in Figure 4.27

These observations infer that there has to be sufficient metal vapor in the contact gap after current zero for a Townsend avalanche to take place. This in turn suggests that for successful current interruption to take place, the contacts at current zero have to have cooled enough to prevent excessive evaporation of metal vapor into the contact gap. Temperature measurements by Donen et al. [51] of a TMF spiral contact interrupting 16 kA (rms) gives a final contact temperature at current zero of 1675K which is similar to that measured by Pieniak et al. [52]. This will give a metal vapor density in the contact gap of less than $10^{20}.m^{-3}$. Poluyanova et al. [53] record the anode temperature at current zero as a function i_{lim} for an AMF contacts: see Figure 4.32. The anode temperature at current zero for the current i_{lim} is 1660K. So, the temperature of the vacuum interrupter's contacts at current zero appears to be the most important parameter for successful interruption of high currents. Also, the post-arc current plays little or no part in determining a vacuum interrupter's performance. As

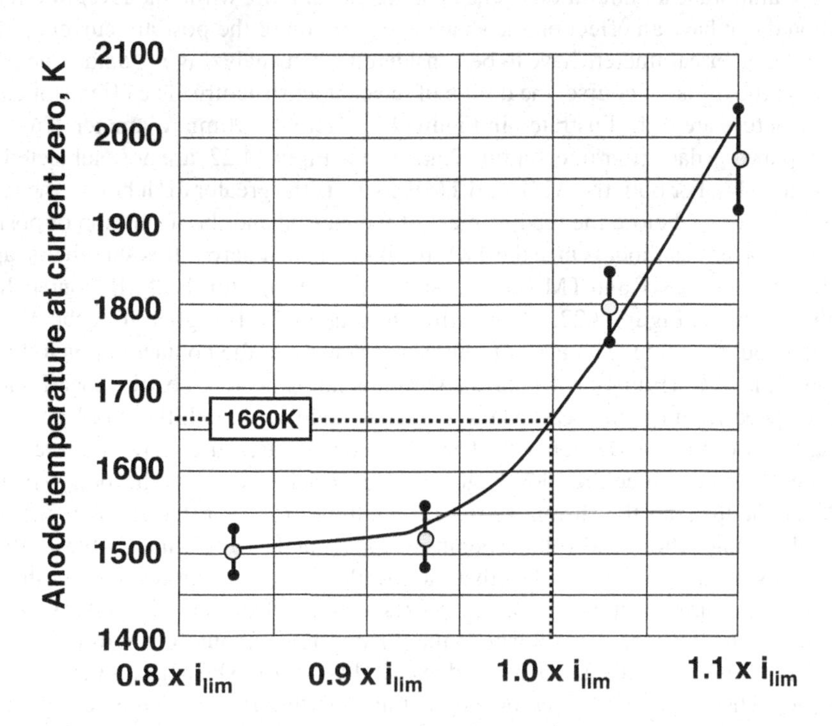

FIGURE 4.32 The anode temperature of the anode for an AMF contact after current zero as a function of i_{lim} [53].

Lindmayer et al. [9] state, there can be several current enhancement mechanisms working together to a level that the circuit cannot be interrupted. These include:

1. The i_{pac} resulting from the ion sheath expansion towards the new anode and the ion current flowing towards the new cathode
2. The emission of electrons from the new cathode liberated by the field at the cathode resulting from $U_R(t)/s(t)$. If there were a residual hot spot on the cathode after the current zero, the electron emission would be enhanced by T–E emission
3. The emission of electrons will also be enhanced by the bombardment of the new cathode by ions traveling from the edge of the space charge sheath
4. If there is a high enough residual metal vapor density after the current zero and/or if the contact surfaces remain at a high enough temperature to continue to evaporate metal vapor, the electrons can begin to ionize this metal vapor. The resulting higher current would also continue the heat input into the contact surfaces and result in the continued evaporation of metal atoms
5. The cooling of the contacts by conduction or from the evaporation of the metal, as well as the dispersion of the residual metal vapor from the contact gap, would limit the effects of this ionization

Again, the structure of the contact plays a major role at high currents. For example, Lindmayer et al. [9] show that the peak of the i_{pac} for the butt contact when it fails to interrupt the 15kA circuit is the same as that observed for both a TMF and an AMF contact structure after a half cycle of arcing at 25kA, where they both successfully interrupt the circuit at current zero. This again emphasizes the importance of these contact structures for distributing the energy from the high current vacuum arc more or less uniformly over the contacts' surfaces.

As Kaumanns [54] has shown, the interruption of high currents can result in somewhat random behavior in the amplitude and duration of the post-arc current. So, while the level of circuit current to be interrupted can have an effect on the shape and duration of the post-arc current pulse as well as its magnitude, other parameters have to be considered when analyzing i_{pac} data. One overwhelming influence on the i_{pac} is, of course, the design of the contact structure. The effects of current level and contact structure are well illustrated in Figure 4.33. Here the 90mm diameter, Cu–Cr contacts used to develop the i_{pac} data after interrupting 25kA (rms), Figure 4.22, are now subjected to a 40kA (rms) vacuum arc [14]. For both the AMF and TMF contacts the greater di/dt before the current zero gives rise to a higher i_{pac} before the readjustment of the plasma and the formation of positive space charge sheath at the new cathode. Once the TRV begins to appear across this sheath, the appearance of the i_{pac} differs for the AMF and TMF contact structures. The i_{pac} for the AMF contact has a shape similar to that shown in Figure 4.22. The maximum value of i_{pac} is higher after the 40kA arc, but its duration is about the same. The larger total charge density in the contact gap is to be expected. However, the same *ds/dt* (Equation 4.7) of the sheath is not expected, i.e., the sheath moves at the same average speed from the new cathode to the new anode after both the 25kA rms arc and the 40kA rms arc between the AMF contacts. The TMF contact structure shows quite a different i_{pac} pulse after the 40kA arc. Once the TRV begins to appear across the sheath, the i_{pac} increases to a value of 10A about 4μs after the current zero, before falling to a low value at about 7μs.

Interestingly enough, the shield voltage begins to appear at the same time for the TMF contact as it does for the AMF contact. This implies that the sheath reaches the anode at about the same time for both contact structures even though the i_{pac} pulses are quite different. Figure 4.21 shows that the vacuum arc between TMF contacts returns to the diffuse mode about 1ms before the current zero for the 25kA rms current. At 40kA rms I would expect this return to be perhaps 0.5ms or less before the current zero. Thus, there could well be a higher temperature region on the new cathode as well as a higher residual metal vapor density between the contacts after current zero. The voltage appearing across the sheath would then be sufficient to liberate electrons from the new cathode by T–E

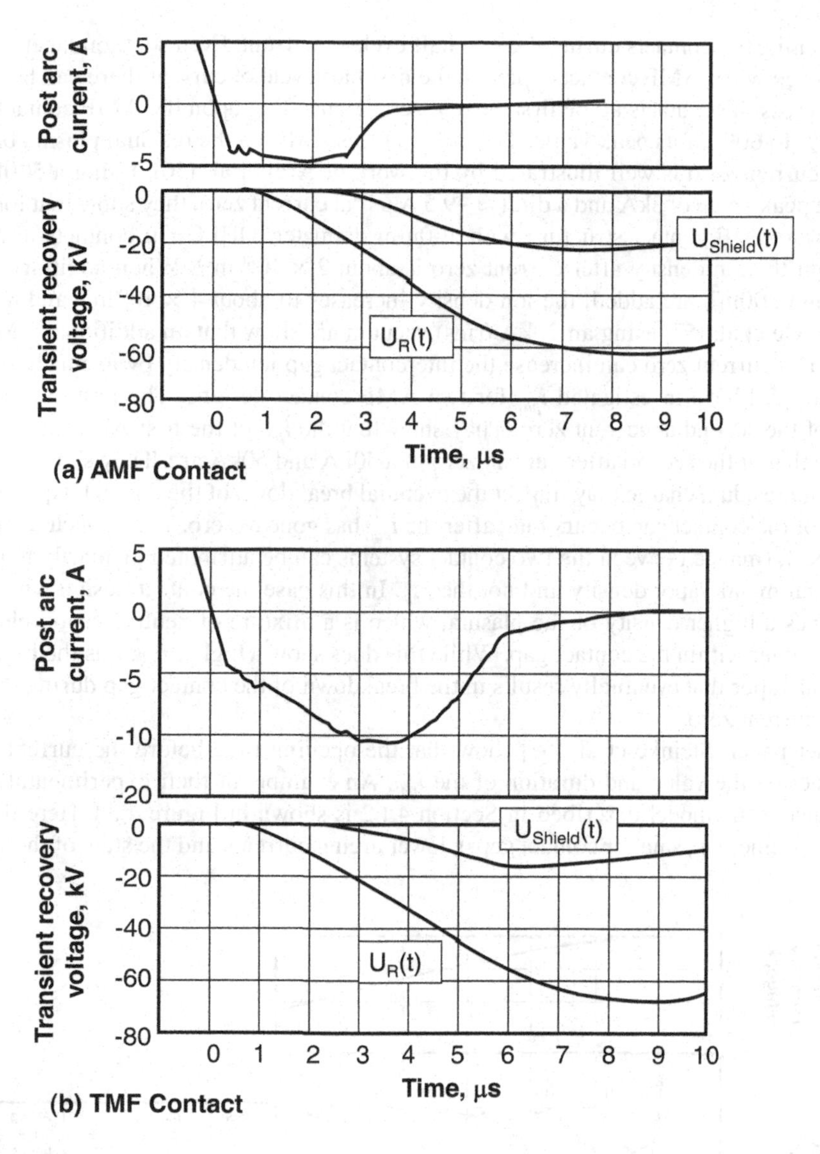

FIGURE 4.33 The post arc current between (a) AMF contacts and (b) TMF contacts from Figure 4.17 after the interruption of a 40kA (rms) current circuit and a 9 ms arcing time [14].

emission. If the metal vapor density is high enough ionization could also take place. The increase in the i_{pac} indicates that both of these phenomena are occurring. However, in this example neither of these enhancement processes is sustained. Once the TRV appears across the full contact gap, the contact surface cools and the residual metal vapor dissipates, the i_{pac} then begins to decrease to a very low value and eventually ceases to exist. It is interesting to note that in this case the much higher i_{pac} does not result in a failure to interrupt the circuit. Both the TMF contact and the AMF contact withstand the full TRV.

As I will present in Chapter 6 in this volume, vacuum circuit breakers that usually have an opening delay of a few current cycles will still have a considerable AMF across the contacts at current zero. This lag in the AMF results from the eddy currents in the AMF contact structure. The AMF across the contact gap after the current zero will keep the residual plasma trapped between the contacts. Most experimental studies of the post current zero phenomena, however, minimize this effect by opening the contacts during the first half cycle of current. Even so, Steinke et al. [55] have shown that even when AMF contacts open on the first half cycle of current there can be a considerable AMF across the contacts at the first current zero. Depending upon the AMF contact structure, it can be 30% to 60% of its peak value. The effect of this AMF on the residual plasma between the contacts at current zero is well illustrated by the work of Arai et al. [56]. Using a 500Hz current pulse with a peak value of 3kA and a di/dt ≈ −9.5 A.μs^{-1}at current zero, they show that for a residual AMF from an AMF contact structure with 100mm diameter, OHFC Cu contacts and a contact gap of 30mm the ion density after current zero is about 2×10^{18} m^{-3}. When additional dc AMFs of 100mT and 200mT are added, the ion density increases to about 4×10^{18}m^{-3} and 8×10^{18}m^{-3}, respectively. Ge et al. [57] using an 20kA (rms) vacuum arc show that an additional AMF pulse of up to 100mT at current zero can increase the intercontact gap ion density by as much as 3.5 times. Steinke et al. [55] have investigated i_{pac} for two AMF contact designs. The first has an AMF 2.2 times that of the second at current zero. They show that the i_{pac} of the first AMF design is always greater than that of the second after current zero of a 50kA and 60kA arc. They show that while the higher i_{pac} and residual charge may impact the eventual breakdown of the contact gap, the dielectric breakdown of the contact gap occurs long after the i_{pac} has gone to zero. They conclude that the difference in performance between the two contact systems can be attributed primarily to differences in the residual metal vapor density and not the i_{pac}. In this case the contact design with the higher AMF confincs a higher density of the plasma, which is a mixture of neutral vapor, electrons and metal ions, longer within the contact gap. While this does show a higher i_{pac}, it is the higher density of the neutral vapor that eventually results in the breakdown of the contact gap during the recovery phase after current zero.

In another paper, Steinke et al. [58] show that the opening time before the current zero has a marked effect on the value and duration of the i_{pac}. An example of their experimental data and a simulation using the model described in Section 4.1.2 is shown in Figure 4.34. Here the effect of shorter arcing time (i.e., smaller contact gaps), lower arcing currents and the state of the vacuum arc

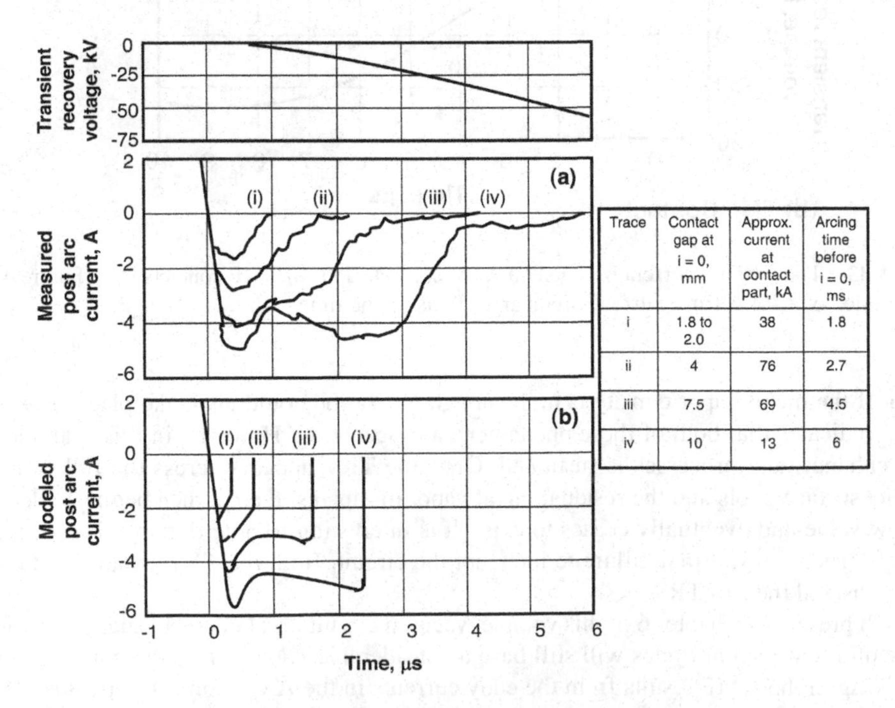

Trace	Contact gap at i = 0, mm	Approx. current at contact part, kA	Arcing time before i = 0, ms
i	1.8 to 2.0	38	1.8
ii	4	76	2.7
iii	7.5	69	4.5
iv	10	13	6

FIGURE 4.34 The post arc current pulse as a function of opening time before current zero for an AMF contact structure, comparing experimental data with a calculation using the sheath model [58].

(i.e., at what stage it is in when becoming fully diffuse) on the i_{pac} is clearly observable. The first i_{pac} peak is proportional to the product of the initial ion density, the initial ion velocity and the effective area of the contact between the residual plasma and the new cathode after the current zero. The increase in the i_{pac} duration is explained by the time it takes for the sheath to cross the contact gap to the new anode. It is interesting to note that for the shortest contact gap and the shortest arcing time, the vacuum arc would most probably still be in the bridge column mode. Thus, even though the i_{pac} has the lowest peak value and has the shortest duration, there would be a high density of metal vapor trapped between the closely space contacts and there would be a high probability of reigniting the arc once the recovery voltage had reached a high enough value. This study also shows that by increasing the RRRV, both the magnitude and the duration of the i_{pac} are increased. Their simulations indicate that this results from secondary electron emission caused by positive ions impacting the new cathode. These data imply that the residual metal vapor between the contacts after current zero is the major cause of arc reignition and not the i_{pac}. For this reason, Dullni et al.'s conclusion [37, 38] that the residual metal vapor and not the residual charge has the greater influence on whether or not a contact gap will recover its dielectric strength after a current zero is valid even though there is plasma trapped by the residual AMF at current zero. This is indeed fortunate: as long as the total density of metal vapor between the contacts remains below about 10^{22} m^{-3}, then the vacuum interrupter will have an excellent chance of interrupting an ac circuit's current. The recovery of the contact gap can be seen in three stages:

1) The residual plasma in the contact gap dominates the first stage after the current zero. It can be described using the sheath model, which is determined by the TRV (i.e., the $dU_R(t)/dt$) and the velocity (ds/dt) of the sheath as it travels from the new cathode to the new anode
2) The vapor density dominates the second stage as the TRV increases after the post arc current has decreased to zero. If the metal vapor density between the contacts is too high a gas-assisted breakdown of the recovering contact gap is possible
3) The third stage is the full recovery of the contact gap to its full design high voltage, withstand value

What does this mean for the vacuum interrupter designer? First of all, when designing a contact structure to interrupt ac currents, it is important to ensure that the vacuum arc reverts to the diffuse mode before the current zero is reached. For low currents, the recovery process is so fast that for all practical ac circuits it is almost impossible not to interrupt it. The sheath model of the contact gap recovery provides a good description of the physical phenomena involved in changes in the residual plasma during the initial stages of the recovery process, but it is not an indicator of the vacuum interrupter's ability to interrupt current. At high currents, if a columnar vacuum arc persists until current zero, the vacuum interrupter will fail to interrupt the ac circuit. This results primarily from the ionization of the increased density of the neutral metal vapor in the intercontact gap by electrons emitted from the hot spots formed on the new cathode.

The TMF and AMF contact structures that have been developed to control the high current vacuum arc have both demonstrated an excellent ability to distribute the vacuum arc's energy over the surface of the contacts. They also ensure the diffuse vacuum arc mode at current zero. In fact, if the diffuse vacuum arc is present at current zero, then the sheath model of the post-arc current is still valid even for high-current vacuum arc [49]. If the contact structure is designed correctly, then the value prior to the current zero of the arcing current seems not to be important: i.e., the contact structure's memory of the high current arc is not the primary driver during the recovery process. Thus, for the vacuum interrupter designer, it is important to size the contact for the expected level of current to be interrupted and to make sure the diffused vacuum arc is present 1 or 2ms before the expected current zero. Figures 3.33 and 4.35 gives examples of the vacuum arc appearance at peak current and then just before current zero for well-designed TMF and AMF contacts.

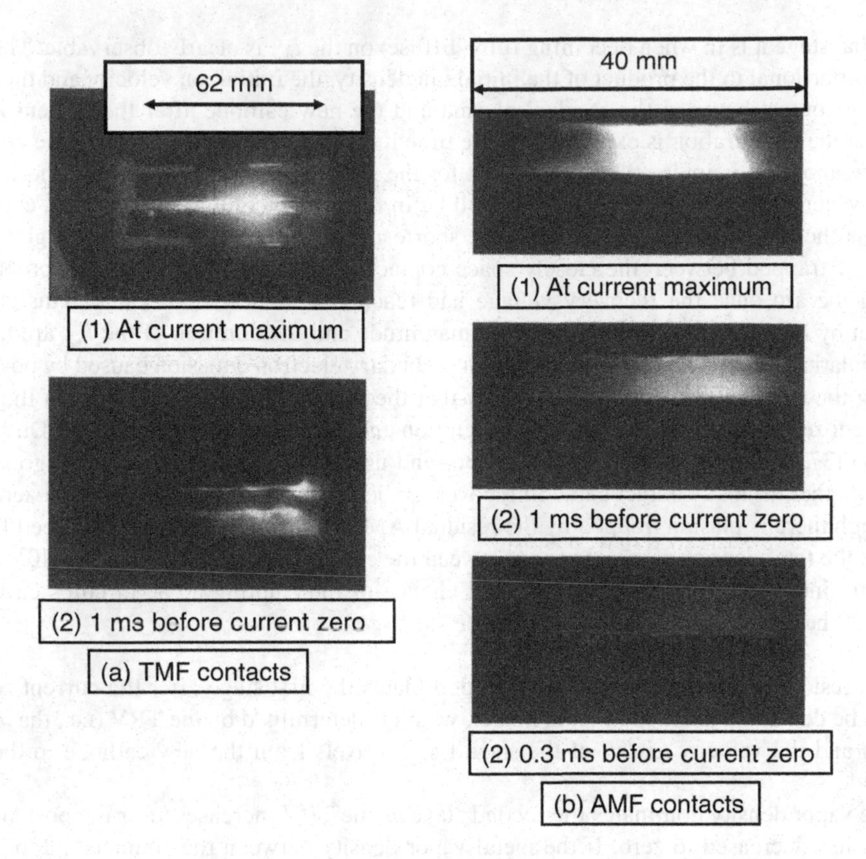

FIGURE 4.35 Photographs of a high current vacuum arc (31.5kA rms) between TMF, Cu–Cr (25 wt.%) contacts and a high current vacuum arc (18kA rms) between and Cu–Cr (20 wt.%) contacts with an AMF contact structure at current maximum and just before current zero.

As Figures 3.30, 3.58, and 4.23 indicate, there is a limit of current that can be interrupted by a contact with a given diameter for both the TMF and AMF designs. The discussion in this Section is relevant to explain this phenomenon. If at current zero the gas density of metal vapor between the contacts is too high, the gap between the contacts will breakdown at a voltage lower than its true vacuum break down level, because electrons crossing the gap will ionize the gas. A metal vapor density can result from two interacting processes on the contact face.

The first process is the time before current zero that the vacuum arc goes into the diffuse mode. The greater this time, the more completely will the metal vapor from the high current arc phase dissipate from the contact gap. It will also permit the contact surfaces to cool down to a temperature low enough that the metal vapor released from the contact surfaces after current zero does not add significantly to the gas density between the contacts. Thus, the probability of the gas density remaining below about 10^{22} m^{-3} will increase.

The second and related process is the high current arc phase. If the heat input into a contact face is too great, it may liberate so much metal vapor that it can still have a high density before the current zero even if the vacuum arc has gone into the diffuse mode. The TMF contact is successful because the high current columnar vacuum arc moves rapidly around the periphery of the contact once the contacts exceed a gap of 3–4mm: see Section 3.3.3 in this volume. As the ac current falls to zero, the motion of the arc roots results in a dearth of metal vapor released and this eventually results in a failure to maintain the high-pressure arc column. This then leads to a transition of the vacuum arc into the diffuse mode and a cooling of the residual arc spots. A good example is presented by Donen et al. [51] who show that during the columnar arc motion between TMF spiral contacts the maximum temperature is 2000C. As the 16kA current reaches zero, the surface temperature of the contacts has dropped to about 1400C. As the arc moves around the contact the arc root regions on the contact surfaces cool once the arc has moved on. If now the current is increased for a given contact diameter, not only does the heat input into the contact surface increase, but also the velocity of the column increases (at least up to about 95kA: see Figure 3.27). Thus, not only is more metal vapor released into the contact gap, but also the arc roots return to their starting point faster, shortening the cooling time of the contact surfaces. Consequently, a high enough current will eventually be reached beyond which a TMF contact of a given diameter will fail to interrupt at a current zero.

At first sight, it would appear that the AMF contact would not have the same limitation discussed above for the TMF contact. Experimental evidence, however, does show an interruption limit for a given contact diameter. This limit results from the nature of the diffuse high current vacuum arc discussed in Sections 2.6.3 and 3.3.4 in this volume. The cathode spots will spread over the contact. If the plasma plumes from these spots do not overlap before they reach the anode, then the heating of the anode will occur over most of the anode's surface. Once the number of cathode spots (which is a function of the current) exceeds a given number for a given contact diameter (i.e., area), the plasma plumes will overlap and a diffuse plasma column will develop. The current flowing through this column will subject it to a magnetic pinch force, which will limit the area where it attaches to the anode. Thus, the heat input to the anode will be increased in this area. If the current is high enough, the heat into this anode spot will also be high enough to permit too high a density of metal vapor at current zero.

Watanabe et al. [59] using AMF contact structures show that for a metal vapor density greater than 10^{20} m^{-3}, the contact gap will fail to interrupt the current. They state that this occurs when the contact surfaces reach a temperature of about 1750K. Table 4.2 shows that the vapor pressure of Cu plus Cr at this temperature is about 33Pa. Niwa et al. [60] indicate that a somewhat higher temperature of 2150K is required. Here Table 4.2 shows that the vapor pressure of Cu plus Cr is now about 2×10^3Pa. At this temperature, Table 4.1 shows that there is a gas density of 7×10^{22} m^{-3} and a small electron mean free path (about 10^{-2}mm). Thus, it is possible for a Townsend avalanche to be established and the vacuum arc to reignite. It can therefore be seen that for both the TMF and AMF contact structures there will be a limited current that can be interrupted for a given contact diameter.

Other factors can influence the density of metal vapor at current zero. One, of course, is the nature of the contact material itself. For example, its thermal conductivity, its electrical conductivity and its vapor pressure at given temperatures will all influence the interruption performance at current zero. A second and perhaps not so obvious factor, is the total volume inside the vacuum interrupter. The metal vapor requires a volume into which it can move away from the contacts themselves. If this volume is limited, it is possible for a higher pressure of metal vapor to exist at current zero. The third effect is the contact opening time, the rate of rise of the recovery voltage and its peak value. All will affect the interruption ability of the butt, TMF and AMF contact structures. This will be discussed further in Chapters 5 and 6 in this volume.

4.3 INTERRUPTION OF AC CIRCUITS WHEN THE CONTACTS OPEN JUST BEFORE CURRENT ZERO

4.3.1 Low Current Vacuum Arcs

Up until now, I have discussed only the interruption of an ac circuit when the vacuum interrupter's contacts are fully open, or at least open by several millimeters at current zero. However, as I stated at the beginning of this chapter, the vacuum interrupter's contacts usually open at random, i.e., there is no preferred place on the ac current cycle where the contacts open. It may therefore be possible for the contacts to open just before the current zero. A typical mechanism may cause the vacuum

TABLE 4.4
Contact Opening Speed 1 ms⁻¹, Peak TRV 20KV in 20μS, 800A (RMS) Current

Time before current zero that the contacts part	Current at contact part 50Hz	60Hz	Contact gap at current zero	Electric Field at the peak of the TRV
1 ms	350A	416A	1mm	2×10^7 V m^{-1}
0.5 ms	177A	212A	0.5mm	4×10^7 V m^{-1}
0.1 ms	36A	43A	0.1mm	2×10^8 V m^{-1}

interrupter's contacts to open at 1ms^{-1}. Let us assume the interrupter is in an inductive circuit where the peak of the TRV is 20kV, and this occurs 20μs after current zero for low currents. When the contacts part just before current zero, the contact gap and the electric field at the peak of the TRV are given for a range of opening times in Table 4.4. These high field values are impressed across a very short contact gap. Trapped in the short contact gap there will be a high-density metal vapor, the residual plasma and hot metal particles. These can all certainly lead a reignition of the vacuum arc sometime during the rise of the TRV [61, 62]. The reignition of the contact gap will most probably result from a Townsend avalanche occurring in the high-pressure metal vapor between the closely spaced contacts [63]. Once the vacuum arc has reignited, it can be expected that the arc would burn until the next ac current zero one-half cycle of current later. After this further half cycle of arcing the contacts in most operating systems would be fully open and the circuit would be interrupted. There are, however, some circuits and some properties of the vacuum interrupter that complicate this simple picture.

4.3.1.1 Low Current Interruption of Inductive Circuits

It has long been recognized that the outstanding ability of the vacuum interrupter to successfully switch high frequency currents can result in some interesting voltage escalation events in an inductive circuit [61–69]. Consider the inductive circuit shown in Figure 4.36, where L is the load inductance, C_L is the load side stray capacitance, L_S is the inductance on the source side, C_S is the source side stray capacitance, ℓ is the small inductance, and r is the resistance in the [vacuum interrupter-C_L–C_S circuit] and VI is the vacuum interrupter. Taking Figure 4.36, I can now describe a possible sequence of events. Suppose VI opens just before the ac current zero. In Figure 4.37(a), there is a chopping current i_C and a corresponding voltage given by:

$$U_C = \pm i_C\left(1-\wp\right)\sqrt{\frac{L}{C_L}} \tag{4.13}$$

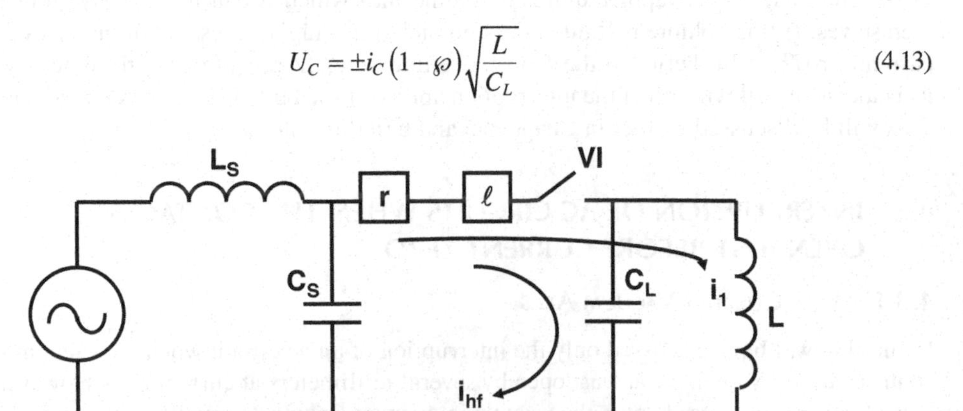

FIGURE 4.36 A schematic of an inductive circuit.

where $\wp$<1 represents certain circuit losses and can be quite significant in limiting the value of U_C [70]. As described in Section 4.1, when there is no current chop (i.e., the ac current goes all the way to the zero value) the TRV now develops whose frequency and peak value depend upon L and C_L once the circuit has been interrupted. In Section 3.2.6, in this volume, I previously presented a discussion of the TRV that appears across the contacts in a L–C–R circuit. Now consider a somewhat simpler and more visual analysis using Figure 4.36 with a chopping current i_C. Also, assume that there are no losses in the system. At the moment that the current chops the voltage on the stray capacitance to ground, C_L, will be close to the supply voltage ($\approx U_{peak}$). At this time there will also be energy stored in the inductance L of $\frac{1}{2} L\, i_C^2$. The current from L will now oscillate in the C_L–L circuit. With the peak value of the voltage resulting from the current chop being U_c, the energy balance will be given by:

$$\tfrac{1}{2} C_L U_C^2 = \tfrac{1}{2} L i_C^2 \tag{4.14}$$

The total voltage on the load side of the vacuum interrupter is now given by:

$$U_L(t) = U_{peak} cos(\omega_L t) + U_C sin(\omega_L t) \tag{4.15}$$

or

$$U_L(t) = U_{peak}\, cos(\omega_L t) \pm i_c \sqrt{\frac{L}{C_L}} sin(\omega_L t) \tag{4.16}$$

where $\omega_L = 1/[L\, C_L]^{1/2}$ and $\wp = 0$. Equation (4.15) is shown in Figure 4.37(a). The interesting consequence of Equation (4.14) is that the TRV term $[U_{peak} cos(\omega_L t)]$ reduces the level of the voltage peak resulting from the chopping current. The peak value of $U_L(t)$ is given by:

$$\tfrac{1}{2} C_L U_{L(peak)}^2 = \tfrac{1}{2} C_L U_{peak}^2 + \tfrac{1}{2} L\, i_C^2 \tag{4.17}$$

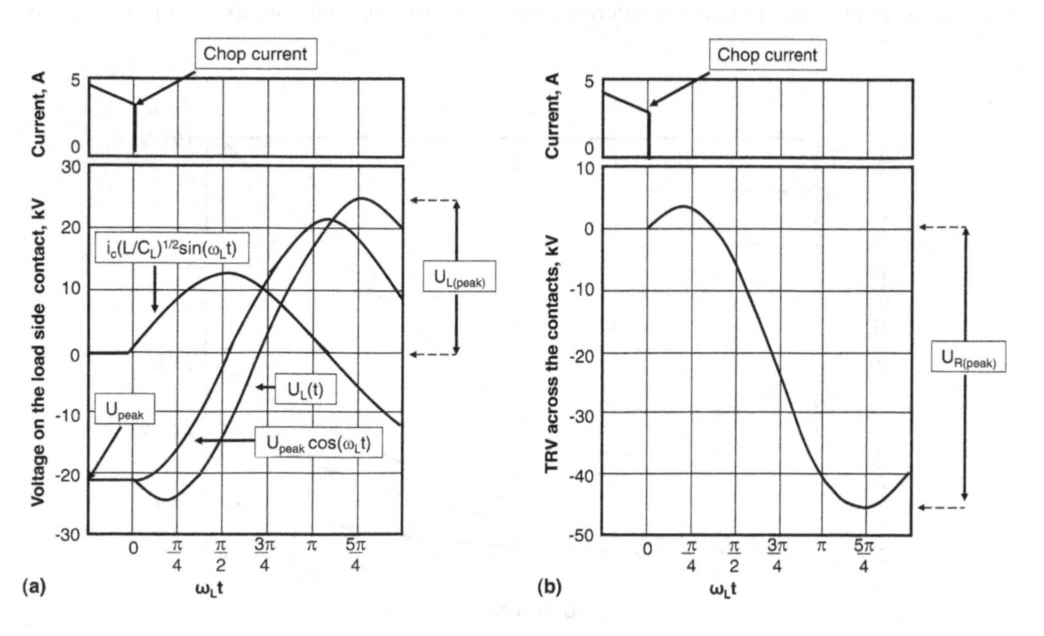

FIGURE 4.37 (a) The voltage $U_L(t)$ on the load side of the vacuum interrupter for the peak of the supply voltage U_{peak} = – 21 kV and i_C = 3 A, L = 12mH, C_L = 800pF, (b) The TRV $U_R(t)$ across the vacuum interrupter's contacts for the peak of the supply voltage U_{peak} = – 21 kV and i_C = 3 A, L = 12mH, C_L = 800pF.

$$U_{L(peak)} = \left[U_{peak}{}^2 + i_C{}^2 \left(L / C_L\right)\right]^{1/2} \tag{4.18}$$

Using Equation (4.13):

$$U_{L(peak)} = \left[U_{peak}{}^2 + U_C{}^2\right]^{1/2} \tag{4.19}$$

The initial TRV across the contacts is:

$$U_R(t) = U_{peak} - \left[U_{peak} \cos(\omega_L t) \pm i_C \left[L / C_L\right]^{1/2} \sin(\omega_L t)\right] \tag{4.20}$$

This TRV is illustrated in Figure 4.37(b). Notice that the development of the TRV across the contacts is delayed for a few microseconds by the presence of the chop voltage. For U_{peak}= 21kV, L = 12.6 mH and C_L = 800 pF a chop current of 3A only increases the peak of the TRV by about 7.7% above the value for a zero-chop current. The percentage increase in the peak value of the TRV as a function of chop current and U_{peak} is shown in Figure 4.38. It is highly dependent upon U_{peak}, but for U_{peak}= 21kV for a 6A chopping current, the peak of the TRV is only increased by 26%. If we assume the usual energy losses (e.g. Equation (4.12)) the peak value of this TRV will be approximately:

$$U_{R(peak)} = 0.85\left[U_{peak} + U_{L(peak)}\right] \tag{4.21}$$

When the contacts open just before a current zero and there is only a small contact gap, we can assume once the TRV reaches a value of U_{B1} (Figure 4.39(b)) a reignition occurs. The current in the load inductance, L, is now i_{L1}. However, when the reignition occurs, a high-frequency i_{hf} current is superimposed on the circuit current i_l that now begins to flow through the contacts; see Figure 4.36 and Figure 4.39(a). This i_{hf} results from the current flowing in the $VI–C_L–C_S$ circuit. It is superimposed on the rising ac current and consists of two frequencies; (a) a very high-frequency in the MHz range resulting from the 'ℓ' and the capacitance of the vacuum interrupter (this current is damped very quickly in a few microseconds) and (b) a second component in the range up to

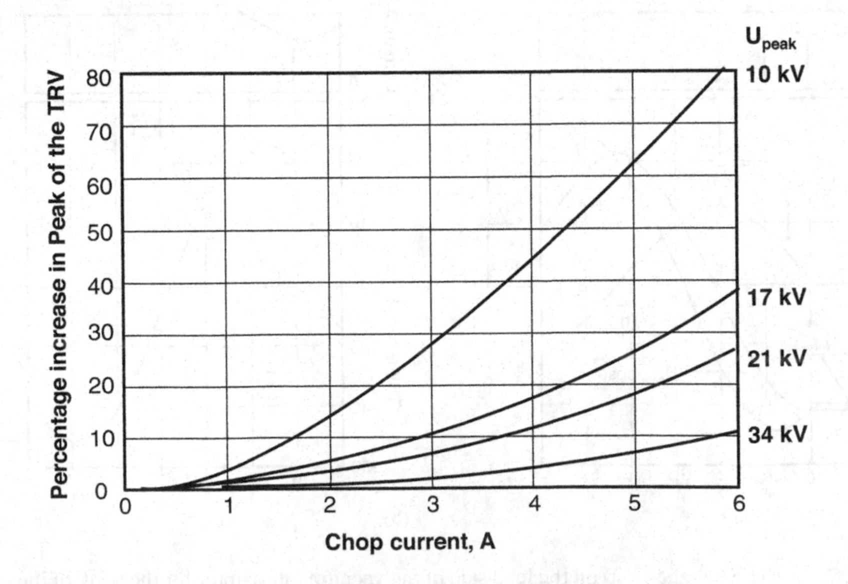

FIGURE 4.38 Percentage increase in the peak of the TRV as a function of the chop current and the peak of the supply voltage U_{peak} = 10kV to 34kV and L = 12mH, C_L = 800pF.

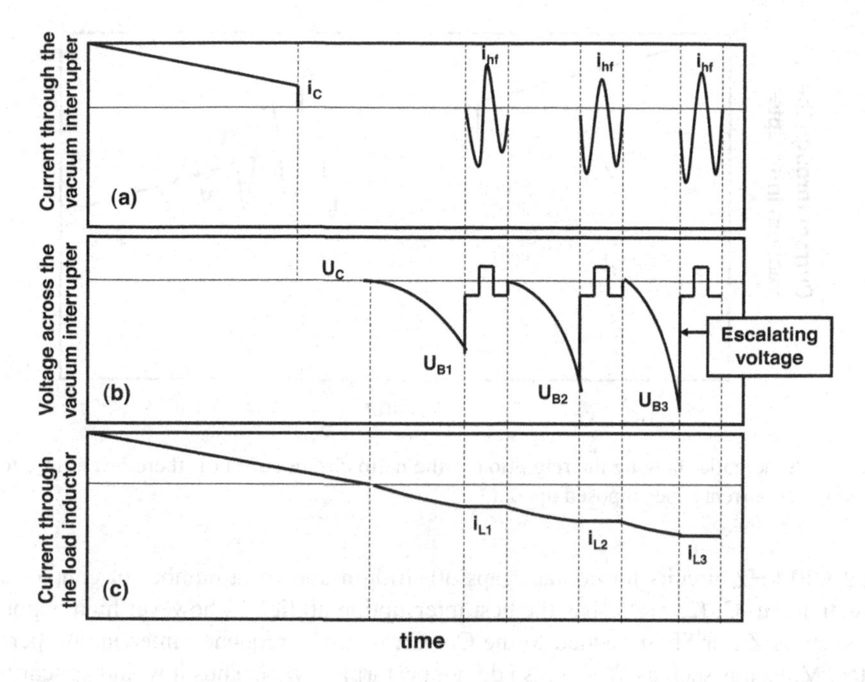

FIGURE 4.39 Schematic of the voltage escalation event.

100 kHz that results from the natural frequency of the L–C_L circuit. Once the i_{hf} begins to flow it is damped. Depending upon the contact material, it can eventually reach a value where it will be interrupted. The change in the load inductance current i_{L1} is small during this flow of high-frequency current. Thus, once the high-frequency current is interrupted, it is reasonable to assume, to a first approximation, the L now has a current i_{L1}. At interruption of the i_{hf}, the L and C_L now impose a new TRV across the contacts and the current into the load inductance, L, increases to a value of i_{L2} as the voltage on C_L now again drives current into it. If now a second reignition occurs at U_{B2}, a high-frequency current i_{hf} is again superimposed on the circuit current flowing through the contacts. Also, the current in L does not increase much above its value i_{L2}. When the i_{hf} is again interrupted, the voltage on C_L drives the current into L to a new value i_{L3}. When a third breakdown occurs at U_{B3}, the sequence will continue. Figure 4.39(b) shows schematically the typical increase in U_B as the contacts continue to open and the gradual increase in the dU/dt as the energy stored in L increases after each breakdown event, Figure 4.39(c). As Murano et al. [68] have shown, however, this voltage escalation does not necessarily sustain itself indefinitely. It can be diminished depending upon how long the i_{hf} continues and once i_{hf} has been interrupted, how long it is before another breakdown. The most important thing to notice is that the voltage escalation is completely independent of the chop event. Some application engineers loosely talk about current chop when they are really referring to a voltage escalation event. Voltage escalation is an interaction between a device that can interrupt very high frequency currents and the circuit in which it finds itself.

What happens if the contact material inside the vacuum interrupter cannot interrupt the high-frequency current? Figure 4.40 illustrates this; the i_{hf} is still superimposed on the circuit current flowing through the vacuum interrupter, but now it is not interrupted and eventually, as the i_{hf} gradually reduces in magnitude and the main circuit current gradually increases in magnitude, no current zeroes are obtained and the vacuum arc burns until the next ac current zero [69]. Here the contacts will have opened to a sufficient gap that the circuit's TRV is easily withstood.

Heyn et al. [71] show that not all contact materials have the same interruption characteristics when switching high-frequency currents. They have investigated the interruption of a 200A current

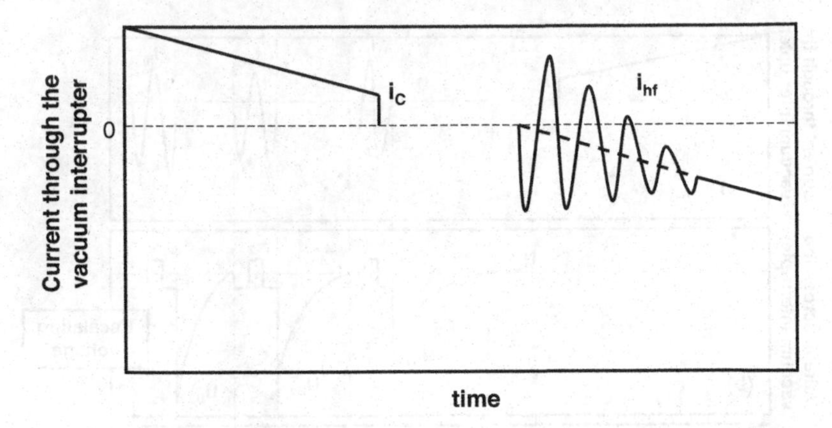

FIGURE 4.40 Schematic showing the reignition of the main circuit current if there is a failure to interrupt the high frequency current superimposed upon it.

in 200 and 600 kHz circuits for contact gaps 30–100μm and for a number of contact materials. They show that Cu–Cr (25 wt.%) has the best interruption ability. If, however high vapor pressure materials such as Zn or Sb are added to the Cu–Cr, its high frequency interruption performance deteriorates. Materials such as W–Cu also do not perform as well. Thus it would appear to be possible to develop so called "low surge" contact materials (i.e., contact materials that will not interrupt high frequency currents) for use in vacuum interrupters to switch inductive circuits where voltage escalation may occur: examples are Ag–WC [72], Co–Ag–Se [73], and Cu–Cr–Bi [74].

4.3.1.2 Low Current Interruption of Capacitive Circuits

Consider the simple capacitive circuit shown in Figure 4.41 as the ac current goes to zero. The effects are shown in Figure 4.42(a), (b), and (c). At current zero, the system voltage lags the current by approximately 90° and thus in this example the voltage that would appear on the source side of the vacuum interrupter is $-U_{peak}$. Also, the capacitor will be fully charged to the same maximum voltage [i.e., $-U_{peak}$] when the circuit is interrupted. The capacitor, now isolated from the source,

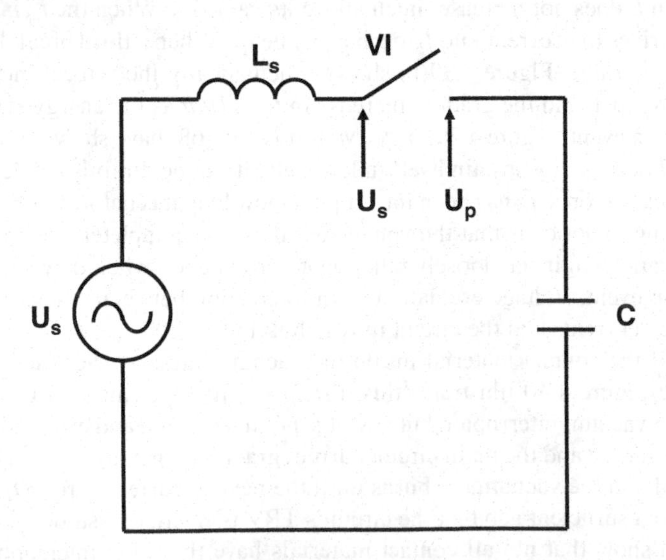

FIGURE 4.41 A schematic of a capacitive circuit.

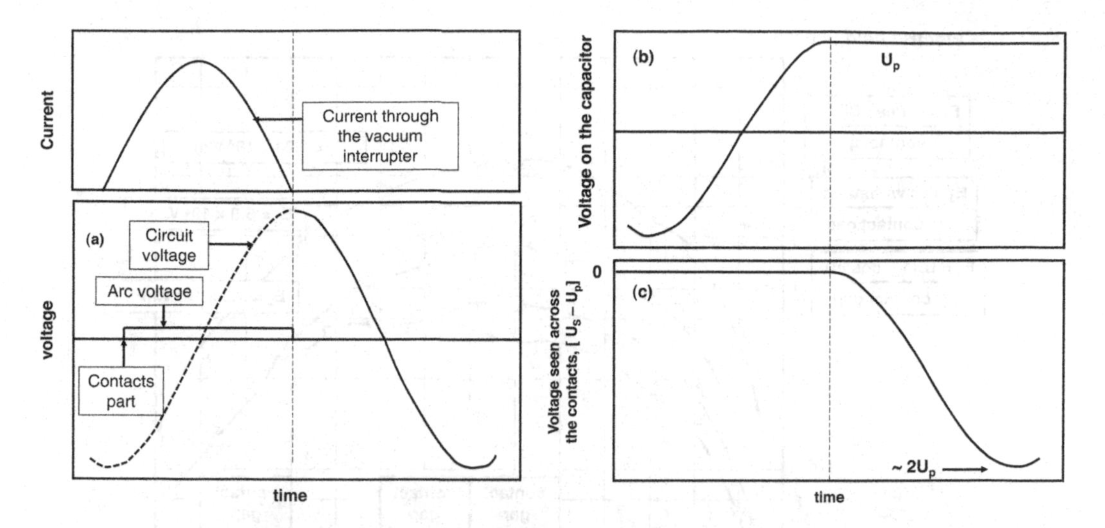

FIGURE 4.42 Schematic to show the transient recovery voltage (TRV) across the open vacuum interrupter contacts after the interruption of ac current in the capacitive circuit.

retains this voltage as is shown in Figure 4.42(b). Hence, the voltage difference across the vacuum interrupter contacts at current zero will itself be zero: see Figure 4.42(c). The voltage on the source side of the vacuum interrupter will continue to follow the ac waveform. This will result in a unidirectional voltage drop across the open contacts that varies from zero volts to about two times the system peak voltage [i.e., about $2U_{peak}$]. As the initial TRV has such a low value and rises only quite slowly, the circuit will easily be interrupted even if the contacts open just immediately before current zero. In this case, it is important for the mechanism designer to open the contacts just enough so that the electric field across the opening contacts is less than a critical value (E_{crit}) at which breakdown will occur, i.e.,

$$\frac{U_{peak}\left[1-cos\left(\omega t\right)\right]}{d\left(t\right)} < E_{crit} \tag{4.22}$$

An example of this calculation is shown in Figure 4.43. Here the vacuum interrupter rated at 24kV, with a peak ac withstand voltage of 50kV(rms.) (or 70.7kV peak), and a gap of 12mm has been designed to accept a maximum gross field (i.e., peak of the 1 minute withstand voltage / contact gap) across the open contacts of about $E_2 = 5.9 \times 10^6$ V.m^{-1}. It has also been designed to accept a BIL pulse peak voltage of 125kV, which would give a momentary gross field of $E_1 = 1.04 \times 10^7$ V.m^{-1}. Shown on this figure are the gross fields experienced by the contacts during the opening. If the maximum field across the contacts were to be less than 5.9×10^6Vm^{-1}, then the opening speed of the contacts would have to exceed 1.5 ms^{-1}. If the maximum field is to be kept below the long-term field across the contacts $E_3 = 3.3 \times 10^6$ V.m^{-1}(for a 3-phase ungrounded system), then the contact opening speed should be greater than 2m.s^{-1}. I shall discuss other aspects of capacitor switching, such as closing a capacitor circuit and the effect of restrikes, in Chapter 5 in this volume.

4.3.2 High Current Interruption

For vacuum contacts opening with high currents, the duration of the bridge column stage can be long enough that at current zero there will be considerable density of metal vapor and plasma trapped between the closely spaced contacts. With these present, it is unlikely that the high-frequency currents, such as those that occur when a vacuum arc in an inductive circuit restrikes, will

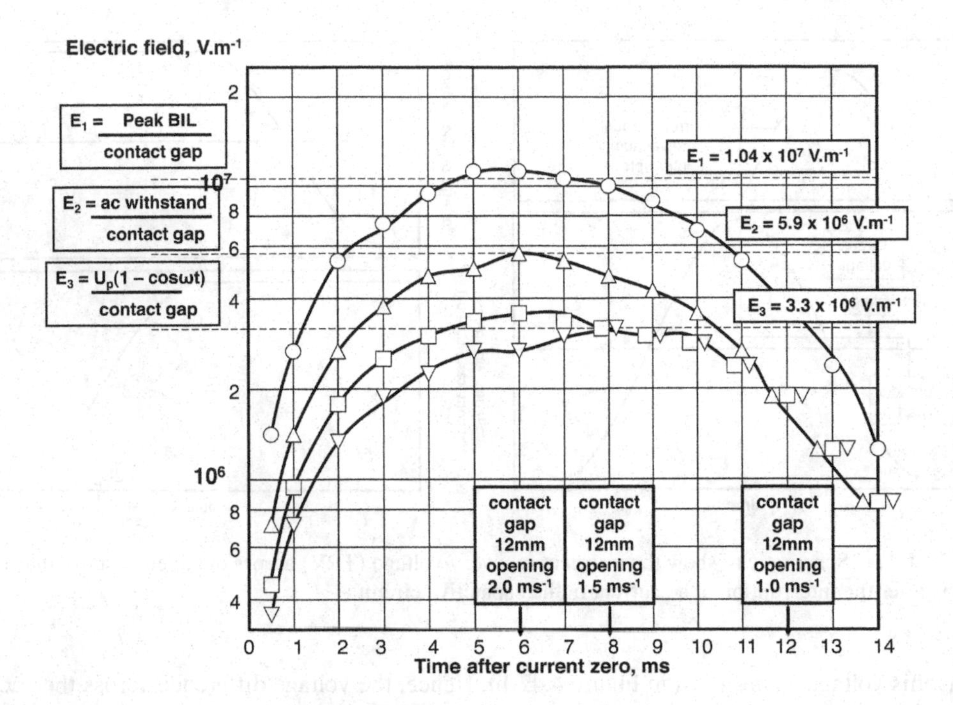

FIGURE 4.43 The gross field (E = {capacitive circuit TRV from Figure 4.42 / contact gap}) as function of time after current zero for contacts opening just before current zero to a 12mm gap in a three phase, ungrounded, 24kV circuit and of contact opening speed (0.5 ms^{-1} {○}, 1 ms^{-1} {△}, 1.5 ms^{-1} {□}, 2 ms^{-1} {▽}); E_1 = (design peak BIL / 12mm), E_2 = (design peak 1 minute withstand voltage / 12mm, E_3 = (peak of the TRV / 12mm).

be interrupted. Thus, the vacuum arc will burn to the next current zero. Telander et al. [75] claim that if the circuit current is above 600A (rms.), the voltage escalation is unlikely to occur. However, I have seen voltage escalation events at currents as high as 2000A (rms.), but it is above this range that the event becomes less and less likely to occur. By the time a circuit current of a few thousand amperes is interrupted, the current at an opening less than 1ms before current zero can be at least a 1–2kA. Here the bridge column arc will either still be in existence at current zero, or the effects of its presence will still be felt. If a restrike occurs, then the number density of the metal vapor and the plasma density between the open contacts will reduce the chance of interrupting the high-frequency currents and another half cycle of the circuit current will develop. At the next current zero the contacts will be fully open, and the contact gap will hold off the resulting TRV.

Betz et al. [76] have investigated the recovery of vacuum arcs for ac currents in the range of 5kA to 10kA for contacts opening at $1ms^{-1}$. They show that for short arcing times (i.e., opening the contacts close to the current zero), because the contact gap is small enough and the vacuum arc is still in the bridge column stage, the contact gap will break down below its full voltage withstand value. Figure 4.44 shows a schematic of the expected performance. Donen et al. [51] also show that if a spiral contact opens less than 2ms before current zero it will not interrupt a 16kA (rms) current. For high short-circuit currents, Slade et al. [77] have investigated the occurrence of the arc burning for an extra half cycle. Their results for one TMF vacuum interrupter style are shown in Table 4.5. In these experiments, voltage escalation is never observed. In this case when the contacts open just before a current zero, a stationary bridge column arc or a stationary constricted column arc will still be in place at current zero and no diffuse arc will be expected to form as the current decreases to zero [78]. Thus, the conditions between the contacts at current zero, the close contact spacing, the increased plasma density, and the increased metal vapor

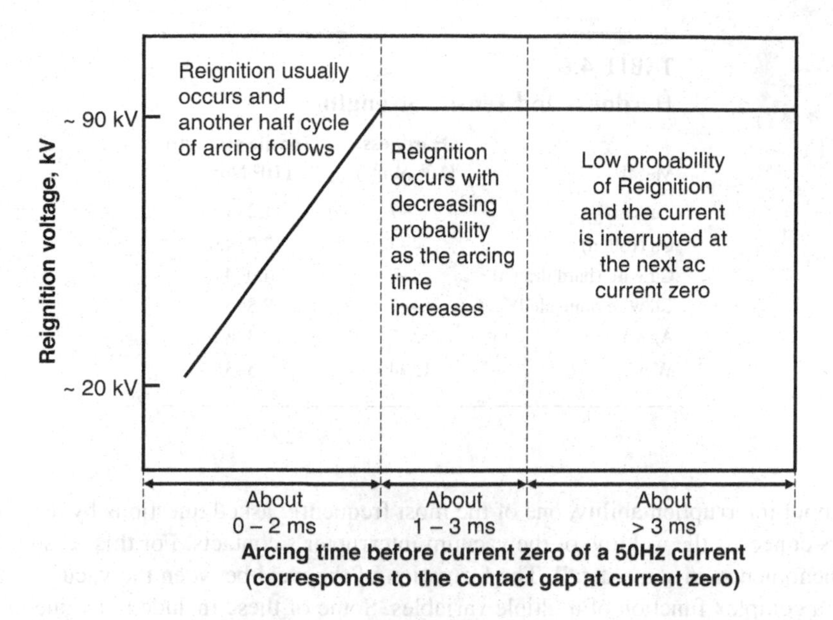

FIGURE 4.44 The expected interruption performance at the first current zero for Cu–Cr contacts subjected to a high, ac current, as a function of the time before current zero the contacts initially open [76].

TABLE 4.5
Probability of Interruption at the First Current Zero as a Function of the Contact Parting Time Before Current Zero for One Style of TMF Contact, Vacuum Interrupter [77]

Experiment	Contact part is:	Time to the first current zero	Probability of interrupting at first current zero
A	Shortly before CZ	0 –3 ms	• Nearly zero for all currents
B	Around current peak	> 3 to 6 ms	• 30% to 80% • 70 to 80% through 36kA • 50% at 40kA • 35% to55% above 40kA
C	Just after one current zero and before current peak	> 6 ms	• Nearly 100% through 40kA • 60 to 85% above 40kA

density will be ideal for the reignition of the arc. The high metal vapor density and the plasma conditions between the contacts will also prevent the interruption of any high frequency currents that are superimposed upon the main circuit current. Thus, the high current vacuum arc will burn to the next current zero where successful interruption of the circuit will take place when the contacts will be open to their full design gap.

4.4 CONTACT WELDING

4.4.1 Introduction

All electrical contacts weld or stick to some extent. The contacts in a vacuum interrupter are no exception. It is only when the strength of the weld prevents the proper operation of the vacuum circuit breaker or vacuum switch that it presents a serious problem. In my experience following

TABLE 4.6
Hardness and Tensile Strength

Metal	Hardness (10^8 Nm^{-2})	Tensile strength (10^8 Nm^{-2})
Cu (cast)	4–7	1.2–1.7
Cu (rolled)	4–7	2.0–4.0
Cu wire (hard drawn)	4–7	4.0–4.6
Cu wire (annealed)	4–7	2.8–3.1
Ag wire	3–7	2.3–3.5
W wire	12–40	15–35

inquiries about interruption ability, one of the most frequently asked questions by users of vacuum interrupters concerns the welding of the vacuum interrupter's contacts. For this reason, I shall discuss the phenomenon in some detail. The formation of the weld between the vacuum interrupter's contacts is a complex function of multiple variables. Some of these include the value of the circuit current, whether or not arcing occurs, whether or not the contacts are closed or are closing, the contact material, the structure of the contacts' surfaces, and the design of the mechanism in which the vacuum interrupter operates. In this section I will develop the general concepts of contact welding. I will also present reference to the analytical analyses of welding phenomena for your further study. It is sufficient for the vacuum interrupter designer, or the user of a vacuum interrupter, to have a general knowledge of what can occur to cause the vacuum interrupter's contacts to weld and to use this knowledge to prevent contact welding from becoming a problem in the switching device. In Chapter 6 in this volume, I will discuss the design of mechanisms used to operate vacuum interrupters and in particular, the criteria to open vacuum interrupter contacts even if welding occurs. The weld force F_W is given by:

$$F_W = \Gamma A_W \tag{4.23}$$

Where Γ is the tensile strength of the material and A_W is the area of the weld. Now as $\Gamma \approx H/3$, where H is the hardness, then:

$$F_W = \frac{HA_W}{3} \tag{4.24}$$

The tensile strength or breaking strength of a material is not an exact number, and is usually given as a range (see, for example, Table 4.6). The elastic limit is always exceeded before the breaking stress is reached. The process of drawing a material into a wire increases its tensile strength; in fact, the finer the wire, the greater its breaking stress. Cold working generally tends to increase the breaking stress of the material. So even with simple equations like Equation (4.23) or Equation (4.24), you can see that for any given contact system there can be a rather large variation in the measured weld force F_W, given the changing micro-structure of the contact surfaces and their change in hardness during their switching life. Contact welding can occur if a high enough current passes through closed contacts and causes the contact spot to melt [79]. Welding can also occur after an arc is initiated between contacts as they close. This arc can result from the electric breakdown of the closing contact gap (see Section 1.6 in this volume), which results in an arc that persists until the contacts finally close. Arcing also can continue as the contacts bounce open once they have initially touched.

4.4.2 Welding of Closed Contacts

4.4.2.1 Cold Welding and Diffusion Welding

In the early development of power vacuum interrupters there was a concern that the two very clean contact surfaces coming together in vacuum would offer an ideal situation for the formation of a strong weld. Indeed, *cold welding* experiments on carefully cleaned metal surfaces in ultra-pure vacuum ($< 10^{-6}$ Pa) showed that a strong bond could develop through normal inter-atomic attraction [80]. There was also a concern that a strong weld would form when the contacts in a vacuum interrupter were together in contact for long time periods with or without the passage of load current. This process, called *diffusion welding* was also demonstrated, again in high vacuum ($< 10^{-6}$ Pa), for Cu, Ti, and Be, at temperatures between 0.1 and 0.2 of their melting points [81]. In a series of papers Maio et al. [82–84] have evaluated the effect of diffusion welding in vacuum for Cu–Cr(30wt.%) contacts with and without a small percentage of Zn and Te at a temperature of 800C. They show that the addition of a small percentage of Te (~100ppm) does reduce the strength of the diffusion weld. The Te atoms become segregated at the interface of the Cu/Cr phases and thus weaken the surface structure of the contact material. The weld strengths measured by Slade [85] for materials under normal load currents, vacuum interrupter pressures and contact material construction were significantly lower than those measured by Maio et al. The impactive opening forces given to the vacuum interrupter by all the usual mechanisms that have been developed easily break these small weld forces. Practical experience since 1967 has shown that for successful vacuum interrupter contact materials fears of cold and diffusion welding were unjustified.

4.4.2.2 Welding Caused by the Passage of High Current

It is possible, however, to form a stronger weld between closed contacts if the current flowing through the microscopic contact spots is high enough to raise them to a temperature where adhesion takes place. This phenomenon is known as *resistance welding.* This has been shown to occur at temperatures as low as the softening temperature (i.e., about one-half of the melting temperature) of contacting metals in a vacuum environment [86]. In fact, spot, roll-spot, seam and projection welding comprise a group of standard resistance welding practices where the passage of a high density current and the resistance of the work parts generate the heat at the joints to be welded [87]. In a series of papers Slade et al. [88–92] have assumed that strong welds will form after the contacting region or regions have reached the contact metal's melting point. In order for melting to occur the threshold welding current i_W must exceed a certain value. This value will depend upon the contact material and the total force, F_T, holding the contacts together. A straightforward model can easily be envisioned: the region of contact melts, the molten contact material flows until the contacting area is large enough to conduct the current in a solidified state, the molten material freezes and a weld is formed. This area of frozen metal is then the weld area A_W given in Equation (4.24).

Let us consider a closed contact system held together with a total force F_T and its initial material hardness and resistivity at temperature T_0 are H_0 and ρ_0, respectively passing a current i for a time period t. I will consider two cases, the first in this section, for a very short pulse of current (~ 10^{-5} to 10^{-2} s) where adiabatic heating of the microscopic contact spots will occur. Then in Section 4.4.4 a second case, where steady state heating is dominant. To illustrate this, a cross section through part of the actual contact region is shown in Figure 4.45. The region affected by adiabatic heating is shown by the dotted lines and can be loosely approximated by a volume $4/3[\pi a^3]$ where a is the average radius of the constriction; see Section 2.1.2 in this volume. For the short duration current pulse, only the volume close to the micro contact spots will be heated beyond the metal's softening point. The softening of the contact spots may cause them to enlarge and even allow other peaks to make contact. However, as Greenwood [93] has shown, if these increased area spots or extra spots occur close to the original contact region, the constriction of the current will not be greatly affected.

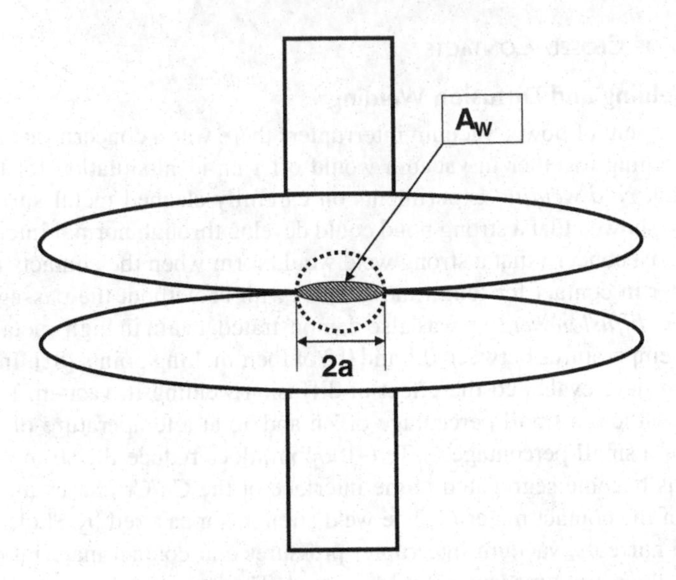

FIGURE 4.45 Closed contacts showing the contact area and the region of adiabatic heating.

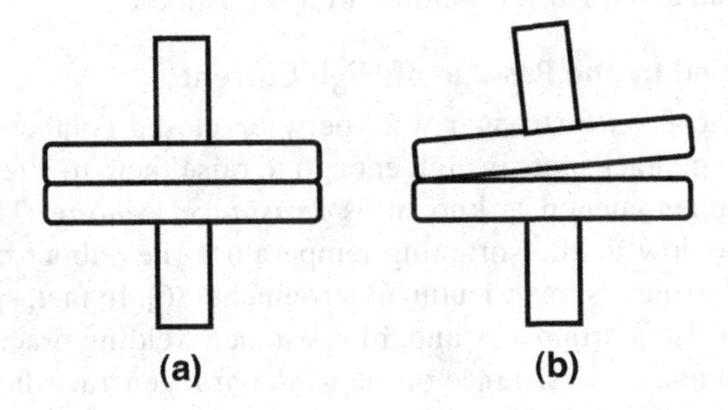

FIGURE 4.46 (a) The assumed mating of the vacuum interrupter contacts with flat face against flat face and (b) tilted contacts making contact along one edge.

A model will be developed to determine the threshold welding current i_W for closed vacuum interrupter contacts with "*n*" regions of contact for a short duration passage of high-current. The model requires the following assumptions:

(a) The contacts make with each contact face against the other and not tilted: See Figure 4.46
(b) The contact though each region of contact is made up of a number of micro contact spots, the contact resistance is determined by an average contact region radius "*a*," see Sections 2.1.1 and 2.1.2 in this volume
(c) The regions of contact are located sufficiently far apart that they can be considered independent of each other [94, 95]; i.e., large area contact regions are similar to separate contact paths in parallel.
(d) The duration of the overload or fault current is short enough to not significantly increase the bulk temperature of the contacts: e.g., an ac half cycle
(e) The duration of the current pulse is long enough for the contact region to reach a high equilibrium temperature: i.e., greater than two or three milliseconds [96–98]
(f) The closed contacts weld when the contact voltage reaches the melting voltage U_m: see Section 2.1.3 in this volume

For the short duration current pulse, only the volume close to the micro contact spots will be heated. Let us assume that the contacts weld when the contact region reaches the melting point. When this occurs, the voltage across the contact region will be the melting voltage U_m: see Section 2.1.3 in this volume. If the threshold welding current is i_W and the contact resistance is R_C then:

$$U_m = i_W \times R_C \tag{4.25}$$

Using the usually accepted value of R_C, Equation (2.4):

$$U_m = i_W \times \frac{\rho}{2}\sqrt{\frac{\pi H}{F_T}} \tag{4.26}$$

where ρ is the contact material's resistivity, H its hardness and F_T is the total force on the closed contacts. In order to determine a value for the threshold welding current i_W the values of ρ and H need to be determined as the temperature in the contact regions increases. Also, the value of an expression for F_T needs to be defined. The total force F_T holding closed switching contacts together is made up of four components:

1. The designer's spring force F
2. The blow-off force F_B resulting from the current flowing into a contact region; see Section 2.1.4 in this volume
3. The Force F_D that results from the passage of current inside the vacuum switch or vacuum breaker design, which can be a blow-off force or a blow-on force [99, 100]
4. The attractive force F_A resulting from the design of the contact structure: e.g. the vacuum interrupter's axial magnetic contacts

Thus:

$$F_T = F - F_B + F_A \pm F_D \tag{4.27}$$

The usual expression for the blow-off force for a single region of contact is given by:

$$F_B = \frac{\mu_0 i^2}{8\pi} log_e\left(\frac{HA}{F_T}\right) \tag{4.28}$$

See Section 2.1.4 in this volume, where i is the instantaneous current, H is the contact material's hardness, A the total area of the contact face and $\mu_0 = 4\pi \times 10^{-7} N.A^{-2}$ is the magnetic permeability in vacuum. Taylor and Slade compiled experimental F_B data for a single region of contact and showed that the blow-off force could be given by [101]:

$$F_B = \beta \times i^2 \tag{4.29}$$

where $\beta = 4.8 \times 10^{-7} N / A^2$; see Figure 2.10. An example of how the current passage through the conductors to the contacts in a switching device affects F_D is given in Figure 4.47. Here an often-used common contact structure is shown. The total force on the moving contact arm and hence on the contact from the passage of current is: (Figure 4.48)

$$F_D = \frac{\mu_o \times L \times i^2}{2\pi(d_1 + d_2)} - \frac{\mu_o \times L \times i^2}{2\pi(d_2)} \tag{4.30}$$

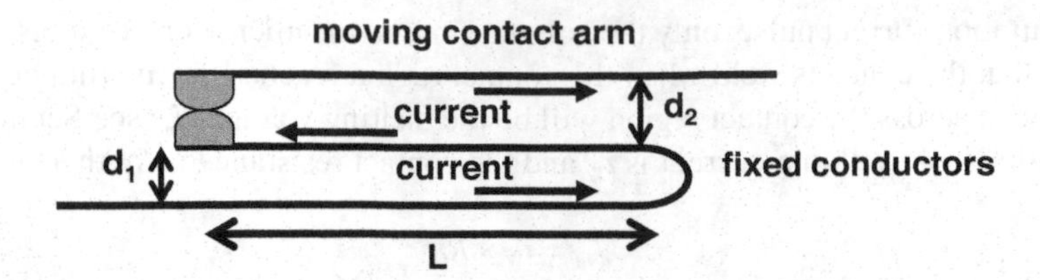

FIGURE 4.47 An example of the effect of the passage of current external to the electrical contacts.

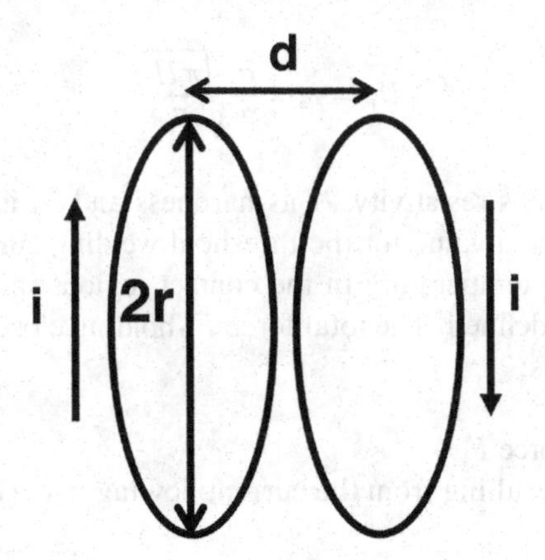

FIGURE 4.48 The current flowing between two coils.

$$F_D = \frac{\mu_o \times L(-d_1)}{2\pi(d_1 + d_2) \times d_2} i^2 \tag{4.31}$$

In general F_D is either a blow-on or a blow-off force [99] and can also be represented by the equation:

$$F_D = \pm k_2 i^2 \tag{4.32}$$

where k_2 is a constant.

Figure 4.48 gives an illustration of the effect of the magnetic field coil behind the contact faces of a vacuum interrupter's magnetic field contact structure. If this simple structure is analyzed for two coils spaced "d" apart, with radius "r" and a peak current "i," the extra holding force is:

$$F_A = \frac{\mu_o \times r \times i^2}{d} \tag{4.33}$$

where F_A is in Newtons when $\mu_0 = 4\pi \times 10^{-7} N.A^{-2}$ and d and r are in meters. Thus in Equation (4.33) F_A can be given by [91, 92]:

$$F_A = k_1 \times i^2 \tag{4.34}$$

where k_1 is a constant. F_A and F_D are forces experienced by the whole contact and do not depend upon the number of contacting regions. The average resistivity for a contact whose region of contact temperature is T_1 close to, but less than the contact's melting temperature is [102]

$$\rho_{T1} = \rho_0 \left(1 + \frac{2}{3}\alpha\left[T_1 - T_0\right]\right) \tag{4.35}$$

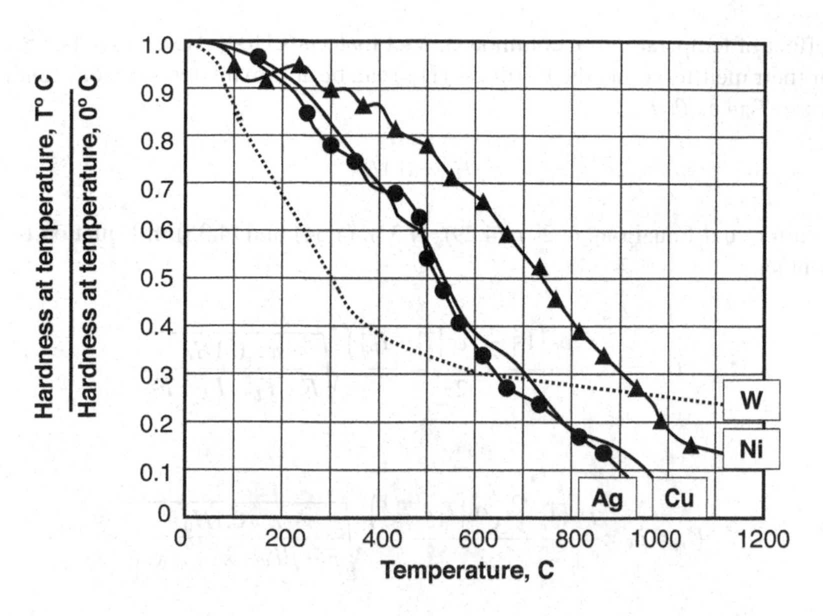

FIGURE 4.49 The effect of temperature on the hardness of common contact materials [104].

TABLE 4.7
Temperature Coefficient of Electrical Resistance for Metals Used in Vacuum Interrupter Contacts

Contact Metal	Temperature coefficient of resistance ($\times 10^{-3}K^{-1}$)
Cu	4.3
Cr	3.0
Ag	4.1
W	4.8
Cu–Cr (25 wt.%)	~ 4.0
Cu–W (90 wt.%)	~ 4.8
Ag–WC (40 wt.%)	~ 4.3
Cu–Bi (0.5 wt.%)	~ 4.3

where ρ_{Tl} is the resistivity of the metal at the contact spot, ρ_0 is the metal's resistivity at ambient temperature (i.e., the contact's bulk temperature) and α is the temperature coefficient of electrical resistance: see Table 4.7. The temperature T_l is chosen because at the contact material's melting point its resistivity can more than double and its hardness has no meaning. For calculations of i_W in this model T_l is the contact material's melting temperature minus 50C. For those contact materials that are a nonalloyed mixture the melting point of the lower melting point component is used. The hardness of the contact spot as it approaches the melting point of the contact metal is greatly reduced from its value at room temperature. Its value is generally given by:

$$H_T = A^* e^{-\frac{B^*}{T}} \tag{4.36}$$

where A^* is the intrinsic hardness at 0K and B^* is the softening coefficient [103]. As with F_B this equation for hardness is not easily used. Instead experimental data can again be used. Figure 4.49

shows the effect of temperature on common contact materials [104]. Here it can be seen that as Cu and Ag near their melting points the Hardness (H_{T1}) can be approximated to be one tenth of its value at temperature, T_0: i.e., $0.1H_0$.

$$H_T = 0.1H_0 \tag{4.37}$$

Using this datum and Equations (4.27), (4.29), (4.32), (4.34) and (4.35) in Equation (4.26) for one region of contact:

$$U_m = i_W \times \frac{\rho_0\left(1 + \frac{2}{3}\alpha\{T_1 - T_0\}\right)}{2}\sqrt{\frac{\pi \times 0.1H_0}{F - F_B + F_A \pm F_D}} \tag{4.38}$$

or

$$U_m = i_W \times \frac{\rho_0\left(1 + \frac{2}{3}\alpha\{T_1 - T_0\}\right)}{2}\sqrt{\frac{\pi \times 0.1H_0}{F - \beta i_W^2 + k_1 i_W^2 \pm k_2 i_W^2}} \tag{4.39}$$

which gives

$$i_W = \frac{2U_m\sqrt{F}}{\left[\left\{\rho_0\left[1 + \frac{2}{3}\alpha\left(T_1 - T_0\right)\right]\right\}^2 \times \pi\left(0.1H_0\right) + 4U_m^2\left(\beta - k_1 \pm k_2\right)\right]^{\frac{1}{2}}} \tag{4.40}$$

where $\beta = 4.8 \times 10^{-7} N/A^2$.

It has been observed that multiple regions of contact can occur in large area contacts found in vacuum interrupters [105], where two or more parallel arcs have been seen as these contacts part. This has been observed for at least 25 years. Figure 4.50 gives an example where seven initial arcs can be seen as the contacts open [106]. In fact, analysis of the contact resistance of vacuum interrupter contacts after arc interruption can best be explained by their having three regions of contact [107].

If there are "n" contact regions, then for the same contact force the total contact area is the sum of the individual areas in each contact region:

$$\pi a^2 = c_1\pi a^2 + c_2\pi a^2 + \ldots + c_n\pi a^2 \tag{4.41}$$

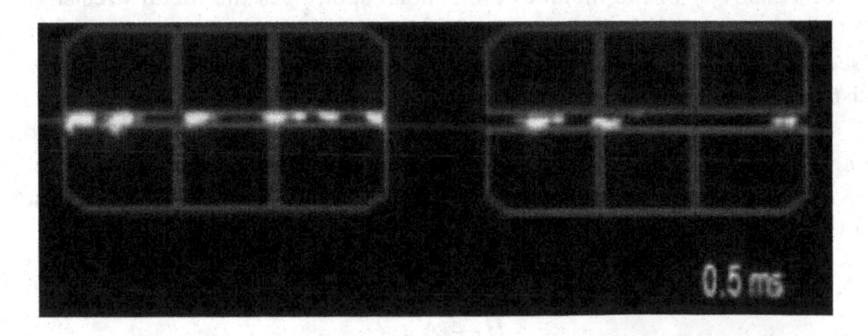

FIGURE 4.50 The opening sequence of an experimental vacuum interrupter with Cu–Cr contacts, showing multiple initial contact spots. The left-hand and right-hand images are taken perpendicular to each other (Figure12 from Reference [106], Courtesy D. Gentsch).

i.e.,

$$\pi a^2 = \sum_1^n c_n \pi a^2 \tag{4.42}$$

Where c_1, c_2 …….. c_n are fractions, i.e., less than 1 and their total equals 1, i.e.,

$$c_1 + c_2 + \ldots + c_n = 1 \tag{4.43}$$

i.e.,

$$\sum_1^n c_n = 1 \tag{4.44}$$

When a fraction of the overload or fault current flows through each contact spot it experiences a blow-off force F_{b1}, F_{b2} …… F_{bn} giving a total blow-off force:

$$F_B = F_{b1} + F_{b2} + \ldots + F_{bn} \tag{4.45}$$

i.e.,

$$F_B = \sum_1^n F_{bn} \tag{4.46}$$

Slade [89] shows that for multiple contact regions:

$$U_m = \frac{i_W}{\sqrt{c_1} + \sqrt{c_2} + \sqrt{c_3}} \times \frac{\rho_0\left(1 + \frac{2}{3}\alpha\left[T_1 - T_0\right]\right)}{2}\sqrt{\frac{\pi(0.1H_0)}{F - \frac{\beta i_W^2}{\left(\sqrt{c_1} + \sqrt{c_2} + \sqrt{c_3}\right)^2} + k_1 i_W^2 \pm k_2 i_W^2}} \tag{4.47}$$

The threshold welding current for "n" regions of contact is:

$$i_W = \frac{2U_m\left(\sum_1^n \sqrt{c_n}\right)\sqrt{F}}{\left[\left\{\rho_0\left[1 + \frac{2}{3}\alpha\left(T_1 - T_0\right)\right]\right\}^2 \pi(0.1H_0) + 4U_m^2\left(\beta - (k_1 \pm k_2)\left\{\sum_1^n \sqrt{c_n}\right\}^2\right)\right]^{1/2}} \tag{4.48}$$

Where $\beta = 4.8 \times 10^{-7} N/A^2$. In a practical world, for a single large contact the values of "c_n" are not usually known. Thus, it is not usually possible to calculate a value of i_W from Equation 4.38. It has been shown [89] that even for a gross range of values different from equal values of "c_n" there is only about a 10% variation in the $\sum_1^n \sqrt{c_n}$ value of contact resistance. In order to estimate a threshold welding current for a vacuum interrupter, each "c_n" can be considered equal. Thus, if $c_1 = c_2 = \ldots = c_n = 1/n$ then $\sqrt{c_1} + \sqrt{c_2} + \ldots + \sqrt{c_n} = \frac{n}{\sqrt{n}}$ so $\sum_1^n \sqrt{c_n} = \sqrt{n}$. This gives:

$$i_W = \frac{2U_m\sqrt{n}\sqrt{F}}{\left[\left\{\rho_0\left[1 + \frac{2}{3}\alpha\left(T_1 - T_0\right)\right]\right\}^2 \pi(0.1H_0) + 4U_m^2\left(\beta - (k_1 \pm k_2)n\right)\right]^{1/2}} \tag{4.49}$$

4.4.3 A Comparison of the Calculated "i_W" with Experimental Values

4.4.3.1 Simple Butt Contacts with One Region of Contact and a Short Current Pulse

Here the closed contact experimental data for i_W developed by Walczuk [108] and by Chrost et al. [109] for one half cycle of 50 Hz current and a contact force of 35N is used. A number of studies have shown that the temperature of the contact spot reaches a high value after the high current has flowed for only a few milliseconds [96–98]. Thus, during an ac half cycle of current or the passage of dc current for a few milliseconds the contacts region will have reached the temperature expected for a given current level. The temperature for the bulk of the contact, however, will not have changed appreciably during this time, so ρ_0 can be assumed to have remained constant An example of their date is shown in Figure 4.51 for a Ag–Cu (3wt.%) contact material. Here the threshold welding current is 4.1kA. Values of i_W can be calculated the using Equation (4.49) for k_1 and k_2 equal to zero and $n = 1$. The values for ρ_0 and H_0 are taken from reference [110]. The major difficulty in using Equation (4.49) is that the H_0 numbers given in the contact material tables have a range that can be as large as a factor of two. Figure 4.52 compares the calculated values of i_W with experimental data developed by Walczuk and by Chrost et al. for a number of contact materials. The interesting observation from Figure 4.52 is that Ag has the highest i_W value. All of the other compounds of Ag and other materials have lower i_W numbers both for the values calculated from Equation (4.49) and from the experimental data. This would have been expected from Equation (4.49), because all the compound contact materials have a greater hardness and a higher resistivity than pure Ag. Figure 4.53 shows a comparison of measured threshold welding currents for Cu–Cr(30wt.%) butt contacts that are plated with a thin layer of Ag [97] with the expected values calculated from Equation (4.49) for one region of contact. Again, the agreement is relatively good considering the uncertainty of the exact value for each contact's hardness.

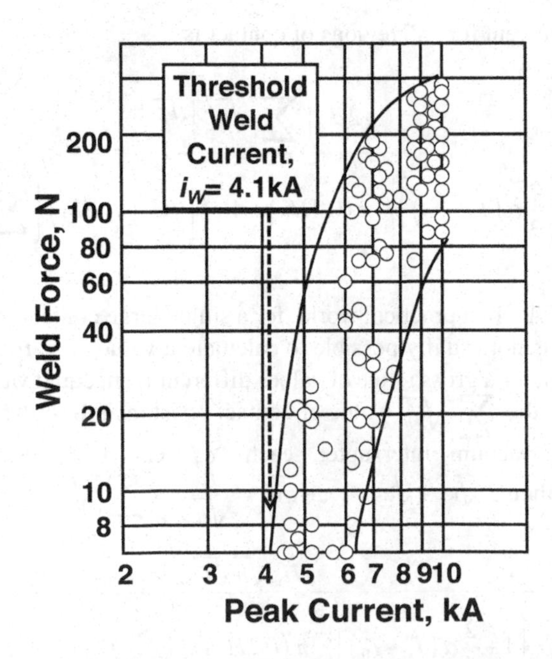

FIGURE 4.51 Dependence of weld strength on current for closed Ag–Cu (3wt.%) contacts for a 35N applied contact force [108].

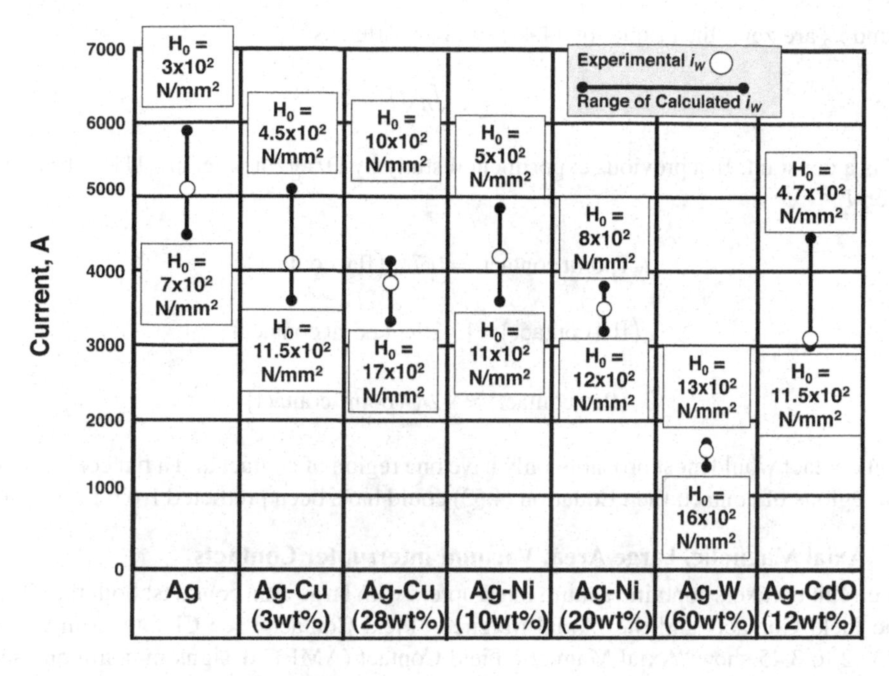

FIGURE 4.52 Comparison of the calculated threshold welding current with experimental values [108, 109] for butt contacts with a 35N contact force.

4.4.3.2 Simple Butt Contacts with More Than One Region of Contact and a Short Current Pulse

As has been discussed above, more than one region of contact has been observed with large area vacuum interrupter contacts. Unfortunately there are no direct experimental determinations of the effect of $n > 1$ on the value of i_W.

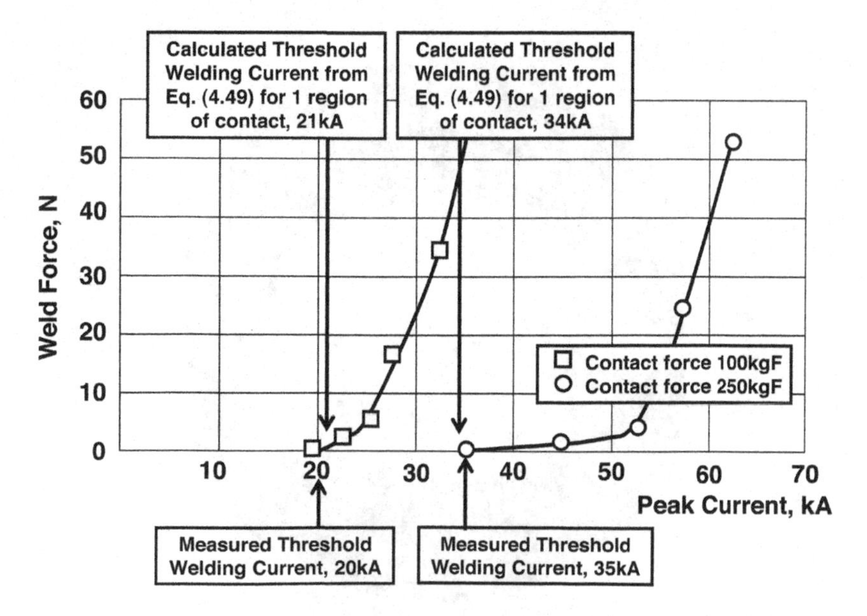

FIGURE 4.53 Comparison of the calculated threshold welding current for one region of contact (Equation (4.49) with experimental values for Cu–Cr(30wt.%) butt contacts plated with a thin layer of Ag [97]

If k_1 and k_2 are zero, then Equation (4.49) can be written as:

$$i_W = C\sqrt{n}\sqrt{F} \tag{4.50}$$

where C is a constant. In a previous experimental study by Dzierzbicki et al. [111]. They stated that for a given F:

$$i_W(\text{point contact}\} = 0.7i_W(\text{flat contact}) \tag{4.51}$$

$$i_W(\text{flat contact}) = 1.429i_W(\text{point contact}) \tag{4.52}$$

i.e. $$i_W(\text{flat contact}\} \approx \sqrt{}2i_W(\text{point contact}\} \tag{4.53}$$

As a point contact would most probably only have one region of contact and a flat contact could well have two regions of contact, then Equation (4.53) could have been predicted from Equation (4.49).

4.4.3.3 Axial Magnetic, Large Area, Vacuum Interrupter Contacts

Vacuum circuit breakers contain vacuum interrupters with large area contacts: both the "Transverse Magnetic Field Contact" and the "Axial Magnetic Field Contact": see Chapter 3 in this volume. Figures 3.42 to 3.45 show "Axial Magnetic Field Contact (AMF)" designs in common use in vacuum interrupters. This contact design consists of relatively flat contact faces and is subjected to high, closed contact forces. As Dullni et al. have shown these contacts usually have at least three regions of contact [107]. When closed, the total contact force for AMF designs is supplied by the design force of the mechanism in which it is housed, the force from atmospheric pressure acting on the vacuum interrupter's bellows and an extra holding force from the magnetic field developed by the current flowing through the coil structure behind the contact faces as illustrated in Figure 4.48 and Equation (4.34). A 38kV vacuum interrupter able to interrupt a fault current of 31.5kA (44.5kA peak) has an axial magnetic field contact 62mm in diameter [91, 92]. For the three-arm coil shown in Figure 3.42(c) and Figure 4.54, the coil thickness $r = 26mm$ and $d = 19mm$ and the current in each

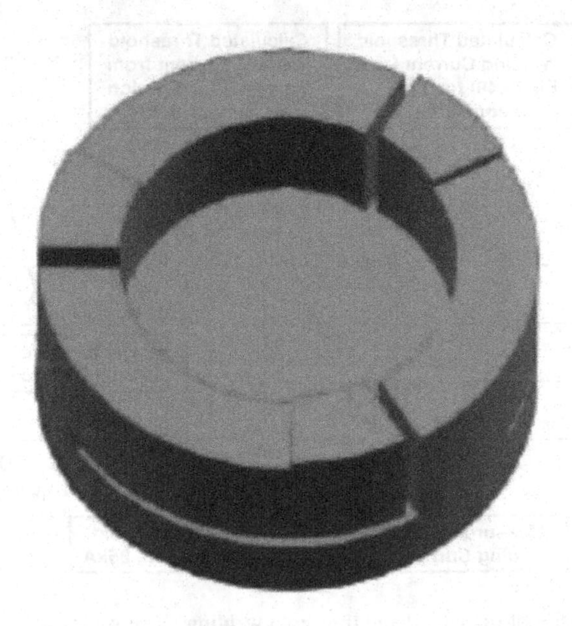

FIGURE 4.54 The three-arm, single coil, axial magnetic field coil for a vacuum interrupter contact.

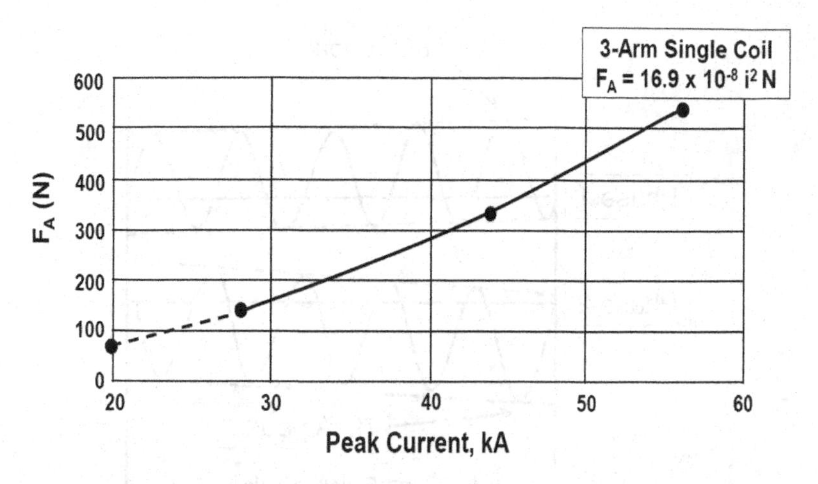

FIGURE 4.55 FEA analysis of the added holding force for a three-arm single magnetic field coil for an AMF vacuum interrupter contact design (91, 92).

arm is $i = 44.5/3$. Substituting in the simple Equation (4.34), the extra holding force for the three-arm coil is 341N ($k_1 = 17.2 \times 10^{-8}$ N/A^2). Slade et al. [87, 88] used a three-dimensional FEA analysis of the three-arm contact structure shown in Figure 3.42(c). Their results of the three-dimensional modeling for a 40kA (56.6kA peak) and 20kA (28.3kA peak), 50 Hz current with 62mm diameter AMF coils gives an additional attractive force for 31.5 kA (46.8kA peak) of 334N and $k_1 = 16.9 \times 10^{-9}$ N/A^2, see Figure 4.55. This is remarkably similar to the values found using Equation (4.43). The attractive force counters the effect of the blow-off force and provides a nontrivial increase in the threshold welding current for a given design contact force. It is now possible to calculate i_W using Equation (4.49) for Cu–Cr contacts with three regions of contact, a diameter of 62mm, $\rho = 33\mu\Omega.mm$, $\alpha = 3.5 \times 10^{-3}$, $n = 3$, $U_m = 4.3V$ $U_m = 0.43V$, and $H_0 = 700N.mm^{-2}$ with a design force of *4020N* and the added force from atmospheric pressure on the vacuum interrupter's bellows of *230N* which gives $F = 4250N$. For a non-AMF contact (i.e., with $k_1 = 0$), $i_W = 46.5kA$, but for the AMF contact $i_W = 51.1kA$ an increase of 4.6kA. This i_W value is lower than the peak asymmetric current for a 31.5kA short circuit current shown in Figure 4.56: see Section 6.6.3 in this volume. The contacts in two of the phases would have formed welds, but those in phase 3 would not. Fortunately, for vacuum circuit breaker manufacturers the certification standards do not require that the contacts not weld. They only require that the contacts open once the current withstand period is over. This requires a mechanism design that will break welds that may occur.

4.4.4 The Model to Determine the Threshold Welding Current for Closed Contacts with "N" Regions of Contact for Passage of Current of 1 to 4 Seconds

The design standards for transmission and distribution vacuum circuit breakers requires them to remain closed and to withstand the passage of high fault currents up to 4 seconds. They then must be able to open and isolate an electrical circuit [112]. These standards assume that these circuit breakers have to coordinate with those breakers closer to the fault and remain closed while the fault is isolated. It has therefore been critical in vacuum circuit breaker design to know the correct contact force required to prevent the contacts from welding during this current withstand period.

One of the certification requirements for a three-phase circuit breaker is the short circuit "withstand test." Here the three phases of the breaker remain closed while (depending upon the certification standard [112]) a short circuit current is passed from between 1s and 4s. The initial peak of this short circuit current will be much greater than the peak value of the symmetrical rms value of the short circuit

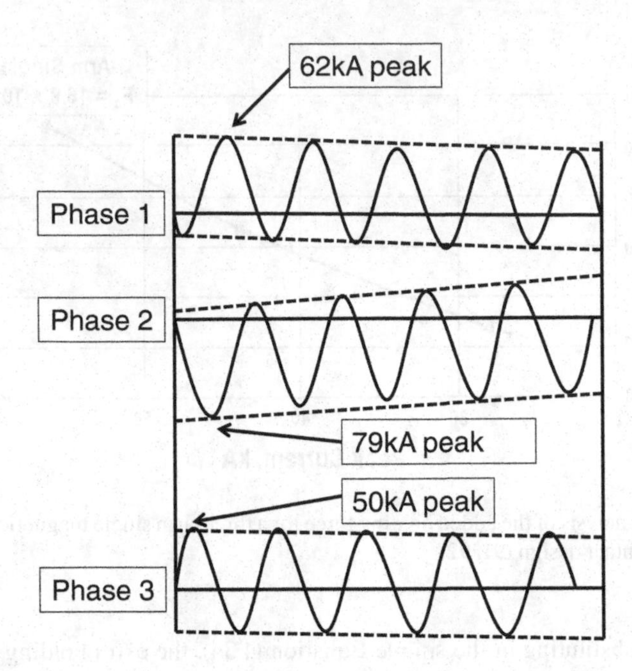

FIGURE 4.56 A 31.5kA, 3-Phase short circuit current showing the initial asymmetric peak current values and the gradual decay to the symmetric peak value of 44.5kA on each of the phases.

current that flows for most of the current withstand period. The initial asymmetrical high current peak values result from the effect of the circuit's inductance not permitting an instantaneous change in current as the short circuit evolves. The result is a decaying dc current that is superimposed upon the symmetrical current value: see Section 6.3 in this volume. For example, a 31.5kA (rms) symmetric current with a peak value of 44.5kA can have a maximum asymmetric peak of 79kA. The peak value of the asymmetric current also depends upon the phase of the current at which the short circuit is initiated: see also Section 6.3 in this volume. As the three current phases are 120^0 with respect to each other only one phase can have the maximum value. This is illustrated in Figure 4.56. Here the maximum value of 79kA can be seen only in phase 2 while the peak values on phases 1 and 3 are 62kA and 50kA, respectively. As the dc component decays in perhaps 5 to 10 cycles, the short circuit current returns to its symmetrical form with the corresponding much lower peak value of 44.5kA.

The assumption used in Equation (4.49) to calculate i_W, that the bulk temperature of the contact remains close to T_0 with the passage of a few ac half cycles of current is no longer valid for current durations of a few seconds. These current, withstand times are usual when high current circuit breakers are subjected to certification testing [112]. With the current passing for a longer time the bulk temperature of the contact would be expected to increase to a temperature T_B. Thus, the bulk resistivity ρ_B would increase to:

$$\rho_B = \rho_0\left[1+\alpha\left(T_B - T_0\right)\right] \tag{4.54}$$

by the end of the passage of the current. Here ρ_0 is the resistivity at the ambient temperature T_0 and α is the temperature coefficient of resistivity. Thus, for "n" points of contact Equation (4.49) becomes:

$$i_W = \frac{2U_m\sqrt{n}\sqrt{F}}{\left[\left[\left\{\rho_0\left[1+\alpha\left(T_B-T_0\right)\right]\left[1+\frac{2}{3}\alpha\left(T_1-T_0\right)\right]\right\}^2 \pi\left(0.1H_0\right)+4U_m^2\left(\beta-\left(k_1\pm k_2\right)n\right)\right]\right]^{1/2}} \tag{4.55}$$

where $\beta = 4.8 \times 10^{-7} N / A^2$. During a longer passage of current, as the bulk temperature of the contact increases, its resistivity will increase and the devisor in Equation (4.55) will increase. Thus, for a constant i_W, F will also have to increase.

4.4.4.1 Closed Large Area Vacuum Interrupter Contacts Passing Fault Currents from 1 to 4 Seconds

Experiments to determine the current level at which welding will occur have been performed on Cu–Cr (25wt.%) vacuum sintered, nonmagnetic field spiral contacts manufactured into vacuum interrupters [79]. The contact shape and dimensions are shown in Figure 4.57. The expected three regions of contact with these large area contacts is also shown. In these experiments the passage of a 20kA (rms) current is adjusted for durations of 0.5s to 4.0s. The region for a low probability of welding for a given contact force is shown in Figure 4.58. The material values for Cu–Cr(25wt.%) contacts are [79]: $(T_T–T_0) = 1000C$, $H_0 = 700$ N.mm^{-2}, $U_m = 0.43V$, $\rho_0 = 3.3 \times 10^{-5}$ Ω.mm and $\alpha = 3.5 \times 10^{-3}$. As expected from Equation (4.55), when the duration of 20kA (rms) current passage increases, the contact force F has to become greater in order to prevent the contacts from welding. Figure 4.58 shows that it takes a total contact force of 1768N to prevent the closed contacts from welding when a current of 20kA (rms) (i.e., 28.28kA peak) passes for 2s, but 2160N when passing the same current for 4s. Using the contact force of 1373N (the specified contact force for this current level [79]) in Equation (4.49), with k_1 and $k_2 = 0$, for a pulse of current of a half cycle and for $n = 1$, a threshold current "i_W" for the contacts to weld is 22.5kA. With more regions of contact the threshold current level increases dramatically (e.g., $n = 3$, $i_W = 28.4kA$). This current level is greater than the peak for a symmetrical current of 20kA. Thus, it can be concluded, that in the experiment there must have been more than one region of contact with these large contacts. In the further analysis the three regions of contact shown in Figure 4.57 will be assumed. From Figure 4.58 at $t = 0.5s$ *to*

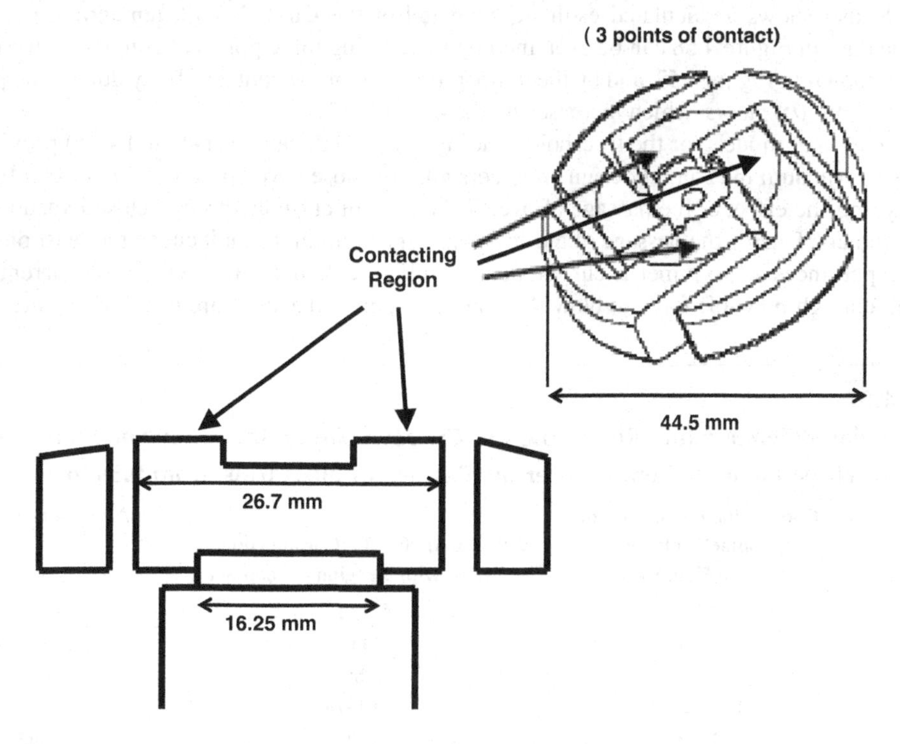

FIGURE 4.57 The large area contact structure (Cu–Cr{25wt.%}) in the vacuum interrupter that was used to develop the closed contact welding data shown in Figure 4.57 [79].

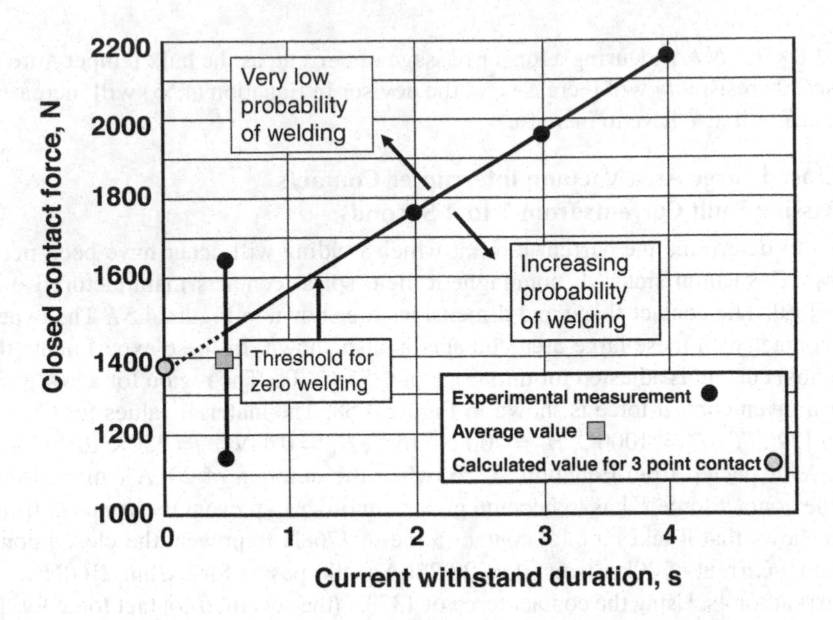

FIGURE 4.58 The probability of Cu–Cr(25wt.%) contacts welding as a function of the total applied contact force "*F*" (the contact spring force + the force on the vacuum interrupter's bellows from atmospheric pressure) and the duration of the 20 kA (rms) current [79].

4s, using Equation (4.44) with $F = (1080–1670 \text{ to } 2160N$ and $i_W = 28.28kA$ the only unknown is $(T_B–T_0)$. Thus, it is possible to calculate values for $(T_B–T_0)$ that would be expected for a passage of 20kA (rms) for *0.5s to 4s* with three points of contact. The calculated values are shown in Table 4.8. Table 4.8 also shows a calculated estimate expected of the Cu–Cr's bulk temperature rise [90]. Thus, the data in Figure 4.58 can be explained by there being three points of contact for the contact structure shown in Figure 4.57. and by the temperature rise of the contact's body during the passage of current from *0.5s to 4s,* which increases its the bulk resistivity.

The analytical models for the threshold welding current (Equations 4.49 and 4.55) provide useful tools for vacuum interrupter design engineers and for those developing vacuum circuit breakers for analyzing the effect of the passage of overload and fault currents through closed vacuum interrupter contacts. During the passage of a three-phase short circuit or fault current at least one of the phases experiences an asymmetric current peak; see Figure 4.56. If this asymmetric current results in a high enough blow-off force to blow the contacts open and a short arc or a floating arc [113] is

TABLE 4.8
The Calculated Temperature Rise of the Cu–Cr (25wt.%) Contact Body Shown in Figure 4.57 with Three Points of Contact after the Passage of 20kA (rms) from 0.5s to 4s

Current duration s	Contact force F to prevent contact welding N, from Figure 4.58	Calculated $(T_B - T_0)$ C using Equation (4.55) with 3 regions of contact	A rough estimate of $(T_B - T_0)$ C [90]
4	2160	132	118–145
3	1964	112	84.5–118
2	1768	89	25–63
0.5	1080–1670	0.1–78	25–30
*10×10^{-3}	1373	0	0.03

* Calculated from Equation (4.49) with $n=3$.

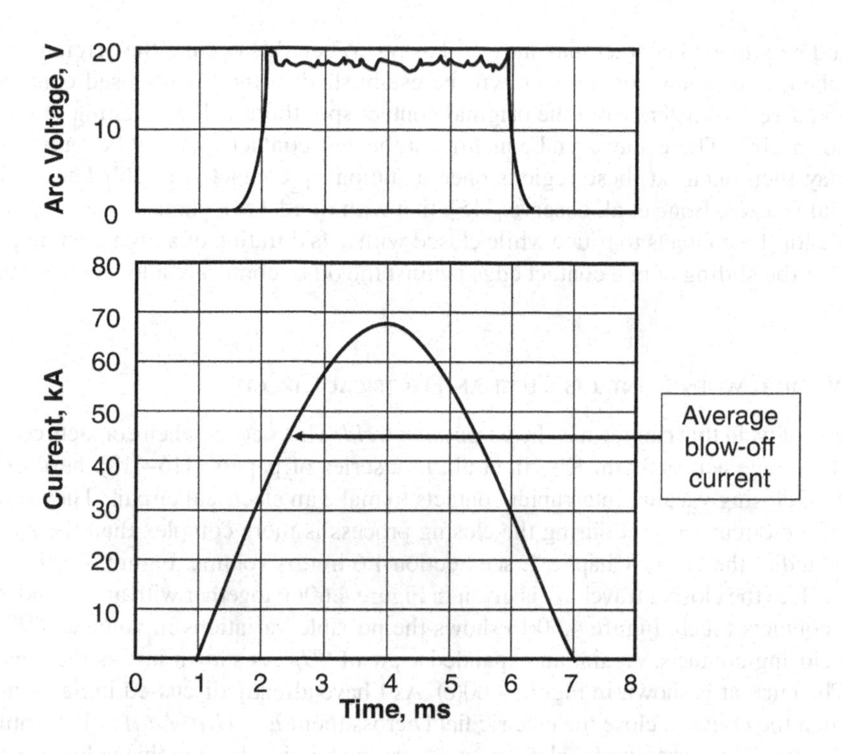

FIGURE 4.59 A current pulse through butt-shaped Ag–W contacts showing the average current where the blow-off force results in the contacts parting and the formation of a short, high-current arc [114].

formed between them, then a strong weld can form once the contacts come back together again. The formation of a strong weld is also highly dependent upon the contact material. Figure 4.59 shows the result of an experiment to determine the ability of Ag–WC contacts to withstand high currents and not form strong welds even when they blow apart. Here the contacts blow apart at 45kA. Even so no welding is observed [114].

4.4.5 Welding of Contacts That Slide

As I will discuss in Section 6.5.2 in this volume, it is possible for the vacuum interrupter's moving contact to slide over the fixed contact. This can occur if care is not taken in the design of the operating mechanism and if a high current passes through the contacts. The magnetic force on the moving terminal resulting from this flow of current can force the moving contact to slip sideways on the surface of the fixed contact. Usually vacuum interrupter contacts are held closed with a relatively high force F. Thus, there will be a related high frictional force $P_{fr} = MF$ (where M if the coefficient of friction) opposing any movement of one contact over the other. For that reason, when a low current flows through the vacuum interrupter contact, sliding is unlikely to occur. At the peak level of a high short circuit current, however, the force holding the contacts together can be severely reduced, that is, when the total contact force F_T is:

$$F_T = F - F_B < 0 \text{ or } F_T = F - \sum_{1}^{n} F_{bn} < 0 \tag{4.56}$$

This in turn will reduce the frictional force that prevents the contacts from sliding. Now, if the force on the moving terminal resulting from the interaction of the high current through it and magnetic flux from the current flow through other parts of the system is high enough, the moving terminal and its attached contact may move sideways. When this occurs, the original contact spot will be broken, and a new contact spot will be established in the final, closed contact position. As the high current transfers from the original contact spot there will be melting of that spot and perhaps even arcing. There may even be melting at the new contact spot as it becomes established. Welding may then occur at these regions once a stationary contact is established again and the molten metal freezes. Bono et al. observe [115], that with tilted TMF contacts (See Figure 4.46(b)), it is possible for the contacts to rotate while closed with a 4s duration of a high-current pulse. They postulate that the sliding of one contact edge against the other contact results in the contacts welding together.

4.4.6 Welding when Contacts Close an Electrical Circuit

Another type of weld that can form is the *percussion weld*. This occurs when contacts come together with a high voltage across them. Kharin et al., in a series of papers [116–118], have explored the complexity of closing vacuum interrupter contacts to make an electrical circuit. They show that the initiation of the circuit current during the closing process is more complex than the simple model that I presented at the end of Chapter 1: see Section 1.6 in this volume. Figure 4.60 illustrates this complexity. Here the closing travel "*d*" is given in Figure 4.60(a) together with an expanded view just before the contacts touch. Figure 4.60(b) shows the possible variations in voltage "U_C" measured across the closing contacts. Again, an expanded view of "U_C" is shown just as the contacts come together. The current is shown in Figure 4.60(c). As I have already discussed in Section 1.6 in this volume, when the contacts close the electric field across them $E = U_C(t) / d(t)$ will eventually reach the critical value E_C for vacuum breakdown to occur; see Section 1.3.2 in this volume. Kharin et al. [118] have shown that a breakdown event does not necessarily result in the formation of a vacuum arc between the contacts and the initiation of the circuit current. This is illustrated in Figure 4.60(b)

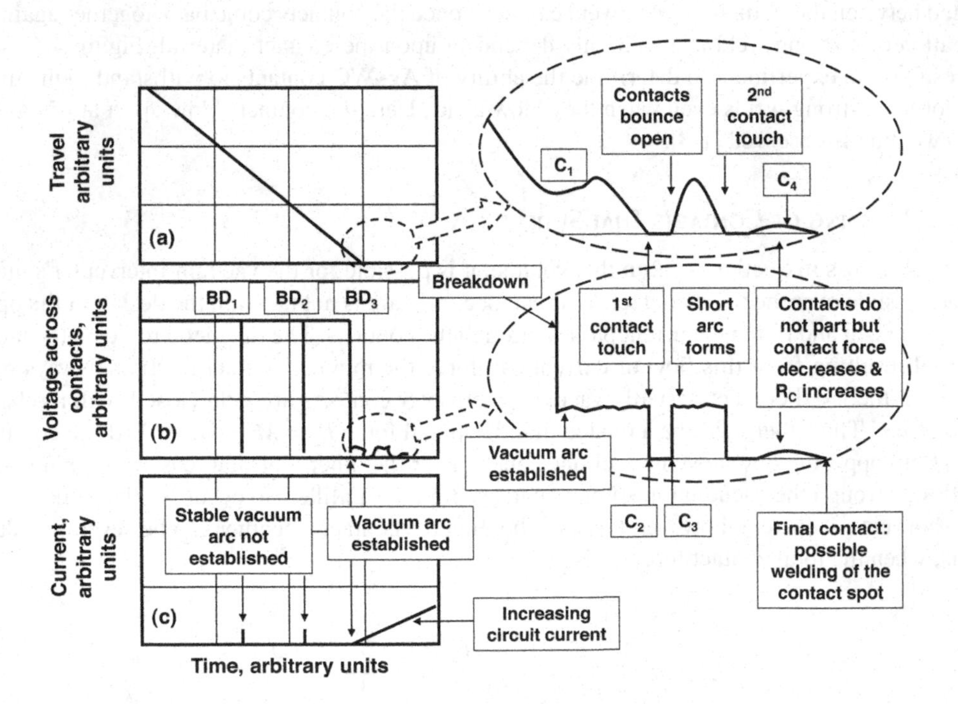

FIGURE 4.60 Schematics of vacuum interrupter contact closing: (a) the travel showing contact bounce, (b) the prestrike breakdown, arcing and arc formation during the contact bouncing and (c) the current flow through the contacts during contact closing [116–118].

and (c); see BD_1 and BD_2. As I discussed in Section 1.3.5 in this volume, the formation of the vacuum arc requires the development of a self-sustaining cathode spot. In this example the breakdowns at BD_1 and BD_2 do not satisfy this criterion; the discharge self-extinguishes and the contact gap recovers its dielectric strength. At BD_3, however, a vacuum arc is established, and the circuit current begins to increase. The current begins at zero and its fastest rate of rise occurs for the symmetric ac current wave, i.e.,

$$i = i_p sin\omega t \tag{4.57}$$

where i_p is the peak current and $\omega = 2\pi f$ and f is the ac frequency. A 2000A(rms) current will only reach 100A after 100µs but will reach about 2000A after 2ms. For a short circuit current of 25kA(rms) the corresponding currents are 1250A and 25kA. Thus, the initial current can be relatively small at C_1; i.e., where the initiation of the vacuum arc first takes place. Even so it may be possible that the pressure [119–121] of the expanding metal vapor from the vacuum arc will result in a slight hesitation in the closing motion. The contacts first touch at C_2. If the point of contact is made in the region of the breakdown arc's roots, the contact will be made on molten metal. When this metal freezes, a weld spot can result. There is, however, a good probability that the final point of contact will not be in the region of the arc's roots. In this case no welding will occur. Thus, it is possible to have a wide range of weld strengths from this first contact touch. The moving contact's kinetic energy will result in it bouncing off the fixed contact a short time later; see C_3. This will break any initial weld that has formed. If the contacts open far enough the molten metal bridge and the bridge column arc that I discussed in Section 2.2 in this volume will form. This time when the contacts again touch there is now a high probability that they will meet on a region of molten metal. When the metal freezes, a weld will form. I have only shown one bounce event in Figure 4.60 as an illustration. The bouncing and arc formation may occur many times before the bouncing finally stops. A possible final motion of the contact is shown at C_4. Here there is not enough energy to part the contacts, but there is enough to reduce the closed contact force. In this regime it is possible for the contact resistance to increase to a value such that the region of contact melts; see Equation 4.26. Again, when contact is finally made, a weld can be formed. The strength of the final weld is usually quite variable from one operation of the contacts to the next. Its value depends upon the fraction of the melted area that forms at the final area of contact.

As I have discussed in Chapter 1 in this volume, the condition of the contact surface is critical for the initiation of a vacuum breakdown of the closing contact gap. Thus, it is advisable, even if minor welding has occurred, to make sure the vacuum interrupter contacts open under a load current. The arc on opening the contacts will go a long way to condition the contacts and leave a smoother contact surface, see Section 1.3.7 in this volume. If this is not possible and the contacts open with no current, then the breaking of welded spots will leave the contact surface with high projections. On closing the contacts, a second time into a high fault current, the pre-strike breakdown might be expected to occur at a longer gap and result in a longer arcing time. This, in turn, may result in a larger molten pool and a stronger weld after the contacts have mated.

In an experiment to demonstrate the effects of closing contacts in vacuum and opening them without current, two contact materials: Cu–Cr (25 wt.%) and W–Cu (10 wt.%) have been compared [122]. These materials are placed in vacuum interrupters, which in turn are placed in a practical vacuum circuit breaker mechanism. The vacuum circuit breaker has a spring mechanism, which is capable of breaking welds with a yield strength of up to about 700KgF. Thus, when the mechanism opens the contacts after a close and latch operation, it means that if a weld has formed and its yield strength is lower than 700KgF, the impactive opening of the mechanism will easily break it. The Cu–Cr contact surfaces are initially conditioned by opening them six times at 8kA (rms) with an arcing time of 16 ms, i.e., about one-half cycle of 30 Hz, ac current.

TABLE 4.9
A Sequence of Five Close and Latch Operations on a 50 kA (PEAK), 30 Hz, AC Current with a No-Load Opening between Each Operation for Cu–Cr(25 wt.%) and W–Cu(10 wt.%) Contacts. The Closed Contact Force is 180Kgf, the Average Closing Speed is 1.17M.s^{-1} and the dc Voltage Across the Contacts During the Closing Operation is 25kV

Close and latch sequence	Measured contact gap at the start of the prestrike arc, mm	Time from the measured start of the prestrike arc to contact touch, ms	Current at contact touch, kA	Did the mechanism break the weld?	β (*n*th op.) /β (1st op.)
Cu–Cr(25wt.%) Contacts					
1	2.31	1.82	17.0	Yes	1
2	2.95	2.37	21.8	Yes	1.27
3	4.36	3.30	29.1	Yes	1.89
4	3.31	2.56	23.4	Yes	1.43
5	6.54	5.49	42.7	No (Yield strength of the weld is 1043 KgF)	2.8
W–Cu (10wt.%) Contacts: Experiment 1					
1	1.43	1.3	12.6	Yes	1
2	4.50	4.0	33.9	Yes	3.15
3	3.51	3.2	28.4	Yes	2.45
4	4.43	4.1	34.6	Yes	3.1
5	2.50	2.5	19.7	Yes	1.75
W–Cu (10wt.%) Contacts: Experiment 2					
1	1.14	0.8	8.6	Yes	1
2	3.79	3.0	27.0	Yes	3.32
3	2.45	2.0	18.6	Yes	2.18
4	3.03	2.5	23.0	Yes	2.66
5	2.26	1.9	17.9	Yes	1.98

These opening operations are performed three times at one polarity and three times at the other. The W–Cu contacts are not subjected to this surface modification procedure. The vacuum circuit breaker is placed in a single phase, high-power capacitor circuit (see Section 6.6.1 in this volume). The circuit is tuned to a 30Hz frequency and a peak current of 50kA. This is similar to the peak asymmetric current wave when closing in on a 25kA (rms) short circuit current in a 50Hz ac circuit. Initially, the capacitor bank is charged to 25kV. Thus, the vacuum interrupter contacts have this dc voltage across them as they close. Before the contacts finally touch a vacuum breakdown of the contact gap, d_B, will occur when the electrical field between the contacts equals a critical value E_C gives by:

$$E_C = \frac{\beta \times U_B}{d_B} \tag{4.58}$$

where β is the enhancement factor ($\beta = \beta_g \times \beta_m$; here β_g is the geometric enhancement factor and β_m is the microscopic enhancement factor: see Section 1.3.2 in this volume) and U_B is the breakdown voltage (25kV in these experiments). Once the vacuum breakdown occurs, a vacuum arc is formed, that continues to burn until contact is established between the two contacts. In these experiments, the final contact always occurs during the passage of the one-half cycle of the 30 Hz current. The circuit breaker then opens with no current flowing. The sequence is repeated five times. Table 4.9 shows the results for a sequence of five such operations. Here you will observe that the first operation in the sequence for both contact materials always occurs at the shortest contact gap and always has the shortest arcing time. Thus, the total energy into the arc roots is much less than it is for subsequent close and latch operations. Because the new contacts surfaces will have a more or less uniform β_m, we can assume that the first breakdown will be in the region of the maximum β_g (see Section 1.4.1): i.e., close to the contacts' edge shown in Figure 4.61(a). Noting that U_B is constant in these experiments, if we assume that E_C is constant for a given closing operation then from Equation (4.58):

$$\frac{E_C(1st-op)}{E_C(nth-op)}=1=\frac{\beta(1st-op)}{\beta(nth-op)}\times\frac{d_B(nth-op)}{d_B(1st-op)} \tag{4.59}$$

i.e.,

$$\frac{\beta_g\beta_m(n^{th}-op)}{\beta_g\beta_m(1^{st}-op)}=\frac{d_B(n^{th}-op)}{d_B(1^{st}-op)}\,or\,\frac{\beta_m(n^{th}-op)}{\beta_m(1^{st}-op)}=\frac{d_B(n^{th}-op)}{d_B(1^{st}-op)}\times\frac{\beta_g(1^{st}-op)}{\beta_g(n^{th}-op)} \tag{4.60}$$

An FEA electric field analysis gives β_g as a function of the contact gap for each contact and shield configuration in their respective vacuum interrupters (e.g., see Figures 1.109–1.111). Table 4.10 shows how the close and latch operation changes the enhancement factor β_g (1st–*n*th) and the *βm* (1st–*n*th) as the experiment progresses. In this Table the possible range of *βm* (nth–op)/ *βm* (1st–op) is given. The nominal value of *βm* (nth–op)/ *βm*(1st–op) results if the breakdown occurs at the contacts' edge each time (Figure 4.61(a)), where the *βg* is the greatest. The maximum value of *βm* (*n*th–op)/ *βm* (1st–op) results when the breakdown occurs in the center of the contacts after first operation: See Figure 4.61(b). For the Cu–Cr contacts, the *βm* (5th–op)/ *βm*(1st–op) is between 2.46 and 3.29 times that of the first operation Thus, while β_g increases perhaps 15.5% as the contact gap increases, the majority of the increase in β will be from β_m as the contact surface is modified by breaking the contact weld spot.

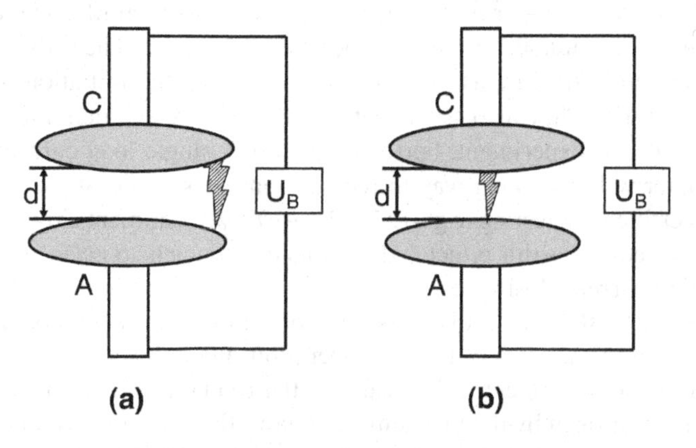

FIGURE 4.61 The prestrike breakdown: (a) occurring at the contacts' edge and (b) occurring at the contacts' center.

TABLE 4.10
The Effect of the Close and Latch Sequence on the Microscopic Enhancement Fact β_m

Close and latch sequence	Gap d mm	d_n / d_1	β_g at contact center	β_g at contact edge	β_g (n^{th} op.) / β_g (1st op.) at contact edge	β_m (n^{th} op.)/ β_m (1st op.) Nominal breakdown at contact edge 1st op. then continued at edge	β_m (n^{th} op.) / β_m (1st op.) Maximum at contact edge 1st op. then continued at center
Cu–Cr(25wt.%) Contacts							
1	2.31	1.00	1.00	1.16	1	1.00	1.00
2	2.95	1.28	1.00	1.19	1.02	1.25	1.48
3	4.36	1.89	1.00	1.26	1.08	1.74	2.19
4	3.31	1.43	1.00	1.21	1.04	1.38	1.67
5	6.54	2.83	1.00	1.34	1.15	2.46	3.29
W–Cu (10wt.%) Contacts: Experiment 1							
1	1.43	1.00	1.00	1.09	1	1.00	1.00
2	4.50	3.15	1.00	1.26	1.15	2.73	3.44
3	3.51	2.45	1.00	1.21	1.10	2.22	2.68
4	4.43	3.10	1.00	1.26	1.15	2.69	3.39
5	2.50	1.75	1.00	1.15	1.05	1.66	1.91
W–Cu (10wt.%) Contacts: Experiment 2							
1	1.14	1.00	1.00	1.08	1	1.00	1.00
2	3.79	3.32	1.00	1.22	1.14	2.93	3.58
3	2.45	2.15	1.00	1.15	1.07	2.01	2.32
4	3.03	2.66	1.00	1.18	1.10	2.43	2.86
5	2.26	1.98	1.00	1.14	1.06	1.88	2.14

On the fifth operation, the pre-strike arcing time is so long and the current at contact touch is so high that a strong weld is formed. It is so strong that the mechanism cannot break it. The force F_W to break the weld is 1045 kgF. Thus, if A_W is the area of the weld and $\Gamma$is the yield strength of the contact material (it equals one-third of the materials hardness $\approx$ 300N.mm^{-2}) then from Equation (4.23) the area of the weld is about 34mm^2, which represents a weld spot diameter of about 6mm.

Nicolle et al. [123], using Cu–Cr and W–Cu contacts, show a similar lengthening of the pre-contact arcing time as the number of closing operations increase. They also show a high-speed video sequence of the pre-strike arc as the contacts close from the initiation of the pre-strike arc to the final contact closure. They also show that the W–Cu contacts maintain a strong resistance to welding throughout their experiment. Fortunately, even a single load current open operation of the vacuum interrupter will go a long way to recondition the surface of the vacuum interrupter's contacts. Also, the certification testing (e.g. IEC 62271–100, Amendment 2) of vacuum contacts into short circuit currents recognizes this effect and permits the contacts to perform load current switching between each high current closing test.

For the W–Cu contacts, the greatest increase in β occurs for the second operation. Even though this increase is also three times that of the first operation, the contact gap, the pre-strike arc time and the current magnitude at contact touch are all smaller than they are for the Cu–Cr contacts. The weld that forms is easily broken by the mechanism. Again, the first close and latch operation modifies the contact surface and greatly enhances β_m. Interestingly, with this contact material, the contact surface modification does not continue to affect the pre-strike breakdown contact gap. Although

β on operations 2–5 are always greater than β (1st–op), they do not continue to increase after the second operation. In fact, in each case, the fifth operation shows that β is smaller than that observed with the second operation. For W–Cu contacts, the mechanism always breaks the weld

It is possible to obtain an estimate of the maximum weld force by making some simple assumptions [124, 125]. Let us consider Figure 4.45. The volume of metal in the contact region that is melted is assumed to be a sphere whose radius (a) is the radius of the weld in area A_W. It is also assumed that almost all the energy from the welding current is used for adiabatic heating of this spherical melted region. The energy W_c to melt the contact region is:

$$W_C = m\left[c_V\left(T_m - T_0\right) + c_L\right] \tag{4.61}$$

where m is the mass of material melted, c_V the specific heat, T_m the melting temperature, T_0 the initial temperature, and c_L the latent heat of fusion. Now if δ is the material density, then $m = [4/3](\pi a^3 \delta)$ and:

$$W_C = \frac{4}{3}\pi a^3 \delta\left[c_V\left(T_m - T_0\right) + c_L\right] \tag{4.62}$$

Now, if the weld area is assumed to be the same as the cross-section of the molten sphere, then:

$$A_W = \pi a^2 \tag{4.63}$$

The weld force F_W is:

$$F_W = \Gamma \pi a^2 \tag{4.64}$$

Using Equations (4.62) and (4.64), eliminating "a,"

$$F_W = K W_C^{2/3} \tag{4.65}$$

where

$$K = \Gamma \pi \left\{\frac{3}{4\pi\delta\left[c_V\left(T_m - T_0\right) + c_L\right]}\right\}^{2/3} \tag{4.66}$$

If all the energy is used to melt the contact spot, then:

$$W_C = \int U_C i(t)\, dt \tag{4.67}$$

where U_C is the voltage measured across the contacts. Thus combining Equation (4.66) with Equation (4.67) gives the maximum weld force that would be expected to occur. Examples of theoretical limits for different materials are given below:

for silver: $F_W = 67 W_C^{2/3}$
for copper: $F_W = 127 W_C^{2/3}$

Figure 4.62 gives an example for Cu and Ag–CdO contacts [126, 127]. Figure 4.63 gives examples for a large number of contact materials [126, 127].

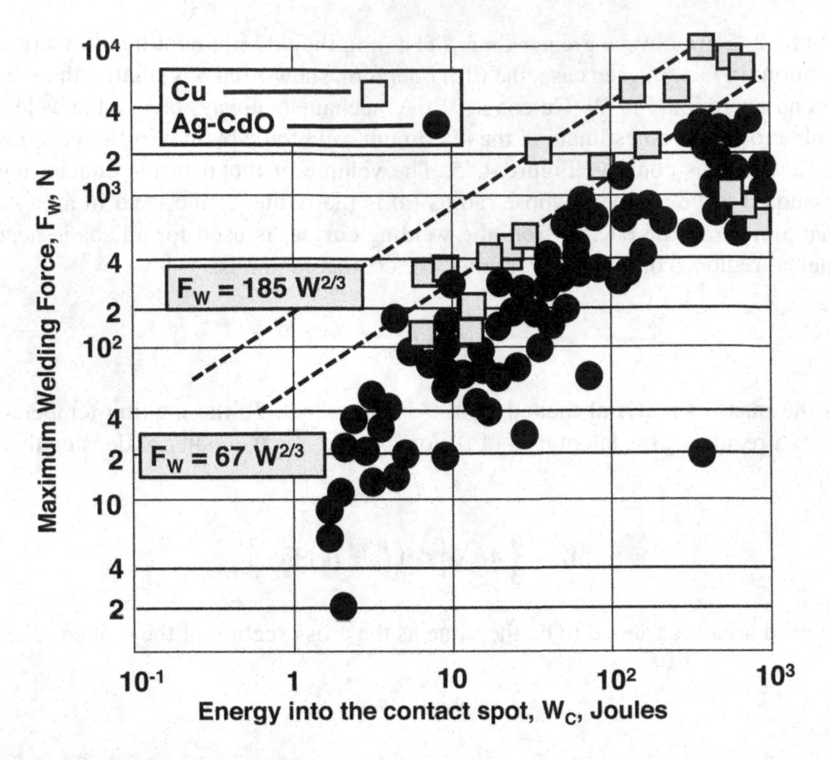

FIGURE 4.62 The weld force as a function of energy into the closed contact spot showing the maximum weld force expected for Cu and Ag–CdO(12wt.%) contacts for high current pilses 30kA –180kA [126, 127].

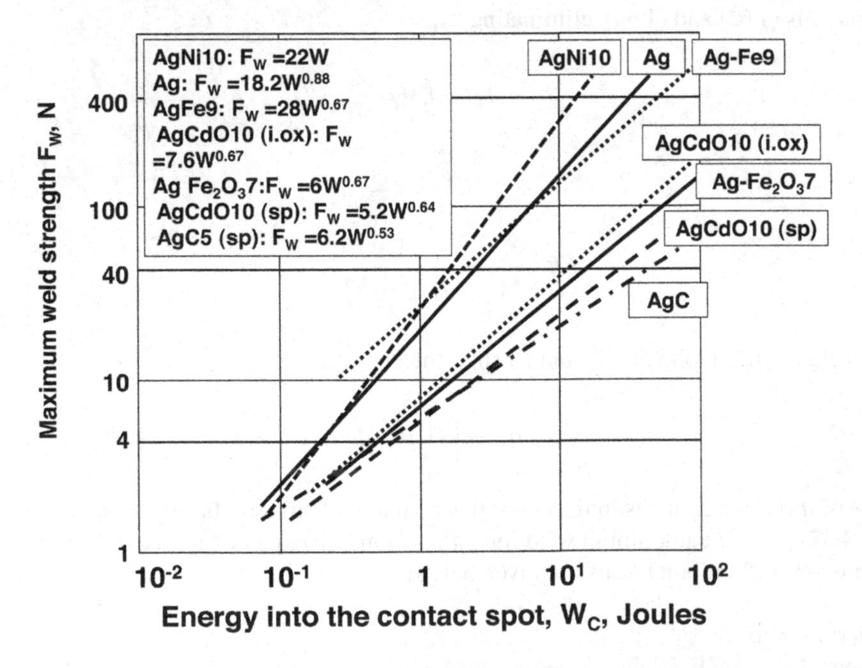

FIGURE 4.63 The maximum weld force for various metals subjected to a single contact bounce [126, 127].

As I have discussed in the beginning of this section, the energy into the contact spot during contact closing is a complex function of:

$$\text{Energy from the breakdown arc} + \text{Energy from the bounce arc}(s) + \text{Energy from contact spot heating} \tag{4.68}$$

$$W_C = \lambda_1 \int_{t(BD)} U_A i(t)dt + \lambda_2 \int_{t(BC)} U_A i(t)dt + \int_{t(C)} U_C i(t)dt \tag{4.69}$$

where λ_1 and λ_2 are fractions less than one, *t(BD)* is the duration of the breakdown arc, *t(BC)* is the bouncing duration, *t(C)* is the time that the contacts are together but with a reduced contact force, U_A is the arc voltage and U_C is the closed contact voltage. If the contacts bounce "*n*" times open, then:

$$\lambda_2 \int_{t(BC)} U_A i(t)dt = \sum_n \lambda_{2(n)} \int_{t(BC)n} U_A i(t)dt \tag{4.70}$$

where *t(BC)n* is the arcing time for the *nth* bounce and $\lambda_{2(n)}$ is the fraction less than one for the *n*th bounce.

Values of F_W less than those calculated by knowing W_C are frequently observed and can be explained by a combination of a number of possible physical effects. For example, if the arc roots on the contacts move, the melted spots may not be exactly opposite each other when the contacts finally come to rest, thus reducing A_W. If the arc duration $t(BD) + \sum_n t(BC)n$ is great enough, the heating of the contact region will not be adiabatic and the effect of heat conduction into the bulk of the contact has to be considered. If the arc is very long, not all the arc energy goes into heating the contact spots: e.g., some is lost by radiation and some is lost by radial conduction of heat. The bounce time of the contacts can be complex with the contacts opening and closing a number of times during one closing operation. Also, during the bouncing the contact spot may vary from bounce to bounce. Thus, the exact value of the energy into the final melt zone is not easy to determine. Finally, the contact surface itself can be different from the bulk metal. We have already seen, for example, in Figure 3.6 that the surface microstructure of Cu–Cr contacts after arcing is quite different from its bulk structure. Thus, the values of T_m and *H(T)* may vary considerably.

In a practical switching device that employs a vacuum interrupter in a well-designed mechanism, it can be seen from voltage or current records whether or not the mechanism will provide enough force (or energy) to break any welds that may form. Indeed, in commercial devices such as vacuum circuit breakers the only record of contact welding is a failure of the breaker to open. From the design point of view, if the device can pass all its certification tests without the contacts forming a strong enough weld to prevent the device from opening, then detailed knowledge of the strength of any welds that may have formed is not necessary. It is only when the mechanism cannot break a weld that knowledge of how contact welding can occur becomes important. In fact, testing of weld strength such as that described above in Figure 4.58 may not be strictly relevant in a practical mechanism. In an ideally brittle-elastic weld, no yielding occurs and the weld will break instantaneously when the opening force exceeds the tensile strength of the metal. An ideally plastic weld, on the other hand, yields at constant force before it finally breaks. Therefore, not only the momentary value of the weld force when it breaks should be considered, but also the rate of rise of the force applied (i.e., the impact). In the weld force example presented above, the force is applied slowly ($\approx 6 \times 10^2$ Ns^{-1}). The separating force for a practical circuit breaker is supplied by the impact of an accelerating mass, which can be as high as (4×10^6 Ns^{-1}). In this case, the fracture energy rather than the tensile strength will be the most important parameter. This aspect of the mechanism's operation has already been well illustrated above in Table 4.9, where weld yield strengths up to 750 KgF are easily ruptured by the mechanism used in those experiments. I shall further discuss the general aspects of mechanism design and the criteria to apply for minimizing the effects of vacuum interrupter, contact welding in Chapter 6, Section 6.5.3 in this volume.

REFERENCES

1. Crawford, F., Edels, H., "The reignition voltage characteristics of freely recovering arcs", *Proceedings IEE*, 107(A), pp. 202–212, April 1960.
2. Farrall, G., "Decay of residual plasma in a vacuum gap after forced extinction of a 250A arc", *Proceedings of the IEEE*, 56(12), pp. 2137–2145, December 1968.
3. Kimblin, C., "Dielectric recovery and shield currents in vacuum interrupters", *IEEE Transactions on Power Apparatus and Systems*, PAS-90, p. 1261, 1971.
4. Lins, G., Paulus, I., Pohl, F., "Neutral copper vapor density and dielectric recovery after forced extinction of vacuum arcs", *IEEE Transactions on Plasma Science*, 17(5), pp. 676–678, October 1989.
5. Lins, G., "Measurement of the neutral copper density around current zero of a 500A vacuum arc using laser-induced fluorescence", *IEEE Transactions on Plasma Science*, PS-13, pp. 577–581, December 1985.
6. Lins, G., "Evolution of copper vapor from the cathode of a diffuse vacuum arc", *IEEE Transactions on Plasma Science*, PS-15(5), pp. 552–556, October 1987.
7. Lins, G., "Measurement of the temperature of evaporating macro particles after current zero of vacuum arcs", *IEEE Transactions on Plasma Science*, 16(4), pp. 433–437, August 1988.
8. Weuffel, M., Gentsch, D., Nikolic, P., "Influence of current interruption operations on internal pressure in vacuum interrupters", *IEEE Transactions on Plasma Science*, 45(8), pp. 2144–2149, August 2017.
9. Lindmayer, M., Unger-Weber, F., "Events associated with interruption of high voltage vacuum arcs", *Proceedings 13th International Symposium on Discharges and Electrical Insulation in Vacuum*, pp. 330 –334, June 1988.
10. Wikening, E., Lindmayer, M., Reininghaus, U., "A test method for 50Hz interruption capability of vacuum interrupter materials", *IEEE Transactions on Dielectrics and Electrical Insulation*, 4(6), pp. 854–856, December 1997.
11. Wang, Z., Geng, Y., Liu, Z., "Stepwise behavior of free recovery processes after diffuse vacuum arc extinction", *IEEE Transactions on Dielectrics and Electrical Insulation*, 19(2), pp. 582–590, April 2012.
12. Farrall, G. W., "Recovery of dielectric strength after current interruption in vacuum", *IEEE Transactions on Plasma Science*, PS-6(4), pp. 360–369, December 1978.
13. Johnson, O. E., Molter, L., "A floating double probe method for measurements in gas discharges", *Physics Review*, 80(1), pp. 58–68, October 1950.
14. Fenski, B., Lindmayer, M., "Post arc currents of vacuum interrupters with radial and axial magnetic field contacts – Measurements and simulations", *Proceedings 19th International Conference on Electric Contact Phenomena*, pp. 259–267, September 1998.
15. Reece, M. P., "The vacuum switch: Part 2, extinction of an ac vacuum arc", *Proceedings IEE*, 110(4), pp. 803–811, April 1963.
16. Andrews, T. G., Varey, R. H., "Sheath growth in a low pressure plasma", *Physics of Fluids*, 14(2), pp. 339–343, February 1971.
17. Holmes, R., Yanabu, S., "Post arc current mechanism in vacuum interrupters", *Journal of Physic Part D: Applied Physics*, 6(10), pp. 1217–1231, 1973.
18. Childs, S. E., Greenwood, A. N., "A model for dc interruption in diffuse vacuum arcs", *IEEE Transactions on Plasma Science*, PS-8(4), pp. 289–294, December 1980.
19. Childs, S. E., Greenwood, A. N., Sullivan, J. S., "Events associated with zero current passage during the rapid commutation of a vacuum arc", *IEEE Transactions on Plasma Science*, PS-11(3), pp. 181–188, September 1983.
20. Huber, E., Weltmann, K. D., Frohlich, K., "Influence of interrupted current amplitude on the post arc current and gap recovery after current zero – Experiment and simulation", *IEEE Transactions on Plasma Science*, 27(4), pp. 930–937, August 1999.
21. Takahashi, S., Arai, K., Morimiya, O., Kaneko, S., Okabe, S., "A PIC-MCC simulation of the high-voltage interruption ability of a vacuum interrupter", *IEEE Transactions on Plasma Science*, 35(4), pp. 912–919, August 2007.

22. Van Lanen, E., Smeets, R., Popov, M., van der Sluis, L., "Vacuum circuit breaker post arc current modelling on theory of Langmuir probes", *IEEE Transactions on Plasma Science*, 35(4), pp. 925–932, August 2007.
23. Solot, A., Jadinian, J., Agheb, E., Hoidalen, H.-K., "Two-dimensional simulation of post arc phenomenon in vacuum interrupter with axial magnetic field", *Proceedings 24th International Symposium on Discharges and Electrical Insulation in Vacuum*, pp. 339–342, September 2010.
24. Mo, Y., Shi, Z., Li, J., Jia, S., Wang, L., "Study of post-arc sheath expansion process with a two-dimensional particle-in-cell model", *Proceedings 28th International Symposium on Discharges and Electrical Insulation in Vacuum*, pp. 239–242, September 2018.
25. Shemshadi, A., "Modeling of plasma dispersion process in vacuum interrupters during post arc interval based on FEM", *IEEE Transactions on Plasma Science*, 47(1), pp. 647–653, January 2019.
26. Lins, G., "Influence of electrode separation on ion density in the vacuum arc", *IEEE Transactions on Plasma Science*, 19(5), pp. 718–724, October 1991.
27. Dullni, E., Schade, E., Gellert, B., "Dielectric recovery of vacuum arcs after strong anode spot activity", *IEEE Transactions on Plasma Science*, PS-15(5), pp. 538–544, October 1987.
28. Carlsson, J., Powis, A., "Large-scale PIC-MCC simulation of low temperature plasmas", *ExB workshop*, Princeton, November 1–2, 2018 https://htx.pppl.gov/exb2018presentations/Friday/7%20Carlsson%20Large-scale%20PIC-MCC%20simulation%20of%20low-temperature%20plasmas.pdf.
29. Frost, L., Browne, T., "Calculation of arc – Circuit interaction". In: Browne, T. (Editor), *Circuit Interruption, Theory and Techniques*, (pub. CRC Press), pp. 187–240, 1984.
30. Frind, G., Carroll, J. J., Goody, C. P., "Recovery times of vacuum interrupters which have stationary anode spots", *IEEE Transactions on Power Apparatus and Systems*, PAS-101(4), pp. 775–781, April 1982.
31. Binnendijk, M., Merck, W., Smeets, R., Watanabe, K., Kaneko, E., "High current interruption in vacuum circuit breakers", *IEEE Transactions on Dielectrics and Electrical Insulation*, 4(6), pp. 836–840, December 1997 [erratum Vol. 5, No. 3, p. 467, June 1998].
32. Schneider, A., Popov, S., Batrkov, A., Sandolache, G., Schellekens, H., "Anode temperature and plasma sheath dynamics of high current vacuum arc after current zero", *IEEE Transactions on Plasma Science*, 41(8), pp. 2022–2028, August 2013.
33. Lee, L. H., Greenwood, A., "Theory for the cathode mechanism in metal vapor arcs", *Journal of Applied Physics*, 32(5), pp. 916–923, 1961.
34. Dakin, T., Luxa, G., Oppermann, G., Vigreux, J., Wind, G., Winkelnkemper, H., "Breakdown of gases in uniform fields: Paschen Curves for nitrogen, air and sulfur hexafluoride", *Electra*, Vol. 32, pp. 61–82, 1971
35. Schönhuber, M., "Breakdown of gas below the Paschen minimum: Basic design data of high voltage equipment", *IEEE Transactions on Power Apparatus and Systems*, PAS-88(2), pp. 100–107, February 1969.
36. Slade, P. G., Taylor, E. D., "Electrical breakdown in atmospheric air between closely spaced (0.2μm - 40μm) electrical contacts", *IEEE Transactions on Components and Packaging Technology*, 25(3), pp. 390–396, September 2002.
37. Dullni, E., Schade, E., "Investigations of high-current interruption of vacuum circuit breakers", *IEEE Transactions on Electrical Insulation*, 28(4), pp. 607–620, August 1993.
38. Schade, E., Dullni, E., "Recovery breakdown strength of a vacuum interrupter after extinction of high-currents", *IEEE Transactions on Dielectrics and Electrical Insulation*, 9(2), pp. 207–215, April 2002.
39. Takahashi, S., Arai, K., Moriyima, O., Hayashi, K., Noda, E., "Laser measurement of copper vapor density after a high-current vacuum arc discharge in an axial magnetic field", *IEEE Transactions on Plasma Science*, 33(5), pp. 1519–1526, October 2005.
40. Wang, Z., Tian, Y., Ma, H., Geng, Y., Liu,Z., "Dielectric recovery strength after vacuum arc extinctions", *Proceedings 26th International Symposium on Discharges and Electrical Insulation in Vacuum*, pp. 333–336, September 2014.
41. Pieniak, T., Kurrat, M., Gentsch, D., "Surface temperature measurement of transverse magnetic field contacts using a thermography camera", *IEEE Transactions on Plasma Science*, 45(8), pp. 2157–2163, August 2017.
42. Gentsch, D., Shang, W., "High speed observations of arc modes and material erosion of TMF and AMF contacts", *IEEE Transactions on Plasma Science*, 33(5), pp. 1605–1610, October 2005.
43. Hakamata, Y., Kurosawa, Y., Natsui, K., Hirasawa, K., "Post arc current of vacuum interrupter after large current interruption", *IEEE Transactions on Power Delivery*, 1(4), pp. 1692–1697, October 1988

44. Yanabu, S., Souma, S., Tamagawa, T., Yamashita, B., Tsutsumi, T., "Vacuum arc under axial magnetic field and its interrupting ability", *Proceedings IEE*, 126(4), pp. 313–320, 1979.
45. Godechot, X., Ernst, U., Dalmazio, L., Kantas, S., Hairour, M., Fadat, N., "Late breakdown phenomena investigation with various kinds of contact material", *Proceedings 24th International Symposium on Discharges and Electrical Insulation in Vacuum*, pp. 206–209, September 2010.
46. Smeets, R., Van Lanen, E., "Short-line fault interruption capability of vacuum circuit breakers", *IEEE Transactions on Plasma Science*, 31(5), pp. 852–858, October 2003.
47. Van Lanen, E., Smeets, R., Popov, M., Van der Sluis, L., "Performance trend and indicator analysis on vacuum circuit breaker current zero measurements", *Proceedings 3rd IEE Conference on Reliability of Transmission and Distribution*, February 2005.
48. Van Lanen, E., Popov, M., Van der Sluis, L., Smeets, R., "Vacuum circuit breaker current zero phenomena", *IEEE Transactions on Plasma Science*, 33(5), pp. 1589–1593, October 2005.
49. Yanabu, S., Satoh, Y., Homma, M., Tamagawa, T., Keneko, E., "Post arc current in vacuum interrupters", *IEEE Transactions on Power Delivery*, PWRD-1(4), pp. 209–214, October 1986.
50. Smeets, R., Van Lanen, E., Popov, M., Van der Sluis, L., "In search for performance indicators of short-circuit current interruption in vacuum", *Proceedings 23rd International Symposium on Discharges and Electrical Insulation in Vacuum*, pp. 79–82, September 2008.
51. Donen, T., Abe, J., Tsukima, M., Takai, Y., Miki, S., Ochi, S., "Temperature measurement and arc rotation observation of spiral-type contact, *Proceedings 27th International Symposium on Discharges and Electrical Insulation in Vacuum*, pp. 255–258, September 2016.
52. Pieniak, T., Kurrat, M., Gentsch, D., "Surface temperature measurement of transverse magnetic field contacts using a thermography camera", *Proceedings 26th International Symposium on Discharges and Electrical Insulation in Vacuum*, pp. 133–136, September 2016.
53. Poluyanova, I., Zabello, K., Logatchev, A., Yakovlev, V., Shkol'nik, S., "Measurements of thermal radiation of anode surface after current zero for range of current levels", *IEEE Transactions on Plasma Science*, 45(8), pp. 2119–2125, August 2017.
54. Kaumanns, J., "Measurements and modeling in the current zero region of vacuum circuit breakers for high current interruption", *IEEE Transactions on Plasma Science*, 25(4), pp. 632–636, August 1997.
55. Steinke, K., Lindmayer, M., "Differences in current zero behavior between bipolar and Quadrapolar AMF contacts", *Proceedings 20th International Symposium on Discharges and Electrical Insulation in Vacuum*, pp. 530–534, July 2002.
56. Arai, K., Morimiya, O., Niwa, Y., "Probe measurement of residual plasma of a magnetically confined high-current vacuum arc", *Proceedings 20th International Symposium on Discharges and Electrical Insulation in Vacuum*, pp. 239–24237, July 2002.
57. Ge, G., Cheng, X., Liao, M., Duan, X., Zou, J., "Vacuum arcs and post arc characteristics of vacuum interrupters with external AMF at current zero", *IEEE Transactions on Plasma Science*, 46(4), pp. 1003–1009, April 2018.
58. Steinke, K., Lindmayer, M., Weltmann, K., "Post-arc currents of vacuum interrupters with axial magnetic field contacts under high current and voltage stress", *Proceedings 19th International Symposium on Discharges and Electrical Insulation in Vacuum*, pp. 475–480, September 2000.
59. Watanabe, K., Sato, J., Kagenaga, K., Somei, H., Homma, M., Kaneko, E., Takahashi, H., "The anode surface temperature of Cu-Cr contacts at the limit of current interruption", *Proceedings 17th International Symposium on Discharges and Electrical Insulation in Vacuum*, pp. 291–295, July 1996.
60. Niwa, Y., Sato, J., Yokokura, K., Kusano, T., Kaneko, E., Ohshima, I., Yanabu, S., "The effect of contact material on temperature and melting of anode surface in vacuum interrupters", *Proceedings 19th International Symposium on Discharges and Electrical Insulation in Vacuum*, pp. 524–527, September 2000.
61. Niayesh, K., "Reignitions in short vacuum arcs after interruption of high frequency currents caused by ion bombardment", *IEEE Transactions on Plasma Science*, 29(1), pp. 69–74, February 2001.
62. Glinkowski, M., Greenwood, A., "Numerical solution of the high current vacuum arc", *S.P.I.E.*, 2259, pp. 153–159, February 1994.
63. Smeets, R., Fu, K., "Townsend type breakdown as a criterion for high frequency vacuum arc interruption at sub-millimeter gaps", *IEEE Transactions on Plasma Science*, 19(5), pp. 767–771, October 1991.
64. Greenwood, A., Jurtz, D., Sopianek, J., "A guide to the application of vacuum circuit breakers", *IEEE Transactions Power Apparatus and Systems*, PAS-90(4), pp. 1589–1597, July/August 1971.
65. Itoh, T., Murai, Y., Ohkura, T., Takami, T., "Voltage escalation in the switching of the motor control circuit by vacuum interrupters", *IEEE Transactions on Power Apparatus and Systems*, PAS-91, pp. 189–1903, 1972.

66. Matsui, Y., Yokoyama, T., Umeya, E., "Reignition current interruption characteristics of the vacuum interrupter", *IEEE Transactions on Power Delivery*, PD-3(4), pp. 1672–1677, October 1988.
67. Greenwood, A., Glinkowski, M., "Voltage escalation in vacuum switching operations", *IEEE Transactions on Power Delivery*, PD-3(4), pp. 1698–1706, 1988.
68. Murano, M., Fujii, T., Nishikawa, M., Nishikawa, S., Okawa, M., "Voltage escalation in interrupting inductive current by vacuum switches", *IEEE Transactions on Power Apparatus and Systems*, PAS-93, pp. 264–270, 1974 and "Three-phase simultaneous interruption in interrupting inductive current using vacuum switches", *IEEE Transaction Power Apparatus and Systems*, PAS-93, pp. 2720–2280, 1974.
69. Helmer, J., Lindmayer, M., "Mathematical modeling of the high frequency behavior of vacuum interrupters and comparison with measured transients in power systems", *Proceedings 17th International Symposium on Discharges and Electrical Insulation in Vacuum*, pp. 323–331, July 1996.
70. Tuohy, E., Parek, J., "Chopping of transformer magnetizing currents, Part 1: Single phase transformers", *IEEE Transactions on Power Apparatus and Systems*, PAS-97(1), pp. 261–268, 1978.
71. Heyn, D., Wilkering, E., "Effect of contact material on the extinction of vacuum arcs under line frequency and high frequency conditions", *IEEE Transactions on Components, Hybrids and Manufacturing Technology*, 14(1), pp. 65–70, March 1991.
72. Kaneko, E., Yokokura, K., Homma, M., Satoh, Y., Okawa, M., Okutomi, I., Oshima, I., "Possibility of high current interruption of vacuum interrupter with low surge contact material: Improved Ag-WC", *IEEE Transactions on Power Delivery*, 10(2), pp. 797–803, April 1995.
73. Kurosawa, Y., Iwasita, K., Watanabe, R., Andoh, H., Takasura, T., Watanabe, H., "Low surge vacuum circuit breakers", *IEEE Transactions on Power Apparatus and Systems*, PAS-104(12), pp. 3634–3642, December 1985.
74. Santilli, V., "Electrical contacts for vacuum interrupter devices", *US Patent 4,743,718*, May 1988.
75. Telander, S., Wilhelm, M., Stamp, K., "Surge limiters for vacuum circuit breaker switchgear", *IEEE Transactions on Industry Applications*, 24(4), pp. 554–559, July/August 1988.
76. Betz, T., Koenig, D., "Fundamental studies on vacuum circuit breaker quenching limits with a synthetic circuit", *IEEE Transactions on Dielectrics and Electrical Insulation*, 4(4), pp. 356–369, August 1997.
77. Slade, P., Smith, R., "The use of vacuum interrupters to control short circuit currents and the probability of interruption at the first current zero after contact part", *Proceedings 9th International Conference on Switching Arc Phenomena*, (IEA, Technical Univ. Lodz, Poland), pp. 54–61, September 2001.
78. Schulman, M., "The behavior of vacuum arcs between spiral contacts with small gaps", *IEEE Transactions on Plasma Science*, 23(6), pp. 915–918, December 1995.
79. Slade, P., Taylor, E., Haskins, R., "Effect of short circuit current duration on the welding of closed contacts in vacuum", *Proceedings 51st IEEE Holm Conference on Electrical Contacts*, pp. 69–74, September 2005.
80. "Adhesion or cold welding of materials in a space environment", *A.S.T.M. STP 431*, American Soc. Testing Materials, 1967.
81. Batzer, T. H., Bunshah, R. F., "Warm-welding of metals in ultra high vacuum", *Journal of Vacuum Science and Technology*, 4(1), pp. 19–28, 1967.
82. Miao, B., Zhang, Y., He, J., Liu, G., Wang, W., Li, G., Wang, X., "The welding tendency of Cu-Cr contact materials in vacuum", *Proceedings 24th International Symposium on Discharges and Electrical Insulation in Vacuum*, pp. 265–268, September 2010.
83. Miao, B., Xie, J., He, J., Liu, G., Wang, W., Wang, X., "Effects of trace Te on the anti-welding property of Cu-30CrTe alloy contact material", *Proceedings 25th International Symposium on Discharges and Electrical Insulation in Vacuum*, pp. 189–192, September 2012.
84. Maio, B., He, J., Liu, G., Wang, W., Wang, X., Liu, K., "Microstructure strength and anti-welding property of Cu-25crTe alloy contact materials", *Proceedings 26th International Symposium on Discharges and Electrical Insulation in Vacuum*, pp. 445–449, October 2014.
85. Slade, P., "An investigation into the factors contributing to welding of contact electrodes in high vacuum", *IEEE Transactions on Parts, Materials and Packaging*, PMP-7(1), pp. 23–33, March 1971.
86. Holm, R., Holm, E., *Electric Contacts, Theory and Application*, (pub. Springer), 4th Edition, pp. 160–164, 2000.
87. Phillips, A., *The Welding Handbook*, (pub. American Welding Soc.), Chapter 30, 1967.
88. Slade, P., "The current level to weld closed contact", *Proceedings 59th IEEE Holm Conference Electrical Contacts*, pp. 123–1128, October 2013.
89. Slade, P., "The threshold welding current for large area closed contacts with two or three points of contact", *Proceedings 60th IEEE Holm Conference on Electrical Contacts*, pp. 7–13, October 2014.

90. Slade, P., "The contact force to prevent large area, closed, vacuum interrupter contacts from welding when passing high fault currents of up to 4 seconds", *Proceedings 59th IEEE Holm Conference, Electrical Contacts*, pp. 93–100, October 2015.
91. Slade, P., Taylor, E., Lawall, A., "Effect of the axial magnetic field structure on the threshold welding current for closed, axial magnetic field vacuum interrupter contacts, *Proceedings 63rd IEEE Holm Conference on Electrical Contacts*, pp. 252–256, 2017.
92. Taylor, E., Lawall, A., Slade, P., "Model for the welding of axial magnetic field vacuum interrupter contacts", *Proceedings 26th International Symposium on Discharges and Electrical Insulation in Vacuum*, pp. 581–584, September 2016.
93. Greenwood, J., "Constriction resistance and the real area of contact", *British Journal of Applied Physics*, 17(12), pp. 1621–1632, December 1966.
94. Kawase, Y., Mori, H., Ito, S., "3-D finite element analysis of electro-dynamic repulsion forces in stationary electric contacts taking into account asymmetric shape", *IEEE Transactions on Magnetics*, 33(2), pp. 1994–1999, 1997.
95. Malucci, R., "The effects of current density variations in power contact interfaces", *Proceedings 57th IEEE Holm Conference on Electrical Contacts*, pp. 55–61, 2011.
96. Walczuk, E., "Resistance and temperature variation of heavy current contacts in welding conditions", *Proceedings 10th International Conerence on Electrical Contacts*, pp. 367–375, August 1980.
97. Chaly, A., Dmitriev, V., Pavleino, M., Pavleino, O., "Experimental research and computer simulation process of pulse heating high current contacts of vacuum interrupters", *Proceedings 24th International Symposium on Discharges and Electrical Insulation in Vacuum*, pp. 418–411, September 2010.
98. Yoshioka, Y., Mano, K., "Theoretical calculation of the transient temperature rise and the welding limit current of point contact due to momentary current flow", *Proceedings 8th International Conference on Electrical Contacts*, pp. 227–232, August 1976.
99. Ferree, J., Petrovic, L., Anheuser, M., "Contact blow-apart forces: Experience in molded case circuit breaker contact systems", *Proceedings 27th International Conference Electrical Contacts*, p. 46, 2014.
100. Barkan, P., "The high current stability limit of electric contacts", Proceedings *11th International Conference on Electrical Contacts*, p. 185, 1982.
101. Taylor, E., Slade, P., "Repulsion or blow-off force between closed contacts carrying current", *Proceeding 62nd IEEE Holm Conference on Electrical Contacts*, pp. 159–162, 2016.
102. Holm, R., Holm, E., op cit., pp. 71–78.
103. Kutty, T., Ravi, K., Ganguly, C., "Studies on hot hardness of Zr and its alloys for nuclear reactors", *Journal of Nuclear Materials*, 265(1–2), pp. 91–99, 1999.
104. Losinsky, M., *High Temperature Metallurgy*, (Pub, Pergamon), 1961.
105. Schulman, M., "Separation of spiral contacts and the motion of vacuum arcs at high ac currents", *IEEE Transactions on Plasma Science*, 21(5), pp. 484–488, October 1993.
106. Lamara, T., Gentsch, D., "Theoretical and experimental investigation of innovative TMF-AMF contacts for high current vacuum arc interruption", *IEEE Transactions on Plasma Science*, 41(8), p. 2048, Fig. 12, August 2013.
107. Dullni, E., Gentsch, D., Shang, W., Delachaux, T., "Resistance increase of vacuum interrupters due to high-current interruptions, *IEEE Transactions on Dielectrics and Electrical Insulation*, 23(1), p. 1, Fig. 12, February 2016.
108. Walczuk, E., "Untersuchung der Schweisseigenschaften von Kontaktwerkstoffen bei geschlossenen Starstromkontakten", *Metall*, 4, pp. 381–384, 1977.
109. Chrost, K., Wojciechowski, S., Walczuk, E., "Properties of directionally solidified silver – copper", *Proceedings 36th IEEE Holm Conference on Electrical Contacts*, pp. 259–264, 1978.
110. Slade, P., *Electrical Contacts: Principles and Applications*, (pub. CRC Press), 2nd Edition, Chapter 24, p. 1195, 2013.
111. Dzierzbicki, S., Walczuk, E., "On welding of closed heavy-current contacts", *Proceedings 3rd International Symposium on Electrical Contact Phenomena*, pp. 243–255, 1966.
112. For example, IEEE C37-Series, IEC62271-Series, GB-JB-DL (the Chinese version of the IEC Standards), GOST (the Russian Standard).
113. Huber, B. F., "An oscilloscope study of the beginning of a floating arc", *Proceedings Holm Seminar on Electrical Contacts*, (pub. IIT), pp. 141–152, November 1967.
114. Smith, R. K., "Private communication".
115. Bono, M., Douchin, J., Sannino, L., Schellekens, H., "Welding on RNF contact during STC test due to rotation phenomena", *Proceedings 28th International Symposium on Discharges and Electrical Insulation in Vacuum*, pp. 539–542, September 2018.

116. Kharin, S., Ghori, Q., "Influence of pre-arcing on the duration of the vacuum arc", *Proceedings 19th International Symposium on Discharges and Electrical Insulation in Vacuum*, pp. 278–285, September 2000.
117. Kharin, S., Nouri, H., Amft, D., Dynamics of electrical contact floating in vacuum", *Proceedings 48th IEEE Holm Conference on Electrical Contacts*, Orlando, USA, IEEE Cat. 02CH37346, pp. 197–205, October 2002.
118. Kharin, S., Nouri, H., Amft, D., Dynamics of arc phenomena at closure of electrical contacts in vacuum circuit breakers,", *IEEE Transactions on Plasma Science*, 33(5), pp. 1576–1581, October 2005.
119. Holmes, A., Slade, P., "Suppression of pip and crater formation during interruption of alternating currents", *IEEE Transactions on Components, Packaging and Manufacturing Techology, Part A*, CPMT-1(1), pp. 59–65, March 1978.
120. Shea, J., DeVault, B., Chien, Y., "Blow open forces on double-break contacts", *IEEE Transactions on Components, Packaging and Manufacturing Technology, Part A*, 17(1), pp. 32–38, March 1994.
121. Zhou, X., Theisen, P., "Investigation of arcing effects during contact blow-open process", *IEEE Transactions on Components and Packaging Technology*, 23(2),pp. 271–277, 2000.
122. Slade, P., Smith, R., "The effect of contact closure in vacuum with fault current on prestrike arcing time, contact welding and field enhancement factor", *Proceedings 53rd IEEE Holm Conference on Electrical Contacts,* pp. 32–36, September 2007.
123. Nicolle, C., Mazzucchi, D., Gauther, J., Gentils, F., "Behavior of CuCr and WCu contacts during making tests at high voltage and high currents", *Proceedings 24th International Symposium on Discharges and Electrical Insulation in Vacuum*, pp. 261–264, September 2010.
124. Bet, M., Souques, G., "Behavior of electrical contacts under current waves of great intensity", *Proceedings 10th International Conference on Electrical Contacts*, pp. 23–34, August 1980.
125. Slade, P., *Electrical Contacts: Principles and Applications*, (pub. CRC Press), 2nd Edition, Chapter 10, pp. 653–656, 2013.
126. Walczuk, E., Boczowski, D., "Computer controlled investigations of the dynamic welding behavior of contact materials", *Proceedings 42nd IEEE Holm Conference on Electrical Contacts*, pp. 11–16, October 1996.
127. Walczuk, E., Boczowski, D., Wojcik, D., "Electrical properties of Ag-Fe and Ag-Fe_2O_3 composite contact materials for low-voltage switchgear", *Proceedings 24th International Conference on Electrical Contacts*, pp. 48–54, June 2008.

5 Application of the Vacuum Interrupter for Switching Load Currents

But to my mind – although I am native here
And to the manner born – it is an custom
More honor'd in the breach than in the observance.

Shakespeare
(Hamlet, Act I, Scene IV)

5.1 INTRODUCTION

At present most vacuum interrupters are used in distribution systems, i.e., electrical circuits with voltages in the range of 3kV to 40.5kV. There is, however, an increasing trend to apply the vacuum interrupter in switches and circuit breakers at 72.5kV and even at transmission circuits greater than 126kV. In spite of this, I will concentrate most of this chapter on the application of vacuum interrupters that connect and disconnect distribution circuits by switching the load current: i.e., less than 4kA. This discussion will, in fact, apply in general to those vacuum interrupters being used in higher voltage circuits. It will also apply to vacuum interrupters being used in low voltage (≤ 1.5kV). The use of vacuum interrupters to protect circuits by interrupting fault or short-circuit currents will be presented in Chapter 6 in this volume. There are some applications at low voltages of less than or equal to 1.5kV, and some at sub transmission voltages (e.g., 72kV) and even at transmission voltages up to 220kV. You can find a detailed analysis of power distribution circuits in a number of sources; see, for example, Fehr [1], Burke [2], and Short [3]. In this chapter, I will present only enough of an overview to enable you to understand the vacuum interrupter's application.

Figure 5.1 shows possible distribution feeder designs. The radial system is the simplest but can be subject to long delays in the restoration of electrical service after a component failure. The primary loop system offers a great improvement, because it provides a two-way feed to each transformer. Any section of the primary can now be isolated, and only those sections on which the fault occurs will have their service interrupted. The primary selective system uses components similar to those used in the primary loop, but they are now arranged in a somewhat more complex manner. Each transformer can select its source and automatic switching is frequently used. The secondary selective system is commonly used in industrial plants. The secondary spot network gives the maximum reliability and operating flexibility. The reliabilities of thc 5 systems are given in Table 5.1 [2].

A load break switch has to be designed and tested under the severest possible requirements that it would see in service for a given circuit voltage and a given circuit power factor. The guesswork on how to design such a switch has been eliminated to a great extent by the certification most manufacturers obtain from independent testing laboratories. These laboratories provide facilities that test load break devices according to a set of standards established by committees of manufacturers and utilities such as the IEEE C37.30 in North America and IEC 60625–1 in Europe. These standards include the switch's dielectric performance, its thermal capabilities, its ability to withstand the passage of fault currents while closed, and its interruption performance. An example is given in Table 5.2.

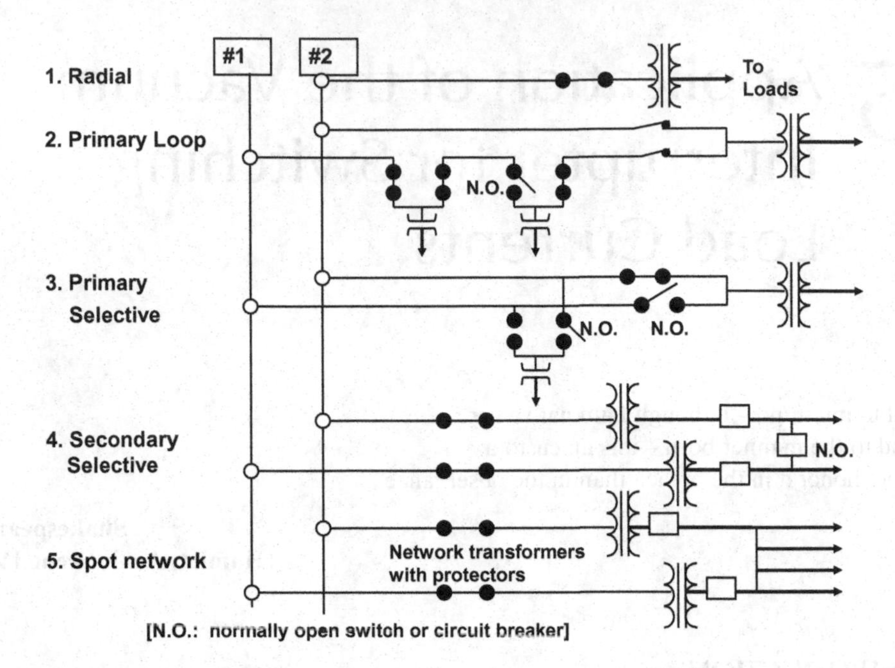

FIGURE 5.1 Examples of distribution networks [2].

TABLE 5.1
Measured Reliability of Different Distribution Systems [2]

Type of system	Radial	Primary auto-loop	Primary selective	Secondary selective	Spot network
Outages/yr	0.3–1.3	0 .4–0.7	0.1–0.5	0.1–0.5	0.02–0.1
Average outage duration, min.	90	65	180	180	180
Momentary interruptions/yr	5–10	10–15	4–8	2–4	0–1

TABLE 5.2
Typical Rating for a Load-Break Switch for Use on an Underground Distribution System

The distribution system

15.5 kV, three-phase grounded wye
95kV BIL insulation system
630 A continuous current
630A load break current
12kA to 25kA symmetrical, fault closing
25A capacitive current interruption (cable charging)
10A inductive current interruption (unloaded transformer)
2000 mechanical life-time operations

Other requirements

- Pad mount: small size, low height
- Submersibles: small size, top operable
- Ability to operate after years of no operation and no maintenance
- Safety in operation both to operating personnel and to general public

5.2 LOAD CURRENT SWITCHING

5.2.1 Switches Used at Distribution Voltages

Use of the vacuum interrupter for load current switching was first introduced in the 1950s [4]. By the 1960's there was an increasing number of papers discussing the use of the vacuum interrupters for load break switches, e.g. [5, 6]. Since that time, the use of vacuum switches on distribution systems has become widespread. In the United States, this has been driven both by the increased use of 15kV distribution systems as well as the increased need for safely operated underground distribution. For the most part, a load break switch must connect and disconnect loads with a relatively high-power factor. The magnitude of load break currents is such that a diffuse vacuum arc is formed as the vacuum interrupter's contacts open. Therefore, as discussed in Chapter 4 in this volume, this operation presents a rather easy duty and one for which the vacuum interrupter is ideally suited. The vacuum interrupter has several additional advantages:

- All arcing is contained within the ceramic envelope
- Its electrical switching life equals or exceeds its mechanical life
- It maintains a constant contact resistance that is within 10–20% of its original value throughout its switching life
- It is essentially maintenance free
- It requires only a very short contact stroke to provide good voltage withstand
- It can be used in very compact configurations
- Vacuum interrupters for more than one phase can be sealed inside one external enclosure
- There is no preferred orientation for successful current interruption: the vacuum interrupters can be applied vertically or horizontally or in any other orientation
- It is possible to incorporate it easily into complex switching systems
- It is easily used in both pad mount and submersible switching systems
- A vacuum interrupter will operate successfully even after long periods (even years) of nonoperation
- It is very safe when being operated, both to the operator and to the general public

Examples of vacuum interrupters used for general load-breaking functions are shown in Figures 5.2 and 5.3. The contact used in the vacuum interrupters in Figure 5.3 usually have a simple butt structure and, because the currents to be interrupted are usually less than 2000A, W–Cu contacts (with the Cu content being in the range 10 wt.% to 25 wt.%) are normally used. The W–Cu contacts for the 72.5kV load-break switch shown in Figure 5.3 are profiled [7, 8] to give a relatively uniform field at the longer contact gaps required to withstand the BIL test voltage. Trinh et al [9] state that the Rogowski profile is dependent on the contact gap. The Bruce profile is a good compromise with negligible edge effects and a constant uniform field region. A simple vacuum switch configuration is shown in Figure 5.4. The mechanism can be a stored energy spring design; see Section 6.5.3 in this volume, which can be manually charged and operated using a "hot stick," or it can be charged using an electric motor drive. The motor's output torque can also be amplified by suitable gearing and then linked to the mechanism to open and close the vacuum interrupter. The mechanism can also be magnetic solenoid that is usually found in contactors: this will be discussed in Section 5.4. It can also be a magnetic actuator that uses permanent magnets; see also Section 6.5.3 in this volume. The major advantage of using a vacuum interrupter at distribution voltages (3kV to 40.5kV) is that the open contact gap is relatively small (2mm to 20mm), so the switch design can be relatively compact. Even at 72kV the contact gap is still only about 30mm.

No matter what type of mechanism is used, the way it attaches to the moving terminal of the vacuum interrupter is critical. The attachment of the mechanism is illustrated schematically in Figure 5.5. In the fully open position, the contact force spring is compressed inside the overtravel component. As the mechanism begins to close the vacuum interrupter's contacts, the overtravel

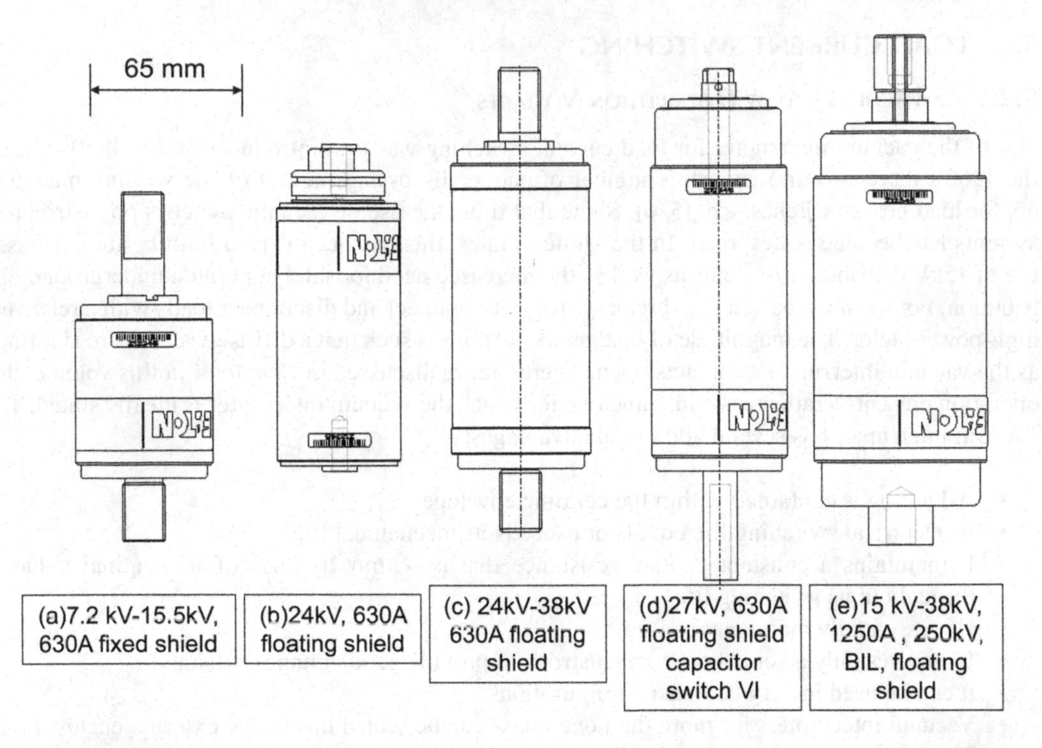

FIGURE 5.2 Load-break vacuum interrupters used in 7.2kV–38kV circuits (courtesy Eaton Corporation).

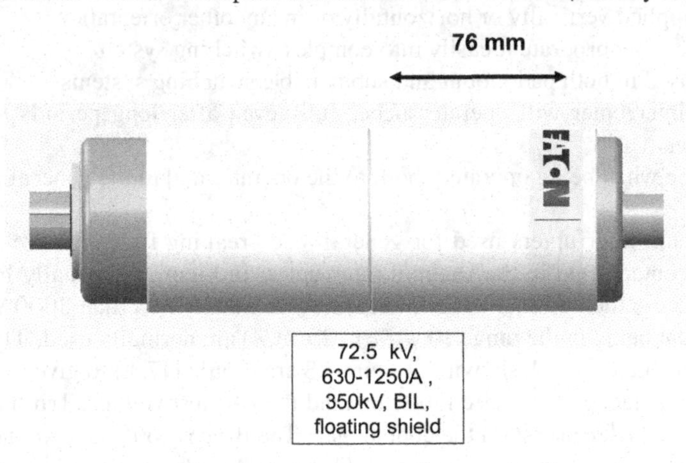

FIGURE 5.3 Load-break vacuum interrupters used in 72.5kV circuits (courtesy Eaton Corporation).

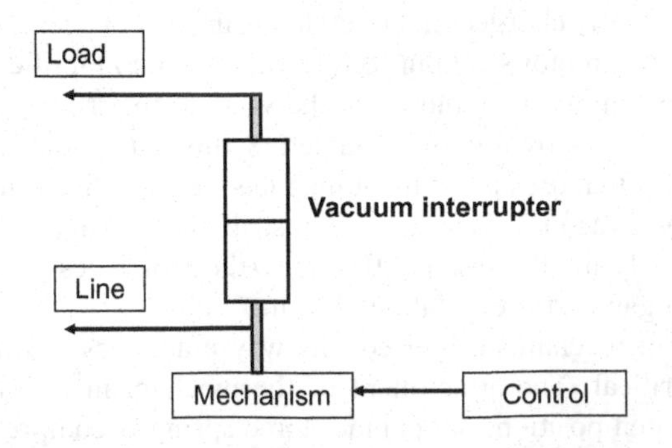

FIGURE 5.4 Schematic of a simple load-break switch using a vacuum interrupter.

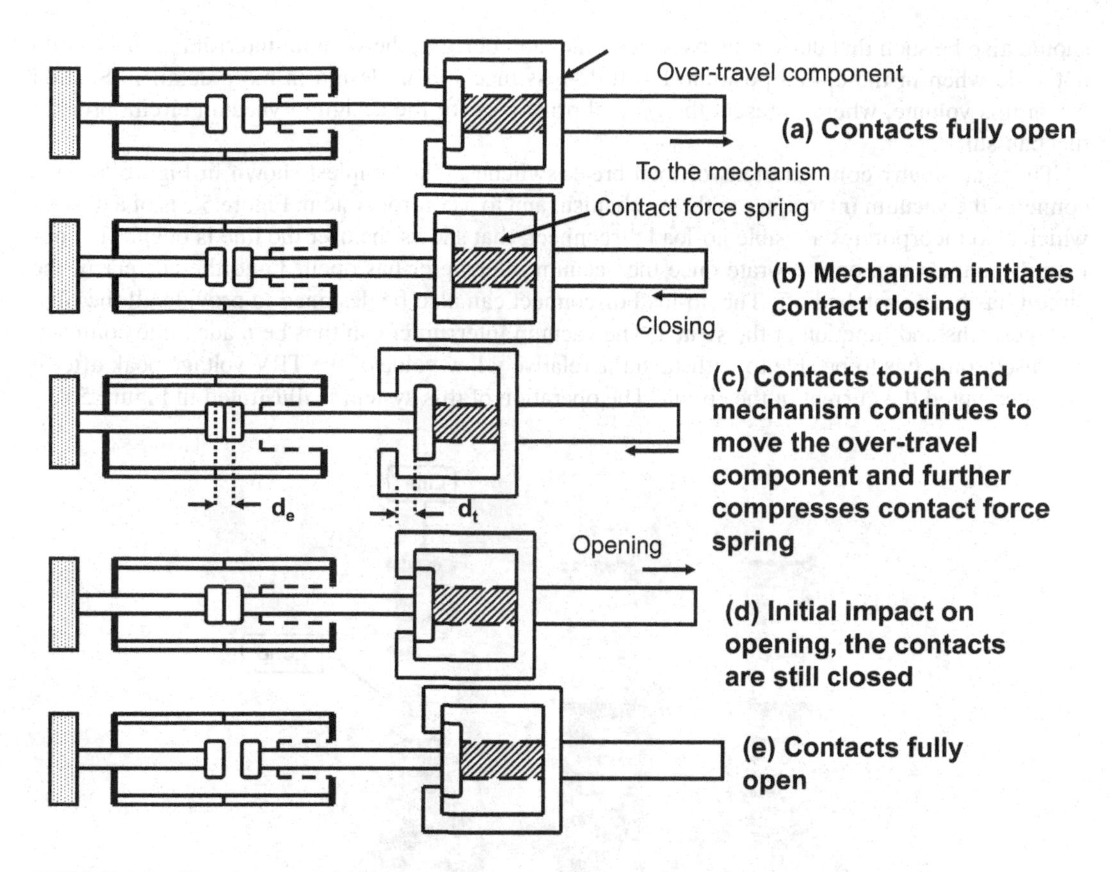

FIGURE 5.5 The overtravel design for operating a vacuum interrupter.

component and the moving terminal move together. Once the contacts touch, the force from the compressed spring provides an immediate minimum contact force. The mechanism continues to move the overtravel component, further compressing the contact force spring. It stops moving once the designed overtravel gap d_t is achieved. When the mechanism begins the opening operation only the mechanism and the overtravel component initially move. It is only when these components have closed the overtravel gap d_t that an impact force is applied to the moving terminal of the vacuum interrupter. As I have discussed in Section 4.4 in this volume, this impact on opening is necessary to break any welds that may have formed between the contacts. It also provides a very high initial acceleration of the contacts so that the full opening speed is obtained just after the contacts begin to part. It is important to design the opening system so that the vacuum interrupter's maximum erosion depth d_e is taken into account. There will be a minimum over travel distance $d_{t\,(min)}$ to provide the required impact given by:

$$d_{t(\text{new})} = d_{t(\text{min})} + d_e \tag{5.1}$$

While the vacuum load break switch only has to interrupt the load current (i.e., usually ≤ 2000A), it usually has to be able to close and latch on to fault currents up to and including the circuits rated short circuit value. Then it must permit the fault current to flow for a specific length of time (perhaps 1 to 3 seconds) until a circuit breaker interrupts it. This requires that the mechanism supply enough contact force to prevent the contacts from blowing open during the passage of the fault current (see Section 2.1.4 in this volume) and to maintain a low enough contact resistance so that the contact spots do not melt and cause the contacts to weld (see Section 4.4.2.2 in this volume). The mechanism should also be such that during the passage of the fault current, the vacuum interrupter contacts do not slide when in the closed position. I will discuss mechanism design in more detail in Section 6.5 in this volume, where I present the general principles for the design of vacuum circuit breaker mechanism.

There are many configurations of load break switches. The simplest shown in Figure 5.6 just connects the vacuum interrupter to the mechanism and to a control system. Figure 5.7 is of a design, which also incorporates a visible no-load disconnect, that shows the user the line is open. This no-load disconnect can only operate once the vacuum interrupter has opened and the current in the circuit has been switched off. The no-load disconnect can also be designed to provide all the high voltage withstand function of the switch. The vacuum interrupter can thus be made quite compact, because it only has to be able to withstand the relatively low value of the TRV voltage peak after it has interrupted the current in the circuit. The operation of this system is illustrated in Figure 5.8.

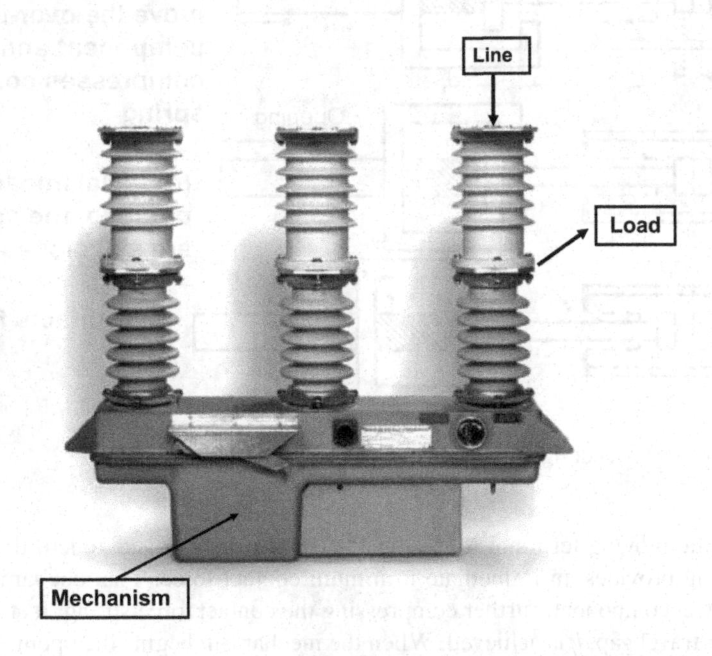

FIGURE 5.6 An example of a practical load break switch [courtesy Joslyn Hi-Voltage].

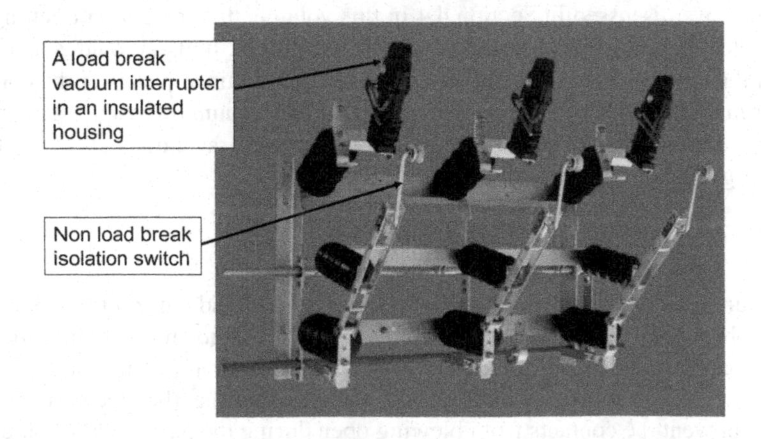

FIGURE 5.7 An example of a practical load break switch that incorporates a series, air insulated, no-load disconnect switch (courtesy Fritz Driescher & Söhne GmbH).

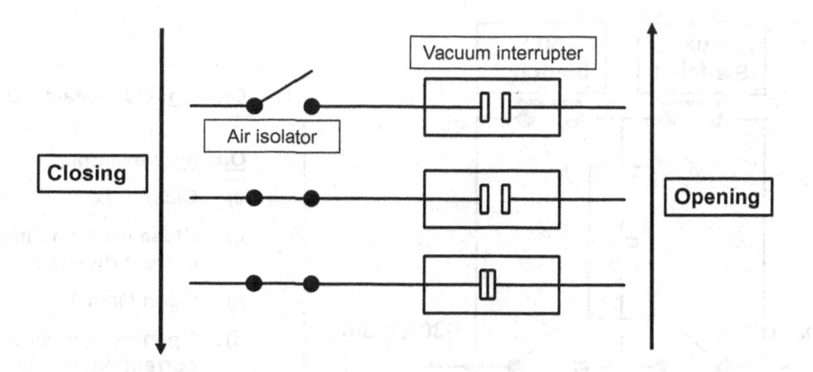

FIGURE 5.8 The operation of a vacuum load-break switch with a series air-break disconnect.

Another concept that has been successful is to only have the vacuum interrupter in the circuit during the load switching operations and have an auxiliary set of contacts to actually carry the load current and to provide the high voltage withstand capability of the device. Figure 5.9 shows a line diagram of this opening and closing sequence. In such a design the vacuum interrupter only has to switch load current and will never see or have to withstand a fault current. Thus, the mechanism operating the vacuum interrupter can be comparatively light. The fault current withstand can be supplied by the auxiliary contacts, which can be designed to have a "blow-on" structure, so that during the passage of the fault current the current itself provides the force to keep the auxiliary contacts closed. When the vacuum interrupter has interrupted the load current, the auxiliary contacts can provide a visible "open" gap as well as provide all the high voltage withstand capability required by the switch. Because the high voltage performance is provided by the auxiliary switch, the vacuum interrupter can again be extremely compact. Again, for normal switching duty (i.e., noncapacitor switching), the maximum voltage it will see will be the peak of the TRV when interrupting an inductive circuit.

For automated switches, a control function can be incorporated into them so that the switch can open and close upon command. This control can be designed to have outputs that tell the user the state of the device and also the state of the circuit the device has switched. It is also possible to maintain

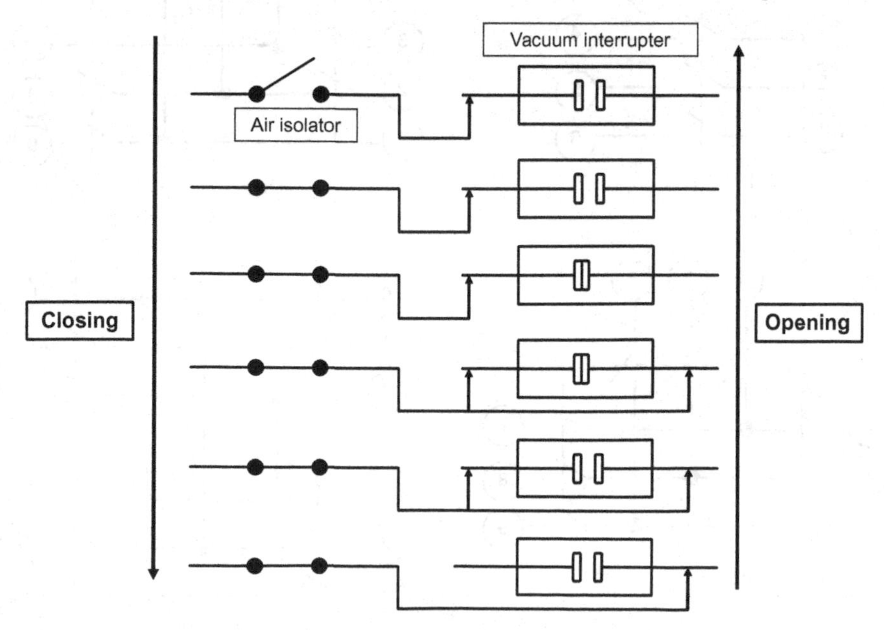

FIGURE 5.9 Example of a vacuum load-break switch with a series no-load disconnect switch, the vacuum interrupter is in the circuit only for switching on and switching off the circuit's current.

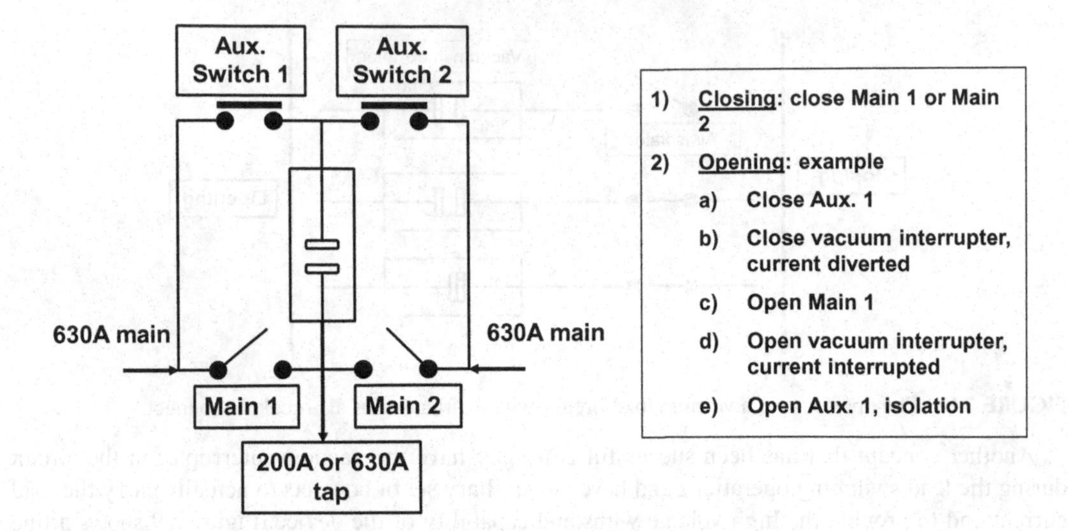

Aux. Switch 1
Aux. Switch 2
630A main
630A main
Main 1
Main 2
200A or 630A tap
1) Closing: close Main 1 or Main 2
2) Opening: example
a) Close Aux. 1
b) Close vacuum interrupter, current diverted
c) Open Main 1
d) Open vacuum interrupter, current interrupted
e) Open Aux. 1, isolation

FIGURE 5.10 A line diagram of a two-way switch with a single tap using a vacuum interrupter to switch the current.

proper coordination with the electrical system's fault-interrupting equipment such as circuit breakers and reclosers. This can, for example, tell the switch to remain closed during the passage of the fault. It can also tell the switch to open at the correct time to isolate the branch of the circuit in which a fault has occurred. Figure 5.10 shows a single line diagram for a two-way switch with a single tap. It is possible to use this type of configuration with various fuse arrangements to provide both switching and protection functions. Figure 5.11 (a–d) give examples of some typical diagrams of three-phase switching schemes.

As the increased distribution requirement of high system reliability and low system downtime has become a necessity for the utility industry, so too the application of sectionalizers using vacuum switches

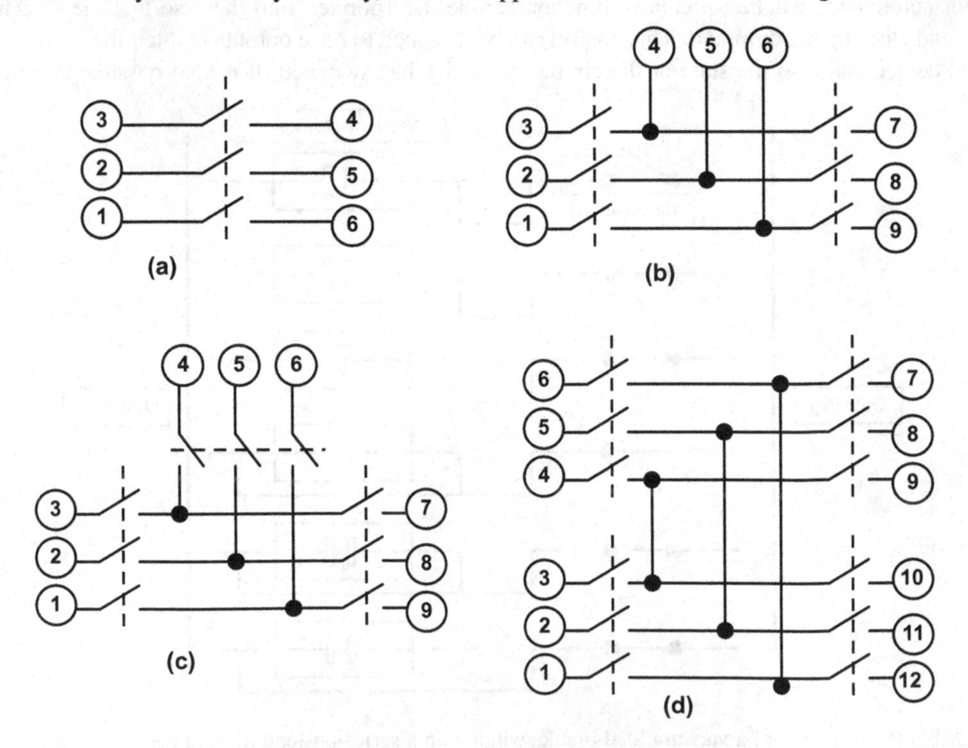

FIGURE 5.11 Examples of four switching schemes using load break switches.

has become more common [6]. A sectionalizer is a switch that is used with an automatic recloser (see Section 6.8.3 in this volume) or a circuit breaker with a reclosing relay, which isolates a faulted section of a line. The sectionalizer does not interrupt the fault current. Its control senses the recloser's operation and opens a circuit when the recloser has opened and has interrupted the fault. Once the sectionalizer has opened and has isolated the faulted section, the rest of the system can be energized again. The sectionalizer can reset itself if the fault in the isolated section is only temporary. It, therefore, provides great flexibility and safety. If the fault is permanent, the fault close-in capability of the vacuum switch greatly simplifies the testing of the circuit. If the fault is still present, the recloser interrupts it while the sectionalizer stays closed. An example for vacuum switches with fault sensing capability together with reclosers for a radial feeder is illustrated in Figure 5.12 [10]. In Figure 5.12 the system could operate as follows:

1. During normal operation, all reclosers, sectionalizers and switches are closed and sensing line current; Figure 5.12 (a)
2. A persistent fault occurs at F, recloser B operates and eventually locks out; see Figure 5.12 (b)
3. In the simplest mode the switches have an internal control that counts the number of close-open operations of the recloser B. When recloser B locks out, the downstream switch B1 opens. In more modern systems a feeder control recognizes that recloser B is open and interrogates switches B1 and B2; Figure 5.12 (c)
4. In Figure 5.12(a), B2 indicates no fault, so the fault must be between B1 and B2. B1 then opens and removes the faulted section. Recloser B is closed and restores power to the unfaulted section

The use of sectionalizers on electrical distribution systems will continue to become more widespread as the demand for system "up-time" becomes more prevalent. Figure 5.13 gives a similar example for circuits tied by a normally open switch.

5.2.2 Switches Used at Transmission Voltages

The use of vacuum interrupters in series to switch high voltage circuits up to 220kV also has a long history [4]. Figure 5.14 shows one successful design in the field. Here the series vacuum interrupters are in the column on the left. The copper conductor joining the two vertical insulator stacks also opens after the vacuum interrupters have interrupted the current. It thus provides a visible air disconnect and also gives good high voltage isolation: e.g.it will satisfy the Lightning Impulse Withstand Voltage (or BIL). When a switch is designed with vacuum interrupters in series, care must be taken to ensure a good distribution of voltage across all the interrupters both during a load switching operation and when they are fully open withstanding the open circuit system voltage. The usual way to achieve this is to place a capacitor, or a capacitor plus a resistor, or a capacitor plus a resistor with a parallel ZnO varistor across each of the vacuum interrupters connected in series. This is illustrated in Figure 5.15. Under ideal conditions, the vacuum interrupters open and close exactly at the same time so any voltage appearing across them is uniformly distributed. However, even under perfect opening conditions the voltage will not be distributed completely uniformly. Consider Figure 5.16. Let C_{VI} be the capacitance of the vacuum interrupters plus any capacitor installed in parallel and let C_g be the stray capacitance from the connecting parts to ground. C_{VI} can be made the same for each unit, but C_g may vary depending upon the exact design of the pole unit. If there are n vacuum interrupters in series, then according to Greenwood [11], the total capacitance to ground, assuming C_g and C_{VI} are constant is:

$$C_{GT} = (n-1)C_g \tag{5.2}$$

and the total series capacitance of the vacuum interrupter is:

$$C_s = \frac{C_{VI}}{n} \tag{5.3}$$

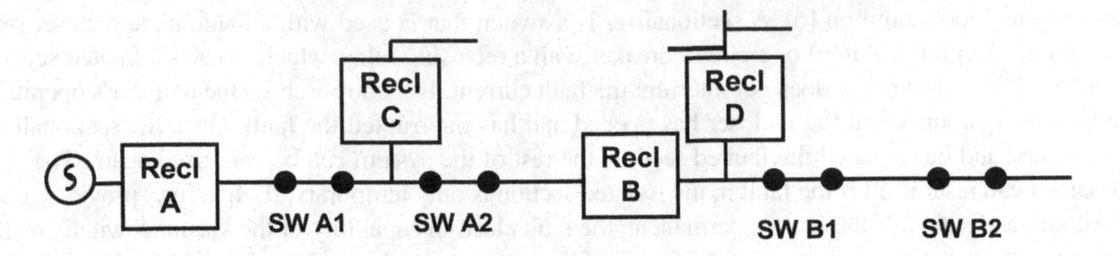

(a) Steady state condition; all elements closed

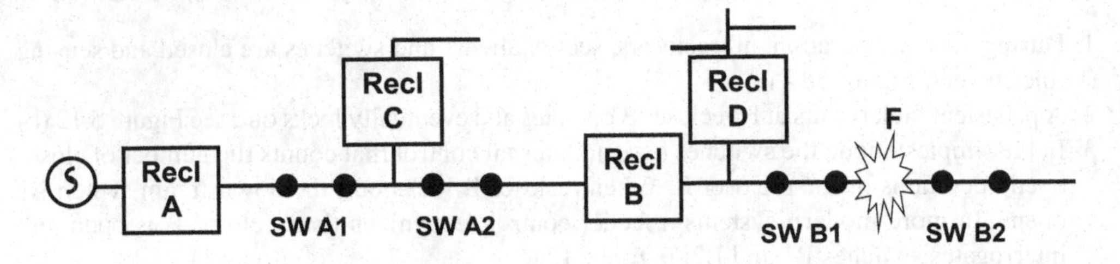

(b) Fault occurs at F and Recloser B clears it

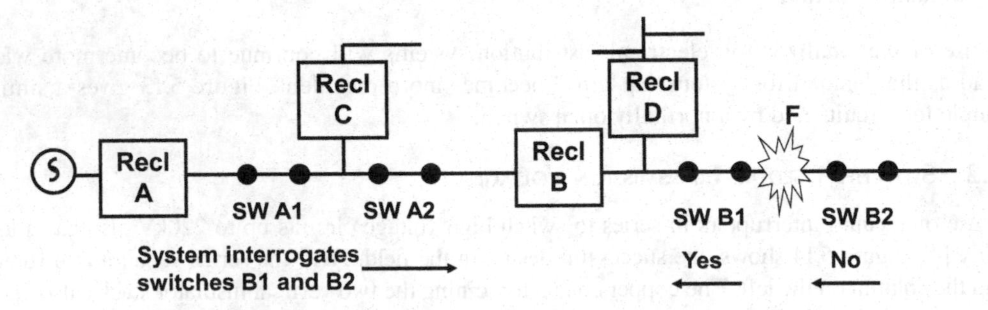

(c) Fault indicators in switches B1 and B2 respond to system interrogation

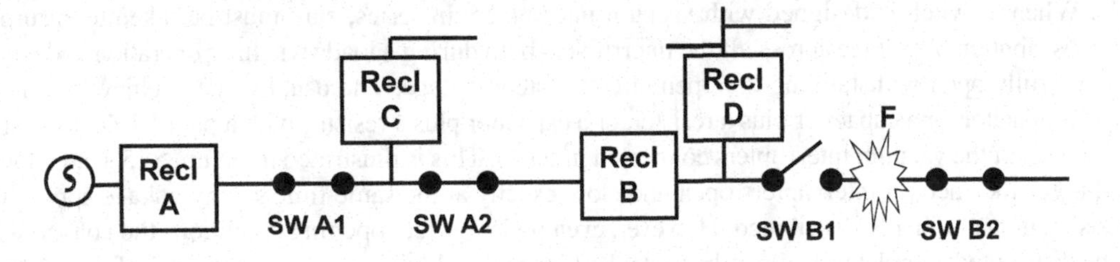

(d) System commands switch B1 to open and Recloser B to close

FIGURE 5.12 An example of using load break sectionalizers together with vacuum reclosers to enhance system up time after a fault on the line [10].

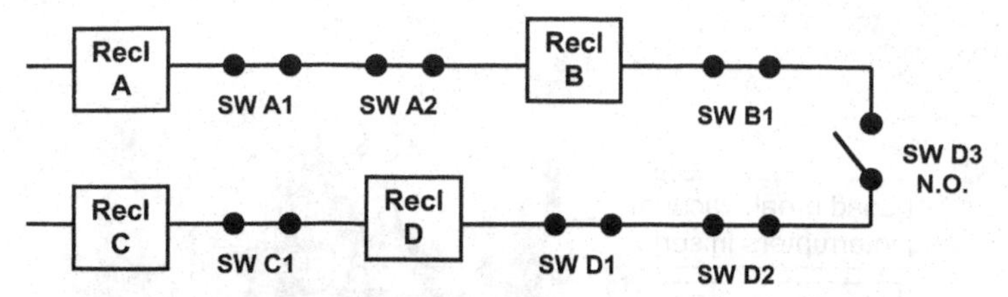

(a) Steady state condition; all elements except switch D3 closed

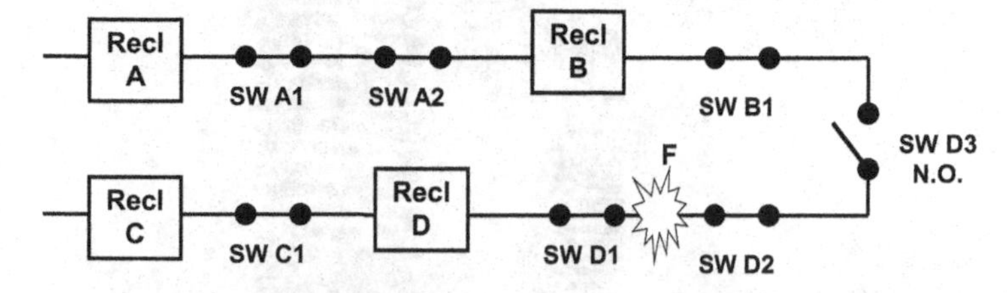

(b) Fault occurs at F and Recloser D clears it

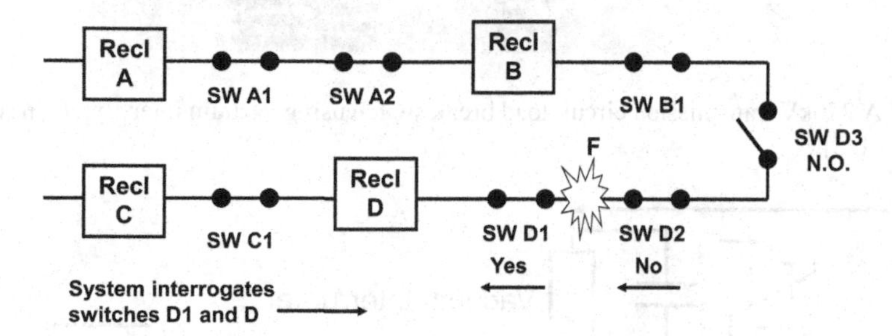

(c) Fault indicator response enables system control to locate the fault between switches D1 and D2

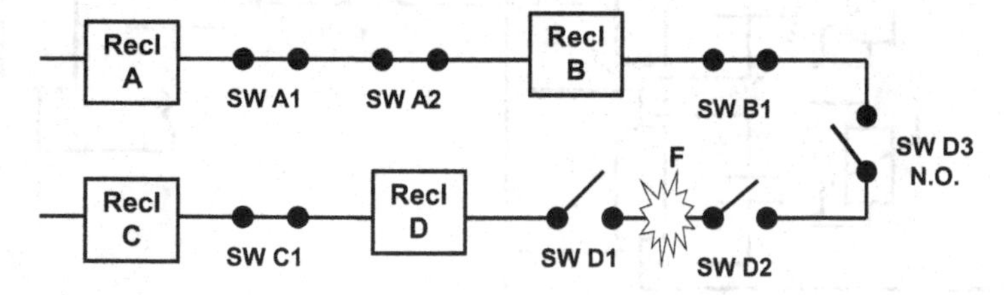

(d) System commands switches D1 and D2 to open and isolate the fault

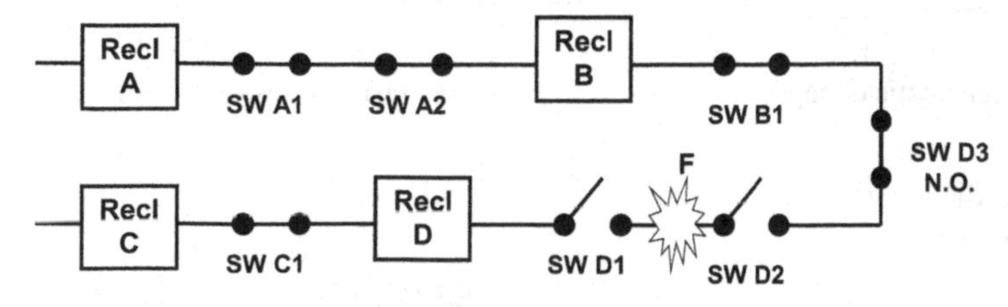

(e) To restore the remainder of the system a close command is sent to switch D3 and the Recloser D

FIGURE 5.13 An example of using load break sectionalizers together with vacuum reclosers in circuits tied with a normally open load break switch [10].

FIGURE 5.14 A 220kV transmission circuit load break switch using vacuum interrupters in series [courtesy Turner Electric LLC].

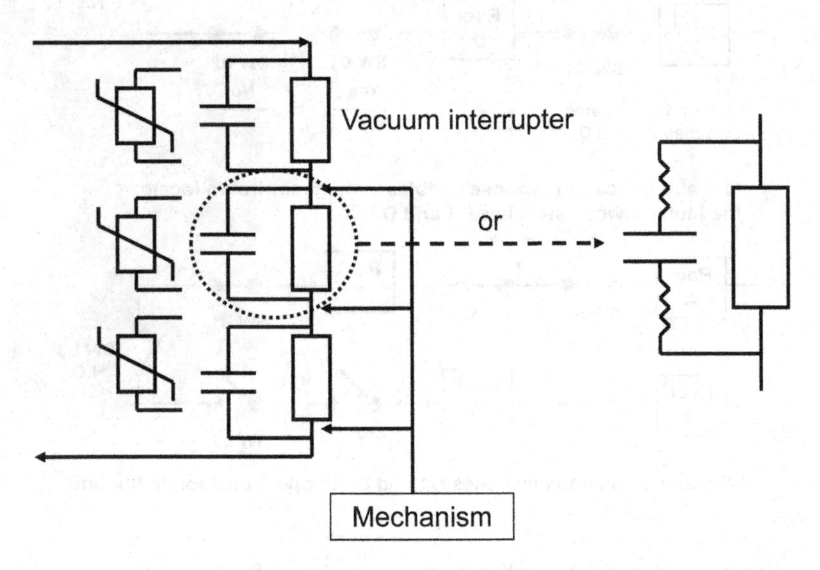

FIGURE 5.15 Schematic of capacitor/resistor/ZnO varistor across series connected vacuum interrupters for voltage grading.

then the voltage distribution is:

$$U(x) = U_a \frac{\sinh\left(\frac{\alpha x}{X}\right)}{\sinh(\alpha)} \tag{5.4}$$

where U_a is the applied voltage across all the series vacuum interrupters, x/X is the fraction of the chain from the line end to the ground end, and $\alpha = [C_{GT}/C_s]^{1/2}$. Suppose there are five units in series, $C_{VI} = 100$ pF $C_g = 10$ pF:

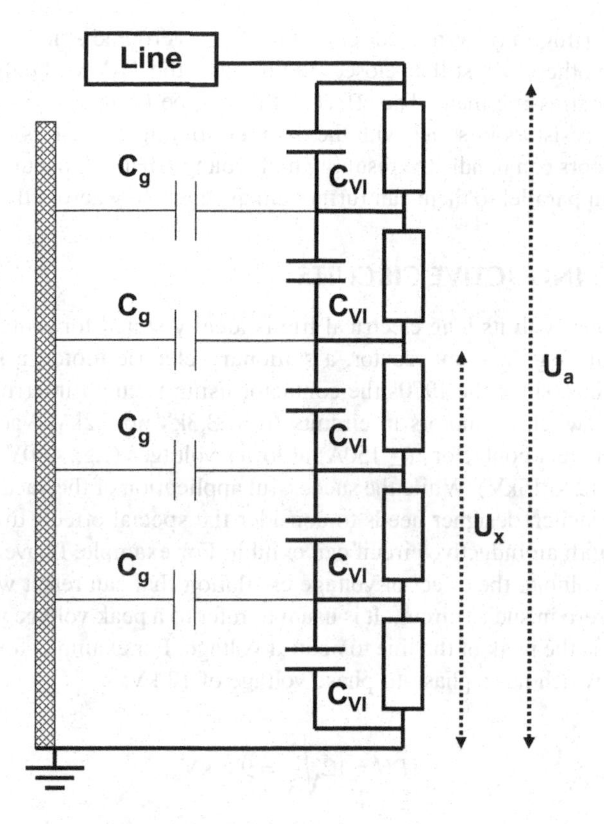

FIGURE 5.16 Example for calculating the voltage grading with capacitors across the series connected vacuum interrupters [11]

$$C_{GT} = 40\,\text{pF};\ Cs = 100/5 = 20\,\text{pF and } \alpha = \left[40/20\right]^{1/2} = 1.414 \tag{5.5}$$

The voltage across the vacuum interrupter closest to the ground would be:

$$U_a \times \frac{\sinh\left(\frac{\alpha}{5}\right)}{\sinh(\alpha)} \cong 0.15U_a \tag{5.6}$$

and the voltage across the vacuum interrupter closest to the line would be:

$$U_a \times \left[1 - \frac{\sinh\left(\frac{4\alpha}{5}\right)}{\sinh(\alpha)}\right] \cong 0.28U_a \tag{5.7}$$

Thus, in this example, the vacuum interrupter closest to the line sees nearly two times the voltage of the one closest to the ground. Also, if the series vacuum interrupters had been designed for a uniform voltage, the distribution across each vacuum interrupter would have been $0.2V_a$, which clearly would have been a problem for the interrupter closest to the line. Increasing the value of the grading capacitor in parallel with each vacuum interrupter reduces this voltage difference. For example, if C_{VI} were 300pF, then the voltage across the vacuum interrupter closest to ground would be $0.18V_a$. In practice you cannot expect that the series connected contacts will always open precisely at the same time. Also, contact parting may even occur close to current zero and some may be open and interrupt the current while others will still be closed. In this case, the TRV will only be impressed across the open contacts overstressing them. The effect of this can be mitigated to some extent by placing noninductive snubber resistors in series with the parallel grading capacitors. Care must be taken to ensure that these resistors can handle the resulting high voltage that will occur across them. Placing a metal oxide varistor in parallel to them can further enhance grading across the vacuum interrupters.

5.3 SWITCHING INDUCTIVE CIRCUITS

The vacuum interrupter with its long electrical life is ideally suited for switching the load current of an inductive circuit; e.g., a shunt reactor, a stationary electric motor, a stalled electric motor and a transformer. Thus, since the 1970s the contactor using vacuum interrupters has become the dominant device for switching motors in circuits from 3.3kV to 7.2kV. Vacuum interrupters are also used in higher current contactors ($\geq$ 150A) at lower voltages (e.g., 440V to 1.5kV) and also at higher voltages (e.g., 12 to 15kV). While the successful application of the vacuum interrupter is now well established, the switch designer needs to consider the special effects that the interaction of a vacuum interrupter with an inductive circuit can exhibit. For example, I have already discussed, in Section 4.3.1 in this volume, the effect of voltage escalation that can result when the contacts part just before a current zero in such a circuit. It is usual to refer to a peak voltage value "per unit value." The 1 per unit (1PU) is the peak of the line to neutral voltage. For example, for a 12 kV, three phase, ungrounded system, which has a phase-to-phase voltage of 12 kV:

$$1PU = 12\sqrt{\frac{2}{3}} = 9.8 \text{ kV} \tag{5.8}$$

5.3.1 Voltage Surges When Closing an Inductive Circuit

When a switch closes an inductive circuit, the voltage across it goes to zero, but reappears instantaneously across circuit elements on each side of it. The voltage will divide in proportion to the surge impedance on each side. If the inductive element is connected to the switch via a cable, the voltage impressed on the load side of the switch will travel down the cable to it. Figure 5.17 illustrates this for a motor connected to the switch. In this example the source impedance is low (i.e., there are a number of cables connected to the bus), thus a high fraction of the source voltage U_{source} travels as a wave to the motor. As the surge impedance of the motor is significantly higher than that of the cable, the reflected wave will have the same sign as the incident wave. In the limit, the voltage wave will double its value when it reaches the motor terminals. This puts severe stress on the first few turns of the motor, especially on the interturn insulation [12, 13]. Of course, in a practical circuit the small inductance of the buswork reduces the steepness of the voltage wave front and the motor capacitance also has an effect. This closing voltage surge occurs whenever an inductive load is connected to the line. It does not depend upon the type of switch used, i.e., it will occur for all switching devices whether they use vacuum, air, oil, or SF_6 as the interruption medium. If the initiation of the circuit closing results from a pre-strike arc between the closing vacuum interrupter contacts, there is the possibility of a high frequency current, which the vacuum interrupter may interrupt. However, as the contact gap in the vacuum interrupter when this occurs will be small and they will continue to close to the final closed contact, the opportunity for voltage escalation is remote.

5.3.2 Voltage Surges When Opening an Inductive Circuit

Disconnecting an inductive component such as a locked-rotor or stalled motor can give rise to the voltage escalation effects discussed in Section 4.3.1 in this volume. Reignitions will certainly occur from time to time if the vacuum interrupter contacts open at random with respect to the ac current wave.

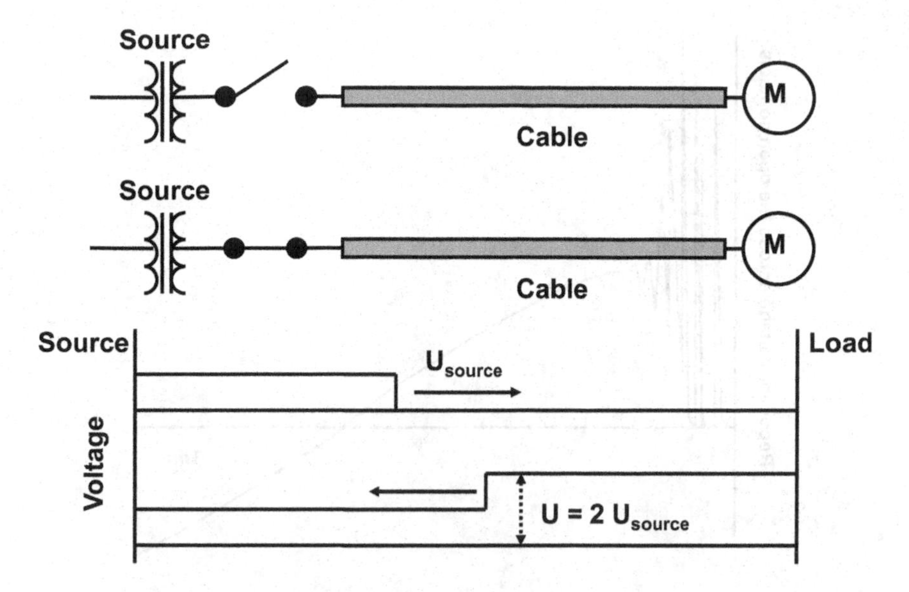

FIGURE 5.17 The voltage surge on closing an inductive circuit.

When a vacuum interrupter opens just before a current zero and interrupts the current, the contact gap can be too small to hold off the TRV imposed across it. This is especially true for the first phase to clear in a three-phase ungrounded circuit or for all three phases in a three-phase grounded circuit. Once the first breakdown or reignition occurs, the stage is set for the multiple reignition process to proceed.

5.3.3 Surge Protection

There are three possible consequences of this reignition process. The first is shown in Figure 5.18. Here the high frequency current superimposed on the ac current wave is not interrupted by the

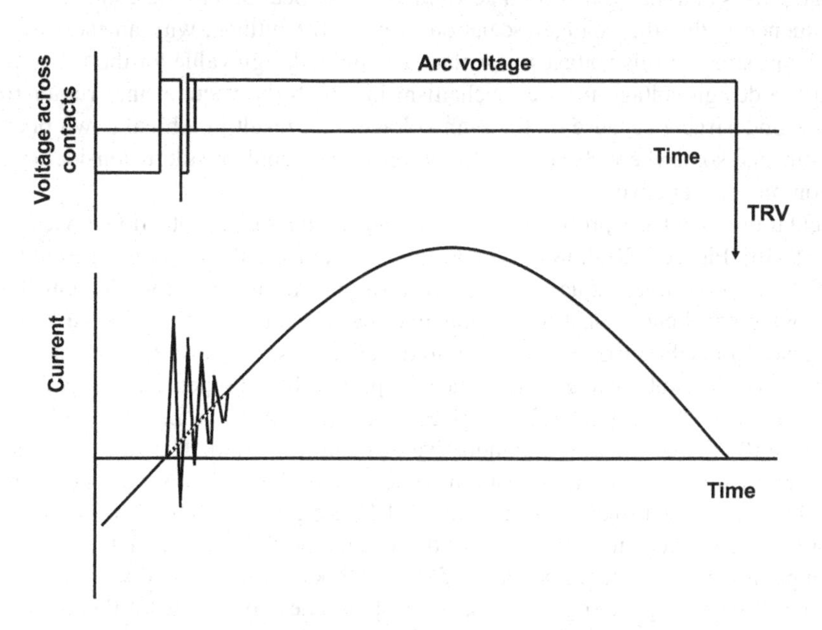

FIGURE 5.18 Example of a restrike on opening an inductive circuit where the high frequency current superimposed on the circuit current is not interrupted.

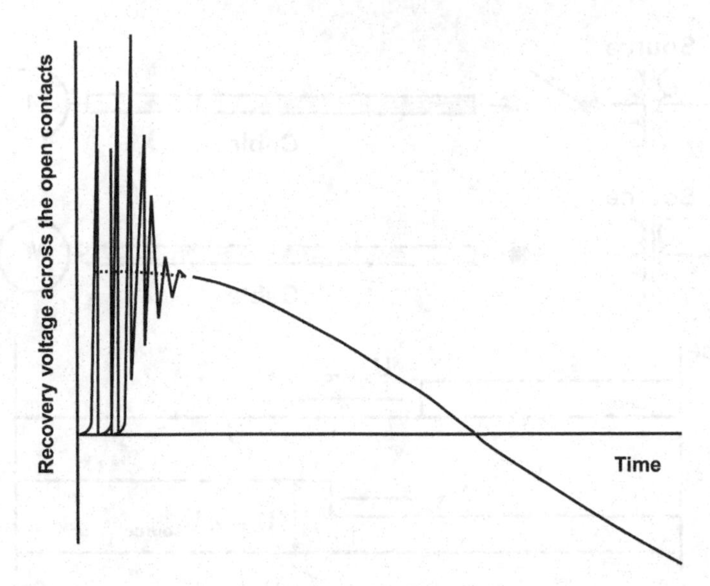

FIGURE 5.19 An example of repetitive restrikes on opening an inductive circuit where the open contact gap eventually withstands the restriking voltage resulting in no further breakdown of the contact gap.

vacuum interrupter. Eventually, as the ac current increases, the high frequency current will no longer have current zeros and the vacuum arc between the contacts will carry the ac current to the next current zero. At that time, the contacts will be fully open and will be able to withstand the voltage that appears across them. A second possible consequence is shown in Figure 5.19. Here the contacts would be somewhat further apart at the first current zero. As the TRV rises the contact gap is such that the electric field allows a breakdown to occur. Now if the contact material interrupts the high frequency current, the voltage escalation events can be initiated. It is possible, however, that the contact gap is sufficient that it will be able to withstand the voltage impressed across it after a few voltage pulses and then the normal ac voltage will appear across the contacts. The third possible consequence is that the voltage escalation event will continue, with an increase of the peak voltages. It is possible for this voltage to escalate above the design value for the vacuum interrupter and beyond the design voltage for the mechanism in which the vacuum interrupter resides. This occurrence is undesirable because (a) the high voltages can result in a breakdown phase-to-phase, or phase-to-ground somewhere else in the circuit; or (b) they could result in considerable damage to the insulation on the inductive element [14].

If the escalation event has a probability of occurring, then a surge protection device is required in the circuit [15, 16]. Figure 5.20 shows examples of such devices. The first arrangement is a resistor-capacitor (R-C) in series that is placed right at the terminals of the inductive element. This has been shown to provide excellent protection to inductive loads from fast voltage wave fronts. The R-C surge suppressor limits the surge voltage magnitude and also slows down the rate of rise of the voltage wave front. These combine to greatly reduce the probability of repetitive surges from occurring. The arrangement shown in Figure 5.20 (a) gives excellent protection, i.e., the R-C combination is connected from the line terminal of the inductive element to ground. The capacitor should be connected between the resistor and ground and the other end of the resistor should be connected to the terminal and the line. The typical capacitor values [17] are given in Table 5.3. These values are for guidance only. A user needs to judge whether or not they provide an adequate safety margin for a particular application. For voltages between 15kV and 38kV, the suggested value is about 0.125μF/phase. Of course, the voltage rating of the capacitor should be appropriate for the line to ground system voltage. There are capacitors especially manufactured for surge protection applications, which have a greater high voltage margin than those manufactured for power factor correction.

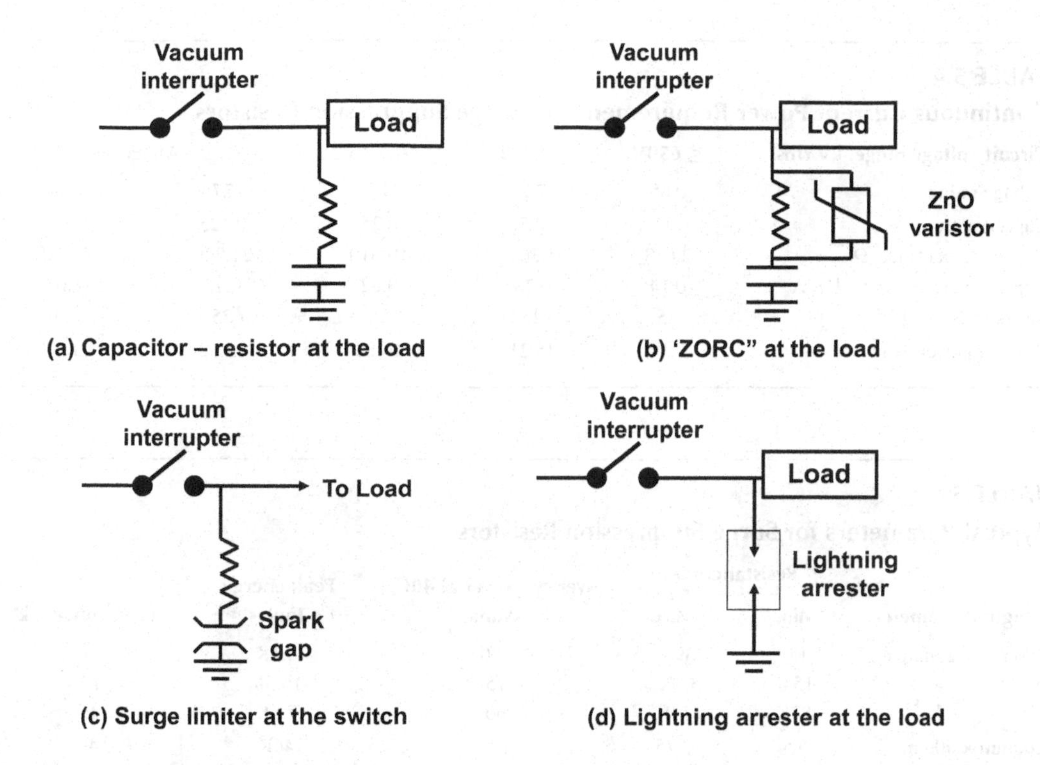

FIGURE 5.20 Possible surge protection schemes to be used at the terminals of a large inductance such as a motor or a transformer.

TABLE 5.3
Typical Capacitance of Surge Protection Capacitors Per Line Terminal of the Inductive Load Connected to Ground [17]

	Rated Circuit Voltage			
	≤ 650V	2.4 kV–7.2 kV	7.2 kV–15 kV	15 kV–38 kV
Capacitance, μF	1.0	0.5	0.25	0.125

The resistor should have an ohmic value, which is about 50% to 200% of the surge impedance of the cable. The resistor (a) absorbs the energy stored in the transient voltage wave and (b) acts as a surge impedance matching device as it is connected in parallel with the much higher surge impedance of the inductive load. The matching of the surge impedance of the cable means that any traveling wave reflections (and related voltage doubling) are significantly reduced. The steady state voltage withstand capability and wattage of the resistor are determined by the steady state capacitor current flow. However, the reactance of the series capacitor is 100 to 1000 times larger than the resistance and hence the normal current flowing through it is relatively small, see Table 5.4. For the fast wave fronts, typical of a voltage escalation event, the capacitor looks like a short circuit. In this case the peak of the voltage surge will be seen across the resistor. Therefore, the resistor must be chosen so that it can withstand the peak of the expected transient voltage pulses. This is the most important parameter in determining the size and design of the resistor. In reality an adequate safety margin in excess of the expected transient should be used. The peak energy absorbed by the resistor

TABLE 5.4
Continuous Current Power Requirements for Surge Suppression Resistors

Circuit voltage range, kV rms.	≤ 650V	2.4–7.2	7.2–15	Above 15	
Voltage rating, kV	0.65	7.2	15	27	38
Capacitance, μF	1	0.5	0.25	0.125	0.125
Capacitor reactance, Ω	2,653	5,305	10,610	21,220	21,220
Capacitor current @ 60 Hz, A	0.14	0.78	0.82	0.73	1.03
Series resistor, Ω	25	25	25	25	25
Resistor power, Watts	0.49	15.21	16.81	13.32	26.52

TABLE 5.5
Typical Parameters for Surge Suppression Resistors

Length & diameter	Resistance, Ω Min.	Resistance, Ω Max.	Average power at 40C, Watts	Peak energy, Joules	Peak voltage, kV
200mm × 25mm	1.0	390	190	2.1K	6
	15.0	3.9K	75	16.5K	45
	4.7K	470K	60	7.5K	15
200mm × 40mm	6.5	1,875	100	46K	45
305mm × 25mm	1.0	680	275	3.2K	10
	25	6.8K	100	27K	75
	8.2K	680K	90	12.5K	25
305mm × 40mm	9.0	2.5K	150	75K	75

during a voltage escalation event can be many times greater than the steady state values given in Table 5.4. Examples of commercial resistors are given in Table 5.5.

Although the best connection of the surge suppressor is directly from the terminals of the inductive element to ground, Pretorius [18] and Pretorius et al. [19] have presented some alternative locations; see Figure 5.21. For example, when it is difficult to attach a surge suppression device right at the terminals a good second choice is to connect a cable to the terminals that has the same surge impedance and voltage rating (e.g., the same type and length) as the cable from the switch to the inductive element. The surge suppressor would then be attached to the end of the cable and then to ground. The high frequency connections from the surge suppression device to the load and to the ground should be as short as possible. Flat conductors no more than a few feet in length are recommended. For example, a wide flexible copper brade, 1cm × 5cm, is a good choice. A round wire is not recommended, because its equivalent inductance at the transient frequencies observed in voltage escalation events can be 200 times that at the usual 50/60 Hz ac power frequencies.

An even more effective surge suppression device is shown in Figure 15.20(b) [18, 19]. Here a ZnO varistor is connected in parallel to the resistor. This further clips the voltage seen across the resistor and also the terminals of the inductance. Other suggestions for surge suppressor devices are a ZnO varistor in series with a spark gap, Figure 5.20(c). This effectively clips the voltage at a given value and permits the contacts to eventually achieve the recovery state shown in Figure 5.18. This type of device can be made to be quite compact and while it does nothing to slow the rate of rise of the voltage pulses, it does not have the disadvantage of a somewhat limited capacitor life. Figure 5.20(d) is another possible method of limiting the peak voltage experienced at the load's terminals. Table 5.6

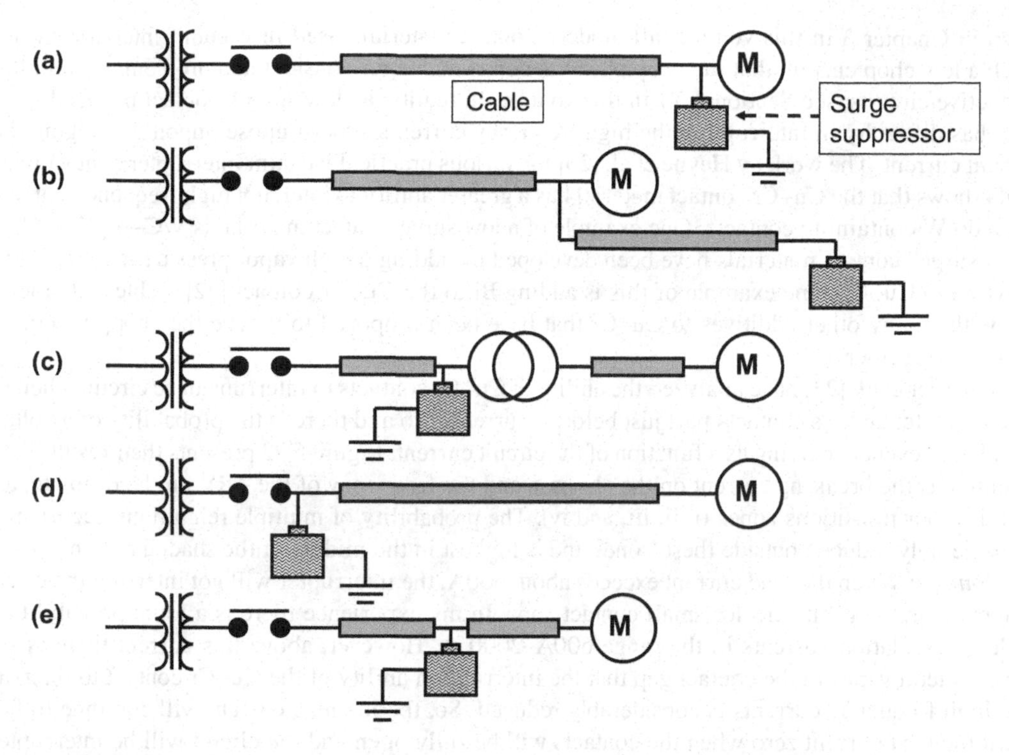

FIGURE 5.21 How to position the surge arrester with respect to the inductive element [18, 19].

TABLE 5.6
The Effect of Surge Suppression When Switching Motors with Vacuum Circuit Breaker Compared to an SF_6 or Air Circuit Breaker

		Vacuum		SF_6/Air	
Operation	**Type and location of surge suppression**	**Voltage at motor, PU**	**Voltage rise time, μs**	**Voltage at motor, PU**	**Voltage rise time, μs**
Switching On	None or arrester	4	0.2–0.5	4	0.2–0.5
	Capacitance at motor	4	3–7	4	3–7
	RC at motor	2	0.2–0.5	2	0.2–0.5
	ZnO-RC at motor	≤ 1	0.2–0.5	≤ 1	0.2–0.5
	ZnO-RC at panel	≤ 2	0.2–0.5	≤ 2	0.2–0.5
Switching Off	None or arrester	4–5	0.2–0.5	4–4.5	0.2–0.5
	Capacitance at motor	6	3–7	4–4.5	3–7
	RC at motor	3	0.2–0.5	2.2–2.8	0.2–0.5
	ZnO-RC at motor	≤ 1.5	0.2–0.5	≤ 1.5	0.1–0.5
	ZnO-RC at panel	≤ 3	0.2–0.5	≤ 3	0.2–0.5

shows the strong effect surge suppression has on suppression of multiple reignition events [18, 19] for vacuum, air and SF_6 circuit breakers.

The voltage escalation phenomenon can be reduced considerably and can even be eliminated by using a "low surge" contact material in the vacuum interrupter. In the past, the term "low surge" contact materials have been erroneously linked with "low chop" contact materials. As we have seen in Chapter 3 in this volume, all modern contact materials used in vacuum interrupters have such a low chop current that the chop phenomenon is no longer considered a problem, even in high inductive circuits; see Section 4.3.1 in this volume. In reality, a "low surge" contact material is one that has difficulty in interrupting the high frequency currents super-imposed upon the reignited ac circuit current. The work by Heyne et al. [20] for various practical vacuum interrupter contact materials shows that the Cu–Cr contact material has a greater ability to interrupt high frequency currents than do W containing contacts. One example of a low surge contact material is WC–Ag [21]. Other "low surge" contact materials have been developed by adding a high vapor pressure material (up to 5 wt.%) to Cu–Cr. One example of this is adding Bi to the Cu–Cr contact [22]. Tables 3.6 and 3.7 show the many other additives to Cu–Cr that have been proposed to reduce the chopping current level even further.

Telander et al. [23] have analyzed the ability of Cu–Cr contacts to interrupt an ac circuit when the vacuum interrupter's contacts part just before a current zero and there is the probability of a voltage escalation event occurring as a function of the circuit current. Figure 5.22 presents their results. This Figure has the breaking current on the abscissa and the frequency of the TRV on the ordinate, and defines four transitions zones (i, ii, iii, and iv). The probability of multiple reignitions occurring is considerably reduced outside these zones and is highest in the middle of the shaded region.

Zone (i): When the load current exceeds about 600A, the interrupter will not interrupt the circuit after the first current zero for small contact gaps. In my experience there is a finite probability of voltage escalation currents in the range 600A–2000 A. However, above this current there is sufficient metal vapor in the contact gap that the interruption ability of the Cu–Cr contact to interrupt the high frequency currents is considerably reduced. So, the main ac current will continue to flow until the next current zero when the contacts will be fully open and the circuit will be interrupted. When the circuit current increases to a few thousand amperes, the event is extremely unlikely and when interrupting a full short circuit current it almost never occurs.

Zone (ii): When the rate of rise of the TRV is less than the rate at which the dielectric strength of even the small contact gap increases, no reignition will take place and hence the voltage escalation event will not begin. Telander's values for this lower TRV boundary depend upon the system voltages, i.e., 900 Hz for 4.16 kV and 6.9 kV (i.e., a rise to a TRV peak ~ 0.28ms), 500 Hz for 13.8 kV (i.e., a rise to a TRV peak ~ 0.5ms) and 250Hz for 23 kV (i.e., a rise to a TRV peak ~ 1ms). This is considerably slower than the rise of the TRV in an inductive circuit of about 10 – 70 μs.

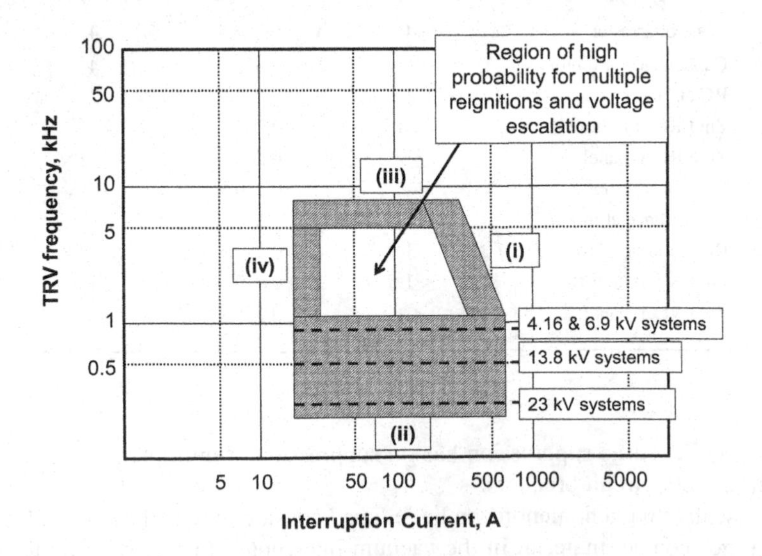

FIGURE 5.22 The region where restriking is probable [23].

A more resistive or capacitive circuit will thus have a much lower occurrence probability of voltage escalation events.

Zone (iii): When the frequency of the TRV is extremely high, e.g., the capacitance of the load is very small, the vacuum interrupter is again not able to interrupt the high frequency current superimposed upon the ac current. This is the example shown in Figure 5.18. An extra half cycle of ac current is seen during which the vacuum interrupter's contacts will be fully open. Circuit interruption will occur at the next current zero.

Zone (iv): At low currents, say below 20A, e.g., magnetizing currents in transformers any over voltages are likely to be limited by iron losses in the transformer.

5.3.4 Switching Three-Phase Inductive Circuits: Virtual Current Chopping

When switching the three-phase inductive circuit shown in Figure 5.23, the first phase to clear (e.g., phase C) will have a recovery voltage similar to that shown in Figure 4.37(b). In fact, damping in the circuit allows its peak voltage to reach about $1.7\sqrt{2}\ U_{(\ell\text{-}n)}$ (where $U_{(\ell\text{-}n)}$ is the line to neutral voltage). The voltages in phases A and B are halfway to the positive peak when phase C interrupts: A decreasing and B increasing. Momentarily there is no voltage between A and B: i.e., the phase-to-phase voltage U_{AB} passes through zero. The currents in these phases are equal and opposite having the instantaneous value of $\sqrt{3}/2 \times i_{peak}$. After phase C has interrupted the currents i_A and i_B must continue to be equal and opposite. As seen in Figure 5.24 they will change slopes when i_C is interrupted and come to zero 90° later. The recovery voltage across phase A and B is split more or less evenly across them.

There is a relatively rare phenomenon (see Figure 5.25) which can occur in this circuit if there is a restrike in phase A before the currents in phases B and C interrupt. It is called *virtual current chopping.* This is not the chopping phenomenon discussed in Sections 2.3.3, 3.2.6, and 4.3.1 in this volume, which depends upon both the properties of the contact material and the vacuum arc. It is instead a property of the particular circuit and the effect of interrupting the transient high frequency current [15, 24]. Figure 5.25 shows an example for contacts opening at t_1 as phase A is approaching zero. When the current in phase A is interrupted at t_2 a fast rising TRV appears across its contacts. If now there is a restrike between these contacts at t_3, a high frequency current superimposed on the main circuit current i_{fa} will flow in phase A; see Section 4.3.1. This high frequency current in phase A will follow the path to ground

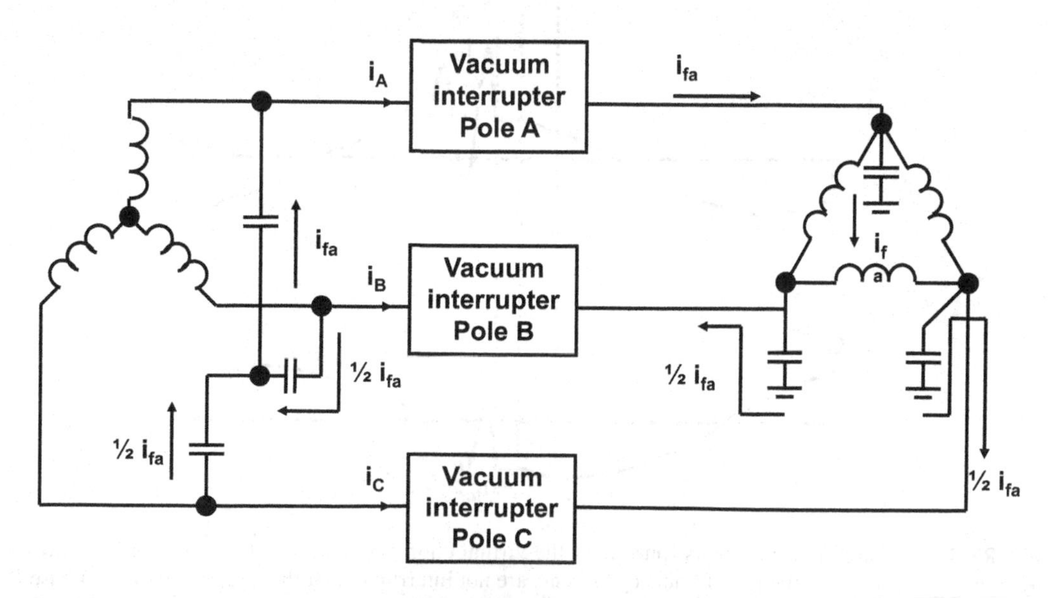

FIGURE 5.23 The circuit to illustrate virtual chop.

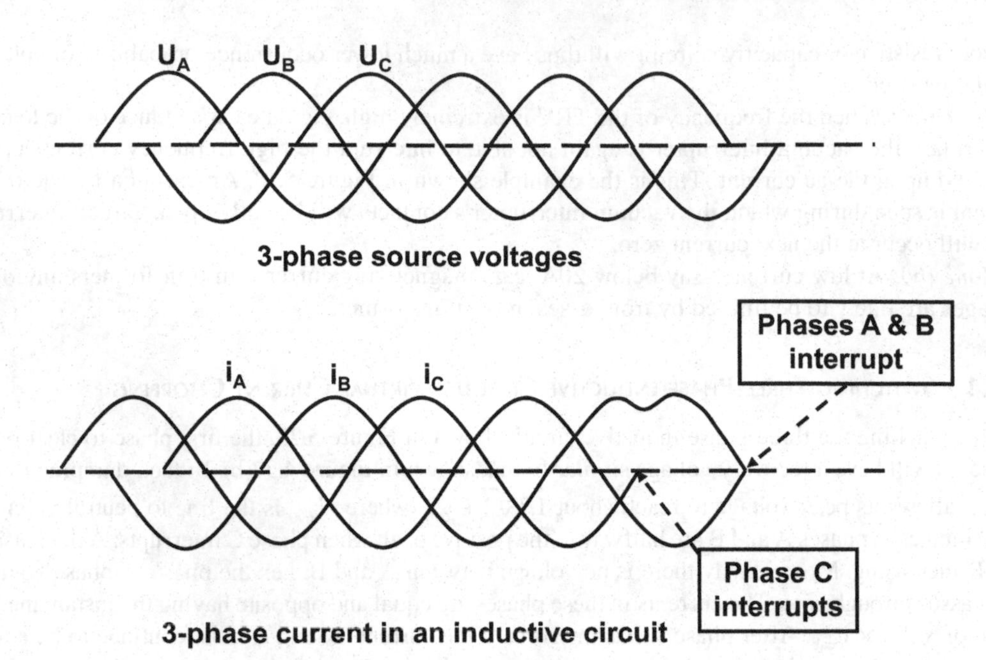

FIGURE 5.24 Three-phase currents in an inductive circuit.

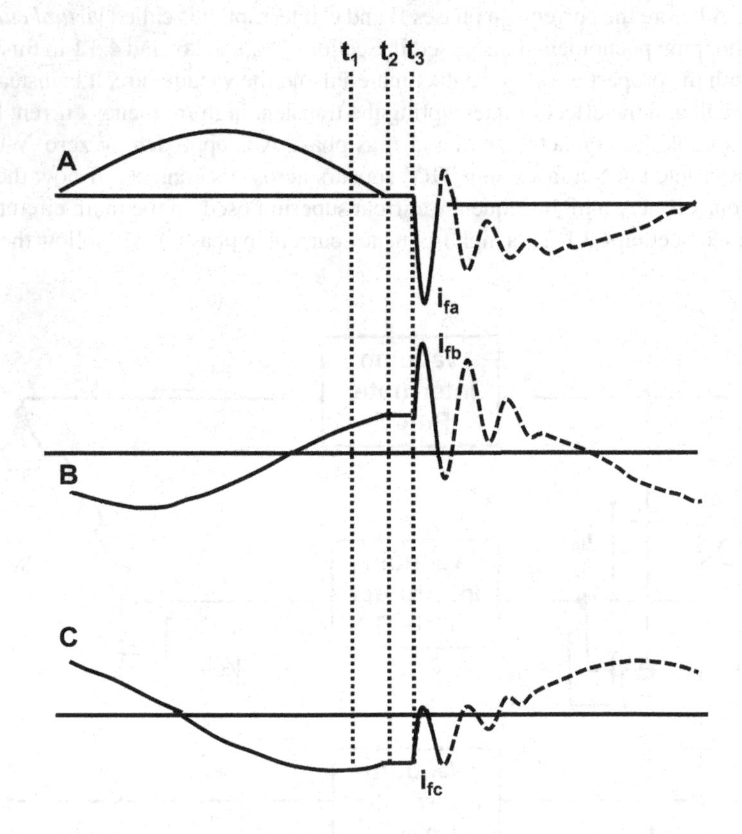

FIGURE 5.25 Three phase currents illustrating the virtual chop phenomenon. The dotted lines show the circuit breaker currents if the high frequency currents are not interrupted. At the next current zero, Phase B, the contacts will be far enough apart that reignition will be unlikely.

through the stray capacitance of the load, as shown in Figure 5.23. If the three-phase system is balanced, i_{fa} divides in two so that $i_{fa}/2$ flows to phases B and C through their respective stray capacitances at the load. These high frequency currents couple back into phase A on the source side of the switch.

When the reignition occurs in phase A, say a few tens of microseconds after phase A initially interrupts the current, the power frequency current (50 or 60 Hz) in phases B and C will be approximately $0.86 \times i_{peak}$. If the magnitude of $i_{fa}/2$ in phases B and C is greater than this value, then the high frequency currents will force the currents in phases B and C to zero, as can be seen in Figure 5.25. This circuit-induced rapid decrease of current to zero is the virtual current chopping phenomenon. The magnitude of this virtually chopped current can be close to the value of the power frequency current: i.e., hundreds of amperes. It is therefore very much higher than the normal level expected as described in Section 2.3.3 in this volume. Virtual current chopping can result in very high overvoltages. The much lower surge impedance, however, affects the resulting surge voltage. In the case of an ungrounded system, neutral phases B and C will be in-phase with opposite polarities. This will result in a line-to-line overvoltage of twice their corresponding line-to-ground overvoltage. An example of virtual chopping is shown in Figure 5.26. The capacitor–resistor combination shown in Figure 5.20(a) when installed phase-to-ground, reduces the effect of virtual chopping. The capacitor coupling between the phases lowers the frequency of the reignition current and also its surge impedance. The resister dissipates the reignition energy and can reduce the travelling wave in the cable systems. The resistor must have a similar surge impedance as the cable for this to be effective. Adding a ZnO varistor across the resistor (Figure 5.20(b)) will limit the level of the surge voltage.

5.3.5 Transformer Switching

A transformer whether loaded or unloaded places a large inductance into a circuit. Switching such a circuit with a vacuum interrupter has been studied since the late 1950's and early 1960's [4, 25, 26]. The previous discussions on voltage transients that can occur when switching such circuits are certainly relevant here; see Section 4.3.1 in this volume and Sections 5.3.2 and 5.3.3. There are two characteristics of this circuit that need to be considered. First is the current level:

1) When switching an unloaded transformer (i.e., one whose secondary windings are not connected to a load) the current is the transformer's magnetizing current. The

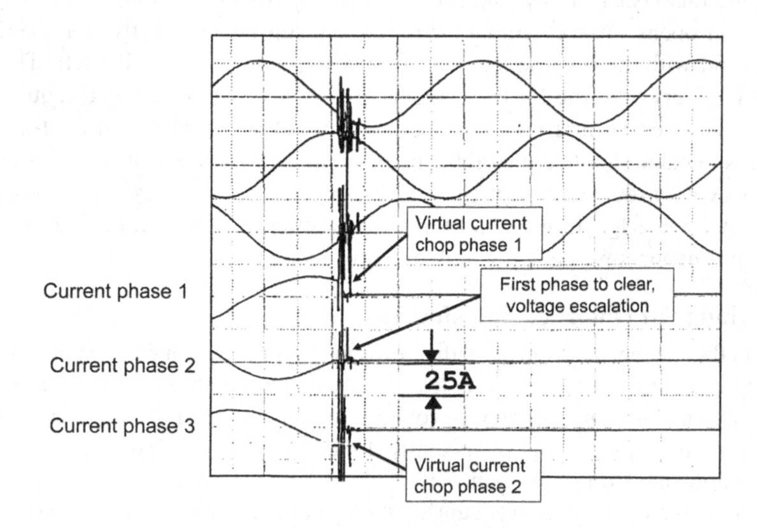

FIGURE 5.26 An example of a three-phase virtual chop event showing the high frequency reignition current on phase 2 and the overvoltages from the virtual chop in phases 1 and 3

kVA of the transformer (defined in Equation (5.9)) determines the magnitude of this current:

$$\text{Transformer kVA} = \frac{U_{(\ell-\ell)} i_L \sqrt{3}}{10^3} \text{ and Transformer MVA} = \frac{U_{(\ell-\ell)} i_L \sqrt{3}}{10^6} \quad (5.9)$$

where $U_{(\ell\text{-}\ell)}$ is the line-to-line voltage and i_L is the transformer's normal load current. As the kVA of a transformer is reduced its capacitance decreases, but its magnetizing inductance increases. Thus, it is possible for the surge impedance to also increase. At the same time, however, the value of the magnetizing current will decrease. Random energizing of unloaded transformers may result in high currents and high overvoltages. These transients may also have harmonic components that can reduce the transformer's life and even cause protective relays to misfire. Brunke et al. [27, 28] and Duan et al. [29] have examined the minimization of the in-rush current effects by controlled switching. Li et al. [30] show that controlled closing can reduce the in-rush current by 80% to 90% compared to uncontrolled random closing. Indeed, Li et al. [31] show that the combination of a closing resistor and controlled switching can significantly reduce the in-rush current to 0.1PU

2) It is a possible to switch off the in-rush current to an unloaded transformer in certain rare occasions. This current is also of low average magnitude but can have high current harmonics superimposed upon it
3) A third current level is the transformer's normal load current. This occurs when switching off a loaded transformer (i.e., a transformer that is connected to its load)

Second, a transformer is generally designed to withstand the high voltage BIL pulse for the circuit in which it finds itself; see Table 1.1.

5.3.5.1 Tap Changers

The use of a vacuum interrupter as a tap changer inside oil filled transformers has been employed since the 1960s. A good review of this application is given by Dohnal et al. [32, 33] The use of vacuum interrupters as tap changers is continuing to be investigated, e.g., Hammer et al. [34]. The fully enclosed arc chamber of the vacuum interrupter proves to be a major advantage. When the vacuum interrupter opens while changing the taps, the resulting arc is fully contained and does not contaminate the oil with the by-products that result from an arc in the oil itself. The low currents involved result in a very long life for the contacts. The low voltages switched require only a small bellows motion. This results in a very long bellows life. One consideration for the vacuum interrupter designer is to make sure that the oil filling the bellows has an unrestricted flow as the bellows chamber changes volume during the opening and closing operations. This application has been very successful. To my knowledge, transformer failures resulting from vacuum interrupters performing the tap changing function have been negligible to nonexistent.

5.3.5.2 Switching Off Unloaded Transformers

As stated above, the magnetizing currents that are switched have a very low value. It is possible for the magnetizing current to have a value below which a vacuum arc can be sustained; see Section 1.3.5, Figure 1.80, and Section 2.3.3 in this volume. Thus, soon after the contacts part the vacuum arc between them will self-extinguish, especially if the magnetizing current is 1A or 2A. The close contact gap will now break down as soon as the voltage across it reaches a high enough value. This will permit a momentary flow of current until the vacuum arc again self-extinguishes and the voltage transient again appears across the opening contacts. This "showering arc" sequence may well be repeated several times until the contact gap is large enough to withstand the transient voltage. For a single-phase transformer these voltage pulses are impressed only on the first few windings of the transformer and a well-designed transformer will have little difficulty withstanding them.

Most distribution transformers, however, are connected in three-phase systems. When three phases are switched the voltage escalation on the first phase to clear can result in the virtual chop phenomenon, discussed in Section 5.3.4, affecting the other two phases. Popov et al. [35, 36] have modeled the peak voltages that can occur at the terminals of a three-phase, unloaded, 13.8kV transformer. They use a magnetizing current less than the average chop current for Cu–Cr contacts and assume that the current chops to zero as soon as the vacuum arc is initiated. Their model includes both the voltage escalation and virtual chop events. Figure 5.27(a) shows the cumulative probability of a given PU voltage at the transformer's terminals (in this case using Equation (5.8), $1PU = 13.8\sqrt{2}/\sqrt{3} = 11.27kV$). The length of the cable joining the switch to the transformer has a major influence on the magnitude of the overvoltages. The capacitance of a cable is typically $250pf.m^{-1}$; thus a 15m cable will add 3.75nF to the system's capacitance. This value may be close to that of the transformer; thus, the total surge impedance of the system will be reduced by $\sqrt{2}$. Figure 5.27(a) [36] shows the probability of overvoltages for a switch connected with a 10m and a 60m cable. Here even the 10m cable shows a maximum overvoltage of only 3.5 PU. This corresponds to 39.4kV, which is far less than the transformer's designed BIL rating of 95kV. In practice transformer losses represent a large damping coefficient that can reduce the modeled values of the overvoltages considerably. Chaly et al. [37] have compared their modeling of these overvoltages with measured values. They show that the reduction factor is a function of the magnetizing current;

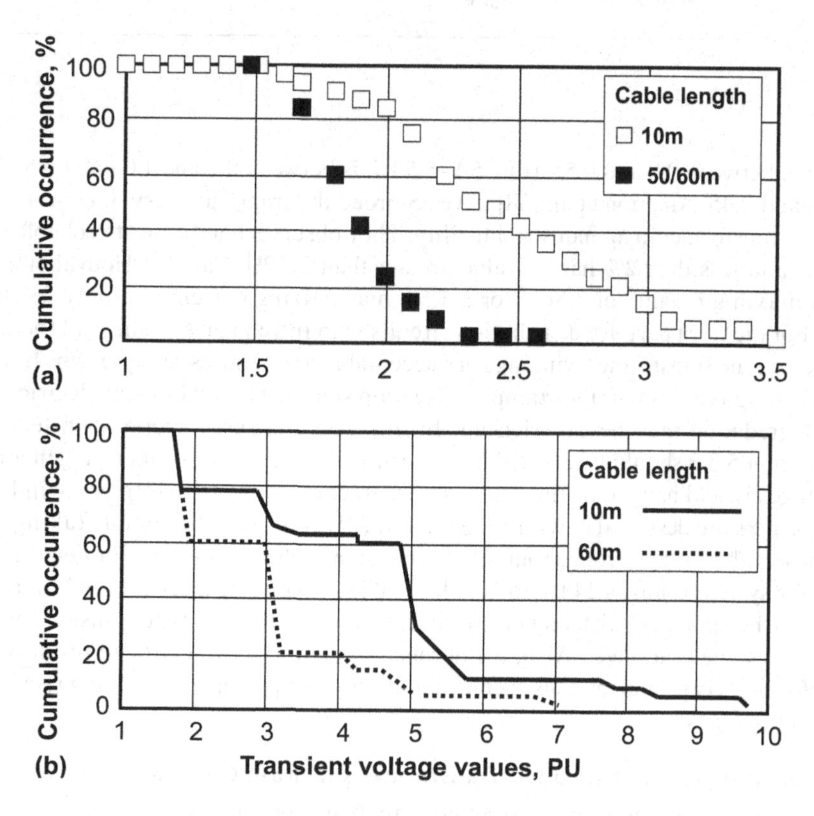

FIGURE 5.27 (a) The probability of observing a PU overvoltage level while switching off the magnetizing current for a 13.8kV transformer as a function of the cable length joining the vacuum interrupter to the transformer and (b) The probability of observing a PU overvoltage level while switching off the magnetizing in-rush current for a 13.8kV transformer as a function of the cable length joining the vacuum interrupter to the transformer [36].

TABLE 5.7
Transformer Protection

	Load	
Switching Device	**Dry-type transformer**	**Oil filled transformer**
Vacuum contactors, 1kV–7.2kV class	Protection not necessary for transformers rated for more than 400kVA. For smaller transformers use surge protection	Protection not necessary
Vacuum circuit breakers and switches, 5kV class	Surge protection should be used, unless the transformer is rated for ≥ 60kV BIL	Protection not necessary for transformers rated more than 300kV. For smaller transformers use surge protection
Vacuum circuit breakers and switches,10kV–40.5kV class	Surge protection should be used unless the transformer is rated for full BIL, e.g., Circuit Voltage — BIL 10kV — 75kV 15kV — 95kV 24kV — 125kV 27kV — 125kV 36kV — 170kV 40.5kV — 185kV	Protection is recommended, but not usually necessary if the transformer is rated for the full BIL voltage. It may be required for transformers rated less than 3MVA. However, RC at the load terminals reduces the peak of the transient voltages and their number

(a) 1A–2A it is between 0.8 and 0.55, (b) 2.5A–5.5A it is between 0.8 and 1.0 and (c) 5.5A–10A it is between 0.8 and 0.45. Ananion et al. [38] have recorded the transient overvoltages on a 2000kVA, 13.8kV/480V, transformer in an industrial facility. They observe that the most probable overvoltage is about 1PU and less than 2% have a value greater than 2.4PU. Van den Heuval et al. [39] have measured a maximum value of 3.5PU for a 1.5A magnetizing current and only 2% have values above 3PU. For the most part, for distribution circuits with infrequent switching of the transformer, these stresses on the transformer windings are acceptable and failures rarely occur. In cases where frequent switching is expected, for example in a pump storage facility or in an electric arc furnace, insulation degradation may be experienced. In this case the surge suppression techniques discussed in Section 5.3.3 should be applied to the terminals of the transformer. In general, however, the overvoltages would have to be unusually severe to damage the transformer's windings. All oil filled transformers are designed to withstand the full BIL voltage of the circuit. Taking values from Table 1.1 these values range from about 7 PU for system voltages less than or equal to 15 kV and about 6 PU for system voltages 24 kV to 40.5 kV. Table 5.7 presents a generalized overview of protection that can be applied to different classes of transformer. For dry-type transformers the design voltages can be somewhat lower. Also, the absence of the dielectric fluid can result in a decrease of the transformer's capacitance, thus increasing its surge impedance and giving rise to somewhat increased overvoltages.

5.3.5.3 Switching Off an Unloaded Transformer's In-Rush Current

The actual occurrence of this event is extremely rare. It may occur as a result of improper relay tripping immediately after initially energizing the transformer or in an electric arc furnace application where frequent switching is likely. Popov et al. [36] have also modeled this case. Their cumulative probability of the occurrence of a given PU overvoltage is shown in Figure 5.27(b). They show that there is now a greater probability of seeing quite high overvoltages. For example, with a 10m cable there is a finite probability of reaching 9.5PU (i.e., 111kV). This value is above the 95kV BIL design voltage. With a 60m cable the peak value is 7.5PU (1.e. 84.5kV). This value is below the 95kV BIL design value. Ohashi et al. [40] show that when the in-rush current is interrupted the overvoltages produced are not high enough to operate the protective lightning arresters. They also observe virtual chopping events. The peak overvoltages that they observe, however, are less than 3 PU. When they apply an RC surge suppressor to the terminals of their transformer all overvoltage events are eliminated. Popov et al. [41] also show that while high overvoltages are certainly possible, their effect is damped by high losses in the transformer's core. The amplitude of the resulting overvoltages is thus not excessive.

5.3.5.4 Switching Off Loaded Transformers

Normally switching off a loaded transformer presents little problem. However, this does depend upon the transformer's design and the load that is connected to the transformer's secondary. One spectacular example of a transformer failure is presented in Reference [42]. In this case there are a number of uninterruptible power supplies with their harmonic filters attached to the secondary side of the transformer. When the vacuum breaker on the primary side is switched the transformer can fail. On examination of one such failure no damage could be observed on the vacuum switch or on the first few windings of the transformer. What the Swindler Team eventually concludes is that an internal resonance results in a high voltage of 65kV being impressed on the internal windings. This voltage could have been as high as 144 times the peak of the 60Hz value and 27 times the BIL value at the winding's mid-point. An RC surge suppressor at the transformer's terminals (5Ω and 0.6μF) connected line-to-ground on the high voltage side of the transformer eliminated the problem. Similar problems have been observed when power semiconductor systems are attached to the secondary of a transformer [43].

Internal breakdown of the internal winding of a transformer as a result of resonance effects can occur from events unrelated to switching using a vacuum interrupter. Hori et al. [44] in an interesting study show how the voltage pulse resulting from a winter lightning strike on an electrical line leading to a transformer can result in the breakdown of the transformer's internal windings. Surge arresters that operated successfully and clipped the peak of the voltage pulse to well below the transformer's BIL design voltage protect the transformer in their study. A Fourier analysis of the voltage pulse shows peak frequencies at 10–15 kHz and at 30–40 kHz. In their analysis these correspond to the resonance frequencies of the internal winding where the breakdown occurs. They show that the resonance results in a peak voltage in that region beyond the design value. When they change the insulation design (see also, for example, [45]) and when they add internal surge suppressers the observed failure mode is eliminated.

Because of these effects and the rare event of transformer failure, a number of review committees (see, for example, [46]), have reached the following conclusions:

1) All switching technologies, vacuum, SF_6, and oil can produce voltage transients at the transformer's terminals
2) Infrequent switching of distribution transformers does not usually result in transformer failure
3) If the transformer is switched frequently, then it is wise to place an RC surge suppressor at the transformer's terminals
4) If a failure occurs as a result of an internal resonance, then an RC surge suppressor at the transformer's terminals will prevent it
5) Considering the low rate of transformer failures, no further dielectric tests to evaluate transformers performance are recommended
6) Problems such as internal resonance should be resolved on a case-by-case basis

5.4 VACUUM CONTACTORS

5.4.1 Introduction

The vacuum contactor is a special type of load break switch that has been developed to primarily switch three-phase electric motors. The switching elements are vacuum interrupters, which are mounted on a frame. They are closed by energizing an electric solenoid and are opened by a spring when the current to the solenoid is switched off. Figure 5.28 illustrates the typical operation. When the solenoid's coil is energized, the opening springs are compressed and the vacuum interrupter contacts begin to close. Once the vacuum interrupter's contacts touch, the solenoid continues to compress the contact force spring; see Figure 5.5. When fully closed, the opening springs are "charged" and the contact force spring provides enough force for the contacts to satisfy all the closed contact requirements. The opening springs accelerate the moving contacts of the vacuum interrupter to the open position once the current to the solenoid coil is switched off. The typical static force requirements for the vacuum contactor are shown in Figure 5.29. The static force must be more than compensated, at least dynamically, by the solenoid's magnetic force. Figures 5.30 and 5.31 show an example of a distribution vacuum contactor and a low voltage vacuum contactor. Vacuum contactors have been developed for use with circuit voltages from 440V to 15kV: typical voltage and continuous current ratings are shown in Table 5.8. The major requirement that sets the vacuum contactor apart from other load break switches is the need for a very long electrical switching life. Low voltage vacuum contactors claim electrical switching lives in excess of 10^6 operations and distribution voltage designs have electrical switching lives in excess of $3x10^5$ operations.

Like all other switching devices, an independent testing laboratory must certify the vacuum contactor to a set standard. In the United States, the standard is the UL 347–1995 standard [47]. This standard has requirements for both with and without coordination with a fault protection fuse. It also has testing considerations for certain overload conditions. Most of the world, however, tests vacuum contactors to IEC Standards: e.g., IEC 60947-1 (low voltage) and IEC 60470 (high voltage) [47]. These standards have a far greater range of specifications to which the contactor can be tested. They also include specifications for a number of possible applications of the vacuum contactor

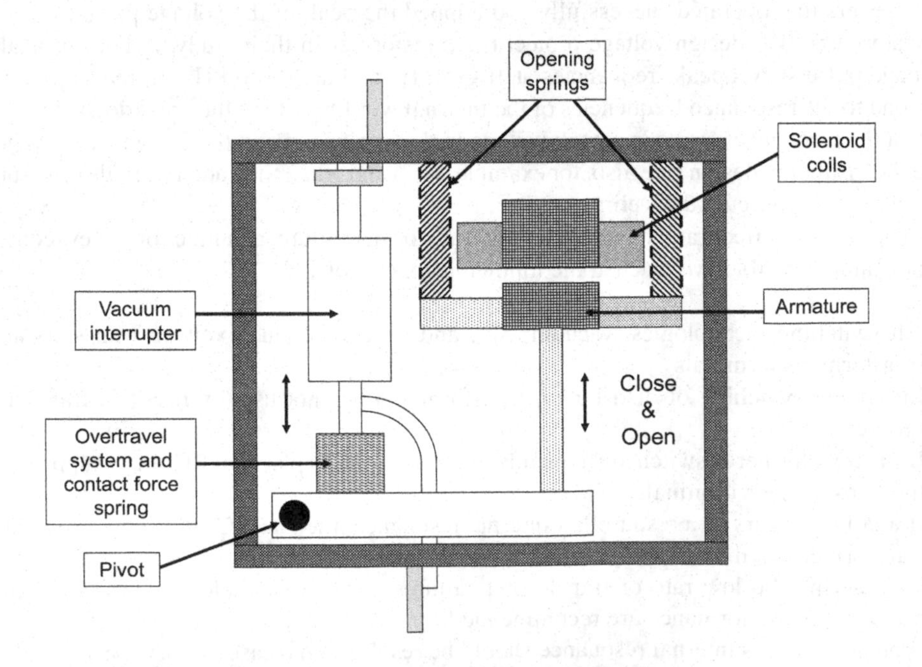

FIGURE 5.28 Schematic diagram of a solenoid operated vacuum contactor.

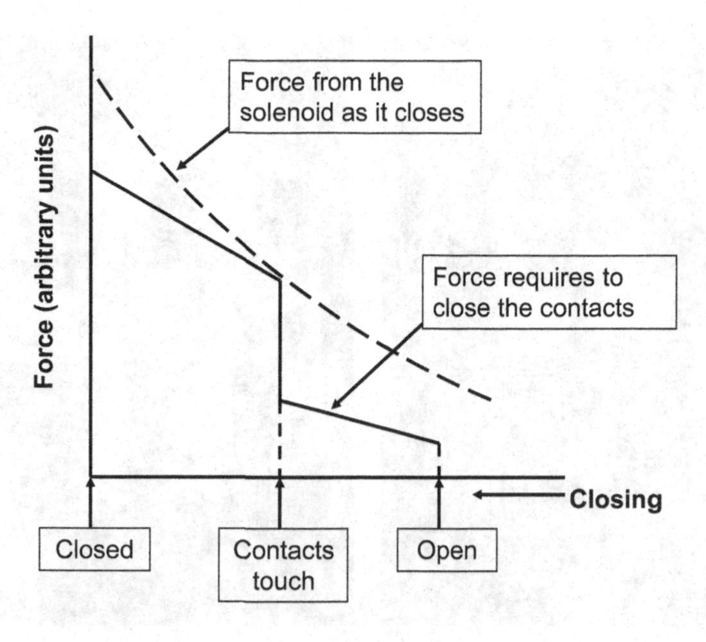

FIGURE 5.29 Generalized force requirements to close a vacuum contactor together with a typical force vs. travel characteristic of a solenoid.

FIGURE 5.30 An example of a medium voltage (7.2 kV) contactor [courtesy of Eaton Corporation].

FIGURE 5.31 Examples of low voltage vacuum contactors (e.g., 440V to 1400V) [courtesy of Eaton Corporation].

TABLE 5.8
Typical Current and Voltages for Vaccuum Contactors

Circuit voltage, V	Continuous current, A	Maximum interruption current for fuse coordination, kA
1) 440–1000	150–210	1.5–2.1
	300–450	3.0–4.5
	600–630	6.0–6.3
	800	8.0
	1000	10.0
	1400	14.0
2) 7,200	400–450	4.0–6.3
	800	8.0–12.0
3) 12,000–15,000	300–400	3.0–4.0

beyond just operating motors. The specifications include a wide range of operating currents; overload currents, frequency of switching operation as well as fuse coordination. Table 5.9 gives a sampling of the extent of the IEC specifications. The special needs of the vacuum interrupter for use in a vacuum contactor place a number of special considerations on its design:

1. The vacuum interrupter must coordinate with the solenoid operation
2. The contact must be designed for the expected electrical life and yet must not be too costly
3. The vacuum interrupter must maintain its voltage withstand capability during its whole life
4. A contact material has to be chosen that ensures the expected the electrical life, maintains its mechanical integrity, maintains a low contact resistance, maintains its high voltage

TABLE 5.9
IEC Categories for Motor Contactor Testing

Category	**Typical application**
AC–1	Noninductive or slight inductive loads, resistance furnaces
AC-2	Starting and plugging–slip ring motors
AC–3	Stating and switching off motors during running–squirrel cage motors
AC–4	Starting, plugging and inching–squirrel cage motors
Class	**Number of operations per hour**
0.01	Up to 1 operation per hour
0.03	Up to 3 operation per hour
0.1	Up to 12 operation per hour
0.3	Up to 30 operation per hour
1	Up to 120 operation per hour (allowable 'off–time' up to 24s)
3	Up to 300 operation per hour (allowable 'off–time' up to 10s)

1) Overload Testing

	Making capacity		**Breaking capacity**			
			Minimum		**Maximum**	
Category	i / i_e	**Power Factor (P.F)**	i_c / i_e	**P.F**	i_c / i_e	**P.F**
AC–1	1.5	95%	0.2	95%	1.5	95%
AC–2	4	65%	0.2	65%	4	65%
AC–3	8	35%	0.2	15%	8	35%
AC–4	10	35%	0.2	15%	8	35%

Note: i_e is the rated current, *ic* and *i* are the test currents

The number of operations for verification of the rated making and breaking capacity:

1) 50 at 85% rated coil voltage (25ops. At minimum break current i_c and 25 ops. at maximum break current)
2) 50 at 110% rated coil voltage
3) Total 100 operations

	Making Capacity		**Breaking capacity**	
Category	i / i_e	**P.F**	i_c / i_e	**P.F**
AC–1	1.5	95%	1.5	95%
AC–2	2.5	65%	2.5	65%
AC–3	6	35%	1 (at 0.17 × circuit voltage)	35%
AC–4	8	35%	6	35%

withstand capability, is resistant to contact welding, and is a low surge material; i.e., it resists voltage escalation

5. The internal design of the shielding has to effectively prevent the deposit of too much metal vapor on to the internal ceramic walls (hence, maintain the voltage withstand performance) for the life of the vacuum interrupter
6. The bellows must be capable of maintaining its performance for the mechanical and electrical life of the vacuum interrupter

5.4.2 Solenoid Operation

There are two issues to be considered when using a solenoid to operate the contactor's vacuum interrupters. The first is preventing any possibility of synchronous opening and closing to the load circuit's current wave. The second is coping with the generally low contact forces that are typical with solenoid mechanism designs. Shea [48] has modeled the effect on the total erosion depth of the contacts (d_e in Figure 5.5) for a three-phase operation as a function of synchronicity to the

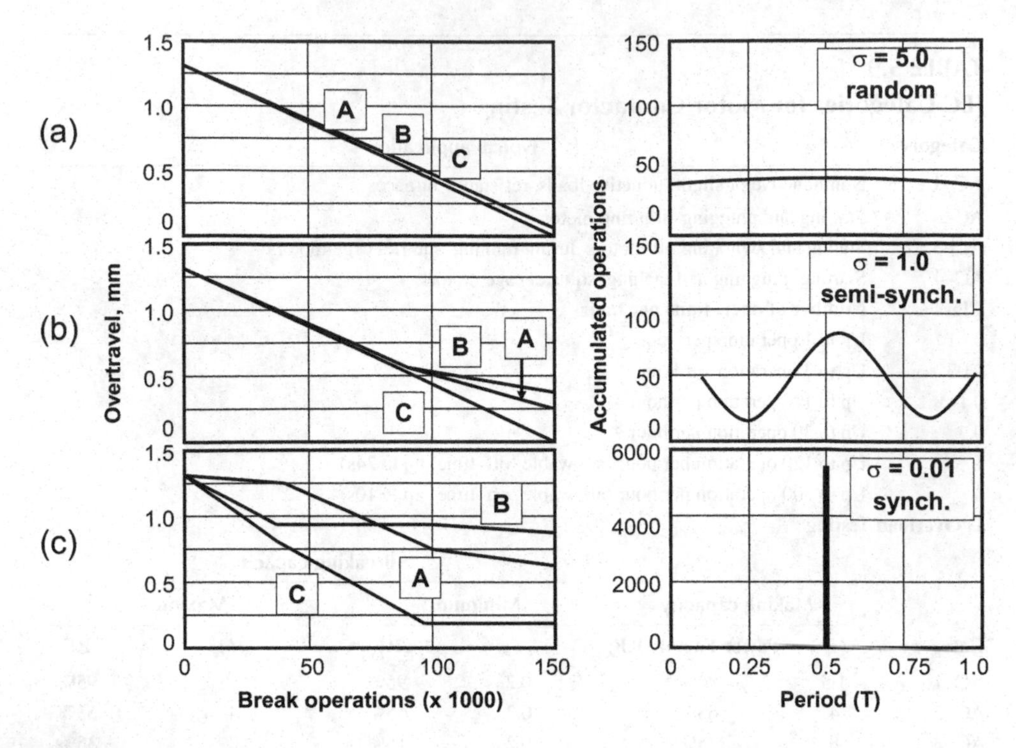

FIGURE 5.32 The effect of (a) random opening, (b) semi-random opening and (c) synchronous opening with respect to the ac current wave on the vacuum interrupter's, contact erosion for each of the three-phases A, B and C) in a vacuum contactor [48].

three-phase ac current. He uses a synchronicity factor σ (where $\sigma = 5.0$ is completely random opening: $\sigma = 1.0$ is when the opening is less random, but reasonably spread over the ac cycle*: and* $\sigma = 0.1$ is for synchronous opening). His results are shown in Figure 5.32. Here it can be seen that the contacts in each phase erode in a completely similar manner when the contact opening is random with respect to the ac current. When the opening is completely synchronous, one phase will eventually erode more rapidly than the other two. In order to achieve a random opening, an ac solenoid cannot be used; instead the solenoid has to be designed to operate with a dc current. Note; if the dc current results from a rectified ac current, then it must be smoothed enough to at least allow the semi-random operation exhibited in Figure 5.32 (b).

The relatively low contact force ($F \approx$ 25–100KgF) gives rise to two concerns. The first is the contact resistance, R_c. We know from Eq (2.4) that $R_c \propto (F)^{-1/2}$. Therefore, the contact force has a considerable influence on the total impedance of the vacuum interrupter: i.e., R_c + the resistance of the fixed and moving terminals. The value of this total impedance must be low enough to allow passage of the load current through the vacuum interrupter without generating an excessive amount of heat. As most of the heat generated must flow from the vacuum interrupter via the Cu terminals and there is generally a limit on the temperature that they are permitted to reach, it is important to minimize the vacuum interrupter's total impedance. Thus, a contact material must be chosen that gives a low R_c even with a low value of the contact force. The second is to ensure that contact welding is minimized during the passage of overload and fault currents. This is especially true when the contactor is placed in series with a circuit-protecting fuse. Here the let-through current can be high enough that the blow-off force (see Section 2.1.4 in this volume) is greater than the contact force. Thus, the contact material must be resistant to welding even in the presence of the arc that will form between the contacts. Fortunately, the WC–Ag contact material combines the properties of a low enough contact resistance with a resistance to welding and thus has found widespread acceptance for vacuum interrupters used in contactors. It is possible to find other materials that also satisfy these requirements such as Cu–Cr–Bi (5 wt.%), but the long-term mechanical strength of this type of material is not as reliable as that of WC–Ag and is not really recommended.

5.4.3 Sizing the Contact

As I have discussed in Chapter 2 in this volume, the arc that develops between the opening contacts of a vacuum interrupter forms and operates in metal vapor eroded from the contacts themselves. In general, for the usual currents switched by contactor vacuum interrupters the diffuse vacuum arc is the norm. The absolute mass erosion, m_e per operation, from such an arc occurs only at the cathode and is given by:

$$m_e = G^* \int_{t_a} i_{arc} dt \tag{5.10}$$

where i_{arc} is the current, t_a is the arc time and G^* is a constant (see Section 2.3.2 and Table 2.8 in this volume). If the contact has a radius r, a density δ, and n is the required number of electrical switching operations then using Equation (5. 10), the total erosion depth d_e will be given by:

$$\frac{m_e n}{\delta} = \pi r^2 d_e \tag{5.11}$$

$$d_e = \frac{m_e n}{\delta \pi r^2} \tag{5.12}$$

For WC–Ag contact, if the current is 400A (rms), $r = 1.35$ cm, $n = 10^6$, $m_e = 2.5 \times 10^{-6}$ g.C^{-1}, $\delta = 13$g. cm^{-3}, and the average charge/operation over the 10^6 operations for a 50Hz current is $400 \times 5 \times 10^{-3}$ C (i.e., assuming a ¼ cycle of acing on the average) then:

$$d_e = 0.7 \text{ cm} \tag{5.13}$$

Thus, the contact would have to be designed with its thickness more than half that of its radius. Fortunately, the above calculation is a gross overestimate of the true erosion of vacuum interrupter contacts. Schulman et al. [49, 50] have shown the true nature of contact erosion from the diffuse vacuum arc in a practical vacuum interrupter system. In their studies they measure the effective erosion of WC–Ag and Cu–Cr butt contacts after switching load currents in the range 400A–600A. In these experiments they ensure that the contacts change polarity after each operation. Their measure of the effective contact erosion thus takes into account the material from the cathode that is deposited on the anode during each operation: this is illustrated in Figure 5.33. Their results for contact gaps of $<g>$ = *4mm and 6mm* fall in the range reported by Rieder et al. [51] for a contact gap $<g>$ = *10mm*, which is based on measured mass loss. The data for the WC–Ag contact materials is shown in Table 5.10. Figure 5.34 shows that the effective mass erosion rate more than doubles as the *contact gap* $<g>$*/contact diameter* $<\varphi>$ increases from 0.1 to 0.3. The effective erosion of 27.1mm, diameter, WC–Ag (50 wt.%) contacts, for example, shows an increase in effective erosion rate as $<g>/<\varphi>$ increases from 0.096 to 0.149, followed by a leveling off for larger $<g>/<\varphi>$.

To interpret the results of this study, it is necessary to consider the spatial distributions of the three components of the eroded cathode material – ions, vapor and molten droplets. The opening sequence of the vacuum contacts complicates the interpretation of the effective contact erosion data. As I have described in Section 2.2 in this volume, a molten metal bridge forms as the contacts begin to part. Daalder [52] has shown that the contribution of this molten metal bridge to the contact

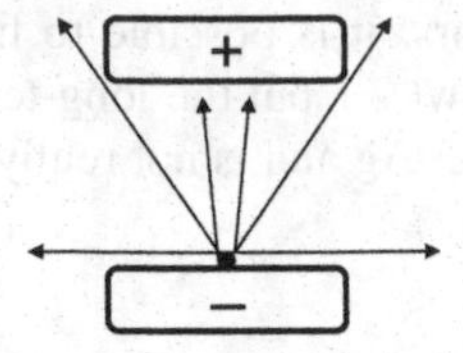
+
−
(a) Absolute cathode erosion from a cathode spot using a
dc current; i.e. the cathode is always the same contact

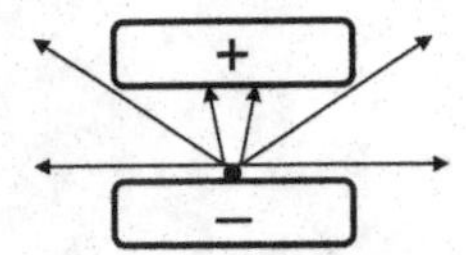
+
−

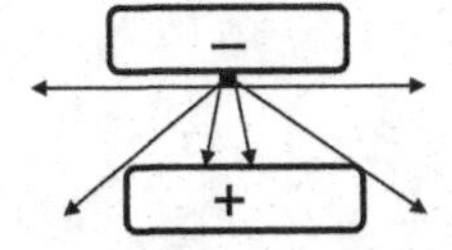
−
+
(b) Effective erosion in a ac switch with contact material form the cathode
spot being deposited on the anode which becomes the new cathode

FIGURE 5.33 Diagrams showing (a) absolute cathode contact erosion for the case of a dc, diffuse vacuum arc where the cathode is always the same contact and (b) the effective contact erosion for the case of an ac diffuse vacuum arc where the contacts change polarity each switching operation.

TABLE 5.10
Effective Erosion Rate for Ag–WC Contacts for Different Contact Gaps, Diameters, and Styles [49]

wt. % WC	Contact style	Open gap, <g> mm	Diameter <Φ> mm	<g> / <φ>	Erosion rate, 10^{-7} cm³/C	Erosion rate, µg/C
60	butt	2.2	23.0	0.096	2.3±0.3	2.9±0.4
50	butt	4	26.9	0.149	6.8±0.9	8.2±1.0
50	butt	6	26.9	0.223	7.0±1.0	8.4±1.2
50	butt	8	27.1	0.295	7.4±1.0	8.9±1.2
50	slotted	4	27.0	0.148	9.7±1.3	11.8±1.5

erosion is negligible. After the rupture of the molten metal bridge, the bridge column arc forms in the vicinity of the molten metal bridge and will last for less than 100 µs for the currents used in these experiments: see Figure 2.16 in this volume. It is only after the collapse of the bridge column arc that the fully diffuse vacuum arc develops. Once it does develop, the anode is a passive current-collecting element. It also receives deposits of metal vapor and metal particles evaporated from the cathode spots. Once the fully diffuse vacuum arc has formed, the contacts will have a gap of about 50 µm. The continued contact opening sequence now has a considerable influence on the measured effective contact erosion. Examples of contact opening sequences are shown in Figure 5.35. Even though the final contact gap varies from ~2.2mm to 10mm, the contacts are considerably closer for a significant proportion of the half cycle. Consider the cases of opening to a gap of 6mm. From i_1 to i_2 the contact gap is less than ~2.2mm, from i_1 to i_3 the contact gap is less than ~4mm, and the effect of the full gap of 6mm is only experienced after the current is past its peak.

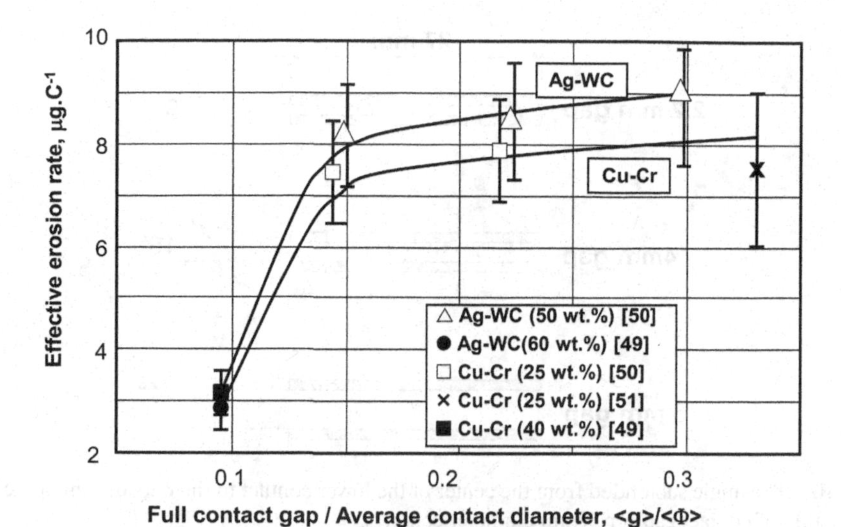

FIGURE 5.34 The effective erosion of Ag–WC and Cu–Cr contacts as a function of the ratio: contact gap <g> / contact diameter <φ> [49–51].

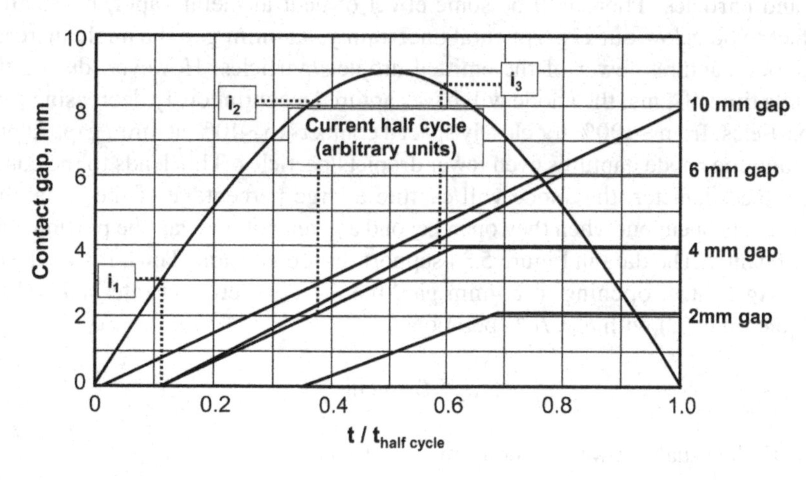

FIGURE 5.35 Travel of opening vacuum interrupter contacts with a speed about 1m.s^{-1} showing the time period that the contact gap is less than the fully open gap.

This varying contact gap greatly influences the effective contact erosion. Consider the effect of contact gap on the effectiveness of the anode contact to collect material eroded from the cathode spots. This discussion can be simplified by using the argument of Heberlein et al. [53] that the angle subtended by the anode from the center of the cathode gives the average effect of the anode for a multiple cathode spot, diffuse, vacuum arc. This is illustrated in Figure 5.36. Once the cathode spots have spread over the contact surface (~0.3 ms after the rupture of the molten metal bridge), the effect of the angle subtended by the anode is experienced. For the first 2.2mm of opening the angle with respect to the cathode plane goes from 0° to ~9°, (and the effective erosion measured is ~10% of the absolute cathode erosion). From 2.2mm to 4mm the angle goes from ~9° to ~16° (and the effective erosion rate is ~22% of the absolute cathode erosion), and so on. We know from Daalder's data [52] and the work by Tuma et al. [54] that the majority of the droplets emitted by the cathode spots are at an angle between the cathode plane and 30°. Thus, during the opening sequence

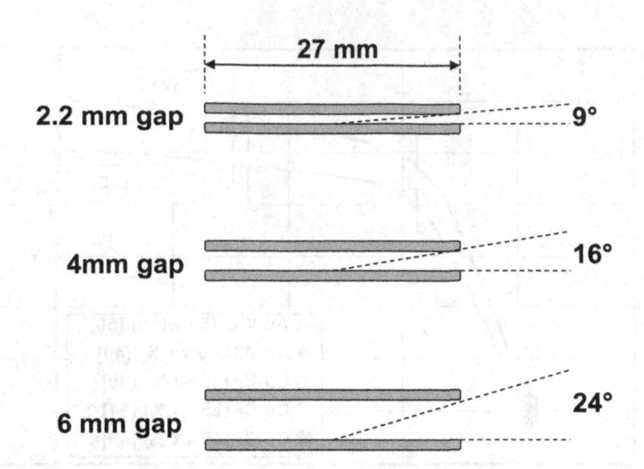

FIGURE 5.36 The angle subtended from the center of the lower contact to the outside rim of the upper contact as a function of contact gap for 27mm diameter contacts.

to 2.2mm, the anode will not only receive the forward directed ions, but it will also capture a large percentage of the metal droplets/particles. At 2.2mm gap, the anode should capture more than half the droplets and particles. There will be some effect of neutral metal vapor, but again we would expect its effect to be below our experimental uncertainty. At 4mm gap the angle increases to ~16°, so now the anode captures fewer of the emitted droplets/particles. However, during the opening sequence up to a time of 4 ms, the anode will have captured a continuously decreasing percentage of the emitted particles, from ~100% for closely spaced contacts to ~10% at 4mm gap. At 6mm gap the angle is ~24° and the anode captures even fewer droplets/particles. This leads to the conclusion that for a given contact diameter, the anode will capture a large percentage of the metal droplets/particles as the contacts open, but when they open beyond a given contact gap the portion collected will decrease significantly. The data in Figure 5.34 support this conclusion. Thus, if we take the 27.1mm diameter WC–Ag contact opening to a 4mm gap, the effective erosion rate is 7×10^{-7} cm^3 C^{-1}, then using Equation (5.12) with $n = 10^6$ operations:

$$d_e = 0.24 \text{ cm} \tag{5.14}$$

which is close to the usual allowed erosion limit of 2 to 3mm.

5.4.4 The Shield

The shield for contactor vacuum interrupters for use at circuit voltages less than or equal to 7.2kV is usually the fixed design: see Figure 3.69(b). In designing the shield special care must be taken to capture the eroded contact material that is not deposited on the anode and to prevent an excessive coating of the material on the inner surfaces of the ceramic. Schulman et al. [50] have shown that even though most of the deposit is within the open contact gap, see Figure 5.37, there is always a small percentage that will be observed far from this region.

5.4.5 The Contact Material

From the above discussion, the WC–Ag material certainly has the properties of: satisfactory erosion, good mechanical strength throughout its electrical life, low probability of welding, and an adequately low contact resistance with the expected contact forces. The work by Kaneko et al. [21] shows that WC–Ag cannot interrupt high frequency currents well. Thus, it serves well as a low surge

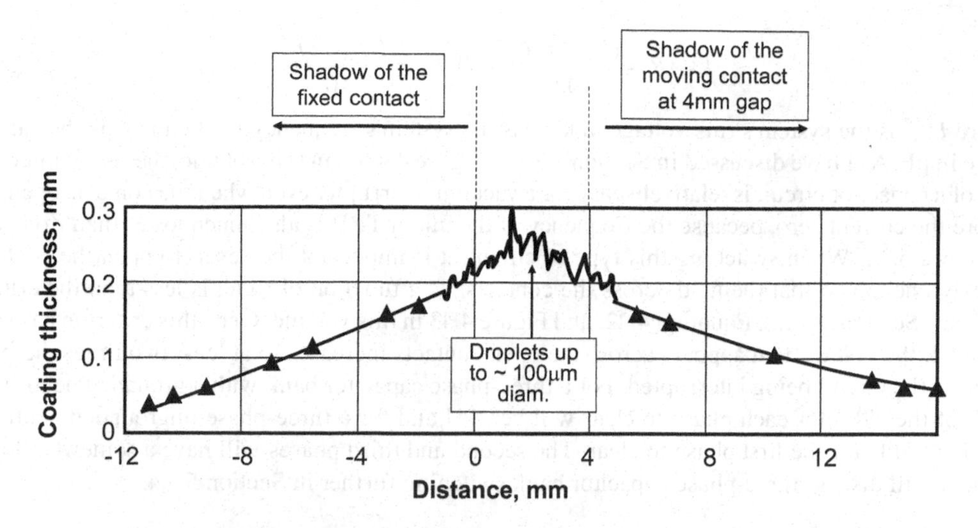

FIGURE 5.37 The shield deposit thickness of material eroded from Ag–WC contacts after switching 600A (rms) for 150,000 operations [49].

contact material; i.e., the phenomenon of voltage escalation (see Section 4.3.1 in this volume) is unlikely to occur when this contact material is used. The major disadvantage of this material is that the current interruption limit for butt contacts is about 4 kA (rms.). So, to interrupt higher currents, internal axial magnetic field contact structures or external magnetic field coils have to be utilized. Even so, it is still difficult to interrupt currents much above 10kA (rms) without increasing the size of the contactor vacuum interrupter considerably. Some manufacturers use Cu–Cr contacts at higher currents, but when they do, they must protect the motor by using the surge protection techniques discussed in Section 5.3.3.

5.5 SWITCHING CAPACITOR CIRCUITS

The capacitor switch is a load break device that can close in on high currents, but only has to open the load currents associated with inserting and disconnecting capacitor banks in a distribution system. It has no capability to interrupt or to provide protection from short circuit and fault currents. This duty has become increasingly necessary as more and more capacitor banks are used for power factor correction in utility distribution systems [55]. The information discussed in this section also applies to vacuum circuit breakers that are installed to switch capacitor circuits. When switching capacitor circuits, they are essentially only interrupting load currents. I will again discuss the use of vacuum circuit breakers for switching capacitor circuits in Section 6.7.2 in this volume. The vacuum circuit breaker does have the advantage of providing protection against short circuits and is increasingly being applied to switch capacitor circuits. In general, however, a switch designed exclusively to perform the capacitor switching function will perform that function better [56]. The vacuum interrupters for this duty are similar to the ones used for load switching shown in Figurea 5.2 and 5.3. Figure 5.2 (d) shows an example of a capacitor switch vacuum interrupter that can be used for three phase system voltages 36 kV to 40.5 kV, where all three phases operate simultaneously. It can also be used in 24 kV to 27 kV circuits where the phases open independently from each other. The main difference with a vacuum interrupter designed for capacitor switching is that the designer must take special care to minimize the geometric enhancement factor (β_g, see Section 1.3.2 in this volume) especially in the design of the contact geometry. The contact material that is used is the usual load break switch material W–Cu. This material has excellent high voltage properties. Experiments using this material have shown it to be excellent for switching capacitors [57]. Capacitor banks are usually designated by their reactive power in kilovars or megavars, which can be defined:

$$kVAR = \frac{U_{rms}^2 2\pi fC}{1000} \quad MVAR = \frac{U_{rms}^2 2\pi fC}{10^6} \tag{5.15}$$

where U_{rms} is the system's rms voltage in kV, f is the system's frequency in Hz and C is the capacitance in µF. As I have discussed in Section 4.3.1.2, Figure 4.42(c) in this volume, the initial interruption of a capacitor circuit is relatively easy for a vacuum interrupter even when the contacts open just before the current zero, because the frequency of the initial TRV is also much lower than Zone (iii) in Figure 5.22. When switching this type of circuit, it is important, however, to open the contacts quickly enough so that the field across the contacts, as a function of time, is less than the critical field: see Section 4.3.1.2, Equation 4.22, and Figure 4.43 in this volume. Once this criterion has been satisfied, the voltage that appears across the open contacts increases to at least two times the peak value of the circuit being interrupted. For a three-phase capacitor bank with a grounded neutral the peak of the TRV for each phase to clear will be 2PU and for a three-phase ungrounded neutral it will be 2.5PU for the first phase to clear. The second and third phases will have a somewhat lower value. I will discuss three-phase capacitor bank switching further in Section 5.5.4.

5.5.1 Inserting a Capacitor Bank

Two cases must be considered when closing a capacitive circuit; see, for example, Fu et al. [58] and McCoy et al. [59]. The first is illustrated in Figure 5.38. Here the switch closes on a single bank whose circuit elements are the inductance of the source L_S, the local inductance in the capacitor's cable L_C (typical values for distribution voltages are 10µH to 30µH) and the capacitance C (of the bank). This is termed *single bank* or *isolated bank* switching. When the contacts are close enough, a prestrike arc is established and a resonant current, superimposed on the ac current flows in the circuit. Initially this resonant current will have a high value (a few kiloamperes) and a high frequency (about one kilohertz). It is given by:

$$i_{R1}(t) = \frac{\left(\sqrt{2}U_{(\ell-n)} \sin \omega t_{cl}\right) \sin \omega_0 t}{Z_C} \tag{5.16}$$

where $\omega = 2\pi f$ (f is the ac frequency 50 Hz or 60Hz), ωt_{cl} is the angle that the switch closes on the ac wave, $U_{(\ell-n)}$ is the line to neutral voltage, $Z_C = (L_T/C)^{1/2}$, $L_T = L_S + L_C$ and $\omega_o = 1/(L_T C)^{1/2}$. If L_T is in µH and C is in µF then $\omega_o = 10^6/(L_T C)^{1/2}$.There will be considerable damping in a practical capacitor circuit, so the sinusoidal current described in Equation (5.16) will decay rapidly and the circuit current will be established in the capacitor circuit. The maximum value of the initial $i_{R1}(t)$ is when the switch closes on the peak of the ac voltage, i.e.,

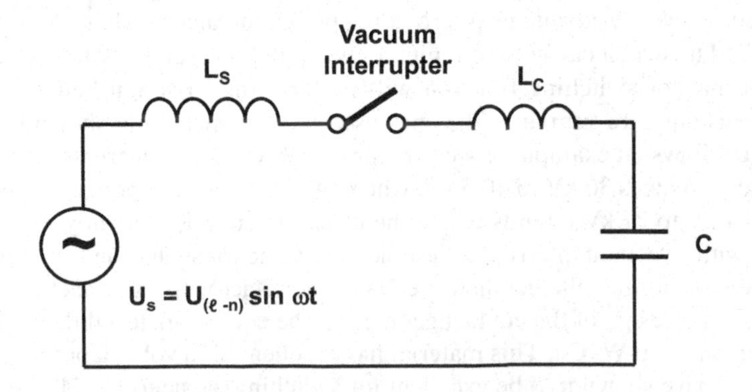

FIGURE 5.38 A single capacitor bank circuit.

$$i_{R1(\text{Max})}(t) = \frac{\sqrt{2}U_{(\ell-n)}\sin\omega_0 t}{Z_C} \tag{5.17}$$

The peak value of this current is thus:

$$i_{R1(\text{peak})} = \frac{\sqrt{2}U_{(\ell-n)}}{Z_C} \tag{5.18}$$

If the system voltage is 15kV, the capacitor bank is 60 μF, L_S is 1mH and L_C is 20μH then the peak of the in-rush current will be:

$$i_{R1(\text{peak})} = \frac{\sqrt{2}\times 15\times 10^3\times\sqrt{60\times 10^{-6}}}{\sqrt{3}\sqrt{1020\times 10^{-6}}} \approx 3\text{ kA},\ \text{frequency} \approx 0.65\text{ kHz} \tag{5.19}$$

This peak current is typical when inserting a single capacitor bank. The data from 220 capacitor banks for 42 utilities is shown in Figures 5.39 and 5.40 given by Bonfanti et al. [60]. These figures

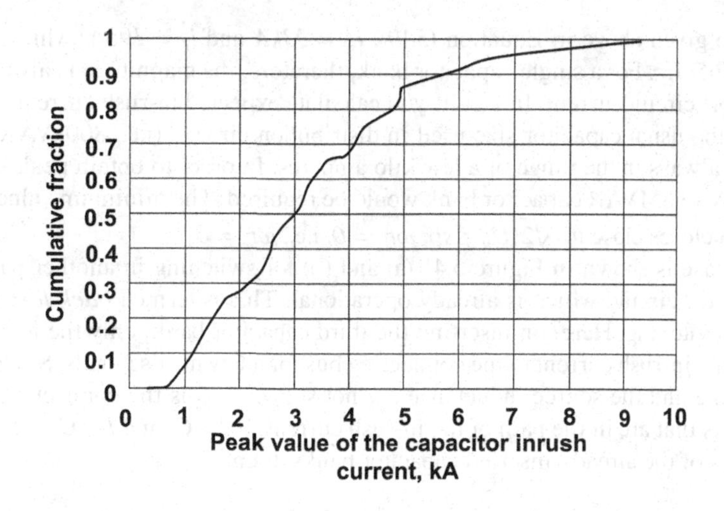

FIGURE 5.39 An example of the cumulative inrush current distribution for a single bank [60].

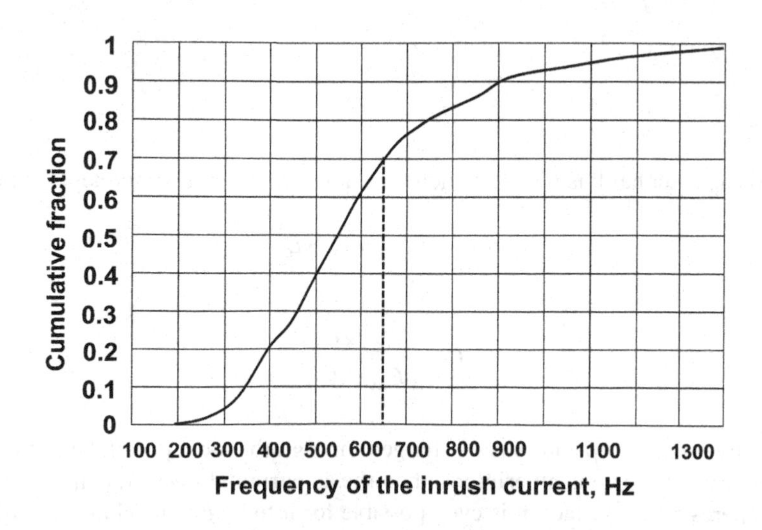

FIGURE 5.40 An example of the cumulative inrush current frequency for a single capacitor bank [60].

present the cumulative distribution of actual peak currents and frequencies measured. The high current peak will be impressed upon the whole system from the terminals of the power supply to the terminals of the capacitor. The source voltage, the circuit inductance, the closing angle and the capacitance determine this in-rush current's magnitude. If we examine Equation (5.18) it is possible to estimate the value of $i_{R1(peak)}$ and the frequency of the in-rush current f_{IR} by using the value of the system's rated short circuit current, i_{sc} and the capacitor bank's rated load current i_L.

$$i_{R1(\text{peak})} = \frac{\sqrt{2U_{(\ell-n)}^2}}{\sqrt{L_T/C}} \approx \sqrt{2}\sqrt{U_{\ell-n}\omega C}\sqrt{\frac{U_{(\ell-n)}}{\omega L_S}} \approx \sqrt{2\left(i_{sc} \times i_L\right)} \tag{5.20}$$

and

$$f_{IR} = f\sqrt{\frac{i_{sc}}{i_L}} \tag{5.21}$$

For the example given above in Equation (5.19), $i_{sc} = 23kA$ and $i_L = 196A$, which gives $i_{R1(peak)} = 3kA$ and $f_{IR} = 0.65kHz$. For a single capacitor bank, therefore, the magnitude is always less than the circuit's full short circuit current. In fact, if you calculate expected in-rush currents for single bank switching with the usual capacitor size used in distribution circuits (i.e., 800kVAR: See Table 7.2 in [3]), they are always in the range of a few kilo amperes. In order to obtain peak in-rush currents greater than 6kA an 8MVAR capacitor bank would be required. The minimum value of $i_{R1}(t)$ occurs when the switch closes close to $\sqrt{2}\, U_{(\ell-n)} \sin \omega t_{cl} = 0$, i.e., $\omega t_{cl} = 0$.

The second case is shown in Figure 5.41 (a) and (b) for switching in another parallel capacitor bank to a capacitor circuit, which is already operational. This is termed *back-to-back* or *bank-to-bank* capacitor switching. Here, on inserting the third capacitor bank, only the local capacitor circuits dominate the in-rush current to the connecting bus for a few milliseconds. So, in Figure 5.41(b) the source voltage and the source inductance are not shown. L_{EQ} is the connecting bus, cable and other inductances that are in the path of the in-rush current. If L_1, C_1 and L_2, C_2 are the inductances and capacitances of the already inserted capacitor banks, then:

$$L_{EQ} = \frac{L_1 \times L_2}{L_1 + L_2} \tag{5.22}$$

And

$$C_{EQ} = C_1 + C_2 \tag{5.23}$$

When the third capacitor bank is inserted, then the equivalent circuit components are in series thus:

$$L_{TP} = L_{EQ} + L_{EX} + L_3 \tag{5.24}$$

$$C_T = \frac{C_{EQ} \times C_3}{C_{EQ} + C_3} \tag{5.25}$$

The charge on the original capacitors is discharged through the local parallel circuit. The resulting high frequency resonant current superimposed on the ac current, however, can be very high, in the tens of kilo amperes range. In fact, it is even possible for it to be greater than the full fault current of the system. This current now only stresses the components in the local parallel capacitor circuit

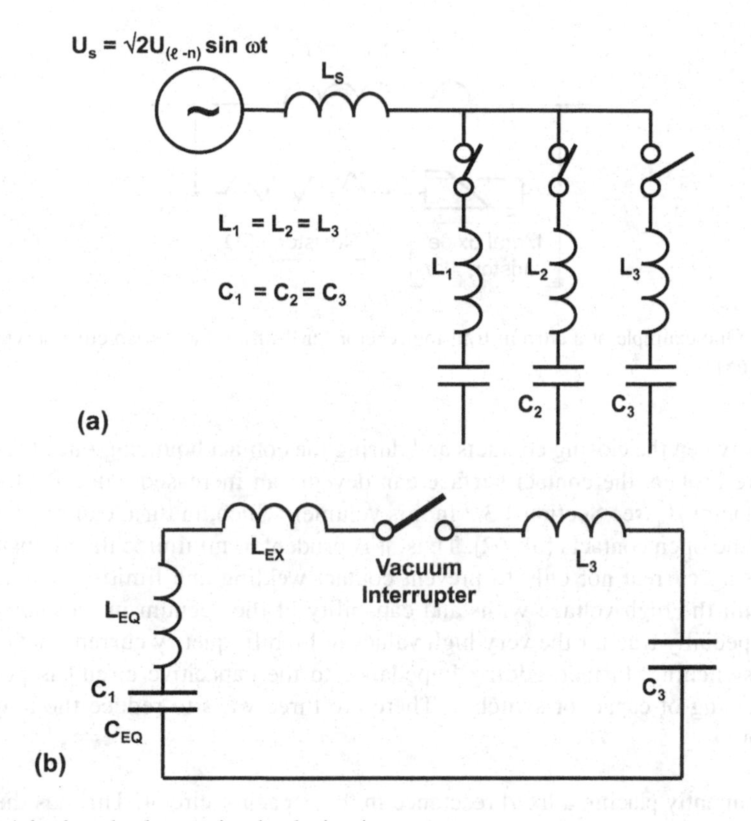

FIGURE 5.41 A back-to-back capacitor bank circuit.

and does not affect the rest of the system. It is given by (remembering that the voltage on the parallel capacitors is $\sqrt{2}U_{(\ell-n)}$ *sin ωt*):

$$i_{R2}(t) = \frac{\left(\sqrt{2}U_{(\ell-n)} \sin \omega t_{cl}\right) \sin \omega_0 t}{Z_C} \tag{5.26}$$

where $\omega_o = 10^6/(L_{TP}C_T)^{1/2}$. If the system voltage is 15kV, each capacitor bank is 60 μF (i.e., $C_1 = C_2 = C_3 = 60\mu F$), if $L_{EX} = 20\mu H$, if and $L_1 = L_2 = L_3 = 20\mu H$ (note: the typical value of the bus inductance between distribution capacitor banks is about 15μH and the inductance of the capacitor banks themselves is about 5μH), then the peak of the in-rush current will be:

$$i_{R2(\text{peak})} = \frac{\sqrt{2 \times 15 \times \sqrt{40}}}{\sqrt{3 \times \sqrt{50}}} \approx 11 \text{ kA, frequency} \approx 3 \text{ kHz} \tag{5.27}$$

The majority of the technical literature on capacitor bank switching is concerned with the effects on the Cu–Cr contact material. This contact material is primarily used in vacuum interrupters that are employed in vacuum circuit breakers. There is very little data on the performance of the Cu–W contact material that is typically used for vacuum interrupters housed in vacuum switches. Therefore, the discussion that follows primarily presents data for the Cu–Cr contact material. In general, I would expect that the data for the Cu–Cr contact material should also apply to the Cu–W contact material.

Upon the insertion of both single bank and back-to-back capacitor banks the initial in-rush current can result in minor welding of the contact spots during the closing operation when a prestrike

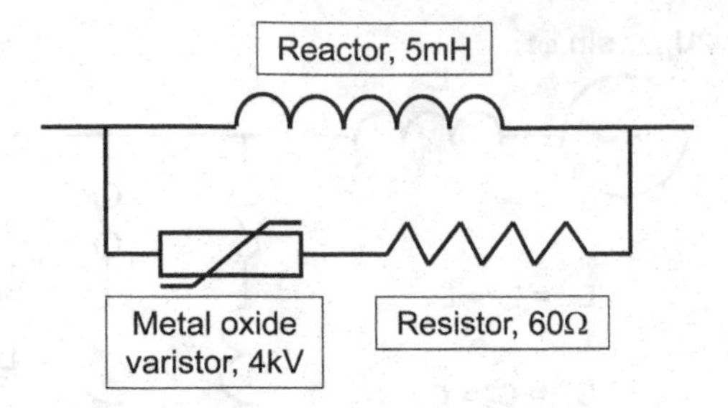

FIGURE 5.42 One example of a current limiting reactor for limiting the inrush current when connecting a capacitor bank [65].

arc develops between the closing contacts and during the contact bouncing once they touch. When these welds are broken, the contact surface can develop an increased value for the microscopic enhancement factor β_m (see Section 1.3.2 in this volume), which, in turn, can result in the vacuum breakdown of the open contacts [61, 62]. Thus, it is prudent to minimize the intensity of the high-frequency closing current not only to prevent contact welding and limiting contact erosion, but also to maintain the high voltage withstand capability of the vacuum interrupter's contacts [63, 64]. This is especially true for the very high values of high frequency current that can result from back-to-back switching. In fact, adding impedance to the capacitive circuit is permitted in the certification testing of capacitor switches. There are three ways to reduce the magnitude of the in-rush current:

1. By permanently placing a fixed reactance in the capacitor circuit. This has the advantage of being easily accomplished. The two minor disadvantages of this solution are that the reactance will increase energy losses in the system and will also reduce the effectiveness of the capacitors. Figure 5.42 illustrates a particularly effective limiting circuit proposed by Sabot et al. [65]
2. By momentarily inserting a current limiting impedance (a resistor or an inductor) just prior to closing the vacuum interrupter when inserting the capacitor. Then, after the in-rush current has decayed to zero, this will bypass or short out the impedance [66]. The operation of such a system is illustrated in Figure 5.43. This effectively overcomes the disadvantages of the permanent reactance. It does, however, introduce a complex switching and coordination function into the system. At transmission voltages the momentary insertion of a current limiting impedance is achieved by using a series air, making switch [67]. The operation of such a system is illustrated in Figure 5.44. It would be possible to incorporate this design into a vacuum switch with the series air-break shown in Figure 5.7
3. By synchronizing the closing of the vacuum interrupter so that it makes permanent contact when the system voltage is close to zero for the single capacitor bank, or the voltage is close to zero on the parallel banks. On a three-phase system this means that each phase must close separately. Therefore, each phase will have its own control and its own mechanism [68]. If all three phases close together the best you can do is to compromise and find the point on the ac wave that gives the lowest value of current below the maximum in-rush current [69]. It would be possible to develop a closing sequence for a three-phase switch such that initially phase-one would close on zero voltage, then on the second operation the phase-two would close on zero voltage and so on. Schellekens et al. [70] discuss the benefits of synchronous capacitor bank closing and the timing of the prestrike breakdown before the contacts touch. They conclude that the scatter of the vacuum interrupter's dielectric behavior limits the in-rush current to about 40% of its possible maximum value. Any synchronous closing scheme also introduces the need for a complex sensing and control

system as well as an accurate mechanism that has little variance in its closing time. I discuss synchronous closing and opening of vacuum circuit breakers used for capacitor switching in Section 6.7.2 in this volume

The seasoning effect of switching the load current while opening the contacts and the subsequent erosion of any microprojections on the cathode's surface by the cathode spots alleviates some of the effects of the weld spots. But, even so, there is a higher probability of restriking when switching back-to-back capacitor banks than when switching a single capacitor bank [71] unless the in-rush current is severely limited.

5.5.2 Disconnecting a Capacitor Bank

As I have previously discussed in Section 4.3.1.2 in this volume, interrupting a capacitor circuit is a relatively easy task for the vacuum interrupter, especially as the currents involved are typically less than 1000A and the resulting vacuum arc is in the diffuse mode. Also, the initial recovery voltage across the contacts is zero and its initial rate of rise is very slow. However, its value reaches from about 2PU to 2.5PU. The resulting voltage appearing across the contacts is unidirectional. That is the voltage will be fully offset as is shown in Figure 5.45. For the most part, a well-designed vacuum capacitor switch will withstand this unidirectional voltage during the time it takes for the charge residing on the capacitor bank to leak away and for the normal bidirectional system ac voltage to be impressed across the open contacts. The initial unidirectional voltage across the open contacts does, however, impress an unusual stress on the withstand ability of the contact gap. For most of a vacuum capacitor switch's operating life it will hold off the unidirectional voltage, but once in a while, a restrike will occur. This phenomenon is not unique to vacuum switching technology. Restriking can also occur with all other switching technologies such as SF_6 and oil. In fact, one study by the Kansas City Power and Light Utility states that there have been so many problems with

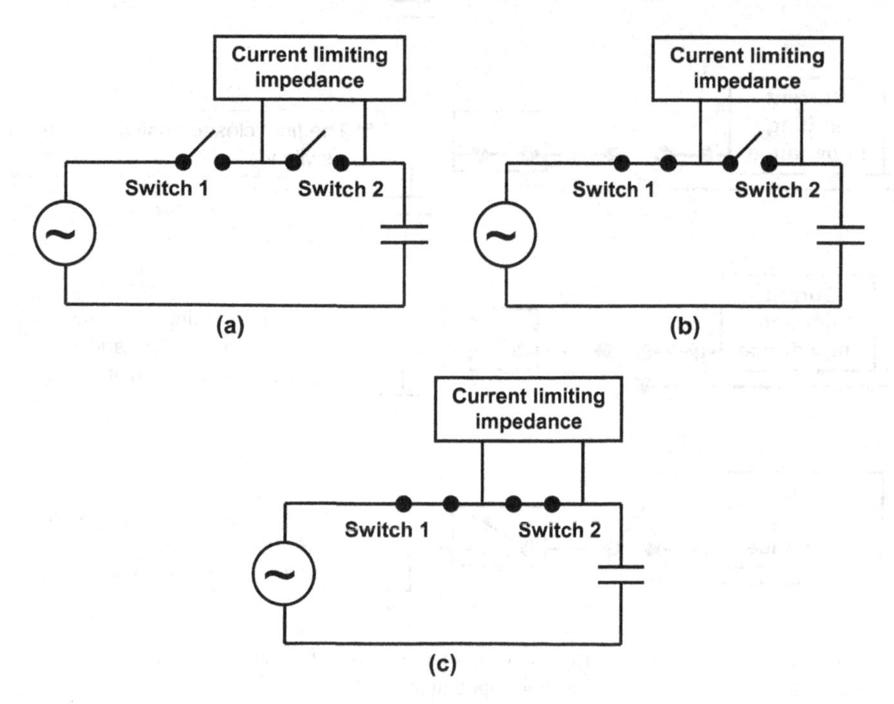

FIGURE 5.43 A scheme for connecting a current limiting impedance for connecting a capacitor bank: (a) both switches open, (b) switch 1 closes while switch 2 remains open, the capacitor bank is connected with a limited inrush current and (c) switch 2 is closed shorting the current limiting impedance.

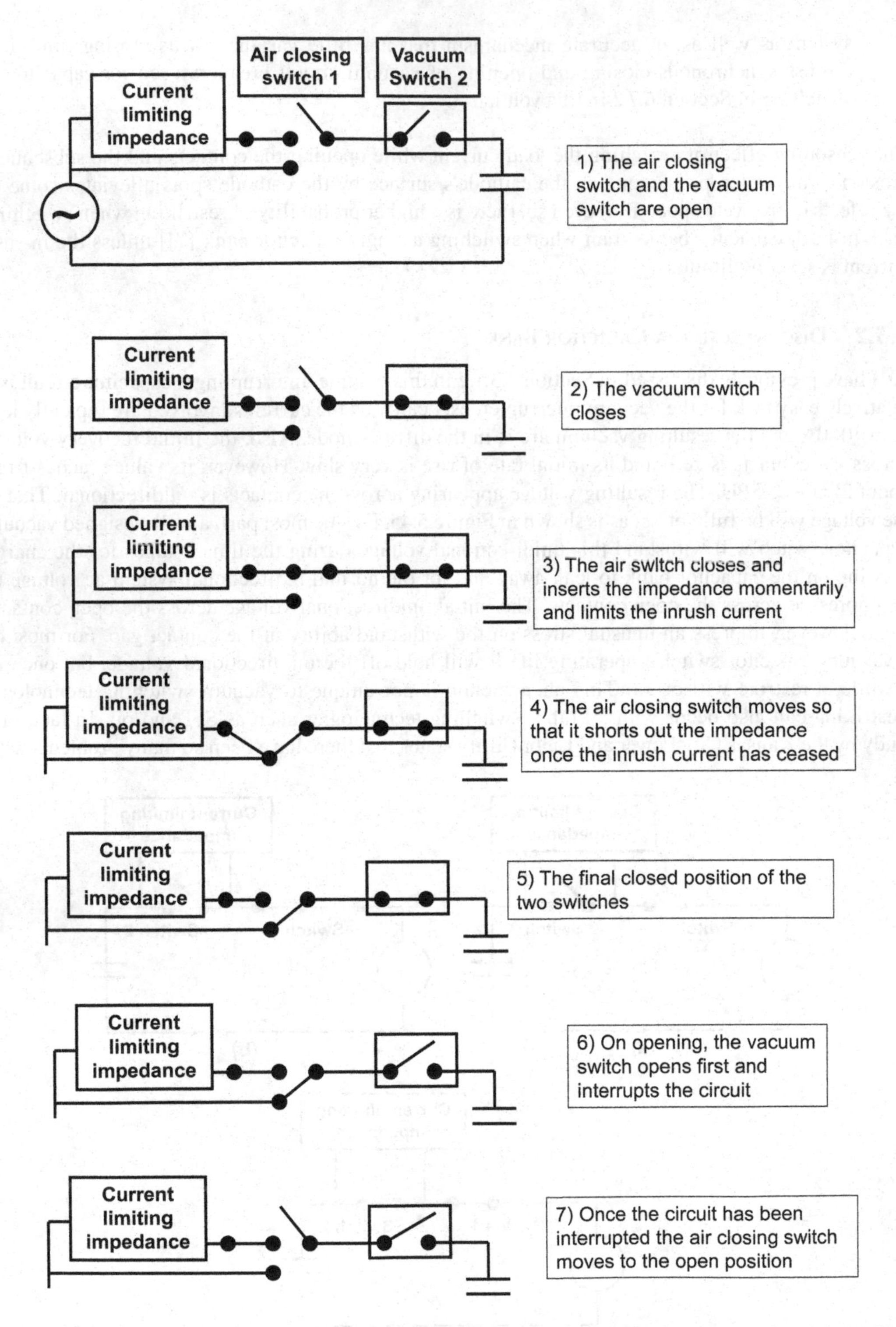

FIGURE 5.44 A schematic showing the use of a vacuum capacitor switch in series with an air, making switch to momentarily insert a current limiting impedance.

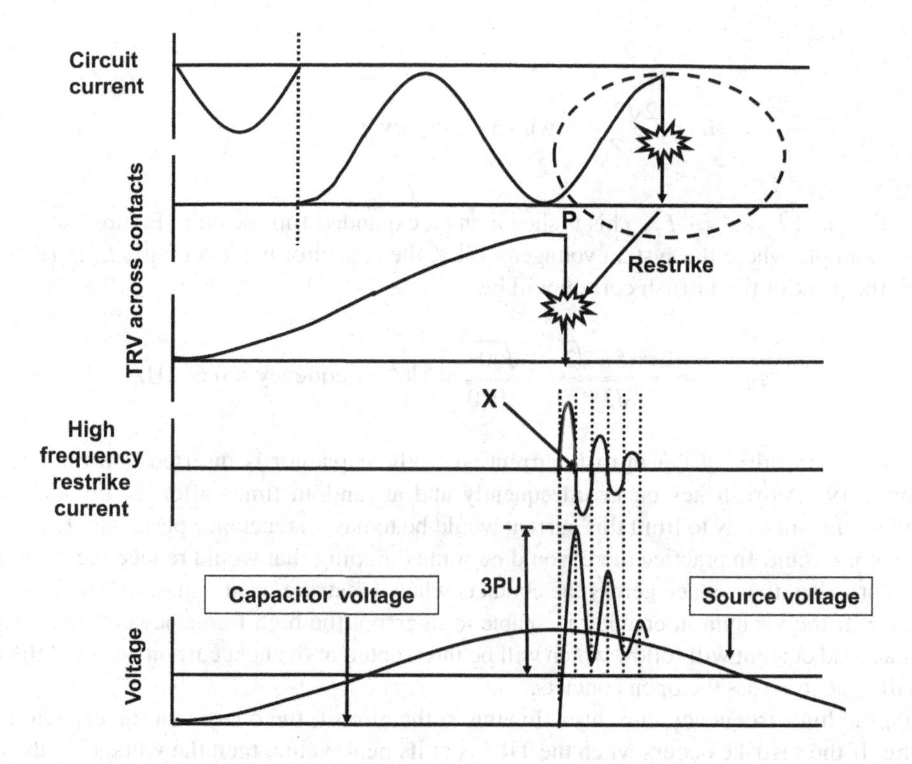

FIGURE 5.45 Schematic showing the effect of a restrike at the capacitor switch when recovery voltage is at its peak value.

their installed base of oil capacitor switches that they are gradually replacing them with vacuum capacitor switches [72].

Let us now consider the effect of a restrike [73–75]. When the restrike occurs, the equations for the transient current are [73]:

$$\sqrt{2}U_{\ell-n}\cos(\omega t) = U_C + L_T\frac{di_{R3}}{dt} \tag{5.28}$$

$$U_C = U_C(0) + \frac{1}{C}\int i_{R3}dt \tag{5.29}$$

where the voltage on the capacitor is U_C and the restriking current is i_{R3}. If we now assume that the restrike occurs at the peak value of the supply voltage $[\sqrt{2}U_{(\ell-n)}]$: i.e., at the peak of the restored voltage, point 'P' in Figure 5.45, and it remains more or less constant during the high frequency flow of current, then:

$$i_{R3} = \left[\sqrt{2}U_{\ell-n} - U_C(0)\right]\times\left(\frac{C}{L_T}\right)^{1/2}\sin(\omega_0 t) \tag{5.30}$$

where $\omega_0 = 1/(L_TC)^{1/2}$. The quantity $[\sqrt{2}U_{(\ell-n)} - U_C(0)]$ is the voltage across the contacts when the restrike occurs $[U_{contacts}]$. Here, for a single grounded capacitor bank its value is $2 \times \sqrt{2}U_{(\ell-n)}$. Thus, the peak value of the in-rush current i_{R3} and its frequency would be:

$$i_{R3} = \left[2\sqrt{2}U_{\ell-n}\right]\times\left(\frac{C}{L_T}\right)^{1/2} \tag{5.31}$$

or

$$i_{R3} = \frac{2\sqrt{2}U_{\ell-n}}{Z_C} \text{ with a frequency } f = \frac{1}{2\pi\left(L_T C\right)^{1/2}} \tag{5.32}$$

$Z_C = (L_T/C)^{1/2}$ and $L_T = L_S + L_C$. This is shown on an expanded time scale in Figure 5.45. Using the previous example where the system voltage is 15kV, the capacitor bank is 60 µF, L_S is 1mH and L_C is 20µH, the peak of the in-rush current will be:

$$i_{R3(\text{peak})} = \frac{2 \times 15 \times \sqrt{2}}{\sqrt{3}} \times \frac{\sqrt{60}}{\sqrt{1020}} \approx 6\,\text{kA}, \text{ frequency} \approx 0.65\,\text{kHz} \tag{5.33}$$

This is twice the value of the in-rush current when the capacitor is inserted into the circuit; see Equation (5.19). As restrikes occur infrequently and at random times after the circuit current is interrupted, the only way to limit this current would be to have a reactance permanently installed in the capacitor circuit. In practice there would be some damping that would reduce the magnitude of this current. Also, the voltage across the contacts when a restrike occurs may well be less than the peak value. If the vacuum interrupter is unable to interrupt the high frequency current, then a half wave of ac load current will follow which will be interrupted at the next current zero and the normal TRV will appear across the open contacts.

While the high frequency current is flowing in the circuit, the charge on the capacitor is also changing. If the restrike occurs when the TRV is at its peak value, then the voltage on the capacitor is:

$$U_C = -\sqrt{2}U_{\ell-n} + 2\sqrt{2}U_{\ell-n}\left(1 - \cos\omega_0 t\right) \tag{5.34}$$

which has a peak value of $+3x\ \sqrt{2}U_{(\ell-n)}$. If the high frequency current is not interrupted, then the charge left on the capacitor will decrease as the current decreases as is shown in Figure 5.45. A vacuum interrupter using Cu–Cr contacts does, however, have a reasonable ability to interrupt high frequency currents of this magnitude if not at the first current zero then certainly later when the value of the current has decayed to a lower value. The worst case would be when the high frequency current is interrupted at the first current zero; "X" in Figure 5.45. It is instructive to consider this case and the potential for voltage escalation in the circuit. At the time the transient current passes through its first zero in Figure 5.46, the transient voltage reaches its peak value of about 3PU and this becomes the new value of voltage residing on the capacitor. Thus, the voltage across the contacts is:

$$U_{\text{contacts}} = 1PU\left(\text{source side}\right) - 3PU\left(\text{capacitor side}\right) = -2PU \tag{5.35}$$

If the contact gap does not breakdown, the source side will continue to follow the ac voltage wave so that half a cycle later there will be −4PU across the contacts. If a second restrike occurs again at this peak voltage a high frequency current will again be initiated whose magnitude will be about twice that given in Equation (5.32). If this current is interrupted at the first high frequency current zero (as is shown in Figure 5.46), the capacitor voltage will now swing to −5PU. Now the load side voltage will increase until 6PU will appear across the vacuum interrupter's contacts. Theoretically the voltage on the capacitor can build up according to a series 1, 3, 5, 7… etc. if the high frequency current is always interrupted at the first current zero and the breakdown always occurs at the maximum voltage across the vacuum interrupter's contacts. As stated before, this is a worst-case sequence.

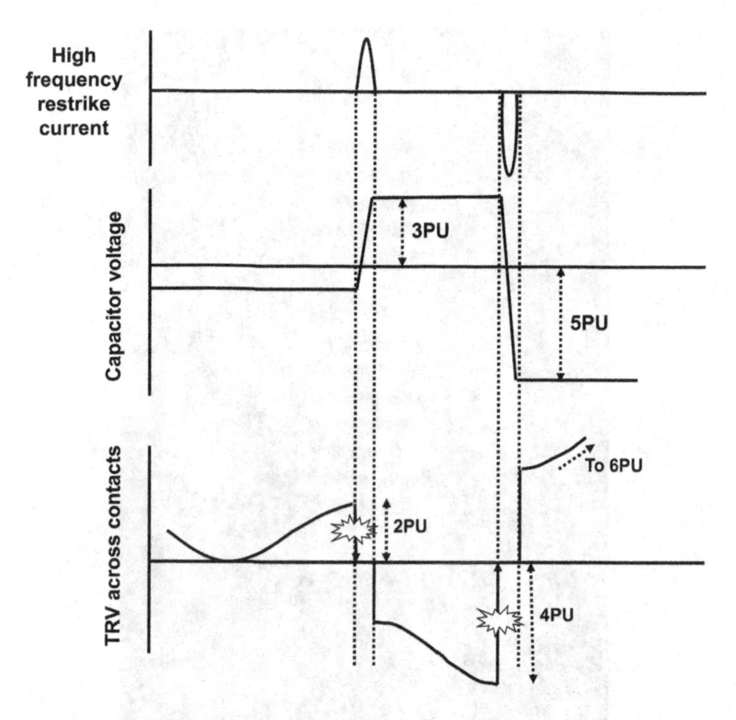

FIGURE 5.46 Schematic of a worst-case scenario when the first restrike at the capacitor switch occurs at the peak of the recovery voltage and the high frequency current is interrupted at its first current zero. It also shows the possible further voltage escalation events.

In practice the probability of such a sequence of restrikes is rather low. First of all, a restrike will not necessarily occur at the peak of the recovery voltage and the high frequency current will not necessarily be interrupted at the first current zero. There is even the possibility for the voltage trapped on the capacitor being reduced if the current is interrupted at the second current zero shown in Figure 5.45. Also, in the above example there is no damping of the current. In a practical circuit, for example, it will be considerably damped to give 60% of the theoretical voltage overshoot [74, 75]. Thus, the voltage escalation will develop more slowly. There is, however, the opportunity to develop high voltages, which can damage the capacitor bank or even the insulation of adjacent electrical equipment. It is also possible that they can initiate phase-to-phase or phase-to-ground faults. Therefore, it is always wise to protect a capacitor bank with a device, which will limit the voltage it may be exposed to [76, 77]. The IEEE red book states [78]:

> *Overvoltage protection should be considered whenever shunt capacitor banks are installed. The possibility of overvoltages from lightning, switching surges, and temporary overvoltages requires a detailed evaluation to determine the duty on arrestors applied to the vicinity of a shunt capacitor bank. Due to the low surge impedance of large high-voltage capacitor banks, it may not be necessary to add arrestor protection beyond that which already exists in the substation. However, additional protection may be needed to protect equipment from overvoltages due to capacitor switching or the switching of lines or transformers in the presence of capacitors. (IEEE Std. C62.22–1991)*

An example of a successful capacitor switch is shown in Figure 5.47. This switch contains a vacuum interrupter with W–Cu contacts. The usual operating mechanism is a charged spring mechanism that can be remotely operated by an electric motor. A lineman can also operate it using an insulated pole. There are some designs that use a solenoid and also magnetic actuators. The device shown in Figure 5.47 has the capability of interrupting a capacitor circuit with a peak recovery voltage of

FIGURE 5.47 A practical capacitor switch using a vacuum interrupter with Cu–W (10wt.%) contacts [courtesy Joslyn Hi-Voltage].

75kV. It can also be used for back-to-back capacitor switching if the peak of the high frequency inrush current is limited to 6kA.

Certification testing for such a switch can be found in IEEE Standards C37.66 – 1969 (R1980) and in IEC Standards 62271-103. The testing sequence for the IEEE standards is given in Table 5.11. As you can see, the test requires 1200 switching operations. It recognizes that the occasional restrike will occur, but to pass the Class A certification, only two are permitted and when they occur, the voltage appearing on the system must be less than 2.5PU.

5.5.3 Switching Three-Phase Capacitor Banks

The preceding description of switching a single capacitor bank serves as a useful introduction to illustrate the closing current transients and the effect of restrikes after interrupting the capacitor circuit. With increasing frequency capacitor banks are being applied to three-phase distribution systems for power factor correction. These three-phase capacitor banks can be switched relatively frequently; e.g., once or twice a day [60]. The effect on a capacitor switch operating such a bank can result in some significant differences and it is important for the vacuum interrupter designer to be familiar with them. First of all, if the neutral of the three-phase capacitor bank and the source neutral are solidly connected (i.e., a grounded capacitor bank), this is then the same as three independent single-phase circuits and the analysis presented in Sections 5.5.1 and 5.5.2 can be applied directly. If, however, the capacitor bank neutral is isolated from the source (i.e., an ungrounded capacitor bank) as is commonly the case, then the transients on connecting and disconnecting the capacitor banks can be somewhat different.

TABLE 5.11
An Example of the IEEE (C37.66) Testing for a Capacitor Switch Used for 15.5KV Outdoor Application

Dielectric performance for a capacitor switch			
	1 minute power frequency withstand tests, kV (rms)		
Rated maximum voltage, V (rms)	**Dry**	**Wet**	**BIL, kV (crest)**
15.5	**50**	**45**	**95**

Current ratings for capacitor switches applied in circuits with a time constant < 45ms

	Power frequency current ratings			High frequency inrush current ratings	
Capacitor switching, A	Symmetrical making, kA (rms)	Peak withstand, kA (peak)	Short time symmetrical withstand, kA (rms)	High frequency transient making current, kA (peak)	Frequency of transient inrush current, kHz
200	10	26	5	10	6
200	12.5	32.5	6.25	10	6
400	10	26	5	10	6
400	12.5	32.5	6.25	20	4
600	10	26	5	10	6
600	12.5	32.5	6.25	24	4

Switching tests; random opening on the ac wave (at least four at max. current and four close to current zero for both polarities)

400 at 100% rated current

400 at 45–50% rated current

400 at 15–20% rated current

1200 switching operations and 10,000 no-load operations

Capacitor switch classifications

Class A: 0.2% probability of restrike in the 1200 operations (i.e., 2 restrikes in 1200 operations): a ‘very low’ probability of restrike

Class B: 2% probability of restrike in the 1200 operations (i.e., maximum of 24 restrikes in 1200 operations): ‘low’ probability of restrike

Class C: greater than 2% probability of restrike in the 1200 operations (i.e., > 25 restrikes in 1200 operations): ‘moderate to high’ probability of restrike

Case 1: The three phases of the capacitor switch close and open at the same time (i.e., synchronous closing and opening)

On closing, the initial in-rush current i_{RI} in each phase will depend upon the point on the ac voltage wave at which the current is initiated. The peak value of the current will be similar to that given in Equation (5.19) for single bank switching and Equation (5.25) for back-to-back capacitor bank switching. On disconnecting a three-phase ungrounded capacitor bank, the interrupter for the first phase to clear (e.g., phase A) will experience a greater recovery voltage across it than will the interrupters in phases B and C [73–75]. The maximum values will be:

Phase A: $2.5 \times \sqrt{2}U_{(\ell-n)}$, occurring 90° after phases B and C interrupt and every 360° afterwards.
Phase B: $(1 + \{\sqrt{3}/2\}) \times \sqrt{2}U_{(\ell-n)}$, occurring 210° after phases B and C interrupt and every 360° afterwards.
Phase C: $(1 + \{\sqrt{3}/2\}) \times \sqrt{2}U_{(\ell-n)}$, occurring 150° after phases B and C interrupt and every 360° afterwards.

Case 2: The three phases of the capacitor switch do not open and close at the same time

In this case, it is possible to control the closing operation such that the closing current transient is minimized [65–68]. On disconnecting a capacitor bank, the peak voltage across a given phase of the switch can be substantially higher than those given above for the synchronous opening case [73, 75]. If, for example, phase A (the first phase to clear) opens while phase B and C remain closed, then the peak value of the voltage across the interrupter in phase A will be 3PU, which will occur 180° after the current zero in that phase. If phases A and B open with phase C closed, then the peak value across the interrupter in phase A will be 4.101 PU. The same voltage will be reached on phase C with phase A closed and phases B and C open. If phases B and A are open and phase C closed, the recovery voltage across phase B will be 3.46 PU. Finally, if phase A is open and phase B is closed, then the recovery voltage across the interrupter in phase C will be 3.46 PU. It is thus advisable when operating a three-phase capacitor switch to make sure the three phases are timed to open as closely as possible to each other. This is not always possible, especially if the switch is designed to be opened manually by a human using a "hot stick." If such a three-phase operation is expected, then the vacuum interrupters must be designed to withstand the much higher recovery voltages presented above. From Greenwood's analysis [73], a restrike in phase A would result in a voltage of −4.134 PU on capacitor B and −5.866 PU on capacitor C if the high frequency current is interrupted at the first current zero. A practical circuit would have some considerable damping. The resulting voltages would be reduced by perhaps 60%. However, high enough voltages might be left that could result in insulation failure.

5.5.4 The Capacitor Switch Recovery Voltage, Late Restrikes, and NSDDs

The peak voltages impressed across the contacts of the vacuum interrupters in the capacitor switch are, in general, much lower than one minute withstand and BIL design voltages. Even so, some electron emission from the cathode would still be expected, because the microscopic enhancement factor, β_m, would increase the electric field at the cathode contact's microprojections: see Section 1.3.2, Equation (1.39) in this volume. Table 5.12 shows the peak values of the voltage across the first phase to clear for an ungrounded capacitor bank compared to a vacuum interrupter's designed one-minute ac withstand voltage and its designed BIL withstand voltage for several common system voltages. It can be seen that the designed voltage withstand values are generally much higher than the voltages impressed after switching capacitor banks. In spite of this fact, restrikes do occasionally occur when switching capacitor circuits. An illustration of the complete sequence of a single phase, back-to-back insertion of a capacitor bank is shown in Figures 5.48–5.50. During the closing of the vacuum interrupter's contacts, Figure 5.48 shows a prestrike arc between the contacts before

TABLE 5.12
Peak Voltages from Capacitor Switching Compared to Vacuum Interrupter Design Voltages

System Voltage (kV)	Maximum Peak Voltage for 3 Phase Capacitor Switching (2.5PU) (kV)	Peak Open Circuit Voltage, 3 Phase Ungrounded System (kV)	BIL Voltage (kV)	Peak, One Minute Withstand Voltage (kV)	Peak Transient Recovery Voltage, First Phase to Clear, 3 Phase Ungrounded System (kV)
12	**24.5**	9.9	75	40	20.6
15	**30.6**	12.2	95	51	28
17.5	**35.7**	14.3	95	54	30
24	**49.0**	19.6	125	71	41.2
27	**55.1**	22	125	85	51
36/38	**77.6**	31	170	113	71

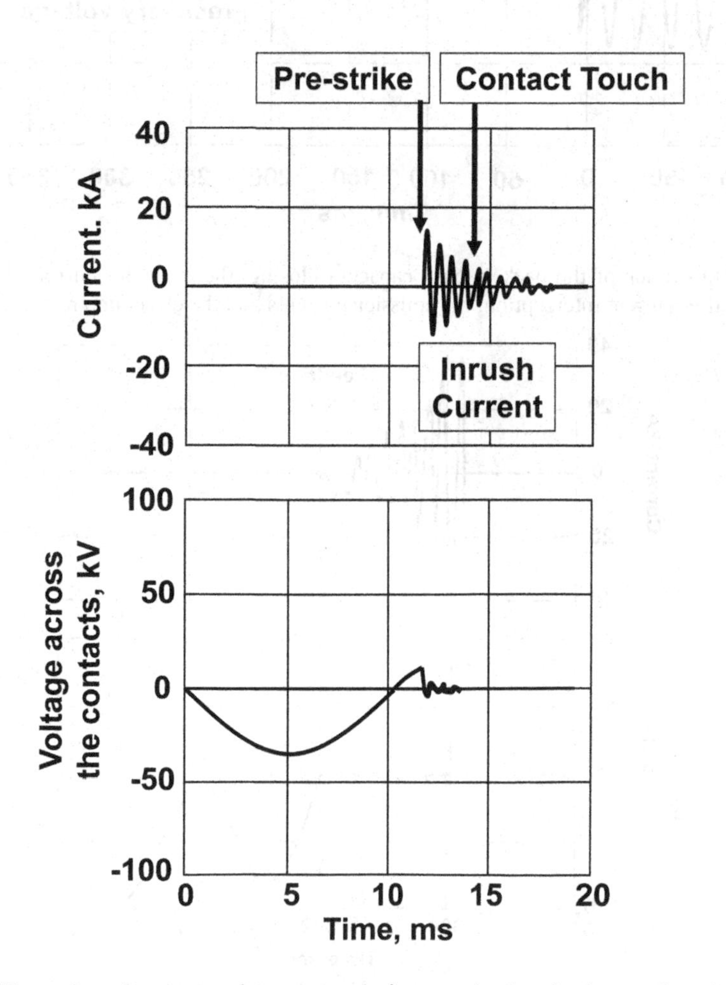

FIGURE 5.48 Illustration of a vacuum interrupter closing on a back to back capacitor circuit showing the high-frequency current as the contacts close and make contact.

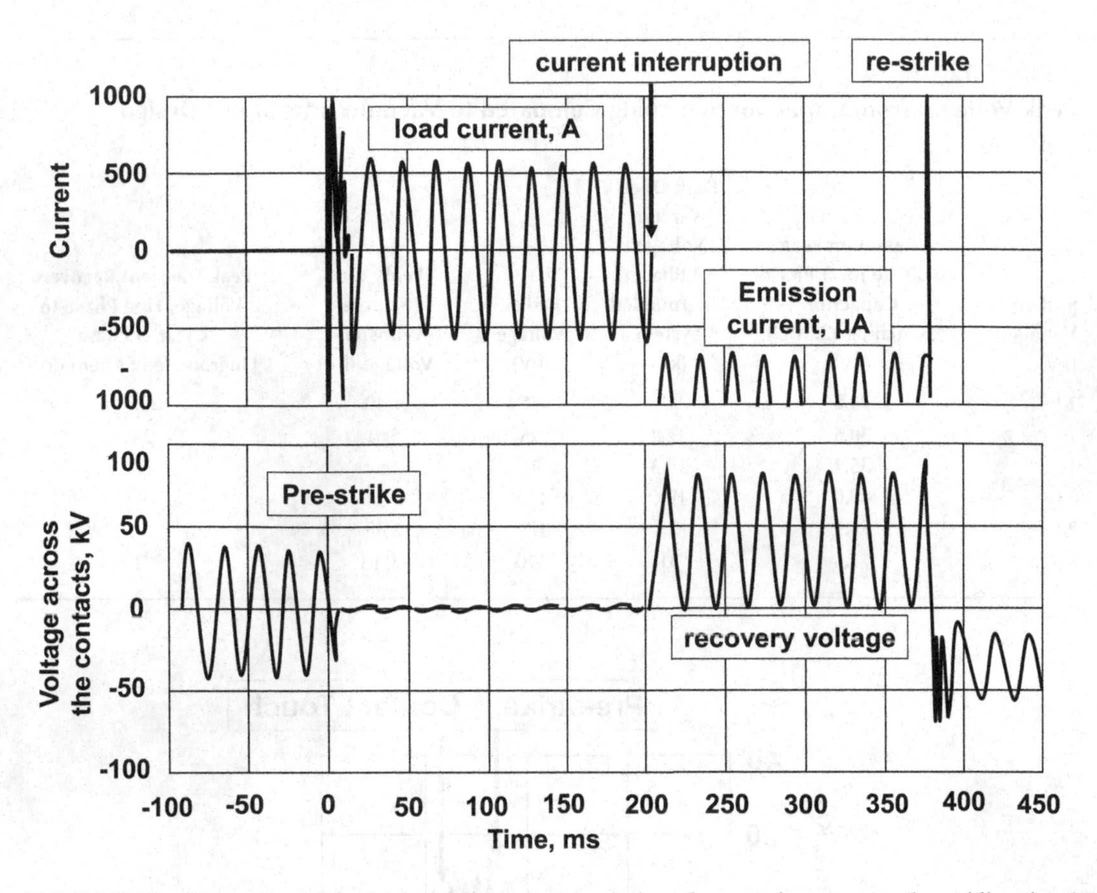

FIGURE 5.49 Illustration of the back to back capacitor closing, the capacitor current, the unidirectional recovery voltage after current interruption, the emission currents and the effect of a restrike.

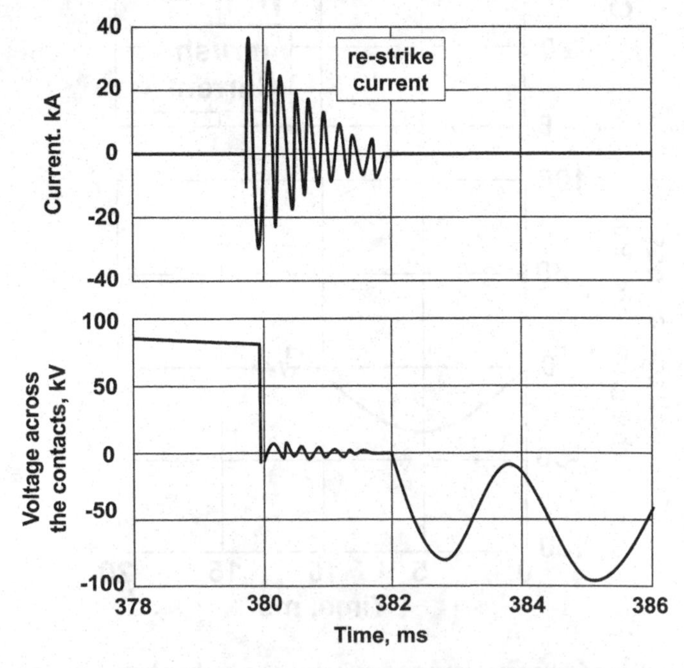

FIGURE 5.50 Illustration of the restrike current decaying and the subsequent recovery voltage across the vacuum interrupter's contacts.

they touch. This initiates the high frequency closing current with a peak of 36kA. The passage of its load current of about 550A, the interruption of the load current, the unidirectional recovery voltage, the emission currents of about 400μA, and a restrike 4 cycles later are shown in Figure 5.49. Figure 5.50 illustrates the restrike current which is not interrupted at the first current zero and the later recovery voltage which is much less that the 4 PU shown in Figure 5.46.

The relevant observations determined during capacitor switching using vacuum interrupters using Cu–Cr contacts are:

1) When opening a vacuum interrupter to disconnect a capacitor bank, a unidirectional alternating voltage is impressed across the open contacts. The peak value of this voltage can range from 2PU to 4PU depending upon the current and the switch's mechanical design
2) The delayed or late restrikes have been observed at random times after current interruption from just a few, recovery voltage cycles to up to 10 seconds later [79–81] The cathode contact during the recovery period produces field emission currents ranging from a few hundred microamperes to a few milli-amperes. These emission currents are maximum at each peak of the unidirectional voltage [79–82]
3) The probability of a delayed restrike occurring increases as the circuit voltage increases
4) A high frequency, high current is initially impressed upon the switch contacts when inserting a capacitor bank. This high current can result in some weld spots on the contact surfaces, which in turn can cause an increase in the enhancement factor and result in higher emission currents [9, 82]
5) The probability of late restrikes increases as the high-frequency in-rush current increases [83]
6) Observations of the open contact gap after a vacuum interrupter has interrupted the capacitor current shows that the delayed restrike initiates with the metal vapor plume from the anode contact [84]
7) The intensity of the emission currents decreases as the number of vacuum interrupter capacitor switching operations increases: See Figure 5.51 [81]
8) When interrupting a typical capacitor bank current, the resulting vacuum arc is in the diffuse mode with multiple cathode spots moving in a retrograde motion: See Section 2.3.1 in this volume. The cathode spots condition the cathode contact [79, 81]

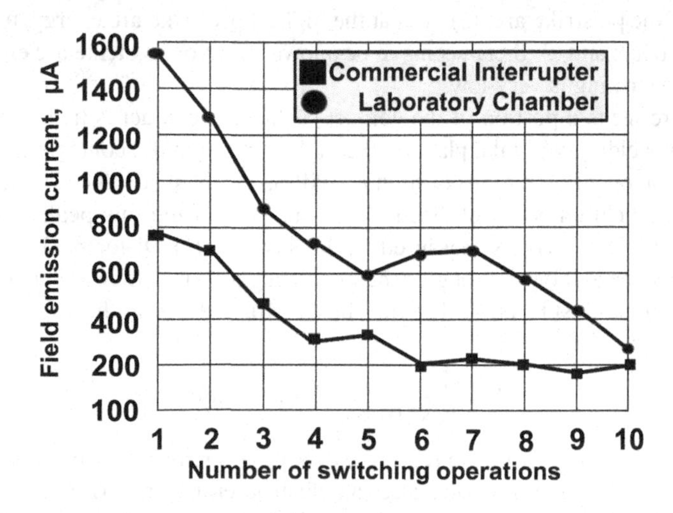

FIGURE 5.51 Decrease in the average emission current as a function of capacitor switching operations (10 × 10 measurements, inrush current 6kA at 1.5 kHz, interrupting current 20A and the peak recovery voltage 75kV) [81].

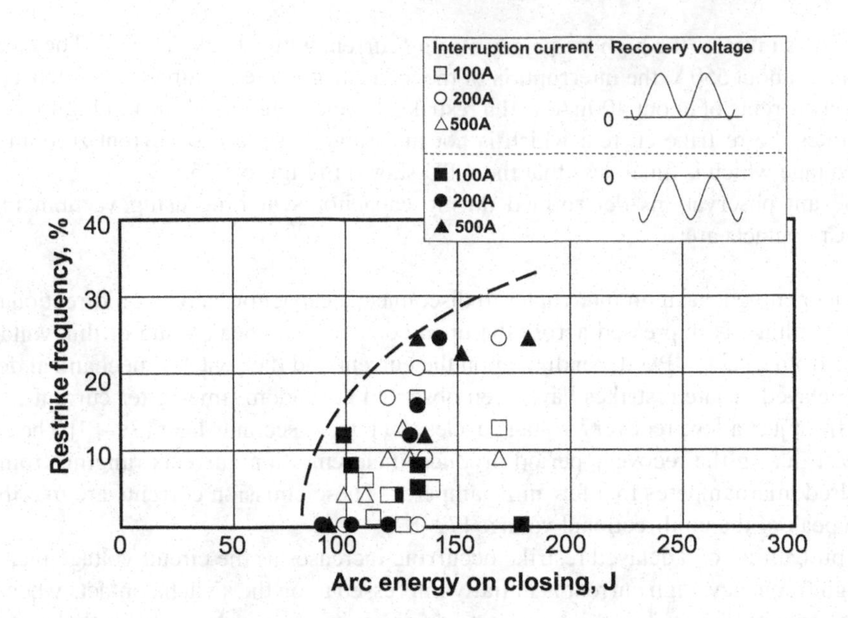

FIGURE 5.52 The probability of restrikes while switching a capacitor bank with Cu–Cr contacts after disconnecting a capacitor bank as a function of the prestrike arc energy measured while connecting the capacitor bank with an inrush current of 6.3kA (peak) as a function of the disconnecting current [79].

9) Microparticles form about 80% of the cathode erosion from the diffuse vacuum arc: See Section 2.3.2 in this volume

Experiments by Juhász et al. [79] show that late restrikes occur any time from the first peak of the recovery voltage up to 10 seconds later. They also show that after interrupting the capacitor circuit an emission current of between 0.1mA and 8mA from the cathode contact is possible. In their experiment they connect a capacitor bank with a high peak making current and they measure the energy in the prestrike arc. They then plot the restrike frequency as a function of the energy of prestrike arc; see Figure 5.52. This figure shows that: (1) the frequency of restrikes is dependent upon the energy of the prestrike arc; (2) even at the highest prestrike arc energy, there is only a 30% probability of a restrike; and (3) there seems to be a lower limit of prestrike arc energy below which the probability of restriking is very low.

The prestrike arc melts a portion of the contact surface. If contact is finally established in this melted region, then welding will take place as the molten metal in the contact region freezes. When the weld is broken on opening, regions of high β_m will be established. These higher β_m regions will become sites for the field emission of electron current even at the low peak of the TRV voltage. Figure 5.53 shows typical currents measured by Juhsász et al. [79] for peak recovery voltage of 40kV. Unfortunately, the authors do not give their contact gap, but as they are operating their experiment in a 24kV circuit it must be about 12mm. Thus, at 40kV the gross field E in the contact gap is:

$$E = \frac{40 \times 10^3}{12 \times 10^{-3}} = 3.3 \times 10^6 \text{ V.m}^{-1} \tag{5.36}$$

From Chapter 1 in this volume it would require a $\beta_g \times \beta_m$ of greater than 10^3 to initiate an immediate breakdown. As can be seen in Figure 5.53, an electron emission current is observed each time the voltage across the contact gap increases close to its maximum value; but the breakdown of the contact gap can occur at any time up to 10 seconds after current interruption. The intensity of this emission depends upon the impressed voltage and the effects of field enhancement at the cathode

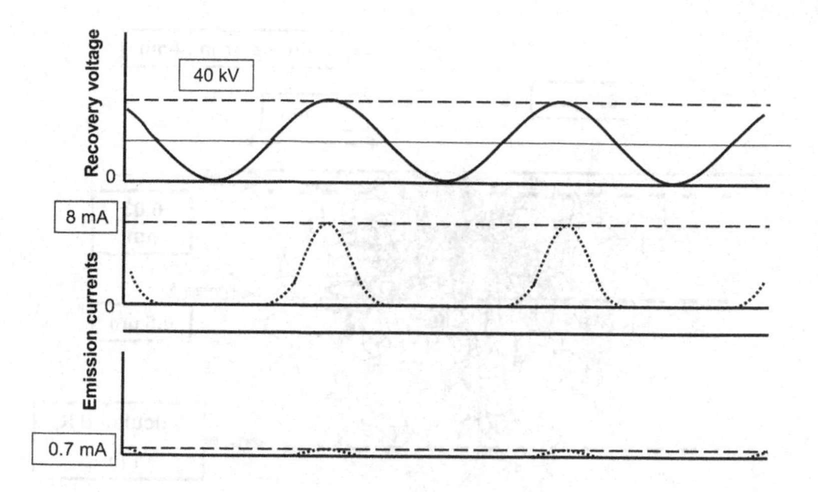

FIGURE 5.53 Examples of measured emission currents between Cu–Cr contacts after connecting a capacitor bank with an inrush current of 6.3kA(peak) and disconnecting it [79].

surface. An example of the emission currents observed during the recovery period after interrupting a capacitor circuit from the moment of current interruption to the development of a delayed restrike is shown in Figure 5.54 from the paper by Koochack-Zadeh et al. [81]. This figure shows a delayed restrike after at the eighth peak of the recovery voltage together with the value of each emission current up to that time. Slade et al. [85] use this data to analyze the possibility that this sequence of emission currents can result in the delayed restrike. They assume for their analysis that there is only one region on the cathode from which the emission current is emitted. When the emission current reaches the anode contact during capacitor switching it has a peak energy of the peak value of the recovery voltage in eV. The electrons will not reside at the anode's surface but will penetrate into the anode. The maximum depth of penetration R_p is (see Section 1.3.3.2 in this volume):

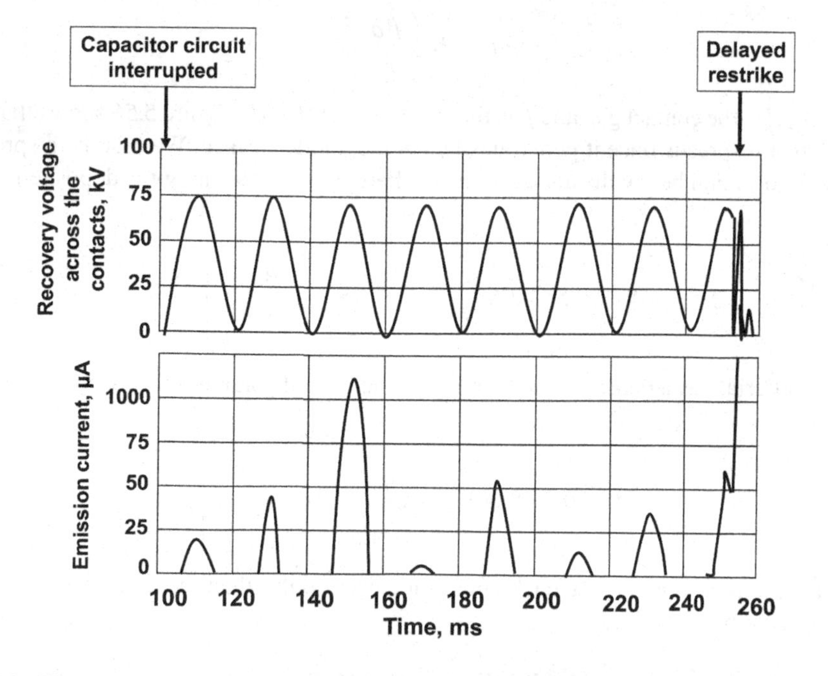

FIGURE 5.54 The recovery voltage, emission currents and delayed restrike after capacitor switching from [81].

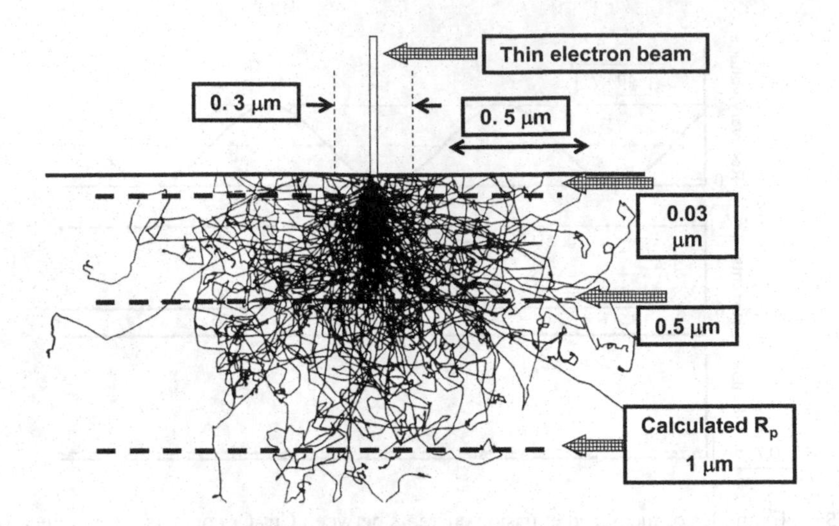

FIGURE 5.55 A Monte Carlo simulation of a thin electron beam incident upon an iron anode.

$$R_p \approx 0.06 \frac{W^{Y}}{\delta} \quad (5.37)$$

where R_p is in µm, W is the electron energy in keV, δ is the material's density in g.cm^{-3}, and Υ ranges from 1.2 to 1.7. [A value of 1.67 is used for these calculations]. Figure 5.55 presents a Monte Carlo simulation of a narrow 20 keV electron beam interacting with an Fe anode. This figure shows that the thin beam expands as it interacts with the metal below its surface. Most of the electron beam's energy is deposited below the anode's surface, in a cylinder whose depth is about $0.5R_p$. The emission current comes from a small region of the cathode given in with radius r_e. The electron emission beam spreads as it crosses the contact gap so when it reaches the anode it has a radius r_a given by:

$$r_a = 2r_e \left(\frac{\beta d}{r_p} \right)^{\frac{1}{2}} \quad (5.38)$$

where $r_p \approx 2r_e$, d is the contact gap and β is the enhancement factor. Figure 5.56 shows that the emission current beam spreads once it penetrates below the anode surface. We assume a spread $n \times r_a$. The volume V_a of metal below the anode surface where most of the energy is deposited is:

$$V_a = 0.5 \times 0.06 \frac{W^{Y}}{\delta} \times \pi \times \left(n \times 2r_e \left(\frac{\beta d}{r_p} \right)^{\frac{1}{2}} \right)^2 \quad (5.39)$$

The contact material has a density δ so the mass of the heated volume is:

$$M_a = 0.5 \times 0.06 W^{Y} \times \pi \times \left(n \times 2r_e \left(\frac{\beta d}{r_p} \right)^{\frac{1}{2}} \right)^2 \quad (5.40)$$

Assuming adiabatic heating during each emission current pulse then the temperature rise T_r *i*nside the anode is given by:

$$I_e \times W \times t_e = c_p \times M_a \times T_r \quad (5.41)$$

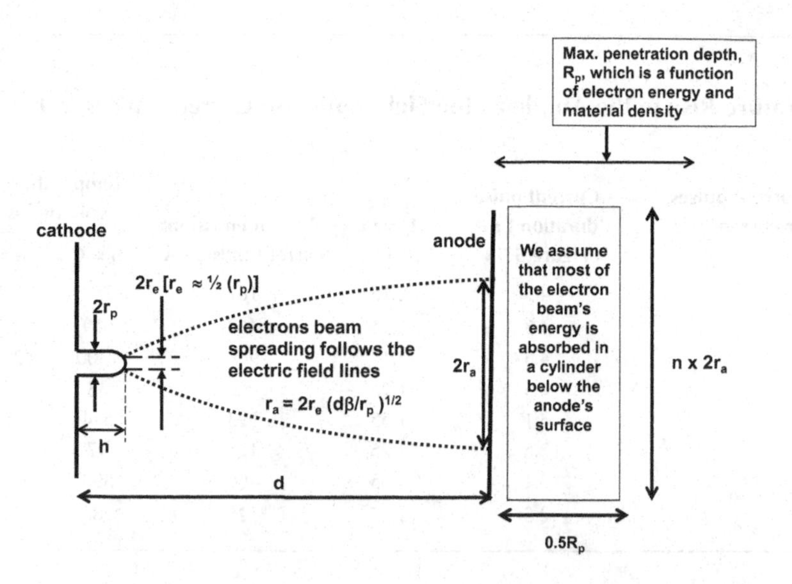

FIGURE 5.56 The spread of the emission current crossing the contact gap and its penetration into the anode.

$$T_r = \frac{I_e \times t_e \times W}{c_p \times M_a} \tag{5.42}$$

where I_e is the emission current, t_e is its average duration, and c_p is the specific heat of the anode material. Using Figure 5.54, I_e is the peak of the emission current, $W = 80kV$, t_e is approximately ½ × the width of the emission current duration. For Cu–Cr (50% Cu) used for the contact material [81], $\delta = 7.5\ g.cm^{-3}$ and $c_p = 0.4$. Using Equations (5.37) and (5.38) and Figure 5.56 we assume that most of the energy from these emission currents will be deposited below the anode's surface to a depth of 12 × 0.5 = 6μm or 6 × 10⁻⁴ cm. The emission current comes from a small area of the cathode given in [80, 81]:

$$A_e = 5.33 \times 10^{-16}\ \text{m}^2 \tag{5.43}$$

with radius r_e:

$$r_e = 13\ nm \tag{5.44}$$

For the emission currents shown in Figure 5.54, $d = 8mm$ and $\beta = 300$, [81], so r_a is:

$$r_a = 26 \times 10^{-9} \left(\frac{300 \times 8 \times 10^{-3}}{26 \times 10^{-9}} \right)^{\frac{1}{2}}\ \text{m} \tag{5.45}$$

$$r_a = 2.5 \times 10^{-2}\ \text{cm} \tag{5.46}$$

The calculated temperature rise T_r for each emission current pulse in Figure 5.54 is shown in Table 5.13, for n = 6, 7, and 8. Using these values of T_r and assuming some temperature drop after each current impulse then Figure 5.57 shows the gradual increase of the anode's temperature in the subsurface. Once the temperature reaches approximately 2200°C, the flux of evaporated coppers atoms could become large enough to lead to a microexplosion and a breakdown [86]. Figure 5.57 shows it is possible that the region below the anode's surface can reach beyond the melting point of the Cu–Cr contact material. When this occurs, metal vapor will erupt from the anode. This vapor

TABLE 5.13
The Temperature Rise in the Anode as the Eight Emission Current Pulses in Figure 5.53 Progress

The order of current pulses after capacitor current interruption	Current pulse duration t ms Figure 5.53	$t_e = ½ t$	Peak of emission current pulse, μA	Temperature rise of heated volume inside anode, C n = 6	n = 7	n = 8
1	4.3	2.15	187	251	209	141
2	4.3	2.15	468	629	523	353
3	4.3	2.15	968	1300	2082	730
4	2.5	1.25	62	38	40	27
5	3.1	1.55	593	574	478	323
6	2.5	1.25	125	97	81	54
7	3.1	1.55	406	393	327	253
8	2.5	1.25	725	585	426	258

will be ionized as it approaches the cathode, breakdown of the contact gap will follow in less than a microsecond, Figure 5.58, and the delayed restrike will be observed: see Section 1.3.5 in this volume. Thus, a compelling argument can be made that the majority of late restrikes periodically observed when switching capacitors are caused by unidirectional recovery voltage and the resulting field emission currents from the cathode and heating of the anode from the high energy electrons. Koochack-Zadeh et al. [80, 81] experimented with Cu–Cr contacts in two interrupter systems: (a) a commercial vacuum interrupter and (b) a laboratory apparatus. They observe that there is a very low

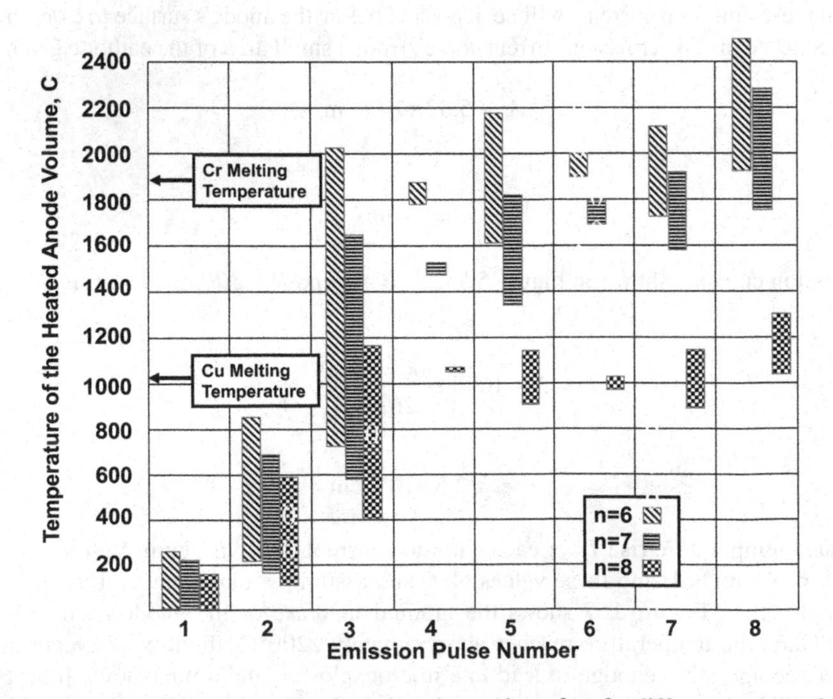

FIGURE 5.57 The calculated temperature rise below the anode surface for different spreads of the emission current inside the anode.

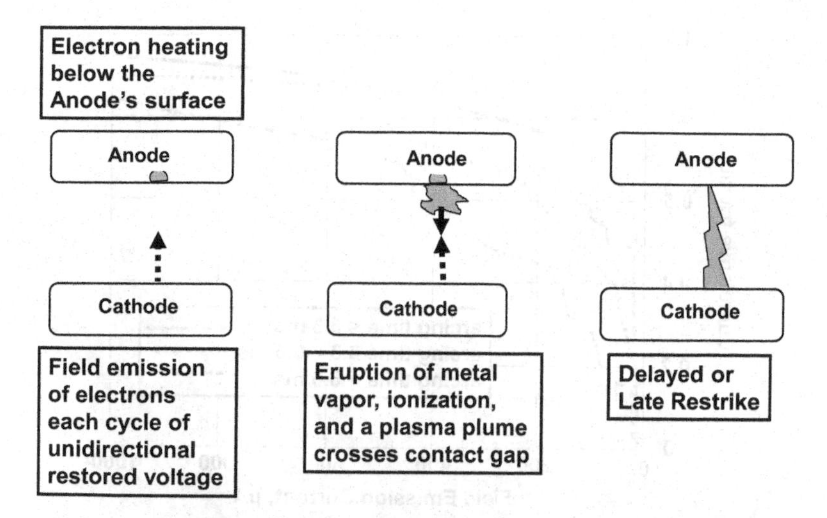

FIGURE 5.58 Sequence from field emission electrons to the delayed or late restrike after capacitor switching.

probability of late restrike using vacuum interrupter (a). In fact, in their experiments no restrikes are seen. Restrikes only occur using the laboratory apparatus (b). The probability of restrikes is 8 in 38 operations for peak emission currents > 0.5mA and 7 in 61 operations for peak values < 0. 5mA.The peak value of the emission currents from (b) were double those from (a). This may have been the result of smoothing the contact Cu–Cr surfaces by the conditioning process during the manufacture of commercial vacuum interrupters.

The randomness of the occurrence of the delayed restrikes can certainly be attributed to the random variability of the emission currents. The authors in [80, 81] observe that the emission currents for both (a) and (b) systems decrease to less than 0.4mA as the number of switching operations increases. This is consistent with the conditioning effect of the cathode spots moving over the contacts from the diffuse vacuum arc. As the emission currents decrease the probability of delayed restrike also decreases. This gives rise to the practical observation "to switch capacitors you need to switch capacitors" [87]. The observation that the delayed restrike initiates with a metal vapor plume from the anode contact [84] can certainly be explained by this analysis. During the interruption of currents below about 1kA the vacuum arc is in the diffuse mode with multiple cathode spots. About 80% of the erosion from these cathode spots is in the form of particles.

While it is possible that some restrikes result from particle impact on the cathode particle breakdown in general is not consistent with the observations presented above. For example: (i) As both (a) and (b) would be expected to produce similar number of particles in each operation why did (a) show no restrikes? (ii) Why did the probability of restriking depend upon the emission current value? (iii) As particle breakdown is usually initiated at the cathode, why do restrikes after capacitor switching initiate at the anode? Juhász et al. [79] observe lower emission currents after interrupting the highest capacitor current of 500A. This again is consistent with the current conditioning effect discussed in Section 1.3.7 in this volume. The increased number of cathode spots from the higher current would have a greater effect on smoothing the cathode surface by current conditioning (see Section 1.3.7.3 in this volume) and thus reducing any projections that may have formed as a result of breaking the contact weld that can form when making contact. Smeets et al. [88] also show the current conditioning effect. The peak emission current is reduced as a function of the arcing time during the interruption of a 400A capacitor circuit: see Figure 5.59.

Unfortunately, Figure 5.54 [80, 81] is the only published data to date that shows a sequence of emission currents leading to an eventual late restrike. Emission currents have been observed by other researchers. Smeets et al. [82, 88] and Sandolache et al. [89] have performed intensive back-to-back capacitor switching tests using a variety of vacuum circuit breakers with 36 kV vacuum

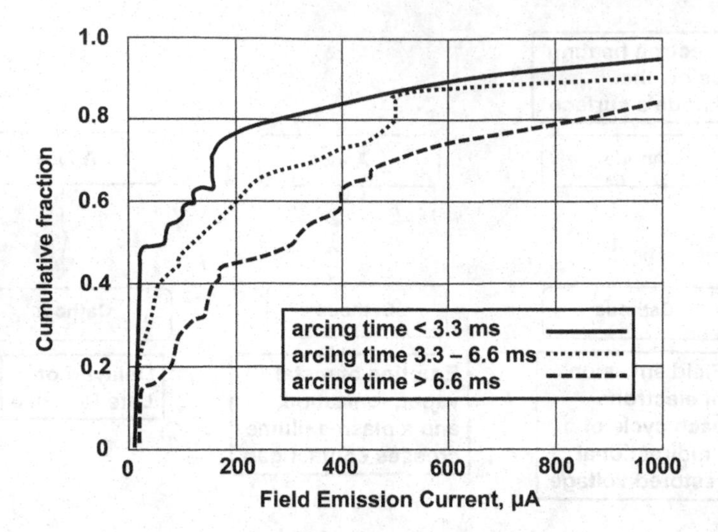

FIGURE 5.59 Cumulative distribution of field emission currents for a 36kV vacuum interrupter making a 4.25kHz, 20kA inrush current during the recovery phase after interrupting the capacitor current of 400A as a function of the interrupting current's duration [82, 88].

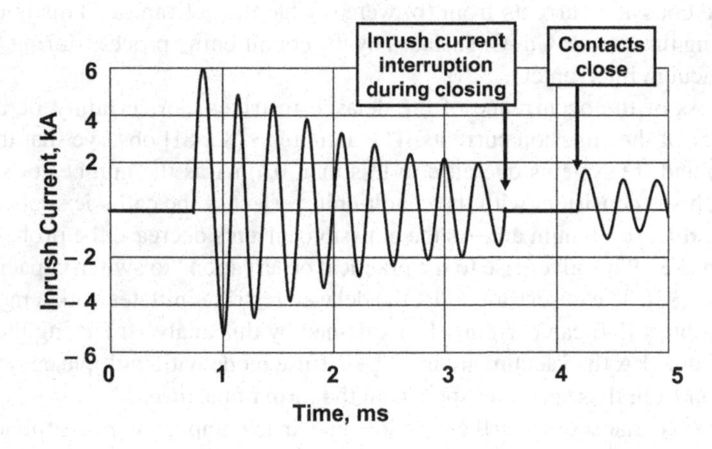

FIGURE 5.60 The decay of the inrush current showing the effect of its interruption as the contacts are closing, but are still apart and its continuation once the contacts close [90].

interrupters. The high-frequency in-rush current used in these tests has an initial peak of 20 kA. The capacitor circuit interruption current is 400A. The measured emission currents observed during the voltage recovery phase have considerable variation. They can range from a few tens of micro-amperes to one or two milli-amperes. The contact bouncing time during contact closure does not appear to affect the capacitor switching performance. Zhang et al. [90] show that the highest current occurs during the prestrike arc as the contacts closing. This prestrike arc may be interrupted as the high-frequency current decays, but once contact is made it continues at the much lower current level: See Figure 5.60. In the experiments by Smeets et al. and Sandolache et al. no correlation is found between the average field emission current and the occurrence of a late restrike.

An example of their observed field emission currents during the recovery phase after switching capacitors is shown in Figure 5.61 [82, 88]. Here the emission currents, after the initial high value, remain at about 300 μA for 20 cycles of the recovery voltage. If only one emission site is assumed,

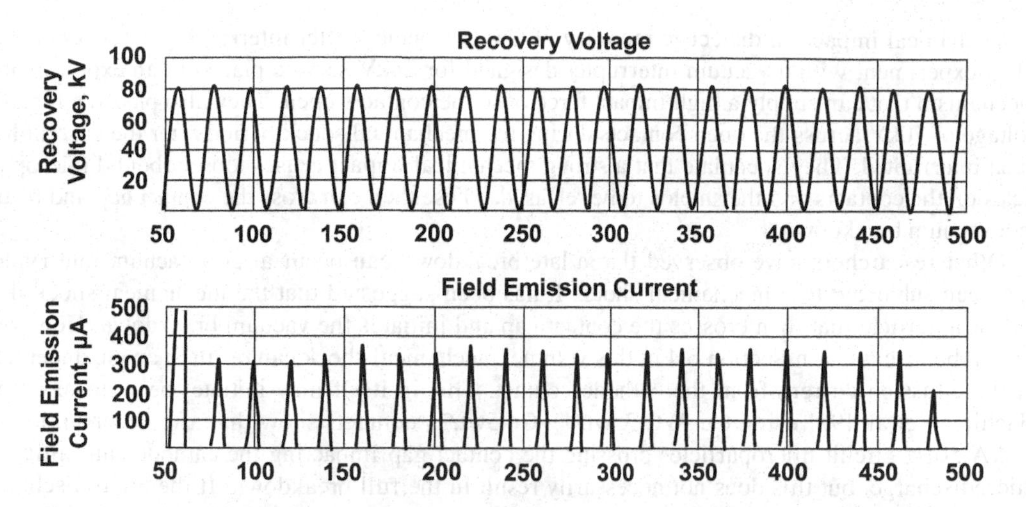

FIGURE 5.61 An example of the field emission current for a 36kV vacuum interrupter making a 4.25kHz, 20kA inrush current during the recovery phase after interrupting the capacitor current of 400A [82, 88].

the analysis presented above would show that the anode would have reached the eruption temperature during about the eighth cycle of recovery voltage. As shown in Figure 1.89 the data shows a best fit if three emission sites are assumed. It is therefore possible that more than one emission site can occur during the vacuum breakdown process. If three emission sites are assumed for the data shown in Figure 5.61, then the average emission current from each site into the anode would now be about 100 μA. The three anode regions receiving the emission currents would now only reach about 1000C at the end of 20 recovery voltage cycles. This temperature is much lower than required for the development of the anode metal vapor plume that initiates the late restrike. So, even though the authors found no direct correlation between the average emission current and the occurrence of the late restrike, they did not take into account the impact of repeated field emission current impulses heating the anode below its surface. Certainly, more analysis of the effect of the repeated emission current pulsed into the anode needs to be undertaken to show that this is indeed a useful criterion for determining the formation of a late restrike.

There is certainly much circumstantial, experimental evidence that points to the importance of the emission current and the formation of the late restrike. For example, there is statistical evidence between the prestriking, high-frequency in-rush current during contact closure and the occurrence of late restrikes during the voltage recovery phase [62, 82]. The in-rush current can result in the formation of minor contact welds. When these welds are broken as the contacts open, high field enhancement sites can form. Donen et al. [83] state the probability of late restrikes increases as the amplitude of the in-rush current increases for Cu–Cr contacts. They explain this by a correlation between the value of the in-rush current and an increase in the effective field emission current.

While Sandolache et al. [89] found no correlation between the average field emission current and occurrence of the late restrike, they postulated that another breakdown process such as particle induced breakdown should be considered. This certainly is a possibility, as the discussions in Sections 1.3.3.3 and 1.3.4 in this volume have shown. Is it, however, the main cause of late restrikes after capacitor switching? When interrupting the typical capacitor circuit currents microparticles are 80% of the main erosion product: see Section 2.3.2 in this volume. Dullni et al. [91] and Schade et al. [92] conclude that microparticles from Cu–Cr contacts play a secondary role in interruption performance. In fact, Dullni et al. [93] go on to conclude that a correlation between the preignition field strengths and the formation of late restrikes suggests that the breakdown is field emission dominate. Indeed, microparticles often come in contact with the cathode without breakdown occurring: see Sections 1.3.3.3 and 1.3.4 in this volume. Be that as it may, Gebel et al. [94] show the effect of mechanical impact on dielectric recovery of Cu–Cr contacts after interrupting a 10kA current. They experiment with a vacuum interrupter designed for 24kV service placed in an experimental mechanism that can supply a high impact force once the contacts open. They also placed a high dc voltage of 70kV across the open contacts during the mechanical shock period after the current has been interrupted. They speculate that a strong mechanical impact causes loosely bound microparticles on the contacts and the shields to be released. These then can cross the contact gap and result in a vacuum breakdown.

Other researchers have observed that a late breakdown can occur after a vacuum interrupter has been subjected to a mechanical shock. It has been suggested that the mechanical shock dislodges a particle that then crosses the contact gap and initiates the vacuum breakdown. However, as will be discussed in Section 6.4 in this volume, mechanical shock can result in an enhancement of the emission current from the cathode contact. This in itself may initiate the restrike event. Hachiman et al. [95] using Cu–WC(35wt.%)–Co(5wt.%) contacts show that after interruption of a 3kA 50Hz circuit microparticles crossing the contact gap impacting the cathode can initiate a microdischarge, but this does not necessarily result in the full breakdown. If the microdischarge is self-limiting (i.e., the charge passed between the contacts is the sum of the charge residing on the contacts resulting from their capacitance added to that resulting from any stray capacitance to ground) and it does not result in the full breakdown of the contact gap and the flow of circuit current, it is termed "a nonsustained disruptive discharge" or NSDD (see Section 6.4 in this volume for a fuller discussion on NSDDs). This type of discharge is sometimes observed in the testing laboratory. It has been determined that an observation of an NSDD during a capacitor switching certification is not a reason to fail the device. In a three-phase ungrounded system, it results in a shift in the neutral voltage, but does not usually lead to any adverse effects [96]. Indeed, in a three-phase ungrounded system, two phases would have to break down at the same time before the circuit current would be allowed to flow again. In a single-phase system or a grounded three-phase system circuit current could, of course, flow if only one phase breaks down and a full vacuum arc is established. However, even in this case the formation of an NSDD does not necessarily lead to the full breakdown of the contact gap, the establishment of a cathode spot, and a reignition of the circuit current.

So, the question remains, what causes the late restrikes after interrupting the capacitor circuit? Microparticle impact at the cathode may result in the late restrike, but there is overwhelming experimental evidence that the field emission currents observed during the recovery phase play a major role its formation. I look forward to further experiments like those of Koochack-Zadeh et al. [81, 82] who show that a series of emission currents results in a late restrike, so that the preliminary hypothesis by Slade et al. [85] can be tested. I would like to **challenge** researchers who have an interest in studying the restrike phenomenon after switching capacitors to determine if the hypotheses of the electron beam causing the eruption of anode metal vapor into the contact gap is valid. This will require experimental work that records the late restrike event visually and equates it with either the field emission current pulses or with a particle impacting the cathode contact. An answer to this question will result in a pivotal paper in the study of capacitor switching using vacuum interrupters.

5.5.5 Switching Cables and Overhead Lines

A cable has a certain amount of capacitance per unit length. Thus, a long length of cable can have a reasonably large capacitance and behave somewhat like a capacitor bank when switched [11, 73, 97, 98]. In this case, however, the capacitance is distributed and is not a lumped capacitor. As I have previously described in Section 5.3, when closing in on an open cable or one with a high inductive load, traveling voltage waves can move down the cable. The point on the ac voltage wave at which the circuit is closed determines the magnitude of these wave fronts. The voltage waves, however, quickly disappear into the power frequency voltage and charging

TABLE 5.14
Preferred Breaking Currents Values for Cable and Line Switching (IEC 62271-100)

Rated Circuit Voltage, kV rms	Rated line-charging breaking current, A rms	Rated cable-charging breaking current, A rms
3.6	10	10
4.76	10	10
7.2	10	10
8.25	10	10
12	10	25
15	10	25
17.5	10	31.5
24	10	31.5
25.8	10	31.5
36	10	50
38	10	50

current. When a cable, with an open, remote end is disconnected, the interrupter switches just the charging current of the cable. For a 15kV circuit this is approximately 1A/km. Thus, the current is small and the testing standards reflect this; see Table 5.14. Because the cable acts similarly to a capacitor, the resulting recovery voltage takes on the form shown in Figure 4.42(c) and Figure 5.45.

Values of peak voltage are given in Table 5.15 for different cable configurations. For a three-phase system with each cable surrounded by its own grounded sheath, the peak TRV will be the same as for a capacitor bank with a grounded neutral; i.e., 2PU. For belted cables (i.e., three cables surrounded by one grounded sheath) the capacitance has a phase-to-phase component as well as a phase-to-neutral component. Overhead lines also have similar characteristics. In this case they both have a combination effect of grounded and ungrounded capacitor banks [73]. The peak TRV for these circuits is thus between 2.2PU to 2.3PU. It is possible for restrikes to occur, but they are unlikely to result in voltage escalation events. In general, load-break switches using load break vacuum interrupters have little problem in passing certification tests for cable switching and line dropping. A restrike during the opening operation could result in voltage surges traveling down the cable, but in my experience voltage escalation rarely, if ever, occurs. Certainly, I have never seen a field failure of a vacuum switch or vacuum breaker when switching a cable or a line. I will discuss the cable and line switching certification testing standards when I again address this subject for vacuum circuit breakers in Section 6.7.3 in this volume.

TABLE 5.15
Peak Voltages for Restrike Free Cable and Capacitor Switching

a) Grounded capacitor banks: 2PU
b) Cables with individual grounded sheaths: 2PU
c) Cables with ungrounded sheaths or overhead lines: 2.2PU to 2.3PU
d) Ungrounded capacitor banks: 2.5PU
e) Non simultaneous 3-phase switching of ungrounded capacitor banks: 2.5PU to 4.1PU

5.6 VACUUM INTERRUPTERS FOR CIRCUIT SWITCHING, CIRCUIT ISOLATION, AND CIRCUIT GROUNDING

5.6.1 Background

There has been considerable experience in using an insulating medium such as air, oil, and SF_6 to isolate the load from an incoming line. Isolation means more than just interrupting the current and then withstanding the resulting TRV and open circuit voltage. To completely isolate the load, the contact gap must also withstand any voltage surges, such as a lightning surge, that may appear on the line as well as completely protect anyone working on the load side of the circuit. In Section 5.2.1, I have already described how a load break vacuum interrupter can be operated in series with an air insulated no-load disconnect switch to provide excellent switching and isolation, see Figures 5.8 to 5.10.

A practical example of a completely enclosed, three-phase vacuum interrupter in series with an air isolation switch is shown in Figure 5.62 [99] The device shown has a rating of 15kV, 800A, and a three-second withstand of 16kA. The drive mechanism connects both the vacuum interrupter and the air isolation switch. Initially, it rapidly opens the vacuum interrupter and then, after a short delay, slowly opens the isolator contacts. The isolator's contacts are coated with a reflective red paint and can easily be observed through the observation window shown. When closing, the air isolation switch closes before the vacuum interrupter. Figure 5.9 gives another example where the vacuum interrupter is inserted into the circuit to only switch the load on and off. The air insulated auxiliary component is used for continuous current flow, overload current withstand, and isolation requirements. As I have discussed in Chapter 1 in this volume, the breakdown of a gap in an insulating medium is mainly dependent upon the medium itself, unlike a vacuum gap where the state of the contacts is the critical component. Thus, an isolation gap in air, oil, or SF_6 can easily be designed to produce excellent isolation performance by making the contact gap larger.

An added advantage of an isolation gap in air is that it can easily be seen. To observe an isolation gap in enclosed systems such as oil or SF_6 a window or perhaps a miniature video camera has to be used.

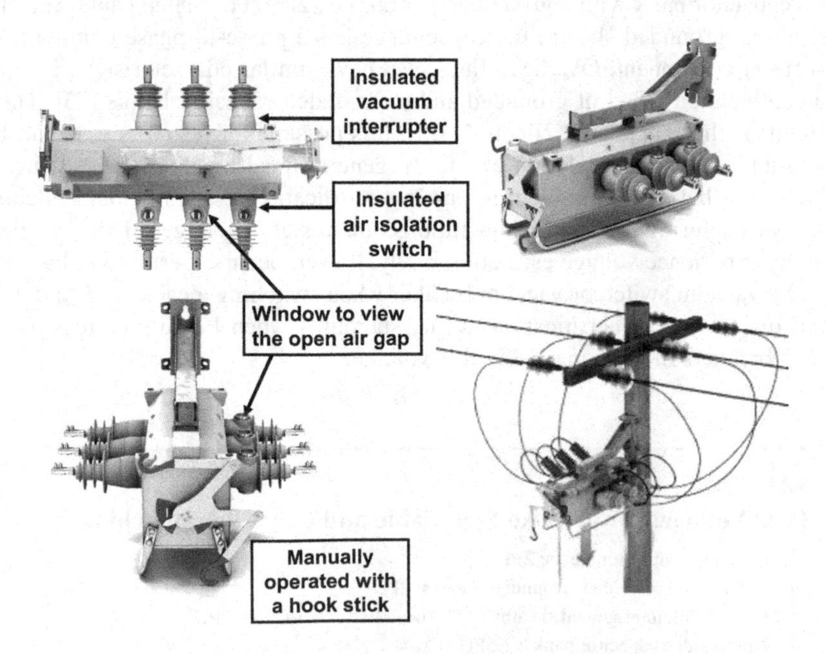

FIGURE 5.62 The VISI-SWITCH a vacuum interrupter and an air isolation switch in series (courtesy of NOJA Power Corporation).

Typically, an isolation gap has a greater BIL level of at least 110% (IEEE Standard) and 115% (IEC Standard) of than the standard level for a particular circuit voltage. Thus for a 15kV circuit (IEEE) where the BIL is 95kV peak, the isolation gap has to be able to withstand a peak value of 105kV. For a 24kV circuit (IEC) where the BIL has a peak value of 125kV the isolation value will be 145kV peak. For the vacuum interrupter, this may not be conservative enough. Thus, for isolation the vacuum interrupter would have to not only satisfy its certification BIL, but it would have to be designed to satisfy a significantly higher level and maintain that high level throughout its operating life.

This places a special operational duty on a vacuum interrupter that is expected to operate in a disconnecting function. First of all, if the load break, vacuum interrupter closes in on a fault current or even if it remains closed as a fault current passes through the contacts, the contacts would experience a small weld; see Sections 4.4.2 and 4.4.4 in this volume. If the contacts then open without current, a small microprojection would form that could decrease the contact gap's high-voltage withstand ability. An example of this can be seen in Table 4.10. This may also result from cold welding if the contacts close with no current. Fortunately, a load break vacuum interrupter usually opens and interrupts its load current. In fact, the latest IEC standards call for the disconnect switch to open with load current after closing in on a fault current. Picot [100] has discussed the potential concern of the long-term high voltage withstand limitations of vacuum disconnect switches and their open circuit protection ability under all operating circumstances. He concludes that excellent isolation from line voltage can easily be achieved using a load break vacuum interrupter. However, true circuit isolation may not always be possible when low probability voltage surges occur; e.g., those resulting from lightning strikes on the line.

For a person to work on the load side of the isolation switch, it is good practice to ground the load side at the switch. To this end, three-position switches have been developed for use in air, oil, and SF_6 [101, 102] that have a grounding position after the switch has opened, interrupted the load current and isolated the load from the line. These switches not only have the high voltage withstand capability, but they are also designed to carry the load current and withstand short circuit currents. These switches have to be certified according to the relevant standards, such as IEC 62271-102. The operation of such a switch is shown schematically in Figure 5.63. It is possible under certain

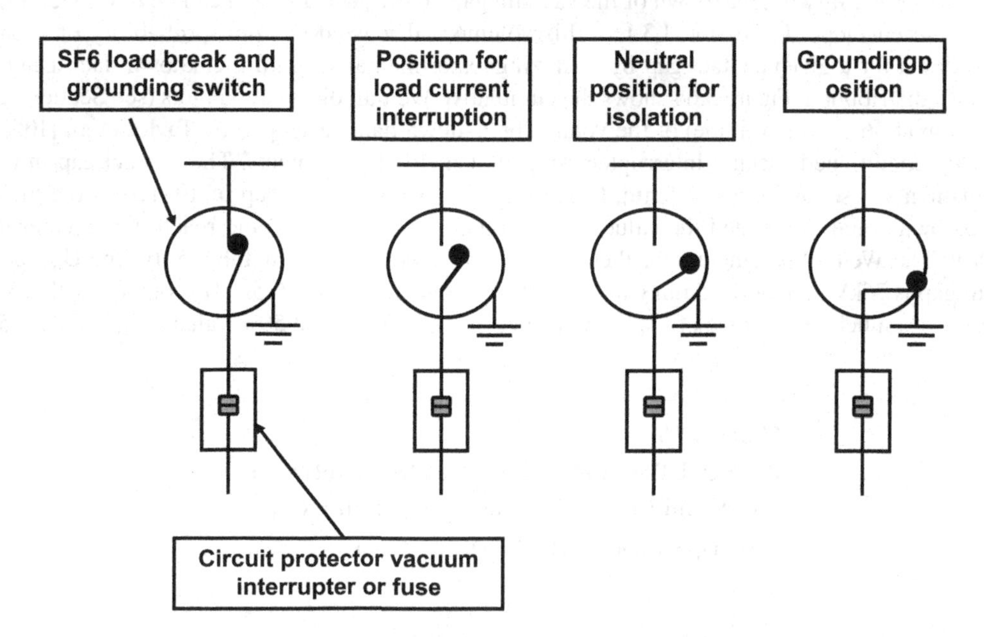

FIGURE 5.63 Schematic of an SF_6 three-position disconnect switch in series with a circuit protection vacuum interrupter.

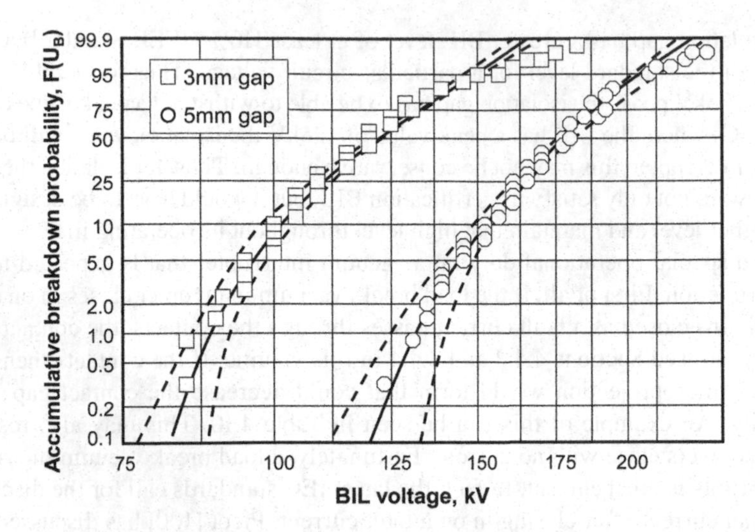

FIGURE 5.64 Weibull cumulative distribution plots of the breakdown voltages as a function of contact gap for Cu–W (10wt.%) contacts in a practical load break vacuum interrupter [105].

circuit conditions that the switch will experience the full short circuit current as the grounding takes place. Therefore, the switch has to be designed to close and latch on the full short circuit current in the ground position. It is relatively easy to design a low cost three-position switch in an insulating medium such as SF_6 and oil. The dielectric constant of these media is such that excellent isolation can be achieved in an economical, compact design. They have been designed into SF_6 gas insulated, compact switchgear and have been successfully used in series with a vacuum interrupter for short circuit protection [103, 104].

The development of a vacuum interrupter to also perform the function of the three-position switch presents some significant challenges. First of all, the vacuum interrupter designer has to consider the probability of breakdown of the vacuum gap before such a switch can be developed inside a vacuum envelope. In Section 1.3.1 in this volume, I discussed the zero probability of vacuum breakdown for a given contact gap by analyzing experimental vacuum breakdown data using the Weibull distribution. Figure 5.64 shows the cumulative Weibull distribution plots (see Section 1.3.1, Equation (1.36) in this volume) of the voltage breakdown data developed by Taylor et al. [105] for a newly conditioned vacuum interrupter using W–Cu (10wt.%) contacts. The contact gap in these experiments is set at 3mm and 5mm, because at the design contact gap of 10–12mm the highest breakdown voltage exceeded the value to which their power supply could reach: i.e., greater than 350kV. The Weibull parameters for these two sets of data are given in Table 5.16. The U_{B0} for the 3mm gap is 76kV and for the 5mm gap is 107kV. The data for Cu–Cr (25wt.%) contacts with a 5mm gap show a much lower value of U_{B0} of 59 kV: i.e., about 59% of the 50% value (U_{B50}). Figure 5.65

TABLE 5.16
WEIBULL Parameters for Breakdown between Cu–W and Cu–Cr Contacts in Vacuum [104]

Contact gap and material	Θ, kV	b	U_{B0}, kV
3mm Cu–W	126	3.2	76
5mm Cu–W	182	4.1	107
5mm Cu–Cr	105	4.1	59

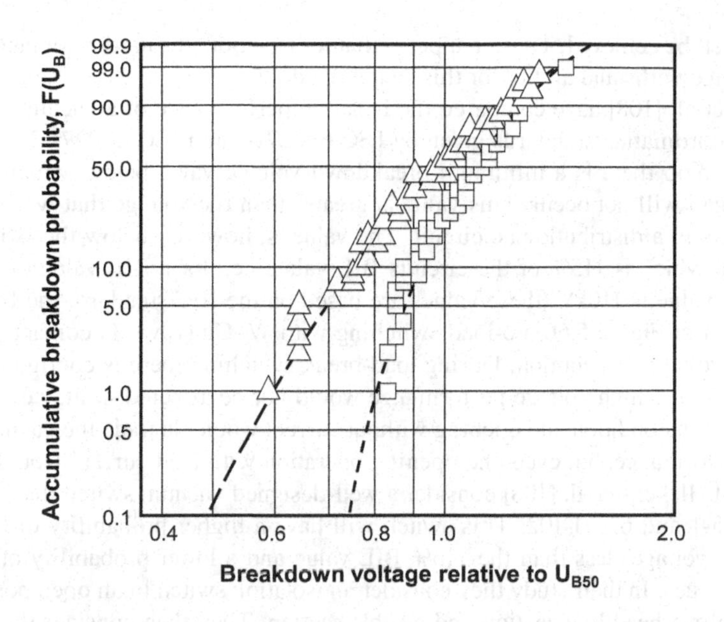

FIGURE 5.65 Weibull cumulative distributions plot of the breakdown voltages of contact for Cu–Cr (25wt.%) contacts: □ newly conditioned contacts and △ after no load switching [106].

shows data developed by Shioiri et al. [106] for newly conditioned Cu–Cr (25wt.%) contacts in vacuum with a 10mm gap and for the same system that has had no-load, close and open operations. The 50% breakdown value (U_{B50}) drops by about 10%, but the U_{B0} for zero probability of breakdown decreases from $0.65U_{B50}$ to $0.41U_{B50}$. Taylor et al. [107] performing the same no-load operations with the W–Cu (10wt.%) contacts show little effect on the high voltage withstand performance of the contacts; see Figure 5.66. Using the data shown in Figures 5.64 and 5.65, if the vacuum interrupter is initially designed and conditioned to 115% of the BIL value, there will be a finite probability of a breakdown for the full lightning surge voltage for Cu–Cr contacts and there will be a finite probability of holding off this voltage surge if W–Cu contacts are used. In this latter case true

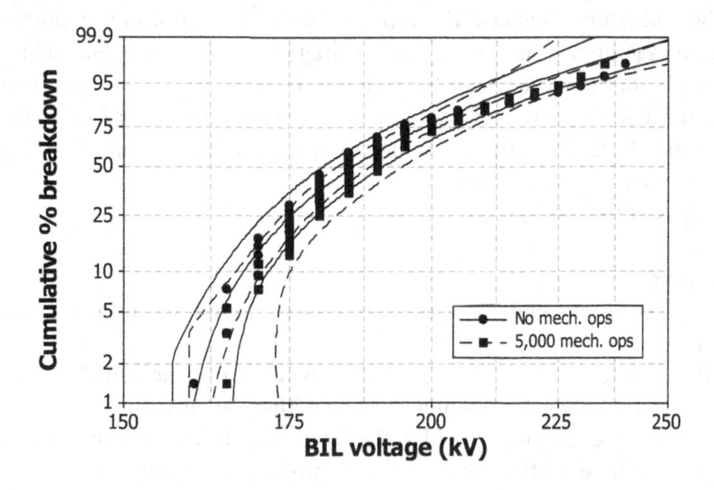

FIGURE 5.66 Weibull cumulative distribution plots of the breakdown voltages as a function of contact gap for Cu–W (10wt.%) contacts in a practical load break vacuum interrupter: ● newly conditioned contacts and ■ after 5000 no load switching operations [107].

isolation may well be achieved. Thus, it appears that for isolation, the harder contact material shows better high voltage withstand ability for this switching duty.

Schellekens et al. [108] have compared the isolation performance of a vacuum interrupter with the insulation coordination standards given by IEC 600271-1 and IEC 600271-2. As we have seen in Figures 5.64–5.66 there is a minimum breakdown voltage value below, which the breakdown of the vacuum gap will not occur. This value is greater than the voltage that would appear across the open contacts in a distribution ac circuit. The value is, however, below the BIL voltage for an isolation switch, which is 115% of the circuits BIL value: i.e., for a BIL value of 95kV the isolation switch BIL value is 110kV (these values are based on the IEC standard, the IEEE standard is 110%). As shown in Figure 5.66, no-load switching with W–Cu (10wt.%) contacts does not affect the breakdown voltage distribution. During load-break switching there is continuous conditioning of the contacts, so the high voltage performance would not be affected: it may even be improved. Closing in on a fault condition and opening without current would degrade the vacuum interrupter's high voltage performance, but even one opening operation with load current would restore it to its former level. Schellekens et al. [108] consider a well-designed vacuum switch that satisfies the IEC standards 60265-1 and 62271-102. This switch will have a higher probability of breakdown than a gas switch for voltages less than the 115% BIL value and a lower probability of breakdown for higher voltage values. In their study they consider an isolation switch in an open position on a junction between an overhead line section and a cable section. They then consider the actual voltages that may occur at an isolation switch as a result of lightning activity on the overhead line section. These values depend upon the distribution of the current magnitude in lightning strikes and the circuit impedance values. Using a Monte Carlo analysis, they show that the mean time to failure of the vacuum isolation switch is about 40% greater than that of an equivalent gas isolation switch. They conclude that a vacuum isolation switch designed to comply with the IEC standards 60265-100 and 62271-102 complies with the insulation coordination given in IEC 600271-1 and 600271-2. In other words, a vacuum isolation switch is at least as reliable as its gas counterpart.

5.6.2 Vacuum Interrupter Design Concepts for Load Switching and for Isolation

The patent literature shows a large number of vacuum interrupter designs that attempt to separate the current switching function from the isolation function. Figure 5.67 shows an example by Crouch et al. [109]. The contacts initially open to the correct distance to interrupt the circuit current. They are then moved to a second position within the fixed shields that are individually connected to each contact and to the end plates. Because these fixed shields have minimal exposure to the vacuum arc they can be manufactured from a good, high voltage withstand material such as stainless steel. Their surfaces can be designed and can be prepared to produce a very low field at the cathode. Hence the field emission electron current that can be produced when a high voltage is impressed across these fixed shields can be kept low enough to minimize the probability of breaking down the vacuum gap; see Section 1.3 in this volume.

The disadvantages of this design are:

(a) it requires two bellows
(b) the operating mechanism becomes quite complex because it has to move two contacts into two positions
(c) there is still a chance that cathode spots will transfer to the fixed shield during the switching operation thereby reducing its high voltage withstand performance
(d) in order to reduce the geometric field at the fixed shields, the design of the vacuum interrupter will be much larger than is necessary to just interrupt the current

A second design that eliminates the need to have two moving contacts is shown in Figure 5.68 [110]. However, in this type of design, the cathode spots will definitely transfer to the fixed shields during

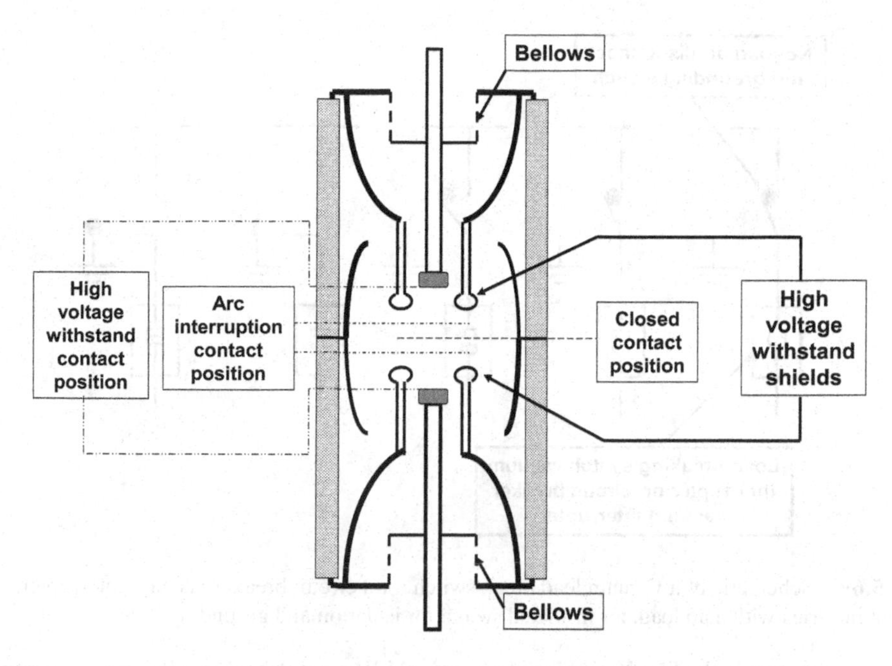

FIGURE 5.67 One vacuum interrupter designed to be used as an isolator [109].

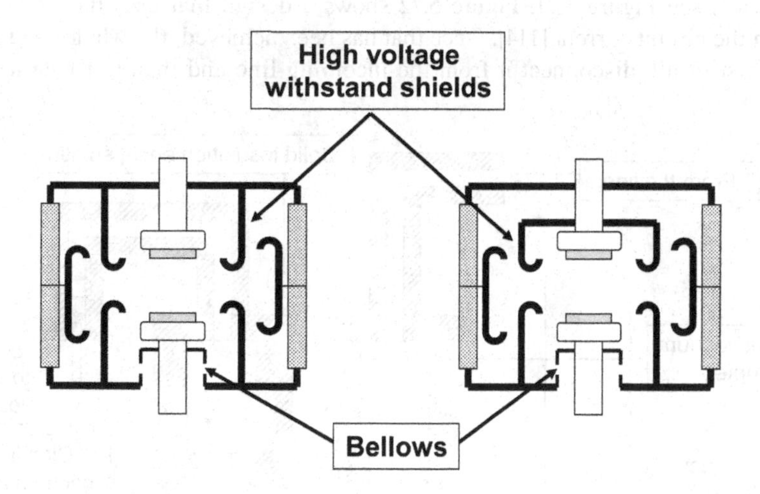

FIGURE 5.68 A second vacuum interrupter designed to be used as an isolator [110].

current interruption and thus reduce their high voltage withstand performance to some extent. Interestingly, in spite of these disadvantages, the patent literature continues to show similar designs. As far as I know, there has been almost no application of this type of vacuum interrupter design to practical load break switches.

5.6.3 Vacuum Interrupter Design for Switching and Grounding

One way of achieving this function is to use a vacuum interrupter in series with a grounding switch; see Figure 5.69. Figure 5.70 shows one modular example by Sato et al. [111] for load switching, circuit protection and grounding. In this design, once the load break switch is opened, the air or SF_6 grounding switch can be closed and the load side is protected. Hodkin et al. [112] have proposed a

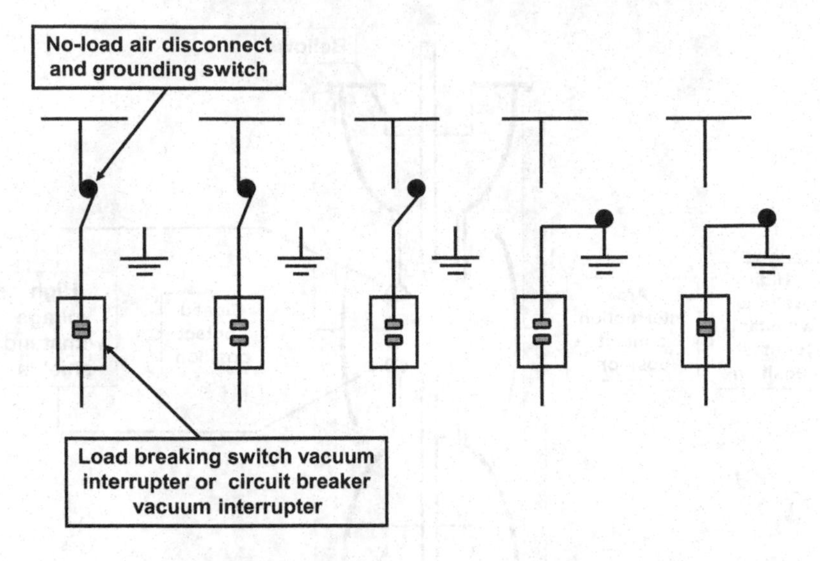

FIGURE 5.69 Schematic of a vacuum load break switch (or a circuit breaker vacuum interrupter) to switch the current in series with a no load, air insulated switch for isolation and grounding.

similar modular approach. The Xiria ring main unit [113] also integrates a vacuum interrupter load break switch, or a vacuum interrupter capable of interrupting fault currents, with an air insulated grounding device; see Figure 5.71. Figure 5.72 shows a design that uses the vacuum interrupter itself to switch the circuit current [114]. Once that has been achieved, the whole vacuum interrupter is rotated to, first of all, disconnect it from the incoming line and then to rotate it further into a

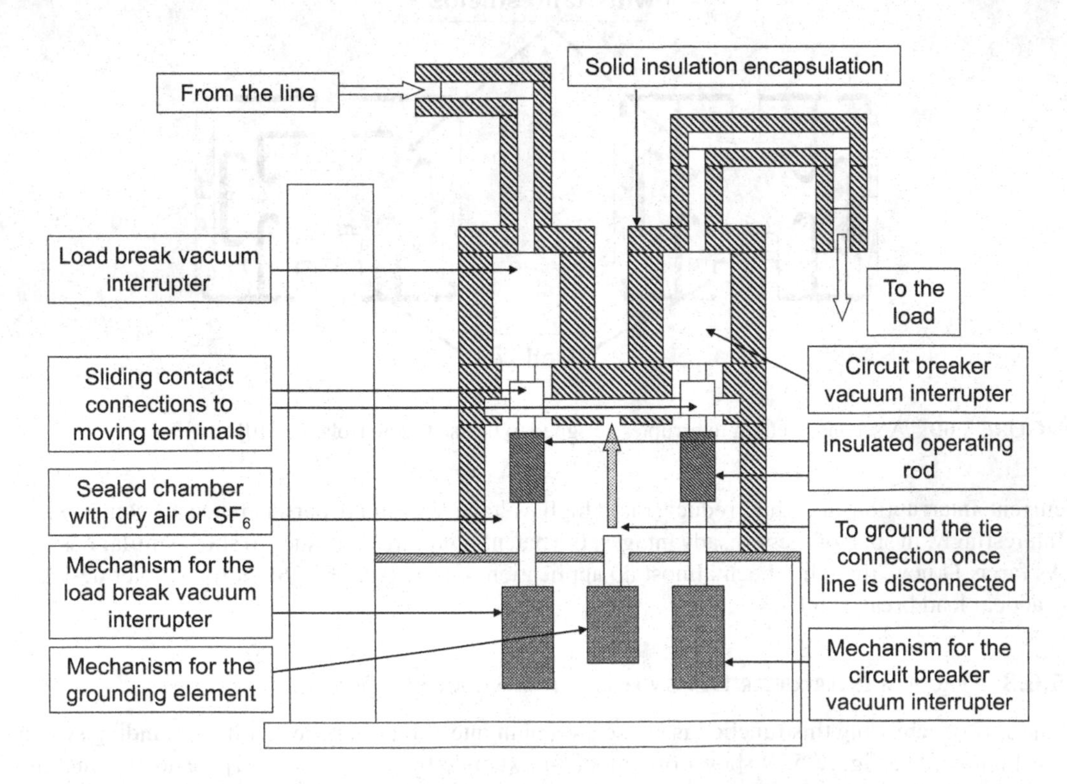

FIGURE 5.70 A schematic of a pole unit using solid insulation with a vacuum interrupter for fault current control and a vacuum interrupter for load current switching together with grounding capability [111].

FIGURE 5.71 Xiria Ring Main Unit with a no load, air insulated switch for isolation and grounding [courtesy Eaton Corporation].

grounding position. Date et al. [115] have proposed a similar concept, but only for disconnecting the vacuum interrupter from the load. The advantage of this type of arrangement is that it is possible to see that the system has been grounded.

Vacuum interrupter designs have been suggested where the switch and the grounding element reside inside the same vacuum envelope. Gentsch et al. [102] have proposed the design shown schematically in Figure 5.73. Here the contacts have two opening positions; the first to interrupt the

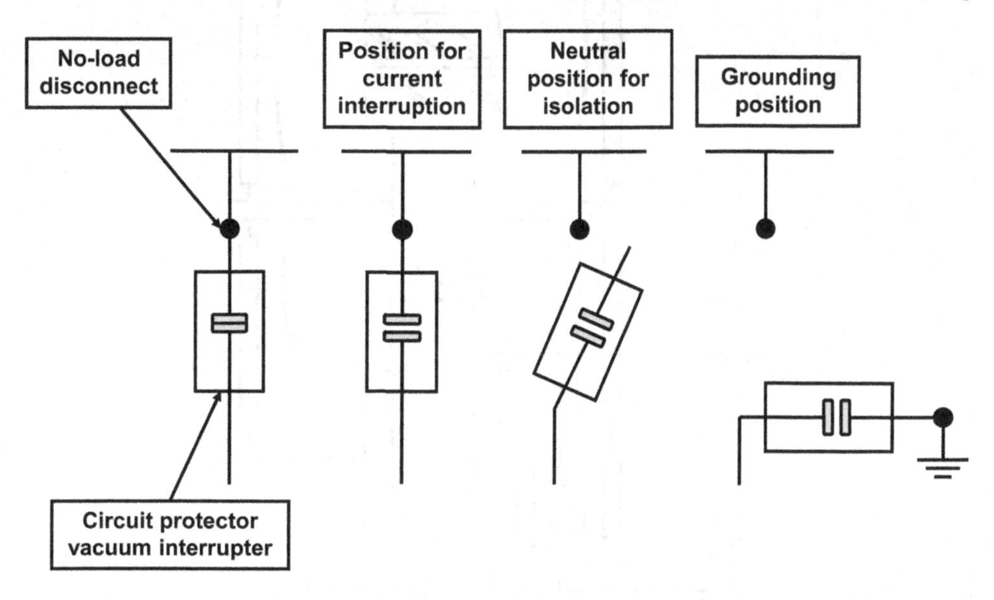

FIGURE 5.72 Schematic of a rotating vacuum interrupter for isolation and grounding [114].

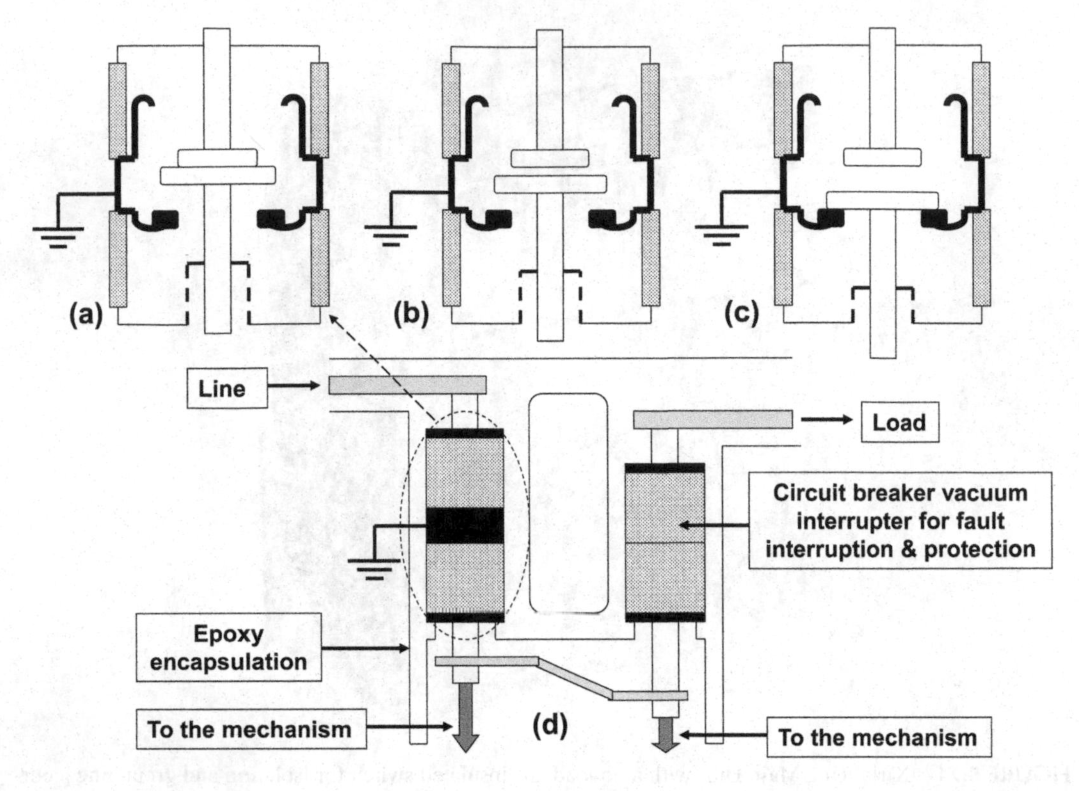

FIGURE 5.73 The load break and grounding function within one vacuum envelope [102].

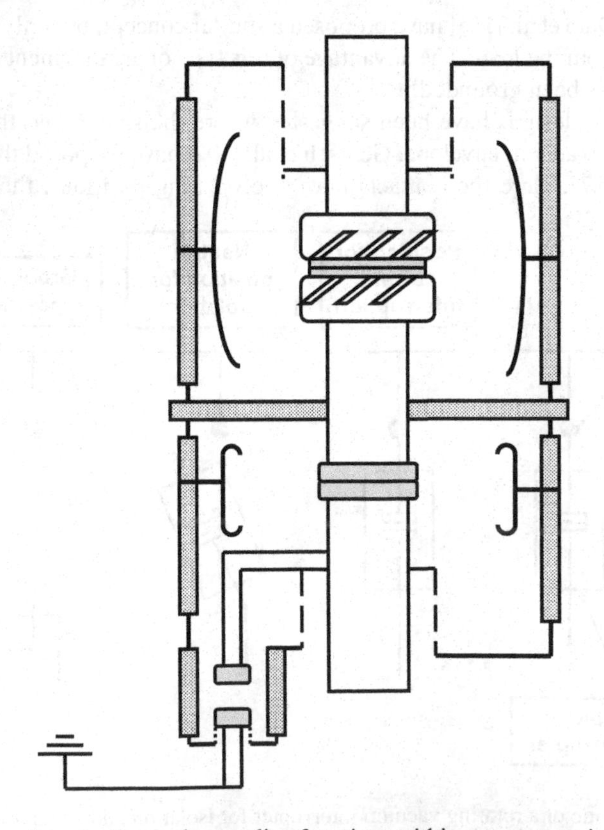

FIGURE 5.74 The circuit protection and grounding functions within one vacuum interrupter [117].

circuit current, and the second to make contact to the grounded shield. Kimblin [116] has proposed a similar design. The disadvantages of this design are:

(a) there is a likelihood that the vacuum arc will transfer to the shield during the interruption function
(b) the internal design has to be capable of sustaining the high voltage withstand between the grounded shield and the contacts even when the contacts are closed
(c) the grounding contacts have to be able to close and latch on the short-circuit current
(d) arcing in one chamber can lead to breakdown in the other chamber

Other designs move the isolating and grounding elements to a separate chamber; see Figure 5.74 [116], but this still has the high voltage withstand design problems discussed above.

5.6.4 Vacuum Interrupter Design for Fault Protection, Isolation, and Grounding

One proposed design for these functions in Renz et al. [118] is illustrated in Figure 5.75. This design has similarities to those shown in Figure 5.73 for the load break vacuum interrupter with a grounding

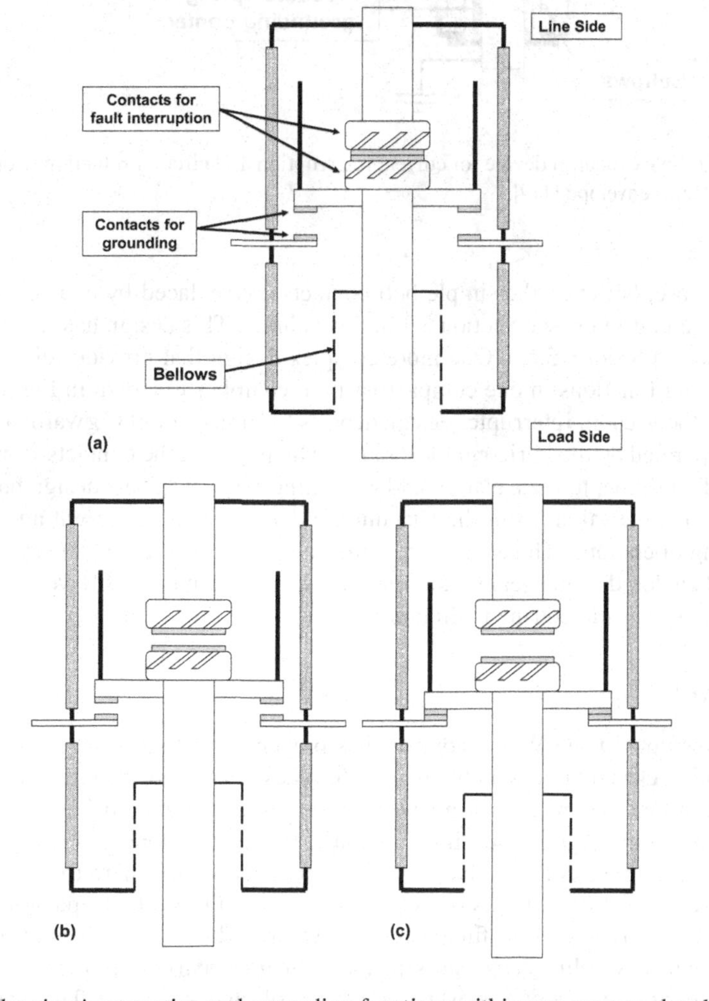

FIGURE 5.75 The circuit protection and grounding functions within one vacuum chamber: (a) closed (b) open to interrupt the circuit current and (c) the grounding position [118].

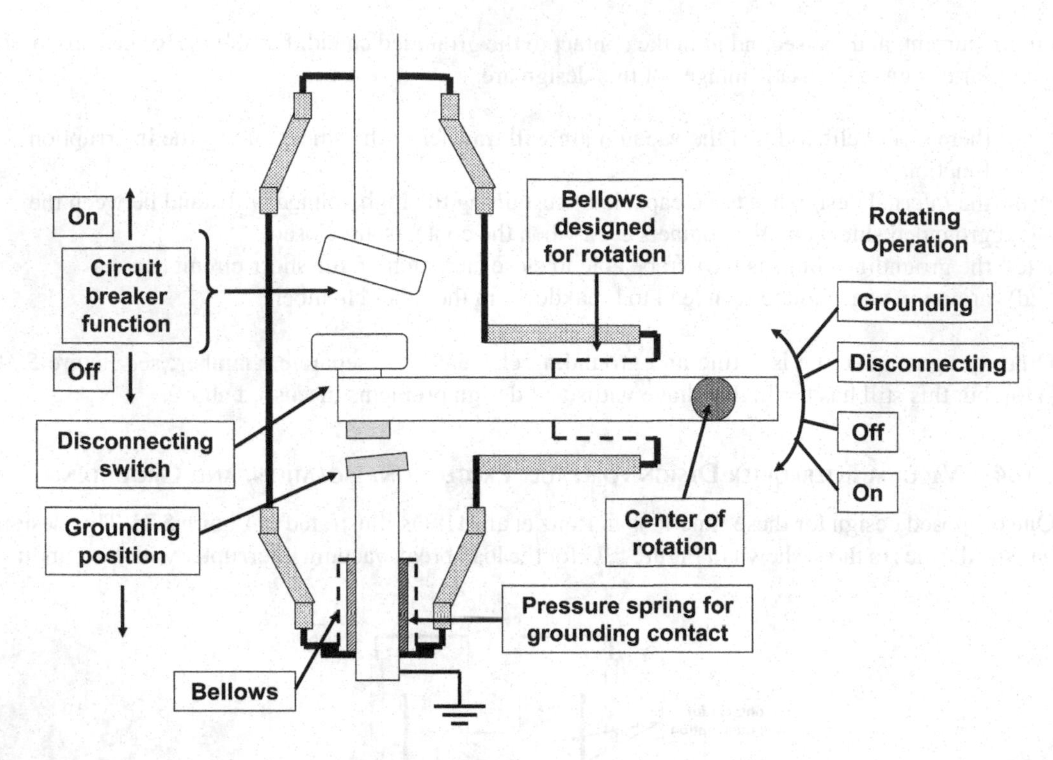

FIGURE 5.76 Complex vacuum device for current interruption and circuit protection, isolation and grounding within one vacuum envelope [119].

element. In this case, however, the simple butt contacts are replaced by contacts capable of interrupting short circuit currents; see Section 3.3 in this volume. This design has similar disadvantages to those discussed in Section 5.6.3. One more complex design that provides circuit breaker, isolation, and grounding functions in one compact vacuum chamber is shown in Figure 5.76, Kajiwara et al. [119]. Here the vacuum interrupter design departs from the straightforward opening operation. The contact is operated by the horizontal lever arm, which rotates the contacts from the on position to the open or off position, to the isolation and grounding positions. This design has the problem of insulating the ground terminal from the vacuum arc during the load current and the short circuit current switching operations. The advantage is the compactness of the whole system. The authors claim that metal enclosed switchgear containing this design can have a 50% volume reduction compared to a typical gas insulated ring main unit.

5.7 SUMMARY

The vacuum interrupter using W–Cu contacts has proven to be highly satisfactory for use as the current interruption element of a load break switch. It has been used since the late 1950s in a wide variety of distribution systems and in many different configurations. It has proven successful in above ground and in underground designs. Its maintenance-free long switching life and its long mechanical life have been a considerable advantage for application in hard to reach locations where replacement would be difficult. It has been successfully used to switch capacitor circuits. Further experimental research is needed to finally resolve whether the late restrikes observed during the recovery phase when switching capacitors result from microparticle impact at the cathode or a series of field emission current pulses over-heating the sub-anode surface. This vacuum interrupter has been successfully applied to cable-connected systems and in systems with overhead lines. It can be applied in locations with limited space and in any orientation without affecting its high voltage or current interruption performance. Future development of the vacuum interrupter for this application needs to focus on researching the factors that will permit a vacuum interrupter design to completely satisfy the isolating function. This will always be a concern, because the operator of such a device can never see the open contact gap. An example of a design that satisfies this requirement is the Visi-Switch shown in Figure 5.62. Thus, it is perhaps advisable that an auxiliary, visibly open, no-load contact will be required until users become familiar and comfortable with a vacuum interrupter isolation switch. The Ag–WC contact material has been successful in vacuum interrupters used for contactors which are used to switch electric motors. By using this contact material, the phenomenon of voltage escalation when switching inductive circuits is prevented.

REFERENCES

1. Fehr, R., *Industrial Power Distribution*, (pub. Wiley), 2016.
2. Burke, J. L., *Power Distribution Engineering*, (pub. Marcel Dekker), 1994.
3. Short, T. A., *Electric Power Distribution*, (pub. CRC Press), 2014.
4. Ross, H. C., "Vacuum switch properties for power switching applications", *AIEE Transactions Part III: Power Apparatus and Systems*, 77, pp. 104–117, April 1958.
5. Curtis, T. E., "A switch for multiple-circuit load break switching with a common interrupter", *IEEE Winter Power meeting, Paper 64-23*, February 1964.
6. Clair, M., Parks, J. E., Larner, A. F., R. A., "Vacuum switching for U.D. protection and sectionalizing", *IEEE Conference Record on Underground Distribution*, pp. 205–212, May 1969.
7. Rogowski, W., "Die elektrische Festigkeit am Rande Des Plattenkondensators. Ein Beitrag zur Theorie der Funkenstrecken um Durchfulhrungen", *Archiv für Electrotechnik*, 12(1), pp. 1–15, 1923.
8. Bruce, F. M., "Calibration of uniform-field spark gaps for high-voltage measurement at power frequencies", *Journal of the Institution of Electrical Engineers-Part II: Power Engineering*, 94(38), pp. 138–149, 1947.
9. Trinh, N., "Electrode design for testing in uniform fields", *IEEE Transactions on Power Apparatus and Systems*, PAS-99(3), pp. 1235–1242, May 1980.
10. Hardtfeldt, G. E., Kumbra, D. G., "A new stored-energy vacuum switch for distribution system automation", *Proceedings IEEE PES Transmission and Distribution Conference Paper 84 T&D 322-4*, April/May 1984.
11. Greenwood, A., *Vacuum Switchgear* (pub. Inst. Electrical Engineers, U.K.), pp. 98–104, 1994.
12. Orace, H., McLaren, P., "Surge voltage distribution in line-end turns of induction motors", *IEEE Transactions on Power Apparatus and Systems*, PAS-104, pp. 1843–1848, 1985.
13. Corrick, K., Thompson, T., "Steep fronted switching voltage transients and their distribution in electrical machine windings, part 2: Distribution of steep-front voltage transients in motor windings", *Proceedings IEE B*, 130, p. 56, 1982.
14. Nassar, O. M., "Motor insulation degradation due to switching surges and surge protection devices", *IEEE Transactions on Energy Conversion*, EC-1(3), pp. 182–189, September 1986.
15. Perkins, J. F., "Evaluation of switching surge over voltages on medium voltage power systems", *IEEE Transactions on Power Apparatus and Systems*, 101(6), pp. 1727–1734, 1982.
16. Perkins, J., Bhasavanich, D., "Vacuum switchgear application study with reference to switching surge protection", *IEEE Transactions on Industry Applications*, IA-19(5), pp. 879–888, September/October 1983.
17. *IEEE Red Book*, ANSI/IEEE Std. 141-1993, Tables 6–7, p. 351.
18. Pretorius, R., "Guide for the application of surge suppressors to medium voltage motors", *Report Issued by the South African Electric Power Coordinating Committee*, August 1992.
19. Pretorius, R., Kane, C., Golubev, A., "A new approach towards surge suppression and insulation monitoring for medium voltage motors and generators", *IEEE Electrical Insulation Conference*, 2009.
20. Heyne, D., Lindmayer, M., Wilkening, E.-D., "Effect of contact material on extinction of vacuum arcs under line frequency and high frequency conditions", *IEEE Transaction CHMT*, 14(1), pp. 65–70, March 1991.
21. Kaneko, E., Yokokura, K., Homma, M., Satoh, Y., Okawa, M., Okutomi, I., Oshima, I., "Possibility of high current interruption of vacuum interrupter with low surge contact material: Improved Ag-WC", *IEEE Transactions on Power Delivery*, 10(2), pp. 797–803, April 1995.

22. Santilli, V., "Electrical contacts for vacuum interrupter devices", *US Patent 4,743,718*, May 1988.
23. Telander, S., Wilheim, M., Stump, K., "Surge limiters for vacuum circuit breaker switchgear", *IEEE Transactions on Industry Applications*, 24(4), pp. 1672–1677, October 1998.
24. Garzon, R., *High Voltage Circuit Breakers: Design and Application*, (Pub. Marcell Dekker), 1997.
25. Lee, T., "The effect of current chopping in circuit breakers on networks and transformers: I – theoretical considerations", *AIEE Transactions Part III: Power Apparatus and Systems*, 79, pp. 535–544, August 1960.
26. Greenwood, A., "The effect of current chopping in circuit breakers on networks and transformers: II – experimental techniques and investigations", *AIEE Transactions Part III: Power Apparatus and Systems*, 79, pp. 545–554, August 1960.
27. Brunke, J., Frohlich, K., "Elimination of transformer inrush currents by controlled switching 1. Theoretical considerations", *IEEE Transactions on Power Delivery*, 16(2), pp. 276–280, April 2001.
28. Brunke, J., Frohlich, K., "Elimination of transformer inrush currents by controlled switching II. Application and performance considerations", *IEEE Transactions on Power Delivery*, 16(2), pp. 281–285, April 2001.
29. Duan, X., Liao, M., Zou, J., "Controlled vacuum circuit breaker for transformer inrush current minimization", *Proceedings 23rd International Symposium on Discharges and Electrical Insulation in Vacuum*, pp. 133–136, September 2008.
30. Li, W., Fang, C., Zhang, B., Xie, P., Ren, X., Luo, Y., "Research on controlled switching in reducing unloaded power transformer inrush current considering circuit breaker's prestrike" Characteristics", *Proceedings 27th International Symposium on Discharges and Electrical Insulation in Vacuum*, pp. 145–148, September 2016.
31. Li, R., Liu, H., Liao, M., Jing, Y., Duan, X., Li, K., "Investigation on transformer inrush current switched by controlled vacuum circuit breaker", *Proceedings 28th International Symposium on Discharges and Electrical Insulation in Vacuum*, pp. 563–566, September 2018.
32. Dohnal, D., Kurth, B., "Vacuum switching, a well proven technology has found its way into resistance-type load tap changers", *Proceedings IEEE Transmission and Distribution Conference*, 2001.
33. Dohnal, D., "On-load tap-changers for power transformers a technical digest". https://www.reinhausen.com/PortalData/1/Resources/tc/research_development/vacuum_technology/PB252_en_Power_Transformers.pdf.
34. Hammer, C., Albrecht, W., Pircher, C., Rehkopf, S., Sachsenhauser, A., Radlinger, K., Stelzer, A., Teichmann, J., Renz, R., Lawall, A., Kosse, S., Hartmann, W., Wenzel, N., "Tap changer and vacuum interrupter for such a tap changer", *European Patent, EP2695177A1*, May 2016.
35. Popov, M., Acha, E., "Over-voltages due to switching off an Unloaded transformer with a vacuum circuit breaker", *IEEE Transactions on Power Delivery*, 14(4), pp. 1317–1326, October 1999.
36. Popov, M., Van der Sluis, L., "Improved calculations for no-load transformer switching loads", *IEEE Transactions on Power Delivery*, 16(3), pp. 401–408, July 2001.
37. Chaly, A., Chalaya, A., "A computer simulation of transformer magnetizing current interruption by a vacuum circuit breaker", *Proceedings 17th International Symposium on Discharges and Electrical Insulation in Vacuum*, Berkeley, USA, IEEE No. 96CH35939, pp. 249–253, July 1996.
38. Ananian, L., Miller, K., Titus, C., Walsh, G., "Field testing of voltage transients associated with vacuum breaker no-load switching of a power transformer in an industrial plant", *Conference Record of the IEEE Conference Industrial and Commercial Power*, Houston, TX, May 1980.
39. Van den Heuval, W., Daalder, J., Boone, M., Wilmes, L., "Interruption of a dry-type transformer by a vacuum circuit breaker", *EUT Report 83-E-141* (ISBN 90-6144-141-2, ISSN 0167-9708), August 1983.
40. Ohashi, H., Mizuno, T., Yanabu, S., "Application of a vacuum circuit breaker to dry-type transformer switching", *Proceedings IEEE Power Engineering Society Winter Meeting, New York, USA, Paper A 76 174 – 3*, pp. 174–173, January 1976.
41. Popov, M., Van der Sluis, "Statistical estimation of reignition over-voltages and probability of reignition when switching transformers and motors with a vacuum circuit breaker", *Proceedings 3rd Conference on Reliability of Transmission and Distribution*, February 2005.
42. Square D/Schneider Electric, "Power and control sleuths track transformer failures to troublesome circuit",, *Power Quality Assurance*, pp. 14–18, November 2000.
43. Voss, G., Mattatia, S., Bajog, G., "Interaction between power electronic components and distribution transformers – risk and mitigation techniques", *CIRED, International Conference on Electricity Distribution*, Turin, Italy, June 2005.
44. Hori, M., Nishioka, M., Ikeda, Y., Noguchi, K., Kajimura, K., Motoyama, H., Kawamura, T., "Internal winding failure due to resonance over-voltage in distribution transformer caused by winter lightning" ", *IEEE Transactions on Power Delivery*, 21(3), pp. 1600–1606, July 2006.

45. Vogelsang, R., Weiers, T., Fröhlich, K., Brütsch, R., "Electrical breakdown in high voltage winding insulations of different manufacturing quantities", *IEEE Dielectrics and Electrical Insulation Magazine*, 22(3), pp. 5–12, May/June 2006.
46. CIGRE Workshop, "Electrical environment of transformers", Paris France, http://www.cigre-se12.org/SC_12/activities.htm, August 2002.
47. ANSI/ UL contactor standard – 347-1999 and the IEC contactor standards – 60947-1 and 60470.
48. Shea, J., "Modeling contact erosion in three phase vacuum contactors", *IEEE Transactions CPMT - Part A*, 21(4), pp. 556–564, December 1998.
49. Schulman, M., Slade, P., Bindas, J., "Effective erosion for selected contact materials in low-voltage contactors", *IEEE Transactions on Components, Packaging, and Manufacturing Technology, – Part A*, 18(2), pp. 329–333, June 1995.
50. Schulman, M., Slade, P., Loud, L., Li, W.-P., "Influence of contact geometry and current on effective erosion of Cu-Cr, Ag-WC, and Ag-Cr, vacuum contact materials", *IEEE Transactions on Components and Packaging Technology*, 22(3), pp. 405–413, September 1999.
51. Rieder, W., Schussek, M., Glatzle, W., Kny, E., "The influence of composition and Cr particle size of Cu-Cr contacts on chopping current, contact resistance and breakdown voltage in vacuum interrupters", *IEEE Transactions on Components, Hybrids, and Manufacturing Technology*, 12(2), pp. 273–283, June 1989.
52. Daalder, J., "Cathode spots and vacuum arcs", *Physica*, 104C(1–2), pp. 91–106, 1981.
53. Heberlein, J. V.R ., Porto, D. R., "The interaction of vacuum arc ion currents and axial magnetic fields", *IEEE Transactions on Plasma Science*, PS-11, pp. 152–159, September 1983.
54. Tuma, D. T., Chen, C. L., Davies, D. K., "Erosion products from the cathode spot region of a copper vacuum arc", *Journal of Applied Physics*, 49(7), pp. 3821–3831, July 1978.
55. For example: "Joslyn VerSaVac capacitor switches", Joslyn Hi–Voltage Corporation, U.S.A. www.joslynhivoltage.com.
56. The Editor, "Capacitor switching needs a switch, not a circuit breaker", *Electrical World*, pp. 43–46, April 1996.
57. Yokokura, K., Matsuda, M., Atsumi, K., Nizazama, T., Sohma, S., Kaneko, E., Ohshima, I., "Capacitor switching capability of vacuum interrupters with Cu-W contacts", *IEEE Transactions on Power Delivery*, 10(2), pp. 804–810, April 1995.
58. Fu, Y., Damstra, G., "Switching transients during energizing capacitor loads by vacuum circuit breaker", *IEEE Transactions on Electrical Insulation*, 28(4), pp. 657–666, August 1993.
59. McCoy, C., Floryancic, B., "Characteristics and measurement of capacitor switching at medium voltage distribution level", *IEEE Transactions on Industry Applications*, 30(6), pp. 1480–1489, November/December 1994.
60. Bonfanti, I., (Convener), "Shunt capacitor bank switching stresses and tes methods (2nd Part)", *Electra* 183, April 1999.
61. Donen, T., Yano, T., Tsukima, M., Miki, S., "The impact of capacitive switching condition on vacuum breakdown phenomena", *Proceedings 26th International Symposium on Discharges and Electrical Insulation in Vacuum*, pp. 137–140, September 2014.
62. Yu,Y., Wang, J., Yang, H., Geng, Y., Liu, Z., "The impact of inrush current on field emission current of vacuum interrupters during capacitive current switching", *Proceedings 26th International Symposium on Discharges and Electrical Insulation in Vacuum*, pp. 145–148, September 2014.
63. Godechot, X., Novak, P., "New trends in capacitor switching", *Proceedings 28th International Symposium on Discharges and Electrical Insulation in Vacuum*, pp. 599–602, September 2018.
64. Yang, H., Geng, Y., Liu, Z., "Capacitive current switching of vacuum interrupters and inrush currents", *Proceedings 25th International Symposium on Discharges and Electrical Insulation in Vacuum*, pp. 228–231, September 2012.
65. Sabot, A., Morin, C., Guillaume, G., Pons, A., Taisne, J., "A unique multipurpose damping circuit for shunt capacitor bank switching", *IEEE Transactions on Power Delivery*, 8(3), pp. 1173–1183, July 1993.
66. Camm, E., "Shunt capacitor overvoltages and a reduction Technique",, Presented at the panel session on "Overvoltages: Analysis and Protection", *IEEE, Power Engineering Society, T&D Conference*, New Orleans, April 1999.
67. For example: "Circuit switcher for capacitor bank switching and protection", www.sandc.com/products/default.asp.
68. For example: "Digital zero voltage closing (ZVC) control for capacitor switching", www.joslynhivoltage.com.

69. Moulaert, G., Couvreur, M., Van Ranst, A., Decock, W., Morant, M., "Application of controlled switching to MV capacitor banks", *Proceedings CIRED Conference*, Brussels Belgium, pp. 1.08.0–1.08.4, 1995.
70. Schellekens, H., Bono, M., "Synchronous capacitor bank switching with vacuum circuit breakers", *Proceedings 28th International Symposium on Discharges and Electrical Insulation in Vacuum*, pp. 567–570, September 2018.
71. Osmokrovic, P., "Influence of switching operation on the vacuum interrupter dielectric strength", *IEEE Transactions on Dielectrics and Electrical Insulation*, 1(2), pp. 340–347, April 1994.
72. Goeckeler, C., "Beyond SCADA: Capacitor automation at KCPL", *Transmission and Distribution World*, pp. 43–48, August 1997.
73. Greenwood, A., *Electrical Transients in Power Systems*, (Pub. Wiley Interscience), 1971.
74. Zaborszky, J., Rittenhouse, J., "Fundamental aspects of some switching overvoltages on power systems", *IEEE Transactions on Power Apparatus and Systems*, Vol. 81, N0. 3, pp. 822–830. April 1962
75. Zaborszky, J., Rittenhouse, J., "Some fundamental aspects of recovery voltages", *IEEE Transactions on Power Apparatus and Systems*, Vol. 82, No. 2, February 1963.
76. Report by the Working Group 3.4.17 of the IEEE Surge Protection Committee, "Surge protection of high voltage shunt capacitor banks on ac power systems survey results and application considerations", *IEEE Transactions on Power Delivery*, 6(3), pp. 1065–1072, July 1991.
77. Report by the Working Group 3.4.17 of the IEEE Surge Protection Committee, "Impact of shunt capacitor banks on substation surge environment and surge arrester applications"„ *IEEE Transactions on Power Delivery*, 11(4), pp. 1798–1809, October 1996.
78. *IEEE Red Book*, ANSI/IEEE Std. 141-1993, Section 6.7.3.6, p. 349.
79. Juhsász, A., Rieder, W., "Capacitative switching with vacuum interrupters", *Proceedings 10th International Conference on Gas Discharges and Their Applications*, pp. 62–65, 1992.
80. Koochack-Zadeh, M., Hinrichsen, V., Kirvenhoven, S., "Measurement of field emission current switching of capacitive current in vacuum", *Proceedings 24th International Symposium on Discharges and Electrical Insulation in Vacuum*, pp. 210–213, September 2010.
81. Koochack-Zadeh, M., Hinrichsen, V., Smeets, R., Lawall, A., Field emission currents in vacuum breakers after capacitor switching", *IEEE Transactions on Dielectrics and Electrical Insulation*, 18(3), pp. 910–917, 2011.
82. Smeets, R., Kuivenhoven, S., Chakraborty, S., Sandolache, G., "Field emission current in vacuum interrupters after large inrush current", *Proceedings 25th International Symposium on Discharges and Electrical Insulation in Vacuum*, pp. 157–160, September 2012.
83. Donen, T., Tsukima, M., Sato, S., Yoshida, T., "Investigation of correlation between vacuum breakdown phenomena and field emission current during shunt capacitor switching", *Proceedings 25th International Symposium on Discharges and Electrical Insulation in Vacuum*, pp. 161–164, September 2012.
84. Niayesh, K., Rager, F., Schacherer, C., "Electrode phenomena before long delayed breakdowns in vacuum after switching capacitor currents", *IEEE Transactions on Plasma Science*, 33, pp. 258–259, April 2002.
85. Slade, P., Taylor, E., "Calculations on the potential role of emission currents on restrikes after capacitor switching using vacuum interrupters", *Proceedings 28th International Symposium on Discharges and Electrical Insulation in Vacuum*, pp. 177–180, September 2018.
86. Wang, L., Jia, S., Yang, D., Liu, K., Su, G., Shi, Z., "Modelling and simulation of anode activity in high-current vacuum arc", *Journal of Physics. Part D: Applied Physics*, 42(14), June 2009.
87. Wayland, P., Former Engineering Manager, Westinghouse Vacuum Interrupter Operation: Private communication.
88. Smeets, R., Wiggers, R., Bannink, H., Kuivenhoven, S., Chakraborty, S., Sandolache, G., "The impact of switching capacitor banks with very high inrush current", *CIGRE Conference (Paris Session), Paper A-201*, pp. 1–12, 2012.
89. Sandolache, G., Chakraborty, S., Gaches, L., Smeets, R., Kuivenhoven, S., Novak, P., Beer, P., "An investigation into late breakdown phenomena during capacitor switching performances in relation with vacuum interrupter design and field emission current", *Proceedings 25th International Symposium on Discharges and Electrical Insulation in Vacuum*, pp. 441–444, September 2012.
90. Zhang, Y., Yang, H., Geng, Y., Liu, Z., Jin, L., "Effect of high-frequency high-voltage impulse conditioning on inrush current interruption of vacuum interrupters", *IEEE Transactions on Dielectrics and Electrical Insulation*, 22(2), pp. 1306–1313, April 2015.

91. Dullni, E., Schade, E., "Investigation of high-current interruption of vacuum circuit breakers", *EEE Transactions on Electrical Insulation*, 28(4), pp. 607–620, August 1993.
92. Schade, E., Dullni, E., "Recovery of breakdown strength of a vacuum interrupter after extinction of high currents", *IEEE Transactions on Dielectrics and Electrical Insulation*, 9(2), pp. 207–213, April 2002.
93. Dullni, E., Shang, W., Gentsch, D., Kleberg, I., Niayesh, K., "Switching capacitive currents and correlation of restrike and pre-ignition behavior", *IEEE Transactions on Dielectrics and Electrical Insulation*, 13(1), pp. 65–71, February 2006.
94. Gebel, R., Hartmann, W., "mechanical shocks as cause of late discharges in vacuum circuit breakers", *IEEE Transactions on Electrical Insulation*, 28(4), pp. 469–472, August 1993.
95. Hachiman, Y., Oshiro, F., Nagayo, Y., Kaneko, E., Inada, Y., Taguchi, Y., Yamano, Y., Maeyama, M., Kitabayashi, Y., Iwabuchi, H., Ejiri, H., Kumada, A., Hidaka, K., "Late dielectric breakdown phenomenon caused by microparticles released after current interruption", *Proceedings 28th International Symposium on Discharges and Electrical Insulation in Vacuum*, pp. 173–175, September 2018.
96. Smeets, R., Lathouwers, A., "Non-sustained disruptive discharges: Test experiences, standardization status and network consequences", *IEEE Transactions on Dielectrics and Electrical Insulation*, 9(2), pp. 195–200, April 2002.
97. Batara, R., Krolikowski, C., Opyda, W., "Switching unloaded medium voltage cable lines by a vacuum interrupter", *Proceedings 19th International Symposium on Discharges and Electrical Insulation in Vacuum (IEEE No. 00CH37041)*, Xi'an, China, pp, pp. 396–402, September 2000.
98. Damstra, G., Fu, Y., "Switching transients in a cable network by vacuum interrupters", *Proceedings 15th International Symposium on Discharges and Electrical Insulation in Vacuum Darmstadt, Germany*, pp. 436–439, 1992.
99. https://www.nojapower.com.au/visi-switch.
100. Picot, P., "Vacuum disconnector: Acceptability issues", *Proceeding 18th CIRED Conference on Electrical Distribution*, Turin, Italy, 2005.
101. Date, K., Wainio, R., "Vacuum interrupter and disconnect function", *US patent 4,105,878*, August 1978.
102. Gentsch, D., Fink, H., Dullni, E., "Three position switch based on vacuum interrupter technology for disconnecting and earthing purposes", *Proceedings 22nd International Symposium on Discharges and Electrical Insulation in Vacuum*, pp. 311–314, September/October 2004.
103. Lav, C., Staley, D., Olsen, T., "Practical design considerations for application of GIS MV Switchgear", *IEEE Transactions on Industry Applications*, 40(5), pp. 1427–1434, September/October 2004.
104. Ormazabal Company Brochure, "Medium voltage switchgear for primary distribution" and "Medium voltage switchgear for secondary distribution networks", *National Book Catalogue Numbers*, SS-188/04 and SS-0032/04, www.ormazabal.com.
105. Taylor, E., Slade, P., "High voltage breakdown performance and circuit isolation capability of vacuum interrupters", *Proceedings 22nd International Symposium on Discharges and Electrical Insulation in Vacuum*, pp. 208–211, September 2006.
106. Shioiri, T., Sasage, K., Kamikawaji, T., Homma, M., Kaneko, E., "Investigation of dielectric breakdown probability distribution for vacuum disconnecting switch", *Proceedings IEEE/PES Transmission and Distribution Conference and Exhibition 2002*, China, vol. 3, pp. 1780–1785, October 2002.
107. Taylor, E., Slade, P., "Private communication."
108. Schellekens, H., Gal, I., Goulielmakis, D., "Vacuum switch-disconnectors: 2. Compliance with insulation coordination", *Proceedings 23rd International Symposium on Discharges and Electrical Insulation in Vacuum*, pp. 153–154, 2008.
109. Crouch, D., Kurtz, D., Sofianek, J., "High-voltage vacuum switch", *US Patent 3,914,568*, October 1975.
110. Shioiri, T., Yamazaki, T., Shin, M., Yokokura, K., Sato, J., "Vacuum switchgear", *European Patent EP 1 005 058*, May 2000.
111. Sato, J., Osamu, S., Nobutaka, K., Miyagawa, M., Ohno, M., Saito, T., Shioiri, T., Makishima, S., Kinoshita, S., *European Patent EP 1 343 233 A2*, September 2003.
112. Hogkin, A., Marshall, T., Davies, N., Theisen, P., Marchand, F., "Modular miniaturized switchgear", *International Patent, WO 01/50561 A1*, July 2001 and "Isolator switch", *International Patent, WO 01/50562 A1*, July 2001.
113. Eaton Electrical (Holec) Company Brochure, "Xiria 24kV Ring Main Unit", www.eatonelectric.com.au/brochures/XiriaEN.pdf.
114. Slade, P., Taylor, E., "Thee position vacuum interrupter disconnect switch providing current interruption, disconnection and grounding" *US Patent 7,186,942 B1*, March, 2007.
115. Date, K., Wainio, R., "Vacuum interrupter and disconnect combination", *US Patent 4,105,878*, August 1978.

116. Kimblin, C., "Multiple contact switch", *US Patent 6,255,615 B1*, July 2001.
117. Kikukawa, S., Kojima, K., Tanimizu, T., Morita, A., Tsuji, M., "Vacuum switch including vacuum measuring devices, switchgear using the vacuum switch, and operation method thereof", *US Patent 6,426,627 B2*, July 2002.
118. Rentz, R., Steinemer, N., "Vacuum interrupter with two contact systems", *US Patent 6,720,515 B2*, April 2004.
119. Kajiwara, S., Watanabe, Y., Tanimizu, T., Morita, A., "Development of 24 kV switchgear with multi-functional vacuum interrupters for distribution", *Hitachi Review*, 49(2), pp. 93–100, 2000.

6 Circuit Protection, Vacuum Circuit Breakers, and Reclosers

I bend, and do not break.

La Fontaine
(Fables, no. 22)

6.1 INTRODUCTION

The first successful application of the vacuum interrupter was for load switching (see the Introduction). It was only in the mid-1950s that the extensive R&D program undertaken by the General Electric Company (United States) began to show that it was possible for vacuum interrupters to reliably interrupt high fault currents. This work led to an initial vacuum interrupter design that could interrupt 12 kA in a 15kV circuit. Since that time there has been continued development of the vacuum interrupter in the United States, Europe and Asia, which has resulted in a wide range of vacuum interrupter designs. Vacuum interrupters have now been successfully applied to protect all distribution circuits from 4.76kV to 40.5kV, with fault currents up to 100kA [1–3]. Vacuum interrupters have also been developed for subtransmission voltages (50kV–84kV) [2, 4–6] and for transmission voltages (110kV–242kV) [7–14].

The circuit breaker using a vacuum interrupter as the current interruption component is universally known as a *vacuum circuit breaker*. A general description of a circuit breaker is:

> *A mechanical switching device capable of making, carrying and interrupting currents under normal circuit conditions and making, carrying for a specified time and interrupting currents under specified abnormal conditions, such as those from a short circuit.*

That is, the vacuum circuit breaker can perform the load switching functions that I presented in Chapter 5 in this volume as well as having the capability of interrupting all fault current levels up to, and including, its maximum rated short circuit current. The vacuum circuit breaker's primary purpose is to protect an electrical circuit from the deleterious effects of fault currents. After the interruption of the fault current, the vacuum circuit breaker's performance must not be compromised. For example, once the cause of the fault current has been identified and eliminated the vacuum circuit breaker must be capable of closing and continuing its safe and reliable performance. When open the vacuum circuit breaker is expected to protect the downstream circuit from voltage surges such as may occur when lightning strikes the upstream side of the electric circuit.

The automatic circuit recloser, developed in the United States in the 1930s, permits the automatic restoration of a circuit once a temporary fault, such as that resulting from a lightning strike, has been removed. This development became necessary when long electric distribution lines were built for rural electrification. When temporary faults occur on these lines, the utility companies require a circuit breaker that will reset itself if the fault proves to be truly temporary or to stay "locked-out" if the fault is permanent. The automatic circuit recloser (or recloser for short) using a vacuum interrupter is called a *vacuum recloser*: a general description is:

> *A self-controlled device for automatically interrupting and resetting an alternating current circuit with a predetermined sequence of opening and reclosing followed by resetting, hold closed or lock-out.*

TABLE 6.1
An Example of the Test Ratings for Certification of a Vacuum Circuit Breaker

Rated Circuit Voltage, kV (rms)	4.76 (IEEE)	7.2 (IEC)	12 (IEC)	15 (IEEE)	24 (IEC)	38 (IEEE)
1-Minute withstand Voltage, kV (rms)	19	20	28	36	50	80
Basic Impulse Voltage, kV (peak)	60	40/60	60/75	95	95/125	150
Normal Rated Load Current, A (rms)	630–3150					
Rated Fault Interruption Current, kA (rms)	≤ 50		≤ 80		≤ 50	
Rated Making Current, kA (peak)	≤ 130	≤ 125	≤ 200	≤ 208	≤ 125	≤ 130
Short Time Fault Current withstand, s	1–3					

The vacuum circuit breaker and the vacuum recloser are designed to pass a series of tests that have been established by committees with representatives from equipment manufacturers, electric utilities, government agencies, testing laboratories, consultants and universities [15, 16]. Typical testing requirements are given in Table 6.1. In North America, the test standards are established by IEEE. Most of the rest of the world uses a variation of the IEC standards. In China GB/DL standards and in Russia the GOST standards are used. For development of a vacuum interrupter and a vacuum circuit breaker these various standards present a challenge, because there can be enough differences between them that one design, which successfully passes the tests for one standard, may not necessarily pass all the tests for another. Fortunately, there is considerable effort worldwide to consolidate these different standards into a set that will be accepted by all equipment manufacturers and electric utilities [17–19].

Once a vacuum circuit breaker or a vacuum recloser has been developed it is tested according to a schedule outlined in the standards at an independent high-power test laboratory [20]. When the device passes these tests, the laboratory certifies that the device has been successful. The manufacturer is then free to sell his device to all customers who will accept that certification. In general, this series of certification tests subjects the device to far greater stresses than it will ever be subjected to in its service life. For example, Burke et al. [21], have studied fault characteristics for widely distributed utilities in the United States with circuits ranging from 4.76kV to 34.5kV and a large range of potential fault circuits. They show that only 1% of fault currents reach 85% of the maximum rated value for the circuit and the most probable value is close to 20%. In my own experience, when examining vacuum interrupters from vacuum circuit breakers that had been installed for 25 years, I see little evidence of interruption duty. The vacuum interrupters exhibit so little stress to their internal structures that they could have easily remained in the field for another 25 years.

In this chapter, I will first discuss the different levels of current that a vacuum circuit breaker has to interrupt, beginning with the standard load currents and continuing with fault current interruption. I will then present a discussion of the transient recovery voltages expected across the open contacts once the fault current has been interrupted. The chapter continues with a general review of vacuum circuit breaker design criteria. It ends with a broad outline of the application of vacuum circuit breakers to a wide variety of ac and dc circuits.

6.2 LOAD CURRENTS

Although a vacuum circuit breaker is relied upon as a circuit protection device and must successfully interrupt a fault when one occurs, the vacuum circuit breaker's most usual function is to remain closed and pass the circuit's normal load current. This load current is usually in the range of 200A to 3150A. There are some applications, for example, generator breakers (see Section 6.8.6) where designs of 6300A and higher are required. The international electrical industry has determined an acceptable guideline for the interval between two levels of continuous current. These follow the R10 series developed by the French engineer Col. Charles Renard in the 1870s. He proposed a set of preferred numbers to be used with the metric system. These then would establish a set of standard intervals that consumers could use to compare one manufacturer's product with another's. In 1952 these numbers became the International Standard ISO 3. Renard's numbers divide the interval from 1 to 10 into 5, 10, 20, or 40 steps. The factor between two consecutive numbers in a Renard series is constant before rounding. This factor is the 5th, 10th, 20th, or 40th root of 10; i.e., 1.58, 1.26, 1.12, or 1.06, respectively. After some rounding, the R10 series used by the electrical industry is: 1, 1.25, 1.6, 2.0, 2.5, 3.15, 4.0, 5.0, 6.3, and 8.0. Thus, the accepted R10 series for continuous current from 100A is:

100A, 125A, 160A, 200A, 250A, 315A, 400A, 500A, 630A, 800A, 1000A, 1250A, 1600A, 2000A, 2500A, 3150A, 4000A, and so on.

and for short circuit current levels:

4kA, 5kA, 6.3kA, 8kA, 10kA, 12.5kA, 16kA, 20kA, 25kA, 31.5kA, 40kA, 50kA, 63kA, 80kA, and so on.

The R10 series is also now accepted in the United States, where the former load current levels were 600A, 1200A, 2000A, and 3000A. Thus, if a user's system has a maximum value of the continuous load current of 600A, then a vacuum circuit breaker designed to pass 630A would be chosen.

Figure 6.1 shows a typical schematic for a vacuum circuit breaker pole structure. The certification standards require that the first connection to the vacuum interrupter have a limit to the temperature it can reach while passing the rated continuous load current. The IEEE standards limit it to 50°C above the ambient temperature for bare Cu and 60°C above ambient temperature for Ag plated Cu [16]. The IEC standards value is 60°C above the ambient temperature [16]. In addition, 115C is the maximum permitted temperature; see Table 6.2. This can present a challenge when a vacuum circuit breaker is applied in circuits where a high ambient temperature such as 50°C–55°C is possible. The circuit breaker designer has to take into account the three contact resistances shown

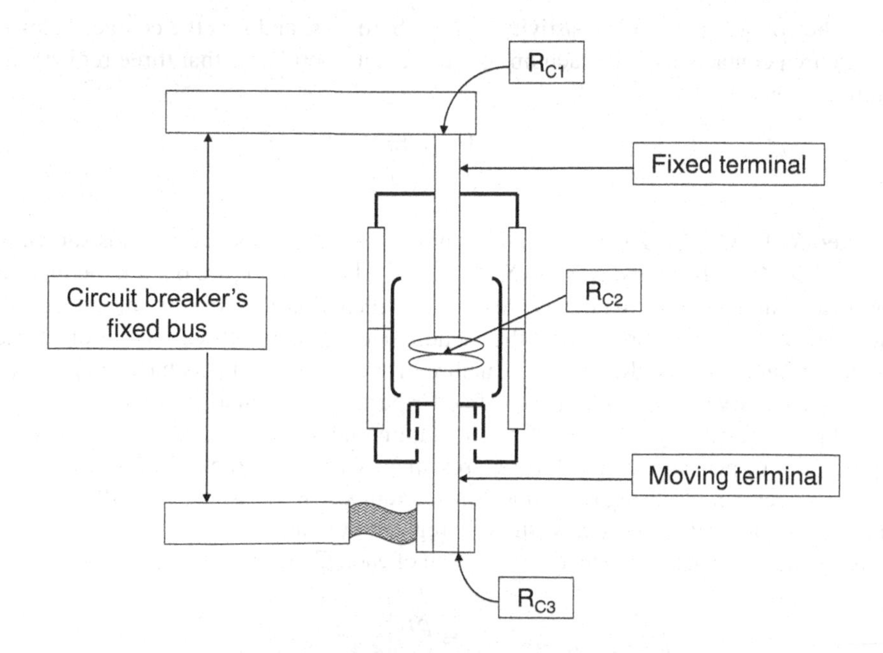

FIGURE 6.1 The connections of the vacuum interrupter to the vacuum circuit breaker's fixed bus.

TABLE 6.2
Temperature Standards for Bolted Joints

The Bolted Joint Material		Maximum Temperature Rise above a 40C Ambient	Maximum Temperature Allowed, C	Standard
Cu against Cu	In air	50	90	IEC & IEEE
	In SF_6	75	115	
	In oil	60	100	
Ag or Ni plated Cu against Ag or Ni plated Cu	In air	75	115	IEC & IEEE
	In SF_6	75	115	
	In oil	60	100	
Sn plated Cu against Sn plated Cu	In air	65	105	IEC & IEEE
	In SF_6	65	105	
	In oil	60	100	
Ag or Ni plated Cu against Cu		50	90	IEEE
Ag or Ni plated Cu against Cu		75	115	IEC

in Figure 6.1; i.e., R_{C1} (the connection to the fixed terminal), R_{C2} (the total vacuum interrupter resistance: (i.e., the contact resistance of the closed contacts R_C plus the resistance of the Cu terminals, R_T, plus the bulk resistance of the contacts), and R_{C3} (the connection between the movable terminal and the circuit breaker's bus).

Two things have to be considered when estimating a vacuum interrupter's contact resistance R_C: (1) the number of regions of contact and (2) the change in the surface structure of the Cu–Cr contact material after interrupting high short circuit currents. The contact resistance can be estimated for one region of contact using Equation (2.4), i.e.,

$$R_C = \frac{\rho_B}{2}\sqrt{\frac{\pi H}{F}} \tag{6.1}$$

Where ρ_B is the contact material's resistivity, H is its hardness, and F is the contact's closing force. For the large area contacts used in vacuum interrupters it is probable that three regions of contact occur. In that case:

$$R_C = \frac{1}{3}\frac{\rho_B}{2}\sqrt{\frac{\pi H}{F/3}} \tag{6.2}$$

The calculated R_C for Cu–Cr (25wt.%) contacts one and three regions of contact is shown in Figure 6.2 for $\rho_B = 3.3 \times 10{-}5$ Ωmm and H =900 Nmm^{-2}. Also shown in Figure 6.2 are the measurements of R_{C2} for a vacuum interrupter with new Cu–Cr contacts and after short circuit interruptions [22]. When the contacts are new it appears as if the vacuum interrupter has three regions of contact. After the short circuit interruptions, the surface structure of the Cu–Cr contacts has changed: see Figure 3.6. The top surface layer now contains very fine Cr particles in a matrix of Cu. It has a thickness of about 100 μm, a resistivity of $\rho_{SL} = 4.3 \times 10^{-5}$ Ωmm and a hardness, H = 2000 Nmm^{-2}. As the layer is quite thin the plating factor will be approximately 1 [23], so the effective resistivity used to calculate R_C will remain ρ_B. Using H = 2000 Nmm^{-2} and $\rho_B = 3.3 \times 10{-}5$ Ωmm the data after short circuit interruption are still consistent with three regions of contact.

For butt contacts or TMF contacts the resistance of each Cu terminal R_T is:

$$R_T = \frac{\rho L}{A} \tag{6.3}$$

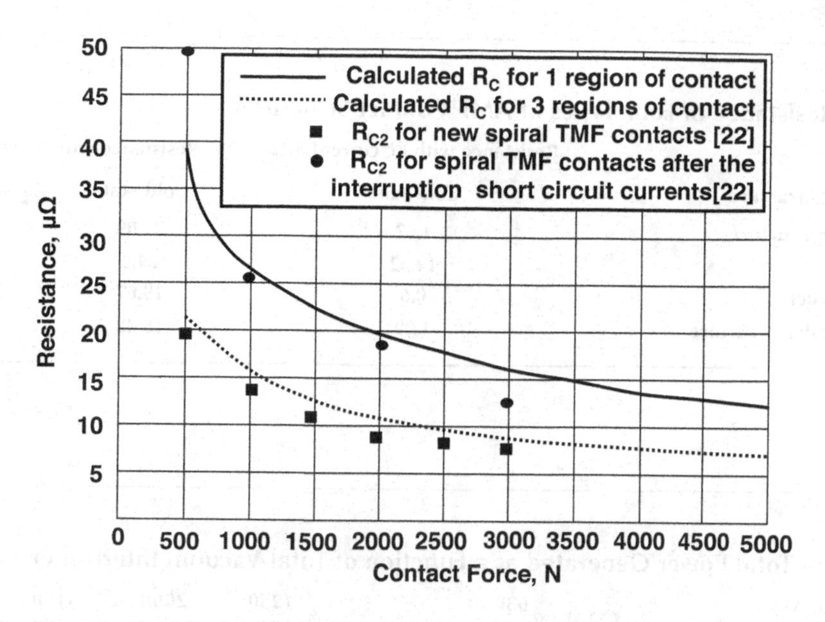

FIGURE 6.2 The calculated R_C for Cu–Cr(25wt.%) contacts with one and three regions of contact and the measured R_{C2} for Cu terminals plus Cu–Cr(25wt.%) spiral TMF contacts before and after short circuit current interruptions [22].

Where the resistivity of the oxygen free, high conductivity Cu is 1.65×10^{-5} Ωmm, L is the terminal's length and A its cross-section area. Its value is typically a few micro-ohms. When axial magnetic field contact structures are included in the vacuum interrupter's design, an added impedance results to the current path. Liu et al. [24] have analyzed four AMF structures that have been suggested for use in 126 kV, single break vacuum interrupters. These are shown in Figure 6.3:

1. A three-arm, two-layer coil [25]
2. ¾ coil [26]

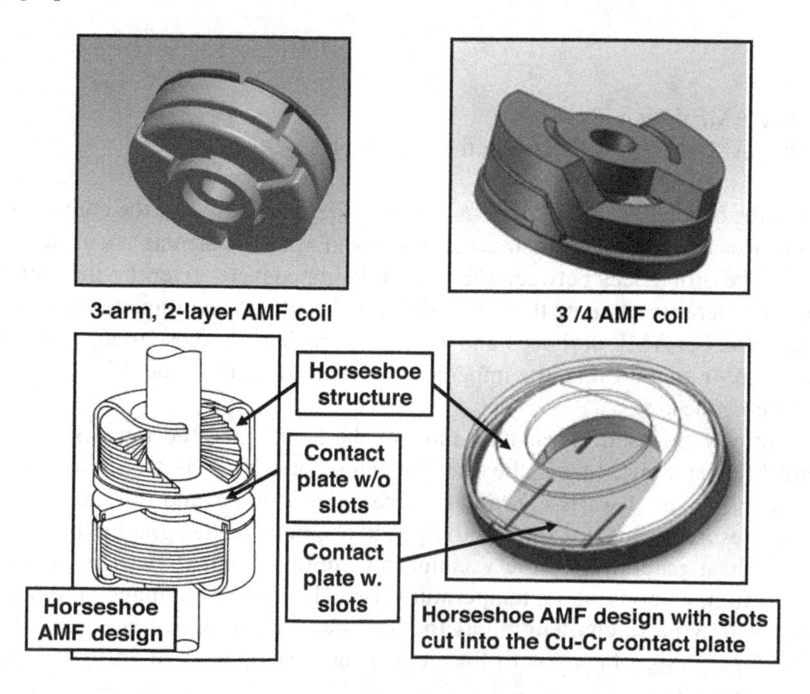

FIGURE 6.3 Four AMF structures.

TABLE 6.3
The Bulk Resistance of Four Types of AMF Contact Structures

AMF Contact Structure	Resistance with dc current, μΩ	Resistance with ac current, μΩ	
	Cold state	Cold State	Hot state
3-arm 2-layer structure	12.2	21.09	30.23
3/4 structure	14.32	24.2	36.93
Horseshoe structure	0.6	19.67	23.98
Slotted Horseshoe structure	1.09	15.48	16.25

TABLE 6.4
Examples of Total Power Generated as a Function of Total Vacuum Interrupter Impedance

Load Current, A	630	1250	2000	3150	4000
Total Vacuum Interrupter Impedance, Ω	Power Generated by Passing Load Current through the Vacuum Interrupter, Watts				
10	4	14	40	99	160
15	6	22	60	149	240
20	8	29	80	198	320
25	10	39	100	248	400
30	12	47	120	298	480
35	14	55	140	347	560
40	16	62	160	397	640

3. Horseshoe AMF design [27]
4. Horseshoe AMF design with slots in the contact plate [28]

They show that the bulk resistance of these AMF structures is affected by the eddy current losses in the contact surfaces and for the horseshoe designs, 3 and 4, eddy currents occur in the iron horseshoe structure. The differences between the ac and dc impedances given by the authors is shown in Table 6.3. It is interesting to note that for AMF coils 1 and 2 the ac impedance is approximately double the dc value. For AMF designs 3 and 4, the ac impedance is close to an order of magnitude times the dc value. In spite of this, the impedance of AMF designs 3 and 4 is lower than 1 and 2, with design 4 being the lowest.

As R_{C2} is entirely enclosed in vacuum, most of the heat generated must be conducted away by the circuit breaker's bus. Inside the vacuum interrupter there is some heat transfer to the vacuum shield and its inner walls via radiation from the contacts and the interior terminals. In spite of this, however, the bus, together with R_{C1} and R_{C3} must be designed to be heat sinks. This will ensure that heat generated by the vacuum interrupter's contacts will be dissipated, at the first connection to it, without undue temperature rise. The thermal management of the pole unit becomes increasingly more challenging as the load current increases: see Table 6.4. It is also becoming a critical design factor with the increasing development of fully encapsulated pole units; see Section 6.5.4.

For the vacuum interrupter designer, the diameter of the terminals is important. Obviously, the passage of 630A load current will require terminals with smaller diameters than a vacuum interrupter designed to carry a load current of 3150A. For a dc current, the resistance of a cylindrical conductor with a resistivity ρ, length L, and radius r is:

$$R_T = \frac{\rho L}{\pi r^2} \tag{6.4}$$

When an ac current of a given frequency f flows through a conductor most of it flows in a layer from the conductor's surface down to a depth from the surface termed the skin depth. The skin depth d_s is given by:

$$d_s = \left[\frac{\rho}{\pi f \mu_0}\right]^{1/2} \tag{6.5}$$

where μ_0 is the magnetic permeability, $4\pi \times 10^{-7}$ H.m^{-1} (Henries per meter). Table 6.5 gives the skin depth for various frequencies. From this table, it can be seen that once the diameter of terminal is more than $2xd_s$ the resistance of a conductor of length $L$is:

$$R_T = \frac{\rho L}{\pi d_s (2r - d_s)} \tag{6.6}$$

Thus, the area determined by d_s will conduct the current through the terminal. The added volume of Cu, however, does provide heat sinking and will contribute to the extraction of heat generated inside the vacuum interrupter. Most vacuum interrupters will be applied to 50Hz or 60Hz ac circuits where the vacuum designer must consider a skin depth of about 10mm. There are, however, circuits that supply considerable harmonics, e.g., ac to dc converters and arc furnaces. These harmonics, even on 50/60 Hz ac circuits, can result in reducing the skin depth thus making them an important design parameter. Therefore, knowledge of the vacuum circuit breaker's application is an important consideration when determining the diameter of the vacuum interrupter's terminals, the size of the breaker's bus and the heat sinking required.

The connection to the fixed terminal is usually a bolted connection. See Figure 6.4(a). In such a connection the force to provide contact and the region of contact is limited to a region close to the bolt's head, or if a washer is used, the region under the washer [29]. The resistance of this joint can be reduced by having multiple contact regions. The pad diagram shown in Figure 6.4(b) shows one often-used design that achieves four regions of contact. Connecting the bus to the moving terminal of the contact presents more of a challenge. There are two principal methods for doing this: (a) using a flexible conductor [30] and; (b) using a sliding [31] or rolling conductor. Examples are shown in Figures 6.5 and 6.6. For the flexible connection, as the load current is increased the number of Cu

TABLE 6.5
The Skin Depth for Current Flowing in a Copper Cylinder (ρ = 1.65×10^{-8} Ω.m)

Frequency, Hz	Skin Depth, $d_s$mm
50	9.2
60	8.4
100	6.5
200	4.6
400	3.2

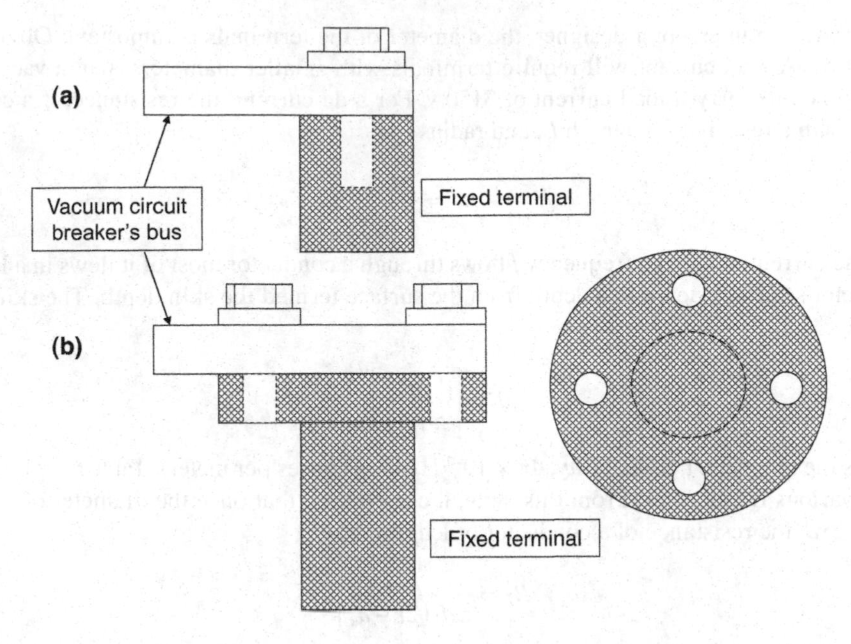

FIGURE 6.4 (a) A single bolted connection to the fixed end of the vacuum interrupter. (b) A multiple bolted joint to the fixed end of the vacuum interrupter using a brazed pad to the fixed terminal.

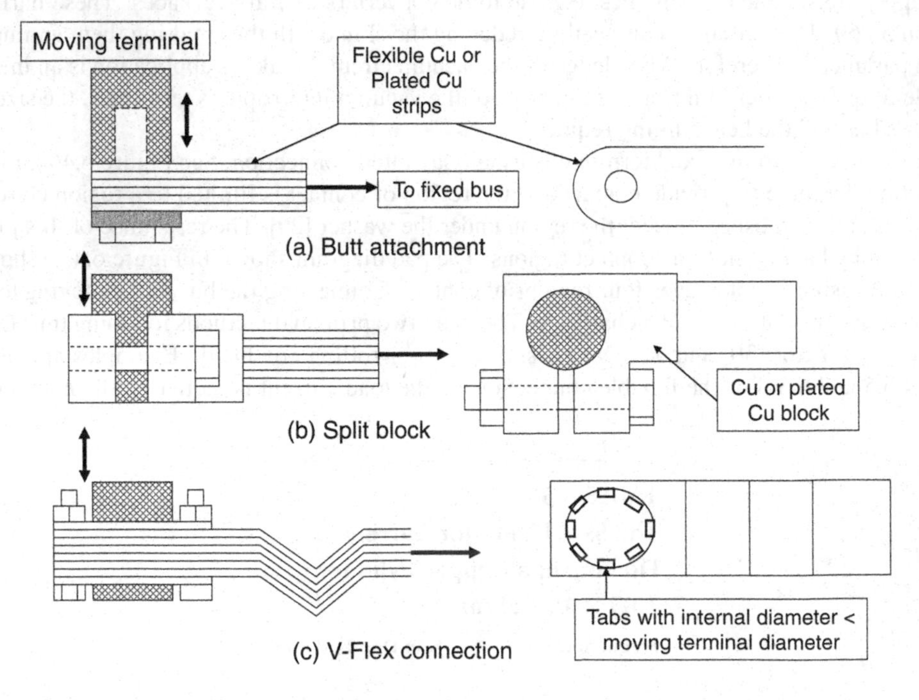

FIGURE 6.5 (a) A flexible connection bolted to the end of the vacuum interrupter's moving terminal. (b) A flexible connection using a split block to connect it to the vacuum interrupter's moving terminal. (c) A V-Flex connection [30].

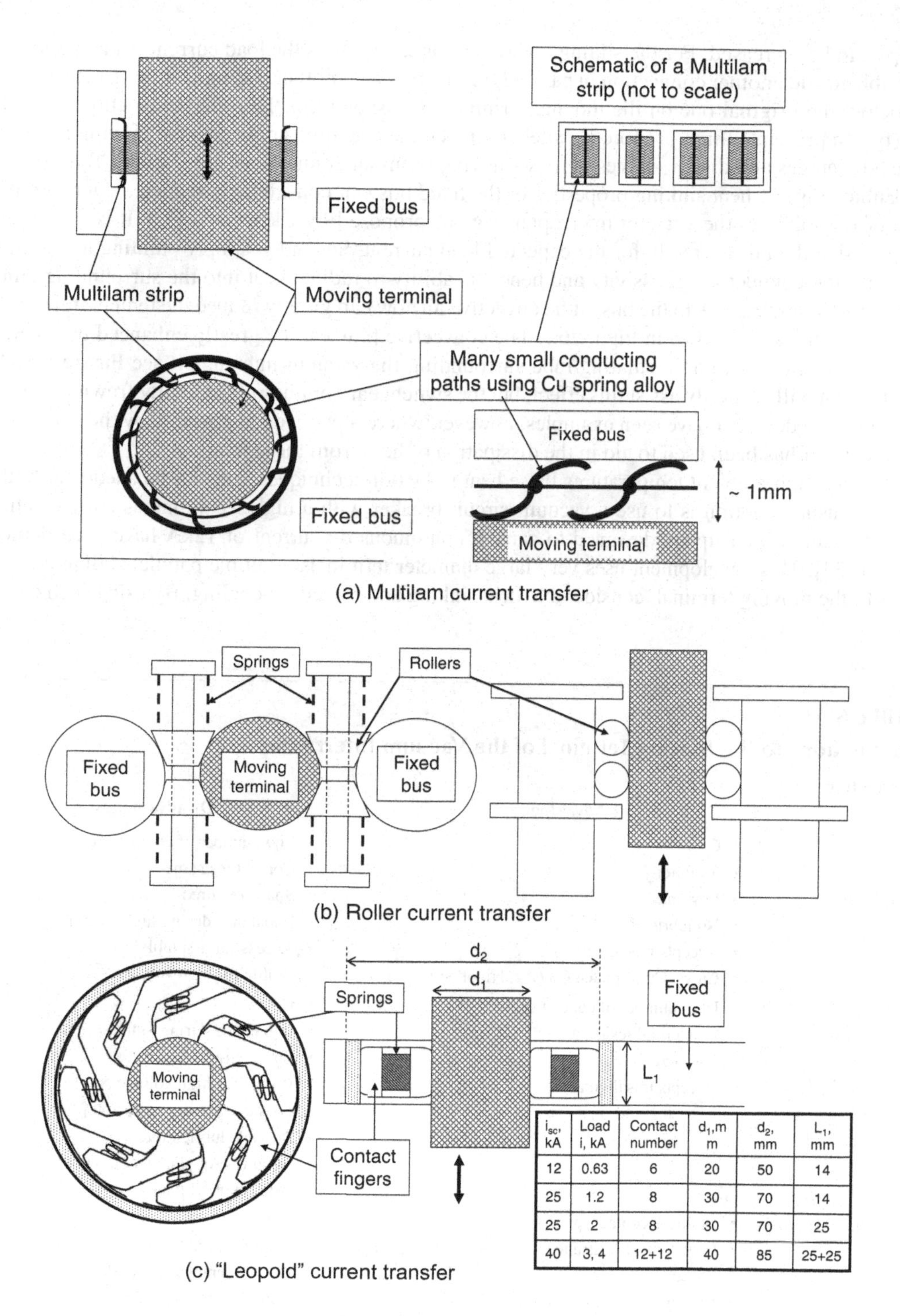

i_{sc}, kA	Load i, kA	Contact number	d_1,mm	d_2, mm	L_1, mm
12	0.63	6	20	50	14
25	1.2	8	30	70	14
25	2	8	30	70	25
40	3, 4	12+12	40	85	25+25

FIGURE 6.6 (a) The "Multilam sliding connection to the vacuum interrupter's moving terminal [31]. (b) The roller connection to the vacuum interrupter's moving terminal. (c) The "Leopold" connection to the vacuum interrupter's moving terminal.

strips can be increased. For the sliding and roller connections, as the load current is increased, it is possible to add another connection in parallel (i.e., by placing another sliding/rolling structure above or below the original one on the moving terminal as is shown for the roller connection in Figure 6.6(b)). In principle three or more parallel sliding contacts can be applied to the moving terminal. The advantages and disadvantages of these moving terminal connections are given in Table 6.6.

Enhancing the heat sinking properties of the fixed bus can considerably reduce the temperature rise of the joints to the vacuum interrupter; e.g., as proposed by Cleaveland [32]. It is first necessary to size the bus correctly for the expected load current. Secondly, simply painting it black can increase the conductor's emissivity and hence its ability to radiate heat into the surrounding ambient. Thirdly, adding fins to the bus with convective airflow can greatly reduce the temperature of the joint. Examples are shown in Figure 6.7. The convective flow can be greatly enhanced by creating a chimney effect with an insulating tube surrounding the vacuum interrupter; see Figure 6.8. An electric fan will certainly assist this effect, but the switchgear community generally frowns upon the use of such a device. I have seen examples, however, where a passive fan powered by the convective airflow itself has been used to aid in the dissipation of heat from these joints.

At very high ambient temperatures these heat reduction techniques may not be adequate. In this case the usual solution is to use a vacuum circuit breaker with a higher continuous current rating. Single vacuum interrupters designed to carry a continuous dc current of 12kA have been demonstrated [33]. This development uses very large diameter terminals, multiple parallel sliding connections to the moving terminal, considerable heat sinking and forced air-cooling. By using such design

TABLE 6.6
Connections to the Moving Terminal of the Vacuum Interrupter

Connection Type	Advantages	Disadvantages
Butt & Split Block	• Cost • Simplicity • Low mass • No lubrication • Accepts misalignment • Cu vs. Cu or plated Cu (Ag, Ni, or Sn)	• Low number of contact spots • Contact resistance • Space required • Imbalance during fault current • Access for assembly • Profile for high voltage
V-Flex [23]	• High number of contact spots • Low initial resistance • No friction • Accepts misalignment	• Cost • Imbalance during fault current • Space required • Profile for high voltage • Possible fretting corrosion [29] • Possible lubrication • Requires Ag plate (should not be used with Sn or Ni plate)
Multilam [24] & Roller & Leopold	• Multiple contact spots • Low contact resistance • Space • Damping during operation • Uniform load during fault current • Blind assembly possible • Possible lower contact loads • Smooth high voltage profile • Blow-on of contacts during faults (Multilam & Leopold)	• Cost • High friction • Good tolerance required • Lubrication required • Possible fretting corrosion [29] • Terminal and contact structure should be Ag plated (should not be used with Sn or Ni plate) • Possible wear will increase contact resistance

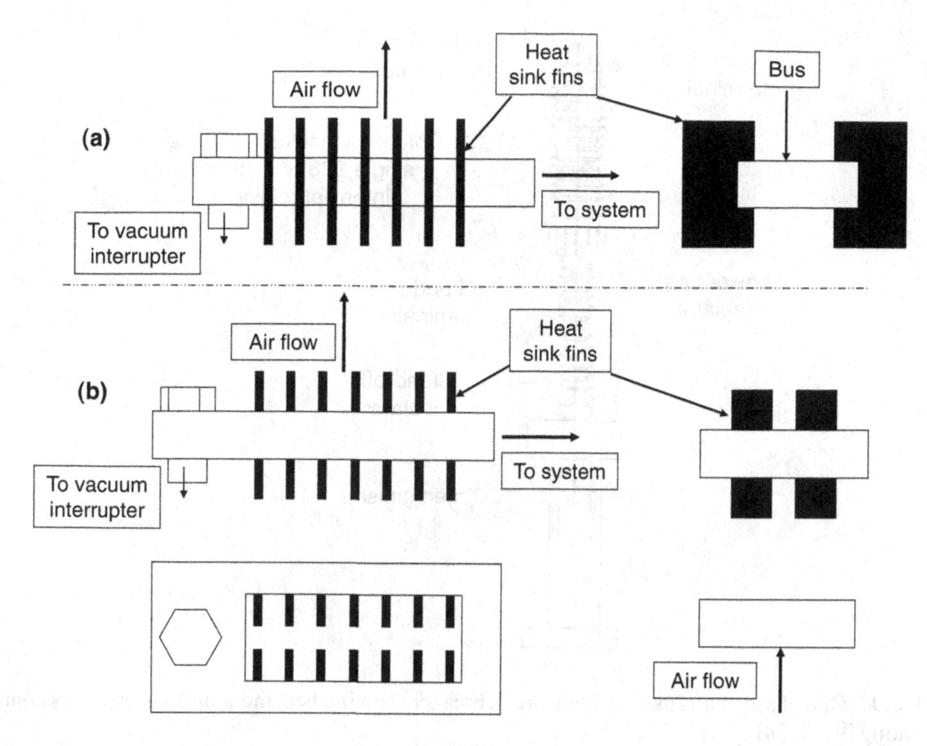

FIGURE 6.7 Examples of heat sink designs for the vacuum circuit breaker's fixed bus.

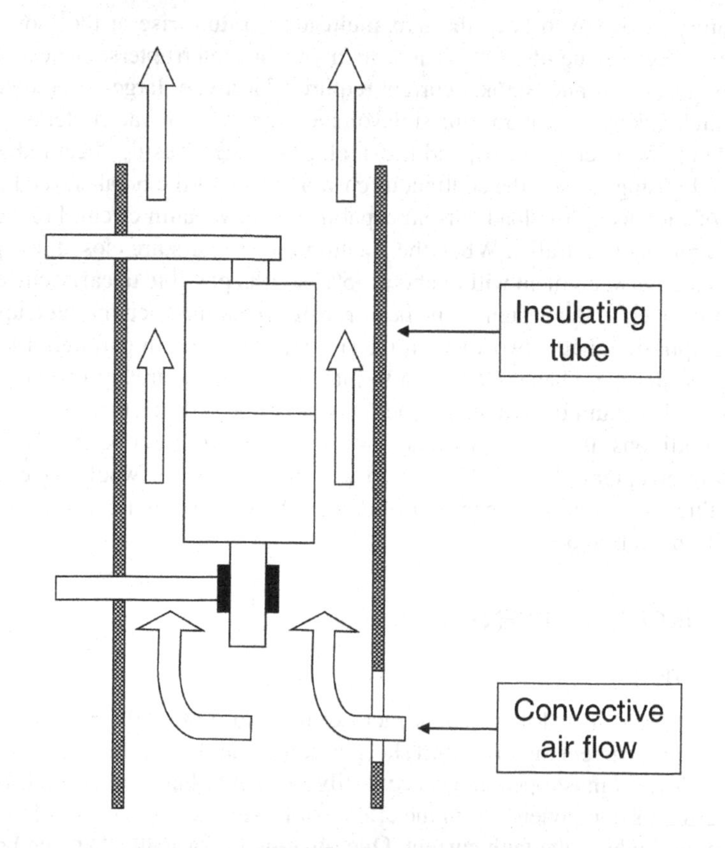

FIGURE 6.8 Enhancement of the convective airflow passed a vacuum interrupter using the chimney effect.

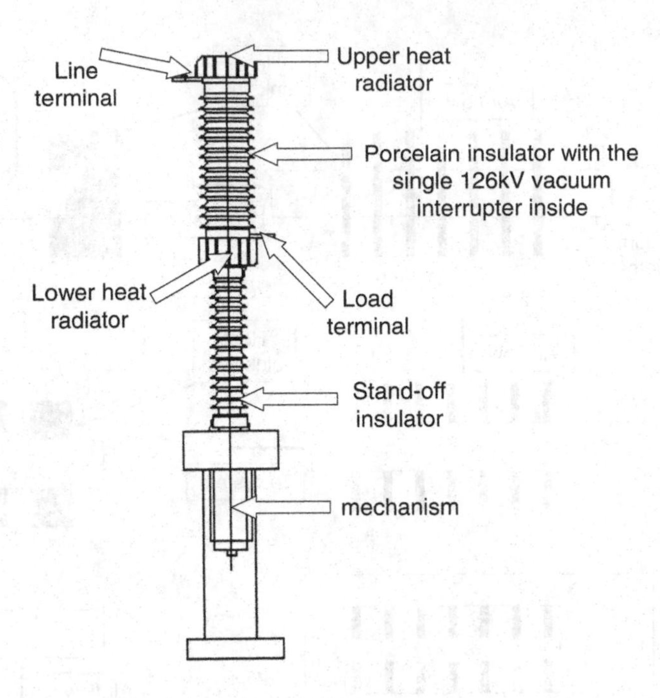

FIGURE 6.9 One phase of a 126kV vacuum circuit breaker showing heat radiators to increase its continuous current rating [9, 10, 24].

concepts, the authors manage to keep the maximum temperature rise at the movable terminal to 55.2C above the ambient value of 22.8C. Three such vacuum interrupters connected in parallel are capable of carrying the continuous 36kA current required for a very large superconducting magnet. For vacuum circuit breakers used at transmission voltages, the vacuum interrupter often resides inside a porcelain insulator. Wang et al. [9, 10] and Liu et al. [24] suggest using "heat radiators" at the top and bottom of the housing to raise the continuous current rating of the breaker, see Figure 6.9.

One method of increasing the load current capability of a vacuum circuit breaker is to connect two vacuum interrupters in parallel. When the vacuum interrupters are closed it is possible for the current to be divided between them within about ±5%, so it is possible to carry currents up to 10kA if close attention is given to the design of the heat transfer from the vacuum interrupter's terminals. Sato et al. [34] emphasize that if two vacuum interrupters are used in parallel, it is important that they open at the same time when called upon to interrupt a short circuit current. Thus, the major concern with parallel vacuum interrupters is the division of the full short circuit current when opening under fault conditions. It is quite probable that the current will not be evenly divided between the two vacuum interrupters [35–37]. Thus, each vacuum interrupter would have to be capable of interrupting a value current greater than half of that of the circuit's fault current capability. I shall discuss this again in Section 6.8.6.

6.3 SHORT CIRCUIT CURRENTS

6.3.1 INTRODUCTION

The vacuum circuit breaker and the vacuum recloser have to protect the circuit to which they are applied. Thus, if a fault current (up to the full short circuit capability of the circuit) is detected, these circuit protection devices must open and successfully interrupt that fault current. In doing so they will protect the circuit's components from the effects of the excessive current. Many circuit parameters affect the magnitude of the fault current. One obvious factor is the distance between the fault and the circuit breaker. Any line impedance can decrease the magnitude of a short circuit current to a value considerably less than the maximum value the circuit is capable of delivering. Other circuit components such as generators, motors and utility feeds can contribute to the level of short circuit current and to its initial asymmetry. For example, induction motors may contribute for a short time while synchronous motors will have a much longer-term effect.

6.3.2 The Short Circuit Current and Asymmetry

A simplified model of the effects of a sudden short circuit on a single-phase ac circuit is illustrated in Figure 6.10. A short circuit at the load side terminals of a vacuum circuit breaker is called a terminal fault, a bus fault, or a bolted fault. Since the impedance on the load side is very low, the current can be very high. In Greenwood's analysis [38] the load is represented by a resistance R and inductance L in series, whose power factor is:

$$cos\varphi = \frac{R}{|Z|} = \frac{R}{\left(R^2 + \omega L^2\right)^{1/2}} \tag{6.7}$$

The short circuit current i_{sc} is initiated when the switch S closes, then:

$$Ri + L\frac{di_{sc}}{dt} = U_m sin\left(\omega t + \theta\right) \tag{6.8}$$

The switch S closes at the ac voltage's phase angle θ. In Greenwood's analysis the short circuit current is:

$$i_{sc}\left(t\right) = \frac{U_m}{\left(R^2 + \omega^2 L^2\right)^{1/2}}\left[sin\left(\omega t + \theta - \varphi\right) + sin\left(\theta - \varphi\right)e^{-\left(\frac{R}{L}\right)t}\right] \tag{6.9}$$

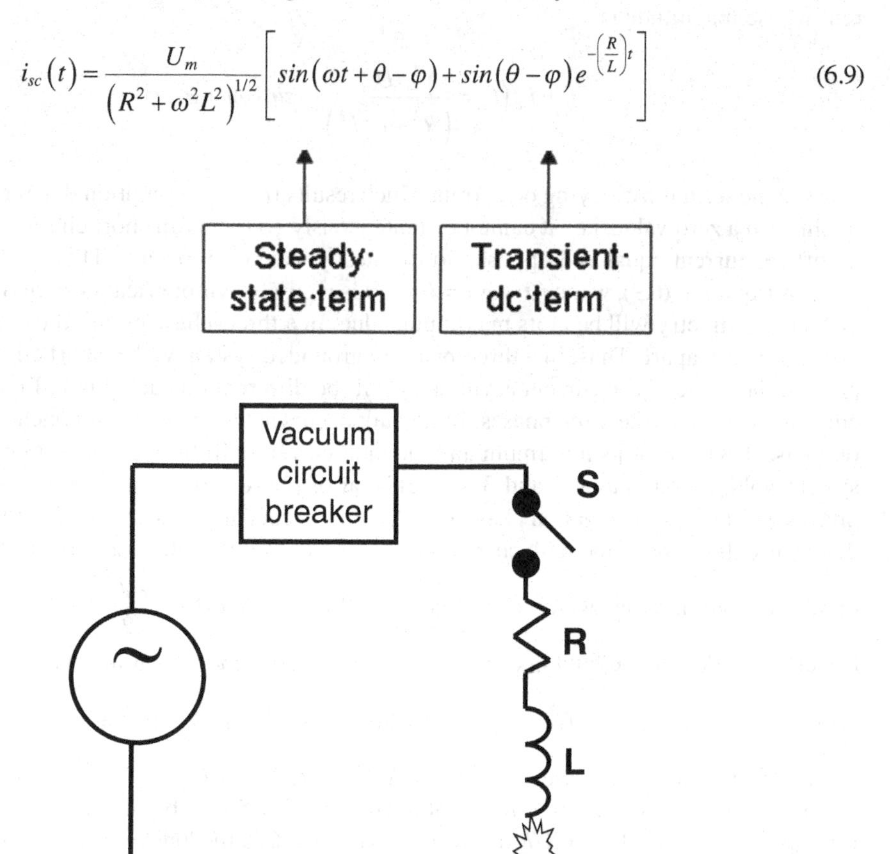

FIGURE 6.10 Circuit for analyzing the fault current level.

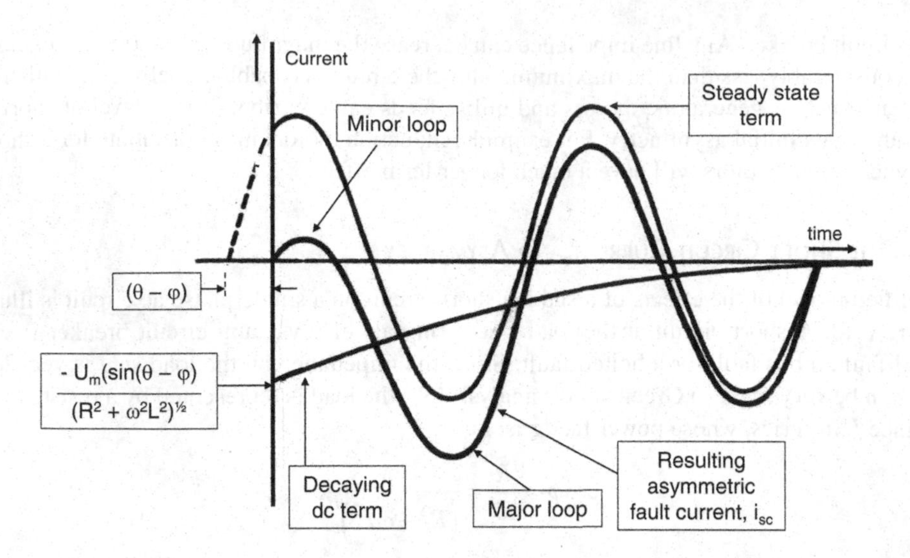

FIGURE 6.11 The short circuit current, showing the steady state term, the dc component and the resulting asymmetric fault current.

Where $tan\varphi = \omega L/R$ and The first term in Equation (6.9) is the steady state symmetrical ac current whose magnitude is:

$$i_{sc}(t) = \frac{U_m}{\left(R^2 + \omega^2 L^2\right)^{1/2}} sin(\omega t + \theta - \varphi) \tag{6.10}$$

The second term is a decaying dc current which results from the condition that the fault current must begin with a zero value; i.e., it cannot instantaneously go to its full short circuit value, but at $t = 0$ a dc off-set current equal and opposite to the steady-state term occurs. This is illustrated in Figure 6.11. In Equation (6.9), when $(\theta - \varphi) = 0$, $i_{sc}(t)$ is a purely symmetrical current, but when $(\theta - \varphi) = \pi/2$ the asymmetry will be at its maximum value. In a three-phase circuit the current zeros in each phase are 120° apart. Thus, in a three-phase ungrounded system with a short circuit across all three phases, the value of the current asymmetry will be different on each phase. Figure 6.12 shows the currents on each of the three phases. If the fault occurs at "A," the dc component of the fault current on phase 1 will be at its maximum and the fault current will have the maximum asymmetry. The system voltages on phases 2 and 3 are half that of phase 1 therefore the dc offsets on those two phases will be less than that on phase 1. If the fault occurs at "B" there will be no dc offset on phase 2, because the value of the fault current's dc component on this phase is zero. Thus, the fault current on phase 2 will have no asymmetry. Phase 1 will have a dc offset $\frac{-\sqrt{3}}{2}$ and phase 3 will have a dc offset $\frac{+\sqrt{3}}{2}$. When the fault occurs at "C" the peak asymmetry is now on phase 3. Phases 1 and 2 will now have similar dc offsets as occurred on phases 2 and 3 at position "A." The dc offset decay term $e^{-\frac{R}{L}t}$ is usually discussed in terms of X/R where $X = \omega L$ (i.e., $e^{-\frac{R}{X}\omega t}$). A usual value for short circuit certification testing is $X/R = 17$ at 60Hz and 14 at 50Hz, but it is possible for some practical circuits to have much larger X/R ratios. The larger the X/R, the longer the decay time of the transient dc current. For example, for $X/R = 12$ a short circuit current will be approximately symmetrical after five cycles, but for $X/R = 22$ it will take about nine cycles. Figure 6.13 shows an example of a close–open operation on a three-phase fault current. It clearly illustrates the different values of

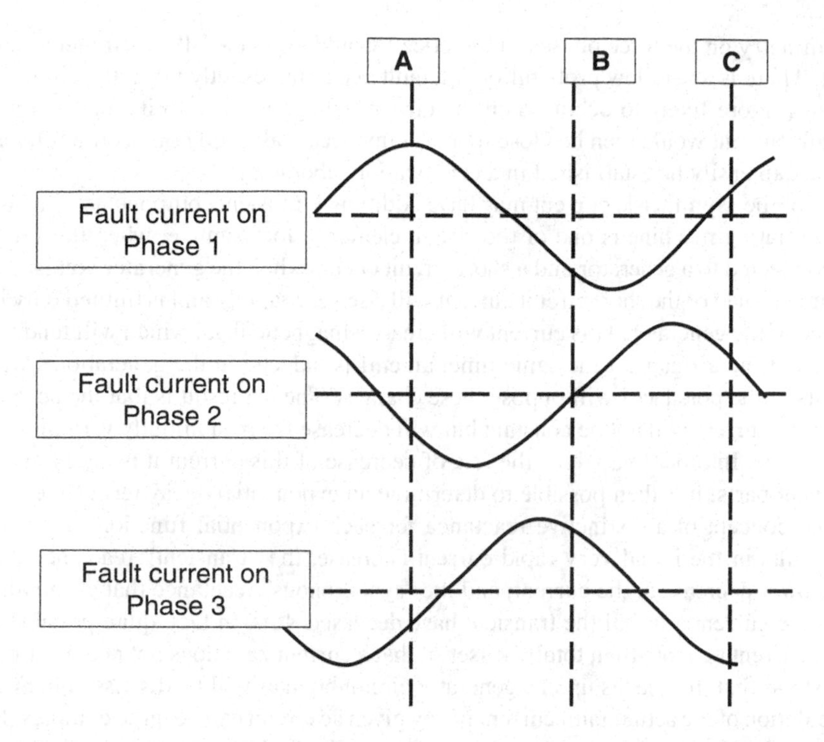

FIGURE 6.12 The currents on the 3-phases to illustrate they are 120° apart.

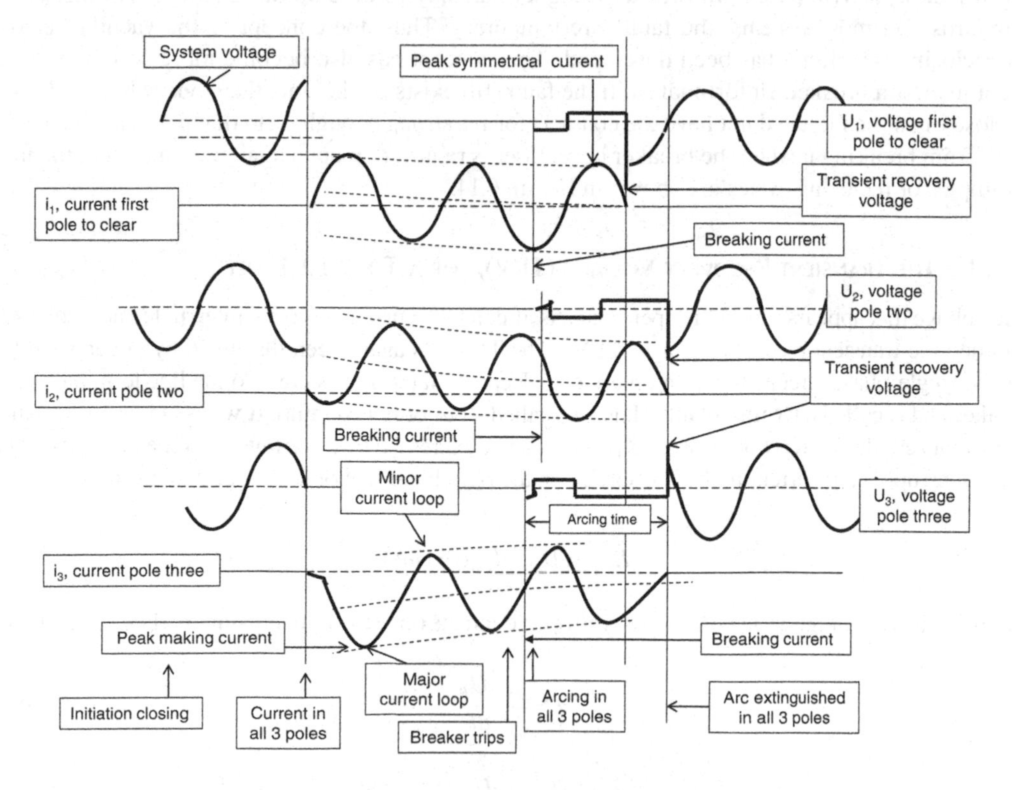

FIGURE 6.13 An example of a close–open operation on a short circuit fault occurring in three-phase ungrounded circuit.

current asymmetry on the three phases. In practice I would expect a fully asymmetric current to be a rare event. There is a very low probability of a fault occurring exactly when the circuit's voltage is zero. In fact, it more likely to occur when the circuit voltage is close to its maximum value. The resulting fault current would then be close to its symmetrical value. Of course, the fully asymmetrical condition can easily be established in a certification laboratory.

The peak of the asymmetric current may have additional transient components. This is especially true when a rotating machine is one of the circuit elements; for example, when the vacuum circuit breaker is connected to a generator and a short circuit occurs when the generator voltage is at its peak value. The magnitude of the short circuit current will rise very rapidly and is limited only by the leakage reactance of the generator. This current will create a magnetic field, which will tend to cancel the magnetic flux at the air gap. At the same time, an emf is induced in the generation's windings and eddy currents in the pole faces will oppose these changes. The net result is that the ac component of the short circuit current will not be constant but will decrease from an initially very high value to the steady state value. In order to analyze the rate of decrease of this current it is necessary to divide it into constituent parts. It is then possible to determine an exponential decay term for each part. This results in the concept of a distinctive reactance for each exponential function: the "subtransient" reactance results in the initial very rapid current decrease; the "transient" reactance produces the somewhat slower decrease in the current; and the "synchronous" reactance that eventually results in the steady state current after all the transient have deceased. It is, in fact, quite possible to have the short circuit current be more than totally offset so that a current zero does not occur for many cycles. The added stress that this places upon a generator circuit breaker will be discussed in Section 6.8.6.

The calculation of the actual fault current in any given ac circuit can be quite complex. Fortunately for the vacuum circuit breaker designer, knowledge of the fault levels in a practical circuit is not required. The standards for the certification testing set the fault levels and the degree of asymmetry. The vacuum interrupter circuit breaker designer has only to develop designs that will meet these standards. On most systems, the faults are temporary. Thus, the concept of the vacuum recloser for reclosing the circuit has been developed. This concept has also become more common for the vacuum circuit breaker. Unfortunately, if the fault still exists on the line, the devices have to be able to close in on the fault and not have the contacts form a strong enough weld that prevents the mechanism from breaking it when the breaker or recloser is required to open again and interrupt the fault. Examples of reclosing cycles are shown in Figure 6.14.

6.3.3 The Transient Recovery Voltage (TRV), for a Terminal Fault

The voltage that appears across the open contacts after a fault current has been interrupted at current zero is called the transient recovery voltage (TRV). The TRV has again been discussed by Greenwood [38] for the single-phase circuit shown in Figure 6.15. Here the fault occurs close to the line terminals of the breaker and is called a terminal fault. (It is also called a bus or bolted fault). If we assume at the moment of current zero the source voltage is at its peak, it can be expressed by $U_m cos\omega t$. After the fault current has been interrupted the current in the L C circuit on the right-hand side of Figure 6.15 is given by:

$$L\frac{di}{dt} + U_R = U_m cos(\omega t) \tag{6.11}$$

where U_R is the voltage appearing across the open contacts once the fault current has been interrupted.

$$i = C\frac{dU_R}{dt} \tag{6.12}$$

$$\frac{d^2U_R}{dt^2} + \frac{U_R}{LC} = \frac{U_m}{LC} cos(\omega t) \tag{6.13}$$

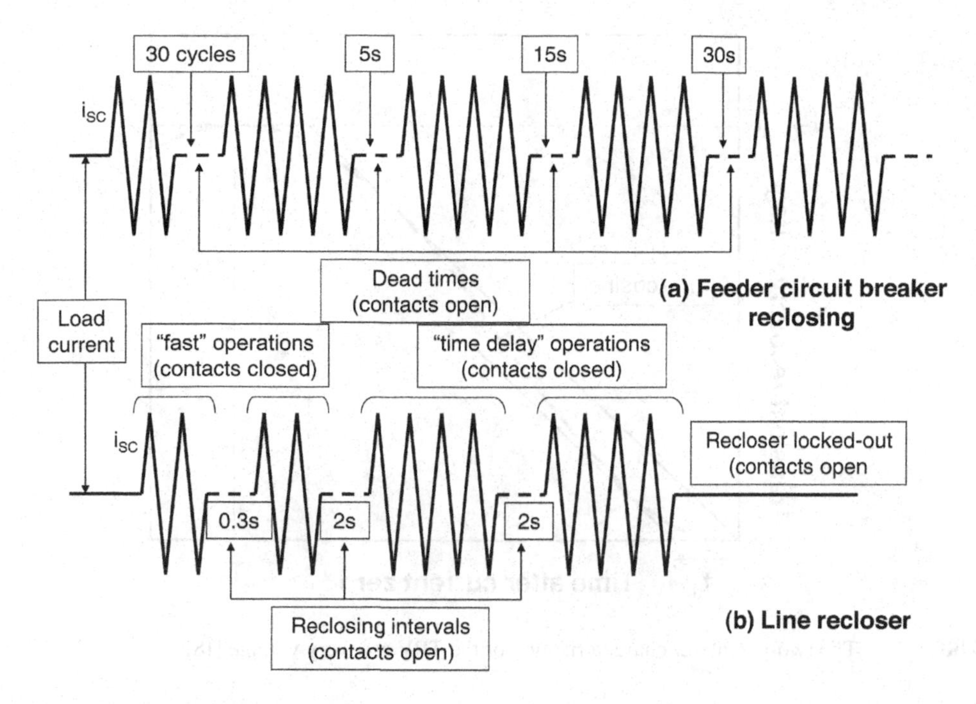

FIGURE 6.14 Possible reclosing cycle for vacuum circuit breakers and vacuum reclosers.

Greenwood's solution for these equations gives:

$$U_R(t) = U_m\left[cos(\omega t) - cos(\omega_0 t)\right] \tag{6.14}$$

where $\omega_0 = 1/(LC)^{1/2}$. Now at current zero $cos(\omega t) = 1$ and remains more or less constant for a few 10s of microseconds, i.e., during the time to the peak value of $U_R(t)$. Thus Equation (6.14) reduces to:

$$U_R(t) = U_m\left[1 - cos(\omega_0 t)\right] \tag{6.15}$$

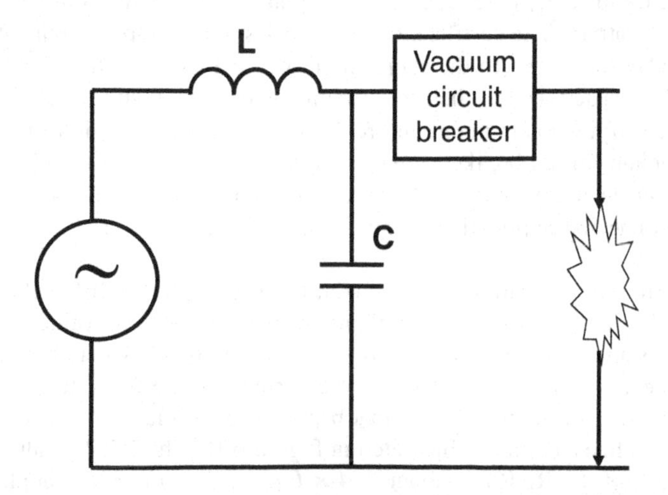

FIGURE 6.15 Circuit for analyzing the transient recovery voltage (TRV) after interrupting the fault current.

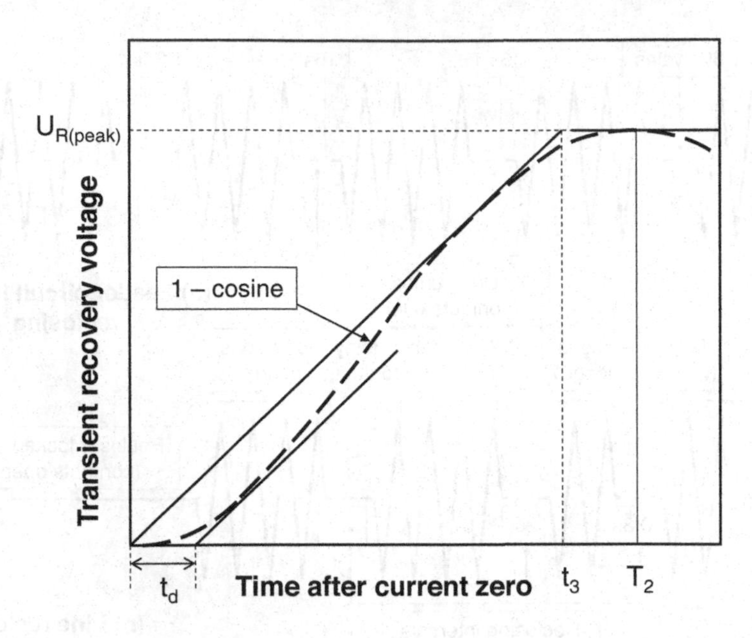

FIGURE 6.16 The two-parameter characterization of the TRV recovery voltage [18].

Although the maximum value of $U_R(t)$ in Equation (6.15) is $2U_m$, damping in the circuit results in lowering the peak value by about 20%–30%; this is illustrated in Figure 6.16. The TRV will oscillate with the frequency $1/[2\pi(LC)^{1/2}]$, but will decay exponentially until there is just the open circuit voltage across the open vacuum interrupter's contacts; see Figure 6.17. Because the value of the circuit voltage $U_m cos\omega t$ will be lower after interrupting an asymmetrical current, $U_R(t)$ will also be lower; this is also illustrated in Figure 6.17. The exact description of the TRV can become quite complicated; see, for example, the IEEE Application Guide [39], and Garzon [40]. Fortunately, the one-minus-cosine TRV wave applies for most of the vacuum circuit breaker applications at circuit voltages 72.5kV and below.

Figure 6.16 shows two ways of characterizing the *(1 – cosine)* TRV. The first method is that formerly used by the IEEE standards where the $U_{R(peak)}$ value (called E_2)) and the time t_p to reach the peak value (called T_2) are measured. The Rate of Rise of the Recovery Voltage (RRRV) is then $U_{R(peak)}/ t_p$. The values for $U_{R(peak)}$ and t_p are based upon system studies, but have the disadvantage of being somewhat arbitrary. The certification laboratories, for example, often have cables connecting the circuit breaker under test to the short circuit generators. Both the cable and the generator can add considerable capacitance to their test circuits, which can result in relatively slow RRRV's. When testing and certifying vacuum circuit breakers for indoor application this does not present a problem. Indoor vacuum circuit breakers are usually connected to the distribution system by cables. For outdoor, line connected applications, however, the RRRV values are about two times those for the indoor, cable-connected applications. Thus, a more formal way of characterizing the TRV is necessary.

In recent times the harmonization of the standards has brought the IEEE characterization close to that of the IEC [18]. This is also shown in Figure 6.16. Here the peak value $U_{R(peak)}$ is again measured, but the time is now t_3. As can be seen from Figure 6.16 this is where the line from the origin and tangential to the TRV curve intersects with the horizontal line from the peak voltage. There is also a delay line shown parallel to this line that begins after a delay time t_d. The TRV curve must lie between these two lines as I have illustrated in Figure 6.16. The RRRV value is now defined as $U_{R(peak)} / t_3$: this value of the RRRV is about $1.14 \times U_{R(peak)}/T_2$. For indoor applications (i.e., cable connected) the RRRV at 15kV is about 0.5 kV.μs^{-1} and for outdoor applications (i.e., line connected)

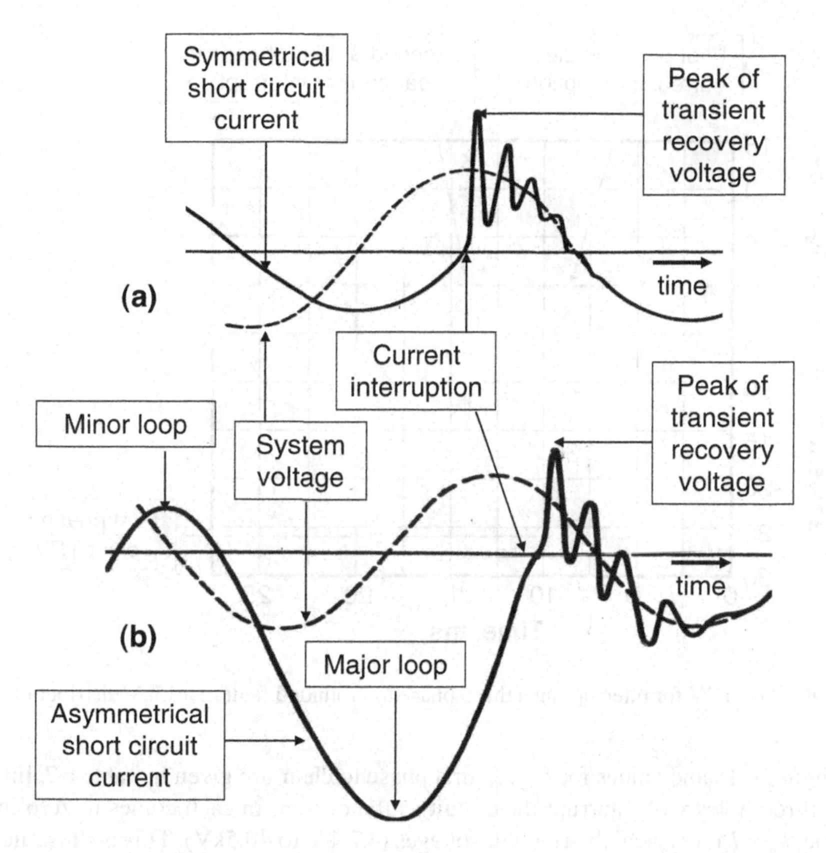

FIGURE 6.17 Illustrating the TRV shape and the TRV decay when interrupting different values of asymmetric current.

it is about 1.0 kV.μs^{-1}. In Sections 6.3.5, 6.3.6, and 6.8.6 I will discuss some special applications where the RRRV is much higher.

6.3.3.1 First Pole-to-Clear Factor

The peak value of the TRV ($U_{R(peak)}$) is defined by [39]:

$$U_{R(peak)} = k_{pp} \times k_{af} \times \frac{\sqrt{2}}{\sqrt{3}} U_{\ell\ell} \tag{6.16}$$

Where $U_{\ell\ell}$ is the system's line-to-line rms voltage, k_{pp} is the "first pole to clear" factor (it is the ratio of the power frequency voltage across the interrupting pole before current interruption in the other poles, to the power frequency voltage occurring across the pole or poles after interruption in all three poles [39]) and k_{af} is a transient amplitude factor that depends whether the circuit breaker is line connected or cable connected [18, 39]. Typical distribution voltage (4.76kV to 40.5kV) values of k_{af} for a 100% short circuit current are 1.4 for cable-connected systems and 1.54 for line-connected systems. The value of k_{af} can increase to 1.7 for a fault level of 10% of the full short circuit current. With a three-phase fault there will be one phase that is the first to interrupt the current. In an ungrounded system $k_{pp} = 1.5$. After the interruption of the first phase in the ungrounded system the fault current is then carried by the other two phases (with opposite sign); see Figure 6.13. Thus, the last two interrupters will be in series when this current is interrupted. Each pole will share the phase-to-phase voltage, so that for each pole $k_{pp} = \frac{\sqrt{3}}{2} = 0.87$. A typical sequence of the TRVs is

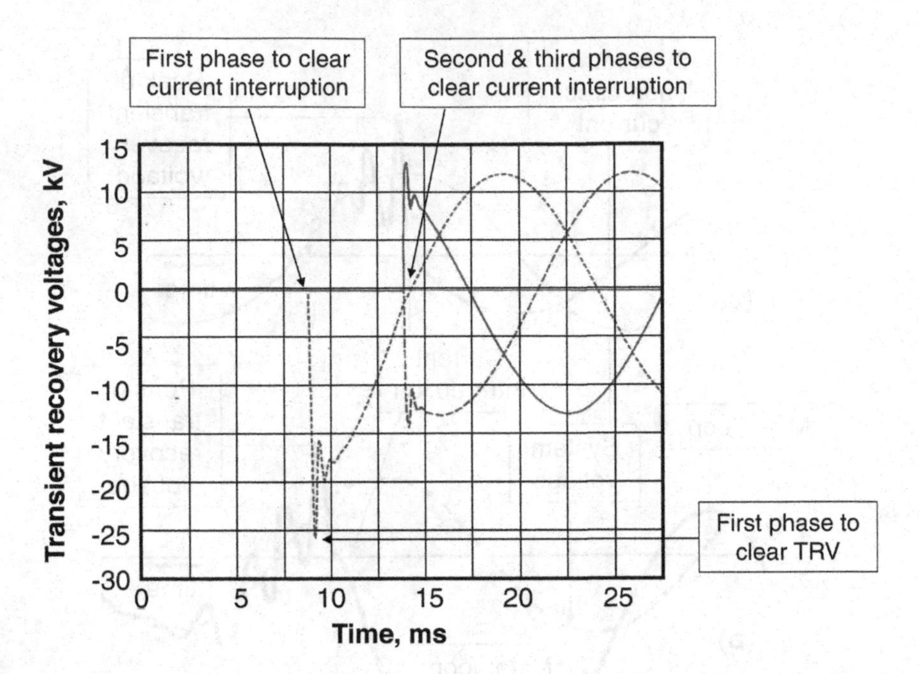

FIGURE 6.18 The TRV for interrupting a three-phase ungrounded fault in a 15kV distribution system.

shown in Figure 6.18 and values for $U_{R(peak)}$ first phase to clear are given in Table 6.7. In a grounded system the three poles will interrupt the circuit as the current in each comes to zero in sequence: for each pole, $k_{pp}=1$ for typical distribution voltages (4.76kV to 40.5kV). This is illustrated in Figure 6.19. You will notice that there is some ground shift after the second and third phases interrupt the current, but this is generally not a significant factor in determining the vacuum circuit breaker's performance. It is only when the system voltage is above 100kV that $k_{pp} = 1.3$.

TABLE 6.7
Expected Peaks for Each Phase of the Transient Recovery Voltages for a 100% Short Circuit Current in a 15kV Line-to-Line System Voltage ($U_{R(peak)} = k_{pp} \times k_{af}$ x$\{\sqrt{2}/\sqrt{3}\} \times U_{\ell\ell}$)

Peak of the TRV, $U_{R(peak)}$	Phase to Clear	Connection	Ungrounded Neutral	Grounded Neutral
	1st	Line connected (k_{af} = 1.54)	(k_{pp} = 1.5) 28.3 kV	(k_{pp} = 1.0) 18.9 kV
		Cable connected (k_{af} = 1.4)	(k_{pp} = 1.5) 25.7 kV	(k_{pp} = 1.0) 17.1 kV
	2nd	Line connected (k_{af} = 1.54)	(k_{pp} = 0.87) 16.4 kV	(k_{pp} = 1.0) 18.9 kV
		Cable connected (k_{af} = 1.4)	(k_{pp} = 0.87) 14.9 kV	(k_{pp} = 1.0) 17.1 kV
	3rd	Line connected (k_{af} = 1.54)	(k_{pp} = 0.87) 16.4 kV	(k_{pp} = 1.0) 18.9 kV
		Cable connected (k_{af} = 1.4)	(k_{pp} = 0.87) 14.9 kV	(k_{pp} = 1.0) 17.2 kV

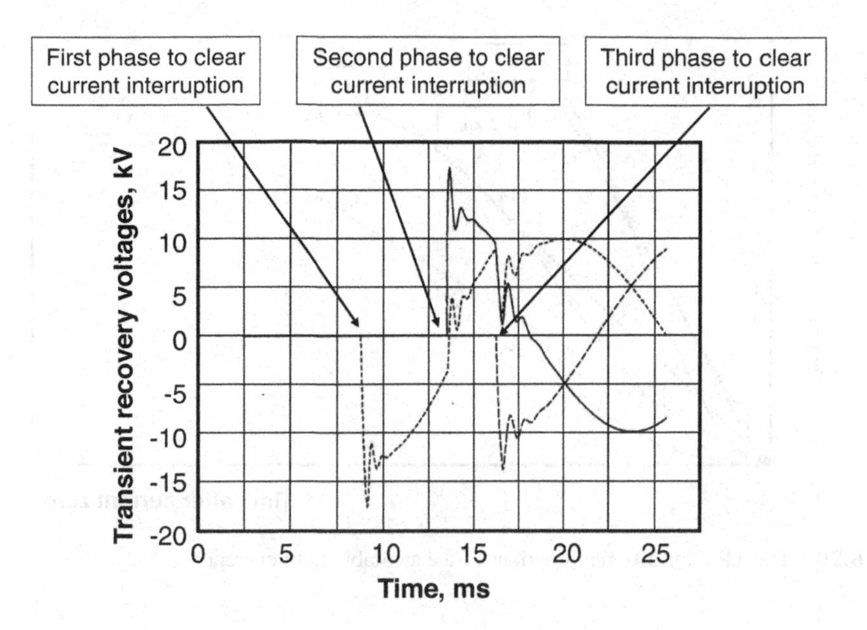

FIGURE 6.19 The TRV for interrupting a three-phase grounded fault in a 15kV distribution system.

Distribution vacuum interrupters (4.76 kV to 40,5 kV) are tested according to IEC and IEEE standards for short circuit interruption on ungrounded circuits with k_{pp} = 1.55 for the first phase to clear. There is a new IEC testing series for effectively grounded circuits where the transformer is connected to ground with an impedance less than the source circuit's impedance: here the k_{pp} = 1.3 [41, 42]. Taylor et al. [43, 44] have tested a vacuum interrupter with TMF contacts (rated 17.5 kV/ 31.5 kA) and another with AMF contacts (rated 40.5 kV/ 31.5 kA) that had both previously passed the IEC ungrounded fault tests with k_{pp} = 1.5. They show that these vacuum interrupters easily passed the effectively grounded fault tests with k_{pp} = 1.3. They also show that a single-phase test is a reasonable representation of a worst-case three-phase test. The substitute test is similar to the T100a test given in reference [45].

The time to the $U_{R(peak)}$, t_p, is about 70μs for the first phase to clear in an indoor, cable connected, three-phase, ungrounded 15kV system at its full (100%) short circuit current. As the magnitude of the fault current decreases from its 100% value, $U_{R(peak)}$ increases and t_p decreases. This is illustrated in Figure 6.20. One explanation for this is that higher short circuit currents are associated with an increasing number of connected lines. Thus, the TRV is not so dependent upon one transformer's natural frequency. During the certification testing of a vacuum circuit breaker, currents lower than its 100% short circuit capability will have to be interrupted. Thus, the vacuum interrupters will have to be able to withstand the higher peak values of the TRV and a faster RRRVs than are shown in Figure 6.20.

One rule of thumb for a vacuum breaker's performance given by Greenwood [46] is:

$$\frac{di_{sc}\left(i_{sc}=0\right)}{dt} \times \frac{dU_R}{dt} = \varepsilon \tag{6.17}$$

where ε is a "figure of merit" for a particular vacuum circuit breaker so that as the short circuit current decreases, *di/dt* at current zero decreases, and the dU_R/dt that the vacuum interrupter can sustain increases. (For example, ε for the 60% of a 40kA fault current in a 15kV circuit from Figure 6.20 would have a value 1.4 × 10^4 A.V.μs^{-2}, but vacuum circuit breakers have ε values exceeding 3 × 10^4 A.V. μs^{-2}.) In fact, Smith [47] has demonstrated an ε value of 14 × 10^4 A.V.μs^{-2} for a 102mm

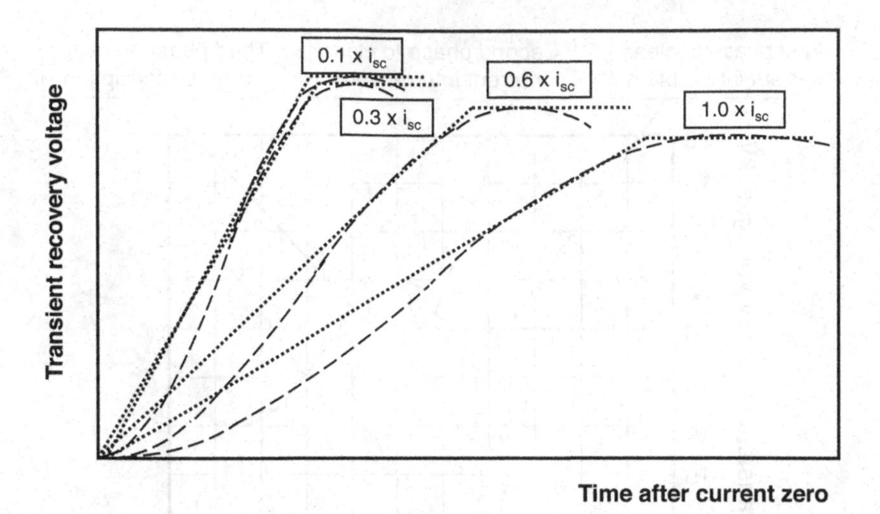

FIGURE 6.20 The TRV for different fractions of the available fault current.

diameter, vacuum interrupter with Cu–Cr TMF contacts. In general, Equation (6.17) is rather conservative for both indoor and outdoor TRV values. Most of the vacuum interrupter designs I have been involved with easily meet the t_p for indoor applications and also for outdoor applications, which have a t_p approximately one half the value required for the indoor cases. Even so, if you examine Figure 6.20 you will find that Equation (6.17) is easily satisfied for the values given for the 100%, 60% and 30% fault current levels. This means that if a well-designed vacuum interrupter can interrupt the full short circuit current, it will most probably easily interrupt short circuit currents of lower magnitude even if the RRRV and the TRV peak increase. The vacuum interrupter designer who does not know where the vacuum interrupter will be applied should design to satisfy the more severe line-connected, outdoor fault interruption whose TRV is faster than will be experienced by a vacuum interrupter switching a similar current for an indoor cable-connected system.

6.3.4 The Terminal Fault Interruption Performance of Vacuum Interrupters

The vacuum interrupter has to interrupt short circuit currents with the asymmetrical component discussed above. Most vacuum circuit breakers have a sensing system that identifies a fault condition and sends a signal to the breaker's mechanism to open the vacuum interrupters. This is illustrated in Figure 6.21, where possible lag times such as the fault sensing time and the delay before the mechanism begins to move are shown. The actual opening of the vacuum interrupter is thus delayed for a few cycles after the initiation of the fault. During this delay, the peak value of the asymmetrical current will have decayed in accordance with the *L/R (or X/R where X = ωL)* ratio of the circuit (see Figure 6.11 and Equation (6.9)). Most vacuum circuit breakers are designed to perform well when opening an asymmetric current with a 50% dc component. Some, with a longer interval between sensing a fault and the contacts parting, are designed for asymmetric current with a 30% dc component, which permits the interrupting performance of their vacuum interrupters to be less robust. Slade et al. [48] have determined the probability of interruption for the first phase to clear for a three-phase ungrounded circuit condition. The data they use is for both symmetrical currents and for currents with a 50% asymmetry. Their data are shown in Figure 6.22 and Table 6.8. For symmetric short circuit currents, if the vacuum interrupter's contacts open too close to a current zero, there is a high probability of the vacuum arc restriking and continuing to burn for another current half cycle. This is expected from the discussion in Section 4.3.2 in this volume on the interruption of the high current vacuum arc. At the current zero of the following current half

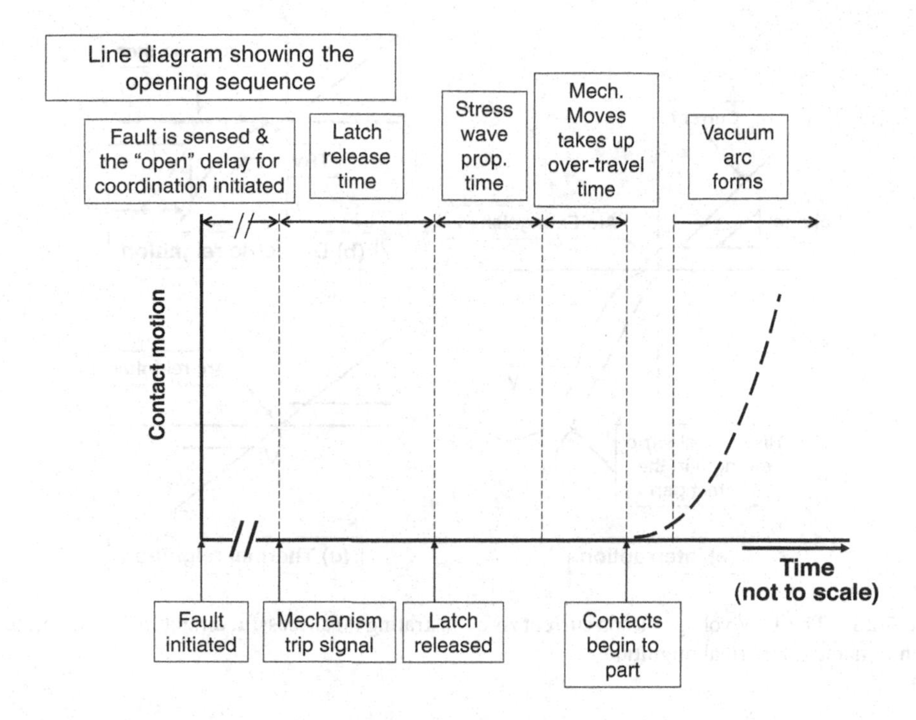

FIGURE 6.21 Possible delays in opening the vacuum circuit breaker.

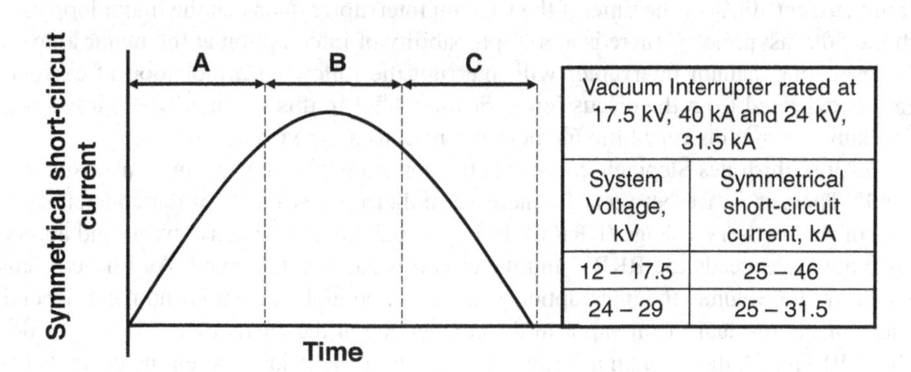

FIGURE 6.22 Contact part intervals for data in Table 6.8 [47].

TABLE 6.8
Probability of Interruption of Symmetrical Short-Circuit Current at the First Current Zero after Contact Part [48]

Arcing time range, Fig 6.22	Contact parting	Time to first current zero	Probability of interruption at the first current zero
A	Just after current zero and before current peak	> 6 ms	• Nearly 100% for $i_{SC} \leq 40$kA • 60–85% for $i_{SC} \geq 40$kA
B	Around current peak	< 6 ms & > 3 ms	• 17–80 % for $i_{SC} \leq 36$kA • 50% for $i_{SC} = 40$ kA • 35– 55% for $i_{SC} \geq 40$kA
C	Just before current zero	< 3 ms	• Nearly zero for all i_{SC}

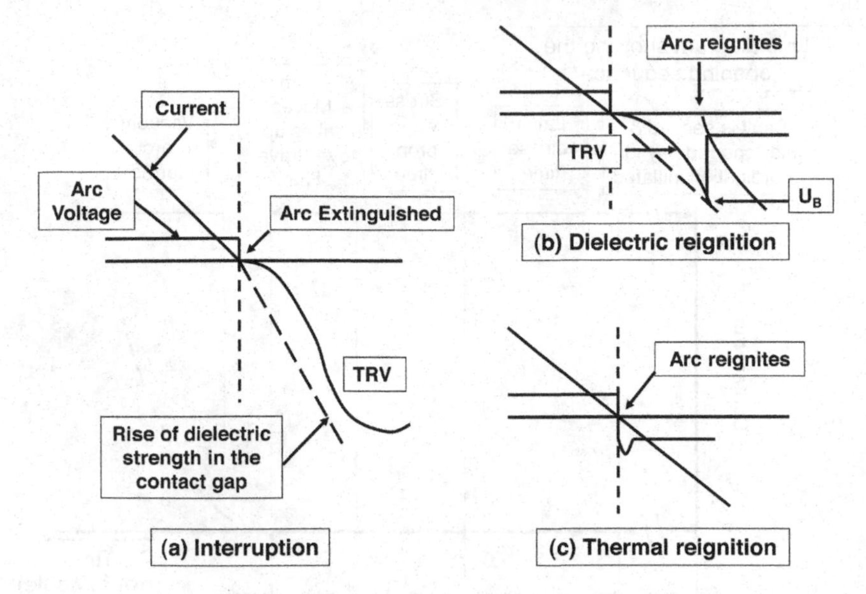

FIGURE 6.23 The TRV voltage after a current zero illustrating (a) successful circuit interruption, (b) dielectric reignition and (c) thermal reignition.

cycle, when the contacts will be fully open, a well-designed vacuum interrupter will interrupt the short circuit current 100% of the time. If the vacuum interrupter opens on the major loop of current, i.e., with the 50% asymmetry, there is a 60% probability of interruption at the major loop's current zero. However, the vacuum interrupter will interrupt the following minor loop of current easily. This again is expected from the discussion in Section 4.2.3 in this volume, where it has been seen that the vacuum arc retains very little memory of previous arcing events.

Figure 6.23(a) illustrates Slepian's concept of the contact gap's recovery in a gas-insulated circuit breaker [49]. There is a race between the increase in dielectric strength of the contact gap and the rate of rise of the recovery voltage (RRRV). In Figure 6.23(a) the dielectric withstand ability of the contact gap always exceeds the RRRV and the circuit is easily interrupted. As I have discussed in Section 4.2.3 in this volume, the interruption process for the high current vacuum arc depends upon the pressure of the residual metal vapor in the contact gap at the current zero. If the gas density is below about 10^{22} n.m^{-3}, the vacuum arc can only reignite if the field between the contacts is greater than the critical value for vacuum breakdown; see Section 1.3.3 in this volume. For example, this condition may exist if the contacts have only just parted at the current zero. When the pressure is greater than about 10 Pa the Paschen curve (see Figures 1.8 and Figure 4.16 in this volume) controls the reignition process. This is illustrated in Figure 6.23(b). Here the residual metal vapor pressure is high enough that once the TRV exceeds the U_B at that pressure, the contact gap will break down and the vacuum arc will reignite allowing the circuit current to pass for another half cycle. This is termed a *Dielectric Reignition* and is the most common type of reignition observed in vacuum interrupters that have exceeded their interruption limit. Another type of reignition that has been described for gas circuit breakers is the *Thermal Reignition*. Here there is little or no observed momentary development of the TRV. At current zero the arc reignites immediately, and the circuit current continues to flow without interruption: this is illustrated in Figure 6.23(c). In this case the new cathode must be capable of supplying the electrons to the arc immediately after the current passes through zero.

An illustration of the two types of reignition in a vacuum interrupter is shown in Figure 6.24. This figure has been recorded in an experiment to explore the interruption limits of a vacuum interrupter with a Cu–Cr (25wt.%), TFM contact designed to interrupt a 25kA (rms) current in a 17.5kV circuit. Here at 31.5kA(rms) the vacuum interrupter cannot interrupt the circuit. During the first half cycle the

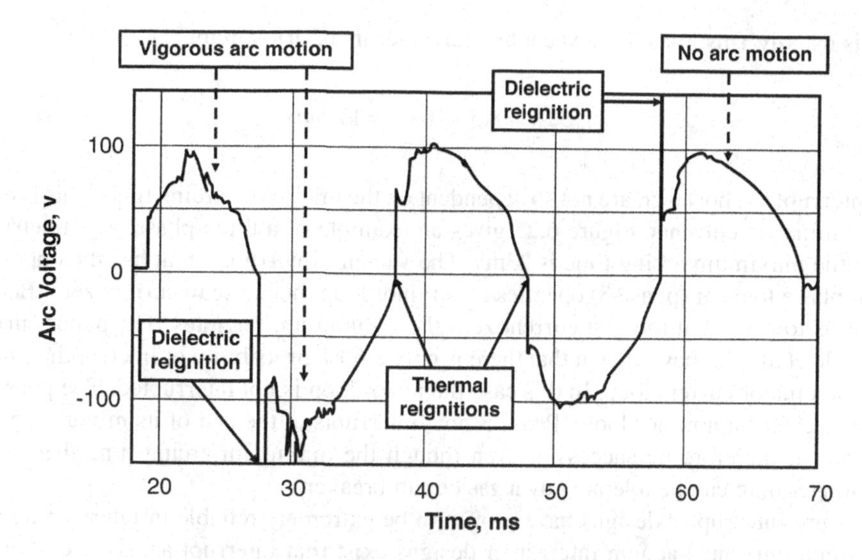

FIGURE 6.24 The voltage across a vacuum interrupter designed to interrupt a 25kA rms fault current tested at 31.5kA rms illustrating both dielectric and thermal reignitions.

arc voltage shows the characteristic expected for a high current, columnar, vacuum arc when it moves readily around the contacts' periphery; see Figures 3.21 and 3.32. At the first current zero the TRV across the contact gap rises rapidly, but the contact gap breaks down and the vacuum arc is reignited. During the second half cycle of arcing the columnar vacuum arc seems to move, but more sluggishly. At current zero there is no sign of the TRV and a thermal reignition is shown. The vacuum arc is stationary during the third half cycle of arcing and again a thermal reignition is observed. Surprisingly at the next current zero the TRV again appears momentarily and a dielectric reignition again occurs.

The concept of a "minimum arcing time" before current interruption can be achieved follows from my discussion on high current interruption in Sections 4.2.3 and 4.3.2 in this volume. The arcing time depends upon how long before a current zero the contacts part. If the contact opening speed is known, the arcing time also gives the contact gap at the current zero. From Section 3.3.3 in this volume, it will also give the state of the high current, columnar, vacuum arc rotation for a TMF contact as well as whether or not the AMF contact has made the high current vacuum arc fully diffuse. Slade et al. [48] show for a vacuum interrupter opening at about 1 ms^{-1} when interrupting a 15 kV, 50Hz fault current using a TMF contact (a) for arcing times below 3 milliseconds, the VI does not interrupt, (b) for arcing times between 3 and 6 milliseconds, the VI interrupts about 75% of the time and (c) for arcing times greater 6 milliseconds, the VI interrupts 95% of the time. Therefore, it is reasonable to conclude that the minimum arcing time for vacuum interrupters with TMF contacts protecting a 50Hz circuit is about 8.5ms and about 7ms for 60Hz circuits. The "minimum arcing time" for AMF contacts will depend upon the time it takes for the high current arc to go diffuse and this, in turn, will depend upon the current at the time of contact parting; see for example Figure 2.73. I would expect that this time would be at least as long as the time for the TMF contacts. The maximum arcing time is a concept that has been developed for gas circuit breakers such as SF6 puffer breakers; see for example, Calvino [50]. These circuit breakers require considerable gas flow at current zero for successful current interruption. There is, however, only a window of time when this flow is effective. If the arcing continues beyond this time current interruption will not occur. The maximum arcing time is usually given by [50]:

$$t_{arc\,max} = \left(t_{arc\,min} - \varepsilon^*\right) + ½\,\text{cycle} \tag{6.18}$$

where ε^* is usually 1ms: thus, for a vacuum interrupter in a 50Hz circuit,

$$t_{arc\ max} = 8.5 + 10 - 1 = 17.5ms. \quad (6.19)$$

Vacuum interrupters, however, are not so dependent on the maximum arcing time. This is especially true for asymmetric currents. Figure 6.25 gives an example of a three-phase, asymmetric current test where the maximum arcing time is 23ms. The vacuum interrupter that has the opportunity to be the first phase to clear (phase-3) operates on a minor loop, but close to current zero. Because the contact gap is too small at the first current zero the vacuum arc reignites. The major current loop follows. Slade et al. [48] have shown that there is only a 60% probability of interrupting the current at the end of a major current loop. In this case the major loop is not interrupted. The phase-2 is the first phase to clear (at a minor loop). Phase 3 now interrupts at the end of its minor loop. Vacuum interrupters can therefore be successful even though the maximum arcing time they sometimes exhibit is longer than can be tolerated by a gas circuit breaker.

The vacuum interrupter designs have proven to be extremely reliable in interrupting all levels of short circuit current. Vacuum interrupter designs exist that interrupt ac, short circuit currents in excess of 80kA (rms) [3], currents with asymmetries up to 75%, and even short circuit currents that have no immediate current zeros [51]. The TMF and AMF contact designs are so effective in distributing the energy from the vacuum arc uniformly over the contact surfaces that the contact erosion from the arc roots is quite uniform. This has enabled the development of vacuum interrupter designs capable of interrupting the full short circuit current 30, 50, and even 100 times while still retaining their ability to satisfy the full 1 minute withstand voltage across the open contacts. Even though they may experience a small increase in contact resistance, they will still be able to pass their specified load current with very little effect on the temperature rise at the terminals. Because the interruption ability of a vacuum interrupter only depends upon the interaction of the

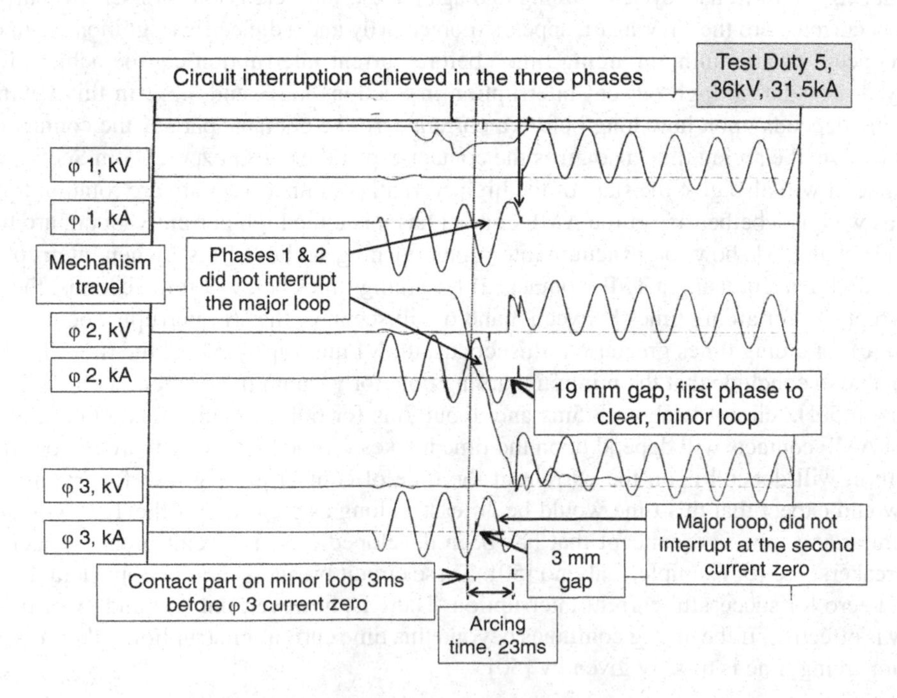

FIGURE 6.25 A three-phase, ungrounded, asymmetric short circuit test illustrating a long arcing time on Phase 3.

vacuum arc with the contact configuration and the contact material, the vacuum interrupter can easily interrupt "developing" faults. The current in this type of fault can begin at one level and can rapidly change to a higher level if the impedance limiting the magnitude of the fault current is suddenly reduced. For example, the scenario in which a tree limb that has fallen onto an overhead line creating a high impedance path to ground suddenly falls off the line, drawing an arc from the line directly to ground.

6.3.5 The Transient Recovery Voltage for Short Line Faults (SLF)

It is possible for the fault to occur some distance from the load terminal of the vacuum circuit breaker. This short distance may be a few hundred meters to a few kilometers. This type of fault is called a *short line fault* (SLF), and it will impose an initially severe TRV across the circuit breaker contacts [41, 45]. The fault condition is illustrated in Figure 6.26(a) for a single-phase circuit. The short circuit current is reduced from $\frac{U_m}{X_s}$ to $\frac{U_m}{(X_s + X_L\lambda)}$; where U_m is the instantaneous voltage driving the short circuit current, λ is the distance from the breaker's terminals to the fault and X_L is the line's reactance per unit length. The voltage is divided on each side of the breaker in proportion to the impedance of the source and the line. This is shown in 6.26(b). The voltage on the line side at the breaker is $\frac{X_L\lambda U_m}{(X_s + X_L\lambda)}$. Thus, the line has a charge distribution, which is at its maximum at the circuit breaker and drops more or less linearly to the fault. This charge cannot remain static; it results in a traveling wave between the breaker's line terminal and the fault where it is reflected. Thus, while the breaker's terminal on the source side of the breaker will experience the relatively

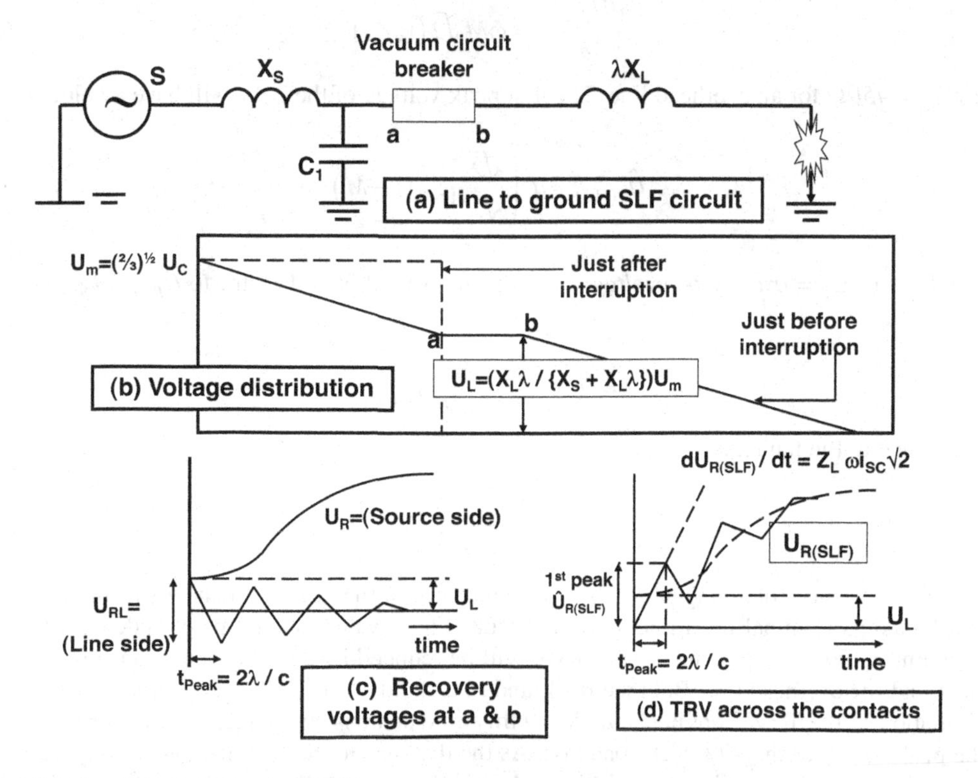

FIGURE 6.26 The short line fault TRV.

slow (*1 – cosine*) TRV of the terminal fault, the line side terminal will see a fast-triangular voltage wave: this is shown in Figure 6.26(c). The resulting TRV across the circuit breaker's terminal will be:

$$U_{R(SLF)} = U_R\left(\text{terminal fault}\left[1-\cos(\omega_0 t)\right]\right) - U_{RL}\left(\text{traveling wave}\right) \tag{6.20}$$

This is illustrated in Figure 6.26(d). The rate of rise of the U_{RL} (traveling wave) is much faster than U_R (terminal fault) and for the initial rate of RRRV it is dominant and is approximately equal to $Z_L \frac{di_{SLF}(i=0)}{dt}$ where Z_L is the surge impedance of the line (about 450Ω). The short circuit current for a SLF, i_{slf}, is reduced from the bus fault short circuit current i_{sc} by a factor:

$$i_{slf} = i_{sc}\left[\frac{X_s}{\lambda X_L + X_s}\right] \tag{6.21}$$

The triangular TRV wave starts just after current zero and reaches its first peak value $\hat{U}_{R(SLF)}$ after t_{peak} (the transit time along the line to the fault and back, $t_{peak} = \frac{2\lambda}{c}$ where c is the speed of light, 300,000 km.s^{-1}: i.e., $i_{(peak)} = 6.67 \times \lambda$ μs when λ is in kilometers). If the SLF current is reduced from the bus fault value i_{sc} by a fraction $M < 1$ then:

$$i_{SLF} = M \times i_{SC} \tag{6.22}$$

Then from [39]

$$\frac{dU_{R(SLF)}}{dt} = \omega M\sqrt{2}\left(i_{sc}Z_L\right) \tag{6.23}$$

where $Z_L = 450\Omega$ for an overhead line. The first peak voltage of the TRV will have a value of:

$$\hat{U}_{R(SLF)} = d^*\left(\frac{\sqrt{2}}{\sqrt{3}}\right)U_C\left(1-M\right) \tag{6.24}$$

where $d^* \approx 1.6$, U_C = *(rms system voltage)*. For line to ground fault the time to $\hat{U}_{R(SLF)}$, t_{peak} is:

$$t_{peak} = \frac{\hat{U}_{R(SLF)}}{dU_{R(SLF)}/dt} \tag{6.25}$$

The distance to the fault λ is:

$$\lambda = \frac{ct_{peak}}{2} \tag{6.26}$$

Table 6.9 gives values of $dU_{R(SLF)}/dt$, $\hat{U}_{R(SLF)}$ and distance of the SLF in kilometers from the vacuum circuit breaker's terminal for a line to ground fault. These values should be considered a theoretical maximum. For example, the initial RRRV will be reduced by the stray capacitance between the breaker and the overhead line. Both the IEEE and IEC standards permit a delay time of 0.2μs before the initiation of the TRV. Looking at Table 6.9 this time delay is an appreciable fraction of the time to the peak value for the 90% SLF condition. As the distance to the fault increases, so i_{slf} decreases because the line impedance increases. This results in a decrease of $dU_{R(SLF)}/dt$, however the first peak

TABLE 6.9
Effect of i_{slf} / i_{sc} on SLF Values and the Distance to Fault

System Voltage, kV(rms)	15.5			27			38		
Line to Ground Voltage, kV(rms)	8.9			15.6			21.9		
M, $\frac{i_{slf}}{i_{sc}}$	$\frac{dU_{R(SLF)}}{dt}$, kV/µs	$\hat{U}_{R(peak)}$ kV	λ, km	$\frac{dU_{R(SLF)}}{dt}$ kV/µs	$\hat{U}_{R(peak)}$ kV	λ, km	$\frac{dU_{R(SLF)}}{dt}$ kV/µs	$\hat{U}_{R(peak)}$ kV	λ, km
0.9	5.4	2.01	0.06	5.4	3.53	0.10	5.4	4.95	0.14
0.75	4.5	5.03	0.17	4.5	8.82	0.29	4.5	12.39	0.41
0.6	3.6	8.05	0.34	3.6	14.47	0.60	3.6	20.31	0.85

value of $\hat{U}_{R(SLF)}$ increases. Thus, there are three parameters that will affect the recovery of the contact gap after current zero as the distance to the fault increases: there is a decrease in the value of the fault current, there is a reduction in the rate of rise of the TRV, but an increase in its value. For most circuit breakers there is a line length, λ, where these effects combine to produce the most difficult SLF interrupting condition. Certification for SLF performance has traditionally been performed for line to ground faults on circuit breakers operating at system voltages ≥ 72.5kV. Since 2004 the IEEE and IEC standards have included SLF tests for outdoor medium voltage circuit breakers for system voltages 15.5kV and higher with potential terminal short circuits of greater than 12.5kA. SLF testing is not performed on indoor, medium voltage circuit breakers, because they are usually connected by cables whose surge impedance ranges from about 50Ω to 75Ω. Thus, from Equation 6.23 the $dU_{R(SLF)}/dt$ will be quite slow, giving the vacuum circuit breaker ample opportunity to recover after interrupting the fault current. Smeets et al. [52] have shown that vacuum interrupters perform well under SLF conditions. The maximum value of $\mathcal{E}$ in Equation (6.17) that can be calculated from the data in Table 6.9 is 6×10^4 A.V.µs^{-2}, which is well below the 14×10^4 A.V.µs^{-2} measured by Smith [47].

6.3.6 TRV FROM TRANSFORMER SECONDARY FAULTS

The application of a vacuum circuit breaker very close to a transformer can result in a terminal fault TRV with a very fast RRRV [53]. The reason for this is that there is only a very small capacitance from the connecting cables or bus duct. Thus, the TRV is driven by the high natural frequency of the transformer. A compilation of the expected frequency of the TRV for the first phase to clear of a three-phase ungrounded short circuit as a function of current is given in the IEEE TRV guide [39]. Figure 6.27 shows the data for a 15kV system. The upper curve represents the median value of the time to reach the peak of the TRV (t_p), a function of short circuit current: i.e., 50% of all transformer faults have a t_p equal to or greater than the values shown. The lower curve presents the 90% data. The black dots and expected lower limit of performance for vacuum interrupters using Cu–Cr (25wt.%) TMF contacts are from data presented by Smith [47]. These data show that a vacuum circuit breaker with these vacuum interrupters will protect transformers from the effects of short circuit currents even though the TRV with its very fast RRRV imposes a severe interruption duty.

6.4 LATE BREAKDOWNS AND NON-SUSTAINED DISRUPTIVE DISCHARGES (NSDDS)

After interrupting an ac current, the vacuum circuit breaker must withstand the TRV that appears across its vacuum interrupter's contacts. In a single phase or a three-phase grounded circuit If it fails

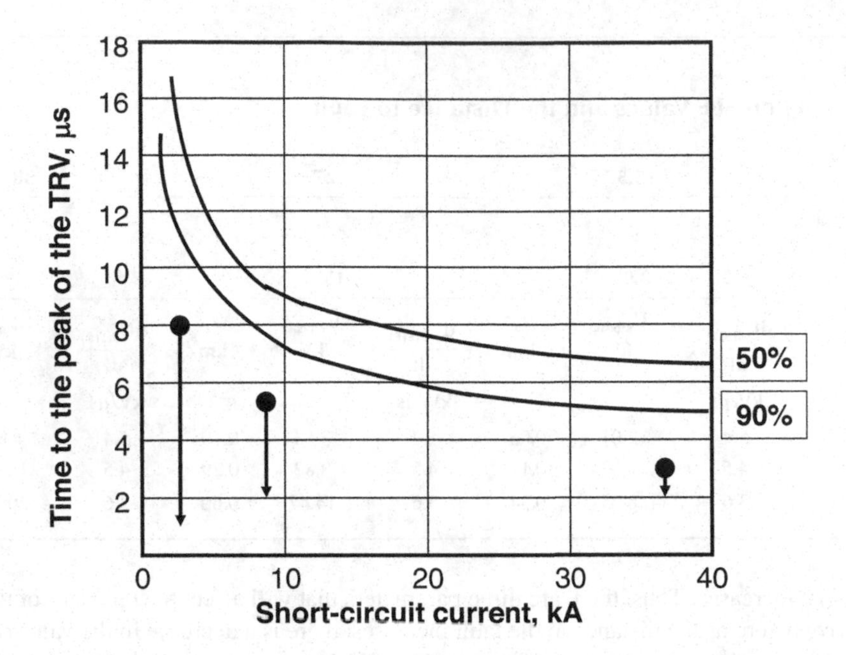

FIGURE 6.27 The time to peak value of the TRV for transformer secondary faults as a function of the fault current also showing performance data for one style of vacuum interrupter [39, 47].

to do this in a time less than or equal to one-quarter cycle of the ac frequency, it is considered a disturbance caused by the TRV and the standards permit it to occur with no reflection on the vacuum circuit breaker's performance. In a three-phase ungrounded circuit if it the first phase to clear fails to do so before all three phases have interrupted, the disturbance is called a *reignition*. If at a later time the contact gap breaks down and permits the circuit current to flow, then the vacuum circuit breaker will have failed the test; this is termed a *restrike*. The vacuum interrupter will certainly interrupt the restrike current at the following current zero, so in reality, even though the vacuum circuit breaker would have failed a certification test, its purpose to protect the circuit would have been satisfied.

In a single-phase fault, or phase-to-ground fault, a discharge between the contacts, which does not result in the restoration of the circuit current, has also been observed. This phenomenon can also be observed when interrupting a three-phase ungrounded circuit. In this case, if the discharge occurs in one-phase only, no circuit current can flow. Even if it occurs in two-phases at the same time, there is still a finite probability that the circuit current flow will not be restored. This breakdown event can occur long after (up to a few seconds) the circuit current has been interrupted. It permits a high frequency current to pass between the open vacuum contacts and is self-extinguishing: i.e., the vacuum contact gap restores itself to its original insulating state with no resumption of the circuit current. This event is sometimes observed during the certification testing of vacuum circuit breakers. In the circuit breakers standards (e.g., IEC, IEEE) these self-restoring, late breakdowns are termed *Non-Sustained Disruptive Discharges* (NSDDs); see, for example, Leusenkamp's review [54]. This term is somewhat pejorative and misleading. The discharge may or may not be disruptive: for the most part it is not. Certainly, the vacuum interrupter has the interesting characteristic that it can restore itself to its insulating state almost immediately after the discharge has occurred. To my mind, the term *Non-Sustained Transient Discharges* (NSTDs) would be a more appropriate term.

Smeets et al. [55] analyzed the occurrence of NSDDs during high power testing of vacuum circuit breakers at the KEMA High-Power Laboratories for the year 1999. Their analysis is shown in Figure 6.28. They conclude that NSDDs occur frequently in greater than 30% of the vacuum circuit

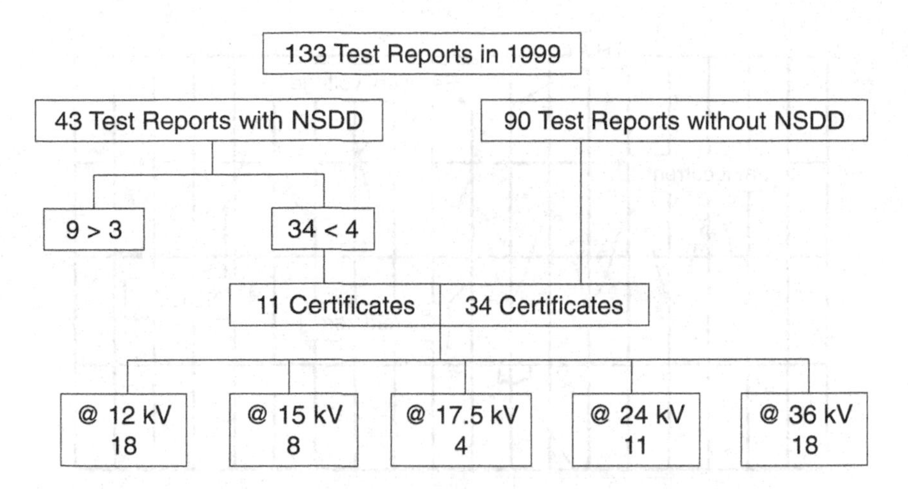

FIGURE 6.28 The occurrence of NSDDs during certification testing at the KEMA Testing Laboratory (Holland) in 1999 [55].

breakers tested. The majority of NSDDs (~ 80%) occurred within 300ms after current interruption. The occurrence of NSDDs is not confined to fault current interruption. As I have discussed in Section 5.5.4 in this volume, they can also occur after interrupting a capacitor circuit's load current. They occur over the whole range of circuit voltages from 12kV through 40.5kV. Smeets et al. have also expanded the definition of the NSDD to include any restrike (or voltage disturbance) that results in a high-frequency current that does not develop into a power frequency current after short circuit interruption. It seems to be unrelated to the current being interrupted; it has been observed after switching load currents [56] and after switching short circuit currents [55]. While this phenomenon has mainly been associated with vacuum interrupters, Smeets et al. [57] state that there is undocumented evidence that it can also occur with SF_6 interrupters.

I have already discussed one example of a self-extinguishing discharge between two open contacts in vacuum in Chapter 1, Section 1.3.3, in this volume: the micro discharge. When such a discharge takes place, it usually results in the flow of a high frequency current between the contacts as local capacitances share charge. It seems likely that an NSDD is initiated at the cathode either from field emission, Xu et al. [58] or by particle impact, Hachiman et al. [59]. If the resulting micro-discharge cannot be sustained an NSDD will result. In single-phase circuits or in three-phase grounded circuits, vacuum interrupters usually interrupt the short duration NSDD before the main circuit current becomes established. In the worst case, if the discharge persists and the power frequency circuit current is restored, this power frequency current will be interrupted at the following current zero. An example of an NSDD in a single-phase circuit is shown in Figure 6.29. Normally, a vacuum circuit breaker is certified in a three-phase ungrounded circuit. In this case, two-phases would have to form an NSDD at the same time and each would have to result in the power frequency current flow for a restrike of this current to occur. Figure 6.30 shows an example of an NSDD occurring on one-phase after the interruption of a three-phase short circuit current and the restoration of the circuit voltage across the open contacts of the vacuum interrupters in an ungrounded system. The momentary collapse of the voltage across one-phase results in an offset of the voltages on each of the other two phases. The amplitude and time constant of this offset depends upon the local values of inductance, capacitance, and resistance. Because the discharge duration is extremely short (perhaps a few 10s of microseconds) and does not usually cause the power frequency current to be reestablished, the impact, if any, of an NSDD in a practical ac circuit is not clear. As I will show in Section 6.7.2, however, an NSDD in one phase may shift the voltages on the other two phases. This can result in a voltage in one of those phases being much higher than the normal circuit voltage.

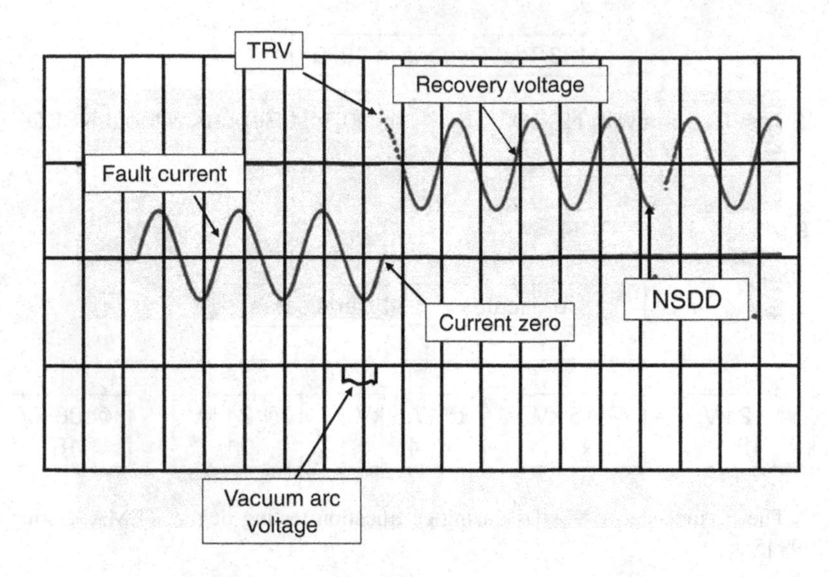

FIGURE 6.29 An example of a NSDD after the interruption of a single-phase fault current.

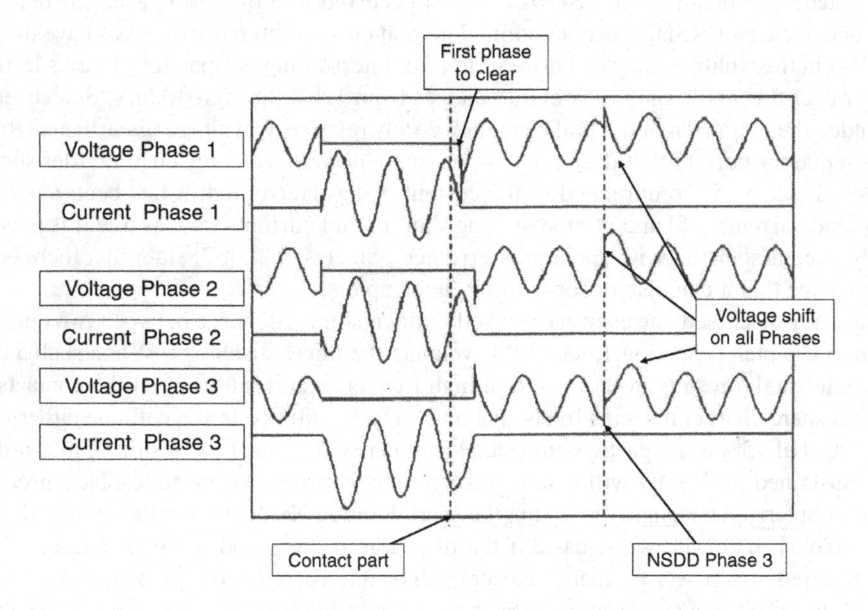

FIGURE 6.30 An example of a NSDD occurring on one phase after the interruption of a three-phase fault in an ungrounded circuit.

The presence of NSDD phenomenon has resulted in some vigorous discussion by the vacuum interrupter and vacuum circuit breaker manufacturing, vacuum circuit breaker testing laboratory and end user communities. At the present time, the consensus seems to be that although NSDDs may occur during the recovery voltage period following a current interruption, their occurrence is not a sign of distress in the switching device under test. Therefore, the number of NSDDs in a testing sequence is of no significance in interpreting the performance of the device. The number only needs to be reported in order to differentiate them from restrikes. This new consensus is quite different from the former arbitrary contention that the vacuum circuit breaker cannot be certified if four or more NSDDs are observed during the entire test duties required by the Standards. The new consensus is entirely reasonable when you remember that vacuum circuit breakers and reclosers have been applied in the field since the late 1960s: this translates into over a few hundred million vacuum interrupter-years of service in medium voltage circuits. To my knowledge, NSDDs, if they have occurred, have not had any detrimental effect.

The initiation process for the NSDD has no definitive physical explanation at present. Indeed, there are conflicting interpretations of the experimental observations:

1) The occurrence of a late breakdown decreases exponentially with time after current interruption: Bernauer et al. [60], Platter et al. [61], Bernauer et al. [62], Schlaug et al. [56], Gebel et al. [63], Smeets et al. [55]
2) Late breakdowns can occur after switching all current levels; i.e., load current levels ~ 1000A, Bernauer et al. [62, 64], to full rated short circuit current [55]
3) The probability of late breakdown occurrence increases as the voltage impressed across the vacuum contact gap increases [56, 64]
4) For a given vacuum interrupter design, if the peak of the restored voltage is less than 0.45 × U_{CDF10}(10th percentile of the Cumulative Density Function, CDF, see Equation 1.34; i.e., the 10th percentile of the cumulative probability of breakdown) of the cold contact gap, then the occurrence of an NSDD is negligible [61, 65]. This U_{CDF10} breakdown voltage most probably will be lower for contacts that have made contact under current and have opened with no current than for those contacts that have performed both the making and breaking functions
5) When the recovery voltage is impressed across the vacuum interrupter after its contacts close with current and then open with no current, field emission currents in the range 0.1mA to 10mA are measured [66]. Higher field emission currents are measured after the contacts close with a current of 44kA than are measured when they close with a current of 6kA. The emission currents decay over time (in typically 10 to 40 minutes) to perhaps half of the initial value. After this, the emission current continues to decay, but very slowly. For the contact gaps typically used by vacuum interrupters in a vacuum circuit breaker, this field emission current fluctuates at random about the average value; see Platter et al. [61, 67]. Once a breakdown occurs with a low current discharge, the emission current decreases appreciably [66]
6) Mechanical shock to the vacuum interrupter seems to increase the probability of breakdown of the open contact gap; see Jüttner [68], Jüttner et al. [69] and [62, 63]. Mechanical shock can increase the emission current considerably [69]
7) NSDD's occur in vacuum interrupters containing Ag–WC, Cu–W, and Cu–Cr contacts. After switching currents in the range 1kA to 8kA each material shows appreciable emission currents when a 20 to 30kV (peak), 50Hz ac voltage is impressed across a 4mm contact gap [70]
8) When opening Cu–Cr contacts with no current, an NSDD is associated with a current pulse (a "current bump") with a peak value up to 10mA and duration of about 5μs [71]. There is no prior electron emission before the occurrence of the "current bump", which is characteristic of the micro-discharge discussed in Section 1.3.3 in this volume
9) As the number of close-open operations with current increases, the occurrence of NSDDs decreases

How should these data be interpreted? As I discussed in Chapter 1 in this volume, the breakdown of a vacuum gap is dependent upon the electric field at the surface of the cathode contacts, which in turn depends upon the microscopic profile of these surfaces. After current interruption and after the post-arc current has fully decreased, there will be a residual emission current resulting from the microscopic field enhancement at the cathode (see Section 1.3.3 in this volume). Thus, the first and second conditions for breakdown to occur are satisfied; an electric field and the presence of electrons (see Sections 1.3.2 and 1.1.3 in this volume). During the initial operation of the breaker, there is considerable shock to the vacuum interrupter. Later, however, the effects of opening will no longer be present. Now some researchers postulate that a mechanical shock will result in the release of a micro particle into the contact gap, which causes a breakdown [54, 56]. These researchers never give an explanation of how the particle initiates a breakdown. As I have discussed at length in Chapter 1 in this volume, I believe that vacuum breakdown is always initiated by the emission current from the cathode [72]. Particles may play a role, but on their own most probably do not initiate the late breakdown phenomenon between the vacuum interrupter contacts.

The research by Jüttner et al. [68, 69] indeed shows that mechanical shock can change the emission characteristic of a cathode surface. It is interesting to note that there can be a considerable delay between a mechanical shock and the ultimate value of the emission current (a few milliseconds, up to seconds [68, 73]). Thus, if a circuit breaker is still vibrating after it has opened, there may well be a change in emission current after a long-time interval. This increased emission current could well be the initiator mechanism for a late breakdown. Jüttner [68, 73] explains that these variations in the field emission current are possible through changes in gas adsorptions on a cathode surface. For example, adsorption of residual gas can change the local current density of the emitter, such that a critical value can be reached. The residual gas adsorbed on the cathode can result from switching current or even from the electron emission current causing adsorption of gas on the anode. Jüttner states that the time constraints for these processes can vary from milliseconds to seconds. A remarkable series of photographs of contacts during a late breakdown event has been published by Ponthenier et al. [74]. They show that the discharge is initiated near a residual hot spot on the cathode contact's surface. Unfortunately, they could not determine what caused the hot spot and whether or not particles play a role in the late breakdown process.

As I have already discussed in Section 5.5.4 in this volume, it is quite possible for the emission current from the cathode to eventually heat the anode to a high enough temperature that a vacuum discharge can be initiated. The peak open circuit voltage across the vacuum interrupter's contact gap is low compared to the peak design voltage given by the BIL level; see Table 1.16. It is generally about one sixth of the BIL voltage peak. Thus, the inter-contact field that would accelerate any particle would result in a less than optimum energy to cause the breakdown of the contact gap; see Sections 1.3.3 and 1.3.4 in this volume. The photographs of the delayed breakdown of the contact gap after current interruption [74] and the electron emission data [66] show that the NSDD results from the vacuum breakdown processes described in Chapter 1; see Figures 1.85 in this volume. The NSDD is always initiated by the electron emission from the cathode. Particles in the contact gap either have little effect, or their role is minor. The discharge is self-limiting and extinguishes, because the cathode does not establish a cathode spot capable of delivering the electrons required to sustain the discharge. For particles to have any effect on the vacuum breakdown process, the field between the fully open contacts would have to be much higher. Even so, I believe that vacuum breakdown of the contact gap is always initiated by the electron emission from the cathode.

Table 1.16 shows the peak value of the open circuit voltage once the initial TRV period is past for load current operation or short circuit current operation. I will discuss the special case for capacitor circuits in Section 6.7.2. From the work by Platter et al. [61] for a 10mm contact gap, the breakdown voltage at the 10th percentile of the Cumulative Density Function, U_{CDF10}, (CDF, see Equation (1.34)); i.e., the 10th percentile of the cumulative probability of breakdown, after making the circuit at 1kA, 2kA, 5kA, 10kA, and 20kA for Cu–Cr (25 Wt.%) is in the range 59kV to 75kV. The minimum $U_{BL(min)}$ voltage for late breakdowns is in the range 47kV (peak) to 70kV (peak) with the worst ratio of $U_{BL(min)} / U_{CDF10}$) being 0.71. In later work, this ratio is reduced to 0.45 [65]. If the average U_{CDF10} of 65kV (peak), see Figure 5.65, is taken then the average value of $U_{BL(min)}$ where late breakdowns are unlikely is 29kV (peak). For a 10mm contact gap this may become critical for circuit voltages greater than 27kV. This seems to agree with observation number 3 given above. Vacuum interrupters designed for circuit voltages of greater than or equal to 36kV, however, will have larger contact gaps and an internal design that would result in a $0.45 \times U_{CDF10}$ being greater than 29kV. Switching load currents (630A and 3150A) will continually condition the contact surfaces (see Section 1.3.7), so perhaps $U_{BL(min)}$ will be modified.

In my experience, the incidence of NSDDs after fault current operations in a wide range of vacuum interrupter designs for interruption tests within their design limit ranges from 0/100 trials to 10/100 trials. The higher level being for either the 36 kV to 40.5 kV level or for interrupter designs with test currents very close to their interruption limit. The incidence of NSDDs observed during certification testing of these vacuum interrupters in practical circuit breakers and reclosers is not common; perhaps 30 NSDDs /year are observed. In the course of my career, which includes experience with more than 20 million vacuum interrupters in the field, none have ever been returned for a problem with late breakdowns. Indeed, Greenwood states [75]:

> *Just because vacuum has the unique ability to burp on occasions to relieve its indigestion, it should not be punished, indeed, it should be commended.*

Indeed, as stated at the beginning of this section, the NSDD resulting from a late breakdown is now considered an interesting phenomenon of the vacuum gap and the contact structure. It is no longer considered a sign of distress in a vacuum circuit breaker and has no significance to the vacuum circuit breaker's performance. It is, therefore, reasonable to accept the performance of a vacuum circuit breaker, even if it does exhibit some level of NSDDs during its certification testing.

6.5 VACUUM CIRCUIT BREAKER DESIGN

6.5.1 INTRODUCTION

The vacuum interrupter is the core of the vacuum circuit breaker. It is the component that performs all the current interruption and voltage withstand functions. It cannot, however, perform these functions without a suitably designed mechanism. In order to obtain the optimum performance from a given vacuum interrupter design, the mechanism design must be perfectly matched to it. When this has been achieved, then an optimal vacuum circuit breaker or recloser is obtained. The details of vacuum circuit breaker design really deserve its own book. Most of these details are kept secret by the companies that employ the vacuum circuit breaker design engineers and are not publicly disclosed. Thus, in this section I will only outline the critical components and the thought processes that must take place in order to match the vacuum interrupter to the vacuum circuit breaker's mechanism.

As I have already stated in the introduction to this chapter, the vacuum circuit breaker is designed to satisfy a given set of operating and performance standards. It is then certified that it can pass this set of standards by an independent testing laboratory. The user can then choose from a range of standard vacuum circuit breakers to protect their electrical system. They will thus choose the standard circuit breaker whose performance (i.e., maximum fault current, system voltage, ambient, etc.) equals or exceeds the calculated worst-case conditions that the electrical system may experience. The vacuum circuit breaker manufacturer will design a range of breakers that will cover the majority of conditions that may be experienced in an electrical system. They may also develop special vacuum circuit breakers that can be employed in exceptional electrical circuits whose performance criteria fall outside the usual circuit breaker standards. In general, the vacuum circuit breaker is developed to satisfy the following specific requirements:

A. The certification standards [16]: e.g., IEEE, IEC, GB/DL, GOST, etc.
    a. Rated power frequency
    b. Rated maximum system voltage
    c. Rated insulation level (see Table 1.1)
        i. Rated Lightning Impulse Withstand Voltage (or Basic Impulse Level)

ii. Rated power frequency withstand voltage
d. Rated normal (or continuous) current
e. Rated short-circuit withstand current
f. Rated short-time, short-circuit current withstand duration
g. Rated short-circuit breaking current
h. Rated dc component of the short-circuit breaking current
i. Rated peak value of the short-circuit making current
j. Rated mechanical endurance
k. Rated electrical switching endurance
 i. At normal current
 ii. At the rated short-circuit breaking current

B. The circuit breaker style:
 a. Fixed
 b. Horizontal withdrawable
 c. Vertical withdrawable

C. The ambient insulation:
 a. Air
 b. SF_6 or SF_6 + gas (e.g., N_2) mixtures
 c. Suitable insulation gases to replace SF_6: see Section 6.5.4
 d. Oil
 e. Solid insulation (e.g., epoxy, polyurethane, thermal plastics, etc.)

D. Specific applications (e.g., generator breakers, locomotive, electric arc furnaces, etc. (see Section 6.8) and vacuum dc circuit breakers (see Section 6.11)

6.5.2 Closed Contacts

The vacuum circuit breaker usually spends most of its life in the closed condition, passing the electrical circuit's load current. In fact, most of the time this continuous current will be significantly less than the rated design value. The usual rated values are given in Table 6.10, which follow the R10 series. The importance of the attachments to the fixed and moving ends of the vacuum interrupter has already been discussed in Section 6.2. One major advantage of the vacuum interrupter is that the contacts are never exposed to the ambient, thus the contact resistance can only change by a small amount as the contact's surface structure and its hardness change during its operating life.

TABLE 6.10
Range of Continuous Load Currents at Different System Voltages and Short Circuit Current Level

System voltage, kV	Symmetric short circuit current, kA (rms)								
	6	12.5	16	20	25	31.5	40	50	63
	Range of continuous currents, A								
4.76–7.2				630, 1250	630, 1250	1250, 2000	1250, 2000	2000, 3150	
12–17.5	200, 400, 630	630	630	630, 1250	630, 1250, 2000	1250, 2000	2000, 2500, 3150	2000, 2500, 3150	3150, 4000, 5000
24–27	200, 400, 630	630	630	630, 1250	1250, 2000	1250, 2000	2000, 2500, 3150	2000, 2500, 3150	
36–40.5			630	1250	1250, 1600, 2000	1250, 2000, 2500	1250, 2000, 2500, 3150		

TABLE 6.11
Suggested Added Contact Force at Various 50 Hz Fault Current Levels

	Added Contact Force, KgF			
Manufacturer	**16 kA (rms) 40 kA (peak)**	**20 kA (rms) 50 kA (peak)**	**25 kA (rms) 63 kA (peak)**	**40 kA (rms) 100 kA (peak)**
Areva (France)	120	140	210	
ABB (Germany)	160	200	220	400
BEL (India)	80	120	200	500
Crompton Grieves (India)	120	175	227	
Eaton Electrical (USA)	100	140	200	410
Hitachi (Japan)	60	97	140	230
Holec (Holland)	100	150	180	480
Lucky Goldstar (South Korea)	68	154	227	367
MELCO (Japan)	125	150	200	
Shanxi (China)	120	160	200	367
Siemens (Germany)	84	120	200	500
Toshiba (Japan)			200	400
A&E Power(Japan)	56	92	138	311

When the peak of the maximum short circuit current (i_{sc} rms) passes through the closed contacts, the blow-off force, F_B (see Section 2.1.4, Equations (2.14), (2.17), and (2.19) in this volume), reduces the total force holding the contacts together. For normal 50Hz circuits, the maximum peak value of the asymmetrical currents is $2.5 \times i_{sc}$ and for normal 60Hz circuits, $2.6 \times i_{sc}$. Thus, the circuit breaker's mechanism must supply a closed contact force that exceeds the maximum possible blow-off force, plus enough force for maintenance of a low enough contact resistance to prevent the contact spot from melting during the passage of the high current (see Sections 2.1.4.2 and 4.4.2 in this volume). Equation (2.27) in this volume gives one rule of thumb for calculating the added contact force for a vacuum interrupter. Table 6.11 gives the values of added contact forces given by a range of vacuum interrupter manufacturers. The total contact force then equals this added contact force plus the force from atmospheric pressure (10^5 Pa) acting on the area of the vacuum interrupter's bellows.

The high value of the closed contact force is also required for the closed contact, high current withstand. Its duration can range from 1s to 4s depending upon which certification standard the vacuum circuit breaker is designed to meet. The need for this specification is for coordination between upstream and downstream circuit breakers in a distribution circuit. This is illustrated in Figure 6.31. If a fault occurs on the load side of VCB4, then it is usual for VCB1 to remain closed unless VCB4 fails to interrupt the current. This explains the need for VCB1 to pass fault current in a closed position for this period of time. The range of 1s to 4s for this withstand time results from the sophistication of the fault current sensing, the relaying and the required coordination of the utility system. A shorter coordination time (i.e., a 1s fault withstand rather than one of 4s duration) places much less stress on the cabling and joints in the distribution system and is therefore more desirable for system maintenance and reliability.

For most vacuum circuit breakers (and load break switches), it is possible to supply the levels of contact force given in Table 6.11 by directly using the contact force spring; see Figure 5.5. It must be understood that the added contact force values given in Table 6.11 are a worst-case condition with only one contact spot. If there are two equally spaced spots, then Kawase et al. [76] show that the new blow-off force F_{B2} is ~ $0.6\ F_B$ and for three spots, F_{B3} is ~ $0.5\ F_B$. These values are somewhat optimistic because the authors do not change the value of "a" in Equation (2.13) in this volume when they calculate for two and three spots. See also the effect of three regions of contact in Equation (6.2) and

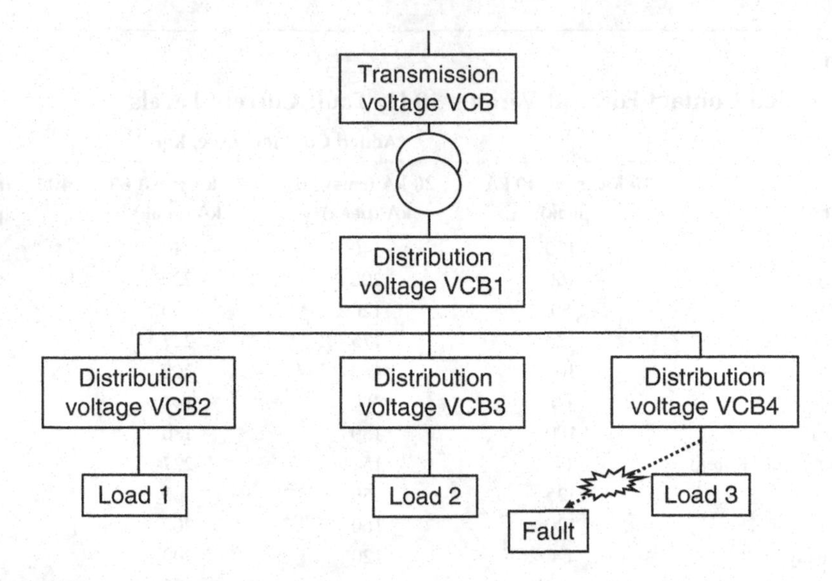

FIGURE 6.31 Coordination between up-stream and down-stream vacuum circuit breakers (VCB)

the effect on closed contact welding in Equation (4.49) in this volume. One would expect a vacuum interrupter contact to usually have three regions of contact in the closed position [22]. There have been contact designs proposed to ensure more than one contact spot for every vacuum interrupter contact closing. One example is given by Banghard et al. [77]. This design has an auxiliary, center, sectored contact that is attached to "springy supports" and is raised a little above the arcing sections of a TMF contact structure. This type of scheme has four problems; (a) at short circuit current levels the columnar arc will initiate at the final point of contact and erode the center region, (b) if the contact spots weld, the springy supports may give enough to prevent the weld from rupturing, (c) vacuum arc erosion can completely change the contact structure, and (d) the springiness of the support structure may not survive the annealing that results from the final braze cycle in the vacuum interrupter's manufacture. The use of AMF contacts reduces the magnitude of the required added contact force because the field coils behind the contact faces attract each other when current flows through them. Unfortunately, because of eddy current effects, the maximum magnetic field may not occur at the maximum current flowing through the coils. Thus, the AMF contacts may only provide a fraction of the total closed contact force needed at the maximum of the asymmetric short circuit current: See for example, Figure 4.55.

Another design that can provide added contact force uses the current flow through the conductors attached to the moving terminal of the vacuum interrupter. Figure 6.32(a) illustrates this principle. Current flowing through the conductor that is held in position by the circuit breaker frame provides a blow-on force F_{BN}. F_{BN} can be adjusted by the gap between the conductors D^* and their length L^*. The disadvantage of this scheme is that the mechanism has to overcome F_{BN} before it can open the vacuum interrupter contacts. A variation of this scheme, suggested by Lane et al. [78], is shown in Figure 6.32(b). Here the conductor loop is constrained by the overtravel box and not by the breaker housing. When the overtravel box is latched in place, the current in the loop again provides the force on the contact. Once the overtravel box is released by the mechanism, the force in the loop now assists in the opening of the contacts. Experiments with this type of arrangement have shown that only quite small closed-contact spring forces are needed for satisfactory operation even at full short circuit currents.

As discussed in Section 4.4.5 in this volume, sideways contact sliding can result in contact welding. Thus, it is important in the vacuum circuit breaker design to prevent the closed contacts from sliding with respect to each other, especially during the passage of fault currents. A high enough closed contact force will provide a frictional force that prevents contact sliding during normal load current operation. During a peak asymmetric current flow, the blow-off force will lower the frictional

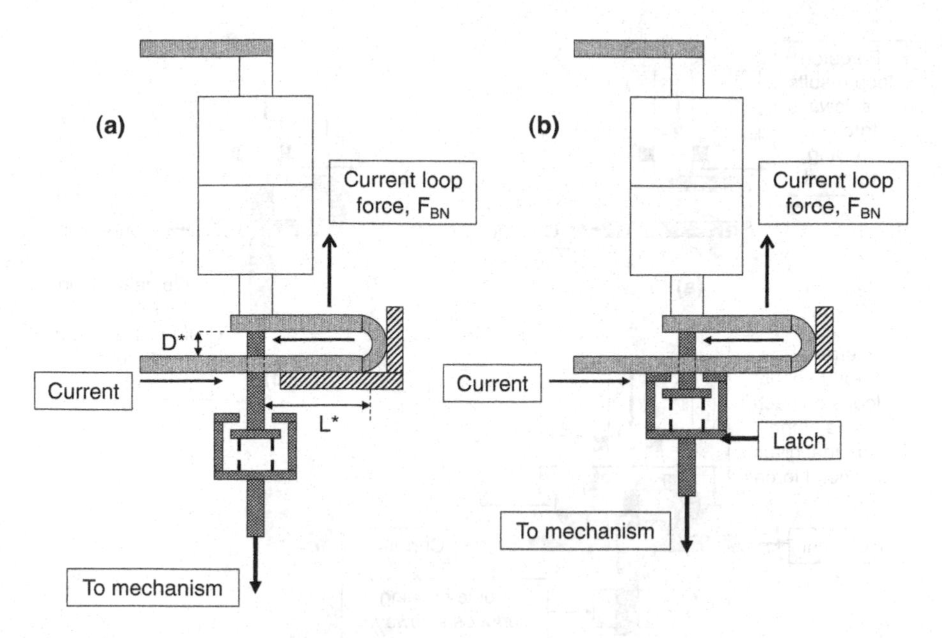

FIGURE 6.32 Blow-on conductor arrangements to provide added contact force during the passage of fault currents through the vacuum interrupter.

force considerably, thus permitting other forces acting on the vacuum interrupter and its terminals to cause one contact to slide against the other. The breaking of one contact spot and the development of another during this process can cause the spots to melt and result in the welding of the two contacts. Figure 6.33(a) shows the forces and constraints acting on a vacuum interrupter mounted vertically in a U-shaped current flow. There is the loop force from the current flow that acts upon the fixed terminal and moving terminals and a restraining force resulting from the frictional force at the contact faces. The guide bushing at the moving terminal and the end of the terminal where it connects to a sliding current transfer device or else a guide bushing on the rod that connects to the

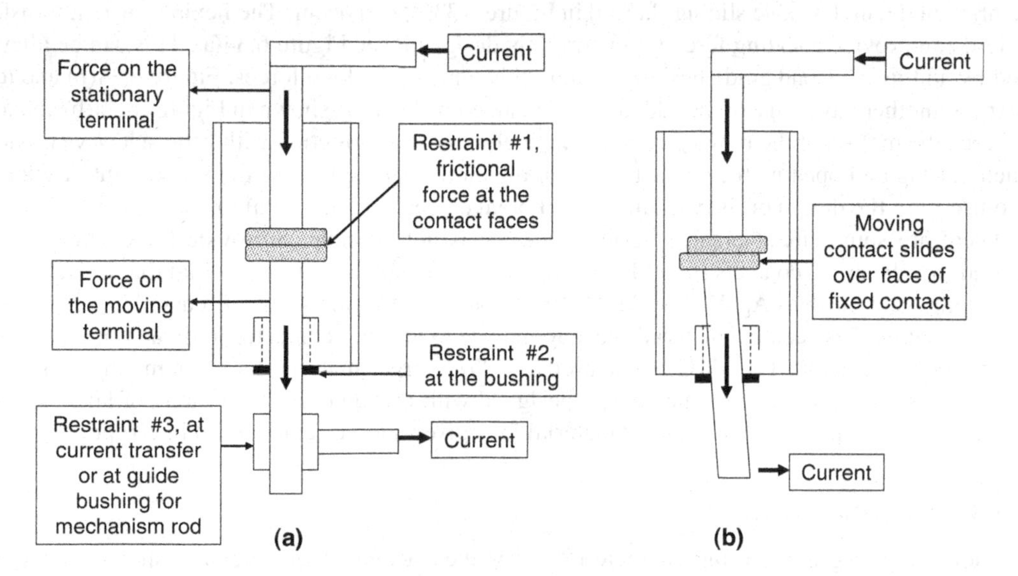

FIGURE 6.33 (a) The forces acting in the vacuum interrupter and its internal structure and the restraints during the passage of fault current. (b) The possible contact sliding motion if the restraints are nor adequate.

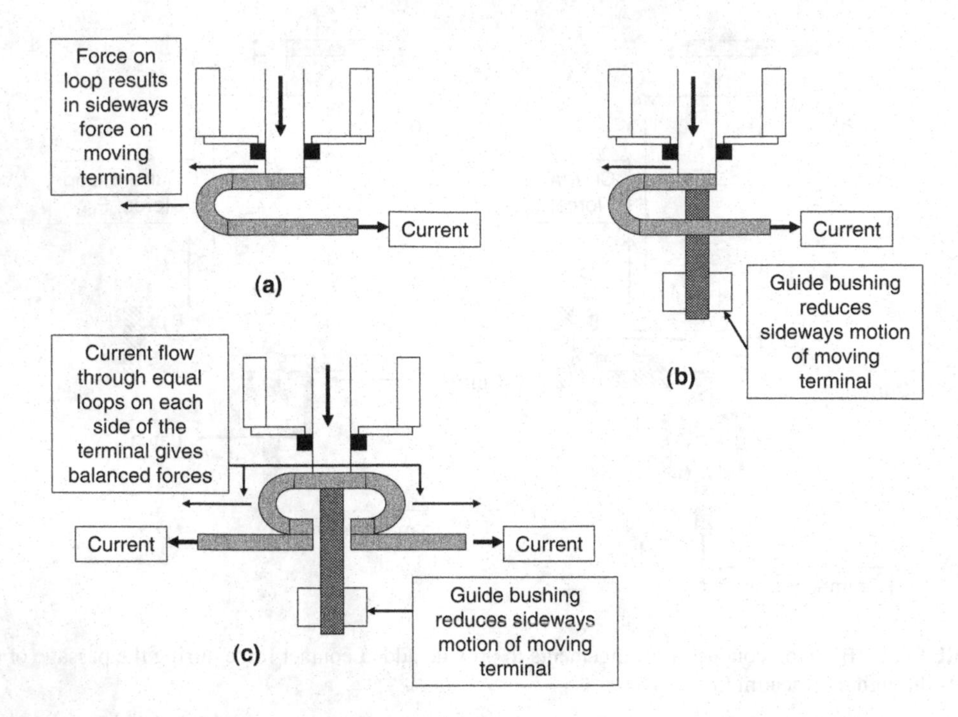

FIGURE 6.34 (a) Loop forces from the flexible connection resulting in contact slipping. (b) A restraint from a guide bushing. (c) Two restraints, one from the guide bushing and the other from balancing the flexible connector loops.

mechanism provide further restraining forces. The Multilam, the Roller and the Leopold current transfer (Figure 6.6) systems also provide an added frictional force on the moving terminal during the passage of fault currents. This, to some extent, limits the effects of blow-off force at the contact's interface. If restraints #2 and #3 in Figure 6.33(a) are sturdy enough to limit sideways motion of the terminal, then contact interface sliding will be prevented. When these restraints are weak or nonexistent, then the undesirable sliding showed in Figure 6.33(b) can occur. The flexible current transfer system can provide a sliding force if not correctly designed; see Figure 6.34(a). This can be alleviated by adding a second guide bushing and/or by balancing the loop forces, Figure 6.34(b) and (c). There is another advantage of the added mechanism guide bushing shown in Figures 6.34(b) and (c). It forces the motion of the moving contact to travel in a linear direction with little sideways motion when closing and opening the contacts. Preventing motion of the base of the vacuum interrupter is also useful in the design of the vacuum current breaker; see Figure 6.35(a) to (c).

There are some circuit breaker designs where it is not possible to provide the contact forces needed to prevent the contacts from blowing open. In this case, a contact material that resists forming strong welds such as Ag–WC or Cu–Cr-Bi can be used. Certainly, as I discussed in Section 5.4 in this volume, these contact materials have found use in vacuum contactor designs. It has periodically been suggested that Ag–WC can be used as a "low surge" material for vacuum circuit breaker applications [79]. The vacuum interrupters designed with this material are usually of larger diameter than those using the Cu–Cr contact material for a given short circuit current interruption ability.

6.5.3 Mechanism Design

The ideal opening and closing characteristics for the vacuum interrupter are shown in Figure 6.36(a). The more practical characteristics are given in Figure 6.36(b). On closing there is an initial acceleration period the length of which is not critical. Then a more or less uniform closing speed

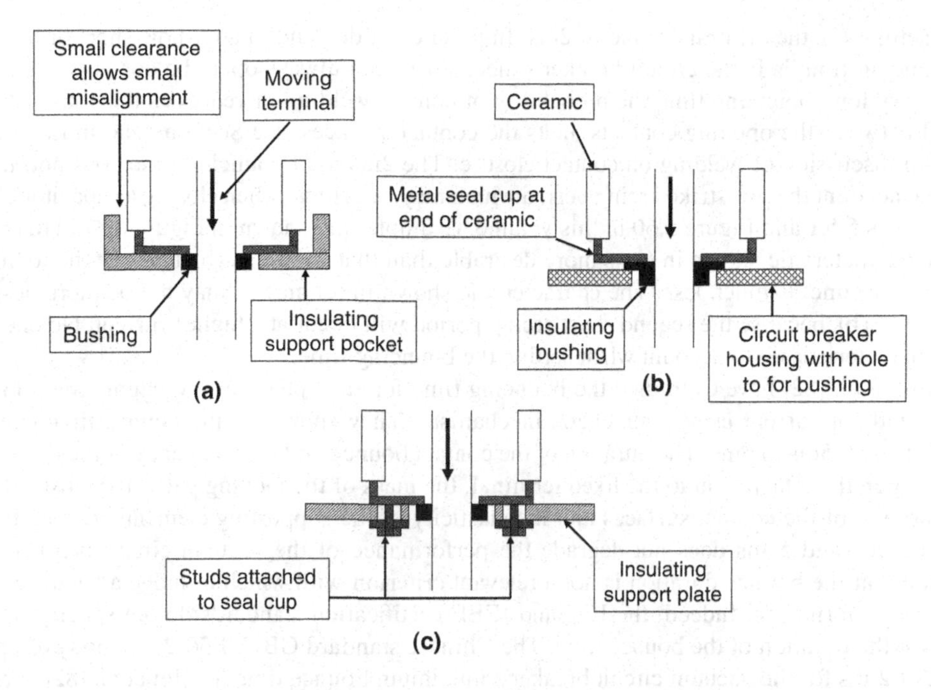

FIGURE 6.35 Restraining the base of the vacuum interrupter.

occurs until the contacts touch. The kinetic energy of the moving contact structure must then be dissipated. This is achieved by oscillations in the vacuum circuit breaker's structure and by bouncing of the contacts. Greenwood [80] and Barkan [81] both give an excellent introduction to designing for reduction of the bouncing period. The vacuum circuit breaker designer must, however, expect some period of bouncing. The accepted total bouncing period specified by most vacuum interrupter

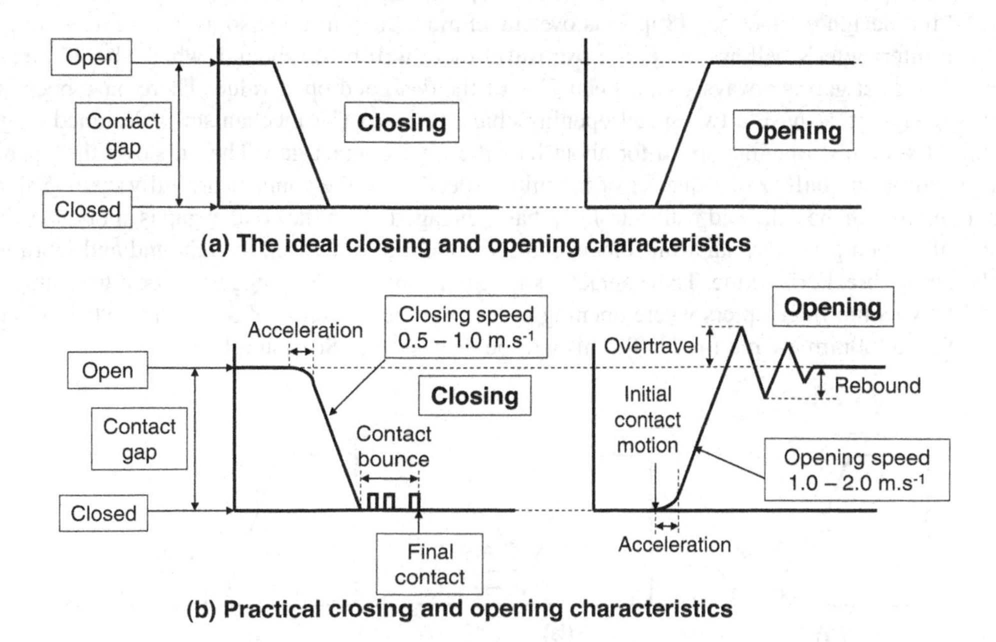

FIGURE 6.36 (a) The ideal opening and closing characteristics for a vacuum circuit breaker pole. (b) The opening and closing characteristics of a practical design.

manufacturers is the arbitrary value of 2ms. In practice, it does not matter how short or how long the bouncing time is if the circuit breaker's mechanism can always open the contacts. Of course, with a very long bouncing time the probability of contact welding increases as the constricted arc formed between the opening contacts melts the contact surfaces; see Section 4.4.6 in this volume for a full discussion of welding on contact closure. The 2ms total bouncing time does not, in fact, take into account the pre-strike arcing period which may be critical when closing a capacitor circuit: see Sections 5.5.1 and Figure 5.60 in this volume. Examples are shown in Figure 6.37(a) to (c). The bounce characteristic shown in (a) is more desirable than that shown in (b), because the total open time (arcing time) is much less. The characteristic shown in (c) may or may not be more desirable than that of (b), because the second short arcing period will occur at a higher current. No one takes these characteristics into account when setting the bouncing time.

Dullni et al. [82] have examined the bouncing time for each phase of a vacuum circuit breaker closing without current using a magnetic mechanism. They show that the bounce time can have periods from 0.5ms to 6ms. The duration of the contact bounce and the frequency of the driving rod depend upon the attachment to the fixed terminal, the mass of the moving parts, the closing speed, the condition of the contact surfaces and the elasticity of the supporting elements. Even a bounce time of 4 ms and 5 ms does not degrade the performance of the vacuum circuit breaker. They conclude that the bounce duration is not a relevant criterion with which to judge a vacuum circuit breaker's performance. Indeed, the IEC and IEEE certification standards do not specify requirements for the duration of the bounce time. The Chinese standard GB 501.50-2006, however, places a limit of 2 ms for the vacuum circuit breaker's maximum bounce time. Dullni et al. [82] correctly state that there is really no technical reason to require this limitation. A statement I certainly agree with! Also, when closing with current any minor welding of the contacts will certainly modify the bounce characteristics.

On opening there will be a period of contact acceleration before the contacts reach a uniform opening speed. For vacuum circuit breakers used in distribution circuits (4.76 kV to 40.5 kV) this is usually between 1m.s^{-1} and 2m.s^{-1}. The mechanism designer must keep this acceleration period very short to ensure the opening characteristic is as close as possible to the ideal shown in Figure 6.36(a). Once the vacuum interrupter reaches the open position the contacts will continue to open beyond the design contact gap [80]. This overtravel must be controlled so as not to overstress the vacuum interrupter's bellows. After the overtravel there will be a rebound, which should be such that the contact gap is always greater than 75% of the designed open value. There have been suggested designs that have a two speed-opening characteristic. The mechanism is designed to give an initial very fast opening speed for about half the total contact gap. Then it slows the opening to a speed of one-half to one-quarter of the initial speed until the contacts are fully open [83]. For example, this can be achieved with a dashpot that is engaged when the contact gap is about one-third of its fully open gap. Such an arrangement could reduce the overtravel, the rebound and vibrations of the circuit breaker's frame. This contact opening concept has been suggested for high voltage (> 72.5 kV) vacuum interrupters where opening speeds can be in excess of 3m.s^{-1} and where contact gaps of up to 60mm are required [10]. This will be discussed in Section 6.10.

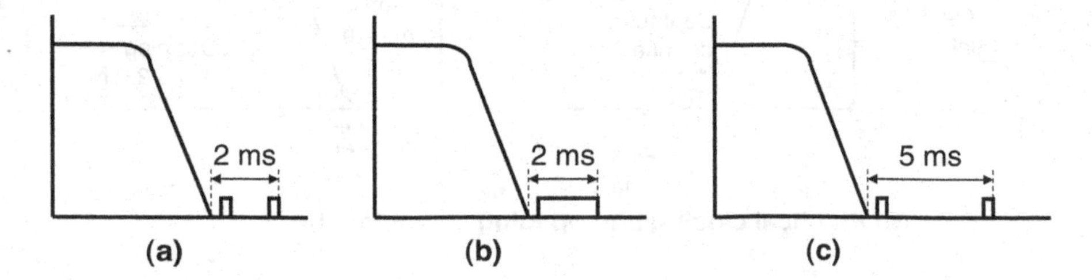

FIGURE 6.37 Examples of bouncing time characteristics.

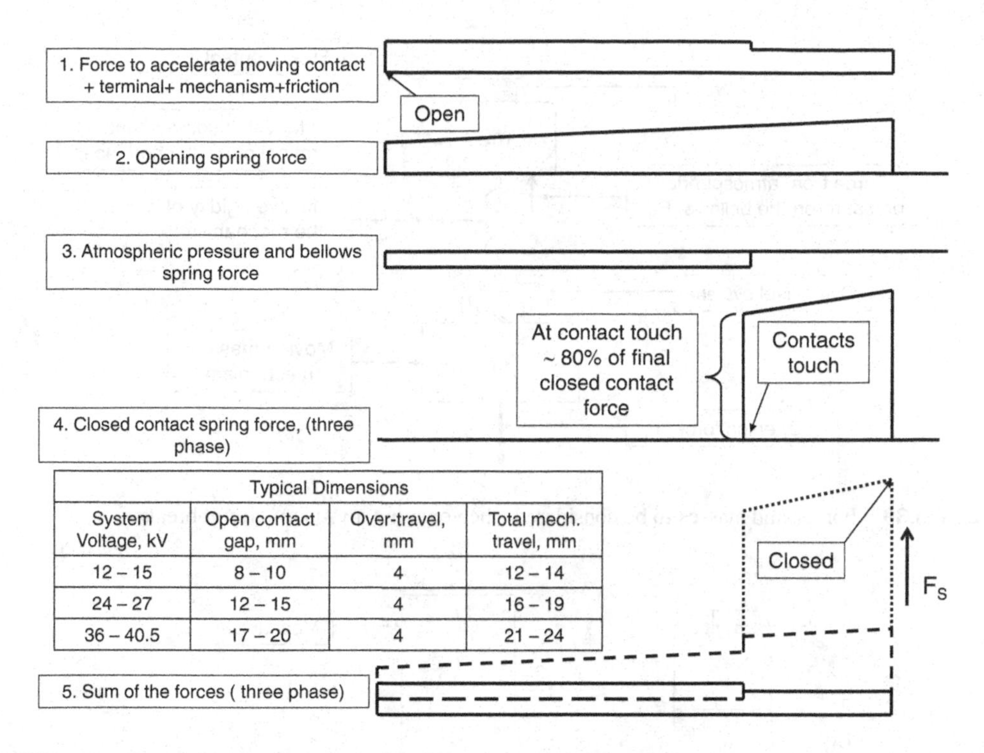

Typical Dimensions			
System Voltage, kV	Open contact gap, mm	Over-travel, mm	Total mech. travel, mm
12 – 15	8 – 10	4	12 – 14
24 – 27	12 – 15	4	16 – 19
36 – 40.5	17 – 20	4	21 – 24

FIGURE 6.38 Forces involved when closing the vacuum circuit breaker.

Figure 6.38 illustrates the forces that must be considered when closing the vacuum circuit breaker. Initially there is the force to overcome friction in the mechanism, accelerate the mechanism and begin to move the vacuum interrupter's contacts. At the same time compression of the opening springs begins. There is some assistance from the atmospheric pressure on the bellows. At contact touch the force on the contacts should be at least 80% of the final contact force. This prevents the contacts from blowing apart if the vacuum circuit breaker is closing-in on a full short circuit current. The contact spring is then compressed to its full value as the overtravel gap is established.

Some welding usually occurs when closing the Cu–Cr contacts in a vacuum interrupter into a fault current greater than 8kA or when they pass a high withstand current (see Section 4.4.2, Figure 4.45 in this volume). Thus, the mechanism must be designed to break these welds when opening the vacuum interrupter's contacts. A general schematic of the closed contacts with a weld, a moving terminal, contact and linkages of mass m_1, and a mechanism with a moving mass m_2 is shown in Figure 6.39. The mechanism open force F_0 moves the mass m_2 through the overtravel gap d_e (see Figure 5.5). At this time, the impact on mass m_1 breaks any weld and the opening characteristic shown in Figure 6.36(b) results. To achieve this, a rule of thumb is that $m_2 \approx 2 \times m_1$. When considering this relationship, it is important to add all the moving linkages attached to m_1 above the contact force spring. Figure 6.40 illustrates some examples.

In designing for m_2 it is also important not to make it too large. The vacuum interrupter's terminals are made from fully annealed oxygen free Cu and are thus relatively soft. With $m_2 \approx 2 \times m_1$ the first 50 to 100 operations of the vacuum circuit breaker will cause the total length of the vacuum interrupter to shorten by about 1mm as the Cu work hardens. The hardness of the Cu now can absorb the impact of m_2 on closing without further shortening. If, however, m_2 is too large the plastic deformation of the Cu will continue and the terminals will continue to shorten, which is undesirable. Thus, there is an optimum value of m_2.

The first and still most commonly used mechanism for vacuum circuit breakers is the *spring mechanism*. Here a charged spring provides a simple and reliable source of energy required to

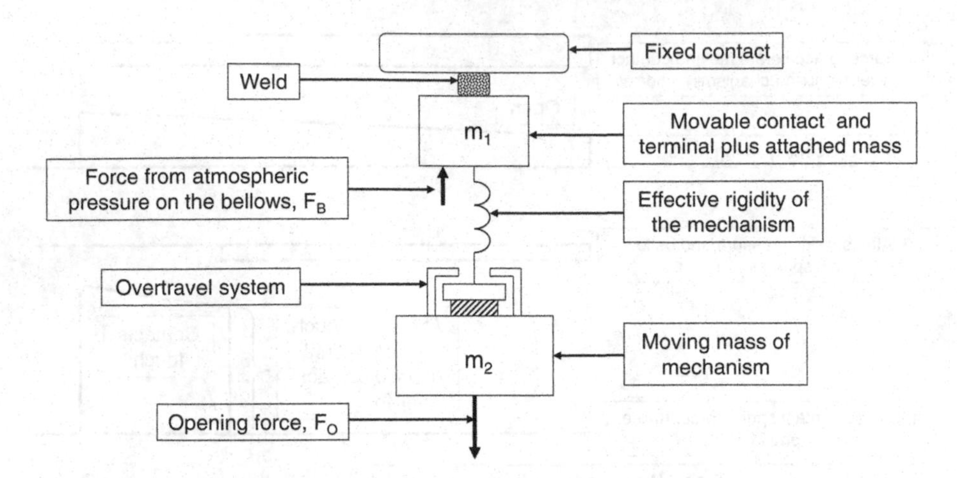

FIGURE 6.39 Forces and masses to be considered when opening a vacuum circuit breaker.

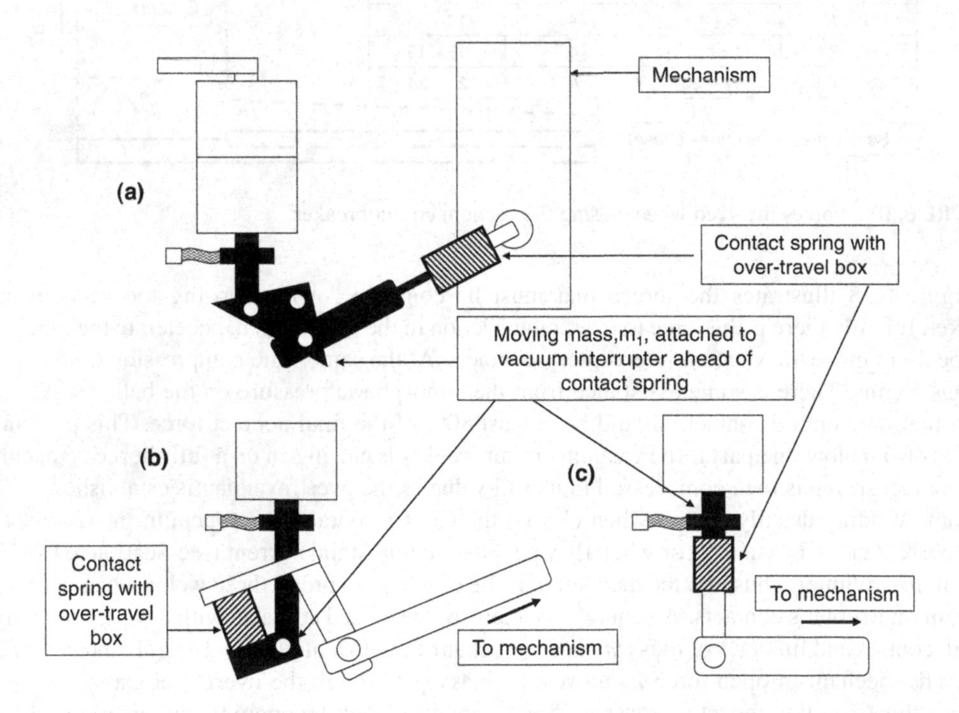

FIGURE 6.40 Examples of the attached mass m_1.

provide the breaker's operating sequences. The closing spring is usually charged remotely by an electric motor, although most vacuum circuit breaker designs also have a manual charging option. The spring mechanism is ideal for operating a vacuum interrupter where high forces are required over short distances. The spring mechanism in its simplest form consists of the closing spring, a charging motor, and a charging ratchet. The spring is held in its fully charged position by a latch. The rest of the mechanism consists of closing cams, closed contact force springs, opening springs, and various toggles and linkages. A closing signal releases the charged closing spring's latch, which allows the closing spring to drive the vacuum interrupter's contacts into the closed and latched position. At the same time the opening springs are fully charged and a trip-open latch is engaged to maintain the mechanism in the closed position. An opening signal releases the trip-open latch, which in turn releases the opening springs. The vacuum interrupters' contacts then open and the vacuum circuit breaker interrupts the circuit. The energy stored in the closing spring is sufficient to close the circuit breaker and to charge the opening springs. Once the circuit breaker is closed the charging motor recharges the closing spring. With the breaker closed and both the closing and opening springs fully charged, the stored energy is sufficient to provide an open-close-open operating sequence. Tripping of the opening latch is usually initiated electrically, although most circuit breaker designs also have a manual trip option, which is generally used only during maintenance. A photograph of a typical spring mechanism is shown in Figure 6.41. A spring mechanism has perhaps 200 parts. Greenwood [84], Barkan [85, 86], and Shi et al. [87] give further information on spring mechanisms. While all spring mechanisms follow similar principles of operation, all manufacturers have their own trade-secret features that are unique to their designs.

Because it is unlikely that the three phases of a vacuum circuit breaker will close at exactly the same time, there will be slight differences in the linkages to the three phases, the contact erosion of each vacuum interrupter and the work hardening of the copper terminals. Yu et al. [88] have studied the effect of 0.5 ms differences in closing time using a spring mechanism. Their data is shown in Table 6.12. The initial impact force is 6 to 20 times the final closed contact force. Also, the first phase to close has the highest impact force. The difference in the impact forces results from the slight nonsynchronous closing caused by slight differences in the contact gaps.

A second mechanism that is increasingly being developed for use in the vacuum circuit breaker and vacuum recloser is the *magnetic mechanism*. The concept certainly predates Lindsay's 1965 Patent [89] that matches the magnetic mechanism to the vacuum interrupter. The magnetic actuator is a flux-transfer system; one embodiment is illustrated in Figure 6.42. It consists of a chamber made from a magnetic material (usually steel), inside which an armature resides also made of magnetic material. Two coils are placed inside this housing, one at the top and one at the bottom. Halfway down the inside surface of the chamber, a high permeability magnetic material (e.g., sintered Nd–Fe–B or sintered Sm–Co) is placed. The housing and/or the armature may have a laminated structure. The upside rest position is shown in Figure 6.42(a). When a current usually supplied by a charged capacitor (in some designs the coil current is supplied from a lithium oxide battery) is discharged through the lower coil (Coil 2), Figure 6.42(b), the magnetic flux transfers to the lower path shown. The force on the armature is now such that it begins to move down. Figure 6.42(c) shows the armature in the lower position. To move the armature back to the upper position, a current pulse is passed through the upper coil (Coil 1), 6.42(d), which transfers the magnetic flux to the upper

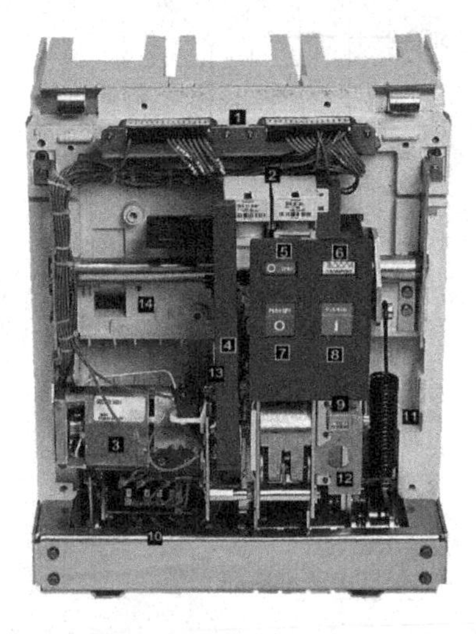

FIGURE 6.41 The spring-operated vacuum circuit breaker (courtesy of Eaton Corporation).

TABLE 6.12
The Impact Forces on Each Phase of a Vacuum Circuit Breaker as a Function of the Design Short Circuit Current Level Compared to the Expected Close Contact Force for That Current Level [88]

Short Circuit Current kA	Impact Force 1st Phase to close, F_1, N	Impact Force 2nd Phase to close. F_2, N	Impact Force 3rd Phase to close, F_3. N	Closed Contact Force, F_C. N
25	21,430	12,482	7,853	1,150
31.5	32,771	21,772	11,601	1,600
40	70,484	60,082	48.021	4,000

Short Circuit Current kA	F_1 / F_C	F_2 / F_C	F_3 / F_C
25	13	10	6
31.5	20	14	7
40	18	15	12

position. This simple design uses far fewer parts than does the spring mechanism and can easily be attached to a vacuum interrupter; see Figure 6.43.

Other magnetic mechanism designs are illustrated in Figure 6.44(a) and (b). Figure 6.44(a) (1) shows the magnetic flux latching the armature in the open position. When a closing signal is received current is supplied to the coil (2). This causes the magnetic flux to transfer to the upper

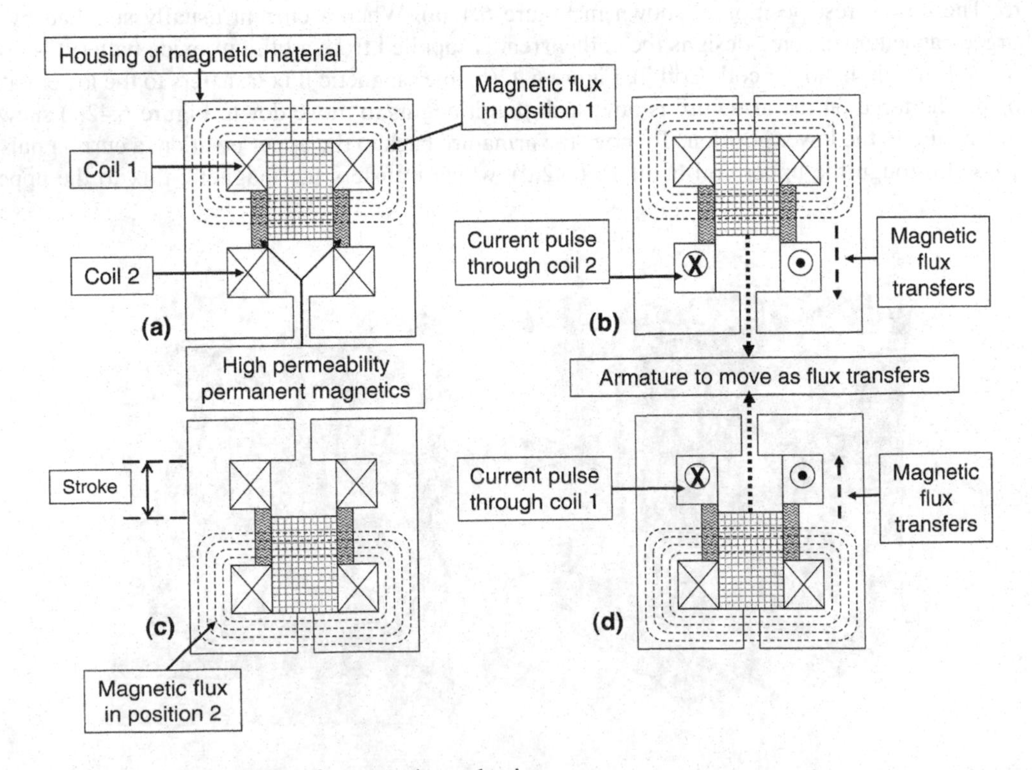

FIGURE 6.42 The operation of a magnetic mechanism.

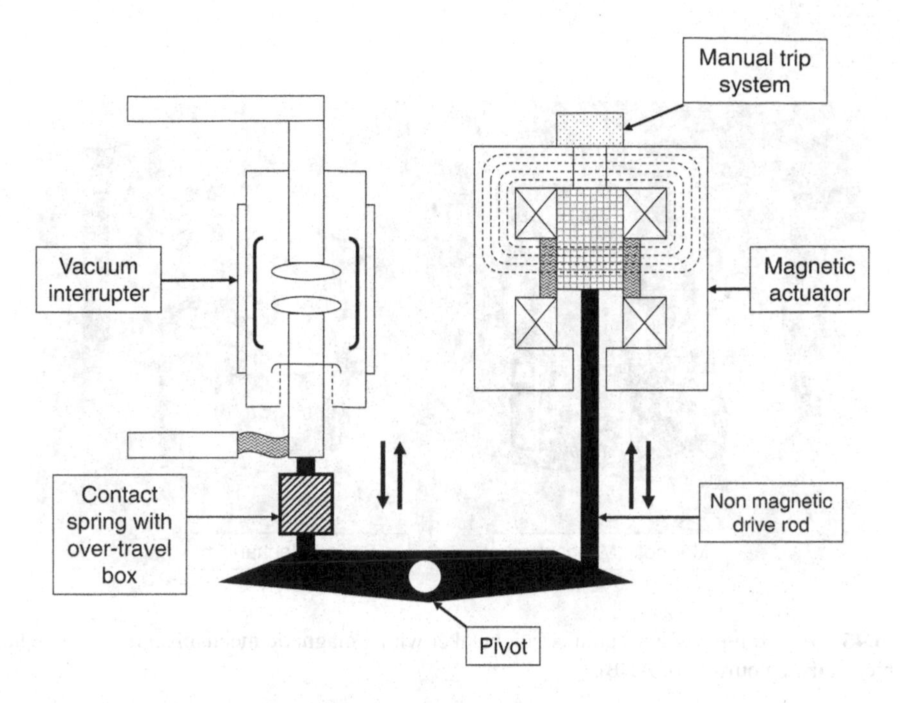

FIGURE 6.43 A schematic of the attachment of the magnetic mechanism to a vacuum interrupter.

position, which in turn results in the armature moving into position (3). When doing so, the contact force spring and the opening spring are compressed. To return the armature into the open position a reverse current is supplied to the coil, which reduces the holding force on the armature allowing the opening spring to drive it as shown (4). In doing so the magnetic flux transfers back to that shown in position (1) latching the armature in the open position. Figure 6.44(b) shows another example with the permanent magnets now placed on the armature. The magnetic mechanism has become

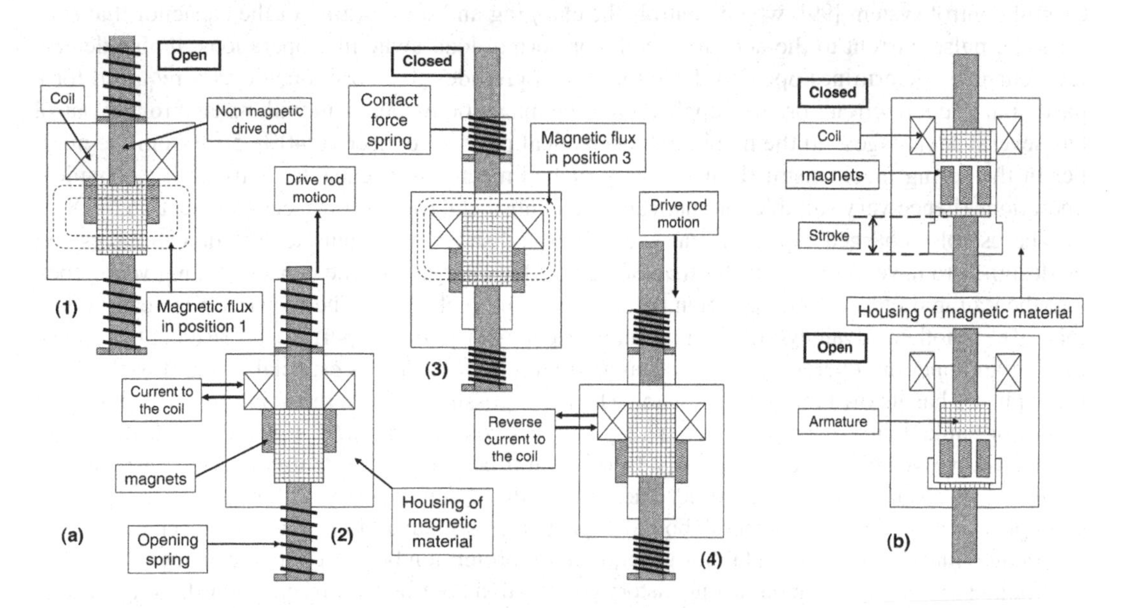

FIGURE 6.44 Other magnetic mechanism designs.

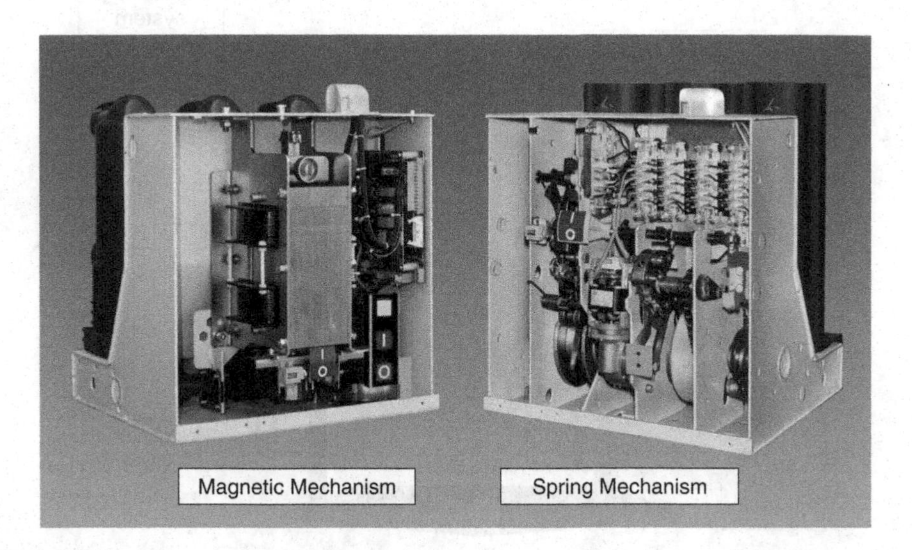

FIGURE 6.45 An example of a vacuum circuit breaker with a magnetic mechanism in the same housing as a spring mechanism (courtesy of ABB).

accepted for recloser application [89, 90] since the mid-1990s. Its general use for vacuum circuit breaker applications is also gaining strength in the twenty-first century [91–98]. Most vacuum circuit breaker manufacturers have developmental designs in process. One practical vacuum circuit breaker design is shown in Figure 6.45.

The magnetic actuator is well suited for the vacuum interrupter where short strokes (8mm–20mm) and large forces (2000–4000N/phase) are required. For this mechanism to operate effectively it must be matched with an electronic capacitor charging and control system. This control system has to be immune from electromagnetic interference resulting from the high currents and high impulse voltages to which the vacuum circuit breaker is exposed. Dullni et al. present a successful control system [99], which controls the charging and discharging of the capacitor that supplies the pulse current to the actuator's coils for normal load switching operations. It also detects fault currents, determines opening delay times, and provides the close–open cycles required for a particular vacuum circuit breaker application. The magnetic mechanism with its electronic control has several advantages: (a) the number of mechanical parts is reduced to 40%–50% from the number in the spring mechanism; (b) it is a very rugged mechanism, capable of 100,000 mechanical operations (hence very suitable for application to control electric arc furnaces (see Section 6.8.8).

The use of a charged capacitor may seem to be a disadvantage, but electrolytic capacitors can be designed to have a life of greater than 30 years. One charging of the capacitor can give an open and then a rapid close–open operation before it must be recharged. The recharge time can be less than 10s. Another advantage of the magnetic mechanism is that it permits precise control of the closing and opening operation of the vacuum circuit breaker. This is especially true if each of the three phases has its own magnetic actuator. Thus, it is possible to have the point-on-wave closing of capacitor banks that I discuss in Sections 5.5.1 in this volume and 6.7.2. I would expect that other point-on-wave switching applications [100,101] would become more common as this mechanism technology becomes more widely used. There is continued development of this mechanism: see, for example Cai et al. [102]. . In general, both the spring mechanism and the magnetic mechanism have advantages and disadvantages, but for general circuit protection both are reliable enough.

Another mechanism that has a long history (invented in the nineteenth century) is finding a use in transmission vacuum circuit breakers (Section 6.10) and in high-voltage dc vacuum circuit breakers

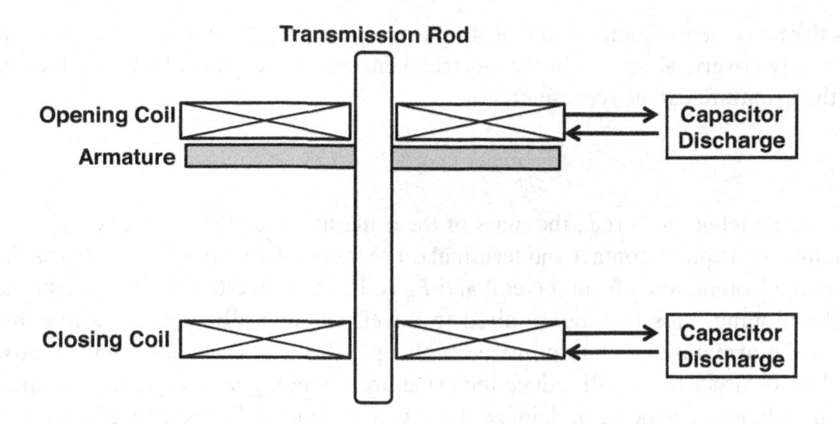

FIGURE 6.46 The Thomson Coil mechanism.

(Section 6.11). This is the *Thomson Coil Mechanism* (T-C) [103]. The T-C mechanism is ideal where a fast reaction time and a rapid opening speed are required. Figure 6.46 shows a schematic of the T-C system. It consists of opening and closing coils that are commonly pancake coils illustrated in Figure 6.47. There is a transmission rod attached to a metal armature which operates the vacuum interrupter. Figure 6.46 shows the armature abutting the spring coil. When a time varying current is passed through the opening coil it produces a time varying magnetic field. This magnetic field induces eddy currents in the armature. The interaction of these eddy currents and the coil's magnetic field produces a very strong repulsive force. The current pulse to the opening and closing coils is supplied by discharging a capacitor bank. The electrical energy E_T stored in the capacitor bank is:

$$E_T = ½CU^2 \tag{6.27}$$

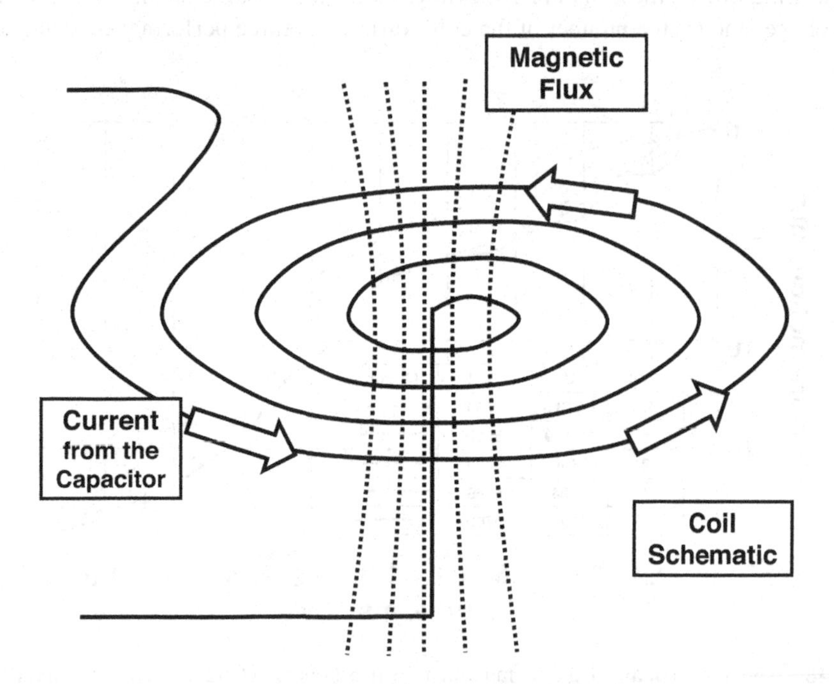

FIGURE 6.47 The pancake-shaped coil.

where C is the bank's capacitance and U is its charging voltage. Bissal at al [104] show that the T-C mechanism only coverts about 5% of the electrical energy to mechanical energy. The initial acceleration of the armature can be represented by:

$$ma = \pm mg + F_L - F_B \tag{6.28}$$

where m is the attached mass (i.e., the mass of the armature, plus the operating rod, plus the mass of the vacuum interrupter's contact and terminal). The value of which is typically less than 10 kilograms. F_L is the Lorenz force from the coil and F_B is the force exerted by the vacuum interrupter's bellows. The moving mass may be required to travel tens of millimeters in a few milliseconds. The mass of the armature can be optimized with care. A thicker armature will improve the T-C efficiency, but its larger mass will reduce the armature's opening acceleration. A thinner armature can result in its bending which can damage the T-C operation and reduce its efficiency. The coil is usually operated using a single pulse from the capacitor bank. Vilchis-Rodriguez et al. [105] have analyzed the performance of a T-C mechanism for a single current pulse and a two-stage current pulse with the same total electrical energy. They find that the performance of the T-C operated by the two stage pulse lags that of the single current pulse. The reason for this is that as the armature separates from the coil during the application of the first current pulse, the electromagnetic force is reduced by the time the second pulse is applied. Figure 6.48 [105] shows the opening displacement for an 8kg total moving mass as a function of time for different capacitor banks with approximately the same electrical energy E_T. It shows that as the capacitance decrease with a corresponding increase in the charging voltage the armature reaction time increases. They observe, however, that this increase of speed and efficiency comes at a cost of the armature flexing. Augustin et al. [103] also show that for a given total moving mass and capacitor energy there is an optimum charging voltage–capacitance combination. Figure 6.49 Illustrates that above the dotted line the mass moves too quickly and below it travels too slowly.

Zhang et al. [106] have analyzed the influence of six design parameters on three performance factors for the T-C mechanism: see Table 6.13. The six parameters are: (1) the moving mass; (2) the radius of the armature; (3) the armature's thickness; (4) the total charge on the capacitor bank; (5) the charging voltage; and (6) the number of the coil's turns. The three performance factors are: (1) the

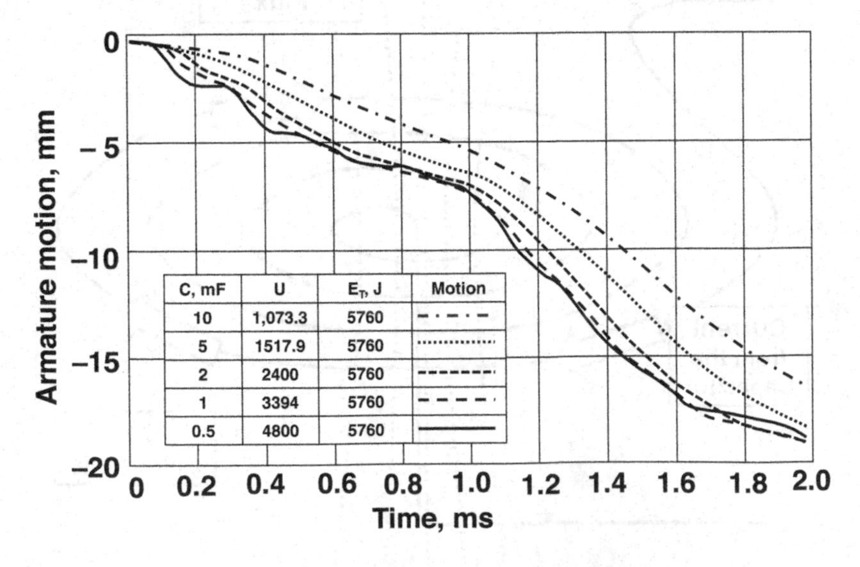

FIGURE 6.48 The motion of an 8 Kg total moving mass in a Thomson Coil mechanism as a function of time for current pulses from capacitor banks with the same total electrical energy [105].

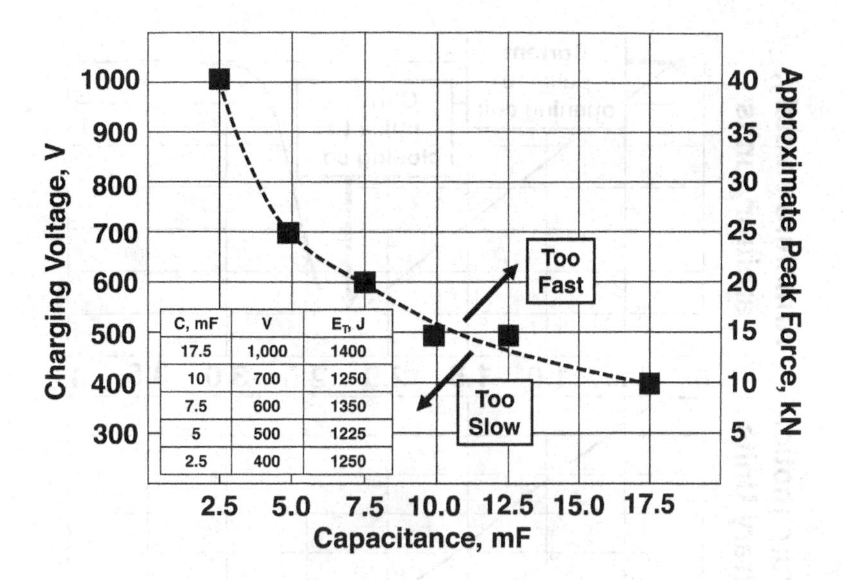

FIGURE 6.49 Optimized capacitor-voltage combinations providing a current pulse to a Thomson coil mechanism to operate a 2.24 Kg moving mass [103].

average opening speed; (2) the maximum opening speed; and (3) the maximum electromagnetic force on the armature. As might be expected, they show that an increase in the moving mass results in a decrease of its average opening speed and also a decrease in its maximum opening speed. The average and maximum opening speeds are directly proportional to the armature's thickness, the initial capacitor voltage and capacitance as well as the number of the coil's turns. It is interesting to note that when the radius of the armature is smaller than that of the coil, the performance factors increase linearly with an increase of the armature's radius. However, when the armature's radius is larger than that of the coil the performance factors do not change.

The armature moves from the opening coil with a high speed. For example, if a vacuum interrupter and operating rod have a total moving mass of 8 kilograms and opens to a 20mm contact gap in 2 ms, then 400kJ of energy has to be absorbed at the end of the travel. It is possible to use the

TABLE 6.13
The Parameters Analyzed in the Operation of a Thomson Coil Mechanism [106]

Design Parameters	Value
Rated contact force	4800 ± 200 N
Total travel	13.9mm
Moving mass	3.5 Kg
Armature radius	90mm
Armature thickness	10mm
Capacitor bank	2,200 μF
Capacitor bank voltage	2000 V
Coil turns	22
Outside radius of the coil	90mm

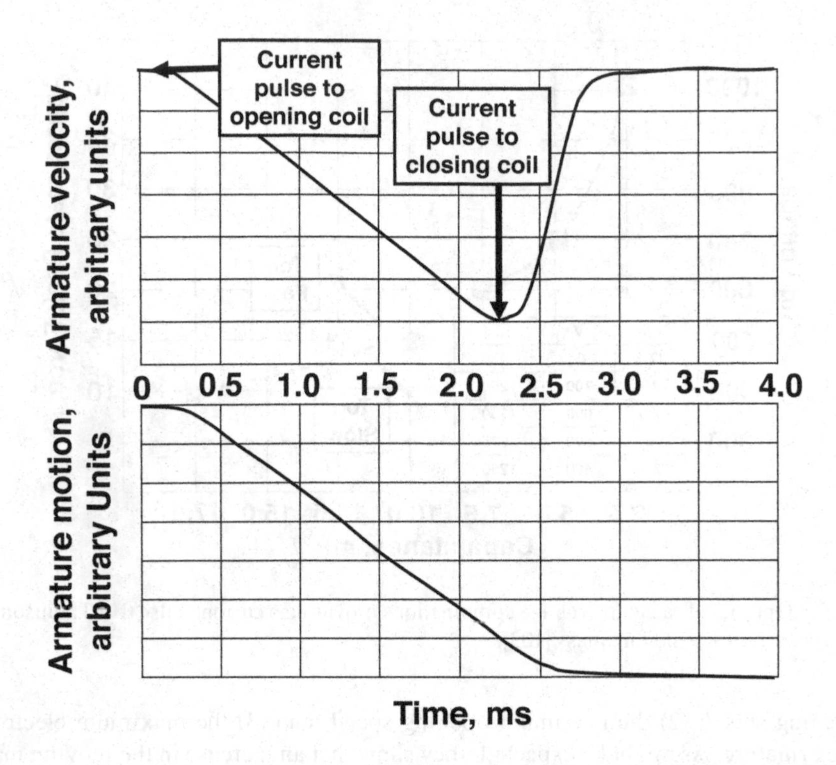

FIGURE 6.50 Active damping of the Thomson Coil armature using the closing coil operating at the end of the armature's stroke [107].

closing coil as an active soft-stop system by activating it just as the armature reaches its fully open position. Figure 6.50 shows an example of the damping of the motion using the closing coil to slow the armature's speed. Vilchis-Rodriguez et al. [107] show a passive system using an air chamber shown in Figure 6.51. The damping is achieved by limiting the flow of air from the gas chamber by controlling the size of the vents. When using the T-C for opening and closing a vacuum interrupter a latching system must also be incorporated once the final open and close are achieved. Figure 6.52 and 6.53 [106] show examples of latches that can be used with an active or a passive damping system.

6.5.4 The Vacuum Interrupter Mounting and Insulation

The vacuum circuit breaker is a three-phase unit with usually one vacuum interrupter per phase. The circuit breaker's design has to take into account both the current path to and from each vacuum interrupter as well as the phase-to-phase distance between the vacuum interrupters. Figure 6.54 gives some typical vacuum interrupter mountings. The "U"-shaped conduction path, Figure 6.54(a) and (b), is the most common pole system. Most vacuum circuit breaker designs have the vertical arrangement of the vacuum interrupter shown in Figure 6.54(a). Care must be taken to ensure that the distance between the bus "D_1" is large enough to satisfy the high voltage withstand and the BIL standards. The current loop in this design produces a repulsion force F_{arc} (see Figure 6.54(e)) on the high current columnar arc (TMF contacts), and on the high current diffuse column (AMF contacts). Therefore, the internal shield in the direction of the U's base will experience severe heating from the vacuum arc plasma when the vacuum interrupter is switching its maximum fault current. The vertical arrangement shown in Figure 6.54(c) is more common when a vacuum circuit breaker is used to retrofit, or replace, the former magnetic air and minimum/bulk oil circuit breaker installations.

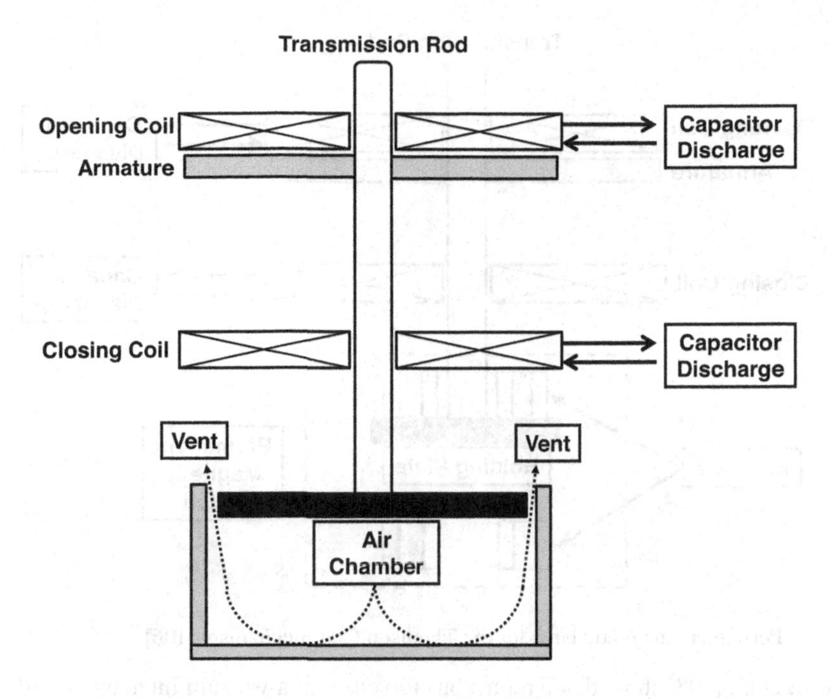

FIGURE 6.51 Passive damping system using an air chamber [107].

The dimensions D_1 and D_2 are once again determined by the high voltage requirements. There is, again, a repulsion force F_{arc} (see Figure 6.54(e)) between the vacuum arc and the return bus, which results in the vacuum arc impinging upon the internal shield. Thus, the gap D_3 must be made as large as possible to minimize the effects of F_{arc}. If D_3 is too small this F_{arc} can be large enough that the vacuum interrupter will not perform up to its designed fault current interruption rating.

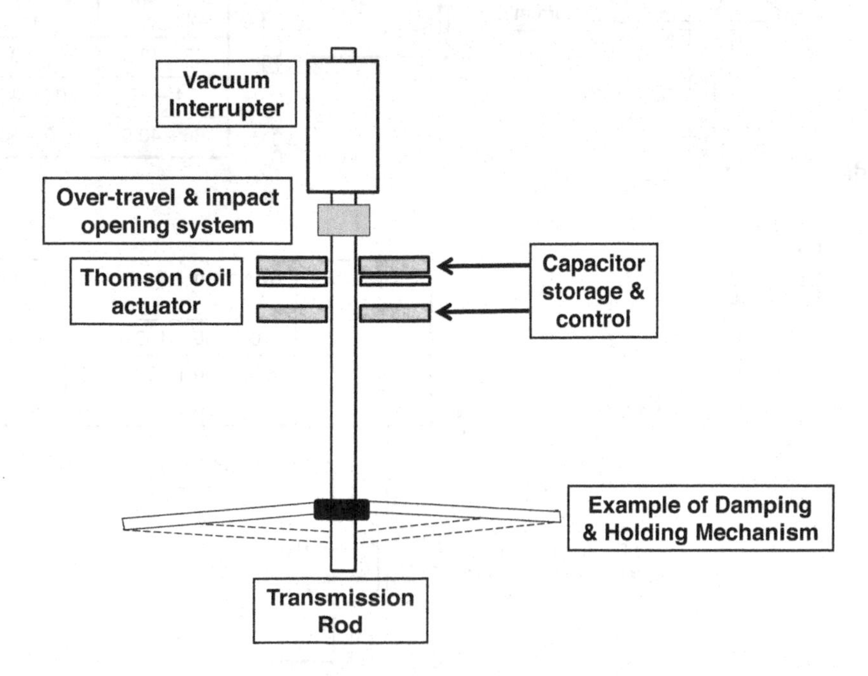

FIGURE 6.52 Schematic of a vacuum interrupter operated using a Thomson Coil mechanism with a damping and latching system.

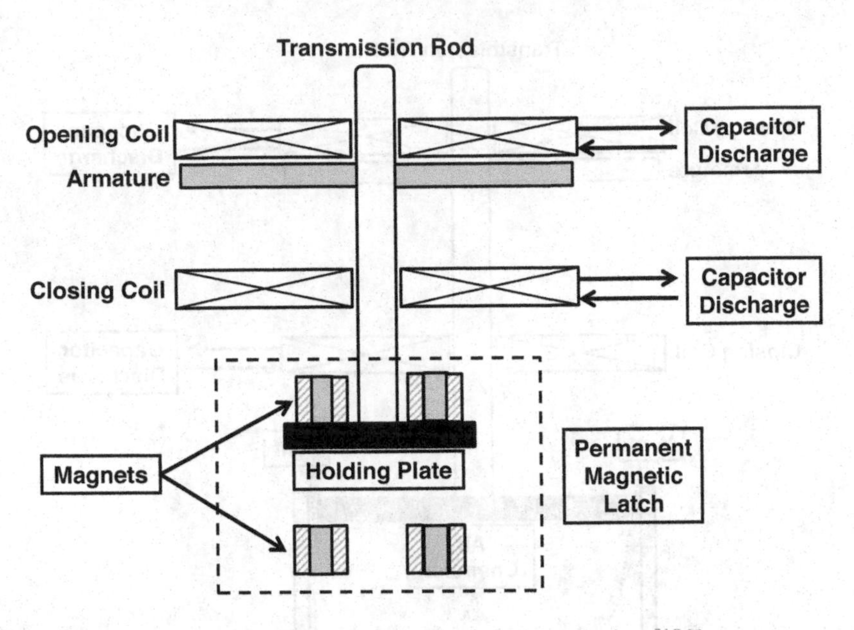

FIGURE 6.53 Permanent magnetic latch for the Thomson Coil mechanism [106].

Poluyanova et al. [108] show that a return bus too close to a vacuum interrupter with AMF contacts can reduce its short circuit performance by as much as 20%. I have also experienced a 20% derating of the short circuit interruption ability in a well-designed vacuum interrupter with TMF contacts when a return bus is too close. When the return bus is moved further away, the rating returns to its full 100% value. Long et al. [109] have evaluated a number of bus configurations and

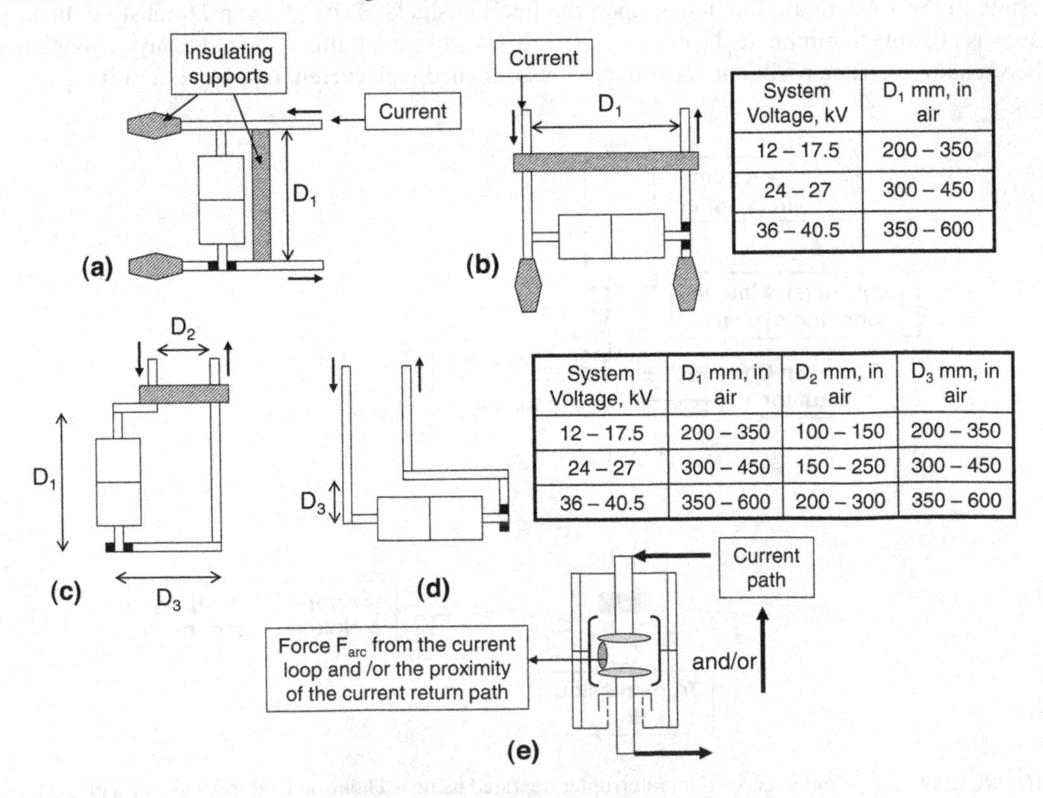

System Voltage, kV	D_1 mm, in air
12 – 17.5	200 – 350
24 – 27	300 – 450
36 – 40.5	350 – 600

System Voltage, kV	D_1 mm, in air	D_2 mm, in air	D_3 mm, in air
12 – 17.5	200 – 350	100 – 150	200 – 350
24 – 27	300 – 450	150 – 250	300 – 450
36 – 40.5	350 – 600	200 – 300	350 – 600

FIGURE 6.54 Vacuum and current loop arrangements in practical vacuum circuit breakers.

show that indeed the bus design affects the value of F_{arc}. They recommend that the bus connection of the vacuum circuit breaker to the power source be taken into account when testing for certification. The arrangement shown in Figure 6.54(d) is seen in some gas-insulated systems (GIS). Again, the same comments apply concerning the distance D_3 of the return bus. Jia et al. [110] and Kong et al. [111] have suggested that the effects of the magnetic fields on the vacuum arc can be mitigated to some extent by surrounding the arc chamber with a magnetic material that will provide a magnetic shield for the vacuum arc.

The pole spacing D_4 is mostly decided by the pole-to-pole insulation required. This is shown in Figure 6.55(a) for an air-insulated design. Wang et al. [112] have studied the effect of the transverse magnetic field from adjacent phases on the high-current plasma distribution between AMF contacts. Their parameters are: (1) two contact diameters, 49mm and 35mm; (2) a phase spacing 210mm; (3) a current of 20 kA (r.m.s). Figure 6.56 represents the observed motion of the inter-contact plasma. The exact motion of the plasma is complicated by the current in each phase being 120° to each other. Under the influence of the external magnetic field the arc column inclines in the Amperian direction. Initially the cathode spots move in a retrograde direction, but as the plasma continues in the Amperian direction the cathode spots begin to form under the plasma in that direction. The deflection of the plasma column is greatly affected by the contact gap and the level of the external magnetic field. Usually, however, at the spacing needed for air insulation, the magnetic fields resulting from the pole spacing do not greatly affect the fault current interruption ability of a well-designed vacuum interrupter. Placing insulating barriers, as shown in Figure 6.55(b) and (c), can reduce pole spacing. An insulating tube, Figure 6.55(d), can even surround the vacuum interrupter. In this case it is possible to lower the temperature of the connections to the vacuum interrupter if the convective flow shown is designed into the structure. Compact vacuum circuit breaker designs are possible for system voltages less than 17.5kV. One example is shown in Figure 6.57 where the width is 180mm. At 24–27kV and 36–40.5kV, air insulated designs require larger pole-to-pole and conductor-to-conductor spacing and the vacuum circuit breakers become much larger. In order to produce more compact designs, the vacuum interrupter poles have to be placed into an insulating medium such as oil [113] or SF_6 [114]. Both oil and SF_6 at 1 atmosphere gauge (i.e., two atmospheres) give a dielectric strength about five times that of air.

One disadvantage of SF_6 is that it is a greenhouse gas [115] and its use is generally being discouraged. To overcome this disadvantage, manufacturers who use SF_6 for insulation have developed hermetically sealed designs that prevent the gas from leaking into the environment. It is possible to reduce the amount of SF_6 and maintain about 86% of its dielectric strength of *~9kV/mm* (at 1 atmosphere) by adding up to 60% N_2 [116], see Table 6.14. Nitrogen by itself is a reasonable dielectric gas with about 39% the dielectric strength of SF_6. Its dielectric strength can be doubled by adding as little as 10% SF_6. The effect of adding SF_6 to a poor dielectric gas such as He can be quite dramatic as can also be seen in Table 6.14. There have been suggestions of using other gases and gas mixtures gases to replace SF_6 One example presented by Widger et al. [117] for use in ring main units (see Section 6.8.4) is a mixture of trifluoroidomethane (C_3F_3I) and carbon dioxide (CO_2). Investigations using $C_3F_3I–CO_2$ ratios of 30% to 70% demonstrate promising insulation properties. The global warming potential of CF_3I is less than 5 compared to that of SF_6. which is 22,500. Zhou et al. [118] have analyzed the dielectric performance of $C_3F_3I–N_2$ gas mixtures using a sphere-to-plane electrode system. Figure 6.58 shows a comparison of the dielectric performance for $CF_3I–N_2$(20/80) with $SF_6–N_2$ (20/80) and pure SF_6. The lightning impulse withstand voltage (BIL) for $CF_3I–N_2$(20/80) is 90% of that for $SF_6–N_2$ (20/80) and 77% of that for pure SF_6. One disadvantage is C_3F_3I's high boiling point of–22C. Other gas mixtures discussed by Guo et al. [119] are the perfluoronitriles (PFN) (e.g., C4-PFN or $(CF_3)_2CFCN$) and perfluoroketons (PFK) (e.g., CF5-PFK or $CF_3COCF(CF_3)_2$) mixed with air, CO_2, and SF_6. Again, these gases also have high boiling points. I expect that the search to find an effective insulating gas or gas mixture to replace SF_6 will continue.

It would be possible to use gas pressures greater than two atmospheres, but the thin hydro-formed bellows used with most vacuum interrupters places a limit on the pressure they can withstand before

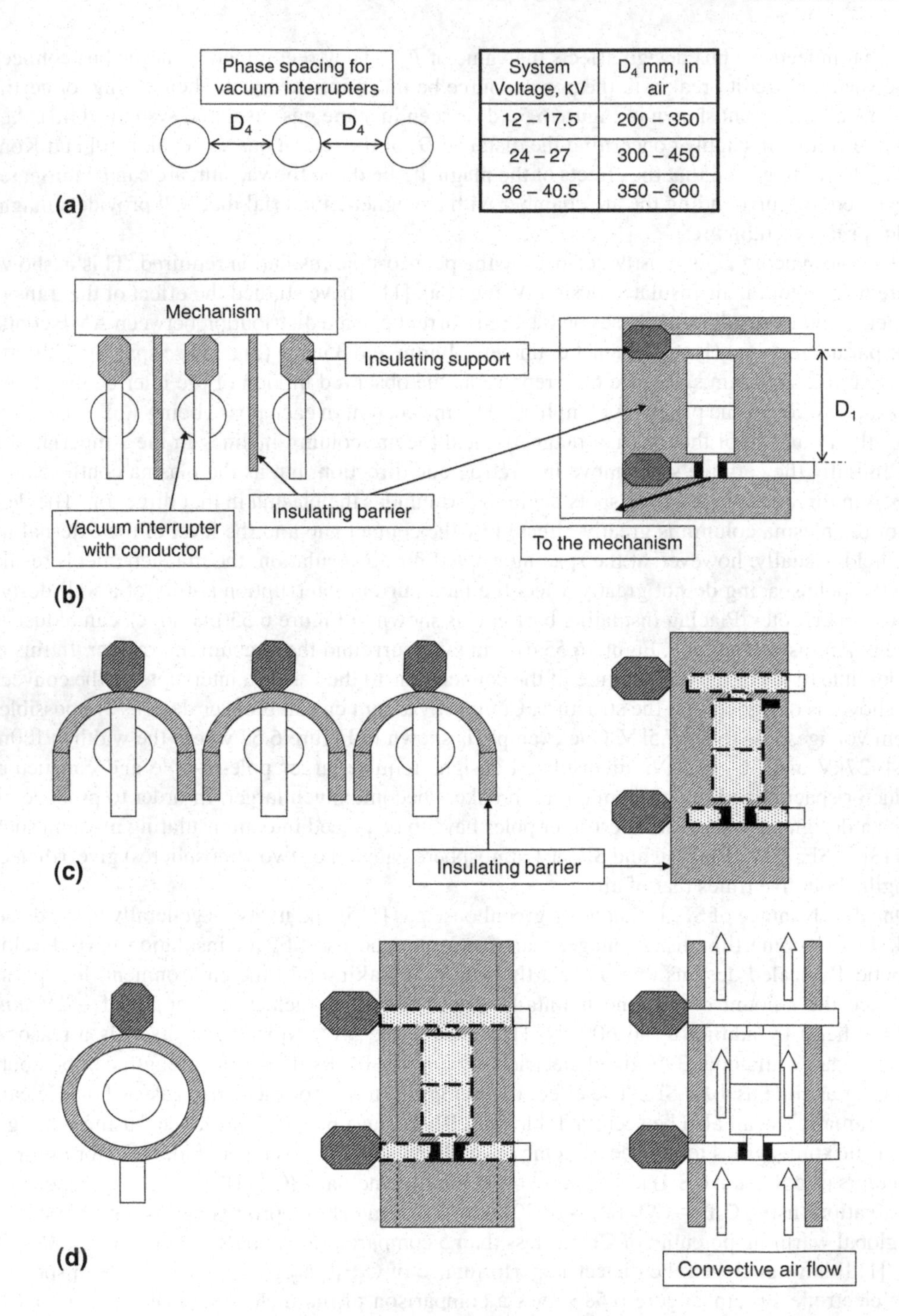

FIGURE 6.55 Pole spacing for three-phase vacuum circuit breakers with and without insulation between phases.

they deform. If a vacuum interrupter is to be used in a high-pressure environment, then a suitably designed bellows must be chosen: see Section 3.4.3 in this volume. Figure 6.59 shows an example of an outdoor breaker with the vacuum interrupter inside a ceramic insulator. The space between the insulator and the vacuum interrupter is filled with an insulating gas. Again, this system must be hermetically sealed to prevent the ingress of moisture from the atmosphere [120]. One way to achieve

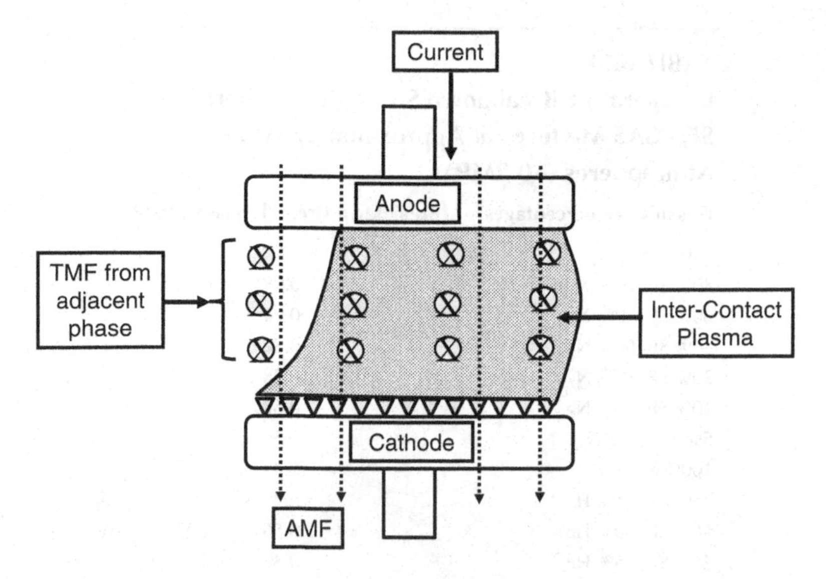

FIGURE 6.56 Schematic of the expected plasma deflection between AMF contacts in a vacuum interrupter caused by TMF from current passing in an adjacent phase in a three-phase vacuum circuit breaker [112].

this is to pour a quick setting liquid insulation, such as polyurethane, into the cavity between the vacuum interrupter and the porcelain bushing. Once this insulation solidifies, no moisture will be able to enter the space. There is an environmental concern with the mineral oil used as insulation. Biodegradable insulating oils would seem to reduce this concern [121]. The use of a vacuum interrupter under oil is an excellent combination because during current interruption the vacuum arc is

FIGURE 6.57 The compact T-VAC vacuum circuit breaker, 17.5kV, 25kA, 1250A (courtesy of Eaton Corporation).

TABLE 6.14
Comparative Breakdown Strength of Different SF_6–GAS Mixtures at Approximately Two Atmospheres (~0.2MPa)

Gas mixture percentages	Breakdown strength relative to SF_6
100% SF_6	1
80% SF_6 20% N_2	0.98
60% SF_6 40% N_2	0.94
40% SF_6 60% N_2	0.86
20% SF_6 80% N_2	0.78
10% SF_6 90% N_2	0.68
5% SF_6 95% N_2	0.56
100% N_2	0.39
75% SF_6 25% He	0.78
50% SF_6 50% He	0.56
25% SF_6 75% He	0.32
15% SF_6 85% He	0.21
10% SF_6 90% He	0.18
5% SF_6 95% He	0.08
100% He	<0.03

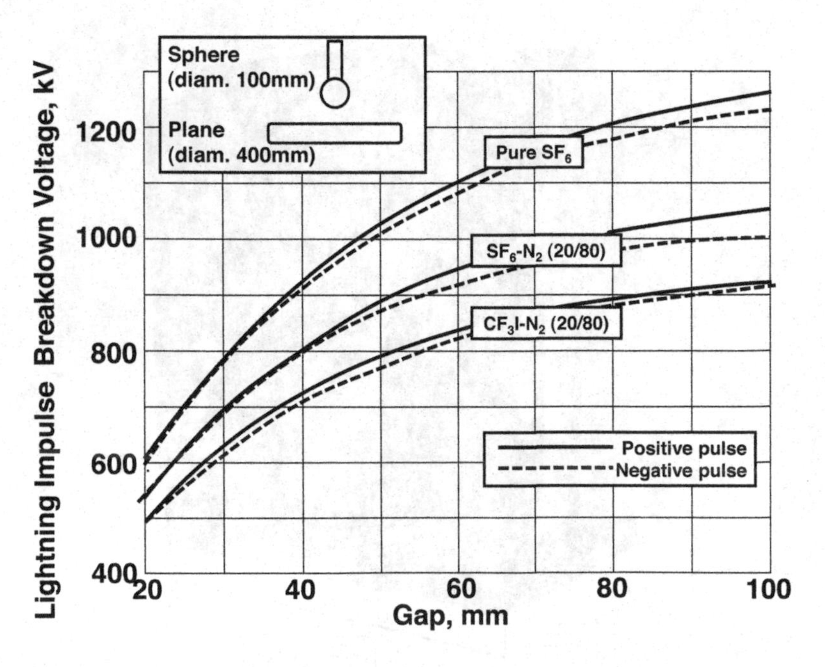

FIGURE 6.58 Lightning impulse breakdown comparison for pure SF_6, an SF_6–N_2 mixture, and a CF_3I–N_2 mixture [118].

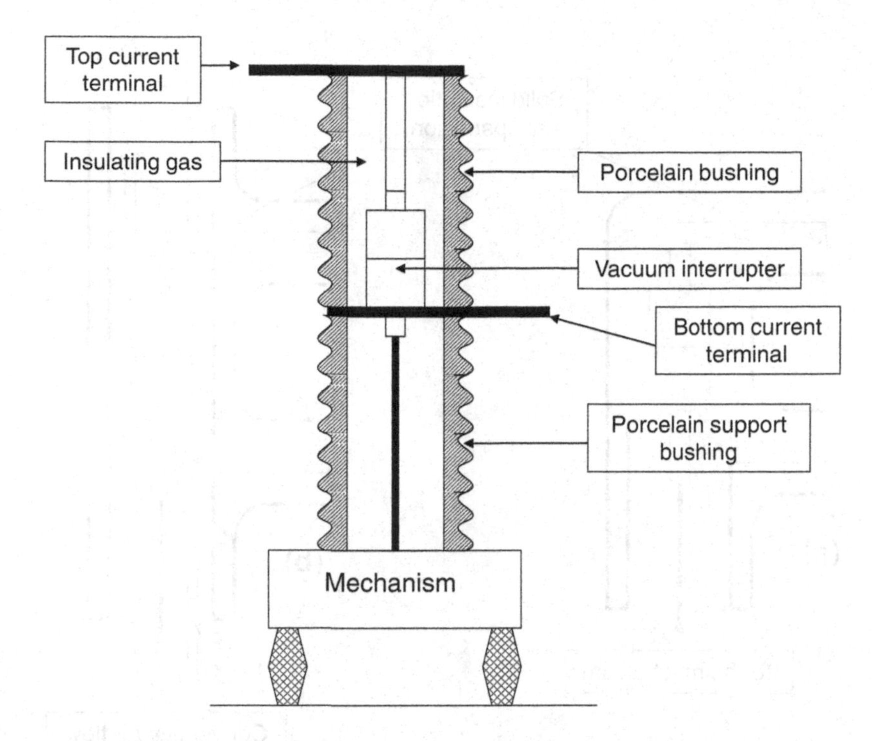

FIGURE 6.59 A schematic of an outdoor, live-tank, vacuum circuit breaker pole using porcelain ceramics.

fully contained inside the interrupter and thus will not 'carbonize' the oil. The oil acts as an efficient damper for oscillations of vacuum interrupter's bellows during opening and closing. This increases the vacuum interrupter's mechanical life. The resulting structure is virtually maintenance free.

Another approach that has been used to develop compact high voltage vacuum circuit breaker designs is to use a solid insulating material such as cyclo-aliphatic epoxy, polyurethane or even thermo-plastics. Cookson et al. [122] give a good introduction to the desired high voltage properties for effective solid insulation. This type of material when molded around the vacuum interrupter and its connecting bus will give a completely encapsulated pole unit. An example is shown in Figure 6.60 (a) [123]. Care must be taken with this system to take into account the differences in thermal expansion coefficients between the insulating material on the copper bus and the vacuum interrupter. It is suggested that a thin layer of pliable material be placed between the vacuum interrupter, the copper and the solid insulation to relieve the mechanical stresses that can arise [124, 125]. Also, it is extremely important that gas bubbles not be trapped in the material during the encapsulation process. The high voltage stress across such bubbles can result in the eventual failure of the dielectric material [122].

Another design concern is the potential for thermal runaway when the heat input from the passage of the current exceeds the heat lost to the insulation and through the bus. Even without a thermal runaway, the design must be such that temperature of the joints to the vacuum interrupter does not exceed the required standards, see Table 6.2. A number of designs have been proposed to overcome these concerns: (a) oversizing the copper bus; (b) the use of a thermally conducting solid electrical insulation; (c) the use of a hollow insulating chimney plus solidly insulated bus (see Figure 6.60(b)) and (d) using a thermally conducting encapsulation material with external heat sinks.

I have discussed the use of finite element analysis to design high voltage properties into the vacuum interrupter in Chapter 1, Section 1.3.2 in this volume. This analysis is ideally suited to the design and spacing of a three-phase vacuum circuit breaker [126]. One interesting design consideration is when a grounded container surrounds the vacuum interrupter. The change in the distribution

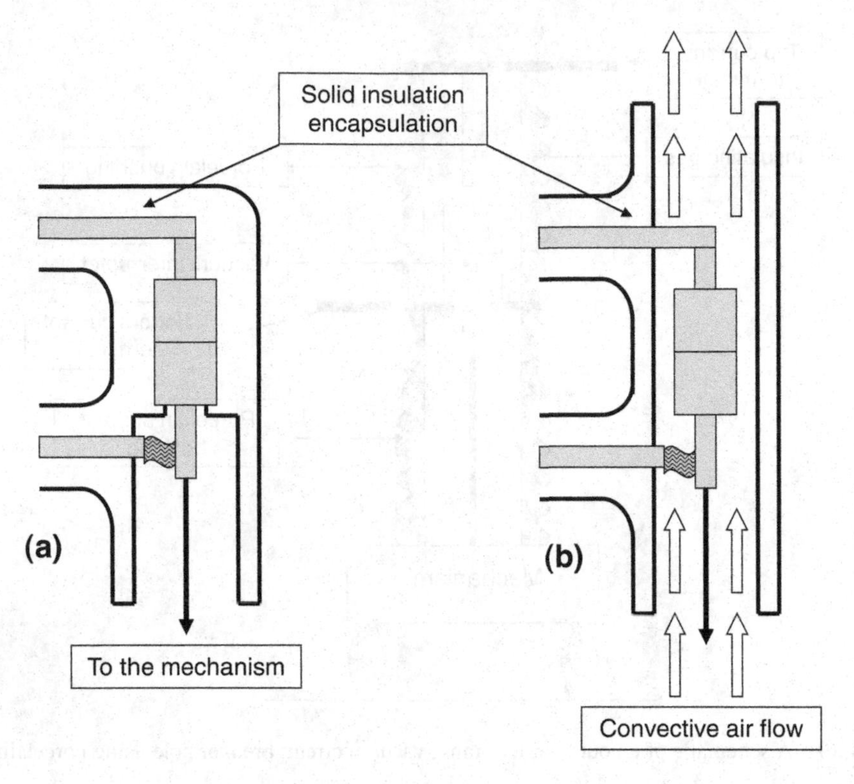

FIGURE 6.60 A fully solid encapsulation of a vacuum circuit breaker pole.

of the equipotential lines is shown in Figure 6.61(a), (b), and (c); Taylor [127]. The redistribution of the equipotential lines when the vacuum interrupter is placed inside the grounded metal tube is illustrated in Figure 6.61(b). This changed configuration also changes the internal field structure of the vacuum interrupter. In order to return it to a semblance of its original structure, metal shields can be connected to each end of the vacuum interrupter, see for example Figure 6.60(c). Now, however, the dielectric strength of the insulation between the vacuum interrupter and the grounded metal wall has to be sufficient to withstand the high fields that can appear across the insulated gap.

6.5.5 The Vacuum Circuit Breaker's Electrical Life

The bellows determines the mechanical life of the vacuum interrupter. The life span of the bellows is a function of the number of its convolutions, its length, its diameter, its travel, and the speed of operation. It can be designed to operate for many tens of thousands of operations for a vacuum circuit breaker application. The electrical life (i.e., the number of close-open operations switching current) of the vacuum circuit breaker is determined by two parameters: (a) the arc erosion of the contacts; and (b) the deposit of metal on the interior walls of the ceramic envelope. The maximum erosion permitted is determined by the minimum overtravel $d_{t(min)}$ (see Section 5.2.1 and Figure 5.5 in this volume), required by the vacuum breaker's mechanism. This linear erosion figure is a rather arbitrary number, but most manufacturers usually set its value at 3mm. The second parameter determines the continued high voltage withstand of the vacuum interrupter when its contacts are open.

When switching load currents (e.g., 200A to 3150A), the vacuum arc is in the diffuse mode following the initial bridge column (see Sections 2.1 and 2.2 in this volume). As I discussed in Section 5.4.3 in this volume, the measured electrical erosion is a function of the total charge passed by the vacuum arc and the ratio of the final contact gap $<g>$ to the diameter of the contact $<\phi>$,

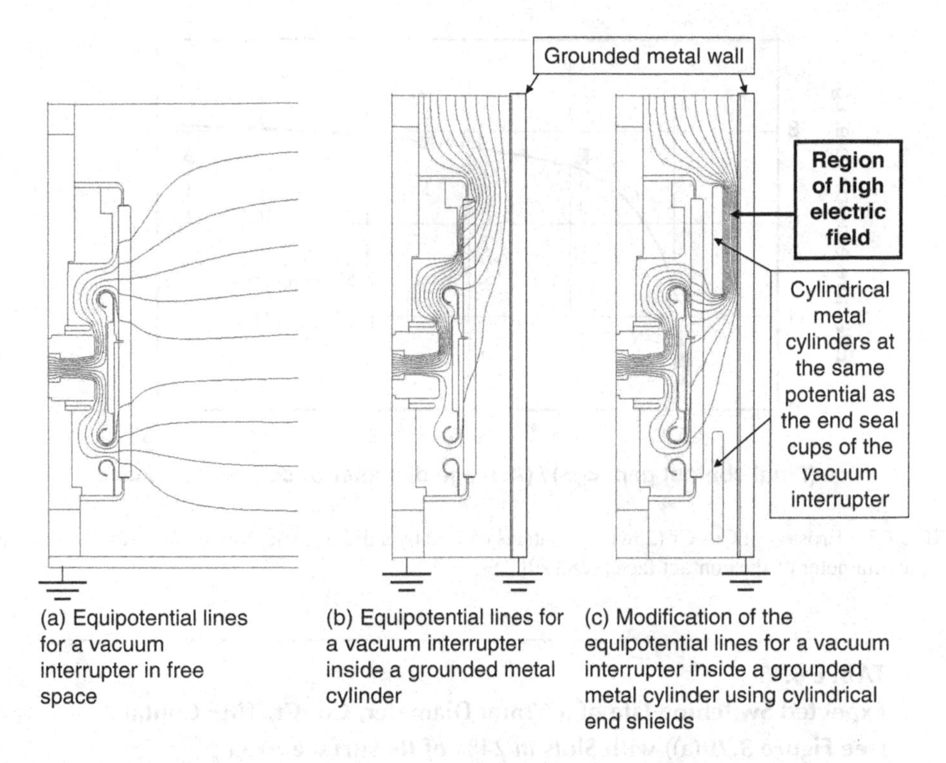

FIGURE 6.61 The effect on the distribution of the equipotential lines inside and outside a vacuum interrupter when a grounded metal cylinder surrounds it.

(i.e., $<g> / <\phi>$). The effect of slots in the contact faces is to increase the effective contact erosion rate as arc eroded contact material passes through them and is not deposited on their surfaces; see Schulman et al. [128]. Their data for Cu–Cr contacts is shown in Figure 6.62 together with one datum point from the work by Rieder et al. [129]. From Table 5.10 the slots that reduce the contact surface area by about 20% increase the erosion rate by about 40%. Thus, if we take a TMF contact of diameter 62mm and a final contact gap of 10mm with slots that reduce the surface capture area by 20%, then from Figure 6.62 the erosion rate for this contact will be ~ 1.4 × 7.5 μg.C^{-1} or ~ 1.4 × 9.1 × 10^{-7} cm^3.C^{-1}. If we assume one-half cycle of arcing (10ms) or one-quarter cycle of arcing (5ms) each switching operation of a 50Hz circuit, then Table 6.15 gives the possible number of electrical operations before the 3mm erosion limit is reached. Schlaug et al. [130] measure only 0.3mm erosion after switching 1250A for 30,000 operations with a contrate cup, TMF contact. If we assume a linear erosion rate this would result in a contact life of about 3 × 10^5 operations to erode the full 3mm. This load current switching life will be similar for an AMF contact with the same arcing surface with slots cut into it to reduce the effects of eddy currents. As you can see from Table 6.15, and from [130] the expected switching life is so long that most manufacturers give the electrical life of a vacuum circuit breaker equal to its mechanical life.

An example of the erosion of a Cu–Cr (25wt.%), TMF contact while switching a fault current of 25kA is shown in Figure6.63. After an initial period, the erosion is more or less linear with the number of operations. The life (i.e., number of operations, n, to reach 3mm of linear erosion) of a TMF contact interrupting fault currents has been shown by Slade et al. [131] to be inversely proportional to the rms value of the short circuit current (i_{sc}) squared, see Figure 6.64:

$$n \propto \frac{1}{i_{sc}^2} \tag{6.29}$$

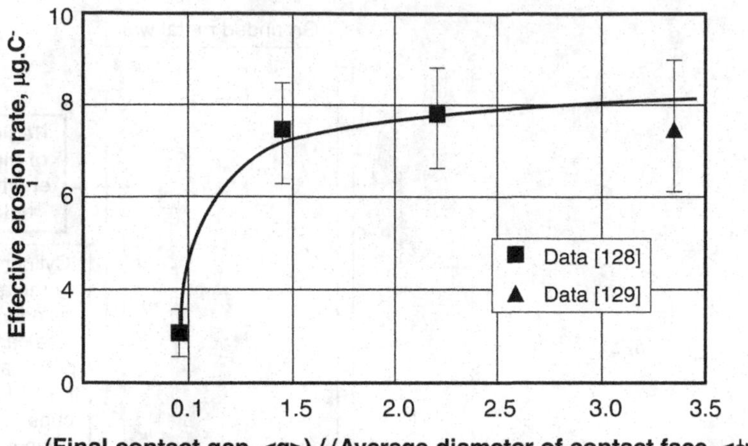

FIGURE 6.62 Erosion of Cu–Cr (25wt.%) contacts caused by a diffuse vacuum arc as a function of the final contact gap/diameter of the contact face [128,129].

TABLE 6.15
Expected Switching Life of a 62mm Diameter, Cu–Cr, TMF Contact (see Figure 3.20(a)) with Slots in 24% of Its Surface Area

Contact Profile

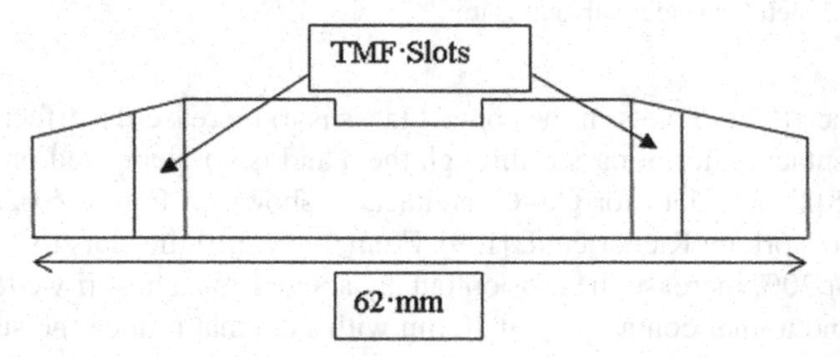

Load Current, A (rms)	Number of operations until contact surface erosion depth reaches 3mm	
	½ cycle of arcing /op.	¼ cycle of arcing /op.
630	9.2×10^5	18.4×10^5
1250	4.6×10^5	9.2×10^5
2000	2.9×10^5	5.8×10^5
2500	2.3×10^5	4.6×10^5
3150	1.8×10^5	3.6×10^5

Thus, if a contact has a life of 100 operations at 25kA (rms) then at 40kA it will have:

$$n(40) = n(25)\frac{25^2}{40^2} = 100\frac{625}{1600} = 39 \tag{6.30}$$

The AMF contact has a much lower erosion rate, see for example Figure 6.65. Figure 6.66 shows a direct comparison of the erosion of TMF and AMF contacts each with a 35mm diameter [131]. The end of electrical life in this case is not the minimum value of the overtravel required, but results

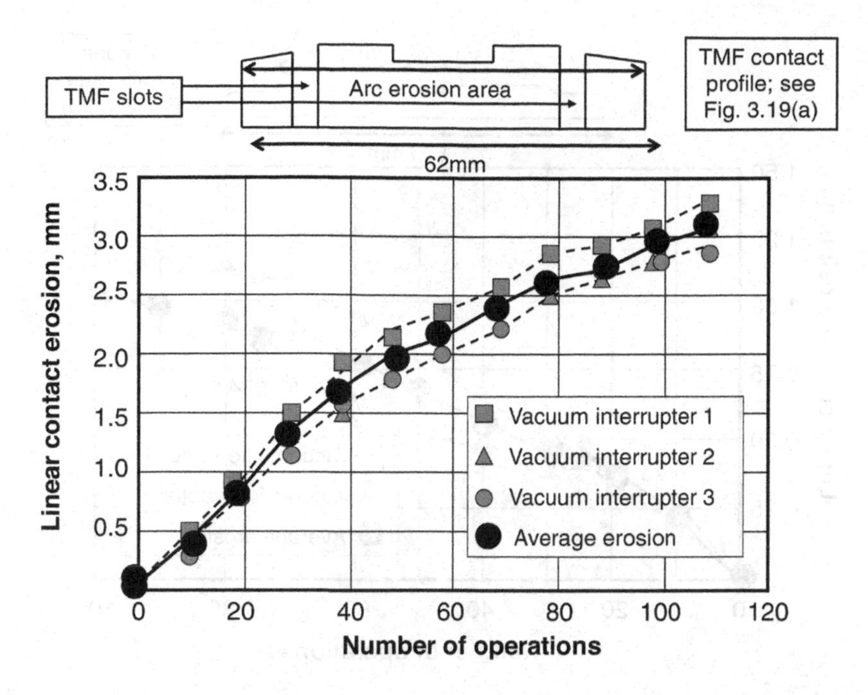

FIGURE 6.63 The linear erosion of a 62mm diameter, TMF contact interrupting a short circuit current of 25 kA (rms).

from the deposit of metal vapor and metal particles eroded from the contacts onto the interior walls of the ceramic, see for example Schellekens et al. [132]. Osmokrovic [133] has operated a 12kV vacuum interrupter with a TMF and AMF, Cu–Cr contacts 100 times at its rated short circuit current of 25 kA (r.m.s). The measured withstand voltage of the 8mm contact gap after each operation for the TMF contact is shown in Figure 6.67. He shows that there is also a gradual decline of the dielectric

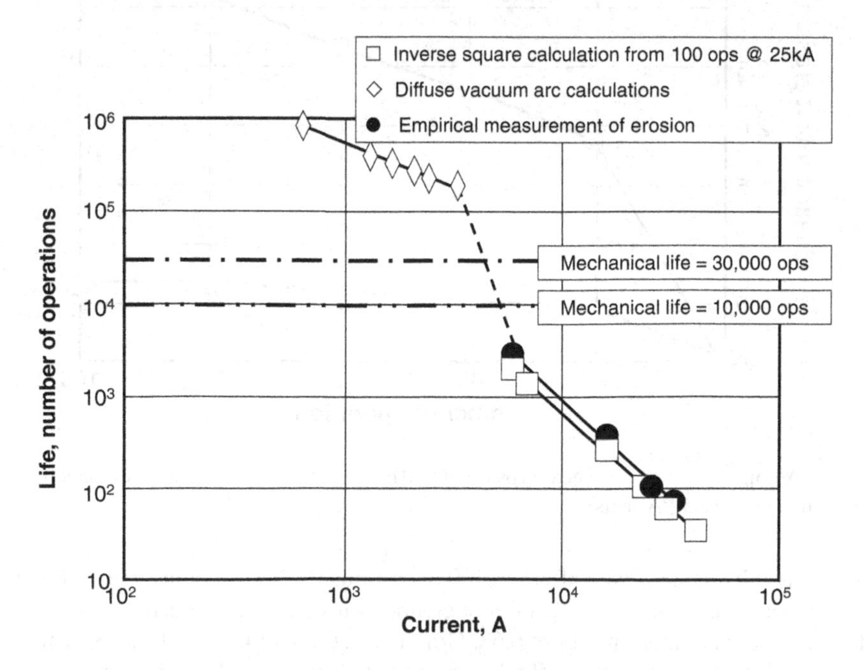

FIGURE 6.64 The dependence of the contact life on the value of the short circuit current for a 62mm diameter, TMF contact.

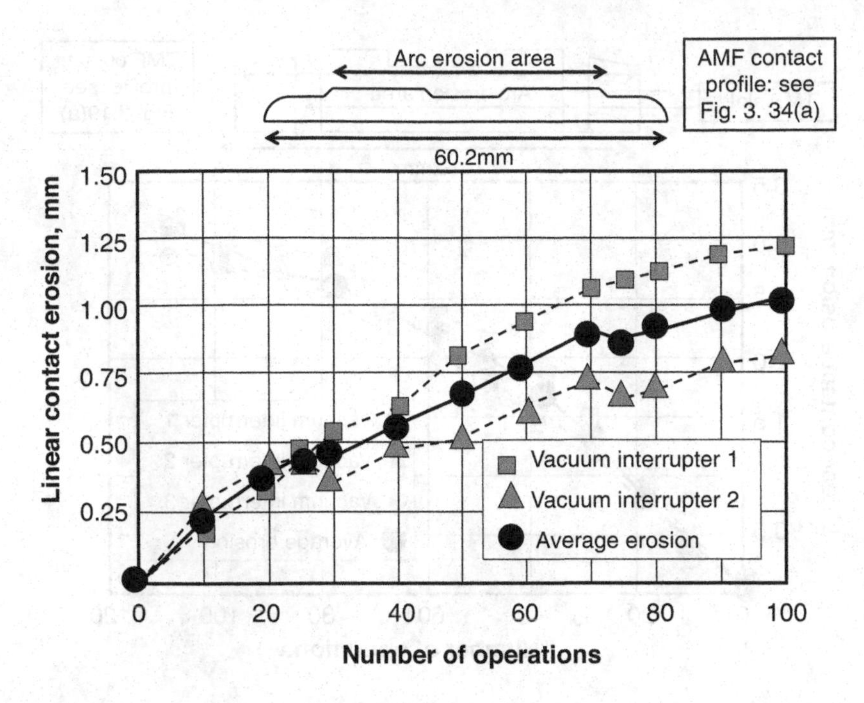

FIGURE 6.65 The linear erosion of a 55mm diameter, AMF contact interrupting a short circuit current of 25 kA (rms).

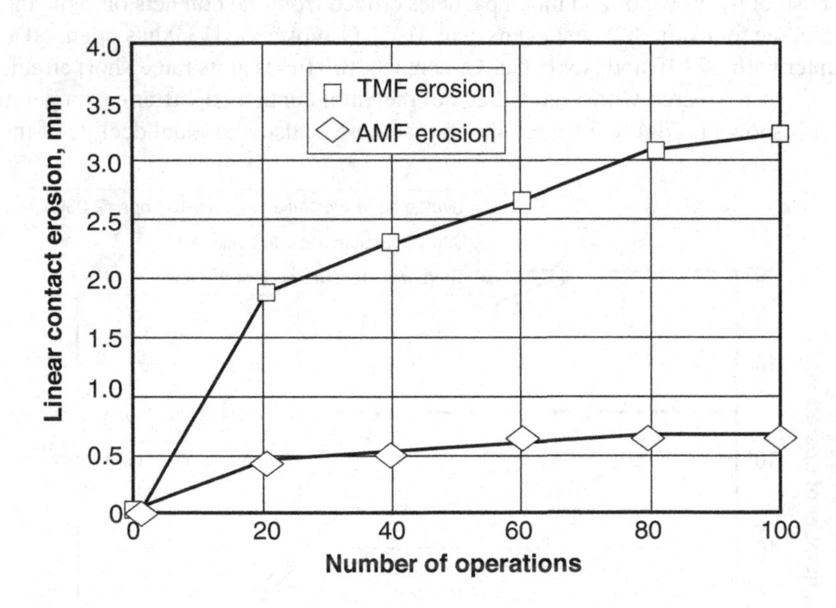

FIGURE 6.66 A comparison of the linear erosion of TMF and AMF 35mm diameter contacts, interrupting a short circuit current of 12.5 kA (rms).

strength for the AMF contact. Even after the 100 operations the TMF contact can still withstand 70 kV across the open contact gap. The vacuum interrupters used in this experiment have nonoverlapping shields that are illustrated in Figure 6.68. Gramberg et al. [134, 135] show that after interrupting 20kA, 12kV with nonoverlapping shields the metal deposit on the exposed, internal ceramic also results in a decline in the lightning impulse withstand voltage. Kurrat et al. [136] measure the

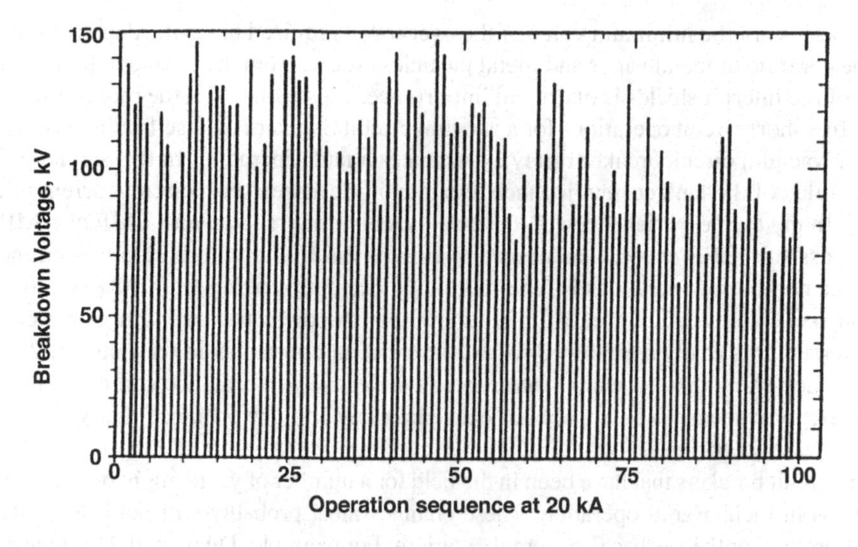

FIGURE 6.67 Variation of the dielectric strength of a vacuum interrupter TMF, Cu–Cr contact as a function of the number of operations interrupting a current of 20 KA [133].

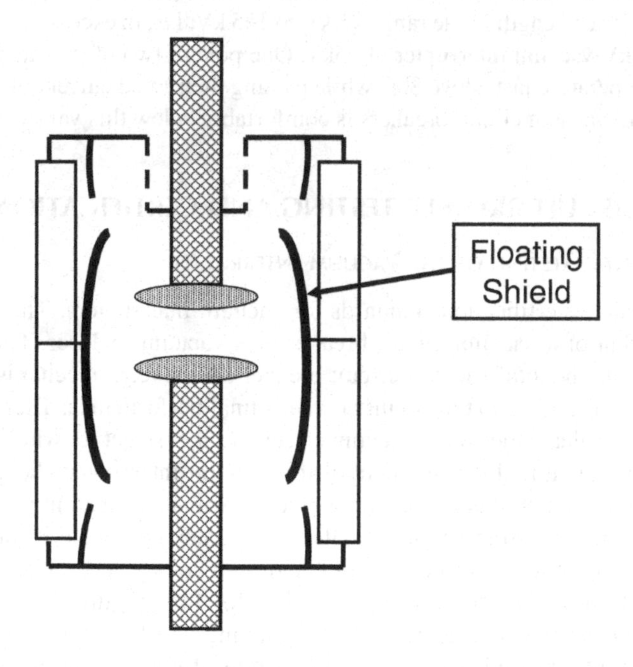

FIGURE 6.68 Schematic of a vacuum interrupter with nonoverlapping shields.

resistance of the deposited metal layer from Cu–Cr contacts on the exposed, interior ceramics after interrupting a current of 15 kA. They show that only when the deposit's thickness is greater than 10 nm will its surface resistance be measurable. However, a minimum layer thickness of 30 nm is necessary for current to flow. They also conclude that the metal deposit gradually reduces the voltage withstand ability of the vacuum interrupter.

In each of these cases, the design of the interior shields is of critical importance. In their experiments, however, Slade et al. [131] show that the high voltage performance of the AMF vacuum interrupter only begins to deteriorate after about 200 short circuit operations. The end of electrical life in this case is not the minimum value of the over travel required but is the deposit on the interior walls of the ceramic of metal vapor and metal particles eroded from the contacts. In this case again, the design of the interior shields is of critical importance. This being said, the requirement of a large number of full short circuit operations for a vacuum circuit breaker is in itself rather arbitrary. In the first place, a vacuum circuit breaker rarely experiences a fault operation. In the second place, when it does, the value of the fault current is much less than its designed short circuit current interruption ability [21]. In my own experience a well-designed vacuum interrupter with TMF or AMF contacts can easily pass the IEEE recloser endurance testing or the IEC circuit breaker endurance testing where a large number of operations is required, but at currents ranging from 10% (i_{sc}) to 100% (i_{sc}), see Section 6.6.3. Indeed, Ernst et al. [137] show that the dielectric strength of a well-designed vacuum interrupter is easily maintained for 50 short circuit operations in the range of 25kA rms to 50kA rms. They show that a vacuum interrupter's dielectric performance can even improve. The vacuum interrupter seems to be capable of exceeding any life performance that would be experienced, in practice, in the field.

Vacuum circuit breakers that have been in the field for a number of years might be expected to show some changes in their overall operation. These changes most probably will not affect their general function of protecting the electrical system they are in. For example, Duan et al. [138] have analyzed 13, 12 kV/31.5 kA vacuum circuit breakers that have been in the field from 5 to 10 years. They have no record of the number of operations or the current levels interrupted. They do show some changes in the circuit breaker's closing and opening speeds. Similarly, Wu et al. [139] show that these vacuum circuit breakers have a dielectric strength in the range 95 kV to 125 kV: i.e., in excess of the lightning impulse requirement for a 12kV vacuum interrupter of 75kV. One pole of two of the vacuum circuit breakers have a maximum temperature just above 80C while passing their rated current of 1250 A. Every other pole for each of the 13 vacuum circuit breakers is comfortably below this value.

6.6 VACUUM CIRCUIT BREAKER TESTING AND CERTIFICATION

6.6.1 Developmental Testing of the Vacuum Interrupter

There are no stand-alone certification standards for vacuum interrupters. The vacuum interrupter is only one component of a vacuum circuit breaker or a vacuum recloser. The interaction of the vacuum interrupter and the total vacuum circuit breaker design (e.g., mechanism, speed of operation, current path, etc.) is critical to the circuit breaker's final performance. Therefore, all the certifications are written to evaluate the whole vacuum circuit breaker structure with vacuum interrupters mounted in it. During the initial development of the vacuum interrupter, the Westinghouse R&D team, in the 1960s, determined that it was impossible to evaluate a vacuum interrupter's final fault interruption performance by extrapolating from its current interrupting ability at low currents. They thus developed a vacuum interrupter high current and high voltage test procedure using the tuned capacitor bank circuit described by Schulman et al. [140, 141]. Figure 6.69 shows a diagram of this tuned capacitor bank circuit together with the making and back-up circuit breakers and the capacitor-resistor circuit for shaping the TRV and its rate of rise. The interruption testing is performed using the symmetrical i_{sc} (rms) current with 14 or 16 operations. The tests are controlled so that the contacts part just after a current zero and at current maximum on the positive current wave, then just after current zero and at current maximum on the negative current wave. This is a single-phase direct test, so if the vacuum interrupter fails to interrupt at a current zero another half cycle of current will occur. An example of a successful interruption is shown in Figure 6.70. Tests are performed at the maximum rated short-circuit current with a TRV equivalent to the first phase to clear for the three-phase ungrounded circuit. Experience since the late 1960s [142] has shown that when testing three vacuum interrupters, if the interruption success is greater than 90%, then a well-designed, three-phase vacuum circuit breaker, or vacuum recloser, using these interrupters will easily be certified in an independent testing laboratory.

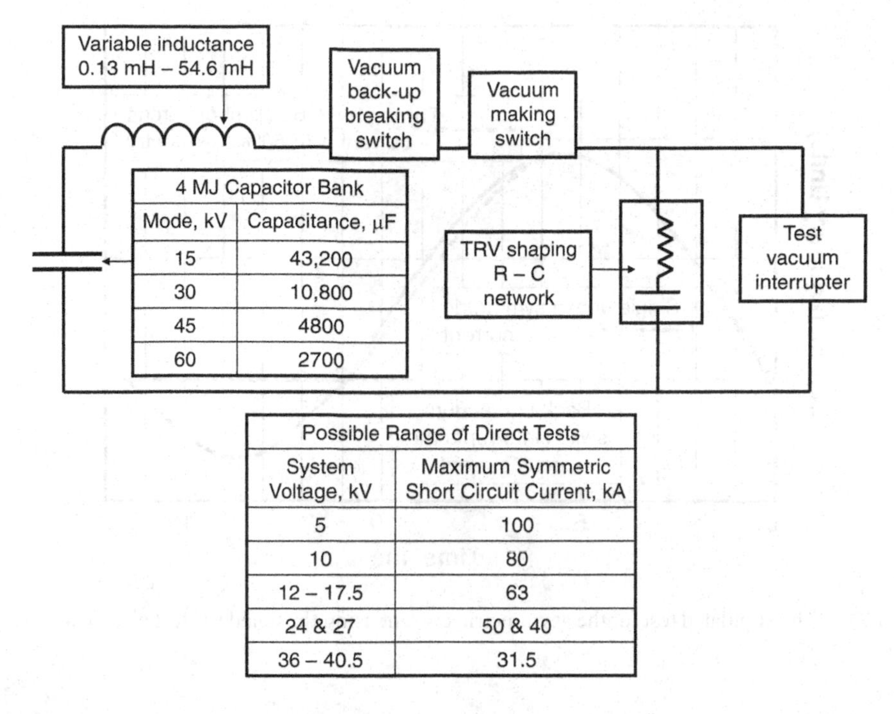

4 MJ Capacitor Bank	
Mode, kV	Capacitance, μF
15	43,200
30	10,800
45	4800
60	2700

Possible Range of Direct Tests	
System Voltage, kV	Maximum Symmetric Short Circuit Current, kA
5	100
10	80
12 – 17.5	63
24 & 27	50 & 40
36 – 40.5	31.5

FIGURE 6.69 A tuned LC circuit for developmental testing of vacuum interrupters [140, 141].

This test circuit can only provide symmetrical currents. To test the asymmetric capability of a vacuum interrupter design, the circuit is tuned to a lower frequency that will give the peak of the asymmetric fault current and the same duration, see Figure 6.71. The vacuum interrupter is then opened five times with the full half cycle and with a quarter cycle of arcing again, using positive and negative waves. The back-up breaker also opens at the same time so the vacuum interrupter being tested only experiences one half cycle of arcing. High voltage tests are performed and the internal

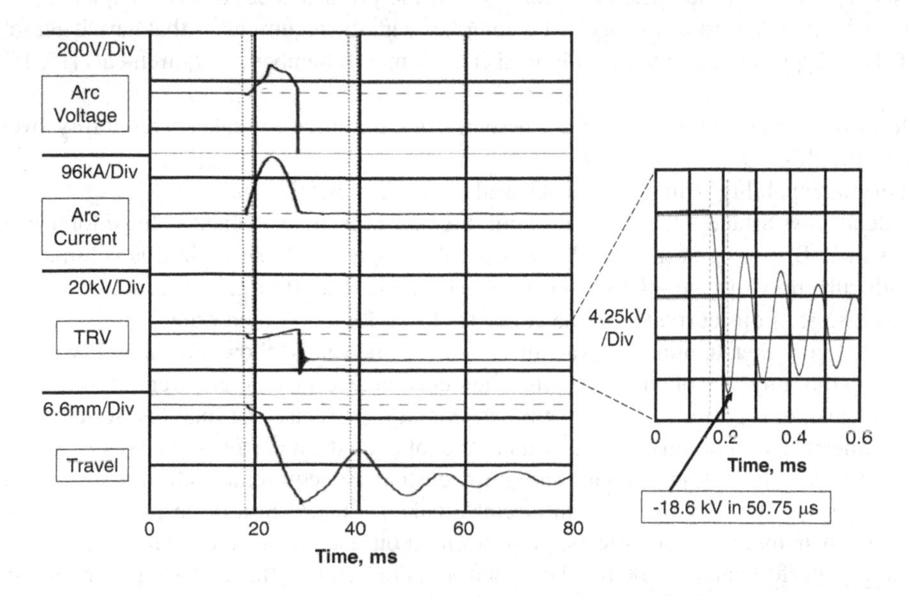

FIGURE 6.70 The result of a successful interruption of the tuned LC circuit shown in Figure 6.69.

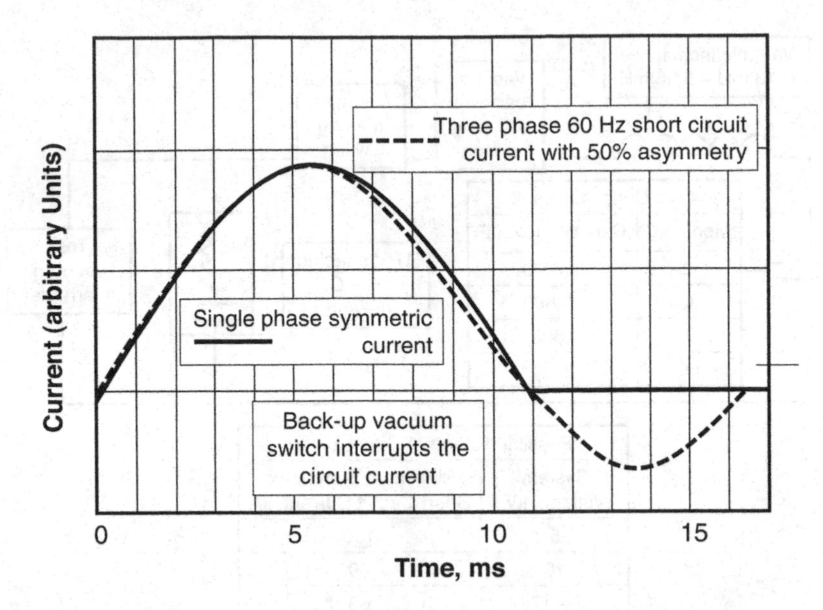

FIGURE 6.71 The simulated test for the asymmetric current using the tuned LC test circuit shown in Figure 6.69.

structure of the vacuum interrupter is then examined. If no severe damage is observed (such as holes burned through the shield), then it is assumed the vacuum interrupter will pass asymmetric current interruption because even if it does not interrupt the major loop of the asymmetric current it will interrupt the following minor loop's much lower value [48, 140]. This assumption has also proved true from experience gained since the late 1960s.

6.6.2 Certification Testing at an Independent High-Power Testing Laboratory

Before the final user accepts a vacuum circuit breaker or vacuum recloser, it must pass a number of "type-tests" at an independent test laboratory. When the product successfully completes these tests then it is certified by the test laboratory that it complies with the requirements that have been set by the standards [16]. Typically, a vacuum circuit breaker must meet a number of requirements [15, 143–145]:

a) Mechanical capabilities – these are essential for the circuit breaker's reliability over its operating life
b) Thermal capability – this I have discussed in detail in Section 6.2
c) Dielectric withstand – the vacuum circuit breaker must ac withstand voltage for 1 minute as well as BIL or lightning impulse withstand voltage; see Table 1.1 in this volume
d) High current withstand – Close and latch on the rated short circuit current and withstand the passage of this current with the contacts closed for a specified time
e) Short circuit interruption – the vacuum circuit breaker must show the capability for interrupting the full range of bus or terminal fault current levels with their attendant TRVs. Most of this testing is performed using symmetrical fault currents. The final test series uses full asymmetric currents and can involve a number of open (O) and close–open (CO) cycles that depend upon the testing standard and/or the customer's requirements. Examples are given in Table 6.16. Usually the vacuum circuit breaker is tested in a three-phase ungrounded circuit. A number of further tests are performed on one phase in a grounded condition to evaluate performance of the breaker when it is applied in a three-phase grounded circuit [44]. For some outdoor applications the ability to interrupt short line faults may be required

TABLE 6.16
Examples Of (Open)–(Close–Open) Cycles

The Standard or Other Requirement	Open (O)–Close (C)–Open (O) Cycles
IEEE C37.04, (a) Normal reclosing	O–15s–CO–3m–CO
(b) Fast reclosing	O–0.3s–CO–3m–CO
(c) Possible derated	O–15s–CO–15s–CO–15s–CO–15s–CO
IEEE C37.60, Recloser	O–fast–CO–slow–CO–slow–CO
	C–1s to 4s –O (difficult test for certification labs)
IEC 6237–100, (a) Normal reclosing	O–3m–CO–3m–CO
(b) Fast reclosing	O–0.3s–CO–3m–CO
(c) Optional	CO–15s–CO

f) Line charging and cable charging – outdoor breakers will be evaluated for line charging and indoor breakers for cable charging. Again, success with these tests will be noted in the certification
g) Capacitor bank switching – some vacuum circuit breaker tests will demonstrate the vacuum circuit breaker's ability to switch single and/or back-to-back capacitor banks. Success with these optional tests will be added to the certification
h) Transformer switching (i.e., unloaded transformer switching, shunt-reactor switching, and out-of-phase switching: see Section 6.8.2) – these tests are not required of standard breakers, but increasingly these tests are performed so that breaker manufacturers can claim the widest application market for their product

Interestingly enough, the certification standards have set levels of performance; see for example Table 6.10. In fact, a vacuum circuit breaker manufacturer when testing and certifying to the IEC standard, can combine load current, short circuit current and system voltage levels to suit the particular design as long as the testing conforms to the set level of system voltages and the R10 series for load current and short circuit currents. If a practical application lies between two set levels, then the user must choose a vacuum circuit breaker that satisfies the higher certification standard. For example, the R10 series for short circuit currents are; 12.5kA, 16kA, 20kA, 25kA, 31.5, kA, 40, kA, 50, kA, 63kA, and 80kA. If a user's circuit has a potential full short current of 36kA, then a vacuum circuit breaker that has been certified at 40kA must be employed. This also has implications for the vacuum interrupter designer. It makes no sense to create a design that can interrupt 55kA at a given system voltage when the certification level is either 50kA or 63kA. The same is true for the circuit voltage levels. For example, the outdoor test level in the USA is 15.5kV, but there are many utility systems that run at 13.8kV. These then must use the vacuum circuit breakers that have been certified at the 15.5kV level.

6.6.3 Fault Current Endurance Testing

The vacuum recloser has to pass the endurance testing as given by the IEEE recloser standards [146]. These involve a series of fault interruption tests and examples are given in Table 6.17. The IEC standards [16] have endurance tests for vacuum reclosers and for vacuum circuit breakers that are also given in Table 6.17. The vacuum interrupter has no difficulty in passing these endurance tests. In fact, the Vacuum ESV Power Recloser is one of the first applications of the vacuum interrupter technology by Westinghouse in the early 1970s [147]. More recently, testing of vacuum circuit breakers for a large number of full fault currents has been a requirement for the Chinese DL standards. Also, some individual utilities require vacuum circuit breaker manufacturers to endurance test their products beyond the IEEE and the IEC requirements. One example is shown in Table 6.17 [130].

TABLE 6.17
Examples of Vacuum Circuit Breaker Fault Current Endurance Standards

IEEE C37.60 Recloser Testing Sequence

System voltage, kV	Fault current i_{sc}, kA	Test operating duty*: 15% – 20% i_{sc}	Test operating duty*: 45% – 55% i_{sc}	Test operating duty*: 90% – 100% i_{sc}	Total number of operations
12, 15, 17.5	2	52	60	18	138
	6	48	60	16	124
	8, 12.5, 16, 20	44	56	16	110
24, 27	8. 10, 12.5, 16, 20				
36, 38	8, 12.5, 16, 20				

* The usual sequence for testing is C–fast–CO–slow–CO at each current level until the number of tests is complete.

IEC 62271–100 preferred vacuum circuit breaker endurance testing sequence for auto-reclosing duty

System voltage, kV	Test current	Operating sequence	Number of required tests
12–36	10% i_{sc}	O	84
		O-0.3s-CO	14
		O-0.3s-CO-t-CO	6
	30% i_{sc}	O	84
		O-0.3s-CO	14
		O-0.3s-CO-t-CO	6
	60% i_{sc}	O	2
		O-0.3s-CO-t-CO	2
	100% i_{sc}	O-0.3s-CO-t-CO	2
Total			214

IEC 62271–100 and other vacuum circuit breaker endurance testing

System Voltage, kV	Test current as % of rated short circuit current	IEC E1	IEC E2-1	IEC E2-3	An example of one utility's specification [130]
12–36	3–13	3	130	3	220
	20–30	3	130	3	160
	50–64	3	8	60	115
	100	7	6	6	50
Total		16	274	72	545

Both of these extra endurance tests beyond the already conservative IEEE and IEC standards are usually not required for vacuum circuit breaker application. They do, however, add an interesting challenge to the vacuum interrupter and vacuum circuit breaker designer. On the downside, they may also add an extra cost factor that is not really required by the ultimate application.

6.7 VACUUM CIRCUIT BREAKERS FOR CAPACITOR SWITCHING, CABLE AND LINE SWITCHING, AND MOTOR SWITCHING

6.7.1 Introduction

The vacuum circuit breaker's primary purpose is to provide protection to a circuit when a fault occurs. Once the fault is detected and the vacuum circuit breaker operates, it must interrupt the fault current with 100% reliability. Both the design of the vacuum interrupter and the design of the vacuum circuit breaker itself are predicated on this one major criterion. In Chapter 5 in this volume,

I discussed vacuum interrupter designs that have been developed especially for switching capacitors and for switching motors with just the load or rated current. These designs use contact materials that have properties that permit them to perform these specialized functions. The contact material for rated current, capacitor switching, vacuum interrupters is W–Cu and for rated current, motor contactor, vacuum interrupters is WC–Ag. Neither of these materials performs well when called upon to interrupt high levels of fault current. The contact material that has shown outstanding reliability for use in circuit breaker vacuum interrupters is Cu–Cr. So much so that all modern vacuum interrupter designs now use the Cu–Cr contact material in one form or another. Because the vacuum circuit breaker performs its circuit protection function with such high reliability and because it also switches load currents with no problem, it is also used for switching capacitor and high voltage motor circuits. The use of a vacuum circuit breaker eliminates the use of an auxiliary protection device such as a fuse. However, when the vacuum circuit breaker is employed to switch capacitors or motors, the user must be aware that Cu–Cr is not the best performing contact material for these functions. That being said, vacuum circuit breakers have been successfully used since the 1970s for both these applications.

6.7.2 Capacitor Switching

Capacitor banks are switched perhaps once or twice a day. One study in 1999 by Bonfanti et al. [148] shows that 60% of all capacitor banks are switched up to 300 times a year and a further 30% are switched up to 700 times a year. Thus, a vacuum circuit breaker rated at 10,000 operations could operate from between 14 and 33 years, while one rated at 30,000 operations would be satisfactory for between 40 and 100 years. The limit of a circuit breaker's mechanical life compared to a vacuum capacitor switch (~ 50,000 operations), would, for the most part, not be a problem. For 90% of all applications, the capacitive current is between 300A and 400A. Also, the closing-in current is limited from 5kA to 6kA with a frequency of 0.9 kHz. Almost 90% of the medium voltage capacitor banks use damping elements to reduce the amplitude of this in-rush current. A more recent survey by Dullni et al. [149] of 34 utilities that use substation, capacitor banks shows the mean capacitor current for all rated voltages (3.6 kV to 550 kV) is between 165A and 320A. Also 58% of the capacitor banks are switched once or twice a day, but 40% are switched less than once a week. About 70% use single capacitor banks and 30% use back to back banks. Interestingly, this survey shows that switches specially developed to switch capacitor banks are rarely applied. Vacuum circuit breakers designed primarily for circuit protection are the most common devices used to switch these capacitor banks. As this duty places undue stress on the vacuum circuit breaker's mechanism regular maintenance is required over a five-year period.

The extensive discussion on capacitor switching presented in Section 5.5 in this volume is also applicable in its entirety to capacitor switching when using a vacuum circuit breaker. The analysis for the in-rush currents and the TRVs for single capacitor bank and back-to-back capacitor switching is the same and so is the effect of a restrike and the potential for voltage escalation. For the most part, a vacuum circuit breaker will only be used to switch three-phase ungrounded capacitor banks, but the three phases will operate at the same time, so the peak of the TRV will be limited to 2.6 PU for the first phase to clear. In order to be certified for capacitor switching, a vacuum circuit breaker has to satisfy a series of switching test duties in standards established by IEEE (C37.04a–2003 and C37.09a–2005) or IEC (62271-100:2003). Both sets of standards assume that there is always a finite probability of a restrike occurring while switching a capacitor back during the vacuum circuit breaker's life. The tests to certify a vacuum circuit breaker for single bank and for back-to-back capacitor switching require careful interpretation of the relevant standards. A simplified synopsis of the IEEE and the IEC switching tests are given in Table 6.18. Both the IEEE and the IEC test duties are divided into two classes: (a) C1 for vacuum circuit breakers with a low probability of restrike, and (b) C2 for vacuum circuit breakers with a very low probability of restrike.

The IEEE has a third class, C0. This level of performance permits a restrike with every operation as long as there is no breakdown across the outside of the vacuum interrupter and there is no

TABLE 6.18
Three-Phase Capacitor Bank Switching Tests for Certification (Single Bank or Back-to-Back Banks)

IEEE: BC0	**SWITCHING TEST SEQUENCE**	**Probability of restrike permitted**
The same as for the BC1 or BC2 tests	The same as for the BC1 or BC2 tests	1 restrike allowed per operation. External flashovers and phase-to-ground flashovers are not permitted
IEEE: BC1	**SWITCHING TEST SEQUENCE**	**Probability of restrike permitted**
a) 24 Open tests: Capacitor load current i_L = 400A b) For back to back switching the in-rush current 20 kA peak @ 4250 Hz	a) 4 O, on one polarity (step 15°) b) 6 O, at minimum arcing time on one polarity c) 4 O, distributed on the other polarity (step 15°) d) 6 O, at minimum arcing time on the other polarity e) Additional tests to make up a total of 24	If no restrikes 0 in 48 (0%) If 1 restrike during the first 24 operations the full test is repeated with no restrikes: 1 in 48 (2.1%)
IEEE: BC2	**SWITCHING TEST SEQUENCE**	**Probability of restrike permitted**
a) 80 Close–open tests: Capacitor load current i_L = 400A b) For back to back switching the in-rush current 20 kA peak @ 4250 Hz	c) 4 CO, on one polarity (step 15°) d) 32 CO, at minimum arcing time on one polarity e) 4 CO, distributed on the other polarity (step 15°) f) 32 CO, at minimum arcing time on the other polarity g) Additional tests to make up a total of 80	If no restrikes 0 in 80 (0%) If 1 restrike during the first 80 operations the full test is repeated with no restrikes: 1 in 160 (0.6%)
IEC: C1	**SWITCHING TEST SEQUENCE**	**Probability of restrike permitted**
c) Capacitor load current i_L = 400A d) For back to back switching the in-rush current 20 kA peak @ 4250 Hz	a) Open 3 times at 60% full short circuit current b) Can perform 10% full short circuit current: optional c) 24 open only at 0.1–0.4 × i_C d) 24 close–open at i_C	If no restrikes 0 in 48(0%) If 1 restrike during the first 48operations the full test is repeated with no restrikes: 1 in 96 (1%)
IEC: C2	**SWITCHING TEST SEQUENCE**	**Probability of restrike permitted**
a) Capacitor load current i_L = 400A b) For back to back switching the in-rush current 20 kA peak @ 4250 Hz	a) Open 3 times at 60% full short circuit current b) Can perform 10% full short circuit current: optional c) 24 open only at 0.1–0.4 × i_C d) 80 close–open at i_C	If no restrikes 0 in 104 (0%) If 1 restrike during the first 104 operations the full test is repeated with no restrikes: 1 in 208 (0.5%)

breakdown to ground. These classes of performance have been decided by the respective standards committees after a great deal of vigorous discussion. In spite of this, however, they are somewhat arbitrary. There is, for example, always a finite probability of passing these test sequences and still having a vacuum circuit breaker whose probability of restriking is higher than would be expected. Thus, there will continue to be a strong debate between vacuum circuit breaker manufacturers and the users of these products on how test laboratory performance equates to successful performance in the field. Dullni et al. [149] state for circuits below 38 kV only 50% of their respondents require a C2 rating. Even so, the C2 class seems to be the required standard rating while, in practice, this is a difficult performance level for a vacuum circuit breaker to achieve. This especially true when switching back to back capacitor banks. This survey differs from that of Bonfanti et al. [148]. In Dullni et al.'s survey, few of the respondents use damping elements to reduce the amplitude of this in-rush current. It is also surprising that less than 25% use surge protection for circuit voltages in the range 15 kV to 24 kV. This does rise to 40% for 27 kV to 38 kV circuits. About 50% of the capacitor circuit are protected by fuses, but these do not protect against overvoltages.

Also, from the discussion in Section 5.5, Equation (5.20), in this volume, the in-rush current for an isolated bank, for example will always be considerably less than a circuit's rated short circuit current. A simple calculation using Equation (5.18) in this volume will convince you that an in-rush current of greater than 6kA would be rather rare. The 20 kA, 4.25 kHz in-rush current required in the standards for back-to-back capacitor bank switching certification (see Table 6.18) is excessive given that some utilities limit this in-rush current to about 6kA. In fact, in order to achieve these values a certification laboratory must use two large capacitor banks with values in excess of 8MVAR (using Equation (5.26) in this volume). The certification laboratories experience the high stress placed on all the joints and bus by the high in-rush current during these tests. I am sure they would recommend to any utility that would listen, limiting this in-rush current is a very good idea. However, Dullni et al. [149] find that use of current limiting reactors is not common. There does appear to be a real disconnect between the real-world application of capacitor banks and the testing requirements given in the standards. For example, in China, all capacitor banks are installed with a current limiting impedance, but as the standards are similar to those shown in Table 6.18, all utilities require certification with the 20kA in-rush current. This places a large burden on vacuum circuit breaker designers and, I suspect, increases the cost of the final product with no real advantage in practice. Going forward, the real standard for certifying a back-to-back capacitor switching vacuum circuit breaker will continue to be debated by vacuum circuit breaker manufacturers and the users of these products.

The in-rush current can result in weld spots on the contact surfaces that will serve as emission sites for electrons, which can result in a late breakdown of the contact gap. Delachaux et al. [150] show that closing an in-rush current of only 5.6kA, 270Hz results in an increase of the local electric field 3.5 times above that of the original contact surface. Indeed, as I discussed in Section 5.5 in this volume from the work of Juhász et al. [151] and Figure 5.52, the energy in the pre-strike arc correlates with the probability of a restrike. Thus, in order to have a better chance of certifying a vacuum circuit breaker for capacitor switching, it is always advisable to limit the in-rush current to below 6kA. There has been an increase in vacuum circuit breaker designs using three mechanisms that can give a point-on-wave closing. This can be used to limit the in-rush current during the closing operation, but it greatly complicates the circuit breaker's design and control system. It can also add significantly to its cost; see for example Fang et al. [152].

Ding et al. [153] demonstrate that a magnetic mechanism (see Section 6.5.3) on each phase with a suitable electronic sensing and control system can provide very precise control over closing and opening a vacuum circuit breaker. Thus, it is possible to use such a system for switching capacitor banks more reliably than using a standard vacuum circuit breaker. Figure 6.72(a) shows the closing sequence. The requirement here is that each phase must close just before the system's voltage zero. This will ensure the lowest in-rush current to each of the vacuum interrupters: see Section 5.5.1 in this volume. The opening of the three phases has to ensure that each vacuum interrupter has a long arcing time. The purpose of this is twofold: first, the long arcing time will assist in conditioning the vacuum interrupters with each operation (see Section 1.3.7 in this volume); second, at current zero the contacts will be far enough apart to withstand the peak voltage (see Section 5.5.2 in this volume). Figure 6.72(b) shows where phase 1, as first phase to clear, should open to have its longest arcing time. Phases 2 and 3 then have the arcing time shown. In this scheme, the control should have each phase in turn be the first phase to clear. This will ensure that the contact erosion will be similar for each vacuum interrupter as the vacuum circuit breaker continues to switch the three-phase capacitor bank.

Dullni et al. [154] show that the probability of successfully certifying a vacuum circuit breaker (with random closing and opening) switching a 24kV capacitor bank increases with contact gap of at

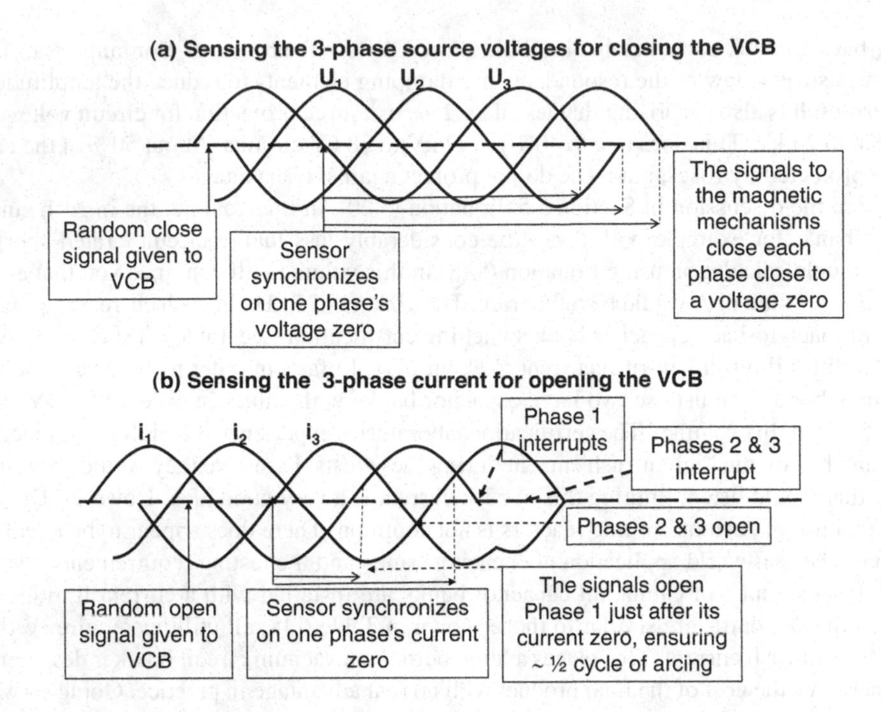

FIGURE 6.72 The synchronous closing and opening of a vacuum circuit breaker with magnetic mechanisms for reliable capacitor switching [152, 153].

least 14mm. They also state that for a 36kV capacitor bank, using two vacuum interrupters in series would provide the highest probability of successful certification. Indeed, Wang et al. [155] show that two 24kV/31.5kA/1250A vacuum interrupters in series closing on an inrush current of 20kA operate successfully for the C2 class during a sequence of 20 close–open operations. In my own experience, capacitor switching for circuit voltages less than or equal to 12kV–17.5kV usually has considerable certification success. As the circuit voltage increases, the probability of restriking also increases, but I have seen successful C2 testing at 36kV when the in-rush current is limited to 6kA.

6.7.2.1 Capacitor Switching and NSDDs

The usual current interrupted in capacitor circuits is a few hundred amperes. A vacuum circuit breaker can easily interrupt this level of current. If the vacuum circuit breaker is opened rapidly enough, the first phase to clear will successfully withstand the peak of the TRV (see Section 5.5 in this volume). Even so, NSDDs are observed when performing this switching operation. The cause of these late discharges seems to be related to the current level and the duration of the pre-strike arc when the vacuum interrupter closes the capacitor circuit [151]. The breaking of the welded spots on the contact surface provides electron emission sites, which can initiate the breakdown process. In fact, Koochack-Zadeh [156] and Smeets [157] measure peak emission currents in the milliampere range long after current interruption with the unidirectional capacitor switching TRV; see Section 5.5.4 in this volume. For a full discussion of the NSDD phenomenon, see Section 6.4

As stated in Section 6.4, an NSDD is no longer considered a failure mode for certification of a vacuum circuit breaker when it is being tested for fault current performance. This criterion is also true for all other vacuum interrupter switching conditions. Thus, when testing a vacuum circuit breaker switching a three-phase ungrounded capacitor bank, NSDDs are counted but are not a reason to deny the circuit breaker a certification. The NSDDs conductive period is usually too short for a restriking of the capacitive current. Indeed, Smeets et al. [57] show this is true even when the discharge receives charge beyond the local capacitance of the vacuum interrupter and some stray

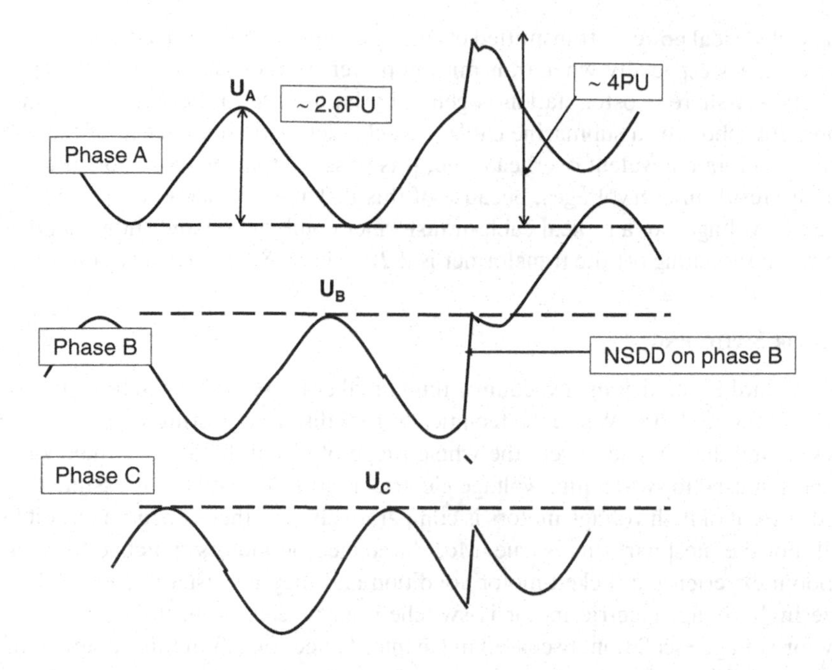

FIGURE 6.73 An example of a NSDD while switching a three-phase ungrounded capacitor bank [55].

capacitance to ground. However, the NSDD can result in a shift of the neutral voltage. Smeets et al. [55] show a good example of a very high voltage generated during a capacitor-switching test; see Figure 6.73. An NSDD on phase B results in a substantial shift in the neutral so that the voltage across the vacuum interrupter in phase A is now increased to about 4PU. Thus, in practice it may be prudent to install surge suppressors or lightning arresters on the capacitor bank to limit such voltage surges if they occur.

6.7.3 Cable Switching and Line Dropping

Again, I have discussed this duty in Chapter 5, Section 5.5.5 in this volume. The discussion there applies equally to testing of vacuum circuit breakers. Table 5.14 shows the preferred values of current for testing these duties. Indoor circuit breakers usually have to satisfy only the cable-switching test while outdoor circuit breakers must satisfy the line-switching test. In my experience, this is usually not a difficult test to satisfy for all medium voltage circuits. Chaly et al. [158] in an analysis of real electrical system networks vigorously argue that the IEC test requirements for certifying vacuum circuit breakers using a capacitor bank for cable and line switching do not match reality. In fact, they state that these tests are not a valid evaluation of a vacuum circuit breaker's true performance for this switching duty. The capacitance of a cable and a line is distributed along the line and there is generally considerable impedance in the cable/line from loads that are still attached to it. Thus, any charge trapped on the cable/line will decay quite quickly and the voltage appearing across the open contacts will rapidly change from the offset capacitor voltage to the normal ac system voltage. Unfortunately, the usual testing method is to simulate the cable/line with a lumped capacitor. There are two problems with this approach: first the capacitor chosen is equivalent to a very long line/cable, which, in practice, is rather rare: and second, the recovery voltage during the test stays fully offset. They suggest that a better approach would be to use an R-C network to more closely simulate the cable/line switching circuit. It is now proposed for the IEC standard (clause 6.111.5.1) to insert a series resistance that is 5% of the "capacitive impedance" to limit the in-rush current.

While most electrical power is transmitted over long distances via overhead lines, there is increasing use of long cables especially when transmitting power from off-shore wind farm power to land substations. The off-shore booster station is generally built close to the wind farm and the raised voltage is brought ashore by a submarine cable. The characteristic impedance of the cable is much smaller than that of an equivalent overhead line. It is possible that switching off the current at the transformer can result in overvoltages, because of this difference. Xiao et al. [159] have modelled the expected overvoltages for a typical cable-transformer combination and show that the maximum overvoltage when switching off the transformer is 2.2PU, but 1.8PU when switching it on.

6.7.4 Motor Switching

The testing standard for certifying a vacuum circuit breaker for reliably switching an electric motor is given in IEC/TR 62271-306. While the technical report discusses testing vacuum circuit breakers for motor switching duty it also covers the whole range of circuit breaker testing. Vacuum circuit breakers are often used to switch high voltage electric motors (≥ 12kV). It is assumed, because the normal rated current of high voltage motors is comparatively low, that a vacuum circuit breaker will operate well. For the most part this is true. Most large electric motors switched by vacuum circuit breakers seldom experience a locked motor condition and they are usually switched infrequently. So, once the high voltage electric motor is switched on, it reaches its full operating speed. The opportunity for voltage escalation discussed in Chapter 5, Section 5.3 in this volume will occur less frequently than would be expected for a vacuum contactor.

Even though there is very limited evidence of its effectiveness [160], it would still be advisable to use surge suppression (e.g., a R-C network to ground; see Figure 5.20) at the terminals of the high voltage motor, especially if one or more of the following conditions are in place:

1. The motor's function is extremely critical
2. The motor is expensive to replace
3. The motor has marginal insulation
4. The motor has frequent aborted starts (i.e., jogging)
5. The circuit has poor power quality

6.8 APPLICATION OF VACUUM CIRCUIT BREAKERS FOR DISTRIBUTION CIRCUITS (4.76 KV TO 40.5 KV)

6.8.1 Indoor Switchgear

Acceptance in the industry of the vacuum circuit breaker developed quite slowly; See the Introduction. The older technologies (the magnetic air breaker, the bulk oil breaker, and the minimum oil breaker [161–166]) having been established and having performed satisfactorily for many years had an entrenched acceptability. Indeed, they performed their principal function of protecting the distribution circuit from the effects of fault currents extremely well and had done so for many years. In the conservative electricity distribution industry, the question asked seems to have been, "Why replace the tried and true technology with a new and untried technology?" Thus, the vacuum circuit breaker had to earn its spurs! Once it had done so and its advantages had been clearly demonstrated, it has gradually become the technology of choice. This has been true not only in new installations, but also in replacing or retrofitting the former technologies [164].

It is instructive to follow my own experience with the Westinghouse/Eaton development of its vacuum circuit breaker. The original Westinghouse design used the magnetic air circuit breaker's mechanism with the vacuum interrupters mounted in place of the contact chambers and the huge arc ceramic chutes [164, 165]. The spring operating mechanism is located under the three vacuum interrupters. The second-generation design used a similar spring mechanism, but this is now mounted in front of the three vacuum interrupters as is shown in Figure 6.41. Later generations followed with more compact designs with smaller footprints and more efficient mechanisms. These developments have resulted in vacuum circuit breakers that can even be stacked two-high in the metal clad and metal enclosed switchgear.

Almost all-indoor, vacuum circuit breakers used for utility, industry, and commercial applications enclose them inside steel housings. In the United States, these housings are termed metal enclosed switchgear or metal clad switchgear [166]. The definition of Metal Clad Switchgear and Metal Enclosed Switchgear from IEEE Standard C37-100 is:

> ***Metal-Enclosed Power Switchgear*** *is a switchgear assembly completely surrounded by a metal case or housing, usually grounded. The enclosure is made from with grounded sheet metal on all sides and top (except for ventilating openings and inspection windows). It contains the power vacuum circuit breaker with buses and connections, and may include control and auxiliary devices. The vacuum circuit breaker may be of the stationary or removable type. When of the removable type, mechanical interlocks are provided to ensure a proper and safe operating sequence. Doors or removable covers provide access to the interior of the enclosure.*
> ***Metal-Clad Switchgear*** *is Metal-Enclosed Power Switchgear with the following features:*
>
> 1) *The vacuum circuit breaker is removable (draw out) arranged with a mechanism for moving it physically between connected and disconnected positions and equipped with self-aligning and self-coupling primary disconnecting devices and disconnectable control wiring.*
> 2) *Major parts of the primary circuit (i.e. the circuit switching or interrupting devices, buses, voltage transformers, and control power transformers) are completely enclosed by grounded metal barriers that have no openings between compartments. Specifically included is a metal barrier in front of or a part of the vacuum circuit breaker to ensure that, when in the connected position, no primary circuit components are exposed by the opening of a door.*
> 3) *All live parts are enclosed within grounded metal compartments.*
> 4) *Automatic shutters that cover primary circuit elements when the removable element is in the disconnected, test, or removed position.*
> 5) *Primary bus conductors and connections are covered with insulating material throughout.*
> 6) *Mechanical interlocks are provided for proper operating sequence under normal operating conditions.*
> 7) *Instruments, meters, relays, secondary control devices, and their wiring are isolated by grounded metal barriers from all primary circuit elements with the exception of short lengths of wire such as at instrument transformer terminals.*
> 8) *The door through which the vacuum circuit breaker is inserted into the housing may serve as an instrument or relay panel and may also provide access to a secondary or control compartment within the housing.*

The metal enclosures are divided into cubicles. Typically, one cubicle will contain the vacuum circuit breaker while another will contain the bus or cables connecting the upstream distribution transformer to the downstream system; see Figure 6.74. The steel enclosures are grounded so that they can be safely approached from all sides. These cubicles contain other components such as current transformers, potential transformers, interlocks, control and tripping sections, etc. The design of the cubicles has to be such that all the electrical joints can pass the expected load currents without overheating and also pass the vacuum circuit breaker's full short circuit current withstand rating. All the clearances inside the cubicles must be such that the high voltage withstand and the BIL requirements are also met. The vacuum circuit breaker, when placed inside the cubicle, is thus fully capable of coordinating with the upstream and downstream distribution systems. It also provides all the protection required of it. A photograph of metal enclosed switchgear is shown in Figure 6.75.

There are two methods of connecting the vacuum circuit breaker to the cubicle's internal bus. One is a fixed connection. Here the vacuum circuit breaker is permanently bolted to the bus so that it is integral with it. Present developments of vacuum circuit breakers increasingly claim that they are

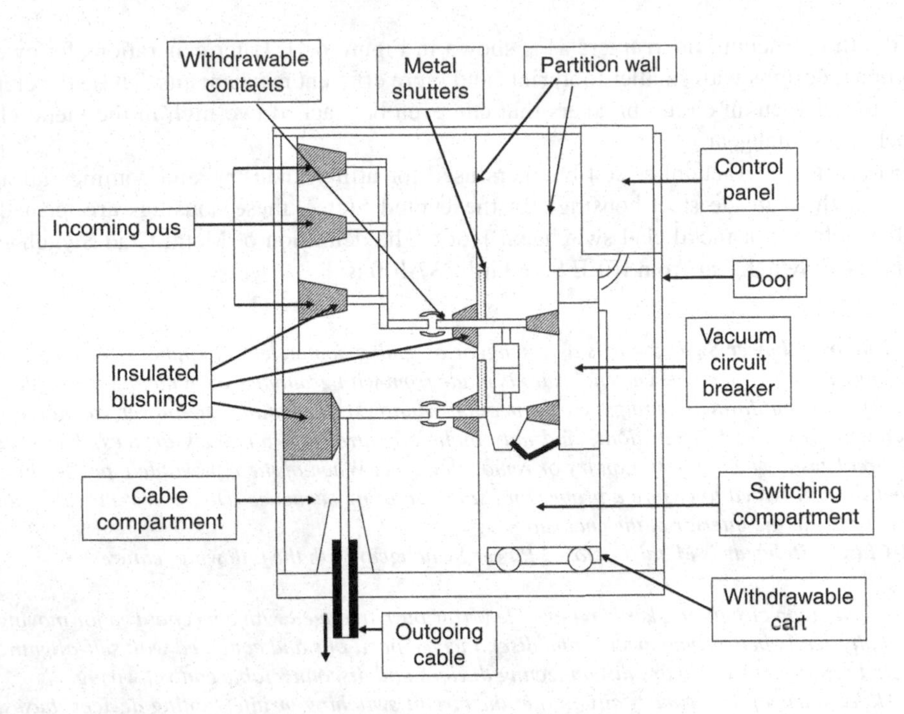

FIGURE 6.74 A diagram of a metal enclosed switchgear enclosure.

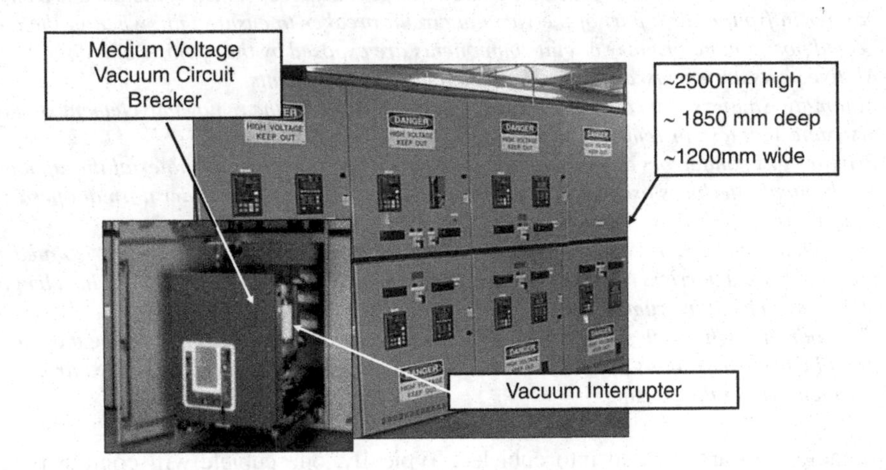

FIGURE 6.75 A photograph of metal enclosed switchgear (courtesy of Eaton Corporation).

"maintenance free." Thus, this type of fixed connection is becoming more common, especially on fully sealed systems; see below. The second method of application is the withdrawable design. Here the vacuum circuit breaker is mounted on wheels (or rollers) or is mounted to a withdrawable frame. Once the vacuum circuit breaker is opened, the vacuum circuit breaker can be withdrawn from its bus connection. As the vacuum circuit breaker is withdrawn, a grounded shutter is deployed that isolates the circuit breaker cubicle from the bus cubicle. It is only then that the door of the cubicle can be opened. After the cubicle door is opened, the vacuum circuit breaker can be rolled from the cubicle for inspection or servicing. Most modern cubicles have a horizontal draw-out. There are some designs that have a vertical, drop-down capability. These are mostly retrofit designs to replace older circuit breaker technologies in existing switchgear. The connection from the vacuum circuit breaker to the bus is a spring-loaded stab type connection. As it is a no-load connection, it is

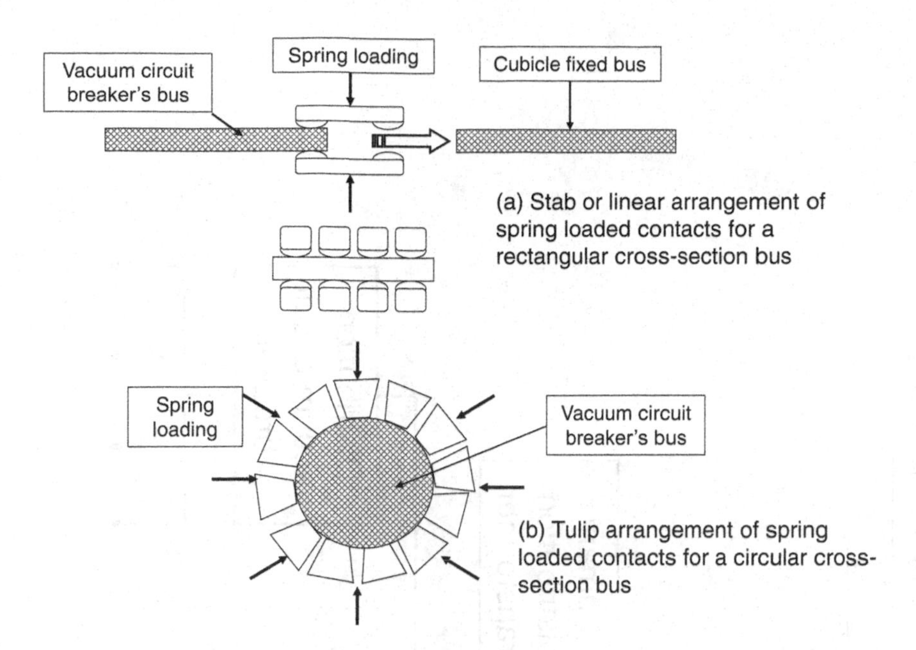

FIGURE 6.76 Stab and tulip connections for the circuit breaker to the switchgear cubicle's bus.

usually manufactured from Ag plated Cu. The two common stab connections, (a) the linear, and (b) the tulip, are illustrated in Figure 6.76. Wang et al. [167] have analyzed the effect of eddy currents and air flow in a switchgear chamber on the temperature rise of vacuum interrupter terminals. They show that while natural convection helps to maintain a reasonable temperature rise at rated continuous current, as would be expected, forced convection does work better.

The metal enclosed/clad switchgear is used on both the primary and secondary distribution systems [168]. A simple illustration is shown in Figure 6.77. The primary distribution system is that portion of the power system between the distribution substation and the distribution transformers. The vacuum circuit breaker in primary switchgear generally must pass higher continuous currents and interrupt higher fault currents. It must also have a large fault current withstand capability in order to coordinate with the secondary distribution switchgear. The secondary distribution system is that portion of the power system between the distribution transformers and the consumer. The secondary system consists of distribution transformers, circuit breakers, secondary circuits, and metering. The medium voltage secondary system used in large industrial and commercial applications are three-phase and can cover the system voltage range from 4.76kV to 38kV. Commercial examples of both types of switchgear can be found on all major manufacturer's websites [169].

Switchgear for system voltages of 15kV and lower are usually air insulated. The bus or cabling needed to connect the vacuum circuit breaker to the upstream and downstream electrical systems determines the minimum volume of the cubicle. In general, the volume required for these connections is such that for voltages 15kV and lower air insulation is quite adequate. At higher voltages the air insulated enclosures become much larger. This has led to the development of hermetically sealed SF_6 insulated cubicles where a one-atmosphere gauge of SF_6 (i.e., two atmospheres of SF_6 in the cubicle) gives about five times the dielectric strength of air. This permits a significant reduction in the size of the cubicle [170]. Another advantage of this sealed cubicle is that it can be used at high altitudes without having to take into account the reduction of the dielectric strength of the ambient air as its pressure decreases; see Section 1.4.4 in this volume. As the concern that SF_6 is a greenhouse gas becomes a major concern, alternative gases and gas mixtures will be introduced to replace SF_6 [117–119]. This will permit the hermetically sealed system to be used. Also, I would

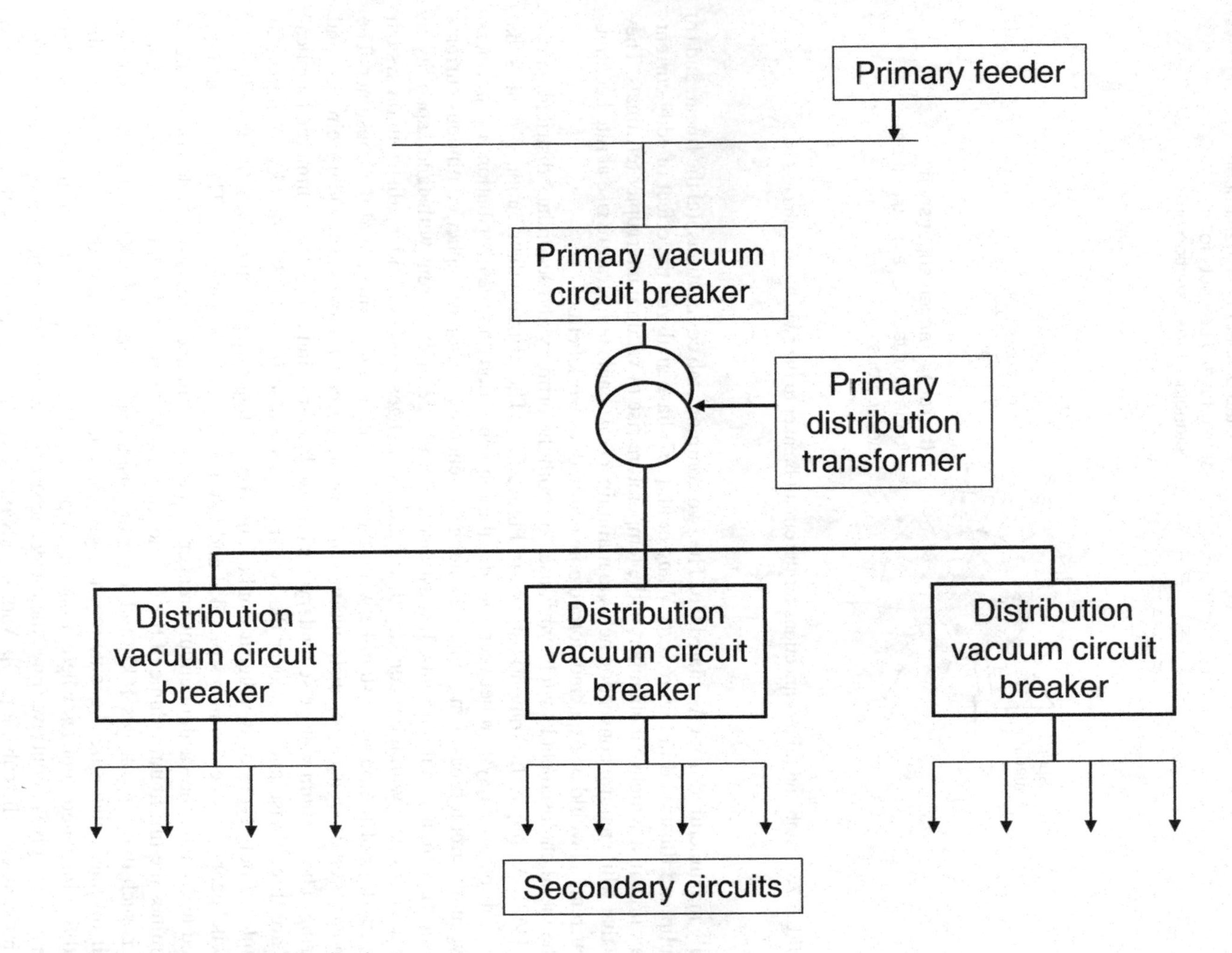

FIGURE 6.77 The primary and secondary distribution systems.

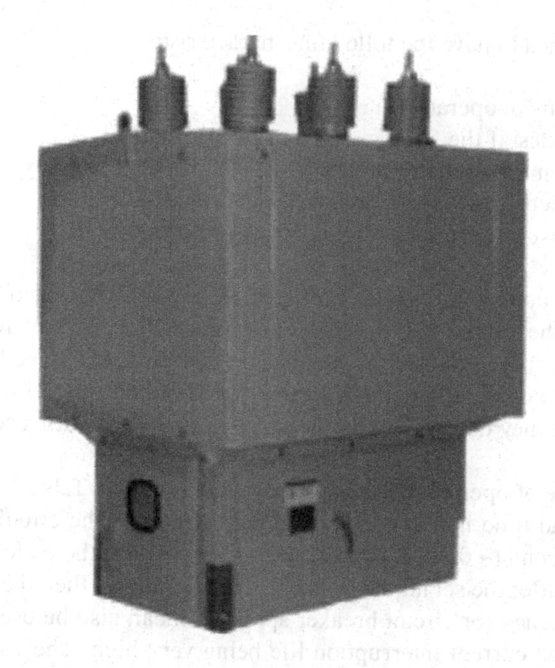

FIGURE 6.78 An example of an outdoor, dead-tank circuit breaker (courtesy of ABB, USA).

expect that in future, switchgear designs with solid insulation will increasingly be applied at the higher distribution voltages.

6.8.2 Outdoor Circuit Breakers

Figure 6.78 shows an example of an outdoor distribution vacuum circuit breaker. Here again, the vacuum circuit breaker is contained within a grounded metal housing. This stand-alone vacuum circuit breaker typically functions as the main distribution circuit breaker at the primary circuit's distribution substation. The circuit breaker's cubicle contains the sensing and control for downstream coordination. The cubicle also contains heaters that prevent moisture build-up on the components inside. These heaters are essential for maintaining the high voltage performance of the circuit breaker and its connecting bus, especially in humid ambients. The standard outdoor circuit breaker usually has reclosing capabilities so that a temporary fault will not result in an extended loss of power to the whole distribution system. Examples again can be found on manufacturer's websites [171].

6.8.3 Vacuum Reclosers

An increasingly important outdoor type of vacuum circuit breaker is the vacuum recloser. The general definition for the recloser can be found in Section 6.1. Nylen [172] presents a comprehensive review of auto-closing. The vacuum interrupter is ideally suited for the recloser application. In fact, it was one of the first practical U.S. applications of the vacuum interrupter in the early 1970s to outdoor medium voltage, line protection [147]. They are increasingly being used in distribution substations in place of circuit breakers. In fact, a 1995 IEEE survey shows that 51% of substation feeder interruption devices at that time in the United States were reclosers [173]. It appears that the recloser is being used more frequently in smaller substations while the circuit breaker is applied in larger ones. With the need to maintain the up-time of distribution circuits and because only about 5% of faults are permanent, the application of the recloser and the auto resetting circuit breaker has become increasingly common.

The recloser is designed to have the following characteristics:

a) Frequent and repetitive operation
b) Rapid reclosing to test if the fault is temporary or permanent
c) It is often placed on a long distribution line where expected fault levels can be quite low
d) A high BIL level across the open contacts is not critical
e) The total cumulative current interruption duty can be high

Reclosers with a continuous load current rating of 50A–200A (i.e., feeder reclosers) have fault interruption ratings in the range 2kA–6.3 kA. The larger reclosers that are usually found in distribution substations can have continuous currents up to 1250A and have fault current ratings up to 25 kA. Reclosers are not usually derated for multiple fault current, close and open operations. Unlike a circuit breaker they do not usually have a close and latch with a subsequent high current withstand rating.

The general sequence of operations is shown in Figure 6.79 and Table 6.17. After the first interruption sequence, a dead time is required for the faulted part of the circuit to recover. Typically, this is a fraction of a second (~ 0.3s). It is possible to then perform the reclosing operation again. If the fault is still present after the set number of reclosing operations, then the recloser locks out. The vacuum interrupter designed for circuit breaker application can also be used as a vacuum recloser with no changes; its fault current interruption life being very high. The vacuum interrupter has a unique ability to manage the reclosing operation and many variations placed by different utilities on the resetting time from a few cycles to a few minutes. It also has the ability to operate against faults for more than three times. The reclosing function is becoming more common as additional automation of the distribution system becomes more common in order to lower its SAIDI (System Average Interruption Duration Index = Sum of all customer interruption durations / Total number of customers served) [174]. Staszesky et al. [175], for example, state that several utilities in the United States are automating their entire distribution systems.

The initial vacuum recloser designs looked very similar to the vacuum circuit breaker shown in Figure 6.78. The air insulated vacuum interrupters with their attendant mechanism, sensing and

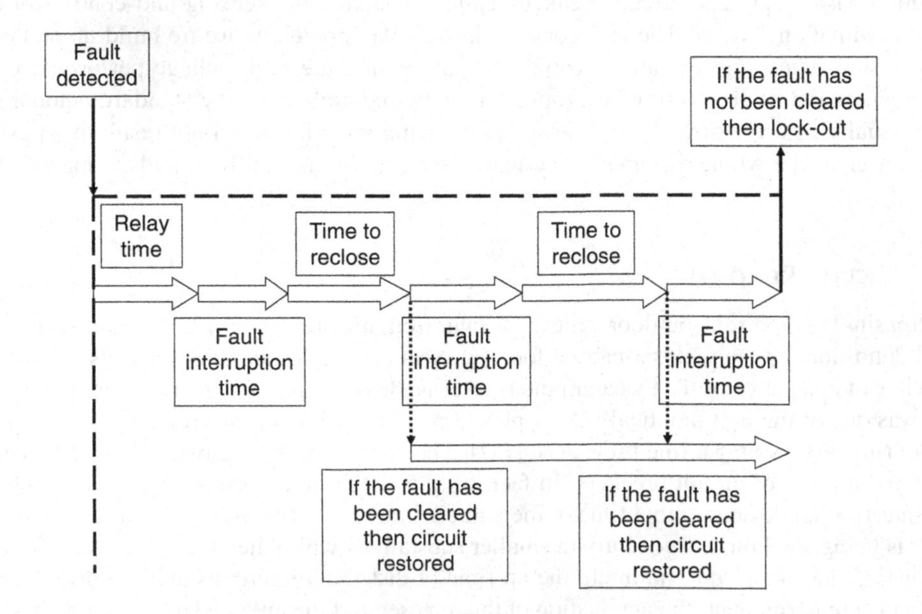

FIGURE 6.79 The operating sequence of a vacuum recloser.

FIGURE 6.80 Pole mounted reclosers (courtesy of FKI Engineering).

control are placed in a large, stand-alone grounded metal cubicle. The vacuum recloser has developed in a number of ways. One example is the pole-mounted unit, an example of which is shown in Figure 6.80. Modern designs use hermetically sealed enclosures with SF_6 insulation [90] or solid encapsulation [176]. In the recent past the reclosers have used a spring mechanism with an electronic or electromechanical control system: the relay for tripping the mechanism being mechanical and the relay control being electronic [177]. The magnetic mechanism with full electronic control entered the market at the end of the twentieth century and is becoming increasingly common. Reclosers are available as three-phase and as single-phase units. The single-phase designs can be used on single-phase taps in place of fuses. There are other designs that can operate each phase independently, so that the recloser only opens the phase on which the fault occurs. This type of operation presents a challenge to the vacuum interrupter designer as well as to the recloser designer. In a three-phase ungrounded system, if the fault is only present on two phases, the second phase to clear will have the full line-to-line voltage across it and thus will see a very high TRV.

The recloser continues to evolve and new designs are being introduced with intelligent controls that record the number of operations, the level of current interrupted and the expected life; see Figure 6.81. These data are then sent to a central control base where they can be used to determine when a recloser is nearing its end-of-life and needs to be replaced. One such data recording and communication system is called SCADA (Supervisory Control and Data Acquisition) [178, 179]. It is an industrial measurement and control system, which consists of a central host and a number

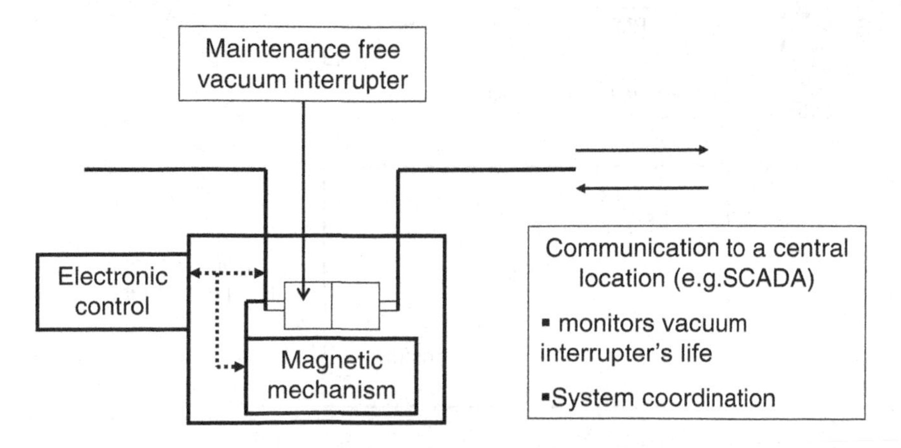

FIGURE 6.81 The remote control and information transfer for recloser maintenance.

of remote data gathering and control units. By using suitable software, it is possible to monitor remotely located field installations such as reclosers and their attendant sectionalizers (see Sections 5.2.1 and Figures 5.12 and 5.13 in this volume). This enables the "host" to have a "supervisory level" control over the remote site. For an electrical system the data sent to the "host" can be used to develop system trends. For example, trend analysis makes it possible to predict which parts of the distribution system will experience an overload condition. This analysis would then enable a central operator to isolate that part of the system. As discussed above, the data could also be used to determine how close a circuit's component is to its end of life and to determine when a timely replacement is in order. SCADA operates in a supervisory mode and does not usually perform the safety or fault interruption functions. This sensing and control is still part of the electronic sensing and electromechanical components of the recloser itself: See, for example, the NOJA website [180]

6.8.4 The Ring Main Unit (RMU) for Secondary Distribution

The RMU has been in continuous development in Europe since the1950s. Figure 6.82(a) shows a typical ring-shaped cable network in which an RMU operates. A vacuum circuit breaker is placed at the feeder substation; i.e., the beginning of the open ring. An RMU is a secondary distribution switching and protection component that is placed around the ring at points where feeds to a low voltage distribution are required. A simple diagram of an RMU is shown in Figure 6.82(b). It consists of two load break switches with grounding capability, a fault interrupter, and a step-down transformer. By using such a system, it is possible to isolate a fault condition at one of the feeders and maintain service to the rest of the system, thus increasing the uptime of the ring. In the RMU's initial embodiment, either fuses or circuit breaker interrupters could be used to provide the protection function. Also, as the fault interruption duty is expected to be less severe than that for a circuit breaker, it does not have to pass the full circuit breaker certification standards. Indeed, the early designs used oil interrupters for this duty [181].

In recent years, more emphasis has been placed upon a more robust fault interruption requirement and thus the vacuum interrupter has been increasingly incorporated into the RMU to perform this

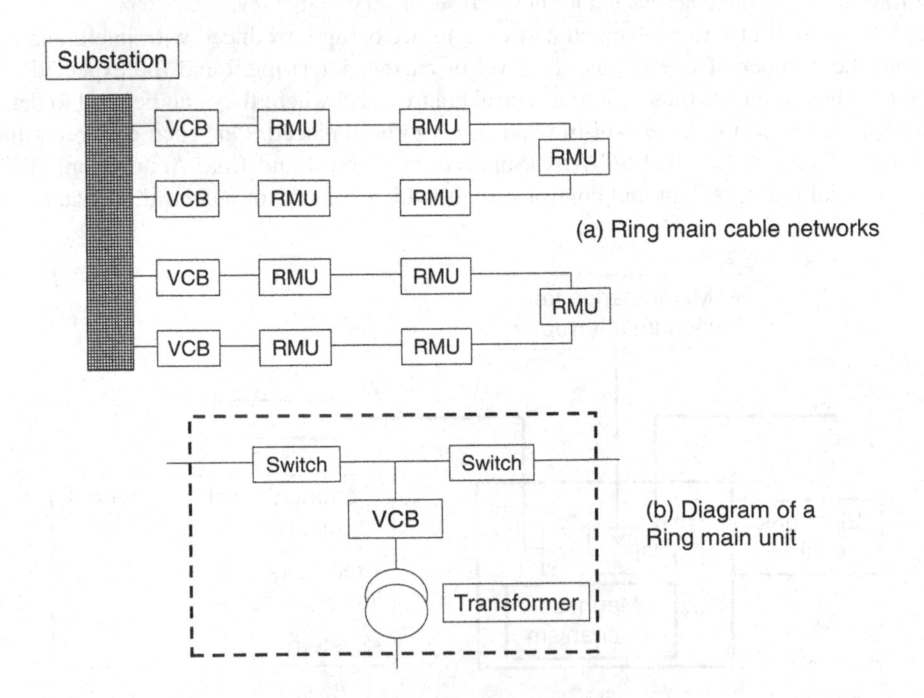

FIGURE 6.82 The Ring Main Unit secondary distribution system.

TABLE 6.19
Nominal Ratings for Ring Main Units (IEC)

Parameters	12 kV system voltage	24 kV system voltage
a) System		
Rated bus current, A	630	630
BIL, kV	75	125
Power frequency withstand voltage, kV	28	50
Rated frequency	50/60	50/60
Short-time withstand current, kA–s	20–1	16–1
Peak withstand current, kA	50	40
b) Vacuum circuit breaker		
Rated current, A	630	630
Short circuit interruption current, kA	20	16
Peak short circuit making current, kA	50	40
DC component	35 or 50	35 or 50
c) Load break switch		
Rated current, A	630	630
Rated breaking current (0.7 pf)	630	630
Peak short circuit making current, kA	50	40

function [181, 182]. RMUs are typically applied to medium voltage circuits in the range 12kV to 24kV. Their performance requirements are shown in Table 6.19. Insulation of RMUs has progressed from oil to SF_6 [183] to solid epoxy insulation [184, 185]. The vacuum interrupter, with its maintenance free operations, is ideally suited to the hermetically sealed SF_6 container and the epoxy encapsulation. Figure 5.71 in this volume showed an example of an RMU with solid encapsulation of the vacuum switches and the vacuum circuit breaker. Thus, its use for this function will continue to expand. In the future, more automated control, in coordination with precise high-speed mechanisms will be introduced [186, 187]. In SF_6-insulated RMUs, the load switch usually uses SF_6 gas as the interruption medium. However, for the solid insulation, a vacuum interrupter load break switch has to be used [184]. As the load current levels increase above 630A, and as the fault interrupter requirements go above 20kA, the vacuum interrupter will gradually become the future interrupter of choice for both RMU functions.

6.8.5 Pad-Mount Secondary Distribution Systems

The pad-mounted secondary distribution system is common in the United States. It is similar in concept to the RMU. It consists of a metal outdoor enclosure that can be attached to a concrete pad and supply power to the secondary distribution system [188]. Inside the metal enclosure are usually load break switches and fault protection devices. A line diagram of one example is given in Figure 6.83. The original designs used fuses for protection. The application of the vacuum interrupter for load switching and for the circuit protection has resulted in quite compact designs insulated with oil or SF_6[188]: see Figure 6.84. The pad-mount switchgear is usually located close to a customer's site and can be rather remote from the central distribution control center. This has also led to the application of the SCADA [178, 179] to these units.

6.8.6 The Generator Vacuum Circuit Breaker

The generator circuit breaker is applied between a generator and a step-up transformer. The connection from the generator to the generator breaker and then to the transformer is usually designed to

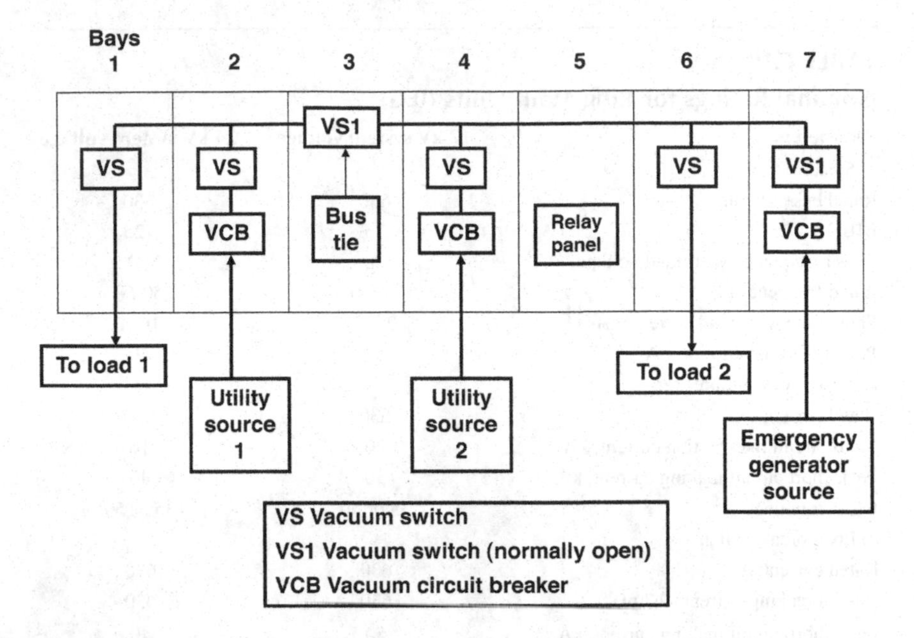

FIGURE 6.83 A complex Pad Mount secondary distribution system with seven bays showing two utility sources, an emergency generator source and a bus-tie connection. All switches, circuit breakers, utility feeds and the generator feed are either computer controlled or by SCADA.

be as short as possible and have as low an impedance as can be arranged. For the most part, these connections are in the form of a low impedance isolated three-phase bus, but for small generators they can be cable connected [189]. A simple diagram for this application is shown in Figure 6.85. It became clear in the late 1960s that a special circuit breaker would have to be developed to give protection for this application. At this time both Delle-Alsthom and Brown Boveri developed generator circuit breaker designs for large generators using an air blast arc interruption technique [190]. By the late 1970s, about 140 of these mammoth designs had been installed worldwide [191].

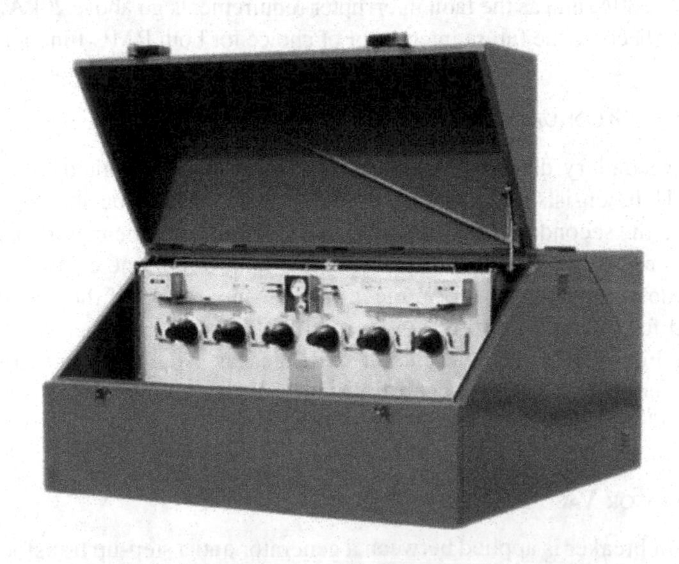

FIGURE 6.84 A Vacuum Pad Mount system (courtesy of Trayer Engineering).

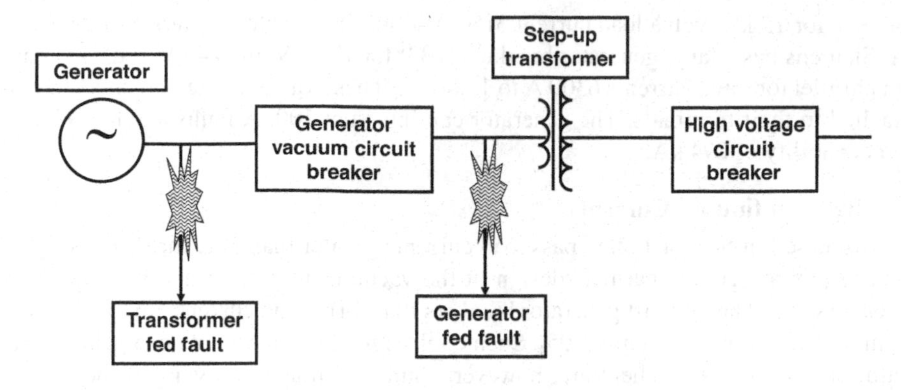

FIGURE 6.85 The positioning of the vacuum generator circuit breaker showing potential fault locations.

The bulk and operating noise of this circuit breaker technology has proven a disadvantage when applied to small generating systems. The need for a more compact and a quieter generator breaker has led to the use of the vacuum interrupter in the development of vacuum generator breakers for these smaller systems. The certification testing of a generator breaker places extreme demands upon both the vacuum interrupters and the circuit breaker's mechanism. A comparison of normal circuit protection requirements and those for a generator breaker given by Long et al. [51] can be seen in Table 6.20. The standards for generator breakers are given by the IEEE [192, 193] and the IEC [194]. From Table 6.20 it can be seen that the vacuum generator breaker has to meet the following performance criteria:

1) High continuous load currents
2) Interrupt generator fed faults (on the RHS of the generator breaker shown in Figure 6.85)
3) Interrupt transformer/system fed faults (on the LHS of the generator breaker show in Figure 6.85)
4) Interrupt fault currents with high X/R values
5) Interrupt fault currents when out-of-phase conditions exist

The air blast generator breakers, for example, are able to carry continuous load currents up to 35kA and interrupt fault currents in the range 250kA–350kA [190, 195]. Within a few years of introducing practical vacuum circuit breakers to protect distribution circuits, there was a thought to use vacuum interrupters as the switching elements in generator breakers. Indeed, while preparing this Section I came across a memo I had written in 1977 to the Westinghouse Power Circuit Breaker Engineering Department where I had optimistically recommended the use of parallel and series combinations of vacuum interrupters to achieve the performance rating of a large generator breaker. This design eventually proved to be impractical.

The research on, and the development of vacuum interrupters for generator breaker applications, however, did continue. Alexander et al. [196] demonstrated that a vacuum interrupter using TMF contacts would make a good candidate for application as a small generator breaker. Kulicke et al. [197] again demonstrated that a vacuum interrupter with TMF contacts that gave an arc voltage in the range 100V–150V greatly ameliorated the effect of the initial high *X/R* experienced by the generator-fed fault. Smith et al. [51, 198] showed a successful compact vacuum generator breaker based upon a standard distribution vacuum circuit breaker design with a spring mechanism. This development used vacuum interrupters with TMF contacts. This 15kV vacuum generator breaker design is capable of 4000A continuous load current and interrupt fault currents of 63kA and 75kA. A photograph is shown in Figure 6.86. The development of generator breakers using vacuum interrupters continues. For example, Gentsch et al. [199] showed a design with vacuum interrupters with

TMF contacts for 15 kV with a load current 3150 A–4000 A capable of interrupting a fault current of 50 kA. Siemens has a large generator breaker (HB3) for 17.5 kV and 24 kV. It uses vacuum interrupters in parallel for rated currents 6300 A to 12,000A. These vacuum interrupters have heroic heat sinks attached to their terminals. The generator can interrupt 100kA faults and has a high-current withstand capability of 274 kA.

6.8.6.1 High Continuous Currents

As I have discussed in Section 6.2, to pass load currents greater than 3kA through a single vacuum interrupter requires a good "thermal" design of the vacuum interrupter and its attachment to the circuit breakers bus. The air blast generator breakers that carry load currents up to 36kA use forced air [200] and water-cooling systems [195]. Such cooling methods have not been generally accepted for vacuum circuit breakers. There are, however, some manufacturers who are beginning to use

TABLE 6.20
Comparison of Requirements (IEEE–United States) for Generator Class Circuit Breakers and for Distribution Class Circuit Breakers

Parameter	Units	Generator Class	Distribution Class
Design & application philosophy		Custom	Standard
Rated system voltage (U_C)	kV	15.75, 28.35	4.76, 8.25, 15, 27, 38
Normal frequency	Hz	60	60
Rated continuous current	kA	3,4,6.3,10, 12.5….	0.63, 0.8, 1, 1.25, 2, 2.5, 3.15
Emergency current if cooling loss	A	Req'd as applicable	Not req'd
Lightning impulse withstand voltage	kV pk	110	95
Power frequency withstand voltage	kV rms	50	36
Short-circuit parameters			
Rated short-circuit current (i_{sc})	kA	63–120…	16–80
Close & latch (making) current	kA pk	$2.74 \times i_{sc}$	$2.6 \times i_{sc}$
Rated short-circuit duty cycle		CO–30 min–CO	O–0.3s–CO–3 min–CO O–15s–CO–3 min–CO
Rated interrupting time	ms (cycles)	Approx. 60 to 90 (4 to 6)	83 max (5)
Rated permissible trip delay	s	0.25	2
Duration of short-time current	s	1	2
(Reference) X/R circuit		50	17
(Reference) Time constant of circuit		133	45
Rate of rise of recovery voltage	kV/μs	3.5	0.2–0.55
Peak of TRV	kV	$1.84 \times U_C$	$1.88 \times U_C$
Time to the peak of TRV	μs	8–15	50–125
Generator-fed asymmetrical fault			Not required
Delayed current zeroes		Must demonstrate	
Relevant dc component		> 100% (110% – 135%)	
Endurance capabilities: operations			
No-load mechanical		1000	5,000–30,000
Rated load current switching		50	500
Rated shot-circuit switching		None specified	16…
Out-of-phase switching capability			
Out-of phase switching current	kA	$50\% \times i_{SC}$	$25\% \times i_{SC}$
Power frequency recovery voltage	kV rms	$1.22 \times U_C$	$1.44 \times U_C$
Rate of rise of recovery voltage	kV/μs	3.3	~ 0.3
Peak of TRV	kV pk	$2.6 \times U_C$	$2.54 \times U_C$
Time to peak of TRV	μs	~15	~100

FIGURE 6.86 A photograph of a 15kV, 75kA, 4000A vacuum generator circuit breaker (courtesy of Eaton Corporation).

forced air-cooling in vacuum switchgear for load currents ≥ 4kA. For load currents in excess of 6kA it is possible to use vacuum interrupters connected in parallel. Using this approach, it is quite possible to use three circuit breakers, one for each phase, with their vacuum interrupters connected in parallel to pass continuous currents up to 15kA. An example of such a design is shown in reference [201]. The total impedances of the closed vacuum interrupters are very close to being equal, i.e., there would be close to equal division of current between the three vacuum interrupters. During fault interruption the variation of arc voltage may be such that the division of fault current may not be equal. So, the design of each vacuum interrupter should be such that it can interrupt currents greater than one third of the available short circuit current.

Also, it is necessary to open the parallel-connected vacuum interrupters on each phase at exactly the same time. Pertsev et al. [36] have demonstrated a fault current division of 40%, 30%, and 20% on three vacuum interrupters connected in parallel with TMF contacts. They also postulate that an AMF contact structure would give a better current division. Unfortunately, a TMF contact is better suited for interrupting generator fed faults. Table 6.21 shows my data for the division for three vacuum interrupters with AMF contacts. Note that adding a small inductance in front of each vacuum interrupter greatly assists in approximately equalizing the division of current. For parallel-connected vacuum interrupters to be effective, all three phases must be opened such that a reasonable arcing period is established in each phase. Also, each vacuum interrupter must be designed to operate without damage if only one of the parallel-connected vacuum interrupters experiences the full fault current arc. For example, if one of the vacuum interrupters opens just before a current zero and a second one just after the second interrupter may have to interrupt the full short circuit current [202].

6.8.6.2 Transformer/System Fed Faults

The fault current on the generator side of the generator breaker is fed through the transformer by the whole system connected to it. The value of this current is limited only by the relatively low

TABLE 6.21
Current Division Between Three Vacuum Interrupters with AMF Contacts Connected in Parallel

	Currents in the three interrupters at 90° and % departure from equal sharing							
Total peak current, kA	Phase A, kA	%	Phase B, kA	%	Phase C, kA	%	Arcing time, ms	Inductance before each interrupter, ¼H
25.6	9.3	9	8.1	–5	8.2	–4	1.5	0
27.2	11.3	25	9.6	6	6.2	–31	7.5	
49.5	24	45	12	–27	13.6	–18	6.1	
51.6	20.5	19	18.1	5	13	–24	8	
80	31.1	16.5	26.9	0.5	22.1	–17	4.3	
24.7	7.9	–4	7.5	–9	9.3	13	3.3	1.5
26.3	9.2	5	8.1	–8	9	3	4	
50.5	16	–5	18.8	11	15.1	–6	8.6	
76.8	28	9	26	–1	22.8	–11	1.2	
78.5	26.9	3	25.9	2	25.7	–2	3.8	

impedance of the transformer and results in very high magnitude current levels. Also, the *X/R* ratio is relatively high (50~100; see Table 6.22 [196]), which gives a slow decay of the dc component of the asymmetrical fault current. For the normal three-cycle vacuum circuit breaker, this means that the peak asymmetric fault current can be 2.74 times the rms symmetrical value. This is greater than the 2.5 (at 50Hz) to 2.6 (at 60Hz) times experienced by normal distribution breakers. The high natural frequency and the very small capacitance of the connections to the generator breaker combine to produce a TRV with a very fast rate of rise (perhaps 3–5 kV.μs^{-1}) [203]. The rate of rise of this TRV can be reduced considerably by placing capacitance to ground at the transformer's terminals. If this capacitance is used to pass the severe certification test, the resulting generator circuit breaker has to be installed with the same capacitance. This test duty is so severe that Dufournet [203] considers it the most difficult one for the generator breaker to satisfy.

TABLE 6.22
A Selection of Industrial Generator Characteristics

				No. of cycles to a current zero after fault initiation	
Generator No.	kV	MVA	X/R	No arc resistance	Arc resistance (mΩ) for 5 cycle breaker rating
1	13.8	116.8	82.9	6	0.2
2	14.4	72.0	110.5	7	5.1
3	14.4	66.8	86.3	8	2.6
3	13.8	51.2	73.5	6	1.3
4	13.8	43.8	49.0	5	0
5	13.8	29.4	106.7	7	4.2
6	14.2	25.5	91.2	6	2.3

6.8.6.3 Generator Fed Faults

When the fault occurs on the transformer side of the generator circuit breaker, only the generator feeds the current through the breaker. Typically, the fault current levels are only about half those of the system-fed fault. However, this fault current can be offset by as much as 110% to 135% of the symmetrical current value resulting in ac current cycles with no current zeros. As discussed in Section 6.3.2, this condition results from the effect of the fault on the generator itself [196, 204, 205]. This means that at least one phase, and frequently two phases, in a three-phase ungrounded fault will pass ac currents with no current zeros for four to eight cycles. Table 6.22 shows characteristics determined for a number of different generators [196]. For all of these generators the *X/R* is much higher than the value of 17, usual for distribution faults. However, the introduction of the small impedance from the arc between open contacts can reduce this value considerably and permit the ac current to pass through a current zero earlier. The effect of the 100V to 150V arc voltage from a vacuum arc between TMF contacts has indeed been shown to provide a very effective impedance that reduces *X/R*, [196, 204, 205]. Figure 6.87 shows the arc voltage for an 80kA (rms) contact operating between 125mm diameter TMF contacts with a contact gap of 10mm, which introduces an impedance of about 1mΩ into the circuit. For a generator-fed fault, the current would be about 40kA(rms) and the impedance would be 2.7mΩ. A simple analysis by Smith et al. [198] that illustrates the effect of introducing this impedance is shown in Figure 6.88.

The vacuum interrupter technology is ideal for operating under conditions of delayed current zeros, because its capability of interrupting current does not depend upon the position of the contacts, it depends only upon a reliable current zero crossing. The TMF contact structure and the internal shield would have only to endure a much longer arcing time than the usual one cycle experienced with normal distribution circuit faults. Because this contact would have been designed for the transformer-fed fault with a 75% dc component, the longer arcing time with the expected one-half of this current value has proven to be readily achieved. Indeed, Slade et al. [206] have demonstrated the effectiveness of TMF contacts for long arcing times. An AMF contact designed for the transformer-fed fault might also be able to endure the long arcing time of the generator-fed fault with its lower value of current. Unfortunately, the AMF contact whose arc voltage is typically 50V would have to experience a much longer time before the first current zero is achieved.

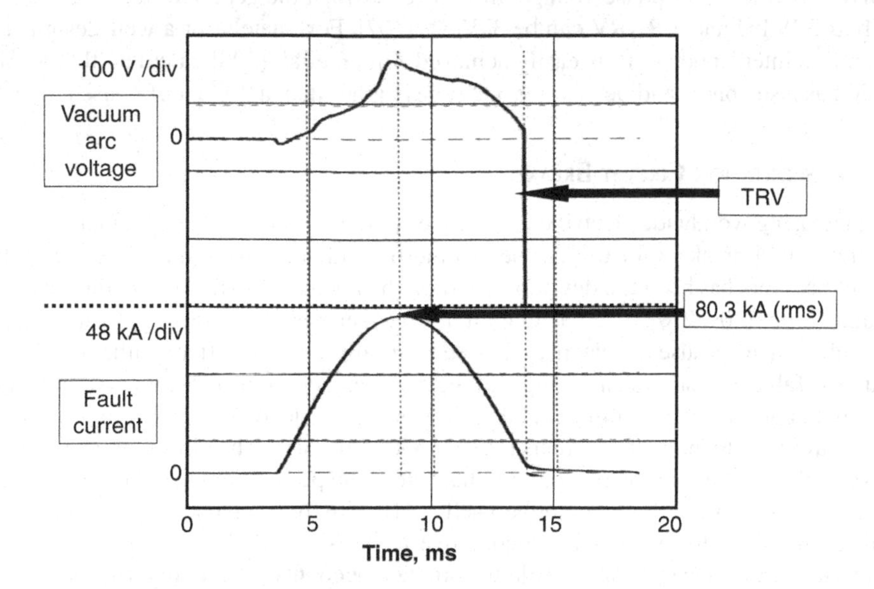

FIGURE 6.87 The arc voltage greater than 100V during the interruption of an 80kA (rms) symmetric fault current.

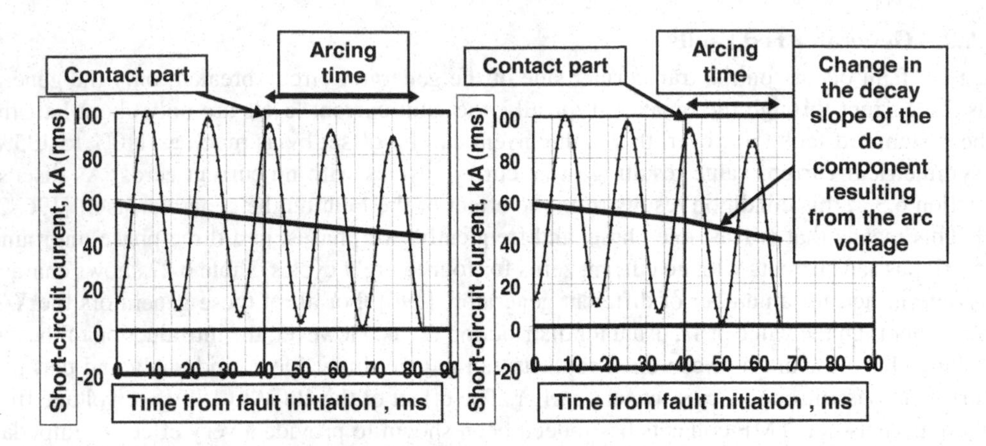

FIGURE 6.88 The decay of the dc component from a generator fault and the effect of an arc voltage of 150V reducing the time before a current zero is established by one full current cycle [198].

6.8.6.4 Out-of-Phase Switching

This third condition can occur if the generator breaker opens with the generator supplying the load and then is reclosed to reconnect the generator to the system. While the generator is disconnected, there is no electrical output to counter the generator's large mechanical energy input. This causes the generator to gradually speed up. Thus, the generator's phase begins to differ from that of the system. When the generator is reconnected, an *out-of-phase* condition may exist. If it is dramatically out-of-phase, a synchronism check relay is employed to prevent reconnection of the generator to the system. However, the standards allow for the possibility of reconnection for an out-of-phase condition of 90 electrical degrees. When the generator is reconnected to the system under this condition, a substantial current will flow in the circuit. This current can be as high as 50% of the rated short circuit of the generator. A current of this magnitude would be interpreted as a fault by the generator breaker's protective relay and a signal will be sent for it to open. After interrupting this current, the TRV is driven by the out-of-phase voltage difference between the generator and the system. It can be as high as 3.18 PU and it RRRV can be $3kV\mu s^{-1}$ [207]. Fortunately for a well-designed vacuum interrupter, this interruption duty is easily achieved. Braun et al. [208] calculate that a 180° out of phase TRV has a steepness and peak value $\sqrt{2}$ times higher than a 90° out of phase angle.

6.8.7 Transportation Circuit Breakers

The ever-increasing worldwide electrification of the railroad has led to the development of specialized vacuum circuit breakers for use on the locomotive and at the trackside. One major driver for these circuit breakers has been the development of high-speed (> 300km/hr.) locomotives in Japan, Europe, and China [209, 210]. The vacuum circuit breaker design for this application presents an interesting challenge because the electrical systems for railroads can differ significantly from country to country. Table 6.23 shows an example of the different ac frequencies and system voltages that are presently being used. A vacuum circuit breaker on top of a locomotive traveling from France to Germany would have to interrupt currents at 50Hz and at 16 ⅔Hz. The circuit breaker on top of the locomotive is placed between the pantograph that brings the power from the overhead catenary and the electrical system inside the locomotive itself [211]. This breaker not only provides protection from fault currents, it also enables the locomotive to be isolated while passing through section-gaps in the overhead catenary line. This latter function accounts for the majority of the switching operations considering these section gaps can occur every 25km to 100km. On a trip from Paris to Marseilles, for example, this circuit breaker would open and close between 10 and 20 times. Thus,

TABLE 6.23
Examples of the Variation in System Voltages and Frequencies for Worldwide Railroads

Country	kV	Hz
USA	15.5	25–60
France	24	50
Germany	15	16⅔
Norway, Sweden	15	16⅔
UK, Denmark	24	50
Taiwan	25	60
China	24	50
Other Systems	1kV and 3kV	dc

the vacuum interrupter used for this circuit breaker application needs to have a mechanical and electrical life in excess of 250,000 operations. This presents a challenge to the designer to incorporate a suitable bellows that can sustain this duty. Tian et al. [212] suggest an alternative method to transfer the catenary from one electric system to another. They present a combination of vacuum circuit breakers and IGBTs (integrated-gate bipolar transistors) that are attached to the power lines themselves. This permits the catenary to cross the gap between the two systems with no arcing and no damage to the catenary. The catenary wire system is an overhead line that stretches for many thousands of kilometers. It is, therefore, an inviting target for lightning strikes [213]. Thus, for both trackside and locomotive vacuum circuit breakers, lightning ipulse withstand voltage (BIL) values of 175kV to 250kV are required. These values are much higher than the 15kV systems outdoor standard of 110kV and the 24kV systems standard of 125kV. Thus, the designer of both the vacuum interrupter and the vacuum circuit breaker must take this into account.

A second type of transportation circuit breaker is gradually being introduced to switch and protect ac circuits with a frequency of a few hundred Hz. These are primarily being applied in shipboard applications. Here the vacuum interrupter designer is challenged with the shorter half-cycle times and the effects of the contacts being close together at the first current zero, even if the contacts begin to part just after a current zero. Thus, in a 400Hz system, vacuum interrupter's contacts in a circuit breaker opening at $1m.s^{-1}$ will be less than 1mm apart at the first current zero. If the vacuum circuit breaker is switching an inductive circuit, surge suppression will most probably have to be used to prevent voltage escalation; see Sections 4.3.1 and 5.3.3.

6.8.7.1 Interrupting Fault Currents at Frequencies Less Than and Greater Than 50/60 Hz

What short circuit rating should you give a vacuum circuit breaker designed to operate at 50/60Hz and at 16⅔Hz? Also, what is the interruption rating of a vacuum interrupter designed for 50/60Hz service when it is installed in a vacuum circuit breaker to protect a 400Hz ac system? When the frequency is decreased the average arcing, time is increased. For example, one symmetrical half-cycle of fault current at 50Hz lasts for 10ms, but at 16⅔Hz it will increase to 30ms. A TMF contact in a 50Hz ac system operating at its full short circuit current will experience a columnar vacuum arc for 8 to 9 ms before reverting to a diffuse vacuum arc 1 to 2 ms before current zero. The same contact in a 16⅔Hz ac system will endure the columnar vacuum arc for 26–27 ms before the diffuse mode takes over just before current zero. As I have discussed in Section 3.3.3 in this volume, this columnar arc moves rapidly around the periphery of the contact. However, at longer arcing times, the columnar vacuum arc will return to its starting position more frequently. Thus, more metal vapor would be expected to remain in the contact gap at current zero. This, in turn, would result in a lower current interruption capability for a given TRV as the system ac frequency decreases. If we assume that the dielectric recovery of the vacuum interrupter after current zero varies inversely with the residual density of the metal vapor in the contact gap after current zero (see Section 4.2.3 in this volume), and if the residual metal vapor density is proportional to the arc energy (i.e., $\frac{i_{sc}^2}{f}$, where i_{sc} is the short circuit current and f is the frequency), then the peak recovery voltage $U_{R(peak)}$ is:

$$U_{R(peak)} \propto \frac{f}{i_{sc}^2} \tag{6.31}$$

Thus, for a given $U_{R(peak)}$:

$$\frac{i_{sc}(f_1)}{i_{sc}(f_2)} = \sqrt{\frac{f_2}{f_1}} \tag{6.32}$$

Figure 6.89 presents experimental data given by Slade et al. [206]. The data show that Equations (6.31) and (6.32) are somewhat conservative but can be useful for estimating a vacuum interrupter's performance at lower frequencies when its performance at 50/60 Hz is known. The interrupting performance of this TMF contact drops from 40 kA at 50Hz to 27 kA at 16⅔Hz (a decrease of 32.5%). The AMF contact structure's performance also drops at a similar rate [206].

What would be expected at frequencies greater than 50/60Hz? As I have discussed in Section 6.3.3, the interruption for a given RRRV is determined by $di_{sc}(i_{sc} = 0)/dt$, which in turn, is given by $-2\pi f\sqrt{2}i_{sc}$; see Equation (4.2) in this volume. Thus, if the frequency increases, one would expect the value of the i_{sc} that can be interrupted to decrease. Experimental data given by Smith [214] for a vacuum interrupter with TMF contacts are shown in Figure 6.90. Interestingly enough, the range between 40Hz and 100Hz seems to be an optimum frequency for fault interruption. Wang et al. [215] show for a vacuum interrupter with AMF contacts a decline in current interrupting ability as the current's frequency increases it goes from 14.7 kA at 400Hz to 7.1 kA at 800Hz.

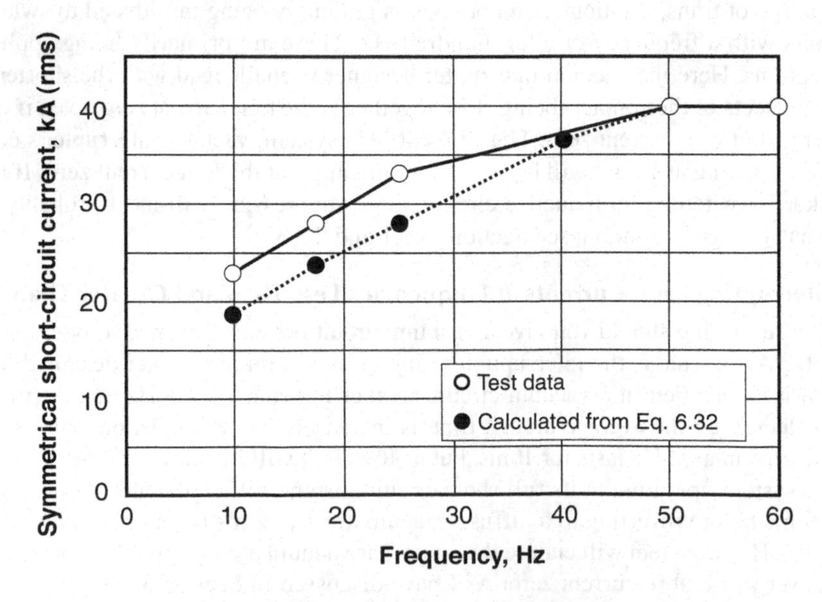

FIGURE 6.89 The reduction in the interruption ability of a vacuum interrupter as the circuit frequency is reduced [206]

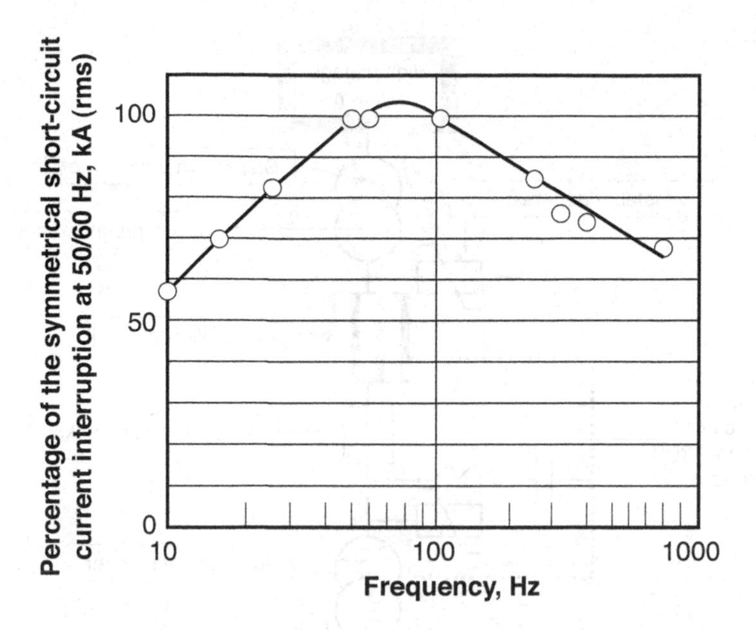

FIGURE 6.90 The interruption performance of vacuum interrupters as a function of the frequency of the electrical system [214].

6.8.8 Switching Electric Arc Furnaces (EAF)

The vacuum circuit breaker is a good choice for reliably switching and protecting an arc furnace circuit. A well-designed vacuum interrupter can handle the effects of the nonlinear and stochastic nature of the EAF load. Here the furnace current and voltage, the real and reactive power, the furnace resistance and reactance can all exhibit sudden and significant variations with time. This is especially true when the furnace is performing one of its primary functions: scrap metal melting [216, 217]. These variations can be controlled to some extent by designing the EAFs electrical system with higher rated secondary transformers. They can also be controlled with special attention being paid to the circuit components such as reactors and capacitors that are used to match the electrical current settings for the different phases of the EAFs heating cycle [218].

The EAF may be switched as much as 100 times a day. The vacuum interrupter can easily handle this duty. A vacuum circuit breaker designed for 30,000 operations will operate reliably for 300 days, or approximately 1 year before a replacement is required. It would be possible to design specialized vacuum circuit breakers to switch and protect EAF systems for more than 250,000 operations; e.g., using vacuum interrupters similar to those on top of the electrical locomotive as discussed above, together with the magnetic mechanism discussed in Section 6.5.3. When interrupting the EAF load current high overvoltages are possible as a result of voltage escalations similar to those discussed in Section 5.3 in this volume. Dangerous resonances, even in transformers with excellent insulation, may be established inside the transformers that can result in their failure (see Section 5.4.5 in this volume). Thus, it is recommended that great attention be paid to adding "heroic" surge suppression to components in an EAF electrical circuit; see for example Hesse et al. [219]. Harmonic filter banks designed for transient, dynamic, and steady-state duties (Dudley et al. [220] and Akdağ et al. [218]), RC surge networks as well as metal oxide varistor lightning arresters can be employed. Figure 6.91 gives an example [220]. The vacuum interrupter has great flexibility in being able to withstand high frequency TRVs as well as in being able to interrupt high frequency currents. It is important, however, that any surges on the feeder and/or line side of the vacuum circuit breaker be limited so that they remain below the breakers' design values. The above discussion strongly suggests that the entire EAF electrical system and not just the vacuum circuit breaker requires careful design. The

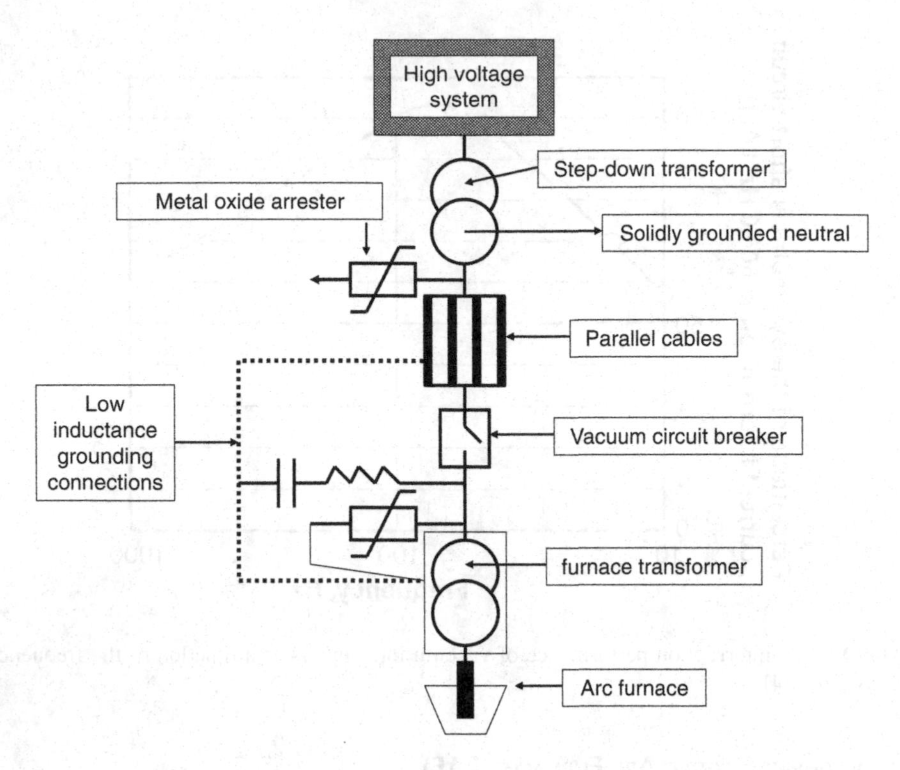

FIGURE 6.91 An example of surge protection for an arc furnace electrical circuit [220].

vacuum circuit breaker can perform the switching and protection functions with great reliability, but it will not perform satisfactorily without careful consideration given to what surge protection is required. An example of this type of systems approach can be found in reference [221].

6.9 VACUUM INTERRUPTERS IN SERIES

There has been a long history of placing vacuum interrupters in series to switch circuits with higher voltage levels than can be achieved by one vacuum interrupter alone. One early example from 1967 is a 132 kV, 15 kA circuit breaker that used seven vacuum interrupters in series [222, 223]. Westinghouse in the early 1970s used two 15 kV vacuum interrupters in series in their commercially successful 38 kV vacuum recloser [147]. The General Electric Corporation (GE) in the late 1960s envisioned the use of series vacuum interrupters in circuit breakers to replace the ubiquitous bulk oil circuit breakers for system voltages 121 kV to 800 kV [224, 225]. Shores et al. [226] stated that the first commercial transmission vacuum circuit breaker will be rated at 242kV/40kA/3000A and would be a dead tank design. The unit would use 40kV/40kA vacuum interrupters in series. Unfortunately, this development was way ahead of its time and while a development prototype was produced, no commercial products were ever sold. Eventually, GE stopped all development work on this product.

As discussed in Section 5.2.2 in this volume, series vacuum interrupters have long been used in circuits up to 242 kV as load break isolation switches. Figure 6.92 shows such a 242kV load break switch with an air-break gap for complete circuit isolation. In the late 1980s, I was involved in the development of an upgrade and retrofit kit for 135kV, 20kA, and 1250A bulk oil breakers using series vacuum interrupters. This very successful development used four 38kV, 40kA vacuum interrupters in series. When installed in the bulk oil breaker's tank it was fully certified to the IEEE C35 standard for 121–145 kV, 40kA, and 2000A interruption, as well as for line dropping and for C2

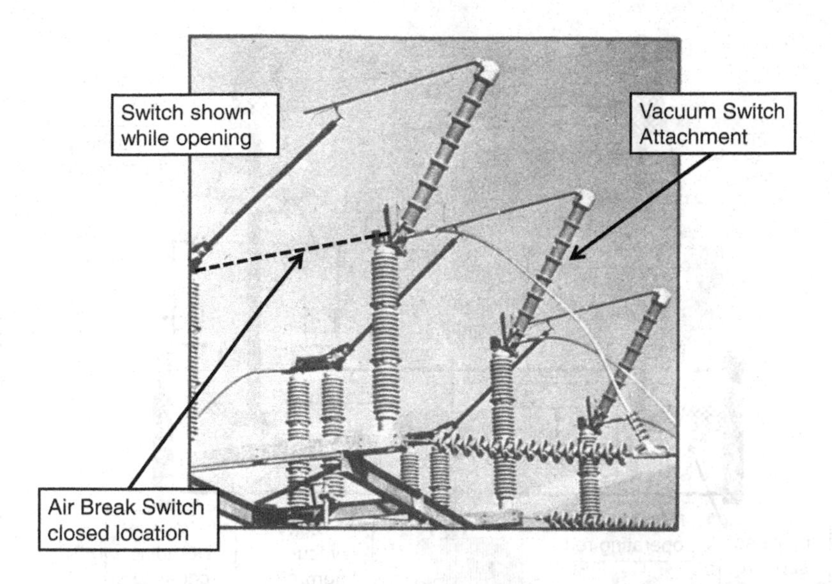

FIGURE 6.92 242 kV load break switch with air isolation using series vacuum interrupters (courtesy Turner Electric Corporation).

capacitor switching [227]. A diagram of this development is shown in Figure 6.93. Unfortunately, Westinghouse T&D business was sold to ABB just as this development was being introduced to U.S. utilities. As ABB was more interested in selling their new SF_6 puffer breakers than this VI retrofit in the existing housing of the bulk oil breakers, the design died a quiet death.

The use of two 38 kV vacuum interrupters in series as a 72.5 kV circuit breaker has been commercially available since the late 1990s. Three 40.5 kV vacuum interrupters in series have been used in the development of a 125kV vacuum circuit breaker [228, 229]. There have also been some

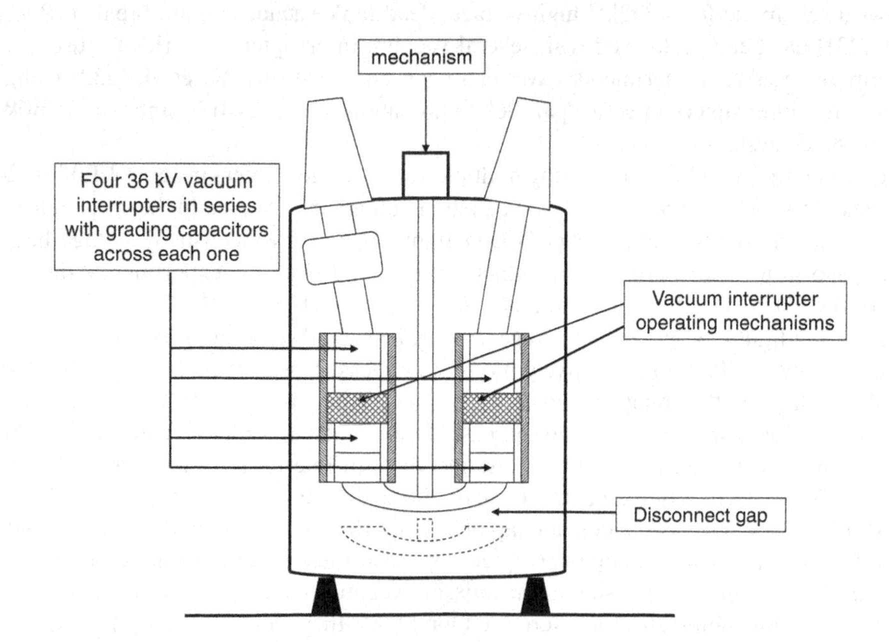

FIGURE 6.93 The vacuum interrupter retrofit and performance upgrade for an outdoor, bulk oil, transmission circuit breaker [227].

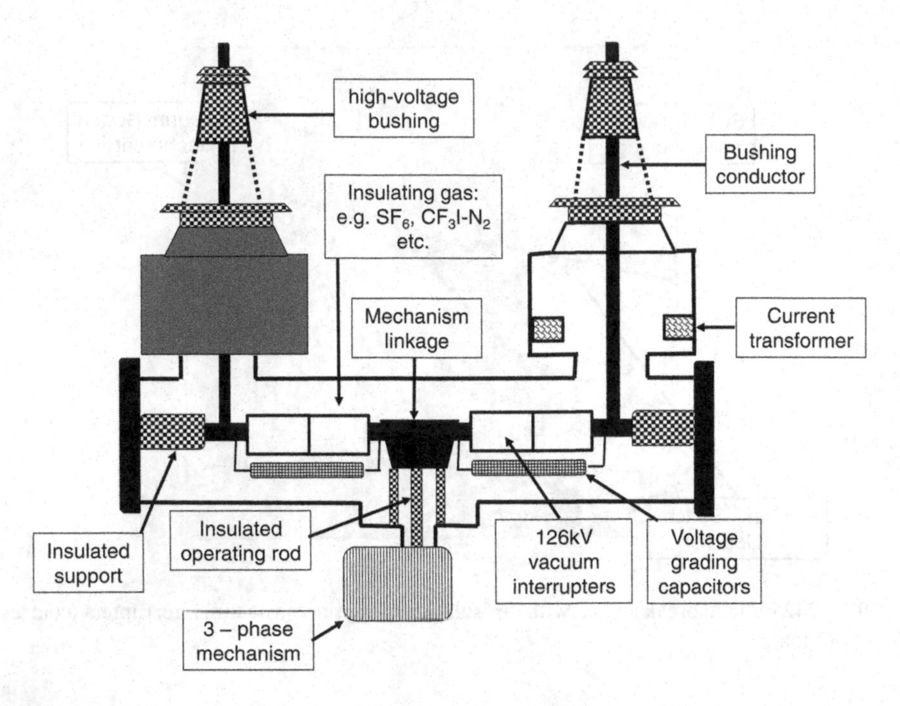

FIGURE 6.94 Schematic of one pole of a 245kV vacuum circuit breaker with two 126 kV vacuum interrupters in series.

concepts using two 72.5kV vacuum interrupters for this service. Okhi [14] showed a design concept for two 126 kV vacuum interrupters in series for a 204kV dead tank vacuum circuit breaker. Falkingham et al. [230] showed a similar concept for 242 kV transmission circuit breakers. Figure 6.94 shows a schematic for a 242kV high voltage, dead tank vacuum circuit breaker. Research by Lui et al. [231] has been performed with several vacuum interrupters in series for use in a 750kV vacuum circuit breaker. Experiments have also been conducted by Yu et al. [232] using 6 × 6, 40.5kV vacuum interrupters in series/parallel combination to successfully interrupt a 60kA short circuit at a 363kV system voltage.

Why consider the complication of using multiple vacuum interrupters in series? Table 6.24 shows that certification standards require that as the system voltage increases, a higher peak TRV voltage, a higher withstand voltage and a higher BIL or lightning impulse withstand voltage be satisfied. Table 6.24 also shows the approximate contact gap required from the experimental data shown in Figure 1.18 (i.e., $U = 38d^{0.64}$kV and also that calculated using $U = 58d^{0.58}$ kV) for a single vacuum gap to satisfy the lightning withstand voltage. As can be seen the contact gap required for distribution voltage (12kV to 40.5kV) is relatively modest; i.e., less than 20mm. The vacuum interrupters in this voltage class are thus relatively compact. As the system voltage increases the contact gap for a single vacuum interrupter will have to increase dramatically in order to withstand the lightning impulse voltage (BIL) level. For example, a single vacuum interrupter for use in a 123 kV circuit with a BIL of 550 kV will require a contact gap of at least 70mm and one for use in a 145 kV circuit with a BIL of 660 kV will need a contact gap of at least 100mm. Along with the increased contact gap other dimensions of the vacuum interrupters for use at these system voltages will also increase. I will discuss the design of these single transmission vacuum interrupters in Section 6.10. If three 40.5 kV vacuum interrupters in series were used for 123kV they would have an increased high voltage withstand as a result of the "performance factor" presented in Equation (1.92), Section 1.4.2 in this volume. Assuming that each of the 40.5 kV vacuum interrupters has a contact gap of 20mm and

TABLE 6.24
Example of the High Voltage Withstand and the BIL Levels as the System Voltage Increases

System Voltage	TRV Peak Voltage (3φ ungrd.), kV	1 minute withstand, kV	BIL, kV	Contact gap (mm) needed for BIL voltage from	
				$U = 58d^{0.58}$ kV	$U = 38d^{0.64}$ kV
12	17	40	75	1.6	3
24	33	71	125	3.8	6
36/40.5	50/56	99/134	150/200	7.5	8/13
72.5	100	175	350	22	50
123	170	280	550	48	65
145	200	390	650	65	84
245	340	650	900	113	140
360	500	755	1175	180	211
420	580	860	1425	250	285
550	760	1130	1550	290	325

the single break 123 kV vacuum interrupter has a contact gap of 60mm, then the performance factor for three vacuum interrupters in series is;

$$P_3^* = 3^{(1-\eta)} \tag{6.33}$$

If $\eta = 0.64$ then $P^*_3 = 1.42$, that means that the three vacuum interrupters in series would have a voltage withstand ability 1.42 times a single vacuum interrupter with the same total contact gap. This also assumes that there is equal voltage division over each of the three vacuum interrupters. The capacitance of a vacuum interrupter of an open vacuum interrupter is:

$$C = \frac{\epsilon_0 A}{d} \tag{6.34}$$

where A is the contact's area, d is the contact gap and ϵ_0 is the permittivity of free space (8.9×10^{-12} $F.m^{-1}$). The open gap capacitance of the 40.5 kV vacuum interrupter with contact diameters 60mm and a contact gap of 20mm is approximately 40pF. Using the analysis in Section 5.22 in this volume for a peak lightning withstand pulse of 550 kV (U_a peak), the vacuum interrupter closest to ground will see a voltage of

$$U_a \times \frac{sinh\left(\frac{a}{3}\right)}{sinh(a)} \cong 0.26U_a \tag{6.35}$$

or 143 kV and the one closest to the line will see:

$$U_a \times \left[1 - \frac{sinh\left(\frac{2a}{3}\right)}{sinh(a)}\right] \cong 0.43U_a \tag{6.36}$$

or 236.5 kV and the middle vacuum interrupter will see:

$$0.31U_a \tag{6.37}$$

or 170.5 kV. So, for this division of voltage across the three vacuum interrupters the performance factor will be somewhat less than given in Equation (6.33). Looking at Figure 6.95 the vacuum interrupter closest to the line would easily withstand 236.5 kV. In spite of this, Shioiri et al. [233] and Giere et al. [234] do show that two vacuum interrupters in series offer greater high voltage withstand and a lower probability of breakdown than one well-designed single gap vacuum interrupter with the same total contact gap in spite of the inequality of the voltage sharing across each vacuum interrupter. Liao et al. [235] show in Figure 6.96 the effect of two and three vacuum interrupters in series. Indeed, Wang et al. [155] show that two vacuum interrupters in series can successfully pass the C2 level, 40.5kV capacitor switching certification. The fault current interruption performance of two vacuum interrupters in series is also improved over one vacuum interrupter at half the circuit voltage [234, 235]. Interestingly enough, an NSDD in one vacuum interrupter does not necessarily result in a breakdown through both vacuum interrupters [236–241]. The other advantage of using series vacuum interrupters is that the vacuum interrupters are more compact, but the disadvantage is that a more complex mechanism is required.

The lightning impulse voltage test is performed over open high vacuum contact gaps. When interrupting short circuit current the discussion in Section 4.2.3 in this volume shows that a post arc current (PAC) can flow during the rise of the transient recovery voltage (TRV). There appears to be no correlation between the value or the duration of the PAC on a single, well designed, vacuum interrupter's ability to interrupt these currents. However, when vacuum interrupters in series interrupt short circuit currents the PAC in each will affect the capacitance of each interrupter and thus affect the voltage division across each one. In spite of this, Liao et al. [238] and Sun et al. [239] show that for a double break circuit breaker the breakdown of one vacuum interrupter after current zero does not necessarily result in failure to interrupt the circuit. The second vacuum interrupter may be able to withstand the total TRV for the time it takes the first vacuum interrupter to recover. Placing

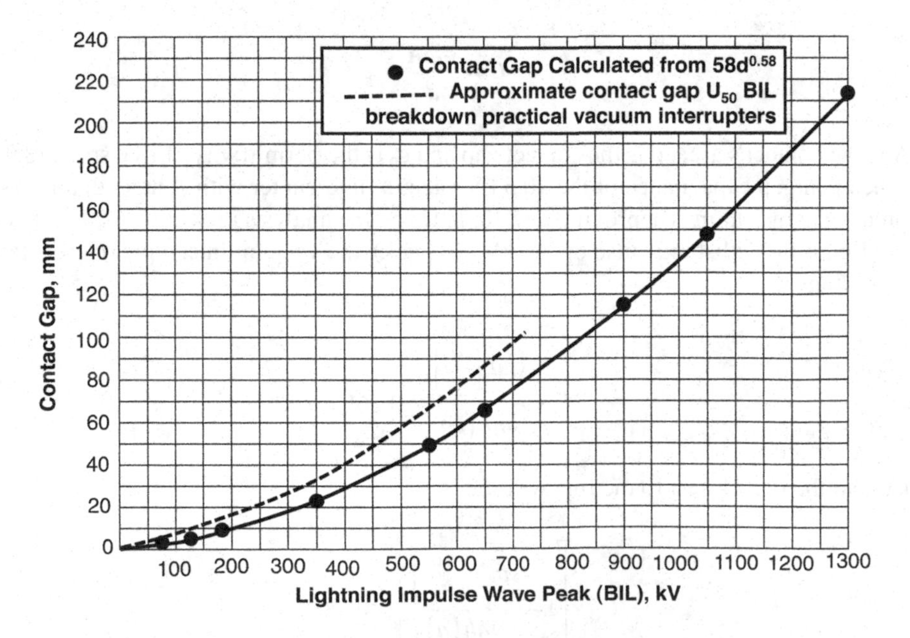

FIGURE 6.95 Contact gap to withstand increasing lightning impulse voltages calculated from $U = 58d^{0.58}$ (Figure 1.18) together with the expected contact gap for practical vacuum interrupters.

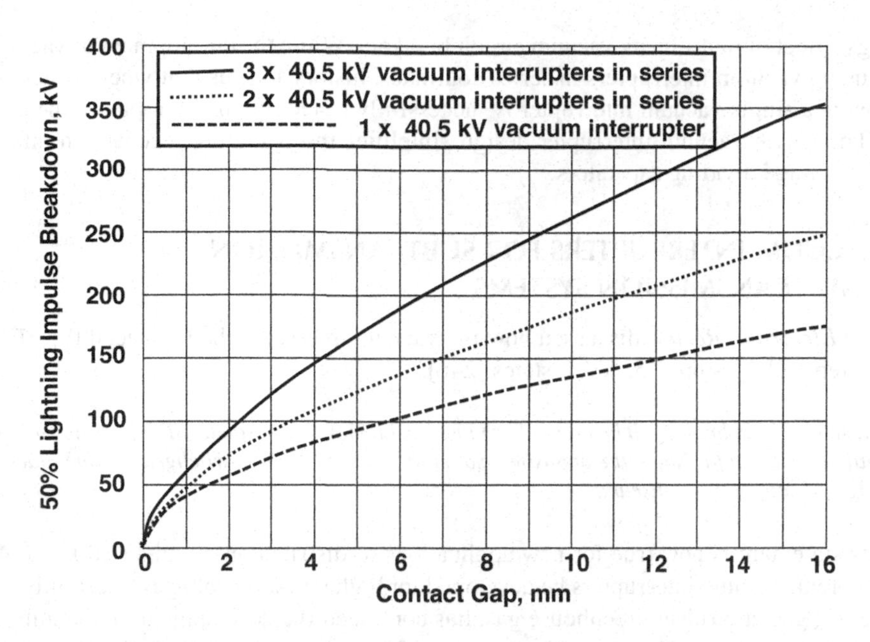

FIGURE 6.96 The 50% Lightning impulse withstand voltage as a function of contact gap for a single vacuum interrupter, two vacuum interrupters in series and three vacuum interrupters in series [229].

a capacitor with a much higher capacitance than the open contact gap capacitance across each vacuum interrupter will force a better distribution of voltage across each vacuum interrupter and will counter the effects of variable PACs: See, for example, Betz et al. [237]. Further investigations by Huang [242], Ge et al. [243] and Liao et al. [244] for double and triple break vacuum circuit breakers show that to overcome the effects of variable PACs the parallel capacitors shown in Figure 6.97 are necessary. For a 126kV vacuum circuit breaker with three series vacuum interrupters Huang et al. [242] show that the capacitors should have a value above 55pF. They also show that above 1000pF no further improvement in the voltage distribution is evident. Each of the above researchers state that it is important for each of the series vacuum interrupters to open synchronously. This does place an extra burden on the engineers developing such a vacuum circuit breaker. Ge et al. [245] review

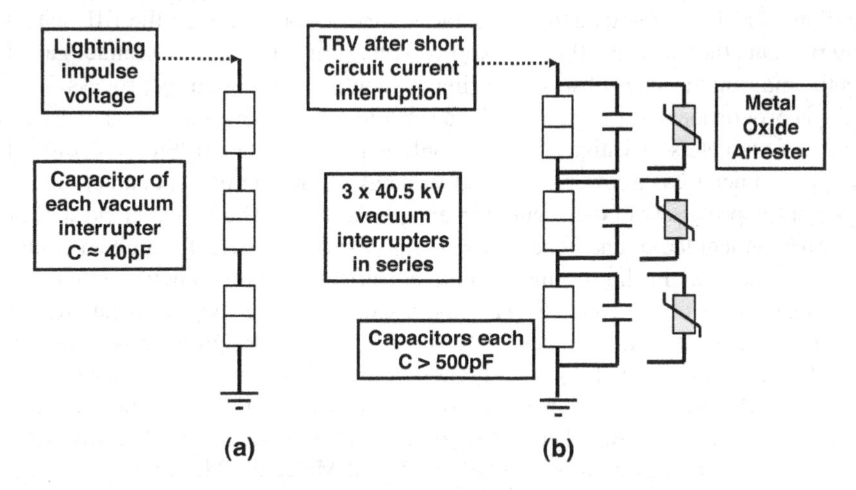

FIGURE 6.97 Series vacuum interrupters without parallel voltage grading capacitors and series vacuum interrupters with voltage grading capacitors and optional parallel metal oxide arresters.

the synergy effect of multi-break vacuum circuit breakers. While the development of vacuum circuit breakers using vacuum interrupters in series continues, history tells us that when a vacuum circuit breaker using a single vacuum interrupter is successfully tested, it quickly replaces the prior series designs. The single vacuum interrupter design simplifies the operating mechanism and does not require the voltage dividing capacitors.

6.10 VACUUM INTERRUPTERS FOR SUBTRANSMISSION AND TRANSMISSION SYSTEMS

In 2011 the *Electrical Review* discussed how the vacuum interrupter had replaced the SF_6 technology for system voltages up to 52 kV. It states [246]:

> *SF_6 is yesterday's technology: it has served its purpose, but now it's obsolete. SF_6 offers no technical or financial benefit – in fact quite the opposite – so let us confine SF_6 MV switchgear to one place where it belongs, of course, in a museum!*

While this statement is now true for new applications to distribution circuits (3.6 kV to 40.5 kV), can single break vacuum interrupters be developed for higher system voltages? Certainly, the desire to replace SF_6 gas, a virulent greenhouse gas, has continued the development of vacuum interrupters for subtransmission and transmission systems. Beginning in the late 1970s development of a 72.5 kV single break vacuum interrupter showed considerable promise: See, for example, Voshall et al. [4]. Now all major vacuum interrupter manufacturers have successfully certified a vacuum interrupter with this voltage rating. As I have discussed in Section 6.9 and shown in Figure 6.95 a single vacuum gap that can withstand the lightning impulse withstand voltage (BIL) has to increase dramatically. Table 6.24 shows that a 72.5kV vacuum interrupter with a BIL of 350 kV requires a 30mm to 40mm contact gap and a 123kV vacuum interrupter requires a 60mm to 70mm contact gap. Also, emission currents from the cathode contact will gain considerable energy from the high voltages shown in Table 6.24. I have discussed the effect of this energy being deposited below the surface of the anode contact in Section 1.3.3 in this volume. Thus, during the design process of higher voltage, single break, vacuum interrupters special attention must be made in the design of all the internal metal surfaces: i.e., the contacts, the shields and the terminals (see Section 1.3.2 in this volume) [10, 247, 248]. Zhang et al. [249, 250] using butt contacts in the transmission style vacuum interrupter shown in Figure 6.98 have determine the contact gaps needed to withstand high BIL voltages. In their excellent papers they have explored the effects of contact gap, the radius of the contact edges, the diameter of the contact and the contact's surface roughness on the BIL withstand level. Figure 6.99 presents the average, 50%, breakdown voltage as a function of contact gap. This graph also shows the maximum values observed during the contact's conditioning process. The withstand value for U_{50} is proportional to the gap $d^{0.64}$. The variation of the withstand voltage, as a function of the three other parameters investigated, shows only small variations in Figure 6.100. The contact with the larger diameter has a lower withstand value as a function of the contact gap. This shows the influence of the proximity of the vacuum interrupter's shield. The closer shield will increase the geometric enhancement factor and hence lower the breakdown voltage for a given contact gap: see Sections 1.3.2.2 and 1.4.1 in this volume. The major effect on the geometric enhancement factor is from the edge radius of the contacts. The macroscopic surface roughness has little significant effect. Of course, the high voltage vacuum interrupter's internal design is greatly facilitated by the use of Finite Element Analysis (FEA) to study the electric fields internal and external to the vacuum interrupter. Using this software, it is possible to determine the optimum shape for a given contact diameter and the effect of the shields and the contact gap: see Section 1.4.1 in this volume. Thus, by using this computer aided design Kato et al. [247] and Ma et al. [248] obtain a good estimate of their high voltage vacuum interrupter's high voltage withstand ability. It is unfortunate that Zhang et al. [249, 250] did not include such an analysis in their papers.

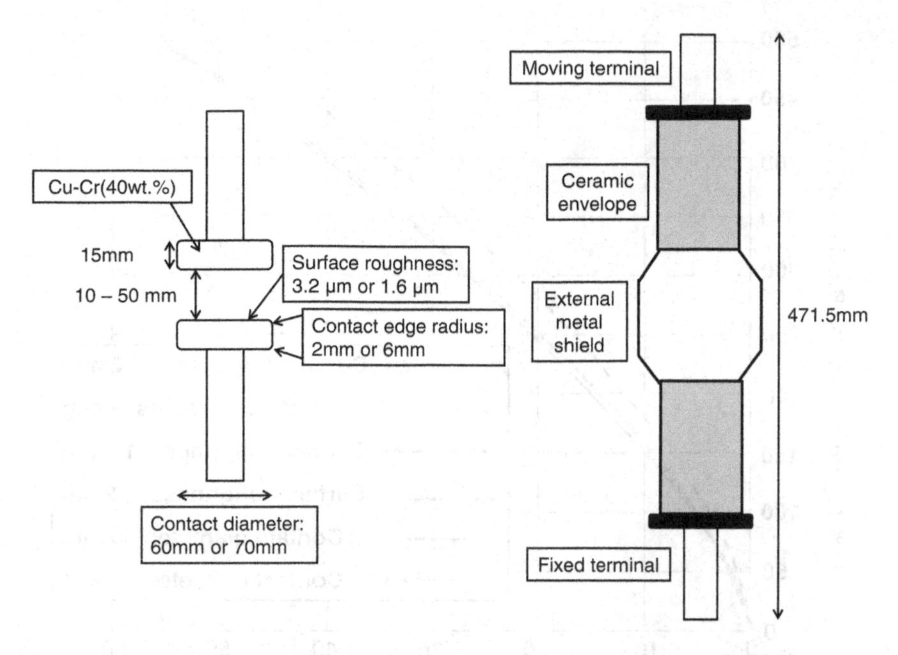

FIGURE 6.98 Schematic of the contact parameters and an outline of the vacuum enclosure for the lightning impulse withstand experiments as a function of contact gap [250].

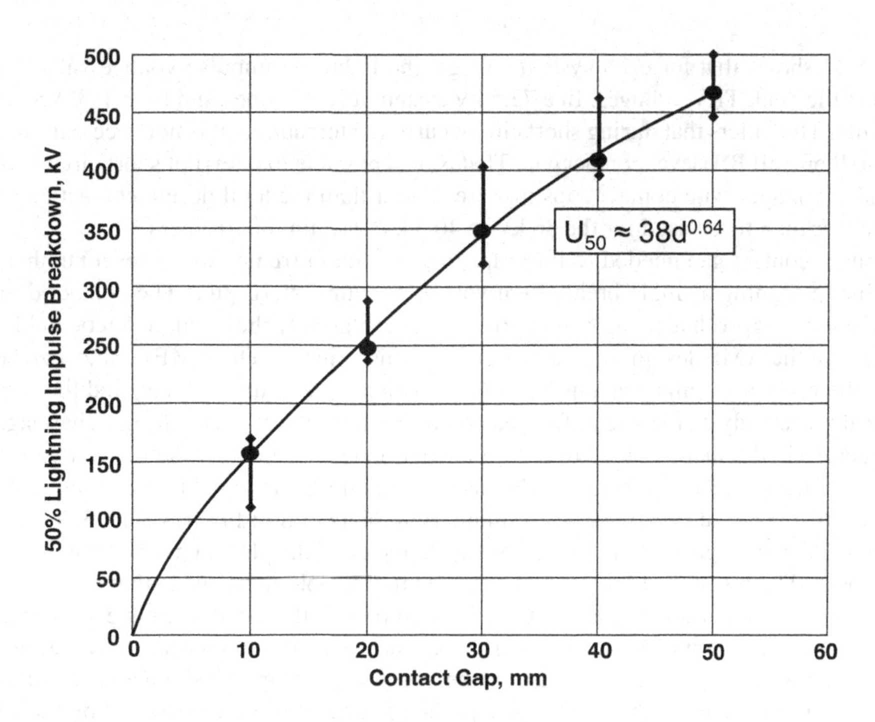

FIGURE 6.99 The average U_{50} as a function of contact gap ● with the range of maximum ◆——◆ during the contacts' conditioning process [250].

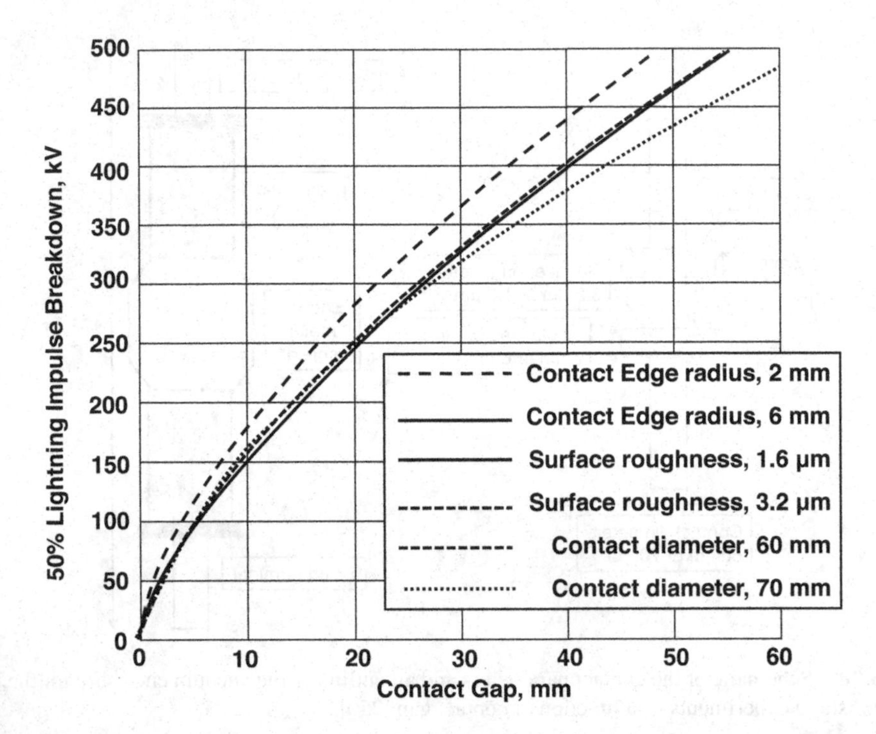

FIGURE 6.100 The effect of contact diameter, the contact edge radius and the contact surface roughness on U_{50} as a function of contact gap [250].

Table 6.24 shows that for each system voltage, the lightning impulse voltage values are much higher than the peak TRV voltages. In a 72.5 kV system it is 3.5 times and for a 145kV system 3.25 times higher. This infers that during short circuit current interruption, it is not necessary to open the contacts to their full BIL level contact gap. That is, it is possible to interrupt short circuit currents at transmission voltages with contact gaps that are shorter than the total design contact gap, but they are still longer than those used for the 36 kV to 40.5 kV, vacuum interrupters.

The longer contact gap needed to interrupt short circuit currents introduces a number of challenges when designing a single break, high voltage vacuum interrupter. The contact design used for single break, high voltage, vacuum interrupters is exclusively the axial magnetic field style (for a discussion of the AMF design, see Section 3.3.4 in this volume). The AMF contact can be shaped to reduce the geometric enhancement factor (see Section 1.3.2 in this volume) and the contact surfaces remain relatively stable even after interrupting fault currents. The design challenge for the AMF structure is the maintenance of a high enough magnetic field as the contact gap increases [251, 252]. As the contact gap increases the intergap, axial magnetic field for a given AMF design decreases. This has been illustrated for a simple Helmholtz coil in Figures 3.49, 3.50, and 3.52 in this volume. As the magnetic field decreases the diameter of the plasma plume from each cathode spot increases. The result, as Figures 2.89 and 2.90 in this volume show, is that the overlap of the plasma plumes occurs at a lower current. In order to reduce the effect of this earlier overlapping the surface contact area will have to increase: i.e., the contact diameter has to increase. This, in turn, will necessitate an overall increase in the vacuum interrupter's diameter. As discussed in Section 2.7.2 in this volume, the overlapping plasma plumes will be constricted by the azimuthal magnetic field flowing in the plasma. If this persists for too long an anode spot will develop and this will affect the vacuum interrupter's high current interruption performance. The contact gap when interrupting the short circuit currents still has to be large enough to withstand the peak of the TRV after current zero. This means that the opening speed of single break, high voltage, vacuum interrupters has to be increased above the 1 ms^{-1} to 2 ms^{-1} common for vacuum interrupters rated less than or equal to 40.5kV [8].

The development high voltage vacuum interrupters has had a long history. Liu et al. [253] even showed a prototype 242 kV single break vacuum interrupter in 2007. The development of the AMF contact for these designs and their effect on the formation and structure of the vacuum arc during the current interruption process has also been the subject of continuing research. Xui et al. [254] give an example of how the magnetic field decreases as the contact gap increases in Figure 6.101 for a 2-arm AMF structure. Both the two-layer, three-arm AMF structure (Figure 3.51 in this volume) and the horseshoe AMF structure (Figure 3.48 in this volume) show promise for maintaining a high enough magnetic field at the higher contact gaps to interrupt short circuit currents up to 40 kA [255]. For successful short circuit current interruption, it is necessary to reduce the time of the vacuum arc's constricted mode to ensure a diffuse structure at the ac current zero. Zhang et al. [256] in an important paper using the two-layer, three-arm AMF structure, show a single break, vacuum interrupter's contact travel when successfully interrupting a 40kA short circuit current in a 126kV circuit. The Cu–Cr(50wt.%) contacts have a diameter of 100mm. Figure 6.102 shows the travel curve for the opening contact. Successful interruptions are observed when the arcing times range from 6.4 ms to 14.6 ms. That is for contact gaps ranging from 22.5mm to 41mm. The maximum magnetic field at 10.4 ms is 484mT and at 17.9 ms it is 384mT. Both of these values are greater than the B_{crit} value given in Equation 2.43 for a 40kA current. This example shows that the contact gap for successful interruption of the 126kV circuit is significantly less than the 60mm needed to withstand the BIL voltage. Li et al. [255] using the horseshoe AMF contact structure open a 102mm diameter Cu–Cr(50wt.%) contact with a speed of 2.4 ms^{-1} to a total contact gap of 60mm. They show for a total arc time of 10 ms a 40kA arc is totally diffuse at the current zero, Again the contact gap at the current zero is about 24mm. They also characterize the vacuum arc appearance for currents ranging from 40kA to 50kA and for contact gaps ranging from 1mm to 60mm. Li et al. [255] show that a minimum arcing time of 7 ms is required for successful interruption of the 126kV circuit.

Single break vacuum interrupters have been developed and available in Japan since the mid-1970s [5, 11, 257]. The original glass envelope 72kV design achieved its low field objectives by being very large; 736mm long and 204mm in diameter. The development of this vacuum interrupter

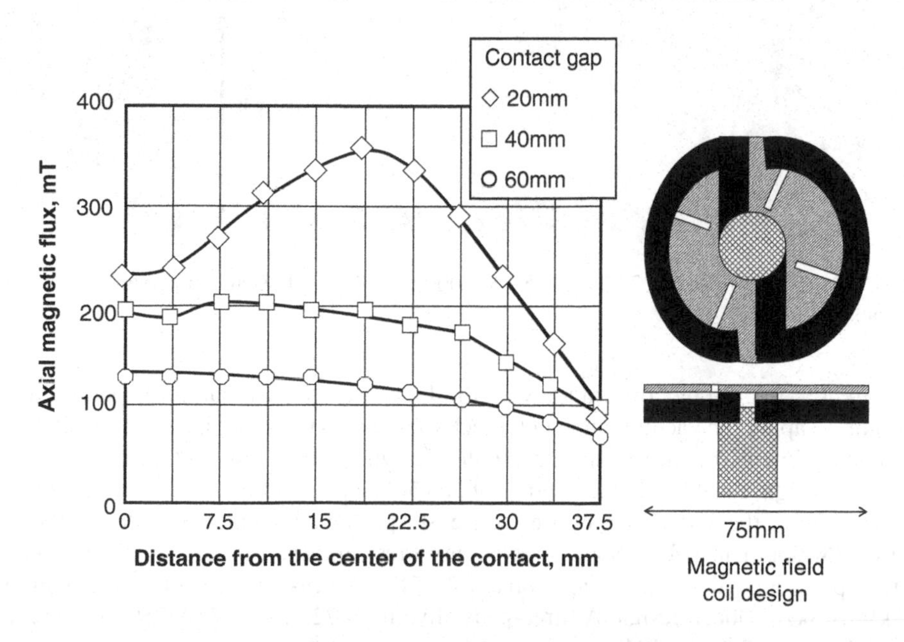

FIGURE 6.101 The drop on mid-gap magnetic flux as the contact gap increases [254].

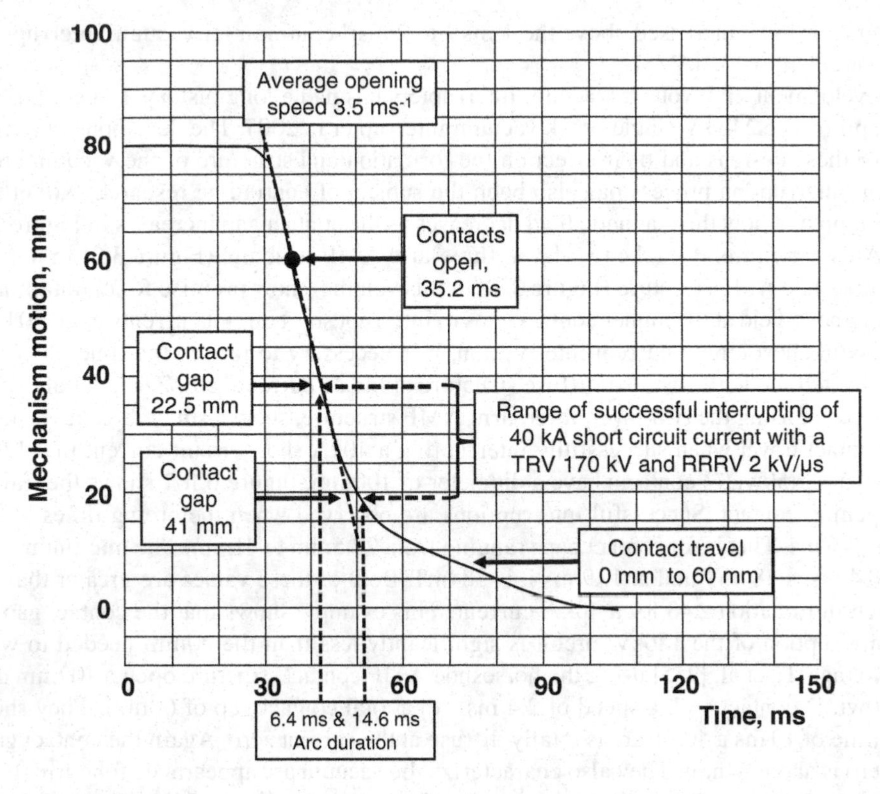

FIGURE 6.102 The mechanism, travel for a 126kV single break vacuum interrupter showring the contact gap and vacuum arc duration for successful interruption of 40 kA asymmetric current [256].

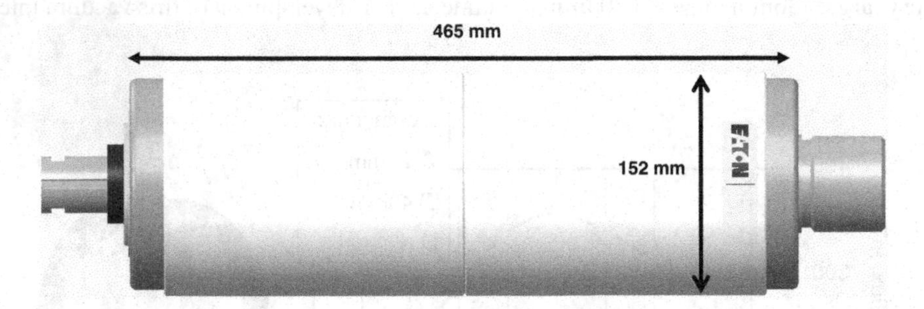

FIGURE 6.103 An example of a 72.5 kV/31.5 kA/ 2000A single break vacuum interrupter (Courtesy Eaton Corporation).

has been greatly assisted by cooperation with the Japanese utilities that have been willing to pay the premium to apply the vacuum interrupter technology into their 72.5kV systems [258]. My own team also experimented with vacuum interrupter designs during this time period and developed a very large design capable of interrupting about 50kA [4]. Unfortunately, in the early 1980s, the development of the much more cost-effective SF_6 puffer interrupters resulted in the cancellation of this development effort. Now in the twenty-first century, all major manufacturers of vacuum interrupters have developed single break, 72.5kV designs. Figure 6.103 shows an example for 72.5kV/31.5kA/2000A design. A three-phase live tank 72.5kV/31.5kA/2500A vacuum circuit breaker is shown in Figure 6.104.

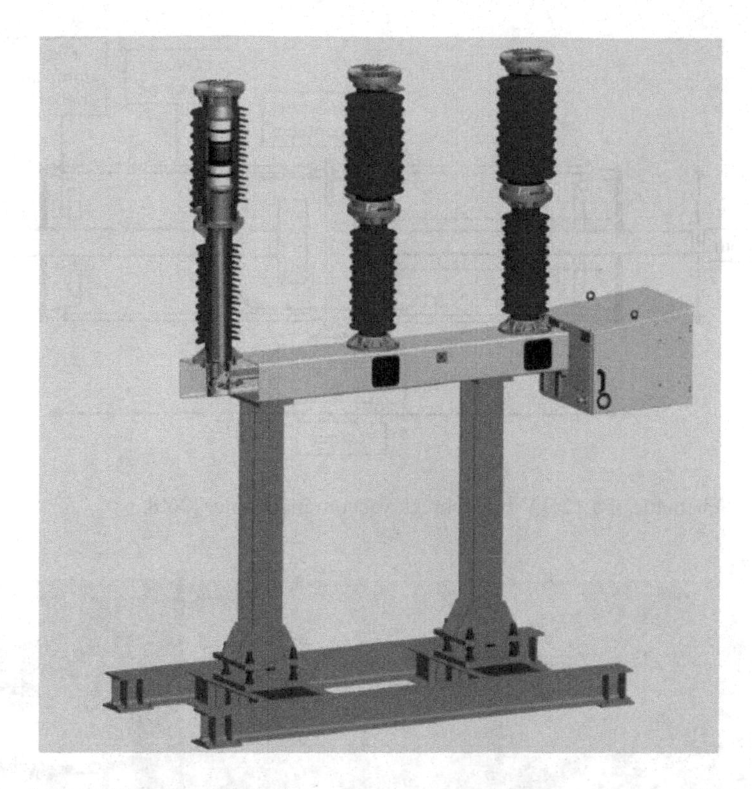

FIGURE 6.104 A 72.5 kV/31.5 kA/200A live tank vacuum circuit breaker (Courtesy Siemens AG).

Continued development of single break, vacuum interrupters has resulted in successful designs for system voltages from 121kV to 170kV. One experimental design shown in 1987 is extremely large [259]. In 2002 Saitou et al. [7] presented a glass envelope design 1000mm long with a diameter of 257mm. Yao et al. [260] presented an example of a vacuum interrupter that satisfied the IEC 62271-100 certification tests for 126kV/40kA/2500A. A schematic of their vacuum interrupter is shown in Figure 6.105. The contact diameter is 100mm which is larger than the 60mm diameter contact used for a 38kV vacuum interrupter capable of interrupting the same current. The open contact gap is 60mm, which is close to what would be expected from Figure 6.95. Again, the larger contacts and the large contact gap results in a large vacuum interrupter. It is interesting to note the bellows at the fixed terminal end. I assume that the breaker design incorporates some kind of shock absorber that cushions the fixed terminal during the closing operation. An example of a live tank 145kV vacuum circuit breaker is shown in Figure 6.106 and a dead tank 145kV switchgear in Figure 6.107. Ryi et al. [12] have demonstrated a single break vacuum interrupter with three shields for 170 kV. It is capable of interrupting a 50kA short circuit current with a peak TRV of 254kV. Its minimum and maximum arcing times are 7.1 ms and 16.8 ms. It also performs well for a short line fault of 42.2kA. Here the minimum and maximum arcing times are 8.4 ms and 16.3 ms. The 2-layer, 3-arm AMF contact design is 120mm in diameter and the contact gap to withstand the BIL of 750 kV would have to be over 110mm. Wang et al. [10] show a 242kV single break design that has a length of 1300mm and a diameter of 269mm. The contacts for 40kA interruption are 140mm and the contact gap is 80mm.

A very comprehensive review by Smeets et al. [261] of the transmission vacuum circuit breaker has been published. It delineates the concerns and hesitations in their application compared to the equivalent SF_6 designs. It shows that the transmission vacuum circuit breaker for the system voltage range 72.5 kV to 145 kV is now a viable product and will be increasingly installed in the future. Vacuum circuit breakers using one 72.5 kV vacuum interrupter have been installed in Japan beginning in the late 1970s. An approximate total of 8300 units 52 kV and above have been installed

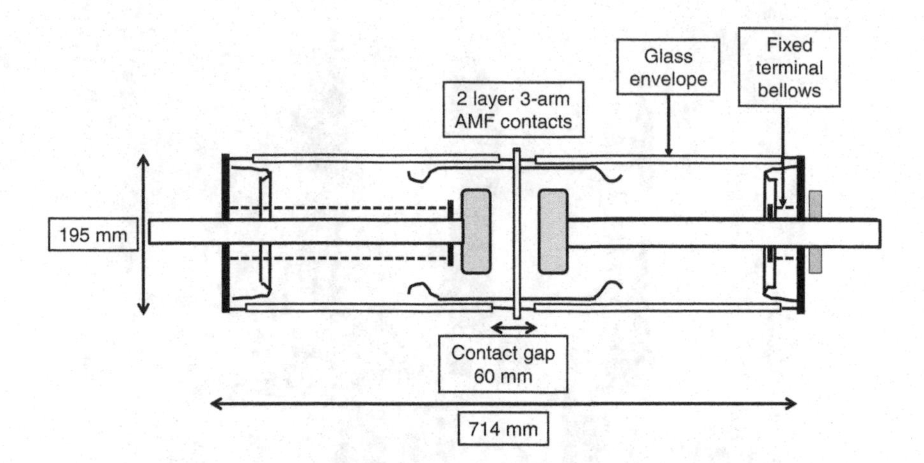

FIGURE 6.105 Schematic of a 126kV single break vacuum interrupter [260].

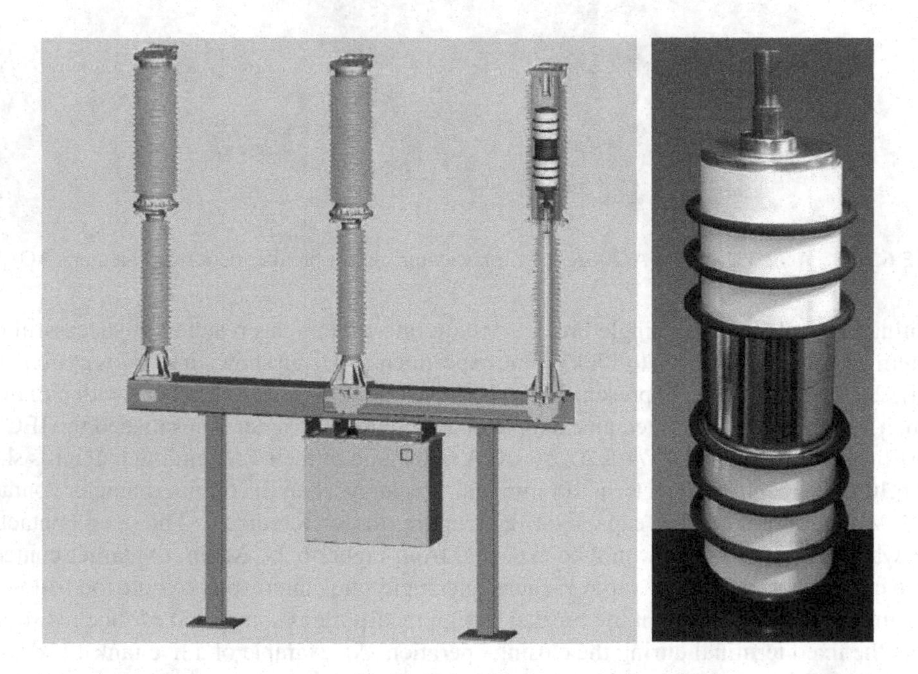

FIGURE 6.106 A 145 kV/40 kA/3150A live tank vacuum circuit breaker (Courtesy Siemens AG).

by five manufacturers [261]. Single break, vacuum interrupters for 121kV to 145kV transmission circuit breakers are increasingly being installed in China and Japan. 145kV vacuum circuit breakers are now being offered in Europe. The development of the high voltage vacuum circuit breaker is currently where the development of 36kV to 40.5kV vacuum circuit breaker was in the 1980s. At present there is a cost premium over equivalent SF_6 designs, but as vacuum interrupter development continues the cost will eventually equalize and eventually become cost competitive. Also, as the concern regarding SF_6 as a greenhouse gas continues the vacuum circuit breaker will offer a viable alternative. The development of single break, high voltage vacuum interrupters continues, Figure 6.108 shows prototype 170 kV and 242 kV single break vacuum interrupters. I believe that in the next 10 to 15 years cost effective designs will become increasingly available and electrical utilities who are traditionally rather conservative will increasingly apply them in their systems.

FIGURE 6.107 A gas insulated, dead tank vacuum switchgear 145 kV/40 kA/ 3150A (Courtesy Siemens AG).

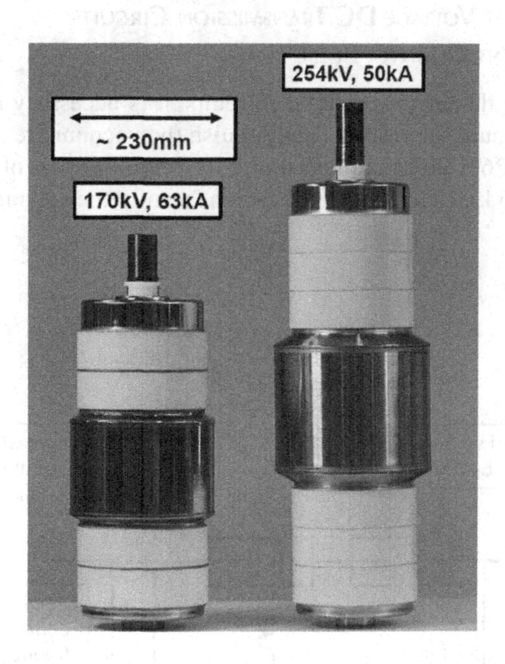

FIGURE 6.108 Prototype 170 kV and 242 kV single break vacuum interrupters (Courtesy Siemens AG).

6.11 SWITCHING DC CIRCUITS USING VACUUM INTERRUPTERS

6.11.1 DC Interruption Using the Natural Vacuum Arc Instability

The vacuum interrupter without external support is really an ac current switching device. Where there is a natural current zero, the vacuum interrupter has a well-proven capability to reliably interrupt that current and to withstand the resulting TRV. In saying this, vacuum interrupters have been used for many years to switch low dc currents (~10A) in circuits with a system voltage of a few kV

[262]. These vacuum interrupters use the instability phenomenon of low current vacuum arcs discussed in Section 1.3.5, Figure 1.84 in this volume. Here it can be seen that a 10A arc between W contacts would self-extinguish in about 1ms. Once this arc goes out the contact gap will be able to withstand recovery voltages up to tens of kilovolts.

6.11.2 DC Current Interruption Using an External Magnetic Field Pulse

A high-current dc switch using a vacuum interrupter has been developed which uses the effect of a pulsed magnetic field transverse to a diffuse vacuum arc, as discussed in Section 2.6.1 in this volume. When a high magnetic field pulse is applied transverse to the diffuse vacuum arc it is possible to develop a voltage of several kilovolts across the open contacts of a vacuum interrupter. This voltage is high enough to divert the dc current into a parallel capacitor/ZnO varistor network and thus create a current zero in the vacuum interrupter [263]. Figure 6.109 shows such an arrangement. A successful metallic return transfer breaker (MRTB) has been developed using this technique. This MRTB interrupts between 200A and 2160A dc against a recovery voltage of 80kV [264]. In spite of extensive research in the 1980s by my Westinghouse R&D team, we were never able to interrupt dc currents greater than about 15kA using this technique. As far as I know, no further development of this technique has been undertaken, although occasional papers are presented on its design; see for example, Bartosik et al. [265] and Alferov et al. [266, 267].

6.11.3 Switching High Voltage DC Transmission Circuits Using a Current Counter Pulse

In order to interrupt high dc currents in HVdc circuits, it is necessary to create an artificial current zero to allow the vacuum interrupter to extinguish the vacuum arc and to withstand the TRV. Greenwood et al. [268, 269] demonstrated the first successful use of a vacuum interrupter to switch dc currents up to 5 kA. They created a current zero in the vacuum interrupter by injecting

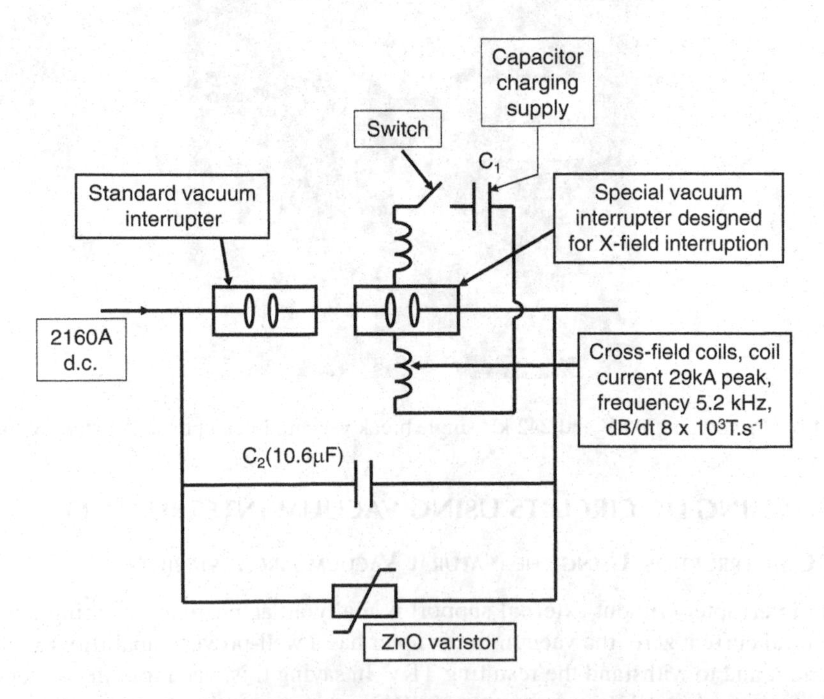

FIGURE 6.109 The metallic return transfer switch, switching 2000 A in an 80kV circuit [263].

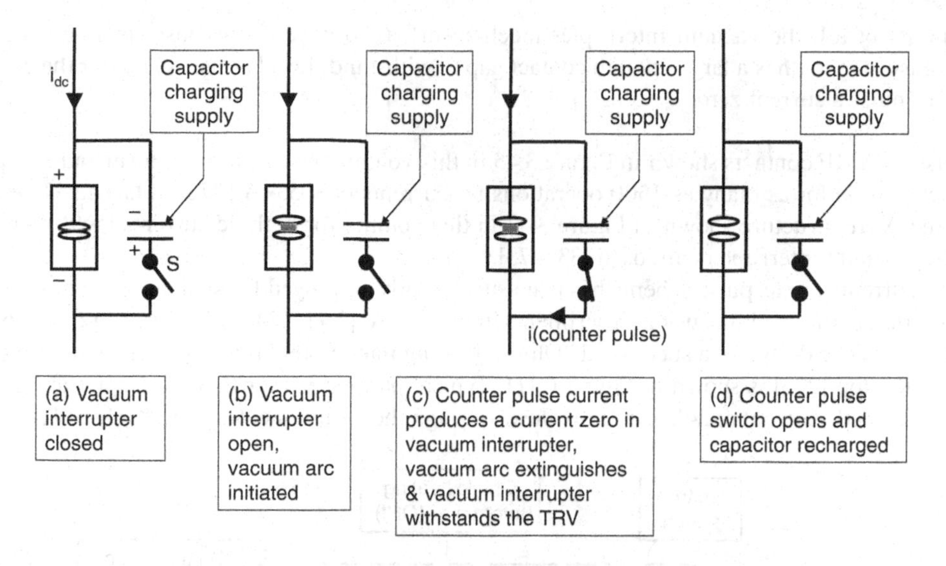

FIGURE 6.110 A diagram of a HVdc circuit breaker using a vacuum interrupter.

an opposing current from a parallel, pre-charged capacitor. Figure 6.110 illustrates the opening sequence for the successful interruption of the dc current. When the dc current has to be interrupted, the vacuum interrupter opens and the vacuum arc is established. Once the contacts reach their fully open position, the "switch" S is closed, which initiates a high-frequency counter current through the vacuum interrupter. Another way of looking at this is that the dc current is commutated from the vacuum interrupter into the capacitance C. Either way, this creates a current zero in the vacuum interrupter, which allows the vacuum arc to extinguish and the dc circuit to be interrupted.

The initial dc circuit breaker counter pulse experiments used vacuum interrupters with TMF contacts. Anderson et al. [270] showed that variability of the vacuum arc voltage for a spiral, TMF, Cu–Bi contacts gave a somewhat erratic performance. The interruption performance was limited to 8kA. Their system would not interrupt currents equal to or greater than 9 kA. The development of vacuum interrupters with AMF contacts or with external AMF field coils and the resulting diffuse vacuum arc with its stable arcvoltage greatly improved the reliability and performance of this method for high current dc interruption. Belda et al. [271] have investigated the dc interruption performance of vacuum interrupters with TMF and AMF contacts designed for ac interruption (ratings: 38 kV/ 2500A/ 31.5 to 40 kA). They tested three dc current levels, 2 kA, 10 kA and 16 kA and a peak TRV of 45 kV. The counter pulse was applied when the vacuum interrupter contacts were about 3.5 mm apart. They show that after the counter pulse the two vacuum interrupters with the AMF contacts showed a more reliable interruption performance than did the vacuum interrupter with the TMF contact. This would have been expected from Anderson et al's. experiments [270]. There are, however, essential criteria to take into account when using a vacuum arc in an AMF:

1) First of all, there is a time lag, which is a function of the current to be interrupted, before the fully diffused vacuum arc is established; see Figure 2.73, Section 2.7.2 in this volume.
2) Second, the strength of the AMF to develop a fully diffuse vacuum arc depends upon the current to be interrupted.
3) Third, too small a contact diameter for a given current will result in a constricted diffuse column, which can affect the recovery performance of the vacuum gap between the open contacts; see Section 4.2.3 in this volume.

4) Last of all, the vacuum interrupter mechanism has to have a very fast opening time to ensure that it has a large enough contact gap to withstand the TRV across it after the counterpulse's current zero.

The use of AMF contacts shown in Figure 3.45 in this volume has ensured uniform and acceptable contact erosion for as many as 1500 operations for currents of ~ 55kA [272]. It has also been used with the AMF structure shown in Figure 3.43 in this volume for high dc currents (> 90kA) where parallel vacuum interrupters are used [33, 273].

The current counterpulse scheme has been successfully employed for switching the dc current in the Ohmic-heating coil for fusion experiments in the U.S.A. [264, 274, 275], Europe [273, 276], and Japan [277]. The design of a successful "Ohmic-heating interrupter" that my team at Westinghouse helped develop [273] is shown in Figure 6.111. Typical parameters for this interrupter are given in Table 6.25. A dc current of 24kA initially flows through the Ohmic-heating coil (OH) and the closed

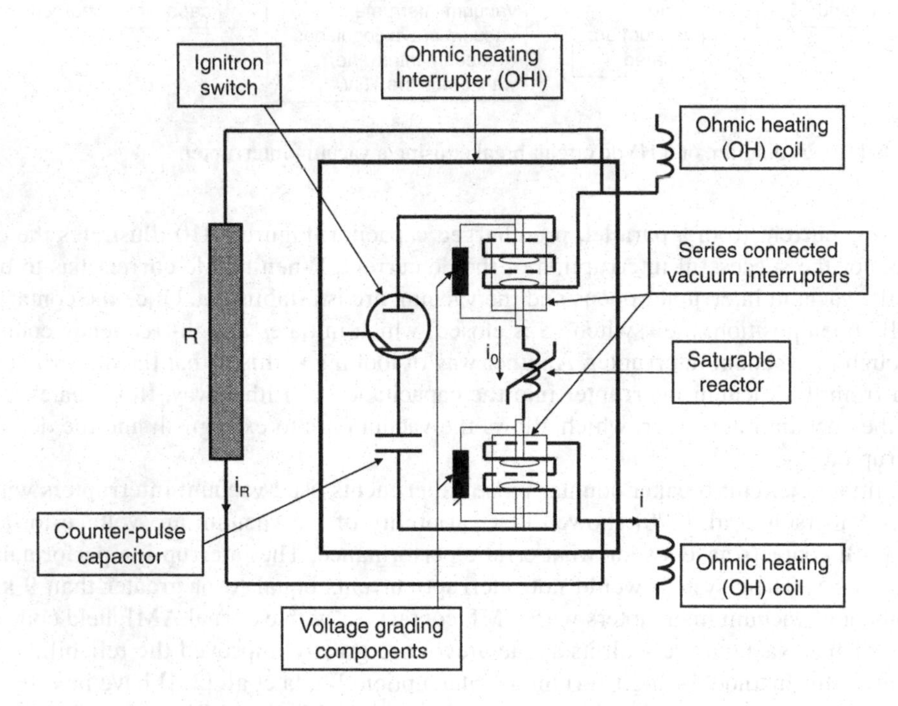

FIGURE 6.111 The Ohmic Heating Coil dc switching system [263, 274].

TABLE 6.25
The Ohmic Heating Interrupter Specifications

Property	Rating
BIL to ground	95 kV
Maximum interruption current	24 kA d.c.
High current withstand rating for the vacuum interrupters	24 kA for 0.1s with a repetition rate of ≤ 300s
Maximum i^2t rating	5.8×10^7 $A^2.s$
$(di/dt)_m$	~ 150 $A.s^{-1}$
$(di/dt)_0$	~ 30 $A.s^{-1}$
Recovery voltage peak, $U_{R(peak)}$	25 kV (+2 kV/–0 kV)
(dU_R/dt)	$5kV.ms^{-1}$–$12\ kV.ms^{-1}$
Recovery voltage withstand time	≥ 15ms

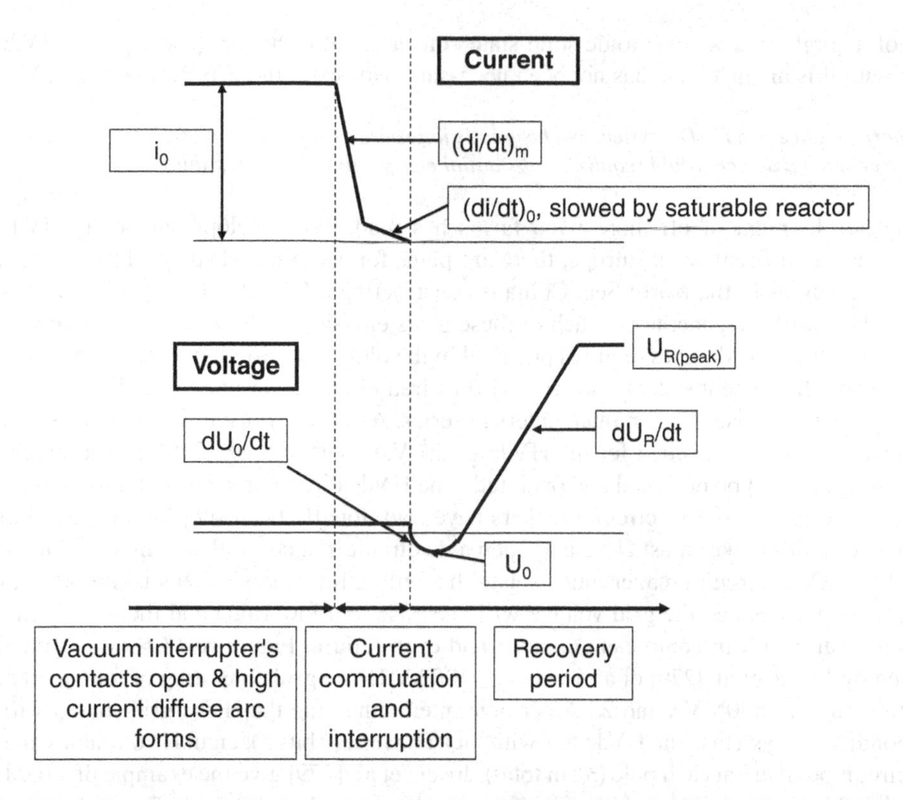

FIGURE 6.112 The current and voltage characteristics when switching the Ohmic Heating Coil dc switch [263].

vacuum interrupters. The plasma striking voltage is created by opening the two vacuum switches and by discharging the counterpulse current from the parallel capacitor. This diverts the current to the energy dissipation resistor *R*. The insertion of this resistor results in a rapid rate of decay of the current, which produces the striking voltage [263, 274, 275]. The current and voltage characteristics are shown in Figure 6.112. A current (i_o) is established in the Ohmic-heating coils. The vacuum interrupters are then opened to their designed contact gap and a high current, diffuse vacuum arc is established. The AMF is supplied by the circuit current opening through field coils external to the vacuum interrupters. The counter pulse from the charged capacitors is only initiated once the high current vacuum arc is in the fully diffuse mode. As the counter pulse timing has to be operated at a very precise time, an ignitron initiated the counter pulse current. During the counter pulse and interruption period, the current in the vacuum interrupters is forced to zero with a high rate of current decrease ($di/dt_m = 0.15\ kA/\mu s$). The inductance of the series saturable reactor shown between the vacuum interrupters becomes large at low currents and the ($di/dt_0 = 0.03\ kA/\mu s$) through the vacuum interrupters slows down just before the current zero. The slower rate of current decrease gives cathode spots from the diffuse vacuum arc a longer time to extinguish. This assists the recovery of the vacuum interrupters when the TRV is imposed across the open contacts once the current goes to zero. At current zero, when the vacuum arc extinguishes, a part of the pre-charge voltage remains as a residual voltage on the counterpulse capacitor C. After interruption, it is discharged in the OH coil. The vacuum interrupters see this voltage as the small negative voltage pulse (U_o). Later, C is charged by the stored energy in the OH coil and this is the TRV (dU_R/dt *to* $U_{R(peak)}$) seen across the open contacts of the vacuum interrupters.

High voltage dc (HVdc) transmission has been used since the 1950s, but in the twenty-first century there has been an increase in planning for its use. At present the vast majority of HVdc systems consist of a single line with ac to dc solid state converters at each end. The use of a HVdc circuit breaker with this arrangement has not been necessary. However, in 1976 Bowles et al. [278] stated:

The performance of a HVDC system will be greatly improved by the addition of a HVdc circuit breaker. However, such a device would require a substantial energy dissipation capability

Now the development of off-shore wind farms has made the development of the HVdc circuit breaker a necessary reality. In Europe, there are plans for a 525 kV, HVdc grid to harvest 200GW from wind turbines in the North Sea. China is constructing a 500 kV, HVdc grid to harvest 5 GW from off-shore turbine generators. Each of these grids envision multiple dc lines in networks. The connection of multiple dc lines requires practical high voltage, dc circuit breakers. Jovcic et al. [279] have reviewed the current state of the art in HVdc circuit breaker development. They show that one viable approach is to use vacuum interrupters in series. As HVdc transmission becomes more available substations will be needed for the HVdc grids. Van Herten et al. [280] give a review of how these substations will be designed and operated. The HVdc circuit breaker will have to pass similar certification tests that HV ac circuit breakers have traditionally been subjected to; see Table 6.26. The HVdc circuit breaker must close and open a dc circuit in a controlled manner. When a fault is detected the HVdc circuit breaker must isolate the faulted line quickly. This is important, because if this line is not isolated the grid voltage will decrease to a low value and the very high currents that occurs can result in component damage and even failure. Figure 6.113 shows a line diagram presented by Jovcic et al. [279] of a five node ± 400 kV HVdc grid consisting of four ac to dc bipole converters rated at 800MVA and 2 kA per pole interconnecting the HVdc grid (buses 1 to 4) with corresponding ac systems, one HVdc bus without a converter (bus 5), eight HVdc cables and sixteen HVdc circuit breakers at each pole (32 in total), Jovcic et al. [279] give the example of a fault linking the number 3 bus to the number 4 bus. This inserts a low impedance at the fault and the arrows show that high currents will flow from other parts of the network. These currents can reach 10 to 30 times the rated current (i.e., ~ 20 kA to 60 kA) in some cables. The HVdc circuit breaker must detect this fault immediately after it is initiated and open to the required contact gap very quickly. This infers that the HVdc grid the HVdc circuit breaker must respond much faster and isolate the fault much quicker than is usual in a traditional ac grid. Once the HVdc circuit breaker isolates the dc current fault, the dc voltage recovers and dc power flows over the remaining HVdc grid. Even so, blocking

TABLE 6.26
Testing for Insulation and Operating Characteristics for HVDC Circuit Breakers

Test Type	Test Objective	Tests
Insulation	Phase to Ground	DC voltage withstand
		Lightning impulse withstand
	Across the HVdc CB contacts	DC voltage withstand
		Lightning impulse withstand
Circuit Breaker Operation	Closed Contact Tests	Current withstand
		Overload current
		Short-time current withstand
	Current Interruption Tests	Rated load current interruption
		Short circuit interruption
		Rated load current making
		Short circuit current making
	Operating Mechanism	Very fast

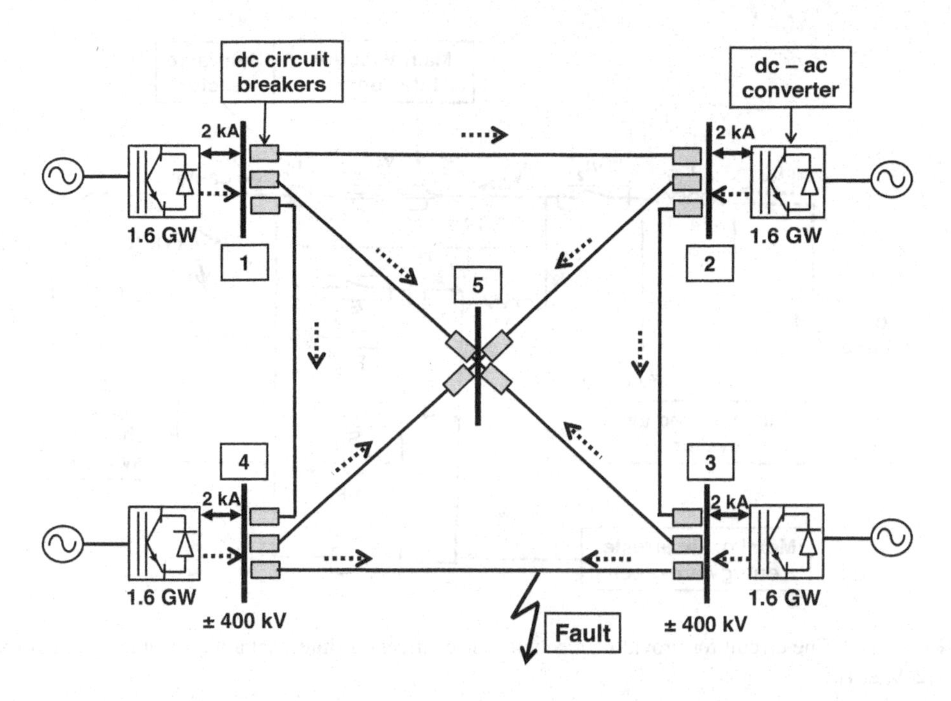

FIGURE 6.113 Schematic showing a five-node dc grid with 4 ac to dc converters, 8 cables, and 16 dc circuit breakers pe pole [279].

of the nearby converter cannot be completely avoided. Some studies indicate that temporary blocking of the converter may occur from 10 ms to 30ms until the faulted line is isolated. The converter can then be unblocked.

A schematic of a practical HVdc circuit breaker is shown in Figure 6.114 [279, 281]. A vacuum interrupter (VI_1) with an AMF contact design is the primary circuit interrupter. This could also be a vacuum interrupter with a butt contact and an external coil around it as shown in Figure .6111. In the closed position, it has a low impedance when passing the dc load current. To open, it is operated by a high-speed mechanism such as the Thomson coil discussed in Section 6.5.3. In this example, the current injection system shows two charged capacitor banks. The second capacitor bank is not required if the HVdc circuit breaker's duty cycle is a single operation which then allows the single capacitor bank time to recharge. When the switch S_{1a} is closed the injected current flows into V_1. S_{1a} or S_{1b} have to be fully closed at a precise time during the rise time of the fault current in V_1. It is possible that these switches could be vacuum interrupters operated by a fast closing mechanism. The disadvantage of using a vacuum interrupter is the fully closed condition has some uncertainty that results from the possibility of a pre-strike before the contacts touch and then contact bouncing after they first touch. Gorman et al. [263] (Figure 6.111) use an ignitron which can be triggered precisely for this duty. It is also possible to use IGBT's with high power ratings to perform the precise closing function or a trigged vacuum gap. A reasonable frequency for the LC closing circuit is 2 kHz to 3 kHz. The resistors R_{1a} and R_{1b} are connected to the capacitor banks' charging supplies. The metal oxide arrester and energy absorber are placed across VI_1 and the current injection system. The vacuum interrupter VI_2 interrupts the residual current at the end of the HVdc fault current interruption cycle. The inductor L_{dc} limits the peak value of the of the fault current and it also reduces its rate of rise. The saturable reactor L_{sr} slows the –di/dt seen by V_1 close to the induced current zero. This is important as it aids V_1 to withstand the fast TRV expected after the dc current is interrupted.

Figure 6.115 shows an example of a HVdc circuit breaker's operation. In Figure 6.115(a) the fault current initiates at t = 2 ms and a trip signal is sent to the mechanism to open V_1 soon

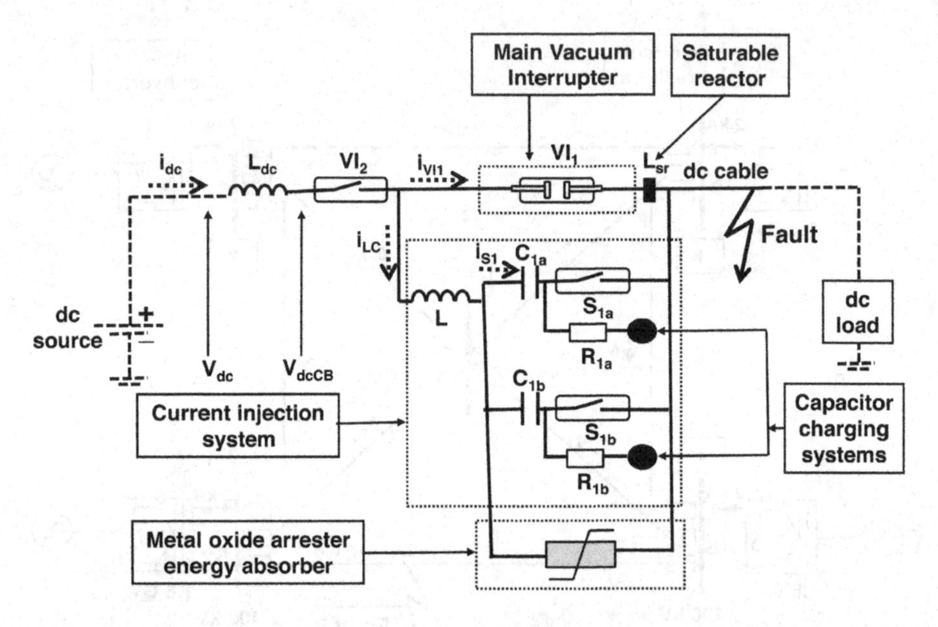

FIGURE 6.114 The circuit for providing a counterpulse current to interrupt a high-voltage, high current dc circuit [279, 281].

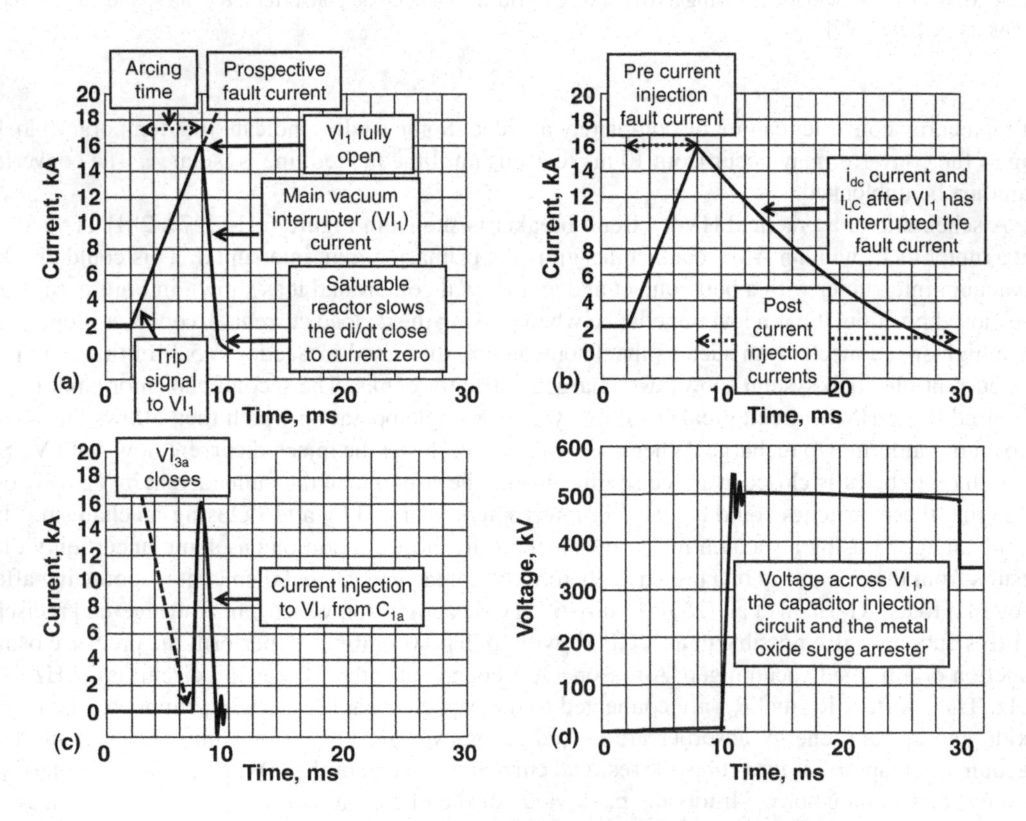

FIGURE 6.115 The dc current interruption process: (a) shows interruption of a dc fault current at 16 kA; (b) shows the stages of the fault current i_{dc} through L_{dc} and the current i_{LC} into the current injection system after the fault current in VI_1 has been interrupted; (c) the current injection from the capacitor C_{1a}; (d) the voltage across VI_1 and the injection circuit limited by the metal oxide arrester and energy absorber.

afterwards. In this example the fault current rises to16 kA in 7 ms after V_1 begins to open. By this time V_1 should be fully open and the vacuum arc controlled by the axial magnetic field will be fully diffuse [263]. Also, at this time S_{1a} is closed and current (see Figure 6.115(c)) flows into V_1 from the charged capacitor bank C_{1a}. The dc current in V_1 rapidly falls to zero and the diffuse vacuum arc extinguishes. The open contact gap then withstands the voltage that appears across it. Figure 6.115(b) shows the current in the circuit at the left-hand side of V_1 before the current injection: i.e., the current in the inductor L_{dc}. After the current injection it shows the current flowing in L_{dc} and through L in the injection circuit. The current i_{LC} flowing into the capacitor bank results in a very high voltage being suddenly impressed across V_1 and the current injection system: see Figure 6.115(d). The metal oxide arrester limits the voltage to 500 kV and absorbs the resulting mega joules of energy. When the current in L_{dc} falls close to zero VI_2 is opened and the HVdc circuit breaker isolates the circuit. Belda et al. [271] show that a well-designed metal oxide arrester will perform well if its temperature is kept below 200Cand the energy injected per volume is limited to less than 200J.cm^{-3}. They also show that for temperatures greater than 200C the metal oxide varistors will be damaged. Different failure modes such as cracks and punctures are observed.

The one vacuum interrupter shown in Figure 6.115 will not withstand a voltage of 500 kV across its open contacts. Practical HVdc circuit breakers for this level of voltage will have to be designed with series vacuum interrupters and series switches shown in Figure 6.116 [281]. The main vacuum interrupters will have to use grading capacitors across each one for the same reason that was discussed in Section 6.9. The HVdc circuit breaker with the current injection technique is far larger and its controls far more complicated than is usual for a vacuum circuit breaker for ac applications. The current injection capacitor bank C and the metal oxide arrester and energy absorber are very large. The opening mechanism for the main vacuum interrupters has to open faster than is usually necessary for an ac circuit breaker. That being said, Tokoyada et al. [282] have successfully tested a HVdc circuit breaker with a single, main vacuum interrupter for a fault current of 16 kA in a 100kV dc circuit. They conclude that a 300/400 kV HVdc circuit breaker should be possible with four main vacuum interrupters in series and a 550/600 kV HVdc circuit breaker with six in series.

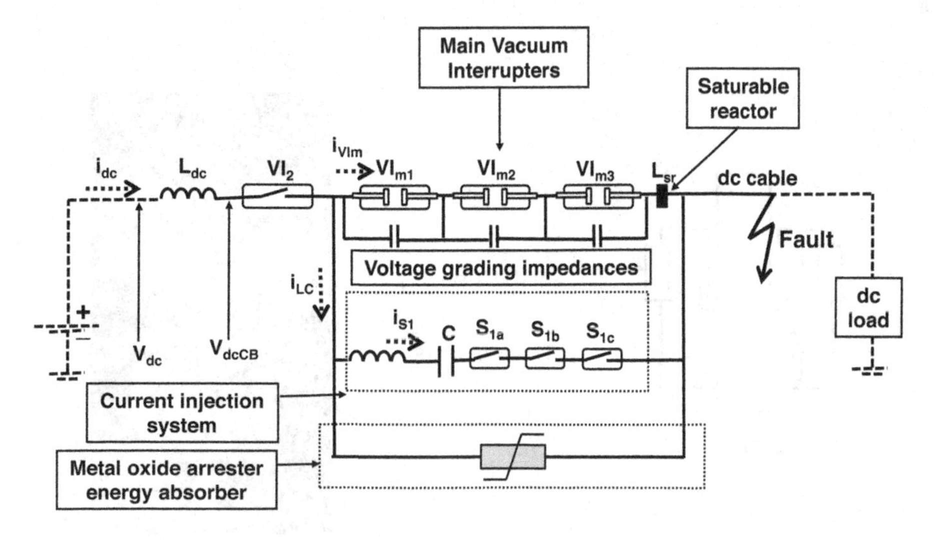

FIGURE 6.116 A diagram of a dc circuit breaker with series vacuum interrupters for use in very high voltage dc circuits.

6.12 DEVELOPMENT OF VACUUM INTERRUPTERS FOR LOW VOLTAGE (< 1000V) CIRCUIT BREAKERS

At low voltages, typically 440V, the dominant circuit breaker design is the low voltage molded case and power breaker using contacts in air. The current interruption is achieved using the arc in air between the opening contacts together with a suitably designed arc chamber [283, 284]. At first sight, it would seem that the vacuum interrupter would have a difficult time competing with the relatively low-cost air-arc interrupting system. There are, however, environments where the "dragon's breath" that the air circuit breakers expel when interrupting high currents is undesirable: e.g., in potentially explosive ambient and fire hazard environments. Also, the enclosed vacuum interrupter contacts have an advantage in chemically aggressive ambients.

The short circuit current interruption ability of a given contact diameter is a function of the system voltage; this is illustrated for a TMF contact in Figure 3.31 in this volume. The design of a low voltage vacuum breaker can be as compact as the equivalent air breaker design. The length of a vacuum interrupter for low voltage can be much shorter than for the medium voltage design. Figure 6.117 shows an example of a possible cross-section. I would expect, however, that for a given fault current there is a minimum internal volume required to ensure that the density of metal vapor at current zero is below that at which a dielectric breakdown will occur even at low peak values of the TRV; see Section 4.2.3 in this volume. Figure 6.118 [285] shows a commercial low voltage vacuum circuit breaker developed by Siemens.

6.13 CONCLUDING SUMMARY

The vacuum interrupter's development and application has surpassed even the most optimistic expectations its early proponents had in the 1960s. It has proved to be extremely reliable. In fact, vacuum interrupters placed in service in the late 1960s are still operating quite satisfactorily and there is no reason to replace them any time soon. The concept of "sealed for life" has been well demonstrated [286–288]. Indeed, the concern of vacuum integrity so frequently voiced when the product was initially introduced has largely disappeared. The vacuum interrupter has earned its well-deserved reputation for maintenance free performance and overall reliability. The vacuum interrupter has been shown to be useful in an ever-widening range of applications. For new installations it continues to achieve a dominant position worldwide at distribution voltages (3.6 kV to 40.5 kV) e.g.:

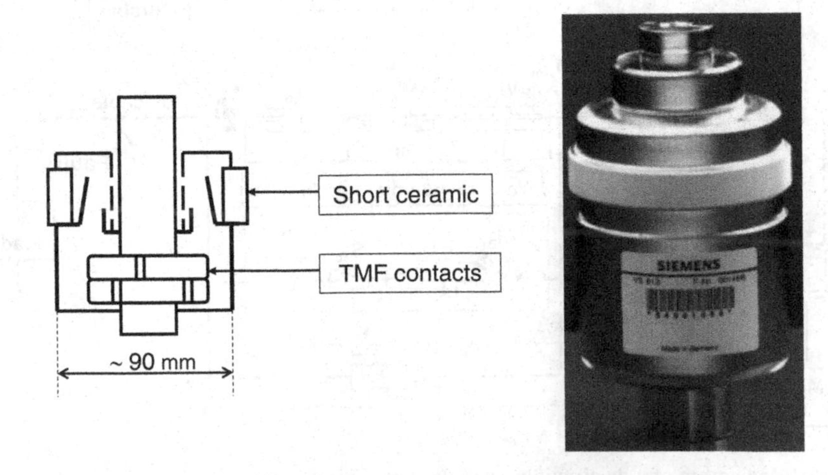

FIGURE 6.117 A low voltage vacuum interrupter design (courtesy Siemens AG).

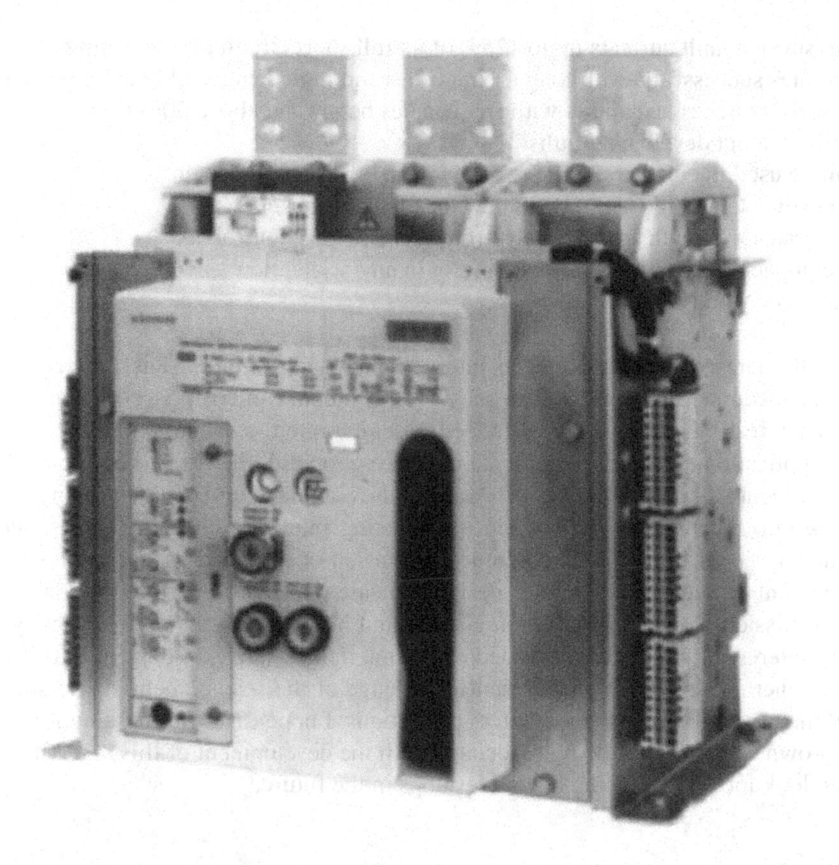

FIGURE 6.118 A low voltage vacuum circuit breaker (courtesy Siemens AG).

- North America and Japan – close to 100%
- Northern Europe – 100%
- Southern Europe – 100%
- China, India, Southeast Asia – strongly vacuum (tending toward 100%)
- South America, Africa, Middle East, Russia – a strong trend towards vacuum interrupter technology (tending toward 90%)

The introduction of 72.5 kV–170 kV vacuum circuit breakers to protect subtransmission and transmission circuits seems to show that these vacuum circuit breakers will eventually garner a significant percentage of these markets as well. As I have presented in Chapters 3, 5, and 6 in this volume, the vacuum interrupter has the following advantages:

1. It is environmentally friendly
2. Its electrical switching life equals, or even exceeds, its mechanical life
3. It can be designed such that its mechanical life can be as high as 10^6 operation
4. Its fault interruption capability exceeds the requirements for recloser certification and for circuit breaker endurance certification. It can be designed to perform 10, 30, 50, 100 and 300 full short circuit interruptions
5. Its reliability enables it to be designed into maintenance free switching systems
6. Its current interrupting ability has no blind spots. A given design can interrupt every current level from below its rated load current (e.g., cable/line switching and unloaded transformer switching) up to its full rated short circuit current

7. It can survive fault currents up to 125% of its full short circuit current rating
8. It operates successfully with long arcing times and high degrees of current asymmetry
9. It reliably switches ac circuits with frequencies below and above 50/60 Hz
10. It will interrupt developing faults
11. It can be used in corrosive, toxic explosive and other harsh ambients
12. It is easily adapted to operate with new mechanism concepts
13. It integrates well with advanced electronic sensing and control
14. It is compact and can operate successfully in any orientation
15. It has a low "total lifetime" cost

As I review the latest proceedings of the International Symposium on Discharges and Electrical Insulation in Vacuum, I am delighted to see that the vacuum interrupter continues to be a fruitful topic for interesting scientific study. Indeed, research into the vacuum arc, vacuum interrupter design and application continues to generate interesting topics to be studied by the next generation of graduate students. I am truly looking forward to the continued unfolding of its mysteries.

As I look into the future, I do not see a competing interruption technology superseding the vacuum interrupter. In fact, I see a continued expansion of the vacuum interrupter's application. The development of vacuum interrupter designs outside the traditional distribution voltage range into subtransmission circuits,transmission circuits and high voltage dc circuits is very exciting. It will be very interesting to see the continued expansion in these areas of design. In this book I have presented a general overview of our present knowledge of the science behind vacuum interrupter design and the range of vacuum interrupter applications. I hope that after reading it you take away some of my own excitement of being associated with the development of this versatile product and that you too look forward to seeing how it develops in the future.

REFERENCES

1. Slade, P. G., "Vacuum interrupters: The new technology for switching and protecting distribution circuits", *IEEE Transactions on Industry Applications*, 33(6), pp. 1501–1511, November/December 1997.
2. Homma, M., Sakaki, M., Kaneko, E., Yanabu, S., "History of vacuum circuit breaker and recent developments in Japan", *IEEE Transactions on Dielectrics and Electrical Insulation*, 13(1), pp. 85–92, February 2006.
3. Sunada, Y., Ito, N., Yanabu, S., Awaji, H. Okumura, Kanai, H.Y., "Research and development on 13.8kV 100kA vacuum circuit breaker with huge capacity and frequent operation", *Proceedings International Conference on Large High Voltage Electric Systems (CIGRE), Paper 13-04*, September 1982.
4. Voshall, R., Kimblin, C., Slade, P., Gorman, J., "Experiments on vacuum interrupters in high voltage 72kV circuits", *IEEE Transactions on Power Apparatus and Systems*, PAS-99(2), pp. 658–666, March/April 1980.
5. Umeya, E., Hisatsune, H., Yanagisawa, H., Sano, T., "A new high voltage interrupter", *IEEE PES Winter Meeting, Paper No. C75099-7*, 1975.
6. Shioiri, T., Homma, M., Miyagawa, M., Kaneko, E., Ohshima, I., "Insulation characteristics of vacuum interrupter for a new 72/84kV GIS", *IEEE Transactions on Dielectrics and Electrical Insulation*, 6(4), pp. 486–490, August 1999.
7. Saitou, H., Ichikawa, H., Nishijima, A., Mataui, Y., Sakai, M., Honma, M., Okubu, M., "Research and development on 145kV/40kA one break vacuum circuit breaker", *IEEE T&D Conference*, Yokohama, Japan, pp. 1465–1468, 2002.
8. Lui, T., Ma, Z., "Design and dynamic simulation of moving characteristics of 126kV vacuum circuit breaker", *Proceedings 21st International Symposium on Discharges and Electrical Insulation in Vacuum*, pp. 307–310, September 2004.
9. Wang, J., "126kV vacuum circuit breaker debuted in China", *Proceedings 22nd International Symposium on Discharges and Insulation in Vacuum*, pp. 1–5, September 2006.
10. Wang, J., Liu, Z., Xiu, S., Wang, Z., Yuan, S., Jin, L., Zhou, H., Yang, R., "Development of high voltage vacuum circuit breaker in China", *Proceedings 22nd International Symposium on Discharges and Electrical Insulation in Vacuum*, pp. 247–252, September 2006.

11. Matsui, Y., Nagatake, K., Takeshita, M., Katsumata, K., "development and technology of high voltage VCB's; brief history and state of the art", *Proceedings 22nd International Symposium on Discharges and Electrical Insulation in Vacuum*, pp. 263–256, September 2006.
12. Ryi, J., Kim, Y.-G., Choi, J., Park, S., "The experimental research of 170kV VCB using single-break vacuum interrupter", *Proceedings 25th International Symposium on Discharges and Electrical Insulation in Vacuum*, pp. 493–496, September 2012.
13. Smeets, R., (Convener), "the impact of the application of vacuum switchgear at transmission voltages", CIHRE Working Group A3.27, 2014.
14. Ohki, Y., "Development of a 204kV vacuum circuit breaker", *IEEE Electrical Insulation Magazine*, 31(1), pp. 44–46, January/February 2015.
15. An excellent general overview of circuit breaker requirements is given by Wagner, C., "Circuit breaker applications", *Circuit Interruption, Theory and Techniques*, Browne, T. Editor, (pub. Marcel Dekker), Chapter 3, pp. 39–134, 1984.
16. For example: IEEE C37 – Series, IEC 62271 – Series, GB-JB-DL (the Chinese version of the IEC Standards), GOST (the Russian Standards).
17. Johnson, B., Dunn, D., Hulett, R., "Seeking global harmony in standards", *IEEE Transactions on Industry Applications Magazine*, p. 14, January/February 2004.
18. Dufournet, D., Montillet, G., "Harmonization of TRVs in ANSI/IEEE and IEC standards for high-voltage circuit breakers rated less than 100kV", *IEEE General Meeting of the Power Engineering Society*, San Francisco, June 2005.
19. Das, J., Mohla, D., "Harmonizing with the IEC", *IEEE Industry Applications Magazine*, pp. 16–26, January/February 2013.
20. There are a number of such certification laboratories around the world, e.g.: KEMA, Holland; CESI, Italy; LAPEM, Mexico; Power-Tech, Canada; Siemens, Germany; EDF, France; MOSKVA, Russia; VUSE, Czech Republic; PSM, U.S.A.; Xi'an, China; Beijing, China.
21. Burke, J., Lawrence, D., "Characteristics of fault currents on distribution systems", *IEEE Transactions on Power Apparatus and Systems*, PAS-103(1), January 1984.
22. Dullni, E., Gentsch, D., Shang, W., "Resistance increase of vacuum interrupters due to high-current interruptions", *IEEE Transactions, on Dielectrics and Electrical Insulation*, 23(1), pp. 1–7, February 2016.
23. Timsit, R., "Electrically conducting layers on a conducting substrate". In: Slade, P. (Editor), *Electrical Contacts: Principles and Applications, 2nd Edition*, (pub. CRC Press), Chapter 1, pp. 22–25, 2014.
24. Liu, S., Wang, Z., Li, H., Li, Y., Geng, Y., Liu, Z., Wang, J., "Temperature-rise performance of 126 kV single-break vacuum circuit breakers with 4 types of AMF contact", *Proceedings 28th International Symposium on Discharges and Electrical Insulation in Vacuum*, pp. 547–550, September 2018.
25. Zhang, Y., Yao, X., Liu, Z., Geng, Y., Liu, P., "Axial magnetic field strength needed for a 126-kV single break vacuum circuit breaker during asymmetrical circuit switching", *IEEE Transactions on Plasma Science*, 41(8), pp. 2034–2042, August 2013.
26. Zhang, X., Wang, X., Guan, Q., Li, M., "A new axial magnetic field contact three-quarters of coil for high-voltage vacuum interrupter and its properties", *Proceedings 27th International Symposium on Discharges and Electrical Insulation in Vacuum*, pp. 541–544, September 2016.
27. Li, H., Wang, Z., Geng, Y., Liu, Z., Wang, J., "Arcing contact gap of a 126-kV horseshoe-type bipolar axial magnetic field vacuum interrupters", *IEEE Transactions on Plasma Science*, 46(10), pp. 3713–3720, October 2018.
28. Liu, Z., et al, "A kind of axial magnetic field contact with high nominal rated current in, vacuum interrupter", *CN Patent, CN107093535A*, 2017.
29. Braunovic, M., "Power connections". In: Slade, P. (Editor), *Electrical Contacts, Principles and Applications, 2nd Edition*, (pub. CRC Press), Chapter 5, pp. 231–373, 2014.
30. Milianowicz, S., "Electrical junction of high conductivity for a circuit breaker or other electrical apparatus", *US Patent 4,376,235*, March 1983 and "Stiff flexible connector for a circuit breaker or other electrical apparatus", *US Patent 4,384,179*, May 1983.
31. Multilam brochure, "Multi-Contact", www.multi-contact.com.
32. Cleaveland, C., "Vacuum type circuit interrupters having heat-dissipating devices associated with the contact structures thereof", *US Patent 4,005,297*, January 1977.
33. Matsukawa, M., Miura, Y., Tsunehisa, T., Kimura, T., "Development of a vacuum switch carrying a continuous current of 36kA dc", *Proceedings 19th International Symposium on Discharges and Electrical Insulation in Vacuum*, pp. 415–418, September 2000.

34. Sato, Y., Arazoe, S., Okuma, H., Takao, N., Yanabu, S., "Increase of current carrying ability of vacuum interrupter", *Proceedings 24th International Symposium on Discharges and Electrical Insulation in Vacuum*, pp. 198–201, September 2010.
35. Tamura, R., Shimoda, R., Kito, Y., Kanai, S., Ikeda, H., Yarabu, S., "Parallel interruption of heavy direct current by vacuum circuit breakers", *IEEE Transactions on Power Apparatus and Systems*, PAS-99(3), pp. 1119–1127, 1980.
36. Pertsev, A., Christjakov, S., Rylskaya, L., Chistjakov, R., "Parallel connection of several vacuum interrupters in circuit breaker pole", *Proceedings 19th International Symposium on Discharges and Electrical Insulation in Vacuum*, pp. 407–410, September 2000.
37. Liu, Z., Wang, J., Wang, Z., "Current transfer process in open operation between parallel breaks of vacuum generator circuit breakers", *Proceedings 19th International Symposium on Discharges and Electrical Insulation in Vacuum*, pp. 403–406, September 2000.
38. Greenwood, A., *Electrical Transients in Power Systems*, (pub. Wiley-Interscience, NY), Chap. 3, pp. 32–47, 1971.
39. IEEE, "IEEE Application guide for transient recovery voltage for AC high-voltage circuit breakers", *IEEE PC37.011/D13*, September 2004.
40. Garzon, R., *High Voltage Circuit Breakers* (pub. Marcel Dekker), Chapter 3 "Transient recovery voltage", pp. 77–109, 1997.
41. "High voltage switchgear and control gear – Part 100 Alternating-Current Circuit-Breakers", *IEC Standard 62271-100*, Edition 2.2, 2017.
42. "Test procedure for AC high-voltage circuit breakers rated on symmetrical current basis amendment 2: To change the description of transient recovery voltage for harmonization with IEC standard 62271-100 and IEEE #C37.096-2010", 2011.
43. Taylor, E., Oemisch, J., Eiselt, M., Hinz, M., "Performance of vacuum interrupters in electrical systems with an effectively earthed neutral", *Proceedings 27th International Symposium on Discharges and Electrical Insulation in Vacuum*, pp. 513–516, September 2016.
44. Taylor, E., Lawall, A., Gentsch, D., "Single-phase short-circuit testing of vacuum interrupters for power systems with an effectively earthed neutral", *Proceedings 28th International Symposium on Discharges and Electrical Insulation in Vacuum*, pp. 579–582, September 2018.
45. "Test procedure for AC high-voltage circuit breakers rated on a symmetrical current basis", *IEEE Standard 37.09-1999 (R2007)*.
46. Greenwood, A., *Vacuum Switchgear*, (pub. IEE UK), p. 62, 1994.
47. Smith, R. K., "Tests show ability of vacuum circuit breaker to interrupt fast transient recovery voltage rates of rise of transformer secondary faults", *IEEE Transactions on Power Delivery*, 10(1), pp. 266–274, January 1995.
48. Slade, P., Smith, R. K., "The use of vacuum interrupters to control short circuit currents and the probability of interruption at the first current zero after contacts part", *Proceedings 9th International Conference on Switching Arc Phenomena*, pp. 54–61, September 2001.
49. Slepian, J., "Extinction of an ac arc", *AIEE Transactions* 47, pp. 1398–1408, 1928.
50. Calvino, B., "Single pressure SF6 circuit breakers". In: Browne, T. (Editor), *Circuit Interruption, Theory and Techniques*, (pub. Marcel Dekker), Chapter 10, pp. 402–404, 1984.
51. Long, W., Smith, R. K., "Are all circuit breakers created equal? Not when it comes to some generator circuits", *Proceedings Water Power Conference*, 2003.
52. Smeets, R., Van der Linder, W., "Current zero measurements of vacuum circuit breakers interrupting short-line faults", *IEEE Transactions on Plasma Science*, 31(5), pp. 852–858, October 2003.
53. Swindler, D., "A case study of the application of medium voltage circuit breakers in adverse transient recovery voltage conditions", *Proceedings IEEE Pulp and Paper Technical Conference*, pp. 107–112, 1995.
54. Leusenkamp, M., "Non-sustained disruptive discharges in vacuum interrupters", *Proceedings 19th International Symposium on Discharges and Electrical Insulation in Vacuum*, pp. 495–498, September 2000.
55. Smeets, R., Lathouwers, A., "Non-sustained disruptive discharges: Test experiences, standardization status, and network consequences", *IEEE Transactions on Dielectrics and Insulation*, 9(2), pp. 194–200, April 2002.
56. Schlaug, M., Falkingham, L., "Non-sustained disruptive discharges (NSDD) – A new investigation method leading to increased understanding of this phenomenon", *Proceedings 19th International Conference on Discharges and Electrical Insulation in Vacuum*, pp. 490–494, September 2000.
57. Smeets, R., Thielens, D., Kirkenaar, R., "The duration of arcing following late breakdown in vacuum circuit breakers", *IEEE Transactions on Plasma Science*, 33(5), pp. 1582–1588, October 2005.

58. Xu, S., Kumada, A., Hidaka, K., Ikeda, H., Kaneko, E., "Observation of pre-discharge phenomena with point-to-plane electrodes in vacuum under ac", *IEEE Transactions, on Dielectrics and Electrical Insulation*, 22(6), pp. 3633–3640, December 2015.
59. Hachiman, Y., Oshiro, F., Nagayo, Y., Kaneko, E., Inada, Y., Taguchi, Y., Yamano, Y., Maeyama, M., Kitabayashi, Y., Iwabuchi, H., Ejiri, H., Kumada, A., Hidaka, K., "Late dielectric breakdown phenomenon caused by microparticles released after current interruption", *Proceedings 28th International Symposium on Discharges and Electrical Insulation in Vacuum*, pp. 173–175, September 2018.
60. Bernauer, C., Rieder, W., Kny, E., "Non-sustained disruptive discharges in vacuum circuit breakers", *Proceedings 13th International Symposium on Discharges and Electrical Insulation in Vacuum*, pp. 338–340, June 1988.
61. Platter, F., Rieder, W., "Long time dielectric behavior of Cu-Cr contacts in vacuum interrupters", *Proceedings 14th International Symposium on Discharges and Electrical Insulation in Vacuum*, Santa Fe (USA), pp. 507–511, 1990.
62. Bernauer, C., Kny, E., Rieder, W., "Restrikes in vacuum circuit breakers within 9s after current interruption", *ETEP* 4(6), pp. 551–555, November/December 1994.
63. Gebel, R., Hartmann, W., "Mechanical shocks as a cause of late discharges in vacuum circuit breakers", *IEEE Transactions on Electrical Insulation*, 28(4), pp. 468–472, August 1993.
64. Bernauer, C., Kny, E., Rieder, W., "Late restrikes in vacuum interrupters", *Proceedings 6th International Conference on Switching Arc Phenomena*, Lodz (Poland), September 1989.
65. Bernauer, C., Kny, E., Rieder, W., "Restrikes in vacuum circuit breakers within 9s after current interruption", *Proceedings 14th International Symposium on Discharges and Electrical Insulation in Vacuum*, pp. 512–516, September 1990.
66. Zalucki, Z., Kutzner, J., "Dielectric strength of a vacuum interrupter contact gap after making current operations", *IEEE Transactions on Dielectrics and Electrical Insulation*, 10(4), pp. 583–589, August 2003.
67. Platter, F., Rieder, W., "Long-range dielectric behavior of vacuum interrupters", *Proceedings 13th International Symposium on Discharges and Electrical Insulation in Vacuum*, pp. 335–337, June 1988.
68. Jüttner, B., "Surface migration as a possible cause for late breakdowns", *Proceedings 18th International Symposium on Discharges and Electrical Insulation in Vacuum*, pp. 408–411, 1998.
69. Jüttner, B., Lindmayer, M., Düring, G., "Instabilities of pre-breakdown currents in vacuum 1: Late breakdowns", *Journal of Physics. Part D: Applied Physics*, 32(19), pp. 2537–2543, 1999.
70. Tanaka, D., Okawa, M., Yanabu, S., "Non-sustained disruptive discharge phenomena after current interruption in vacuum", *Proceedings 22nd International Symposium on Discharges and Electrical Insulation in Vacuum*, pp. 280–283, September 2006.
71. Khvorost, J., Baturin, A., Sheshin, E., "The role of emission properties on non –sustained disruptive discharge (NSDD) evolution", *Proceedings 22nd International Symposium on Discharges and Electrical Insulation in Vacuum*, pp. 79–82, September 2006.
72. See Chapter 1, Section 1.3.5, Figure 1.85 and References [157, 158].
73. Jüttner, B., "Possible role of field electron emission for late breakdowns in vacuum switches", *Elektrotechick und Informationstechnick (e & i)*, 107, pp. 115–117, 1990.
74. Ponthenier, J., Giraud, D., Schellekens, H., "Visualization of late breakdown in vacuum", *Proceedings 19th International Symposium on Discharges and Electrical Insulation in Vacuum*, pp. 499–503, September 2000.
75. Greenwood, A., *Vacuum Switchgear*, (pub. IEE UK), p. 253.
76. Kawase, Y., Mori, H., Ito, S., "3-D finite element analysis of electrodynamics repulsion forces in stationary electric contacts taking into account asymmetric shape", *IEEE Transactions on Magnetics*, 33(2), pp. 1994–1999, March 1997.
77. Banghard, J., Hahn, M., Hartman, W., Renz, R., "Contact arrangement for a vacuum interrupter", U.S. Patent 6,674,039 B1, January 2004.
78. Lane, S., "Electrical circuit breakers", *U.S. Patent 6,140,894*, October 2000.
79. Kaneko, E., Yokokura, K., Homma, M., Satoh, Y., Okawa, M., Okutomi, I., Oshima, I., "Possibility of high current interruption of vacuum interrupter with low surge contact material: Improved Ag-WC", *IEEE Transactions on Power Delivery*, 10(2), pp. 797–803, April 1995.
80. Greenwood, A., op cit 51, pp. 146–149, 1994.
81. Barkan, P., "A study of the contact bounce phenomena", *IEEE Transactions on Power Apparatus and Systems*, PAS-86, pp. 231–240, 1967.
82. Dullni, E., Zhao, S.-F., "Bouncing phenomena of vacuum interrupters", *Proceedings 24th International Symposium on Discharges and Electrical Insulation in Vacuum*, pp. 463–466, September 2010.

83. Kusserow, J., Renz, R., "Method for opening the contact gap of a vacuum interrupter", *US Patent 2004/0124178 A1*, July, 2004.
84. Greenwood, A., op cit 51, pp. 152–162.
85. Barkan, P., McGarrity, R., "A spring actuated cam-follower system: Design theory and experimental results", *Transactions on ASME, Journal of Engineering for Industry*, 87(3), pp. 279–286, 1965. Series B.
86. Barkan, P., "The mechanical phenomena of circuit breakers". In: Lee, T. H. (Editor), *Physics and Engineering of High Power Switching Devices*, (pub. MIT Press, Cambridge, MA), pp. 461–536, 1975.
87. Shi, F., Lin, X., Xu, J., "Theoretical analysis of mechanical characteristics of vacuum circuit breaker with spacing operating mechanism", *Proceedings 21st International Symposium on Discharges and Electrical Insulation in Vacuum*, pp. 438–441, September/October 2004.
88. Yu, L., Uppalapati, B., Leusenkamp, M., Bao, J., Chen, R., "Non-synchronous impact phenomena of closing operation in medium voltage vacuum circuit breakers", *Proceedings 26th International Symposium on Discharges and Electrical Insulation in Vacuum*, pp. 293–296, September 2014.
89. Lindsay, W., "Electromagnetic actuator", *U.S. Patent 3,218,409*, November 1965.
90. See, for example, the FKI Switchgear's GVR Recloser, www.fki-eng.com.
91. McKean, B., Kenworthy, D., "Bistable magnetic actuator", *World Patent Application WO 95/07542*, 1994.
92. McKeen, B., Reuber, C., "Magnets and vacuum, a perfect match", *Proceedings IEE Trends in Distribution Switchgear*, (IEE Conf. Pub. No. 459), London (UK), pp. 73–79, November 1998.
93. Dullni, E., Fink, H., Hőrner, G., Leonhardt, G., Reuber, C., "Totally maintenance-free: New vacuum circuit breaker with permanent magnetic actuator", *Elektriziätswirtschafft*, 21, Jg. 96, pp. 1208–1212, 1997.
94. Dullni, E., "A vacuum circuit breaker with permanent magnet actuator for frequent operations", *Proceedings 18th International Symposium on Discharges and Electrical Insulation in Vacuum*, pp. 688–691, 1998.
95. Ma, S., Wang, J., "Research and design of permanent magnetic actuator for high voltage vacuum circuit breaker", *Proceedings 20th International Symposium on Discharges and Electrical Insulation in Vacuum*, pp. 487–490, July 2002.
96. Lin, X., Gao, H., Cai, Z., "Magnetic field calculation and dynamic behavior analysis of the permanent magnetic actuator", *Proceedings 19th International Symposium on Discharges and Electrical Insulation in Vacuum*, pp. 532–535, September 2000.
97. Kato, K., Matsumoto, Y., Matsuo, K., Homma, M., "The development of low-surge-type VCB with a balanced-type magnetic actuator", *Proceedings 22nd International Symposium on Discharges and Electrical Insulation in Vacuum*, pp. 470–473, September 2006.
98. Huang, Z., Duan, X., Zou, J., Chen, N., "A permanent magnetic actuator with scheduled stroke curve for vacuum circuit breakers", *Proceedings 24th International Symposium on Discharges and Electrical Insulation in Vacuum*, pp. 162–165, September 2010.
99. Dullni, E., Fink, H., Reuber, C., "A vacuum circuit breaker with permanent magnetic actuator and electronic control", (pub. CIRED, Nice (France)), 1999.
100. Waldron, M. (Convenor) et al, "Controlled switching of HVAC circuit breakers: Benefits and economic aspects", *Report of CIGRE Working Group A3.07*, January 2004.
101. Waldron, M. (Convenor) et al, "Controlled switching of HVAC circuit breakers: Guidance for further applications including unloaded transformer switching, load and fault interruption and circuit breakers uprating", *Report of CIGRE Working Group A3.07*, January 2004.
102. Cai, Z., Ma, S., Wang, J., "An approach of improved permanent magnet actuator of vacuum circuit breaker", *Proceedings 23rd International Symposium on Discharges and Electrical Insulation in Vacuum*, pp. 165–168, September 2008.
103. Augustin, T., Magnusson, J., Parekh, M., Garcia, M., Nee, H.-P., "System design of fast actuator in DC applications", *Proceedings 28th International Symposium on Discharges and Electrical Insulation in Vacuum*, pp. 527–530, September 2018.
104. Bissal, A., Magnusson, J., Engdahl, G., "Comparison of two ultra-fast actuator concepts", *IEEE Transactions on Magnetics*, 48(11), pp. 3315–3318, October 2012.
105. Vilchis-Rodriguez, D., Shuttleworth, R., Smith, A., Barnes, M., "Performance of high-power Thomson coil actuator excited by a current pulse train", *The Journal of Engineering*, 2019(17), pp. 3937–3941, 2019.
106. Zhang, L., Yang, K., Liu, Z., Geng, Y., Wang, J., "Influence of design parameters on electro-magnetic repulsion mechanism performance", *Proceedings 27th International Symposium on Discharges and Electrical Insulation in Vacuum*, pp. 545–548, September 2016.

107. Vilchis-Rodriguez, D., Shuttleworth, R., Smith, A., Barnes, M., "A comparison of damping techniques for the soft-stop of ultra-fast linear actuators for HVDC breaker applications", *9th International Conference on Power Electronics, Machines and Drives*, April 2018.
108. Poluyanova, I., Bugayov, V., "Interrupting capability of the AMF electrodes in the external Transverse field", *Proceedings 25th International Symposium on Discharges and Electrical Insulation in Vacuum*, pp. 244–247, September 2012.
109. Long, R. W., Calhoun, G., "High magnetic fields due to current flowing in the main circuit conductors of circuit breaker conversions can reduce short circuit switching performance of circuit breaker elements", *Proceedings PDS 12th Annual Electrical System Reliability and Energy Conference*, August 1999.
110. Jia, S., Fu, J., Yan, J., Li, H., Wang, J., "A kind of magnetic shield for vacuum interrupters", *Proceedings 18th International Symposium on Discharges and Electrical Insulation in Vacuum*, pp. 480–483, 1998.
111. Kong, G., Wei, J., Liang, J., Wang, H., Li, X., "Study on the transverse burning action to contact shield during high-current vacuum interruptions in a solid insulated switchgear", *Proceedings 27th International Symposium on Discharges and Electrical Insulation in Vacuum*, pp. 535–538, September 2016.
112. Wang, L., Qin, K., Hu, L. Zhang, X., Jia, S., "Numerical simulation of vacuum arc behavior considering action of adjacent phases in vacuum circuit breakers", *IEEE Transaction on Plasma Science* 45(5), pp. 859–874, May 2017.
113. See for example the Trayer 2000 Series Vacuum Fault Interrupter (VFI), www.trayer.com .
114. Masaki, N., Matsuzawa, K., Yoshida, T., Ohshima, I., "72/84 kV cubicle style SF6 gas – insulated switchgear with vacuum circuit breaker", *Proceedings IEE 2nd International Conference on Developments in Distribution Switchgear*, IEE publication # 261, 1981.
115. Christophorou, L. G., Olthoff, J. K., Brunt, R. J., "Sulfur hexafluoride and the electric power industry", *IEEE Electrical Insulation Magazine*, 13(5), pp. 20–24, September/October 1997.
116. Christophorou, L., Olthoff, J., Green, D., "Gases for electrical insulation and arc interruption: Possible present and future alternatives to pure SF_6", *National Institute of Standards and Technology, Technical Note 1425*, November 1997.
117. Widger, P., Haddad, A., Griffiths, H., "Breakdown performance of vacuum circuit breakers using alternative CF_3I-CO_2 insulation gas mixture", *IEEE Transactions on Dielectrics and Electrical Insulation*, 23(1), pp. 14–21, February 2016.
118. Zhou, B., Tan, D., Xue, J., Cai, F., Xiao, D., Lightning impulse withstand performance of CF_3I-N_2 gas mixture for 252 kV GIL Insulation,", *IEEE Transactions on Dielectrics and Electrical Insulation*, 26(4), pp. 1190–1196 , August 2019.
119. Guo, Z., Li, X., Li, B., "Dielectric properties of C5-PFK mixtures as a possible SF_6 substitute for MV power equipment", *IEEE Transactions on Dielectrics and Electrical Insulation*, 26(1), pp. 129–136, February 2019.
120. Baghavan, V., Sathish, D., "Problems of moisture ingress in porcelain clad vacuum circuit breakers", *Proceedings 5th International Seminar on Switchgear and Control Gear*, SWICON 2000, Hyderabad (India), paper, IV-31 to IV-40, November 2000.
121. Oommen, T. V., Claiborne, C. C., "Electrical transformers containing insulation fluids comprising high oleic acid oil compositions", *U. S. Patent 5,949,017*, September 1999 and "An agriculturally based biodegradable dielectric fluid", *Proceedings IEEE T&D Conference*, New Orleans, LA, April 1999.
122. Cookson, A., Mandleconn, L., Woolton, R., Roach, F., "Dielectric properties of circuit breakers". In: Browne, T. E. (Editor), *Circuit Interruption Theory and Techniques*, (pub. Marcel Dekker), Chapter 8, pp. 310–325, 1984.
123. For example, the "INNOVAC SYS 3.6–24 kV", www.eatonelectric.com.au.
124. Leusenkamp, M., Hilderink, J., Lenstra, K., "Field calculations on epoxy resin insulated vacuum interrupters", *Proceedings 17th International Symposium on Discharges and Electrical Insulation in Vacuum*, pp. 1065–1069, July 1996.
125. Bestel, E., "Encapsulated vacuum interrupter and method of making same", *US Patent 5,917,167*, June 1999.
126. Cao, Y., Liu, X., Wang, E., Wang, L., "Numerical analysis on electric field of three-phase outdoor vacuum circuit breaker", *Proceedings 19th International Symposium on Discharges and Electrical Insulation in Vacuum*, pp. 131–134, September 2000.
127. Taylor, E., "Private communication".
128. Schulman, M., Slade, P., Loud, L., Li, W.-P., "Influence of contact geometry and current on effective erosion of Cu-Cr, Ag-WC, and Ag-Cr, vacuum contact materials", *IEEE Transactions on Components and Packaging Technology*, 22(3), pp. 405–413, September 1999.

129. Rieder, W., Schussek, M., Glatzle, W., Kny, E., "The influence of composition and Cr particle size of Cu-Cr contacts on chopping current, contact resistance and breakdown voltage in vacuum interrupters", *IEEE Transactions on Components, Hybrids, and Manufacturing Technology*, 12(2), pp. 273–283, June 1989.
130. Schlaug, M., Dalmazio, L., Ernst, U., Godechot, X., "Electrical life of vacuum interrupters", *Proceedings 22nd International Symposium on Discharges and Insulation in Vacuum*, pp. 177–180, September 2006.
131. Slade, P., Smith, R. K., "Electrical switching life of vacuum circuit breaker interrupters", *Proceedings 52nd IEEE Holm Conference on Electrical Contacts*, pp. 32–37, September 2006.
132. Schellekens, H., Battandier, J., "Metal vapor deposition on ceramic insulators and vacuum interrupter lifetime prediction", *Proceedings 19th International Symposium on Discharges and Electrical Insulation in Vacuum*, pp. 119–122, September 2002.
133. Osmokrovic, P., Lazarevic, Z., Irreversibility of dielectric strength of vacuum interrupters after switching operations,", *European Transactions on Electrical Power*, 7(2), pp. 129–135, February 2007.
134. Gramberg, I., Kurrat, M., Gentsch, D., "Investigations on the behavior of vacuum circuit breakers after switching operations", *Proceedings 24th International Symposium on Discharges and Electrical Insulation in Vacuum*, pp. 269–272, September 2010.
135. Gramberg, I., Kurrat, M., Gentsch, D., "Investigations of copper chrome coatings on vacuum circuit breaker ceramics by electron probe microanalysis and electric field simulation", *IEEE Transactions on Plasma Science*, 1(8), pp. 2074–2080, August 2013.
136. Kurrat, M., Hilbert, M., "Surface conductivity investigations of deposition layers on VCB ceramics", *Proceedings 28th International Symposium on Discharges and Electrical Insulation in Vacuum*, pp. 115–118, September 2018.
137. Ernst, U., Cheng, K., Godechot, X., Schlaug, M., "Dielectric performance of vacuum interrupters after switching", *Proceedings 22nd International Symposium on Discharges and Insulation in Vacuum*, pp. 17–20, September 2006.
138. Duan, X., Wang, T., Ye, H., Feng, D., Fan, M., Xiu, S., "Experiment an simulation research on mechanical characteristic of vacuum circuit breaker", *Proceedings 27th International Symposium on Discharges and Electrical Insulation in Vacuum*, pp. 565–568, September 2016.
139. Wu, S., Sun, Z., Duan, X., Mao, L., Wang, T., Xiu, S., "Experimental research on characteristic parameters of 12kV vacuum circuit breakers", *Proceedings 28th International Symposium on Discharges and Electrical Insulation in Vacuum*, pp. 663–666, September 2018.
140. Schulman, M. B., Smith, R. K., "Vacuum arc research expands interpretation of test data from vacuum interrupters", *Proceedings 13th International Conference on Electricity Distribution, CIRED*, Brussels (Belgium), May 1995.
141. Schulman, M. B., Smith, R. K., "Better interpretation of vacuum interrupter tests", *IEEE Industry Applications Magazine*, pp. 18–27, March/April 1999.
142. Slade, P., "The use of binomial distribution in the analysis of the pass-fail test on circuit breakers and high voltage installations", *Proceedings IEEE Power Engineering Society Summer Meeting and EHV/UHV Conference*, Vancouver (Canada), July 1973.
143. Garzon, R., *High Voltage Circuit Breaker Design and Application*, (pub. Marcel Dekker, NY), Chapters 7–9, 1997.
144. Fahrnkopf, D., Vallo, D., "Interruption testing". In: Browne, T., (Editor), *Circuit Interruption Theory and Techniques*, (pub. Marcel Dekker, NY), Chapter 17, 1984.
145. Smeets, R., "Testing and certification of vacuum switchgear", IEEE Tutorial on the Vacuum Switchgear, IEEE Cat. No. 99TP135-0, 1999.
146. ANSI/IEEE C 37.60 Recloser Standards.
147. http://optimalcontrol.net/user_manuals/Westinghouse/Reclosers/index.html.
148. Bonfanti, I., et al, "Shunt capacitor bank switching stresses and test methods", *Electra* 183, pp. 13–41, 1999.
149. Dullni, E., Baum, B., Desmond, D., Heinrich, C., "The performance of in-service shunt capacitor switching devices as investigated by CIGRE WG A3.38", *25th International Conference on Electricity Distribution*, Paper No. 1312, June 2019.
150. Delachaux, T., Rager, F., Gentsch, D., "Study of vacuum circuit breaker performance and weld formation for different drive closing speeds for switching capacitive current", *Proceedings 24th International Symposium on Discharges and Electrical Insulation in Vacuum*, pp. 241–244, September 2010.
151. Juhász, A., Rieder, W., "Capacitive switching with vacuum interrupters", *Proceedings 10th International Conference on Gas Discharges and Their Applications*, pp. 62–65, 1992.

152. Fang, C., Zou, J., Duan, X., Ding, F., "Statistical characteristic analysis and an adaptive control of synchronous vacuum switches", *Proceedings 21st International Symposium on Discharges and Electrical Insulation in Vacuum*, pp. 350–353, September/October 2004.
153. Ding, F., Duan, X., Zou, J., Liao, M., "Controlled switching of shunt capacitor banks with vacuum circuit breaker", *Proceedings 22nd International Symposium on Discharges and Electrical Insulation in Vacuum*, pp. 447–450, September 2006.
154. Dullni, E., Shang, W., Gentsch, D., Kleberg, I., Niayesh, K., "Switching of capacitive currents and the correlation of restrike and pre-ignition behavior", *IEEE Transactions on Dielectrics and Electrical Insulation*, 13(1), pp. 65–71, February 2006.
155. Wang, T., Yan, J., Zhai, X., Liu, Z., Geng, Y., Chen, Y., "Research on 40.5kV two-break vacuum circuit breaker for switching capacitor banks", *Proceedings 24th International Symposium on Discharges and Electrical Insulation in Vacuum*, pp. 647–650, September 2016.
156. Koochack-Zadeh, M., Hinrichsen, V., Smeets, R., Lawall, A., Field emission currents in vacuum breakers after capacitor switching,", *IEEE Transactions on Dielectrics and Electrical Insulation*, 18(3), pp. 910–917, 2011.
157. Smeets, R., Kuivenhoven, S., Chakraborty, S., Sandolache, G.," Field emission current in vacuum interrupters after large inrush current", *Proceedings 25th International Symposium on Discharges and Electrical Insulation in Vacuum*, pp. 157–160, September 2012.
158. Chaly, A., Poluyanova, I., "Relevancy of IEC requirements related to switching cable and line charging currents for medium voltage circuit breakers (VCB)", *Proceedings 22nd International Symposium on Discharges and Electrical Insulation in Vacuum*, pp. 455–457, September 2006.
159. Xiao, Y., Fang, C., Li, W., Zhang, B., Ren., X., Luo, Y., "Modeling of high frequency transients of vacuum circuit breaker switching transformers in offshore wind parks", *Proceedings 27th International Symposium on Discharges and Electrical Insulation in Vacuum*, pp. 453–456, September 2016.
160. Jackson, D., "Survey of failures of surge protection capacitors and arresters on ac rotating machines; report by working group 3.4.9 of surge protection devices committee", *IEEE Transactions on Power Delivery*, 4(3), pp. 1725–1730, July 1989.
161. Browne, T. (Editor), *Circuit Interruption, Theory and Techniques*, (pub. Marcel Dekker, USA), Chapter 11, "Magnetic air circuit breakers", 1984.
162. Flurscheim, C. (Editor), *Power Circuit Breaker Theory and Design* (pub. IEE UK), 1982; Amer, D., "Oil circuit breakers", Chapter 4, pp. 125–188; Morton, J., "Air-break circuit breakers", Chapter 5, pp. 189–214; Gonek, S., "Air-blast circuit breaker", Chapter 6, 235–302.
163. Lythall, R., *The JSP Switchgear Book*, (pub. JSP Books, UK), 7th Edition. 4th Impression, 1980; "Bulk oil breakers", Chapter 12, pp. 375–417; "Small-oil-volume circuit breakers", Chapter 13, pp. 418–432; "Heavy duty air break circuit breaker", Chapter 14, pp. 433 – 465; "Air blast circuit breaker", Chapter 15, pp. 466–499.
164. Bowen, J., Burse, T., "Medium voltage replacement breaker projects", *IEEE Transactions on Industry Applications*, 38(2), pp. 584–595, March/April 2002.
165. Browne, T. op cit [114], Fig. 12.12, p. 473.
166. Telander, S., "AC switchgear", Chapter 14, published in *Switchgear and Control Handbook, 3r Edition*, Smeaton, R., Ubert, H., (Editors), (pub. McGraw- Hill), 1997.
167. Wang, L., Wang, L., Li, X., Lin., J., Zheng, W., Jia., S., "Multi physical field simulation of medium voltage switchgear and optimal design", *Proceedings 27th International Symposium on Discharges and Electrical Insulation in Vacuum*, pp. 457–460, September 2016.
168. Westinghouse Distribution Book (pub. Westinghouse Electric Corporation), 1st edition, 5th printing, 1965; Lokay, H. "Primary and secondary distribution", Chapter 4, pp. 109–148.
169. For example: www.eatonelectrical.com, www.ormazabal.com, www.abb.com, and similar sites for Siemens, Areva, Schneider, Toshiba, Meidensha etc.
170. Lav, C., Staley, D., Olsen, T., "Practical design considerations for application of GIS MV switchgear", *IEEE Transactions on Industry Applications*, 40(5), pp. 1427–1434, September/October 2004.
171. For example: www.abb.com, (USA) and similar sites for other vacuum circuit breaker manufacturers.
172. Nylén, R., "Auto-reclosing", *ASEA Journal*, 52(6), pp. 127–132, 1979.
173. IEEE Working Group on Distribution Protection "Distribution line protection practices – industry survey results",, *IEEE Transactions on Power Delivery*, 10(1), pp. 176–186, January 1995.
174. Short, T., *Electric Power Distribution Handbook*, (pub. CRC Press), Chapter 9, Reliability, pp. 441–478, 2004.
175. Staszesky, D., Craig, D., Befus, C., "Advanced feeder automation is here", *IEEE Power and Energy Magazine*, pp. 56–63, September/October 2005.

176. For example: www.fki-eng.com.
177. For example, go to the Cooper web site, the G&W web site, the Joslyn web site, the Nulec web site, NOJA web site etc.
178. SCADA: "Supervisory control and data acquisition": http://ref.web.cern.ch/ref/cern/CNC/2000/003/SCADA.
179. Boyer, S., *Supervisory Control and Data Acquisition*, (pub. Instrumentation and Control Soc.), 2nd edition, 2002.
180. https://www.nojapower.com.au (click on "News" then "White Paper").
181. Michel, A., Hollingwood, G., "Ring main distribution witnessing the evolutionary process", *Proceedings IEE Conference on Developments in Distribution Switchgear,* May 1986.
182. Pikkert, A., Schoonenberg, G., "Innovative ring main unit design: Safety and availability aspects", *17th International Conference on Electricity Distribution (CIRED), Session 1,* Paper No. 6, May 2003.
183. For example: http://www.schneider-electric.co.uk, "The "ringmaster" RMU.".
184. See, for example, Eaton, Holec, "Xiria", RMU, www.eatonelectrical.com and www.xiria.nl.
185. Hou, C., Li, C., Han, Q., Gao, Y., Cao, Y., "Insulation analysis and improvement of 12kV solid-Insulated ring main units", *Proceedings 27th International Symposium on Discharges and Electrical Insulation in Vacuum*, pp. 441–444, September 2016.
186. Harris, T., Rigden, D., Morton, D., "Ring main distribution switchgear – Latest trends in use, control, and operation", *Proceedings IEE Conference on Trends in Distribution Switchgear,* London (UK), pp. 66–70, November 1994.
187. Torben, S., Gulbrandsen, T., "The intelligent ring main unit", *Proceedings 5th IEE International Conference on Trends in Distribution Switchgear,* London (UK), pp. 167–172, November 1998.
188. For example: www.trayer.com; www.sandc.com.
189. Colclaser, R., Reckleff, J., "Investigation of inherent transient recovery voltage for generator circuit breakers", *IEEE Winter Power Meeting, Paper 83,* January/February 1983.
190. Scruggs, E., Weinrich, H., "Generator breakers for large nuclear, fossil fired, and pump storage power stations", *Proceedings ASME/IEEE Joint Power Generation Conference*, September 1973.
191. Friedlander, G., "What's new in power plants?", *Electrical World*, pp. 46–49, July 15, 1978.
192. IEEE C37.013-1997, "Standards for ac high-voltage generator breakers rated on a symmetrical current basis", December 1997.
193. IEEE PC37.013a/D3, "Standard for ac high-voltage generator circuit breakers rated on symmetrical current basis – Amendment 1: Supplement for use with generators rated 10-100 MVA", October 2002.
194. IEC/IEEE 62271-37-013:2015 High-voltage switchgear and controlgear - Part 37-013: Alternating-current generator circuit-breakers.
195. Burckhardt, P., Schweiger, K., Skorobala, L., "BBC generator breakers in pumped storage", *Brown Boveri Review*, pp. 336–342, May 1979.
196. Alexander, G., Urbanek, J., "Interruption of asymmetrical fault current with a medium voltage switchgear breaker", *Proceedings of the American Power Conference*, pp. 688–693, 1983.
197. Kulicke, B., Schramm, H., "Application of vacuum circuit breaker to clear faults with delayed current zeros", *IEEE Transactions on Power Delivery*, 3(4), pp. 1714–1723, October 1988.
198. Smith, R. K., Long, W., Burmingham, D., "Vacuum interrupters for generator circuit breakers, they're not just for distribution circuit breakers anymore", *Proceedings 17th International Conference on Electricity Distribution (CIRED)*, Barcelona (Spain), May 2003.
199. Gentsch, D., Goettlich, S., Wember, M., Lawall, A., Anger, N., Taylor, E., "Interruption performance at frequency 50 or 60Hz for generator breaker equipped with vacuum interrupters", *Proceedings 27th International Symposium on Discharges and Electrical Insulation in Vacuum*, pp. 429–432, September 2014.
200. Thuries, E., Jeanjean, R., Collod, M., "Control of temperature rises in generator circuit breakers", *Proceedings International Conference on Large High Voltage Systems (CIGRE)*, September 1982.
201. Greenwood, A., *Vacuum Switchgear,* (pub. IEE, London), Fig. 10.24, p. 239, 1994.
202. Sato, Y., Arazoe, S., Okumo, H., Takao, N., Yanabu, S., "Increase of current carrying ability of vacuum interrupter", *Proceedings 24th International Symposium on Discharges and Electrical Insulation in Vacuum*, pp. 198–201, September 2010.
203. Dufournet, D., "Transient recovery voltage requirements for system source fault interruption by small generator circuit breakers", *IEEE Transactions on Power Delivery*, 17(2), pp. 474–478, April 2002.
204. Schramm, H., "Abschalten generatornaker Kurzschlüsse beim Ausbleiben von Strom-Nulldurchgängen", *ETZ-a*, Bd. 89, M.2, pp. 149–153, 1977.
205. Wagner, C., op cit [9], Table 3.6, pp. 65–82.

206. Slade, P., Smith, R. K., "A comparison of short circuit performance using transverse magnetic field contacts and axial magnetic field contacts in low frequency circuits with long arcing times", *Proceedings 21st International Symposium on Discharges and Electrical Insulation in Vacuum*, pp. 337–341, September/October 2004.
207. Ruoss, E., Kolarik, P., "A new IEEE/ANSI standard for generator circuit breakers", *IEEE Transactions on Power Delivery*, 10(2), pp. 811–816, April 1995.
208. Braun, D., Koeppl, G., "Transient recovery voltages during switching under out-of-phase conditions", *International Conference on Power System Transients*, 2003.
209. Steimel, A., "Electric railway traction in Europe", *IEEE Industry Applications Magazine*, pp. 6–17, November/December 1996.
210. Kurz, H., "Rolling across Europe's vanishing frontiers", *IEEE Spectrum*, Vol. 36, No. 2, pp. 44–49, February 1999.
211. Vernon, D., "An improved design of vacuum circuit breakers for ac traction vehicles", *GEC Review*, 1(2), pp. 89–94, 1985.
212. Tian, Y., Wang, S., Zhang, L., Dong, E., Yu, G., Li, Y., "Theory and application research of hybrid circuit breaker in ground-switching auto-passing neutral section system", *Proceedings 28th International Symposium on Discharges and Electrical Insulation in Vacuum*, pp. 603–606, September 2018.
213. He, J., Han, S., Cho, H., "Lightning over voltage protection of ac railroad vehicles by polymeric arrestors", *IEEE Transactions on Power Delivery*, 14(4l), pp. 1304–1310, October 1999.
214. Smith, R. K., "Private communication".
215. Wang, J., Wu, J., Zhu, L., "Properties of intermediate-frequency vacuum arc under axial magnetic field", *IEEE Transactions on Plasma Science*, 37(8), pp. 1477–1483, August 2009.
216. Lavers, J., Bringer, P., "Real time measurement of electric arc furnace disturbances and parameter variations", *IEEE Transactions on Industry Applications*, IA-22(4), pp. 568–576, August 1986.
217. Lavers, J., Danai, B., "Statistical analysis of electric arc furnace parameter variations", *Proceedings IEE, C*, 132(2), pp. 82–93, 1985.
218. Akdağ, A., Cadirçi, I., Nalcaçi, E., Ermiş, M., Tadakuma, S., "Effects of main transformer replacement on the performance of an electric arc furnace system", *IEEE Transactions on Industry Applications*, 36(2), pp. 649–657, March/April 2000.
219. Hesse, I., Schultz, W., "Switching arc furnace transformers in the medium voltage range", *Proceedings 13th International Conference on Electricity Distribution (CIRED)*, Brussels (Belgium), 1995.
220. Dudley, R., Fellers, C., Bonner, J., "Special design considerations for filter banks in arc furnace installations", *IEEE Transactions on Industry Applications*, 33(1), pp. 226–233, January/February 1997.
221. The "SIMELT and the "SIVAC-X" systems: http://www.siemens.com/metals.
222. Vacuum circuit breaker at 132kV,", *Engineering*, p. 996, June 1967.
223. Reece, M., "Improved electric circuit breaker comprising vacuum switches", *British Patent 1,149,413*, April 1969.
224. GE Brochure No. 701b, "Type VIB – vacuum circuit breakers, 121kV through 800kV", December 1973.
225. Kurtz, D., Sofianek, J., Crouch, D., "Vacuum interrupters for high voltage transmission circuit breakers", *IEEE Power Engineering Society Winter Meeting*, Paper No. C 75 054-2, January 1975.
226. Shores, R., Phillips, V., "High voltage vacuum circuit breakers", *IEEE Transactions on Power Apparatus and Systems*, PAS – 94, pp. 1821–1831, 1975.
227. Slade, P., Voshall, R., Wayland, P., Bamford, A., McCracken, G., Yeckley, R., Spindle, H., "The development of a vacuum interrupter retrofit for the upgrading and life extension of 125kV – 145kV oil circuit breakers", *IEEE Transactions on Power Delivery*, 6(3), pp. 1124–1131, July 1991.
228. Wu, G., Ruan, J., Huang, D., Shu, S., "Voltage distribution characteristics of multiple-break vacuum circuit breakers", *Proceedings 24th International Symposium on Discharges and Electrical Insulation in Vacuum*, pp. 186–189, September 2010.
229. Huang, D., Shu, S., Ruan, J., "Transient recovery voltage distribution ratio voltage sharing measure of double and triple break vacuum circuit breakers", *IEEE Transaction Packaging and Manufacturing Technology*, 6(4), pp. 545–552, April 2016.
230. Falkingham, L. K. W., Cheng, K., Molan, W. *Proceedings 26th International Symposium on Discharges and Electrical Insulation in Vacuum*, pp. 497–500, September 2016.
231. Liu, D., Wang, J., Xiu, S., Liu, Z., Wang, Z., Yang, R., "Research on 750kV circuit breaker composed of several vacuum interrupters in series", *Proceedings 21st International Symposium on Discharges and Electrical Insulation in Vacuum*, pp. 315–318, September/October 2004.
232. Yu, X., Yang, F., Li, X., ai, S., Huang, Y., Fan, Y., Du, W., "Static sharing design of a sextuple-break 363kV vacuum circuit breaker", *Energies*, 12, 2019.

233. Shioiri, T., Niwa, Y., Kamikawaji, T., Homma, M., "Investigation of dielectric breakdown probability distribution for double breaker vacuum circuit breaker", *Proceedings 20th International Symposium on Discharges and Electrical Insulation in Vacuum*, pp. 323–326, July 2002.
234. Giere, S., Kärner, H., Knobloch, H., "Dielectric strength of double and single break vacuum interrupters", *IEEE Transactions on Dielectrics and Electrical Insulation*, 8(1), pp. 43–47, February 2001.
235. Liao, M., Duan, X., Cheng, X., Zou, J., "Dielectric property of vacuum circuit breaker with multiple breaks based on fiber-controlled vacuum interrupter modules", *Proceedings 24th International Symposium on Discharges and Electrical Insulation in Vacuum*, pp. 170–173, September 2010.
236. Betz, T., König, D., "Influence of grading capacitors on the breaking capability of two vacuum interrupters in series", *Proceedings 18th International Symposium on Discharges and Electrical Insulation in Vacuum,* Eindhoven (Holland), pp. 679–683, 1998.
237. Betz, T., "Simulation of reignition process of vacuum circuit breakers in series", *IEEE Transactions on Dielectrics and Electrical Insulation*, 4(4), pp. 370–373, August 1997.
238. Liao, M., Zou, J., Duan, X., "Analysis on dynamic dielectric recovery and statistical property of vacuum circuit breakers with multiple breaks", *Proceedings 20th International Symposium on Discharges and Electrical Insulation in Vacuum*, pp. 602–605, July 2002.
239. Sun, L., Wang, Z., Zhai, X., Geng, Y., Liu, Z., "The reignition characteristics of double vacuum interrupters in series after interrupting vacuum arcs", *Proceedings 26th International Symposium on Discharges and Electrical Insulation in Vacuum*, pp. 401–404, September 2014.
240. Fugel, T., König, D., "Switching and transient phenomena in a series design of two vacuum circuit breakers", *Proceedings 21st International Symposium on Discharges and Electrical Insulation in Vacuum*, pp. 399–402, September/October 2004.
241. Fugel, T., König, D., "Switching performance of two 24kV vacuum interrupters in series", *IEEE Transactions on Dielectrics and Electrical Insulation*, 9(2), pp. 164–168, April 2002.
242. Huang, D., Wu, G., Ruan, J., "Study on static and dynamic voltage sharing design of a 126 kV modular triple-break vacuum circuit breaker", *IEEE Transactions on Plasma Science*, 43(8), pp. 2694–2702, August 2015.
243. Ge, G., Liao, M., Duan, X., Cheng, X., Zhao, Y., Liu, Z., Zou, J., "Experimental investigation into synergy of vacuum circuit breaker with double break", *IEEE Transactions on Plasma Science*, 44(1), pp. 79–84, January 2016.
244. Liao, M., Duan, X., Huang, Z., Zou, J., "Influence of the AMF arc control on voltage distribution of double-break VCBs", *IEEE Transactions on Plasma Science*, 44(10), pp. 2455–2461, August 2018.
245. Ge, G., Cheng, X., Liao, M., Duan, X., Zou, J., "A review of the synergy effect in multibreak VCBs", *IEEE Transactions on Plasma Science*, 47(1), pp. 671–679, January 2019.
246. "Feature switchgear –SF_6 – Yesterday's technology", *Electrical Review*, March 2011.
247. Kato, K., Okubo, H., "Optimization of electrode contour for improvement of insulation performance of high voltage vacuum circuit breaker", *Proceedings 22nd International Symposium on Discharges and Insulation in Vacuum*, pp. 21–24, September 2006.
248. Ma, S., Li, X., Wang, J., "Electric field optimization of the 72kV high voltage vacuum interrupter", *Proceedings 21st International Symposium on Discharges and Electrical Insulation in Vacuum*, pp. 391–394, September/October 2004.
249. Zhang, Y., Liu, Z., Geng, Y., Yang, L., Wang, J., "Lightning impulse voltage breakdown Characteristics of vacuum interrupters with 10 to 50mm contact gaps", *Proceedings 24th International Symposium on Discharges and Electrical Insulation in Vacuum*, pp. 44–47, September 2010.
250. Zhang, Y., Liu, Z., Geng, Y., Yang, L., Wang, J., "Lightning impulse voltage breakdown Characteristics of vacuum interrupters with 10 to 50mm contact gaps", *IEEE Transactions on Dielectrics and Electrical Insulation*, 18(6), pp. 2123–2129, December 2011.
251. Wang, Z., Wang, J., "Theoretical research and design on high voltage vacuum interrupters with long electrode gap", *Proceedings 17th International Symposium on Discharges and Electrical Insulation in Vacuum*, pp. 258–262, July 1996.
252. Hartman, W., Haas, W., Rõmheld, M., Wenzel, N., "AMF vacuum arcs at large contact separation", *Proceedings 21st International Symposium on Discharges and Electrical Insulation in Vacuum*, pp. 450–453, September/October 2004.
253. Liu, Z., Wang, J., Xiu, S., Wang, Z., Yuan, S., Jin, L., Zhou, H., Yang, R., "Development of high-voltage vacuum circuit breakers in China", *IEEE Transactions on Plasma Science*, 35(4), pp. 856–865, January 2007.
254. Xiu, S., Pang, L., Wang, J., Lin, J., He, G., "Analysis of axial magnetic field electrode applied to high voltage vacuum interrupters", *Proceedings 22nd International Symposium on Discharges and Insulation in Vacuum*, pp. 317–320, September 2006.

255. Li, H., Wang, Z., Geng, Y., Liu, Z., Wang, J., "Arcing contact gap of a 126-kV horseshoe-type bipolar axial magnetic field vacuum interrupters", *IEEE Transactions on Plasma Science*, 46(10), pp. 3712–3720, October 2018.
256. Zhang, Y., Yao, X., Liu, Z., Geng, Y., Liu, P., "Axial magnetic field strength needed for a 126-kV single-break vacuum circuit breaker during asymmetrical current switching", *IEEE Transactions on Plasma Science*, 41(8), pp. 2034–2042, August 2013.
257. Umeya, E., Yanagisawa, H., "Vacuum interrupters", *Meiden Review*, pp. 3–11, Series 45, 1975.
258. Himi, H., Takashima, T., Shinmon, Y., "A high voltage vacuum circuit breaker for the "66kV super clad substation", *IEEE Power Engineering Society Winter Meeting*, Paper C75117-7, January 1975 and *Meiden Review*, pp. 35–40, 1975.
259. Okawa, M., Yanabu, S., Tamagawa, T., Okubo, H., Kaneko, E., Aiyoshi, T., "Development of vacuum interrupters with high interrupting capacity", *IEEE Transactions on Power Delivery*, PWRD 2(3), pp. 805–809, July 1987.
260. Yao, X., Wang, J., Geng, Y., Yan, J., Liu, Z., Yao, J., Liu, P., "Development and type test of a single-break 126-kV/40-kA–2500-A vacuum circuit breaker", *IEEE Transactions on Power Delivery*, 31(1), pp. 182–190, February 2016.
261. Smeets, R., (Convenor) 'the impact of the application of vacuum switchgear at transmission voltages", *CIGE Working Group*, A3(27, July), 2014.
262. See for example the Ross Engineering HV DC relays; www.rossengineeringcorp.com.
263. Gorman, J., Kimblin, C., Voshall, R., Wien, R., Slade, P., "The interaction of vacuum arcs with magnetic fields and applications", *IEEE Transactions on Power Apparatus and Systems*, PAS-102(2), pp. 257–265, February 1983.
264. Courts, A., Vithayathil, J., Hingorani, N., Porter, J., Gorman, J., Kimblin, C., "A new dc breaker used as a metallic return transfer switch", *IEEE Transactions on Power Apparatus and Systems*, PAS-101(10), pp. 4112–4120, October 1982.
265. Bartosik, M., Jasiulewicz, S., "Theoretical basis of the design of magnetic field generator for dc vacuum circuit breakers", *IEEE Transactions on Plasma Science*, 25(4), pp. 647–651, August 1997.
266. Alferov, D., Yevsin, D., Londer, Y., "Studies of the stable stage of the electric arc at the contact separation in a vacuum gap with a transverse magnetic field", *IEEE Transactions on Plasma Science*, 35(4), pp. 953–958, August 2007.
267. Alferov, D., Belkin, G., Yevsin, D., "DC vacuum arc extinction in a transverse axisymmetric magnetic field", *IEEE Transactions, on Plasma Science*, 37(8), pp. 1433–1437, August 2009.
268. Greenwood, A., Lee, T., "Theory and application of the commutation principle for h.v.d.c. circuit breakers", *IEEE Transactions on Power Apparatus and Systems*, PAS-91(4), pp. 1570–1574, 1972.
269. Greenwood, A., Barkan, P., Kracht, W., "H.V.D.C. vacuum circuit breakers", *IEEE Transactions on Power Apparatus and Systems*, PAS-91(4), pp. 1575–1588, 1972.
270. Anderson, J., Carroll, J., "Applicability of a vacuum interrupter as the basic switch element in H.V.D.C. breakers", *IEEE Transactions on Power Apparatus and Systems*, 97(5), pp. 1893–1900, September/October 1978.
271. Belda, N., Smeets, R., Nijman, R., "Experimental investigation of electrical stresses on the main components of HVDC circuit breakers", *IEEE Transactions on Power Delivery*, early access, DOi 10.1109/TPWRD.2020.2979934, 2020
272. Benfatto, I., DeLorenzi, A., Maschio, A., Weigand, W., Timment, H., Weyer, H., "Life tests on vacuum switches breaking 50kA unidirectional current", *IEEE Transactions on Power Delivery*, 6(2), pp. 824–832, April 1993.
273. Kito, Y., Kaneko, E., Miyamae, K., Shimada, R., "Current sharing between two vacuum circuit breakers in parallel operation for large direct current interruption", *Electrical Engineering in Japan*, 99(5), pp. 67–75, 1979.
274. Kimblin, C., Slade, P., Gorman, J., Voshall, R., "Vacuum interrupters applied to pulse power switching", *Proceedings 3rd IEEE International Pulse Power Conference*, pp. 440–443, June 1981.
275. Bellomo, P., Calpin, J., Cussel, R., Zuvers, H., "Plasma striking voltage production", *Proceedings 9th IEEE Symposium on Engineering Problems of Fusion Research*, pp. 1427–1430, 1981.
276. Benfatto, I., Maschio, A., Feilcke, F., Rösher, D., Weyer, H., "The RFX energy transfer units", *Proceedings 13th IEEE Symposium on Fusion Engineering*, Knoxville (USA.), pp. 1384–1387.
277. Schimada, R., Tani, K., Kishimoto, H., Tamura, S., Ikeda, H., Tamagawa, T., Yanabu, S., Matsushita, T., "Synthetic test methods of hi-direct-current circuit breaker", *Proceedings IEE*, 126(10), pp. 965–970, October 1979.

278. Bowles, J. Vaughan, L., Hingorani, N., "Specification of HVDC circuit breakers for different system applications", *International Conference on Large High Voltage Electric Systems*, Paper 13-09, August/September 1976
279. Jovcic, D., Tang, G., Pang, H., "Adopting circuit breakers for high-voltage dc networks", *IEEE Power and Energy Magazine*, 17(3), pp. 82 *EEE Power and Energy Magazine*, 17(4), pp. 56–66, July/Aug. 82–93, May/June 2019.
280. Van Hertem, D., Leterme, W., Chaffey, G., Abedrabbo, M., "Substations for future HVDC grids", *IEEE Power and Energy Magazine*, 17(4), pp. 56–66, July/August 2019.
281. Shi, Z., Zhang, Y., Jia, S., Song, X., Wang, L., Chen, M., "Design and numerical investigation of a HVDC vacuum switch based on artificial current zero", *IEEE Transactions on Dielectrics and Electrical Insulation*, 22(1), pp. 135–141, February 2015.
282. Tokoyoda, S., Inagaki, T., Kamimae, R., Tahata, K., Kamei, K., Minagawa, T., Yoshida, D., Ito, H., "Development of EHV DC circuit breaker with current injection", *CIGRE-IEC-Conference on EHV and UHV (AC & DC)*, Session A3, Paper 2-1, April 2019.
283. Lee, A., Slade, P., "Molded case, low voltage circuit breakers". In: Browne, T. E., (Editor), *Current Interruption, Theory and Technique*, (pub. Marcel Dekker, NY), pp. 527–565, 1984.
284. Lindmayer, M., "Medium to high current switching". In: Slade, P. G. (Editor)., *Electrical Contacts, Principles and Applications*, (pub. Marcel Dekker), pp. 627–680, 1999.
285. Fink, H., Renz, R., "Future trends in vacuum application", *Proceedings 20th International Symposium on Discharges and Electrical Insulation in Vacuum*, pp. 25–29, July 2002.
286. Slade, P., Li, W., Mayo, S., Smith, R., Taylor, E., "The Vacuum interrupter, the high reliability component of distribution switches, circuit breakers and contactors", *International Conference on Reliability of Electrical Components & Electrical Contacts*, March 2007.
287. Renz, R., Gentsch, D., Fink, H., Slade, P., Schlaug, M., "Vacuum interrupters, sealed for life", *19th International Conference on Electrical Distribution*, May 2007.
288. Taylor, E., Lawall, A., Gentsch, D., "Long term integrity of vacuum interrupters", *Proceedings 26th International Symposium on Discharges and Electrical Insulation in Vacuum*, pp. 433–436, September 2014.

Author Index

A

Abe, J., 355, 361, 368
Abe, K., 55
Abe, N., 264, 266, 267, 181
Abedrabbo, M, 594
Abplanalp, M., 179, 186
Acha, E., 425, 426
Acharya, V., 207, 299
Afanas'ev, V., 279
Agarwal, M., 188
Agheb, E., 333
Ai, S., 578
Aiyoshi, T., 264, 266, 587
Akdağ, A., 575
Akira, N., 264, 266
Aksenov, I., 162
Alexander, G., 567, 570, 571
Alferov, D., 29, 182, 282, 590
Almeida, N., 154, 156
Alpert, D., 31
Altcheh, L., 26
Althoff, F., 250, 251
Altimani, T., 204, 277
Alyoshi, T., 300
Amer, D., 556
Amft, D., 227
Amft, D., 386
An, Z., 33, 86
Ananian, L., 426
Ananthakrishnan, S., 3
Anderko, K., 226
Anders, A., 85, 86, 88, 92, 149, 153, 154, 158, 159, 160, 162, 169, 191, 192
Anders, S., 154, 160, 169, 191, 192
Anderson, J., 591
Andlauer, R, 143
Andoh, H., 242, 366
Andrews, L, 281
Andrews, T. G., 333, 335
Anger, N., 567
Anheuser, M., 373
Aoki, K., 58, 59, 60, 69, 84, 89
Aplanalp, M., 205, 206
Arai, K., 333, 335 344, 358
Aranaga, S., 277, 297
Arazoe, S., 492, 569
Asari, N., 33, 281,
Ascher, H., 27
Atsumi, K., 437
Augustin, T., 529, 531
Awaji, H., 481, 506

B

Baghavan, V., 536
Bajog, G., 427
Balachanndra, T. C., 51
Balasubramaman, G., 16, 17
Ballat, J., 87, 88, 89, 90, 93, 95, 233
Bamford, A., 233, 577, 581
Banghard, J., 518
Bannink, H., 459, 460, 461
Bao, J., 525, 526
Barbour, J., 68
Barengolts, S., 154, 160, 162
Barino, Y., 279
Barkan, P., 144, 145, 221, 289, 573, 521, 525, 590
Barnes, M., 530, 532
Barrio, S., 277
Bartosik, M., 590
Batara, R., 462
Batrakov, A., 33, 56, 68, 71, 78, 92, 154, 160, 162, 171, 172, 173, 174, 206, 227, 340
Battandier, J., 543
Baturin, A., 513
Batzer, T. H., 371
Baum, B., 553
Baus, S., 141
Beavis, L., 301
Beer, P., 459, 460, 461
Befus, C., 174
Behrens, V., 223, 224
Beilis, I, 153, 154, 155, 156, 158, 171, 182, 204
Bektas, S., 16, 97
Belda, N., 591, 297
Belkin, G., 182, 590
Bellomo, P., 592, 593
Ben Jemma, N., 165
Bender, H. G., 84, 85
Benfatto, I., 592
Benilov, L, 154, 155
Benilov, M., 154, 155
Bennett, W. H., 53
Bereza, A., 292
Berkey, W., xx
Bernauer, C., 51, 519
Beroual, A., 18, 19
Berthon, B., 91
Bestel, E., 539
Bestel, F., 23, 264, 266, 273
Bet, M., 391
Betz, T., 368, 369, 580, 581
Bhasavanich, D., 414
Bi, D., 281
Bindas, J., 433, 434, 435, 436, 437
Binnendijk, M., 337, 339, 341, 352
Biondi, M., 38, 39, 45, 46, 47, 53, 55, 65, 71, 72
Bissal, A., 530
Bochkarev, M., 71, 149
Boczowski, D., 391, 392
Boehm, M., 233
Böhm, H., 100
Bolongeat-Mobleu, R., 264, 287
Bolotov, A., 149, 161
Bonfanti, I., 439, 448, 551, 53

Böning, M., 229, 230, 231
Bonner, J., 575, 576
Bono, M., 386, 442
Boone, M., 426
Borisova, T, 56, 57
Bouwmeester, C., 300
Bowen, J., 556
Bowles, J., 594
Boxman, R. L., xxiii, xxiv, 24, 161, 171, 186, 204
Boyer, S., 563, 565
Braun, D., 572
Braunovic, M., 487, 490
Breedis, J.F., 225
Bringer, P., 575
Brown, I., 158, 159, 162, 191, 192
Brown, K., 221
Browne, T. E., 335, 482, 539, 548, 556, 598
Bruce, F. M., 403
Bruning, A., 27
Brunke, J., 424
Brunt, R. J., 21, 535
Brütsch, R., 427
Buchenauer, C., 15, 16
Bugaev, A., 158, 159, 160
Bugayov, V., 534
Bulanchuk, O. N., 53
Bunshah, R. F., 371
Burckhardt, P., 567, 568
Burger, E, 221
Burke, J. L., 401, 402, 482, 646
Burmingham, D., 286, 287, 567, 572
Burse, T., 556
Byon, E., 159

C

Cadirçi, I., 575
Cai, F., 535, 538, 559
Cai, Z., 528
Calhoun, G., 534
Calpin, J., 592, 593
Calvino, B., 505
Camm, E., 442
Campbell, L., 16, 17
Cao, Y., 539, 565
Carlsson, J., 335
Carroll, J. J., 246, 337, 338, 339, 352, 591
Chabrier, J., 80, 145
Chaffey, G., 594
Chakraborty, S., 91, 453, 459, 460, 461, 54
Chalaya, A., 425
Chaly, A., 182, 196, 201, 203, 278, 279, 372, 378, 379, 425, 555
Chandrasekharan, M., 291
Charbonnier, F. M., 66
Chatterjee, N., 18
Chatterton, P. A., 30, 31, 38, 46, 53, 56, 57, 60, 61, 65
Chaudhari, S., 91
Chen, B, 204, 205
Chen, C, 154, 162, 485
Chen, M., 595, 596, 597
Chen, N., 528
Chen, R., 525, 526
Chen, S., 274
Chen, Y., 146, 554, 580
Cheng, K., 546, 578
Cheng, L., 110
Cheng, S., 98, 270, 277
Cheng, X, 358
Cheng, X., 580, 581
Cherry, S., 287, 300
Chien, Y., 186, 387
Child, C. D., 68
Childs, S. E., 333
Chistjakov, R., 492, 569
Cho, H., 573
Choi, J., 271, 273, 289
Choi, J., 481, 587
Christjakov, S., 492
Christophorou, L. G., 21, 535
Chrost, K., 378, 379
Claiborne, C. C., 21, 537
Clair, M., 403, 409
Cleaveland, C., 490
Cline, C., 292
Cobine, J. D., xxii, 12, 16, 17, 31, 75, 77, 156, 157, 166, 221, 300, 231
Colclaser, R., 566
Collod, M., 568
Conybear, J., 294
Cookson, A., 539
Cornick, K., 242, 414
Courts, A., 590, 592
Couvreur, M., 442
Cox, B., 33
Craggs, J. D., 8
Craig, D., 174
Cranberg, L., 65
Crawford, F., 325
Cross, J. D., 75, 76, 81, 82
Crouch, D, xxii, 468, 469, 576
Cunha, M., 154
Curtis, T. E., 403
Cussel, R., 592, 593
Czarnecki, L., 166

D

Daalder, J., 154, 155, 169, 426, 433, 435
Dai, L., 282
Daibo, A., 281
Dakin, T., 13, 340
Dalmazio, L, 351, 352, 541, 549, 550
Dampilon, B., 227
Damstra, G. C., 31, 63, 300, 438, 462
Dan, S., 262
Danai, B., 575
Das, J., 482
Date, K., 465, 471
Davies, D. K., 38, 39, 45, 46, 47, 53, 55, 65, 71, 72, 73, 154, 162, 435
Davies, N., 469
Davis, W., xx
Decock, W., 442
Dehonova, S., 227
Del Rio, L., 277, 287, 297
Delachaux, T, 140, 179, 186, 205, 206, 233, 253, 256, 376, 380, 553

DeLorenzi, A., 592
Deng, J., 229
Desmond, D., 553
DeVault, B., 15, 387
Devisme, M.-F., 227, 232
Dhahbi-Megriche, N., 18, 19,
Diamond, W. T., 39, 40, 41, 43, 44, 61, 63, 74, 114
Dickinson, M. R, 192
Dimitrescu, G, 253
Ding, B., 225, 226
Ding, C., 166, 167, 234, 241, 242
Ding, F., 553
Ding, S., 226
Djakov, B., 153
Djogo, G., 75, 76, 81, 82
Djurabekova, F., 63, 82, 83
Dmitriev, V., 372, 378, 379
Dohnal, D., 424
Dolan, W., 68
Donen, T., 55, 56, 57, 66, 355, 361, 368, 442, 453, 461
Dong, E., 573
Doremieux, J., 148
Douchin, J., 386
Dougal, R. A., 16, 285
Du, W., 578
Duan, X., 104, 358, 424, 528, 546, 553, 554, 580, 581
Dubrovskaya, E, 91, 94
Dudley, R., 575
Duffy, S., 27
Dufournet, D., 482, 498, 570
Dullni, E., 140, 229, 230, 231, 234, 242, 252, 253, 254, 255, 334, 338, 339, 341, 342, 343, 345, 359, 376, 380, 461, 465, 471, 472, 484, 518, 522, 528, 553
Dunham, B., 286, 287
Dunn, D, 482
Duracov, V., 227
Düring, G., 56, 513, 514
Dyke, W. P., 68
Dzierzbicki, S., 376, 380

E

Eastham, D. A., 53, 65
Ebeling, W., 149, 161
Edels, H., 165, 325
Einstein, P., 95
Eiselt, M., 501
Ejiri, H., 55, 56, 57, 66, 462, 511
Ellis, D., 233
Emelyanov, A., 71, 74, 78
Emelyanov, E., 71, 74, 78
Emmerich, W., 247
Emtage, P., 182
Engdahl, G., 530
Enholm, O.A., xix
Ermiş, M., 575
Ernst, U., 351, 352, 541, 546, 549, 550
Erzhi, W., 98
Escholz, O., xx

F

Fabian, R., 294
Fadat, N., 351, 352
Fahrnkopf, D., 548
Falkingham, L., 287, 288, 300, 511, 513, 578
Fan, J, 300
Fan, M., 546
Fan, X., 104, 300
Fan, Y., 578
Fang, C., 424, 553, 556
Fang, D., 181
Farish, O., 16, 97,
Farrall, G. A., 16, 30, 31, 34, 56, 63, 65, 75, 77, 86, 166, 325, 330
Fedun, V. I., 53
Fehr, R., 401
Feilbach, A., 231
Feilcke, F., 592
Feinberg, B., 162
Feldman, L. C., 47, 48, 49
Fellers, C., 575, 576
Feng, D., 248, 249, 546, 554
Fenski, B., 264, 266, 273, 281, 282, 332, 333, 346, 347, 348, 352, 356, 357
Ferrario, B., 295
Ferree, J., 373
Filip, G., 234, 236
Fink, H., 93, 227, 280, 306, 465, 471, 472, 528, 598
Fischer, B., 23
Fischer, O., 33
Floryancic, B., 438
Flurscheim, C., 556
Foosnaes, J., 192
Förster, A., 149, 161
Fowler, R., 38
Franke, S, 160, 171, 172, 186, 205, 206
Frants, O., 149, 161
Frey, P., 230, 232
Friedlander, G., 566
Frind, G., 246, 337, 338, 339, 352
Fritz, O, 186, 253, 256
Fröhlich, K., 43, 44, 74, 333, 352, 424, 427
Frontzek, F., 300
Frost, L., 335
Fu, J., 280, 535
Fu, K., 362
Fu, Y., 438, 462
Fugal, T., 299, 303, 580
Fujii, H., 63, 64, 65
Fujii, T., 362
Fukai, T., 227
Fursey, G. N., 66, 67, 78

G

Gal, I., 468
Galonska, M., 192
Galvin, J., 162
Gang, Li, 226
Ganguly, C., 375
Gao, H., 528
Gao, Y., 565
Garcia, M., 529, 531
Garzon, R., 3, 6, 421, 498, 548
Gauther, J., 390
Gauyacq, J., 80
Ge, G., 358, 581

Gebel, R., 461, 513
Geire, S., 116, 117, 118
Geisselman, M., 16, 97
Gellert, B., 229, 231, 233, 334, 338, 339
Geneguard, P., 247
Geng, Y., 33, 41, 53, 63, 82, 83, 92, 98, 107, 108, 116, 117, 118, 119, 169, 170, 171, 198, 200, 201, 207, 208, 209, 227, 268, 271, 272, 273, 283, 330, 336, 344, 346, 442, 460, 461, 485, 485, 486, 492, 530, 534, 554, 580, 582, 583, 584, 585, 586, 587, 588
Gentils, F., 390
Gentsch, D., 93, 116, 119, 122, 140, 186, 188, 223, 224, 227, 230, 231, 233, 249, 250, 253, 255, 256, 261, 262, 282, 283, 299, 303, 306, 326, 329, 345, 346, 355, 376, 380, 461, 465, 472, 484, 501, 518, 544, 548, 553, 567, 598
Gerdien, xx
Gere, S., 98, 99
Gherendi, F., 253
Ghori, Q., 386
Giere, S., 580
Giorgi, T., 295
Giraud, D., 514
Glatzle, W., 242, 433, 438, 541
Glinkowski, M., 231, 280, 281, 362
Gnyusov, S, 227
Goa, J. C., 18
Godechot, X., 91, 299, 351, 352, 442, 541, 546, 549, 550
Goeckeler, C., 445
Goettlich, S., 567
Goldsmith, S., xxiii, 24
Golubev, A., 418, 419
Gonek, S., 556
Goody, C. P., 246, 337, 338, 339, 352
Gorlt, K., 231
Gorman, J.G., 176, 177, 182, 184, 185, 233, 264, 266, 267, 268, 481, 506, 582, 586, 590, 592, 593, 595, 597
Gortschakow, S., 171, 172, 186, 205, 205
Gorur, R. S., 18, 19
Gosh, P. S., 18
Goulielmakis, D., 468
Gramberg, I., 544
Green, D., 535
Greenwood, A, xxii, 24, 80, 156, 166, 231, 286, 287, 281, 333, 340, 362, 409, 413, 423, 445, 450, 462, 463, 493, 496, 501, 515, 521, 522, 525, 569, 590
Greenwood, J., 138, 139, 140, 141, 371
Greenwood, T., xx
Griffiths, H., 535, 559
Grohs, C., 250
Gruntjes, R., 300
Gruszka, H., 92
Guan, Q., 271, 283, 485
Gubanski, S., 23
Guillaume, G., 442, 450
Gulbrandsen, T., 565
Gundlach, H., 188, 189, 191, 201, 202, 204
Guo, H., 226
Guo, Z., 535, 559

H

Haas, W., 253, 584
Hachiman, Y., 462, 511
Hackham, R., 26
Haddad, A., 535, 559
Hahn, M., 518
Hairour, M., 299, 351, 352
Hakamata, Y., 348, 349, 350, 353
Hallal, M. P., 69,
Hammons, T., 3
Han, Q, 565
Han, S, 573
Hansen, M., 226
Hao, M., 281
Hardtfeldt, G. E., 409, 410, 411
Harris, T., 565
Hartmann, W., 154, 155, 156, 186, 187, 253, 424, 461, 513, 518, 584
Hasegawa, K., 87
Haskins, R., 162, 163, 370, 383, 384
Hässler, H., 226, 231
Hauf, U., 231
Hauh, R., 148
Hauner, F., 227
Hayakawa, N, 87
Hayashi, K., 344
Hayashi, T., 92
Hayes, W.C., xx
Hazel, T., 3, 6
He, G., 116, 117, 118, 264, 266, 268, 277, 281, 283, 585
He, J., 110, 233, 371, 573
Heberlein, J., 155, 162, 176, 177, 182, 184, 185, 188, 194, 195, 196, 199, 208
Heil, B., 93
Heilmaier, M., 229, 230, 231, 232, 233
Heimback, M., 227, 264, 266, 280, 281, 282
Heinemeyer, R., 57, 58, 61, 300
Heinrich, C., 553
Heinz, T., 116, 117, 118, 119, 122
Helmer, J., 362, 435
Hemachander, M., 207
Henken, K., 125, 249, 250
Henon, A., 204, 227, 232, 277
Hesse, I., 575
Heylan, A., 13
Heyn, D., 231, 232, 365, 420
Hidaka, K., 41, 42, 55, 56, 57, 66, 70, 71, 74, 462, 511
Hidemitsu, T., 264, 266
Hilbert, M., 544
Hildebrandt, S., 295, 301
Hilderink, J., 22, 262, 280, 286, 287, 539
Hill, J., 231
Himi, H., 586
Hingorani, N., 590, 592, 594
Hinrichsen, V., 231, 232, 453, 454, 455, 458, 459, 462, 554
Hinz, M., 501
Hirasawa, K., 348, 349, 350, 353
Hirose, H., 28
Hisatsune, H., 481, 585
Hizal, M., 16, 97
Hochstrasser, M., 250
Hogkin, A., 469
Hoidalen, H.-K., 333

Hollinger, R., 192
Hollingwood, G., 564, 565
Holm, E., 137, 138, 139, 140, 142, 143, 146, 371, 374
Holm, R., 137, 138, 139, 140, 142, 143, 146, 371, 374
Holmes, F. A., 182, 184, 185, 236, 238, 239, 240, 241, 387
Holmes, R, 165, 188, 333, 335
Homma, M., 16, 26, 27, 58, 59, 60, 63, 69, 84, 89 97, 166, 223, 233, 275, 278, 279, 280, 282, 289, 291, 352, 353, 354, 359, 361, 366, 420, 436, 467, 481, 529, 528, 580
Honda, M., 63, 84, 99
Hong, Y., 226
Honig, T, 223, 224
Hori, M., 427
Horn, F., 221
Hőrner, G., 528
Hou, C., 565
Howe, F., 300
Hoyaux, M., 149, 151, 153, 180
Hu, C., 271, 273
Hu, D., 262
Hu, L., 535, 537
Hu, Y., 281
Huang, D., 577, 581
Huang, X., 171
Huang, Y., 578
Huang, Z., 300, 528, 581
Huber, B. F., 146, 384
Huber, E., 333, 352
Hudda, F. G., 56
Hulett, R., 482
Humpert, C., 93
Hundstad, R., 247

I

Ibuki, K., 3
Ichikawa, H., 87, 89, 91, 299, 481, 581
Ichikawa, T., 33, 70, 71, 74
Ikeda, H., 41, 42, 70, 71, 74, 492, 511, 592
Ikeda, Y., 427
Imani, M., 171
Inada, Y., 70, 71, 74, 462, 511
Inagaki, T., 597
Inagawa, Y., 87, 89, 91
Isono, H., 40
Ito, H., 597
Ito, N., 481, 506
Ito, S., 97, 372, 511
Ito, T., 262
Itoh, T., 362
Ivanov, V, 282
Iwabuchi, H, 462, 511
Iwai, A., 87, 98
Iwasita, K., 242, 366
Izcara, J., 277, 297

J

Jackson, D., 556
Jadinian, J, 333
Jäger, K., 232
Jaghel, D., 291
Jai, S., 227
Jaitly, N., 285
Janiszewski, J., 178, 262
Jasiulewicz, S., 590
Jeanjean, R., 568
Jennings, J.E., xxi
Jia, S., 50, 63, 153, 171, 181, 182, 188, 190, 192, 193, 198, 199, 201, 202, 204, 205, 229, 280, 333, 355, 457, 535, 537, 550, 559 595, 596, 597
Jiang, S., 262
Jiang, Y., 171
Jin, L., 33, 86, 102, 460, 481, 492, 522, 582, 585, 587
Jing, Y., 424
Johnson, B., 482
Johnson, D., 33, 56, 56
Johnson, O. E, 331
Jovcic, D., 594, 595
Juhász, A., 453, 454, 455, 459, 553, 554
Jurtz, D., 362
Juttner, B., 42, 43, 56, 85, 86, 88, 74, 77, 78, 79, 80, 81, 90, 92, 93, 153, 154, 155, 158, 165, 169, 513, 514

K

Kadish, A., 15, 16
Kagenaga, K., 275, 361
Kageyama, K., 298, 299
Kahl, B., 102, 103
Kajimura, K., 427
Kajiwara, S., 474
Kaljatsky, I., 63
Kamble, D., 291
Kamei, K., 597
Kamikawaji, T., 26, 27, 28, 51, 52, 53, 64, 86, 91, 99, 291, 467, 580
Kamimae, R., 597
Kanai, H.Y., 481, 506
Kanai, S., 492
Kanai, Y., 291
Kane, C., 418, 419
Kaneko, E., 26, 27, 28, 41, 42, 51, 52, 53, 64, 86, 91, 99, 223, 233, 241, 264, 266, 275, 276, 289, 291, 337, 339, 341, 352, 353, 354, 359, 361, 366, 420, 436, 437, 462, 467, 481, 511, 520, 587, 592
Kaneko, S., 333, 335
Kantas, S., 351, 352
Kapin, A., xx, 154, 158, 162
Kärner, H. C., 43, 44, 74, 84, 85, 102, 103, 580
Kashiwagi, Y., 225
Kassirov, G., 63
Kassubek, F., 179
Kasture D., 27
Kasyanov, A., 149, 161
Kato, K., 87, 89, 91, 97, 528, 582
Kato, M., 225
Katsumata, K., 481, 585
Kaufmann, H., 155
Kaumanns, J., 356
Kawakubo, Y., 264, 266, 267, 181, 275
Kawamura, T., 427
Kawase, Y., 140, 372, 517
Keider, M., 194, 195, 196, 199, 204, 208
Keil, A., 31, 32
Kenworthy, D., 528
Kesaev, I., 154, 155

Khakpour, A, 160, 171, 172
Kharin, S., 386
Khoroshikh, V. M., 162
Khromoi, Y., 181
Khvorost, J., 513
Kikuchi, R., 70, 71, 74
Kikuchi, Y., 55
Kikukawa, S., 471
Kim, Y.-G., 271, 273, 289, 481, 587
Kimblin C., 154, 160, 162, 169, 171, 180, 182, 184, 185, 191, 233, 267, 325, 473, 481, 506, 582, 586, 590, 592, 593, 595, 597
Kimura, T., 490, 592
Kinoshita, S., 469, 470
Kippenberg, H., 226, 232
Kirkenaar, R., 511, 554
Kirvenhoven, S., 453, 454, 455, 458, 459, 462
Kishimoto, H., 592
Kitabayashi, Y., 462, 511
Kitakizaki, K., 225
Kitamura, T., 300
Kito, Y., 492, 592
Kleberg, I., 461, 553
Klink, N., 230, 232
Klinski-Wetzel, K., 229, 232
Knobloch, H., 580
Kny, E., 51, 242, 433, 438, 513, 541, 542
Kobayashi, S., 42, 87
Kobayashi, T., 300
Koenig, D., 368, 369
Koeppl, G., 572
Kohl, W.H., 285
Kojima, A., 40
Kojima, H., 58, 59, 60, 69, 84, 87, 89, 98
Kojima, K., 471
Kokura, K., 55
Kolarik, P., 572
Kolyada, Y., E., 53
Komatsu, H., 274, 275
Kong, G., 172, 207, 208, 209, 227, 535
König, D., 43, 44, 57, 58, 61, 74, 87, 88, 89, 90, 93, 95, 105, 107, 233, 300, 580, 581
Koochack-Zadeh, M., 453, 454, 455, 458, 459, 462, 554
Koren, P, 186
Korolev, Y., 149, 161
Koshiro, K., 227, 229, 230, 232, 233, 250
Kosse, S, 154, 424
Kouahou, T., 148
Koulik, Y., 171
Kowalczyk, M., xxi
Koyama, K., 33, 34, 63, 64, 65
Kozyrev, A., 162
Kracht, W., 590
Kraft, V. V., 56, 57
Kranjec, P., 38
Kraus, A., 223, 224
Krinberg, I., 158, 159, 160, 192
Krock, R., 221
Krolikowski, C., 462
Kudo, T., 148
Kuhl, W., 231
Kuivenhoven, S., 453, 459, 460, 461, 554
Kulicke, B., 567
Kulkarni, S., 91, 207, 281, 299
Kumada, A., 41, 42, 55, 56, 57, 66, 70, 71, 74, 462, 511
Kumara S., 23
Kumbra, D. G., 409, 410, 411
Kurosawa, Y., 242, 264, 266, 267, 181, 348, 349, 350, 353, 466
Kurrat, M., 98, 99, 186, 188, 253, 255, 261, 262, 345, 348, 349, 350, 353, 355, 466, 544
Kurth, B., 424
Kurtz, D., xxiii, 468, 469, 576
Kurz, H., 572
Kusano, T., 166, 276, 278, 279, 280, 282, 361
Kusserow, J., 522
Kutty, T., 375
Kutzner, J., 513
Kyritsakis, A., 63, 82, 83

L

Lafferty, J., xxiv, 221, 291
Lagotzky, S., 86
Lamara, T., 23, 249, 250, 282, 376
Lamarsh, J. R., 47, 117
Landl, V., 149, 161
Lane, S., 518
Larner, A. F., 403, 409
Latham, R. V., xxiv, 30, 32, 38, 42, 45, 46, 53, 54, 55, 66
Lathouwers, A, 462, 510, 511, 513, 555
Laux, M., 154, 162
Lav, C., 466, 559
Lavers, J., 575
Lawail, A., 116, 117, 118, 119, 122, 141, 154, 306, 371, 374, 380, 381, 424, 453, 454, 455, 458, 459, 462, 501, 548, 554, 567, 598
Lawrence, D., 482, 546
Lazarevic, Z., 543, 545
Leblanc, T., 143
Lee, A., 186, 598
Lee, D., 31
Lee, L., 156
Lee, L. H., 340
Lee, T, 423, 590
Lee, T. H., xxii, 80, 221, 231, 286, 287
Lei, P., 283
Lenstra, K., 22, 267, 268, 539
Leonhardt, G., 528
Leterme, W., 594
Leusenkamp, M., 22, 233, 260, 262, 510, 525, 626, 539
Li, B., 535, 559
Li, C., 565
Li, G, 169, 170, 371
Li, H., 198, 200, 201, 268, 485, 486, 492, 535, 585
Li, J., 333, 335
Li, K., 424
Li, M., 248, 249, 250, 271, 282, 283, 485
Li, P., 226
Li, R., 260, 262, 424
Li, S., 33
Li, W., 16, 17
Li, W., 282
Li, W-P., 153, 162, 163, 166, 233, 237, 239, 241, 260, 282, 302, 303, 304, 306, 360, 424, 433, 435, 436, 437, 541, 542, 556, 598
Li, X., 274, 535, 559, 578, 582
Li, Y., 53, 63, 82, 83, 485, 492, 573

Li, Yu., 227
Liang, C., 300
Liang, J., 535
Liao, M., 104, 358, 424, 553, 580, 581
Licheng, W., 98
Lietz, A., 227
Liljestrand, L., 234, 242
Lin, F., 282
Lin, J., 264, 266, 268, 283, 559, 585
Lin, R., 229
Lin, X., 525, 528
Lindell, E., 234, 242
Lindmayer, M., 43, 44, 56, 74, 85, 86, 88, 92, 164, 165, 166, 186, 223, 224, 231, 232, 253, 255, 261, 262, 264, 266, 273, 281, 282, 326, 328, 332, 333, 341, 346, 347, 348, 352, 354, 356, 357, 358, 362, 420, 513, 514, 525, 528, 598
Lins, G., 325, 326, 328, 334, 336
Little, R. P., 33
Litvinov, E., 30
Liu, C., 192, 193, 198, 199
Liu, D., 578
Liu, G., 66, 226, 233, 249, 371
Liu, H., 424
Liu, J., 53
Liu, K., 50, 63, 171, 233, 371
Liu, P., 271, 272, 273, 283, 485, 585, 586, 587
Liu, S., 198, 200, 201, 485, 492
Liu, S-C, 29
Liu, X., 539
Liu, Z., 33, 41, 53, 92, 98, 102, 107, 108, 116, 117, 118, 119, 169, 170, 171, 172, 198, 200, 201, 207, 208, 209, 221, 226, 227, 249, 250, 260, 264, 267, 268, 271, 272, 273, 277, 280, 281, 283, 300, 330, 336, 344, 346, 442, 460, 461, 481, 485, 486, 492, 522, 530, 534, 554, 578, 580, 581, 582, 583, 584, 585, 586, 587, 588
Llewellyn-Jones, F., 8, 11, 12, 139, 142, 148
Logachev, A., 150, 151, 152, 196, 201, 203, 279, 279, 355
Lokay, H, 559
Londer, Y., 182, 590
Long, R. W., 506, 534, 567, 572, 576
Long, Y., 249, 250
Losinsky, M., 375, 376
Loud, L., 16, 17, 162, 163, 433, 435, 436, 437, 541, 542
Lui, G., 226
Lui, T., 481, 585
Lunev, V., 162
Luo, Y., 424, 556
Luxa, G., 13, 340
Lyman, E., 31
Lysniak, M., 160
Lythall, R., xxi, 556

M

Ma, H., 171, 207, 208, 344, 346
Ma, S., 528, 582
Ma, Z., 481, 585
MacGill, R. A., 162, 192
Maeyama, M., 462, 511
Magnusson, J., 529, 530, 531
Maier, W., 15, 16
Makishima, S., 469, 470
Malakhovsky, S., 279
Malluci R, 33, 140, 372
Mandelelbrot, B., 161
Mandleconn, L., 539
Mano, K., 372, 378
Mao, L., 546, 554
Marchand, F., 469
Marshall, T., 469
Martin, E. E., 66, 68
Martin, P., xxiv, 161
Maruyama, X. K., 69,
Masaki, N., 21
Masaki, N., 535
Maschio, A, 592
Masilamani, H., 91
Mason, R., xx
Mataui, Y., 481, 581, 585
Matsuda, M, 231, 437
Matsui, Y., 274, 275, 289, 362
Matsukawa, M., 490, 592
Matsumoto, K., 63, 99
Matsumoto, Y., 528
Matsuo, K., 528
Matsuoka, S., 55, 66, 70, 71, 74
Matsushita, T., 592
Matsuzawa, K., 21, 535
Mattatia, S., 427
Mauro, R., 231
Mayer, J. W., 47, 48, 49
Mayo, S., 264, 267, 286, 287, 306, 598
Mazurek, B., 63, 74, 75, 114
Mazurova, L., 56, 57
Mazzucchi, D., 390
McCoy, C., 438
McCracken, G, 233, 577, 581
McGarrity, R., 525
McKean, B., 528
McLaren, P., 414
McQuin, N., 3
McShane, C. P., 21
Meek, J. M., 8
Mendenhall, M., xx
Meng, Z., 262
Menon, M. M., 53
Menzel, K, 179, 205, 206
Merck, W., 337, 339, 341, 352
Merk, W., 300
Merl, W. A., 31, 32
Mesyats, G., xxiv. 30, 34, 35, 71, 74, 77, 78, 79, 81, 153, 154, 155, 160, 162, 169
Methling, R., 171, 172, 186, 205, 206
Miao, B., 226, 233, 371
Michal, R., 186, 230
Michel, A., 564, 565
Miki, S., 355, 361, 368, 442
Milianowicz, S., 487, 488
Miller, H. C., xx, xxiii, 16, 38, 86, 162, 169, 196, 259, 285
Miller, K., 426
Millikan, R., xx
Minagawa, T., 597
Mitchell, G., 173, 174, 175, 247
Miura, Y., 490, 592
Miyagawa, M., 289, 469, 470, 481
Miyamae, K., 592

Miyazaki, F., 87, 89, 91
Mizuno, T., 427
Mizutani, H., 188, 189
Mo, Y., 333, 335
Mohla, D., 482
Molan, W., 578
Möller, K, 43, 44, 74
Molter, L, 331
Montillet, G., 482, 498
Moore, D.M., xx
Morant, M., 442
More, S., 291
Mori, H., 140, 372, 517
Morimiya, O., 188, 189
Morin, C., 442, 450
Morita, A., 471, 474
Moriyima, O., 333, 335, 344, 358
Morton, J., 556, 565
Mościck-Grzesiak, H., 57, 58, 61, 62, 92
Motoyama, H., 427
Moulaert, G., 442
Müller, F., 229, 230, 231, 232, 233, 250
Müller, G., 86
Müller, R., 226, 227, 233
Murai, Y., 362
Murano, M., 264, 266, 362

N

Nagabhushana, G. R., 51, 63, 84
Nagai, H., 70, 71, 74
Nagatake, K., 481, 585
Nagayo, Y., 462, 511
Nahemow, M., 148, 149
Nakajima, Y., 221
Nalcaçi, E., 575
Nash, W., 264, 266
Nassar, O. M., 414
Natsui, K., 348, 349, 350, 353
Naya, E., 225
Nee, H.-P., 529, 531
Nemchinsky, V., 181, 188, 189, 191
Niak, P., 293
Niayesh, K, 453, 459 362, 461, 553
Nicolle, C., 299, 390
Niedermann, P., 33
Niemeyer, K., 237
Nijman, R., 591, 297
Nikolaev, A., 158, 162
Nikolic, P., 326, 329
Nishijima, A., 289
Nishijima, A., 481, 581
Nishikawa, M., 362
Nishikawa, S., 362
Nishimura, R., 58, 59, 60, 69, 84, 89
Nishioka, M., 427
Nitu, C., 253
Niwa, Y., 252, 253, 276, 278, 279, 280, 281, 278, 279, 280, 358, 361, 580
Nizazama, T., 437
Nobutaka, K., 469, 470
Noda, E., 344
Noda, Y., 225, 227
Noe, S., 100
Noguchi, K., 427
Nordheim, L., 38
Norris, A., 3
Nouri, H., 386
Novak, P., 442, 459, 460, 461
Nowak, A., 63, 74, 75, 114
Nylén, R., 561

O

Ochi, S., 355, 361, 368
Oemisch, J., 501
Ohashi, H., 427
Ohira, K., 87
Ohki, Y., 481, 578
Ohkura, T., 362
Ohno, M., 469, 470
Ohshima, I., 21, 26, 27, 28, 51, 52, 53, 63, 64, 84, 86, 91, 99, 231, 241, 276, 278, 279, 280, 289, 461, 481, 535
Okabe, N., 291
Okabe, S., 333, 335
Okawa, M, 51, 52, 53, 64, 86, 91, 98, 99, 166, 223, 233, 300, 362, 366, 420, 436, 513, 520, 587
Oks, E., 158, 159, 160, 162, 192
Okubo, H., 58, 59, 60, 69, 84, 87, 89, 91, 95 96, 97, 98, 99, 102, 289, 582, 587
Okubu, M., 481, 581
Okuma, H., 492, 569
Okumura, 481, 506
Okumura, H., 63, 84, 99, 264, 266
Okumura, M., 225
Okutomi, I., 366, 420, 436, 520
Okutomi, T, 87, 89, 166, 223
Olive, S., 227, 232, 299
Olsen, T., 466, 559
Olthoff, J. K., 21, 33, 91, 94, 535
Onishchenko, I. N., 53
Ookura, T., 262
Oommen, T. V., 21, 537
Ootaka, M., 61
Oppermann, G., 13, 340
Opyda, W., 462
Orace, H., 414
Osamu, S., 469, 470
Osawa, Y., 300
Oshima, I., 223, 282, 366, 420, 436, 520
Oshiro, F., 462, 511
Osmokrovic, P., 28, 29, 91, 108, 443, 543, 545
Otobe, K., 231

P

Padalka, V., 162
Palad, R., 252, 253
Pang, H., 594, 595
Pang, L., 264, 266, 268, 585
Papillon, A., 86, 91, 232
Parashar, R., 283
Parek, J., 363
Parekh, M., 529, 531
Park, S., 271, 273, 289, 481, 587
Parks, J. E., 403, 409
Paschen, F., 13

Paul, B.-J., 264, 266, 268
Paulus, I., 165, 261, 325, 326, 328, 336
Paulzagade, D., 299
Pavelescu, D., 253
Pavleino, M., 372, 378, 379
Peach, N., xxii
Pearson, J., xxi
Pegel, I., 68
Penning, F., 298, 299
Perkins, J. F., 414, 421
Persky, N., 181
Pertsev, A., 492, 569
Petrovic, L., 373
Pfieffer, W., 16, 97
Phillips, A., 371, 381
Phillips, V., xxiii, 576
Piccoz, D., 143
Picot, P., 204, 227, 232, 299, 465
Pieniak, T., 186, 188, 345, 355
Pietsch, C., 237
Pikkert, A., 565
Pilsinger, G, 100
Plakensteiner, A., 250
Platter, F., 513, 514
Plessl, A., 229
Plyutto, A., xx, 154, 158, 162
Pohl, F., 325, 326, 328, 336
Pokrovskaya-Soboleva, A., 56, 57
Poluyanova, I., 278, 355
Poluyanova, I., 534, 555
Pons, A., 442, 450
Ponthenier, J., 514
Popov, M., 333, 335, 352, 355, 425, 426, 427
Popov, S, 71, 78, 160, 171, 172, 173, 174, 206, 227, 340
Porter, J., 221, 590, 592
Porto, D. R., 154, 155, 162, 188, 435
Poulussen, H., 300
Powers, L., 27
Powis, A., 335
Pretorius, R., 418, 419
Proskurovskii, D., 30, 33, 56, 68, 71, 78, 92
Puchkarev, V., 149, 153
Pursch, H., 42, 34, 85, 86, 88, 92, 93, 153, 154, 162, 164, 165

Q

Qian, Z., 198, 199
Qin, K., 535, 537

R

Radtke, R., 149, 161
Rager, F., 233, 453, 459, 553
Rajan, S., 281
Rajhans, R., 281, 299
Ramesh, M., 23
Rankin, W., xx
Ravi, K., 375
Reckleff, J., 566
Reece, M.P., xxii, 221, 240, 249, 333, 576
Reeves, R., 300
Reininghaus, U., 87, 88, 89, 90, 93, 229, 329
Ren, X., 424, 556
Rengier, H., 31
Renz, R., 116, 119, 122, 264, 266, 268, 283, 306, 424, 473, 518, 522, 598
Rettenmaier, T., 232, 233
Reuber, C., 528
Rich, J., 247
Rieder, W., 43, 44, 51, 74, 234, 236, 242, 433, 438, 453, 454, 455, 459, 513, 514, 541, 542, 553, 554
Rigden, D., 565
Ritskaya, L., 85
Rittenhouse, J., 445, 447, 450
Roach, F., 539
Roach, J.F., 264, 266, 268
Robinson, A.A., 225
Robiscoe, R., 15, 16
Robson, A., 192
Rogowski, W., 31, 403
Rohrbach, F., 32, 86
Rolle, S., 227
Romheld, M, 186, 187, 253, 584
Rondeel, W., 188, 189, 191, 192
Rong, M., 172, 280
Rosenkrans, B., 286, 287
Rösher, D, 592
Ross, H. C., xxi, 403, 409, 423
Rowe, S., 173, 174, 206, 232
Ruan, J., 577, 581
Ruby, L., 38
Ruoss, E., 572
Rusteberg, C., 85, 86, 88, 92, 164, 165
Rüttel, R., 231
Ryi, J., 481, 587
Rylskaya, L., 231, 492, 569
Ryu, J., 271, 273, 289
Ryzhkov, V., xx, 154, 158, 162

S

Sabot, A., 442, 450
Sachteben, J., 154, 162
Saeger, K.E., 230, 232
Safin, R., 3, 6
Safonova, T., 71, 74, 78
Saito, H., 87, 274, 275
Saito, T., 469, 470
Saito, Y., 42, 87, 98
Saitoh, H., 289
Saitou, H., 481, 581
Sakaguchi, W., 252, 253, 281
Sakahi, M., 87, 89, 91, 289
Sakai, M, 481, 581
Sana, A., 274, 275
Sanders, D., xxiv, 161
Sandolache, G., 173, 174, 206, 340, 453, 459, 460, 461, 554
Sannino, L., 386
Sano, T., 481, 585
Santilli, V., 366, 420
Sasage, K., 252, 253, 467
Satau, H., 274, 275
Sathish, D., 536
Sato, J., 233, 276, 282, 461, 468, 469, 470
Sato, K., 33
Sato, S., 33, 34, 53, 64, 65, 453, 461
Sato, Y., 492, 569

Satoh, K., 231
Satoh, Y., 223, 352, 353, 354, 359, 366, 420, 435, 520
Savage, M., 56
Savkin, K., 158, 162
Schacherer, C., 453, 459
Schächter, L., 16, 97
Schade, E., 186, 229, 231, 233, 253, 254, 255, 256, 334, 338, 339, 341, 342, 343, 345, 359, 461
Schellekens, H., 88, 204, 225, 227, 232, 247, 264, 267, 269, 273, 274, 275, 276, 277, 280, 286, 287, 300, 340, 386, 442, 468, 514, 543
Schimada, R., 592
Schlaug, M., 306, 511, 513, 541, 546, 549, 550, 598
Schlenk, W., 231
Schmidt, B., 100
Schmidt, H., 105, 107
Schmoelzer, T., 233
Schneider, A., 91, 94, 172, 173, 174, 206, 227, 340
Schneider, H., 247, 248, 286, 287
Schnettler, A., 93
Schönhuber, M., 13, 340, 341
Schoonenberg, G., 565
Schramm, H., 97, 287, 289, 567, 571
Schreiner, H., 226
Schulman, M. B., 179, 180, 194, 195, 196, 199, 200, 204, 208, 254, 256, 258, 259, 260, 261, 273, 274, 275, 276, 305, 368, 376, 433, 434, 435, 436, 437, 541, 542, 546, 547, 548
Schulpen, F., 154, 169
Schultz, W., 575
Schulz-Gulde, E., 162
Schussek, M., 234, 236, 242, 433, 438, 541
Schwagner, A., xxi, 14, 31, 50, 62, 75, 80, 95, 96, 137, 138, 142, 146, 147, 148, 149, 150, 151, 152, 153, 162, 163, 165, 166, 179, 180, 182, 184, 185, 186, 194, 195, 196, 198, 199, 200, 208, 220, 221, 225, 229, 233, 237, 239, 241, 246, 256, 267, 286, 287, 302, 303, 304
Schwaiger, A., 250
Schweiger, K., 567, 568
Schwirzke, F., 69,
Scruggs, E., 566, 567
Seco, M., 297
Seki, T., 87, 89, 278, 279, 280, 282
Sekimori, Y., 281
Sekisov, F., 63
Sekiya, K., 158
Sellappan, A., 291
Selzer, A., xxiii, 24
Serdyuk, Y., 23
Serikov, I., 71, 74, 78
Shang, W., 140, 230, 253, 254, 255, 264, 266, 267, 269, 280, 281, 282, 283, 346, 376, 380, 461, 484, 518, 553
Shanker, P., 91
Sharpe, R., 56
Shea, J., 14, 84, 387, 431, 432
Shemshadi, A., 333
Shemyakin, L., 149, 161
Sheshin, E., 513
Shi, F., 525
Shi, W, 229, 300
Shi, Z, 50, 63, 153, 171, 181, 182, 188, 190, 192, 193, 198, 199, 201, 204, 280, 295, 333, 335, 357, 595, 596, 597
Shimoda, R., 492, 592
Shin, M., 468, 469
Shinmon, Y., 586
Shioiri, T., 26, 27, 28, 33, 51, 52, 53, 58, 59, 60, 63, 64, 69, 70, 71, 74, 84, 86, 89, 91, 95, 96, 98, 99, 289, 467, 468, 469, 470, 481, 580
Shkol'nik, S., 162, 182, 196, 201, 203, 355
Shmelev, D., 154, 160, 162, 186, 253, 256
Shol'nik, S., 278, 279
Shores, R., xxiii, 576
Short, T., 174
Short, T. A., 401
Shu, S., 577
Shümann U., 98, 99
Shuttleworth, R., 530, 532
Sidorov, V., 29, 282
Siemroth, P., 42, 43, 93, 154, 162
Simon, R., 233
Singh, 23
Siodla, K., 57, 58, 62
Skorobala, L., 567, 568
Slade, P. G., xxi, 14, 31, 50, 62, 75, 80, 95, 96, 137, 138, 142, 146, 147, 148, 149, 150, 151, 152, 153, 162, 163, 165, 166, 179, 180, 182, 184, 185, 186, 194, 195, 196, 198, 199, 200, 208, 220, 221, 225, 229, 233, 237, 239, 241, 246, 256, 267, 286, 287, 302, 303, 304, 306, 341, 368, 369, 370, 371, 373, 374, 374, 378, 380, 381, 383, 384, 387, 391, 433, 434, 435, 436, 437, 455, 462, 467, 470, 471, 481, 484, 487, 490, 502, 505, 506, 541, 542, 545, 546, 548, 571, 574, 577, 581, 582, 586, 590, 592, 593, 595, 597, 598
Slepian, J., xx, 504
Slivkov, I. N., 57, 65
Smeaton, R., 557
Smeets, R., 154, 166, 167, 168, 169, 241, 300, 333, 335, 337, 339, 341, 352, 355, 362, 453, 454, 455, 458, 459, 460, 461, 462, 481, 509, 510, 511, 513, 548, 554, 555, 587, 588, 591, 597
Smith, A., 530, 532
Smith, R.K., 96, 107, 233, 260, 264, 265, 306, 368, 369, 385, 387, 501, 502, 505, 506, 509, 510, 541, 542, 545, 546, 547, 548, 567, 571, 572, 574, 575, 598
Smith, S., 247, 249
Snowdon, A., 143
Sodeyama, H., 264, 266
Sofianek, J., xxiii, 468, 469, 576
Sohma, S, 188, 189, 437
Solot, A., 333
Somei, H., 233, 361, 275, 278, 279, 280
Sone, M., 40, 41
Song, X., 182, 192, 193, 198, 199, 595, 596, 597
Sopianek, J., 362
Sorensen, R., xx
Souma, S, 349
Souques, G., 391
Spaedtke, P, 192
Spatami, C., 80
Spindle, H., 577, 581
Spindle, R., 233
Srinivasa, K. V., 63, 84

Srivastava, K. D., 53
Staley, D., 466, 559
Stamp, K., 368
Starlinger, A., 27
Staszesky, D., 174
Steiler, C., 116, 117, 118, 119, 122
Steimel, A., 572
Steinemer, N., 473
Steinke, K., 282 385
Stelzer, A., 424
Storey, B, 295
Störi, H., 92
Stoving, P.N., 273
Strayer, R. W., 66
Streater, A., 247
Stuchenkov, V. M., 56, 57
Stump, K., 420
Su, G., 50, 63, 171, 457
Sudarshan, T. S., 16, 285
Sugarwara, T, 188, 189
Sugawara, H., 264, 266, 267, 181
Sullivan, J. S., 333
Summer, R., 23
Sun, H., 104, 289
Sun, J., 66
Sun, L., 53
Sun, L., 580
Sun, W., 204, 205
Sun, Z., 249
Sun, Z., 546
Sunada, Y., 481, 506
Sundararajan, R., 18, 19
Sutterlin, R.-P., 179, 186, 205, 206
Suzuki, N., 227
Swanson, L. W., 66
Swindler, D., 509
Sysun, S., 181

T

Tadakuma, S., 575
Taguchi, Y., 462, 511
Tahata, K., 597
Taisne, J., 442, 450
Takaaki, F., 264, 266
Takahashi, E., 40, 41
Takahashi, H., 26, 27, 63, 84, 275
Takahashi, S., 333, 335, 344
Takahashi, T., 87
Takai, Y., 355, 361, 368
Takami, T., 362, 362
Takao, N., 492, 569
Takashima, T., 586
Takasuma, T., 242
Takasura, T, 366
Takeshita, M., 481, 585
Talento, J., 221
Tamagawa, T., 223, 233, 349, 352, 353, 354, 359, 587, 592
Tamura, R., 492
Tamura, S., 592
Tan, D., 535, 538, 559
Tanaka, D., 513
Tanberg, R., xx
Tang, G., 594, 595
Tani, K., 592
Tanimizu, T., 471, 474
Tateyama, C., 70, 71, 74
Taylor, E., 14, 28, 29, 30, 35, 36, 37, 49, 50, 96, 97, 100, 101, 116, 117, 118, 119, 122, 141, 146, 162, 163, 166, 194, 195, 196, 197, 198, 199, 208, 237, 239, 241, 273, 275, 306, 341, 370, 371, 373, 374, 380, 381, 383, 384, 455, 462, 466, 467, 470, 471, 501, 540, 548, 567, 598
Teate, Ph., 80, 143
Teichmann, J., 186, 187, 253, 424
Teillet-Billy, D., 80
Telander, S., 368, 420, 557
Temborius, S., 223, 224
Tenitskiy, P., 150, 151, 152
Ter Hennepe, J., 286, 287
Theissen, P, 145, 387, 469
Thielens, D., 511, 554
Thomas, J., 281
Thomas, R., 233, 302, 303, 304
Thompson, T., 414
Thuries, E., 568
Tian, Y., 171, 344, 346, 573
Timment, H., 592
Timsit, R. S., 31, 138, 139, 140, 484
Titus, C., xxii, 426
Tokoyoda, S., 597
Tomaschke, H., 31
Torben, S., 565
Toya, H., 92
Triaire C, 91
Tricarico, C., 23
Trinh, N., 403, 453
Trolan, J., 68
Tsuda, H., 264, 266, 267, 181
Tsuji, M., 471
Tsukima, M, 55, 355, 361, 368, 442, 453, 461
Tsunehisa, T., 490, 592
Tsuruta, K., 61, 158
Tsutsumi, T., 95, 96, 291, 300, 349
Tsventoukh, M., 160
Tuma, D. T., 72, 73, 154, 162, 435
Tuoky, E., 246, 363
Turner, C, 175
Turner, H., 175

U

Ubert, H., 557
Ubiennykh, B., 227
Uchiyama, K, 278, 279, 280
Uhrlandt, D., 160, 171, 172, 198, 200, 201
Uimanov, I., 70, 71, 78, 154, 162
Umeya, E., xxiii, 362, 481, 585
Unger-Weber, F., 85, 86, 88, 92, 326, 341, 352, 354, 356
Uppalapati, B, 525, 526
Urbanek, J., 567, 570, 571
Usui, N., 300
Utsumi, T., 32, 33, 45

V

Vallejo, H., 297
Vallo, D., 548

Van den Heuval, W., 426
Van der Linder, W., 509
Van der Sluis, L., 333, 335, 352, 355, 425, 426, 427
Van Hertem, D., 594
Van Lanen, E., 333, 335, 352, 355
Van Ranst, A., 442
Varey, R. H., 333, 335
Varneckes, V., 231
Vaughan, L., 594
Venna, K., 97, 287, 289
Vernon, D., 572
Vigreux, J., 13, 340
Vilchis-Rodriguez, D., 530, 532
Vinaricky, E., 31, 32
Vine, J., 95
Vithayathil, J., 590, 592
Vogelsang, R., 427
Von Engel, A., 9
Voshall, R, 182, 184, 185, 181, 233, 247, 264, 267, 278, 289, 481, 506, 577, 581, 582, 586, 590, 592, 593, 595, 597
Voss, G., 427
Vries, L. M., 31, 63
Vykhodtsev, A., 150, 151, 152

W

Wagner, C., 482, 548, 571
Wainio, R., 465, 471
Walczak, K., 300
Walczuk, E., 175, 372, 378, 379, 380, 391, 392
Waldron, M., 528
Walsh, G., 426
Wang, C., 153, 181, 182, 188, 190
Wang, D, 172, 280
Wang, E., 539
Wang, H., 535
Wang, J, 33, 41, 53, 98, 102, 169, 170, 171, 198, 200, 201, 208, 209, 221, 225, 226, 227, 264, 266, 267, 268, 271, 272, 273, 274, 277, 280, 281, 283, 283, 289, 442, 461, 481, 485, 486, 492, 522, 528, 530, 534, 535, 574, 578, 582, 583, 584, 585, 587, 588
Wang, L, 50, 63, 171, 181, 182, 188, 190, 192, 193, 199, 201, 204, 205, 229, 280, 333, 335, 457, 535, 537, 539, 559, 595, 596, 597
Wang, Q., 277, 280, 281
Wang, S., 573
Wang, T., 249, 250, 546, 554, 580
Wang, W., 169, 170, 233, 371
Wang, X., 169, 170, 226, 233, 249, 250, 260, 271, 282, 283, 371, 485
Wang, Y., 282
Wang, Z., 53, 63, 82, 102, 171, 198, 200, 201, 264, 266, 267, 268, 280, 281, 289, 330, 336, 344, 346, 481, 485, 486, 492, 522, 578, 580, 582, 584, 585
Watanabe, G., 158
Watanabe, H., 242, 366
Watanabe, K., 233, 275, 278, 279, 280, 337, 339, 341, 352, 361
Watanabe, R., 242, 366
Watanabe, Y., 474
Watson, W.G., 225
Wayland, P.O., 233, 264, 266, 268, 287, 459, 577, 581
Wei, J., 535
Wei, L, 3
Weibull, W. A., 27
Weiers, T., 427
Weigand, W., 592
Wein, R., 185
Weinrich, H., 566, 567
Weltmann, K.-D., 160, 171, 333, 352, 358
Wember, M, 567
Wen, H., 289
Wenkai, S., 280, 281
Wenze, N., 116, 117, 118, 119, 122, 154, 156, 424
Wenzel, N., 584
West, M., 162
Wethekam, S., 116, 117, 118, 119, 122
Weuffel, M., 326, 329
Weyer, H., 592
Whyte, W., 292
Widger, P., 535, 559
Wiehl, G., 295, 301
Wien, R., 590, 592, 593, 595, 597
Wiggers, R., 459, 460, 461
Wilhelm, M., 368, 420
Wilkening, E.D., 231, 232, 329, 365, 420
Williamson, J., 139
Wilmes, L., 426
Wilson, A. H., 142
Wilson, N., 301
Wind, G., 13, 340
Winkelnkemper, H., 13, 340
Wojciechowski, S., 378, 379
Wolf, C., 253, 255, 261, 262
Wood, A., 221
Wood, R. W., 30
Wooton, R.E., 278, 539
Wu, B., 181, 182, 188, 190
Wu, G., 577, 581
Wu, J., 574
Wu, S., 546, 554
Wu, W., 232

X

Xia, X., 262
Xiao, D., 535, 538, 559
Xiao, Y., 556
Xiaoming, L., 98
Xie, P., 424
Xiu, S., 102, 221, 227, 248, 249, 250, 260, 264, 266, 268, 481, 492, 522, 546, 554, 578, 582, 585
Xu, H., 226
Xu, J., 525
Xu, N, 32, 41, 42, 86
Xu, S., 41, 42, 511
Xu, X., 33, 86
Xue, J., 535, 538, 559
Xue, X., 207
Xui, S., 249, 250

Y

Yakovlev, V., 355
Yamada, M., 300

Yamamoto, A., 87, 89, 166, 282
Yamano, Y., 33, 97, 462, 511
Yamashita, B., 349
Yamazaki, T., 468, 469
Yan, J., 116, 117, 118, 119, 271, 272, 273, 283, 535, 554, 580, 587, 588
Yan, W., 53
Yanabu, S., 126, 28, 95, 96, 98, 99, 66, 167, 223, 231, 233, 234, 241, 242, 264, 266, 276, 333, 335, 349, 352, 353, 354, 359, 361, 427, 481, 492, 506, 513, 569, 597, 592
Yanagisawa, H., xxiii, 481, 585
Yang, D., 50, 63, 171, 188, 457
Yang, F., 578
Yang, H., 41, 53, 442, 460, 461
Yang, K., 530, 534
Yang, L., 98, 582, 583, 584
Yang, R., 102, 481, 492, 522, 578, 582, 587, 585
Yang, Z., 225
Yano, T., 442
Yao, J., 271, 272, 273, 283, 587, 588
Yao, X., 191, 192, 207, 226, 271, 273, 283, 485, 585, 586, 587, 588
Yarabu, S., 492
Ye, H, 546, 554
Yeckley, R., 233, 577, 581
Yen, Y. T., 72, 73
Yevsin, D., 182, 580
Yi, X., 169, 170, 207, 208
Yokokura, K., 26, 28, 223, 231, 276, 278, 361, 366, 420, 436, 437, 468, 469, 520
Yokoyama, T., 362
Yoshida, D., 597
Yoshida, H, 63, 84
Yoshida, T., 21, 453, 461, 535
Yoshihara, T., 227
Yoshihiko, M., 264, 266
Yoshioka, N., 227
Yoshioka, Y., 372, 378
Yu, G., 158, 162, 573
Yu, K., 191, 192
Yu, L., 271, 273, 525, 526
Yu, X., 578
Yu, Y., 41, 53, 442, 461
Yu, A., 71
Yuan, B., 248, 249, 250, 260,
Yuan, F., 232
Yuan, S., 102, 481, 492, 522, 582, 585, 587
Yungdong, C., 98
Yushkov, G., 158, 159, 160, 162

Z

Zabello, K., 203, 278, 279, 355
Zaborszky, J., 445, 447, 450
Zalucki, Z., 178, 513
Zang, Y, 226
Zdanuk, E., 221
Zemskov, 71
Zhai, X., 207, 554, 580
Zhang, B., 424, 556
Zhang, C., 225, 226
Zhang, H., 225, 226
Zhang, J., 282
Zhang, J. G., 18
Zhang, L., 204, 205, 530, 534, 573
Zhang, S., 116, 117, 118, 119, 226
Zhang, W., 207
Zhang, X., 171, 264, 271, 273, 277, 282, 283, 300, 485, 535, 537
Zhang, Z, 169, 170, 207, 208
Zhao, F., 268
Zhao, L., 249, 250
Zhao, Li., 248
Zhao, S.-F., 522
Zhao, Y., 581
Zheng, J., 249
Zheng, W., 559
Zheng, Y., 264, 281
Zhoa, Y., 226
Zhoa, Z., 289
Zhou, B., 535, 538, 559
Zhou, F., 226
Zhou, H., 102, 481, 492, 522, 582, 585, 587
Zhou, X., 146, 387
Zhou, Z., 63, 82, 83
Ziomek, W., 57, 58, 61, 62
Zou, H., 289
Zou, J, 104, 110, 358, 424, 528, 553, 574, 580, 581
Zulucki, Z., 262
Zuvers, H., 592, 593

Subject Index

A

Advantages of vacuum interrupters, xix, 403, 599
AMF (axial magnetic field)
  see, Vacuum arc in axial magnetic field
AMF contacts,
  see, Axial magnetic field contact structures
Anode flares, 78–84
Anode phenomena, vacuum breakdown, 44–51
  electron beam radius at anode, 45, 46
  electron beam heating below anode surface, 47–51
  electron beam penetration, 47, 48
  microparticle source, 46, 47
  surface heating, 46
  time to boiling, 49, 51
  time to melting, 49, 50
  anode flares, 78–84
Anode spot, 169–174, 206–207
  columnar vacuum arc at current zero, 174
  Type I and Type II anode spots, 171–173
  axial magnetic field contacts, 206–210
  overview and review, 209, 210
Appearance diagrams,
  axial magnetic field contacts, 275
  butt contacts, 176–178
  development of 176, 177
  spiral and contrate cup contacts, 258–260
  TMF contacts opening at peak ac fault current ½ cycle, 259
  TMF contacts opening with full ac fault current ½ cycle, 258
  TMF contacts opening with small contact gap, 260
  transition vacuum arc, 180
Arc formation
  see, Vacuum arc formation
Arc vacuum
  see, Vacuum arc
Axial magnetic field contact structures 262–273
  advantages and disadvantages, 262–273
  appearance diagram, 275
  arc energy distribution, 192, 205–208, 274, 275
  arc voltage for high current diffuse arc, 197–204
  cathode spot distribution, 205–207, 274
  closed contact impedance, 262–269, 385, 486
  comparison with transverse magnetic field structures, 284
  contact designs, 263–274
  contact gap effect, 265, 270–272, 485
  design parameters, 265
  eddy currents, 265, 272, 273
  electrical switching life, 544, 545
  external coil, 262–264
  finite element analysis, 273
  high current diffuse column vacuum arc, 274
  high current diffuse vacuum arc, 274
  high current diffuse vacuum arc energy distribution, 346
  interruption current as f(contact diameter), 274–278
  magnetic field distribution, 268–273, 278–281, 585
  magnetic materials use, 269, 278–281
  multiple spot vacuum arc, 273
  opening sequence, transition to diffuse vacuum arc, 194–202
  rod structure, 282
  slots, 265, 266, 268, 272, 273, 278, 280
  vacuum arc modes, 273, 274

B

Basic impulse level
  see, BIL
Bellows, 289–291
  anti-twist bushing, 290
  long stroke, 291
  mechanical life, 290
  seam-welded hydro-formed, 289
  seamless hydro-formed, 289
  squirming, 291
  twist, effect of, 290
  welded washers, 289
BIL (Basic Impulse Level), 3–6
  see also, LIWV (Lighting Impulse Withstand Voltage)
  chopped wave, 6, 86
Blow-off force, 143–148, 373
  butt contacts, 145
  contact melting, 147, 148, 373–376
  required contact force, 145, 146, 373
  welding, closed contacts, 371, 373
Bolted faults
  see, Short circuit currents
Boltzmann's constant, 142, 341
Breakdown across a ceramic cylinder, 16–21
  see also, Creepage
Breakdown
  see, Electrical breakdown in gas and Electrical breakdown in vacuum
Bus faults
  see, Short circuit currents
Butt contacts, 246, 247
  see also, Circuit interruption ac circuits < 2kA
  appearance diagram, 176–179
  arc erosion, 173–176, 431–436
  interruption, 224, 232, 323–345
  large area design problem, 247
  load break switch contacts, 403
  motor contactor switch contacts, 433

C

Cable switching, 462, 463, 555
  distributed capacitance, 463
  peak recovery voltages, 463
  testing currents, 463
  criticism of certification circuits, 555
Capacitor switching, 437–462, 550–555
  see also, Cable switching and Overhead line switching

see also, Circuit interruption (capacitive), small contact gaps
circuit breaker, 231, 550–555
contact material, 437, 441, 551
contact welding, 442, 451, 454, 462
disconnecting capacitor banks, 443–448
electron emission, 452–462
inrush current, 438–443
inrush current limiters, 442, 443
inserting a parallel capacitor bank, 440, 441
inserting a single capacitor bank, 438–440
kVAR and MVAR definition, 438
NSDD, 462
see also, Non sustained disruptive discharge
see also, NSDD fault current switching
over-voltage protection, 447
recovery voltage, 366, 367, 450
restriking, 445–447
restriking resulting from electron emission, 450–462
restriking probability as f(prestrike arc energy), 451
standards, 448, 449
synchronous circuit breaker operation, 442, 553
testing, 449, 551–554
three phase capacitor banks, 448–450
vacuum interrupter design, 437
voltage escalation, 445–447
welding and β_m, 450–459
Cathode flares, 71–73, 78–84
Cathode spot erosion, 162–164
see also, Contact erosion
absolute erosion, 162–164, 192, 454
axial magnetic field effect, 191, 192
effective erosion, 454, 455
ions, 158–161
metal particles, 161–163
neutral metal vapor, 161–163
Cathode spot, 149–161
see also, Cathode spot erosion
see also, Contact erosion
see also, Current chop
cathode fall, 156–158, 161, 162
cells per spot, 154
crater size, 154
current change effect, 165–167, 330
current per spot, 153, 154
current per spot f(axial magnetic field), 188–190
electron emission, 156
electron energy, 162, 330
erosion, 162–164
explosive formation, 154
fractal characteristic, 141
high pressure gradient/electron-ion friction theory, 156, 157
ion angular distribution, 162–164
ion charge state, 156–162
ion emission, 156–151
ion energy, 156–165, 330
ion energy, pressure gradient effect, 156–158
ion Mach number, 159–161
ion velocity, 159, 160, 162
model, 156–161
plasma plume, 162–164
potential hump theory, 161, 162
random walk, 154, 155
retrograde motion, 154–156, 330
splitting, 154, 155
threshold spot current f(axial magnetic field), 188–190
type I and type II, 153
Ceramic body, 283–285
advantages, 284
metallization, 284, 285
Child's law, 68, 158, 273
permittivity in vacuum, 273
Chopping current
see, Current chop
Circuit breaker applications
see also, Generator vacuum circuit breaker
see also, Switching de circuits
circuit breaker connection to switchgear bus, 559
electric arc furnace (EAF), 575
electric transit circuit frequencies, 573
metal clad switchgear definition, 557
metal enclosed switchgear definition, 557
outdoor circuit breakers, 561
pad mount secondary distribution systems, 565, 566
parallel connection, 492, 569, 570
primary distribution, 559, 560
reclosers, see Reclosers
ring main units, 471, 564, 565
secondary distribution, 559, 560
shipboard circuit breakers, 573–575
switchgear, 557–561
transmission circuits (72 kV–242 kV), 582–589
transportation breakers on locomotive, 573
transportation breakers trackside, 573
Circuit breaker design distribution circuits, 515–546
see also, Circuit breaker load current
see also, Generator vacuum circuit breaker
see also, NSDD
see also, Welding closed contacts
see also, Welding closing contacts
blow off force, 517
bolted connection to vacuum interrupter, 488
cable switching, 555, 556
capacitor switching, 551–554
certification, 546–549
closed contact force, 517
closing characteristics, 386, 521–523
closing forces, 523
conditioning installed vacuum interrupters, 105–111
contact bounce, 386, 521
contact force from current loop at moving contact, 519
contact material, 224
contact material low surge, 223, 433, 520
contact sliding prevention, 520–521
convective flow pole design, 491, 536
coordination, 517, 518
definition, 481
electrical life, fault current, 541–546
electrical life, load current, 543
endurance, testing, 549
fault current interruption probability, 482
fault currents, see Short circuit currents
field effects with grounded cylinder surrounding vacuum interrupter, 541
flexible connection to vacuum interrupter, 488, 490
heat sinks, 490–492
interruption delays, 503

interruption performance as f(opening time before ac current zero), 502–505
line dropping, 555
load current, 482–492
low voltage (<600V circuits), 598
magnetic mechanism, 525–528
motor switching, 556
moving mass requirements, 523, 524
NSDD capacitor switching, 554, 555
NSDD fault current switching, 509–515
opening characteristics, 503, 521
opening force, 523, 524
overtravel, 403, 405
parallel connected poles, 492, 568
pole spacing, 532–537
prevention of contact sliding, 519, 520
reclosing cycles, 496, 497
short circuit currents, *see* Short circuit currents
sliding connection to vacuum interrupter, 394
spring mechanism, 423–425
stab connection to switchgear bus, 559
standards, 482, 484, 515, 516
synchronous operation for capacitor switching, 553, 554
temperature rise at joints, 483–486
terminal (or bus. bolted) fault, 492–507
test ratings, 482, 515, 516
transient recovery voltage, *see* Short circuit currents
tulip connection to switchgear bus, 559
vacuum interrupter insulation, 535, 536
vacuum interrupter mounting, 534–535

Circuit breaker design sub-transmission and transmission circuits, 576–590
see also, Circuit breaker design distribution circuits 12 1–145kV
121 kV to 171kV designs, 582–589
242kV future designs, 589
72 kV designs, 582, 586, 587
AMF contact design 269, 270–272, 485, 486
contact gap requirements, 582–586
contact opening speed, 585, 586
high voltage requirements, 579
series interrupters, 576–582

Circuit breaker load current 482–492, 516
see also, Constriction resistance
connection advantages and disadvantages, 490
connections to vacuum interrupter, 487–490
heat sinks, 490–492
parallel connected poles, 492, 568
power as f(vacuum interrupter impedance), 486
R 10 series of load currents, 483, 516
Renard series, *see* R 10 series
Renard, Charles, 483
skin depth, 487
temperature rise at joints, 483–486

Circuit interruption ac circuits < 2kA, 263– 275
see also, Post arc current
see also, Transient recovery voltage (TRY)
application of dc circuit interrupters, 329
effect of transient recovery voltage, 333–337
effect of getters on internal pressure, 329
electron mean free path as f(pressure and temperature), 327
free recovery measurement, 324–330
gas density as f(pressure and temperature), 327
metal vapor density after current interruption, 326–328
metal vapor mean free path as f(pressure and temperature), 327
plasma redistribution at ac current zero, 331, 332
post arc current model, 333–337
reignition butt contacts, 224, 232
rate of change of current at ac current zero, 323, 324
vapor pressure of metals as f(temperature), 331

Circuit interruption ac circuits 2 kA–12kA, 337–345
see also, Post arc current
see also, Transient recovery voltage (TRV)
anode temperature at current zero, 340
critical *pd* and *nd* values, 341–344
free recovery measurement, 337–345
influence of metal drop lets, 230, 345
metal vapor density influence on current interruption, 340–344
reignition types, 389
role of Paschen curve, 341–344
Weil–Dobke circuit measurement, 337

Circuit interruption high currents ac circuits TMF and AMF contacts, 345–361
see also, Post arc current
see also, Short circuit currents
see also, Transient recovery voltage (TRY)
contact opening just before ac current zero, 361–369
contact temperature influence at current zero, 205, 206, 340, 360, 361, 365
distribution of arc energy, 345
factors preventing current interruption, 356
interruption as f(contact diameter and circuit voltage), 255, 256, 276, 277, 349
metal vapor density influence on current interruption, 356, 359–361, 504
performance comparison TMF and AMF contacts, 284
post arc current, 332–335, 347–355
transition to diffuse vacuum arc before current zero TMF contacts, 345, 346, 360

Circuit interruption (capacitive), small contact gaps, 367–368
contact opening just before ac current zero, 367
electric field as f(contact opening speed), 368
TRV, 366, 367

Circuit interruption (inductive), small contact gaps, 362–366
see also, Inductive switching
contact opening just before ac current zero, 362
effect of current chop, 362–364
reignition, 364–366, 420, 421
TRV, 362–364
voltage escalation, 364, 365

Circuit interruption dc circuits, 589–597
AMF contact structure advantage for counter-pulse circuits, 591
counter-pulse circuit, 590–593, 595–587
criteria required for AMF vacuum arc, 591, 592
criteria for metal oxide surge arrester, 597
current and voltage at counter pulse interruption, 595–597
low current (few amperes), 589, 590
practical operation with fusion reactors, 590–593
pulsed transverse magnetic field circuit, 590
series interrupters, 597

Clean room, 292–294
activity effect, 294

classifications, 293
design, 292
Columnar vacuum arc, 173–179
anode and cathode jet columns, 177–179
anode spot after current zero, 173, 174
appearance diagram, 176, 177
arc temperature, 179
bridge column arc, 176
constricted column, 176–179
diffuse column arc, 176–178
energy balance, 175
erosion, 173–176
plasma jet column, 177, 178
arc pressure ac currents 10kA–30kA (peak), 178, 179
arc temperature ac currents 10kA–30kA (peak), 178, 179
Conditioning of assembled vacuum circuit breakers, 105–113
Conditioning, 26, 27, 86–94, 304
BIL conditioning vacuum circuit breakers, 100–105
contact surface after conditioning, 93, 94
current conditioning, 91, 92
deconditioning, 95, 96
other conditioning processes, 92
puncture, 94, 95
spark condition ac voltage, 87–90
spark conditioning using BIL pulse, 88–90
vacuum interrupter manufacture, 304
Constriction resistance, 137–141
a-spots, 138, 140
contact, mechanical area, 138–140
contact resistance, one region of contact, 139, 140
contact resistance, 'n' regions of contact, 140, 376, 377
contact surface topology, 32, 138
contact nominal area, 138, 139, 372
function of temperature, 374
hardness as f(temperature), 375
hardness relation, 139, 141
load relation, 138–140, 373
micro topography, 32, 138
plastic deformation, 138–140
stability in vacuum, 220
temperature of
see Voltage–Temperature relation
Contact breakdown parameters, closing contacts
see also, Electrical breakdown in vacuum
see also, Anode phenomena, vacuum breakdown
see also, Welding closing contacts
see also, Conditioning
contact closing prestrike arc, 125, 386, 452
contact bounce, 386, 521, 522
micro topography, 32
Contact definition, 137–139
Contact erosion, 153, 161–164, 174, 175, 192, 432
absolute cathode erosion, 161–164, 192, 433–436
columnar vacuum arc, 174, 175
contactors, 432
diffuse vacuum arc in axial magnetic field, 192
diffuse vacuum arc, 153, 161–164
effect of contact gap, 435, 436
effective cathode erosion, 433–436, 542
electrical life, fault current, 542–544
electrical life, load current, 542–544
metal particles, 161–164, 436
transition vacuum arc, 179, 180, 443
Contact materials, vacuum interrupters, 220–234
advantages and disadvantages, 245
application, 244
capacitor switches, 427
chopping current, 235, 236
copper and copper-based materials, 221
see also, Copper Bismuth
interruption with butt contacts, 224, 232, 323–345
low surge, 366, 420, 436, 520
material properties, 222, 223, 225, 229, 230, 243
material properties assessment, 243
performance needs, 222
refractory plus good conductor materials, 221–224
see also, Copper Tungsten
see also, Copper Tungsten Carbide
see also, Silver Tungsten Carbide
see also, Tungsten Copper
reignitions, 224, 366
semi- refractory plus good conductor materials, 224–233
see, Copper Chromium
vapor pressure of metals as f(temperature), 331
weld resistant, 222, 385, 432, 520
Contact mechanical area, 138, 139
see also, Constriction resistance
contact surface topography, 32, 138
Contact structures, 198–230
see, Axial magnetic field contact structures
see, Butt contacts
see, Transverse magnetic field contact structures
comparison, 284
Contact welding
see Welding
Contact, nominal area, 138, 139, 372
see also, Blow-off force
Contact, true area of, 138, 140
see also, Voltage–Temperature relation
Contactor
see, Vacuum contactors
Contour wave ceramic, 21
Contrate cup contacts,
see Transverse magnetic field contact structures
Conversion of pressure units, 24
Copper Bismuth, 221
current chop, 234, 236
history, 221
material properties, 223
percentage bismuth, 221
weld resistance, 221
Copper Chromium, 224–233
additives, 225, 233
circuit breakers, 224
current chop, 235, 241
droplet formation during arcing, 229–231
effect of impurities, 232
gas content, 228–232
interruption with butt contacts, 231, 232
manufacturing techniques, 226, 227
material properties, 229, 230
multilayered, 227
optimum material ratio, 238
percentage copper, 233
phase diagram, 226, 227
surface change after arcing, 227, 228, 232, 233

- Copper Tungsten Carbide, 221–224
  - current chop, 223, 235
  - interruption with butt contacts, 224
  - material properties, 223
  - percentage copper, 223
  - reignition with butt contacts, 224
  - weld resistance, 221
- Cranberg's Hypothesis
  - *see,* Microparticles
- Creepage, 17–21
  - breakdown models, 19
  - ceramic design, 20, 21
  - contamination, 18
  - creepage distance, 19, 20
  - standards, 19, 20
  - surface conductivity, 19, 20
- Crown of thorns impact structure, 54
- Current chop, 165–169, 224, 233–242, 362–364
  - *see also,* Virtual current chop
  - current at chop, pure metals, 166, 234, 235
  - current at chop, metal mixtures, 166
  - current change effect, 167, 168
  - effect of switching, 234, 236
  - effect of circuit surge impedance, 236–241
  - effect on TRV, 362–364
  - final chop event, 168–169
  - first instability, 167, 168
  - other arc media (air, SF6 and oil), 165, 242
  - practical measurement of, 238–241
  - statistical distribution, 240–242
  - vacuum arc stability at low currents, 77, 166, 590
  - virtual current chopping, 421–423, 425
- Current interruption
  - *see,* Circuit interruption (capacitive), small contact gaps
  - *see,* Circuit interruption (inductive), small contact gaps
  - *see,* Circuit interruption ac circuits < 2kA
  - *see,* Circuit interruption ac circuits 2 kA–l 2kA
  - *see,* Circuit interruption ac circuits high currents TMF and AMF contacts
  - *see,* Circuit interruption de circuits

D

- Deconditioning, 93, 95, 96, 107–113
- Diffuse vacuum arc, 152–165
  - *see also,* Cathode spot
  - anode spot, 169–174
  - cathode spot, 153–161
  - inter-contact plasma, 161–165
- Distribution systems, 401, 402
  - primary loop, 401, 402
  - primary selective, 401, 402
  - radial, 401, 402
  - reliability, 402
  - secondary selective, 401, 402
  - secondary spot, 401, 402

E

- Electric field for vacuum breakdown, 31–37
  - breakdown site distribution, 42, 43
  - critical field, 31, 63
  - geometric enhancement factor, 34–37
  - microparticle effects, 51–56, 63–66
  - microscopic enhancement factor, 31–34
- Electrical breakdown at high altitudes, 113, 114
- Electrical breakdown in gas, 6–17
  - *see also,* external vacuum interrupter high voltage design
  - altitude effect, 113, 114
  - arc formation, 8
  - dissociation, 8
  - elastic collision, 8
  - electric field, 7, 8
  - electron-gas interaction, 7–15
  - electron avalanche, 10
  - excitation and relaxation, 8
  - formative time lag, 15, 16
  - inelastic collision, 8, 9
  - ionization, 9
  - ionization efficiency, 10
  - ionization potential, 9
  - mean free path, electron, 11, 327
  - mean free path, gas, 11, 327
  - mean free path, molecule, 11
  - metal-cylinder sandwich, 16, 17
  - necessary conditions, 15
  - Paschen's law, 12–15
  - statistical time lag, 15
  - Townsend breakdown, 10–12
  - Townsend ionization coefficients, 10–12
  - triple point, 16, 17
  - voltage collapse, 22
- Electrical breakdown in vacuum, 23–114
  - *see also,* Anode phenomena, vacuum breakdown
  - *see also,* Electric field for vacuum breakdown
  - *see also,* Enhancement factors
  - *see also,* Field emission
  - *see also,* Internal vacuum interrupter high voltage design
  - *see also,* Microdischarges *see also* Microparticles
  - *see also,* Microprojections *see also* Prebreakdown
  - *see also,* Vacuum arc formation, breakdown
  - *see also,* Vacuum breakdown voltage
  - absorbed power into contacts, 63
  - anode heating, 47–51
  - anode thermal conductivity, 51
  - breakdown sequence, 78
  - breakdown site distribution, 41, 42
  - breakdown voltage as f(contact gap), 25, 26
  - breakdown voltage as f(pressure), 26
  - breakdown voltage series gaps, 103, 104, 554, 578–582
  - breakdown voltage statistics, 27–31
  - closing contacts, 125, 386
  - conditioning, 26, 27, 86–94
  - contact surface topography, 32
  - critical electric field, 31
  - electric field, 31–37
  - enhancement factors, 31–37
  - field emission current, 38–44
  - insulating inclusions, 56, 57
  - location of metal vapor in contact gap, 63, 70–72, 75, 79, 83
  - microdischarges, 56–62
  - microparticles, 51–57
  - microparticle formation at anode, 47
  - microprojections, 32–34

models of breakdown time, 78–84
necessary conditions, 15
prebreakdown effects, 38–62
rate of rise of voltage pulse, 74, 104–107
spatial light distribution in contact gap, 63, 70–72, 75, 79, 83
statistics, 27–31
time to breakdown, 84–86
transition to the vacuum arc, 62–84
triple point, 97
vacuum breakdown voltage, 25–31
voltage collapse time, 79–83
voltage across vacuum interrupters in series, 103, 104, 409, 412–414, 578–582
voltage pulse width, 74, 104–107
Weibull distribution, 27–31
work function values, 31
Electron avalanche, 10–13
Electron emission
see, Field emission
Electron gas collision processes, 7–13
Encapsulation, 21–23
criteria, 23
cycloaliphatic epoxy, 22, 540
polyurethane, 22
porcelain housing plus insulating medium, 22, 492, 539
porcelain housing moisture ingress, 536
shrink wrap, 22
silicon rubber, 22
thermoplastics, 22
Enhancement factor (β), 31–37
see also, Electrical breakdown in vacuum
see also, Internal vacuum interrupter high voltage design
(β_g) geometric enhancement factor, 34–37, 96–103
(β_m) microscopic enhancement factor, 31–34
calculated (β_m), 32
conditioning effect (β), 86–94
contact area effect (β), 96–100
contact closing, no current (β), 96
contact gap and diameter with floating shield (β_g), 100–103
contact gap and diameter with no floating shield (β_g), 35–37
contact surface topography, 32
design for (β_g), 100–103
effect of contact welding on (β_m), 93, 387–390
effect of microparticle on (β_m), 32, 33
surface effects and (β_m), 33
Environmental advantages of vacuum interrupters, 306
Erosion
see Contact erosion
External vacuum interrupter high voltage design, 17, 21–23

F

FEA (finite element analysis), 35–37, 95, 97, 273, 541
electric field, 35–37, 95, 97
magnetic field, 273
Field emission, 38–44, 450–462
see also, Fowler–Nordheim equation
annealing effect, 40, 41
Child's law, 68, 82, 184
current density at breakdown, 66–69
currents at vacuum breakdown, 39
electron beam radius at anode, 45–50
electron beam spread, 45
function of prestrike arc energy, 454
Fowler–Nordheim equation, 38, 39
interpretation problems, 39–42
Nottingham effect, 66
number of sites, 39–41
penetration into anode, 47–50
space charge limitation, 68–70
temperature rise of emission site, 67
work function values, 31
X-radiation, 43, 44, 114–125
Finite element analysis,
see, FEA
Fowler–Nordheim equation, 38–39
field sensitivity, 45—46
interpretation problems, 48
space charge limitation, 69–70
work function sensitivity, 39, 40, 68–70
work function values, 31

G

Gas insulation,
see also, SF_6 insulation
breakdown strength of gas mixtures, 538
potential gases to replace SF_6, 535
SF_6, 535–538
SF_6 mixed with N_2 or He, 538
Generator vacuum circuit breaker, 565–572
continuous currents, 567–570
generator fed faults, 571, 572
high dc asymmetric fault currents, 571, 572
history, 567
out of phase switching, 572
parallel operation, 569, 570
performance comparison generator and distribution circuit breakers, 568
performance criteria, 567
standards, 568
TMF contact structure superior to AMF structure, 571
transformer/system fed faults, 569, 570
X/R high values, 570
Geometric enhancement factor (β_g),
see, Enhancement factor (β)

H

High voltage standards
see, Standards high voltage
History, xix–xxiv
books, xxiv
copper–bismuth material, 221
copper–chromium material, 225
generator circuit breakers, 566
manufacturing improvements, 304–306
market share, xxii
patents, xx
publications, xxiii
size reduction, xxii, 220
time-line, xxiv

I

Inter-contact plasma, 141–144
  see also, Diffuse vacuum arc
  electron energy, 162
  ion angular distribution, 163, 164
  ion current, 162
  ion energy, 162–164
  ion energy distribution f(current), 165
  ion velocity, 159, 160
  overlapping plasma plumes, 164
  plasma plume, 164
Internal vacuum interrupter high voltage design, 96–105
  see also, Finite element analysis
  component area effect, 97, 98
  design for (β_g), 97, 98
  effect of shield, 97–103
  FEA (finite element analysis), 35–37, 95, 97, 541
  field strength model, 97, 98
  roughness of surfaces, 32, 97
  shield radius, 103
  three phase effect, 535, 537
  vacuum interrupter inside a grounded container, 539–541
Ionization in gas, 6–17
Ionization potential, 9
Isolation switches, 464–474
  design concepts for fault protection and grounding, 470–473
  design concepts for switching and grounding, 464, 465, 470–474
  design concepts for vacuum interrupters, 464, 469
  encapsulation, 470
  SF_6 3-postion switch, 469–471
  standards, 468
  threshold breakdown voltage, 466–468
  vacuum interrupter after no-load switching, 468
  vacuum interrupter with new Cu–W contacts, 466, 467
  vacuum isolation compared with gas isolation, 468
  Weibull distribution vacuum breakdown, 466, 467

L

Leopold sliding connection, 489
Lightning Impulse Withstand Voltage
  see, LIWV
LIWV (Lightning Impulse Withstand Voltage), 3–6
Load break switches, 403–414
  air isolator in series with, 406–408, 412
  cable switching, 462, 468
  capacitor switching, 437–462
  contact material, 436
  isolation switches, 464–474
  mechanisms, 403–407
  overtravel, 405
  rating, 403
  sectionalizers, 408–411
  series operation, 409, 412, 413
  standards, 401, 468
  switching schemes, 407–411
  transmission voltages, 409, 412–414
  vacuum interrupter advantages, 403
  vacuum interrupters for, 404
Lorenz constant, 142
Low current vacuum arc stability, 77

M

Mach number, ion, 159
Magnetic field, 181–209
  see, Axial magnetic field contact structures
  see, Transverse magnetic field contact structures
  see, Vacuum arc in transverse magnetic field
  see, Vacuum arc in axial magnetic field
  axial magnetic field, 187–209
  transverse magnetic field, 181–187
Magnetron pressure testing, 298–300
Manufacture, 283–304
  assembly, 291–296
  batch production, 296-390
  clean room, 292–294
  cleaning, 294
  conditioning, 304
  gas species liberated in vacuum furnace, 303, 304
  leak rates, 300–302
  magnetron pressure gauge, 298–300
  one step assembly, 293
  pressure testing, 298–300
  residual gas analysis, 298, 303, 304
  sub-assembly then final assembly, 293–296
  tubulation exhaust, 291
  vacuum furnace, 292, 296–298
  vacuum furnace temperature-time ramp, 297
Mean free path, 11, 327
  electron as f(pressure and temperature), 327
  metal vapor as f(pressure and temperature), 327
Mechanisms, 403–405, 520–532
  circuit breaker, 520–532
  contact bounce, 522
  closing force sequence, 523
  impact force during closing, 526
  load break switches, 403–405
  magnetic mechanism, 525–528
  overtravel, 405–405
  solenoid mechanism, 403, 428–433
  spring mechanism, 523–525
  Thomson coil mechanism, 528–532
  vacuum contactor, 403, 428–433
Microdischarges, 56–62
  charge passed, 59, 61
  experimental observations, 27, 58
  formation hypotheses, 60, 61
  inception voltage and breakdown voltage, 62
  inception voltage, 58
Microparticles, 51–57
  acceleration in electric field, 53, 54
  bouncing, 55–57
  breakdown with rough contacts, 63–66
  breakdown with smooth contacts, 63–66
  charge, 53
  Cranberg's hypothesis, 65
  distribution on contacts after machining, 52
  energy to cause breakdown, 65
  energy, 53
  from diffuse vacuum arc, 154, 162–164, 433–437
  impact effects, 54–55
  nonmetallic particles, 56, 57
  velocity, 54, 55
Microprojections, 31–34
  annealing, 41

(β_m) microscopic enhancement factor, 31–37
capacitor inrush current effect, 452–455, 461
explosive rupture, 68, 74, 75, 78, 79
modeling, 32
thermal stability, 67, 68
welding effect, 485–490
Microscopic enhancement factor (β_m),
see, Enhancement factor (β)
Molten metal bridge, 130
contact melting voltage, 142–143
duration as f(current), 152
duration bridge column arc, 152
formation, 147, 148
rupture, 148
transition to bridge column arc, 149–152
transition to pseudo arc, 149
Motor switching,
see, Switching inductive circuits
see, Vacuum contactors
Multilam sliding connection, 489

N

Non sustained disruptive discharge,
see, NSDD
NSDD, 462, 510–515, 554, 555
certification lab experience, 511
circuit breaker capacitor switching, 554, 555
circuit breaker, 510–515
contact closing, 386
electron emission, 454, 513
experimental observations, 513
field experience on circuit breaker performance, 513
load switch capacitor switching, 462
mechanisms of initiation, 513
NSDD, non-sustained transient discharge, 512
to burp is to be commended, 515

O

Overhead line switching, 462, 463
criticism of certification circuits, 555
distributed capacitance, 463
testing currents, 463
Overtravel, 403–405

P

Paschen's law, 12–14, 342
air and SF6, 14
non applicable contact gaps, 14
minimum breakdown voltage, 14
minimum pressure times gap, 14
calculation of, 13
interruption of vacuum arc, 330–345
Penning discharge, 298–300
Magnetron gauge, 299
Post arc current, 332–336, 339, 347–359
post arc current model, 333–337
butt vs. AMF contacts, 353
copper content of copper chromium contacts, 354
effect of di/dt at current zero, 350
effect of peak current, 350
excess metal vapor, 351, 354–357
lack of correlation with interruption, 352
Prebreakdown, 38–62
anode phenomena, 44–51
field emission from cathode, 38–44, 63
microdischarges, 56–62
microparticles, 51–57
Pressure ranges in vacuum technology, 24
conversion of units, 24
electron mean free path, 11, 327
gas mean free path, 11, 327
monolayer formation, 11
other features, 11
particle number density, 11, 327
vapor pressure of metals as f(temperature), 331
Pseudo arc, 149
Puncture, 94, 95

R

Rate of change of current at ac current zero, 323, 324
Rate of rise of recovery voltage (RRRV), 321–324, 496–510
circuit breaker terminal faults, 496–501
generator faults, 571, 572
short line faults, 507–509
transformer secondary faults, 509, 510
Recloser, 481
definition, 481
design characteristics, 561–564
endurance testing, 561
operation sequence, 410, 411, 562
use with sectionalizers, 410, 411
Reignition, 337–345, 351
dielectric, 339, 504, 505
inductive circuit, 362–365
thermal, 339, 504, 505
Residual gas analysis, 303, 304

S

SCADA, supervisory control and data acquisition, 565, 566
Sectionalizers use with reclosers, 410, 411
Series vacuum interrupters, 103, 104, 409, 412–413, 576–582
breakdown voltage, 103, 104, 581
transmission load break switch, 412
transmission circuit breaker, 576–582
voltage distribution, 409, 412–414, 579–581
SF_6 insulation, 535–538
see also, Gas insulation
dielectric strength 14, 535
greenhouse gas, 535
potential replacement gases, 535
SF_6 with $CF_3I–N_2$, 538
SF_6 with N_2 and He, 538
Shields, 286–289
attachment, 286, 287
ceramic, 288
effect on (β_g), 96–103
fixed, 286
floating, 287
three shields, 289
Short circuit currents, 351, 492–496

see also Circuit interruption high ac cur rents TMF and AMF contacts
see also Generator breaker
see also NSDD
AMF and TMF contact energy distribution, 231, 277, 506
asymmetry, 492–496
de component, 492–496
dielectric reignition, 339, 504, 505
figure of merit, 501
first pole to clear factor, 499–502
generator fed faults, 571, 572
high frequency (>60 Hz), 573–575
interruption performance as f(opening time before ac current zero), 503
long arcing times, 506
low frequency (<*50* Hz), 573–575
minimum and maximum arcing time, 505, 506
NSDD after fault current interruption, 509–515
parallel pole operation, 568, 570
performance comparison TMF and AMF contacts, 284
R 10 series of terminal fault currents, 516, 549
reclosing cycles, 410, 411, 562
short line faults, 507–509
terminal (or bus, bolted) faults, 493–486
thermal reignition, 399, 504, 505
three phase asymmetry, 493–496
transformer secondary faults, 509, 510
transformer system fed faults, 569, 570
TRV after fractional faults, 502
TRV after short line faults, 507–509
TRV after terminal fault, 496–502
X/R definition, 494
Short line faults
see Short circuit currents
Silver Tungsten Carbide, 221–225
current chop, 2254, 235
interruption with butt contacts, 224
material properties, 223
percentage silver, 223
reignition with butt contacts, 224
vacuum contactors, 23, 244, 245, 432
weld resistance, 223, 244, 245, 385
Spiral contacts
see, Transverse magnetic field contact structures
Standards
capacitor switches, 447–449, 551, 552
circuit breaker, 3, 483, 515, 516, 549, 550
contactors, 428, 431
creepage, 19, 20
generator breakers, 567
high voltage, 3, 483
isolation switches, 468
load break switches, 402, 468
reclosers, 550
Surge protection, 415–421
effect of, 419, 420
electric motors, 419
positioning, 419
protection schemes, 417
surge capacitors, 417
surge resistors, 418
transformers, 487
Switching dc circuits
see, Circuit interruption de circuits
Switching inductive circuits, 414–428
see also, Circuit interruption (inductive), small contact gaps
see also, Transformer switching
see also, Vacuum contactors
multiple reignitions, 362–366
surge capacitors, 317
surge protection for electric motors, 419
surge protection positioning, 419
surge protection schemes, 417
surge resistors, 418
TRV, 363, 374
virtual current chopping, 442, 443
voltage escalation, 362–366
voltage surges on closing, 414, 415
voltage surges on opening, 414–416

T

Temperature coefficient of resistance, 308
Terminal faults
see, Short circuit currents
Testing
capacitor switch, 449
certification testing vacuum circuit breakers, 546–550
circuit breaker endurance testing, 549, 550
circuit breaker capacitor switching, 551–554
free recovery method, 325–330, 337, 342
generator breakers, 568
load break switch, 401, 402
out of phase switching, 572
recloser endurance testing, 550
three poles connected in parallel, 570
vacuum contactors, 431
vacuum interrupter development designs, 546–548
Weil-Dobke circuit, 329, 330
TMF (transverse magnetic field)
see Vacuum arc in transverse magnetic field
TMF contacts
see Transverse magnetic field contact structures
Townsend ionization coefficients, 10–12
Transformer secondary faults
see Short circuit currents
Transformer switching, 423–427
interrupting inrush current, 426, 427
kVA and MVA definition, 424
loaded transformer, 427
overvoltages, 425
surge protection, 427
tap changers 424
TRV (transformer secondary faults), 509, 510
unloaded transformer, 424–427
virtual current chopping, 425
Transient recovery voltage (TRV), 323, 324, 363, 364, 498, 502
effect of current chop, 363, 364
first pole to clear factor, 499–501
rate of rise of recovery voltage (RRRV), 323, 324
TRV after short line fault, 507–509
TRV after terminal (or bus, bolted) fault, 496–502
TRY capacitive circuit, 367
TRV inductive circuit, 322, 323, 363
TRV resistive circuit, 321, 322

TRV transformer secondary faults, 509, 510
two parameter characterization, 498
Transition vacuum arc, 152, 165, 179–181, 197–201
appearance diagram, 280
axial magnetic field, 197–201
contact erosion, 180, 200
current range, 180
pressure, 180
Transverse magnetic field contact structures, 247–262
advantages and disadvantages, 284
Amperian motion (columnar arc), 185–187
appearance diagram, 258–261
large contact gaps, 258, 259
small contact gaps, 260
arc energy distribution, 231, 346, 506
arc revolutions, 254, 255
arc voltage characteristics, 251, 261
comparison with axial magnetic field structures, 284
contrate cup contact s, 248–251
effect on multiple cathode spots, 182–184
drag forces, 186, 187
duration stationary columnar arc as contacts open, 250–252, 258–263
electrical switching life, 540–546
force on columnar arc, 185, 186, 252
generator circuit breaker use, 571
interruption current as f(contact diameter and circuit voltage), 256
interruption current as f(contact diameter), 255, 256
minimum arc length for columnar arc motion, 186, 250–252
single cathode spot retrograde motion, 181, 182
spiral contacts, 257–251
transition to diffuse arc, 251, 252, 257, 261
velocity of columnar arc between contrate cup contacts, 261, 262
velocity of columnar arc between spiral contacts, 252–254, 257
TRV, *see* Transient recovery voltage
Tungsten Copper, 221–224
additives to lower current chop, 223, 224
capacitor switches, 437
current chop, 224, 226
interruption with butt contacts, 224, 232
load current switches, 403
material properties, 225
percentage copper, 225
reignition with butt contacts, 224, 232
weld resistance, 223

V

Vacuum arc
see, Columnar vacuum arc
see, Diffuse vacuum arc
see, Transition vacuum arc
see, Vacuum arc in axial magnetic field
see, Vacuum arc in transverse magnetic field Vacuum arc formation,
see, Vacuum arc formation, breakdown
see, Vacuum arc formation, closing contacts
see, Vacuum arc formation, opening contacts
Vacuum arc formation, breakdown, 62–86
anode flares, 70, 71, 75, 78–83
breakdown sequence, open contacts, 78
cathode flares, 71, 75, 87–83
closing contacts, 11, 126, 386–390
current increase, 63, 76, 78–81, 83
current density at breakdown, 66–69
explosive rupture of field emission site, 70, 71, 75–83
field emission current, 38–44
location of metal vapor in contact gap, 63, 70–73, 75, 83
models, 79–84
space charge limitation, 68–70
spatial light distribution in contact gap, 63, 70, 71, 75, 79, 83
temperature rise of emission site, 67
time to breakdown, 84–86
voltage collapse time and number of emission sites, 81, 82
voltage collapse time, 76
voltage collapse, 76, 79–84
voltage pulse width, 70, 71, 84, 85, 107
Vacuum arc formation, closing contacts, 125, 126, 386–390
Vacuum arc formation, opening contacts, 147–152
bridge column arc, 149–152, 176
duration bridge column arc f(current), 152
duration molten metal bridge f(current), 152
molten metal bridge formation and rupture, 147–149
pressure decline in bridge column arc f(time), 151
pseudo arc phase, 149
Vacuum arc in axial magnetic field (AMF), 160–171, 221–223, 283
low current, diffuse vacuum arc, 187–193
arc voltage, 189–191
average life-time 20A cathode spot f(AMF), 189, 190
cathode erosion rate, 191, 192
cathode spot trajectory, 193
cathode spot velocity, 189, 192, 193
electron helix, 187
electron temperature, 192
floating shield potential, 191
ion charge, 192
ion current, 191
overlapping plasma plumes, 191, 192
plasma plume radius, 188
Robson angle, 192, 193
single cathode spot current f(AMF), 188–190
threshold cathode spot current f(AMF), 188–190
high current vacuum arc, 193–209
anode spot formation, 206–209, 274
anode temperature for 15kA(rms) current half cycle, 205
anode temperature for 15kA (rms) current half cycle f(AMF), 206
anode temperature for half cycle currents 10kA to 20kA (rms) AMF, 180mT
appearance diagram, 202, 275
arc voltage, 201–206
arc voltage for high current diffuse arc, 274, 347
bridge column duration, 196
cathode spot dispersion, 204, 205
contact gap, 197–201
critical axial field, 194–196, 208
eddy currents and AMF phase shift, 272, 273, 257, 258

high current column vacuum arc, 274
high current diffuse vacuum arc, 274
model, 203, 204
multiple spot vacuum arc, 273
opening sequence, 194–201
overlapping plasma plumes, 206–209
transition to high current diffuse vacuum arc, 194–201
threshold anode spot formation f(AMF, contact diameter and gap), 194–201
Vacuum arc in transverse magnetic field (TMF), 185–187, 247–262, 346, 347, 571
columnar vacuum arc, 185–187, 247–261
Amperian motion, 185, 186, 247, 252
arc voltage of moving column, 251, 261, 346, 347, 571
drag forces, 186, 187
force on columnar arc, 185, 186, 252
minimum arc length, 186, 250–252
velocity, spiral contacts, 252–257
velocity, contrate cup contacts, 261, 262
diffuse vacuum arc, 182–185
cathode spot alignment, 182–185
Child's law, 184
de current interruption, 183, 184, 590
Hall field, 182
pulsed magnetic field, 182, 590
retrograde motion of a single cathode spot, 181, 182
Vacuum arc interruption,
see, Circuit interruption
Vacuum arc stability at low currents, 75, 77, 165–169, 188–190
Vacuum breakdown formation of vacuum arc
see, Vacuum arc formation, breakdown
Vacuum breakdown initial processes,
see Prebreakdown
Vacuum breakdown voltage, 25–31
ac breakdown voltage as a f(contact gap), 74
anode tensile, strength, 63
anode thermal conductivity, 63
anode specific heat, 49
anode vapor pressure, 63
BIL breakdown voltage as a f(contact gap), 74
conditioning effect, 26, 27, 86–94
Cranberg's hypothesis, 68
critical electric field, 31, 63
de breakdown as a f(contact gap), 25, 26, 65, 579–584
deconditioning effect, 88, 89, 93
effect of contact roughness, 32, 34, 64
emission current density, 66–68
function of onset voltage, 44
function of pressure, 26
insulating inclusions, 57
microdischarge inception voltage, 62
microparticle effect, 64–66
probability distributions, 27–31
rate of rise of voltage pulse, 72–74, 78, 82, 105, 107
series vacuum interrupters, 103, 104, 579–581
statistics, 27–31 466, 467
step-up measurement method, 28
threshold breakdown voltage, 27–31
up-and-down measurement method, 28
voltage pulse width, 72–74, 78, 92, 105, 107
Weibull distribution, 27–30, 464, 467
Vacuum breakdown
see, Electrical breakdown in vacuum
Vacuum breakdown, series vacuum interrupters, 103, 104, 579–581
Vacuum circuit breaker
see Circuit breaker design distribution circuits
Vacuum contactors, 428–437
absolute cathode erosion, 162–164, 433, 434
contact erosion effect of contact gap, 434–436
contact mate rial, 221–225, 431
contact sizing, 433–436
current and voltage ranges, 430
design criteria, 430, 431
design, 428–431
effective cathode erosion, 433–436
IEC categories, 431
low voltage, 430
mechanism, 428, 429, 431–433
medium voltage, 349, 430
metal particles, 436, 437
shield, 436
solenoid operation, 428, 429, 431–433
standards, 431
testing, 431
welding resistant contact material, 432, 433
Vacuum furnace, 293–298
gas species during temperature ramp, 298
temperature-time ramp, 297
Vacuum interrupter design, 283–291
see also, Axial magnetic field contact. structures
see also, Bellows
see also, Butt contacts
see also, Contact materials, vacuum interrupters
see also, Internal vacuum interrupter high voltage design
see also, Load break switches
see also, Manufacture
see also, Transverse magnetic field contact structures
active metal braze, 285
anti-twist bushing, 290, 291
bellows, 289–291
brazing material, 295
ceramic body, 283–286
advantages, 284
as a shield, 288
metallization, 284, 285
contact structures, 246–251, 264–273
contact materials, 220–233
electrical life, 540–546
FEA (finite element analysis), 35–37, 95, 97, 273, 541
getter, 295, 296
mechanical life, 290
reliability, 300–302, 306, 598
shield, 286–289
floating, 286–289
fixed, 286, 436
attachment, 286, 287
ceramic, 288
three shields, 289
Vacuum interrupter isolation switches
see, Isolation switches
Vacuum interrupter manufacture,
see, Manufacture
see, vacuum interrupter design
Vacuum interrupter reliability, 300–302, 306, 598

leak rate assessment, 300–302
life testing using high voltage withstand test, 301, 302
electrical switching life, 540–546
mechanical life, 290
Vacuum load break switches
see, Capacitor switching
see, Contactors
see, Inductive switching
see Isolation switches
see, Load break switches
see, Sectionalizers use with reclosers
Vacuum recloser
see, Recloser
Virtual current chopping, 421–425
Voltage–Temperature relation, 141–145
Boltzmann's constant, 142
Lorenz constant, 142
melting voltage and temperature, 142, 143, 146, 147
opening contacts, 148
softening voltage and temperature, 142, 143
Wiedemann–Franz law, 142
Voltage escalation, 262–266, 420, 421
surge suppression, 415–421

W

Weibull distribution, 27–30, 466, 467
after load switching, 407
application, 27 = 30, 466, 467
CFO (cumulative distribution function), 27
characteristic or scale parameter, 23, 466
function of contact gap, 466
PDF (probability density function), 27
shape factor, 27, 467
three parameter distribution, 27, 28, 466, 467
threshold value, 27, 28, 466, 467
two parameter distribution, 27, 28
Welding, 369–394
Welding closed contacts, 304–311
AMF contact structure, 380, 381
blow off force, 373
calculated threshold welding current, 376, 377
cold welding, 371
constriction resistance as f(temperature), 374
contact force, 144, 373, 517
contact sliding prevention, 519–521
diffusion welding, 371
effect on contact resistance, 144
effect of 1 to 4 seconds passage of current, 381–384
effect of sliding, 385, 386, 519–521
hardness, 373–365
mechanism requirements, 393, 524
resistance welding, 371
temperature coefficient of resistance, 375
tensile strength, 370
weld area, 370, 372, 391
welding current one region of contact, 373–376
welding current 'n' regions of contact, 378, 379
weld resistant contact material, 221–224, 385, 432, 433
Welding closing contacts, 386–381
bouncing, 386, 393, 521, 522
capacitor inrush current, 438–441, 553
contact force, 517
effect on (β_m), 386–390
maximum weld force, 391–393
mechanism requirements, 393, 524
percussion weld, 386
prestrike arcing, 386–390, 393
weld area, 370, 372, 391
Welding, contacts that slide, 385, 386
Wiedemann–Franz Law, 142
Withstand voltage ac (1 minute), 3, 4

X

X-ray emission, 114–125
absorbed X-ray dose, 117
background radiation dose, 116
Bremsstrahlung X-rays, 114, 115
characteristic X-rays, 114
conditions for no detectable X-ray emission, 115
conditions for X-ray emission, 114
definition of units, 115, 116
effect on vacuum breakdown, 124, 125
effect of contact conditioning, 119
effect of contact of contact gap, 118
expected X-rays emission from high voltage vacuum interrupters, 120–124
radiation dose limits, 116
shielding, 124
X-ray dose f(applied voltage), 118, 120–123